FROM MOLECULES
TO NETWORKS

ELSEVIER science & technology books

ELSEVIER

 Companion Web Site:

http://elsevierdirect.com/companions/9780123741325

From Molecules to Networks, **Second Edition**
by John H. Byrne and James L. Roberts

Resources for Professors:

- **All figures from the book available as PowerPoint slides**
- **Links to web sites carefully chosen to supplement the content of the textbook**

ELSEVIER

TOOLS FOR ALL YOUR TEACHING NEEDS
textbooks.elsevier.com

ACADEMIC
PRESS

FROM MOLECULES TO NETWORKS

AN INTRODUCTION TO CELLULAR AND MOLECULAR NEUROSCIENCE

Second Edition

JOHN H. BYRNE

June and Virgil Waggoner Chair and Chairman
Department of Neurobiology and Anatomy
The University of Texas Medical School at Houston
Houston, Texas

JAMES L. ROBERTS

Department of Pharmacology
Center for Biomedical Neuroscience
Sam and Ann Barshop Institute for Aging and Longevity Studies
Audie Murphy Veterans Medical Center
University of Texas Health Science Center
San Antonio, Texas

AMSTERDAM • BOSTON • HEIDELBERG • LONDON
NEW YORK • OXFORD • PARIS • SAN DIEGO
SAN FRANCISCO • SINGAPORE • SYDNEY • TOKYO
Academic Press is an imprint of Elsevier

ELSEVIER

Academic Press is an imprint of Elsevier

30 Corporate Drive, Suite 400, Burlington, MA 01803, USA
525 B Street, Suite 1900, San Diego, California 92101-4495, USA
84 Theobald's Road, London WC1X 8RR, UK

This book is printed on acid-free paper. ⊗

Library of Congress Cataloging-in-Publication Data
From molecules to networks : an introduction to cellular and molecular
neuroscience / editors, John H. Byrne, James L. Roberts. — 2nd ed.
      p. ; cm.
   Includes bibliographical references and index.
   ISBN 978-0-12-374132-5 (hardcover : alk. paper) 1. Molecular neurobiology.
2. Cytology.   3. Neurons. I. Byrne, John H. II. Roberts, James Lewis, 1951–
[DNLM: 1. Nervous System—cytology.   2. Nerve Net.
3. Neurobiology—methods. WL 102 F9309 2009]
QP356.2.F76 2009
611'.0188—dc22

                                                                2008029618

British Library Cataloguing-in-Publication Data
A catalogue record for this book is available from the British Library.

ISBN: 978-0-12-374132-5

For information on all Academic Press publications
visit our Web site at www.books.elsevier.com

Printed in China
09   10   11   12   9   8   7   6   5   4   3   2   1

Table of Contents

Contributors

Douglas A. Baxter (181, 413) Department of Neurobiology and Anatomy, The University of Texas Medical School at Houston, Houston, TX

Scott Brady (19) Department of Anatomy and Cell Biology, University of Illinois at Chicago, Chicago, IL

Peter Brophy (19) Department of Preclinical Veterinary Sciences, University of Edinburgh, Edinburgh, Scotland, UK

John H. Byrne (111, 181, 413, 469, 489, 539) Department of Neurobiology and Anatomy, The University of Texas Medical School at Houston, Houston, TX

Luz Claudio (1) Department of Community and Preventive Medicine, Mount Sinai School of Medicine, New York, NY

David R. Colman (19) Director's Office, Montreal Neurological Institute, Montreal, QC, Canada

Mark R. Cookson (609) Neurogenetics Laboratory, NIH, National Institute of Aging, Bethesda, MD

Ariel Y. Deutch (267, 301) Department of Psychiatry, Vanderbilt Medical Center, Vanderbilt University, Nashville, TN

Gerald A. Dienel (49) Department of Neurology, University of Arkansas for Medical Sciences, Little Rock, AR

Andrea Giuffrida (301) Department of Pharmacology, The University of Texas Health Science Center at San Antonio, San Antonio, TX

P. John Hart (609) Department of Biochemistry, The University of Texas Health Science Center, San Antonio, TX

Patrick R. Hof (1) Department of Neuroscience, Mount Sinai School of Medicine, New York, NY

Lily Yeh Jan (159) University of California, San Francisco, San Francisco, CA

Yuh Nung Jan (159) University of California, San Francisco, San Francisco, CA

Grahame Kidd (1) Department of Neurosciences, Cleveland Clinic Lerner Research Institute, Cleveland, OH

James J. Knierim (513) Department of Neurobiology and Anatomy, The University of Texas Medical School at Houston, Houston, TX

Dimitri M. Kullmann (217) Department of Clinical Neurology, Institute of Neurology, Queen's Square, University College London, London, England, UK

Kevin S. LaBar (539) Center for Cognitive Neuroscience, Duke University, Durham, NC

Joseph E. LeDoux (539) Center for Neural Science, New York University, New York, NY

James R. Lundblad (391) Division of Endocrinology, Diabetes, and Clinical Nutrition, School of Medicine, Oregon Health Sciences University, Portland, OR

David A. McCormick (133) Department of Neurobiology, Yale University School of Medicine, New Haven, CT

Bruce J. Nicholson (445) Department of Biochemistry, The University of Texas Health Science Center at San Antonio, San Antonio, TX

Esther A. Nimchinsky (1) Department of Radiology, Mount Sinai School of Medicine, New York, NY

James L. Roberts (301, 391, 609) Department of Pharmacology, The University of Texas Health Science Center at San Antonio, San Antonio, TX

Robert H. Roth (267) Department of Psychiatry, Yale University School of Medicine, New Haven, CT

Juan C. Saez (445) Departamento de Ciencias Fisiologicas, Pontificia Universidad Catolica de Chile, Santiago, Chile

Glenn E. Schafe (539) Department of Psychology, Yale University, New Haven, CT

Howard Schulman (359) PPD Biomarker Discovery Sciences, LLC, Menlo Park, CA

Thomas L. Schwarz (217) Department of Neurology, Children's Hospital, Harvard University, Boston, MA

Gordon M. Shepherd (111, 489) Department of Neurobiology, Yale University School of Medicine, New Haven, CT

Paul D. Smolen (413) Department of Neurobiology and Anatomy, The University of Texas Medical School at Houston, Houston, TX

Randy Strong (609) Department of Pharmacology, The University of Texas Health Science Center, San Antonio, TX

J. David Sweatt (539) Department of Neurobiology, University of Alabama at Birmingham, Birmingham, AL

Richard F. Thompson (539) Neuroscience Research Institute, University of Southern California, Los Angeles, CA

Bruce D. Trapp (1) Department of Neuroscience, Cleveland Clinic Foundation, Cleveland, OH

M. Neal Waxham (321) Department of Neurobiology and Anatomy, The University of Texas Medical School at Houston, Houston, TX

David Matthew Young (159) University of California, San Francisco, San Francisco, CA

Robert S. Zucker (217) Neurobiology Division, Department of Molecular and Cell Biology, University of California, Berkeley, Berkeley, CA

Preface to the Second Edition

The second edition contains substantial improvements over the first edition. All chapters have been updated to include recent developments in the field, and major revisions have been done on the chapters on *Energy Metabolism in the Brain*, *Molecular Properties of Ion Channels*, *Gap Junctions*, and *Learning and Memory*. In addition, this edition features two new chapters, *Information Processing in Neural Networks* and *Molecular and Cellular Mechanisms of Neurodegenerative Disease*. Although the first edition covered biochemical and gene networks in significant detail, little was included on neural networks. It is the neural networks in the brain that collect and process information about the external world and about the internal state of the body and generate motor commands. Therefore, an understanding of these networks is essential to understanding the brain and also helps to put the cellular and molecular processes in perspective. However, discussing all of the brain systems is beyond the scope of a text book on cellular and molecular neuroscience. Rather, our goal is to describe the principles of operation of neural networks and the key circuit motifs that are common to many networks. The second new chapter reports on the progress in the last 20 years on elucidating the cellular and molecular mechanisms underlying brain disorders This chapter focuses specifically on amyotrophic lateral sclerosis (ALS), Parkinson disease, and Alzheimer's disease, and the progress that has been made and the strategies that have been used to study and treat the disorders. The fact that all three diseases are associated with neuronal loss, albeit in different brain regions and with different neurotransmitter groups, suggests that there may be common aspects to the degenerative process.

We are once again extremely grateful to Johannes Menzel at Elsevier for his unfading support and encouragement throughout the project. Thanks also to Clare Caruana, Meg Day, Kristi Gomez, Kirsten Funk, Megan Wickline, and members of the production staff. Special thanks to Lorenzo Morales, the graphic artist on the project, who did an outstanding job of creating many of the illustrations in the second edition and restyling all the illustrations for consistency among chapters. He also designed the cover illustration.

John H. Byrne
James L. Roberts

Preface to the First Edition

The past twenty years have witnessed an exponential increase in the understanding of the nervous system at all levels of analyses. Perhaps the most striking developments have been in the understanding of the cell and molecular biology of the neuron. The field has moved from treating the neuron as a simple black box that added up impinging synaptic input to fire an action potential to one in which the function of nerve cells involves a host of biochemical and biophysical processes that act synergistically to process, transmit and store information. In this book, we have attempted to provide a comprehensive summary of current knowledge of the morphological, biochemical, and biophysical properties of nerve cells. The book is intended for graduate students, advanced undergraduate students, and professionals. The chapters are highly referenced so that readers can pursue topics of interest in greater detail. We have also included material on mathematical modeling approaches to analyze the complex synergistic processes underlying the operation and regulation of nerve cells. These modeling approaches are becoming increasingly important to facilitate the understanding of membrane excitability, synaptic transmission, as well gene and protein networks. The final chapter in the book illustrates the ways in which the great strides in understanding the biochemical and biophysical properties of nerve cells have led to fundamental insights into an important aspect of cognition, memory.

We are extremely grateful to the many authors who have contributed to the book, and the support and encouragement during the two past years of Jasna Markovac and Johannes Menzel of Academic Press. We would also like to thank Evangelos Antzoulatos, Evyatar Av-Ron, Diasinou Fioravanti, Yoshihisa Kubota, Rong-Yu Liu, Fred Lorenzetii, Riccardo Mozzachiodi, Gregg Phares, Travis Rodkey, and Fredy Reyes for help with editing the chapters.

John H. Byrne
James L. Roberts

Cellular Components of Nervous Tissue

Patrick R. Hof, Esther A. Nimchinsky, Grahame Kidd, Luz Claudio, and Bruce D. Trapp

Several types of cellular elements are integrated to constitute normally functioning brain tissue. The neuron is the communicating cell, and many neuronal subtypes are connected to one another via complex circuitries, usually involving multiple synaptic connections. Neuronal physiology is supported and maintained by neuroglial cells, which have highly diverse and incompletely understood functions. These include myelination, secretion of trophic factors, maintenance of the extracellular milieu, and scavenging of molecular and cellular debris from it. Neuroglial cells also participate in the formation and maintenance of the blood–brain barrier, a multicomponent structure that is interposed between the circulatory system and the brain substance and that serves as the molecular gateway to brain tissue.

NEURONS

The neuron is a highly specialized cell type and is the essential cellular element in the central nervous system (CNS). All neurological processes are dependent on complex cell–cell interactions between single neurons and/or groups of related neurons. Neurons can be categorized according to their size, shape, neurochemical characteristics, location, and connectivity, which are important determinants of that particular functional role of the neuron in the brain. More importantly, neurons form circuits, and these circuits constitute the structural basis for brain function. *Macrocircuits* involve a population of neurons projecting from one brain region to another region, and *microcircuits* reflect the local cell–cell interactions within a brain region. The detailed analysis of these macro- and microcircuits is an essential step in understanding the neuronal basis of a given cortical function in the healthy and the diseased brain. Thus, these cellular characteristics allow us to appreciate the special structural and biochemical qualities of a neuron in relation to its neighbors and to place it in the context of a specific neuronal subset, circuit, or function.

Broadly speaking, therefore, there are five general categories of neurons: inhibitory neurons that make local contacts (e.g., GABAergic interneurons in the cerebral and cerebellar cortex), inhibitory neurons that make distant contacts (e.g., medium spiny neurons of the basal ganglia or Purkinje cells of the cerebellar cortex), excitatory neurons that make local contacts (e.g., spiny stellate cells of the cerebral cortex), excitatory neurons that make distant contacts (eg., pyramidal neurons in the cerebral cortex), and neuromodulatory neurons that influence neurotransmission, often at large distances. Within these general classes, the structural variation of neurons is systematic, and careful analyses of the anatomic features of neurons have led to various categorizations and to the development of the concept of cell type. The grouping of neurons into descriptive cell types (such as chandelier, double bouquet, or bipolar cells) allows the analysis of populations of neurons and the linking of specified cellular characteristics with certain functional roles.

General Features of Neuronal Morphology

Neurons are highly polarized cells, meaning that they develop distinct subcellular domains that subserve different functions. Morphologically, in a typical neuron, three major regions can be defined: (1) the cell body (*soma* or *perikaryon*), which contains the nucleus and the major cytoplasmic organelles; (2) a variable number of dendrites, which emanate from the perikaryon and ramify over a certain volume of gray matter and which differ in size and shape, depending on the neuronal type; and (3) a single axon, which extends, in most cases, much farther from the cell body than the dendritic arbor (Fig. 1.1). Dendrites may be spiny (as in pyramidal cells) or nonspiny (as in most interneurons), whereas the axon is generally smooth and emits a variable number of branches (collaterals). In vertebrates, many axons are surrounded by an insulating myelin sheath, which facilitates rapid impulse conduction. The axon terminal region, where contacts with other cells are made, displays a wide range of morphological specializations, depending on its target area in the central or peripheral nervous system.

The cell body and dendrites are the two major domains of the cell that receive inputs, and dendrites play a critically important role in providing a massive receptive area on the neuronal surface. In addition, there is a characteristic shape for each dendritic arbor, which can be used to classify neurons into morphological types. Both the structure of the dendritic arbor and the distribution of axonal terminal ramifications confer a high level of subcellular specificity in the localization of particular synaptic contacts on a given neuron. The three-dimensional distribution of dendritic arborization is also important with respect to the type of information transferred to the neuron. A neuron with a dendritic tree restricted to a

particular cortical layer may receive a very limited pool of afferents, whereas the widely expanded dendritic arborizations of a large pyramidal neuron will receive highly diversified inputs within the different cortical layers in which segments of the dendritic tree are present (Fig. 1.2) (Mountcastle, 1978). The structure of the dendritic tree is maintained by surface interactions between adhesion molecules and, intracellularly, by an array of cytoskeletal components (microtubules, neurofilaments, and associated proteins), which also take part in the movement of organelles within the dendritic cytoplasm.

An important specialization of the dendritic arbor of certain neurons is the presence of large numbers of dendritic spines, which are membranous protrusions. They are abundant in large pyramidal neurons and are much sparser on the dendrites of interneurons (see following text).

The perikaryon contains the nucleus and a variety of cytoplasmic organelles. Stacks of rough endoplasmic reticulum are conspicuous in large neurons and, when interposed with arrays of free polyribosomes, are referred to as *Nissl substance*. Another feature of the perikaryal cytoplasm is the presence of a rich cytoskeleton composed primarily of neurofilaments and microtubules. These cytoskeletal elements are dispersed in bundles that extend from the soma into the axon and dendrites.

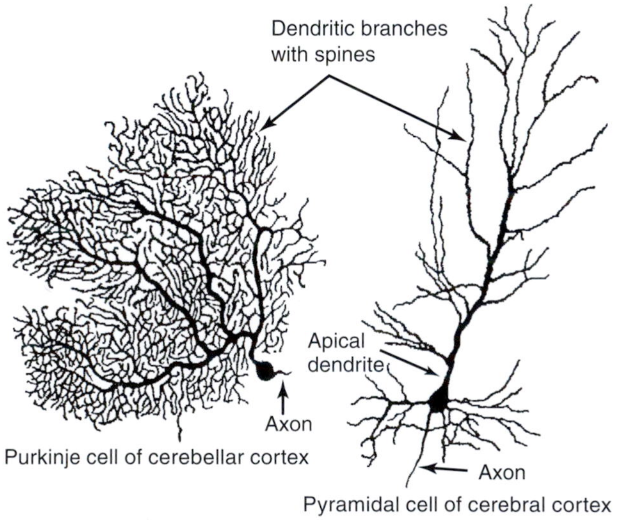

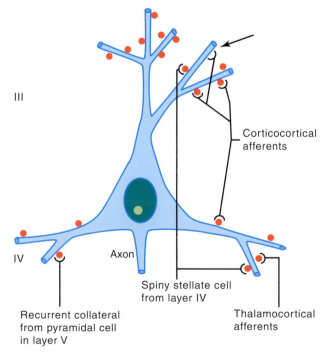

FIGURE 1.1 Typical morphology of projection neurons. (Left) A Purkinje cell of the cerebellar cortex and (right) a pyramidal neuron of the neocortex. These neurons are highly polarized. Each has an extensively branched, spiny apical dendrite, shorter basal dendrites, and a single axon emerging from the basal pole of the cell.

FIGURE 1.2 Schematic representation of four major excitatory inputs to pyramidal neurons. A pyramidal neuron in layer III is shown as an example. Note the preferential distribution of synaptic contacts on spines. Spines are labeled in red. Arrow shows a contact directly on the dendritic shaft.

Whereas dendrites and the cell body can be characterized as domains of the neuron that receive afferents, the axon, at the other pole of the neuron, is responsible for transmitting neural information. This information may be primary, in the case of a sensory receptor, or processed information that has already been modified through a series of integrative steps. The morphology of the axon and its course through the nervous system are correlated with the type of information processed by the particular neuron and by its connectivity patterns with other neurons. The axon leaves the cell body from a small swelling called the *axon hillock*. This structure is particularly apparent in large pyramidal neurons; in other cell types, the axon sometimes emerges from one of the main dendrites. At the axon hillock, microtubules are packed into bundles that enter the axon as parallel fascicles. The axon hillock is the part of the neuron where the action potential is generated. The axon is generally unmyelinated in local circuit neurons (such as inhibitory interneurons), but it is myelinated in neurons that furnish connections between different parts of the nervous system. Axons usually have higher numbers of neurofilaments than dendrites, although this distinction can be difficult to make in small elements that contain fewer neurofilaments. In addition, the axon may be extremely ramified, as in certain local circuit neurons; it may give out a large number of recurrent collaterals, as in neurons connecting different cortical regions, or it may be relatively straight in the case of projections to

subcortical centers, as in cortical motor neurons that send their very long axons to the ventral horn of the spinal cord. At the interface of axon terminals with target cells are the synapses, which represent specialized zones of contact consisting of a presynaptic (axonal) element, a narrow synaptic cleft, and a postsynaptic element on a dendrite or perikaryon.

Synapses and Spines

Synapses

Each synapse is a complex of several components: (1) a *presynaptic element*, (2) a *cleft*, and (3) a *postsynaptic element*. The presynaptic element is a specialized part of the presynaptic neuron's axon, the postsynaptic element is a specialized part of the postsynaptic somatodendritic membrane, and the space between these two closely apposed elements is the cleft. The portion of the axon that participates in the axon is the *bouton*, and it is identified by the presence of synaptic vesicles and a presynaptic thickening at the active zone (Fig. 1.3). The postsynaptic element is marked by a postsynaptic thickening opposite the presynaptic thickening. When both sides are equally thick, the synapse is referred to as *symmetric*. When the postsynaptic thickening is greater, the synapse is *asymmetric*. Edward George Gray noticed this difference, and divided synapses into two types: *Gray's type 1* synapses are symmetric and have variably

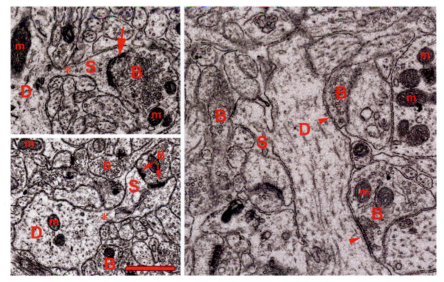

FIGURE 1.3 Ultrastructure of dendritic spines (S) and synapses in the human brain. Note the narrow spine necks (asterisks) emanating from the main dendritic shaft (D) and the spine head containing filamentous material, and the cisterns of the spine apparatus particularly visible in the lower panel spine. The arrows on the left panels point to postsynaptic densities of asymmetric excitatory synapses (arrows). The apposed axonal boutons (B) are characterized by round synaptic vesicles. A perforated synapse is shown on the lower left panel. The panel at right shows two symmetric inhibitory synapses (arrowheads) on a large dendritic shaft (D). In this case the axonal boutons (B) contain some ovoid vesicles compared to the ones in asymmetric synapses. The dendrites and axons contain numerous mitochondria (m). Scale bar = 1 μm. Electron micrographs courtesy of Drs. S.A. Kirov and M. Witcher (Medical College of Georgia), and K.M. Harris (University of Texas – Austin).

shaped, or pleomorphic, vesicles. *Gray's type 2* synapses are asymmetric and have clear, round vesicles. The significance of this distinction is that research has shown that, in general, Gray's type 1 synapses tend to be inhibitory, while Gray's type 2 synapses tend to be excitatory. This correlation greatly enhanced the usefulness of electron microscopy in neuroscience.

In cross section on electron micrographs, a synapse looks like two parallel lines separated by a very narrow space (Fig. 1.3). Viewed from the inside of the axon or dendrite, it looks like a patch of variable shape. Some synapses are a simple patch, or *macule*. Macular synapses can grow fairly large, reaching diameters over 1 μm. The largest synapses have discontinuities or holes within the macule and are called *perforated synapses* (Fig. 1.3). In cross section, a perforated synapse may resemble a simple macular synapse or several closely spaced smaller macules.

The portion of the presynaptic element that is apposed to the postsynaptic element is the *active zone*. This is the region where the synaptic vesicles are concentrated and where, at any time, a small number of vesicles are docked and presumably ready for fusion. The active zone is also enriched with voltage gated calcium channels, which are necessary to permit activity-dependent fusion and neurotransmitter release.

The synaptic cleft is truly a space, but its properties are essential. The width of the cleft (∼20 μm) is critical because it defines the volume in which each vesicle releases its contents, and therefore, the peak concentration of neurotransmitter upon release. On the flanks of the synapse, the cleft is spanned by adhesion molecules, which are believed to stabilize the cleft.

The postsynaptic element may be a portion of a soma or a dendrite, or rarely, part of an axon. In the cerebral cortex, most Gray's type 1 synapses are located on somata or dendritic shafts, while most Gray's type 2 synapses are located on dendritic spines, which are specialized protrusions of the dendrite. A similar segregation is seen in cerebellar cortex. In nonspiny neurons, symmetric and asymmetric synapses are often less well separated. Irrespective of location, a postsynaptic thickening marks the postsynaptic element. In Gray's type 2 synapses, the postsynaptic thickening (or postsynaptic density, PSD), is greatly enhanced. Among the molecules that are associated with the PSD are neurotransmitter receptors (e.g., NMDA receptors) and molecules with less obvious function, such as PSD-95.

Spines

Spines are protrusions on the dendritic shafts of some types of neurons and are the sites of synaptic contacts, usually excitatory. Use of the silver impregnation techniques of Golgi or of the methylene blue used by

Ehrlich in the late nineteenth century led to the discovery of spiny appendages on dendrites of a variety of neurons. The best known are those on pyramidal neurons and Purkinje cells, although spines occur on neuron types at all levels of the central nervous system. In 1896, Berkley observed that terminal axonal boutons were closely apposed to spines and suggested that spines may be involved in conducting impulses from neuron to neuron. In 1904, Santiago Ramón y Cajal suggested that spines could collect the electrical charge resulting from neuronal activity. He also noted that spines substantially increase the receptive surface of the dendritic arbor, which may represent an important factor in receiving the contacts made by the axonal terminals of other neurons. It has been calculated that the approximately 20,000 spines of a pyramidal neuron account for more than 40% of its total surface area (Peters *et al.*, 1991).

More recent analyses of spine electrical properties have demonstrated that spines are dynamic structures that can regulate many neurochemical events related to synaptic transmission and modulate synaptic efficacy. Spines are also known to undergo pathologic alterations and have a reduced density in a number of experimental manipulations (such as deprivation of a sensory input) and in many developmental, neurologic, and psychiatric conditions (such as dementing illnesses, chronic alcoholism, schizophrenia, trisomy 21). Morphologically, spines are characterized by a narrower portion emanating from the dendritic shaft, the neck, and an ovoid bulb or head, although spine morphology may vary from large mushroom-shaped bulbs to small bulges barely discernable on the surface of the dendrite. Spines have an average length of ∼2 μm, but there is considerable variability in their dimensions. At the ultrastructural level (Fig. 1.3), spines are characterized by the presence of asymmetric synapses and contain fine and quite indistinct filaments. These filaments most likely consist of actin and α- and β-tubulins. Microtubules and neurofilaments present in dendritic shafts do not enter spines. Mitochondria and free ribosomes are infrequent, although many spines contain polyribosomes in their neck. Interestingly, most polyribosomes in dendrites are located at the bases of spines, where they are associated with endoplasmic reticulum, indicating that spines possess the machinery necessary for the local synthesis of proteins. Another feature of the spine is the presence of confluent tubular cisterns in the spine head that represent an extension of the dendritic smooth endoplasmic reticulum. Those cisterns are referred to as the *spine apparatus*. The function of the spine apparatus is not fully understood but may be related to the storage of calcium ions during synaptic transmission.

Specific Examples of Different Neuronal Types

Inhibitory Local Circuit Neurons

Inhibitory Interneurons of the Cerebral Cortex A large variety of inhibitory interneuron types is present in the cerebral cortex and in subcortical structures. These neurons contain the inhibitory neurotransmitter γ-aminobutyric acid (GABA) and exert strong local inhibitory effects. Their dendritic and axonal arborizations offer important clues as to their role in the regulation of pyramidal cell function. In addition, for several GABAergic interneurons, a subtype of a given morphologic class can be defined further by a particular set of neurochemical characteristics. Interneurons have been extensively characterized in the neocortex and hippocampus of rodents and primates, but they are present throughout the cerebral gray matter and exhibit a rich variety of morphologies, depending on the brain region as well as on the species studied.

In the neocortex and hippocampus, the targets and morphologies of interneuron axons are most usefully classified into morphological and functional groups. For example, *basket cells* have axonal endings surrounding pyramidal cell somata (Somogyi *et al.*, 1983) and provide most of the inhibitory GABAergic synapses to the somas and proximal dendrites of pyramidal cells. These cells are also characterized by certain biochemical features in that the majority of them contain the calcium-binding protein parvalbumin, and cholecystokinin appears to be the most likely neuropeptide in large basket cells.

Chandelier cells have spatially restricted axon terminals that look like vertically oriented "cartridges," each consisting of a series of axonal boutons, or swellings, linked together by thin connecting pieces. These neurons synapse exclusively on the axon initial segment of pyramidal cells (this cell is also known as *axoaxonic cell*), and because the strength of the synaptic input is correlated directly with its proximity to the axon initial segment, there can be no more powerful inhibitory input to a pyramidal cell than that of the chandelier cell (Freund *et al.*, 1983; DeFelipe *et al.*, 1989).

The double bouquet cells are characterized by a vertical bitufted dendritic tree and a tight bundle of vertically oriented varicose axon collaterals (Somogyi and Cowey, 1981). There are several subclasses of double bouquet cells based on the complement of calcium-binding protein and neuropeptide they contain. Their axons contact spines and dendritic shafts of pyramidal cells, as well as dendrites from non-pyramidal neurons.

Inhibitory Projection Neurons

Medium-Sized Spiny Cells These neurons are unique to the striatum, a part of the basal ganglia that comprises the caudate nucleus and putamen. Medium-sized spiny cells are scattered throughout the caudate nucleus, and putamen and are recognized by their relatively large size compared with other cellular elements of the basal ganglia, and by the fact that they are generally isolated neurons. They differ from all others in the striatum in that they have a highly ramified dendritic arborization radiating in all directions and densely covered with spines. They furnish a major output from the caudate nucleus and putamen and receive a highly diverse input from, among other sources, the cerebral cortex, thalamus, and certain dopaminergic neurons of the substantia nigra. These neurons are neurochemically quite heterogeneous, contain GABA, and may contain several neuropeptides and the calcium-binding protein calbindin. In Huntington disease, a neurodegenerative disorder of the striatum characterized by involuntary movements and progressive dementia, an early and dramatic loss of medium-sized spiny cells occurs.

Purkinje Cells Purkinje cells are the most salient cellular elements of the cerebellar cortex. They are arranged in a single row throughout the entire cerebellar cortex between the molecular (outer) layer and the granular (inner) layer. They are among the largest neurons and have a round perikaryon, classically described as shaped "like a chianti bottle," with a highly branched dendritic tree shaped like a candelabrum and extending into the molecular layer, where they are contacted by incoming systems of afferent fibers from granule neurons and the brainstem. The apical dendrites of Purkinje cells have an enormous number of spines (more than 80,000 per cell). A particular feature of the dendritic tree of the Purkinje cell is that it is distributed in one plane, perpendicular to the longitudinal axes of the cerebellar folds, and each dendritic arbor determines a separate domain of cerebellar cortex (Fig. 1.1). The axons of Purkinje neurons course through the cerebellar white matter and contact deep cerebellar nuclei or vestibular nuclei. These neurons contain the inhibitory neurotransmitter GABA and the calcium-binding protein calbindin. Spinocerebellar ataxia, a severe disorder combining ataxic gait and impairment of fine hand movements, accompanied by dysarthria and tremor, has been documented in some families and is related directly to Purkinje cell degeneration.

Excitatory Local Circuit Neurons

Spiny Stellate Cells Spiny stellate cells are small multipolar neurons with local dendritic and axonal arborizations. These neurons resemble pyramidal cells in that they are the only other cortical neurons with large numbers of dendritic spines, but they differ from pyramidal neurons in that they lack an elaborate apical dendrite. The relatively restricted dendritic arbor of these neurons is presumably a manifestation of the fact that they are high-resolution neurons that gather afferents to a very restricted region of the cortex. Dendrites rarely leave the layer in which the cell body resides. The spiny stellate cell also resembles the pyramidal cell in that it provides asymmetric synapses that are presumed to be excitatory, and is thought to use glutamate as its neurotransmitter (Peters and Jones, 1984).

The axons of spiny stellate neurons have primarily intracortical targets and a radial orientation, and appear to play an important role in forming links among layer IV, the major thalamorecipient layer, and layers III, V, and VI, the major projection layers. The spiny stellate neuron appears to function as a high-fidelity relay of thalamic inputs, maintaining strict topographic organization and setting up initial vertical links of information transfer within a given cortical area (Peters and Jones, 1984).

Excitatory Projection Neurons

Pyramidal Cells   All cortical output is carried by pyramidal neurons, and the intrinsic activity of the neocortex can be viewed simply as a means of finely tuning their output. A pyramidal cell is a highly polarized neuron, with a major orientation axis perpendicular (or orthogonal) to the pial surface of the cerebral cortex. In cross section, the cell body is roughly triangular (Fig. 1.1), although a large variety of morphologic types exist with elongate, horizontal, or vertical fusiform, or inverted perikaryal shapes. Pyramidal cells are the major excitatory type of neurons and use glutamate as their neurotransmitter. A pyramidal neuron typically has a large number of dendrites that emanate from the apex and form the base of the cell body. The span of the dendritic tree depends on the laminar localization of the cell body, but it may, as in giant pyramidal neurons, spread over several millimeters. The cell body and dendritic arborization may be restricted to a few layers or, in some cases, may span the entire cortical thickness (Jones, 1984).

In most cases, the axon of a large pyramidal cell extends from the base of the perikaryon and courses toward the subcortical white matter, giving off several collateral branches that are directed to cortical domains generally located within the vicinity of the cell of origin (as explained later). Typically, a pyramidal cell has a large nucleus, and a cytoplasmic rim that contains, particularly in large pyramidal cells, a collection of granular material chiefly composed of *lipofuscin*. Although all pyramidal cells possess these general features, they can also be subdivided into numerous classes based on their morphology, laminar location, and connectivity with cortical and subcortical regions (Fig. 1.4) (Jones, 1975).

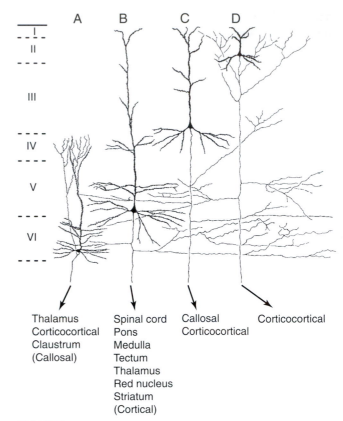

FIGURE 1.4   Morphology and distribution of neocortical pyramidal neurons. Note the variability in cell size and dendritic arborization, as well as the presence of axon collaterals, depending on the laminar localization (I–VI) of the neuron. Also, different types of pyramidal neurons with a precise laminar distribution project to different regions of the brain. Adapted from Jones (1984).

Spinal Motor Neurons   Motor cells of the ventral horns of the spinal cord, also called α motoneurons, have their cell bodies within the spinal cord and send their axons outside the central nervous system to innervate the muscles. Different types of motor neurons are distinguished by their targets. The α motoneurons innervate skeletal muscles, but smaller motor neurons (the γ motoneurons, forming about 30% of the motor neurons) innervate the spindle organs of the muscles. The α motor neurons are some of the largest neurons in the entire central nervous system and are characterized by a multipolar perikaryon and a very rich cytoplasm that renders them very conspicuous on histological preparations. They have a large number of spiny dendrites that arborize locally within the ventral horn. The α motoneuron axon leaves the central nervous system through the ventral root of the peripheral nerves. Their distribution in the ventral horn is not random and corresponds to a somatotopic representation of the muscle groups of the limbs and axial musculature (Brodal, 1981). Spinal motor neurons use acetylcholine as their neurotransmitter. Large motor neurons are severely affected in lower motor neuron disease, a neurodegenerative disorder characterized by progressive muscular weakness that affects, at first, one or two limbs but involves more and more of the body musculature, which shows signs of wasting as a result of denervation.

Neuromodulatory Neurons

Dopaminergic Neurons of the Substantia Nigra
Dopaminergic neurons are large neurons that reside mostly within the pars compacta of the substantia nigra and in the ventral tegmental area. A distinctive feature of these cells is the presence of a pigment, *neuromelanin*, in compact granules in the cytoplasm. These neurons are medium-sized to large, fusiform, and frequently elongated. They have several large radiating dendrites. The axon emerges from the cell body or from one of the dendrites and projects to large expanses of cerebral cortex and to the basal ganglia. These neurons contain the catecholamine-synthesizing enzyme *tyrosine hydroxylase*, as well as the monoamine dopamine as their neurotransmitter. Some of them contain both calbindin and calretinin. These neurons are affected severely and selectively in Parkinson disease—a movement disorder different from Huntington disease and characterized by resting tremor and rigidity—and their specific loss is the neuropathologic hallmark of this disorder.

NEUROGLIA

The term *neuroglia,* or "nerve glue," was coined in 1859 by Rudolph Virchow, who erroneously conceived of the neuroglia as an inactive "connective tissue" holding neurons together in the central nervous system. The metallic staining techniques developed by Ramón y Cajal and del Rio-Hortega allowed these two great pioneers to distinguish, in addition to the ependyma lining the ventricles and central canal, three types of supporting cells in the CNS: oligodendrocytes, astrocytes, and microglia. In the peripheral nervous system (PNS), the Schwann cell is the major neuroglial component.

Oligodendrocytes and Schwann Cells Synthesize Myelin

Most brain functions depend on rapid communication between circuits of neurons. As shown in depth later, there is a practical limit to how fast an individual bare axon can conduct an action potential. Organisms developed two solutions for enhancing rapid communication between neurons and their effector organs. In invertebrates, the diameters of axons are enlarged. In vertebrates, the myelin sheath (Fig. 1.5) evolved to permit rapid nerve conduction.

Axon enlargement accelerates the rate of conduction of the action potential in proportion to the square root of axonal diameter. Thus small axons conduct at slower rates than larger ones. The largest axon in the invertebrate kingdom is the squid giant axon, which is about the thickness of a mechanical pencil lead. This axon conducts the action potential at speeds of 10–20 m/s. As the axon mediates an escape reflex, firing must be rapid if the animal is to survive. Bare axons and continuous conduction obviously provide sufficient rates of signal propagation for even very large invertebrates, such as the giant squid, and many human axons also remain bare. However, in the human brain with 10 billion neurons, axons cannot be as thick as pencil lead, otherwise human heads would weigh 100 pounds or more.

Thus, along the invertebrate evolutionary line, the use of bare axons imposes a natural, insurmountable limit— a constraint of axonal size—to increasing the processing capacity of the nervous system. Vertebrates, however, get around this problem through evolution of the myelin sheath, which allows 10- to 100-fold increases in conduction of the nerve impulse along axons with fairly minute diameters.

In the central nervous system, myelin sheaths (Fig. 1.6) are elaborated by oligodendrocytes. During brain development, these glial cells send out a few cytoplasmic processes that engage adjacent axons and form myelin around them (Bunge, 1968). Myelin consists of a long sheet of oligodendrocyte plasma membrane, which is spirally wrapped around an axonal segment. At the end of each myelin segment, there is a bare portion of the axon, the node of Ranvier. Myelin segments are thus called "internodes." Physiologically, myelin has insulating properties such that the action potential can "leap" from node to node and therefore does not have to be regenerated continually along the axonal segment that is covered by the myelin membrane sheath. This leaping of the action potential from node to node allows axons with fairly small diameters to conduct extremely rapidly (Ritchie, 1984) and is called "saltatory" conduction.

Because the brain and spinal cord are encased in the bony skull and vertebrae, CNS evolution has promoted compactness among the supporting cells of the CNS. Each oligodendrocyte cell body is responsible for the construction and maintenance of several myelin sheaths (Fig. 1.6), thus reducing the number of glial cells required. In both PNS and CNS myelin, cytoplasm is removed between each turn of the

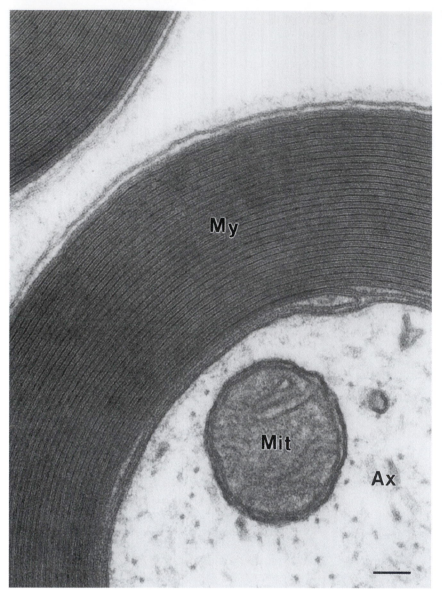

FIGURE 1.5 An electron micrograph of a transverse-section through part of a myelinated axon from the sciatic nerve of a rat. The tightly compacted multilayer myelin sheath (My) surrounds and insulates the axon (Ax). Mit, mitochondria. Scale bar: 75 nm.

myelin, leaving only the thinnest layer of plasma membrane. Due to protein composition differences, CNS lamellae are approximately 30% thinner than in PNS myelin. In addition, there is little or no extracellular space or extracellular matrix between the myelinated axons passing through CNS white matter. Brain volume is thus reserved for further expansion of neuronal populations.

Peripheral nerves pass between moving muscles and around major joints and are routinely exposed to physical trauma. A hard tackle, slipping on an icy sidewalk, or even just occupying the same uncomfortable seating posture for too long can painfully compress peripheral nerves and potentially damage them. Thus, evolutionary pressures shaping the PNS favor

robustness and regeneration rather than conservation of space. Myelin in the PNS is generated by Schwann cells (Fig. 1.7), which are different from oligodendrocytes in several ways. Individual myelinating Schwann cells form a single internode. The biochemical composition of PNS and CNS myelin differs, as discussed in following text. Unlike oligodendrocytes, Schwann cells secrete copious extracellular matrix components and produce a basal lamina "sleeve" that runs the entire length of myelinated axons. Schwann cell and fibroblast-derived collagens prevent normal wear-and-tear compression damage. Schwann cells also respond vigorously to injury, in common with astrocytes but unlike oligodendrocytes. Schwann cell growth factor secretion, debris removal by Schwann cells after injury,

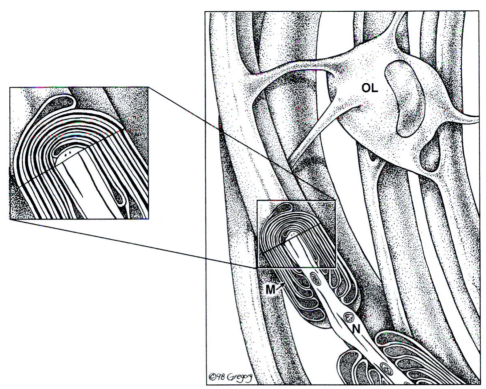

FIGURE 1.6 An oligodendrocyte (OL) in the central nervous system is depicted myelinating several axon segments. A cutaway view of the myelin sheath is shown (M). Note that the internode of myelin terminates in paranodal loops that flank the node of Ranvier (N). (Inset) An enlargement of compact myelin with alternating dark and light electron-dense lines that represent intracellular (major dense lines) and extracellular (intraperiod line) plasma membrane appositions, respectively.

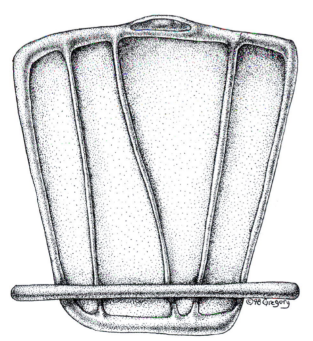

FIGURE 1.7 An "unrolled" Schwann cell in the PNS is illustrated in relation to the single axon segment that it myelinates. The broad stippled region is compact myelin surrounded by cytoplasmic channels that remain open even after compact myelin has formed, allowing an exchange of materials among the myelin sheath, the Schwann cell cytoplasm, and perhaps the axon as well.

and the axonal guidance function of the basal lamina are responsible for the exceptional regenerative capacity of the PNS compared with the CNS.

The major integral membrane protein of peripheral nerve myelin is protein zero (P0), a member of a very large family of proteins termed the *immunoglobulin gene superfamily.* This protein makes up about 80% of the protein complement of PNS myelin. Interactions between the extracellular domains of P0 molecules expressed on one layer of the myelin sheath with those of the apposing layer yield a characteristic regular periodicity that can be seen by thin-section electron microscopy (Fig. 1.5). This zone, called the *intraperiod line,* represents the extracellular apposition of the myelin bilayer as it wraps around itself. On the other side of the bilayer, the cytoplasmic side, the highly charged P0 cytoplasmic domain probably functions to neutralize the negative charges on the polar head groups of the phospholipids that make up the plasma membrane itself, allowing the membranes of the myelin sheath to come into close apposition with one another. In electron microscopy, this cytoplasmic apposition is a bit darker than the intraperiod line and is termed the *major dense line.* In peripheral nerves, although other molecules are present in small quantities in compact myelin and may have important functions, compaction (i.e., the close apposition of membrane surfaces without intervening cytoplasm) is accomplished solely by P0–P0 interactions at both extracellular and intracellular (cytoplasmic) surfaces.

Curiously, P0 is present in the CNS of lower vertebrates such as sharks and bony fish, but in terrestrial vertebrates (reptiles, birds, and mammals), P0 is limited to the PNS. CNS myelin compaction in these higher organisms is subserved by proteolipid protein (PLP) and its alternate splice form, DM-20. These two proteins are generated from the same gene, both span the plasma membrane four times, and they differ only in that PLP has a small, positively charged segment exposed on the cytoplasmic surface. Why did PLP–DM-20 replace P0 in CNS myelin? Manipulation of PLP and P0 content of CNS myelin established an axonotrophic function for PLP in CNS myelin. Removal of PLP from rodent CNS myelin altered the periodicity of compact myelin and produced a late-onset axonal degeneration (Griffiths *et al.*, 1998). Replacing PLP with P0 in rodent CNS myelin stabilized compact myelin but enhanced the axonal degeneration (Yin *et al.*, 2006). These and other observations in primary demyelination and inherited myelin diseases have established axonal degeneration as the major cause of permanent disability in diseases such as multiple sclerosis.

Myelin membranes also contain a number of other proteins such as the myelin basic protein, which is a major CNS myelin component, and PMP-22, a protein that involved in a form of peripheral nerve disease. A large number of naturally occurring gene mutations can affect the proteins specific to the myelin sheath and cause neurological disease. In animals, these mutations have been named according to the phenotype that is produced: the shiverer mouse, the shaking pup, the rumpshaker mouse, the jimpy mouse, the myelin-deficient rat, the quaking mouse, and so forth. Many of these mutations are well characterized and have provided valuable insights into the role of individual proteins in myelin formation and axonal survival.

Astrocytes Play Important Roles in CNS Homeostasis

As the name suggests, astrocytes are star-shaped, process-bearing cells distributed throughout the central nervous system. They constitute from 20 to 50% of the volume of most brain areas. Astrocytes come in many shapes and forms. The two main forms, protoplasmic and fibrous astrocytes, predominate in gray and white matter, respectively (Fig. 1.8). Embryonically, astrocytes

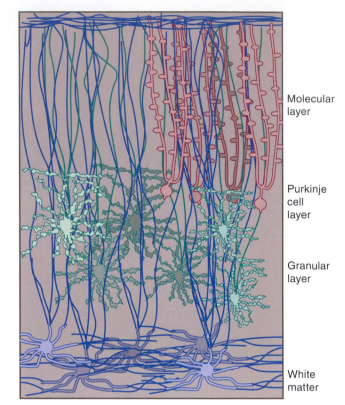

Molecular layer

Purkinje cell layer

Granular layer

White matter

FIGURE 1.8   The arrangement of astrocytes in human cerebellar cortex. Bergmann glial cells are in red, protoplasmic astrocytes are in green, and fibrous astrocytes are in blue.

develop from radial glial cells, which transversely compartmentalize the neural tube. Radial glial cells serve as scaffolding for the migration of neurons and play a critical role in defining the cytoarchitecture of the CNS (Fig. 1.9). As the CNS matures, radial glia retract their processes and serve as progenitors of astrocytes. However, some specialized astrocytes of a radial nature are still found in the adult cerebellum and the retina and are known as *Bergmann glial cells* and *Müller cells*, respectively.

Astrocytes "fence in" neurons and oligodendrocytes. Astrocytes achieve this isolation of the brain parenchyma by extending long processes projecting to the pia mater and the ependyma to form the glia limitans, by covering the surface of capillaries and by making a cuff around the nodes of Ranvier. They also ensheath synapses and dendrites and project processes to cell somas (Fig. 1.10). Astrocytes are connected to each other by gap junctions, forming a syncytium that allows ions and small molecules to diffuse across the brain parenchyma. Astrocytes have in common unique cytological and immunological properties that make them easy to identify, including their star shape, the glial end feet on capillaries, and a

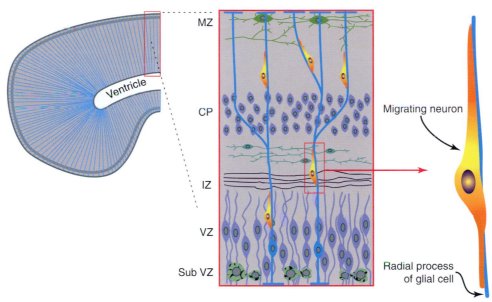

FIGURE 1.9 Radial glia perform support and guidance functions for migrating neurons. In early development, radial glia span the thickness of the expanding brain parenchyma. (Inset) Defined layers of the neural tube from the ventricular to the outer surface: VZ, ventricular zone; IZ, intermediate zone; CP, cortical plate; MZ, marginal zone. The radial process of the glial cell is indicated in blue, and a single attached migrating neuron is depicted at the right.

unique population of large bundles of intermediate filaments. These filaments are composed of an astroglial-specific protein commonly referred to as *glial fibrillary acidic protein* (GFAP). S-100, a calcium-binding protein, and glutamine synthetase are also astrocyte markers. Ultrastructurally, gap junctions (connexins), desmosomes, glycogen granules, and membrane orthogonal arrays are distinct features used by morphologists to identify astrocytic cellular processes in the complex cytoarchitecture of the nervous system.

For a long time, astrocytes were thought to physically form the blood–brain barrier (considered later in this chapter), which prevents the entry of cells and diffusion of molecules into the CNS. In fact, astrocytes are indeed the blood–brain barrier in lower species. However, in higher species, astrocytes are responsible for inducing and maintaining the tight junctions in endothelial cells that effectively form the barrier. Astrocytes also take part in angiogenesis, which may be important in the development and repair of the CNS. However, their role in this important process is still poorly understood.

Astrocytes Have a Wide Range of Functions

There is strong evidence for the role of radial glia and astrocytes in the migration and guidance of neurons in early development. Astrocytes are a major source of extracellular matrix proteins and adhesion molecules in the CNS; examples are nerve cell–nerve cell adhesion molecule (N-CAM), laminin, fibronectin, cytotactin, and the J-1 family members janusin and tenascin. These molecules participate not only in the migration of neurons but also in the formation of neuronal aggregates, so-called nuclei, as well as networks.

Astrocytes produce, *in vivo* and *in vitro*, a very large number of growth factors. These factors act singly or in combination to selectively regulate the morphology, proliferation, differentiation, survival, or all four, of distinct neuronal subpopulations. Most of the growth factors also act in a specific manner on the development and functions of astrocytes and oligodendrocytes. The production of growth factors and cytokines by astrocytes and their responsiveness to these factors is a major mechanism underlying the developmental function and regenerative capacity of the CNS. During neurotransmission, neurotransmitters and ions are released at high concentration in the synaptic cleft. The rapid removal of these substances is important so that they do not interfere with future synaptic activity. The presence of astrocyte processes around synapses positions them well to regulate neurotransmitter uptake and inactivation (Kettenman and Ransom, 1995). These possibilities are consistent with the presence in astrocytes of transport systems for many neurotransmitters. For instance, glutamate reuptake is performed mostly by astrocytes, which convert glutamate into glutamine

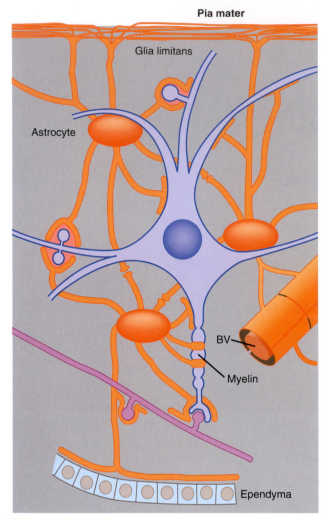

FIGURE 1.10 Astrocytes (in orange) are depicted *in situ* in schematic relationship with other cell types with which they are known to interact. Astrocytes send processes that surround neurons and synapses, blood vessels, and the region of the node of Ranvier and extend to the ependyma, as well as to the pia mater, where they form the glial limitans.

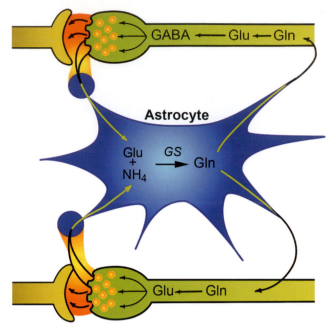

FIGURE 1.11 The glutamate–glutamine cycle is an example of a complex mechanism that involves an active coupling of neurotransmitter metabolism between neurons and astrocytes. The systems of exchange of glutamine, glutamate, GABA, and ammonia between neurons and astrocytes are highly integrated. The postulated detoxification of ammonia and the inactivation of glutamate and GABA by astrocytes are consistent with the exclusive localization of glutamine synthetase in the astroglial compartment.

and then release it into the extracellular space. Glutamine is taken up by neurons, which use it to generate glutamate and γ-aminobutyric acid, potent excitatory and inhibitory neurotransmitters, respectively (Fig. 1.11). Astrocytes contain ion channels for K^+, Na^+, Cl^-, HCO_3, and Ca^{2+}, as well as displaying a wide range of neurotransmitter receptors. K^+ ions released from neurons during neurotransmission are soaked up by astrocytes and moved away from the area through astrocyte gap junctions. This is known as "spatial buffering." Astrocytes play a major role in detoxification of the CNS by sequestering metals and a variety of neuroactive substances of endogenous and xenobiotic origin.

In response to stimuli, intracellular calcium waves are generated in astrocytes. Propagation of the Ca^{2+} wave can be visually observed as it moves across the cell soma and from astrocyte to astrocyte. The generation of Ca^{2+} waves from cell to cell is thought to be mediated by second messengers, diffusing through gap junctions. Because they develop postnatally in rodents, gap junctions may not play an important role in development. In the adult brain, gap junctions are present in all astrocytes. Some gap junctions have also been detected between astrocytes and neurons. Thus, they may participate, along with astroglial neurotransmitter receptors, in the coupling of astrocyte and neuron physiology.

In a variety of CNS disorders—neurotoxicity, viral infections, neurodegenerative disorders, HIV, AIDS, dementia, multiple sclerosis, inflammation, and trauma—astrocytes react by becoming hypertrophic and, in a few cases, hyperplastic. A rapid and huge upregulation of GFAP expression and filament formation is associated with astrogliosis. The formation of reactive astrocytes can spread very far from the site of origin. For instance, a localized trauma can recruit astrocytes from as far as the contralateral side, suggesting the existence of soluble factors in the mediation process. Tumor necrosis factor (TNF) and ciliary neurotrophic factors (CNTF) have been identified as key factors in astrogliosis.

Microglia Are Mediators of Immune Responses in Nervous Tissue

The brain has traditionally been considered an "immunologically privileged site," mainly because the blood–brain barrier normally restricts the access of immune cells from the blood. However, it is now known that immunological reactions do take place in the central nervous system, particularly during cerebral inflammation. Microglial cells have been termed the *tissue macrophages* of the CNS, and they function as the resident representatives of the immune system in the brain. A rapidly expanding literature describes microglia as major players in CNS development and in the pathogenesis of CNS disease.

The first description of microglial cells can be traced to Franz Nissl (1899), who used the term *rod cell* to describe a population of glial cells that reacted to brain pathology. He postulated that rod-cell function was similar to that of leukocytes in other organs. Cajal described microglia as part of his "third element" of the CNS—cells that he considered to be of mesodermal origin and distinct from neurons and astrocytes (Ramón y Cajal, 1913).

Del Rio-Hortega (1932) distinguished this third element into microglia and oligodendrocytes. He used silver impregnation methods to visualize the ramified appearance of microglia in the adult brain, and he concluded that ramified microglia could transform into cells that were migratory, ameboid, and phagocytic. Indeed, a hallmark of microglial cells is their ability to become reactive and to respond to pathological challenges in a variety of ways. A fundamental question raised by del Rio-Hortega's studies was the origin of microglial cells. Some questions about this remain even today.

Microglia Have Diverse Functions in Developing and Mature Nervous Tissue

On the basis of current knowledge, it appears that most ramified microglial cells are derived from bone marrow–derived monocytes, which enter the brain parenchyma during early stages of brain development. These cells help phagocytose degenerating cells that undergo programmed cell death as part of normal development. They retain the ability to divide and have the immunophenotypic properties of monocytes and macrophages. In addition to their role in remodeling the CNS during early development,

microglia secrete cytokines and growth factors that are important in fiber tract development, gliogenesis, and angiogenesis. They are also the major CNS cells involved in presenting antigens to T lymphocytes. After the early stages of development, ameboid microglia transform into the ramified microglia that persist throughout adulthood (Altman, 1994).

Little is known about microglial function in the healthy adult vertebrate CNS. Microglia constitute a formidable percentage (5–20%) of the total cells in the mouse brain. Microglia are found in all regions of the brain, and there are more in gray than in white matter. The neocortex and hippocampus have more microglia than regions like the brainstem or cerebellum. Species variations have also been noted, as human white matter has three times more microglia than rodent white matter.

Microglia usually have small rod-shaped somas from which numerous processes extend in a rather symmetrical fashion. Processes from different microglia rarely overlap or touch, and specialized contacts between microglia and other cells have not been described in the normal brain. Although each microglial cell occupies its own territory, microglia collectively form a network that covers much of the CNS parenchyma. Because of the numerous processes, microglia present extensive surface membrane to the CNS environment. Regional variation in the number and shape of microglia in the adult brain suggests that local environmental cues can affect microglial distribution and morphology. On the basis of these morphological observations, it is likely that microglia play a role in tissue homeostasis. The nature of this homeostasis remains to be elucidated. It is clear, however, that microglia can respond quickly and dramatically to alterations in the CNS microenvironment.

Microglia Become Activated in Pathological States

"Reactive" microglia can be distinguished from resting microglia by two criteria: (1) change in morphology and (2) upregulation of monocyte–macrophage molecules (Fig. 1.12). Although the two phenomena generally occur together, reactive responses of microglia can be diverse and restricted to subpopulations of cells within a microenvironment. Microglia not only respond to pathological conditions involving immune activation but also become activated in neurodegenerative conditions that are not considered immune mediated. This latter response is indicative of the phagocytic role of microglia. Microglia change their morphology and antigen expression in response to almost any form of CNS injury.

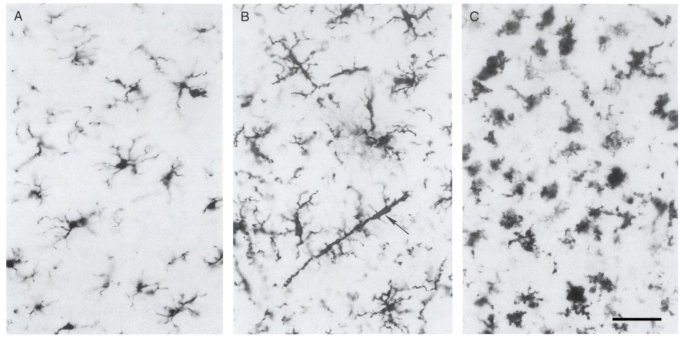

FIGURE 1.12 Activation of microglial cells in a tissue section from human brain. Resting microglia in normal brain (A). Activated microglia in diseased cerebral cortex (B) have thicker processes and larger cell bodies. In regions of frank pathology (C), microglia transform into phagocytic macrophages, which can also develop from circulating monocytes that enter the brain. Arrow in B indicates rod cell. Sections stained with antibody to ferritin. Scale bar = 40 μm.

CEREBRAL VASCULATURE

Blood vessels form an extremely rich network in the central nervous system, particularly in the cerebral cortex and subcortical gray masses, whereas the white matter is less densely vascularized (Fig. 1.13) (Duvernoy *et al.*, 1981). There are distinct regional patterns of microvessel distribution in the brain. These patterns are particularly clear in certain subcortical structures that constitute discrete vascular territories and in the cerebral cortex, where regional and laminar patterns are striking. For example, layer IV of the primary visual cortex possesses an extremely rich capillary network in comparison with other layers and adjacent regions (Fig. 1.13). Interestingly, most of the inputs from the visual thalamus terminate in this particular layer. Capillary densities are higher in regions containing large numbers of neurons and where synaptic density is high. Progressive occlusion of a large arterial trunk, as seen in stroke, induces an ischemic injury that may eventually lead to necrosis of the brain tissue. The size of the resulting infarction is determined in part by the worsening of the blood circulation through the cerebral microvessels. Occlusion of a large arterial trunk results in rapid swelling of the capillary endothelium and surrounding astrocytes, which may reduce the capillary lumen to about one-third of its normal diameter, preventing red blood cell circulation and oxygen delivery to the tissue. The severity of these changes subsequently determines the time course of neuronal necrosis, as well as the possible recovery of the surrounding tissue and the neurological outcome of the patient. In addition, the presence of multiple microinfarcts caused by occlusive lesions of small cerebral arterioles may lead to a progressively dementing illness, referred to as *vascular dementia*, affecting elderly humans.

The Blood–Brain Barrier Maintains the Intracerebral Milieu

Capillaries of the central nervous system form a protective barrier that restricts the exchange of solutes between blood and brain. This distinct function of brain capillaries is called the *blood–brain barrier* (Fig. 1.14) (Bradbury, 1979). Capillaries of the retina have similar properties and are termed the *blood–retina barrier*. It is thought that the blood–brain and blood–retina barriers function to maintain a constant

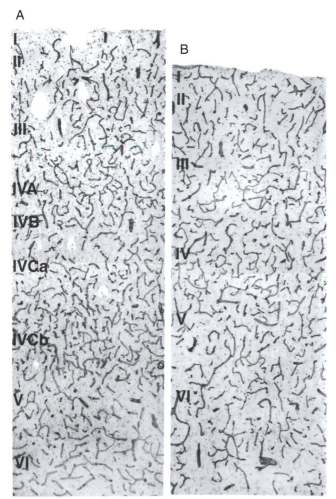

FIGURE 1.13 Microvasculature of the human neocortex. (A) The primary visual cortex (area 17). Note the presence of segments of deep penetrating arteries that have a larger diameter than the microvessels and run from the pial surface to the deep cortical layers, as well as the high density of microvessels in the middle layer (layers IVCa and IVCb). (B) The prefrontal cortex (area 9). Cortical layers are indicated by Roman numerals. Microvessels are stained using an antibody against heparan sulfate proteoglycan core protein, a component of the extracellular matrix.

intracerebral milieu, so that neuronal signaling can occur without interference from substances leaking in from the bloodstream. This function is important because of the nature of intercellular communication in the CNS, which includes chemical signals across intercellular spaces. Without a blood–brain barrier, circulating factors in the blood, such as certain hormones, which can also act as neurotransmitters, would interfere with synaptic communication. When the blood–brain barrier is disrupted, edema fluid accumulates in the brain. Increased permeability of the blood–brain barrier plays a central role in many neuropathological conditions, including multiple sclerosis, AIDS, and childhood lead poisoning, and

may also play a role in Alzheimer disease. The cerebral capillary wall is composed of an endothelial cell surrounded by a very thin (about 30 nm) basement membrane or basal lamina. End feet of perivascular astrocytes are apposed against this continuous basal lamina. Around the capillary lies a virtual perivascular space occupied by another cell type, the pericyte, which surrounds the capillary walls. The endothelial cell forms a thin monolayer around the capillary lumen, and a single endothelial cell can completely surround the lumen of the capillary (Fig. 1.14). A fundamental difference between brain endothelial cells and those of the systemic circulation is the presence in the brain of interendothelial tight junctions, also known as *zonula occludens*. In the systemic circulation, the interendothelial space serves as a diffusion pathway that offers little resistance to most blood solutes entering the surrounding tissues. In contrast, blood–brain barrier tight junctions effectively restrict the intercellular route of solute transfer. The blood–brain barrier interendothelial junctions are not static seals; rather they are a series of active gates that can allow certain small molecules to penetrate. One such molecule is the lithium ion, used in the control of manic depression.

Another characteristic of endothelial cells of the brain is their low transcytotic activity. Brain endothelium, therefore, is by this index not very permeable. It is of interest that certain regions of the brain, such as the area postrema and periventricular organs, lack a blood–brain barrier. In these regions, the perivascular space is in direct contact with the nervous tissue, and endothelial cells are fenestrated and show many pinocytotic vesicles. In these brain regions, neurons are known to secrete hormones and other factors that require rapid and uninhibited access to the systemic circulation.

Because of the high metabolic requirements of the brain, blood–brain barrier endothelial cells must have transport mechanisms for the specific nutrients needed for proper brain function. One such mechanism is the glucose transporter isoform 1 (GLUT1), which is expressed asymmetrically on the surface of blood–brain barrier endothelial cells. In Alzheimer disease, the expression of GLUT1 on brain endothelial cells is reduced. This reduction may be due to a lower metabolic requirement of the brain after extensive neuronal loss. Other specific transport mechanisms on the cerebral endothelium include the large neutral amino acid carrier-mediated system that transports, among other amino acids, L-3,4-dihydroxyphenylalanine (L-dopa), used as a therapeutic agent in Parkinson disease. Also on the surface of blood–brain barrier endothelial cells are transferrin receptors that allow the transport of iron into specific areas of the brain. The amount of iron that is transported into the

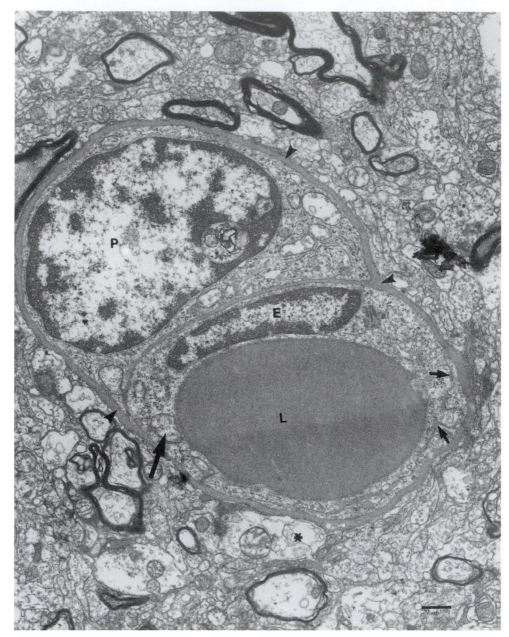

FIGURE 1.14 Human cerebral capillary obtained at biopsy. Blood–brain barrier (BBB) capillaries are characterized by the paucity of trans-cytotic vesicles in endothelial cells (E), a high mitochondrial content (large arrow), and the formation of tight junctions (small arrows) between endothelial cells that restrict the transport of solutes through the interendothelial space. The capillary endothelium is encased within a basement membrane (arrowheads), which also houses pericytes (P). Outside the basement membrane are astrocyte foot processes (asterisk), which may be responsible for the induction of BBB characteristics on the endothelial cells. L, lumen of the capillary. Scale bar = 1 μm. From Claudio *et al.* (1995).

various areas of the brain appears to depend on the concentration of transferrin receptors on the surface of endothelial cells of that region. Thus, the transport of specific nutrients into the brain is regulated during physiological and pathological conditions by blood–brain barrier transport proteins distributed according to the regional and metabolic requirements of brain tissue.

In general, disruption of the blood–brain barrier causes perivascular or vasogenic edema, which is the accumulation of fluids from the blood around the blood vessels of the brain. This is one of the main features of multiple sclerosis. In multiple sclerosis, inflammatory cells, primarily T cells and macrophages, invade the brain by migrating through the blood–brain barrier and attack cerebral elements as if these elements were foreign antigens. It has been observed by many investigators that the degree of edema accumulation causes the neurological symptoms experienced by people suffering from multiple sclerosis.

Studying the regulation of blood–brain barrier permeability is important for several reasons. Therapeutic treatments for neurological disease need to be able to cross the barrier. Attempts to design drug delivery systems that take therapeutic drugs directly into the brain have been made by using chemically engineered carrier molecules that take advantage of receptors such as that for transferrin, which normally transports iron into the brain. Development of an *in vitro* test system of the blood–brain barrier is of importance in the creation of new neurotropic drugs that are targeted to the brain.

References

Altman, J. (1994). Microglia emerge from the fog. *Trends Neurosci.* **17**, 47–49.

Bradbury, M. W. B. (1979). *"The Concept of a Blood-Brain Barrier,"* pp. 381–407. Wiley, Chichester, UK.

Brodal, A. (1981). *"Neurological Anatomy in Relation to Clinical Medicine,"* 3rd ed. Oxford University Press, New York.

Bunge, R. P. (1968). Glial cells and the central myelin sheath. *Physiol. Rev.* **48**, 197–251.

Claudio, L., Raine, C. S., and Brosnan, C. F. (1995). Evidence of persistent blood-brain barrier abnormalities in chronic-progressive multiple sclerosis. *Acta Neuropathol.* **90**, 228–238.

DeFelipe, J., Hendry, S. H. C., and Jones, E. G. (1989). Visualization of chandelier cell axons by parvalbumin immunoreactivity in monkey cerebral cortex. *Proc. Natl. Acad. Sci. USA* **86**, 2093–2097.

del Rio-Hortega, P. (1932). Microglia. *In* "Cytology and Cellular Pathology of the Nervous System" (W. Penfield, ed.), Vol. 2, pp. 481–534. Harper (Hoeber), New York.

Duvernoy, H. M., Delon, S., and Vannson, J. L. (1981). Cortical blood vessels of the human brain. *Brain Res. Bull.* **7**, 519–579.

Freund, T. F., Martin, K. A. C., Smith, A. D., and Somogyi, P. (1983). Glutamate decarboxylase-immunoreactive terminals of Golgi-impregnated axoaxonic cells and of presumed basket cells in synaptic contact with pyramidal neurons of the cat's visual cortex. *J. Comp. Neurol.* **221**, 263–278.

Griffiths, I., Klugmann, M., Anderson, T., Yool, D., Thomson, C., Schwab, M. H., Schneider, A., Zimmermann, F., McCulloch, M., Nadon, N., and Nave, K. A. (1998). Axonal swellings and degeneration in mice lacking the major proteolipid of myelin. *Science.* **280**, 1610–1613.

Jones, E. G. (1984). Laminar distribution of cortical efferent cells. *In* "Cellular Components of the Cerebral Cortex" (A. Peters and E. G. Jones, eds.), **Vol. 1**, pp. 521–553. Plenum, New York.

Jones, E. G. (1975). Varieties and distribution of non-pyramidal cells in the somatic sensory cortex of the squirrel monkey. *J. Comp. Neurol.* **160**, 205–267.

Kettenman, H., and Ransom, B. R., eds. (1995). "Neuroglia." Oxford University Press, Oxford.

Mountcastle, V. B. (1978). An organizing principle for cerebral function: The unit module and the distributed system. *In* "The Mindful Brain: Cortical Organization and the Group-Selective Theory of Higher Brain Function" (V. B. Mountcastle and G. Eddman, eds.), pp. 7–50. MIT Press, Cambridge, MA.

Nissl, F. (1899). Über einige Beziehungen zwischen Nervenzellenerkränkungen und gliösen Erscheinungen bei verschiedenen Psychosen. *Arch. Psychol.* **32**, 1–21.

Peters, A., and Jones, E. G., eds. (1984). "Cellular Components of the Cerebral Cortex," Vol. 1. Plenum, New York.

Peters, A., Palay, S. L., and Webster, H. de F. (1991). "The Fine Structure of the Nervous System: Neurons and Their Supporting Cells," 3rd ed. Oxford University Press, New York.

Ramón y Cajal, S. (1913). Contribucion al conocimiento de la neuroglia del cerebro humano. *Trab. Lab. Invest. Biol.* **11**, 255–315.

Ritchie, J. M. (1984). Physiological basis of conduction in myelinated nerve fibers. *In* "Myelin" (P. Morell, ed.), pp. 117–146. Plenum, New York.

Somogyi, P., and Cowey, A. (1981). Combined Golgi and electron microscopic study on the synapses formed by double bouquet cells in the visual cortex of the cat and monkey. *J. Comp. Neurol.* **195**, 547–566.

Somogyi, P., Kisvárday, Z. F., Martin, K. A. C., and Whitteridge, D. (1983). Synaptic connections of morphologically identified and physiologically characterized basket cells in the striate cortex of cat. *Neuroscience* **10**, 261–294.

Yin, X., Baek, R. C., Kirschner, D. A., Peterson, A., Fujii, Y., Nave, K. A., Macklin, W. B., and Trapp, B. D. (2006). Evolution of a neuroprotective function of central nervous system myelin. *J. Cell Biol.* **172**, 469–478.

Suggested Readings

Brightman, M. W., and Reese, T. S. (1969). Junctions between intimately apposed cell membranes in the vertebrate brain. *J. Cell Biol.* **40**, 648–677.

Broadwell, R. D., and Salcman, M. (1981). Expanding the definition of the BBB to protein. *Proc. Natl. Acad. Sci. USA* **78**, 7820–7824.

Fernandez-Moran, H. (1950). EM observations on the structure of the myelinated nerve sheath. *Exp. Cell Res.* **1**, 143–162.

Gehrmann, J., Matsumoto, Y., and Kreutzberg, G. W. (1995). Microglia: Intrinsic immuneffector cell of the brain. *Brain Res. Rev.* **20**, 269–287.

Kimbelberg, H., and Norenberg, M. D. (1989). Astrocytes. *Sci. Am.* **26**, 66–76.

Kirschner, D. A., Ganser, A. L., and Caspar, D. W. (1984). Diffraction studies of molecular organization and membrane interactions in myelin. *In* "Myelin" (P. Morell, ed.), pp. 51–96. Plenum, New York.

Lum, H., and Malik, A. B. (1994). Regulation of vascular endothelial barrier function. *Am. J. Physiol.* **267**, L223–L241.

Rosenbluth, J. (1980). Central myelin in the mouse mutant shiverer. *J. Comp. Neurol.* **194**, 639–728.

Rosenbluth, J. (1980). Peripheral myelin in the mouse mutant shiverer. *J. Comp. Neurol.* **194**, 729–753.

Subcellular Organization of the Nervous System: Organelles and Their Functions

Scott Brady, David R. Colman, and Peter Brophy

Cells have many features in common, but each cell type also possesses a functional architecture related to its unique physiology. In fact, cells may become so specialized in fulfilling a particular function that virtually all cellular components may be devoted to it. For example, the machinery inside mammalian erythrocytes is completely dedicated to the delivery of oxygen to the tissues and the removal of carbon dioxide. Toward this end, this cell has evolved a specialized plasma membrane, an underlying cytoskeletal matrix that molds the cell into a biconcave disk, and a cytoplasm rich in hemoglobin. Modification of the cell machinery extends even to the discarding of structures such as the nucleus and the protein synthetic apparatus, which are not needed after the red blood cell matures. In many respects, the terminally differentiated, highly specialized cells of the nervous system exhibit comparable commitment—the extensive development of subcellular components reflects the roles that each plays.

The neuron serves as the cellular correlate of information processing and, in aggregate, all neurons act together to integrate responses of the entire organism to the external world. It is therefore not surprising that the specializations found in neurons are more diverse and complex than those found in any other cell type. Single neurons commonly interact in specific ways with hundreds of other cells—other neurons, astrocytes, oligodendrocytes, immune cells, muscle, and glandular cells. This chapter defines the major functional domains of the neuron, describes the subcellular elements that compose the building blocks of these domains, and examines the processes that create and maintain neuronal functional architecture.

AXONS AND DENDRITES: UNIQUE STRUCTURAL COMPONENTS OF NEURONS

Neurons and glial cells are remarkable for their size and complexity, but they do share many features with other eukaryotic cells (Peters *et al.*, 1991) As discussed in Chapter 1, the perikaryon, or cell body, contains a nucleus and its associated protein synthetic machinery. Most neuronal nuclei are large and typically contain a preponderance of euchromatin. This is consistent with the need to create and maintain a large cellular volume. Because protein synthesis must be kept at a high level just to maintain the neuronal extensions, transcription levels in neurons are generally high. In turn, the wide variety of different polypeptide constituents associated with cellular domains in a neuron requires that a large number of different genes be transcribed constantly.

When specific mRNAs have been synthesized and processed, they move from the nucleus into a subcellular region that can be termed the *translational cytoplasm* comprising cytoplasmic ("free") and membrane-associated polysomes, the intermediate compartment of the smooth endoplasmic reticulum, and the Golgi complex. The constituents of translational cytoplasm are thus associated with the synthesis and processing of proteins. Neurons in particular have relatively large amounts of translational cytoplasm to accommodate a high level of protein synthesis. This protein synthetic machinery is arranged in discrete intracellular "granules" termed *Nissl substance* (Fig. 2.1) after the histologist who first discovered these structures in the

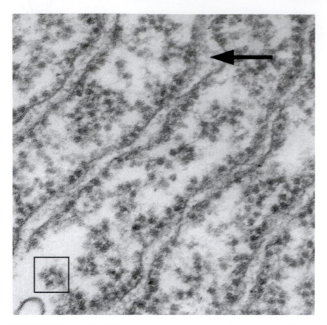

FIGURE 2.1 The "Nissl body" in neurons is an array of cytoplasmic-free polysomal rosettes (boxed) interspersed between rows of rough endoplasmic reticulum (RER) studded with membrane-bound ribosomes. Nascent polypeptide chains emerging from the ribosomal tunnel on the RER are inserted into the lumen (arrow), where they may be processed before transport out of the RER. The relationship between the polypeptide products of these "free" and "bound" polysome populations in the Nissl body, an arrangement that is unique to neurons, is unknown.

nineteenth century. The Nissl substance is actually a combination of stacks of rough endoplasmic reticulum (RER), interposed with rosettes of free polysomes (Fig. 2.1). This arrangement is unique to neurons, and its functional significance is unknown. Most, but by no means all, proteins used throughout the neuron are synthesized in the perikaryon. During or after synthesis and processing, proteins are packaged into membrane-limited organelles, incorporated into cytoskeletal elements, or remain as soluble constituents of the cytoplasm. After proteins have been packaged appropriately, they are transported to their sites of function.

With a few exceptions, vertebrate neurons have two discrete functional domains or compartments, the axonal and the somatodendritic compartments, each of which encompasses a number of microdomains (Fig. 2.2). The axon is perhaps the most familiar functional domain of a neuron and is classically defined as the cellular process by which a neuron makes contact with a target cell to transmit information, providing a conducting structure for transmitting the action potential to a synapse, a specialized subdomain for transmission of a signal from neuron to target cell (neuron, muscle, etc.), most often by release of appropriate neurotransmitters. Consequently, most axons

end in a presynaptic terminal, although a single axon may have many (hundreds or even thousands in some cases) presynaptic specializations known as *en passant synapses* along its length. Characteristics of presynaptic terminals are presented in greater detail later.

The axon is the first neuronal process to differentiate during development. A typical neuron has only a single axon that proceeds some distance from the cell body before branching extensively. Usually the longest process of a neuron, axons come in many sizes. In a human adult, axons range in length from a few micrometers for small interneurons to a meter or more for large motor neurons, and they may be even longer in large animals (such as giraffes, elephants, and whales). In mammals and other vertebrates, the longest axons generally extend approximately half the body length.

Axonal diameters also are quite variable, ranging from 0.1 to 20 μm for large myelinated fibers in vertebrates. Invertebrate axons grow to even larger diameters, with the giant axons of some squid species achieving diameters in the millimeter range. Invertebrate axons reach such large diameters because they lack the myelinating glia that speed conduction of the action potential. As a result, axonal caliber must be large to sustain the high rate of conduction needed for the reflexes that permit escape from predators and capture of prey. Although axonal caliber is closely regulated in both myelinated and nonmyelinated fibers, this parameter is critical for those organisms that are unable to produce myelin.

The region of the neuronal cell body where the axon originates has several specialized features. This domain, called the *axon hillock,* is distinguished most readily by a deficiency of Nissl substance. Therefore, protein synthesis cannot take place to any appreciable degree in this region. Cytoplasm in the vicinity of the axon hillock may have a few polysomes but is dominated by the cytoskeletal and membranous organelles that are being delivered to the axon. Microtubules and neurofilaments begin to align roughly parallel to each other, helping to organize membrane-limited organelles destined for the axon. The hillock is a region where materials either are committed to the axon (cytoskeletal elements, synaptic vesicle precursors, mitochondria, etc.) or are excluded from the axon (RER and free polysomes, dendritic microtubule-associated proteins). The molecular basis for this sorting is not understood. Cytoplasm in the axon hillock does not appear to contain a physical "sizing" barrier (like a filter) because large organelles such as mitochondria enter the axon readily, whereas only a small number of essentially excluded structures such as polysomes are occasionally seen only in the initial segment of the axon and not in the axon proper. An exception to this general rule is during development

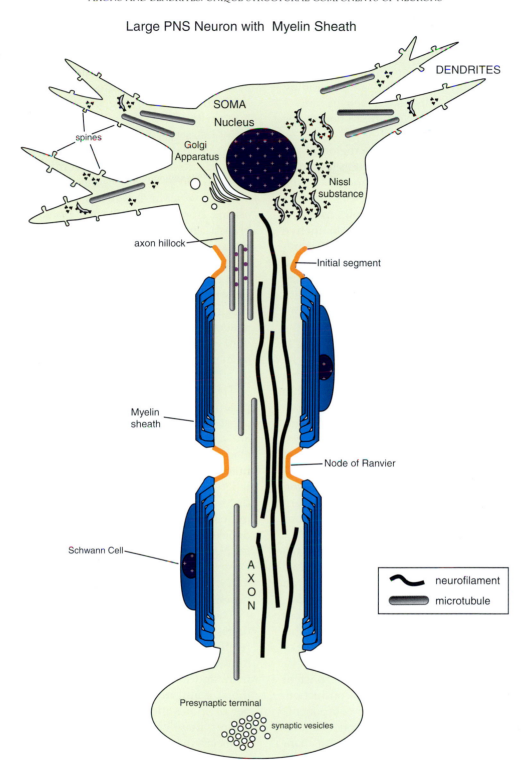

FIGURE 2.2 Basic elements of neuronal subcellular organization. The neuron consists of a soma, or cell body, in which the nucleus, multiple cytoplasm-filled processes termed *dendrites,* and the (usually single) axon are placed. The neuron is highly extended in space; one with a cell body of the size shown here might maintain an axon several miles in length! The unique shape of each neuron is the result of a cooperative interplay between plasma membrane components and cytoskeletal elements. Most large neurons in vertebrates are myelinated by oligodendrocytes in the CNS and by Schwann cells in the PNS. The compact wraps of myelin encasing the axon distal to the initial segment permit rapid conduction of the action potential by a process termed *saltatory conduction* (see Chapter 4).

when local protein synthesis does take place at the axon terminus or growth cone. In the mature neuron, the physiological significance of this barrier must be considerable because axonal structures are found to accumulate in this region in many neuropathologies, including those due to degenerative diseases (such as amyotrophic lateral sclerosis) and to exposure to neurotoxic compounds (such as acrylamide).

The initial segment of the axon is the region of the axon adjacent to the axon hillock. Microtubules generally form characteristic fascicles, or bundles, in the initial segment of the axon. These fascicles are not seen elsewhere. The initial segment and, to some extent, the axon hillock also have a distinctive specialized plasma membrane. Initially, the plasmalemma was thought to have a thick electrondense coating actually attached to the inner surface of the membrane, but this dense undercoating is in reality separated by 5–10 nm from the plasma membrane inner surface and has a complex ultrastructure. Neither the composition nor the function of this undercoating is known. Curiously, the undercoating is present in the same regions of the initial segment as the distinctive fasciculation of microtubules, although the relationship is not understood. The plasma membrane is specialized in the initial segment and axon hillock in that it contains voltage-sensitive ion channels in large numbers, and most action potentials originate in this domain. The molecular composition of the axon initial segment is very similar to that of the node of Ranvier; however, evidence is growing that the mechanisms that govern the assembly of these components in the two locations are distinct.

Ultimately, axonal structure is geared toward the efficient conduction of action potentials at a rate appropriate to the function of that neuron. This can be seen from both the ultrastructure and the composition of axons. Axons are roughly cylindrical in cross section with little or no taper. As discussed later, this diameter is maintained by regulation of the cytoskeleton. Even at branch points, daughter axons are comparable in diameter to the parent axon. This constant caliber helps ensure a consistent rate of conduction. Similarly, the organization of membrane components is regulated to this end. Voltage-gated ion channels are distributed to maximize conduction. Sodium channels are distributed more or less uniformly in small nonmyelinated axons, but are concentrated at high density in the regularly spaced unmyelinated gaps, known as *nodes of Ranvier*. An axon so organized will conduct an action potential or train of spikes long distances with high fidelity at a defined speed. These characteristics are essential for maintaining the precise timing and coordination seen in neuronal circuits.

Nodes of Ranvier in myelinated fibers are flanked by paranodal axoglial junctions composed of the axolemmal proteins Caspr/Paranodin and Contactin and the glial isoform of neurofascin, Nfasc155. There has been considerable debate about the role of axoglial junctions in assembling the node of Ranvier, but, at least in the PNS, the nodal isoform of Neurofascin, Nfasc186, seems to be the crucial molecule that allows NrCAM, beta-IV spectrin, ankyrinG, and sodium channels to form a nodal complex. (Sherman and Brophy, 2005).

Most vertebrate neurons have multiple dendrites arising from their perikarya. Unlike axons, dendrites branch continuously and taper extensively with a reduction in caliber in daughter processes at each branching. In addition, the surface of dendrites is covered with small protrusions, or spines, which are postsynaptic specializations. Although the surface area of a dendritic arbor may be quite extensive, dendrites in general remain in the relative vicinity of the perikaryon. A dendritic arbor may be contacted by the axons of many different and distant neurons or innervated by a single axon making multiple synaptic contacts.

The base of a dendrite is continuous with the cytoplasm of the cell body. In contrast to the axon, Nissl substance extends into dendrites, and certain proteins are synthesized predominantly in dendrites. There is evidence for the selective placement of some mRNAs in dendrites as well (Steward, 1995). For example, whereas RER and polysomes extend well into the dendrites, the mRNAs that are transported and translated in dendrites are a subset of the total neuronal mRNA, deficient in some mRNA species (such as neurofilament mRNAs) and enriched in mRNAs with dendritic functions (such as microtubule-associated protein, MAP2, mRNAs). Also, certain proteins appear to be targeted, postsynthesis, to the dendritic compartment as well.

The shapes and complexity of dendritic arborizations may be remarkably plastic. Dendrites appear relatively late in development and initially have only limited numbers of branches and spines. As development and maturation of the nervous system proceed, the size and number of branches increase. The number of spines increases dramatically, and their distribution may change. This remodeling of synaptic connectivity may continue into adulthood, and environmental effects can alter this pattern significantly. Eventually, in the aging brain, there is a reduction in complexity and size of dendritic arbors, with fewer spines and thinner dendritic shafts. These changes correlate with changes in neuronal function during development and aging.

As defined by classical physiology, axons are structural correlates for neuronal output, and dendrites constitute the domain for receiving information. A neuron without an axon or one without dendrites might therefore seem paradoxical, but such neurons do exist. Certain amacrine and horizontal cells in the vertebrate retina have no identifiable axons, although they do have dendritic processes that are morphologically distinct from axons. Such processes may have both pre- and postsynaptic specializations or may have gap junctions that act as direct electrical connections between two cells. Similarly, the pseudounipolar sensory neurons of dorsal root ganglia (DRG) have no dendrites. In their mature form, these DRG sensory neurons give rise to a single axon that extends a few hundred micrometers before branching. One long branch extends to the periphery, where it may form a sensory nerve ending in muscle spindles or skin. Large DRG peripheral branches are myelinated and have the morphological characteristics of an axon, but they contain neither pre- nor postsynaptic specializations.

The other branch extends into the central nervous system, where it forms synaptic contacts. In DRG neurons, the action potential is generated at distal sensory nerve endings and then transmitted along the peripheral branch to the central branch and the appropriate central nervous system (CNS) targets, bypassing the cell body. The functional and morphological hallmarks of axons and dendrites are listed in Table 2.1.

Summary

Neurons are polarized cells that are specialized for membrane and protein synthesis, as well as for conduction of the nerve impulse. In general, neurons have a cell body, a dendritic arborization that is usually located near the cell body, and an extended axon that may branch considerably before terminating to form synapses with other neurons.

TABLE 2.1   Functional and Morphological Hallmarks of Axons and Dendrites[a]

Axons	Dendrites
With rare exceptions, each neuron has a single axon	Most neurons have multiple dendrites arising from their cell bodies
Axons appear first during neuronal differentiation	Dendrites begin to differentiate only after the axon has formed
Axon initial segments are distinguished by a specialized plasma membrane containing a high density of ion channels and distinctive cytoskeletal organization	Dendrites are continuous with the perikaryal cytoplasm, and the transition point cannot be distinguished readily
Axons typically are cylindrical in form with a round or elliptical cross section	Dendrites usually have a significant taper and small spinous processes that give them an irregular cross section
Large axons are myelinated in vertebrates, and the thickness of the myelin sheath is proportional to the axonal caliber	Dendrites are not myelinated, although a few wraps of myelin may occur rarely
Axon caliber is a function of neurofilament and microtubule numbers with neurofilaments predominating in large axons	The dendritic cytoskeleton may appear less organized, and microtubules dominate even in large dendrites
Microtubules in axons have a uniform polarity with plus ends distal from the cell body	Microtubules in proximal dendrites have mixed polarity, with both plus and minus ends oriented distal to the cell body
Axonal microtubules are enriched in tau protein with a characteristic phosphorylation pattern	Dendritic microtubules may contain some tau protein, but MAP2 is not present in axonal compartments and is highly enriched in dendrites
Ribosomes are excluded from mature axons, although a few may be detectable in initial segments	Both rough endoplasmic reticulum and cytoplasmic polysomes are present in dendrites, with specific mRNAs being enriched in dendrites
Axonal branches tend to be distal from the cell body	Dendrites begin to branch extensively near the perikaryon and form extensive arbors in the vicinity of the perikaryon
Axonal branches form obtuse angles and have diameters similar to the parent stem	Dendritic branches form acute angles and are smaller than the parent stem
Most axons have presynaptic specializations that may be *en passant* or at the ends of axonal branches	Dendrites are rich in postsynaptic specializations, particularly on the spines that project from the dendritic shaft
Action potentials are usually generated at the axon hillock and conducted away from the cell body	Dendrites may generate action potentials, but more commonly they modulate the electrical state of perikaryon and initial segment
Traditionally, axons are specialized for conduction and synaptic transmission, i.e., neuronal output	Dendritic architecture is most suitable for integrating synaptic responses from a variety of inputs, i.e., neuronal input

[a]Neurons typically have two classes of cytoplasmic extensions that may be distinguished using electrophysiological, morphological, and biochemical criteria. Although some neuronal processes may lack one or more of these features, enough parameters can generally be defined to allow unambiguous identification.

PROTEIN SYNTHESIS IN NERVOUS TISSUE

Both neurons and glial cells have strikingly extended morphologies. Protein and lipid components are synthesized and assembled into the membranes of these cell extensions through pathways of membrane biogenesis that have been elucidated primarily in other cell types. However, some adaptations of these general mechanisms have been necessary, due to the specific requirements of cells in the nervous system. Neurons, for example, have devised mechanisms for ensuring that the specific components of the axonal and dendritic plasma membranes are selectively delivered (targeted) to each plasma membrane subdomain.

The distribution to specific loci of organelles, receptors, and ion channels is critical to normal neuronal function. In turn, these loci must be "matched" appropriately to the local microenvironment and specific cell–cell interactions. Similarly, in myelinating glial cells during the narrow developmental window when the myelin sheath is being formed, these cells synthesize sheets of insulating plasma membrane at an unbelievably high rate. To understand how the plasma membrane of neurons and glia might be modeled to fit individual functional requirements, it is necessary to review the progress that has been made so far in our understanding of how membrane components and organelles are generated in eukaryotic cells.

There are two major categories of membrane proteins: integral and peripheral. Integral membrane proteins, which include the receptors for neurotransmitters (e.g., the acetylcholine receptor subunits) and polypeptide growth factors (e.g., the dimeric insulin receptor), have segments that are either embedded in the lipid bilayer or are bound covalently to molecules that insert into the membrane, such as those proteins linked to glycosyl phosphatidylinositol at their C termini (e.g., Thy–1). A protein with a single membrane-embedded segment and an N terminus exposed at the extracellular surface is said to be of "type I," whereas "type II" proteins retain their N termini on the cytoplasmic side of the plasma membrane. Peripheral membrane proteins are localized on the cytoplasmic surface of the membrane and do not traverse any membrane during their biogenesis. They interact with membranes either by means of their associations with membrane lipids or the cytoplasmic tails of integral proteins, or by means of their affinity for other peripheral proteins (e.g., platelet-derived growth factor receptor-Grb2-Sos-Ras complex). In some cases, they may bind electrostatically to the polar head groups of the lipid bilayer (e.g., myelin basic protein).

Integral Membrane and Secretory Polypeptides Are Synthesized *de Novo* in the Rough Endoplasmic Reticulum

The subcellular destinations of integral and peripheral membrane proteins are determined by their sites of synthesis. In the secretory pathway, integral membrane proteins and secretory proteins are synthesized in the rough endoplasmic reticulum, whereas the mRNAs encoding peripheral proteins are translated on cytoplasmic "free" polysomes, which are not membrane associated but which may interact with cytoskeletal structures.

The pathway by which secretory proteins are synthesized and exported was first postulated through the elegant ultrastructural studies on the pancreas by George Palade and colleagues (Palade, 1975). Pancreatic acinar cells were an excellent choice for this work because they are extremely active in secretion, as revealed by the abundance of their RER network, a property they share with neurons. Nissl deduced, in the nineteenth century, that pancreatic cells and neurons would be found to have common secretory properties because of similarities in the distribution of the Nissl substance (Fig. 2.3).

Pulse–chase radioautography has revealed that in eukaryotic cells newly synthesized secretory proteins move from the RER to the Golgi apparatus, where the proteins are packaged into secretory granules and transported to the plasma membrane across which they are released by exocytosis. Pulse–chase studies in neurons reveal a similar sequence of events for proteins transported into the axon. Unraveling of the detailed molecular mechanisms of the pathway began with the successful reconstitution of secretory protein biosynthesis *in vitro* and the direct demonstration that, very early during synthesis, secretory proteins are translocated into the lumen of RER vesicles, prepared by cell fractionation, termed *microsomes*. A key observation here was that the fate of the protein was sealed as a result of encapsulation in the lumen of the RER at the site of synthesis. This cotranslational insertion model provided a logical framework for understanding the synthesis of integral membrane proteins with a transmembrane orientation.

The process by which integral membrane proteins are synthesized closely follows the secretory pathway, except that integral proteins are of course not released from the cell, but instead remain bound to cellular membranes. Synthesis of integral proteins begins with synthesis of the nascent chain on a polysome that is not yet bound to the RER membrane (Fig. 2.4). Emergence of the N terminus of the nascent protein from the protein-synthesizing machinery allows a

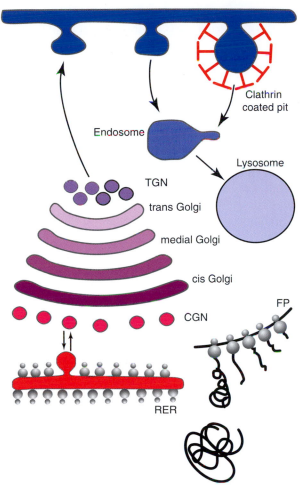

FIGURE 2.3 The secretory pathway. Transport and sorting of proteins in the secretory pathway occur as they pass through the Golgi before reaching the plasma membrane. Sorting occurs in the *cis*-Golgi network (CGN), also known as the *intermediate compartment,* and in the *trans*-Golgi network (TGN). Proteins exit from the Golgi at the TGN. The default pathway is the direct route to the plasma membrane. Proteins bound for regulated secretion or transport to endosomes are diverted from the default path by means of specific signals. In endocytosis, one population of vesicles is surrounded by a clathrin cage and destined for late endosomes.

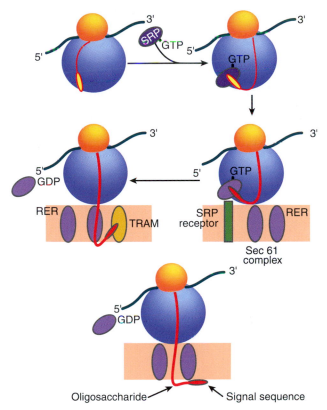

FIGURE 2.4 Translocation of proteins across the rough endoplasmic reticulum (RER). Integral membrane and secretory protein synthesis begins with partial synthesis on a free polysome not yet bound to the RER. The N terminus of the nascent protein emerges and allows a ribonucleoprotein, signal recognition particle (SRP), to bind to the hydrophobic signal sequence and prevent further translation. Translation arrest is relieved once the SRP docks with its receptor at the RER and dissociates from the signal sequence in a GTP-dependent process. Once protein synthesis resumes, translocation occurs through an aqueous pore termed the *translocon,* which includes the translocating chain associating membrane protein (TRAM). The signal sequence is removed by a signal peptidase in the RER lumen.

ribonucleoprotein, a signal recognition particle (SRP), to bind to an emerging hydrophobic signal sequence and prevent further translation (Walter and Johnson, 1994). Translation arrest is relieved when SRP docks with its cognate receptor in the RER and dissociates from the signal sequence in a process that requires GTP. Synthesis of transmembrane proteins on RER is an extremely energy-efficient process. The passage of a fully formed and folded protein through a membrane is thermodynamically formidably expensive; it is infinitely "cheaper" for cells to thread amino acids, in tandem, through a membrane during initial protein synthesis. Protein synthesis then resumes, and the emerging polypeptide chain is translocated into the

RER membrane through a conceptualized "aqueous pore" termed the *translocon.*

A few polypeptides deviate from the common pathway for secretion. For example, certain peptide growth factors, such as basic fibroblast growth factor and ciliary neurotrophic factor, are synthesized without signal peptide sequences but are potent biological modulators of cell survival and differentiation. These growth factors appear to be released under certain conditions, although the mechanisms for such release are still controversial. One possibility is that release of these factors may be associated primarily with cellular injury.

Two cotranslational modifications are commonly associated with the emergence of the polypeptide on the luminal face of the RER. First, an N-terminal hydrophobic signal sequence that is used for insertion into the RER is usually removed by a signal peptidase.

Second, oligosaccharides rich in mannose sugars are transferred from a lipid carrier, dolichol phosphate, to the side chains of asparagine residues (Kornfeld and Kornfeld, 1985). The asparagines must be in the sequence N X T (or S), and they are linked to mannose sugars by two molecules of *N*-acetylglucosamine. The significance of glycosylation is not well understood, and furthermore, it is not a universal feature of integral membrane proteins: some proteins, such as the proteolipid proteins of CNS myelin, neither lose their signal sequence nor become glycosylated.

In general, however, for the vast majority of polypeptides destined for release from the cell (secretory polypeptides), an N-terminal "signal sequence" first mediates the passage of the protein into the RER and is cleaved immediately from the polypeptide by a signal peptidase residing on the luminal side of the RER. For proteins destined to remain as permanent residents of cellular membranes (and these form a particularly important and diverse category of plasma membrane proteins in neurons and myelinating glial cells), however, many variations on this basic theme have been found. Simply stated:

1. Signal sequences for membrane insertions need not be only N-terminal; those that lie within a polypeptide sequence are not cleaved.
2. A second type of signal, a "halt" or "stop" transfer signal, functions to arrest translocation through the membrane bilayer. The halt transfer signal is also hydrophobic and usually flanked by positive charges. This arrangement effectively stabilizes a polypeptide segment in the RER membrane bilayer.
3. The sequential display in tandem of insertion and halt transfer signals in a polypeptide as it is being synthesized ultimately determines its disposition with respect to the phospholipid bilayer, and thus its final topology in its target membrane. By synthesizing transmembrane polypeptides in this way, virtually any topology may be generated.

Newly Synthesized Polypeptides Exit from the RER and Are Moved through the Golgi Apparatus

When the newly synthesized protein has established its correct transmembrane orientation in the RER, it is incorporated into vesicles and must pass through the Golgi complex before reaching the plasma membrane (Fig. 2.3). For membrane proteins, the Golgi serves two major functions: (1) it sorts and targets proteins, and (2) it performs further post-translational modifications, particularly on the oligosaccharide chains that were added in the RER. Sorting takes place in the *cis*-Golgi network (CGN) and in the *trans*-Golgi network (TGN), whereas sculpting of oligosaccharides is primarily the responsibility of the *cis*-, *medial*-, and *trans*-Golgi stacks. The TGN is a tubulovesicular network wherein proteins are targeted to the plasma membrane or to organelles.

The CGN serves an important sorting function for proteins entering the Golgi from the RER. Because most proteins that move from the RER through the secretory pathway do so by default, any resident endoplasmic reticulum proteins must be restrained from exiting or returned promptly to the RER from the CGN should they escape. Although no retention signal has been demonstrated for the endoplasmic reticulum, two retrieval signals have been identified: a Lys-Asp-Glu-Leu or KDEL sequence in type I proteins and the Arg-Arg or RR motif in the first five amino acids of proteins with a type II orientation in the membrane. The KDEL tetrapeptide binds to a receptor called *Erd 2* in the CGN, and the receptor–ligand complex is returned to the RER. There may also be a receptor for the N arginine dipeptide; alternatively, this sequence may interact with other components of the retrograde transport machinery, such as microtubules.

Movement of proteins between Golgi stacks proceeds by means of vesicular budding and fusion (Rothman and Wieland, 1996). The essential mechanisms for budding and fusion have been shown to require coat proteins (COPs) in a manner that is analogous to the role of clathrin in endocytosis. Currently, two main types of COP complex, COPI and COPII, have been distinguished. Although both have been shown to coat vesicles that bud from the endoplasmic reticulum, they may have different roles in membrane trafficking. Coat proteins provide the external framework into which a region of a flattened Golgi cisternae can bud and vesiculate. A complex of these COPs forms the coatamer (coat protomer) together with a p200 protein, AP-1 adaptins, and a family of GTP-binding proteins called *ADP-ribosylation factors* (ARFs). Immunolocalization of one of the coatamer proteins, β-COP, predominantly to the CGN and *cis*-Golgi, indicates that these proteins may also take part in vesicle transport into the Golgi (Fig. 2.5). The function of ARF is to drive the assembly of the coatamer and therefore vesicle budding in a GTP-dependent fashion. Dissociation of the coat is triggered when hydrolysis of the GTP bound to ARF is stimulated by a GTPase-activating protein (GAP) in the Golgi membrane. The cycle of coat assembly and disassembly can continue when the replacement of GDP on ARF by GTP is catalyzed by a guaninenucleotide exchange factor (GEF).

Fusion of vesicles with their target membrane in the Golgi apparatus is believed to be regulated by a series

A

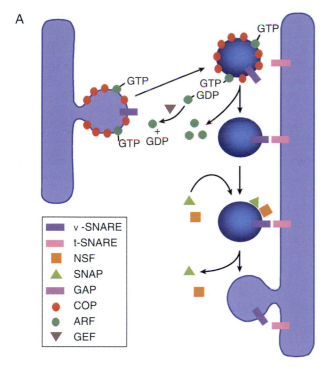

■	v -SNARE
■	t-SNARE
■	NSF
▲	SNAP
■	GAP
●	COP
●	ARF
▼	GEF

B

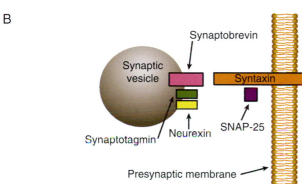

FIGURE 2.5 **A.** General mechanisms of vesicle targeting and docking in the ER and Golgi. Assembly of coat proteins (COPs) around budding vesicles is driven by ADP-ribosylation factors (ARFs) in a GTP-dependent fashion. Dissociation of the coat is triggered by hydrolysis of GTP bound to ARF stimulated by a GTPase-activating protein (GAP) in the Golgi membrane. The cycle of coat assembly and disassembly can continue when replacement of GDP on ARF by GTP is catalyzed by a guanine nucleotide exchange factor (GEF). Fusion of vesicles with target membrane in the Golgi is regulated by a series of proteins, N-ethylmaleimide-sensitive factor (NSF), soluble NSF attachment proteins (SNAPs), and SNAP receptors (SNAREs), which together assist vesicle docking with target membrane. SNAREs on vesicles (v-SNAREs) are believed to associate with corresponding t-SNAREs on target membrane. **B.** Mechanisms of vesicle targeting and docking in the synaptic terminal. The synaptic counterpart of v-SNARE is synaptobrevin (also known as VAMP), and syntaxin corresponds to t-SNARE. SNAP-25 is an accessory protein that binds to syntaxin. Synaptotagmin is believed to be the Ca^{2+}-sensitive regulatory protein in the complex that binds to syntaxin. Neurexins appear to have a role in conferring Ca^{2+} sensitivity to these interactions.

of proteins, N-ethylmaleimide-sensitive factor (NSF), soluble NSF attachment proteins (SNAPs), and SNAP receptors (SNAREs), which together assist the vesicle in docking with its target membrane. The emerging view is that complementary SNAREs on membranes destined to fuse, e.g., synaptic vesicles and the presynaptic membrane, are fundamentally responsible for driving membrane fusion. In addition, Rabs, a family of membrane-bound GTPases, act in concert with their own GAPs, GEFs, and a cytosolic protein that dissociates Rab–GDP from membranes after fusion, called *guanine-nucleotide dissociation inhibitor*. Rabs are believed to regulate the action of SNAREs, the proteins directly engaged in membrane–membrane contact prior to fusion. The tight control necessary for this process and the importance of ensuring that vesicle fusion takes place only at the appropriate target membrane may explain why eukaryotic cells contain so many Rabs, some of which are known to specifically take part in the internalization of endocytic vesicles at the plasma membrane (Fig. 2.3).

Exocytosis of the neurotransmitter at the synapse must occur in an even more finely regulated manner than endocytosis. The proteins first identified in vesicular fusion events in the secretory pathway (namely NSF, SNAPs, and SNAREs or closely related homologues) appear to play a part in the fusion of synaptic vesicles with the active zones of the presynaptic neuronal membrane (Fig. 2.5) (Jahn and Scheller, 2006).

Originally a distinction was made between so-called v-SNAREs and t-SNAREs, reflecting their different locations in the donor and acceptor compartments. An example of specificity is the fact that the synaptic counterpart of v-SNARE is synaptobrevin, also known as *vesicle-associated membrane protein* (VAMP), and syntaxin corresponds to t-SNARE. VAMP does not facilitate fusion with endocytotic vesicle compartments. SNAP-25 is an accessory protein that binds to syntaxin. In the constitutive pathway, such as between the RER and Golgi apparatus, assembly of the complex at the target membrane promotes fusion. However, at the presynaptic membrane, Ca^{2+} influx is required to stimulate membrane fusion. Synaptotagmin is believed to be the Ca^{2+}-sensitive regulatory protein in the complex that binds syntaxin. Neurexins appear to have a role in regulation as well, because, in addition to interacting with synaptotagmin, they are the targets of black widow spider venom (α)-latrotoxin, which deregulates the Ca^{2+}-dependent exocytosis of the neurotransmitter. However, a superficially disturbing lack of specificity in the ability of other membrane-bound SNAREs to complex indicates that much remains to be learned about the regulation of SNARE-mediated membrane fusion.

When comparing secretion in slow-releasing cells, such as the pancreatic (β)-cell, and neurotransmitter release at the neuromuscular junction, two differences stand out. First, the speed of neurotransmitter release is much greater both in release from a single vesicle and in total release in response to a specific signal. Releasing the contents of a single synaptic vesicle at a mouse neuromuscular junction takes from 1 to 2 ms, and the response to an action potential involving the release of many synaptic vesicles is over in approximately 5 ms. In contrast, releasing the insulin in a single secretory granule by a pancreatic (β)-cell takes from 1 to 5 s, and the full release response may take from 1 to 5 min. A 103- to 105-fold difference in rate is an extraordinary range, making neurotransmitter release one of the fastest biological events routinely encountered, but this speed is critical for a properly functioning nervous system.

A second major difference between slow secretion and fast secretion is seen in the recycling of vesicles. In the pancreas, secretory vesicles carrying insulin are used only once, and so new secretory vesicles must be assembled *de novo* and released from the TGN to meet future requirements. In the neuron, the problem is that the synapse may be at a distance of 1 m or more from the protein synthetic machinery of the perikaryon, and so newly assembled vesicles even traveling at rapid axonal transport rates (see later) may take more than a day to arrive. Now, the number of synaptic vesicles released in 15 min of constant stimulation at a single frog neuromuscular junction has been calculated to be on the order of 105 vesicles, but a single terminal may have only a few hundred vesicles at any one time. These measurements would make no sense if synaptic vesicles had to be replaced constantly through new synthesis in the perikaryon, as is the case with insulin-carrying vesicles. The reason that these numbers are possible is that synaptic vesicles are taken up locally by endocytosis, refilled with neurotransmitter, and reutilized at a rate fast enough to keep up with normal physiological stimulation levels. This takes place within the presynaptic terminal, and evidence shows that these recycled synaptic vesicles are used preferentially. Such recycling does not require protein synthesis because the classical neurotransmitters are small molecules, such as acetylcholine, or amino acids, such as glutamate, that can be synthesized or obtained locally. Significantly, neurons have fast and slow secretory pathways operating in parallel in the presynaptic terminal (Sudhof, 2004). Synapses that release classical neurotransmitters (acetylcholine, glutamate, etc.) with these fast kinetics also contain dense core granules containing neuropeptides (calcitonin gene-related peptide, substance P, etc.) that are comparable to the secretory granules of the pancreatic (β)-cell. These are used only once because neuropeptides are produced from large polypeptide precursors that must be made by protein synthesis in the cell body. The release of neuropeptides is relatively slow; as is the case in endocrine release, neuropeptides serve primarily as modulators of synaptic function. The small clear synaptic vesicles containing the classic neurotransmitters can in fact be depleted pharmacologically from the presynaptic terminal, while the dense core granules remain. These observations indicate that even though fast and slow secretory mechanisms have many similarities and may even have common components, in neurons they can operate independent of one another.

Proteins Exit the Golgi Complex at the *trans*-Golgi Network

Most of the N-linked oligosaccharide chains acquired at the RER are remodeled in the Golgi cisternae, and while the proteins are in transit, another type of glycosyl linkage to serine or threonine residues through N-acetylgalactosamine can also be made. Modification of existing sugar chains by a series of glycosidases and the addition of further sugars by glycosyl transferases occur from the *cis* to the *trans* stacks. Some of these enzymes have been localized to particular cisternae. For example, the enzymes (β)-1,4-galactosyltransferase and (α)-2,6-sialyltransferase are concentrated in the *trans*-Golgi. How they are retained there is a matter of some debate. One idea is that these proteins are anchored by oligomerization. Another view is that the progressively rising concentration of cholesterol in membranes more distal to the ER in the secretory pathway increases membrane thickness, which in turn anchors certain proteins and causes an arrest in their flow along the default route.

The default or constitutive pathway seems to be the direct route to the plasma membrane taken by vesicles that bud from the TGN (Fig. 2.3). This is how, in general, integral plasma membrane proteins reach the cell surface. Proteins bound for regulated secretion or for transport to endosomes and from there to lysosomes are diverted from the default path by means of specific signals. It has been assumed that the sorting of proteins for their eventual destination takes place at the TGN itself. However, recent analyses of the three-dimensional structure of the TGN have provoked a revision of this view. These studies have shown that the TGN is tubular, with two major types of vesicles that bud from distinct populations of tubules. The implication is that sorting may already have occurred in the *trans*-Golgi prior to the protein's arrival at the TGN. One population of vesicles consists of those surrounded by the familiar clathrin cage, which are destined for late endosomes. The other population appears to be coated in a lacelike structure, which may prove to be made from the elusive coat protein required for vesicular transport to the plasma membrane. The *β*-COP protein and related coatomer proteins active in more proximal regions of the secretory pathway are absent from the TGN.

Endocytosis and Membrane Cycling Occurs in the *trans*-Golgi Network

Two types of membrane invagination occur at the surface of mammalian cells and are clearly distinguishable by electron microscopy. The first type is a caveola, which has a threadlike structure on its surface made of the protein caveolin. Caveolae mediate the uptake of small molecules and may also have a role in concentrating proteins linked to the plasma membrane by the glycosylphosphatidylinositol anchor. Demonstration of the targeting of protein tyrosine kinases to caveolae by the tripeptide signal MGC (Met-Gly-Cys) also suggests that caveolae may function in signal transduction cascades.

The other type of endocytic vesicle at the cell surface is that coated with the distinctive meshwork of clathrin triskelions. The triskelion comprises three copies of a clathrin heavy chain and three copies of a clathrin light chain (Maxfield and McGraw, 2004). The ease with which these triskelions can assemble into a cage structure demonstrates how they promote the budding of a vesicle from a membrane invagination. Clathrin binds selectively to regions of the cytoplasmic surface of membranes that are selected by adaptins. The AP-2 complex, which is primarily active at the plasma membrane, consists of 100-kDa α and β subunits and two subunits of 50 and 17 kDa each. AP-1 complexes localize to the TGN and have γ and subunits of 100 kDa together with smaller polypeptides of 46 and 19 kDa. Adaptins bind to the cytoplasmic tails of membrane proteins, thus recruiting clathrin for budding at these sites.

A further component of the endocytic complex at the plasma membrane is the GTPase dynamin, which seems to be required for the normal budding of coated vesicles during endocytosis. Dynamins are a family of 100-kDa GTPases found in both neuronal and nonneuronal cells that may interact with the AP-2 component of a clathrin-coated pit (Murthy and De Camilli, 2003). Oligomers of dynamin form a ring at the neck of a budding clathrin-coated vesicle,

and GTP hydrolysis appears to be necessary for the coated vesicle to pinch off from the plasma membrane. The existence of a specific neuronal form of dynamin (dynamin I) may be a manifestation of the unusually rapid rate of synaptic vesicle recycling.

The primary function of clathrin-coated vesicles at the plasma membrane is to deliver membrane proteins, together with any ligands bound to them, to the early endosomal apparatus.

Regulation of membrane cycling in the endosomal compartment is likely to include the Rab family of small GTP-binding proteins. Indeed, each stage of the endocytic pathway may have its own Rab protein to ensure efficient targeting of the vesicle to the appropriate membrane. Rab6 is believed to have a role in transport from the TGN to endosomes, whereas Rab9 may regulate vesicular flow in the reverse direction. In neurons, Rab5a has a role in regulating the fusion of endocytic vesicles and early endosomes and appears to function in endocytosis from both somatodendritic domains and the axon. The association of the protein with synaptic vesicles in nerve terminals, attached presumably by means of its isoprenoid tail, also suggests that early endosomal compartments may have a role in the packaging and recycling of synaptic vesicles.

The Lysosome Is the Target Organelle in Several Inherited Diseases That Affect the Nervous System

Lysosomes were first isolated and characterized as a distinct organelle fraction bounded by a single membrane and separable from mitochondria by differential and sucrose gradient centrifugation. Because of their high content of acid hydrolases, the classic view is that lysosomes are organelles of terminal degradation. Indeed, the latency of hydrolase activity before membrane permeabilization by agents such as nonionic detergents has been used biochemically as a measure of the purity and intactness of lysosomal preparations. However, in addition to their well-established function in lipid and protein breakdown, tubulovesicular lysosomes may overlap in sorting functions with early endosomes, particularly during antigen processing in macrophages.

Inherited deficiencies in lipid metabolism in the lysosome often have particularly devastating consequences on the nervous system because of the abundance of the lipid-rich membrane myelin. Metachromatic leukodystrophy is an autosomal recessive disease caused by a deficiency in arylsulfatase α activity, which is also responsible for degrading the myelin lipid cerebroside sulfate (sulfatide). Oligodendrocytes accumulate sulfatide in metachromatic granules, causing severe disruption of myelination. Peripheral nerve myelination is also affected, as are other organs that normally contain much lower amounts of sulfatide, such as the kidney, the liver, and the endocrine system. Krabbe disease, or globoid cell leukodystrophy, is also a dysmyelinating disease in which there is an almost complete lack of oligodendrocytes and therefore myelin, caused by a deficiency in the β-galactosidase responsible for hydrolyzing galactocerebroside to ceramide and galactose. Galactocerebroside is particularly abundant in myelin, constituting about 25% of myelin lipid. Mice that lack galactocerebroside have the ability to assemble multilamellar myelin; however, this myelin does not support adequate nerve conduction, neither is it stable. Unlike metachromatic leukodystrophy, Krabbe disease is limited to the CNS and peripheral nervous system (PNS). However, why a buildup of galactocerebroside should prove particularly toxic to oligodendrocytes is not entirely clear.

How are proteins destined to operate in lysosomes targeted to these organelles? Soluble lysosomal hydrolase enzymes acquire a phosphorylated mannose on their oligosaccharide chains by a two-step process in the Golgi apparatus. This mannose 6-phosphate label is recognized by specific mannose 6-phosphate receptors, which carry the proteins to late endosomes. In contrast, lysosomal membrane proteins are targeted by means of cytoplasmic tail signals that contain either leucine or tyrosine of type LJ or type YXXJ or NXXY, where J is any hydrophobic amino acid. The LJ signal seems to be essential for efficient delivery directly to endosomes, whereas the second type of signal seems to be more important in the recovery of proteins destined for lysosomes from the plasma membrane. The majority of lysosomal membrane proteins have the YXXJ but do not have the LJ signal. The implication of these observations is that many of the lysosomal membrane proteins make their way to lysosomes from the TGN endosomes through the plasma membrane.

What are the receptors for the type LJ or the type YXXJ or NXXY motifs at the TGN? Because transport from the TGN to the endosomes occurs in clathrin-coated vesicles, the proteins that link such vesicles to membranes, the adaptins, may play a role. The weight of the evidence suggests that AP-1 recognizes the LJ sequence, whereas AP-2 identifies the YXXJ and NXXY motifs. Once the ligands are bound, these adaptins would direct transport of their respective ligands to the endosomes from the TGN or through the plasma membrane, respectively. At present, it is not clear how proteins in the late endosome, such as mannose 6-phosphate receptors that cycle back to the Golgi, are sorted from those whose ultimate destination is a lysosome.

How Are Peripheral Membrane Proteins Targeted to Their Appropriate Destinations?

Peripheral membrane proteins are synthesized in the same type of free polysome in which the bulk of the cytosolic proteins are made. However, the cell must ensure that these membrane proteins are sent to the plasma membrane rather than allowed to attach in a haphazard way to other intracellular organelles. The fact that a complex machinery has evolved to ensure the correct delivery of integral membrane proteins suggests that some equivalent targeting mechanism must exist for proteins that attach to the cytoplasmic surface of the plasma membrane. Such proteins are translated on "free" polysomes, but these polysomes are associated with cytoskeletal structures and are not distributed uniformly throughout the cell body. In a number of cases, mRNAs that encode soluble cytosolic proteins are concentrated in discrete regions of the cell, resulting in a local accumulation of the translated protein close to the site of action. For some peripheral membrane proteins, this is the plasma membrane.

Evidence that this mechanism might operate in peripheral membrane protein synthesis came from studies showing biochemically and by *in situ* hybridization that mRNAs encoding the myelin basic proteins are concentrated in the myelinating processes that extend from the cell body of oligodendrocytes (Colman *et al.*, 1982). As in oligodendrocytes, Schwann cells also transport MBP mRNA by microtubule-based transport, which also appears to require specialized cytoplasmic channels called *Cajal bands* (Court *et al.*, 2004).

Myelin basic protein may be a special case because of its very strong positive charge and consequent propensity for binding promiscuously to the negatively charged polar head groups of membrane lipids. Nevertheless, the facts that actin mRNAs are localized to the leading edge of cultured myocytes and mRNA for the microtubule-associated protein MAP2b is concentrated in the dendrites of neurons suggest that targeting by local synthesis is more common than originally thought. This mechanism is probably less important for peripheral membrane proteins that associate with the cytoplasmic surface of the plasma membrane by means of strong specific associations with proteins already located at the membrane because such proteins would act as specific receptors. Because only selected cytoplasmic mRNAs are localized to the periphery, the process is specific. However, no mRNAs are localized exclusively to the periphery, and a significant fraction is typically localized proximal to the nucleus in a region rich with the translational and protein-processing machinery of the cell (the Nissl substance or translational cytoplasm).

Cytoplasmic Proteins Are Also Compartmentalized

Membrane-bound organelles are the most familiar form of compartmentation in cells, but cytoplasmic regions of the cell containing metabolic compartments exist as well. Regions of the neuronal or glial cytoplasm may have highly specialized polypeptide compositions that are important for function. For example, the neuronal phosphoprotein synapsin is highly enriched in presynaptic terminals, where it participates in the localization and targeting of synaptic vesicles. Similarly, calmodulin and the glycolytic enzyme aldolase have been localized in muscle cells to the region of the I band, where they are thought to facilitate coupling of ATP production to contractility.

As mentioned earlier, cytoplasmic proteins are synthesized on cytoplasmic polysomes, termed "free" polysomes to reflect an absence of underlying ER membrane, even though they may be restricted to specific domains of the cell cytoplasm. This restriction is particularly obvious in the neuronal perikaryon, where both cytoplasmic polysomes and membrane-associated polysomes are concentrated in areas near the nucleus and Golgi complex. In addition, cytoplasmic polysomes containing specific mRNAs may be localized to certain regions of the cell, such as the proximal dendrite (those encoding the microtubule-associated protein MAP2) and the processes of oligodendrocytes (those encoding myelin basic protein). In contrast, the protein synthetic machinery of the polysome appears to be effectively excluded from the mature axon. Therefore, cytoplasmic polysomes are representative of cytoplasmic compartmentation for proteins and nucleic acids.

In most cases, localized cytoplasmic proteins interact with cytoskeletal structures in the cytoplasm (see next section), but macromolecular complexes that form in order to make a cellular process more efficient or free from error have been described. Evidence exists that glycolytic enzymes of neurons and muscle cells may be organized in a labile complex that facilitates energy metabolism, but the existence of such complexes remains controversial.

Perhaps the best-characterized cytoplasmic macromolecular complex is the proteasome, which is a large protein complex (2×10^6 Da, sedimenting as a 20S particle) that contains several distinct enzymatic activities, including catalytic sites for both ubiquitin-dependent and ubiquitin-independent proteolysis (Hochstrasser, 1995). Ubiquitin is a small, highly conserved polypeptide that is added covalently to cytoplasmic proteins targeted for degradation. The catalytic core of the proteasome is a barrel-shaped structure formed by four heptameric stacked rings, but additional proteins (about 16 polypeptides) may interact with the 20S core to form a larger 26S particle. Because proteasomes constitute the primary cytoplasmic pathway for protein degradation (i.e., nonlysosomal pathways), they serve a number of important physiological functions, including regulation of cell proliferation and processing of antigens for presentation. In the nervous system, however, proteasomes are likely to be most important for homeostasis, allowing turnover of cytoplasmic polypeptides at specific sites so that the elaborate cellular extensions of neurons and glia may be maintained.

Cytoplasmic proteins may also be compartmentalized effectively by post-translational modification. Two types of modification may be particularly important for this kind of compartmentalization. Local activation of kinases can lead to the phosphorylation of proteins in specific domains of the neuron. For example, the reversible phosphorylation of synapsin in the presynaptic terminal appears to be responsible for the targeting of synaptic vesicles to the terminal and for the mobilization of vesicles during prolonged stimulation. An impressive variety of cytoplasmic protein kinases that may be selectively activated to modify serines or threonines presented in distinctive consensus sequences have been described. Distinct from these serine or threonine kinases, a number of other kinases that specifically modify tyrosines can be found in the brain. In some cases, the tyrosine kinase is linked directly to a membrane-spanning receptor and phosphorylates cytoplasmic proteins in the vicinity of the receptor after activation. Completing the cycle of phosphorylation and dephosphorylation are a number of phosphatases with varying specificities. The properties and physiological roles for kinases and phosphatases are discussed in greater detail later.

A second common post-translational modification of cytoplasmic proteins is the addition of carbohydrate moieties. Whereas modification of membrane-associated proteins in the Golgi complex proceeds by the addition of complex carbohydrates through N-linkages on selected asparagines, glycosylated cytoplasmic proteins have simpler carbohydrates added through O-linkages to serine or threonine hydroxyls. This modification was first recognized as a feature of many nuclear proteins and components of the nuclear membrane, but subsequent studies showed that a number of cytoplasmic proteins also have O-linked carbohydrates. Unlike phosphorylation, relatively little is known about the functional significance of cytoplasmic glycosylation. Remarkably, however, serines and threonines subject to O-linked glycosylation would also be good sites for phosphorylation by various kinases. This congruence raises the possibility that glycosylation and phosphorylation of some cytoplasmic proteins may serve complementary functions.

Summary

Membrane biogenesis and protein synthesis in neurons and glial cells are accomplished by the same mechanisms that have been worked out in great detail in other cell types. Integral membrane proteins are synthesized in the rough endoplasmic reticulum, and peripheral membrane proteins are products of cytoplasmic-free ribosomes that are found in the cell sap. For transmembrane proteins and secretory polypeptides, synthesis in the RER is followed by transport to the Golgi apparatus, where membranes and proteins are sorted and targeted for delivery to precise intracellular locations. It is likely that the neuron and glial cell have evolved additional highly specialized mechanisms for membrane and protein sorting and targeting because these cells are so greatly extended in space, although these additional mechanisms have yet to be fully described. The basic features of the process of secretion, which includes neurotransmitter delivery to presynaptic terminals, are beginning to be understood as well. The key features of this process are apparently common to all cells, including yeast, although the neuron has developed certain specializations and modifications of the secretory pathway that reflect its unique properties as an excitable cell.

CYTOSKELETONS OF NEURONS AND GLIAL CELLS

The cytoskeleton of eukaryotic cells is an aggregate structure formed by three classes of cytoplasmic structural proteins: microtubules (tubulins), microfilaments (actins), and intermediate filaments. Each of these elements exists concurrently and independently in overlapping cellular domains. Most cell types contain one or more examples of each class of cytoskeletal structure, but there are exceptions. For example, mature mammalian erythrocytes contain no microtubules or intermediate filaments, but they do have highly specialized actin cytoskeletons. Among cells of the nervous system, the oligodendrocyte is unusual in that it contains no cytoplasmic intermediate filaments. Typically, each cell type in the nervous system has a unique complement of cytoskeletal proteins that are important for the differentiated function of that cell type.

Although the three classes of cytoskeletal elements interact with each other and with other cellular structures, all three are dynamic structures rather than passive structural elements. Their aggregate properties form the basis of cell morphologies and plasticity in the nervous tissue. In many cases, the cytoskeleton is biochemically specialized for a particular cell type, function, and developmental stage. Each type of cytoskeletal element has unique functions essential for a functional nervous system.

Microtubules Are an Important Determinant of Cell Architecture

Microtubules are near-ubiquitous cytoskeletal components in eukaryotes (Hyams and Lloyd, 1994). They play key roles in intracellular transport, are a primary determinant of cell morphology, form the structural correlate of the mitotic spindle, and are the functional core of cilia. Microtubules are very abundant in the nervous system, and tubulin subunits of microtubules may constitute >10% of total brain protein. As a result, many fundamental properties of microtubules were defined with microtubule protein from brain extracts. However, neuronal microtubules have biochemical specializations to meet the unique demands imposed by neuronal size and shape.

Intracellular transport and generation of cell morphologies are the most important roles played by microtubules in the nervous system. In part, this comes from their ability to organize cytoplasmic polarity. Microtubules *in vitro* are dynamic, polar structures with plus and minus ends that correspond to the fast- and slow-growing ends, respectively. In contrast, both stable and labile microtubules can be identified *in vivo*, where they help define both microscopic and macroscopic aspects of intracellular organization in cells. Microtubule organization, stability, and composition are all highly regulated in the nervous system.

By electron microscopy, microtubules appear as hollow tubes 25 nm in diameter and can be hundreds of micrometers in length in axons. Microtubule walls typically comprise 13 protofilaments formed by a linear arrangement of globular subunits. Globular subunits in microtubule walls are heterodimers of α- and β-tubulin, with a variety of microtubule-associated proteins (MAPs) binding to microtubule surfaces.

Neuronal microtubules are remarkable for their genetic and biochemical diversity. Multiple genes exist for both α- and β-tubulins. These genes are expressed differentially according to cell type and developmental stage. Some genetic isotypes are expressed ubiquitously, whereas others are expressed only at specific times in development, in specific cell types, or both. Most tubulin genes are expressed in nervous tissue, and some are enriched or specific to neurons. Specific tubulin isotypes prepared in a pure form vary in assembly kinetics and ability to bind ligands. However, when more than one isotype is expressed in a single cell, such as a neuron, they coassemble into microtubules with mixed composition.

The most common post-translational modifications of tubulins are tyrosination–detyrosination, acetylation–deacetylation, and phosphorylation. The first two are intimately linked to assembled microtubules, but little is known about physiological functions for any tubulin modification. Most α-tubulin isotypes are synthesized with a Glu-Tyr dipeptide at the C terminus (Tyr-tubulin), but the tyrosine is removed by tubulin carboxypeptidase after incorporation into a microtubule, leaving a terminal glutamate (Glutubulin). Microtubules assembled for a longer time are enriched in Glu-tubulin, but when Glu-tubulin-enriched microtubules are disassembled, liberated α-tubulins are rapidly retyrosinated by tubulin tyrosine ligase. The tyrosination state of α-tubulin does not affect assembly–disassembly kinetics *in vitro*, but detyrosination may affect interactions of microtubules with other cellular structures. Concurrent with detyrosination, α-tubulins can be subject to a specific acetylation. Tubulin acetylation was first described in flagellar tubulins, but this modification is widespread in neurons and many other cell types. Acetylase acts preferentially on α-tubulin in assembled microtubules, so long-lived or stable microtubules tend to be acetylated, but the distribution of microtubules rich in acetylated tubulin may not be identical to that of Glu-tubulin. Acetylated α-tubulin is rapidly deacetylated upon microtubule disassembly, but acetylation does not alter microtubule stability *in vitro*.

Tubulin phosphorylation involves β-tubulin and may be restricted to an isotype expressed preferentially in neurons and neuron-like cells. Various kinases can phosphorylate tubulin *in vitro*, but the endogenous kinase is unknown. Effects of phosphorylation on assembly are unknown, but phosphorylation is upregulated during neurite outgrowth. As with α-tubulin modifications, the physiological role of phosphorylation on neuronal β-tubulin has yet to be determined. Other post-translational modifications have been reported, but their significance and distribution in the nervous system are not well documented.

The biochemical diversity of microtubules is increased through association of different MAPs with different populations of microtubules (Table 2.2).

TABLE 2.2   Major Microtubule Proteins and Microtubule Motors in Mammalian Brain

	Location and Function
Tubulins	
Tubulins α-and β-tubulins	Neurons, glia, and nonneuronal cells except mature mammalian erythrocytes. Multigene family with some genes expressed preferentially in brain, whereas others are ubiquitous. Primary structural polypeptides of microtubules.
γ-tubulin	Present near microtubule-organizing center in all microtubule-containing cells. Needed for nucleation of microtubules
Microtubule-associated proteins (MAPs)	
MAP1a/1b	Widely expressed in neurons and glia, including both axons and dendrites; developmentally regulated phosphoproteins.
MAP2a/2b, MAP2c	Dendrite-specific MAPs. The smaller MAP2c is regulated developmentally, becoming restricted to spines in adults, whereas 2a and 2b are major phosphoproteins in adult brain.
LMW tau HMW tau	Tau proteins are enriched in axons with a distinctive phosphorylation pattern. A single tau gene is alternatively spliced to give multiple isoforms.
Microtubule-severing proteins	
Katanin	Enriched at the microtubule-organizing center and thought to be important in the release of microtubules for transport into axons and dendrites.
Motor proteins	
Kinesins (kinesin-1s, kinesin-2s, kinesin-3s, and others)	Kinesin-1s are plus-end directed motors associated with membrane-bound organelles and moving them in fast axonal transport. The other members of the kinesin family are a diverse set of motor proteins with a kinesin-related motor domain and varied tails. Many are regulated developmentally, and some are mitotic motors, restricted to dividing cells.
Axonemal dynein	A set of minus-end-directed microtubule motors associated with cilia and flagella, such as ependymal cells.
Cytoplasmic dynein	Cytoplasmic forms may be involved in the axonal transport of either organelles or cytoskeletal elements.

The significance of microtubule diversity is incompletely understood but may include functional differences as well as variations in assembly and stability. In particular, MAP composition may define specific neuronal domains. For example, MAP2 is restricted to dendritic regions of the neuron, whereas tau proteins are modified differentially in axons. Similarly, oligodendrocyte progenitors transiently express a novel MAP2 isoform with an additional microtubule-binding repeat; i.e., 4-repeat MAP2c or MAP2d. This MAP is in cell bodies but not in processes, suggesting that MAP2d might have a role distinct from its capacity to bundle microtubules (Vouyiouklis and Brophy, 1995).

MAPs in nervous tissue fall into two heterogeneous groups: tau proteins and high molecular weight MAPs. Tau proteins have been of intense interest because post-translationally modified tau proteins are the primary constituents of neurofibrillary tangles in the brains of Alzheimer patients. Tau proteins are primarily neuronal MAPs, although tau may be found outside neurons as well. Tau binds to microtubules during assembly–disassembly cycles with a constant stoichiometry and promotes microtubule assembly and stabilization. Tau exists in a number of molecular weight isoforms expressed differentially in different regions of the nervous system and developmental stage. For example, tau proteins in the adult CNS are typically 60–75 kDa, whereas PNS axons contain a higher molecular mass tau of 100 kDa. Different isoforms of tau protein are generated from a single mRNA by alternative splicing, and additional heterogeneity is produced by phosphorylation.

High molecular weight MAPs are a diverse group of largely unrelated proteins found in various tissues, some of which are brain specific. All have molecular masses >1300 kDa and form side arms protruding from microtubule surfaces. Many MAPs may participate in microtubule assembly and cytoskeletal organization. Traditionally, high molecular weight MAPs comprise five polypeptides: MAPs 1a, 1b, 1c, 2a, and 2b. MAP2 proteins are closely related and located primarily in dendrites. In contrast, the polypeptides known as MAP1 are unique polypeptides with little sequence homology. MAPs 1a and 1b are expressed widely and regulated developmentally. MAPs 1a, 1b, and 2 are all thought to play important roles in stabilizing and organizing the microtubule cytoskeleton.

In most cell types, cytoplasmic microtubules are dynamic, although stable microtubule segments are found in all cells. In nonneuronal cells, such as astrocytes and other glia, microtubules are typically anchored in centrosomal regions that serve as microtubule-organizing centers. As a result, their cytoplasmic microtubules are oriented with plus ends at the cell periphery. The biochemistry of microtubule-organizing centers is not fully understood, but they contain a novel tubulin subunit, γ-tubulin, which functions as a microtubule-nucleating protein. In contrast, dendritic and axonal microtubules of neurons are not continuous with a microtubule-organizing center, so alternate mechanisms must exist for their stabilization and organization. The situation is further complicated because dendritic and axonal microtubules differ in both composition and organization. Both axonal and dendritic microtubules are nucleated at the microtubule-organizing center but are subsequently released for delivery to the appropriate compartment. The release of microtubules from the microtubule-organizing center appears to involve the microtubule-severing protein, katanin (Baas, 2002). Surprisingly, axonal and dendritic compartments are not equivalent. First, dendritic and axonal MAPs differ in both identity and phosphorylation state. Second, microtubule orientation in axons has the plus end distal similar to other cell types, but microtubules in dendrites may exhibit both polarities. Finally, dendritic microtubules are less likely to be aligned with one another and are less regular in their spacing. As a result, dendritic diameters taper, whereas axons have a constant diameter as one proceeds away from the cell body.

Stabilization of axonal and dendritic microtubules is essential because of the volume of cytoplasm and the distance from sites of protein synthesis for tubulin. A common side effect of one class of antineoplastic drugs, the vinca alkaloids, underscores the importance of microtubule stability in axons. Vincristine and other vinca alkaloids act by destabilizing spindle microtubules, but dosage must be monitored carefully to prevent development of peripheral neuropathies due to loss of axonal microtubules. Microtubules play critical roles in both dendritic and axonal function, so mechanisms to ensure their proper extent and organization exist.

Axonal microtubules contain a particularly stable subset of microtubule segments resistant to depolymerization by antimitotic drugs, cold, and calcium. Stable microtubule segments are biochemically distinct and may constitute more than half of the axonal tubulin. Stable domains in microtubules may serve to regulate the axonal cytoskeleton by nucleating and organizing microtubules as well as stabilizing them. The biochemical basis of microtubule stability is not well understood but may include post-translational modification of tubulins, presence of stabilizing proteins, or both. Relatively little is known about regulation of dendritic microtubules, but local synthesis of MAP2 in dendrites may play a role.

Microfilaments and the Actin-Based Cytoskeleton Are Involved in Intracellular Transport and Cell Movement

The actin cytoskeleton is universal in eukaryotes, although microfilaments are most familiar as thin filaments in skeletal muscle. Microfilaments (Table 2.3) play critical roles in contractility for both muscle and nonmuscle cells. Actin and its contractile partner myosin are particularly abundant in nervous tissue relative to other nonmuscle tissues. In fact, one of the earliest descriptions of nonmuscle actin and myosin was in brain. In neurons, microfilaments are most abundant in presynaptic terminals, dendritic spines, growth cones, and subplasmalemmal cortex. Although concentrated in these regions, microfilaments are present throughout the cytoplasm of neurons and glia as short filaments (4–6 nm in diameter and 400–800 nm long).

Multiple actin genes exist in both vertebrates and invertebrates. Four α-actin human genes have been cloned, each expressed specifically in a different muscle cell type (skeletal, cardiac, vascular smooth, and enteric smooth muscle). In addition, two nonmuscle actin genes (β- and γ-actin) are present in humans. β-actin and γ-actin genes are expressed ubiquitously and are abundant in nervous tissue. The functional significance of different genetic isotypes is not clear because actins are highly conserved. Across the range of known actin sequences, amino acids are identical at approximately two of three positions. Even the positions of introns

within different actin genes are highly conserved across species and genes. Despite the high degree of conservation, differences in distribution of specific isotypes within a single neuron are seen. For example, β-actin may be enriched in growth cones. The prominent actin bundles seen in some nonneuronal cells in culture are not characteristic of neurons, and most neuronal microfilaments are less than 1 μm in length.

Many microfilament-associated proteins are found in nervous tissue (myosin, tropomyosin, spectrin, α-actinin, etc.), but less is known about their distribution and normal function in neurons and glia. Myosins and myosin-associated proteins are considered in the section on molecular motors, but multiple categories of actin-binding proteins exist (Table 2.3). Monomer actin-binding proteins such as profilin and thymosins are abundant in the developing brain and thought to help regulate assembly of microfilaments by sequestering actin monomers, which may be mobilized rapidly in response to appropriate signals. For example, phosphatidylinositol 4,5-bisphosphate causes the actin–profilin complex to dissociate, freeing monomer for explosive microfilament assembly. This may play a role in growth cone motility, where actin assembly is critical for filopodial extension.

Several proteins have been identified that cap microfilaments, serving to anchor them to other structures or regulate microfilament length. The ezrin–radixin–moesin gene family encodes barbed-end capping proteins that are concentrated at sites where the microfilaments meet the plasma membrane, suggesting a role in anchoring microfilaments or linking them to extracellular components through membrane proteins. They are prominent components of nodal and paranodal structures in nodes of Ranvier. A mutation in a member of this family expressed in Schwann cells, merlin or schwannomin, is responsible for the human disease neurofibromatosis type 2. Development of numerous tumors with a Schwann cell lineage in neurofibromatosis type 2 suggests that this microfilament-binding protein acts as a tumor suppressor.

Whereas some membrane proteins interact directly with microfilaments in the membrane cytoskeleton, others interact with the actin cytoskeleton through intermediaries. Proteins such as spectrin (fodrin), α-actinin, and dystrophins cross-link, or bundle, microfilaments, giving rise to higher order complexes. Spectrin is enriched in the cortical membrane cytoskeleton and is thought to have a role in localization of integral membrane proteins such as ion channels and receptors. Dystrophin is the best-known member of a family of proteins that appear to be essential for clustering of receptors in muscle and nervous tissue. A mutation in dystrophin is responsible

TABLE 2.3   Selected Proteins of the Microfilament Cytoskeleton in Brain

Actins
α-actin (smooth muscle)

β-actin and γ-actin (neuronal and nonneuronal cells)

Actin monomer-binding proteins
Profilin

Thymosin 4 and 10

Capping proteins
Ezrin/radixin/moesin

Schwannomin/merlin

Gelsolin and other microfilament-severing proteins
Gelsolin

Villin

Cross-linking and bundling proteins
Spectrin (fodrin)

Dystrophin, utrophin, and related proteins

α-actinin

Tropomyosin
Myosins I, II, V, VI, VII

for Duchenne muscular dystrophy. Positioning of integral membrane proteins on the cell surface is an essential function of the actin-rich membrane cytoskeleton, acting in concert with a class of proteins that contain the protein-binding module, the PDZ domain.

Members of the gelsolin family have multiple activities. They not only cap the barbed end of a microfilament, but also sever microfilaments and can nucleate microfilament assembly. Severing-capping proteins may be critical for reorganizing the actin cytoskeleton. The Ca^{2+} dependence of gelsolin-severing activity may provide a mechanism for altering the membrane cytoskeleton in response to Ca^{2+} transients. Other second messengers, such as phosphatidylinositol 4,5-bisphosphate, may also regulate gelsolin function, suggesting interplay between different classes of actin-binding proteins such as gelsolin and profilin. Oligodendrocytes are the only nonneural cells in the CNS that express significant amounts of the actin-binding and microfilament-severing protein gelsolin.

Proteins with other functions may interact directly with actin or actin microfilaments. For example, some membrane proteins, such as epidermal growth factor receptor, bind actin microfilaments directly, which may be important in anchoring these components at a particular location on the cell surface. Other cytoskeletal structures also interact with microfilaments. Both MAP2 and tau microtubule-associated proteins can interact with microfilaments *in vitro* and may mediate interactions between microtubules and microfilaments. Finally, the synaptic vesicle-associated phosphoprotein, synapsin I, has a phosphorylation-sensitive interaction with microfilaments that may be important for targeting and storage of synaptic vesicles in the presynaptic terminal (Murthy and De Camilli, 2003). Many of these interactions were defined by *in vitro*-binding studies, and their physiological significance is not always established. The presence of actin as a major component of both pre- and postsynaptic specializations, as well as in growth cones, gives the actin cytoskeleton special significance in the nervous system (Murthy and De Camilli, 2003). The enrichment of the microfilament cytoskeleton at the plasma membrane makes the cytoskeletal components most responsive to changes in the local external environment of the neuron. Microfilaments also play a critical role in positioning receptors and ion channels at specific locations on neuronal surfaces. Although we emphasize enrichment of the microfilament cytoskeleton at the plasma membrane, microfilaments are also abundant in the deep cytoplasm. The microfilaments are best regarded as a uniquely plastic component of the neuronal

cytoskeleton that plays a critical role in local trafficking of cytoskeletal and membrane components.

Intermediate Filaments Are Prominent Constituents of Nervous Tissue

Intermediate filaments appear as solid, ropelike fibrils from 8 to 12 nm in diameter that may be many micrometers long (Lee and Cleveland, 1996). Intermediate filament proteins constitute a superfamily of five classes with expression patterns specific to cell type and developmental stage (Table 2.4). Type I and type II intermediate filament proteins are keratins, hallmarks of epithelial cells. Keratins are not associated with nervous tissue and will not be considered further. In contrast, all nucleated cells contain type V intermediate filament proteins, nuclear lamins. Lamins are the most evolutionarily divergent of intermediate filament genes, with regard to both intron–exon distribution and polypeptide domain structure. Cytoplasmic intermediate filaments in the nervous system are all either type III or type IV.

Type III intermediate filaments are a diverse family that includes vimentin (characteristic of fibroblasts and embryonic cells, including embryonic neurons) and glial fibrillary acidic protein (GFAP, a marker for astrocytes and Schwann cells). Type III

TABLE 2.4   Intermediate Filament Proteins of the Nervous System

Class and Name	Cell Type
Types I and II	
Acidic and basic keratins	Epithelial and endothelial cells
Type III	
Glial fibrillary acidic protein	Astrocytes and nonmyelinating Schwann cells
Vimentin	Neuroblasts, glioblasts, fibroblasts, etc.
Desmin	Smooth muscle
Peripherin	A subset of peripheral and central neurons
Type IV	
NF triplet (NFH, NFM, NFL)	Most neurons, expressed at highest level in large myelinated fibers
α-internexin	Developing neurons, parallel fibers of cerebellum
Nestin	Early neuroectodermal cells
	The most divergent member of this class; some have classified it as a sixth type
Type V	
Nuclear lamins	Nuclear membranes

intermediate filament subunits are typically 45–60 kDa with a conserved rod domain and relatively small gene-specific amino- and carboxy-terminal sequences. As a result, type III intermediate filament subunits form smooth filaments without side arms. Type III polypeptides can form homopolymers or coassemble with other type III intermediate filament subunits.

Type III intermediate filament proteins in the nervous system are typicaly restricted to glia or embryonic neurons. Vimentin is abundant in many cells during early development, including both glioblasts and neuroblasts. Some Schwann cells and astrocytes also contain vimentin. Curiously, mature oligodendrocytes do not have intermediate filaments; an exception to the general rule that metazoan cells contain all three classes of cytoskeletal structures. Oligodendrocyte precursors do, however, express vimentin and may express GFAP transiently.

Peripherin is one type III intermediate filament protein unique to neurons. Peripherin has a characteristic expression during development and regeneration in specific neuronal populations and may be coexpressed with type IV neurofilament proteins. It can coassemble with type IV neurofilament subunits both *in vitro* and *in vivo*, where it can substitute for the low molecular weight neurofilament subunit (NFL). However, whether coassembly is generally the case is not known. Unlike type IV intermediate filaments, intermediate filaments made from type III subunits tend to disassemble more readily under physiological conditions. Thus, the presence of type III intermediate filament subunit proteins may produce more dynamic structures, which could be important during development or regeneration.

Neuronal intermediate filaments typically have side arms that limit packing density, whereas glial intermediate filaments lack side arms and may be very tightly packed. Neuronal intermediate filaments have an unusual degree of metabolic stability, which makes them well suited to the role of stabilizing and maintaining neuronal morphology. Due to this stability, the existence of neurofilaments was recognized long before much was known about their biochemistry or function. Neurofilaments were seen in early electron micrographs, and many traditional histological procedures to visualize neurons were based on a specific reaction of silver and other metals with neurofilaments.

Most neuronal intermediate filaments have three distinct subunits present in varying stochiometries, all type IV polypeptides. Apparent molecular mass for neurofilament subunits vary widely across species, but mammalian forms are typically a triplet ranging from 180–200 kDa for the high molecular weight subunit (NFH), from 130–170 kDa for the medium subunit (NFM), and from 60–70 kDa for NFL. Neurofilament triplet proteins are each encoded by a separate type IV intermediate filament gene, which has a characteristic domain structure that can be recognized in both primary sequence and gene structure. Type IV genes are typically expressed only in neurons, although Schwann cells in damaged peripheral nerves may also transiently express NFM and NFL. Neurofilament polypeptides were initially identified from axonal transport studies. Neurofilament subunits are highly phosphorylated in axons, particularly NFM and NFH. In humans and some other species, NFH has >50 repeats of a consensus phosphorylation site at its carboxy terminus, and levels of NFH phosphorylation indicate that most are phosphorylated *in vivo*. This high level of phosphorylation in neurofilament tail domains is a distinctive characteristic of neurofilaments.

A second motif characteristic of neurofilaments is the presence of a glutamate-rich region in the tail adjacent to the core rod domain. This glutamate region has particular significance for neuroscientists because it appears to be the basis for reaction of the classic neurofibrillary silver stains for neurons. These stains were introduced in the late nineteenth century and used extensively by histologists and neuroanatomists from Ramon y Cajal's time to the present. The molecular basis of neurofibrillary stains was unknown until 1968, when F. O. Schmitt showed that neurofibrils were formed by neurofilaments. Remarkably, the ability of neurofilament subunits to react with silver histological stains is retained even after separation in gel electrophoresis for neurofilaments from organisms as diverse as human, squid, and the marine fanworm, *Myxicola*. Conservation of this glutamate-rich domain suggests both an important functional role and early divergence of neurofilaments from the other intermediate filament families.

Neurofilaments and neurofilament triplet proteins play a critical role in determining axonal caliber. As noted earlier, neurofilaments have characteristic side arms, unique among intermediate filaments. Although all three subunits contribute to the neurofilament central core, side arms are formed only by carboxy-terminal regions of NFM and NFH. Phosphorylation of NFH and NFM side arms alters charge density on the neurofilament surface, repelling adjacent similarly charged neurofilaments. Although cross bridges between neurofilaments are often noted, direct studies of interactions between neurofilaments provide little evidence of stable cross-links between neurofilaments or between neurofilaments and other cytoskeletal structures. The high density of surface charge due to phosphorylation of neurofilaments makes it difficult to imagine a stable interaction between neurofilaments and other structures of like charge. However, dynamic interactions between neurofilaments and cellular structures or proteins may be critical for neurofilament function and metabolism.

Altered expression levels of neurofilament subunits or mutations in neurofilament genes are associated with some neuropathologies. Disruption of neurofilament organization is a hallmark of pathology for many degenerative diseases of the nervous system, particularly those affecting large myelinated axons such as those of spinal motor neurons such amyotrophic lateral sclerosis. Overexpression of normal NFH or expression of some mutant NFL genes in transgenic mouse models leads to the accumulation of neurofilaments in the cell body and proximal axon of spinal motor neurons, similar to those seen in amyotrophic lateral sclerosis and related motor neuron diseases. Similarly, an early indicator of neuropathies due to neurotoxins such as acrylamide and hexanedione is accumulation of neurofilaments in either proximal or distal regions of axons. However, the question of whether neurofilament defects are a primary event in pathogenesis or reflect an underlying metabolic pathology remains unclear.

Another type IV intermediate filament gene expressed only in neurons is α-internexin. Unlike the triplet, α-internexin is expressed preferentially early in development and disappears from most neurons during maturation. Intermediate filaments with α-internexin do persist in some adult neurons, such as the branched axons of granule cells in the cerebellar cortex. Although α-internexin can coassemble with neurofilament triplet subunits, it also forms homopolymeric filaments. The primary sequence of α-internexin has features in common with NFL and NFM that are thought to confer assembly properties distinct from other type IV intermediate filaments.

The final intermediate protein expressed in the nervous system is nestin, which is seen transiently during early development. Nestin is expressed in neurons, Schwann cells, and oligodendrocyte progenitors, which appear late in the development of the embryonic nervous system. Remarkably, nestin is expressed almost exclusively in ectodermal cells after commitment to the neuroglial lineage but prior to terminal differentiation. At 1250 kDa, nestin is the largest intermediate filament subunit and the most divergent in sequence. Several distinctive features lead some to classify nestin as a sixth type of intermediate filament, whereas others group it with type IV genes. Relatively little is known about

assembly properties of nestin *in vivo* or physiological functions of nestin filaments in neuroectodermal cells.

How Do the Various Cytoskeletal Systems Interact?

Each class of cytoskeletal structures may be found without the others in some cellular domains, but all three classes—microtubules, microfilaments, and intermediate filaments—coexist in many domains and inevitably interact. These interactions are typically dynamic, rather than through stable cross-links to one another. As mentioned earlier, microtubules and neurofilaments have highly phosphorylated side arms projecting from their surfaces. The high density of negative surface charge tends to repel structures with a like charge and rigidify microtubules and neurofilaments, affecting axon diameter.

The growth cone is a unique neuronal domain with distinctive cytoskeletal organization, such as longer microfilaments in filopodia, and neurofilaments are excluded from growth cones, typically extending no further than the growth cone neck. In contrast, microtubules and microfilaments play complementary roles in growth cones. Microfilaments are critical in sprouting but less critical for elongation. Disrupting microtubules in distal neurites inhibits neurite elongation but does not affect sprouting.

Summary

The intracellular framework giving shape to neurons and glia is the cytoskeleton, a complicated set of structures and their associated proteins. These organelles are also responsible for intracellular movement of materials and, during development, for cell migration and plasma membrane extension within nervous tissue.

MOLECULAR MOTORS IN THE NERVOUS SYSTEM

Until 1985, our knowledge of molecular motors in vertebrate cells of any type was restricted to myosins and flagellar dyneins. Myosins were identified in nervous tissue, but functions were uncertain. The preponderance of evidence indicated that fast axonal transport was microtubule based, so there was considerable interest in cytoplasmic dyneins, but initial studies failed to find a functional cytoplasmic dynein. However, a better understanding of molecular motors in the nervous system has now emerged, largely through studies on axonal transport (Brady and Sperry, 1995; Hirokawa and Takemura, 2005).

Myosins and dyneins can be distinguished pharmacologically by their differential susceptibility to inhibitors of ATPase activity, but the spectrum of inhibitors active against fast axonal transport fails to match properties of either myosin or dynein. The most striking difference between inhibitor effects on axonal transport and on myosin or dynein motors was seen with a nonhydrolyzable analog of ATP. Adenylyl-imidodiphosphate (AMP-PNP) is a weak competitive inhibitor of both myosin and dynein, requiring a 10- to 100-fold excess of analog. In contrast, both anterograde and retrograde axonal transport stop within minutes of AMP-PNP perfusion into isolated axoplasm, even in the presence of stoichiometric concentrations of ATP. Organelles moving in both directions freeze in place and remain attached to microtubules. AMP-PNP weakens interactions of myosin with microfilaments and of dynein with microtubules, but stabilizes binding of membrane-bound organelles to microtubules. Thus, effects of AMP-PNP indicated that axonal transport of membrane-bound organelles involved another type of motor, distinct from both myosins and dyneins.

The effects of AMP-PNP both demonstrated the existence of a new type of motor protein and provided a basis for identifying its constituent polypeptides. Binding of this ATPase to microtubules should be increased by AMP-PNP and decreased by ATP. Polypeptides meeting this criterion were soon identified and the new mechanochemical ATPase was named kinesin, based initially on an ability to move microtubules across glass coverslips as plus-end directed motor. Studies soon established that kinesin was a microtubule-activated ATPase with minimal basal activity. This combination of ATPase activity and motility *in vitro* confirmed it as the first member of a new class of microtubule-based motor, the kinesins (Brady and Sperry, 1995; Hirokawa and Takemura, 2005).

Kinesins are now known to comprise >40 different genes in at least 14 subfamilies, all with a highly conserved motor domain that includes ATP- and microtubule-binding domains (Miki *et al.*, 2005). Multiple members of the kinesin superfamily are expressed in both adult and developing brains. Many kinesin family members are associated with mitosis, although some of these are also in postmitotic neurons. Members of kinesin families 1–6 are implicated in various neuronal functions ranging from transport of membrane-bounded organelles to translocation of microtubules in dendrites, and others may also have functions in the nervous system. This proliferation of motor proteins has dramatically altered the questions being asked about motor function in the brain. Most kinesins move toward the plus end of microtubules, but some move toward the minus end, increasing the number of potential functions that kinesin family members might serve, including a role in transport of cytoskeletal structures. Studies continue to identify new functions for members of the kinesin superfamily expressed in neurons or glial cells.

Kinesin-1, the founding member of the kinesin superfamily, remains the most abundant class of kinesin expressed in brain and other tissues, leading to an extensive characterization of its biochemical, pharmacological, immunochemical, and molecular properties. Electron microscopic and biophysical analyses reveal kinesin as a long, rod-shaped protein, approximately 80 nm in length. Kinesin-1 is a heterotetramer with two heavy chains (115–130 kDa) and two light chains (62–70 kDa). Localization of antibodies specific for kinesin subunits by high-resolution electron microscopy of brain kinesin indicates that two heavy chains arranged in parallel form the heads and much of the shaft, whereas light chains are localized to the fan-shaped tail region (Fig. 2.6).

ATP-binding and microtubule-binding domains of kinesin are in the heavy chain head regions. Axonal microtubules are oriented with plus ends distal from the cell body, so anterograde transport would require a motor that moves organelles toward the plus-end direction. Three different genes for kinesin-1 heavy chain are expressed in neurons, one of which is neuron specific, along with two light chain genes. Kinesin-1 appears associated with a variety of membrane-bound organelles, including synaptic vesicles or their precursors, mitochondria, and endosomes. Mechanisms for associating specific kinesin-1 isoforms with specific neuronal cargoes are incompletely understood. In the case of kinesin, the interaction is thought to involve both kinesin light chains and the carboxy termini of the heavy chains. Remarkably, mutations in the neuron-specific isoform of kinesin-1 can lead to a form of hereditary spastic paraplegia, an adult onset neurodegenerative disease of motor neurons.

An indirect result of the discovery of kinesin was the long sought cytoplasmic form of dynein, previously identified as a high molecular weight MAP in brain called MAP1c. Both cytoplasmic dynein and kinesin-1 can be isolated from brain by incubation of microtubules with nucleotide-free soluble extracts. Both are bound to microtubules under these conditions and released by ATP. MAP1c dynein moved microtubules *in vitro* with a polarity opposite that seen with kinesin and was identified as a two-headed cytoplasmic dynein using both structural and biochemical criteria. Dynein heavy chains are also a gene family

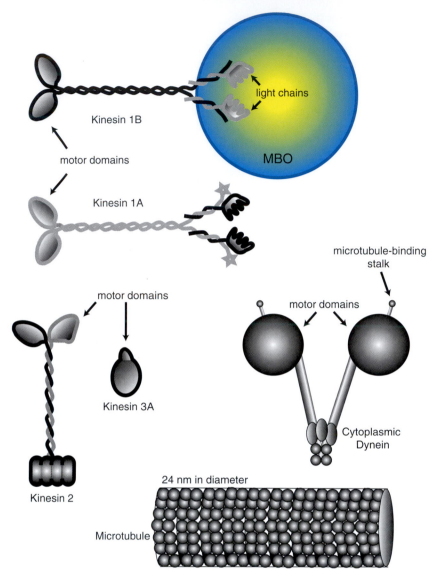

FIGURE 2.6   Examples of microtubule motor proteins in the mammalian nervous system. The first microtubule motor identified in nervous tissue was a kinesin-1 but showed three kinesin-1 genes, including a neuron-specific form (kinesin-1A). Motor domains are well conserved by tail domains and appear to be specialized for interaction with various targets, such as different membrane-bound organelles. After the sequence of the kinesin heavy chain was established, the presence of additional genes that contained sequences homologous to the motor domain of kinesin was soon recognized. The molecular organization of these various motor proteins is diverse, including monomers (kinesin-3A), trimers (kinesin-2), and tetramers (ubiquitous and neuron-specific kinesins). Cytoplasmic dynein may interact with membrane-bound organelles and cytoskeletal structures. Genetic methods have established that there may be >40 kinesin-related genes and 16 dynein heavy chains genes in a single organism.

with 14 flagellar dyneins and two cytoplasmic dyneins (Pfister *et al.*, 2006). Some, but not all, functions of cytoplasmic dynein also involve another complex of polypeptides known as dynactin (Schroer, 2004).

Cytoplasmic dyneins are a 40-nm-long complex of molecular mass 1.6×10^6 Da, that include two heavy chains as well as multiple intermediate and light chains (Figs. 2.6). Nonneuronal cells show immunoreactivity for dynein on mitotic spindles, and a punctate pattern of immunoreactivity present in interphase cells is thought to be dynein bound to membrane-bounded organelles. Dyneins are widely thought to be the motor for retrograde fast axonal transport but are also implicated as motors for slow axonal transport (Baas and Buster, 2004). As with kinesin-1, partial loss of cytoplasmic dynein function leads to degeneration of motor neurons, reflecting the importance of dyneins in neuronal function (Levy and Holzbaur, 2006).

Myosins from muscle were the first molecular motors identified, but research in nonmuscle myosins has increased the number of myosins expressed in humans to some 40 different genes grouped in 18 different subfamilies (Berg *et al.*, 2001). As with kinesins, all myosins share considerable homology in their motor domains but diverge widely in other domains. Many nonmuscle myosins are expressed in neurons and glia (Brown and Bridgman, 2004). Myosins play critical roles in neuronal growth and development, as well as in specialized cells such as sensory hair cells of the cochlea and vestibular organs (Gillespie and Cyr, 2004).

The most familiar myosins are myosin II (Fig. 2.7), forming the thick filaments of smooth and skeletal muscle but also present in nonmuscle cells. Myosin II heavy chains form a dimer that may interact with other myosin II dimers to form bipolar filaments. In tissue culture, many cells contain bundled actin microfilament stress fibers with a characteristic sarcomeric distribution of myosin II, but stress fibers are not apparent in neurons and glia *in situ*. However,

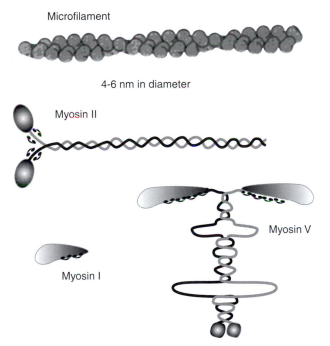

Microfilament

4-6 nm in diameter

Myosin II

Myosin V

Myosin I

FIGURE 2.7 Examples of myosin motor proteins found in mammalian brain. Myosin heavy chains contain the motor domain, whereas light chains regulate motor function. Myosin II was the first molecular motor characterized biochemically from skeletal muscle and brain. Genetic approaches have now defined >15 classes of myosin, many of which are found in brain. Myosin II is a classic two-headed myosin forming thick filaments in nonmuscle cells. Myosin I motors have single motor domains but may interact with actin microfilaments or membranes. Myosin V has multiple binding sites for calmodulin that act as light chains. Mutations in other classes of myosin have been linked to deafness. Myosins I, II, and V have been detected in growth cones as well as in mature neurons.

bipolar thick filaments assembled from myosin II dimers can be isolated from nervous tissues. Although brain myosin II was one of the first nonmuscle myosins to be described, relatively little is known about myosin II function in neurons. Many cellular contractile events in nonneuronal cells, such as the contractile ring in mitosis, involve myosin II.

Myosin I proteins have a single, smaller heavy chain that does not form filaments but possesses a homologous actin-activated ATPase domain and has been purified from neural and neuroendocrine tissues (Fig. 2.7). Some myosin I motors have the ability to interact directly with membrane surfaces, which may generate movements of plasma membrane components or intracellular organelles. Mammals have at least three myosin I genes, and multiple forms are in brain. For example, myosin Ic is in cochlear and vestibular hair cell stereocilia and plays a key role in mechanotransduction (Gillespie and Cyr, 2004).

The mouse mutation *dilute*, which affects coat color, results from mutations in a myosin V gene, which is distinct from both myosins I and II (Fig. 2.7). Coat color changes in *dilute* mouse are due to ineffective pigment delivery to developing hairs by dendritic pigment cells, but *dilute* mutants also have complex neurological deficits, including seizures in early adulthood that may lead to death. The specific cellular localization and function of myosin V motors in neurons remain unclear. Genes for myosin VI and VIIa are implicated in some forms of congenital deafness but are expressed in both brain and other tissues. Myosin VI is the gene responsible for Snell's Walzer deafness, and myosin VIIa is associated with Usher syndrome type 1B, a human disease involving both deafness and blindness. Both of these myosins are expressed in cochlear and vestibular hair cells but exhibit a different localization from each other and from myosin Ic.

The diversity of brain myosins and their distinctive localization suggests that the various myosins may have narrowly defined functions. However, relatively little is known about specific neuronal functions for most myosins despite intensive study of myosins in the nervous system. Myosins likely play roles in growth cone motility, synaptic plasticity, and even neurotransmitter release. The axonal transport of myosin II–like proteins was described, but relatively little progress has been made in defining functions of myosin II in the mature nervous system. Even less is known about myosin I in the nervous system beyond their role in hair cell function. There are few instances in our knowledge of neuronal function in which we fully understand the role played by specific molecular motors, but members of all three classes are abundant in nervous tissue. Proliferation of different motor molecules and isoforms suggests that some physiological activities may require multiple classes of motor molecules.

Summary

The concept is now firmly in place that neurons and glial cells, like other cells, contain multiple molecular motors responsible for moving discrete populations of molecules, particles, and organelles through intracellular compartments. The complex morphologies and diverse functional interactions of neurons mean that motor proteins and their regulation play a critical role in the nervous system.

BUILDING AND MAINTAINING NERVOUS SYSTEM CELLS

The functional architecture of neurons comprises many specializations in cytoskeletal and membranous components. Each of these specializations is dynamic, constantly changing and being renewed at a rate determined by the local environment and cellular metabolism. Axonal transport processes represent a key to understanding neuronal dynamics and provide a basis for exploring neuronal development, regeneration, and neuropathology. Recent advances provide insight into the molecular mechanisms underlying axonal transport and its role in both normal neuronal function and pathology.

Slow Axonal Transport Moves Soluble Proteins and Cytoskeletal Structures

Slow axonal transport has two major components, both representing movement of cytoplasmic constituents (Fig. 2.8). Cytoplasmic elements in axonal transport move at rates comparable to the rate of neurite elongation. Slow component a (SCa) is movement of cytoskeletal elements, primarily neurofilaments and microtubules. SCa rates typically range from 0.1–1 mm/day, and newly synthesized cytoskeletal proteins may take >1000 days to reach the end of a meter-long axon. Slow component b (SCb) is a complex and heterogeneous rate component, including hundreds of distinct polypeptides from cytoskeletal proteins such as actin (and sometimes tubulin) to soluble enzymes of intermediary metabolism (i.e., glycolytic enzymes). SCb moves at 2–4 mm/day and is the rate-limiting component for nerve growth or regeneration.

The coordinated movement of neurofilament and microtubule proteins provided strong evidence for the "structural hypothesis." For example, in pulse-labeling experiments, labeled neurofilament proteins move as a bell-shaped wave with little or no trailing of neurofilament protein (Baas and Buster, 2004). Neurofilament stability under physiological conditions indicates that soluble neurofilament subunit pools are negligible, so coherent transport of neurofilament triplet proteins implied a transport complex, i.e. neurofilaments. Similarly, coordinate transport of tubulin and MAPs made sense only if microtubules move, because MAPs do not interact with unpolymerized tubulin. The simplest explanation is that neurofilaments and microtubules move as discrete cytological structures, but this idea was controversial for many years.

Development of fluorescently tagged neurofilament or microtubule subunits and methods for visualizing these structures in living cells resolved this issue by documenting movements of individual microtubules and neurofilaments in neurites of cultured neurons (Brown, 2003). Direct observations of individual microtubule or neurofilament segments indicated that they move down axons as assembled polymers. Video images of fluorescently tagged microtubules or neurofilaments reveal discontinuous movements, with long pauses punctuated by brief, rapid translocations at 1–2 m/sec. Due to long pauses, average rates are 2–3 orders of magnitude slower than instantaneous velocities (Baas and Buster, 2004; Brown, 2003). Remarkably, dynein plays a major role in slow axonal transport of microtubules and neurofilaments (He et al., 2005).

Studies on transport of neurofilament proteins indicated that little or no degradation occurs until neurofilaments reach nerve terminals, where they are degraded rapidly. Comparable results were obtained with microtubule proteins. Differential metabolism appears to be a key to targeting of cytoplasmic and cytoskeletal proteins. Proteins with slow degradative rates accumulate, reaching higher steady-state levels. Altering degradation rates changes that steady-state concentration, so enrichment of actin in presynaptic terminals is due to slower turnover of actin than neurofilaments and tubulin. As a result, inhibiting calpain causes neurofilament accumulation in terminals. Differential turnover may involve specific proteases or post-translational modifications that affect susceptibility to degradation. Regardless, cytoplasmic proteins are degraded in the distal axon and do not return in retrograde axonal transport.

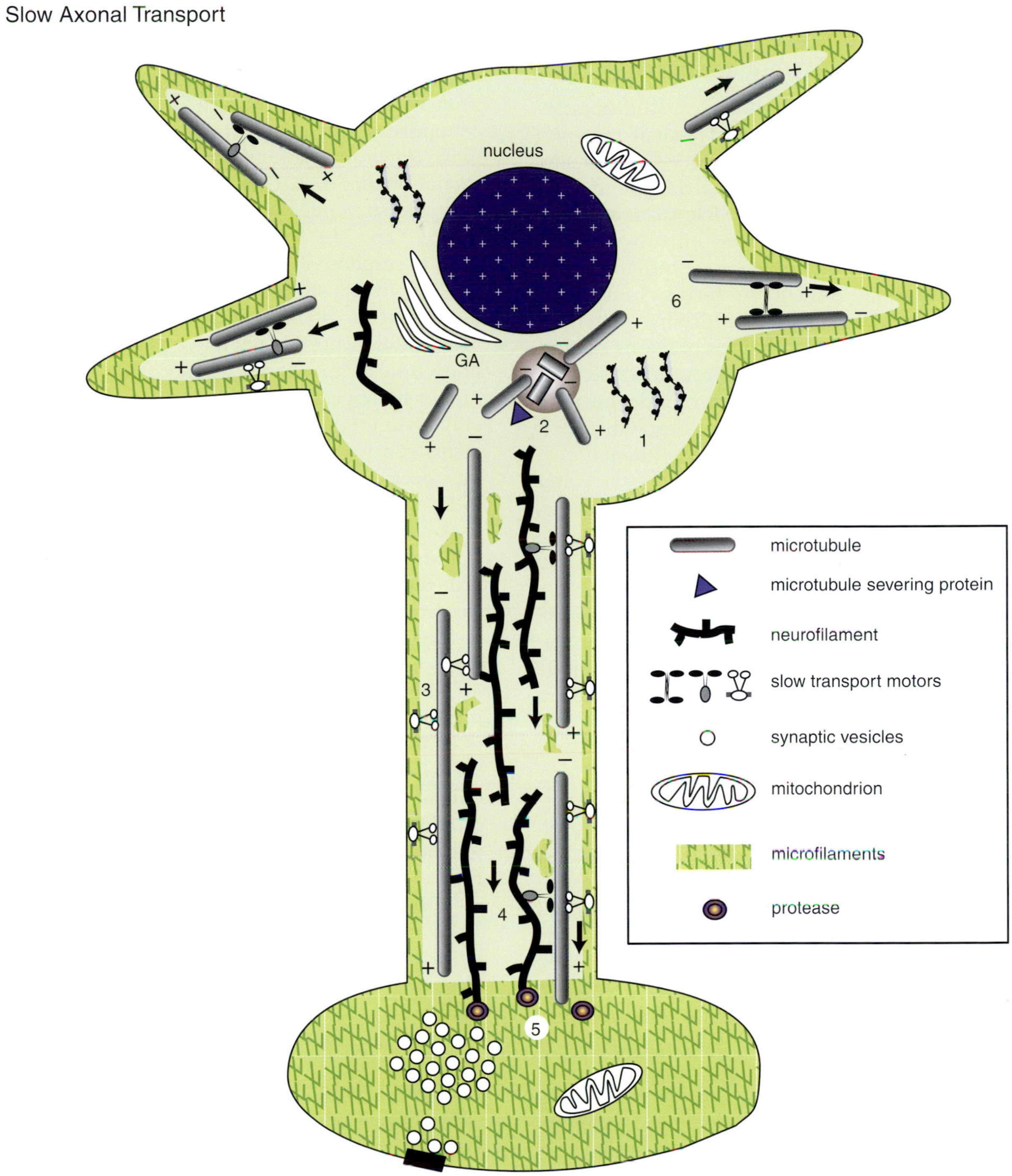

Slow Axonal Transport

FIGURE 2.8 Slow axonal transport represents the delivery of cytoskeletal and cytoplasmic constituents to the periphery. Cytoplasmic proteins are synthesized on free polysomes and organized for transport as cytoskeletal elements or macromolecular complexes (1). Microtubules are formed by nucleation at the microtubule-organizing center near the centriolar complex (2) and then released for migration into axons or dendrites. Slow transport appears to be unidirectional with no net retrograde component. Studies suggest that cytoplasmic dynein may move microtubules with their plus ends leading (3). Neurofilaments may move on their own or may hitchhike on microtubules (4). Once cytoplasmic structures reach their destinations, they are degraded by local proteases (5) at a rate that allows either growth (in the case of growth cones) or maintenance of steady-state levels. The different composition and organization of cytoplasmic elements in dendrites suggest that different pathways may be involved in delivery of cytoskeletal and cytoplasmic materials to dendrites (6). In addition, some mRNAs are transported into dendrites, but not into axons.

Fast Axonal Transport Is the Rapid Movement of Membrane Vesicles and Their Contents over Long Distances Within a Neuron

Early biochemical and morphological studies established that material moving in fast axonal transport was associated with membrane-bound organelles (Fig. 2.9) (Brady, 1995). Mitochondria, membrane-associated receptors, synaptic vesicle proteins, neurotransmitters, and neuropeptides all move in fast anterograde transport. Many cargoes moving down axons in anterograde transport return by retrograde transport (Kristensson, 1987). In addition, exogenous materials taken up in distal regions of axons may be moved back to the cell body by retrograde transport (Fig. 2.9). Exogenous materials in retrograde transport include neurotrophins, such as nerve growth factor, and viral particles invading the nervous system.

Electron microscopic analysis of materials accumulated at a ligation or crush demonstrated that organelles moving in the anterograde direction were morphologically distinct from those moving in the retrograde direction (Tsukita and Ishikawa, 1980). Consistent with ultrastructural differences, radiolabel and immunocytochemical studies indicate quantitative and qualitative differences between anterograde and retrograde moving material. These differences indicate that processing or repackaging events must occur for turnaround in axonal transport. Both proteases and kinases may play a role in turnaround processing for retrograde transport.

Biochemical and morphological approaches resulted in a detailed description of materials moved in fast axonal transport but were not suitable for identifying molecular motors for axonal transport. Methods that permitted direct observation of organelle movements and precise control of experimental conditions were required. Development of video-enhanced contrast (VEC) microscopy allowed characterization of bidirectional movement of membrane-bounded organelles in giant axons from the squid *Loligo pealeii* (Brady, 1995). Years before, studies showed that axoplasm could be extruded from the giant axon as an intact cylinder. VEC microscopic analysis of axoplasm revealed that fast axonal transport continued unabated in isolated axoplasm for hours despite lacking a plasma membrane or other permeability barriers. Combining VEC microscopy with isolated axoplasm, complemented by biochemical and pharmacological approaches, permitted rigorous dissection of mechanisms for fast axonal transport and led to the discovery of kinesin molecular motors (Brady and Sperry, 1995) as well as allowing characterization of regulatory mechanisms associated with fast axonal transport.

How Is Axonal Transport Regulated?

The diversity of polypeptides in each axonal transport rate component and the coherent movement of proteins with very different molecular weights is a conundrum: How can so many different polypeptides move down the axon as a group? Rate components of axonal transport move as discrete waves, each with a characteristic rate and a distinctive composition (Figs. 2.8 and 2.9). The structural hypothesis was formulated in response to such observations. The hypothesis is deceptively simple: Axonal transport represents movement of discrete cytological structures. Proteins in axonal transport do not move as individual polypeptides. Instead, they move as part of a cytological structure or in association with a cytological structure. The only assumption made is that a limited number of elements can interact directly with transport motors, so transported material must be packaged appropriately to be moved. Different rate components result from packaging of transported material into distinct cytological structures. In other words, membrane-associated proteins move as membrane-bounded organelles (vesicles, etc.), whereas tubulins move as microtubules.

Kinesin-1 isoforms appear to be the major (but not sole) motors for fast anterograde movement of membrane-bounded organelles such as vesicles and mitochondria. Similarly, cytoplasmic dynein appears to be the motor for fast retrograde transport of membrane-bounded structures. However, cytoplasmic dynein is also involved in the anterograde transport of microtubules in slow axonal transport. Regulation of motor proteins is needed to assure that appropriate levels of axonal and synaptic components are delivered where needed in the neuron.

Because synthesis of proteins occurs at some distance from many functional domains of a neuron, transport to distal regions of a neuron is necessary, but not sufficient, for proper function. Specific materials must also be delivered to proper sites of utilization and not left in inappropriate locations. For example, synaptophysin has no known function in axons or cell body, so it must be delivered to a presynaptic terminal along with other components necessary for regulated neurotransmitter release. The traditional picture places the presynaptic terminal at the axon end. Such images imply that synaptic vesicles need only move along axonal microtubules until reaching their ends in the presynaptic terminal. However, many CNS synapses are not at axon ends. Many terminals may be located sequentially along a single axon, making *en passant* contacts with multiple target cells. Targeting of synaptic vesicles then becomes a more complex problem, and targeting ion channels

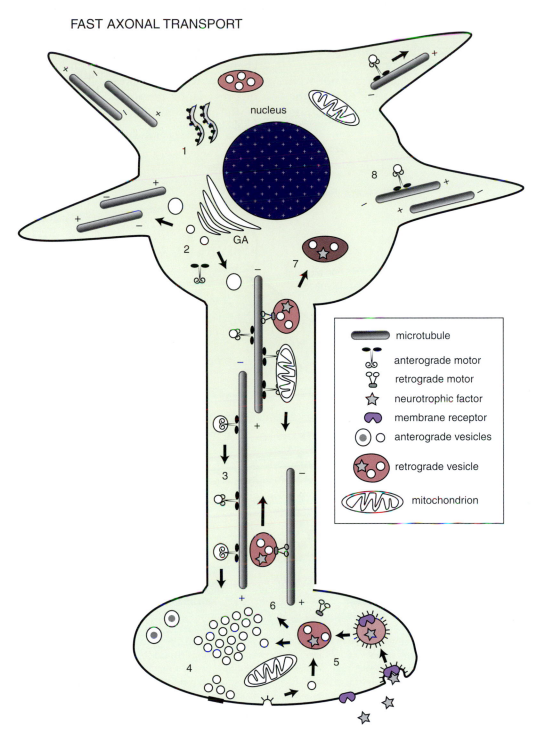

FIGURE 2.9 Fast axonal transport represents transport of membrane-associated materials, having both anterograde and retrograde components. For anterograde transport, most polypeptides are synthesized on membrane-bound polysomes, also known as *rough endoplasmic reticulum* (1), and then transferred to the Golgi for processing and packaging into specific classes of membrane-bound organelles (2). Proteins following this pathway include both integral membrane proteins and secretory polypeptides in the vesicle lumen. Cytoplasmic peripheral membrane proteins such as kinesins are synthesized on free polysomes. Once vesicles are assembled and appropriate motors associate with them, they move down the axon at a rate of 100–400 mm per day (3). Different membrane structures are delivered to different compartments and may be regulated independently. For example, dense core vesicles and synaptic vesicles are both targeted for presynaptic terminals (4), but release of vesicle contents involves distinct pathways. After vesicles merge with the plasma membrane, their protein constituents are taken up in coated vesicles via the receptor-mediated endocytic pathway and delivered to a sorting compartment (5). After proper sorting into appropriate compartments, membrane proteins are either committed to retrograde axonal transport or recycled (6). Retrograde moving organelles are morphologically and biochemically distinct from anterograde vesicles. These larger vesicles have an average velocity about half that of anterograde transport. The retrograde pathway is an important mechanism for delivery of neurotrophic factors to the cell body. Material delivered by retrograde transport typically fuses with cell body compartments to form mature lysosomes (7), where constituents are recycled or degraded. However, neurotrophic factors and neurotrophic viruses act at the level of the cell body and escape this pathway. Vesicle transport also occurs into dendrites (8); less is known about this process.

to nodes of Ranvier or other appropriate sites on the neuronal surface is equally challenging.

Although specific details of targeting are not well understood, a simple model for targeting of synaptic vesicle precursors or ion channels serves to illustrate how such targeting may occur (Fig. 2.10). A local change in the balance between kinase and phosphatase activity in a subdomain like a node of Ranvier or presynaptic terminal can lead to phosphorylation of the motor protein on a vesicle carrying a cargo

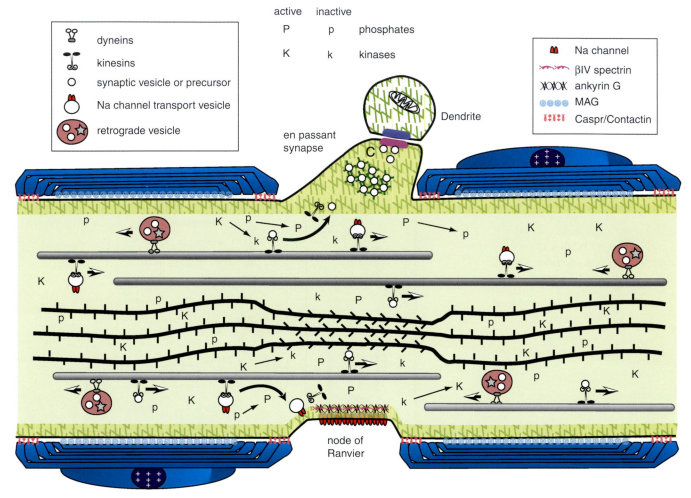

FIGURE 2.10   Axonal dynamics in a myelinated axon from the peripheral nervous system (PNS). Axons are in a constant flux with many concurrent dynamic processes. This diagram illustrates a few of the many dynamic events occurring at a node of Ranvier in a myelinated axon from the PNS. Axonal transport moves cytoskeletal structures, cytoplasmic proteins, and membrane-bound organelles from the cell body toward the periphery (from left to right). At the same time, other vesicles return to the cell body by retrograde transport (retrograde vesicle). Membrane-bound organelles are moved along microtubules by motor proteins such as the kinesins and cytoplasmic dyneins. Each class of organelles must be directed to the correct functional domain of the neuron. Synaptic vesicles must be delivered to a presynaptic terminal to maintain synaptic transmission. In contrast, organelles containing sodium channels must be targeted specifically to nodes of Ranvier for saltatory conduction to occur. Cytoskeletal transport is illustrated by microtubules (rods in the upper half of the axon) and neurofilaments (bundle of ropelike rods in the lower half of the axon) representing the cytoskeleton. They move in the anterograde direction as discrete elements and are degraded in the distal regions. Microtubules and neurofilaments interact with each other transiently during transport, but their distribution in axonal cross sections suggests that they are not stably cross-linked. In axonal segments without compact myelin, such as the node of Ranvier or following focal demyelination, a net dephosphorylation of neurofilament side arms allows the neurofilaments to pack more densely. Myelination is thought to alter the balance between kinase (K indicates an active kinase; k is an inactive kinase) and phosphatase (P indicates an active phophatase; p is an inactive phosphatase) activity in the axon. Most kinases and phosphatases have multiple substrates, suggesting a mechanism for targeting vesicle proteins to specific axonal domains. Local changes in the phosphorylation of axonal proteins may alter the binding properties of proteins. The action of synapsin I in squid axoplasm suggests that dephosphorylated synapsin cross-links synaptic vesicles to microfilaments. When a synaptic vesicle encounters the dephosphorylated synapsin and actin-rich matrix of a presynaptic terminal, the vesicle is trapped at the terminal by inhibition of further axonal transport, effectively targeting the synaptic vesicle to a presynaptic terminal. Similarly, a sodium channel-binding protein may be present at nodes of Ranvier in a high-affinity state (i.e., dephosphorylated). Transport vesicles for nodal sodium channels (Na channel vesicle) would be captured upon encountering this domain, effectively targeting sodium channels to the nodal membrane. Interactions between cells could in this manner establish the functional architecture of the neuron.

targeted to that domain. Thus, phosphorylation of kinesin-1 carrying Na channels at the node of Ranvier would allow delivery of Na channels to the nodal membrane. Evidence for such a mechanism exists for delivery of membrane proteins to growth cones (Morfini *et al.*, 2002) and other domains. A number of kinases have been identified that can regulate kinesin and/or dynein function. Significantly, many of these kinases are misregulated in neurodegenerative diseases such as Alzheimer, Parkinson, and Huntington disease, raising the possibility that axonal transport is disrupted in these diseases.

Although such models are speculative, they satisfy criteria that any mechanism for targeting to specific neuronal subdomains must address. Specifically, mechanisms must act locally because distances to cell body can be great and the number of targets is large. There must be some means to connect a targeting signal to the external microenvironment, such as a glial or muscle cell. Finally, there must be a way of distinguishing subdomains. Thus, synaptic vesicles will not be delivered to nodes of Ranvier, and voltage-gated sodium channels for nodes are not targeted to presynaptic terminals. Careful segregation of different organelles and proteins to different domain of a neuron suggests that highly efficient targeting mechanisms do exist.

Summary

A well-studied feature of the neuron is the phenomenon of axonal transport, which moves in both anterograde and retrograde directions. Axonal transport is responsible for delivery of both membrane-associated and cytoplasmic materials from the cell body to distant parts of the neuron, membrane retrieval and circulation, and uptake of materials from presynaptic terminals and dendrites as well as their delivery to the cell soma. Precise molecular mechanisms by which anterograde and retrograde transport are targeted within an individual dendrite or axon are still being defined.

Neurons and glial cells have unusually large cell volumes enclosed within extensive plasma membrane surfaces. Nature has evolved a number of "universal" mechanisms in other systems and adapted them for the special needs of nervous tissue cells. The synthesis and packaging of components, and in particular proteins, destined for cytoplasmic organelles and cell surface subdomains engage general and evolutionarily conserved molecular mechanisms and pathways that are employed in single-cell yeasts as well as in cells in complex nervous tissue.

Once synthesized and sorted, most intracellular organelles (vesicles destined for axonal or dendritic domains, mitochondria, cytoskeletal components) must be distributed, and targeted, to precise intracellular locations. Because they are so extended in space and exhibit exceptionally complex functional architecture, neurons and glial cells have adapted and developed to a high degree mechanisms that operate to distribute components within all cells. In neurons, movement of materials within the axon has been the central focus of most studies. The motors, cargoes, and regulation of axonal transport are now understood in some measure at the molecular level.

References

Baas, P. W. (2002). Microtubule transport in the axon. Int Rev Cytol *212*, 41-62.

Baas, P. W., and Buster, D. W. (2004). Slow axonal transport and the genesis of neuronal morphology. J Neurobiol *58*, 3-17.

Berg, J. S., Powell, B. C., and Cheney, R. E. (2001). A millennial myosin census. Mol Biol Cell *12*, 780-794.

Brady, S. T. (1995). A kinesin medley: Biochemical and functional heterogeneity. Trends in Cell Biol *5*, 159-164.

Brady, S. T., and Sperry, A. O. (1995). Biochemical and functional diversity of microtubule motors in the nervous system. Curr Op Neurobiol *5*, 551-558.

Brown, A. (2003). Live-cell imaging of slow axonal transport in cultured neurons. Methods Cell Biol *71*, 305-323.

Brown, M. E., and Bridgman, P. C. (2004). Myosin function in nervous and sensory systems. J Neurobiol *58*, 118-130.

Colman, D. R., Kreibich, G., Frey, A. B., and Sabatini, D. D. (1982). Synthesis and incorporation of myelin polypeptides into CNS myelin. J Cell Biol *95*, 598-608.

Court, F. A., Sherman, D. L., Pratt, T., Garry, E. M., Ribchester, R. R., Cottrell, D. F., Fleetwood-Walker, S. M., and Brophy, P. J. (2004). Restricted growth of Schwann cells lacking Cajal bands slows conduction in myelinated nerves. Nature *431*, 191-195.

Gillespie, P. G., and Cyr, J. L. (2004). Myosin-1c, the hair cell's adaptation motor. Annu Rev Physiol *66*, 521-545.

He, Y., Francis, F., Myers, K. A., Yu, W., Black, M. M., and Baas, P. W. (2005). Role of cytoplasmic dynein in the axonal transport of microtubules and neurofilaments. J Cell Biol *168*, 697-703.

Hirokawa, N., and Takemura, R. (2005). Molecular motors and mechanisms of directional transport in neurons. Nat Rev Neurosci *6*, 201-214.

Hochstrasser, M. (1995). Ubiquitin, proteasomes, and the regulation of intracellular protein degradation. Curr Op Cell Biol *7*, 215-223.

Hyams, J. S., and Lloyd, C. W., eds. (1994). Microtubules. New York: Wiley-Liss.

Jahn, R., and Scheller, R. H. (2006). SNAREs—engines for membrane fusion. Nat Rev Mol Cell Biol *7*, 631-643.

Kornfeld, R., and Kornfeld, S. (1985). Assembly of asparagine-linked oligosaccharides. Annu Rev Biochem *54*, 631-664.

Kristensson, K. (1987). Retrograde transport of macromolecules in axons. Annu Rev Pharmacol Toxicol *18*, 97-110.

Lee, M. K., and Cleveland, D. W. (1996). Neuronal Intermediate Filaments. Ann Rev Neurosci *19*, 187-217.

Levy, J. R., and Holzbaur, E. L. (2006). Cytoplasmic dynein/dynactin function and dysfunction in motor neurons. Int J Dev Neurosci 24, 103-111.

Maxfield, F. R., and McGraw, T. E. (2004). Endocytic recycling. Nat Rev Mol Cell Biol 5, 121-132.

Miki, H., Okada, Y., and Hirokawa, N. (2005). Analysis of the kinesin superfamily: insights into structure and function. Trends Cell Biol 15, 467-476.

Morfini, G., Szebenyi, G., Elluru, R., Ratner, N., and Brady, S. T. (2002). Glycogen synthase kinase 3 phosphorylates kinesin light chains and negatively regulates kinesin-based motility. EMBO Journal 23, 281-293.

Murthy, V. N., and De Camilli, P. (2003). Cell biology of the presynaptic terminal. Annu Rev Neurosci 26, 701-728.

Palade, G. (1975). Intracellular aspects of the process of protein synthesis. Science 189, 347-358.

Peters, A., Palay, S. L., and Webster, H. D. (1991). *The Fine Structure of the Nervous System: Neurons and Their Supporting cells*, 3rd ed. New York, NY: Oxford University Press.

Pfister, K. K., Shah, P. R., Hummerich, H., Russ, A., Cotton, J., Annuar, A. A., King, S. M., and Fisher, E. M. (2006). Genetic analysis of the cytoplasmic dynein subunit families. PLoS Genet 2, e1.

Rothman, J. E., and Wieland, F. T. (1996). Protein sorting by transport vesicles. Science 272, 227-234.

Schroer, T. A. (2004). Dynactin. Annu Rev Cell Dev Biol 20, 759-779.

Sherman, D. L., and Brophy, P. J. (2005). Mechanisms of axon ensheathment and myelin growth. Nat Rev Neurosci 6, 683-690.

Steward, O. (1995). Targeting of mRNAs to subsynaptic microdomains in dendrites. [Review]. Curr Opin Neurobiol 5, 55-61.

Sudhof, T. C. (2004). The synaptic vesicle cycle. Annu Rev Neurosci 27, 509-547.

Tsukita, S., and Ishikawa, H. (1980). The movement of membranous organelles in axons. Electron microscopic identification of anterogradely and retrogradely transported organelles. J Cell Biol 84, 513-530.

Vouyiouklis, D. A, and Brophy, P. J. (1995). Microtubule-associated proteins in developing oligodendrocytes: transient expression of a MAP2c isoform in oligodendrocyte precursors. J Neurosci Res 42, 803-817.

Walter, P., and Johnson, A. E. (1994). Signal sequence recognition and protein targeting to the endoplasmic reticulum membrane. Annu Rev Cell Biol 10, 87-119.

Energy Metabolism in the Brain

Gerald A. Dienel

The brain is a complex, heterogeneous organ in which many pathways for its input and output are organized in a somatotopic manner. Changes in the activities of the neural circuits and networks involved in specific functions therefore govern demand for energy at a local level, i.e., shifts in cellular activities in the pathways in which signaling rises and falls cause parallel changes in the rate of metabolism to generate ATP and rate of blood flow to deliver fuel (*functional activity* Box 3.1). Brain energy metabolism is a dynamic, highly regulated process that is governed, in large part, by moment-to-moment interactions among brain cells that process information from sensory and cognitive activities and to direct functions of the body. Cellular energy demands, consumption of oxygen and glucose, and blood flow change simultaneously in response to variations in cellular activity, and these processes are therefore often described as "coupled." The cell types contributing to functional metabolic activity are sometimes considered as a "neurovascular unit," consisting primarily of neurons, astrocytes, and endothelial cells.

Energy budget. Information processing is the hallmark of brain function and, as schematically illustrated in Figure 3.1, there is a close temporal–spatial relationship between functional activity in brain and rates of glucose utilization and blood flow. The largest component of brain work involves consumption of ATP to pump Na^+ and K^+ across membranes. A recent estimate of the brain's energy budget by Attwell and Laughlin (2001) calculated that Na^+, K^+-ATPase activity accounts for most of the ATP turnover related to excitatory signaling activity and predicted that about 10% of the signaling energy in rodent brain is used to maintain somal resting potentials, 47% to support action potentials, 34% consumed in postsynaptic structures to restore membrane ionic gradients, and smaller fractions assigned to other

processes in neurons and astrocytes. The numerical values assigned to these processes are likely to be updated as more is learned about cellular oxidative activity, neuron–astrocyte interactions, and the contributions of astrocytes and inhibitory neurons to overall brain energetics (Hyder *et al.*, 2006; Hertz *et al.*, 2007; Gjedde, 2007; Nicholls, 2007; Nicholls *et al.*, 2007). Defense against oxidative stress arising from generation of reactive oxygen species (ROS) by various enzymes and the electron transport chain involves the glutathione system and consumes NADPH produced by the pentose phosphate shunt pathway (Fig. 3.1). A much smaller fraction of the brain's energy requirements is consumed over a much longer timescale for so-called "housekeeping" activities. In spite of the pedestrian implications of this name, these "housekeeping" energy-requiring processes are essential for brain function and include mRNA, protein, lipid, and organelle turnover; axonal transport; and biosynthetic activities associated with neurotransmitters and other compounds that are synthesized *de novo* in the brain. Neuronal signaling events involving fast excitatory action potentials, receptor activation, ion channel opening and closing, and inhibitory transmitter actions take place on a timescale of <10 milliseconds (Korf and Gramsbergen, 2007). Astrocytes surrounding working synapses must also be quickly activated during excitatory neurotransmission because they control glutamate and K^+ levels in extracellular fluid (Hertz *et al.*, 2007). Sensory perception occurs within a time frame of 0.5 to 2 sec, similar to that of synaptic vesicle filling. Local rates of cerebral blood flow (CBF) are regulated by products of metabolism, but recent evidence indicates that the hemodynamic response is driven by glutamate-mediated signaling processes that involve both neurons and astrocytes and their interactions with endothelial cells in the blood vessels (Attwell and

BOX 3.1

CONCEPTS RELATED TO FUNCTIONAL METABOLISM

Functional metabolic activity. Metabolic rate associated with a specific brain function or pathway. In normal brain, metabolic rates rise when functional activity increases (e.g., during sensory or cognitive stimulation) and fall with decrements in activity (e.g., sensory deprivation or anesthesia). Glucose and oxygen consumption may, but need not, change by the same proportion under different conditions. During activation of normal, normoxic subjects, most studies have observed greater increases in blood flow and glucose utilization compared to oxygen consumption.

Resting or baseline activity. Activity of brain cells during conditions in which no specific stimulus is given to the subject. Resting activity and the associated rates of blood flow and metabolism can be difficult to define due to the unknown interactions of the subject with the specific experimental environment (e.g., paying attention to visual, auditory, or other sensory cues or information) and the level of "stress" that a subject may experience.

Brain activation. Responses to a specific stimulus or experimental condition. Brain work involves mainly ion pumping to maintain and restore ionic gradients across cellular membranes and to control the levels of compounds in extracellular fluid. Increased rates of these ATP-requiring processes stimulate metabolism specifically in the activated pathways.

Glucose (Glc) utilization. The cerebral metabolic rate (CMR) of glucose (CMR_{glc}) denotes the overall rate of glucose consumption by all pathways. Under normal steady-state conditions (Box 3.2), the rate of any step in the glycolytic pathway equals glucose utilization, whereas the rate of the steps in the TCA cycle are twice those of glycolysis due to formation of two pyruvate per glucose. CMR_{glc} is generally assayed at the hexokinase step with radiolabeled deoxyglucose (DG); the rate of phosphorylation of DG is converted to rate of glucose utilization by taking into account the kinetic differences for transport and phosphorylation between DG and glucose; approximately two glucose are utilized per deoxyglucose (Sokoloff *et al.*, 1977).

Oxygen utilization. CMR_{O2} denotes the overall rate of oxygen consumption by all pathways. Most oxygen is consumed via the electron transport chain to generate ATP in mitochondria, but there are a number of other enzymes (e. g., monoamine oxidase, mixed-function oxidases) that utilize oxygen as a substrate and contribute to CMR_{O2}. Global rates of oxygen consumption can be determined by measuring blood flow and arteriovenous differences for oxygen; local CMR_{O2} can be measured by positron emission tomography by determination of metabolism of $^{15}O_2$.

Glycolytic metabolism. Metabolism of glucose or glycogen via the glycolytic pathway to pyruvate by reactions that are not dependent on oxygen consumption. Glycolysis generates a net 2 ATP per glucose converted to 2 pyruvate. Glycolytic flux increases during hypoxic or anoxic conditions, with increased production of lactate from pyruvate in order to regenerate NAD^+ from the $NADH + H^+$ generated by the glyceraldehyde-3-phosphate dehydrogenase reaction.

Oxidative metabolism. Metabolism of pyruvate, keto acids, monocarboxylic acids, fatty acids, and amino acids via the tricarboxylic acid cycle. Oxidative metabolism is linked to the consumption of oxygen and generation of ATP via the electron transport chain. This pathway can be used for generation of energy and for net *de novo* biosynthesis of amino acids from glucose, which also consumes oxygen and generates ATP (see text).

"Coupling" of blood flow and metabolism at a local level. Rates of blood flow and glucose utilization in different structures in the brain are highly correlated and generally change in parallel during different physiological states, leading to the notion of coupling. Regulation of local rates of blood flow by cellular activity and metabolism involves release of substances that regulate vascular diameter and therefore blood flow. Because activities of neurons, astrocytes, and endothelial cells are involved in major aspects of local brain function, metabolism, and blood flow, these cell types together are sometimes collectively referred to as a *neurovascular unit*.

Iadecola, 2002; Takano *et al.*, 2006; Gordon *et al.*, 2007; Iadecola and Nedergaard, 2007). Changes in blood flow can be initiated within hundreds of milliseconds and persist for seconds or longer after a brief stimulus; notably, the tissue volume with increased blood flow response is considerably greater than that of the activated cells. Thus, the highest fraction of the brain's energy production is devoted to the fastest processes, and linkage of neuronal and astrocytic signaling to control of blood flow is essential to maintain an adequate and continuous supply of fuel to the brain.

Fueling brain work. Glucose and oxygen are the primary and obligatory fuels for brain, and other substrates cannot substitute for the continuous supply of glucose

Functional Activity Increases ATP Demand, CMR$_{glc}$, and CBF

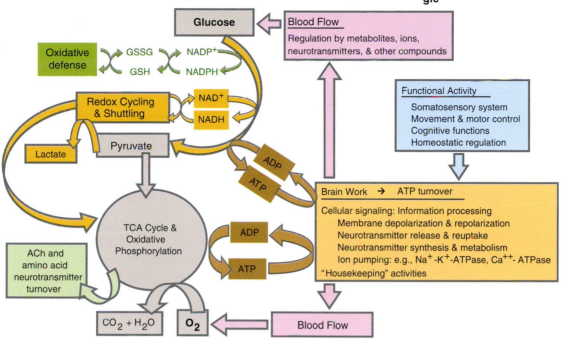

FIGURE 3.1 **Relationships among brain function, metabolism, and blood flow.** The predominant energy-requiring brain activities involve restoration of ionic gradients across cellular membranes in conjunction with cellular signaling and information transfer. Compounds derived from cellular activity and metabolism regulate local rates of blood flow to deliver the obligatory fuel, glucose and oxygen, to the brain. Glucose is nearly completely oxidized via the glycolytic and oxidative (tricarboxylic acid, TCA, cycle) pathways. Metabolism of glucose generates ATP and precursors for *de novo* synthesis of glucose-derived neurotransmitters, e.g., acetylcholine (Ach), glycine, D-serine, and glutamate, within the brain. Redox (oxidation-reduction) cofactors are also produced and can be used to generate ATP or lactate, defend against oxidative stress, and participate in biosynthetic reactions. (Adapted from Dienel, 2002.)

from blood (Clarke and Sokoloff, 1999). Under *resting* conditions (Box 3.1) the stoichiometry of oxygen to glucose utilization is close to the theoretical maximum of 6.0, which corresponds to the complete oxidation of glucose: 1 glucose + 6O$_2$ → 6CO$_2$ + 6 H$_2$O. The ratio of cerebral metabolic rates (CMR) for these two substrates, CMR$_{O2}$/CMR$_{glc}$, in resting human and animal brain is usually found to be slightly lower than 6.0, e.g., in the range of 5.5–5.8, indicating that some of the glucose carbon is used in nonoxidative biosynthetic reactions and a small amount of lactate is released to the cerebral venous blood. Also, some oxygen is consumed by enzymes, such as mixed-function oxidases, that are not components of the oxidative pathways of glucose metabolism. The respiratory coefficient for brain, the ratio of CO$_2$ released from brain to O$_2$ taken up, is approximately 1, indicating that carbohydrate is the primary brain fuel; oxidation of lipid consumes oxygen in steps that do not result in proportionate CO$_2$ production. Alternative fuels are used by brain during development or abnormal conditions, such as prolonged starvation, but these compounds are not capable of sustaining brain function in the normal adult (Roberts, 2007).

Typical concentrations of oxygen and glucose in blood of the normal rat are shown in Table 3.1. Note that the concentrations of oxygen and glucose in whole blood are similar, but extraction fraction for oxygen is 4.5 times higher than that of glucose; nearly half of the oxygen that traverses the brain's microvasculature is extracted and consumed, whereas only 10% of the glucose is removed from the same blood sample by the brain. Negative arteriovenous difference for lactate indicates a small net efflux corresponding to about 6% of the glucose taken up into brain, based on two lactate being equivalent to one glucose. Measurement of arteriovenous (A-V) differences across the brain and rates of cerebral blood flow (CBF) enable calculation of the global cerebral metabolic rate using the Fick principle, CMR = CBF (A-V). In the example provided in Table 3.1, CMR$_{O2}$ and CMR$_{glc}$ are 4.13 and 0.75 µmol g^{-1} min^{-1}, respectively, and the ratio of their rates of their utilization (calculated from the ratio of arteriovenous differences for each substrate, since blood flow cancels out) was 6.1. There is no reservoir for oxygen in brain, whereas there is a small pool of unmetabolized glucose in brain, along with the reserve glucose stored as glycogen

TABLE 3.1 Blood and Brain Metabolites and Global Metabolic Rates During Rest and Brain Activation

Variable	Before Activation	During Sensory Stimulation
Arterial blood concentration (mmol L^{-1})		
Oxygen	8.15 ± 0.59	7.98 ± 0.73
Glucose	6.81 ± 0.82	7.81 ± 1.32**
Lactate	0.50 ± 0.30	1.96 ± 0.43**
Arteriovenous difference (A-V) (mmol L^{-1})		
$(A-V)_{O2}$	3.75 ± 0.56	2.97 ± 0.56
Mean O_2 extraction fraction $[(A-V)_{O2}/A_{O2}]$	0.46	0.37
$(A-V)_{glucose}$	0.68 ± 0.20	0.60 ± 0.06**
Mean Glc extraction fraction $[(A-V)_{glc}/A_{glc}]$	0.10	0.077
$(A-V)_{lactate}$	-0.08 ± 0.06	0.02 ± 0.04**
Cerebral blood flow (CBF) (mL $\cdot$ $g^{-1} \cdot$ min^{-1})	1.1	1.8
Ratios of (A-V) differences (mmol L^{-1}/mmol L^{-1})		
Oxygen/glucose	6.1 ± 1.1	5.0 ± 1.1**
Lactate/glucose	-0.14 ± 0.14	0.02 ± 0.08**
Metabolic rate (μmol g^{-1} min^{-1})		
Mean CMR_{O2} (CBF x $[A-V]_{O2}$)	4.13	5.35
Mean CMR_{glc} (CBF x $[A-V]_{glc}$)	0.75	1.08
Brain metabolite concentration (μmol g^{-1})		
Glucose	2.8 ± 0.5	3.1 ± 0.5
Lactate	1.0 ± 0.5	1.9 ± 0.4*
Glycogen	6.1 ± 1.4	5.4 ± 1.3

Note: Blood metabolite levels and arteriovenous differences were determined in samples of blood taken from the femoral artery and superior sagittal sinus of normal conscious rats; brain metabolites were determined in cortex from funnel-frozen brain. Values are means $\pm$ SD for 39 rats during rest and 8 during activation (*p <0.05, **p <0.01 compared to resting condition). Ratios of arteriovenous differences were calculated for each animal, not from means for all animals. Because blood flow was measured in different groups of animals, mean values are calculated for CMR_{glc} and CMR_{O2}. Data from Madsen *et al.* (1999).

(Table 3.1). The *half-life* (Box 3.2) of glucose in resting brain is 1.2–1.6 min (Savaki *et al.*, 1980), and the major carbohydrate pool (glucose + glycogen + 0.5 lactate) is about 9.4 μmol glucose equivalents g^{-1}. At the normal rate of glucose utilization of about 0.75 μmol g^{-1} min^{-1} all of this fuel would be consumed within 12.5 min if none were provided from blood. Thus, an insulin overdose that severely reduces blood glucose level or a vascular blockage that reduces or eliminates blood flow to the brain can lead to energy failure, loss of consciousness or function, and perhaps cell death, depending on the severity of decrement, extent of tissue involvement, and duration of the insult.

Stimulation of brain activity. During *brain activation* (Box 3.1) induced by generalized sensory stimulation of the animal by brushing of the body and whiskers with soft paintbrushes, blood glucose and lactate levels rose due to glycogenolysis and muscular activity, respectively (Table 3.1). Because plasma lactate rose above that in brain, there was a small net increase in lactate uptake and brain lactate level also increased. Extraction of oxygen and glucose tended to fall somewhat, but because blood flow increased considerably, the rates of utilization of both oxygen and glucose increased. The disproportionate rise in glucose compared to oxygen consumption caused the CMR_{O2}/ CMR_{glc} ratio (Box 3.1) to fall from 6.1 to 5.0 during stimulation. This phenomenon, sometimes called *aerobic glycolysis*, is commonly observed in many (but not all) brain activation studies in humans and experimental animals (e.g., Fox *et al.*, 1988; Madsen *et al.*, 1995, 1999), but the biochemical and cellular basis of utilization of glucose in excess of oxygen even in the presence of adequate levels of oxygen in blood and increased rates of delivery of oxygen to the brain is not understood (reviewed by Dienel and Cruz, 2004; Dalsgaard, 2006). The cellular basis of the metabolic shifts induced by brain activation is an active research area.

To summarize, global measurements of arteriovenous differences in early studies in humans and experimental animals provided the first line of evidence for the major substrates taken up into brain and released from brain, leading to identification of major substrates and metabolic products. Quantitative cerebral metabolic rates were made possible by the work of Kety and colleagues who developed the first quantitative assays to measure cerebral blood flow by determination of rates of clearance of inert gas (Kety and Schmidt, 1948).

Metabolic flow diagrams (Fig. 3.1) illustrate the flux of glucose through the glycolytic and oxidative pathways to end products, but it is important to recognize that conversion of glucose to CO_2 is not a direct process due to the many side and exchange reactions that mix incoming carbon atoms with those of endogenous metabolite pools. As will become evident later in the chapter, this characteristic of metabolism is important for functional imaging and biochemical and magnetic resonance spectroscopic studies of brain metabolism. Due to the high yield of ATP from the oxidative phosphorylation reactions in mitochondria, respiration generates most of the ATP in brain cells, with a smaller fraction produced by the glycolytic pathway. The highest energy yield

BOX 3.2

SUBSTRATE LEVELS, ENZYMES, AND TRANSPORTERS

Metabolite concentration. The concentration of a compound in the intracellular or extracellular fluid or total tissue concentration is the net result of all transport (carrier-mediated and diffusion) and metabolic processes that contribute to its formation and removal. At steady state the overall concentration of a metabolite is stable even though individual molecules are continuously entering and leaving the metabolic pool. Because sum of all input and output rates determines concentration of a metabolite, concentration itself does not reflect flux of any single pathway and changes in concentration need not be in the same direction as changes in the flux of any specific process.

Metabolite pool. The quantity of a compound in a tissue, cell, or subcellular compartment. Pool is sometimes used interchangeably with concentration but the concept of a pool often infers localization or compartmentation, since more than one pool can contribute to the cellular content of a compound. For example, (1) the cytoplasmic and mitochondrial NAD/NADH pools are segregated due to the impermeability of the inner mitochondrial membrane to the pyridine nucleotides, and (2) the glutamate pool that is the precursor for glutamine synthesis is located in astrocytes and is a small fraction of the total glutamate pool, which is mainly neuronal.

Turnover time and half-life. In a simple example of a first-order reaction where $x \rightarrow y$, the turnover time of a metabolite or pool is the reciprocal of the rate constant, k, for a reaction in which the rate of change of compound x is proportional to its concentration, $dx/dt = k[x]$, and $k = (dx/dt)/[x]$. This situation applies to enzymatic reactions in which the substrate concentration is much lower than the K_m so that the initial reaction rate varies with substrate concentration (see Fig. 3.3C). In this case, turnover time or time constant can be estimated by dividing the metabolite concentration by pathway flux rate. Turnover time is the time required to metabolize the quantity of the intermediate in that pool at a constant rate. Due to the exponential decline of the rate of an irreversible first-order reaction, not all molecules in the pool are metabolized even though the pool "turns over." The half-time or half-life is the time required for the concentration to fall by 50%, and the integrated rate equation becomes $kt = 2.3 \log [x]_{t=0}/[x]_{t=t}$. Thus, when x is half gone, $kt = 2.3\log(1/0.5) = 0.69$, and the half-life, $t_{1/2} = 0.69/k$. Thus, at 5 and 6 half-times, 94 and 99% of the pool are consumed, respectively, if there is no substrate replenishment. Note that the expressions for turnover time and half-life are more complex for reversible reactions and for second-order reactions involving two substrates.

Labile metabolite. Any metabolite that is rapidly consumed or produced when normal metabolism is disrupted, e.g., when tissue is sampled and extracted for analysis. Labile metabolites include glucose, glycogen, glycolytic and TCA cycle intermediates, pyridine nucleotides, signaling compounds, and high-energy compounds and their metabolites. More stable compounds include most but not all amino acids, proteins, and lipids. Rapid enzyme-inactivation procedures are required for determination of the levels of labile metabolites.

Enzyme or transporter amount or activity. In a simple case when an enzyme catalyzes a one-step reaction with a single substrate and follows Michaelis–Menten kinetics, the rate of the reaction can vary with the amount of enzyme, the substrate concentration, and affinity for the substrate, according to the following relationship: $v = V_{max} S/(K_m + S)$, where v is the velocity of the reaction, V_{max} is the maximal velocity, S is substrate concentration, and K_m is the concentration of substrate at which the velocity is half-maximal. Note that when the substrate concentration is much higher than K_m, $v = V_{max}$ and the enzyme is "saturated" with substrate; because further increases in S do not alter the rate the activity is maximal (Fig. 3.3C). However, when S is approximately equal to K_m the velocity of the reaction varies as S changes above and below K_m. A similar expression can be derived for transport reactions, $v = T_{max} S/(K_t + S)$, where T_{max} is maximal transport capacity and K_t is analogous to K_m. Enzyme activities assayed under optimal conditions in the test tube are proportional to enzyme amount, which is expressed in terms of V_{max} (maximal rate per unit tissue or protein, e.g., $\mu mol^{-1} min^{-1}$ gram tissue^{-1} or mg protein^{-1}. *In vitro* assays of enzymatic activity are measures of enzyme amount or capacity when determined under optimal conditions; they do not represent *in vivo* activities or fluxes through that enzymatic step. Histochemical assays based on enzyme reactions, immunostaining of tissue, and Western blots are *representations* of enzyme location, relative amount, or capacity, not actual *in vivo* activity related to pathway flux.

Flux. The overall rate of a specific enzymatic reaction or of a pathway composed of more than one enzyme. In a multistep pathway the overall flux is governed by the slowest, or rate limiting, step. Under steady-state conditions the rate of each step in a multistep pathway is

BOX 3.2   (continued)

equal to the flux through that pathway. Determinants of flux include the amounts and activities of enzymes controlling the nonequilibrium or rate-limiting reactions; metabolite concentrations, transport, and diffusion; concentrations of enzyme regulators (activators and inhibitors); coenzyme availability; and pH. *In vivo reaction rates*, not the concentrations of their pathway constituents, provide information about the *dynamic state* of metabolism.

Steady state. Steady state is the condition under which metabolic rates and metabolite concentrations are constant, and glucose utilization rates (glycolytic and/or TCA cycle) can be linked to overall energy production. In contrast, under non-steady-state conditions, rates may transiently rise or fall to and from the baseline level during activating or depressing conditions, and some of the carbon flowing through a pathway may be used for biosynthetic reactions or some may be produced from other compounds (e.g., glycogen).

Physiological relationships among mRNA level, protein amount, and protein function. Changes in mRNA level are not necessarily reflected by corresponding alterations in protein amount, and variations in protein amount (capacity, amount or *in vitro* activity) need not be reflected by parallel shifts in the *in vivo* activity of the protein or metabolic pathway flux due to regulatory controls. Some examples from the literature included in the "Metabolomics, Transcriptomics, and Proteomics" section illustrate

the following points: (1) function-dependent changes in mRNA and protein levels can be discordant and also differ from shifts in pathway activity; (2) the phenotype of a gene knockout animal may arise, in part, from compensatory changes in the expression of hundreds of genes involved in pathways or processes seemingly unrelated to the gene of interest; and (3) marked upregulation of enzyme levels by genetic engineering need not alter either pathway flux or levels of metabolites in the pathway. Thus, caution must be applied to interpretation of studies that examine only one aspect of the biological range from gene to mRNA to protein to metabolite to function. Single time points are not sufficient to characterize functional responses to changes in gene expression because induction or suppression of enzymes by various approaches can have effects that vary with response time, duration of response, direction of response, and magnitude of response. In various mammalian species, substrate and enzyme levels are more similar to each other, whereas metabolic rates differ markedly and are scaled to brain (and body) weight. Thus, within normal variations, metabolic fluxes are not indicated by the concentrations of substrates or of the amount of enzymes. Also, enzyme amounts generally enable considerable excess catalytic capacity compared to pathway flux, so large decrements in enzyme amount or activity can be tolerated before biochemical and functional changes become evident (e.g., in genetic diseases such as phenylketonuria).

---

from glucose is obtained when the NADH generated by glycolysis is shuttled to the mitochondria for oxidation via the electron transport chain. Nevertheless, specific processes in brain cells may preferentially utilize ATP generated by glycolysis or require compounds generated by nonoxidative metabolism, and overall metabolism of glucose rises disproportionately more than oxygen consumption during *brain activation* (Table 3.1, Box 3.1). Pathways that branch from the main glycolytic pathway are also important and participate in defense against oxidative stress (pentose phosphate shunt pathway), as precursors for glucose storage as glycogen and for synthesis of glycoprotein precursors and amino acids, including neuromodulators. Branch pathways linked to the tricarboxylic acid (TCA) cycle are essential for neurotransmitter turnover, and oxidative metabolism is closely linked to turnover of acetylcholine, which is synthesized from glucose via citrate, as well as to

de novo synthesis of glutamate, glutamine, GABA, and aspartate. Because the energy reservoirs in brain are very limited, brain function is dependent on a continuous supply of oxygen and glucose to satisfy local demand for ATP to power brain work, for excitatory and inhibitory neurotransmitter synthesis to signal from cell to cell, and for defense against reactive oxidative species produced by working brain (Fig. 3.1).

To summarize, co-registration of changes in functional activity, metabolism, and blood flow enables the use of markers of metabolism or flow to evaluate brain function as important tools for imaging and quantifying brain function and activity in living subjects under normal and pathophysiological conditions. Understanding the basis for functional-imaging signals requires a detailed knowledge of the pathways of brain energy metabolism, cellular specialization that is unique to or critical for brain function, and the involvement of energy metabolism in neurological diseases or disorders. Because metabolic reactions and mechanisms and their control are common to many tissues, readers are referred to biochemistry texts for details not covered in this chapter. For detailed coverage of many topics presented in this chapter, interested readers are encouraged to consult the reference books listed at the end of the chapter, which along with some recent reviews cited in the text, provide broad access to the brain energy metabolism literature.

MAJOR PATHWAYS OF BRAIN ENERGY METABOLISM

The brain requires uninterrupted delivery of glucose and oxygen from blood to generate the energy required to sustain consciousness. Glucose also serves as the carbon source for biosynthesis of essential compounds within brain. Minor substrates that can be metabolized by brain tissue and maintain near-normal ATP levels *in vitro* do not enter brain in sufficient quantities to satisfy the energy or biosynthetic demands of mature brain.

Fuel Delivery to Brain

Blood–brain barrier. Tight junctions between adjacent endothelial cells in the cerebral vasculature form the blood–brain barrier (see monograph by Davson and Siegel, 1996), which restricts entry of water-soluble or aqueous-phase compounds into brain from blood. This selective permeability is a critical function, since blood contains many neuroactive compounds such as glutamate and catecholamines, and intricate control of cell-to-cell signaling in brain would be severely disrupted if material could readily enter the brain. Thus, transporters are required to facilitate movement of essential materials across the blood–brain barrier (Fig. 3.2). Early studies by Oldendorf and colleagues clearly demonstrated that glucose and neutral amino acids had high blood–brain barrier permeability, whereas transit from blood to brain of acidic amino acids, biogenic amines, and short-chain mono- and dicarboxylic acids was highly restricted (Oldendorf, 1981). Although immature brain cells in tissue culture and in slices of adult brain are capable of metabolizing many compounds, these metabolites are not transferred into brain in quantities sufficient to sustain the brain's high-energy demands. The consequence of restricted entry into brain of critical amino

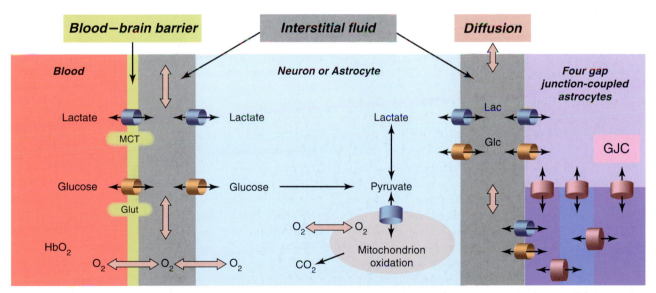

FIGURE 3.2 Transporters govern passage of hydrophilic compounds across the blood–brain barrier. Glucose transporters (GLUTs) are present in high quantities in the endothelial cells in the cerebral microvasculature; the 55 kDA isoform of GLUT1 catalyzes the bidirectional, facilitative transport of glucose (Glc) in blood across the blood–brain barrier, whereas the 45 kDA GLUT1 and GLUT3 deliver glucose to glia and neurons, respectively. The isoforms of monocarboxylic acid transporter (MCT) are also preferentially located in different cell types, MCT1 in endothelial cells, MCT1 and 4 in astrocytes, and MCT2 in neurons. MCTs have broad substrate specificity and can transport lactate (Lac), pyruvate, ketone bodies, and short-chain fatty acids across membranes. Note that an MCT is required for pyruvate transport into mitochondria. Astrocytic but not neuronal MCT(s) preferentially transport acetate as substrate. Astrocytes are extensively coupled via gap junctions to form a large syncytium through which many small molecules (< ~1k Daltons) can quickly diffuse (GJC denotes gap junctional communication). Metabolites can also diffuse throughout brain via the perivascular, interstitial, and cerebrovascular fluids. HbO$_2$ denotes oxygen bound to hemoglobin (Hb) in blood, and free oxygen can readily diffuse across the blood–brain barrier and cellular membranes.

acids and glucose-derived neurotransmitters is that they must be synthesized from glucose within brain, thereby closely linking energy metabolism with neurotransmitter synthesis and degradation within brain.

Glucose transporters. In adult mammalian brain, only glucose has a sufficiently high number of transporters (GLUT1) to facilitate an adequate flux of carbon fuel into brain (Fig. 3.2), and from extracellular fluid into brain cells by transporter isoforms that are preferentially localized in different cell types (e.g., GLUT1 in astrocytes and GLUT3 in neurons; reviewed by Simpson *et al.*, 2007). Maximal glucose transport capacity (T_{max}) is approximately three times maximal glucose utilization capacity (V_{max}) (Holden *et al.*, 1991) (see Box 3.2, paragraph on enzyme activity). In contrast to glucose and other minor fuels that enter brain in small quantities under normal circumstances (e.g., lactate, ketone bodies, fatty acids), oxygen diffuses across the blood–brain barrier and cellular membranes.

The pathways for distribution of glucose within the brain are poorly understood, and the fraction that enters astrocytic endfeet that surround cerebral capillaries compared to that entering interstitial fluid and diffusing through extracellular spaces remains to be established. Diffusion of small molecules through the perivascular and interstitial fluid is restricted by the tortuosity of extracellular fluid space. Cellular membranes present barriers to transport into brain cells, but compounds can rapidly move to and from cells via transporters because glucose and lactate transport across the cellular membranes is much faster than metabolism (Holden *et al.*, 1991; Dienel and Hertz, 2001; Simpson *et al.*, 2007). Astrocytes are extensively coupled via gap junctions to form a large, extensive intracellular syncytium formed by thousands of astrocytes in brain *in vivo*, contrasting the low-level (typically 10–15 cells) gap junctional coupling of cultured astrocytes (Ball *et al.*, 2007). Compounds taken up into astrocytes can diffuse extensively from astrocyte to astrocyte and along perivascular space through the gap junction–coupled endfeet (Fig. 3.2; Ball *et al.*, 2007). Flow of perivascular fluid within brain is powered by arterial pulsations and is capable of quickly distributing compounds in extracellular space throughout the brain within minutes (Rennels *et al.*, 1985). The *maximal capacity for glucose transport* (T_{max}, Box 3.2) across the microvascular membranes is calculated to be considerably higher than that across neuronal and astrocytic plasma membranes (about 260, 35, and 5 nmol per 10^6 cells per min, respectively); taking into account the kinetic parameters of the glucose transporters (transporter protein number, K_t [Box 3.2] and number of transport cycles per transporter per sec [k_{cat}]), neurons are calculated to have a nine-fold higher capacity for glucose transport than astrocytes at 5 mM glucose (Simpson *et al.*, 2007).

Glucose transporters are facilitative, and glucose moves down its concentration gradient from arterial plasma to interstitial fluid to intracellular fluid. Thus, brain glucose level passively follows changes in plasma (Fig. 3.3A). Normal rats have a brain-to-plasma ratio for glucose of approximately 0.22 for arterial plasma glucose concentrations over the normo- and hyperglycemic (about 8–25 mM) range; within this wide range, brain glucose rises and falls in proportion to that in plasma, and the brain-to-plasma ratio is constant (Figs. 3.3A and 3.3B). When the steady-state level of glucose in plasma is reduced below about 8 mM in a graded manner, the brain-to-plasma distribution ratio progressively falls because glucose delivery is increasingly unable to match demand (Fig. 3.3B). Brain hexokinase has a K_m for glucose of about 0.05 mM, and the enzyme is saturated with substrate in normoglycemic conditions (inset, Fig. 3.3C). However, during hypoglycemia, hexokinase becomes progressively more unsaturated as brain glucose level falls below about 0.5 mM, causing the rate of glucose phosphorylation to fall (Fig. 3.3C). Glucose concentration is normally relatively uniform throughout the brain, as shown by autoradiographic studies using the radiolabeled, nonmetabolizable glucose analog, 3-O-methylglucose; local changes in brain glucose level during activation or during depression of brain metabolism can be detected with labeled methylglucose, but the changes are relatively small (see later, Fig. 3.14). To summarize, glucose supply is not limiting for brain metabolism; supply matches demand in activated brain when glucose utilization rates are stimulated by several fold, although there may be brief transients at the onset or termination of stimuli before compensatory processes respond to change in demand.

Monocarboxylic acid transporters (MCT). MCTs are facilitative transporters with broad substrate specificity that cotransport one H^+ ion along with each monocarboxylic acid (reviewed by Halestrap and Price, 1999). In the adult rodent (postnatal day 21 and older), different MCT isoforms are preferentially, but not exclusively, enriched in brain cell types; MCT1 is present in endothelial cells, MCT1 and MCT4 are mainly in astrocytes, whereas MCT2 is mainly in neurons. The K_m values of these isoforms for lactate are in the range of about 0.7 mM, 3–5 mM, and 15–30 mM for MCT2, 1, and 4, respectively (Halestrap and Price, 1999; Hertz and Dienel, 2005; Simpson *et al.*, 2007; and references cited in these reviews), indicating that the neuronal isoform is nearly saturated under normal physiological levels of lactate (Table 3.1),

A   Brain glucose concentration varies with plasma glucose level

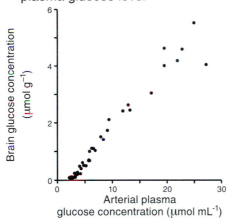

B   Glucose distribution ratio falls during hypoglycemia

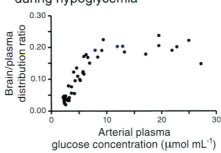

C   Hexokinase becomes unsaturated during severe hypoglycemia

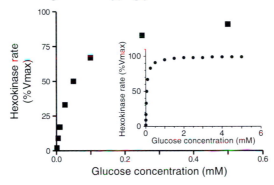

FIGURE 3.3   **Brain glucose concentration during hyperglycemic, normoglycemic, and hypoglycemic conditions.** Steady-state concentration of brain glucose (A) and brain-to-plasma glucose concentration ratio (B) as functions of arterial plasma glucose concentration (panels A and B are plotted from data of Dienel *et al.*, 1991). Under normal and hyperglycemic conditions (plasma glucose range about 8–25 mM), the mean glucose brain-to-plasma distribution ratio is about 0.22, but this ratio falls as plasma glucose is clamped at progressively lower levels because glucose delivery does not match the metabolic demand for glucose. Brain hexokinase has a K_m for glucose of about 0.05 mM. Hexokinase is saturated with substrate under normal conditions (brain glucose concentration >1 mM) (inset, panel C), but the velocity of the hexokinase reaction, calculated using simple Michaelis–Menten kinetics (v = V_{max} S/[K_m + S], where v is the reaction velocity, V_{max} = maximal velocity, S = substrate concentration, and K_m is the value for S at which v = 0.5 V_{max}), falls progressively as intracellular glucose concentration falls below 0.5 mM (C).

whereas flux across the endothelial and astrocytic membranes will rise and fall with lactate concentration within the normal physiological range. An important finding is that the substrate specificity for unidentified astrocytic MCT isoform(s) is the basis for preferential uptake of acetate into astrocytes, not neurons (Waniewski and Martin, 1998). The cell-type specificity of acetate uptake and metabolism enables its use as an astrocyte "reporter molecule" to study astrocytic oxidative metabolism, glutamate–glutamine cycling between astrocytes and neurons, and astrocytic responses to brain activation and pathology; metabolic compartmentation and cellular aspects of metabolism are discussed in more detail in following text.

Blood-borne lactate, acetate, and ketone bodies are minor fuels for normal adult brain due to low metabolite levels in normal arterial plasma and low monocarboxylic acid transporter levels in the microvasculature of adult brain. However, in the suckling mammal, the levels of MCT1 isoform at the blood–brain barrier is about 10-fold higher than in the adult to facilitate utilization of ketone bodies (β-hydroxybutyrate and acetoacetate) and lactate; after weaning, the capacity for monocarboxylic acid transport across the blood–brain barrier falls markedly (Cremer, 1982). The level of MCT1 in the microvasculature can adapt to pathophysiological conditions, and during prolonged starvation or feeding of a ketotic diet the

MCT levels increase, thereby enabling greater use by the brain of ketone bodies in blood (Nehlig, 1996, 2004; Yudkoff *et al.*, 2007; Prins, 2008).

Recent studies in exercising humans have shown that intense muscular activity generates increased amounts of lactate in blood to a range where blood lactate level rises from <1 mM to 7–15 mM during maximal exercise and exceeds that in brain (range: ~0.5–1 μmol g wet weight^{-1} during rest to ~2 during activation) (reviewed by Dalsgaard, 2006). Higher blood-to-brain lactate levels lead to inward facilitative transport of lactate (along with one H^+ ion that is cotransported with each monocarboxylic acid) into brain where it is oxidized and contributes to reducing the calculated ratio of $CMR_{O2}/CMR_{glucose+lactate}$. In normal adult rats, labeled lactate and glucose inserted into the interstitial fluid via microdialysis probes are oxidized by both neurons and astrocytes in substantial amounts (Zielke *et al.*, 2007). Use of endogenously generated lactate as a cellular fuel is controversial and will be discussed in more detail in following text.

Glycolytic Pathway: Generation of Pyruvate from Glucose

Glycolysis is a highly regulated series of reactions that convert the six-carbon glucose (Glc) to pyruvate, a triose, by steps that do not require oxygen (Fig. 3.4). The first committed (irreversible) step of the pathway is catalyzed by hexokinase and consumes 1 ATP to generate glucose-6-phosphate (Glc-6-P), which traps the glucose molecule inside the cell. Glc-6-P regulates hexokinase activity by feedback inhibition, and ATP also inhibits the reaction. Glc-6-P is in equilibrium with Glc-1-P via the phosphoglucomutase reaction and with fructose-6-P (Fru-6-P) via the phosphoglucoisomerase reaction. These three compounds are readily interconvertible and constitute a pool that can be considered as a major "branch point." Thus, Glc-6-P has three major fates: continuation down the glycolytic pathway, entry into the pentose phosphate shunt pathway, and storage as glycogen via Glc-1-P and UDP-glucose (Fig. 3.4). Fru-6-P is also the precursor for synthesis of monosaccharides and complex carbohydrates that are key components of glycoproteins and glycolipids, including mannose, glucosamine, and sialic acid. The pentose shunt pathway is present in all brain cells and regulated mainly by availability of NADP; this pathway generates NADPH for biosynthetic reactions and oxidative defense plus five-carbon precursors for nucleic acid biosynthesis, particularly in dividing cells. Brain glycogen is located mainly, but not exclusively, in astrocytes. Glycogen is the major energy reserve in brain, and it is actively used during normal physiological activity.

Synthesis of fructose-1,6-bisphosphate (Fru-1,2-P_2) by phosphofructo-1-kinase (PFK) consumes a second ATP and is the committed step for glycolysis. PFK is a very highly regulated enzyme, with many activators and inhibitors that together convey to the enzyme information related to the cellular energy demand and the "status" of downstream pathways (Passonneau and Lowry, 1964). Cleavage of Fru-1,6-P_2 by aldolase generates two trioses, dihydroxyacetone-P and glyceraldehyde-3-P, and these three compounds form another readily interchangeable pool.

Oxidation of glyceraldehyde-3-P to 1,3-diphosphoglycerate by glyceraldehyde-3-P dehydrogenase requires the cofactor NAD^+ and generates NADH + H^+. The next enzymatic step catalyzed by 2-P-glyceratekinase generates 1 ATP for each of the two three-carbon molecules formed from glucose, and energy required to "activate" the hexose is now "replaced." The 3-P-glycerate is then converted via 2-P-glycerate to phosphoenolpyruvate (PEP). Note that 3-P-glycerate is the precursor for 3-P-hydroxypyruvate, from which L-serine is synthesized (Fig. 3.4). L-serine is then the precursor for glycine and the one-carbon pool that is essential for the methyl-transfer reactions. Also, L-serine is the precursor for D-serine, a neuromodulator in astrocytes. PEP is converted to pyruvate by the action of pyruvate kinase to produce 1 ATP per PEP. Thus, the net ATP yield for direct conversion of glucose to two pyruvate via glycolytic pathway is 2 ATP, with the generation of 2NADH, which must be oxidized to NAD^+ for glycolytic flux to continue. Pyruvate also has alternative metabolic fates (Fig. 3.4). If NADH is not oxidized via shuttle systems (see following text) as fast as it is generated, lactate dehydrogenase converts pyruvate to lactate and regenerates NAD^+. Pyruvate can also be transaminated to form alanine, and in astrocytes pyruvate is a substrate for pyruvate carboxylase, the enzyme required for CO_2 fixation and anaplerotic reactions. Under normal resting conditions when the ratio of oxygen to glucose utilization is close to 6, nearly all the pyruvate is oxidized via the TCA cycle.

The concentration of each of the glycolytic intermediates in brain is very low, and taken together, the total amount of all intermediates, approximately 0.4 μmol g^{-1}, is in the range of 15–25% of the brain glucose concentration (Table 3.2). Due to the small pool sizes of these compounds and the high glucose utilization rate (~0.7 μmol g^{-1} min^{-1}, Table 3.1), the *turnover rates* (i.e., divide metabolite concentration by pathway flux rate; Box 3.2) of glycolytic intermediates are much higher than that of brain glucose. The total concentration of NAD^+ + NADH is also low and oxidation-reduction (redox) cycling of these

Glycolytic Pathway: Branch Points and Major Regulators

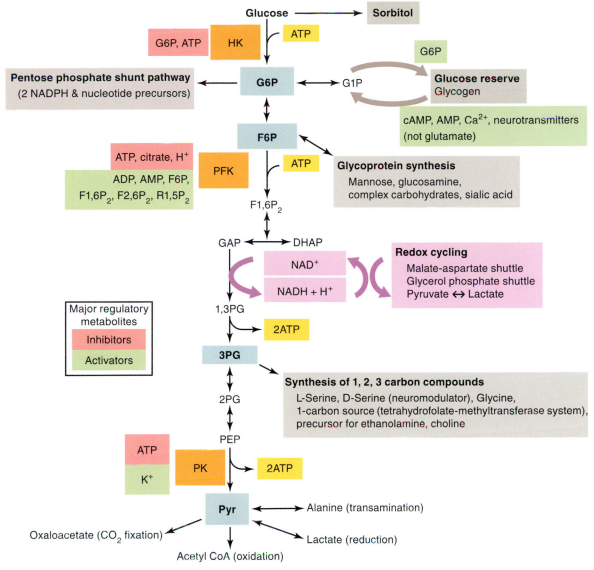

FIGURE 3.4 **Conversion of glucose to pyruvate and use of glucose carbon for storage and biosynthetic reactions.** Glucose is the obligatory carbon source for brain, and most is converted to pyruvate during normal resting conditions. However, glucose also supplies precursors for a number of other critical pathways with diverse and essential roles in brain structure and function. Regulation of the glycolytic pathway involves feedback inhibition and allosteric modulation by metabolites involved in different pathways, thereby integrating control with overall metabolic, biosynthetic, and redox economies of the cell. Major inhibitors or activators of key glycolytic enzymes are indicated in red or green boxes, respectively, adjacent to the enzymes. Abbreviations: G6P, glucose-6-phosphate (P); HK, hexokinase; G1P, glucose-1-P; F6P, fructose-6-P; F1,6P$_2$, fructose-1,6-bisphosphate; F2,6P$_2$, fructose-2,6-bisphosphate; R1,5P$_2$, ribose-1,5-bisphosphate; PFK, phosphofructo-1-kinase; DHAP, dihydroxyacetone-P; GAP, glyceraldehyde-3-P; 1,3PG, 1,3-diphosphoglycerate; 3PG, 3-phosphoglycerate; 2PG, 2-phosphoglycerate; PEP, phosphoenolpyruvate; Pyr, pyruvate.

cofactors must be rapid to sustain a high glycolytic rate, i.e., it must be twice CMR$_{glc}$, or 1.4 μmol g^{-1} min^{-1}, since two trioses are formed per glucose consumed. The calculated cytoplasmic [NAD$^+$]/[NADH] and [NADP$^+$]/NADPH] ratios in freeze-blown brains are about 670 and 0.12, respectively, and the ratio of [ATP]/[ADP][HPO$_4^{2-}$] in cytoplasm is

370 M^{-1}, indicating a high phosphorylation state in brain (Veech *et al.*, 1973).

Gluconeogenesis. Reversal of the glycolytic pathway to synthesize glucose or glycogen from triose precursors (Fig. 3.4) can occur in brain, but it is considered to be a minor pathway. Thus, label from ^{14}C-*labeled* (radiolabeled tracers, Box 3.3) lactate, alanine, aspartate, and

TABLE 3.2 Approximate Concentrations of Metabolites in Rodent Brain[1,3]

Glycolytic Pathway[1]		TCA Cycle Intermediates[1]		Amino Acids[2]		Energy Reserves and High-Energy Metabolites[1,3]	
Metabolite	Concentration (μmol g^{-1})	Metabolite	Concentration (μmol g^{-1})	Metabolite	Concentration (μmol g^{-1})	Metabolite	Concentration (μmol g^{-1})
Glucose (Glc)	1.5-2.5	Acetyl CoA	0.001-0.005	Glutamate	11.6	Glucose-1-P	0.004-0.01
Glucose-6-P	0.06-0.2	CoA	0.003	Taurine	6.6	UDP-Glc	0.1
Fructose-6-P	0.01-0.02	Citrate	0.25-0.35	N-Acetylaspartate	5.6	Glycogen	1.5-12
Fructose-1,6-P$_2$	0.01-0.1	Isocitrate	0.02	Glutamine	4.5	Creatine-P	4.7-4.9
Dihydroxyacetone-P	0.01-0.03	α-Ketoglutarate	0.15-0.2	Aspartate	2.6	Creatine	5.6-6.1
3-P-Glycerate	0.04-0.05	Succinate	0.45-0.7	GABA	2.3	ATP	2.5-3
Phosphoenolpyruvate	0.004-0.005	Fumarate	0.07	Glycine	0.68	ADP	0.4-0.6
Pyruvate	0.05-0.2	Malate	0.35-0.45	Alanine	0.65	AMP	0.03-0.07
Total (Glc-6-P to Pyr)	**~0.4**	Oxaloacetate	0.004-0.007	Serine	0.98	P$_i$	2.0-2.7
		Total	**~1.5**	Threonine	0.66		
Lactate	0.5-1.5			Lysine	0.21		
				Arginine	0.11		
NAD	0.20-0.3			Histidine	0.05		
NADH	0.03-0.1			Leucine	0.05		
NADP	0.005-0.02			Isoleucine	0.02		
NADPH	0.003-0.02			Valine	0.07		
				Phenylalanine	0.05		
				Tyrosine	0.07		
				Proline	0.08		
				Methionine	0.04		
				Ornithine	0.02		
				Tryptophan	0.003		
				Glutathione	2.6		
				Total	**~40**		
				Total (Glu + Gln + Asp + GABA)	**~21**		

Note: Metabolite levels are approximate ranges of values from rat, mouse, or guinea pig brain taken from cited sources. Note that brain glucose level varies as a function of arterial plasma glucose level (Fig. 3.3A), and the normal rat brain:plasma ratio for glucose is about 0.2 over the normo- and hyperglycemic range (Fig. 3.3B). Concentrations of some metabolites vary with animal handling and history (e.g., glycogen) have a large range. Estimates of the total tissue concentrations in each category were calculated using the mean values of the indicated ranges.

[1]Data from Siesjö (1978); Veech (1980); McIlwain and Bachelard (1985).

[2]From the mean values of many studies that were compiled by Clarke et al. (1989).

[3]Data from Siesjö (1978); Cruz and Dienel (2002); McIlwain and Bachelard (1985); Duffy et al. (1975).

BOX 3.3

USE OF RADIOLABELED TRACERS IN BIOLOGY

Nuclide. Atom characterized by atomic number (number of protons in nucleus) and atomic mass (number of protons plus neutrons).

Isotope. Nuclide with same number of protons but different number of neutrons. The chemical properties are the same but the mass is different; a preceding superscript indicates mass.

Radioactive and stable isotopes. A radioactive atom undergoes spontaneous disintegration with emission of radiation, e.g., ^{14}C. A stable isotope does not decay to generate another compound, e.g., ^{13}C. Autoradiographic and positron emission tomographic (PET) studies use many different radiolabeled compounds, whereas magnetic resonance spectroscopic (MRS) studies use stable isotopes.

Activity. Number of nuclear disintegrations per unit time; the standard international unit is the Becquerel (Bq) = 1 disintegration per sec (dps). A commonly used unit of activity is the Curie (Ci), which equals 2.22×10^{12} disintegrations per min (dpm); 1 μCi = 2.22×10^6 dpm.

Specific activity. Ratio of activity to molar quantity (e.g., dpm/mole or nCi/μmol). Knowledge of specific activity of a precursor and counts in a compound derived from the precursor allows calculation of the amount of a labeled compound (e.g., dpm in unknown x mole/dpm of precursor = moles unknown). A similar expression, isotopic enrichment, is used in MRS studies to describe relative proportion of an isotope (see Fig. 3.18B).

Precursor–product relationships. The specific activity or isotopic enrichment of a precursor in a metabolic pathway is always higher than that of the product due to dilution of the label by the quantity of unlabeled material in each of the successive metabolic pools through which a compound is metabolized. For example: (1) If the specific activity of glutamate is 1 mCi/mmol glutamate, and if 1 mmol glutamate is converted to glutamine in a reaction mixture that contains 1 mmol glutamine, the specific activity of glutamine will be 0.5, i.e., 1 mCi in glutamine/(1 mmol unlabeled glutamate + 1 mmol unlabeled glutamine). (2) If [1-^{14}C]glucose is converted to lactate, the maximal specific activity of [^{14}C]lactate is half that of glucose because only one of the two lactate formed from glucose will be labeled (see Fig. 3.18A).

Interpretation of radiolabeling studies is often complicated by metabolic compartmentation and insufficient knowledge of the specific activities of important but small precursor pools. For example, an apparent decrease in a pathway flux can be caused by dilution of the precursor-specific activity even if there is no flux change. Thus, if unlabeled lactate were added to a tissue

along with labeled glucose, the equilibrative steps of lactate–pyruvate interconversion would quickly reduce the specific activity of pyruvate, thereby reducing the amount of labeled pyruvate entering the TCA cycle. In this example, reduced label accumulation into TCA cycle–derived amino acids (e.g., glutamate) might be incorrectly interpreted as lower oxidative metabolism of glucose when in fact the apparent labeling was due to dilution of the pyruvate-specific activity (that was not measured and appropriately taken into account). Increased glycogenolysis could also cause dilution of downstream labeled metabolites derived from labeled glucose, thereby reducing the apparent label incorporation into astrocytic amino acid pools.

Integrated specific activity. Integral of specific activity over time (time–activity integral).

Beta-emitter. Decay releases an electron (beta particle). For example, ^{14}C decays to produce an electron plus nitrogen. Beta emitters commonly used in biological studies include ^{3}H, ^{14}C, ^{35}S, ^{32}P, and ^{45}Ca. These compounds can be assayed by detection of the electron by liquid scintillation counting or autoradiography using X-ray film or imaging plates.

Positron emitter. Decay releases a positron that travels a short distance and then annihilates with an electron to produce simultaneously two photons emitted at a 180° angle; these events are assayed with pairs of coincidence detectors. Examples include ^{18}F, ^{11}C, and ^{15}O, which are commonly used in labeled compounds in positron-emission tomographic (PET) studies (Phelps, 2004).

Physical half-life. The time for half of the atoms to decay. For example, ^{14}C, 5730 years; ^{3}H, 12.3 years; ^{18}F, 109 min; ^{11}C, 20 min; ^{15}O, 2 min. Note that isotopes used in PET assays generally have short half-lives, thereby reducing exposure of the subject to radiation. The relative stability of ^{14}C and ^{3}H facilitates broad utilization of these isotopes in biological and biochemical studies.

Biological half-life. The time for half of the dose to be eliminated; biological half-life depends on the compound and its metabolic and excretory fates.

Tracer amount or quantity. The dose of a radiolabeled compound is usually so small that it does not change the concentration of the endogenous compound and does not alter fluxes; the higher the specific activity, the lower the molar amount of the tracer to achieve the same labeling sensitivity. Introduction of tracer amounts of a radiolabeled compound allows specific enzymatic steps, pathways, and fluxes to be studied *in vitro*, in cultured cells, and *in vivo*.

glutamate can be incorporated into glycogen in cultured astrocytes (Hamprecht and Dringen, 1995). This process involves pyruvate carboxylase to carry out the ATP-requiring, CO_2-fixation step to convert pyruvate to oxaloacetate, followed by decarboxylation to generate PEP, then the action of PEP carboxykinase, fructose-1,6-phosphatase to generate Glc-1-P that can be incorporated into glycogen. A net of 6 ATP is required to convert two molecules of lactate to Glc-6-P, and one more is needed to incorporate the Glc-6-P into glycogen.

Redox Shuttling and Recycling Mechanisms: Oxidation of Pyruvate or Conversion to Lactate

NADH generated in the cytosol cannot directly cross the inner mitochondrial membrane to be oxidized via the electron transport chain (see following text), and alternative processes must be used to regenerate NAD^+. When the CMR_{O2}/CMR_{glc} ratio is close to 6, almost all the glucose is oxidized and reducing equivalents are nearly quantitatively transferred into the mitochondria, mainly by the malate–

aspartate shuttle (MAS), and perhaps also by the glycerol–phosphate shuttle. The first step of the malate–aspartate shuttle (Fig. 3.5A) is reduction of oxaloacetate to generate malate, which enters the mitochondrial matrix in exchange for α-ketoglutarate. The malate is oxidized in the mitochondria to form NADH, which enters the electron transport chain at complex I (see Fig. 3.8), with the subsequent generation of ATP. The oxaloacetate is transaminated to form aspartate that leaves the mitochondrion via the aspartate–glutamate carrier, an electrogenic process driven by cotransport of H^+ with glutamate; oxaloacetate is then regenerated in the cytosol by transamination (reviewed by LaNoue et al., 2007). The glycerol–phosphate shuttle (McKenna et al., 2006) involves cycling of dihydroxyacetone phosphate and glycerol-P (Fig. 3.5B) across the inner membrane and transfer of reducing equivalents to Coenzyme Q, with a lower ATP yield than the MAS (Fig. 3.8). Another potential mechanism for redox shuttling proposed by Cerdán and colleagues (2006) involves transcellular metabolite transfer in which

A
Regeneration of NAD⁺ via Malate-Aspartate Shuttle or LDH

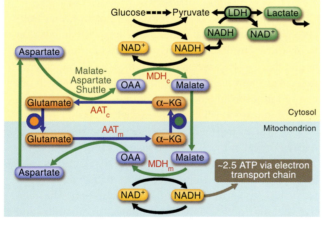

B
Regeneration of NAD⁺ via Glycerol-Phosphate Shuttle

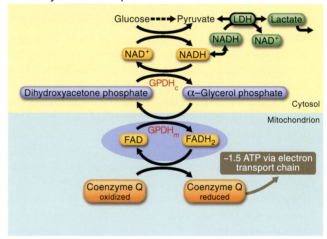

FIGURE 3.5 **Redox shuttle systems transfer reducing equivalents from cytosol to mitochondrial electron transport chain.** NAD^+ can be regenerated by one of three processes: conversion of pyruvate to lactate (A), the malate–aspartate shuttle (A), or the glycerol-phosphate shuttle (B). Note the different ATP yields from the three processes. Lactate formation removes pyruvate from the metabolic pathway of the cell, and NADH oxidation yields no energy. The MAS generates the highest yield of ATP due to entry into the electron transport chain at complex I, whereas the glycerol-P shuttle enters the electron transport chain at the Coenzyme Q step (see Fig. 3.8). The carrier proteins involved in the malate–aspartate shuttle are the aspartate–glutamate carrier, which is an electrogenic process due to cotransport of H^+ with glutamate, and the malate-α-ketoglutarate carrier. The malate–aspartate shuttle is the major shuttle system in brain (McKenna et al., 2006; LaNoue et al., 2007). Panel A is reproduced from Fig. 4B of Dienel and Cruz, 2004, with permission of Elsevier. Note that the mixing of carbon between the mitochondrial and cytosolic pools via the malate–aspartate shuttle serves another function that is critical for metabolic studies of brain. Oxidative metabolism of labeled glucose causes labeling of the tricarboxylic acid (TCA) cycle intermediates, including oxaloacetate and α-ketoglutarate. By means of transaminase-mediated exchange reactions, label is transferred to the mitochondrial glutamate and aspartate amino acid pools. A key aspect of dilution of label in the large cytoplasmic amino acid pools is shuttling of keto and amino acids across the mitochondrial membrane. Without this shuttling, label would be more rapidly lost via the decarboxylation reactions in the TCA cycle. Abbreviations: OAA, oxaloacetate; αKG, α-ketoglutarate; LDH, lactate dehydrogenase; MDH, malate dehydrogenase; AAT, aspartate aminotransferase; GPDH, glycerol-P dehydrogenase. The subscripts c and m denote the cytosolic and mitochondrial enzymes, respectively.

lactate moves from one cell to another where it is converted to pyruvate, which is then shuttled back to the originating cell and oxidized; this mechanism can bypass the requirement for an intracellular redox shuttle in one cell type.

When the rate of glycolysis exceeds the redox shuttle capacity, pyruvate is converted to lactate by lactate dehydrogenase (LDH). This reaction enables the glycolytic pathway to operate under anaerobic conditions, with the release of lactate from the cell; coupling of glycolysis to LDH activity has the advantage of sustaining a high flux rate at the cost of release of the triose carbon from the cell. Because (1) glucose is rarely rate limiting under normal conditions and (2) glucose and lactate transport are facilitative (i.e., driven by concentration gradients and not energy requiring), the energetic cost of release of lactate by brain for use by other body tissues is negligible. However, as discussed in following text, lactate efflux has consequences for metabolic imaging using labeled glucose as a tracer because lactate carries all the labeled carbon atoms of glucose (except those lost via decarboxylation reactions in the pentose shunt pathway).

Early studies demonstrated the predominance of the malate–aspartate shuttle compared to the glycerol–phosphate shuttle system in brain (reviewed by Siesjö, 1978), but recent reports suggest the possibility of considerable differences in the capacities of redox shuttle systems in neurons and astrocytes. This is a very important issue related to understanding metabolic labeling in these cell types and their contributions to oxidative metabolism and brain energetics. The finding that the aspartate–glutamate carrier is poorly detectable by immunocytochemical procedures in astrocytes compared to neurons in adult brain (Ramos et al., 2003) was unexpected because magnetic resonance spectroscopic (MRS) studies have shown substantial oxidative metabolism of glucose in astrocytes in brain in vivo. Also, astrocytes contain pyruvate carboxylase, which confers to astrocytes anaplerotic capability for net synthesis of TCA-cycle-derived amino acids from glucose (Yu et al., 1983; Shank et al., 1985; Öz et al., 2004). Direct evidence for oxidative metabolism of glucose in mature astrocytes that were acutely isolated from adult mice comes from ^{13}C-labeling patterns of TCA-cycle-derived metabolites after incubation of the cells with [U-^{13}C]glucose (Lovatt et al., 2007). Entry of glucose-derived pyruvate into the TCA cycle would not occur without a working redox shuttle mechanism, i.e., if there were no shuttle, all pyruvate must be converted to lactate to maintain glycolysis (Figs. 3.4, 3.5). The quantitative contributions of astrocytic oxidative metabolism to brain energetics has been a difficult,

unresolved issue, and emerging data suggest that the contributions of astrocytes to overall brain energy metabolism is greater than previously recognized (Hertz et al., 2007). For example, the recent finding of mitochondria in perisynaptic astrocytic processes (Lovatt et al., 2007) raises the possibility that these thin structures may not be exclusively glycolytic as suggested by their narrow diameter that may exclude larger-diameter organelles, including mitochondria (Hertz et al., 2007). Thus, glutamate uptake into perisynaptic astrocytic structures may stimulate astrocytic oxidative metabolism of glutamate in addition to glucose in conjunction with excitatory glutamatergic neurotransmission and glutamate–glutamine cycling.

Glycogen: Glucose Storage and Mobilization During Physiological Activity and Energy Crisis

Incorporation of glucose into a macromolecule minimizes cellular osmotic changes, and glycogen, along with the enzymes involved in its synthesis and degradation, is widely distributed throughout the brain. Glycogen is mainly localized in the soma, perivascular endfeet, and fine processes of astrocytes, but glycogen granules are also present in some large neurons in the brain stem and spinal cord. The concentration of glycogen in brain historically has been found to be rather low, in the range of about 1–3 μmol glucosyl units g^{-1} (note: tissue glycogen concentration actually refers to the glucose equivalents [i.e., glucosyl units] released from glycogen when it is assayed in tissue extracts by enzymatic degradation), and glycogen had been considered to be a minor energy store since its level was similar to that of glucose (Table 3.2). However, recent assays using carefully handled animals and tissue fixation and extraction methods procedures that prevent enzymatic degradation of labile metabolites (Box 3.2) found much higher glycogen levels in unstimulated rat brain, ranging from 5 to 12 μmol glucosyl units g^{-1} (Cruz and Dienel, 2002). Glycogen utilization during whisker stimulation of the rat (Swanson et al., 1992) and large compensatory increases in utilization of blood-borne glucose during whisker stimulation when glycogenolysis is blocked (Dienel et al., 2007) indicate that both net consumption of glycogen and glycogen turnover increase during normal physiological activity of astrocytes. A recent, novel finding is that glycogen is involved in taste-aversion learning and memory formation in the newborn chick (Gibbs et al., 2006). In the preceding studies, glycogen turnover was determined under normoglycemic conditions, indicating that activity-evoked glycogenolysis is not due to generalized hypoglycemia in brain and suggesting that glycogen turnover plays an important role at a local level. Because glycogen phosphorylase is located throughout the subcellular compartments of astrocytes, soma, endfeet, and the fine processes that surround synapses (Peters *et al.*, 1991; Hamprecht *et al.*, 2005), one possibility is that glycogen serves as a fuel "buffer" for specific *metabolic pools* (Box 3.2) during times of abrupt increases in energy demand associated with excitatory neurotransmission. Increased levels of extracellular potassium and catecholamine neurotransmitters (e.g., adrenaline, serotonin, histamine), but *not* extracellular *glutamate* (Sorg and Magistretti, 1991), are capable of stimulating glycogenolysis in cultured cells and brain slices, thereby linking astrocyte glycogen turnover to synaptic activity (Hertz *et al.*, 2007). High levels of glutamate cause glycogen levels to rise in cultured astrocytes, ^{13}C-MRS studies have shown that glutamate can be converted to lactate, and ^{14}C labeling studies have shown incorporation of label from lactate into glycogen, so it is conceivable that some glutamate can be converted to glycogen under specific conditions.

The reactions involved in glycogen synthesis and degradation are tightly regulated, and transfer of glucosyl groups to and from glycogen interfaces with the glycolytic pathway at the glucose-6-P step (Fig. 3.4). Glucose-6-P is in equilibrium via the phosphoglucomutase reaction with glucose-1-P, which is about 7% of the level of glucose-6-P. Glucose-1-P is next converted to uridine diphosphoglucose (UDP-glucose) with the release of pyrophosphate, and the glucosyl moiety of UDP-glucose is then added to glycogen in an α-1,4-glycosidic linkage to a growing amylose chain by glycogen synthase, a highly regulated enzyme. Note that epimerization of UDP-glucose forms UDP-galactose, which is an important precursor for complex carbohydrates. The phosphorylated form of glycogen synthetase is dependent (D form) on glucose-6-P as an activator, whereas the unphosphorylated form is independent (I form) of glucose-6-P. Phosphorylation of the I form to the D form is stimulated by 3′,5′-cyclic AMP (cAMP). Glycogen phosphorylase is also very highly regulated and exists in the a (active) and b (inactive) forms; interconversion of these forms is governed by phosphorylase b kinase and phosphorylase a phosphatase. The phosphorylase b form is activated by AMP and inorganic phosphate (P_i) (i.e., during low-energy conditions) and inhibited by ATP and glucose-6-P (i.e., when energy levels are adequate). cAMP stimulates a protein kinase to phosphorylate the phosphorylase b kinase, which is also activated by Ca^{2+}; these steps generate phosphorylase a and stimulate glycogenolysis to form glucose-1-P. Thus, regulation of glycogen synthesis and degradation can be fine-tuned by means of intracellular metabolic

regulators and coordinated with (1) local neuronal activity by means of K^+ uptake from extracellular fluid and (2) local and global neuronal activity via neurotransmitter receptors. For example, the noradrenergic system from the locus coeruleus has widespread innervation throughout the brain, and β-adrenergic receptors on astrocytic membranes are potent stimulators of glycogenolysis. Finally, astrocytic glycogenolysis is also regulated by oxidative stress. Increased demand for generation of NADPH via the pentose phosphate shunt pathway is triggered by exposure of cultured astrocytes to reactive oxygen species and hydrogen peroxide, and glycogen is quickly degraded (Dringen *et al.*, 2007).

Cultured astrocytes degrade glycogen to pyruvate, which is converted to lactate and released to the tissue culture medium, contrasting the liver, which converts glycogen to glucose-6-P that is dephosphorylated, followed by release of "free" glucose to blood (reviewed by Hamprecht and Dringen, 1995). A number of studies of glucose-6-phosphatase activity in brain *in vivo* and in cultured astrocytes *in vitro* by the Sokoloff laboratory failed to detect significant phosphatase activity, consistent with the release of lactate instead of glucose from astrocytes. There have, however, been some reports of activity of this phosphatase enzyme in brain cells, but its physiological significance remains to be established. Thus, a major difference between the fate of glycogen in brain compared to liver is astrocytic metabolism of glycogen to generate to pyruvate–lactate, whereas liver glycogen is used for whole-body glucose homeostasis.

Under *"resting" conditions* (Box 3.1), the rate of glycogenolysis in brain *in vivo* is quite slow, on the order of 0.01 µmol g^{-1} min^{-1}, and the glycogen molecule is relatively stable. However, during physiological stimulation, glycogenolysis rates rise by a factor of 6–50 to 0.06–0.5 µmol g^{-1} min^{-1}, and it increases even more (to 0.2–2.6 µmol g^{-1} min^{-1}) during pathophysiological conditions with very high ATP demand, e.g., seizures, anoxia, or ischemia (Dienel and Cruz, 2006). Note that the highest rates of glycogenolysis are more than three times the resting rate of utilization of blood-borne glucose (Table 3.1). Thus, rapid activation of glycogenolysis by physiological activity, stressful handling conditions, and actions of neurotransmitters probably caused the low levels of brain glycogen found in many studies—sensory stimulation and handling prior to tissue sampling are sufficient to stimulate degradation of this labile metabolite.

The glycolytic energy yield from glycogen can be higher or lower than that from glucose, depending on how the calculation is made. Conversion of one glucosyl unit from *preformed glycogen* to two pyruvate *yields 3 ATP* because glycogen phosphorylase is a phosphoryl transferase that forms glucose-1-P from glycogen plus P_i, thereby bypassing the initial ATP required by the hexokinase reaction. Thus, metabolism of preformed glycogen has a 50% higher ATP yield than glycolysis of blood-borne glucose (Fig. 3.4). On the other hand, if one takes the viewpoint that there is *turnover of glycogen*, the accounting must include the hexokinase and UDP-glucose steps for synthesis of glycogen from glucose. Two ATP are consumed for conversion of glucose to glycogen to glucose-6-P, and another ATP is needed to form fructose-1-6-P_2, for a net consumption of 3 ATP per glucose cycled through glycogen. Because glycolysis has a net yield of 4 ATP, the overall gain of shunting glucose through glycogen is only 1 ATP per glucose. The advantage of this process is that the glycogen can be stored when fuel is readily available and rapidly degraded upon demand, without requiring the initial input of ATP.

To sum up, the finding of high levels of glycogen in brain and increased glycogen turnover during brain activation has stimulated interest in the contributions of glycogen to astrocytic energetics. The specific physiological functions of glycogen during brain activation and the fate of the carbon derived from glycogen during normal brain activity remain to be established, but during a hypoglycemic or ischemic energy crisis glycogen clearly helps prolong the duration of brain and nerve function and minimize cellular damage.

Pentose Phosphate Shunt Pathway: Oxidative Defense and Biosynthesis

The pentose phosphate shunt pathway (Fig. 3.6) has two major roles, provision of NADPH that is utilized in biosynthetic reactions and oxidative defense, and generation of five-carbon intermediates that are precursors

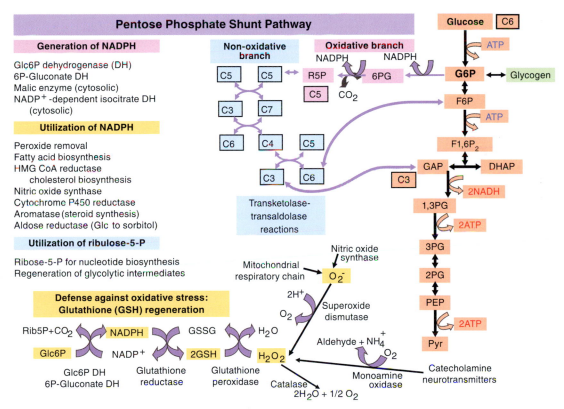

FIGURE 3.6 **The pentose phosphate shunt pathway provides substrates for oxidative defense, biosynthetic reactions, and nucleotide biosynthesis.** The oxidative component of the pathway generates 2 NADPH + 2 H^+ in successive oxidation reactions starting with glucose-6-P and forming 6-P-gluconate (6PG), then ribulose-5-P (R5P) + CO_2. Thus, the six-carbon glucose (denoted as C6 in a box next to glucose) is converted to a C5 intermediate, which by means of interconversions catalyzed by transketolases and transaldolases can regenerate C6 and C3 glycolytic intermediates. The pentose shunt pathway is not the only source of NADPH, but it is likely to be the major supplier due to its activation by oxidative stress and exposure of cells or tissue to peroxides. NADPH is necessary for a variety of biosynthetic reactions, some of which are highly active during brain growth and maturation (e.g., lipid biosynthesis) and some that are involved in biosynthesis of neuroactive compounds, e.g., nitric oxide synthase, as well as in degradation of catecholamine neurotransmitters (monoamine oxidase). An important function of the NADPH is its role as a cofactor in the glutathione reductase and peroxidase systems to eliminate hydrogen peroxide that is produced by various cellular reactions. Note that glucose-6-P can be derived from blood-borne glucose and from glycogen in astrocytes. Abbreviations for compounds in the glycolytic pathway are as in Fig. 3.4; GSH, reduced glutathione; GSSG, oxidized glutathione.

for nucleic acids (Baquer *et al.*, 1988; Dringen *et al.*, 2007). Both these functions are particularly important in developing brain when lipid biosynthesis and cell division are most active. In adult brain the flux through the pentose shunt pathway is approximately 5% of the rate of glucose utilization, but brain tissue has a huge excess capacity that is revealed by incubation of brain slices with an artificial electron acceptor, phenazine methosulfate, which stimulates the pathway by 20- to 50-fold (Hothersall *et al.*, 1979). Flux through the pentose shunt pathway is also stimulated by addition of catecholamine neurotransmitters to brain slices, presumably due to formation of H_2O_2 by monoamine oxidase, as well as by exposure of cells to H_2O_2 or other peroxides that are substrates for glutathione peroxidases (Fig. 3.6). Thus the predominant function of this pathway is likely to serve different purposes in developing compared to adult brain.

There are two major aspects of the pentose shunt pathway, the oxidative branch and the nonoxidative branch (Fig. 3.6). The oxidative branch consists of two sequential steps that convert glucose-6-P to ribulose-5-P. The first, catalyzed by glucose-6-P dehydrogenase (Glc-6-P DH), is the flux-regulating step and forms NADPH plus an unstable intermediate, 6-phosphogluconolactone, which spontaneously hydrolyzes to form 6-P-gluconate. The concentrations of glucose-6-P, $NADP^+$, and NADPH in brain tissue are low (Table 3.2), and the $NADP^+$/NADPH ratio is approximately 0.01 (Veech *et al.*, 1973). NADPH is a competitive inhibitor of glucose-6-P DH, indicating that consumption of NADPH and formation of $NADP^+$ provide the required substrate for the reaction that is dependent on continuous supply of glucose-6-P that can be derived from blood-borne glucose or glycogen. 6-P-gluconate and $NADP^+$ are the substrates for the second step, oxidative decarboxylation, which releases carbon 1 of glucose as CO_2. This reaction forms the basis for the most widely used assay for pentose shunt activity, comparison of the rate of formation of $^{14}CO_2$ from [1-^{14}C]glucose compared to [6-^{14}C]glucose. The nonoxidative branch of the pentose shunt pathway involves interconversions of intermediates via transketolase and transaldolase reactions that can regenerate fructose-6-P and glyceraldehyde-3-P, so that if carbon is not used for nucleic acid biosynthesis it is returned to the glycolytic pathway. Transketolase is a thiamine pyrophosphate (vitamin B_1)-dependent enzyme, and, along with pyruvate and α-ketoglutarate dehydrogenases of the tricarboxylic acid cycle, it is affected by thiamin deficiency (beriberi). Severe thiamin deficiency affects selective areas of the central nervous system, even though all the enzymes affected are present in all cell types. Conversion by phosphopentose

isomerase of ribulose-5-P to ribose-6-P forms the precursor for 5-phosphoribosyl-1-pyrophosphate (PRPP), which is the starting point for *de novo* synthesis of purine ribonucleotides.

To sum up, the overall reaction glucose-6-P plus 2 $NADP^+$ generates ribulose-5-P + CO_2 + 2 NADPH + 2 H^+. Rearrangement of the five-carbon intermediates that are not utilized for biosynthesis via the nonoxidative reactions returns carbon to the glycolytic pathway. In astrocytes, the pentose phosphate shunt pathway is fueled by glucose and glycogen, whereas it is fueled only by glucose in neurons.

The Tricarboxylic Acid (TCA) Cycle: ATP Production and Biosynthetic Activity

Energy generation. The TCA cycle in brain has two major functions, generation of energy and biosynthesis of amino acids and other compounds. After entry into mitochondria via a monocarboxylic acid transporter, pyruvate is converted to acetyl coenzyme A (CoA) by pyruvate dehydrogenase (PDH), a highly regulated thiamine-dependent multienzyme complex. The net reaction results in the oxidative decarboxylation of carbons 3 and 4 of glucose, which are released as CO_2. PDH has a K_m for pyruvate of about 0.05 mM, which is approximately equal to the concentration of pyruvate in brain tissue. Two products, acetyl CoA and NADH, are competitive inhibitors of the PDH reaction. PDH exists in an inactive, phosphorylated form and an active, dephosphorylated form; the protein kinase that renders PDH inactive is inhibited by ADP, pyruvate, CoASH, and NAD^+, whereas it is activated by acetyl CoA and NADH. The protein phosphatase is activated by Mg^{2+} and Ca^{2+}.

Condensation of acetyl CoA with oxaloacetate to generate citrate (Fig. 3.7) is catalyzed by citrate synthase, a regulated enzyme that is inhibited by ATP, NADH, and succinyl CoA and stimulated by ADP. Citrate is converted to isocitrate by aconitase, and the first oxidative decarboxylation step in the cycle is carried out by isocitrate dehydrogenase to generate NADH + H^+ and α-ketoglutarate. This enzyme requires Mg^{2+} or Mn^{2+} and is stimulated by ADP and Ca^{2+}. α-Ketoglutarate dehydrogenase then converts α-ketoglutarate to succinyl CoA in a thiamine-dependent reaction, releasing the second CO_2 in the pathway and generating NADH + H^+; this enzyme can be inhibited by ATP, GTP, NADH, and succinyl CoA. The next step, succinyl CoA synthetase, forms succinate, releases CoASH, and generates GTP. Succinate dehydrogenase oxidizes succinate to fumarate, with conversion of FAD to $FADH_2$. Because FAD is fluorescent, this step can be used to monitor mitochondrial redox status by fluorescence

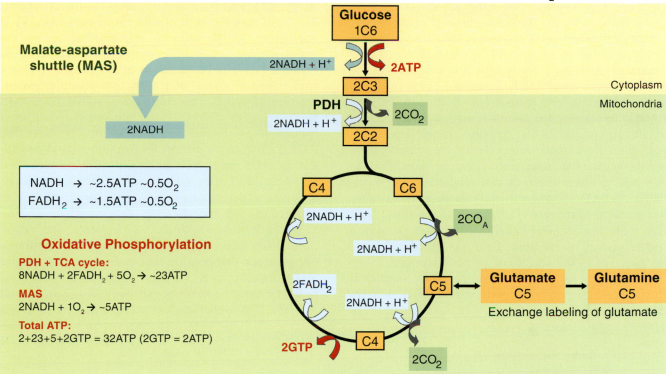

Complete Oxidation of Glucose Produces 32 ATP and Consumes 6 O$_2$

FIGURE 3.7 **Oxidation of pyruvate via the tricarboxylic acid cycle.** Pyruvate is transported from the cytosol to mitochondria along with H$^+$ by a monocarboxylic acid transporter. Oxidative decarboxylation of pyruvate by pyruvate dehydrogenase generates acetyl coenzyme A (CoA) and releases CO$_2$ corresponding to the 3 and 4 positions of glucose. This two-carbon compound (denoted C2 in a box) condenses with oxaloacetate, a four-carbon (C4) catalytic component of the TCA cycle by the action of citrate synthase. Complete oxidation of acetyl CoA releases CO$_2$ via the isocitrate dehydrogenase (C6→C5 step) and α-ketoglutarate dehydrogenase (C5→C4 step) reactions; since two acetyl CoA are generated from glucose, a total of four CO$_2$ are produced by the TCA cycle. Note that for each turn of the cycle, a two-carbon compound enters and two one-carbon compounds are released; there is no net synthesis of any intermediate. Labeling of the glutamate, glutamine, and aspartate pools can occur by equilibrative reactions that shuttle compounds to and from the reactions of the TCA cycle, thereby leading to label dilution and pool labeling. (Note: The transamination reaction involving oxaloacetate (C4) and aspartate (C4) is not shown in the figure.) This characteristic of intermediary metabolism slows the release of CO$_2$ from labeled glucose, since the route to the decarboxylation reactions is not direct. A total of eight NADH, two FADH$_2$ and two GTP are also generated from one glucose via the TCA cycle, for a net energy yield equivalent to about 32 ATP. Note that the lower yield of ATP compared to previous estimates of 36 in the earlier literature is due to adjustments made for H$^+$ leakage across the inner mitochondrial membrane and the electrogenic cost of transport of glutamate via the malate–aspartate shuttle and pyruvate via the MCT. Reproduced with permission from Macmillan Publishers Ltd. Hertz L, Peng L, Dienel GA. 2007. Energy metabolism in astrocytes: high rate of oxidative metabolism and spatiotemporal dependence on glycolysis/glycogenolysis. J Cereb Blood Flow Metab. 27:219–249.

microscopy during brain activation, and this application will be discussed in following text in more detail (see Fig. 3.13). Succinate dehydrogenase is inhibited by malonate, a three-carbon dicarboxylic acid that has been a useful compound for studies of the TCA cycle. Fumarase adds water to fumarate to form L-malate, which is a substrate to regenerate oxaloacetate with formation of the third NADH + H$^+$ of the cycle per acetyl CoA. The activities of TCA cycle dehydrogenases can be stimulated by changes in calcium levels that are evoked by altered functional activity (e.g., Huang *et al.*, 1994).

To sum up, the TCA cycle is a highly regulated "regenerative" process in which two carbons enter the cycle as acetyl CoA and two carbons are released as CO$_2$; there is no net gain or loss of carbon with each full turn of the cycle (Fig. 3.7). The carbon atoms entering the

cycle corresponding to carbons 2 and 5 of glucose are lost on the second turn of the cycle, whereas carbons 1 and 6 of glucose are released on the third turn of the cycle. Note that transaminase exchange reactions at the α-ketoglutarate and oxaloacetate steps can transfer label from intermediates in the cycle to the large glutamate, glutamine, and aspartate pools (Table 3.2), thereby delaying release of labeled carbon atoms. Incorporation of label from glucose into TCA-cycle-derived amino acids is the basis for magnetic resonance spectroscopic analysis of metabolic fluxes in brain (Rodrigues and Cerdán, 2007). The concentrations of all TCA cycle intermediates are quite low, totaling about 1.5 μmol g^{-1} (Table 3.2). Under normal resting conditions, the overall rate of the cycle is twice that of glycolysis due to formation of two pyruvate molecules from each glucose.

Electron transport chain and oxidative phosphorylation. Only one high-energy compound is formed per turn of the TCA cycle, i.e., the substrate-level phosphorylation of GDP to form GTP at the succinyl CoA synthetase step (Fig. 3.7). Most of the ATP generated via glucose oxidation reactions is recovered from oxidation of the 3 NADH and one FADH$_2$ generated per turn of the cycle. Electrons are transferred from NADH and FADH$_2$ to oxygen, the terminal acceptor, via a series of enzyme complexes that constitute the electron transport chain. Protons are pumped from the mitochondrial matrix across the inner mitochondrial membrane to generate an electrochemical potential across the membrane, and ATP synthase uses the electrochemical proton gradient to drive ATP synthesis (Papa *et al.*, 2007; Fig. 3.8). The NADH-linked dehydrogenase has a higher energy yield than the FAD-linked dehydrogenases due to the additional protons pumped across the membrane by complex I compared to complex II (FAD-linked succinate dehydrogenase that delivers electrons to Coenzyme Q). The active state of

respiration requires a continuous supply of oxygen, ADP, and P$_i$ and the rate is much higher than the "resting" respiratory rate. If all the cytoplasmic NADH is transferred to the electron transport chain by the malate–aspartate shuttle, an estimated 32 ATP are generated by the combined action of the TCA cycle, electron transport chain, and oxidative phosphorylation reactions (Fig. 3.7). This value is somewhat lower than early estimates of 36 ATP per glucose due to proton leakage across the mitochondrial membrane. Note that ADP is a required substrate for respiration, and generation of ADP from ATP by brain work provides the "catalytic link" (i.e., ATP turnover) between cellular activities and respiration because phosphorylation is coupled to oxidation.

Biosynthetic roles of the TCA cycle generation of neurotransmitters. Glucose provides the carbon for synthesis of numerous compounds required for brain structure and function, and various biosynthetic reactions branch from the glycolytic pathway (Fig. 3.4), as well as in the TCA cycle (Fig. 3.9A). The TCA-cycle-derived

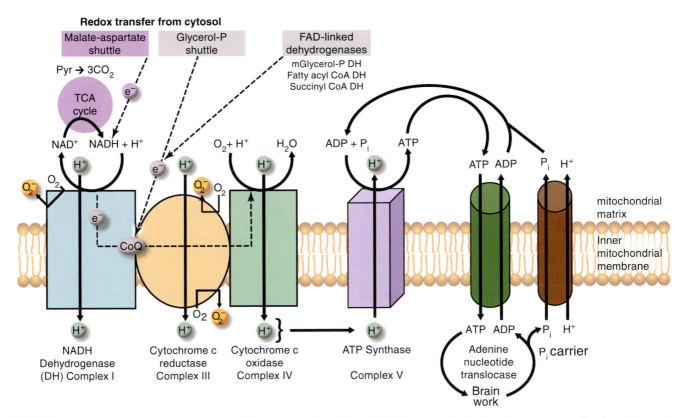

FIGURE 3.8 **Electron transport chain and oxidative phosphorylation.** NADH generated via the malate–aspartate shuttle and from the NAD-linked oxidative reactions in the TCA cycle enters the electron transport chain at complex I, whereas the FAD-linked reactions enter at the Coenzyme Q (CoQ) step, with a lower net ATP yield. Translocation of H$^+$ from the mitochondrial matrix at three steps, complex I, complex III, and complex IV, produces a protein gradient that is coupled to ATP synthesis by ATP synthase (complex V). The adenine nucleotide translocase and phosphate carrier are necessary to transfer ADP and inorganic phosphate (P$_i$) into the mitochondrial matrix and transfer ATP from the matrix to cytoplasm. A continuous supply of ADP is necessary for maximal ATP synthesis; respiration in the presence of ADP (state 3 respiration) is much higher than in the absence of ADP (state 4 respiration). Adapted from Papas *et al.*, 2007, with the kind permission of Springer Science and Business Media.

A Glutamate synthesis and Glu-Gln cycle

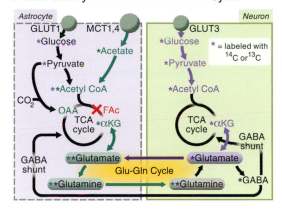

B Astrocytic inhibitor fluoroacetate (FAc) impairs synaptic transmission

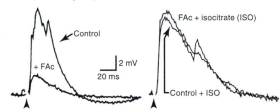

C Fluoroacetate depletes ATP in glia

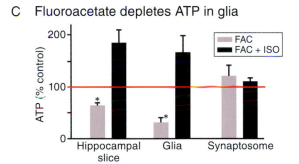

D Metabolic fate of glutamate in cultured astrocytes

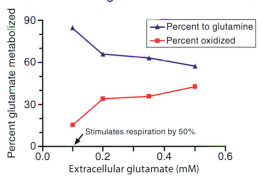

FIGURE 3.9 Anaplerotic reactions are required for net synthesis of glutamate, glutamine, and aspartate and for normal synaptic trans-mission. (A) Net synthesis of TCA-cycle-derived amino acids requires the ATP-dependent CO_2 fixation reaction catalyzed by pyruvate carboxylase to generate oxaloacetate (OAA) from pyruvate. Condensation of the OAA with a second molecule of pyruvate forms a "new" molecule of citrate, a six-carbon compound that, after decarboxylation, can generate a "new" molecule of glutamate, glutamine, aspartate, or GABA. Pyruvate carboxylase is located in astrocytes thereby conferring this cell type with the capacity for generation of important TCA-cycle-derived amino acids within the brain. Because the astrocytic monocarboxylic acid transporters (MCTs) have a substrate specificity for transport of acetate, labeled acetate preferentially labels the glutamine pool in brain, whereas metabolic labeling by glucose preferentially labels the large glutamate pool that is located in neurons (see text). When given at low doses, fluoroacetate (FAc), like acetate, is preferentially transported into astrocytes where it is metabolized via the TCA cycle to generate fluorocitrate, an inhibitor of aconitase that blocks the astro-cytic TCA cycle by preventing conversion of citrate to isocitrate. The GABA shunt is the pathway by which GABA is returned to the TCA cycle following transamination to succinate semialdehyde and oxidation by succinate semialdehyde dehydrogenase; this can take place in neurons and astrocytes. See text for discussion of the glutamate (Glu)–glutamine (GLN) cycle between astrocytes and neurons. Modified from Fig. 8C of Hertz *et al.*, 2007, with the permission of Nature Publishing Group. (B) Acute treatment of hippocampal brain slices fluoroacetate (FAc) quickly interferes with synaptic transmission. Left panel: Intracellular postsynaptic potentials (PSPs) evoked by a stimulating electrode in the stratum radiatum and recorded in the pyramidal neurons in the CA1 region of the hippocampus in the presence or absence of 100 μM FAc for 30 min. The FAc effects on evoked potentials are dose dependent, with IC$_{50}$ values in the range of 40–65 μM, with regional differences in dose sensitivity within the hippocampus (Keyser and Pellmar, 1994, 1997). Right panel: Inclusion of 3 mM isocitrate (ISO), which enters the astrocytic TCA cycle after the metabolic block by 100 μM FAc, restores the synaptic potential (reproduced from Keyser and Pellmar, 1994, with permission of Wiley-Liss, Inc., a subsidiary of John Wiley & Sons, Inc.). Glutamine also attenuated the FAc-induced depression of evoked potentials (Keyser and Pellmar, 1994). (C) Incubation of hippocampal slices, C6 glioma, or a crude synaptosomal fraction from the CA1 region of hippocampal slices in artificial cerebrospinal fluid plus 1 mM FAc depleted ATP levels in the slices and glioma cells but not in the synap-tosomal fraction, and ATP levels were restored when incubated in 1 mM FAc + 3 mM isocitrate (ISO). Note the substantial contribution of the FAc sensitive TCA cycle to the ATP content of the slice and glioma preparations. (Reproduced from Keyser and Pellmar, 1994, with permis-sion of Wiley-Liss, Inc., a subsidiary of John Wiley & Sons, Inc.). (D) Glutamate has two major metabolic fates, conversion to glutamine, which is exclusively an astrocytic process due to the cellular localization of glutamine synthetase in astrocytes, and oxidation via the TCA cycle after transamination or oxidative deamidation. Astrocytes metabolize glutamate by both pathways in a glutamate-concentration-dependent man-ner; the higher the extracellular glutamate the smaller the fraction converted to glutamine and the greater the fraction oxidized via the TCA cycle (plotted from data of McKenna *et al.*, 1996; modified from Figure 2 from Dienel and Cruz, 2006). The 50% stimulation of oxygen consumption by glutamate (100 μM) in cultured astrocytes that can be blocked by ouabain (plotted from data from Eriksson *et al.*, 1995) is con-sistent with stimulation of glutamate oxidation and respiration to help dispose of excess glutamate and to generate ATP to fuel Na$^+$, K$^+$-ATPase activity. Note that the catalytic nature of the TCA cycle (Fig. 3.8) requires that a four-carbon molecule, malate, exit from the TCA cycle and then be converted to pyruvate that must then re-enter the cycle to convert all the glutamate carbon to CO_2, a process called *pyruvate recy-cling* (see text).

biosynthetic reactions directly link oxidative energy generation with neurotransmitter turnover in cholinergic and amino acid transmitter pathways. For example, synthesis of acetylcholine from glucose requires the action of the mitochondrial pyruvate dehydrogenase in neurons; the acetyl CoA is transferred from mitochondria to the cytosol where choline and choline acetyltransferase are localized, but the pathways of the transfer process remain to be established (Joseph and Gibson, 2007). Synthesis of acetylcholine is closely linked to glucose metabolism, and, even though less than 1% of the glucose is used for synthesis of acetylcholine, the cholinergic pathway is especially sensitive to mild hypoglycemia and mild hypoxia, both of which are associated with reduced acetylcholine turnover and decrements in acetylcholine levels.

The neuroactive amino acids are highly restricted from passage across the blood–brain barrier (Oldendorf and Szabo, 1976), yet their concentrations in brain are substantial (Table 3.2). The high brain-to-plasma concentration ratios for GABA (300), aspartate (300), glutamate (~150), and glutamine (10) (McIlwain and Bachelard, 1985) indicate considerable net synthesis of these amino acids within the brain. Because the TCA cycle cannot support net synthesis of four- or five-carbon compounds from acetyl CoA, another reaction, CO_2 fixation, is required to generate oxaloacetate and form a "new" four-carbon backbone from pyruvate. This anaplerotic reaction carried out by pyruvate carboxylase is predominantly localized in astrocytes (Yu et al., 1983; Shank et al., 1985; Öz et al., 2004). Thus, astrocytes have the ability to carry out de novo synthesis of the molecules that are used as precursors for excitatory and inhibitory amino acid neurotransmitters (Fig. 3.9A). The oxaloacetate can undergo a transamination reaction to form a "new" molecule of aspartate, or it can condense with acetyl CoA derived from a second pyruvate to produce citrate, which continues through the TCA cycle to generate α-ketoglutarate. A transamination reaction "withdraws" this new five-carbon compound from the TCA cycle as glutamate. In astrocytes, the glutamate is converted to glutamine by the action of glutamine synthetase, an astrocyte-specific, ATP-dependent enzyme (Martinez-Hernandez et al., 1977; Norenberg and Martinez-Hernandez, 1979). Glutamine can then be released to extracellular fluid and taken up by neurons, where it is converted to glutamate by the enzyme glutaminase. In excitatory neurons, the glutamate is packaged into synaptic vesicles, released as a neurotransmitter, taken up into astrocytes via sodium-dependent glutamate transporters, and either oxidized or converted to glutamine.

This cyclic process involving shuttling of glutamine and glutamate among astrocytes and neurons is called the glutamate–glutamine cycle (Fig. 3.9A).

Glutamine is also a good precursor of GABA, an inhibitory neurotransmitter. Inhibitory neurons convert glutamine to glutamate, which is then decarboxylated by glutamate decarboxylase to form GABA (Fig. 3.9A). After its release to the synaptic cleft, GABA can be taken up by neurons and astrocytes and undergo a transamination reaction by GABA transaminase to generate succinate semialdehyde. Next, oxidation by succinate semialdehyde dehydrogenase forms succinate that is oxidized in the TCA cycle. Because the transfer of carbon from α-ketoglutarate to glutamate to GABA to succinate bypasses some of the TCA cycle reactions, this pathway is referred to as the GABA shunt (Fig. 3.9A).

Neurotoxins have been particularly useful in metabolic studies of astrocyte–neuron interactions, and compounds that have preferential effects on astrocytes (e.g., methionine sulfoximine) or neurons (e.g., 3-nitropropionic acid, aminooxyacetic acid) are particularly useful for ^{13}C-metabolic studies of compartmentation (Sonnewald et al., 2007). The gliotoxin fluoroacetate, when given in low doses, is preferentially transported via a monocarboxylic acid transporter (MCT) into astrocytes where it is metabolized by their TCA cycle to generate fluorocitrate, a potent inhibitor of aconitase (Muir and Clarke, 1986; Hassel et al., 1997) (Fig. 3.9A). Fluorocitrate blocks astrocytic TCA cycle flux, interferes with synaptic transmission (Fig. 3.9B), and reduces glial but not synaptosomal ATP levels (Fig. 3.9C). Blockade of the TCA cycle by fluoroacetate can be circumvented by providing the downstream metabolite, isocitrate, which restores evoked potentials and ATP levels (Fig. 3.9B, C).

The pyruvate carboxylase reaction consumes one ATP and is stimulated by K^+, which is taken up from the synaptic cleft by astrocytes during neuronal activation thereby linking the anaplerotic reactions and the astrocytic TCA cycle flux with neuronal signaling activity (Hertz et al., 2007). If all the reducing equivalents produced from glucose are transferred to the electron transport chain via redox shuttle systems, then synthesis of glutamate from glucose is an energy-producing process, with a net yield of 11 ATP and consumption of 2 O_2. This yield contrasts the production of 32 ATP and consumption of 6 O_2 by complete oxidative metabolism of glucose. Under steady-state conditions, glutamate level is constant and rates of de novo synthesis and complete oxidative degradation of glutamate are equal.

Oxidation of other compounds via the TCA cycle. A variety of compounds that are taken up from blood or turnover in brain are metabolized via the TCA cycle, including ketone bodies, fatty acids, and amino acids, but in adult brain these compounds are not important energy sources compared to glucose, which is continually supplied to brain from the blood. Suckling mammals have high capacity to oxidize the ketone bodies acetoacetate and β-hydroxybutyrate (in addition to glucose) due to high levels of the monocarboxylic acid transporters at the blood–brain barrier and the necessary metabolic enzymes (e.g., β-hydroxybutyrate dehydrogenase); after weaning both transport and metabolic capacity of ketone bodies are markedly down-regulated, but the transporters and oxidative enzymes can be induced by prolonged starvation (Nehlig, 1996, 2004). Fatty acid oxidation is high in cultured astrocytes, whereas it is very low in cultured neurons (Edmond, 1992). Beta-oxidation of fatty acids generates acetyl CoA plus NADH, which feeds into the electron transport chain at complex I (Fig. 3.8), thereby producing ATP and consuming oxygen without generating equivalent amounts of CO_2. Complete oxidation of carbohydrates, such as glucose, consumes 6 O_2 and produces 6 CO_2, yielding a *respiratory quotient* (ratio of CO_2 produced to O_2 consumed) of one; the respiratory quotient for fatty acids and ketone body oxidation is less than one. Assays of arteriovenous differences across the brain for oxygen and CO_2 demonstrate that the respiratory quotient for brain is close to one, indicating that carbohydrate is the primary fuel for normal brain. Thus, fatty acids, amino acids, and ketone bodies taken up from blood into normal adult brain are not significant substrates for brain energy metabolism. However, during development or prolonged starvation when ketone bodies are consumed in much higher amounts, the respiratory quotient is less than one.

Neutral amino acids are readily taken up into brain via amino acid carriers (Oldendorf and Szabo, 1976; Oldendorf, 1981), and these compounds are readily oxidized via the TCA cycle. The branched chain essential amino acids (valine, isoleucine, leucine) have similar degradative pathways, involving transamination to form a keto acid, followed by oxidation reactions that feed into the TCA cycle at different points. Valine produces propionyl CoA, which enters the TCA cycle at the succinyl CoA step after a CO_2 fixation reaction, and isoleucine generates acetyl CoA plus propionyl CoA. In contrast, degradation of leucine generates acetyl CoA plus acetoacetate. Due to their participation in transamination reactions, these branched chain amino acids are thought to serve as ammonia carriers in transcellular shuttle systems that are required to sustain the glutamate–glutamine cycle (Hutson *et al.*, 1998; LaNoue *et al.*, 2007).

Pyruvate recycling. As mentioned previously, glutamate, glutamine, GABA, and aspartate are continuously degraded via the TCA cycle after entry of the carbon skeleton into the cycle at the α-ketoglutarate, succinyl CoA, or oxaloacetate steps. However, the TCA cycle is a catalytic process and only two carbons are removed by decarboxylation reactions per turn of the cycle. Thus, oxidative metabolism of the four-carbon backbone of these amino acids requires its exit from the cycle and re-entry of pyruvate into the TCA cycle followed by its oxidative metabolism (i.e., recycling of pyruvate that was initially used to generate the four- or five-carbon compound via CO_2 fixation). Thus, complete oxidative degradation of the amino acids occurs when malate leaves the mitochondria and undergoes oxidative decarboxylation by the cytosolic malic enzyme to produce NADPH, CO_2, and cytosolic pyruvate. Some of this pyruvate can be converted to lactate and released from the cell. Re-entry of pyruvate into mitochondria generates 3 CO_2 and results in a distinct labeling pattern of TCA-cycle-derived amino acids that can be detected by ^{13}C-magnetic resonance spectroscopy. Complete oxidation of glutamate generates 20 ATP and consumes 4 O_2, and, in conjunction with synthesis of glutamate from glucose described previously, the net yield (31 ATP) of glutamate turnover (synthesis and degradation) is calculated to be nearly the same as complete oxidation of glucose (32 ATP).

Glutamate released by excitatory neurons and taken up into astrocytes has two major fates, conversion to glutamine and oxidation via the TCA cycle (McKenna, 2007). The higher the extracellular glutamate level the greater the fraction oxidized, and at 0.5 mM glutamate cultured astrocytes process similar proportions by these two pathways (McKenna *et al.*, 1996; Fig. 3.9D). The glutamate levels in Fig 3.9D represent the lower end of the range of glutamate transients that occur in the synaptic cleft; glutamate level reaches a peak of several millimolar before it is quickly transported into astrocytes (Bergles and Jahr, 1997). Extracellular glutamate (0.1 mM) also stimulates oxygen consumption in cultured astrocytes by 50% (Fig. 3.9D) and this rise in respiration is blocked by ouabain, directly linking Na^+,K^+-ATPase activity to astrocytic oxidative metabolism. Extracellular glutamate can also inhibit glucose utilization in cultured astrocytes (by 20% at 0.5 mM [Qu *et al.*, 2001] and by about 50% at 1 mM [Swanson *et al.*, 1990]) consistent with the utilization of glutamate by astrocytes as an oxidative energy source.

Glutamate has also been found to stimulate glucose utilization and lactate release in cultured astrocytes (Pellerin and Magistretti, 1994; Takahashi et al., 1995), but these findings have been controversial because several other laboratories find no change in glucose utilization or in lactate production during exposure of cultured astrocytes to glutamate (reviewed by Dienel and Cruz, 2004, 2006). The basis for the apparent discrepancy among cultured astrocytes grown in different laboratories remains to be established, but culture conditions and developmental plasticity are likely to influence the oxidative capability of astrocytes. For example, astrocytes grown in typical culture media containing 22 mM glucose oxidize glucose and lactate *half as well as those grown in 2 mM glucose* (Abe et al., 2006). Normal rat brain glucose levels are about 2 mM (Tables 3.1 and 3.2; Fig. 3.3) and the 11-fold higher glucose level commonly used in tissue culture studies exceeds the glucose content of severe diabetic rat brain by a factor of 4–5. Thus, a myriad of complications associated with chronic hyperglycemia (e.g., oxidative stress, glycation reactions, increased flux into the polyol pathway, and disruption of various signaling pathways) could be expected to have an impact on metabolism and functions of cultured cells grown in exceedingly high glucose levels. While cultured cells have many advantages, particularly in mechanistic studies, pathway fluxes are likely to be governed by many factors, and extrapolation of results from cultured cells to adult brain *in vivo* must be made with caution.

Summary

Glucose carbon enters the energy and biosynthetic pathways via the hexokinase step, where phosphorylation traps the molecule in the cell where it was initially metabolized. The glycolytic pathway is regulated at three major sites, hexokinase, phosphofructokinase, and pyruvate kinase, and fluxes from the major branch point metabolites are tightly controlled. Glycogen turnover is governed by neurotransmitters, second messengers, and phosphorylation reactions, and flux into the pentose shunt pathway governed mainly by NADP availability. Pyruvate dehydrogenase and key enzymes composing the TCA cycle are regulated by Ca^{2+} and metabolites related to the energy status of the cell. Regulatory mechanisms have an "integrative" effect to coordinate and satisfy the overall demands for the cell for energy production, biosynthetic activity, and neurotransmitter turnover. Because glycolysis can be rapidly upregulated, the lactate dehydrogenase reaction serves as a "release valve," in which pyruvate can be converted to lactate to regenerate NAD^+ so that glycolysis can continue. Because lactate formation and transport are equilibrative processes and because glucose is rarely limiting in normal brain, partial metabolism of glucose to lactate may be preferentially used to support specific processes, even though the energy yield is much lower than that of the TCA cycle and electron transport chain.

SUBSTRATES, ENZYMES, PATHWAY FLUXES, AND COMPARTMENTATION

The metabolic machinery of brain and its components are very heterogeneous in terms of distribution, composition, and capacity at the regional, cellular, and subcellular level. The *steady-state concentration* of any metabolite, mRNA, or protein (Box 3.2; Table 3.2) is the net result of all processes that generate and remove the compound. Enzyme activities assayed *in vitro* under optimal conditions represent the capacity for catalytic activity (i.e., *enzyme activity* is proportional to its amount, Box 3.2), but activities are generally in great excess of the *in vivo* fluxes through the steps catalyzed by the enzymes (Fig. 3.10). Thus, determination of *concentrations*, *amounts*, or *in vitro activities* of biological compounds represents a "*static*" *description* of brain function, indicative of capacity. On the other hand, assays of *in vivo fluxes* or *pathway rates* under different conditions represent *dynamic activities* of cells. Thus, changes in amount with development, aging, and disease reflect altered capacity but do not necessarily indicate changes in flux or the actual magnitude of flux changes. There are many examples in which enzyme amounts are reduced by 90% or more under pathophysiological conditions, with no adverse physiological effects; conversely increasing enzyme amounts may, but need not, be reflected by altered rates due to metabolic regulation of key steps in the pathways (Fell, 1996). However, if the catalyzed process is substrate concentration driven, increased capacity can reflect proportionate changes in flux, depending on the substrate level in the system of interest. For example, changes in enzyme or transporter level will reflect differences in reaction or transport rate if the enzyme or transporter is not saturated, but not if the substrate level is well above the K_m. Thus, it is important to understand the relationships among concentrations, activities, and fluxes, particularly under pathophysiological conditions when levels of biomarkers are frequently used as diagnostic indicators; elucidation of function–metabolism relationships has been driven by technological advances to assay processes of interest *in vitro* and *in vivo*.

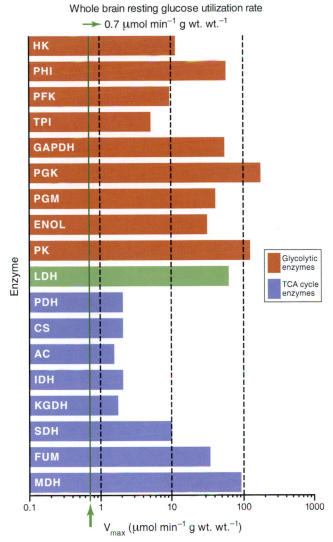

Whole brain resting glucose utilization rate
→ 0.7 μmol min⁻¹ g wt. wt.⁻¹

FIGURE 3.10 **Enzyme activities exceed *in vivo* glucose utilization rate.** Comparison of maximal activities (V_{max}) of glycolytic and TCA cycle enzymes determined *in vitro* to overall glucose utilization rate in brain *in vivo*. (Data from tabulations of McIlwain and Bachelard, 1985.) Note the semi-log scale on the abscissa. Lactate dehydrogenase activity is about 5–6 times higher than hexokinase, and all glycolytic enzyme activities are much higher than those of the TCA cycle. Because metabolism of glucose generates two pyruvate per glucose, the rate of the TCA cycle is twice that of glucose utilization (denoted by the green horizontal line). Abbreviations: HK, hexokinase; PHI, phosphohexose isomerase; PFK, phosphfructo-1-kinase; TPI, triose phosphate isomerase; GAPDH, glyceraldehyde-3-P dehydrogenase (DH); PGK, phosphoglycerate kinase; PGM, phosphoglucomutase; ENOL, enolase; PK, pyruvate kinase; LDH, lactate DH; PDH, pyruvate DH; CS, citrate synthase; AC, aconitase; IDH, isocitrate DH; KGDH, α-ketoglutarate DH; SDH, succinate DH; FUM, fumarase; MDH, malate DH.

Metabolite Levels

Oliver Lowry's laboratory pioneered the field of "enzyme histochemistry" and metabolite analysis by (1) devising precise, sensitive, and quantitative microanalytical techniques with which the maximal activities of as many as nine enzymes and many metabolites could be measured in an extract of a single dissected neuron, (2) developing quantitative fluorescent assays and cycling reactions that enabled accurate determination of 10^{-15} to 10^{-17} moles of specific compounds, and (3) evaluating changes in metabolic flux from changes in metabolite levels in a "closed system" under different experimental conditions (Lowry, 1990; Passonneau and Lowry, 1993). These studies showed that many compounds of interest are very *labile* (Box 3.2) because they are consumed within seconds of the initial postmortem ischemic interval (Lowry *et al.*, 1964; see later, Fig. 3.22). Thus, inadequate procedures to inactivate enzymes yield low levels for glycogen, glucose, glycolytic and TCA cycle intermediates, and high-energy phosphates. Advances in analysis of labile metabolites required development of techniques to quickly stop metabolism, e.g., freeze-blowing (Veech *et al.*, 1973), funnel-freezing (Ponten *et al.*, 1973), and microwave fixation (Stavinoha *et al.*, 1973; Medina *et al.*, 1975) as well as appropriate procedures for tissue extraction and animal handling. Because metabolism is closely integrated with neurotransmitter activity, changes with behavioral state, sensory stimulation, stress due to handling, placement in an apparatus for tissue sampling, anesthesia, and other factors can substantially influence the concentrations of metabolites of interest. For example, glycogenolysis can be stimulated by activation of β-adrenergic receptors on astrocytes, and the adrenergic innervation of the brain from the locus coeruleus, a structure involved in attention and orienting responses, is extensive. Thus, stimuli that disrupt ongoing behavior or activity and generate reorienting behavior elicit strong responses from locus coeruleus neurons, and activation of this pathway could be expected to reduce glycogen levels throughout the brain. Also, glycogen phosphorylase is quickly activated in response to postmortem ischemia, even during decapitation and freezing in liquid nitrogen, and slow inactivation of enzymes in frozen tissue powders during extraction can lead to substantial loss of glycogen and other compounds, with increases in the level of lactate. Thus, metabolic activity *in situ* (pre- and postmortem) and *in vitro* can affect the measured levels of labile metabolites and signaling compounds, and interpretation of results needs to take into account the physiological and biochemical procedures employed with tissue sampling.

Enzyme Levels: Metabolic Capacity or Potential

Maximal activity (V_{max}) of an enzyme assayed under optimal conditions *in vitro* is proportional to enzyme amount and therefore reflects catalytic capacity or potential. The maximal activities of enzymes in the glycolytic and oxidative pathways of glucose metabolism assayed *in vitro* are generally much higher than the overall *in vivo* rate of brain glucose utilization (i.e., about 0.7 µmol g^{-1} min^{-1}; Fig. 3.10), indicating that *metabolic capacities* of the enzymes (Box 3.2) greatly exceed the usual fluxes through the steps they catalyze. For example, it can be seen from Fig. 3.10 that only about 5% of the maximal hexokinase activity is sufficient to sustain the overall rate of glucose utilization in brain *in vivo*. Thus, the brain has a large reserve to meet demands of extreme conditions that is evident during bicuculline-induced seizures where glucose and oxygen utilization rates increase two- to four-fold (Siesjö, 1978). Relative amounts of enzymes in different brain structures, cells, and subcellular organelles are often assessed by immunological and histochemical assays in tissue slices and Western blots; these types of assays report the relative capacity of an enzyme to carry out its function but do not describe the actual rates catalyzed by the enzyme in the living tissue. Nevertheless, enzyme assays are very useful to identify compensatory responses to altered physiological demand and abnormal conditions, as well as the processes that regulate pathways and their interactions.

The large excess capacity of glycolytic compared to TCA cycle enzymes suggests that substantial decrements in the anaerobic metabolic capacity are less likely to have an impact on normal metabolic fluxes than deficits in mitochondrial enzymes (Fig. 3.10), which correlate with deficits in energy metabolism, biosynthetic activity, and cognition in disease states (Bubber *et al.*, 2005). For example, in human and rodent brain, the maximal activity of the α-ketoglutarate dehydrogenase complex is one of the lowest of the enzymes of energy metabolism in brain and it is considered to be one of the especially vulnerable enzymes. This enzyme complex is particularly sensitive to damage by reactive oxygen species (ROS), and deficiencies in its activity occur in many neurodegenerative diseases, including Alzheimer disease (Sheu and Blass, 1999; Gibson *et al.*, 2005).

Levels of Metabolites vs. Pathway Flux: Magnitude or Direction of Change

Substrates of the enzymes in the major energy-producing pathways are present in low concentrations, and, with the exception of glycogen, there is no significant energy reserve in brain (Table 3.2). Although substrate concentrations can change during different physiological states, they need not reflect the magnitude or direction of change in pathway fluxes. In fact, large flux changes can occur while metabolite levels

are relatively constant or change in directions that, in the absence of sufficient information about the system, appear to contrast the flux change. For example, sensory stimulation of the normal conscious rat by brushing of the body and whiskers causes arterial plasma glucose and lactate levels to rise because the animals move around in response to the brushing (Table 3.1). Brain glucose levels at the end of a 10 min stimulation interval rise in proportion to that in plasma (as expected from equilibrative transport), demonstrating that glucose influx across the blood–brain barrier matched the overall increase in metabolic demand during brain activation (Dienel *et al.*, 2007). Whisker stimulation also increased local rates of glucose utilization (CMR$_{glc}$) in the somatosensory cortex by about 30%, and brain tissue lactate level rose two-fold, from about 1 to 2 µmol g^{-1} (Fig. 3.11). Although the *percentage increase* in lactate concentration is *high*, the *net change* (1 µmol g^{-1}) in the *quantity* of glucose-derived carbon in the enlarged lactate pool is *small*, corresponding to *<5% of the flux* from glucose

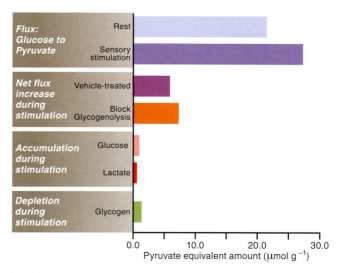

FIGURE 3.11   Metabolite concentrations do not provide information about pathway flux. The amount of glucose converted to pyruvate during rest and generalized sensory stimulation was calculated from *in vivo* assays of the rate of glucose utilization and expressed in pyruvate equivalents (2 pyruvate = 1 glucose or 1 glucosyl unit of glycogen). The net increase in pyruvate formed in somatosensory (whisker barrel) cortex was enhanced by prior inhibition of glycogenolysis, indicating a role for rapid glycogen turnover *in vivo* during sensory stimulation of normal, normoglycemic rats. Because rats moved during stimulation, which involved gentle brushing of the body and whiskers, the plasma glucose and lactate levels rose; the increase in brain glucose concentration was in proportion to that in plasma. Although brain lactate level doubled, the net increase was a tiny fraction of the glucose consumed during the same time interval. Lactate accumulation was also considerably smaller than net depletion of brain glycogen. (Plotted from data of Dienel *et al.*, 2007.)

to pyruvate during the stimulus interval (Fig. 3.11). This sensory stimulation procedure also caused a net consumption of glycogen even though overall brain glucose levels were not below normal, consistent with a role of glycogen in normal physiological activity, not just as an emergency energy reserve (Fig. 3.11). An intriguing finding was that inhibition of glycogenolysis prior and during the whisker stimulation increased glucose utilization by another 17%, suggesting that about half the additional glucose consumed in somatosensory cortex is involved in glycogen turnover (Dienel *et al.*, 2007).

In the studies described previously, the net increase in lactate concentration was much smaller than the stimulus-increased rise in pyruvate formation, the net consumption of glycogen, and the net rise in glucose utilization evoked by glycogenolysis blockade (Fig. 3.11). If only glucose and lactate concentrations were measured, one might incorrectly conclude that glucose utilization was reduced causing brain glucose levels to rise, and that the tissue became "glycolytic," even though there was only a very small net change in lactate level that could arise from various pathways. Changes in tissue concentration do not identify the origin or fate of the metabolite, and lactate level might be explained by contributions from the rise in plasma lactate, increased glycogenolysis, and perhaps a rise in pyruvate level that is reflected by higher lactate level due to mass action and the lactate dehydrogenase equilibrium. As emphasized by Veech (1980), changes in the concentrations of specific glycolytic and TCA cycle intermediates cannot, except under unusual circumstances, be equated with changes in flux through the pathway; it is necessary to consider contributions of many conditions that affect the level of the metabolite.

The combination of tissue low levels of metabolic intermediates and high glucose utilization rate means that there must be fast turnover of the glycolytic and TCA cycle intermediates. In the simplest case, using the mean overall the rate of glucose utilization of 0.7 μmol g^{-1} min^{-1} (Table 3.1) and the tissue concentration of a hypothetical intermediate of 0.1 μmol g^{-1} (compare to Table 3.2) the *turnover time* (Box 3.2) can be estimated by dividing the metabolite concentration by pathway flux rate, or 0.1/0.7 = 0.14 min or 8 sec. Note that a triose (a three-carbon intermediate in the glycolytic pathway) or TCA cycle metabolite that has a concentration of 0.1 μmol g^{-1} would have a turnover time of 4 sec because two triose and two acetyl CoA molecules are generated from each glucose. *Turnover of pools* or molecules is expressed as *half-time* or *half-life*, the time required for half of the pool to be replaced (Box 3.2).

Under steady-state conditions the concentrations of all metabolites in a pathway are constant, and the rate of each step in the pathway is equal to the flux through the entire pathway. In contrast, under non-steady-state conditions, part of the flux may be used to establish new levels of intermediates by pool filling or depletion. Steady state is critical for the conduct of *in vivo* metabolic assays to measure fluxes because the steady-state assumption enables simpler solutions of rate equations, i.e., rates of change of levels of metabolic intermediates can be set equal to zero. The high flux and low pool size also facilitate rapid changes in the concentrations of the various pathway intermediates compared to the large pools of amino acids. Labeling of the large TCA cycle amino acid pools with radioactive or stable isotopes is used to calculate rates of oxidative metabolism and trafficking of intermediates between different metabolic pools located in different cell types, as for example cycling of glutamate and glutamine between neurons and astrocytes (Fig. 3.9A).

Heterogeneity and Compartmentation: Glutamate Metabolism as an Illustrative Example

Mitochondrial heterogeneity. Isolation of crude mitochondrial fractions in the late 1960s by the Whittaker and De Robertis groups provided a means to separate "free" mitochondria (derived from the soma of all brain cells) from synaptosomes, which are pinched-off nerve endings that contain cytosol and mitochondria. Ultracentrifugation studies using continuous and discontinuous gradients to separate mitochondria in conjunction with analysis of their enzyme composition and substrate-supported respiration clearly demonstrated differences between synaptic and nonsynaptic mitochondria, among synaptic mitochondria, and among mitochondria isolated from different brain regions (reviewed by Lai and Clark, 1989; Clark and Lai, 1989). There is not only biochemical heterogeneity of mitochondria (Sonnewald *et al.*, 1998); these organelles also differ morphologically (Perkins and Ellisman, 2007), as well as functionally. Mitochondrial functions differ during development, with preferential expression of specific mitochondrial enzymes during development and in the mature brain. For example, β-hydroxybutyrate dehydrogenase activity is high prior to weaning and falls rapidly thereafter, whereas other activities rise during maturation.

Radiolabeling studies identify "large and small" glutamate pools. Early studies of metabolism in brain during the 1960s and 1970s examined the fate of radiolabeled substrates and developed the concept of metabolic compartmentation involving at least two different functional TCA cycles in brain that do not directly communicate with each other (see monographs edited by Balázs and Cremer, 1972; Berl et al., 1975; and the references cited therein). Metabolic compartmentation was first firmly established for glutamate metabolism on the basis of differences in labeling of glutamate and glutamine from various labeled precursors. In brief, when [^{14}C]glucose was injected intravenously or intraperitoneally and brain amino acids purified and counted to determine their extent of ^{14}C-labeling, the *specific activity* of glutamate (Box 3.3) was higher than the specific activity of its product, glutamine (Fig. 3.9A). This is the expected *precursor-product relationship* because as carbon derived from tracer amounts of [^{14}C]glucose sequentially traverses the glycolytic and TCA cycle pathways, then enters the glutamate pool and finally the glutamine pools. The label is continuously diluted with unlabeled carbon as it flows into successive metabolic pools, and the specific activity of the true precursor is always higher than that of the product.

On the other hand, when [^{14}C]acetate was used as the precursor, the specific activity of glutamine was unexpectedly higher than that of its obligatory precursor. This finding was explained by the existence of two "independent" TCA cycles, one associated with a "large" glutamate pool that is predominantly labeled by glucose, the other associated with a "small" glutamate pool that is the true precursor for glutamine that is preferentially labeled by acetate and other compounds. The large and small glutamate pools are segregated *in vivo,* but when the tissue is homogenized and amino acids are extracted, all the labeled and unlabeled glutamate in the various tissue pools is combined, thereby reducing the apparent specific activity of the true glutamate precursor pool for glutamine; this pool would, in fact, have a higher specific activity than glutamine. Subsequent studies in many laboratories using different experimental approaches clearly demonstrated that the "small" glutamate pool is located in astrocytes and the "large" pool is neuronal. Note that glucose is also oxidized in astrocytes but trapping of the label is achieved by dilution in glutamate, which is mainly neuronal. The basis for the preferential uptake of acetate into astrocytes was found to be transport (Waniewski and Martin, 1998), and acetate and its toxic analog, fluoroacetate, have been very important tools to study oxidative and biosynthetic metabolism in astrocytes, glutamate–glutamine cycling, and astrocyte–neuron interactions.

Biosynthetic capability and ammonia detoxification. Studies of cellular specialization have revealed high enrichment or exclusive localization of specific enzymes in different cell types in brain, as would be expected with the specialized functions of the major classes of brain cells (Wiesinger, 1995). Neurons are endowed with the capability to synthesize and package various neurotransmitters (including acetylcholine from glucose). Astrocytes are highly enriched in the capability to synthesize the neuromodulator D-serine from glucose (Fig. 3.4), as well as the capability for *de novo* synthesis of glutamate and aspartate (Fig. 3.9) due to their enrichment with pyruvate carboxylase (Yu et al., 1983; Shank et al., 1985). Glutamine synthetase is an astrocytic enzyme (Martinez-Hernandez et al., 1977), enabling astrocytes to detoxify ammonia entering brain from blood and convert glutamate to a nonexcitatory compound, glutamine, that is released to the interstitial fluid where it can be taken up by neurons. Phosphate-activated glutaminase is enriched in neurons, serving to convert glutamine into glutamate prior to its entry into the neurotransmitter pool. Glutamine is also a good precursor for GABA (Sonnewald et al., 1993), and inhibitory neurons contain glutamate decarboxylase for the synthesis of GABA from glutamate. The glutamate–glutamine cycle (Fig. 3.9) involves synthesis, trafficking, signaling, and degradation of glutamate (Hertz and Zielke, 2004; Hertz et al., 2007), as well as shuttling of ammonia among neurons and astrocytes via branched chain amino acids (Hutson et al., 1998).

Ammonia and ammonium ion are rapidly interchangeable, and free ammonia readily diffuses across lipid membranes into brain. Astrocytic glutamine synthetase has a critical role in the trapping of blood-borne ammonia (Cooper and Plum, 1987). In patients with liver disease, removal of ammonia produced by gut bacteria by the liver is severely impaired and blood ammonia levels rise, contributing to the progression of hepatic encephalopathy which is characterized, in part, by very high brain glutamine levels (Butterworth, 2006). Inhibitors of reactions in the glutamate–glutamine cycle, methionine sulfoximine (an inhibitor of glutamine synthetase) and fluoroacetate, are important tools in studies of astrocyte–neuron interactions.

Summary

The brain is a complex, heterogeneous organ that is very difficult to study due to cellular specialization of major functions and metabolic activity. Early studies of brain function and metabolism relied on global methods (e.g., arteriovenous differences), and these approaches provided valuable information by identifying glucose and oxygen as the major brain fuel, compounds that enter brain from blood, and metabolites released from brain. However, changes in local rates of functional activity are not detectable unless high-resolution methods are used. Microanalytical and cell culture methods enabled analysis of the composition and capabilities of different cell types *in vivo* and *in vitro,* and radiolabeling studies set the stage for development of *in vivo* methods to study brain function in living subjects. Glutamate metabolism is an important, instructive example of complexities of compartmentation of functional metabolism and integration of neurotransmitter turnover with generation of energy.

IMAGING FUNCTIONAL METABOLIC ACTIVITY IN LIVING BRAIN

Functional activity in brain involves electrical and molecular events that transmit signals from cell to cell within specific networks and anatomical pathways, and our current understanding of functional metabolism in brain has been advanced by development of quantitative local methods to measure blood flow, blood oxygen levels, oxygen consumption, glucose utilization, the fate of specific atoms in labeled precursors, and metabolite levels. Noninvasive technologies employing different approaches (nuclear, optical, magnetic resonance, thermal, acoustic, fluorescence imaging) and various tracers (inert gases, radiolabeled compounds, compounds labeled with stable isotopes, voltage-sensitive dyes, bioluminescence, and endogenous compounds) have been particularly important for a wide variety of *in vivo* brain-imaging studies of molecular, metabolic, physiologic, and cognitive processes in normal and disease states (Dienel, 2006).

Determination of Local Rates of Blood Flow and Metabolism in Brain

Blood flow. Quantitative autoradiographic determination of local rates of blood flow and glucose utilization using *radioactive tracers* (Box 3.3) provided high sensitivity and appropriate spatial resolution to measure flow and metabolism in all regions of the brain. Kety and colleagues (Landau *et al.*, 1955) used a radioactive gas, trifluoromethane, to establish activity-dependent changes in blood flow in the cat, and

Sokoloff's group developed the use of a less volatile tracer with high blood–brain barrier permeability (Sakurada *et al.*, 1978). In brief, the experimental procedure involves a pulse intravenous injection of the tracer and collection of timed arterial samples of blood from which the *time-activity integral* (Box 3.3) in blood is calculated. The brain tissue is sampled, cut into thin sections, and exposed to X-ray film along with calibrated radioactivity standards; local optical densities are measured and converted to tissue concentration by use of the standard curve for that film; and then blood flow rates calculated. Blood flow assays measuring unidirectional uptake of a labeled tracer must be of short duration to prevent efflux of significant amounts of tracer from tissue to blood during the assay interval, and the tracer must have high lipid permeability so that its accumulation in brain tissue is blood flow dependent. The advantage of these *in vivo* procedures is that the tracers label all regions of the brain simultaneously and assay of local tissue concentrations of the tracer enables comparisons between activated and nonactivated pathways in the same brain. Development of the use of calibrated ^{14}C standards along with each X-ray film transformed qualitative pictures and relative changes (e.g., structures of interest normalized to one other structure) to fully quantitative autoradiographic assays, a procedure that had a high impact on many fields that used autoradiographic detection methods. Blood flow is often used to identify activated brain structures, particularly in brief experiments, but blood flow and the oxygen content of blood in the brain's microvasculature are not measures of metabolism; flow and metabolism do not always change proportionately.

Glucose utilization: radiotracer techniques. Early studies of glucose utilization rates in brain *in vivo* used labeled glucose as a precursor but were limited by the loss of labeled products at various stages of metabolism of glucose at rates that vary with the position of the label and the state of the subject. For example, label is released from the one position via the pentose phosphate shunt pathway, from carbons 3 and 4 at the pyruvate dehydrogenase step, from positions 2 and 5 during the second turn of the TCA cycle, and from positions 1 and 6 during the third turn of the TCA cycle; label is also lost when lactate is released from brain (Fig. 3.12). Due to difficulties associated with complete accounting of precursor and products of labeled glucose in brain *in vivo*, radiolabeled glucose has been most useful for assessment of specific aspects of glucose transport and metabolism *in vitro* using cultured cells, brain slices, and preparations of whole homogenates and subcellular

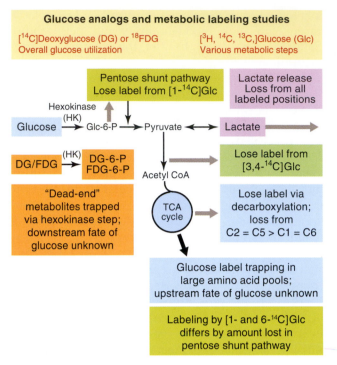

Glucose analogs and metabolic labeling studies

[14C]Deoxyglucose (DG) or 18FDG
Overall glucose utilization

[3H, 14C, 13C,]Glucose (Glc)
Various metabolic steps

Pentose shunt pathway
Lose label from [1-14C]Glc

Lactate release
Loss from all
labeled positions

Hexokinase
(HK)

Glucose → Glc-6-P → Pyruvate ↔ Lactate

DG/FDG (HK) DG-6-P
FDG-6-P

Acetyl CoA

Lose label from
[3,4-14C]Glc

"Dead-end"
metabolites trapped
via hexokinase step;
downstream fate of
glucose unknown

TCA
cycle

Lose label via
decarboxylation;
loss from
C2 = C5 > C1 = C6

Glucose label trapping in
large amino acid pools;
upstream fate of glucose unknown

Labeling by [1- and 6-14C]Glc
differs by amount lost in
pentose shunt pathway

FIGURE 3.12 **Metabolic imaging using labeled glucose and its analogs.** Various steps of glucose metabolism can be measured in different preparations by using differentially labeled glucose because label from glucose is lost at different steps. The glucose analogs 2-deoxy-D-glucose (DG) and 2-fluoro-2-deoxy-D-glucose (FDG) compete with glucose for transport and for metabolism by hexokinase. Their major metabolites, DG-6-P and FDG-6-P, respectively, are trapped intracellularly; hexokinase activity and glucose utilization rates can be calculated from labeled product accumulation (see text and Sokoloff *et al.*, 1977). Modified from Fig. 1 of Dienel and Cruz, 2008; with permission of Wiley-Blackwell Publishers.

fractions derived from brain. As discussed in following text, determination of rates of incorporation of [13C] from [13C]glucose into amino acids is used to calculate rates of oxidative metabolism of glucose.

A major advance in metabolic assays was development of the use of 2-deoxy-D-[14C]glucose (DG), a glucose analog, as a tracer to determine local rates of glucose utilization in brain *in vivo* (Sokoloff *et al.*, 1977). Sokoloff recognized that because (1) DG competes with glucose for transport across the blood–brain barrier and into brain cells and for phosphorylation by hexokinase and (2) DG is not metabolized via glycolysis beyond the hexokinase step (i.e., it is a "dead-end" metabolite, Fig. 3.12), [14C]DG can be given in *tracer doses* (Box 3.3) and used to measure a single reaction, the hexokinase step, in all regions of the living brain simultaneously (Fig. 3.12 and Fig. 3.13). Based on the kinetic differences between glucose and DG for transport and phosphorylation, approximately two molecules of glucose are phosphorylated for each

molecule of [14C]DG that is phosphorylated. Thus, [14C]DG-6-P (and its derivatives—there is some incorporation into other phosphorylated compounds and macromolecules) is trapped in the cells where phosphorylated and quantitatively retained for a reasonable amount of time (30–60 min), and accumulation of labeled products of [14C]DG serves as a "meter" to register the amount of glucose consumed at a local level.

Shortly after development of the autoradiographic [14C]DG method, the procedure was extended for use in human and primate brain by synthesis of a positron-emitting analog [18F]2-fluoro-2-deoxy-D-glucose ([18F]FDG) (Phelps *et al.*, 1979). This tracer has had broad application in positron emission tomographic (PET) studies of brain, particularly in studies of the progression and treatment of human neurodegenerative diseases. [18F]FDG is also widely used for detection and localization of brain tumors, as well as tumors throughout the body, which are readily detected as metabolic "hot spots" due to their high glycolytic metabolism compared to surrounding normal tissue. Interested readers are referred to reviews by Sokoloff (1986, 1996a) for details of the DG method and its broad application to many fields, including physiology, pharmacology, cognition, psychology, behavior, and pathophysiology and to the autobiographical perspectives of Seymour Kety (1996) and Louis Sokoloff (1996b). Since these pioneering studies to devise local quantitative methods to measure blood flow and glucose utilization, a wide variety of compounds labeled with positron-emitting isotopes have been developed for *in vivo* studies of gene expression, protein, phospholipid, and neurotransmitter turnover, receptor assays, transporters, and other processes (Phelps *et al.*, 1986; Phelps, 2004).

Tracking of carbon atoms through metabolic pathways: stable isotopes and magnetic resonance spectroscopy. Assays of the hexokinase step with deoxyglucose quantify the amount of glucose consumed by brain but do not provide any information regarding the downstream fate of glucose carbon (Figs. 3.12 and 3.13). The complexities of compartmentation of metabolism were established by tracer labeling studies using [14C]- and [3H]-labeled precursors, but the analytical procedures are time-consuming and labor-intensive and only one time point can be obtained from each subject; many reviews describing the history of this field are in the monographs edited by Bálazs and Cremer (1972) and Berl *et al.* (1975). The next major advance in the field was the use of stable isotopes, magnetic resonance spectroscopy (MRS), and metabolic modeling to evaluate fluxes in metabolic pathways of interest in diverse preparations,

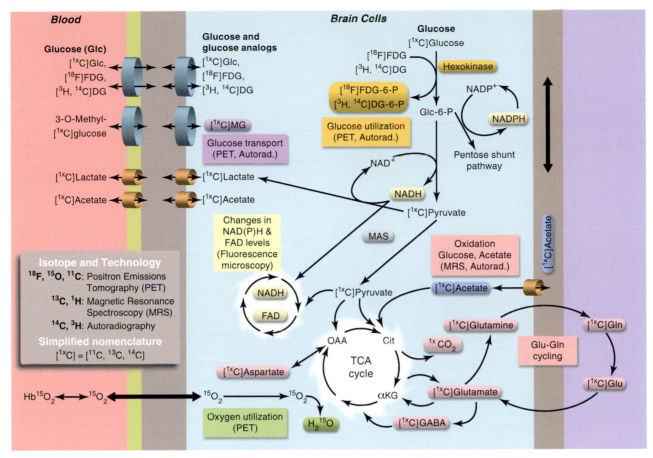

FIGURE 3.13 **Metabolic brain imaging takes advantage of transport and metabolism of specific labeled precursors.** By use of appropriate precursors labeled with beta- or positron-emitting radioactive or stable isotopes (see Box 3.3), specific aspects of transport, hexokinase activity, pentose phosphate shunt activity, glycolytic and oxidative metabolism, and redox state can be assayed in living brain under different experimental conditions using different technologies, positron emission tomography (PET), magnetic resonance spectroscopy (MRS), autoradiography, and fluorescence microscopy. For example, the nonmetabolizable glucose analog, methylglucose, is used to assay blood–brain barrier and cellular membrane transport and the steady state glucose concentration. The glucose analogs DG and FDG are used to assay hexokinase activity and calculate glucose utilization rates. Glucose, lactate, and acetate labeled with radioactive or stable isotopes are used to assess rates of oxidative pathways in neurons and astrocytes, decarboxylation reactions (e.g., pentose shunt or pyruvate dehydrogenase; see Fig 3.12), or glutamate–glutamine cycling. Labeled oxygen is used for *in vivo* assays of cerebral respiration, and labeled water is used to assay cerebral blood flow. Changes in the redox state of endogenous compounds (NAD(P)H and FAD) are assayed by fluorescence microscopy. For simplicity, isotopes of carbon are sometimes denoted by the superscript 1x to collectively represent [11]C-, [13]C-, and ^{14}C-labeled compounds that can be used to evaluate similar pathways using different technologies (see text for details).

ranging from cell cultures to human brain. MRS has the advantages of no radiation exposure, determination of time courses in each subject, and longitudinal studies in the same subject; disadvantages are lower sensitivity compared to radiolabeling studies and immobilization of the head during the assay procedure, which requires use of anesthesia in experimental animals.

Pioneering work in the Shulman laboratory demonstrated the high potential for MRS with ^{31}P and ^{13}C studies in animal brain (reviewed by Pritchard and Shulman, 1986), and initial applications of this *in vivo* approach to human brain included determination of brain glucose levels (Gruetter *et al.*,

1992) and assays of incorporation of ^{13}C from glucose into C4 of glutamate (Rothman *et al.*, 1992). ^{13}C-labeled tracers were also used in other laboratories to explore the complexity of brain energy metabolism, metabolite trafficking, astrocyte–neuron interactions, and metabolic compartmentation (Badar-Goffer *et al.*, 1990, 1992; Cerdan *et al.*, 1990; Kanamori *et al.*, 1991; Brand *et al.*, 1992), and numerous subsequent MRS studies have greatly expanded these research areas (Rodriques and Cerdán, 2007; Zwingmann and Leibfritz, 2007; see Figs. 3.17 and 3.18, later). Metabolic modeling of brain metabolic activity and energy transfer within cells are very important research areas that are continually evolving to accommodate new data, refined mathematical approaches, and revised concepts describing cell–cell interactions (Hyder *et al.*, 2006; Gjedde, 2007; Saks *et al.*, 2007; Shestov *et al.*, 2007).

Changes in redox state: fluorescence microscopy. Optical signals are generated by activity-dependent electrolyte- and metabolism-driven changes in properties of intrinsic molecules, and these signals can be used to probe real-time changes in tissue metabolic activity (Villringer and Chance, 1997). For example, reduction of NAD and NADP generates the respective fluorescent compounds, NADH and NADPH (often denoted as NAD(P)H because the absorption and fluorescence spectra are similar), whereas oxidation of the mitochondrial flavoprotein, $FADH_2$, yields a fluorescent product FAD (Chance *et al.*, 1962). Thus, spatial–temporal changes in cytoplasmic and mitochondrial oxidation state can be assayed by fluorescence microscopy (Fig. 3.13) under a variety of experimental conditions using biological samples of varying complexity, ranging from isolated mitochondria to the intact animal.

The pioneering studies by Chance and colleagues demonstrated the capability of real-time, continuous fluorescent assays of redox changes in living cells and brain tissue during anoxia and other conditions (e.g., Chance and Thorell, 1959; Chance *et al.*, 1962), and early work in other laboratories assessed NAD(P)H fluorescence changes in isolated brain mitochondria, brain slices, and intact cerebral cortex (1) to evaluate effects of oxygen and ADP levels on mitochondrial redox state, (2) to reveal biphasic responses to electrical stimulation (i.e., rapid oxidation followed by a slower reduction phase with an overshoot above the initial baseline level) that were related to changes in ions and signaling molecules, and (3) to characterize responses of intact brain tissue to electrical stimulation and seizure activity (Jöbsis *et al.*, 1971; Rosenthal and Jöbsis, 1971; Lipton, 1973).

An advantage of functional studies that combine electrophysiological and fluorescence techniques is that various fluorescent indicators can be loaded into tissue samples, and interactions among cellular activity, energy metabolism, ion concentration changes, and mitochondrial function evoked by different electrical stimuli or pharmacological challenge can be simultaneously evaluated. For example, NAD(P)H signals in hippocampal organotypic slice cultures from 7–8 day rats were shown to be closely correlated to increased extracellular K^+ concentration and mitochondrial Ca^{2+} concentration (Kann *et al.*, 2003). In hippocampal slices from mature brain the electrical stimulus-induced biphasic response of NAD(P)H fluorescence was shown to be an "inverted match"

of the FAD fluorescence that did not reflect mitochondrial Ca^{2+} dynamics; both components of the biphasic NAD(P)H response were abolished by blockade of ionotropic glutamate receptors and were therefore attributed mainly to postsynaptic neuronal activity, not presynaptic neuronal activity or astrocytic glutamate uptake (Shuttleworth *et al.*, 2003). High-resolution, two-photon fluorescence imaging distinguished the temporal profiles of NADH fluorescence in astrocytes and neurons after electrical stimulation, and the early oxidative change was ascribed to dendrites (on the basis of blockade of postsynaptic kainate/AMPA receptors by CNQX) whereas the delayed reductive overshoot colocalized with astrocytes (Kasischke *et al.*, 2004). Follow-up studies to identify the basis for the apparently discrepant results in the preceding reports employed a variety of biochemical and pharmacological approaches to evaluate the basis of the reductive "overshoot" component of the NAD(P)H response and found that responses to brief stimuli used in the Shuttleworth study were eliminated by ionotropic glutamate receptor antagonists, whereas the response to the longer stimulus used in the Kasischke study was partially resistant to these drugs and was not blocked by a glutamate transport inhibitor; with short stimuli, the reductive NAD(P)H response arises from postsynaptic activation, not astrocytic glycolysis stimulated by glutamate uptake (Brennan *et al.*, 2006; Brennan *et al.*, 2007). The impact of postsynaptic activity of evoked metabolic responses is also emphasized by the results of Caesar *et al.* (2008), who found that electrical stimulus-induced increases in extracellular lactate level, blood flow, CMR_{glc}, and CMR_{O2} in the cerebellum of the anesthetized rat were eliminated by treatment with CNQX, which does not block glutamate-stimulated glycolysis in cultured astrocytes (Pellerin and Magistretti, 1994) or glutamate uptake into astrocytes (Duan *et al.*, 1999). Thus, brain activation-induced lactate generation in this intact animal preparation is dissociated from astrocytic glutamate transport; the lactate arises from postsynaptic activity.

Flavoprotein fluorescence imaging is also emerging as a useful tool to evaluate dynamic, local changes in mitochondrial activity in brain slices and *in vivo* during electrical and physiological stimulation. For example, forepaw or hindpaw stimulation elicits FAD autofluorescence changes in somatosensory cortex that correlate with evoked changes in field potentials in the tissue, and transcranial imaging through the intact skull of anesthetized mice enabled tonotopic mapping of cortical responses to acoustic stimuli of different frequencies (Shibuki *et al.*, 2007). Pharmacological studies showed that spatial patterns of excitatory and inhibitory activity in the cerebellum could be tracked by FAD fluorescence, and stimulation of parallel fibers causes biphasic FAD responses, a rapid initial oxidation followed by slower reduction that is ascribed to neurons and glia, respectively; the signals are tightly coupled to neuronal activity and are primarily postsynaptic in origin due to blockade by CNQX (Reinert et al., 2007). To sum up, autofluorescence responses of NAD(P)H and FAD reveal major contributions of postsynaptic activity to stimulus-evoked metabolism in various brain regions. Interpretation of changes in concentrations of redox compounds involves assessment of the relevant fluxes that underlie the redox changes and the effects of tissue preparation (e.g., brain slices have been subjected to ischemia and mechanical damage, and anesthesia suppresses conscious and metabolic activity). The high spatial and temporal resolution of fluorescence imaging, in conjunction with electrophysiological, pharmacological, and metabolic assays, opens the door for new approaches to studies of the cellular basis of images generated by functional metabolism (Fig. 3.13).

Function–Metabolism Relationships

Carbohydrates, amino acids, carboxylic acids, alcohols, aldehydes, ketones, lipids, nucleic acids, steroids, and other compounds labeled with *beta-emitting isotopes* (Box 3.3), especially ^{14}C and ^3H, as well as ^{35}S, ^{32}P, and ^{33}P, have been used for over 50 years to study metabolic reactions and pathways in biochemical and autoradiographic studies. *Positron-emitting isotopes* (^{18}F, ^{15}O, ^{11}C, and ^{13}N) have been incorporated into a wide variety of tracers used for positron emission tomographic (PET) studies, whereas *stable isotopes* (e.g., ^1H, ^{13}C, ^{31}P, ^{15}N) are used in MRS studies. Some of the specific aspects of blood–brain barrier transport and brain metabolism that can be evaluated by means of these technologies are illustrated in Figure 3.13. Assays can be designed to determine quantitatively uptake across the blood–brain barrier, metabolism, or concentrations of endogenous compounds or of exogenous compounds labeled with radioactive or stable isotopes. For example, the nonmetabolizable glucose analog, 3-O-methylglucose, is useful for biochemical, autoradiographic, and PET assays of unidirectional glucose transport across the blood–brain barrier and into cultured cells. Because the steady-state brain-to-plasma ratio for methylglucose is proportional to brain glucose concentration, methylglucose can also be used to measure local tissue glucose concentration in brain. DG and FDG are metabolized to the hexose-6-P step and measure the first (irreversible) step of glucose utilization, whereas metabolism of $^{15}O_2$ measures the terminal step of the electron transport chain, and $[^{15}O]H_2O$ is useful for PET assays of cerebral blood flow in humans. Note that metabolites of ^{13}C-labeled glucose and acetate are incorporated into metabolites generated by the TCA cycle, and these precursors are particularly useful to measure oxidative metabolism, metabolite cycling, neurotransmitter trafficking, and compartmentation by MRS. Finally, fluorescence microscopy can be used to detect changes in the fluorescence intensity of NADH and FAD with brain function, and evaluate changes in redox status generated during activating or pathophysiological conditions (Fig. 3.13). Taken together, autoradiographic, PET, MRS, and fluorescence microscopic studies of metabolism are complementary, each with its strengths and weaknesses, and can be used together to investigate complex issues in brain function and disease.

Glucose concentration and utilization. Local rates of glucose utilization (CMR_{glc}) are very heterogeneous whereas brain glucose concentration is fairly uniform, indicating that under steady-state conditions glucose delivery matches its rate of utilization over a wide range of normal rates. CMR_{glc} is highest in gray matter (range: about 0.50–1.20 μmol g^{-1} min^{-1}) and lowest in white matter (range: about 0.15–0.25 μmol g^{-1} min^{-1}), and rates in various gray matter structures in different anatomical pathways can differ by a factor of 2 or more (Sokoloff et al., 1977). For example, CMR_{glc} varies throughout the cerebral cortex, as well as in stations in the same pathway (e.g., auditory pathway) (Table 3.3). Furthermore, regional and laminar differences are readily detectable in $[^{14}C]DG$ autoradiographs of the cerebral cortex, hippocampus, caudate-putamen, and thalamus (Fig. 3.14A and 14B, top panels). In contrast, $[^{14}C]$methylglucose autoradiographs (Fig. 3.14A, B, lower panels) have relatively uniform optical densities, demonstrating the similar concentration of glucose throughout the brain; note the relatively small glucose concentration differences between gray and white matter structures, contrasting the much larger differences in metabolic rate (compare top and bottom panels in Fig. 3.14A, B). Focal activation of metabolism by a penicillin-induced seizure doubles cortical CMR_{glc} at the application site and slightly depresses local glucose level, whereas focal suppression of brain activity by barbital depresses CMR_{glc} by 50% at the application sites and causes a small increase in local glucose content. Thus, specific, focal changes in metabolism throughout the brain due to altered physiological activity or pharmacological treatment are readily detected and quantifiable, and at the level of autoradiographic resolution, glucose supply and demand are closely matched over a four-fold range of CMR_{glc}.

TABLE 3.3 Regional Metabolic Responses to Unilateral Broadband Click Stimulus

Region of Interest	Activated (Right) Hemisphere	Contralateral (Left) Hemisphere	Right–Left Ratio
	Glucose utilization (μmol 100 g^{-1} min^{-1})		
Auditory structures			
Superior olive	87.3 ± 23.6	80.2 ± 25.4	1.11 ± 0.17
Lateral lemniscus	92.0 ± 29.4 [a]	63.6 ± 13.6	1.41 ± 0.25
Inferior colliculus	119.8 ± 36.9 [a]	70.8 ± 10.2	1.66 ± 0.34
Medial geniculate	63.9 ± 12.0	57.9 ± 9.5	1.11 ± 0.13
Auditory cortex	67.9 ± 5.0	65.0 ± 4.3	1.05 ± 0.08
Other structures			
Visual cortex	51.5 ± 7.6	52.4 ± 7.5	0.98 ± 0.04
Sensory cortex	71.7 ± 2.0	71.6 ± 4.6	1.00 ± 0.05
Sensorimotor cortex	70.5 ± 5.2	71.7 ± 3.9	0.98 ± 0.03
Thalamus	62.4 ± 6.1	63.1 ± 5.3	0.99 ± 0.02
Caudate	69.7 ± 5.3	71.9 ± 4.7	0.97 ± 0.03
Genu corpus callosum[c]	17.0 ± 7.0	16.4 ± 6.9	1.04
Cerebellar white matter[c]	24.2 ± 7.2	24.2 ± 4.9	1.0
	[2-^{14}C]Acetate minimal net uptake coefficient (mL 100 g^{-1} min^{-1})		
Auditory structures			
Superior olive	5.9 ± 1.5	5.7 ± 1.5	1.04 ± 0.05
Lateral lemniscus	6.0 ± 1.3 [b]	5.1 ± 1.4	1.18 ± 0.10
Inferior colliculus	6.1 ± 1.2 [b]	5.3 ± 1.0	1.15 ± 0.03
Medial geniculate	5.5 ± 1.9 [a]	5.3 ± 1.8	1.03 ± 0.01
Auditory cortex	5.6 ± 1.4	5.5 ± 1.5	1.02 ± 0.04
Other structures			
Visual cortex	5.3 ± 2.1	5.4 ± 2.1	0.99 ± 0.02
Sensory cortex	5.4 ± 1.1	5.4 ± 1.2	1.01 ± 0.02
Sensorimotor cortex	5.6 ± 1.3	5.6 ± 1.4	1.00 ± 0.03
Thalamus	4.9 ± 2.3	4.8 ± 2.3	1.03 ± 0.02
Caudate	4.9 ± 2.6	4.9 ± 2.7	0.99 ± 0.01
Genu corpus callosum[c]	2.1 ± 0.3	2.2 ± 0.3	0.95
Cerebellar white matter[c]	2.1 ± 0.5	2.1 ± 0.5	1.0

Note: The acoustic stimulus (~88 dB) was initiated 10 min prior to intravenous [^{14}C]tracer injection into conscious rats kept in a sound-insulated environment. Glucose utilization was assayed with the routine [^{14}C]deoxyglucose method. The net acetate uptake coefficient was assayed during a 5 min experiment and calculated by dividing the ^{14}C level in the region of interest by the time–activity integral for total ^{14}C in arterial plasma. Plasma acetate levels in a parallel group of rats averaged 0.9 μmol/ml, and net acetate uptake coefficients can be converted to μmol 100 g^{-1} min^{-1} by multiplying by 0.9. Values are mean ± SD (n = 5); statistically significant right–left differences were identified by the paired t test ([a]P <0.05, [b]P <0.01). The right–left ratios are means of values calculated for each structure in each rat, not ratios of the mean values for structures in each hemisphere. Data from Cruz *et al.* (2005) except for the values for two white matter structures from normal unstimulated rats from [c]Cetin *et al.* (2003).

The cellular basis of autoradiographic images of glucose utilization has been of intense interest for more than 30 years because routine ^{14}C-autoradiographic assays do not have cellular resolution. Studies to identify metabolically labeled cells have been hindered by serious technical difficulties, particularly the quantitative retention of labeled metabolites in cells and tissue during histological processing to identify cell type. Label losses during tissue processing can be as high as 90% (Sharp, 1976a,b; Durham *et al.*, 1981; Duncan *et al.*, 1987a,b), and even in the most recent studies with about 50% label retention (Nehlig *et al.*, 2004) or *in vivo* assays using a fluorescent glucose analog (Itoh *et al.*, 2004), the best estimates suggest that overall glucose utilization rates in astrocytic and neuronal cell bodies are similar

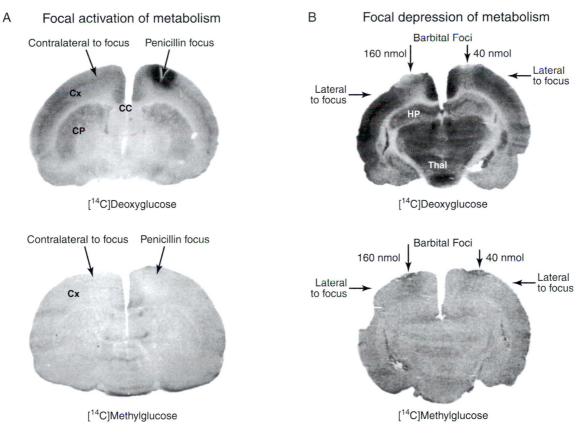

FIGURE 3.14 **Autoradiographic assays of local rates of glucose utilization and glucose concentration.** A focal seizure was induced by topical application of penicillin to the cerebral cortex (A) and focal anesthesia resulted from topical application of 40 or 160 nmol of barbital (B). Local rates of glucose utilization (CMR$_{glc}$) were assayed with [^{14}C]deoxyglucose and local glucose concentration assayed with [^{14}C] methylglucose. In brief, a pulse intravenous bolus of [^{14}C]deoxyglucose is given to the animal, 45 min later the brain is sampled, frozen, cut into serial coronal 20 μm-thick sections, and exposed to X-ray film along with calibrated ^{14}C standards; the darker the region, the higher the ^{14}C concentration and the higher the metabolic rate or glucose level. Similar procedures were used for the [^{14}C]methylglucose assays except that a programmed infusion was used to establish steady-state conditions for the tracer. Note that in the untreated tissue CMR$_{glc}$ is very heterogeneous, whereas glucose concentrations are relatively uniform throughout the brain. CMR$_{glc}$ increased two-fold at the site of the seizure (dark spot in the cerebral cortex in the upper figure in panel A), and barbital anesthesia reduced CMR$_{glc}$ (light spots in cerebral cortex in the upper figure in panel B) by 36% and 52% at the 40 and 160 nmol barbital sites, respectively. The steady-state brain-to-plasma ratio for [^{14}C] methylglucose, which varies with brain glucose concentration, fell 13% at the seizure focus and rose 14–19% in the barbital foci. (Reproduced from Nakanishi *et al.*, 1996, with permission of Nature Publishing Group.) Abbreviations: Cx, cerebral cortex; CC, corpus callosum (white matter); CP, caudate-putamen; Hip, hippocampus; Thal, thalamus.

under resting conditions. Specific subsets of neurons and astrocytes are, however, more metabolically active than others during resting conditions and accumulate high levels of [^{3}H]DG compared to other cells (Duncan *et al.*, 1987a,b, 1990). During swimming or rotation paradigms increased neuronal trapping of labeled DG occurs in activated structures (Sharp, 1976a,b). Increased neuronal glucose utilization during *in vivo* activation is consistent with the large excess capacity of synaptosomes to carry out glycolysis and respiration; synaptosomal glycolytic rate increased 10-fold for at least 30 min and respiration rose 6-fold after treatment of synaptosomes with FCCP (carbonylcyanide p-trifluoromethoxyphenylhydrozone) to uncouple respiration and ATP synthesis

(Kauppinen and Nicholls, 1986). In spite of many attempts to improve the cellular resolution of *in vivo* autoradiographic studies, the available data still do not fully account for the metabolic contributions of major cell types. Label accumulation in the cell bodies of astrocytes and neurons is most readily identified and tallied according to cell type, and neuronal perikarya have lower rates of glucose utilization than the neuropil. The energetic contributions of neuronal pre- and postsynaptic processes and the fine filopodia of astrocytes remain to be established, particularly during brain activation.

Imaging focal activation with different metabolic tracers. Presentation of a single tone stimulus to a subject activates the auditory pathway in a tonotopic manner, and different groups of cells in stations of the auditory processing pathway respond by increasing signaling and rates of glucose utilization. Focal activation of bands of cells in the inferior colliculus of the rat (Fig. 3.15, top left panel) is readily detectable with the routine [^{14}C]DG method using a 45 min

experimental period (the longer assay interval takes advantage of the metabolic stability of [^{14}C]DG-6-P in brain tissue and minimizes the effects of uncertainties in the true values of rate constants used to estimate (1) the unmetabolized [^{14}C]DG in brain and (2) the brain time–activity integral on the calculated rate of

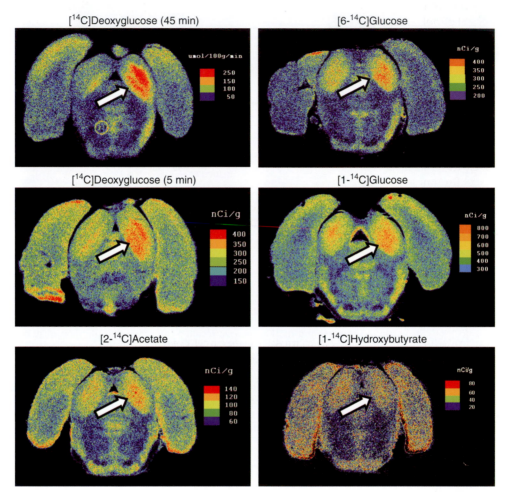

FIGURE 3.15 **Registration of focal activation with different metabolic tracers.** Stimulation of the auditory pathway of the rat with a single tone increases cellular activity in a tonotopic manner, such that specific groups of cells in the stations of the auditory pathway respond to that tone by increasing their signaling and metabolic activity. Tonotopic bands of increased CMR$_{glc}$ identify the regions within the right inferior colliculus (arrows) that respond to a unilateral acoustic stimulus, and these bands are most readily detected with the routine 45 min [^{14}C] deoxyglucose (DG) method (top left panel) but are also detectable within 5 min with [^{14}C]DG (middle left panel), in spite of the higher background due to unmetabolized precursor. The small activation band in left inferior colliculus in the 45 min [^{14}C]DG assay is probably due to crossover of fibers causing some activation in the contralateral hemisphere. The tonotopic bands are modestly detectable with [6-^{14}C]glucose (top right panel) but not with [1-^{14}C]glucose (middle right panel) or [2-^{14}C]acetate (bottom left panel). These findings suggest that most of the products of glucose metabolism produced by activation over and above those during rest are quickly lost from the activated inferior colliculus by label release and decarboxylation reactions (see Cruz et al., 2007, and Dienel and Cruz, 2008, for detailed discussion of these findings and their implications). Because a broadband acoustic stimulus that activates most of the inferior colliculus does increase labeling by [2-^{14}C]acetate (Cruz et al., 2005) and generates a [^{14}C]acetate image similar to that obtained by monotonic acoustic stimulation (that does not register the tonotopic bands; bottom left panel), it is likely that labeled amino acids derived from the astrocytic TCA cycle spread within the colliculus and may also be released from the colliculus; [^{14}C]glutamine spreading does increase during acoustic activation (Cruz et al., 2007). Note that [1-^{14}C]hydroxybutyrate does not register activation of the right inferior colliculus by a broadband stimulus (Cruz et al., 2005). The color-coding scales are not the same in the images obtained from different labeling assays: CMR$_{glc}$ is shown only for the 45 min DG assay; all other panels represent local tissue tracer concentrations. Adapted from Dienel and Cruz, 2008, and Dienel et al., 2007.

glucose utilization). Focal activation can also be imaged in 5 min [^{14}C]DG experiments (Fig. 3.15, middle left panel), but the higher background due to the higher proportion of unmetabolized [^{14}C]DG blunts the relative increase, and calculated metabolic rates are not as accurate as desired due to a higher impact of estimates of values for rate constants. On the other hand, tonotopic bands are barely detectable with [6-^{14}C]glucose (Fig. 3.15, top right panel), and they are not detected with [1-^{14}C]glucose or [2-^{14}C]acetate (Fig. 3.15, bottom left and middle right panels). A broadband acoustic stimulus causes robust increases in glucose utilization throughout the activated inferior colliculus, but functional activation is not registered at all with β-hydroxybutyrate (Fig. 3.15, bottom right panel). These findings are of considerable interest, since glucose and β-hydroxybutyrate label the "large" glutamate pool, whereas acetate labels the small glutamate pool in astrocytes. Although acetate and β-hydroxybutyrate are transported into brain from blood by the same monocarboxylic acid carrier, MCT1, they have different metabolic fates in brain. Acetate does register broadband acoustic activation, whereas the ketone body does not (Cruz et al., 2005). These discrepant labeling patterns are ascribed to rapid spreading and release of metabolites of glucose by various pathways, including decarboxylation by the pentose shunt pathway, release of labeled lactate, spreading via gap junctions, and rapid elimination of labeled products that correspond to most of the additional glucose consumed during activation compared to rest (Cruz et al., 2007).

Calculated rates of glucose utilization in different brain regions during broadband acoustic stimulation of conscious rats illustrate the pathway-specific effects of a physiological stimulus on utilization of glucose by all brain cells and by oxidation of acetate by astrocytes; structures within the auditory pathway respond to different extents, whereas nonauditory pathways have stable rates of glucose utilization (Table 3.3). Note the low metabolic rates in white matter structures (corpus callosum and cerebellar white) compared to gray matter, the large range of CMR$_{glc}$ throughout brain (5- to 7-fold in the resting and activated hemisphere, respectively), the more modest range (2.6- to 2.9-fold) of acetate oxidation, and the large stimulus-induced response of CMR$_{glc}$ (66%) compared to astrocytic oxidative metabolism (18%) (Table 3.3). These findings illustrate the large dynamic range of responsiveness of glucose utilization during physiological activity and demonstrate that oxidative metabolism in astrocytes increases during sensory stimulation. Astrocytes are not simply glycolytic, as is sometimes assumed based on tissue culture studies; they also depend on oxidative metabolism to generate ATP.

Flow-metabolism "coupling." The notion of close coupling of blood flow and metabolism arose from the high correlation between local rates of blood flow and CMR$_{glc}$ in different brain structures (Fig. 3.16A). This concept was further strengthened by demonstration that (1) GLUT1 glucose transporter density and capillary density are also highly correlated with local rates of glucose utilization (Fig. 3.16B) and that (2) regional GLUT1 and MCT1 densities are correlated (Fig. 3.16C). Thus, the physical structure of the microvasculature, rates of blood flow, transporter capacity, and metabolic rate are coordinated at a local level so that fuel supply can satisfy local energy demand. Low activity in white matter is associated with low flow and low transport capacity, whereas high-activity structures are endowed with high flow and transport capacity. Signaling activities of neurons and astrocytes generate compounds that regulate vasodilation and govern rates of blood flow at a local level (Iadecola and Nedergaard, 2007; Gordon et al., 2007).

Whiskers are an extremely important tactile system in many animals, and the representation of whiskers in the sensory cortex is correspondingly high, similar to the high representation of the hands and mouth in somatosensory cortex of primates. Each whisker is represented by a specific group of cells in sensory cortex that involves all lamina from dorsal to ventral cortex and corresponds to a "barrel." Modulation of the activity of the whisker-to-barrel pathway by anesthesia during rest and sensory stimulation is illustrated in Figure 3.16D. Note the progressive rise in CMR$_{glc}$ as the activities of cells in four stations in the pathway increase from the resting anesthetized, activated anesthetized, conscious resting, and conscious activated states. CMR$_{glc}$ rises during the change of state from anesthesia to conscious and from rest to stimulation in each state, but the magnitude of CMR$_{glc}$ is much lower during anesthesia than the corresponding value in the conscious animal, and the local rate of cerebral blood flow changes in proportion to metabolic rate. Because the CMR$_{glc}$ is differentially suppressed in each structure by anesthesia during rest and activation, the data in Figure 3.16D can also be used to estimate the "cost of consciousness" in stations of the auditory pathway.

Neurotransmission–Metabolism Relationships

Glucose and oxygen requirements. The flux of carbon derived from glucose is an integral aspect of neurotransmitter turnover, especially the cycling of glutamate, glutamine, and GABA (Schousboe and Waagepetersen, 2007). However, the TCA-cycle-derived

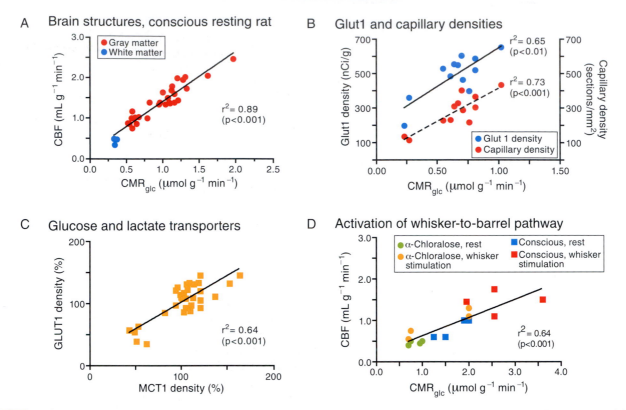

FIGURE 3.16 **Relationships among blood flow, glucose utilization, metabolite transport, and capillary density during rest and brain activation.** (A) Local rates of blood flow correlate with CMR_{glc} in the same rat brain structure; note the low flow–metabolism rates in white compared to gray matter (plotted from data of Sokoloff *et al.*, 1977, and Sakurada *et al.*, 1978). (B) Capillary and glucose transporter GLUT1 densities in adult rat brain are correlated with CMR_{glc} (plotted from data of Zeller *et al.*, 1997). (C) Local glucose and lactate transport capacities are correlated in adult rat brain (plotted from data of Maurer *et al.*, 2004). (D) Flow–metabolism relationships in four stations of the whisker-to-barrel pathway of the rat during rest and whisker stimulation in α-chloralose-anesthetized and conscious rats (plotted from data of Nakao *et al.*, 2001).

amino acid transmitters are not the only systems dependent on glycolytic and oxidative metabolism of glucose and on adequate levels of oxygen. The neuromodulators D-serine and glycine are generated from the glycolytic pathway, and both the catecholaminergic and cholinergic pathways are linked to tissue oxygen content and oxidative metabolism (Fig. 3.17). Catecholamine neurotransmitter synthesis and degradation is highly dependent on the activity of mixed-function oxidases (tyrosine hydroxylase, tryptophan hydroxylase, dopamine-β-hydroxylase, and monoamine oxidase) that consume molecular oxygen (Fig. 3.17). Generation of acetylcholine from mitochondrial pyruvate (Fig. 3.17) is closely coupled to oxidative metabolism of glucose, even though only 1% of the glucose is used to synthesize acetylcholine (Joseph and Gibson, 2007).

Neuronal utilization of glycolysis and glycolytic ATP. During the past decade, considerable attention has been placed on astrocytic glycolysis and lactate production, but neurons also depend on the glycolytic pathway, not only for energy but also to sustain neurotransmission. For example, glycolysis provides the energy for vesicular packaging of neurotransmitters (Fig. 3.17), an integral aspect of synaptic signaling. The driving force for the vesicular glutamate transporter (VGLUT) is the proton gradient inside the vesicle that is generated by H^+-ATPase. Ueda and colleagues found high enrichment of glyceraldehyde-3-P dehydrogenase and 3-phosphoglycerate kinase on synaptic vesicles and showed that impairment of glycolysis reduces vesicular glutamate content and lowers glutamate release from synaptosomes (Ueda and Ikemoto, 2007). These studies demonstrate functional linkage of these glycolytic enzymes to the vesicular proton pump and indicate that glycolytically derived ATP, not mitochondrially derived ATP, is preferred by the vesicular proton pump; the working model of this process portrays cycling of a local pool of ATP and ADP near the vesicle membrane surface. Ongoing work in the Ueda laboratory indicates that vesicular loading of other neurotransmitters uses a mechanism similar to that for glutamate.

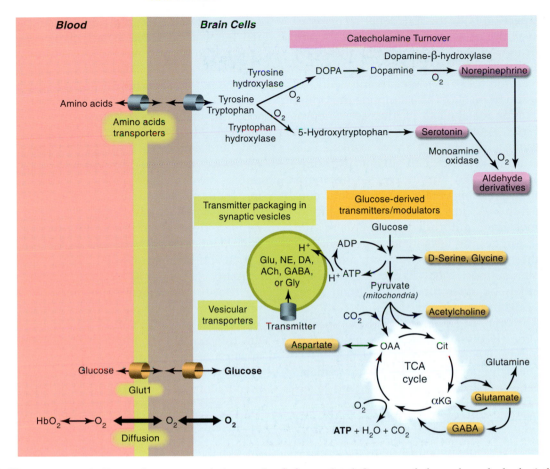

FIGURE 3.17 **Energy metabolism and neurotransmission are closely inter-related.** Oxygen and glucose have dual roles in brain, serving as fuel for generation of ATP and as substrates for enzymes that synthesize neurotransmitters and other compounds required by brain. Oxygen not only serves as the electron acceptor in the electron transport chain, it also is an important substrate for mixed-function oxidases that synthesize and degrade catecholamine neurotransmitters. Glucose-derived neurotransmitters are synthesized via the glycolytic and oxidative pathways, and ATP generated from both processes is essential for critical processes involved in neurotransmission, e.g., packaging of neurotransmitters in synaptic vesicles and maintenance and restoration of transmembrane ionic gradients.

A role for glyceraldehyde-3-phosphate dehydrogenase (GAPDH) in neurotransmission is not limited to excitatory glutamatergic activity because this enzyme can also serve as a GABA$_A$ receptor kinase. The GABA$_A$ postsynaptic receptor is a chloride channel that mediates fast synaptic inhibition by causing hyperpolarization. GAPDH co-localizes with the α1 subunit of the GABA$_A$ receptor and is proposed to become autophosphorylated by locally produced glycolytic ATP, and then this phosphoryl group is transferred to the receptor thereby modulating the receptor-mediated responses to GABA; all the glycolytic machinery to sustain this process is localized in postsynaptic densities (Laschet *et al.,* 2004 and cited references). A deficiency in the GABA$_A$ receptor phosphorylation has been reported in human epileptogenic tissue, and recent work suggests that abnormal GABAergic inhibition may link glycolytic metabolism and enzymes with seizure activity (Laschet *et al.,* 2007).

Thus, neuronal glycolysis has important roles in both presynaptic and postsynaptic structures.

The ability of alternative substrates (e.g., fructose, mannose, pyruvate, lactate) to replace glucose and maintain neurotransmission has been a field of intense interest since the 1960s. Early studies observed that electrically evoked neuronal population spikes in hippocampal brain slices were frequently *not maintained* by alternative substrates even when the ATP concentration exceeded 80% of control, indicating that high energy level is not sufficient to sustain synaptic function (summarized in Fig. 4 of Dienel and Hertz, 2005; also see the related text in that review). A detailed series of studies by Okada and colleagues showed that the specific conditions employed during slice preparation are critical to the outcome of substrate substitution studies. For example, the speed of slice preparation and other conditions have a significant impact on both the functional recovery of slices and the ability of various substrates to support neuronal function; major factors include adenosine, glutamate, and calcium homeostasis (reviewed by Okada and Lipton, 2007). More recent work in cultured cells and brain slices also supports important roles for glycolytic metabolism of glucose in neurotransmission (e.g., Allen *et al.*, 2005; Bak *et al.*, 2006), and synaptosomes do, in fact, have considerable excess glycolytic (as well as oxidative) capacity (Kauppinen and Nicholls, 1986). An unexpected finding is that about half the vesicle-containing presynaptic boutons in axons from CA3 to CA1 neurons in the hippocampus do not have mitochondria (Shepherd and Harris, 1998), suggesting glycolytic dependence of function in these structures. To sum up, alternative substrates can support neuronal energetics and evoked potentials or other functions in brain slices and cultured cells under defined conditions for a certain amount of time, but neurons do have specific requirements for glycolytic and oxidative metabolism of glucose for a number of important processes.

Metabolic modeling of energy supply and demand based on kinetic properties, distributions, and levels of glucose and lactate transporters led Simpson and colleagues (2007) to conclude that glucose uptake and metabolism and generation of lactate transients are mainly neuronal. The likelihood of neuronal generation of lactate during brain activation is strongly supported by a recent study by Caesar *et al.* (2008), who showed that postsynaptic neuronal activation governs the activity-dependent increases in CMR_{glc} and lactate concentration. The stimulus-induced rise in glucose consumption and the increase in extracellular lactate level in cerebellum of anesthetized rats were both blocked by CNQX (Caesar *et al.*, 2008). Because glutamate would be continuously taken up into cerebellar astrocytes during CNQX treatment, the *in vivo* stimulus-induced glycolytic activity is not due to astrocytic glutamate uptake. In fact, activity-dependent stimulation of blood flow and CMR_{O2} in various rat brain structures is dependent on postsynaptic glutamate AMPA or NMDA receptor activity in many studies (see references cited by Caesar *et al.*, 2008), and postsynaptic neuronal activation involving mitochondrial dynamics has been linked to NAD(P)H transients in hippocampal slices (Shuttleworth *et al.*, 2003; Brennan *et al.*, 2006, 2007). Together, these findings emphasize the importance of inclusion of neuronal glucose utilization and lactate production into modeling of functional metabolism; astrocytic glutamate uptake is not necessarily the major or the only factor governing lactate generation during brain activation *in vivo* (see

following text). To sum up, energy supply is an essential but not exclusive aspect of glucose metabolism required to support neurotransmission.

Na^+, K^+-ATPase activation and astrocyte-to-neuron lactate shuttling. Maintenance and re-establishing ionic gradients, mainly by Na^+, K^+-ATPase, is considered to be the major energy expenditure in brain (Attwell and Laughlin, 2001). Ionic pumping increases in neurons and astrocytes during brain activation but the metabolic reactions that provide energy for the pumps have not been definitively identified, contrasting some peripheral tissues in which the Na^+, K^+-ATPase has been shown to use preferentially ATP derived from glycolysis (Paul *et al.*, 1979; Mercer and Dunham, 1981; Lynch and Paul, 1983, 1985; Lynch and Balaban, 1987).

Astrocytic Na^+, K^+-ATPase activity rises during excitatory neurotransmission because glutamate is removed from the synaptic cleft by its cotransport into astrocytes along with sodium. The source of ATP to fuel this transporter is an important but controversial issue arising from apparently discrepant results obtained in different preparations of cultured astrocytes. Some preparations respond to glutamate by increased glycolysis and lactate production (e.g., Pellerin and Magistretti, 1994; Takahashi *et al.*, 1995), whereas others observe no increase in glucose utilization or lactate production (Hertz *et al.*, 1998; Peng *et al.*, 2001; Tables 5 and 6 in Dienel and Cruz, 2004; Fig. 3 in Dienel and Cruz, 2006). Furthermore, glutamate exposure markedly stimulates oxygen consumption in cultured astrocytes (Eriksson *et al.*, 1995) and it increases the fraction of glutamate oxidized compared to that converted to glutamine (Fig. 3.9D; McKenna *et al.*, 1996); these findings are consistent with possibility of oxidation of some of the glutamate taken up by astrocytes to produce ATP for the Na^+, K^+-ATPase. Cultured cells are very useful for studies of regulation, mechanisms, and development, but metabolic fluxes can be influenced by factors that are present only in intact tissue (e.g., maturation of astrocytes within a complex environment containing other cell types), and translation of metabolic findings from pure cultures to intact brain must be made with caution.

The lactate generated by glutamate-evoked glycolysis in cultured astrocytes was proposed by Pellerin and Magistretti (1994) to be taken up and oxidized by neurons, leading to the notion of an astrocyte-to-neuron lactate shuttle. While there is no question that some lactate may be generated by astrocytes and consumed by neurons, unresolved questions include *how much* lactate compared to glucose is actually used by neurons *in vivo* during brain activation in conscious subjects, *what cells actually produce the lactate*, and under what *activating* circumstances is lactate *shuttled and oxidized locally*. There is considerable evidence that both neurons and astrocytes can produce and oxidize lactate and glucose *in vitro* and *in vivo,* and many factors (e.g., local metabolic demand and differences in redox shuttle capacity of different cell types) have a high impact on *local* glycolytic and oxidative activities in brain *in vivo*. It is likely that some lactate generated during activation is consumed in brain, but most of the label corresponding to the additional glucose consumed during activation over and above that used during rest appears to be quickly released from the activated tissue. *Quantitative assessment* of lactate trafficking within and from brain is an important aspect of understanding the biochemical and cellular basis of functional activation of glucose utilization and of metabolic brain images obtained during activation and diseases that involve mitochondrial deficits. A comprehensive model of cell–cell metabolic interactions needs to accommodate and explain the major findings from *in vivo* studies, including regional and stimulus-dependent changes in the oxygen–glucose ratio, local metabolic demand that enhances glycogenolysis, the contributions of pre- and postsynaptic neuronal and astrocytic activity to oxidative metabolism and lactate generation, dissociation of lactate formation from glutamate transport in the cerebellum, and failure of labeled metabolites of [6- and 1-^{14}C]glucose to be retained in the activated tissue (Fig. 3.15). Because these issues are technically difficult to address in brain of conscious subjects *in vivo*, the notion of lactate shuttling has been controversial and key aspects of the model have been challenged and debated. This model has evolved over time and focused attention on metabolite trafficking and the roles of astrocytes in brain energetics and brain imaging. Interested readers are referred to reviews that evaluate evidence for and against the lactate shuttle model and cite the literature relevant to the broad, complex aspects of this field (Magistretti *et al.*, 1999; Chih *et al.*, 2001; Chih and Roberts, 2003; Dienel and Cruz, 2004; Hertz, 2004; Korf, 2006; Hyder *et al.*, 2006; Gjedde, 2007; Hertz *et al.*, 2007; Pellerin *et al.*, 2007; Dienel and Cruz, 2008).

Astrocytic glycogenolysis. Uptake of K$^+$ from extracellular fluid into astrocytes stimulates glycogenolysis in a concentration-dependent manner (Hof *et al.*, 1988), and many neurotransmitters (with the striking exception of glutamate [Sorg and Magistretti, 1991]) also activate glycogenolysis, thereby linking astrocytic glycolytic activity to local and more distant neuronal-signaling activity (Hertz *et al.*, 2007). The ultimate fate of the glycogen degraded during brain activation is also unknown; some may be converted to lactate and released from the cells and some may be oxidized. The recent evidence showing high glycogen turnover and a fall in the oxygen to glucose utilization ratio suggests that overall glycolysis increases during brain activation, accompanied by lactate release. Note that if all the lactate generated from glucose within the activated tissue were locally oxidized, the oxygen–glucose ratio would not fall below the resting value, and if glycogen were locally oxidized, the oxygen–glucose ratio could even increase above the theoretical maximum of 6 due to utilization of endogenous carbohydrate.

Cellular glucose oxidation rates and turnover of excitatory and inhibitory amino acid neurotransmitters. Magnetic resonance spectroscopic (MRS) studies are particularly useful for studies of the metabolic fate of specific atoms in labeled molecules and have made major contributions to our understanding of metabolic processes and compartmentation in living brain. Many brain metabolites, including glucose and lactate, are difficult to detect by ^{13}C-MRS due to low sensitivity, and considerable emphasis has been placed on the relationship of the glutamate–glutamine cycle to excitatory neurotransmission and rates of oxidative metabolic in neurons and astrocytes. As shown in Figure 3.18A, metabolism of glucose that is labeled in carbon 1 (i.e., [1-^{13}C]glucose where the ^{13}C-labeled carbon atom is indicated in red) labels carbon 3 of pyruvate, carbon 2 of acetyl CoA, and, during the first turn of the TCA cycle, it labels carbon 4 of glutamate; the same labeling pattern is obtained with [6-^{13}C]glucose. Glucose metabolism labels both neurons and astrocytes, but because most of the brain tissue glutamate is located in neurons, dilution of ^{13}C into the glutamate pool mainly reflects neuronal metabolism. [2-^{13}C]acetate (indicated in green circles) is preferentially metabolized via the astrocytic TCA cycle, and it labels carbon 2 of acetyl CoA as well as carbon 4 of glutamate and glutamine (Fig. 3.18A). In contrast, entry of ^{13}C derived from [1-^{13}C]glucose into the astrocytic TCA cycle via the anaplerotic pyruvate carboxylase reaction (blue circles) labels C2 of oxaloacetate and glutamate.

MRS analysis of metabolic compartmentation in astrocytes and neurons in rat brain *in vivo* is illustrated in Figure 3.18B, which shows that C4 of glutamate is quickly and highly labeled by [1,6-^{13}C$_2$] glucose, whereas labeling of C4 of glutamine lags and does not attain the same extent of enrichment.

A   Labeling glutamate and glutamine via different pathways in neurons and astrocytes

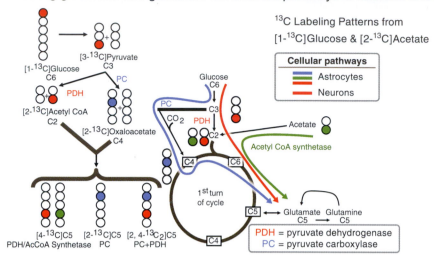

B   MRS spectra and isotopic enrichment after in vivo labeling of rat brain

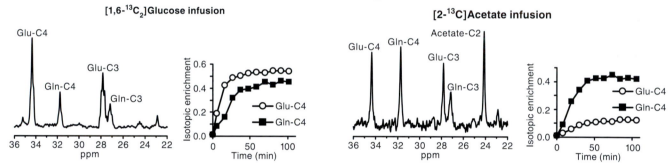

FIGURE 3.18 **Cellular compartmentation of metabolism: labeling patterns of glutamate and glutamine.** (A) Magnetic resonance spectroscopy (MRS) has the advantage that labeled carbons can be identified and the extent of their labeling quantified in brain *in vivo*, thereby allowing real-time assessment of metabolic pathway fluxes and metabolite shuttling between astrocytes and neurons. Metabolism of a glucose molecule that is labeled with ^{13}C in carbon 1 (red) causes labeling of (1) carbon 3 of one of the two pyruvate molecules generated from glucose, (2) carbon 2 of *one* of the two molecules of acetyl Coenzyme A (CoA), and (3) carbon 4 of glutamate during the first turn of the TCA cycle; the same labeling pattern is obtained with $[6-^{13}C]$glucose. Labeling patterns resulting from subsequent turns of the TCA cycle are more complex and can be detected with current technology. When $[1-^{13}C]$glucose is metabolized via the pyruvate carboxylase reaction in astrocytes, carbon 2 of oxaloacetate becomes labeled (blue), labeling C2 of glutamate. Metabolism of $[2-^{13}C]$acetate by astrocytes causes labeling of C4 of glutamate (green). (B) MRS measurements of metabolic labeling by $[1,6-^{13}C_2]$glucose and $[2-^{13}C]$acetate of metabolites in rat brain *in vivo*, determined at 9.4 Tesla in a localized volume of 9x5x9 mm^3 that encompasses most of the brain. The first and third panels from the left show representative spectra at *isotopic steady state* (see Boxes 3.2 and 3.3). Peak intensities reflect ^{13}C concentrations, which are the products of total concentration times isotopic enrichment. Note the strikingly different ratios of glutamate (Glu)-C4 to glutamine (Gln)-C4 peak intensities obtained with $[1,6-^{13}C_2]$glucose and $[2-^{13}C]$acetate, which directly reflect metabolic compartmentation. The second and fourth panels from the left show the time courses of ^{13}C label incorporation into Glu-C4 and Gln-C4 in the rat brain *in vivo* during infusion of $[1,6-^{13}C_2]$glucose (second panel) and $[2-^{13}C]$acetate (fourth panel). With ^{13}C-glucose, Glu-C4 labels more rapidly and reaches higher isotopic enrichment than Gln-C4, consistent with glucose being metabolized predominantly in neuronal TCA cycle. Isotopic enrichment after 120 min is about 54% for Glu-C4 and 45% for Gln-C4. Assuming glutamate and glutamine concentration to be 10 mM and 4 mM, respectively, this corresponds to ^{13}C concentrations of 5.4 mM for Glu-C4 and 1.8 mM for Gln-C4, consistent with a 3:1 peak ratio on the NMR spectrum (first panel). In contrast, with ^{13}C-acetate, Gln-C4 labels more rapidly and reaches higher isotopic enrichment than Glu-C4, consistent with acetate being metabolized predominantly in astrocytic TCA cycle where most glutamine is located. Isotopic enrichment after 120 min is about 42% for Gln-C4 and only 12% for Glu-C4. Glutamate labeling is heavily diluted by unlabeled glucose metabolized in neurons. Assuming glutamate and glutamine concentration to be 10 mM and 4 mM, respectively, ^{13}C concentrations at isotopic steady state are 1.6 mM for Gln-C4 and 1.2 mM for Glu-C4, consistent with similar peak intensities for both signals on the NMR spectrum (third panel). The figures in panel B were kindly provided by Dinesh Deelchand and Pierre-Gilles Henry, Center for Magnetic Resonance Research, University of Minnesota, Minneapolis.

In contrast, infusion of [2-^{13}C]acetate preferentially labels C4 of glutamine, with a much slower incorporation into C4 of glutamate. Reversal of the labeling patterns with the two labeled substrates is explained by differential labeling of the large glutamate pool by glucose and the small glutamate pool by acetate. Note the long time (about 50 min) required to reach *isotopic steady state* (Box 3.3). Model-dependent analysis of the temporal profiles of labeling of different carbon atoms of glutamate, glutamine, and aspartate is used to calculate glucose oxidation rates in neurons and astrocytes, glutamate–glutamine cycling, and anaplerotic rate (Gruetter *et al.*, 2001; Hyder *et al.*, 2006). Current estimates of TCA cycle rates in astrocytes and neurons are about 20–30% and 70–80%, respectively, of the total glucose oxidation rate, and pyruvate carboxylase is about 10% of the oxidative flux (e.g., see Table 1 in Hertz *et al.*, 2007).

Summary

Brain imaging relies mainly on metabolic fluxes that govern the levels, labeling, or redox state reporter molecules, and recent studies in many interrelated fields using different combinations of technologies have made considerable progress in understanding function-induced shifts in metabolic activity. Brain activation is a very complex process that can vary with stimulus intensity and duration, physiological state of the subject (conscious or anesthetized), and the anatomical pathway. Signaling by neural cells is heterogeneously stimulated with varying temporal and spatial profiles causing changes in blood flow and metabolic rates to satisfy increased metabolic demand. Restoration of ionic gradients is considered to be the major energy-consuming process, and recent work has brought attention to the contributions of postsynaptic cells during activation. The cellular basis of increased metabolic activity and the pathway fluxes preferentially stimulated during brain activation are not fully understood due to technical limitations of *in vivo* studies. It has been argued that if (1) neurons account for most of the oxidative metabolism of glucose, (2) similar amounts of glucose are consumed by astrocytes and neurons, and (3) the ratio of oxygen to glucose consumption is close to six, then there is probably some lactate shuttling and oxidation within brain. While this is most likely to occur under resting conditions, it is important to take into account the spatial–temporal limitations of currently available data with respect to quantifying somal metabolic activity compared to that in the pre- and postsynaptic neuronal elements and fine filopodial processes of astrocytes.

PATHOPHYSIOLOGICAL CONDITIONS DISRUPT ENERGY METABOLISM

Derangement of energy metabolism in brain arises from many causes, including inadequate nutrition, genetic diseases involving brain transporters and enzymes, neurodegenerative diseases, traumatic brain injury, mitochondrial dysfunction, diseases involving other organ systems that result in imbalances that seriously affect brain development and function, and diseases that affect oxygen and glucose delivery to brain (Table 3.4). Because cardiovascular disease, pulmonary disease, and diabetes are becoming increasingly important as mid-life diseases that can have profound effects on brain function, this section will focus on some selected consequences of hypoglycemia (low glucose levels), hypoxia (low oxygen levels), and ischemia (no blood flow, therefore depriving tissue of both oxygen and glucose) on metabolite levels that reflect major shifts in overall energy status. Impairment of mitochondrial energy metabolism and its relationship to cell death is a vital aspect of energetics of brain injury and disease, but review of this extensive, complex topic is beyond the scope of this chapter; interested readers are referred to recent reviews (Nicholls *et al.*, 2007; Soane *et al.*, 2007a,b; Wieloch *et al.*, 2007).

Hypoglycemia

Compensatory regulatory mechanisms in peripheral organs involving the endocrine system, mobilization of glycogen, and use of endogenous compounds to synthesize glucose (e.g., amino acids from muscle) tightly control blood glucose levels. For this reason, overt hypoglycemia is rare under normal circumstances. However, with the increase in prevalence of diabetes and the possibility of excessive insulin administration, hypoglycemia is a potentially serious problem for diabetic patients because the supply of glucose to the brain depends on the gradient from blood to brain (Fig. 3.3A). When glucose supply cannot satisfy demand, the brain-to-plasma glucose concentration ratio falls (Fig. 3.3B), and when the brain glucose level falls below about 0.5 µmol g^{-1} (in the rat) hexokinase becomes unsaturated, causing the rate of glucose utilization to fall progressively as glucose level is reduced (Fig. 3.3C). As the severity of hypoglycemia increases, the level of consciousness shifts from lethargy to stupor and to coma, and these transitions are characterized by progressive changes in EEG to the point of isoelectricity or a flat EEG trace.

TABLE 3.4 Examples of Conditions or Disorders Involving Brain Energy Metabolism Pathways

Category	Compound, Reaction, or Processes Affected	Characteristics	Consequences
Nutrition	Thiamine (vitamin B$_1$)	Deficiency impairs transketolase and α-ketoglutarate DH activities	Impaired pentose phosphate shunt and TCA cycle activities; selective regional brain damage
	Pyridoxal phosphate	Deficiency impairs transaminase and decarboxylase activities	Many reactions affected, including synthesis of neurotransmitters
	Biotin	Deficiency impairs carboxylation reactions	Anaplerotic reactions affected; developmental delay, seizures, death if not treated with supplements
Transport	Glucose	GLUT1 deficiency	Reduced glucose transport into brain; seizures, developmental delay
Anaplerosis	CO$_2$ fixation	Pyruvate carboxylase deficiency	Reduced anaplerosis, affecting TCA-cycle-derived amino acids; severe mental retardation or death
Fuel delivery	Glucose	Hypoglycemia (low glucose)	Rate of glucose utilization (CMR$_{glc}$) falls; oxidative metabolism of endogenous metabolites rises; progressive decline in level of consciousness with time and severity, with risk of brain damage and death
	Oxygen	Hypoxia, anoxia (low or no oxygen)	CMR$_{O2}$ is impaired but may be maintained by compensatory mechanisms (e.g., HIF pathway); glycolysis increases; mental capability falls with risk of brain damage and death
	Glucose and oxygen	Ischemia (severe reduction or blockade of blood flow; little or no delivery of glucose and O$_2$)	Unconscious within seconds; selective neuronal death or necrotic infarction of affected tissue, depending on severity, duration, and restoration of blood flow
Mitochondrial defects	Oxidative metabolism	Deficiencies in pyruvate DH and α-ketoglutarate DH	Common phenotype in many diseases that can arise for many reasons. Oxidative deficiencies are present in Alzheimer's and Huntington diseases, stroke, motor neuron disease.
		Mutations in mitochondrial or nuclear DNA have been identified for all respiratory chain complexes and ATP synthase	Deficiencies cause encephalopathy, lactic acidosis, movement disorders, death
Amino acid, fatty acid, and glycogen metabolism	Many pathways	Major effects are in peripheral pathways, causing secondary consequences in brain	Phenylketonuria, maple syrup urine disease, hepatic disease, homocystinuria, urea cycle defects, and fatty acid oxidation defects involve the central nervous system with serious consequences, including mental retardation

The energy status in brain is maintained at normal levels until a threshold is reached at which phosphocreatine (PCr) is consumed to buffer ATP levels; PCr levels then quickly fall and creatine (Cr) levels rise (Fig. 3.19A). During this interval, brain ATP, ADP, and AMP levels are nearly constant, but once a critical point is reached energy failure is rapid and nearly complete; there is an abrupt fall in ATP concentration that is accompanied by increases in ADP and AMP concentrations (Fig. 3.19B). During the hypoglycemic span when ATP levels are maintained at normal levels, endogenous substrates in brain are consumed, and the tissue contents of glycogen and the amino acids that can be oxidized via the TCA cycle are progressively depleted (Siesjö, 1978). A particular problem with diabetic patients is that they often do not recognize the signs and symptoms of hypoglycemia (Table 3.5, top), a condition called *hypoglycemic unawareness* that puts the patient at risk for serious accidents during the stages of impairment of consciousness. Increased cerebral blood flow is one of the compensatory responses to severe hypoglycemia, and oxygen consumption is maintained at about 80% of normal until late stages of a hypoglycemic episode, reflecting the continuous oxidation of endogenous brain metabolites (Table 3.5, top). Administration of glucose is required to bring a subject out of hypoglycemic coma,

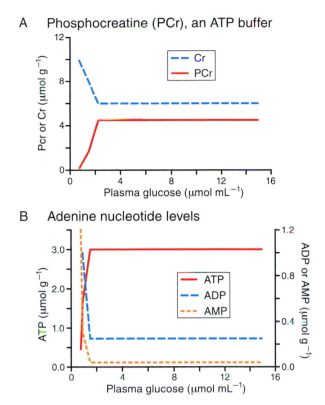

A Phosphocreatine (PCr), an ATP buffer

B Adenine nucleotide levels

FIGURE 3.19 Energy failure occurs during severe hypoglyce-mia. (A) Phosphocreatine (PCr) is a high-energy "buffer" for ATP; its concentration in brain is stable over a wide range of plasma glucose concentrations (3–16 mM). However, when plasma glucose level (2–3 mM) is too low to sustain an adequate supply of glucose to brain (compare to Fig. 3.3), PCr levels fall and creatine (Cr) concentration rises. (B) ATP concentration is maintained constant over a slightly larger range than PCr due to buffering. Energy failure is abrupt, ATP is quickly consumed, and ADP and AMP levels rise. (Plotted from data of Lewis *et al.*, 1974.)

and, although consciousness may be restored relatively quickly, the time required for recovery is considerably longer than might be anticipated because endogenous compounds must be resynthesized from glucose to restore their normal levels via the anaplerotic reactions.

Hypoxia

Hypoxia can be experienced during normal and pathophysiological conditions. An inadequate level of oxygen in the inspired air is frequently experienced by persons going to high altitude, and patients with chronic respiratory and cardiovascular diseases may not have sufficient oxygen levels in their blood. Reducing the oxygen level in inspired air from the normal sea level value of 21% causes progressive deterioration of mental function, with compensatory increases in blood flow to deliver more oxygen to the brain (Table 3.5, bottom). One of the earliest changes during an *acute*

hypoxic episode is reduced neurotransmitter turnover. Mild hypoxia (graded reductions in inspired oxygen level from 30 to 15 and 10%) causes an increase in blood flow that is sufficient to maintain normal rates of oxygen consumption (Fig. 3.20A). Glucose utilization also increases during hypoxia, with greater production of lactate (Fig. 3.20B). Acetylcholine levels are stable during this interval, but acetylcholine synthesis falls by more than 50% (Fig. 3.20C). Thus, acetylcholine synthesis is closely correlated with oxidative metabolism of glucose, and when glucose oxidation is impaired, acetylcholine synthesis rates fall proportionately (Fig. 3.20D). Biogenic amine synthesis is also extremely sensitive to reduced oxygen content, and hydroxylation of tryptophan (Fig. 3.17) falls by nearly half before lactate levels rise appreciably (Fig. 3.20E). During graded hypoxia, the supply of glucose from blood is not limiting and brain ATP levels are preserved by buffering by PCr and by increased glycolysis to cause a large increase in tissue lactate concentration; hypoxic energy failure begins when the arterial oxygen content falls below about 35 mm Hg (Fig. 3.20F). To summarize, synthesis of specific neurotransmitters is much more sensitive to hypoxia than concentrations of these transmitters or levels of energy metabolites. Increased glycolysis compensates for deficits in oxidative metabolism over a limited range of graded hypoxia, and lactate levels increase markedly. ATP concentration is buffered by PCr until a critical supply–demand energy threshold is reached and ATP levels fall.

The brain and other body tissues have an oxygen-sensing mechanism that mediates adaptation to chronic hypoxia. The hypoxia-inducible factor-1 (HIF-1) is a transcription factor that exists in two isoforms, one of which serves as a sensor to assay tissue oxygen levels and controls the coordinated expression of genes that increase capillary density, glucose transporters, and glycolytic enzymes (LaManna *et al.*, 2007). In brief, oxygen-dependent hydroxylation reactions result in the continuous degradation of the HIF-1α subunit by proteosomes (Fig. 3.21). Inadequate levels of oxygen cause HIF-1α to accumulate to a high enough level that it forms a heterodimer with HIF-1β. This complex binds to the hypoxia response element, thereby stimulating graded increases in mRNA and protein levels of factors that substantially increase glycolytic capacity, glucose transport, and vascular density in hypoxic tissue. To summarize, the HIF system carries out a compensatory program to adapt to hypoxia by increasing glucose delivery and metabolic capability (Fig. 3.21). The effects of this system are evident in cultured astrocytes exposed to a low oxygen atmosphere for 8 hours; lactate dehydrogenase and pyruvate kinase activities rise about three- and four-fold, respectively, and glucose utilization doubles during a

TABLE 3.5 Influence of Hypoglycemia and Hypoxia on Human Physiology and Brain Function

Substrate Level	Physiological or Mental Status[1,2]	CBF[3] (ml 100 g^{-1} min^{-1})	CMR$_{O_2}$[3] (ml O$_2$ 100 g^{-1} min^{-1})
Arterial Plasma Glucose[1] [mmol L^{-1} (mg dl^{-1})]			
3.9-6 (70-110)	Normal fasting range	54	3.3
4.5 (81)	Decrease insulin secretion		
3.6-3.8 (65-68)	Counterregulatory hormone responses (increase secretion of glucagon, epinephrine, growth hormone, cortisol)		
3.0 (54)	Symptoms of hypoglycemia become manifest: anxiety, palpitations, hunger, tremor, sweaty, dizzy, weak		
2.6 (47)	Cognitive dysfunction threshold: difficulty speaking and thinking, blurred vision		
1.7 (30)	Mild confusional state and delirium		
1.1 (20)	Cognitive failure, stupor, seizures	61	2.6
<0.6 (10)	Deep coma and ultimately death	63	1.9
%O$_2$ in inspired air[2] (estimated altitude in feet)			
21 (sea level)	Normal	54	3.3
18 (4,000)	Delayed dark adaptation		
16-15 (7,500-9,000)	Impaired complex learning; reading test errors		
14 (11,000)	Hyperventilation		
13 (13,000)	Short-term memory impairment		
14-11 (11,000-17,000)	Acute mountain sickness: headache, poor concentration and test performance		
11 (17,000)	Loss of critical judgment		
10 (19,000)	Increase CBF 35% (no change in global CMR$_{O2}$)	73	3.3
9 (21,500)	Increase CBF 70% (no change in global CMR$_{O2}$)		
8-6 (24,000-31,000)	Loss of consciousness		

Note that the endocrine changes in response to reduced blood glucose levels occur well before cognitive changes. Increasing the duration of exposure to any specific level of hypoglycemia below 3 mM can lead to progressive decline in brain function. During hypoxia (10% O$_2$), the arteriovenous difference for oxygen decreases, but blood flow increases and CMR$_{O2}$ remains normal even though brain function is impaired.
[1]From data reported or compiled by Ferrendelli (1974).
[2]Data from Siesjö et al. (1974). Note that compensatory changes associated with acclimatization at altitude (e.g., increased red blood cell number, angiogenesis, and blood flow) help to maintain normal brain function.
[3]From compilation by McIlwain and Bachelard (1985).

metabolic challenge (Marrif and Juurlink, 1999). Thus, brain cells can adapt to the levels of essential fuel and nutrients in their environment, and glycolytic capacity of cultured cells can be up- or down-regulated by the oxygen level in the culture medium.

Ischemia

Ischemia is the blockade of blood flow to the brain, thereby eliminating the supply of oxygen and glucose and preventing removal of metabolic by-products. Ischemia can be a local event, arising from a blood clot or vascular damage or it can be a global event secondary to a heart attack. Oxygen depletion prevents oxidative metabolism of amino acids and glucose depletion

results in the immediate and rapid consumption of endogenous glucose and glycogen, with generation of large amounts of lactate in the brain. PCr levels fall immediately after onset of ischemia (Fig. 3.22A), followed quickly by decreases in levels of ATP, glucose, and glycogen (Fig. 3.22B); energy failure is complete within about 60 sec and it results in loss of consciousness. During decapitation ischemia, lactate levels rise up to about 10 μmol g^{-1} (Fig. 3.22B), whereas during hypoxia the brain lactate level reaches almost twice this level (Figs. 3.20E and 3.20F) due to the continuous supply of glucose from blood during hypoxia and the restricted transport of lactate from brain across the blood–brain barrier to blood (i.e., lactate transport is limited by the low levels of MCT1 in adult brain).

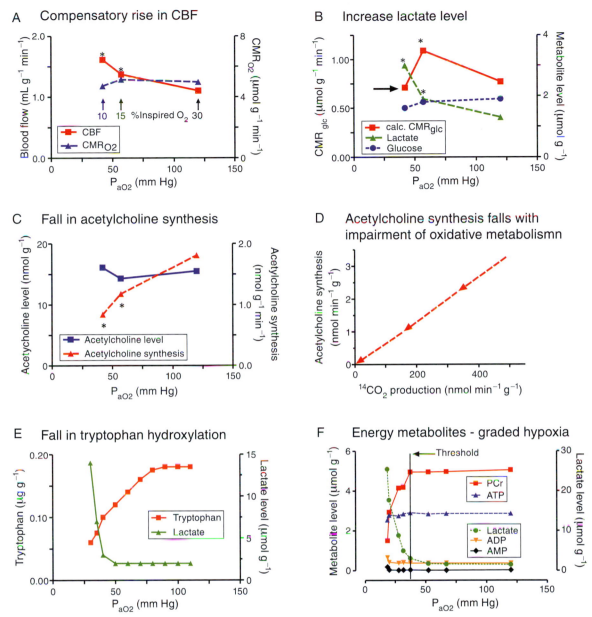

FIGURE 3.20 **Compensatory metabolic and blood flow responses and high sensitivity of neurotransmitter turnover to hypoxia.** Mild hypoxia, a 50% reduction in percent of inspired oxygen (from 21% at sea level to 10%), is associated with (A) a compensatory rise in blood flow that helps maintain the rate of oxygen consumption in brain (i.e., by the Fick principle, $CMR_{O2} = CBF \times (A-V)_{O2}$), (B) an increase in glucose utilization and lactate production while brain glucose levels are stable (glucose delivery is able to match demand), and (C) a fall in the synthesis but not concentration of acetylcholine (panels A–C are plotted from data of Gibson and Duffy, 1981). In panel B, glucose utilization rate was calculated by dividing the quantity of label recovered in metabolites by the specific activity of [U-^{14}C]glucose. The fall in *calculated* rate when inspired oxygen was reduced from 15 to 10% (arrow, panel B) probably reflects efflux of labeled metabolites of glucose from brain (e.g., lactate), not a fall in the true rate of glucose utilization, which should have increased in parallel with the level of lactate accumulation. (D) The rate of acetylcholine synthesis is very sensitive to hypoxia, and, under a variety of conditions, it is strongly correlated with glucose oxidation, reflected by $^{14}CO_2$ production from labeled glucose (plotted from data of Joseph and Gibson, 2007). (E) Tryptophan synthesis is also highly sensitive to graded hypoxia (see Fig. 3.17), and its rate falls well before upregulation of glycolytic metabolism and increased tissue lactate levels (plotted from data of Davis *et al.*, 1973). (F) The apparent threshold for a hypoxia-induced change in energy metabolism is about P_{AO2} of 37 mm Hg, when lactate levels begin to rise and PCr levels begin to fall. Only when the PCr concentration has fallen by more than 50% does ATP level begin to fall and ADP and AMP concentrations increase (plotted from data of Siesjö and Nilsson, 1971).

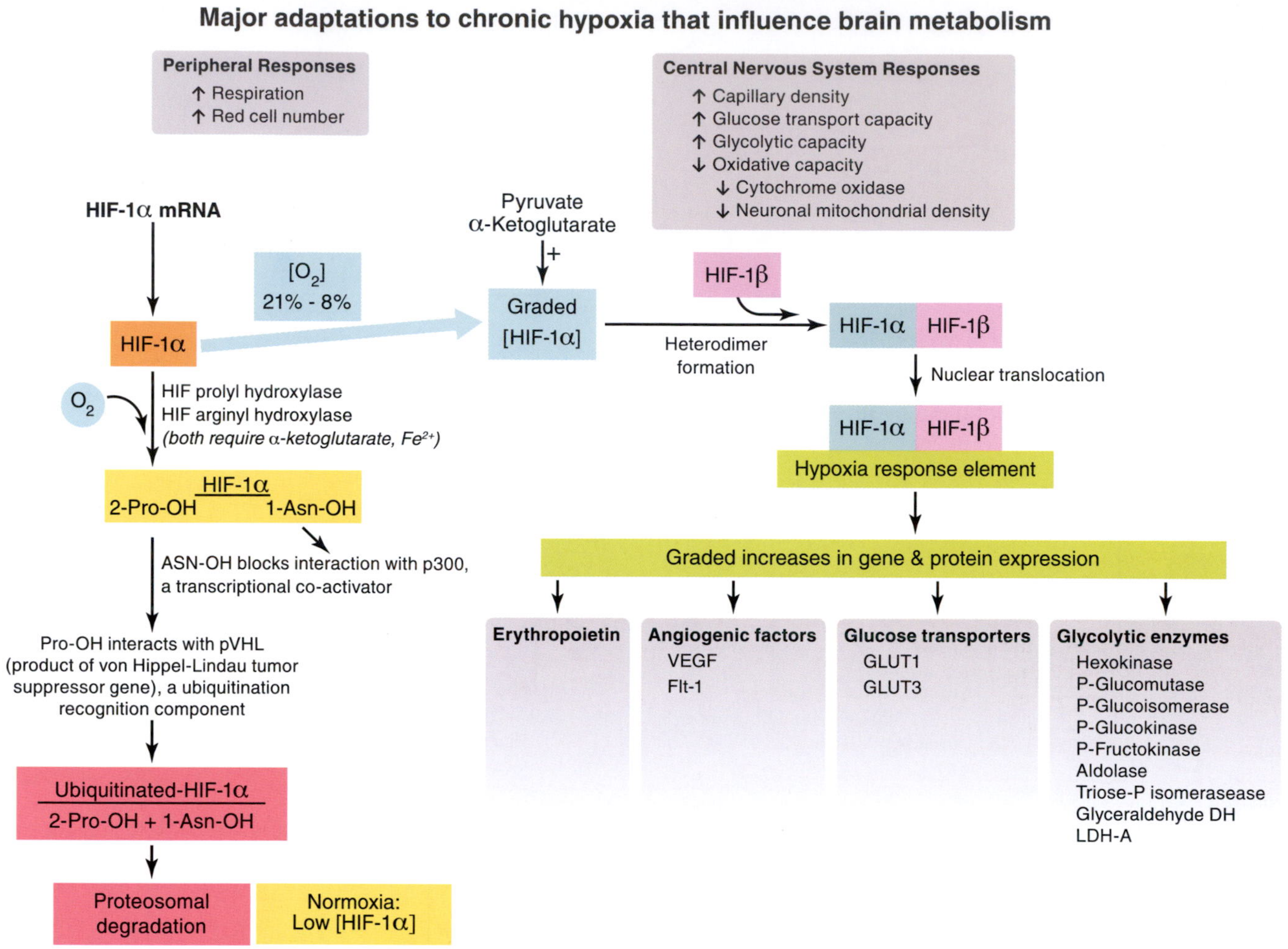

Major adaptations to chronic hypoxia that influence brain metabolism

Peripheral Responses
↑ Respiration
↑ Red cell number

Central Nervous System Responses
↑ Capillary density
↑ Glucose transport capacity
↑ Glycolytic capacity
↓ Oxidative capacity
  ↓ Cytochrome oxidase
  ↓ Neuronal mitochondrial density

HIF-1α mRNA

[O_2]
21% - 8%

Pyruvate
α-Ketoglutarate
+

HIF-1β

HIF-1α

Graded
[HIF-1α]

Heterodimer formation → HIF-1α HIF-1β

O_2

HIF prolyl hydroxylase
HIF arginyl hydroxylase
(both require α-ketoglutarate, Fe^{2+})

Nuclear translocation

HIF-1α
2-Pro-OH   1-Asn-OH

HIF-1α HIF-1β

ASN-OH blocks interaction with p300, a transcriptional co-activator

Hypoxia response element

Pro-OH interacts with pVHL (product of von Hippel-Lindau tumor suppressor gene), a ubiquitination recognition component

Graded increases in gene & protein expression

Erythropoietin	Angiogenic factors	Glucose transporters	Glycolytic enzymes
	VEGF	GLUT1	Hexokinase
	Flt-1	GLUT3	P-Glucomutase
			P-Glucoisomerase
			P-Glucokinase
			P-Fructokinase
			Aldolase
			Triose-P isomerasease
			Glyceraldehyde DH
			LDH-A

Ubiquitinated-HIF-1α
2-Pro-OH + 1-Asn-OH

Proteosomal degradation

Normoxia:
Low [HIF-1α]

FIGURE 3.21 **An oxygen detection system regulates gene expression during chronic hypoxia.** The transcription factor, hypoxia-inducible factor-1 (HIF-1) exists in α and β isoforms and can form a heterodimer that binds to the hypoxia response element in the nucleus and induces transcription of a number of genes that enable adaptation to hypoxia. HIF-1α is hydroxylated on proline (Pro) and asparagine (Asn) moieties by enzymes that require oxygen; these post-translational modifications influence transcriptional activity of HIF-1α and act as recognition sites for ubiquination that leads to proteosome-mediated degradation. Thus, at normal oxygen levels, HIF-1α is continuously degraded and its levels are low; with progressive reduction in oxygen level HIF-1α accumulates, forms heterodimers, and stimulates gene and protein expression, thereby increasing the brain's capacity for blood flow, transport, and carbohydrate metabolism (adapted from LaManna *et al.*, 2007, with the kind permission of Springer Science and Business Media.)

Depending on the duration of ischemia, tissue damage can be selective or severe. In the hippocampus, CA1 pyramidal neurons are most vulnerable to duration of ischemia, but the progression of cell death takes place over several days after a 30 min transient global ischemic episode in the rat; small-to-medium-sized neurons in the caudate also exhibit selective vulnerability, but they die within 24 h of reperfusion. Prolonged ischemia results in necrotic cell death, in which all cell types in the flow-compromised tissue are killed over a relatively short time course. Brief energy failure is not sufficient to kill brain cells, and many of the deleterious processes that irreversibly damage vulnerable neurons (excitotoxicity, oxidative damage, loss of calcium homeostasis, etc.) take place

during the reperfusion interval. Abnormal NMDA receptor activation, dysregulation of calcium homeostasis, and impairment of mitochondrial ATP generation are considered to be major factors in ischemic neuronal death (Nicholls *et al.*, 2007). Many therapeutic strategies have been tested for postischemic neuroprotection but none have had much success in humans (Ginsberg, 2008, and cited reviews).

Summary

Brain can compensate for inadequate supplies of oxygen and glucose over a varied range of fuel deficits, depending on severity and duration. Energy levels can be maintained by increased glycolysis when oxygen

A   High energy metabolites during ischemia

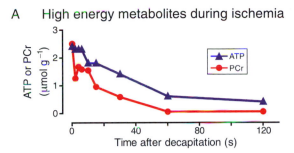

B   Carbohydrate levels during ischemia

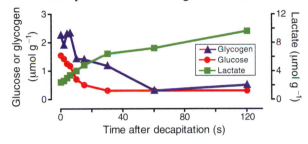

FIGURE 3.22   **Ischemia deprives brain of blood flow, blocking delivery of oxygen and glucose and removal of metabolic by-products.** (A) High-energy metabolites are nearly depleted within 1 min after onset of ischemia in the adult mouse brain; PCr levels fall faster than ATP levels. (B) Glucose and glycogen are quickly consumed, leading to a large increase in tissue lactate concentration. (Plotted from data of Lowry et al., 1964.)

levels are reduced and by upregulation of glucose transporters and glycolytic enzymes via the HIF system, but it is likely that turnover of neurotransmitters dependent on oxygen supply (e.g., catecholamines) and oxidative metabolism (acetylcholine) is impaired. Low glucose levels can also be tolerated to a certain extent, but once hexokinase becomes unsaturated, glucose utilization rates will fall, and when the critical threshold is reached, energy failure is abrupt and complete. Ischemia causes rapid energy failure and loss of consciousness within about a minute of blood flow stoppage. Recovery from energy failure involves several stages, restoration of ionic gradients, generation of high-energy compounds, and resynthesis of compounds consumed when fuel supplies were inadequate.

ROLES OF NUTRIENTS AND METABOLITES IN REGULATION OF SPECIFIC FUNCTIONS AND OVERALL METABOLIC ECONOMY

Emphasis on local control of pathway flux by regulation of the activities of rate-limiting enzymes by metabolic intermediates, second messengers, and energy-related compounds has been a hallmark of biochemistry for decades. However, nutrients, metabolites, and cofactors can also regulate gene expression and have key roles in the integration of the local and global metabolic economy of the cell, tissue, and entire organism. These emerging concepts are currently not well developed in brain tissue but are anticipated to attract attention of neuroscientists in the future.

Novel Roles for Nutrients in Brain Function: Signaling, Gene Expression, and Memory

Glucose is generally regarded as mainly an energy fuel, but glucose itself, its metabolites (mimicked by an analog 2-deoxyglucose-6-phosphate), and downstream products (e.g., glucosamine) are known to regulate transcription of genes, particularly in pancreatic beta cells that release insulin and help to regulate plasma glucose levels, and in other tissues (e.g., glucose-regulated genes) (Scott et al., 1998; Vaulont et al., 2000; Allagnat et al., 2005; Minn et al., 2006). Cofactors, such as NADH and NADP(H), that are integral components of the glycolytic and pentose shunt pathways, not only function in oxidation-reduction (redox) reactions but also have many other important functions (Berger et al., 2004; Pollak et al., 2007), including central nervous system roles in (1) DNA repair via PARP-1 (poly[ADP-ribose]polymerase-1), a nuclear enzyme involved in DNA repair that can cause NAD depletion and inhibition of glycolysis (Alano and Swanson, 2006; Kauppinen and Swanson, 2007) and (2) in calcium signaling after NAD release from a cell and its conversion by ectoenzymes to signaling molecules, nicotinic acid adenine dinucleotide phosphate (NAADP) (Heidemann et al., 2005) or cyclic ADP ribose (cADPR) (Verderio et al., 2001). Furthermore, 2-deoxyglucose has anticonvulsant properties and it blocks seizure-induced expression of brain-derived neurotrophic factor (BDNF) and its receptor TrkB; glycolytic inhibition by deoxyglucose is thought to lower NADH level, thereby increasing formation of a repressor complex on target genes that is destabilized by NADH, which acts to enhance BDNF and TrkB transcription and increase neuronal excitability (Garriga-Canut et al., 2006). Finally, glucose administration can influence cognitive processes in aged rodents and humans by facilitating learning during memory task activities (McNay and Gold, 2002; Salinas and Gold, 2005). Thus, glucose, glucose metabolites, and redox compounds are multifunctional, providing energy, interfacing with cellular damage prevention and repair, regulating genes in pathways relevant to nutritional status, improving cognition in old subjects, and providing signals for detection by various technologies.

AMP-Activated Protein Kinase (AMPK) Signaling Pathway: A Metabolic Sensor

Signaling and integration of local and global metabolic economy (Fig. 3.23) is the function of the AMP-activated protein kinase system (Lindsley and Rutter, 2004; Hardie *et al.*, 2006; Schimmack *et al.*, 2006; Winder *et al.*, 2006; Witters *et al.*, 2006). Cycling of ATP and ADP takes place with biological work, and the cytoplasmic ATP pool is buffered by creatine-phosphate (P-Cr) (Fig. 3.23). When ATP consumption exceeds its production, ADP accumulates and the myokinase (adenylate kinase) reaction converts 2 ADP to ATP + AMP. This is a very sensitive amplification system that reports the "energy charge" of a cell because the AMP–ATP

ratio varies with the square of the ADP–ATP ratio (Fig. 3.23). AMP is an allosteric regulator of phosphofructokinase and glycogen phosphorylase (Fig. 3.4), and AMP also signals *local* energy status by activating the AMPK system by three mechanisms: (1) stimulation of LKB1, a kinase that phosphorylates Thr^{172} on the alpha subunit of AMPK to form AMPK-P, (2) allosteric activation of the AMPK gamma subunit, and (3) inhibition of dephosphorylation of AMPK-P (Hardie *et al.*, 2006; Fig. 3.23). On the other hand, *tissue and organism status* are integrated via receptor-mediated changes in intracellular calcium level to activate a calmodulin kinase kinase (CaMKK) to form AMPK-P (Hardie *et al.*, 2006; Fig. 3.23). Phosphorylated AMPK is allosterically

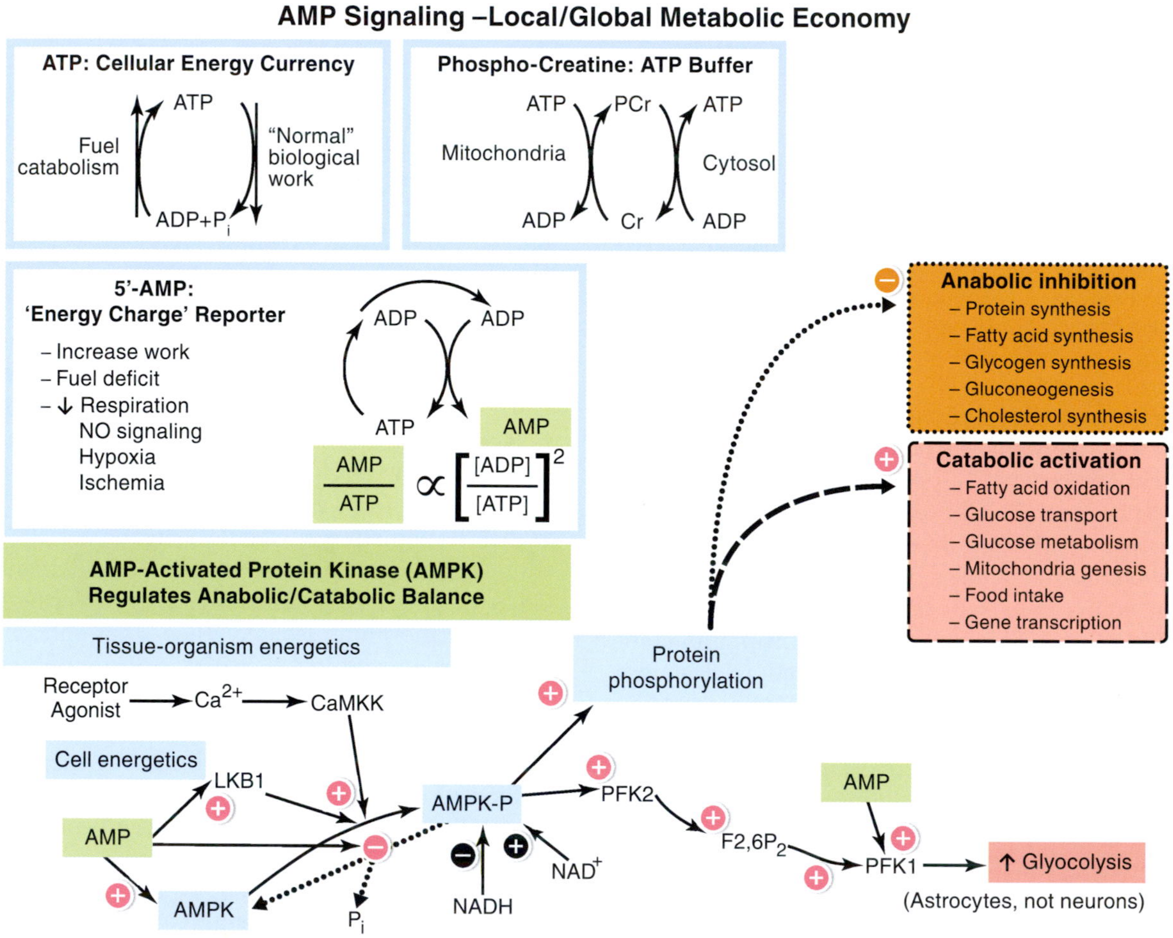

FIGURE 3.23   Some roles of AMP in regulation of the metabolic economy of the cell. ATP is the currency for cellular work; its production is highly regulated and its concentration is buffered by phosphocreatine (top left and right boxes). Under certain conditions when ADP accumulates in tissue, the adenylyl kinase reaction can convert 2 ADP into ATP + AMP, so that AMP levels, which are normally quite low, act to amplify small changes in ATP concentration (middle left box). Thus, AMP not only has a regulatory role in the glycolytic and glycogenolytic pathways (Fig. 3.4), it also has a key role in regulating the overall metabolic balance by anabolic and catabolic processes (see pathways in lower portion of the figure). Phosphorylation of AMP-activated protein kinase (AMPK) is stimulated by increased levels of AMP by several pathways as well as by inhibition of the phosphatase. Redox cofactors also modulate the activity of AMPK-P, which acts via protein phosphorylation to inhibit anabolic (energy-requiring reactions) and stimulate catabolic (energy-producing reactions) processes. In astrocytes but not neurons, AMPK-P phosphorylates phospho-2-fructokinase to stimulate production of 2,6-bisphosphofructose, an activator of phosphofructo-1-kinase, which enhances glycolysis.

modulated by the redox cofactors, NAD and NADH (Rafaeloff-Phail *et al.*, 2004), and the AMP-activated protein kinase regulates anabolic–catabolic balance by enhancing ATP-generating pathways by regulating many steps ranging from reaction fluxes to gene transcription, and by inhibiting ATP-consuming biosynthetic pathways (Fig. 3.23). Thus, the AMPK system is a very sensitive energy sensor at a cellular and organismal level that detects metabolic stress as increased AMP concentration. Other nutrient-sensing systems (e.g., mammalian target of rapamycin [mTOR], PAS kinase, and hexosamine) are also important, particularly in peripheral tissues (Lindsley and Rutter, 2004) that have been more extensively studied compared to brain.

The AMPK system has not been extensively studied in brain tissue and its metabolic effects in brain cells are not well understood, but it has major roles in control of feeding behavior and control of body weight due to regulation of AMPK activity in hypothalamic neurons by hormones, peptides, and nutrients (Ramamurthy and Ronnett, 2006; Xue and Kahn, 2006). AMPK is widely distributed in brain (Turnley *et al.*, 1999), and cellular expression in adult brain tissue appears to be mainly neuronal with undetectable levels in glial cells (Ramamurthy and Ronnett, 2006). These findings contrast results from other studies in cultured astrocytes where AMPK activity was three-fold higher than in cultured cortical neurons; the astrocytic AMPK system was activated by hypoxia and was associated with inhibition of lipid and cholesterol synthesis and increased ketogenesis from palmitate (Blázquez *et al.*, 1999). AMPK activation stimulates production of a potent allosteric activator of phosphofructo-1-kinase (i.e., fructose-2,6-P_2) by stimulating a kinase (6-phosphofructo-2-kinase); this process stimulates glycolytic flux (Fig. 3.23). The fructose-2,6-P_2 flux-activating system is present in cultured astrocytes, not cultured neurons; it conveys resistance to hypoxia–anoxia and is responsive to nitric oxide signaling, since nitric oxide binding to cytochrome c reversibly impairs respiration (Almeida *et al.*, 2004; Moncada and Bolaños, 2006). Because AMP levels are increased in ischemic tissue, the stimulation of energy-producing pathways by activated AMPK might be neuroprotective. Phosphorylated AMPK does substantially increase in neurons during middle cerebral artery occlusion, but it has deleterious effects in the mouse stroke model; pharmacological activation of AMPK increases the extent of stroke damage, whereas its inhibition is neuroprotective (McCullough *et al.*, 2005; Li *et al.*, 2007). To sum up, the AMPK system in the central nervous system has profound effects on whole-body energy homeostasis and it is likely to have important roles in pathophysiological conditions.

METABOLOMICS, TRANSCRIPTOMICS, AND PROTEOMICS

Systems biology studies are increasingly important because they describe complex profiles of metabolites, mRNAs, and proteins in cells and tissues under different conditions. Characterization of relationships among expression of different molecular species during development and aging, under different physiological conditions, and responses to genetic manipulation is particularly useful for development of biomarkers for diseases, assessment of therapeutic efficacies of drugs, and understanding perturbations of regulatory systems after genetic engineering; these are emerging topics in brain energy metabolism.

Metabolomics

Metabolomics deals with assessment of the metabolic profile of a living cell, tissue, organism, or biological fluid to characterize and distinguish between normal and altered states (Box 3.4). Goals and applications of the field include discovery and identification of biomarkers, mechanisms of drug-induced toxicity, efficacy of drugs at preclinical and clinical stages of development, and clinical disease diagnosis and prognosis. Because many factors influence metabolite levels and pathway fluxes (genetics, environment, temperature, nutrition and diet, age and gender, hormones, estrus stage, health status, diurnal and seasonal effects, physical fitness, stress, gut microflora), a systems approach is required. This analytical process involves successive stages: (1) obtain *raw data* for the metabolic signature via one or more technologies (e.g., high-pressure liquid chromatography, gas chromatography, capillary electrophoresis, magnetic resonance spectroscopy, mass spectrometry, etc.) in which the individual peaks do not need to be identified; (2) use *data reduction procedures* to digitize complex spectra from each sample; (3) employ *pattern recognition techniques* (e.g., principle component analysis) to map different samples, (4) carry out *statistical comparisons* of patterns; and (5) develop *databases*. For example, fingerprint analysis (Box 3.4) is a very powerful approach that can readily classify and distinguish biofluid and tissue samples. This approach is capable of distinguishing urine samples from human, rat, rabbit, and mouse, as well as distinguishing mouse strain, rodent gender, stage of estrus cycle, age, and whether the urine was collected during the day or night (Bollard *et al.*, 2005).

Anatomical brain regions have specialized functions and neurotransmitter pathways, and metabolomic analysis of brain tissues is an emerging field

---

BOX 3.4

METABOLOMICS

Definitions of metabolomics vary with specialty disciplines and can be confusing due to different nomenclatures in a rapidly growing field. One set of simple definitions (Fiehn, 2002; Nicholson *et al.*, 2002; Goodacre *et al.*, 2004; Griffin, 2004) is as follows.

Metabolome. A set of metabolites synthesized by an organism, which can be defined on all levels of complexity for organisms, tissues, cells, cell components, or fluids; the conditions of the sample need to be exactly specified.

Metabolomics. Comprehensive analysis of metabolites under a specific set of conditions in which there is a direct connection between genetic activity, protein activity, and the metabolic activity itself.

Metabolite fingerprinting. Classification of samples according to biological relevance or origin by high throughput analytical methods in which unseparated samples are analyzed and individual compounds are not identified. Fingerprinting does not aim to identify individual components; rather; it quickly classifies samples on the basis of their components.

Metabonomics. The quantitative measurement of the multivariate metabolic responses of multicellular systems to pathophysiological stimuli or genetic modification; metabonomics deals with "classifying samples, understanding biochemical mechanisms, identifying biomarkers, quantitatively analyzing concentrations and fluxes, probing molecular dynamics and interactions" (Nicholson *et al.*, 2002).

Metabolic profiling. Analysis of a group of metabolites associated with a specific pathway.

Metabolite target analysis. Restricted assays of metabolites related to a specific enzyme system affected by a perturbation, as in pharmaceutical studies.

Metabolite biomarker. A molecular indicator of pathology or a specific "state" that might be useful for diagnostic purposes or monitoring of effects of therapeutic intervention.

---

that will make major contributions to characterization of normal brain tissue and altered brain function during pathophysiological and drug-treatment conditions (Kristal *et al.*, 2007). For example, brain regions can be identified by differences in their biochemical composition profiles (Tsang *et al.*, 2005) and the effects of metabotropic glutamate receptor agonists and antagonists on cortical samples could be separated from control and from each other by ^{13}C-MRS labeling and metabolomics analysis (Rae *et al.*, 2005). Animals with inflammatory lesions in brain during experimental multiple sclerosis can be identified by metabolic profiling of urine and pattern recognition analysis (Griffin *et al.*, 2004a). Metabolic phenotyping readily identifies regional metabolic perturbations in brain of transgenic mice with spinocerebellar ataxia (Griffin *et al.*, 2004b), and applications of the metabolomic approach include identification of serum biomarkers for patients with motor neuron disease.

Transcriptomics

Analysis of mRNA expression in cultured brain cells and tissue under different conditions provides important information regarding changes in gene expression arising from different developmental stages, physiological conditions, or genetic manipulations that influence the cell's capacity to synthesize specific proteins. For example, a recent transcriptome analysis of acutely isolated astrocytes, neurons, and oligodendrocytes during early mouse brain development (postnatal day 1–30) generated a database composed of more than 20,000 genes and revealed distinct differences among the three major brain cell types (Cahoy *et al.*, 2008). Many studies have examined changes in levels of small numbers of specific mRNAs related to proteins of interest, and results from some of these studies underscore the important observation that *up- or down-regulation of mRNA levels need not be correlated with changes in protein levels*, implying translational regulation. Also, *mRNA and protein* levels may change *disproportionately or in different directions*, and even if there are large changes in *protein concentration, the altered capacity may, but need not, be accompanied by changes in levels of relevant metabolites or fluxes* through the pathway in which enzyme levels are altered (Box 3.2).

During water deprivation, the hypothalamic–neurohypophysial system is activated due to dehydration, and greater secretion of vasopressin and oxytocin increases the activity of magnocellular neurons that regulate water balance; higher rates of neuronal firing are associated with increased glucose utilization by about 2.6-fold. One aspect of the compensatory response to water deprivation is increased levels of glucose transporter proteins, GLUT1, which is expressed in glia and endothelial cells, and GLUT3, which is expressed in neurons. Notably, GLUT1 *mRNA levels fell* by about 43% after 3 days of water deprivation, whereas the GLUT1 *protein level increased* 28%; during rehydration, the GLUT1 *protein level fell* toward normal, and its *mRNA level increased*. GLUT3 mRNA increased by about 74%, and the GLUT3 protein level increased 40%; both normalized during rehydration (Kohler-Stec *et al.*, 2000). During prolonged seizures in 10–21-day-old rats GLUT1 and GLUT3 *mRNA levels* are quickly *upregulated* but there are *no changes* in levels of the respective glucose *transporter proteins* (Nehlig *et al.*, 2006). Thus, mRNA levels, enzymatic catalytic activity, and pathway flux can be independently regulated and may change in opposing directions or by varying magnitude.

Extensive analysis of brain transcripts relevant to energy metabolism is rare, and findings should be accompanied by studies to verify the functional metabolic capacity suggested by the presence or absence of specific mRNAs. For example, analysis of mRNA levels in astrocytes and nonastrocytic cells acutely isolated from mature murine brain showed distinct enrichments of mRNA that were generally in accordance with previously established distributions of metabolic enzymes among astrocytes and neurons (Lovatt *et al.*, 2007). This study also reported the presence of the mRNAs for proteins involved in shuttling of redox equivalents from cytosol to mitochondria in astrocytes (Fig. 3.5), and used ^{13}C-MRS labeling studies to demonstrate the capability of mature, adult isolated astrocytes to oxidize glucose (Lovatt *et al.*, 2007). This important finding indirectly verifies the existence of redox shuttle(s) in astrocytes in adult brain, a topic that is currently debated due to the low levels of immunoreactive aspartate–glutamate carrier protein (a critical component of the malate–aspartate shuttle) in adult murine astrocytes compared to neurons; both cell types express this carrier in neonatal mice (LaNoue *et al.*, 2007).

Genetic Engineering and Metabolism

Knock-in, knock-out, or mutation. Modification, removal, or insertion of genes is a very powerful tool to evaluate regulation, signaling, and function, but the phenotype arising from genetic manipulation can arise, in part, from secondary effects in many seemingly unrelated pathways. For example, astrocytes are extensively linked together via gap junctional channels composed of connexin (Cx) proteins (Cx 43, Cx 32, and Cx 26), and these intercellular pores enable passage from cell to cell of many small

(i.e., molecular weight less than about 1 kDalton) biological compounds, as well as various fluorescent dyes used to assay gap junctional trafficking. Array analysis of gene expression changes in connexin-43 knockout mice revealed up- and down-regulation of 252 mRNAs that encode for proteins in many pathways; 35% were down-regulated and 65% were up-regulated, and of the altered genes, 29% were related to transcription; 19% to energy and metabolism; 9% to cell junctions, adhesion, and extracellular matrix; 9% to cell signaling; 9% to transport; 8% to cell cycle; 7% to unknown functions; and the remaining 9% to two categories, organelle genetics and cytoskeletal proteins (Iacobas *et al.*, 2003, 2004). These findings demonstrate that deletion of a single gene can elicit large, widespread compensatory changes in gene expression in many functional systems that do not appear to be directly related to the function of the deleted gene. It is notable that a substantial fraction of the genes involved in compensatory responses to elimination of Cx 43 is related to metabolic and energy-related functions, suggesting that (1) metabolite trafficking via the astrocytic syncytium has an important role in metabolic activities of the cells, and (2) the presence of other connexins (Cx 32 and Cx 26) in the Cx 43-null astrocytes is not sufficient to carry out the functions of Cx 43. Gene expression linkage patterns in connexin-null brain suggest that regulation of networks of transcriptomes is likely to contribute to altered phenotypes when specific functions are missing (Spray and Iacobas, 2007).

Pathway modulation. Genetic engineering approaches are often used to construct microorganisms that can mass-produce biological compounds of interest. Even when genes are successfully inserted and enzymes are overproduced, a frequent observation is that metabolite levels and metabolic fluxes are unchanged. For example, increasing the activities in yeast of eight enzymes involved in glycolysis and ethanol production by about 4- to 14-fold had no effect on concentrations of major metabolites or ethanol formation, in spite of enhanced levels of all the rate-limiting enzymes hexokinase, phosphofructokinase, and pyruvate kinase (Schaaff *et al.*, 1989). Selective increases in enzyme expression of a rate-limiting enzyme such as phosphofructokinase *generally do not alter glycolytic flux* under a variety of conditions in different organisms, including mammalian cell lines (Urbano *et al.*, 2000). Little is known about metabolic flux changes in brain cells in response to "engineered" variations in enzyme amount, but regional and cell-type differences in amounts of mRNA and enzyme protein have been reported, with the *inference* that differences in capacity are reflected by differences in flux. However, this may not be the case, and metabolic demand (e.g., brain work and ADP generation from ATP) is an important aspect of control of metabolic rates. Metabolic control is a very complex process that involves interactions of many pathways and requirements for energy and metabolites (Fell, 1996).

Metabolic scaling across species. Whole-body metabolic rate in mammals varies inversely with the size, and relationships are approximately linear when plotted on log–log graphs. The smallest animals have the highest metabolic rates, and the metabolic cost of carrying out various activities by small mammals is greater than that for large animals (see monographs by Schmidt-Nielsen, 1972, 1984). Similarly, brain metabolic rate scales with brain size, with larger brains having lower metabolic rates (Karbowski, 2007). Large brains appear to reduce signaling delays and cost in various ways, including increasing axonal fiber volume to enhance conduction velocity and reduce energy expenditure; over the range of brain size from the shrew to monkey, calculated metabolic cost and measured brain glucose utilization rates fall by a factor of 10 as brain weight increases over three orders of magnitude (Wang *et al.*, 2008). Within various primate species, the ratio of glia-to-neurons in area 9L of frontal cortex correlates with brain weight (log–log scale), and comparison of glial density to neuronal density in this brain region in humans and 18 arthropoid primate species reveals that human glial cell (astrocytes plus oligodendrocytes, excluding microglia) density is unexpectedly high (Marino, 2006; Sherwood *et al.*, 2006). Thus, cognitive function, relative densities of major brain cell types, and cellular specialization (e.g., dendritic arborization, axonal length and diameter, etc.) may be linked to the energetics of neuronal signaling. Species differences in scaling of metabolism, structure, and function raise the interesting possibility that similar genetic manipulations in small (mice) compared to large (primates) subjects may have a different, size-dependent impact on functional metabolism.

Proteomics

The entire set of proteins in a cell, organ, or organism composes the proteome, and proteomic analysis often focuses on protein profiles or proteins related to specific functions, subcellular fractions, organelles, or cells. Proteomic analysis of brain has been recently reviewed by Maurer and Kuschinsky (2007), with coverage of techniques, analysis of energy metabolism proteins, and studies of diseases. Proteins are commonly separated by two-dimensional gel electrophoresis (separation by isoelectric point and molecular weight) after extraction of samples of interest, but other separation procedures are also employed. Proteins can be analyzed by image analysis after staining and identified by various procedures, including amino acid sequencing, mass spectrometry, and immunoassays. *Proteomic profiling* is similar in principle to metabolite profiling, and differential levels of proteins under specific conditions are described. *Functional proteomics* has a somewhat different focus related to protein or enzyme activity, protein–protein actions, post-translational modifications (e.g., phosphorylation, glycosylation, sulfation, proteolytic cleavage). An interesting example of proteomic analysis is the effect of 17β-estradiol on brain mitochondrial protein expression and function. Estradiol treatment altered levels of 66 proteins out of the 500 spots detected by image analysis by at least two-fold, with upregulation of 42% and down-regulation of 58%; the levels and activities of key mitochondrial enzymes were increased, and these changes were associated with increases in levels of specific mRNAs and in respiratory activity assayed in isolated mitochondria, as well as reduced free radical formation (Nilsen *et al.*, 2007). Thus, estradiol has widespread effects on mitochondrial protein levels and function that are consistent with its neuroprotective effects.

High-throughput protein analysis is likely to be very useful in characterization of complex disorders of brain function, particularly when used in conjunction with other systems approaches. For example, an integrative study using parallel metabolomic, transcriptomic, and proteomic analyses of postmortem human brain tissue from control subjects and schizophrenic patients identified specific alterations that are related to energy metabolism and oxidative stress (Prabakaran *et al.*, 2004). Analysis of altered proteins, mRNAs, and metabolite levels was consistent with abnormal mitochondrial function, upregulation of oxidative stress response systems, and down-regulation of glycolysis and oxidative metabolism components. These attributes were sufficient to differentiate controls from about 90% of the schizophrenic patients.

Multiple uses of a single protein is called *gene sharing*, and a fascinating aspect of some metabolic enzymes is their capability for multifunctional roles in which the protein fulfills specific and very different functions in the cell or organism. For example, the crystalline proteins account for the vast majority of the water-soluble proteins in the transparent lens, and specific metabolic enzymes are major components of the crystallin lens in a species-dependent manner. In the duck, bird, and crocodile, lactate dehydrogenase has a dual function as the epsilon-crystallin lens protein, whereas α-enolase is the same as tau-crystallin; in cephalopods (squid and octopus), the S-crystallin lens protein is equivalent to glutathione S-transferase (Piatigorsky, 2003). Transketolase, an enzyme in the pentose phosphate shunt pathway, is an abundant component in the mammalian cornea (Sax *et al.,* 2000). The key glycolytic enzyme, glyceraldehyde-3-phosphate dehydrogenase, has multiple functions unrelated to its enzymatic role in glycolysis (Fig. 3.4), including membrane fusion, vesicular secretory transport, DNA replication and repair, transcriptional regulation of histone gene expression, and post-translational modification during hyperglycemic stress (Sirover, 2005). Thus, glyceraldehyde phosphate dehydrogenase is not simply a "housekeeping" protein, and other metabolic enzymes fulfill important functions unrelated to their enzymatic activities.

occur when oxygen and glucose levels are inadequate. During abnormal states, brain ATP concentrations are maintained within the normal range until a critical threshold is reached; subsequent energy failure is often abrupt and complete, with loss of consciousness. Recovery from transient energy failure requires restoration of ionic gradients and levels of high-energy compounds, resynthesis of endogenous compounds consumed during the episode, and repair of damage. Evaluation of disease states and pharmacological treatment paradigms is often monitored by metabolic brain imaging, and emerging systems biology approaches are powerful procedures to characterize and identify tissue and biofluid samples. Major advances in brain energy metabolism are technology driven, and multimodal approaches are making important contributions to understanding functional metabolism in brain.

SUMMARY

Brain energy metabolism is a very complex process involving specialized functions of neurons, astrocytes, the vasculature, and other brain cells. Glucose and oxygen are the primary and obligatory fuels for brain, but minor substrates are also consumed to varying extents under different developmental, physiological, and pathophysiological states. The physical structure of the brain and many regulatory systems ensure that local capacities for fuel delivery, energy generation, and functional activity are closely matched under normal and activating conditions. Ion pumping is the major energy-consuming process in brain, and neuronal activity governs ATP demand and glucose utilization at a local level. Because functional activity and metabolic rates are closely correlated, pathways and compounds involved in metabolism are commonly used for brain-imaging or spectroscopic studies using different technologies and a wide range of experimental approaches and systems. Metabolic fluxes are heterogeneous throughout the brain at a regional, cellular, and subcellular level, and molecular trafficking can be measured in real-time *in vivo.* The cellular contributions to brain energetics and brain images are not firmly established, but specific neurons in the brain stem can control the overall metabolic economy of the body and feeding behavior. Glucose utilization is integrated with pathways of glucose-derived neurotransmitter turnover, which is much more sensitive to metabolic disruption than metabolite or transmitter concentrations. The brain has no significant energy reserves except glycogen, and progressive decrements in cognitive activity

ACKNOWLEDGMENTS

I thank Dr. Joseph LaManna, Case Western Reserve University School of Medicine, for critical review of this chapter.

References

Monographs

Balázs R, Cremer JE, (Eds.). 1972. Metabolic Compartmentation in the Brain. John Wiley & Sons: New York.

Berl S, Clarke DD, Schneider D, (Eds.). 1975. Metabolic Compartmentation and Neurotransmission: Relation to Brain Structure and Function. Plenum Press: New York

Blass JP, McDowell FH, (Eds.). 1999. Oxidative/Energy Metabolism in Neurodegenerative Disorders. Ann New York Academy of Sciences: vol. 893.

Boulton AA, Baker GB, Butterworth RF, (Eds.). 1989. Neuromethods 11. Carbohydrates and Energy Metabolism. Humans Press: Clifton, NJ.

Cryer PE. 1997. Hypoglycemia: Pathobiology, Diagnosis, and Treatment. Oxford University Press: New York.

Davson H, Segal MB. 1996. Physiology of the CSF and Blood-Brain Barriers. CRC Press: Boca Raton.

Dermietzel R, Spray DC, Nedergaard M, (Eds.). 2006. Blood-Brain Barriers. From Ontogeny to Artificial Interfaces. Wiley-VCH Verlag: Weinheim.

Dwyer D. 2002. Glucose Metabolism in the Brain. International Review of Neurobiology, vol. 51. Academic Press: Amsterdam.

Edvinsson L, Krause DN, (Eds.). 2002. Cerebral Blood Flow and Metabolism, 2nd Ed. Lippincott Williams & Wilkins: Philadelphia.

Fell D, Ed. 1996. Understanding the Control of Metabolism. Portland Press: Colchester.

Ferguson SJ, Nicholls DG. 2002. Bioenergetics 3. Academic Press: London.

Gibson GE, Dienel GA, (Eds.). 2007. Brain Energetics. Integration of Molecular and Cellular Processes, Handbook of Neurochemistry and Molecular Biology, 3rd Ed. Lajtha A., Series Editor. Springer-Verlag: Berlin.

Gibson GE, Ratan RR, Beal MF, (Eds.). 2008. Mitochondria and Oxidative Stress in Neurodegenerative Disorders. Annals of the New York Academy of Sciences (in press).

Hertz L, Ed. 2003. Non-Neuronal Cells of the Nervous System: Function and Dysfunction. Advances in Molecular and Cell Biology, vol. 31-1 part I: Structure, Organization, Development and Regeneration; vol. 31-2 part II: Biochemistry, Physiology and Pharmacology; vol. 31-3 part III: Pathological Conditions. Elsevier: Amsterdam.

McIlwain H, Bachelard HS.1985. Biochemistry and the Central Nervous System, 5th Ed. Churchill Livingstone: Edinburgh.

Passonneau JV, Lowry OH. 1993. Enzymatic Analysis. A Practical Guide. Humana Press: Totowa, NJ.

Passonneau JV, Hawkins RA, Lust WD, Welsh FA (Eds.). 1980. Cerebral Metabolism and Neural Function. Williams & Wilkins: Baltimore.

Phelps ME, 2004. PET: Molecular Imaging and Its Biological Applications. Springer: New York.

Phelps ME, Mazziotta JC, Schelbert HR (Eds.). 1986. Positron Emission Tomography and Autoradiography: Principles and Applications for the Brain and Heart. Raven Press: New York

Schmidt-Nielsen K. 1972. How Animals Work. Cambridge University Press: New York.

Schmidt-Nielsen K. 1984. Scaling. Why Is Animal Size So Important? Cambridge University Press: New York.

Siegel GJ, Albers RW, Brady ST, Price DL (Eds.). 2006. Basic Neurochemistry. Molecular, Cellular, and Medical Aspects, 7th Ed. Elsevier: Amsterdam.

Siesjö BK. 1978. Brain Energy Metabolism. Wiley-Interscience, John Wiley & Sons: Chichester.

Literature References

Abe T, Takahashi S, Suzuki N. 2006. Oxidative metabolism in cultured rat astroglia: effects of reducing the glucose concentration in the culture medium and of D-aspartate or potassium stimulation. J. Cereb. Blood Flow Metab. 26:153–160.

Alano CC, Swanson RA. 2006. Players in the PARP-1 cell-death pathway: JNK1 joins the cast. Trends Biochem Sci. 31:309–311.

Allagnat F, Martin D, Condorelli DF, Waeber G, Haefliger JA. 2005. Glucose represses connexin36 in insulin-secreting cells. J Cell Sci. 118:5335–5344.

Allen NJ, Káradóttir R, Attwell D. 2005. A preferential role for glycolysis in preventing the anoxic depolarization of rat hippocampal area CA1 pyramidal cells. J Neurosci. 25:848–859.

Almeida A, Moncada S, Bolanos JP. 2004. Nitric oxide switches on glycolysis through the AMP protein kinase and 6-phosphofructo-2-kinase pathway. Nat Cell Biol. 6:45–51.

Attwell D, Iadecola C. 2002. The neural basis of functional brain imaging signals. Trends Neurosci. 25:621–625.

Attwell D, Laughlin SB. 2001. An energy budget for signaling in the grey matter of the brain. J Cereb Blood Flow Metab. 21:1133–1145.

Badar-Goffer RS, Bachelard HS, Morris PG. 1990. Cerebral metabolism of acetate and glucose studied by ^{13}C-n.m.r. spectroscopy. A technique for investigating metabolic compartmentation in the brain. Biochem J. 266:133–139.

Badar-Goffer RS, Ben-Yoseph O, Bachelard HS, Morris PG. 1992. Neuronal-glial metabolism under depolarizing conditions: 13C-n.m.r.study. Biochem J. 282:225–230.

Bak LK, Schousboe A, Sonnewald U, Waagepetersen HS. 2006. Glucose is necessary to maintain neurotransmitter homeostasis during synaptic activity in cultured glutamatergic neurons. J Cereb Blood Flow Metab. 26:1285–1297.

Ball KK, Gandhi GK, Thrash J, Cruz NF, Dienel GA. 2007. Astrocytic connexin distributions and rapid, extensive dye transfer via gap junctions in the inferior colliculus: implications for [(14)C]glucose metabolite trafficking. J Neurosci Res. 85:3267–3283.

Baquer NZ, Hothersall JS, McLean P. 1988. Function and regulation of the pentose phosphate pathway in brain. Curr Top Cell Regul. 29:265–289.

Berger F, Ramirez-Hernandez MH, Ziegler M. 2004. The new life of a centenarian: signalling functions of NAD(P). Trends Biochem Sci. 29:111–118.

Bergles DE, Jahr CE. 1997. Synaptic activation of glutamate transporters in hippocampal astrocytes. Neuron 19:1297–1308.

Blázquez C, Woods A, de Ceballos ML, Carling D, Guzmán M. 1999. The AMP-activated protein kinase is involved in the regulation of ketone body production by astrocytes. J Neurochem. 73:1674–1682.

Bollard ME, Stanley EG, Lindon JC, Nicholson JK, Holmes E. 2005. NMR-based metabonomic approaches for evaluating physiological influences on biofluid composition. NMR Biomed. 18:143–162.

Brand A, Engelmann J, Leibfritz D. 1992. A ^{13}C NMR study on fluxes into the TCA cycle of neuronal and glial tumor cell lines and primary cells. Biochimie. 74:941–948.

Brennan AM, Connor JA, Shuttleworth CW. 2006. NAD(P)H fluorescence transients after synaptic activity in brain slices: predominant role of mitochondrial function. J Cereb Blood Flow Metab. 26:1389–1406.

Brennan AM, Connor JA, Shuttleworth CW. 2007. Modulation of the amplitude of NAD(P)H fluorescence transients after synaptic stimulation. J. Neurosci. Res. 85:3233–3243.

Bubber P, Haroutunian V, Fisch G, Blass JP, Gibson GE. 2005. Mitochondrial abnormalities in Alzheimer brain: mechanistic implications. Ann Neurol. 57:695–703.

Butterworth RF. 2006. Metabolic encephalopathies. In: Basic Neurochemistry: Molecular, Cellular, and Medical Aspects, 7th Ed. Siegel GJ, Albers RW, Brady ST, Price DL (Eds.). Academic Press: San Diego, pp. 593–602.

Caesar K, Hashemi P, Douhou A, Bonvento G, Boutelle MG, Walls AB, Lauritzen M. 2008. Glutamate receptor dependent increments in lactate, glucose and oxygen metabolism evoked in rat cerebellum in vivo. J Physiol. 586:1337–1349.

Cahoy JD, Emery B, Kaushal A, Foo LC, Zamanian JL, Christopherson KS, Xing Y, Lubischer JL, Krieg PA, Krupenko SA, Thompson WJ, Barres BA. 2008. A transcriptome database for astrocytes, neurons, and oligodendrocytes: a new resource for understanding brain development and function. J Neurosci. 28:264–278.

Cerdan S, Künnecke B, Seelig J. 1990. Cerebral metabolism of $[1,2-^{13}C_2]$ acetate as detected by in vivo and in vitro with ^{13}C NMR. J Biol Chem. 265:12916–12926.

Cerdán S, Rodrigues TB, Sierra A, Benito M, Fonseca LL, Fonseca CP, García-Martín ML. 2006. The redox switch/redox coupling hypothesis. Neurochem Int. 48:523–530.

Cetin N, Ball K, Gokden M, Cruz NF, Dienel GA. 2003. Effect of reactive cell density on net [2-^{14}C]acetate uptake into rat brain: labeling of clusters containing GFAP$^+$- and lectin$^+$-immunoreactive cells. Neurochem Int. 42:359–374.

Chance B, Cohen P, Jöbsis F, Schoener B. 1962. Intracellular oxidation-reduction states in vivo. Science. 137:499–508.

Chance B, Thorell B. 1959. Localization and kinetics of reduced pyridine nucleotide in living cells by microfluorometry. J Biol Chem. 234:3044–3050.

Chih CP, Roberts Jr EL. 2003. Energy substrates for neurons during neural activity: a critical review of the astrocyte-neuron lactate shuttle hypothesis. J Cereb Blood Flow Metab. 23:1263–1281.

Chih CP, Lipton P, Roberts EL, Jr. 2001. Do active cerebral neurons really use lactate rather than glucose. Trends Neurosci. 24:573–578.

Clark JB, Lai JCK. 1989. Glycolytic, tricarboxylic acid and related enzymes in brain. In: Neuromethods 11. Carbohydrates and Energy Metabolism. Boulton AA, Baker GB, Butterworth RF (Eds.). Humans Press: Clifton, NJ, pp. 233–281.

Clarke DD, Lajtha AL, Maker HS. 1989. Intermediary metabolism. In: Basic Neurochemistry: Molecular, Cellular, and Medical Aspects, 4th Ed. Siegel GJ, Agranoff BW, Albers RW, Fisher SK, Uhler MD (Eds.). Raven Press: New York, pp. 541–564.

Clarke DD, Sokoloff L. 1999. Circulation and energy metabolism of the brain. In: Basic Neurochemistry. Molecular, Cellular and Medical Aspects, 6th Ed. Siegel GJ, Agranoff BW, Albers RW, Fisher SK, Uhler MD (Eds.). Lippincott-Raven: Philadelphia, pp. 637–669.

Cooper AJ, Plum F. 1987. Biochemistry and physiology of brain ammonia. Physiol Rev. 67:440–519.

Cremer JE. 1982. Substrate utilization and brain development. J Cereb Blood Flow Metab. 2:394–407.

Cruz NF, Dienel GA. 2002. High brain glycogen levels in brains of rats with minimal environmental stimuli: implications for metabolic contributions of working astrocytes. J Cereb Blood Flow Metab. 22:1476–1489.

Cruz NF, Lasater AS, Zielke HR, Dienel GA. 2005. Activation of astrocytes in brain of conscious rats during acoustic stimulation: acetate utilization in working brain. J Neurochem. 92:934–947.

Cruz NF, Ball KK, Dienel GA. 2007. Functional imaging of focal brain activation in conscious rats: impact of [(14)C]glucose metabolite spreading and release. J Neurosci Res. 85:3254–3266.

Dalsgaard MK. 2006. Fuelling cerebral activity in exercising man. J Cereb Blood Flow Metab. 26:731–750.

Davis JN, Carlsson A, MacMillan V, Siesjö BK. 1973. Brain tryptophan hydroxylation: dependence on arterial oxygen tension. Science. 182:72–74.

Dienel GA. 2002. Energy generation in the central nervous system. In: Cerebral Blood Flow and Metabolism, 2nd Ed. Edvinsson L, Krause DN (Eds.). Lippincott Williams & Wilkins: Philadelphia, pp. 140–161.

Dienel GA. 2006. Functional brain imaging. In: Blood-Brain Barriers. From Ontogeny to Artificial Interfaces, vol. 2. Dermietzel R, Spray DC, Nedergaard M (Eds.). Wiley-VCH Verlag: Weinheim, pp. 551–599.

Dienel GA, Cruz NF. 2003. Neighborly interactions of metabolically activated astrocytes in vivo. Neurochem Int. 43:339–354.

Dienel GA, Cruz NF. 2004. Nutrition during brain activation: does cell-to-cell lactate shuttling provide sweet and sour food for thought? Neurochem Int. 45:321–351.

Dienel GA, Cruz NF. 2006. Astrocyte activation in working brain: energy supplied by minor substrates. Neurochem Int. 48:586–595.

Dienel GA, Cruz NF. 2008. Imaging brain activation: simple pictures of complex biology. Annals New York Acad. Sci. (in press).

Dienel GA, Hertz L. 2001. Glucose and lactate metabolism during brain activation. J Neurosci Res. 66:824–838.

Dienel GA, Hertz L. 2005. Astrocytic contributions to bioenergetics of cerebral ischemia. Glia. 50:362–388.

Dienel G, Cruz N, Mori K, Holden J, Sokoloff L. 1991. Direct measurement of the lambda of the lumped constant of the deoxyglucose method in rat brain: determination of lambda and lumped constant from tissue glucose concentrations or equilibrium brain: plasma distribution ratio for methylglucose. J Cereb Blood Flow Metab. 11:25–34.

Dienel GA, Ball KK, Cruz NF. 2007. A glycogen phosphorylase inhibitor selectively enhances local rates of glucose utilization in brain during sensory stimulation of conscious rats: implications for glycogen turnover. J Neurochem. 102:466–478.

Dringen R, Hoepken H, Minich T, Ruedig C. 2007. Pentose phosphate pathway and NADPH metabolism. In: Brain Energetics. Integration of Molecular and Cellular Processes. Gibson GE, Dienel GA (Eds.). Handbook of Neurochemistry and Molecular Biology, 3rd Ed. Lajtha A, series Ed. Springer-Verlag: Berlin, pp. 41–62.

Duan S, Anderson CM, Stein BA, Swanson RA. 1999. Glutamate induces rapid upregulation of astrocyte glutamate transport and cell-surface expression of GLAST. J Neurosci. 19:10193–10200.

Duffy TE, Howse DC, Plum F. 1975. Cerebral energy metabolism during experimental status epilepticus. J Neurochem. 24:925–934.

Duncan GE, Stumpf WE, Pilgrim C, Breese GR. 1987a. High resolution autoradiography at the regional topographic level with [^{14}C]2-deoxyglucose and [^{3}H]2-deoxyglucose. J Neurosci Methods. 20:105–113.

Duncan GE, Stumpf WE, Pilgrim C. 1987b. Cerebral metabolic mapping at the cellular level with dry-mount autoradiography of [^{3}H]2-deoxyglucose. Brain Res. 401:43–49.

Duncan GE, Kaldas RG, Mitra KE, Breese GR, Stumpf WE. 1990. High activity neurons in the reticular formation of the medulla oblongata: a high-resolution autoradiographic 2-deoxyglucose study. Neurosci. 35:593–600.

Durham D, Woolsey TA, Kruger L. 1981. Cellular localization of 2-[^{3}H]deoxy-D-glucose from paraffin-embedded brains. J Neurosci. 1:519–526.

Edmond J. 1992. Energy metabolism in developing brain cells. Can J Physiol Pharmacol. 70 Suppl:S118–S129.

Eriksson G, Peterson A, Iverfeldt K, Walum E. 1995. Sodium-dependent glutamate uptake as an activator of oxidative metabolism in primary astrocyte cultures from newborn rat. Glia. 15:152–156.

Ferrendelli JA. 1974. Cerebral utilization of nonglucose substrates and their effect in hypoglycemia. Res Publ Assoc Nerv Ment Dis. 53:113–123.

Fiehn O. 2002. Metabolomics: the link between genotypes and phenotypes. Plant Molecular Biology. 48:155–171.

Fox PT, Raichle ME, Mintun MA, Dence C. 1988. Nonoxidative glucose consumption during focal physiologic neural activity. Science. 241:462–464.

Garriga-Canut M, Schoenike B, Qazi R, Bergendahl K, Daley TJ, Pfender RM, Morrison JF, Ockuly J, Stafstrom C, Sutula T, Roopra A. 2006. Deoxy-D-glucose reduces epilepsy progression by NRSF-CtBP-dependent metabolic regulation of chromatin structure. Nat Neurosci. 9:1382–1387.

Gibbs ME, Anderson DG, Hertz L. 2006. Inhibition of glycogenolysis in astrocytes interrupts memory consolidation in young chickens. Glia. 54:214–222.

Gibson GE, Duffy TE. 1981. Impaired synthesis of acetylcholine by mild hypoxic hypoxia or nitrous oxide. J Neurochem. 36:28–33.

Gibson GE, Blass JP, Beal MF, Bunik V. 2005. The alpha-ketoglutarate-dehydrogenase complex: a mediator between mitochondria and oxidative stress in neurodegeneration. Mol Neurobiol. 31:43–63.

Ginsberg MD. 2008. Neuroprotection for ischemic stroke: past, present and future. Neuropharmacology (in press). doi:10.1016/j.neuropharm.2007.12.007.

Gjedde A. 2007. Coupling of brain function to metabolism: evaluation of energy requirements. In: Brain Energetics. Integration of Molecular and Cellular Processes. Gibson GE, Dienel GA (Eds.). Handbook of Neurochemistry and Molecular Biology, 3rd Ed. Lajtha A., series Ed. Springer-Verlag: Berlin, pp. 343–400.

Goodacre R, Vaidyanathan S, Dunn WB, Harrigan GG, Kell DB. 2004. Metabolomics by numbers: acquiring and understanding global metabolite data. Trends Biotechnol. 22:245–252.

Gordon GR, Mulligan SJ, MacVicar BA. 2007. Astrocyte control of the cerebrovasculature. Glia. 55:1214–1221.

Griffin JL. 2004. Defining a metabolic phenotype. Phil Trans R Soc Lond B. 359:857–871.

Griffin JL, Anthony DC, Campbell SJ, Gauldie J, Pitossi F, Styles P, Sibson NR. 2004a. Study of cytokine induced neuropathology by high resolution proton NMR spectroscopy of rat urine. FEBS Lett. 568:49–54.

Griffin JL, Cemal CK, Pook MA. 2004b. Defining a metabolic phenotype in the brain of a transgenic mouse model of spinocerebellar ataxia 3. Physiol Genomics. 16:334–340.

Gruetter R, Novotny EJ, Boulware SD, Rothman DL, Mason GF, Shulman GI, Shulman RG, Tamborlane WV. 1992. Direct measurement of brain glucose concentrations in humans by 13C NMR spectroscopy. Proc Natl Acad Sci USA. 89:1109–1112.

Gruetter R, Seaquist ER, Ugurbil K. 2001. A mathematical model of compartmentalized neurotransmitter metabolism in the human brain. Am J Physiol Endocrinol Metab. 281:E100–E112.

Halestrap AP, Price NT. 1999. The proton-linked monocarboxylate transporter (MCT) family: structure, function and regulation. Biochem J. 343:281–299.

Hamprecht B, Dringen R. 1995. Energy metabolism. In: Neuroglia. Kettenmann H, Ransom BR (Eds.). Oxford University Press: New York, pp. 473–487.

Hamprecht B, Verleysdonk S, Wiesinger H. 2005. Enzymes of carbohydrate and energy metabolism. In: Neuroglia, 2nd Ed. Kettenmann H, Ransom BR (Eds.). Oxford University Press: New York, pp. 202–215.

Hardie DG, Hawley SA, Scott JW. 2006. AMP-activated protein kinase: development of the energy sensor concept. J Physiol. 574:7–15.

Hassel B, Bachelard H, Jones P, Fonnum F, Sonnewald U. 1997. Trafficking of amino acids between neurons and glia in vivo. Effects of inhibition of glial metabolism by fluoroacetate. J Cereb Blood Flow Metab. 17:1230–1238.

Heidemann AC, Schipke CG, Kettenmann H. 2005. Extracellular application of nicotinic acid adenine dinucleotide phosphate induces Ca2+ signaling in astrocytes in situ. J Biol Chem. 42:35630–35640.

Hertz L. 2004. The astrocyte-neuron lactate shuttle: a challenge of a challenge. J Cereb Blood Flow Metab. 24:1241–1248.

Hertz L, Dienel GA. 2005. Lactate transport and transporters: general principles and functional roles in brain cells. J Neurosci Res. 79:11–18.

Hertz L, Zielke HR. 2004. Astrocytic control of glutamatergic activity: astrocytes as stars of the show. Trends Neurosci. 27:735–743.

Hertz L, Swanson RA, Newman GC, Marrif H, Juurlink BH, Peng L. 1998. Can experimental conditions explain the discrepancy over glutamate stimulation of aerobic glycolysis? Dev Neurosci. 20:339–347.

Hertz L, Peng L, Dienel GA. 2007. Energy metabolism in astrocytes: high rate of oxidative metabolism and spatiotemporal dependence on glycolysis/glycogenolysis. J Cereb Blood Flow Metab. 27:219–249.

Hof PR, Pascale E, Magistretti PJ. 1988. K+ at concentrations reached in the extracellular space during neuronal activity promotes a Ca2+-dependent glycogen hydrolysis in mouse cerebral cortex. J Neurosci. 8:1922–1928.

Holden JE, Mori K, Dienel GA, Cruz NF, Nelson T, Sokoloff L. 1991. Modeling the dependence of hexose distribution volumes in brain on plasma glucose concentration: implications for estimation of the local 2-deoxyglucose lumped constant. J Cereb Blood Flow Metab. 11:171–182.

Hothersall JS, Baquer N, Greenbaum AL, McLean P. 1979. Alternative pathways of glucose utilization in brain: changes in the pattern of glucose utilization in brain during development and the effect of phenazine methosulfate on the integration of metabolic routes. Arch Biochem Biophys. 198:478–492.

Huang HM, Toral-Barza L, Sheu KF, Gibson GE. 1994. The role of cytosolic free calcium in the regulation of pyruvate dehydrogenase in synaptosomes. Neurochem Res. 19:89–95.

Hutson SM, Berkich D, Drown P, Xu B, Aschner M, LaNoue KF. 1998. Role of branched-chain aminotransferase isoenzymes and gabapentin in neurotransmitter metabolism. J Neurochem. 71:863–874.

Hyder F, Patel AB, Gjedde A, Rothman DL, Behar KL, Shulman RG. 2006. Neuronal-glial glucose oxidation and glutamatergic-GABAergic function. J Cereb Blood Flow Metab. 26:865–877.

Iacobas DA, Urban-Maldonado M, Iacobas S, Scemes E, Spray DC. 2003. Array analysis of gene expression in connexin-43 null astrocytes. Physiol Genomics. 15:177–190.

Iacobas DA, Scemes E, Spray DC. 2004. Gene expression alterations in connexin null mice extend beyond the gap junction. Neurochem Int. 45:243–250.

Iadecola C, Nedergaard M. 2007. Glial regulation of the cerebral microvasculature. Nat Neurosci. 10:1369–1376.

Itoh Y, Abe T, Takaoka R, Tanahashi N. 2004. Fluorometric determination of glucose utilization in neurons in vitro and in vivo. J Cereb Blood Flow Metab. 24:993–1003.

Jöbsis FF, O'Connor M, Vitale A, Vreman H. 1971. Intracellular redox changes in functioning cerebral cortex. I. Metabolic effects of epileptiform activity. J Neurophysiol. 34:735–749.

Joseph J, Gibson G. 2007. Coupling of neuronal function to oxygen and glucose metabolism through changes in neurotransmitter dynamics as revealed with aging, hypoglycemia, and hypoxia. In: Brain Energetics. Integration of Molecular and Cellular Processes. Gibson GE, Dienel GA (Eds.). Handbook of Neurochemistry and Molecular Biology, 3rd Ed. Lajtha A, series Ed. Springer-Verlag: Berlin, pp. 297–320.

Kanamori K, Ross BD, Farrow NA, Parivar F. 1991. A 15N-NMR study of isolated brain in portacaval-shunted rats after acute hyperammonemia. Biochim Biophys Acta. 1096:270–276.

Kann O, Schuchmann S, Buchheim K, Heinemann U. 2003. Coupling of neuronal activity and mitochondrial metabolism as revealed by NAD(P)H fluorescence signals in organotypic hippocampal slice cultures of the rat. Neuroscience. 119:87–100.

Karbowski J. May 9, 2007. Global and regional brain metabolic scaling and its functional consequences. BMC Biol. 5:18.

Kasischke KA, Vishwasrao HD, Fisher PJ, Zipfel WR, Webb WW. 2004. Neural activity triggers neuronal oxidative metabolism followed by astrocytic glycolysis. Science. 305:99–103.

Kauppinen TM, Swanson RA. 2007. The role of poly(ADP-ribose) polymerase-1 in CNS disease. Neuroscience. 145:1267–1272.

Kauppinen RA, Nicholls DG. 1986. Synaptosomal bioenergetics. The role of glycolysis, pyruvate oxidation and responses to hypoglycaemia. Eur J Biochem. 158:159–165.

Kety SS, 1996. Seymour S. Kety. In: The History of Neuroscience in Autobiography, vol. 1. Squire LR, Ed. Soc Neurosci: Washington, D.C., pp. 382–413.

Kety SS, Schmidt CF. 1948. The nitrous oxide method for the quantitative determination of cerebral blood flow in man: theory, procedure, and normal values. J Clin Invest. 27:476–483.

Keyser DO, Pellmar TC. 1994. Synaptic transmission in the hippocampus: critical role for glial cells. Glia. 10:237–243.

Keyser DO, Pellmar TC. 1997. Regional differences in glial cell modulation of synaptic transmission. Hippocampus. 7:73–77.

Koehler-Stec EM, Li K, Maher F, Vannucci SJ, Smith CB, Simpson IA. 2000. Cerebral glucose utilization and glucose transporter expression: response to water deprivation and restoration. J Cereb Blood Flow Metab. 20:192–200.

Korf J. 2006. Is brain lactate metabolized immediately after neuronal activity through the oxidative pathway? J Cereb Blood Flow Metab. 26:1584–1586.

Korf J, Gramsbergen JB. 2007. Timing of potential and metabolic brain energy. J Neurochem. 103:1697–1708.

Kristal B, Kaddurah-Daouk R, Beal M, Matson W. 2007. Metabolomics: concepts and potential neuroscience applications. In: Brain Energetics. Integration of Molecular and Cellular Processes. Gibson GE, Dienel GA, (Eds.). Handbook of Neurochemistry and Molecular Biology, 3rd Ed. Lajtha A., series Ed. Springer-Verlag: Berlin, pp.889–912.

Lai JCK, Clark JB. 1989. Isolation and characterization of synaptic and nonsynaptic mitochondria from mammalian brain. In: Neuromethods 11. Carbohydrates and Energy Metabolism. Boulton AA, Baker GB, Butterworth RF, (Eds.). Humans Press: Clifton, NJ, pp. 43–98.

LaManna JC, Pichiule P, Chavez JC. 2007. Genetics and gene expression of glycolysis. In: Brain Energetics. Integration of Molecular and Cellular Processes. Gibson GE, Dienel GA (Eds.). Handbook of Neurochemistry and Molecular Biology, 3rd Ed. Lajtha A., series Ed. Springer-Verlag: Berlin, pp. 771–811.

Landau WH, Freygang WH, Rowland LP, Sokoloff L, Kety SS. 1955. The local circulation in the living brain: values in the unanesthetized and anesthetized cat. Trans Am Neurol Assoc. 80:125–129.

LaNoue KF, Carson V, Berkich DA, Hutson SM. 2007. Mitochondrial/cytosolic interactions via metabolite shuttles and transporters. In: Brain Energetics. Integration of Molecular and Cellular Processes. Gibson GE, Dienel GA, (Eds.). Handbook of Neurochemistry and Molecular Biology, 3rd Ed. Lajtha A., series Ed. Springer-Verlag: Berlin, pp. 589–616.

Laschet JJ, Minier F, Kurcewicz I, Bureau MH, Trottier S, Jeanneteau F, Griffon N, Samyn B, Van Beeumen J, Louvel J, Sokoloff P, Pumain R. 2004. Glyceraldehyde-3-phosphate dehydrogenase is a GABAA receptor kinase linking glycolysis to neuronal inhibition. J Neurosci. 24:7614–7622.

Laschet JJ, Kurcewicz I, Minier F, Trottier S, Khallou-Laschet J, Louvel J, Gigout S, Turak B, Biraben A, Scarabin JM, Devaux B, Chauvel P, Pumain R. 2007. Dysfunction of GABAA receptor glycolysis-dependent modulation in human partial epilepsy. Proc Natl Acad Sci USA. 104:3472–3477.

Lewis LD, Ljunggren B, Ratcheson RA, Siesjö BK. 1974. Cerebral energy state in insulin-induced hypoglycemia, related to blood glucose and to EEG. J Neurochem. 23:673–679.

Li J, Zeng Z, Viollet B, Ronnett GV, McCullough LD. 2007. Neuroprotective effects of adenosine monophosphate-activated protein kinase inhibition and gene deletion in stroke. Stroke. 38:2992–2999.

Lindsley JE, Rutter J. 2004. Nutrient sensing and metabolic decisions. Comp Biochem Physiol B Biochem Mol Biol. 139:543–559.

Lipton P. 1973. Effects of membrane depolarization on nicotinamide nucleotide fluorescence in brain slices. Biochem J. 136:999–1009.

Lovatt D, Sonnewald U, Waagepetersen HS, Schousboe A, He W, Lin JH, Han X, Takano T, Wang S, Sim FJ, Goldman SA, Nedergaard M. 2007. The transcriptome and metabolic gene signature of protoplasmic astrocytes in the adult murine cortex. J Neurosci. 27:12255–12266.

Lowry OH. 1990. How to succeed in research without being a genius. Annual Rev Biochem. 59:1–27.

Lowry OH, Passonneau JV. 1964. The relationships between substrates and enzymes of glycolysis in brain. J Biol Chem. 239:31–42.

Lowry OH, Passonneau JV, Hasselberger FX, Schulz DW. 1964. Effect of ischemia on known substrates and cofactors of the glycolytic pathway in brain. J Biol Chem. 239:18–30.

Lynch RM, Balaban RS. 1987. Coupling of aerobic glycolysis and Na+-K+-ATPase in renal cell line MDCK. Am J Physiol. 253:C269–C276.

Lynch RM, Paul RJ. 1983. Compartmentation of glycolytic and glycogenolytic metabolism in vascular smooth muscle. Science. 222:1344–1346.

Lynch RM, Paul RJ. 1985. Energy metabolism and transduction in smooth muscle. Experientia. 41:970–977.

Madsen PL, Hasselbalch SG, Hagemann LP, Olsen KS, Bülow J, Holm S, Wildschiødtz G, Paulson OB, Lassen NA.1995. Persistent resetting of the cerebral oxygen/glucose uptake ratio by brain activation: evidence obtained with the Kety-Schmidt technique. J Cereb Blood Flow Metab. 15:485–491.

Madsen PL, Cruz NF, Sokoloff L, Dienel G. 1999. Cerebral oxygen/glucose ratio is low during sensory stimulation and rises above normal during recovery: excess glucose consumption during stimulation is not accounted for by lactate efflux from or accumulation in brain tissue. J Cereb Blood Flow Metab. 19:393–400.

Magistretti PJ, Pellerin L, Rothman DL, Shulman RG. 1999. Energy on demand. Science. 283:496–497.

Marino L. 2006. Absolute brain size: did we throw the baby out with the bathwater? Proc Natl Acad Sci USA. 103:13563–13564.

Marrif H, Juurlink BH. 1999. Astrocytes respond to hypoxia by increasing glycolytic capacity. J Neurosci Res. 57:255–260.

Martinez-Hernandez A, Bell K, Norenberg M. 1977. Glutamine synthetase: glial localization in brain. Science. 195:1356–1358.

Maurer MH, Kuschinsky W. 2007. Proteomics. In: Brain Energetics. Integration of Molecular and Cellular Processes. Gibson GE, Dienel GA, (Eds.). Handbook of Neurochemistry and Molecular Biology, 3rd Ed. Lajtha A., series Ed. Springer-Verlag: Berlin, pp. 738–769.

Maurer MH, Canis M, Kuschinsky W, Duelli R. 2004. Correlation between local monocarboxylate transporter 1 (MCT1) and glucose transporter 1 (GLUT1) densities in the adult rat brain. Neurosci Lett. 355:105–108.

McCullough LD, Zeng Z, Li H, Landree LE, McFadden J, Ronnett GV. 2005. Pharmacological inhibition of AMP-activated protein kinase provides neuroprotection in stroke. J Biol Chem. 280:20493–20502.

McKenna MC. 2007. The glutamate-glutamine cycle is not stoichiometric: fates of glutamate in brain. J Neurosci Res. 85:3347–3358.

McKenna MC, Sonnewald U, Huang X, Stevenson J, Zielke HR. 1996. Exogenous glutamate concentration regulates the metabolic fate of glutamate in astrocytes. J Neurochem. 66:386–393.

McKenna MC, Waagepetersen HS, Schousboe A, Sonnewald U. 2006. Neuronal and astrocytic shuttle mechanisms for cytosolic-mitochondrial transfer of reducing equivalents: current evidence and pharmacological tools. Biochem Pharmacol. 71:399–407.

McNay EC, Gold PE. 2002. Food for thought: fluctuations in brain extracellular glucose provide insight into the mechanisms of memory modulation. Behav Cogn Neurosci Rev. 1:264–280.

Medina MA, Jones DJ, Stavinoha WB, Ross DH. 1975. The levels of labile intermediary metabolites in mouse brain following rapid tissue fixation with microwave irradiation. J Neurochem. 24:223–227.

Mercer RW, Dunham PB. 1981. Membrane-bound ATP fuels the Na/K pump. J Gen Physiol. 78:547–568.

Minn AH, Couto FM, Shalev A. 2006. Metabolism-independent sugar effects on gene transcription: the role of 3-O-methylglucose. Biochemistry. 45:11047–11051.

Moncada S, Bolaños JP. 2006. Nitric oxide, cell bioenergetics and neurodegeneration. J Neurochem. 97:1676–1689.

Muir D, Berl S, Clarke DD. 1986. Acetate and fluoroacetate as possible markers for glial metabolism in vivo. Brain Res. 380:336–340.

Nakanishi H, Cruz NF, Adachi K, Sokoloff L, Dienel GA. 1996. Influence of glucose supply and demand on determination of brain glucose content with labeled methylglucose. J Cereb Blood Flow Metab. 16:439–449.

Nakao Y, Itoh Y, Kuang TY, Cook M, Jehle J, Sokoloff L. 2001. Effects of anesthesia on functional activation of cerebral blood flow and metabolism. Proc Natl Acad Sci USA. 98:7593–7598.

Nehlig A. 2004. Brain uptake and metabolism of ketone bodies in animal models. Prostaglandins Leukot Essent Fatty Acids. 70:265–275.

Nehlig A. 1996. Respective roles of glucose and ketone bodies as substrates for cerebral energy metabolism in the suckling rat. Dev Neurosci. 18:426–433.

Nehlig A, Rudolf G, Leroy C, Rigoulot MA, Simpson IA, Vannucci SJ. 2006. Pentylenetetrazol-induced status epilepticus up-regulates the expression of glucose transporter mRNAs but not proteins in the immature rat brain. Brain Res. 1082:32–42.

Nehlig A, Wittendorp-Rechenmann E, Lam CD. 2004. Selective uptake of [14C]2-deoxyglucose by neurons and astrocytes: high-resolution microautoradiographic imaging by cellular ^{14}C-trajectography combined with immunohistochemistry. J Cereb Blood Flow Metab. 24:1004–1014.

Nicholls D. 2007. Bioenergetics. In: Brain Energetics. Integration of Molecular and Cellular Processes. Gibson GE, Dienel GA (Eds.). Handbook of Neurochemistry and Molecular Biology, 3rd Ed. Lajtha A., series Ed. Springer-Verlag: Berlin, pp. 3–16.

Nicholls DG, Johnson-Cadwell L, Vesce S, Jekabsons M, Yadava N. 2007. Bioenergetics of mitochondria in cultured neurons and their role in glutamate excitotoxicity. J Neurosci Res. 85:3206–3212.

Nicholson JK, Connelly J, Lindon JC, Holmes E. 2002. Metabonomics: a platform for studying drug toxicity and gene function. Nat Rev Drug Discov. 1:153–161.

Nilsen J, Irwin RW, Gallaher TK, Brinton RD. 2007. Estradiol in vivo regulation of brain mitochondrial proteome. J Neurosci. 27:14069–14077.

Norenberg MD, Martinez-Hernandez A. 1979. Fine structural localization of glutamine synthetase in astrocytes of rat brain. Brain Res. 161:303–310.

Okada Y, Lipton P. 2007. Glucose, oxidative energy metabolism, and neural function in brain slices: glycolysis plays a key role in neural activity. In: Brain Energetics. Integration of Molecular and Cellular Processes. Gibson GE, Dienel GA (Eds.). Handbook of Neurochemistry and Molecular Biology, 3rd Ed. Lajtha A., series Ed. Springer-Verlag: Berlin, pp. 17–39.

Oldendorf WH. 1981. Clearance of radiolabeled substrates by brain after arterial injection using a diffusible internal standard. In: Research Methods in Neurochemistry, vol. 5. Marks N, Rodnight R, (Eds.). Plenum Publishing Corp.: New York, pp. 91–112.

Oldendorf WH, Szabo J. 1976. Amino acid assignment to one of three blood-brain barrier amino acid carriers. Am J Physiol. 230:94–98.

Öz G, Berkich DA, Henry PG, Xu Y, LaNoue K, Hutson SM, Gruetter R. 2004. Neuroglial metabolism in the awake rat brain: CO2 fixation increases with brain activity. J Neurosci. 24:11273–11279.

Papa S, Petruzzella V, Scacco S. 2007. Electron transport. Structure, redox-coupled protonmotive activity, and pathological disorders of respiratory chain complexes. In: Brain Energetics. Integration of Molecular and Cellular Processes. Gibson GE, Dienel GA, (Eds.). Handbook of Neurochemistry and Molecular Biology, 3rd Ed. Lajtha A, series Ed. Springer-Verlag: Berlin, pp. 93–118.

Passonneau JV, Lowry OH. 1964. The role of phosphofructokinase in metabolic regulation. Adv Enzyme Regulation. 2:265–274.

Paul RJ, Bauer M, Pease W. 1979. Vascular smooth muscle: aerobic glycolysis linked to sodium and potassium transport processes. Science. 206:1414–1416.

Pellerin L, Magistretti PJ. 1994. Glutamate uptake into astrocytes stimulates aerobic glycolysis: a mechanism coupling neuronal activity to glucose utilization. Proc Natl Acad Sci. USA. 91:10625–10629.

Pellerin L, Bouzier-Sore AK, Aubert A, Serres S, Merle M, Costalat R, Magistretti PJ. 2007. Activity-dependent regulation of energy metabolism by astrocytes: an update. Glia. 55:1251–1262.

Peng L, Swanson RA, Hertz L. 2001. Effects of L-glutamate, D-aspartate, and monensin on glycolytic and oxidative glucose metabolism in mouse astrocyte cultures: further evidence that glutamate uptake is metabolically driven by oxidative metabolism. Neurochem Int. 38:437–443.

Perkins G, Ellisman MH. 2007. Mitochondrial architecture and heterogeneity. In: Brain Energetics. Integration of Molecular and Cellular Processes. Gibson GE, Dienel GA, (Eds.). Handbook of Neurochemistry and Molecular Biology, 3rd Ed. Lajtha A, series Ed. Springer-Verlag: Berlin, pp. 261–295.

Peters A, Palay SL, Webster H de F. 1991. The Fine Structure of the Nervous System. Neurons and Their Supporting Cells, 3rd Ed. Oxford University Press: Oxford.

Phelps ME, Huang SC, Hoffman EJ, Selin C, Sokoloff L, Kuhl DE. 1979. Tomographic measurement of local cerebral glucose metabolic rate in humans with (F-18)2-fluoro-2-deoxy-D-glucose: validation of the method. Ann Neurol. 6:371–388.

Piatigorsky J. 2003. Crystallin genes: specialization by changes in gene regulation may precede gene duplication. J Struct Funct Genomics. 3:131–137.

Pollak N, Dolle C, Ziegler M. 2007. The power to reduce: pyridine nucleotides—small molecules with a multitude of functions. Biochem J. 402:205–218.

Pontén U, Ratcheson RA, Salford LG, Siesjö BK. 1973. Optimal freezing conditions for cerebral metabolites in rats. J Neurochem. 21:1127–1138.

Prabakaran S, Swatton JE, Ryan MM, Huffaker SJ, Huang JT, Griffin JL, Wayland M, Freeman T, Dudbridge F, Lilley KS, Karp NA, Hester S, Tkachev D, Mimmack ML, Yolken RH, Webster MJ, Torrey EF, Bahn S. 2004. Mitochondrial dysfunction in schizophrenia: evidence for compromised brain metabolism and oxidative stress. Mol Psychiatry. 9:684–697.

Prichard JW, Shulman RG. 1986. NMR spectroscopy of brain metabolism in vivo. Annu Rev Neurosci. 9:61–85.

Prins ML. 2008. Cerebral metabolic adaptation and ketone metabolism after brain injury. J Cereb Blood Flow Metab. 28:1–16.

Qu H, Eloqayli H, Unsgard G, Sonnewald U. 2001. Glutamate decreases pyruvate carboxylase activity and spares glucose as energy substrate in cultured cerebellar astrocytes. J Neurosci Res. 66:1127–1132.

Rae C, Moussa Cel-H, Griffin JL, Bubb WA, Wallis T, Balcar VJ. 2005. Group I and II metabotropic glutamate receptors alter brain cortical metabolic and glutamate/glutamine cycle activity: a ^{13}C NMR spectroscopy and metabolomic study. J Neurochem. 92:405–416.

Rafaeloff-Phail R, Ding L, Conner L, Yeh WK, McClure D, Guo H, Emerson K, Brooks H. 2004. Biochemical regulation of mammalian AMP-activated protein kinase activity by NAD and NADH. J Biol Chem. 279:52934–52939.

Ramamurthy S, Ronnett GV. 2006. Developing a head for energy sensing: AMP-activated protein kinase as a multifunctional metabolic sensor in the brain. J Physiol. 574:85–93.

Ramos M, del Arco A, Pardo B, Martínez-Serrano A, Martínez-Morales JR, Kobayashi K, Yasuda T, Bogónez E, Bovolenta P, Saheki T, Satrústegui J. 2003. Developmental changes in the Ca^{2+}-regulated mitochondrial aspartate-glutamate carrier aralar1 in brain and prominent expression in the spinal cord. Brain Res Dev Brain Res. 143:33–46.

Reinert KC, Gao W, Chen G, Ebner TJ. 2007. Flavoprotein autofluorescence imaging in the cerebellar cortex in vivo. J Neurosci Res. 85:3221–3232.

Rennels ML, Gregory TF, Blaumanis OR, Fujimoto K, Grady PA. 1985. Evidence for a "paravascular" fluid circulation in the mammalian central nervous system, provided by the rapid distribution of tracer protein throughout the brain from the subarachnoid space. Brain Res. 326:47–63.

Roberts, Jr. EL 2007. The support of energy metabolism in the central nervous system with substrates other than glucose. In: Brain Energetics. Integration of Molecular and Cellular Processes. Gibson GE, Dienel GA, (Eds.). Handbook of Neurochemistry and Molecular Biology, 3rd Ed. Lajtha A, series Ed. Springer-Verlag: Berlin, pp. 137–179.

Rodrigues T, Cerdán S. 2007. The cerebral tricarboxylic acid cycles. In: Brain Energetics. Integration of Molecular and Cellular Processes. Gibson GE, Dienel GA, (Eds.). Handbook of Neurochemistry and Molecular Biology, 3rd Ed. Lajtha A, series Ed. Springer-Verlag: Berlin, pp.63–91.

Rosenthal M, Jöbsis FF. 1971. Intracellular redox changes in functioning cerebral cortex. II. Effects of direct cortical stimulation. J Neurophysiol. 34:750–762.

Rothman DL, Novotny EJ, Shulman GI, Howseman AM, Petroff OA, Mason G, Nixon T, Hanstock CC, Prichard JW, Shulman RG. 1992. ^{1}H-[^{13}C] NMR measurements of [4-^{13}C]glutamate turnover in human brain. Proc Natl Acad Sci USA. 89:9603–9606.

Saks VA, Vendelin M, Aliev MK, Kekelidze T, Engelbrecht J. 2007. Mechanisms and modeling of energy transfer between intracellular compartments. In: Brain Energetics. Integration of Molecular and Cellular Processes. Gibson GE, Dienel GA, (Eds.). Handbook of Neurochemistry and Molecular Biology, 3rd Ed. Lajtha A, series Ed. Springer-Verlag: Berlin, pp. 815–860.

Sakurada O, Kennedy C, Jehle J, Brown JD, Carbin GL, Sokoloff L. 1978. Measurement of local cerebral blood flow with iodo[^{14}C] antipyrine. Am J Physiol. 234:H59–H66.

Salinas JA, Gold PE. 2005. Glucose regulation of memory for reward reduction in young and aged rats. Neurobiol Aging. 26:45–52.

Savaki HE, Davidsen L, Smith C, Sokoloff L. 1980. Measurement of free glucose turnover in brain. J Neurochem. 35:495–502.

Sax CM, Kays WT, Salamon C, Chervenak MM, Xu YS, Piatigorsky J. 2000. Transketolase gene expression in the cornea is influenced by environmental factors and developmentally controlled events. Cornea. 19:833–841.

Schaaff I, Heinisch J, Zimmermann FK. 1989. Overproduction of glycolytic enzymes in yeast. Yeast. 5:285–290.

Schimmack G, Defronzo RA, Musi N. 2006. AMP-activated protein kinase: role in metabolism and therapeutic implications. Diabetes Obes Metab. 8:591–602.

Schousboe A, Waagepetersen HS. 2007. GABA: homeostatic and pharmacological aspects. Prog Brain Res.160:9–19.

Scott DK, O'Doherty RM, Stafford JM, Newgard CB, Granner DK. 1998. The repression of hormone-activated PEPCK gene expression by glucose is insulin-independent but requires glucose metabolism. J Biol Chem. 273:24145–24151.

Shank RP, Bennett GS, Freytag SO, Campbell GL. 1985. Pyruvate carboxylase: an astrocyte-specific enzyme implicated in the replenishment of amino acid neurotransmitter pools. Brain Res. 329:364–367.

Shank RP, Leo GC, Zielke HR. 1993. Cerebral metabolic compartmentation as revealed by nuclear magnetic resonance analysis of D-[1-^{13}C]glucose metabolism. J Neurochem. 61:315–323.

Sharp FR. 1976a. Relative cerebral glucose uptake of neuronal perikarya and neuropil determined with 2-deoxyglucose in resting and swimming rat. Brain Res. 110:127–139.

Sharp FR. 1976b. Rotation induced increases of glucose uptake in rat vestibular nuclei and vestibulocerebellum. Brain Res. 110:141–151.

Shepherd GM, Harris KM. 1998. Three-dimensional structure and composition of CA3→CA1 axons in rat hippocampal slices: implications for presynaptic connectivity and compartmentalization. J Neurosci. 18:8300–8310.

Sherwood CC, Stimpson CD, Raghanti MA, Wildman DE, Uddin M, Grossman LI, Goodman M, Redmond JC, Bonar CJ, Erwin JM, Hof PR. 2006. Evolution of increased glia-neuron ratios in the human frontal cortex. Proc Natl Acad Sci USA. 103:13606–13611.

Shestov AA, Valette J, Ugurbil K, Henry PG. 2007. On the reliability of (13)C metabolic modeling with two-compartment neuronal-glial models. J Neurosci Res. 85:3294–3303.

Sheu KF, Blass JP. 1999. The alpha-ketoglutarate dehydrogenase complex. Ann NY Acad Sci. 893:61–78.

Shibuki K, Hishida R, Kitaura H, Takahashi K, Tohmi M. 2007 Coupling of brain function and metabolism: endogenous flavoprotein fluorescence imaging of neuroal activities by local changes in energy metabolism. In: Brain Energetics. Integration of Molecular and Cellular Processes. Gibson GE, Dienel GA, (Eds.). Handbook of Neurochemistry and Molecular Biology, 3rd Ed. Lajtha A, series Ed. Springer-Verlag: Berlin, pp. 321–342.

Shuttleworth CW, Brennan AM, Connor JA. 2003. NAD(P)H fluorescence imaging of postsynaptic neuronal activation in murine hippocampal slices. J. Neurosci. 23:3196–3208.

Siesjö BK, Nilsson L. 1971. The influence of arterial hypoxemia upon labile phosphates and upon extracellular and intracellular lactate and pyruvate concentrations in the rat brain. Scand J Clin Lab Invest. 27:83–96.

Siesjö BK, Johannsson H, Ljunggren B, Norberg K. 1974. Brain dysfunction in cerebral hypoxia and ischemia. Res Publ Assoc Res Nerv Ment Dis. 53:75–112.

Simpson IA, Carruthers A, Vannucci SJ. 2007. Supply and demand in cerebral energy metabolism: the role of nutrient transporters. J Cereb Blood Flow Metab. 27:1766–1791.

Sirover MA. 2005. New nuclear functions of the glycolytic protein, glyceraldehyde-3-phosphate dehydrogenase, in mammalian cells. J Cell Biochem. 95:45–52.

Soane L, Kahraman S, Kristian T, Fiskum G. 2007a. Mechanisms of impaired mitochondrial energy metabolism in acute and chronic neurodegenerative disorders. J Neurosci Res. 85:3407–3415.

Soane L, Solenski N, Fiskum G. 2007b. Mitochondrial mechanisms of oxidative stress and apoptosis. In: Brain Energetics. Integration of Molecular and Cellular Processes. Gibson GE, Dienel GA, (Eds.). Handbook of Neurochemistry and Molecular Biology, 3rd Ed. Lajtha A, series Ed. Springer-Verlag: Berlin, pp. 703–734.

Sokoloff L. 1986. Cerebral circulation, energy metabolism, and protein synthesis: general characteristics and principles of measurement. In: Positron Emission Tomography and Autoradiography: Principles and Applications for the Brain. Phelps M, Mazziotta J, Schelbert H, (Eds.). Raven Press: New York, pp. 1–71.

Sokoloff L. 1996a. Cerebral metabolism and visualization of cerebral activity. In: Comprehensive Human Physiology, vol. 1. Gregor R, Windhorst U, (Eds.). Springer-Verlag: Berlin, 579–602.

Sokoloff L. 1996b. Louis Sokoloff. In: The History of Neuroscience in Autobiography, vol. 1. Squire LR, Ed. Soc Neurosci: Washington, D.C., pp. 454–497.

Sokoloff L, Reivich M, Kennedy C, Des Rosiers MH, Patlak CS, Pettigrew KD, Sakurada O, Shinohara M. 1977. The [^{14}C]deoxyglucose method for the measurement of local cerebral glucose utilization: theory, procedure, and normal values in the conscious and anesthetized albino rat. J Neurochem. 28:897–916.

Sonnewald U, Hertz L, Schousboe A. 1998. Mitochondrial heterogeneity in the brain at the cellular level. J Cereb Blood Flow Metab. 18:231–237.

Sonnewald U, Syversen T, Schousboe A, Waagepetersen H, Aschner M. 2007. Actions of toxins on cerebral metabolism at the cellular level. In: Brain Energetics. Integration of Molecular and Cellular Processes. Gibson GE, Dienel GA, (Eds.). Handbook of Neurochemistry and Molecular Biology, 3rd Ed. Lajtha A, series Ed. Springer-Verlag: Berlin, pp. 569–585.

Sonnewald U, Westergaard N, Schousboe A, Svendsen JS, Unsgård G, Petersen SB. 1993. Direct demonstration by [^{13}C]NMR spectroscopy that glutamine from astrocytes is a precursor for GABA synthesis in neurons. Neurochem Int. 22:19–29.

Sorg O, Magistretti PJ. 1991. Characterization of the glycogenolysis elicited by vasoactive intestinal peptide, noradrenaline and adenosine in primary cultures of mouse cerebral cortical astrocytes. Brain Res. 563:227–233.

Spray DC, Iacobas DA. 2007. Organizational principles of the connexin-related brain transcriptome. J Membr Biol. 218:39–47.

Stavinoha WB, Weintraub ST, Modak AT. 1973. The use of microwave heating to inactivate cholinesterase in the rat brain prior to analysis for acetylcholine. J Neurochem. 20:361–371.

Swanson RA, Yu AC, Chan PH, Sharp FR. 1990. Glutamate increases glycogen content and reduces glucose utilization in primary astrocyte culture. J Neurochem. 54:490–496.

Swanson RA, Morton MM, Sagar SM, Sharp FR. 1992. Sensory stimulation induces local cerebral glycogenolysis: demonstration by autoradiography. Neuroscience. 51:451–461.

Takahashi S, Driscoll BF, Law MJ, Sokoloff L. 1995. Role of sodium and potassium ions in regulation of glucose metabolism in cultured astroglia. Proc Natl Acad Sci USA 92:4616–4620.

Takano T, Tian GF, Peng W, Lou N, Libionka W, Han X, Nedergaard M. 2006. Astrocyte-mediated control of cerebral blood flow. Nat Neurosci. 9:260–267.

Takano T, Tian GF, Peng W, Lou N, Lovatt D, Hansen AJ, Kasischke KA, Nedergaard M. 2007. Cortical spreading depression causes and coincides with tissue hypoxia. Nat Neurosci. 10:754–762.

Tsang TM, Griffin JL, Haselden J, Fish C, Holmes E. 2005. Metabolic characterization of distinct neuroanatomical regions in rats by magic angle spinning 1H nuclear magnetic resonance spectroscopy. Magn Reson Med. 53:1018–1024.

Turnley AM, Stapleton D, Mann RJ, Witters LA, Kemp BE, Bartlett PF. 1999. Cellular distribution and developmental expression of AMP-activated protein kinase isoforms in mouse central nervous system. J Neurochem. 72:1707–1716.

Ueda T, Ikemoto A. 2007. Cytoplasmic glycolytic enzymes. Synaptic vesicle-associated glycolytic ATP-generating enzymes: coupling to neurotransmitter accumulation. In: Brain Energetics. Integration of Molecular and Cellular Processes. Gibson GE, Dienel GA, (Eds.). Handbook of Neurochemistry and Molecular Biology, 3rd Ed. Lajtha A, series Ed. Springer-Verlag: Berlin, pp.241–259.

Urbano AM, Gillham H, Groner Y, Brindle KM. 2000. Effects of overexpression of the liver subunit of 6-phosphofructo-1-kinase on the metabolism of a cultured mammalian cell line. Biochem J. 352:921–927.

Vaulont S, Vasseur-Cognet M, Kahn A. 2000. Glucose regulation of gene transcription. J Biol Chem. 275:31555–31558.

Veech RL. 1980. Freeze-blowing of brain and the interpretation of the meaning of certain metabolite levels. In: Cerebral Metabolism and Neural Function. Passonneau JV, Hawkins RA, Lust WD, Welsh FA, (Eds.). Williams and Wilkins: Baltimore, pp. 34–41.

Veech RL, Harris RL, Veloso D, Veech EH. 1973. Freeze-blowing: a new technique for the study of brain in vivo. J Neurochem. 20:183–188.

Verderio C, Bruzzone S, Zocchi E, Fedele E, Schenk U, De Flora A, Matteoli M. 2001. Evidence of a role for cyclic ADP-ribose in calcium signalling and neurotransmitter release in cultured astrocytes. J Neurochem. 78:646–657.

Villringer A, Chance B. 1997. Non-invasive optical spectroscopy and imaging of human brain function. Trends Neurosci. 20:435–442.

Wang SS, Shultz JR, Burish MJ, Harrison KH, Hof PR, Towns LC, Wagers MW, Wyatt KD. Functional trade-offs in white matter axonal scaling. 2008. J Neurosci. 28:4047–4056.

Waniewski RA, Martin DL. 1998. Preferential utilization of acetate by astrocytes is attributable to transport. J Neurosci. 18:5225–5233.

Wieloch T, Mattiasson G, Hansson MJ, Elmér E. 2007. Mitochondrial permeability transition in the CNS: composition, regulation, and pathophysiological relevance. In: Brain Energetics. Integration of Molecular and Cellular Processes. Gibson GE, Dienel GA (Eds.). Handbook of Neurochemistry and Molecular Biology, 3rd Ed. Lajtha A, series Ed. Springer-Verlag: Berlin, pp. 667–702.

Wiesinger H. 1995. Glia-specific enzyme systems. In: Neuroglia. Kettenmann H, Ransom BR (Eds.). Oxford University Press: New York, pp. 488–499.

Winder WW, Taylor EB, Thomson DM. 2006. Role of AMP-activated protein kinase in the molecular adaptation to endurance exercise. Med Sci Sports Exerc. 38:1945–1949.

Witters LA, Kemp BE, Means AR. 2006. Chutes and ladders: the search for protein kinases that act on AMPK. Trends Biochem Sci. 31:13–16.

Xue B, Kahn BB. 2006. AMPK integrates nutrient and hormonal signals to regulate food intake and energy balance through effects in the hypothalamus and peripheral tissues. J Physiol. 574:73–83.

Yu ACH, Drejer J, Hertz L, Schousboe A, 1983. Pyruvate carboxylase activity in primary cultures of astrocytes and neurons. J Neurochem 41:1484–1487.

Yudkoff M, Daikhin Y, Melø TM, Nissim I, Sonnewald U, Nissim I. 2007. The ketogenic diet and brain metabolism of amino acids: relationship to the anticonvulsant effect. Annu Rev Nutr. 27:415–430.

Zeller K, Rahner-Welsch, S, Kuschinsky W. 1997. Distribution of Glut1 glucose transporters in different brain structures compared to glucose utilization and capillary density of adult rat brains. J Cereb Blood Flow Metab. 17:204–209.

Zielke HR, Zielke CL, Baab PJ, Tildon JT. 2007. Effect of fluorocitrate on cerebral oxidation of lactate and glucose in freely moving rats. J Neurochem. 101:9–16.

Zwingmann C, Leibfritz D. 2007. Glial–neuronal shuttle systems. In: Brain Energetics. Integration of Molecular and Cellular Processes. Gibson GE, Dienel GA, (Eds.). Handbook of Neurochemistry and Molecular Biology, 3rd Ed. Lajtha A, series Ed. Springer-Verlag: Berlin, pp. 197–238.

Electrotonic Properties of Axons and Dendrites

John H. Byrne and Gordon M. Shepherd

The functional operations of neurons are the neural basis of behavior. In order to understand those operations, we need to understand how the different parts of the neuron interact. In this chapter we begin by considering how electrical current spreads.

Neurons characteristically have elaborate dendritic trees arising from their cell bodies and single axons with their own terminal branching patterns (see Chapters 1 and 2). With this structural apparatus, neurons carry out five basic functions (Fig. 4.1):

1. Generate intrinsic activity (at any given site in the neuron through voltage-dependent membrane properties and internal second-messenger mechanisms).
2. Receive synaptic inputs (mostly in dendrites, to some extent in cell bodies, and in some cases in axon hillocks, initial axon segments, and axon terminals).
3. Integrate signals by combining synaptic responses with intrinsic membrane activity (in dendrites, cell bodies, axon hillocks, and initial axon segments).
4. Encode output patterns in graded potentials or action potentials (at any given site in the neuron).
5. Distribute synaptic outputs (from axon terminals and, in some cases, from cell bodies and dendrites).

In addition to synaptic inputs and outputs, neurons may receive and send nonsynaptic signals in the form of electric fields, volume conduction through the extracellular environment of neurotransmitters and gases, and release of hormones into the bloodstream.

TOWARD A THEORY OF NEURONAL INFORMATION PROCESSING

A fundamental goal of neuroscience is to develop quantitative descriptions of these functional operations and their coordination within the neuron that enable the neuron to function as an integrated information processing system. This is the necessary basis for testing experiment-driven hypotheses that can lead to realistic empirical computational models of neurons, neural systems, and networks and their roles in information processing and behavior.

Toward these ends, the first task is to understand how activity spreads. To do this for a single process, such as the axon, is difficult enough; for the branching dendrites it becomes extremely challenging; and for the interactions between the two even more so. It is no exaggeration to say that the task of understanding how intrinsic activity, synaptic potentials, and active potentials spread through and are integrated within the complex geometry of dendritic trees to produce the input–output operations of the neuron is one of the main frontiers of molecular and cellular neuroscience.

This chapter begins with the passive properties of the membrane underlying the spread of most types of neuronal activity. Chapter 17 then considers the active membrane properties that contribute to more complex types of information processing, particularly the types that take place in dendrites. Together, the two chapters provide an integrated theoretical framework for understanding the neuron as a complex

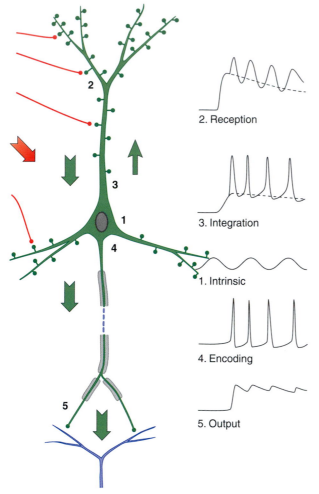

FIGURE 4.1 Nerve cells have four main regions and five main functions. Electrotonic potential spread is fundamental for coordinating the regions and their functions.

Our understanding of electrotonic properties arose in the nineteenth century from a merging of the study of current spread in nerve cells and muscle with the development of cable theory for long-distance transmission of electric current through cables on the ocean floor. The electrotonic properties of neurons are therefore often referred to as *cable properties*. Electrotonic theory was first applied mathematically to the nervous system in the late nineteenth century for spread of electric current through nerve fibers. By the 1930s and 1940s, it was applied to simple invertebrate (crab and squid) axons—the first steps toward the development of the Hodgkin–Huxley equations (Chapters 5 and 7) for the action potential in the axon.

Mathematically it is impractical to apply cable theory to complex branching dendrites, but in the 1960s Wilfrid Rall showed how this problem could be solved by the development of computational compartmental models (Rall, 1964, 1967, 1977; Rall and Shepherd, 1968). These models have provided the basis for a theory of dendritic function (Segev *et al.*, 1995). Combined with mathematical models for the generation of synaptic potentials and action potentials, they provide the basis for a complete theoretical description of neuronal activity.

A variety of software packages now makes it possible for even a beginning student to explore functional properties and construct realistic neuron models. These tools are all freely accessible on the Web (see NEURON; GENESIS; ModelDB; SNNAP). We therefore present modern electrotonic theory within the context of constructing these compartmental models. Exploration of these models will aid the student greatly in understanding the complexities that are present in even the simplest types of passive spread of current in axons and dendrites.

information processing system. Both draw on other chapters for the specific properties—membrane receptors (Chapters 6 and 11), synaptically gated membrane channels (Chapter 16), intrinsic voltage-gated channels (Chapters 5–7), and second-messenger systems (Chapter 12)—that mediate the operations of the neuron.

BASIC TOOLS: CABLE THEORY AND COMPARTMENTAL MODELS

Slow spread of neuronal activity is by ionic or chemical diffusion or active transport. Our main interest in this chapter is in rapid spread by electric current. What are the factors that determine this spread? The most basic are *electrotonic properties*.

SPREAD OF STEADY-STATE SIGNALS

Modern Electrotonic Theory Depends on Simplifying Assumptions

The successful application of cable theory to nerve cells requires that it be based as closely as possible on the structural and functional properties of neuronal processes. The problem confronting the neuroscientist is that processes are complicated. As discussed in Chapters 1 and 2, a segment of axon or dendrite contains a variety of molecular species and organelles, is bounded by a plasma membrane with its own complex structure and irregular outline, and is surrounded by myriad of neighboring processes (see Fig. 4.2A).

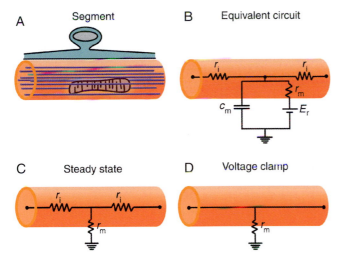

FIGURE 4.2   Steps in construction of a compartmental model of the passive electrical properties of a nerve cell process. (A) Identification of a segment of the process and its organelles. (B) Abstraction of an equivalent electrical circuit based on the membrane capacitance (c_m), membrane resistance (r_m), resting membrane potential (E_r), and internal resistance (r_i). (C) Abstraction of the circuit for steady-state electrotonus, in which c_m and E_r can be ignored. (D) The space clamp used in voltage-clamp analysis reduces the equivalent circuit even further to only the membrane resistance (r_m), usually depicted as membrane conductances (g) for different ions. In a compartmental modeling program, the equivalent circuit parameters are scaled to the size of each segment.

Describing the spread of electric current through such a segment therefore requires some carefully chosen simplifying assumptions, which allow the construction of an *equivalent circuit* of the electrical properties of such a segment. These are summarized in Box 4.1. Understanding them is essential for describing electrotonic spread under the different conditions that the nervous system presents.

Electrotonic Spread Depends on the Characteristic Length (Space Constant)

We begin by using the assumptions in Box 4.1 to represent a segment of a process by electrical resistances: an internal resistance r_i connected to the r_i of the neighboring segments and through the membrane resistance r_m to ground (see Fig. 4.2B). Let us first consider the spread of electrotonic potential under steady-state conditions (Fig. 4.2C). In standard cable theory, this is described by

$$V = \frac{r_m}{r_1} \cdot \frac{d^2 V}{dx^2}. \qquad (4.1)$$

This equation states that if there is a steady-state current input at point $x = 0$, the electrotonic potential (V) spreading along the cable is proportional to the second derivative of the potential (d^2V) with respect

---

BOX 4.1

BASIC ASSUMPTIONS UNDERLYING CABLE THEORY

1. **Segments are cylinders**. A segment is assumed to be a cylinder with constant radius.

   This is the simplest assumption; however, compartmental simulations can readily incorporate different geometrical shapes with differing radii if needed (Fig. 4.2B).

2. **The electrotonic potential is due to a change in the membrane potential**. At any instant of time, the "resting" membrane potential (E_r) at any point on the neuron can be changed by several means: injection of current into the cell, extracellular currents that cross the membrane, and changes in membrane conductance (caused by a driving force different from that responsible for the membrane potential). Electric current then begins to spread between that point and the rest of the neuron, in accord with

$$V = V_m - E_r$$

   where V is the electrotonic potential and V_m is the changed membrane potential.

   Modern neurobiologists recognize that the membrane potential is rarely at rest. In practice, "resting" potential means the membrane potential at any given instant of time other than during an action potential or rapid synaptic potential.

3. **Electrotonic current is ohmic**. Passive electrotonic current flow is usually assumed to be ohmic, i.e., in accord with the simple linear equation

$$E = IR,$$

   where E is the potential, I is the current, and R is the resistance.

   This relation is largely inferred from macroscopic measurements of the conductance of solutions having the composition of the intracellular medium, but is rarely measured directly for a given nerve process. Also largely untested is the likelihood that at the smallest dimensions (0.1 μm diameter or less), the processes and their internal organelles may acquire submicroscopic electrochemical

BOX 4.1   (continued)

properties that deviate significantly from macroscopic fluid conductance values; compartmental models permit the incorporation of estimates of these properties.

4. **In the steady state, membrane capacitance is ignored.** The simplest case of electrotonic spread occurs from the point on the membrane of a steady-state change (e.g., due to injected current, a change in synaptic conductance, or a change in voltage-gated conductance) so that time-varying properties (transient charging or discharging of the membrane) due to the membrane capacitance can be ignored (Fig. 4.2C).

5. **The resting membrane potential can usually be ignored.** In the simplest case, we consider the spread of electrotonic potential (V) relative to a uniform resting potential (E_r) so that the value of the resting potential can be ignored. Where the resting membrane potential may vary spatially, V must be defined for each segment as

$$V = E_m - V_r.$$

6. **Electrotonic current divides between internal and membrane resistances.** In the steady state, at any point on a process, current divides into two local resistance paths: further within the process through an internal (axial) resistance (r_i) or across the membrane through a membrane resistance (r_m) (see Fig. 4.2C).

7. **Axial resistance is inversely proportional to diameter.** Within the volume of the process, current is assumed to be distributed equally (in other words, the resistance across the process, in the Y and Z axes, is essentially zero). Because resistances in parallel sum reciprocally to decrease the overall resistance, axial current (I) is inversely proportional to the cross-sectional area ($I \propto \frac{1}{A} \propto \frac{1}{\pi^2}$); thus, a thicker process has a lower overall axial resistance than a thinner process. Because the axial resistance (r_i) is assumed to be uniform throughout the process, the total cross-sectional axial resistance of a segment is represented by a single resistance,

$$r_i = R_i/A,$$

where r_i is the internal resistance per unit length of cylinder (in ohms per centimeter of axial length), R_i is the specific internal resistance (in ohms centimeter, or ohm cm), and A ($= \pi r^2$) is the cross-sectional area.

The internal structure of a process may contain membranous or filamentous organelles that can raise the effective internal resistance or provide high-conductance submicroscopic pathways that can lower it. In voltage-clamp experiments, the space clamp eliminates current through r_i, so that the only current remaining is through r_m, thereby

permitting isolation and analysis of different ionic membrane conductances, as in the original experiments of Hodgkin and Huxley (Fig. 4.2D; see also Chapter 7).

8. **Membrane resistance is inversely proportional to membrane surface area.** For a unit length of cylinder, the membrane current (i_m) and the membrane resistance (r_m) are assumed to be uniform over the entire surface. Thus, by the same rule of the reciprocal summing of parallel resistances, the membrane resistance is inversely proportional to the membrane area of the segment so that a thicker process has a lower overall membrane resistance. Thus,

$$r_m = R_m/c,$$

where r_m is the membrane resistance for unit length of cylinder (in ohm cm of axial length), R_m is the specific membrane resistance (in ohm cm^2), and c ($= 2\pi r$) is the circumference. For a segment, the entire membrane resistance is regarded as concentrated at one point; i.e., there is no axial current flow within a segment but only between segments (see Fig. 4.2C).

Membrane current passes through ion channels in the membrane. The density and types of these channels vary in different processes and indeed may vary locally in different segments and branches. These differences are incorporated readily into compartmental representations of the processes.

9. **The external medium along the process is assumed to have zero resistivity.** In contrast with the internal axial resistivity (r_i), which is relatively high because of the small dimensions of most nerve processes, the external medium has a relatively low resistivity for current because of its relatively large volume. For this reason, the resistivity of the paths either along a process or to ground is generally regarded as negligible, and the potential outside the membrane is assumed to be everywhere equivalent to ground (see Fig. 4.2C). This greatly simplifies the equations that describe the spread of electrotonic potentials inside and along the membrane.

Compartmental models can simulate any arbitrary distribution of properties, including significant values for extracellular resistance where relevant. Particular cases in which external resistivity may be large, such as the special membrane caps around synapses on the cell body or axon hillock of a neuron, can be addressed by suitable representation in the simulations. However, for most simulations, the assumption of negligible external resistance is a useful simplifying first approximation.

BOX 4.1 *(continued)*

10. **Driving forces on membrane conductances are assumed to be constant**. It is usually assumed that ion concentrations across the membrane are constant during activity.

   Changes in ion concentrations with activity may occur, particularly in constricted extracellular or intracellular compartments; these changes may cause deviations from the assumptions of constant driving forces for the membrane currents, as well as the assumption of uniform E_r. For example, accumulations of extracellular K^+ may change local E_r, and intracellular accumulations of ions within the tiny volumes of spine heads may change the driving force on synaptic currents. These special properties are easily included in most compartmental models.

11. **Cables have different boundary conditions**. In classical electrotonic theory, a cable such as one used for long-distance telecommunication is very long and can be considered of infinite length (one customarily assumes a semi-infinite cable with $V = 0$ at $x = 0$ and only positive values of length x). This assumption carries over to the application of cable theory to long axons, but most dendrites are relatively short. This imposes boundary conditions on the solutions of the cable equations, which have very important effects on electrotonic spread.

   In highly branched dendritic trees, boundary conditions are difficult to deal with analytically but are readily represented in compartmental models.

   Gordon M. Shepherd

---

to distance and the ratio of the membrane resistance (r_m) to the internal resistance (r_i) over that distance. The steady-state solution of this equation for a cable of infinite extension for positive values of x gives

$$V = V_0 e^{-x/\lambda}, \qquad (4.2)$$

where lambda is defined as the square root of r_m/r_i (in centimeters) and V_0 is the value of V at $x = 0$.

Inspection of this equation shows that when $x = \lambda$, the ratio of V to V_0 is $e^{-1} = 1/e = 0.37$. Thus, lambda is a critical parameter defining the length over which the electrotonic potential spreading along an infinite cable decays (is attenuated) to a value of 0.37 of the value at the site of the input. It is referred to as the *characteristic length* (also known as the *space constant* or *length constant*) of the cable. The higher the value of the specific membrane resistance (R_m), the higher the value of r_m for that segment, the larger the value for λ, and the greater the spread of electrotonic potential through that segment (Fig. 4.3). Specific membrane resistance (R_m) is thus an important variable in determining the spread of activity in a neuron.

Most of the passive electrotonic current may be carried by K^+ "leak" channels, which are open at "rest" and largely responsible for holding the cell at its resting potential. However, as mentioned earlier, many cells or regions within a cell are seldom at "rest" but are constantly active, in which case electrotonic current is carried by a variety of open channels. Thus, the effective R_m can vary from values of less than 1000 Ω cm^2 to more than 100,000 Ω cm^2 in different neurons and in different parts of a neuron. Note

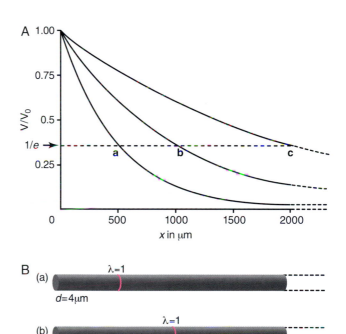

FIGURE 4.3 Dependence of the space constant governing the spread of electrotonic potential through a nerve cell process on the square root of the ratio between the specific membrane resistance (R_m) and the specific internal resistance (R_i). (A) Potential profiles for processes with three different values of λ. (B) Dotted lines represent the location of λ on each of the three processes.

that lambda varies with the square root of R_m, so a 100-fold difference in R_m translates into only a 10-fold difference in lambda.

Conversely, the higher the value of the specific internal resistance (R_i), the higher the value of r_i for that segment, the smaller the value of λ, and the less the spread of electrotonic potential through that segment (see Fig. 4.3). Traditionally, the value of R_i has been believed to be in the range of approximately 50–100 Ω cm based on muscle cells and the squid axon. In mammalian neurons, estimates now tend toward a value of 200 Ω cm. This limited range may suggest that R_i is less important than R_m in controlling passive current spread in a neuron. The square-root relation further reduces the sensitivity of λ to R_i. However, as noted in assumption 7 in Box 4.1, the membranous and filamentous organelles in the cytoplasm may alter the effective R_i. The presence of these organelles in very thin processes, such as distal dendritic branches, spine stems, and axon preterminals, may thus have potentially significant effects on the spread of electrotonic current through them. Furthermore, the relative significance of R_i and R_m depends greatly on the length of a given process, as will be seen shortly.

Electrotonic Spread Depends on the Diameter of a Process

The space constant (λ) depends not only on the internal and membrane resistance but also on the diameter of a process (Fig. 4.4). Thus, from the relations between r_m and R_m, and r_i and R_i, discussed in the preceding section,

$$\lambda = \sqrt{\frac{r_m}{r_i}} = \sqrt{\frac{R_m}{R_i} \cdot \frac{d}{4}}. \tag{4.3}$$

Neuronal processes vary widely in diameter. In the mammalian nervous system, the thinnest processes are the distal branches of dendrites, the necks of some dendritic spines, and the cilia of some sensory cells; these processes may have diameters of only 0.1 μm or less (the thinnest processes in the nervous system are approximately 0.02 μm). In contrast, the thickest processes in the mammal are the largest myelinated axons and the largest dendritic trunks, which may have diameters as large as 20 to 25 μm. This means that the range of diameters is approximately three orders of magnitude (1000-fold). Note, again, that the relation to λ is the square root; thus, over a 10-fold difference in diameter, the difference in λ is only about 3-fold (Fig. 4.4).

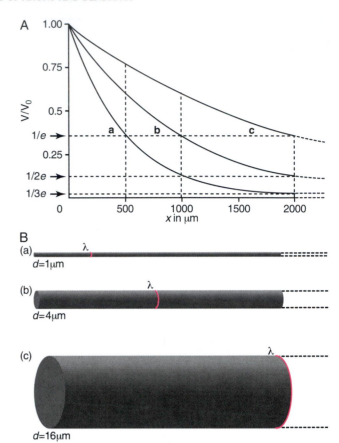

FIGURE 4.4 Dependence of the space constant on the square root of the diameter of the process. (A) Potential profiles for processes with three different diameters but fixed values of R_i and R_m. (B) The three axon profiles in A. Note that to double λ, the diameter must be quadrupled.

Electrotonic Properties Must Be Assessed in Relation to the Lengths of Neuronal Processes

Application of classical cable theory to neuronal processes assumes that the processes are infinitely long (assumption 11 in Box 4.1). However, because neuronal processes have finite lengths, the length of a given process must be compared with λ to assess the extent to which λ accurately describes the actual electrotonic spread in that process. One of the largest processes in any nervous system, the squid giant axon, has a diameter of approximately 1 mm. R_m for this axon has been estimated as 600 Ω cm^2 (a very low value compared to most values of R_m in mammals), and R_i as approximately 80 Ω cm, the value of Ringer solution (note that the very large diameter is counter-balanced by the very low R_m). Putting these values into Eq. (4.3) gives a λ of approximately 4.3 mm.

The real length of the giant axon is several centimeters; to relate real length to characteristic length, we define *electrotonic length* (L) as

$$L = x/\lambda \qquad (4.4)$$

Thus, if $x = 30$ mm, then $L = 30$ mm$/4.3$ mm $= 7$. The electrotonic potential decays to a small percentage of the original value by only three characteristic lengths (see Fig. 4.4), so for this case the assumption of an infinite length is justified. In contrast to axons, dendritic branches have lengths that are usually much shorter than three characteristic lengths. In dendrites, therefore, the branching patterns come to dominate the extent of potential spread. We discuss the methods for dealing with these branching patterns later in this chapter.

A reason often given for why the nervous system needs action potentials is that they overcome the severe attenuation of passively spreading potentials that occurs over the considerable lengths required for transmission of signals by axons. This applies to the long axons of projection neurons, but not necessarily to shorter axons and their collaterals. Recent studies have in fact revealed that excitatory synaptic potentials in the soma may spread through the axon to reach terminal boutons onto nearby cells; the variable amount of synaptic depolarization thus acts as an analog signal to modify the digital signaling carried by the axonal action potentials. This mechanism has been shown in the mossy fiber terminals of dentate granule cells onto CA3 pyramidal cells in the hippocampus (Alle and Geiger, 2006), and in the axon terminals of layer 5 pyramidal neurons onto neighboring cells in the cerebral cortex (Shu *et al.*, 2006). The combined analog and digital signaling is computationally more powerful than digital signaling alone.

A reverse situation is seen in the retina, where a particular type of horizontal cell has elaborate branches of both its dendrites and its terminal axon, interconnected by a long thin axon. Physiological studies have shown that each branching system processes different properties of the visual signal, but they do not interact, because the axon has passive properties that give it a short length constant. This enables one cell to provide two distinct input–output processing systems (Nelson *et al.*, 1975). Never underestimate the ingenuity of the nervous system!

Summary

Passive spread of electrical potential along the cell membrane underlies all types of electrical signaling in the neuron. It is thus the foundation for understanding the interactive substrate whereby the neuron can generate, receive, integrate, encode, and send signals.

Electrotonic spread shares properties with electrical transmission through electrical cables; the mathematical study of cable transmission has put these properties on a quantitative basis. The theoretical basis for extension of cable theory to complex dendritic trees has been developed in parallel with compartmental modeling methods for simulating dendritic signal processing.

Cable theory depends on a number of reasonable simplifying assumptions about the geometry of neuronal processes and current flow within them. Steady-state electrotonus in dendrites depends on passive resistance of the membrane and of the internal cytoplasm and on the diameter and length of a nerve process.

SPREAD OF TRANSIENT SIGNALS

Electrotonic Spread of Transient Signals Depends on Membrane Capacitance

Until now, we have considered only the passive spread of steady-state inputs. However, the essence of many neural signals is that they change rapidly. In mammals, fast action potentials characteristically last from 1 to 5 ms, and fast synaptic potentials last from 5 to 30 ms. How do the electrotonic properties affect spread of these rapid signals?

Rapid signal spread depends not only on all the factors discussed thus far but also on the membrane capacitance (c_m), which is due to the lipid moiety of the plasma membrane. Classically, the value of the specific membrane capacitance (C_m) has been considered to be 1 μF cm^{-2}. However, a value of 0.6–0.75 μF cm^{-2} is now preferred for the lipid moiety itself, with the remainder being due to gating charges on membrane proteins (Jack *et al.*, 1975).

The simplest case demonstrating the effect of membrane capacitance on transient signals is that of a single segment or a cell body with no processes. This is a very unrealistic assumption, equivalent to the single node of neural network models, but a simple starting point. In the equivalent electrical circuit for a neural process, the membrane capacitance is placed in parallel with ohmic components of the membrane conductance, and the driving potentials for ion flows through those conductances (see Fig. 4.2B). Again neglecting the resting membrane potential, we take as an example the injection of a current step into a soma; in this case, the time course of the current spread to ground is described by the sum of the capacitative and resistive current (plus the input current, I_{pulse}):

$$C\frac{dV_m}{dt} + \frac{V_m}{R} = I_{pulse}. \tag{4.5}$$

Rearranging,

$$RC\frac{dV_m}{dt} + V_m = I_{pulse} \cdot R \tag{4.6}$$

where $RC = \tau$ (τ is the time constant of the membrane).

The solution of this equation for the response to a step change in current (I) is

$$V_m(T) = I_{pulse}R(1 - e^{-T}) \tag{4.7}$$

where $T = t/\tau$.

When the pulse is terminated, the decay of the initial potential (V_0) to rest is given by

$$V_m(T) = V_\infty e^{-T}. \tag{4.8}$$

These "on" and "off" transients are shown in Fig. 4.5. The significance of tau is shown in the diagram; it is the time required for the voltage change across the membrane to reach $1/e = 0.37$ of its final value (i.e., the change from V_∞ back to the resting potential). This time constant of the membrane defines the transient voltage

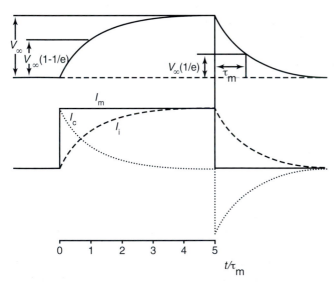

FIGURE 4.5 The equivalent circuit of a single isolated compartment responds to an injected current step by charging and discharging along a time course determined by the time constant, τ. In actuality, because nerve cell segments are parts of longer processes (axonal or dendritic) or larger branching trees, the actual time courses of charging or discharging are modified. V_∞, steady-state voltage in response to the current pulse; I_m (I_{Pulse}), injected current applied to membrane; I_c, current through the capacitance; I_i, (also referred to as I_L) current through the ionic leak conductance; τ_m, membrane time constant. From Jack et al. (1975).

response of a segment of membrane to a current step in terms of the electrotonic properties of the segment. It is analogous to the way that the length constant defines the spread of voltage change over distance.

A Two-Compartment Model Defines the Basic Properties of Signal Spread

These spatial and temporal cable properties can be combined in a two-compartment model (Shepherd, 1994) that can be applied to the generation and spread of any arbitrary transient signal (Fig. 4.6).

In the simplest case, current is injected into one of the compartments, as in an electrophysiological experiment. Positive charge injected into compartment A attempts to flow outward across the membrane, partially opposing the negative charge on the inside of the lipid membrane (the charge responsible for the negative resting potential), thereby depolarizing the membrane capacitance (C_m) at that site. At the same time, the charge begins to flow as current across the membrane through the resistance of the ionic membrane channels (R_m) that are open at that site. The proportion of charge divided between C_m and R_m determines the rate of charge of the membrane, i.e., the membrane time constant, τ. However, charge also starts to flow through the internal resistance (R_i) into compartment B, where the current again divides between capacitance and resistance. The charging (and discharging) transient in compartment A departs from the time constant of a single isolated compartment, being faster because of the impedance load (e.g., current sink) of the rest of the cable (represented by compartment B). Thus, the time constant of the system no longer describes the charging transient in the system because of the conductance load of one compartment on another. The system is entirely passive; the response to a second current pulse sums linearly with that of the first.

This case is a useful starting point because an experimenter often injects electrical currents in a cell to analyze nerve function. However, a neuron normally generates current spread by means of localized conductance changes across the membrane. In Fig. 4.6, consider such a change in the ionic conductance for Na^+, as in the initiation of an action potential or an excitatory postsynaptic potential, producing an inward positive current in compartment A. The charge transferred to the interior surface of the membrane attempts to follow the same paths followed by the injected current just described by opposing the negativity inside the membrane capacitance, crossing the membrane through the open membrane channels to ground, and spreading through the internal resistance to the next compartment, where the charge flows are similar.

FIGURE 4.6 The equivalent circuit of two neighboring compartments or segments (A and B) of an axon or dendrite shows the pathways for current spread in response to an input (injected current or increase in membrane conductance) at segment A. See text for full explanation.

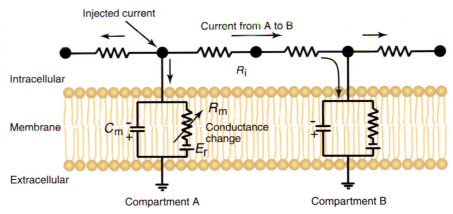

Thus, the two cases start with different means of transferring positive charge within the cell, but from that point the current paths and the associated spread of the electrotonic potential are similar. The electrotonic current that spreads between the two segments is referred to as the *local current*. The charging transient in compartment A is faster than the time constant of the resting membrane; this difference is due both to the conductance load of compartment B (as in the injected current case) and to the fact that the imposed conductance increase in compartment A reduces the time constant of compartment A (by reducing effective R_m). This illustrates a critical point first emphasized by Wilfrid Rall (1964): changes in membrane conductance alter the system so that it is no longer a linear system, even though it is a passive system. Thus, passive electrotonic spread is not so simple as most people think! Nonlinear summation of synaptic responses is further discussed later in this chapter.

Summary

In addition to the properties underlying steady-state electrotonus, passive spread of *transient* potentials depends on the membrane capacitance. Initiation of electrotonic spread by intracellular injection of a transient electrical current pulse produces an electrotonic potential that spreads by passive local currents from point to point. It is more attenuated in amplitude than the steady-state case as it spreads along an axon or dendrite due to the low-pass filtering action of the membrane capacitance. Simultaneous current pulses at that site or other sites produce potentials that add linearly because the passive properties are invariant. However, transient conductance changes, as in synaptic responses, generate electrotonic potentials that do not sum linearly because of the nonlinear interactions of the conductances.

ELECTROTONIC PROPERTIES UNDERLYING PROPAGATION IN AXONS

Impulses Propagate in Unmyelinated Axons by Means of Local Electrotonic Currents

We next apply our knowledge of electrotonic current properties to propagation of an action potential in an unmyelinated axon, i.e., one that is not surrounded by myelin or other membranes that restrict the spread of extracellular current. Details on the ionic mechanisms of the nerve impulse can be found in Chapters 5 and 7. The local current spreading through the internal resistance to the neighboring compartment enables the action potential to propagate along the membrane of the axon. The rate of propagation is determined by both the passive cable properties and the kinetics of the action potential mechanism.

Each of the cable properties is relevant in specific ways. For brief signals such as the action potential, C_m is critical in controlling the rate of change of the membrane potential. R_m is a parameter that can vary widely. Thus, each of these parameters must be assessed in order to understand the exquisite effects of passive variables on the rates of impulse propagation in axons.

A high value of R_m, for example, forces current farther along the membrane, increasing the characteristic length and consequently the spread of electrotonic potential, as we have seen; however, at the same time, it increases the membrane time constant, thus slowing the response of a neighboring compartment to a rapid change. Increasing the diameter of the axon lowers the effective internal resistance of a compartment, thereby also increasing the characteristic length, but without a concomitant effect on the time constant. The conduction rate of any given axon depends on the particular combination of these properties (Rushton, 1951; Ritchie, 1995). For example, in the squid giant axon, the very large diameter (as large as 1 mm) promotes rapid impulse propagation; the very low value of R_m (600 Ω cm^2) lowers the time constant (promoting rapid current spread) but also decreases the length constant (limiting the spatial extent of current spread).

The effects of these passive properties on impulse velocity also depend on other factors. For example, on the basis of the cable equations, we can show that the conduction velocity should be related to the square root of the diameter (Rushton, 1951). However, the density of Na$^+$ channels in fibers of different diameters is not constant; thus, the binding of saxitoxin molecules, for example, to Na$^+$ channels varies greatly with diameter, from almost 300 μm^{-2} in the squid axon to only 35 μm^{-2} in the garfish olfactory nerve (Ritchie, 1995). Thus, both active and passive properties must be assessed in order to understand a particular functional property.

Myelinated Axons Have Membrane Wrappings and Booster Sites for Faster Conduction

The evolution of larger brains to control larger bodies and more complex behavior required communication over longer distances within the brain and body. This requirement placed a premium on the ability of axons to conduct impulses as rapidly as possible. As noted in the preceding section, a direct way of increasing the rate of conduction is by increasing the diameter, but larger diameters mean fewer axons within a given space, and complex behavior must be mediated by many axons. Another way of increasing the rate of conduction is to make the kinetics of the impulse mechanism faster; i.e., make the rate of increase in Na$^+$ conductance with increasing membrane depolarization faster. The Hodgkin–Huxley equations (Chapter 7) for the action potential in mammalian nerves in fact have this faster rate.

As we have seen, the rapid spread of local currents is promoted by an increase in R_m but is opposed by an associated increase in the time constant. What is needed is an increase in R_m with a concomitant decrease in C_m. This is brought about by putting more resistances in series with the membrane resistance (because resistances in series add) while putting more capacitances in series with the membrane capacitance (capacitances in series add as the reciprocals, much like resistances in parallel, as noted earlier). The way the nervous system does

this is through a special satellite cell called a *Schwann cell*, a type of glial cell. As described in Chapters 1 and 2, Schwann cells wrap many layers of their plasma membranes around an axon. The membranes contain special constituents and together are called *myelin*. Myelinated nerves contain the fastest conducting axons in the nervous system. A general empirical finding known as the *Hursh factor* (Hursh, 1939) states that the rate of propagation of an impulse along a myelinated axon in meters per second is six times the diameter of the axon in micrometers. Thus, the largest axons in the mammalian nervous system are approximately 20 μm in diameter, and their conduction rate is approximately 120 m s^{-1}, whereas the thin myelinated axons of about 1 μm in diameter have conduction rates of approximately 5 to 10 m s^{-1}.

As discussed in Chapter 1, myelinated axons are not myelinated along their entire length; at regular intervals (approximately 1 mm in peripheral nerves), the myelin covering is interrupted by a node of Ranvier. The node has a complex structure. The density of voltage-sensitive Na$^+$ channels at the node is high (10,000 μm^{-2}), whereas it is very low (20 μm^{-2}) in the internodal membrane. This difference in density means that the impulse is actively generated only at the node; the impulse jumps, so to speak, from node to node, and the process is therefore called *saltatory conduction*. A myelinated axon therefore resembles a passive cable with active booster stations.

In rapidly conducting axons the impulse may extend over considerable lengths; e.g., in a 20-μm-diameter axon conducting at 120 m s^{-1}, at any instant of time an impulse of 1 ms duration extends over a 120 mm length of axon, which includes more than 100 nodes of Ranvier. It is therefore more appropriate to conceive that the impulse is generated simultaneously by many nodes, with their summed local currents spreading to the next adjacent nodes to activate them.

The specific membrane resistance (R_m) at the node is estimated to be only 50 Ω cm^2, due to a large number of open ionic channels at rest. This value of R_m reduces the time constant of the nodal membrane to approximately 50 microseconds, which enables the nodal membrane to charge and discharge quickly, aiding rapid impulse generation greatly. For axons of equal cross-sectional area, myelination is estimated to increase the impulse conduction rate 100-fold.

In all axons, a critical relation exists between the amount of local current spreading down an adjacent axon and the threshold for opening Na$^+$ channels in the membrane of the adjacent axon so that propagation of the impulse can continue. This introduces the notion of a *safety factor*, i.e., the amount by which the electrotonic potential exceeds the threshold for activating the impulse. The safety factor must protect against a wide range of operating conditions, including adaptation (during high-frequency firing), fatigue, injury, infection, degeneration, and aging.

Normally, an excess of local current ensures an adequate margin of safety against these factors. In the squid axon, the safety factor ranges from 4 to 5. In myelinated axons, an exquisite matching between internodal electrotonic properties and nodal active properties ensures that the electrotonic potential reaching a node has an adequate amplitude and the node has sufficient Na^+ channels to generate an action potential that will spread to the next node. The safety factors for myelinated axons range from 5 to 10. Thus, the interaction of passive and active properties underlies the safety factors for impulse propagation in axons. Similar considerations apply to the orthodromic spread of signals in dendritic branches and the back-propagation of action potentials from the axon hillock into the soma and dendrites.

Theoretically, the conduction velocity, space constant, and impulse wavelength of myelinated fibers vary with fiber diameter (Rushton, 1951; Ritchie, 1995), as indeed is indicated in the aforementioned Hursh factor. This difference between myelinated and unmyelinated fibers in their dependence on diameter is thus related to the scaling of the internodal length. At approximately 1 μm in diameter, the Hursh factor breaks down; at less than 1 μm in diameter, there is an advantage, all other factors being equal, for an axon to be unmyelinated. However, myelinated axons are found down to a diameter of only 0.2 μm, which has been correlated with shorter internodal distances (Waxman and Bennett, 1972). Thus, conduction velocity in myelinated nerve depends on a complex interplay between passive and active properties.

Summary: Passive Spread and Active Propagation

Impulses propagate continuously through unmyelinated fibers because the local currents spread directly to neighboring sites on the membrane. The rate of propagation is directly determined by the electrotonic properties of the fiber. In myelinated axons, the impulse propagates discontinuously from node to node. The electrotonic properties of both the nodal and internodal regions determine not only the rate of impulse propagation but also the safety factor for impulse transmission.

Here and in Chapter 17, it will contribute to clarity to distinguish between passive spread and active propagation (see Box 4.2).

ELECTROTONIC SPREAD IN DENDRITES

Dendrites are the main neuronal compartment for the reception of synaptic inputs. The spread of synaptic responses through the dendritic tree depends critically on the electrotonic properties of the dendrites. Because dendrites are branching structures, understanding the rules governing dendritic electrotonus and the resulting integration of synaptic responses in dendrites is much more difficult than understanding the rules of simple spread in a single axon.

BOX 4.2

ELECTROTONIC POTENTIALS SPREAD, ACTION POTENTIALS PROPAGATE

It is important to distinguish between passive and active spread of potentials, which is helped by using different terms. Based on common dictionary definitions, *spreading* has a more general meaning of distributing something (in this case a current or potential) over an area or along an object. It applies specifically to passive electronic "spread" and to the local circuit currents that spread before an action potential, and can also be used in a general way to refer to spread of the action potential itself. *Conduction* also has a general meaning in the electrical sense. In contrast, *propagating* refers specifically to the action potential, because it carries the dictionary meaning of spreading by sequential active processes of reproducing oneself, which is what an action potential does along an axon or dendrite.

These distinctions of meaning as applied to nervous conduction date from the work of Wilfrid Rall in the 1960s and continue to be useful.

Dendritic Electrotonic Spread Depends on Boundary Conditions of Dendritic Termination and Branching

As noted earlier, compared with axons, dendrites are relatively short, and their length becomes an important factor in assessing their electrotonic properties. Consider, in the mammalian nervous system, a moderately thin dendrite of 1 μm (three orders of magnitude smaller than the squid axon) that has a typical R_m of $60,000\,\Omega\,cm^2$ (two orders of magnitude larger than that of the squid axon) and an R_i of $240\,\Omega\,cm$ (three times the squid value). Inserting these values into the equation for characteristic length [Eq. (4.3)] gives a λ of approximately 790 μm. This illustrates that lambda tends to be relatively long in comparison with the actual lengths of the dendrites; in other words, because of the relatively high membrane resistance, the electrotonic spread of potentials is relatively effective within a dendritic branching tree.

This essential property underlies the integration of signals in dendrites. The effective spread immediately leads to a second property. The assumption of infinite length no longer holds; dendritic branches are bounded by their terminations, on the one hand, and the nature of their branching, on the other. These are termed *boundary conditions*. The spread of electrotonic potentials is therefore exquisitely sensitive to the boundary conditions of the dendrites.

This problem is approached most easily by considering two extreme types of termination of a dendritic branch. First, consider that at $x = \lambda$, which we will define as point *a*, the branch ends in a sealed end with infinite resistance. In this case, the axial component of the current can spread no farther and must therefore seek the only path to ground, which is across the membrane of the cylinder. This current is added to the current already crossing the membrane; in the equation for Ohm's law ($E = IR$), I is increased, giving a larger E. The membrane will thus be more depolarized up to the terminal point *a*; in fact, near point *a*, axial current is negligible and almost all the current is across the membrane, which amounts to a virtual space clamp (Fig. 4.7). If at point *a* the infinite resistance is replaced by the more realistic assumption of an end that is sealed with surface membrane, only a small amount of current crosses this membrane and attenuation of electrotonic potential is only slightly greater. Infinite resistance is therefore a useful approximation for assessing the effects of a sealed end on electrotonic spread in a terminal dendritic branch.

At the other extreme, consider that at point *a* a small dendritic branch opens out into a very large conductance. Examples are, in the extreme, a hole in the membrane; less extreme are a very small dendritic branch on a large soma and a small twig or spine on a large dendritic branch. Recall that large processes sum their

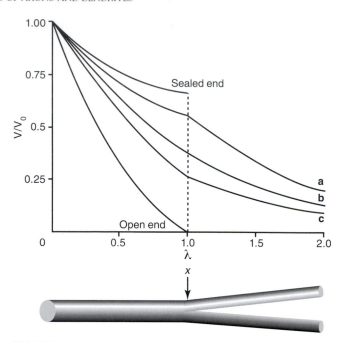

FIGURE 4.7 The spread of electrotonic potential through a short nerve cell process such as a dendritic branch is governed by the space constant and by the size of the branches; the latter imposes a boundary condition at the branch point. Curves a–c represent a range of realistic assumptions about the sizes of the branches relative to the size of the stem, together with the limiting conditions of an open circuit (corresponding to an infinite conductance load) and a closed circuit (corresponding to a sealed end).

resistances in parallel, which gives low current density and small voltage changes. Therefore, a current spreading through the high resistance of a small branch into a large branch encounters a very low resistance. For steady-state current spread, this situation is referred to as a *large conductance load*; for a transient current, we refer to it as a *low impedance* (which includes the effect of the membrane capacitance).

This introduces the key principle of *impedance matching* between interacting compartments, an important principle generally in biological systems. In our example, an impedance mismatch exists between the high-impedance thin branch and the lower impedance thick branch. This mismatch reduces any voltage change due to the current and, in the extreme, effectively clamps the membrane to the resting potential (E_r) at that point. The electrotonic potential is thus attenuated through the branch much more rapidly than would be predicted by the characteristic length (see Fig. 4.7). This does not invalidate λ as a measure of electrotonic properties; rather, it means that, as with the time constant, each cable property must be assessed within the context of the size and branching of the dendrites.

All the different types of branching found in neuronal dendrites lie between these two extremes, with a corresponding range of boundary conditions at

$x = a$. Consider a segment of dendrite that divides into two branches at $x = a$. We can appreciate intuitively that the amount of spread of electrotonic potential into the two branches will be governed by the factors just considered. One possibility is that the two branches have very small diameters, so their input impedance is higher than that of the segment; in this case, the situation will tend toward the sealed end case (Fig. 4.7, top trace). In contrast, the segment may give rise to two very fat branches, so the situation will tend toward the large conductance load case (Fig. 4.7, bottom trace).

For many cases of dendritic branching, the input impedance of the branches is between the two extremes (see Fig. 4.7, traces a–c), providing for a reasonable degree of impedance matching between the stem branch and its two daughter branches. This situation thus approximates the infinite cylinder case, in which by definition the input impedance at one site matches that at its neighboring site along the cylinder. The general rules for impedance matching at branch points were worked out by Rall (1959, 1964, 1967), who showed that the input conductance of a dendritic segment varies with the diameter raised to the 3/2 power. There is electrotonic continuity at a branch point equivalent to the infinitely extended cylinder if the diameter of the segment raised to the 3/2 power equals the sum of the diameters raised to the 3/2 power of all the daughter branches. An idealized branching pattern that satisfies this rule is shown in Fig. 4.8. When the branching tree reduces to a single chain of compartments, as in this case, it is called an "equivalent cylinder." When the branching pattern departs from the d 3/2 rule, the compartment chain is referred to as an "equivalent dendrite" (Rall and Shepherd, 1968).

Dendritic Synaptic Potentials Are Delayed and Attenuated by Electrotonic Spread

We are now in a position to assess the effects of cable properties on the time course of the spread of synaptic potentials through dendritic branches and trees. Consider in Fig. 4.8 the case of recording from a soma while delivering a brief excitatory synaptic conductance change to different locations in the dendritic tree. The response to the nearest site is a rapidly rising synaptic potential that peaks near the end of the conductance change and then decays rapidly toward baseline. When the input is delivered to the middle of the chain of compartments, the response in the soma begins only after a delay, rises more slowly, reaches a much lower peak (which is reached after the end of the conductance change in the soma), and decays slowly toward baseline. For input to the terminal compartment, the voltage delay at the soma is so long that the response has

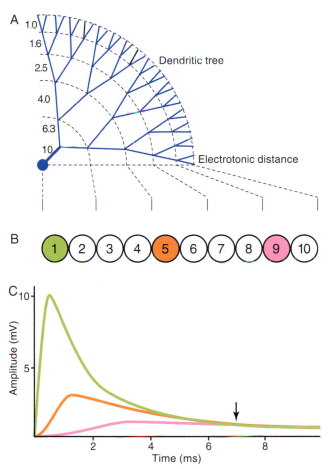

FIGURE 4.8 The spread of electrotonic potentials is accompanied by a delay and an attenuation of amplitude. (A) Dendritic diameters (left) satisfy the d 3/2 rule so that the tree can be portrayed by an equivalent cylinder. An excitatory postsynaptic potential (EPSP) is generated in compartment 1, 5, or 9 (B) while recordings are made from compartment 1. (C) Short latency, large amplitude, and rapid transient response in compartment 1 at the site of input, as well as the later, smaller, and slower responses recorded in compartment 1 for the same input to compartments 5 and 9. Despite the initial differences in time course, the responses converge at the arrow to decay together. Based on Rall (1967).

scarcely started by the end of the conductance change in the distal dendrite; the response rises slowly to a delayed (several milliseconds) and prolonged plateau that subsides very slowly (see Fig. 4.8).

Although the synaptic potentials thus decrease in amplitude as they spread, the rate of electrotonic spread can be calculated in terms of the half-amplitude at any point. If distance is expressed in units of λ, and time in units of τ, then for spread through a semi-infinite cable, we have the simple equation (Jack et al., 1975)

$$\text{Velocity} = 2\frac{\lambda}{\tau}. \qquad (4.9)$$

Thus, if we ignore boundary effects, for a 10 µm process in which $\lambda = 1500$ µm and $\tau = 10$ ms, the velocity of spread would be 0.3 m s^{-1}, or 300 µm s^{-1}. It can be seen that electrotonic spread can be relatively fast over short distances within a dendritic tree but is very slow in comparison with impulse transmission for an axon of this diameter (60 m s^{-1}). Thus, both the severe decrement and the slow velocity make passive spread by itself ineffective for transmission over long distances.

These general rules of delay and attenuation govern the passive spread of all transient potentials in dendritic branches and trees. As a rule of thumb, spread within one space constant (see the decrement between compartments 1 and 5 in Fig. 4.8) mediates relatively effective linkage for rapid signal integration, whereas spread over one or two space constants (see the decrement between compartments 1 and 9 in Fig. 4.8) is limited to slower background modulation. In real dendrites, these limitations are often overcome through boosting the signals at intermediate sites by voltage-gated properties (see Chapter 17).

The spread of electrotonic potential from a point of input involves the *equalization of charge* on the membrane throughout the system. After cessation of the input, a time is reached when charge has become equalized and the entire system is equipotential; from this time on, the remaining electrotonic potential decays equally at every point in the system. This time is indicated by the vertical arrow in Fig. 4.8C. Before this time, the decaying transients are governed by equalizing time constants, indicating electrotonic spread, which can be identified by "peeling" on semilogarithmic plots of the potentials (Rall, 1977). After this time, the decay of electrotonic potential is governed solely by the membrane time constant. In experimental recordings of synaptic potentials, the overall electrotonic length of the dendritic system, considered as an "equivalent cylinder" or "equivalent dendrite" (see earlier discussion), can be estimated from measurements of the membrane time constant and the equalizing time constants. The electrotonic lengths of the dendritic trees of many neuron types lie between 0.3 and 1.5.

What is the spread of the postsynaptic potential throughout the system when a synaptic input is delivered to only a single terminal dendritic branch (Fig. 4.9) (Rall and Rinzel, 1973, Rinzel and Rall, 1974)? Let us begin by considering a steady-state potential?

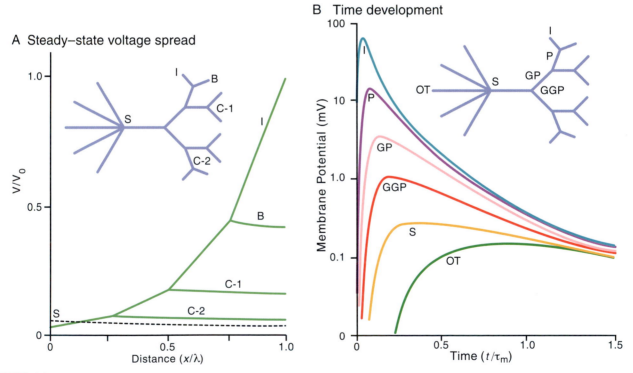

FIGURE 4.9 Electrotonic spread from a single small dendritic branch. (A) For steady-state input (*I*) delivered to the dendritic branch, the electrotonic potential (*V*), relative to the initial potential (*V*₀) at the site of input, spreads from the distal branch through the dendritic tree, with a large decrement into the parent branch (due to the large conductance load) but a small decrement into neighboring branches B, C–1, and C–2 (due to the small conductance loads). The resulting potential in the soma (*S*) is much reduced, as is the response to the same input delivered directly to the soma (because of the low input resistance at the soma and the large conductance load of the dendritic tree). The dashed line indicates the response when the same amount of current is injected into the soma. (B) For transient input (*I*) to a distal branch, transient electrotonic potentials decrease sharply in amplitude and are delayed and slower as they spread toward the soma through the parent (P), grandparent (GP), and great-grandparent (GGP) branches, eventually reaching the soma (*S*) and output trunk (OT). Modified from Segev (1995) based on Rall and Rinzel (1973) and Rinzel and Rall (1974).

Two main factors are involved. First, in the terminal branch, both the effective membrane resistance and the internal resistance are very large; hence, the branch has a very high input resistance, which produces a very large voltage change for any given synaptic conductance change. Balanced against this high input resistance is a second factor: the small branch has a very large conductance load on it because of the rest of the dendritic tree. As a result, there is a steep decrement in the electrotonic potential spreading from the branch through the tree to the cell body (see Fig. 4.9A). For comparison, a direct input to the soma produces only a small potential change there because of the relatively very low input resistance at that site.

For a transient synaptic input, a third factor—membrane capacitance—must be taken into account. The small surface area of a terminal branch has little capacitance, so the amplitude of a transient response differs little from a steady-state response in the branch. However, in spreading out from a small process (such as a distal dendritic twig or spine), the transient synaptic potential is attenuated by the impedance mismatch between the process and the rest of the dendritic tree. Spread of the transient through the dendritic tree is attenuated further by the need to charge the capacitance of the dendritic membrane and is slowed by the time taken for the charging.

The amount of slowing is so precise that the relative distance of a synapse in the dendritic tree from the soma can be calculated from experimental measurements in the soma of the time to peak of the recorded synaptic potential (Rall, 1977; Johnston and Wu, 1995). For these reasons, the peak of a synaptic potential transient spreading from distal dendrites toward the soma may be severely attenuated, several-fold more than for the case of steady-state attenuation. This is often referred to as the *filtering* effect of the cable properties. However, the integrated response (the area under the transient voltage) is approximately equivalent to the steady-state amplitude, indicating that there is only a small loss of total charge (see Fig. 4.9B).

DYNAMIC PROPERTIES OF PASSIVE ELECTROTONIC STRUCTURE

Electrotonic Structure of the Neuron Changes Dynamically

These considerations show that, compared with the anatomical structure of a dendritic system, which is relatively fixed over short periods of time, the electrotonic structure continually shifts over time, producing complex effects on signal integration. The

effects reflect different relations between the electrotonic and signaling properties, such as the direction of signal spread, inhomogeneities in passive properties, rates of signal transfer, and interactions between synaptic or active conductances, to name a few. The effects can be illustrated in graphic fashion for the entire soma–dendritic system by taking a stained neuron and modifying its size according to its electrotonic properties. This is termed a *morphoelectrotonic transform* (MET) or *neuromorphic transform*.

We illustrate three types of neuromorphic transforms, beginning with the direction of signal spread. Figure 4.10 illustrates a CA1 hippocampal pyramidal cell in which a comparison is made between spread of a signal from the soma to the dendrites (voltage out, V_{out}) with spread from the dendrites to the soma (voltage in, V_{in}). On the left (Fig. 4.10A) is the stained neuron, with its long many-branched apical dendrite and shorter basal dendrites and their branches. Figure 4.10B2 illustrates an electrotonic representation of the neuron for signals spreading from the distal dendrites toward the soma. There is severe decrement from each distal branch (cf. Fig. 4.9) so that apical and basal dendritic trees have electrotonic lengths of approximately 3 and 2, respectively. By comparison, Figure 4.10B1 illustrates an electrotonic representation of this neuron for a signal

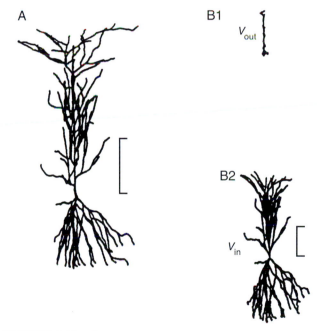

FIGURE 4.10 The electrotonic structure of a neuron varies with the direction of spread of signals. (A) Stained CA1 pyramidal neuron. (B) Electrotonic transform of the stained morphology for the case of a voltage spreading away from the cell body (B1, V_{out}) and toward the cell body (B2, V_{in}). Calibration bar, 1 electrotonic length. Reproduced with permission from Carnevale, N. T., Tsai, K. Y., Claiborne, B. J., and Brown, T. H. (1997). Comparative electronic analysis of three classes of rat hippocampal neurons. *J. Neurophysiol.* **78**, 703–720.

spreading from the soma to the dendrites. The basal dendrites have shrunk to almost nothing, indicating that they are nearly isopotential. This feature emerges because the branches are relatively short compared with their electrotonic lengths and because the sealed end boundary condition greatly reduces the decrement of electrotonic potential through them (cf. Fig. 4.7). The apical dendrite has shrunk to an electrotonic length of approximately 1. Thus, distal synaptic responses decay considerably in spreading all the way to the soma, which active properties help to overcome, as we shall see in Chapter 17, whereas signals at the soma "see" a relatively compact dendritic tree. This would, for example, be the case for a back-propagating action potential.

The analysis in Figure 4.10 applies to spread of steady-state or very slowly changing signals. What about spread of rapid signals? We have seen that membrane capacitance makes the dendrites act as a low-pass filter, further reducing rapid signals. The electrotonic transforms can assess this effect, as shown in Figure 4.11. On the left, the electrotonic representation of a pyramidal neuron is shown for a slow (100 Hz) current injected in the soma. The form is similar to that of the cell in Figure 4.10, with tiny, virtually isopotential basal dendrites and a longer apical dendritic tree of electrotonic length of approximately 1.5. By comparison, a rapid (500 Hz) signal is severely attenuated in spreading into the dendrites, as shown by the basal dendrites with L of approximately 1 and the apical dendritic tree electrotonic lengths of 4–5. Thus, a somatic action potential could back-propagate into the basal dendrites rather effectively but would require active properties to invade very far into the

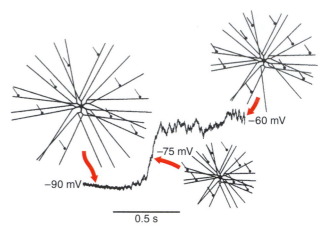

FIGURE 4.12   The electrotonic structure of a neuron can vary with shifts in the resting membrane potential. In this medium spiny cell, the electrotonic transform varies with the resting membrane potential, which in turn reflects the combination of resting voltage-gated K^+ currents and excitatory synaptic currents. See text. From Wilson, C.J. (1998). Basal ganglia. *In* "The Synaptic Organization of the Brain" (G.M. Shepherd, ed.), 5th Ed., pp. 361–414. Oxford Univ. Press, New York. Reproduced with permission fom Oxford University Press.

apical dendrites. There is direct evidence for these properties underlying back-propagating action potentials in apical dendrites (Chapter 17).

The electrotonic structure of a neuron is not necessarily fixed, but may vary under synaptic control. Our final example is shown in Figure 4.12 for the case of a medium spiny cell in the basal ganglia. During low levels of resting excitatory synaptic input, the electrotonic transform of this cell type is relatively large (left) because of the action of a specific K^+ current (known as I_h) in the dendrites that hold them relatively hyperpolarized (see arrow at –90 mV). When synaptic excitation increases, the K^+ current is deactivated, reducing the membrane conductance and thereby increasing the input resistance of the cell; the dendritic tree becomes more compact electrotonically (middle) so that synaptic inputs are more effective in activating the cell. As the cell responds to the synaptic excitation, the resulting depolarization activates other K^+ currents, which expand the electrotonic structure again (right). This example illustrates how cable properties and voltage-gated properties interact to control the integrative actions of the neuron.

Synaptic Conductances in Dendrites Tend to Interact Nonlinearly

Dynamic interactions also occur between synaptic conductances. It is often assumed that synaptic responses sum linearly, but we have already noted that this is not generally true. In an electrical cable, responses to simultaneous current inputs sum linearly (they show "superposition") because the cable

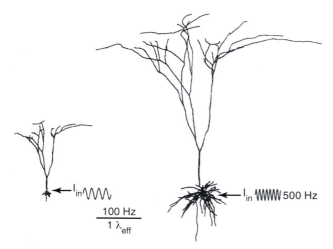

FIGURE 4.11   The electrotonic structure of a neuron varies with the rapidity of signals. (Left) Electrotonic transform of a pyramidal neuron in response to a sinusoidal current of 100 Hz injected into the soma (i.e., this is an example of V_{out}). (Right) Electrotonic transform of same cell in response to 500 Hz. Calibration, 1 electrotonic length. See text. From Zador *et al.* (1995).

properties remain invariant at all times. However, as noted in relation to Figure 4.6, synaptic responses in real neurons generate current by means of changes in the membrane conductance at the synapse, which alters the overall membrane resistance of that segment and with it the input resistance, thereby changing the electrotonic properties of the whole system. As pointed out by Rall (1964), excitatory and inhibitory conductance changes involve "a change in a conductance which is an element of the system; the system itself is perturbed; the value of a constant coefficient in the linear differential equation is changed; hence the simple superposition rules do not hold."

This effect is illustrated by the two-compartment model of Figure 4.6. Consider a synaptic input to compartment A, which decreases the membrane resistance of that compartment. Now consider a simultaneous synaptic input to compartment B, which has the same effect on the membrane resistance of that compartment. The internal current flowing between the two compartments encounters a much lower impedance and hence has much less effect on the membrane potential than would have been the case for current injection. The integration of these two responses therefore gives a smaller summed potential than the summation of the two responses taken individually. This effect is referred to as *occlusion*. In essence, each compartment partially short-circuits the other through a larger conductance load, thus reducing the combined response.

These properties mean that, as noted earlier, synaptic integration in dendrites in general is not linear even for purely passive electrotonic properties. The farther apart the synaptic sites, the fewer the interactions between the conductances, and the more linear the summation becomes (see Fig. 4.13). These nonlinear properties of passive dendrites, combined with the nonlinear properties of voltage-gated channels at local sites on the membrane, contribute to the complexity of signal processing that takes place in dendrites, as will be discussed in Chapter 17. As we shall see, dendritic spines affect these nonlinear properties.

Significance of Active Conductances in Dendrites Depends on Their Relation to Cable Properties

In electrophysiological recordings from the cell body, dendritic synaptic responses often appear small and slow (cf. Fig. 4.8). However, at their sites of origin in the dendrites, the responses tend to have a large amplitude (because of the high input resistances of the thin distal dendrites) and a rapid time course (because of the small membrane capacitance) (cf. Fig. 4.9). These properties have important implications for the

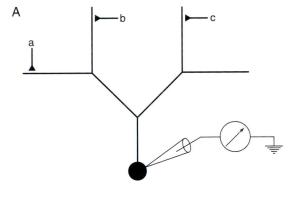

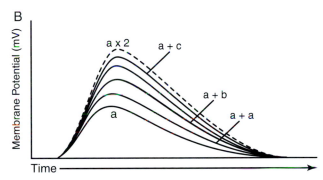

FIGURE 4.13 Schematic diagram of a dendritic tree to illustrate graded effects of nonlinear interactions between synaptic conductances. (A) Three sites of synaptic input (a–c) are shown, with a recording site in the soma. (B) The voltage response (*V*) is shown for the response to a single input at a, the theoretical linear summation for two inputs at a (a × 2), and the gradual reduction in summation from c to a due to increasing shunting between the conductances. See text. From Shepherd and Koch (1990).

signal processing that takes place in dendrites. In particular, the fact that distal dendrites contain sites of voltage-gated channels means that local integration, local boosting, and local threshold operations can take place. These most distal responses need spread no farther than to neighboring local active sites to be boosted by these sites; thus, a rapid integrative sequence of these actions ultimately produces significant effects on signal integration at the cell body. These properties will be considered further in Chapter 17.

In addition to their role in local signal processing, the cable properties of the neuron are also important for (1) controlling the spread of synaptic potentials from the dendrites through the soma to the site of action potential initiation in the axon hillock initial segment and (2) back-propagation of an action potential into the soma–dendritic compartments, where it can activate dendritic outputs and interact with the active properties involved in signal processing. These properties are discussed further in Chapter 17.

Dendritic Spines Form Electrotonic and Biochemical Compartments

The rules governing electrotonic interactions within a dendritic tree also apply at the level of a spine, the smallest process of a nerve cell. A spine may vary from a bump on a dendritic branch to a twig to a lollipop-shaped process several micrometers long (Fig. 4.14). A dendritic spine usually receives a single excitatory synapse; an axonal initial segment spine characteristically receives an inhibitory synapse.

Dendritic spines receive most of the excitatory inputs to pyramidal neurons in the cerebral cortex and to Purkinje cells in the cerebellum, as well as to a variety of other neuron types, so an understanding of their properties is critical for understanding brain function (Shepherd, 1996; Araya *et al.*, 2006; Alvarez and Sabatini, 2007). As with the whole dendritic tree, one begins with their electrotonic properties. Given the rules we have built earlier in this chapter, by simple inspection of spine morphology as shown in Figure 4.14, we can postulate several distinctive features that may have important functional implications (see Box 4.3).

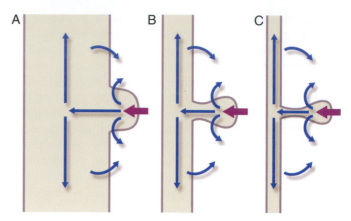

FIGURE 4.14 Diagrams illustrating different types of spines and current flows generated by a synaptic input. (A) Stubby spine arising from a thick process. (B) Moderately elongated spine from a medium diameter branch. (C) Spine with a long stem originating from a thin branch. Parallel considerations apply to diffusion between the spine head and dendritic branch. Modified from Shepherd (1974).

In addition to its electrotonic properties, the spine may have interesting biochemical properties. The same cable equations that govern electrotonic properties also have their counterparts in describing the diffusion of substances (as well as the flow of heat).

BOX 4.3

SOME BASIC ELECTROTONIC PROPERTIES OF DENDRITIC SPINES

a. **High input resistance**. The smaller the size and the narrower the stem, the higher the input resistance; this property gives a large amplitude synaptic potential for a given synaptic conductance. Such a large depolarizing EPSP can have powerful effects on the local environment within the spine.

b. **Low total membrane capacitance.** The small size also means a small total membrane capacitance, implying that synaptic (and any active) potentials may be rapid; this means that spines on dendrites can potentially be involved in rapid information transmission.

c. **Increases in total dendritic membrane capacitance**. Although the membrane capacitance of an individual spine is small, the combined spine population increases the total capacitance of its parent dendrite. This property increases the filtering effect of the dendrite on transmission of signals through it.

d. **Decrement of potentials spreading from the spine.** An impedance mismatch exists between the spine head and its parent dendrite; this means that potentials spreading from the spine to the dendrite will suffer considerable decrement unless there are active properties of the dendrite or of neighboring spines to boost the signal.

e. **Ease of potential spread into the spine.** The other side of the impedance mismatch is that membrane potential changes within the dendrite spread into the spine with little decrement; thus, the spine tends to follow the potential of its dendrite, except for the transient large-amplitude responses to its own synaptic input. This property means that a spine can serve as a *coincidence detector* for nearby synaptic responses or for an action potential backpropagating into the dendritic tree.

f. **Linearization of synaptic integration**. The spine necks increase the anatomical and electrotonic distance between the spine synapses, thereby decreasing the interactions between their conductances, producing more linear superposition of the postsynaptic responses.

Gordon M. Shepherd

Thus, as already noted, accumulations of only small numbers of ions are needed within the tiny volumes of spine heads to change the driving force on an ion species or to affect significant changes in the concentrations of subsequent second messengers. This interest is intensifying, as the ability to image ion fluxes, such as for Ca^{2+}, and to measure other molecular properties of individual spines' increases with new technology such as two-photon microscopy. The interpretation of those results for the integrative properties of the neuron will require considerations in the bio-chemical domain that parallel those discussed in the electrotonic domain. The range of properties and possible functions of spines are discussed further in Chapters 17 and 19.

Summary

In addition to membrane properties, the spread of electrotonic potentials in branching dendritic trees is dependent on the boundary conditions set by the modes of branching and termination within the tree. In general, other parts of the dendritic tree constitute a conductance load on activity at a given site; the spread of activity from that site is determined by the impedance match or mismatch between that site and the neighboring sites. Rules governing these impedance relations have been worked out relative to the case in which the sum of the daughter branch diameters raised to the 3/2 power is equal to that of the parent branch, in which case the system of branches is an "equivalent cylinder," resembling a single continuous cable. This provides a starting point in analyzing synaptic integration, which can be adapted for different types of branching patterns in terms of "equivalent dendrites."

Synchronous synaptic potentials in several branches spread relatively effectively through most dendritic trees. Responses in individual branches may be relatively isolated because of the decrement of passive spread and require local active boosting for effective communication with the rest of the tree. Passive spread can be characterized in terms of several measures, including characteristic length of the equivalent cylinder. There is scaling within individual branches, such that electrotonic spread in finer branches is relatively effective over their shorter lengths. Integration of synaptic potentials in passive dendrites is fundamentally nonlinear because of interactions between the synaptic conductances. The rules for electrotonic spread in dendrites are the basis for understanding the contributions of active properties of dendrites (see Chapter 17).

RELATING PASSIVE TO ACTIVE POTENTIALS

We can now begin to gain insight into the relation between passive and active potentials in a neuron. We consider a model, the olfactory mitral cell, in which we apply the principles of this chapter and look forward to the principles underlying active properties in Chapter 17.

A basic problem is to understand the factors that decide where the action potential will be initiated with different levels of excitatory or inhibitory inputs. The possible sites are anywhere from the axon through the soma to the most distal dendrites. The mitral cell is advantageous for this analysis (1) because all the excitatory synaptic input is through olfactory nerve terminals that make their synapses on the distal dendritic tuft and (2) because the primary dendrite that connects the tuft to the cell body is an unbranched cylinder. Applying depolarizing current to distal dendrite or soma, the experimental findings were counterintuitive: with weak distal inputs the action potential initiation site is far away, in the axon (Fig. 4.15A1), but with increasing excitation it shifts to the distal dendrite (Fig. 4.15A2) (Chen et al., 1997). How can the weak response spread so far passively, and why does not it excite the active dendrites along the way? Electrotonic spread is the key to the answer.

This is much too complex a problem to solve in your head or with "back of the envelope" calculations. The only effective method is a realistic computational simulation. A compartmental model of the mitral cell was therefore constructed, with Na^+ and K^+ conductances scaled to the structure of the mitral cell. Fitting of computed with experimental responses was carried out under stringent constraints, with minimization of eight simultaneous simulations (distal and soma recording sites, distal and soma sites of excitatory current input, strong and weak levels of excitation) (Shen et al., 1999).

We will analyze the active properties in Chapter 17; here we focus on fitting the passive properties. Two steps were essential. First, each experimental recording began with a period of passive charging of the mitral cell membrane (c in Fig. 4.15A,B). The superimposed tracings show that the model gave a very accurate simulation, even when the charging was long lasting (A1, continuous line, weak stimulation). This fit was critical for giving the correct latency of action potential initiation. Second, the longitudinal spread of passive current between the axon and the distal dendrite was calculated. This model showed

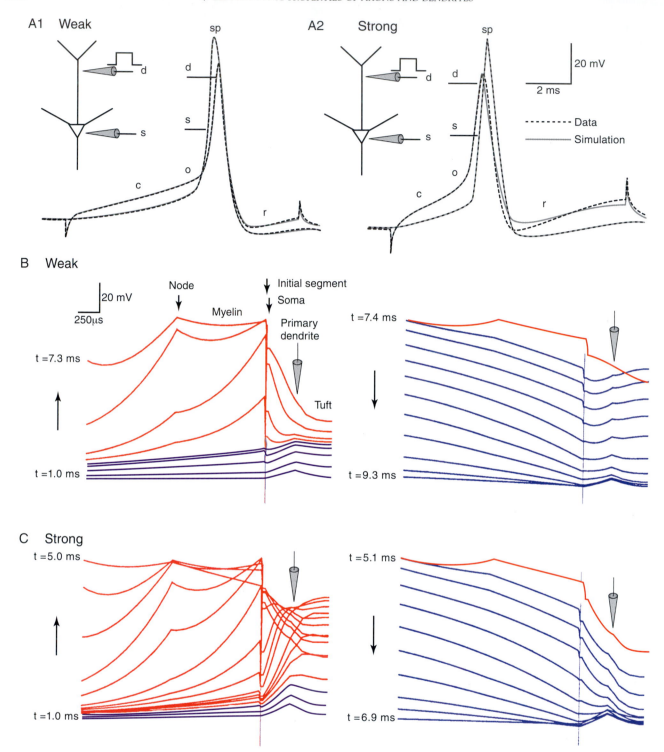

FIGURE 4.15   Interactions of passive and active potentials in the olfactory mitral cell. (A) Insets show diagrams of a mitral cell with recording sites at soma (s) and distal dendrite (d). Traces show fitting of experimental and computed responses to weak (A1) and strong (A2) depolarizing currents injected into the distal primary dendrite. Note the nearly exact superposition of experimental (solid lines) and computed (dashed lines) responses. (B) Longitudinal distribution of membrane potential changes at successive times after the weak distal dendritic excitation. (C) Longitudinal distribution of membrane potential changes at successive times after the strong distal dendritic excitation. Blue lines, predominantly passively generated potentials; red lines, predominantly actively generated potentials. d, dendrite; s, soma; c, passive charging; o, onset of action potential; sp, spike peak; r, recovery. See text. Reproduced with permission from Shen, G., Chen, W.R., Midtgaard, J., Shepherd, G.M., and Hines, M.L. (1999). Computational analysis of action potential initiation in mitra cell soma and dendrites based on dual patch recordings. *J. Neurophysiol.* **82**, 3006–3020.

that with weak distal excitation (Fig. 4.15B), the electrotonic current spread with a shallow gradient from the site of injection along the dendrite to the axon (bottom traces); the action potential arose first in the axon because of the much higher density of Na^+ channels there compared with the dendrite. However, with strong distal excitation (Fig. 4.15C) the direct depolarization of the less excitable distal dendrite led the weaker electrotonic depolarization of the more excitable axon, and dendritic action potential initiation occurred first. These features can be explored online at senselab.med.yale.edu/modeldb.

The computational simulations thus show precisely how the interactions of passive and active potentials control the sites of action potential initiation in the neuron. This is a model for the complex integrative properties of the neuron, which are explored further in Chapter 17.

References

Alle, H., and Geiger, J.R. (2006). Combined analog and action potential coding in hippocampal mossy fibers. *Science* **311**, 1290-1293.

Alvarez, V. A., and Sabbatini, B. L. (2007). Anatomical and physiological plasticity of dendritic spines. *Annu. Rev. Neurosci.* **30**, 79-97.

Araya, R., Eisenthal, K.B., and Yuste, R. (2006). Dendritic spines linearize the summation of excitatory potentials. *Proc. Natl. Acad. Sci. USA* **103**, 18799-18804.

Carnevale, N. T., Tsai, K.Y., Claiborne, B. J., and Brown, T. H. (1997). Comparative electrotonic analysis of three classes of rat hippocampal neurons. *J. Neurophysiol.* **78**, 703-720.

Chen, W. R., Midtgaard, J., and Shepherd, G. M. (1997). Forward and backward propagation of dendritic impulses and their synaptic control in mitral cells. *Science* **278**, 463-467.

Hursh, J. B. (1939). Conduction velocity and diameter of nerve fibers. *Am. J. Physiol.* **127**, 131-139.

Jack, J. J. B., Noble, D., and Tsien, R. W. (1975). "Electrical Current Flow in Excitable Cells." Oxford Univ. Press (Clarendon), London.

Johnston, D., and Wu, S. M. S. (1995). "Foundations of Cellular Neurophysiology." MIT Press, Cambridge.

Nelson, R., Lutzow, A.V., Kolb, H., and Gouras, P. (1975). Horizontal cells in cat retina with independent dendritic systems. *Science* **189**, 137-139.

Rall, W. (1959). Branching dendritic trees and motoneuron membrane resistivity. *Exp. Neurol.* **1**, 491-527.

Rall, W. (1964). Theoretical significance of dendritic trees for neuronal input-output relations. *In* "Neural Theory and Modelling" (R. F. Reiss, ed.), pp. 73-97. Stanford Univ. Press, Stanford, CA.

Rall, W. (1967). Distinguishing theoretical synaptic potentials computed for different soma-dendritic distributions of synaptic input. *J. Neurophysiol.* **30**, 1138-1168.

Rall, W. (1977). Core conductor theory and cable properties of neurons. *In* "The Nervous System, Cellular Biology of Neurons" (E. R. Kandel, ed.), Vol. 1, pp. 39-97. Am. Physiol. Soc., Bethesda, MD.

Rall, W., and Rinzel, J. (1973). Branch input resistance and steady attenuation for input to one branch of a dendritic neuron model. *Biophys. J.* **13**, 648-688.

Rall, W., and Shepherd, G. M. (1968). Theoretical reconstruction of field potentials and dendrodendritic synaptic interactions in olfactory bulb. *J. Neurophysiol.* **3**(6), 884-915.

Rinzel, J., and Rall, W. (1974). Transient response in a dendritic neuron model for current injected at one branch. *Biophys. J.* **14**, 759-790.

Ritchie, J. M. (1995). Physiology of axons. *In* "The Axon, Structure, Function, and Pathophysiology" (S. G. Waxman, J. D. Kocsis, and P. K. Stys, eds.), pp. 68-69. Oxford Univ. Press, New York.

Rushton, W. A. H. (1951). A theory of the effects of fibre size in medullated nerve. *J. Physiol.* (Lond.) **115**, 101-122.

Segev, I. (1995). Cable and compartmental models of dendritic trees. *In* "The Book of Genesis" (J. M. Bower and D. Beeman, eds.), pp. 53-82. Springer-Verlag (Telos), New York.

Segev, I., Rinzel, J., and Shepherd, G. M. (eds.) (1995). "The Theoretical Foundation of Dendritic Function." MIT Press, Cambridge, MA.

Shen, G., Chen, W. R., Midtgaard, J., Shepherd, G. M., and Hines, M. L. (1999). Computational analysis of action potential initiation in mitral cell soma and dendrites based on dual patch recordings. *J. Neurophysiol.* **82**, 3006-3020.

Shepherd, G. M. (1974). "The Synaptic Organization of the Brain." Oxford Univ. Press, New York.

Shepherd, G. M. (1994). "Neurobiology." Oxford University Press, New York.

Shepherd, G. M. (1996). The dendritic spine: A multifunctional integrative unit. *J. Neurophysiol.* **75**, 2197-2210.

Shepherd, G. M., and Koch, C. (1990). Dendritic electrotonus and synaptic integration. *In* "The Synaptic Organization of the Brain" (G. M. Shepherd, ed.), 3rd Ed., pp. 439-574. Oxford Univ. Press, New York.

Shu, Y., Hasenstab, A., Duque, A., Yu, Y., and McCormick, D. A. (2006). Modulation of intracortical synaptic potentials by presynaptic somatic membrane potential. *Nature* **444**, 761-765.

Waxman, S. G., and Bennett, M. V. L. (1972). Relative conduction velocities of small myelinated and nonmyelinated fibres in the central nervous system. *Nature, New Biol.* **238**, 217.

Wilson, C. J. (1998). Basal ganglia. In "The Synaptic Organization of the Brain" (G. M. Shepherd, ed.), 5th Ed., pp. 361-414. Oxford Univ. Press, New York.

Zador, A. M., Agmon-Snir, H., and Segev, I. (1995). The morphoelectrotonic transform: A graphical approach to dendritic function. *J. Neurosci.* **15**, 1169-1682.

Membrane Potential and Action Potential

David A. McCormick

The communication of information between neurons and between neurons and muscles or peripheral organs requires that signals travel over considerable distances. A number of notable scientists have contemplated the nature of this communication through the ages. In the second century AD, the great Greek physician Claudius Galen proposed that "humors" flowed from the brain to the muscles along hollow nerves. A true electrophysiological understanding of nerve and muscle, however, depended on the discovery and understanding of electricity itself. The precise nature of nerve and muscle action became clearer with the advent of new experimental techniques by a number of European scientists, including Luigi Galvani, Emil Du Bois-Reymond, Carlo Matteucci, and Hermann von Helmholtz, to name a few (Brazier, 1959, 1988). Through the application of electrical stimulation to nerves and muscles, these early electrophysiologists demonstrated that the conduction of commands from the brain to muscle for the generation of movement was mediated by the flow of electricity along nerve fibers.

With the advancement of electrophysiological techniques, electrical activity recorded from nerves revealed that the conduction of information along the axon was mediated by the active generation of an electrical potential, called the *action potential*. But what precisely was the nature of these action potentials? To know this in detail required not only a preparation from which to obtain intracellular recordings but also one that could survive *in vitro*. The squid giant axon provided precisely such a preparation, as was first demonstrated by J. Z. Young in 1936 (Young, 1936). Many invertebrates contain unusually large axons for the generation of escape reflexes; large axons conduct more quickly than small ones and so the response time for escape is reduced (see Chapter 4). The squid possesses an axon approximately 0.5 mm in diameter, large enough to be impaled by even a coarse micropipette (Fig. 5.1). By inserting a glass micropipette filled with a salt solution into the squid giant axon, Alan Hodgkin and Andrew Huxley demonstrated in 1939 that axons at rest are electrically polarized, exhibiting a resting membrane potential of approximately −60 mV inside versus outside (Hodgkin and Huxley, 1939; Hodgkin, 1976). In the generation of an action potential, the polarization of the membrane is removed (referred to as *depolarization*) and exhibits a rapid swing toward, and even past, 0 mV (Fig. 5.1B). This depolarization is followed by a rapid swing in the membrane potential to more negative values, a process referred to as *hyperpolarization*. The membrane potential following an action potential typically becomes even more negative than the original value of approximately −60 mV. This period of increased polarization is referred to as the *afterhyperpolarization* or the *undershoot*.

The development of electrophysiological techniques to the point that intracellular recordings could be obtained from the small cells of the mammalian nervous system revealed that action potentials in these neurons are generated through mechanisms similar to that of the squid giant axon (Brock *et al.*, 1952; Buser and Albe–Fessard, 1953; Tasaki *et al.*, 1954; Phillips, 1956).

It is now known that action potential generation in nearly all types of neurons and muscle cells is accomplished through mechanisms similar to those first detailed in the squid giant axon by Hodgkin and Huxley. In this chapter, we consider the cellular

A

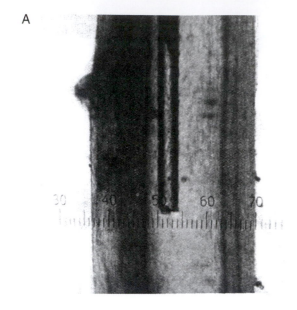

B

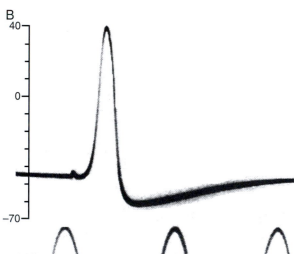

FIGURE 5.1   Intracellular recording of the membrane potential and action potential generation in the squid giant axon. (A) A glass micropipette, about 100 μm in diameter, was filled with seawater and lowered into the giant axon of the squid after it had been dissected free. The axon is about 1 mm in diameter and is transilluminated from behind. (B) One action potential recorded between the inside and the outside of the axon. Peaks of a sine wave at the bottom provided a scale for timing, with 2 ms between peaks. Reproduced with permission from Macmillan Publishers Ltd. Hodgkin, A.L., and Huxley, A.F. (1939). Action potentials recorded from inside a nerve fiber. *Nature (London)* **144,** 710–711.

mechanisms by which neurons and axons generate a resting membrane potential and how this membrane potential is briefly disrupted for the purpose of propagation of an electrical signal, the action potential. The following chapters describe in more detail the molecular properties of ion channels (Chapter 6) and the quantitative analysis and dynamic properties of action potentials (Chapter 7).

THE MEMBRANE POTENTIAL

The Membrane Potential Is Generated by the Differential Distribution of Ions

Through the operation of ionic pumps and special ionic buffering mechanisms, neurons actively maintain precise internal concentrations of several important ions, including Na^+, K^+, Cl^-, and Ca^{2+}. The mechanisms by which they do so are illustrated in Figure 5.2 and Figure 5.3. The intracellular and

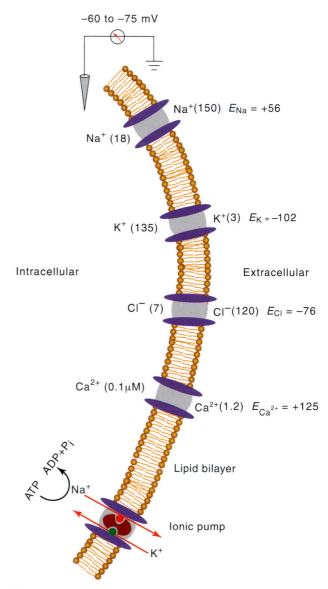

FIGURE 5.2   Differential distribution of ions inside and outside plasma membrane of neurons and neuronal processes, showing ionic channels for Na^+, K^+, Cl^-, and Ca^{2+}, as well as an electrogenic Na^+–K^+ ionic pump (also known as Na^+, K^+-ATPase). Concentrations (in millimoles except that for intracellular Ca^{2+}) of the ions are given in parentheses; their equilibrium potentials (E) for a typical mammalian neuron are indicated.

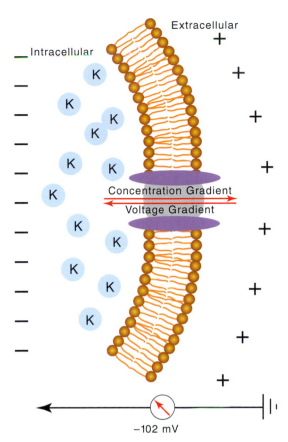

Concentration Gradient

Voltage Gradient

−102 mV

FIGURE 5.3 The equilibrium potential is influenced by the concentration gradient and the voltage difference across the membrane. Neurons actively concentrate K⁺ inside the cell. These K⁺ ions tend to flow down their concentration gradient from inside to outside the cell. However, the negative membrane potential inside the cell provides an attraction for K⁺ ions to enter or remain within the cell. These two factors balance one another at the equilibrium potential, which in a typical mammalian neuron is −102 mV for K⁺.

TABLE 5.1 Ion Concentrations and Equilibrium Potentials

	Inside (mM)	Outside (mM)	Equilibrium Potential (mV)
Squid giant axon			
Na^+	50	440	+55
K^+	400	20	−76
Cl^-	40	560	−66
Ca^{2+}	0.4 mM	10	+145
Mammalian neuron			
Na^+	18	145	+56
K^+	140	3	−102
Cl^-	7	120	−76
Ca^{2+}	100 nM	1.2	+125

K^+ inside glial cells, neurons, and axons results in a tendency for K^+ ions to diffuse down their concentration gradient and leave the cell or cell process (see Fig. 5.3). However, the movement of ions across the membrane also results in a redistribution of electrical charge. As K^+ ions move down their concentration gradient, the intracellular voltage becomes more negative, and this increased negativity results in an electrical attraction between the negative potential inside the cell and the positively charged K^+ ions, thus offsetting the outward flow of these ions. The membrane is selectively permeable; that is, it is impermeable to the large anions inside the cell, which cannot follow the potassium ions across the membrane. At some membrane potential, the "force" of the electrostatic attraction between the negative membrane potential inside the cell and the positively charged K^+ ions exactly balances the thermal "forces" by which K^+ ions tend to flow down their concentration gradient (see Fig. 5.3). In this circumstance, it is equally likely that a K^+ ion exits the cell by movement down the concentration gradient as it is that a K^+ ion enters the cell owing to the attraction between the negative membrane potential and the positive charge of this ion. At this membrane potential, there is no net flow of K^+ (the same number of K^+ ions enter the cell as leave the cell per unit time) and these ions are said to be in equilibrium. The membrane potential at which this occurs is known as the *equilibrium potential*. (See Box 5.1 for the calculation of the equilibrium potential.)

To illustrate, let us consider the passive distribution of K^+ ions in the squid giant axon as studied by Hodgkin and Huxley. The K^+ concentration inside the squid giant axon $[K^+]_i$ is about 400 mM, whereas that outside the axon $[K^+]_o$ is about 20 mM (Table 5.1).

extracellular concentrations of Na^+, K^+, Cl^-, and Ca^{2+} differ markedly (see Fig. 5.2 and Table 5.1); K^+ is actively concentrated inside the cell, and Na^+, Cl^-, and Ca^{2+} are actively extruded to the extracellular space. However, this does not mean that the cell is filled only with positive charge; anions (denoted A^-) to which the plasma membrane is impermeant are also present inside the cell and almost balance the high concentration of K^+. The osmolarity inside the cell is approximately equal to that outside the cell.

Electrical and Thermodynamic Forces Determine the Passive Distribution of Ions

Ions tend to move down their concentration gradients through specialized ionic pores, known as *ionic channels,* in the plasma membrane. Through simple laws of thermodynamics, the high concentration of

---

BOX 5.1

THE NERNST EQUATION

The equilibrium potential is determined by (1) the concentration of the ion inside and outside the cell, (2) the temperature of the solution, (3) the valence of the ion, and (4) the amount of work required to separate a given quantity of charge. The equation that describes the equilibrium potential was formulated by a German physical chemist named Walter Nernst in 1888:

$$E_{ion} = RT/zF \ln \{[ion]_o/[ion]_i\}.$$

Here, E_{ion} is the membrane potential at which the ionic species is at equilibrium, R is the gas constant (8.315 joules per Kelvin per mole), T is the temperature in Kelvin ($T_{Kelvin} = 273.16 + T_{Celsius}$), F is Faraday's

constant (96,485 coulombs per mole), z is the valence of the ion, and $[ion]_o$ and $[ion]_i$ are the concentrations of the ion outside and inside the cell, respectively. For a monovalent, positively charged ion (cation) at room temperature (20°C), substituting the appropriate numbers and converting natural log (ln) into log base 10 (log) results in the equation

$$E_{ion} = 58.2 \log \{[ion]_o/[ion]_i\};$$

at a body temperature of 37°C, the Nernst equation is

$$E_{ion} = 61.5 \log \{[ion]_o/[ion]_i\}.$$

Because $[K^+]_i$ is greater than $[K^+]_o$, potassium ions tend to flow down their concentration gradient, taking positive charge with them. The equilibrium potential (at which the tendency for K^+ ions to flow down their concentration gradient is exactly offset by the attraction for K^+ ions to enter the cell because of the negative charge inside the cell) at a room temperature of 20°C can be determined by the Nernst equation as

$$E_K = 58.2 \log(20/400) = -76 \text{ mV}.$$

Therefore, at a membrane potential of –76 mV, K^+ ions have an equal tendency to flow either into or out of the axon. The concentrations of K^+ in mammalian neurons and glial cells differ considerably from that in the squid giant axon, which is adapted to live in seawater (see Table 5.1). By substituting 3.1 mM for $[K^+]_o$ and 140 mM for $[K^+]_i$ in the Nernst equation, with mammalian body temperature, $T = 37°C$, we obtain

$$E_K = 61.5 \log(3.1/140) = -102 \text{ mV}.$$

Movements of Ions Can Cause Either Hyperpolarization or Depolarization

In mammalian cells, at membrane potentials positive to –102 mV, K^+ ions tend to flow out of the cell. Increasing the ability of K^+ ions to flow across the membrane, that is, increasing the conductance of the membrane to K^+ (g_K), causes the membrane potential to become more negative, or hyperpolarized, owing to the exit of positively charged ions from inside the cell (Fig. 5.4).

At membrane potentials negative to –102 mV, K^+ ions tend to flow into the cell; increasing the membrane conductance to K^+ causes the membrane potential to become more positive, or depolarized, owing to the flow of positive charge into the cell. The membrane potential at which the net current "flips" direction is referred to as the *reversal potential*. If the channels conduct only one type of ion (e.g., K^+ ions), then the reversal potential and the Nernst equilibrium potential for that ion coincide (see Fig. 5.4A). Increasing the membrane conductance to K^+ ions while the membrane potential is at the equilibrium potential for K^+ (E_K) does not change the membrane potential, because no net driving force causes K^+ ions to either exit or enter the cell. However, this increase in membrane conductance to K^+ decreases the ability of other species of ions to change the membrane potential, because any deviation of the potential from E_K increases the drive for K^+ ions either to exit or to enter the cell, thereby drawing the membrane potential back toward E_K (see Fig. 5.4B). This effect is known as a "shunt" and is important for some effects of inhibitory synaptic transmission.

The exit from and entry into the cell of K^+ ions during generation of the membrane potential give rise to a curious problem. When K^+ ions leave the cell to generate a membrane potential, the concentration of K^+ changes both inside and outside the cell. Why does this change in concentration not alter the equilibrium potential, thus changing the tendency for K^+ ions to flow down their concentration gradient? The reason is that the number of K^+ ions required to leave the cell to achieve the equilibrium potential is quite

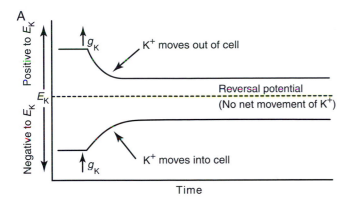

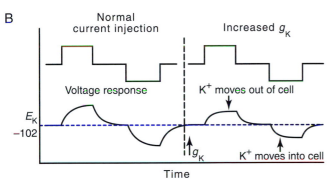

FIGURE 5.4 Increases in K⁺ conductance can result in hyperpolarization, depolarization, or no change in membrane potential. (A) Opening K⁺ channels increases the conductance of the membrane to K⁺, denoted g_K. If the membrane potential is positive to the equilibrium potential (also known as the reversal potential) for K⁺, then increasing g_K will cause some K⁺ ions to leave the cell, and the cell becomes hyperpolarized. If the membrane potential is negative to E_K when g_K is increased, then K⁺ ions enter the cell, making the inside more positive (more depolarized). If the membrane potential is exactly E_K when g_K is increased, then there is no net movement of K⁺ ions. (B) Opening K⁺ channels when the membrane potential is at E_K does not change the membrane potential; however, it reduces the ability of other ionic currents to move the membrane potential away from E_K. For example, a comparison of the ability of the injection of two pulses of current, one depolarizing and one hyperpolarizing, to change the membrane potential before and after opening K⁺ channels reveals that increases in g_K noticeably decrease the responses of the cell.

small. For example, if a cell were at 0 mV and the membrane suddenly became permeable to K⁺ ions, only about 10–12 mol of K⁺ ions per square centimeter of membrane would move from inside to outside the cell in bringing the membrane potential to the equilibrium potential for K⁺. In a spherical cell of 25 μm diameter, this would amount to an average decrease in intracellular K⁺ of only about 4 μM (e.g., from 140 to 139.996 mM). However, there are instances when significant changes in the concentrations of K⁺ may occur, particularly during the generation of pronounced activity, such as an epileptic seizure. During the occurrence of a tonic–clonic generalized (grand mal) seizure, large numbers of neurons discharge throughout the cerebral cortex in

a synchronized manner. This synchronous discharge of large numbers of neurons significantly increases the extracellular K⁺ concentration, by as much as a couple of millimoles, resulting in a commensurate positive shift in the equilibrium potential for K⁺ (Hotson et al., 1973; Prince et al., 1973). This shift in the equilibrium potential can increase the excitability of affected neurons and neuronal processes and thus promote the spread of the seizure activity. Fortunately, the extracellular concentration of K⁺ is tightly regulated and kept at normal levels through uptake by glial cells as well as by diffusion through the fluid of the extracellular space (Kuffler and Nicholls, 1966).

As is true for K⁺ ions, each of the membrane-permeable species of ions possesses an equilibrium potential that depends on the concentration of the ions inside and outside the cell. Thus, equilibrium potentials may vary between different cell types, such as those found in animals adapted to live in saltwater versus mammalian neurons (see Table 5.1). In mammalian neurons, the equilibrium potential is approximately +56 mV for Na⁺, approximately –76 mV for Cl⁻, and about +125 mV for Ca²⁺ (see Table 5.1 and Fig. 5.2). Thus, increasing the membrane conductance to Na⁺ (g_{Na}) through the opening of Na⁺ channels depolarizes the membrane potential toward +56 mV; increasing the membrane conductance to Cl⁻ brings the membrane potential closer to –76 mV; and finally increasing the membrane conductance to Ca²⁺ depolarizes the cell toward +125 mV.

Na⁺, K⁺, and Cl⁻ Contribute to the Determination of the Resting Membrane Potential

If a membrane is permeable to only one ion and no electrogenic ionic pumps are operating (see next section), then the membrane potential is necessarily at the equilibrium potential for that ion. At rest, the plasma membrane of most cell types is not at the equilibrium potential for K⁺ ions, indicating that the membrane is also permeable to other types of ions. For example, the resting membrane of the squid giant axon is permeable to Cl⁻ and Na⁺, as well as K⁺, owing to the presence of ionic channels that not only allow these ions to pass but also are open at the resting membrane potential. Because the membrane is permeable to K⁺, Cl⁻, and Na⁺, the resting potential of the squid giant axon is not equal to E_K, E_{Na}, or E_{Cl} but is somewhere within these three. A membrane permeable to more than one ion has a steady-state membrane potential whose value is between those of the equilibrium potentials for each of the permeant ions (Box 5.2) (Goldman, 1943; Hodgkin and Katz, 1949).

BOX 5.2

THE GOLDMAN–HODGKIN–KATZ EQUATION

An equation developed by Goldman (1943) and later used by Hodgkin and Katz (1949) describes the steady-state membrane potential for a given set of ionic concentrations inside and outside the cell and the relative permeabilities of the membrane to each of those ions:

$$V_m = RT/F \ln \left\{ \left(p_K[K^+]_o + p_{Na}[Na^+]_o + p_{Cl}[Cl^-]_i \right) / \left(p_K[K^+]_i + p_{Na}[Na^+]_i + p_{Cl}[Cl^-]_o \right) \right\}.$$

The relative contribution of each ion is determined by its concentration differences across the membrane and the relative permeability (p_K, p_{Na}, p_{Cl}) of the membrane to each type of ion. If a membrane is permeable to only one ion, then the Goldman–Hodgkin–Katz equation reduces to the Nernst equation. In the squid giant axon, at resting membrane potential, the permeability ratios are

$$p_K : p_{Na} : p_{Cl} = 1.00 : 0.04 : 0.45.$$

The membrane of the squid giant axon, at rest, is most permeable to K^+ ions, less so to Cl^-, and least permeable to Na^+. (Chloride appears to contribute considerably less to the determination of the resting potential of mammalian neurons.) These results indicate that the resting membrane potential is determined by the resting permeability of the membrane to K^+, Na^+, and Cl^-. In theory, this resting membrane potential may be anywhere between E_K (e.g., –76 mV) and E_{Na} (+55 mV). For the three ions at 20°C, the equation is

$$Vm = 58.2 \log \left\{ (1.20 + 0.04 \cdot 440 + 0.45 \cdot 40)/(1 \cdot 400 + 0.04 \cdot 50 + 0.45 \cdot 560) \right\} = -62 mV.$$

This suggests that the squid giant axon should have a resting membrane potential of –62 mV. In fact, the resting membrane potential may be a few millivolts hyperpolarized to this value through the operation of the electrogenic Na^+–K^+ pump.

Different Types of Neurons Have Different Resting Potentials

Intracellular recordings from neurons in the mammalian CNS reveal that different types of neurons exhibit different resting membrane potentials. Indeed, some types of neurons do not even exhibit a true "resting" membrane potential; they spontaneously and continuously generate action potentials even in the total lack of synaptic input. In the visual system, intracellular recordings have shown that the photoreceptor cells of the retina—the rods and cones—have a membrane potential of approximately –40 mV at rest and are hyperpolarized when activated by light (Tomita, 1965). Cells in the dorsal lateral geniculate nucleus, which receive axonal input from the retina and project to the visual cortex, have a resting membrane potential of approximately –70 mV during sleep and –55 mV during waking (Hirsch *et al.*, 1983; Jahnsen and Llinas, 1984a,b), whereas pyramidal neurons of the visual cortex have a resting membrane potential of about –75 mV (McCormick *et al.*, 1985). Presumably, the resting membrane potentials of different cell types in the central and peripheral nervous system are highly regulated and are functionally important. For example, the depolarized membrane potential of photoreceptors presumably allows the

membrane potential to move in both negative and positive directions in response to changes in light intensity. The hyperpolarized membrane potential of thalamic neurons during sleep (–70 mV) dramatically decreases the flow of information from the sensory periphery to the cerebral cortex (Livingstone and Hubel, 1981; Steriade and McCarley, 2005), presumably to allow the cortex to be relatively undisturbed during sleep, and the 20 mV membrane potential between the resting potential and the action potential threshold in cortical pyramidal cells may permit the subthreshold computation and integration of multiple neuronal inputs in single neurons (see Chapters 4, 16, and 17).

Ionic Pumps Actively Maintain Ionic Gradients

Because the resting membrane potential of a neuron is not at the equilibrium potential for any particular ion, ions constantly flow down their concentration gradients. This flux becomes considerably larger with the generation of electrical and synaptic potentials, because ionic channels are opened by these events. Although the absolute number of ions traversing the plasma membrane during each action potential or synaptic potential may be small in individual cells, the collective influence of a large neural network of cells, such as in the brain, and the presence of ion fluxes even at rest can substantially change the distribution of ions inside and outside neurons. As described in Chapter 3, cells have solved this problem with the use of active transport of ions against their concentration gradients. The proteins that actively transport ions are referred to as *ionic pumps*, of which the Na^+–K^+ pump is perhaps the most thoroughly understood (Hodgkin and Keynes, 1955; Skou, 1957, 1988; Thomas, 1972). The Na^+–K^+ pump is stimulated by increases in the intracellular concentration of Na^+ and moves Na^+ out of the cell while moving K^+ into it, achieving this task through the hydrolysis of ATP (see Fig. 5.2). Three Na^+ ions are extruded for every two K^+ ions transported into the cell. Owing to the unequal transport of ions, the operation of this pump generates a hyperpolarizing electrical potential and is said to be electrogenic. The Na^+–K^+ pump typically results in the membrane potential of the cell being a few millivolts more negative than it would be otherwise.

The Na^+–K^+ pump consists of two subunits, α and β, arranged in a tetramer $(\alpha\beta)_2$. The α subunit has a molecular mass of about 100 kDa and six hydrophobic regions capable of forming transmembrane helices (Mercer, 1993; Horisberger *et al.*, 1991). The β subunit is smaller (about 38 kDa) and has only one hydrophobic membrane-spanning region. The Na^+–K^+ pump is believed to operate through conformational changes that alternatively expose a Na^+ binding site to the interior of the cell (followed by the release of Na^+) and a K^+ binding site to the extracellular fluid (see Fig. 5.2). Such a conformation change may be due to the phosphorylation and dephosphorylation of the protein.

The membranes of neurons and glia contain multiple types of ionic pumps, used to maintain the proper distribution of each ionic species important for cellular signaling (Pedersen and Carafoli, 1987; Läuger, 1991). Many of these pumps are operated by the Na^+ gradient across the cell, whereas others operate through a mechanism similar to that of the Na^+–K^+ pump (i.e., the hydrolysis of ATP). For example, the calcium concentration inside neurons is kept at very low levels (typically 50–100 nM) through the operation of both types of ionic pumps as well as special intracellular Ca^{2+} buffering mechanisms. Ca^{2+} is extruded from neurons through both a Ca^{2+}–Mg^{2+}-ATPase and a Na^+–Ca^{2+} exchanger. The Na^+–Ca^{2+} exchanger is driven by the Na^+ gradient across the membrane and extrudes one Ca^{2+} ion for each Na^+ ion allowed to enter the cell.

The Cl^- concentration in neurons is actively maintained at a low level through the operation of a chloride–bicarbonate exchanger, which brings in one ion of Na^+ and one ion of HCO_3^- for each ion of Cl^- extruded (Reithmeier, 1994; Thompson *et al.*, 1988).

Intracellular pH also can markedly affect neuronal excitability and is therefore tightly regulated, in part by a Na^+–H^+ exchanger that extrudes one proton for each Na^+ allowed to enter the cell.

Summary

The membrane potential is generated by the unequal distribution of ions, particularly K^+, Na^+, and Cl^-, across the plasma membrane. This unequal distribution of ions is maintained by ionic pumps and exchangers. K^+ ions are concentrated inside the neuron and tend to flow down their concentration gradient, leading to a hyperpolarization of the cell. At the equilibrium potential, the tendency of K^+ ions to flow out of the cell is exactly offset by the tendency of K^+ ions to enter the cell owing to the attraction of the negative potential inside the cell. The resting membrane is also permeable to Na^+ and Cl^- and therefore the resting membrane potential is approximately -75 to -40 mV, in other words, substantially positive to E_K.

THE ACTION POTENTIAL

An Increase in Na^+ and K^+ Conductance Generates Action Potentials

Hodgkin and Huxley not only recorded the action potential with an intracellular microelectrode (see Fig. 5.1) but also went on to perform a remarkable series of experiments that qualitatively and quantitatively explained the ionic mechanisms by which the action potential is generated (Hodgkin and Huxley, 1952a–d; Hodgkin *et al.*, 1952). As mentioned earlier, these investigators found that during the action potential, the membrane potential of the cell rapidly overshoots 0 mV and approaches the equilibrium potential for Na^+. After generation of the action potential, the membrane potential repolarizes and becomes more negative than before, generating an afterhyperpolarization. Cole and Curtis (1939) had previously shown that these changes in membrane potential during the generation of the action potential are associated with a large increase in conductance of the plasma membrane. But to what does the membrane become conductive to generate the action potential? The prevailing hypothesis was that there was a nonselective increase in conductance causing the negative resting potential to increase toward 0 mV. Since publication of the experiments of Overton in 1902, the action potential had been known to depend on the presence of extracellular Na^+.

BOX 5.3

THE VOLTAGE-CLAMP TECHNIQUE

In the voltage-clamp technique, two independent electrodes are inserted into the squid giant axon: one for recording the voltage difference across the membrane and the other for intracellularly injecting the current (Fig. 5.5). These electrodes are then connected to a feedback circuit that compares the measured voltage across the membrane with the voltage desired by the experimenter. If these two values differ, then current is injected into the axon to compensate for this difference. This continuous feedback cycle, in which the voltage is measured and current is injected, effectively "clamps" the membrane at a particular voltage. If ionic channels were to open, then the resultant flow of ions into or out of the axon would be compensated for by the injection of positive or negative current into the axon through the current-injection electrode. The current injected through this electrode is necessarily equal to the current flowing through the ionic channels. It is this injected current that is measured by the experimenter. The benefits of the voltage-clamp technique are two-fold. First, the current injected into the axon to keep the membrane potential "clamped" is necessarily equal to the current flowing through the ionic channels in the membrane, thereby giving a direct measurement of this current. Second, ionic currents are both voltage and time dependent; they become active at certain membrane potentials and do so at a particular rate. Keeping the voltage constant in the voltage clamp allows these two variables to be separated; the voltage dependence and the kinetics of the ionic currents flowing through the plasma membrane can be directly measured.

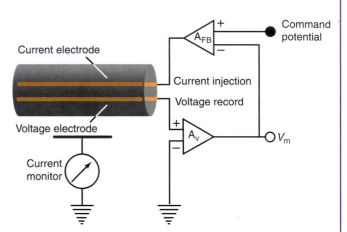

FIGURE 5.5 Voltage-clamp technique. The voltage-clamp technique keeps the voltage across the membrane constant so that the amplitude and time course of ionic currents can be measured. In the two-electrode voltage-clamp technique, one electrode measures the voltage across the membrane while the other injects current into the cell to keep the voltage constant. The experimenter sets a voltage to which the axon or neuron is to be stepped (the command potential). Current is then injected into the cell in proportion to the difference between the present membrane potential and the command potential. This feedback cycle occurs continuously, thereby clamping the membrane potential to the command potential. By measuring the amount of current injected, the experimenter can determine the amplitude and time course of the ionic currents flowing across the membrane.

Reducing the concentration of Na^+ in the artificial seawater bathing the axon resulted in a marked reduction in the amplitude of the action potential. On the basis of these and other data, Hodgkin and Katz proposed that the action potential is generated through a rapid increase in the conductance of the membrane to Na^+ ions. Quantitative proof of this theory was lacking, however, because ionic currents could not be observed directly. The development of the voltage-clamp technique by Cole (1949) at the Marine Biological Laboratory in Massachusetts resolved this problem and allowed quantitative measurement of the Na^+ and K^+ currents underlying the action potential (Box 5.3).

Hodgkin and Huxley used the voltage-clamp technique to investigate the mechanisms of generation of the action potential in the squid giant axon. Neurons have a threshold for the initialization of an action potential of about −45 to −55 mV. Increasing the voltage from −60 to 0 mV produces a large, but transient, flow of positive charge into the cell (known as *inward current*). This transient inward current is followed by a sustained flow of positive charge out of the cell (the *outward current*). By voltage-clamping the cell and substituting different ions inside or outside the axon or both, Hodgkin, Huxley, and colleagues demonstrated that the transient inward current is carried by Na^+ ions flowing into the cell and the sustained outward current is mediated by a sustained flux of K^+ ions moving out of the cell (Fig. 5.6) (Hodgkin and Huxley, 1952a–d; Hodgkin *et al.*, 1952; Hille, 1977).

The Na^+ and K^+ currents (I_{Na} and I_K, respectively) can be blocked, allowing each current to be examined in isolation (see Fig. 5.6B) (see also Chapter 7). Tetrodotoxin (TTX), a powerful poison found in the puffer fish *Spheroides rubripes* (Kao, 1966), selectively blocks

A   **Ion replacement**

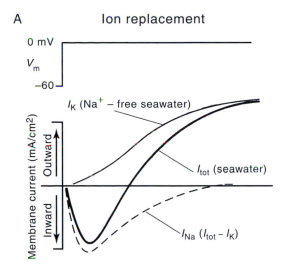

B   **Pharmacological blockade**

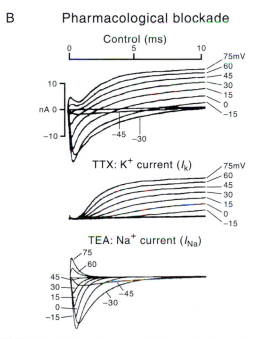

FIGURE 5.6 Voltage-clamp analysis reveals the ionic currents underlying action potential generation. (A) Increasing the potential from –60 to 0 mV across the membrane of the squid giant axon activates an inward current followed by an outward current. If the Na^+ in seawater is replaced by choline (which does not pass through Na^+ channels), then increasing the membrane potential from –60 to 0 mV results in only the outward current, which corresponds to I_K. Subtracting I_K from the recording in normal seawater illustrates the amplitude–time course of the inward Na^+ current, I_{Na}. Note that I_K activates more slowly than I_{Na} and that I_{Na} inactivates with time. (B) These two ionic currents can also be isolated from one another through the use of pharmacological blockers. (1) Increasing the membrane potential from –45 to +75 mV in 15 mV steps reveals the amplitude–time course of the inward Na^+ and outward K^+ currents. (2) After the block of I_{Na} with the poison tetrodotoxin (TTX), increasing the membrane potential to positive levels activates I_K only. (3) After the block of I_K with tetraethylammonium (TEA), increasing the membrane potential to positive levels activates I_{Na} only. (A) from Hodgkin and Huxley (1952); part (B) from Hille (1977).

voltage-dependent Na^+ currents (the puffer fish remains a delicacy in Japan and must be prepared with the utmost care by the chef). Using TTX, one can selectively isolate I_K and examine its voltage dependence and time course (see Fig. 5.6B).

Armstrong and Hille (1972) and others demonstrated that tetraethylammonium (TEA) is a useful pharmacological tool for selectively blocking I_K (see Fig. 5.6B). The use of TEA to examine the voltage dependence and time course of the Na^+ current underlying action-potential generation (see Fig. 5.6B) reveals some fundamental differences between Na^+ and the K^+ currents. First, the inward Na^+ current activates, or "turns on," much more rapidly than does the K^+ current (giving rise to the name "delayed rectifier" for this K^+ current). Second, the Na^+ current is transient; it inactivates, even if the membrane potential is maintained at 0 mV (see Fig. 5.6A). In contrast, the outward K^+ current, once activated, remains "on" as long as the membrane potential is clamped to positive levels; that is, the K^+ current does not inactivate; it is sustained. Remarkably, from one experiment, we see that the Na^+ current both rapidly activates and inactivates, whereas the K^+ current only slowly activates. These fundamental properties of the underlying Na^+ and K^+ channels allow the generation of action potentials.

Hodgkin and coworkers (Hodgkin and Huxley, 1952a–d; Hodgkin et al., 1952) proposed that the K^+ channels possess a voltage-sensitive "gate" that opens by the depolarization and closes by the subsequent repolarization of the membrane potential. This process of "turning on" and "turning off" the K^+ current came to be known as *activation* and *deactivation*. The Na^+ current also exhibits voltage-dependent activation and deactivation (see Fig. 5.6), but the Na^+ channels also become inactive despite maintained depolarization. Thus, the Na^+ current not only activates and deactivates but also exhibits a separate process known as *inactivation*, whereby the channels become blocked even though they are activated. The removal of this inactivation is achieved by relief of the depolarization and is a process known as *deinactivation*. Thus, the Na^+ channels possess two voltage-sensitive processes: activation–deactivation and inactivation–deinactivation. The kinetics of these two properties of Na^+ channels are different: inactivation takes place at a slower rate than activation.

The functional consequence of the two mechanisms is that Na^+ ions are allowed to flow across the membrane only when the channel is activated but not inactivated. Accordingly, Na^+ ions do not flow at resting membrane potentials, because the activation gate is closed (even though the inactivation gate is not). On depolarization, the activation gate opens, allowing Na$^+$ ions to flow into the cell. However, this depolarization also results in closure (at a slower rate) of the inactivation gate, which then blocks the flow of Na$^+$ ions. On repolarization of the membrane potential, the activation gate once again closes and the inactivation gate once again opens, preparing the axon for generation of the next action potential (Fig. 5.7) (see also Chapter 7, Fig. 7.6). Depolarization allows ionic current to flow by virtue of activation of the channel. The rush of Na$^+$ ions into the cell further depolarizes the membrane potential and more Na$^+$ channels become activated, forming a positive feedback loop that rapidly (within 100 μs or so) brings the membrane potential toward E_{Na}. However, the depolarization associated with the generation of the action potential also inactivates Na$^+$ channels, and, as a larger and larger percentage of Na$^+$ channels become inactivated, the rush of Na$^+$ into the cell diminishes. This inactivation of the Na$^+$ channels and activation of K$^+$ channels result in repolarization of the action

potential. This repolarization deactivates the Na$^+$ channels. Then, the inactivation of the channel is slowly removed, and the channels are ready, once again, for generation of another action potential (see Fig. 5.7).

By measuring the voltage sensitivity and kinetics of these two processes, activation–deactivation and inactivation–deinactivation of the Na$^+$ current, as well as activation–deactivation of the delayed rectifier K$^+$ current, Hodgkin and Huxley generated a series of mathematical equations (see Chapter 7 for details) that quantitatively describe the generation of the action potential (calculation of the propagation of a single action potential required an entire week of cranking a mechanical calculator). According to these early experimental and computational neuroscientists, the action potential is generated as follows. Depolarization of the membrane potential increases the probability of Na$^+$ channels being in the activated, but not yet inactivated, state. At a particular membrane potential, the resulting inflow of Na$^+$ ions tips the balance of the net ionic current from outward to inward (remember that depolarization also increases K$^+$ and Cl$^-$ currents by moving the membrane potential away from E_K and E_{Cl}). At this membrane potential, known as the *action potential threshold* (typically about –55 mV), the movement of Na$^+$ ions into the cell depolarizes the axon and opens more Na$^+$ channels, causing yet more depolarization of the membrane; repetition of this process yields a rapid, positive feedback loop that brings the axon close to E_{Na}. However, even as more and more Na$^+$ channels are becoming activated, some of these channels are also inactivating and, therefore, no longer conducting Na$^+$ ions. In addition, the delayed rectifier K$^+$ channels also are opening, owing to the depolarization of the membrane potential, and allowing positive charge to exit the cell. At some point, close to the peak of the action potential, the inward movement of Na$^+$ ions into the cell is exactly offset by the outward movement of K$^+$ ions out of the cell. After this point, the outward movement of K$^+$ ions dominates, and the membrane potential is repolarized, corresponding to the fall of the action potential. The persistence of the K$^+$ current for a few milliseconds following generation of the action potential generates the afterhyperpolarization. During this afterhyperpolarization, which is lengthened by the membrane time constant, the inactivation of the Na$^+$ channels is removed, preparing the axon for generation of the next action potential (see Fig. 5.7).

The occurrence of an action potential is *not* associated with substantial changes in the intracellular or extracellular concentrations of Na$^+$ or K$^+$, as we saw earlier for the generation of the resting membrane potential. For example, the generation of a

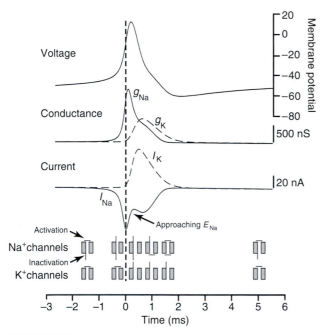

FIGURE 5.7 Generation of the action potential is associated with an increase in membrane Na$^+$ conductance and Na$^+$ current followed by an increase in K$^+$ conductance and K$^+$ current. Before action potential generation, Na$^+$ channels are neither activated nor inactivated (illustrated at the bottom of the figure). Activation of the Na$^+$ channels allows Na$^+$ ions to enter the cell, depolarizing the membrane potential. This depolarization also activates K$^+$ channels. After activation and depolarization, the inactivation particle on the Na$^+$ channels closes and the membrane potential repolarizes. The persistence of the activation of K$^+$ channels (and other membrane properties) generates an afterhyperpolarization. During this period, the inactivation particle of the Na$^+$ channel is removed and the K$^+$ channels close. From Huguenard and McCormick (1994).

single action potential in a 25 μm diameter hypothetical spherical cell should increase the intracellular concentration of Na^+ by only approximately 6 μM (from about 18 to 18.006 mM). Thus, the action potential is an electrical event generated by a change in the distribution of charge across the membrane and not by a marked change in the intracellular or extracellular concentration of Na^+ or K^+.

Action Potentials Typically Initiate in the Axon Initial Segment and Propagate Down the Axon and Backward through the Dendrites

Neurons have complex morphologies including dendritic arbors, a cell body, and typically one axonal output that branches extensively (See Chapter 1). In many cells, all these parts of the neuron are capable of independently generating action potentials. The activity of most neurons is dictated by barrages of synaptic potentials generated at each moment by a variable subset of the thousands of synapses impinging upon the cell's dendrites and soma. Where then is the action potential initiated? In most cells, each action potential is initiated in the initial portion of the axon, known as the *axon initial segment* (Coombs *et al.*, 1957; Stuart *et al.*, 1997; Shu *et al.*, 2007). The initial segment of the axon has the lowest threshold for action potential generation because it typically contains a moderately high density of Na^+ channels and it is a small compartment, which is easily depolarized by the in-rush of Na^+ ions. Once a spike is initiated (e.g., about 30–50 microns down the axon from the cell body in cortical pyramidal cells), this action potential then propagates orthodromically down the axon to the synaptic terminals, where it causes release of transmitter, as well as antidromically back through the cell body and into the cell's dendrites, where it can modulate intracellular processes. See Chapter 17 for additional discussion of propagation of action potentials in dendrites.

Refractory Periods Prevent "Reverberation"

The ability of depolarization to activate an action potential varies as a function of the time since the last generation of an action potential, owing to the inactivation of Na^+ channels and the activation of K^+ channels. Immediately after generation of an action potential, another action potential usually cannot be generated regardless of the amount of current injected into the axon. This period corresponds to the absolute refractory period and is mediated largely by the inactivation of Na^+ channels. The relative refractory period occurs during the action potential afterhyperpolarization and follows the absolute

refractory period. The relative refractory period is characterized by a requirement for the increased injection of ionic current into the cell to generate another action potential and results from the persistence of the outward K^+ current. The practical implication of refractory periods is that action potentials are not allowed to "reverberate" between the soma and the axon terminals.

The Speed of Action Potential Propagation Is Affected by Myelination

Axons may be either myelinated or unmyelinated. Invertebrate axons or small vertebrate axons are typically unmyelinated, whereas larger vertebrate axons are often myelinated. As described in Chapters 1 and 4, sensory and motor axons of the peripheral nervous system are myelinated by specialized cells (Schwann cells) that form a spiral wrapping of multiple layers of myelin around the axon (Fig. 5.8). Several Schwann cells wrap around an axon along its length; between the ends of successive Schwann cells are small gaps (nodes of Ranvier). In the central nervous system, a single oligodendrocyte, a special type of glial cell, typically ensheathes several axonal processes (Bunge, 1968).

In unmyelinated axons, the Na^+ and K^+ channels taking part in action potential generation are distributed along the axon, and the action potential propagates along the length of the axon through local depolarization of each neighboring patch of membrane (see Chapter 4), causing that patch of membrane also to generate an action potential (Fig. 5.8). In myelinated axons, on the other hand, the Na^+ channels are concentrated at the nodes of Ranvier (Ritchie and Rogart, 1977). The generation of an action potential at each node results in the depolarization of the next node and subsequently the generation of an action potential with an internode delay of only about 20 μs, referred to as *saltatory conduction* (from the Latin *saltare*, "to dance"). Growing evidence indicates that, near the nodes of Ranvier and underneath the myelin covering, K^+ channels may play a role in determining the resting membrane potential and the repolarization of the action potential. A cause of some neurological disorders, such as multiple sclerosis and Guillain–Barré syndrome, is the demyelination of axons, resulting in a block of conduction of the action potentials (see Chapter 1).

Ion Channels Are Membrane-Spanning Proteins with Water-Filled Pores

The generation of ionic currents useful for the propagation of action potentials requires the movement of

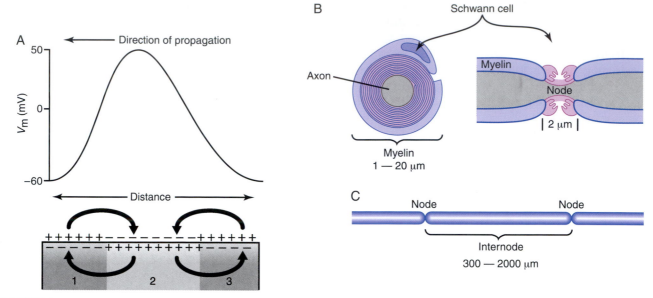

FIGURE 5.8 Propagation of the action potential in unmyelinated and myelinated axons. (A) Action potentials propagate in unmyelinated axons through the depolarization of adjacent regions of membrane. In the illustrated axon, region 2 is undergoing depolarization during the generation of the action potential, while region 3 has already generated the action potential and is now hyperpolarized. The action potential will propagate further by depolarizing region 1. (B) Vertebrate myelinated axons have a specialized Schwann cell that wraps around them in many spiral turns. The axon is exposed to the external medium at the nodes of Ranvier (Node). (C) Action potentials in myelinated fibers are regenerated at the nodes of Ranvier, where there is a high density of Na$^+$ channels. Action potentials are induced at each node through the depolarizing influence of the generation of an action potential at an adjacent node, thereby increasing the conduction velocity.

significant numbers of ions across the membrane in a relatively short time. The rate of ionic flow during the generation of an action potential is far too high to be achieved by an active transport mechanism and results instead from the opening of ion channels. Although the existence of ionic channels in the membrane has been postulated for decades, their properties and structure have only recently become known in detail. The powerful combination of electrophysiological and molecular techniques, and, most recently, X-ray crystallography, has greatly enhanced the knowledge of the structure–function relations of ionic channels (Catterall, 1995, 2000a,b; Jiang et al., 2002a,b; Yellen, 2002; Gouaux and MacKinnon, 2005) (Box 5.4).

Various neural toxins were particularly useful in the initial isolation of ionic channels. For example, three subunits (α, $\beta 1$, $\beta 2$) of the voltage-dependent Na$^+$ channel were isolated with the use of a derivative of a scorpion toxin (Beneski and Catterall, 1980; Catterall, 2000b). The α-subunit of the Na$^+$ channel is a large glycoprotein with a molecular mass of 270 kDa, whereas the $\beta 1$ and $\beta 2$ subunits are smaller polypeptides of molecular masses 39 and 37 kDa, respectively (Fig. 5.9). The α subunit, of which there are at least nine different isoforms, is the building block of the water-filled pore of the ionic channel, whereas the β subunits have some other role, such as in the regulation or structure of the native channel.

The α subunit of the Na$^+$ channel contains four internal repetitions (see Fig. 5.9B). Hydrophobicity analysis of these four components reveals that each contains six hydrophobic domains that may span the membrane as an α-helix. Of these six membrane-spanning components, the fourth (S4) has been proposed to be critical to the voltage sensitivity of the Na$^+$ channels. Voltage-sensitive gating of Na$^+$ channels is accomplished by the redistribution of ionic charge ("gating charge") in the Na$^+$ channel (Armstrong, 1992). Positive charges in the S4 region may act as voltage sensors such that an increase in the positivity of the inside of the cell results in a conformational change of the ionic channel. In support of this hypothesis, site-directed mutagenesis of the S4 region of the Na$^+$ channel to reduce the positive charge of this portion of the pore also reduces the voltage sensitivity of activation of the ionic channel (Catterall, 2000b). See Chapter 6 for additional discussion.

The mechanisms of inactivation of ionic channels have been analyzed with a combination of molecular and electrophysiological techniques. The most convincing hypothesis is that inactivation is achieved by a block of the inner mouth of the aqueous pore. Ionic channels are inactivated without detectable movement of ionic current through the membrane; thus inactivation is probably not directly gated by changes in the membrane potential alone. Rather, inactivation

BOX 5.4

ION CHANNELS AND DISEASE

Cells cannot survive without functional ion channels. It is therefore not surprising that an ever-increasing number of diseases have been found to be associated with defective ion channel function. There are a number of different mechanisms by which this may occur.

1. Mutations in the coding region of ion channel genes may lead to gain or loss of channel function, either of which may have deleterious consequences. For example, mutations producing enhanced activity of the epithelial Na^+ channel are responsible for Liddle syndrome, an inherited form of hypertension, whereas other mutations in the same protein that cause reduced channel activity give rise to hypotension. The most common inherited disease in Caucasians is also an ion channel mutation. This disease is cystic fibrosis (CF), which results from mutations in the epithelial chloride channel, known as CFTR. The most common mutation, the deletion of a phenylalanine at position 508, results in defective processing of the protein and prevents it from reaching the surface membrane. CFTR regulates chloride fluxes across epithelial cell membranes, and this loss of CFTR activity leads to reduced fluid secretion in the lung, resulting in potentially fatal lung infections.

2. Mutations in the promoter region of the gene may cause under- or overexpression of a given ion channel.

3. Other diseases result from defective regulation of channel activity by cellular constituents or extracellular ligands. This defective regulation may be caused by mutations in the genes encoding the regulatory molecules themselves or defects in the pathways leading to their production. Some forms of maturity-onset diabetes of the young (MODY) may be attributed to such a mechanism. ATP-sensitive potassium (K-ATP) channels play a key role in the glucose-induced insulin secretion from pancreatic β cells, and their defective regulation is responsible for two forms of MODY.

4. Autoantibodies to channel proteins may cause disease by down-regulating channel function—often by causing internalization of the channel protein itself. Well-known examples are myasthenia gravis, which results from antibodies to skeletal muscle acetylcholine channels, and Eaton–Lambert myasthenic syndrome, in which patients produce antibodies against presynaptic Ca channels.

5. Finally, a number of ion channels are secreted by cells as toxic agents. They insert into the membrane of the target cell and form large nonselective pores, leading to cell lysis and death. The hemolytic toxin produced by the bacterium *Staphylococcus aureus* and the toxin secreted by the protozoan *Entamoeba histolytica*, which causes amebic dysentery, are examples.

Natural mutations in ion channels have been invaluable in studying the relationship between channel structure and function. In many cases, genetic analysis of a disease has led to cloning of the relevant ion channel. The first K channel to be identified (Shaker), for example, came from the cloning of the gene that caused *Drosophila* to shake when exposed to ether. Likewise, the gene encoding the primary subunit of a cardiac potassium channel (KVLQT1) was identified by positional cloning in families carrying mutations that caused a cardiac disorder known as *long QT syndrome* (see later). Conversely, the large number of studies on the relationship between Na channel structure and function have greatly assisted our understanding of how mutations in Na channels produce their clinical phenotypes.

Many diseases are genetically heterogeneous, and the same clinical phenotype may be caused by mutations in different genes. Long QT syndrome is a relatively rare inherited cardiac disorder that causes abrupt loss of consciousness, seizures, and sudden death from ventricular arrhythmia in young people. Mutations in three different genes, two types of cardiac muscle K channels (HERG and KVLQT1) and the cardiac muscle sodium channel (SCN1A), give rise to long QT syndrome. The disorder is characterized by a long QT interval in the electrocardiogram, which reflects the delayed repolarization of the cardiac action potential. As might therefore be expected, the mutations in the cardiac Na channel gene that cause long QT syndrome enhance the Na current (by reducing Na channel inactivation), while those in the potassium channel genes cause loss of function and reduce the K current.

Mutations in many different types of ion channels have been shown to cause human diseases. In addition to the examples listed previously, mutations in water channels cause nephrogenic diabetes insipidus; mutations in gap junction channels cause Charcot–Marie–Tooth disease (a form of peripheral neuropathy) and hereditary deafness; mutations in the skeletal muscle Na channel cause a range of disorders known as the *periodic paralyses*; mutations in intracellular Ca-release channels cause malignant hyperthermia (a disease in which inhalation

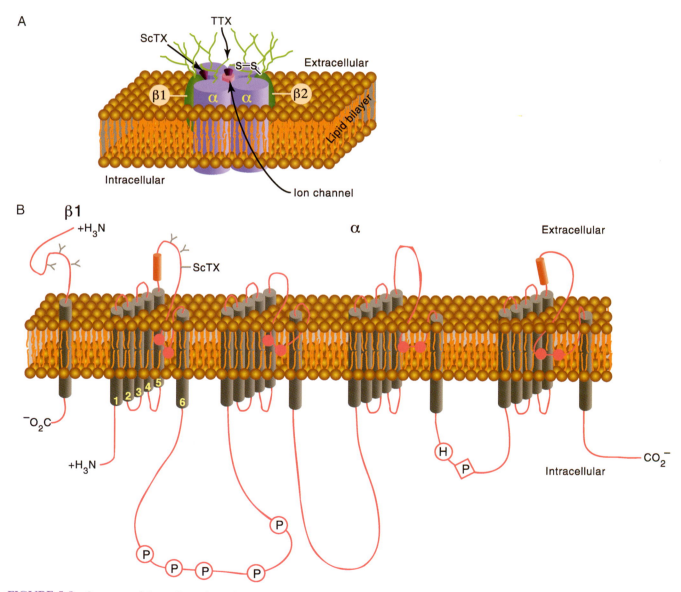

FIGURE 5.9 Structure of the sodium channel. (A) Cross section of a hypothetical sodium channel consisting of a single transmembrane α subunit, which contains four membrane-spanning components (See Part B), in association with a β1 subunit and a β2 subunit. The α subunit has receptor sites for α-scorpion toxin (ScTX) and tetrodotoxin (TTX). (B) Primary structures of α and β1 subunits of sodium channel illustrated as transmembrane folding diagrams. Cylinders represent probable transmembrane α-helices.

may be triggered or facilitated as a secondary consequence of activation. Site-directed mutagenesis or the use of antibodies has shown that the part of the molecule between regions III and IV may be allowed to move to block the cytoplasmic side of the ionic pore after the conformational change associated with activation (Vassilev *et al.*, 1988, 1989; Stuhmer *et al.*, 1989). Additional information on the molecular properties of voltage-gated ion channels is provided in Chapter 6.

Neurons of the Central Nervous System Exhibit a Wide Variety of Electrophysiological Properties

The first intracellular recordings of action potentials in mammalian neurons by Sir John Eccles and colleagues revealed a remarkable similarity to those of the squid giant axon and gave rise to the assumption that the electrophysiology of neurons in the CNS was really rather simple: when synaptic potentials brought the membrane potential positive to action potential threshold, action potentials were produced through an increase in Na^+ conductance followed by an increase in K^+ conductance, as in the squid giant axon. The assumption, therefore, was that the complicated patterns of activity generated by the brain during the resting, sleeping, or active states were brought about as an interaction of the very large numbers of neurons present in the mammalian CNS (Brock *et al.*, 1952; Eccles, 1957). However, intracellular recordings of invertebrate neurons revealed that different cell types exhibit a wide variety of different electrophysiological behaviors, indicating that neurons may be significantly more complicated than the squid giant axon (Arvanitaki and Chalazonitis, 1961; Alving, 1968; Jackelet, 1989). Elucidation of the basic electrophysiology and synaptic physiology of different types of neurons and neuronal pathways within the mammalian CNS was facilitated by the *in vitro* slice technique, in which thin ($\sim$0.5 mm) slices of brain can be maintained for several hours. Intracellular recordings from identified cells revealed that neurons of the mammalian nervous system, similar to those of invertebrate networks, can generate complex patterns of action potentials entirely through intrinsic ionic mechanisms and without synaptic interaction with other cell types. For example, Rodolfo Llinás and colleagues discovered that Purkinje cells of the cerebellum can generate high-frequency ($>$ 200 Hz) trains of Na^+- and K^+-mediated action potentials interrupted by Ca^{2+} spikes in the dendrites (Llinás and Sugimori, 1980a,b), whereas a major afferent to these neurons, the inferior olivary cell, can generate

rhythmic sequences of broad action potentials only at low frequencies ($<$ 15 Hz) through an interaction between various Ca^{2+}, Na^+, and K^+ conductances (Llinás and Yarom, 1981a,b) (Fig. 5.10). These *in vitro* recordings confirmed a major finding obtained with earlier intracellular recordings *in vivo*: each morphologically distinct class of neuron in the brain exhibits distinct electrophysiological features (Llinás, 1988). Just as cortical pyramidal cells are morphologically distinct from cerebellar Purkinje cells, which are distinct from thalamic relay cells, the electrophysiological properties of each of these different cell types also are markedly distinct.

Although no uniform classification scheme has been formulated in which all the different types of neurons of the brain can be classified, a few characteristic patterns of activity seem to recur. The first general class of action potential generation is characterized by those cells that generate trains of action potentials one spike at a time. The more prolonged the depolarization of these cells, the more prolonged their discharge. The more intensely these cells are depolarized, the higher the frequency of action potential generation. This type of relatively linear behavior is typical of brain stem and spinal cord motor neuron functioning in muscle contraction. A modification of this basic pattern of "regular firing" is characterized by the generation of trains of action potentials that exhibit a marked tendency to slow down in frequency with time, a process known as *spike frequency adaptation*. Examples of cells that discharge in this manner are cortical and hippocampal pyramidal cells (Madison and Nicoll, 1984; McCormick *et al.*, 1985; Pennefather *et al.*, 1985).

In addition to these regular firing cells, many neurons in the central nervous system exhibit the intrinsic propensity to generate rhythmic bursts of action potentials (Fig. 5.10) (see also Chapter 7). Examples of such neurons are thalamic relay neurons, inferior olivary neurons, and some types of cortical and hippocampal pyramidal cells (Llinás and Yarom, 1981a,b; Jahnsen and Llinás, 1984a,b; Wang and McCormick, 1993). In these cells, clusters of action potentials can occur together when the membrane is brought above firing threshold. These clusters of action potentials are typically generated through the activation of specialized Ca^{2+} currents that, through their slower kinetics, allow the membrane potential to be depolarized for a sufficient period to result in the generation of a burst of regular Na^+- and K^+-dependent action potentials (discussed in the next section).

Yet another general category of neurons in the brain comprises cells that generate relatively short duration ($<$ 1 ms) action potentials and can discharge at relatively

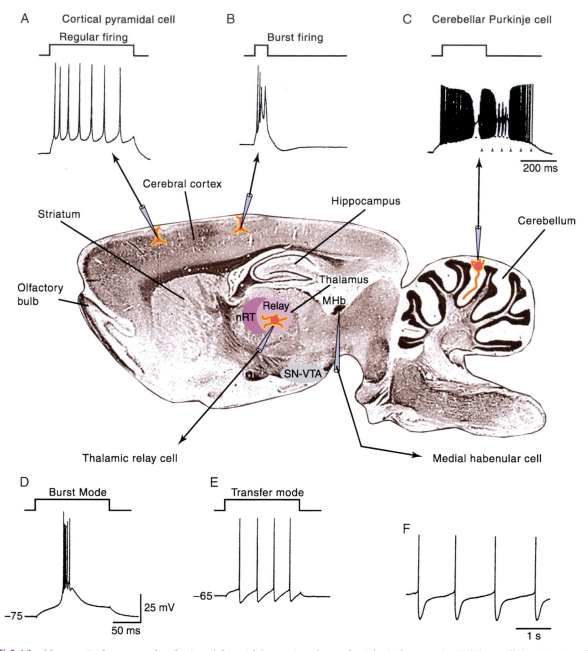

FIGURE 5.10 Neurons in the mammalian brain exhibit widely varying electrophysiological properties. (A) Intracellular injection of a depolarizing current pulse in a cortical pyramidal cell results in a train of action potentials that slow down in frequency. This pattern of activity is known as "regular firing." (B) Some cortical cells generated bursts of three or more action potentials, even when depolarized only for a short period. (C) Cerebellar Purkinje cells generate high-frequency trains of action potentials in their cell bodies that are disrupted by the generation of Ca^{2+} spikes in their dendrites. These cells can also generate "plateau potentials" from the persistent activation of Na^+ conductances (arrowheads). (D) Thalamic relay cells may generate action potentials either as bursts (D) or as tonic trains (E) of action potentials owing to the presence of a large low-threshold Ca^{2+} current. (F) Medial habenular cells generate action potentials at a steady and slow rate, in a "pacemaker" fashion.

high frequencies (> 300 Hz). Such electrophysiological properties are often found in neurons that release the inhibitory amino acid γ-aminobutyric acid (Llinás and Sugimori, 1980a,b) (see Fig. 5.10) including some types of interneurons in the cerebral cortex, thalamus, and hippocampus (Schwartzkroin and Mathers, 1978; McCormick et al., 1985; Pape and McCormick, 1995).

Finally, the last general category of neurons consists of those that spontaneously generate action potentials at relatively low frequencies (e.g., 1–10 Hz). This type of electrophysiological behavior is often associated with neurons that release neuromodulatory transmitters, such as acetylcholine, norepinephrine, serotonin, and histamine (Vandermaelen and Aghajanian, 1983; Williams *et al.*, 1984; Reiner and McGeer, 1987). Neurons that release these neuromodulatory substances often innervate wide regions of the brain and appear to set the "state" of the different neural networks of the CNS, in a manner similar to the modulation of the different organs of the body by the sympathetic and parasympathetic nervous systems (McCormick, 1992; Steriade and McCarley, 2005).

Each of these unique intrinsic patterns of activity in the nervous system is due to the presence of a distinct mixture and distribution of different ionic currents in the cells. As in the classic studies of the squid giant axon, these different ionic currents have been characterized, at least in part, with voltage-clamp and pharmacological techniques, and the basic electrophysiological properties have been replicated with computational simulations (Belluzzi and Sacchi, 1991; McCormick and Huguenard, 1992; Huguenard and McCormick, 1994) (see Fig. 5.7, Fig. 5.12, and Chapter 7).

Neurons Have Multiple Active Conductances

The search for the electrophysiological basis of the varying intrinsic properties of different types of neurons of vertebrates and invertebrates revealed a wide variety of ionic currents. Each type of ionic current is characterized by several features: (1) the type of ions conducted by the underlying ionic channels (e.g., Na^+, K^+, Ca^{2+}, Cl^-, or mixed cations), (2) their voltage and time dependence, and (3) their sensitivity to second messengers. In vertebrate neurons, two distinct Na^+ currents have been identified and six distinct Ca^{2+} currents and more than seven distinct K^+ currents are known (Table 5.2 and Fig. 5.11). This is a minimal number, as these currents are formed from a much greater pool of channel subunits. The following sections briefly review these classes of ionic currents and their ionic channels, relating them to the different patterns of behavior mentioned earlier for neurons in the mammalian CNS.

TABLE 5.2 Neuronal Ionic Currents

Current	Description	Function
Na^+ currents		
$I_{Na,t}$	Transient; rapidly activating and inactivating	Action potentials
$I_{Na,p}$	Persistent; noninactivating	Enhances depolarization; contributes to steady-state firing
Ca^{2+} currents		
I_T, low threshold	Transient; rapidly inactivating; threshold negative to −65mV	Underlies rhythmic burst firing
I_L, high threshold	Long-lasting; slowly inactivating; threshold around −20 mV	Underlies Ca^{2+} spikes that are prominent in dendrites
I_N	Neither; rapidly inactivating; threshold around −20 mV	Underlies Ca^{2+} spikes that are prominent in dendrites
I_P	Purkinje; threshold around −50 mV	Contributes to the generation of dendritic Ca^{2+} spikes in Purkinje cells
K^+ currents		
I_K	Activated by strong depolarization	Repolarization of action potential
I_C	Activated by increases in $[Ca^{2+}]_i$	Action potential repolarization and interspike interval
I_{AHP}	Slow afterhyperpolarization; sensitive to increases in $[Ca^{2+}]$	Slow adaptation of action potential discharge; the block of this current by neuromodulators enhances neuronal excitability
I_A	Transient; inactivating	Delayed onset of firing; lengthens interspike interval; action potential repolarization
I_M	Muscarine sensitive; activated by depolarization	Contributes to spike frequency adaptation; the block of this noninactivating current by neuromodulators enhances neuronal excitability
I_h	Depolarizing (mixed cation) current that is activated by hyperpolarization	Contributes to rhythmic burst firing and other rhythmic activities
$I_{K,leak}$	Contributes to neuronal resting membrane potential	Block of this current by neuromodulators can result in a sustained change in membrane potential

Inward currents

Outward currents

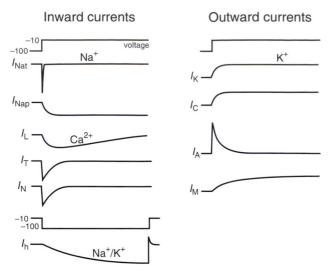

FIGURE 5.11 Voltage dependence and kinetics of different ionic currents in the mammalian brain. Depolarization of the membrane potential from –100 to –10 mV results in the activation of currents entering or leaving neurons. Stepping from –10 to –100 mV can also activate a mixed Na$^+$/K$^+$ current known as I_h.

Na$^+$ Currents Are Both Transient and Persistent

Depolarization of many different types of vertebrate neurons results not only in the activation of the rapidly activating and inactivating Na$^+$ current (I_{Nat}) underlying action potential generation but also in the rapid activation of a Na$^+$ current that does not inactivate and is therefore known as the "persistent" Na$^+$ current (I_{Nap}) (Hotson et al., 1979; Stafstrom et al., 1982; Llinás, 1988; Alzheimer et al., 1993). The threshold for activation of the persistent Na$^+$ current is typically about –65 mV, that is, below the threshold for the generation of action potentials. This property gives this current the interesting ability to enhance or facilitate the response of the neuron to depolarizing, yet subthreshold, inputs. For example, synaptic events that depolarize the cell activate I_{Nap}, resulting in an extra influx of positive charge and therefore a larger depolarization than otherwise would occur. Likewise, hyperpolarizations may result

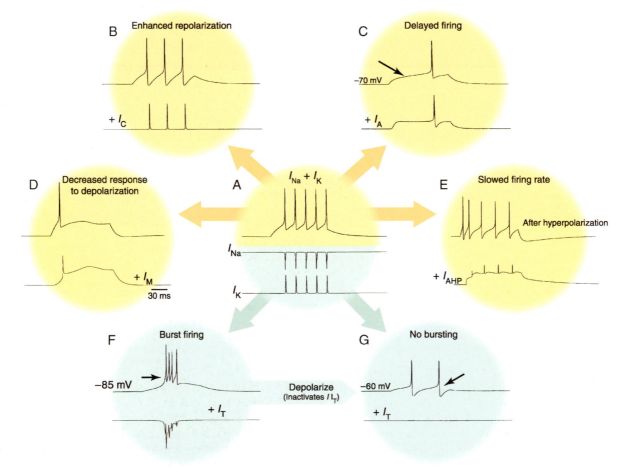

FIGURE 5.12 Simulation of the effects of the addition of various ionic currents to the pattern of activity generated by neurons in the mammalian CNS. (A) The repetitive impulse response of the classic Hodgkin–Huxley model (voltage recordings above, current traces below). With only I_{Na} and I_K, the neuron generates a train of five action potentials in response to depolarization. Addition of I_C (B; along with a Ca^{2+} current to activate IC) enhances action potential repolarization. Addition of I_A (C) delays the onset of action potential generation. Addition of I_M (D) decreases the ability of the cell to generate a train of action potentials. Addition of I_{AHP} (E; along with a Ca^{2+} current to activate IAHP) slows the firing rate and generates a slow afterhyperpolarization. Finally, addition of the transient Ca^{2+} current I_T results in two states of action potential firing: (F) burst firing at –85 mV and (G) tonic firing at –60 mV. From Huguenard and McCormick (1994).

in deactivation of I_{Nap}, again resulting in larger hyperpolarizations than would otherwise occur. In this manner, the persistent Na$^+$ current may play an important regulatory function in the control of the functional responsiveness of the neuron to synaptic inputs and may contribute to the dynamic coupling of the dendrites to the soma.

Persistent activation of I_{Nap} may also contribute to another electrophysiological feature of neurons, namely, the generation of plateau potentials (Llinás, 1988). A *plateau potential* refers to the ability of many different types of neurons to generate, through intrinsic ionic mechanisms, a prolonged (from tens of milliseconds to seconds) depolarization and action potential discharge in response to a short-lasting depolarization (see Fig. 5.10C). One can wonder whether such plateau potentials contribute to persistent firing in neurons during the performance of visual memory tasks, as has been found in some types of neurons in the prefrontal neocortex and superior colliculus of behaving primates (Goldman-Rakic, 1995).

K$^+$ Currents Vary in Their Voltage Sensitivity and Kinetics

Potassium currents that contribute to the electrophysiological properties of neurons are numerous and exhibit a wide range of voltage-dependent and kinetic properties (Jan and Jan, 1990; Storm, 1990; Johnston and Wu, 1995; Coetzee *et al.*, 1999; Yellen, 2002) (see also Chapter 7 for additional details). Perhaps the simplest K$^+$ current is that characterized by Hodgkin and Huxley: this K$^+$ current, I_K, rapidly activates on depolarization and does not inactivate (see Fig. 5.11). Other K$^+$ currents activate with depolarization but also inactivate with time. For example, the rapid activation and inactivation of I_A give this current a transient appearance (see Fig. 5.11), and I_A is believed to be important in controlling the rate of action potential generation, particularly at low frequencies (Connor and Stevens, 1971a,b) (Fig. 5.12). Like the Na$^+$ channel, I_A channels are inactivated by the plugging of the inner mouth of the pore through the movement of an inactivation particle (Hoshi *et al.*, 1990; Zagotta *et al.*, 1990; Yellen, 2002).

Another broad class of K$^+$ channel consists of those that are sensitive to changes in the intracellular concentration of Ca^{2+} (Blatz and Magleby, 1987; Latorre *et al.*, 1989). These K$^+$ currents are collectively referred to as I_{KCa}, with two examples being the K$^+$ currents that underlie slow afterhyperpolarizations following repetitive action potential discharge, I_{AHP}, and a fast K$^+$ current that helps repolarize action potentials, I_C (see Fig. 5.11). Still other K$^+$ channels are not only activated by voltage but also modulated by activation of various modulatory neurotransmitter receptors, such as the M current (see Fig. 5.11). By investigating the ionic mechanisms by which the release of acetylcholine from preganglionic neurons in the brain results in prolonged changes in the excitability of neurons of the sympathetic ganglia, Brown and Adams (1980) discovered a unique K$^+$ current (I_M) that slowly (over tens of milliseconds) turns on with depolarization of the neuron (see Fig. 5.12D). The slow activation of this K$^+$ current results in a decrease in the responsiveness of the cell to depolarization and, therefore, regulates how the cell responds to excitation. This K$^+$ current, like I_{AHP}, is reduced by the activation of a wide variety of receptors, including muscarinic receptors, for which it is named. Reduction of I_M results in a marked increase in responsiveness of the affected cell to depolarizing inputs and again may contribute to the mechanisms by which neuromodulatory systems control the state of activity in cortical and hippocampal networks (Nicoll, 1988; Nicoll *et al.*, 1990; McCormick, 1992).

Between these classic examples of K$^+$ currents are a variety of other types that have not been fully characterized, including K$^+$ currents that vary from one another in their voltage sensitivity, kinetics, and response to various second messengers. Molecular biological studies of voltage-sensitive K$^+$ channels, first done in *Drosophila* and later in mammals, have revealed a large number of genes that generate K$^+$ channels. The voltage-gated K$^+$ channel subunits are contained within nine distinct subfamilies: Kv1-9 (reviewed in Coetzee *et al.*, 1999). These genes generate a wide variety of different K$^+$ channels due not only to the large number of genes involved, but also to alternative RNA splicing, gene duplication, and other post-translational mechanisms. Functional expression of different K$^+$ channels reveals remarkable variation in the rate of inactivation, such that some are rapidly inactivating (A-current-like), whereas others inactivate more slowly. Finally, some K$^+$ channels do not inactivate, such as I_K. One of the largest subfamilies of K$^+$ channels comprises those that give rise to the resting membrane potential, so-called "leak channels." Interestingly, these channels appear to be opened by gaseous anesthetics, indicating that the hyperpolarization of central neurons is a major component of general anesthesia. Recently, MacKinnon and colleagues have succeeded in crystallizing different types of K$^+$ channels, leading to a great leap in our knowledge of their structure and how they function, including the mechanisms by which channels are opened by voltage and ligands (Jiang *et al.*, 2002a,b; Gouaux and MacKinnon, 2005). It is now clear that each type of neuron in the nervous system is likely to contain a unique set of functional voltage-sensitive

K^+ channels, perhaps selected, modified, and placed in particular spatial locations in the cell in a manner to facilitate the unique role of that cell type in neuronal processing.

Ca^{2+} Currents Control Electrophysiological Properties and Ca^{2+}-Dependent Second-Messenger Systems

Ionic channels that conduct Ca^{2+} are present in all neurons. These channels are special in that they serve two important functions. First, Ca^{2+} channels are present throughout the different parts of the neuron (dendrites, soma, synaptic terminals) and contribute greatly to the electrophysiological properties of these processes (Llinás, 1988; Regehr and Tank, 1994; Johnston et al., 1996). Second, Ca^{2+} channels are unique in that Ca^{2+} is an important second messenger in neurons, and entry of Ca^{2+} into the cell can affect numerous physiological functions, including neurotransmitter release, synaptic plasticity, neurite outgrowth during development, and even gene expression.

On the bases of their voltage sensitivity, their kinetics of activation and inactivation, and their ability to be blocked by various pharmacological agents, Ca^{2+} currents can be separated into at least four distinct categories, three of which are I_T ("transient"), I_L ("long lasting"), and I_N ("neither") (Carbonne and Lux, 1984; Nowycky et al., 1985), illustrated in Fig. 5.11. A fourth, I_P, is found in the Purkinje cells of the cerebellum, as well as in many different cell types of the CNS (Llinás et al., 1992). Calcium channels are formed from at least 10 different α subunits as well as a variety of β and γ subunits, indicating that an even greater number of Ca^{2+} currents are yet to be characterized (Tsien et al., 1991; Ertel et al., 2000).

Neurons Possess Multiple Subtypes of High-Threshold Ca^{2+} Currents

High-voltage-activated Ca^{2+} channels are activated at membrane potentials more positive than approximately −40 mV and include the currents I_L, I_N, and I_P. The L-type calcium currents exhibit a high threshold for activation (about −10 mV) and give rise to rather persistent, or long-lasting, ionic currents (see Fig. 5.11). Dihydropyridines, Ca^{2+} channel antagonists, are clinically useful for their effects on the heart and vascular smooth muscle (e.g., for the treatment of arrhythmias, angina, and migraine headaches) and selectively block L-type Ca^{2+} channels (Stea et al., 1995; Bean, 1989). In contrast with I_L, I_N is not blocked by dihydropyridines; rather it is selectively blocked by a toxin found in Pacific cone shells ($\bar{\omega}$-conotoxin-

GVIA). The N-type Ca^{2+} channels have a threshold for activation of about −20 mV, inactivate with maintained depolarization, and are modulated by a variety of neurotransmitters. In some cell types, I_N has a role in the Ca^{2+}-dependent release of neurotransmitters at presynaptic terminals (Wheeler et al., 1994). The P-type calcium channel is distinct from N and L types in that it is not blocked by either dihydropyridines or $\bar{\omega}$-conotoxin-GVIA but is blocked by a toxin ($\bar{\omega}$-agatoxin-IVA) present in the venom of the funnel web spider (Llinás et al., 1992; Stea et al., 1995). This type of calcium channel activates at relatively high thresholds and does not inactivate. Prevalent in Purkinje cells as well as other cell types, as mentioned earlier, the P-type Ca^{2+} channel participates in the generation of dendritic Ca^{2+} spikes, which can strongly modulate the firing pattern of the neuron in which it resides (see Fig. 5.10C).

Collectively, the high-threshold-activated Ca^{2+} channels contribute to the generation of action potentials in mammalian neurons. The activation of Ca^{2+} currents adds somewhat to the depolarizing part of the action potential, but, more importantly, these channels allow Ca^{2+} to enter the cell, and this has the secondary consequence of activation of various Ca^{2+}-activated K^+ currents (Latorre et al., 1989) and protein kinases (see Chapters 12 and 18). As mentioned earlier, the activation of these K^+ currents modifies the pattern of action potentials generated in the cell (see Figs. 5.10 and 5.12). High-threshold Ca^{2+} channels are similar to the Na^+ channel in that they are composed of a central $\alpha 1$ subunit that forms the aqueous pore and several regulatory or auxiliary subunits. As in the Na^+ channel, the primary structure of the $\alpha 1$ subunit of the Ca^{2+} channel consists of four homologous domains (I–IV), each containing six regions (S1–S6) that may generate transmembrane α-helices. Genes for at least 10 different Ca^{2+} channel α subunits have been cloned and are separated into three families (CaV1, CaV2, CaV3). The properties of the products of these genes indicate that I_L is likely to correspond to the CaV1 subfamily, whereas I_N corresponds to CaV2.2 and I_T is formed by the CaV3 subfamily (Bean, 1989; Catterall, 2000a).

Low-Threshold Ca^{2+} Currents Generate Bursts of Action Potentials

Low-threshold Ca^{2+} currents (see Fig. 5.11) often take part in the generation of rhythmic bursts of action potentials (see Figs. 5.10 and 5.12). The low-threshold Ca^{2+} current is characterized by a threshold for activation of about −65 mV, which is below the threshold for generation of typical Na^+–K^+-dependent action potentials (–55 mV). This current inactivates with maintained depolarization. Owing to these properties, the role of low-threshold Ca^{2+} currents differs markedly from that of the high-threshold Ca^{2+} currents. Through activation and inactivation of the low-threshold Ca^{2+} current, neurons can generate slow (about 100 ms) Ca^{2+} spikes, which can result, owing to their prolonged duration, in the generation of a high-frequency "burst" of short-duration Na^+– K^+ action potentials (see Fig. 5.10 and Box 5.5) (Llinás and Jahnsen, 1982).

In the mammalian brain, this pattern is especially well exemplified by the activity of thalamic relay neurons; in the visual system, these neurons receive direct input from the retina and transmit this information to the visual cortex. During periods of slow wave sleep, the membrane potential of these relay neurons is relatively hyperpolarized, resulting in the removal of inactivation (deinactivation) of the low-threshold Ca^{2+} current. This deinactivation allows these cells to spontaneously generate low-threshold Ca^{2+} spikes and bursts of from two to five action potentials (Fig. 5.13) (McCormick and Pape, 1990). The large number of thalamic relay cells bursting during sleep in part gives rise to the spontaneous synchronized activity that early investigators were so surprised to find during recordings from the brains of sleeping animals (Steriade *et al.*, 1993). It has even proved possible to maintain one of the sleep-related brain rhythms (spindle waves) intact in slices of thalamic tissue maintained *in vitro*, owing to the generation of this rhythm by the interaction of a local network of thalamic cells and their electrophysiological properties (von Krosigk *et al.*, 1993).

BOX 5.5

JELLYFISH—WHAT A NERVE!

Research on jellyfish provides intriguing insight into how the properties and distribution of ion channels within a nerve membrane can affect the behavior of the whole animal. *Aglantha digitale* can swim slowly when feeding or quickly if escaping predators just through the "behavior" of a single muscle sheet coupled to a simply organized nervous system.

The jellyfish does this through an unusual form of signaling. Each "giant" motor nerve axon not only has voltage-dependent sodium channels and three types of potassium channels, but also crucial T-type calcium channels. *Aglantha* motor axons are unusual because they develop two entirely different propagating action potentials (Mackie and Meech, 1985). The T-type calcium channels contribute to a low-amplitude calcium-dependent spike that propagates along the motor axon without gaining amplitude or decrementing in the way that electrotonic potentials do (Meech and Mackie, 1995). The motor axon makes direct synaptic contact with the muscle epithelium that makes up the bell of the jellyfish and so the propagating calcium spike induces the weak contractions responsible for propulsion during the regular slow swimming the animal performs when feeding.

Aglantha lives in the colder waters of the world at a depth of about 100 m. Studied in their natural habitat, they are seen to avoid predators by generating an altogether stronger form of swimming. In the laboratory this "escape" swimming can be reproduced by stimulating vibration-sensitive receptors at the base of the bell of the animal (Arkett *et al.*, 1988). The strong synaptic depolarization that this stimulus induces in each of the eight giant motor axons drives its membrane potential beyond the peak of the calcium spike and induces a full-sized sodium action potential. As the sodium spike propagates more rapidly than the slow swim calcium spike, there is a coordinated contraction of the body wall that drives the animal forward.

Sodium and calcium spikes like those seen in *Aglantha* have been recorded from a variety of sites in the mammalian CNS (Llinás and Yarom, 1981b; Llinás and Jahnsen, 1982). However, unlike in *Aglantha*, the peak of the calcium spike always exceeds the threshold of the sodium spike and the two impulses form a single complex signal. Patch-clamp analysis of *Aglantha* axons has revealed a family of potassium channels that are responsible for setting thresholds and repolarizing each of the two different impulses. Each potassium channel class has an identical unitary conductance and appears to be organized in a mosaic fashion over the surface of the axon (Meech and Mackie, 1993). Sodium and T-type calcium channels are clustered together into well-defined "hot spots." George Mackie and I have suggested that the mosaic organization facilitates the turnover of ion channels; channels inserted into the membrane in clusters age together and are eliminated together.

Robert W. Meech

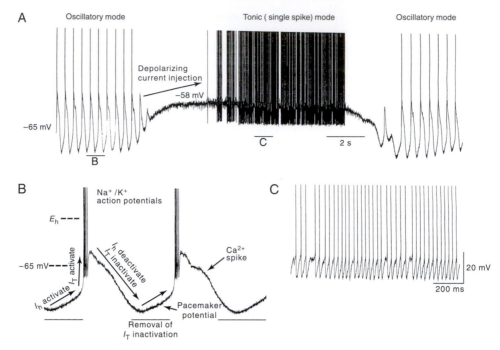

FIGURE 5.13 Two different patterns of activity generated in the same neuron, depending on membrane potential. (A) The thalamic neuron spontaneously generates rhythmic bursts of action potentials owing to the interaction of the Ca^{2+} current I_T and the inward "pacemaker" current I_h. Depolarization of the neuron changes the firing mode from rhythmic burst firing to tonic action potential generation in which spikes are generated one at a time. Removal of this depolarization reinstates the rhythmic burst firing. This transition from rhythmic burst firing to tonic activity is similar to that which occurs in the transition from sleep to waking. (B) Expansion of detail of rhythmic burst firing. (C) Expansion of detail of tonic firing. From McCormick and Pape (1990).

The transition to waking or the period of sleep when dreams are prevalent (rapid eye movement sleep) is associated with a maintained depolarization of thalamic relay cells to membrane potentials ranging from about –60 to –55 mV. The low-threshold Ca^{2+} current is inactivated and therefore the burst discharges are abolished. In this way, the properties of a single ionic current (I_T) help to explain in part the remarkable changes in brain activity taking place in the transition from sleep to waking (Fig. 5.13).

Low-threshold Ca^{2+} channels were cloned and shown to have some similarities to other Ca^{2+} channels (Perez-Reyes *et al.*, 1998). Evidence suggests that some antiepileptic drugs may exert their therapeutic actions through a reduction in I_T. This is especially true of the drugs useful in the treatment of generalized absence (petit mal) seizures, which are known to rely on the thalamus for their generation (Coulter *et al.*, 1990).

Hyperpolarization-Activated Ionic Currents Are Involved in Rhythmic Activity

In most types of neurons, hyperpolarization negative to approximately –50 mV activates an ionic current, known as I_h, that conducts both Na^+ and K^+ ions (see Fig. 5.11). This current typically has very slow kinetics, turning on with a time constant on the order of tens of milliseconds to seconds. Because the channels underlying this current allow the passage of both Na^+ and K^+ ions, the reversal potential of I_h is typically about –35 mV—between E_{Na} and E_K. Because this current is activated by hyperpolarization below approximately –60 mV, it is typically dominated by the inward movement of Na^+ ions and is therefore depolarizing. For what purpose could neurons use a depolarizing current that activates when the cell is hyperpolarized? A clue comes from cardiac cells in which this current, known as I_f for "funny," is important for determining heart rate (DiFrancesco, 2005). Activation of I_f results in a slow depolarization of the membrane potential between adjacent cardiac action potentials. The more that I_f is activated, the faster the membrane depolarizes between beats and, therefore, the sooner the threshold for the next action potential is reached and the next beat is generated. In this manner, the amplitude, or sensitivity to voltage, of I_f can modify heart rate. Interestingly, the sensitivity of I_f to voltage is adjusted by the release of noradrenaline and acetylcholine; the activation of β-adrenoceptors by noradrenaline increases I_f and therefore increases the heart rate, whereas the activation of muscarinic receptors decreases I_f, thereby decreasing the heart rate (DiFrancesco *et al.*, 1989; DiFrancesco, 2005). This continual adjustment of I_f results from a "push–pull" arrangement between adrenergic and muscarinic cholinergic receptors and is mediated by the adjustment of intracellular levels of cyclic AMP. Indeed, the cloning of H channels revealed that their structure is similar to that of cyclic nucleotide-gated channels (Ludwig *et al.*, 1998).

Could I_h play a role in neurons similar to that of I_f in the heart? Possibly. Synchronized rhythmic oscillations in the membrane potential of large numbers of neurons, in some respects similar to those of the heart, are characteristic of the mammalian brain. Oscillations of this type are particularly prevalent in thalamic relay neurons during some periods of sleep, as mentioned earlier. Intracellular recordings from these thalamic neurons reveal that they often generate rhythmic "bursts" of action potentials mediated by the activation of a slow spike that is generated through the activation of the low-threshold, or transient, Ca^{2+} current, I_T (McCormick and Pape, 1990; McCormick and Huguenard, 1992) (see Fig. 5.13). Between the low-threshold Ca^{2+} spikes is a slowly depolarizing membrane potential generated by activation of the mixed Na^+–K^+ current I_h, as with I_f in the heart. The amplitude, or voltage sensitivity, of I_h controls the rate at which the thalamic cells oscillate, and, as with the heart, this sensitivity is adjusted by the release of modulatory neurotransmitters (see Fig. 5.13). In a sense, the thalamic neurons are "beating" in a manner similar to that of the heart.

Summary

An action potential is generated by the rapid influx of Na^+ ions followed by a slightly slower efflux of K^+ ions. Although the generation of an action potential does not disrupt the concentration gradients of these ions across the membrane, the movement of charge is sufficient to generate a large and brief deviation in the membrane potential. Action potentials are typically initiated in the axon initial segment, and the propagation of the action potential along the axon allows communication of the output of the cell to its distal synapses. Neurons possess many different types of ionic channels in their membranes, allowing complex patterns of action potentials to be generated and complex computations to occur within single neurons.

References

Alving, B. O. (1968). Spontaneous activity in isolated somata of *Aplysia* pacemaker neurons. *J. Gen. Physiol.* **51**, 29–45.

Alzheimer, C., Schwindt, P. C., and Crill, W. E. (1993). Modal gating of Na$^+$ channels as a mechanism of persistent Na$^+$ current in pyramidal neurons from rat and cat sensorimotor cortex. *J. Neurosci.* **13**, 660–673.

Arkett, S., Mackie, G. O., and Meech, R. W. (1988). Hair-cell mechano-reception in the jellyfish *Aglantha digitale*. *J. Exp. Biol.* **135**, 329–342.

Armstrong, C. M. (1992). Voltage-dependent ionic channels and their gating. *Physiol. Rev.* **72**(Suppl.), 5–13.

Armstrong, C. M., and Hille, B. (1972). The inner quaternary ammonium ion receptor in potassium channels of the node of Ranvier. *J. Gen. Physiol.* **59**, 388–400.

Arvanitaki, A., and Chalazonitis, N. (1961). Slow waves and associated spiking in nerve cells of *Aplysia*. *Bull. Inst. Oceanogr. Monaco* **58**, 1–15.

Bean, B. P. (1989). Classes of calcium channels in vertebrate cells. *Annu. Rev. Physiol.* **51**, 367–384.

Belluzzi, O., and Sacchi, O. (1991). A five-conductance model of the action potential in the rat sympathetic neurone. *Prog. Biophys. Mol. Biol.* **55**, 1–30.

Beneski, D. A., and Catterall, W. A. (1980). Covalent labeling of protein components of the sodium channel with a photoactivable derivative of scorpion toxin. *Proc. Natl. Acad. Sci. USA* **77**, 639–643.

Blatz, A. L., and Magleby, K. L. (1987). Calcium-activated potassium channels. *Trends Neurosci.* **11**, 463–467.

Brazier, M. A. B. (1959). The historical development of neurophysiology. In "Handbook of Physiology" (J. Field, Ed.), Sect. 1, Vol. 1, pp. 1–58. Am. Physiol. Soc., Washington, D.C.

Brazier, M. A. B. (1988). "A History of Neurophysiology in the 19th Century." Raven Press, New York.

Brock, L. G., Coombs, J. S., and Eccles, J. C. (1952). The recording of potentials from motoneurones with an intracellular electrode. *J. Physiol. (London)* **117**, 431–460.

Brown, D. A., and Adams, P. R. (1980). Muscarinic suppression of a novel voltage sensitive K$^+$ current in a vertebrate neurone. *Nature (London)* **283**, 673–676.

Bunge, R. P. (1968). Glial cells and the central myelin sheath. *Physiol. Rev.* **48**, 197–251.

Buser, P., and Albe-Fessard, D. (1953). Premiers resultats d'une analyse l'activite electrique du cortex cerebral du Chat par micro-electrodes intracellulaires. *C. R. Hebd. Seances Acad. Sci.* **236**, 1197–1199.

Carbonne, E., and Lux, H. D. (1984). A low voltage-activated, fully inactivating Ca channel in vertebrate sensory neurones. *Nature (London)* **310**, 501–502.

Catterall, W. A. (1995). Structure and function of voltage-gated ion channels. *Annu. Rev. Biochem.* **64**, 493–531.

Catterall, W. A. (2000a). Structure and regulation of voltage-gated Ca^{2+} channels. *Annu. Rev. Cell Dev. Biol.* **16**, 521–555.

Catterall, W. A. (2000b). From ionic currents to molecular mechanisms: The structure and function of voltage-gated sodium channels. *Neuron* **26**, 13–25.

Coetzee, W. A., Amarillo, Y., Chui, J., Chow, A., Lau, D., McCormack, T., Moreno, H., Nadal, M.S., Ozaita, A., Pountney, D., Saganich, M., Vega-Saenz de Miera, E., Rudy, B. (1999). Molecular diversity of K$^+$ channels. *Ann. N.Y. Acad. Sci.* **868**, 233–285.

Cole, K. S. (1949). Dynamic electrical characteristics of the squid axon membrane. *Arch. Sci. Physiol.* **3**, 253–258.

Cole, K. S., and Curtis, H. J. (1939). Electric impedance of the squid giant axon during activity. *J. Gen. Physiol.* **22**, 649–670.

Connor, J. A., and Stevens, C. F. (1971a). Voltage clamp studies of a transient outward membrane current in gastropod neural somata. *J. Physiol. (London)* **213**, 21–30.

Connor, J. A., and Stevens, C. F. (1971b). Prediction of repetitive firing behaviour from voltage clamp data on an isolated neurone soma. *J. Physiol. (London)* **213**, 31–53.

Coombs, J.S., Curtis, D.R., Eccles, J.C. (1957). The interpretation of spike potentials of motoneurons. *J. Physiol.* **139**, 198–231.

Coulter, D. A., Huguenard, J. R., and Prince, D. A. (1990). Differential effects of petit mal anticonvulsants and convulsants on thalamic neurones: Calcium current reduction. *Br. J. Pharmacol.* **100**, 800–806.

DiFrancesco, D., Ducouret, P., and Robinson, R. B. (1989). Muscarinic modulation of cardiac rate at low acetylcholine concentrations. *Science* **243**, 669–671.

DiFrancesco, D. (2005). Physiology and pharmacology of the cardiac pacemaker ("funny") current. *Pharmacol. Ther.* **107**, 59–79.

Eccles, J. C. (1957). "The Physiology of Nerve Cells." Johns Hopkins Univ. Press, Baltimore, MD.

Ertel, E. A., Campbell, K. P., Harpold, M. M., Hofmann, F., Mori, Y., Perez-Reyes, E., Schwartz, A., Snutch, T. P., Tanabe, T., Birnbaumer, L., Tsien, R. W., and Catterall, W. A. (2000). Nomenclature of voltage-gated calcium channels. *Neuron* **25**, 533–535.

Goldman, D. F. (1943). Potential, impedance, and rectification in membranes. *J. Gen. Physiol.* **27**, 37–60.

Goldman-Rakic, P. S. (1995). Cellular basis of working memory. *Neuron* **14**, 477–485.

Gouaux, E., and MacKinnon, R. (2005) Principles of selective ion transport in channels and pumps. *Science* **310**, 1461–1465.

Hille, B. (1977). Ionic basis of resting potentials and action potentials. In "Handbook of Physiology" (E. R. Kandel, Ed.), Sect. 1, **Vol. 1**, pp. 99–136. Am. Physiol. Soc, Bethesda, MD.

Hirsch, J. C., Fourment, A., and Marc, M. E. (1983). Sleep-related variations of membrane potential in the lateral geniculate body relay neurons of the cat. *Brain Res.* **259**, 308–312.

Hodgkin, A. L. (1976). Chance and design in electrophysiology: An informal account of certain experiments on nerve carried out between 1934 and 1952. *J. Physiol. (London)* **263**, 1–21.

Hodgkin, A. L., and Huxley, A. F. (1939). Action potentials recorded from inside a nerve fiber. *Nature (London)* **144**, 710–711.

Hodgkin, A. L., and Huxley, A. F. (1952a). Currents carried by sodium and potassium ions through the membrane of the giant axon of *Loligo*. *J. Physiol. (London)* **116**, 449–472.

Hodgkin, A. L., and Huxley, A. F. (1952b). The components of membrane conductance in the giant axon of *Loligo*. *J. Physiol. (London)* **116**, 473–496.

Hodgkin, A. L., and Huxley, A. F. (1952c). The dual effect of membrane potential on sodium conductance in the giant axon of *Loligo*. *J. Physiol. (London)* **116**, 497–506.

Hodgkin, A. L., and Huxley, A. F. (1952d). A quantitative description of membrane current and its application to conduction and excitation in nerve. *J. Physiol. (London)* **117**, 500–544.

Hodgkin, A. L., and Katz, B. (1949). The effect of sodium ions on the electrical activity of the giant axon of the squid. *J. Physiol. (London)* **108**, 37–77.

Hodgkin, A. L., and Keynes, D. (1955). Active transport of cations in giant axons from *Sepia* and *Loligo*. *J. Physiol. (London)* **128**, 28–60.

Hodgkin, A. L., Huxley, A. F., and Katz, B. (1952). Measurement of current–voltage relations in the membrane of the giant axon of *Loligo*. *J. Physiol. (London)* **116**, 424–448.

Horisberger, J.-D., Lemas, V., Kraehenbuhl, J.-P., and Rossier, B. C. (1991). Structure–function relationship of Na,K-ATPase. *Annu. Rev. Physiol.* **53**, 565–584.

Hoshi, T., Zagotta, W. N., and Aldrich, R. W. (1990). Biophysical and molecular mechanisms of Shaker potassium channel inactivation. *Science* **250**, 533–538.

Hotson, J. R., Prince, D. A., and Schwartzkroin, P. A. (1979). Anomalous inward rectification in hippocampal neurons. *J. Neurophysiol.* **42**, 889–895.

Hotson, J. R., Sypert, G. W., and Ward, A. A. (1973). Extracellular potassium concentration changes during propagated seizures in neocortex. *Exp. Neurol.* **38**, 20–26.

Huguenard, J., and McCormick, D. A. (1994). "Electrophysiology of the Neuron." Oxford Univ. Press, New York.

Jackelet, J. W. (1989). "Neuronal and Cellular Oscillators." Dekker, New York.

Jahnsen, H., and Llinás, R. (1984a). Electrophysiological properties of guinea-pig thalamic neurons: An *in vitro* study. *J. Physiol. (London)* **349**, 205–226.

Jahnsen, H., and Llinás, R. (1984b). Ionic basis for the electroresponsiveness and oscillatory properties of guinea-pig thalamic neurons *in vitro*. *J. Physiol. (London)* **349**, 227–247.

Jan, L. Y., and Jan, Y. N. (1990). How might the diversity of potassium channels be generated? *Trends Neurosci.* **13**, 415–419.

Jiang, Y., Lee, A., Cadene, M., Chalt, B. T., and MacKinnon, R. (2002b) The open pore conformation of potassium channels. *Nature* **417**, 523–526.

Jiang, Y., Lee, A., Chen, J., Cadene, M., Chait, B. T., and MacKinnon, R. (2002a) Crystal structure and mechanism of calcium-gated potassium channel. *Nature* **417**, 515–522.

Johnston, D., and Wu, S. M.-S. (1995). "Foundations of Cellular Neurophysiology." MIT Press, Cambridge, MA.

Johnston, D., Magee, J. C., Colbert, C. M., Cristie, B. R. (1996). Active properties of neuronal dendrites. *Annu. Rev. Neurosci.* **19**, 165–186.

Kao, C. T. (1966). Tetrodotoxin, saxotoxin and their significance in the study of excitation phenomena. *Pharmacol. Rev.* **18**, 997–1049.

Kuffler, S. W., and Nicholls, J. G. (1966). The physiology of neuroglia cells. *Ergeb. Physiol.* **57**, 1–90.

Latorre, R., Oberhauser, A., Labarca, P., and Alvarez, O. (1989). Varieties of calcium-activated potassium channels. *Annu. Rev. Physiol.* **51**, 385–399.

Läuger, P. (1991). "Electrogenic Ion Pumps." Sinauer, Sunderland, MA.

Livingstone, M. S., and Hubel, D. H. (1981). Effects of sleep and arousal on the processing of visual information in the cat. *Nature (London)* **291**, 554–561.

Llinás, R. R. (1988). The intrinsic electrophysiological properties of mammalian neurons: Insights into central nervous system function. *Science* **242**, 1654–1664.

Llinás, R., and Jahnsen, H. (1982). Electrophysiology of mammalian thalamic neurones *in vitro*. *Nature (London)* **297**, 406–408.

Llinás, R., and Sugimori, M. (1980a). Electrophysiological properties of *in vitro* Purkinje cell somata in mammalian cerebellar slices. *J. Physiol. (London)* **305**, 171–195.

Llinás, R., and Sugimori, M. (1980b). Electrophysiological properties of *in vitro* Purkinje cell dendrites in mammalian cerebellar slices. *J. Physiol. (London)* **305**, 197–213.

Llinás, R., and Yarom, Y. (1981a). Electrophysiology of mammalian inferior olivary neurones *in vitro*: Different types of voltage-dependent ionic conductances. *J. Physiol. (London)* **315**, 569–584.

Llinás, R., and Yarom, Y. (1981b). Properties and distribution of ionic conductances generating electroresponsiveness of mammalian inferior olivary neurones *in vitro*. *J. Physiol. (London)* **315**, 569–584.

Llinás, R., Sugimori, M., Hillman, D. E., and Cherksey, B. (1992). Distribution and functional significance of the P-type, voltage-dependent Ca^{2+} channels in the mammalian nervous system. *Trends Neurosci.* **15**, 351–355.

Ludwig, A., Zong, X., Jeglitsch, M., Hofmann, F., and Biel, M. (1998). A family of hyperpolarization-activated mammalian cation channels. *Nature* **393**, 587–591.

Mackie, G. O., and Meech, R. W. (1985). Separate sodium and calcium spikes in the same axon. *Nature (London)* **313**, 791–793.

Madison, D. V., and Nicoll, R. A. (1984). Control of repetitive discharge of rat CA1 pyramidal neurons *in vitro*. *J. Physiol. (London)* **354**, 319–331.

McCormick, D. A. (1992). Neurotransmitter actions in the thalamus and cerebral cortex and their role in neuromodulation of thalamocortical activity. *Prog. Neurobiol.* **39**, 337–388.

McCormick, D. A., and Huguenard, D. A. (1992). A model of the electrophysiological properties of thalamocortical relay neurons. *J. Neurophysiol.* **68**, 1384–1400.

McCormick, D. A., and Pape, H.-C. (1990). Properties of a hyperpolarization-activated cation current and its role in rhythmic oscillation in thalamic relay neurones. *J. Physiol. (London)* **431**, 291–318.

McCormick, D. A., Connors, B. W., Lighthall, J. W., and Prince, D. A. (1985). Comparative electrophysiology of pyramidal and sparsely spiny neurons of the neocortex. *J. Neurophysiol.* **54**, 782–806.

Meech, R. W., and Mackie, G. O. (1993). Potassium channel family in giant motor axons of *Aglantha digitale*. *J. Neurophysiol.* **69**, 894–901.

Meech, R. W., and Mackie, G. O. (1995). Synaptic events underlying the production of calcium and sodium spikes in motor giant axons of *Aglantha digitale*. *J. Neurophysiol.* **74**, 1662–1669.

Mercer, R. W. (1993). Structure of the Na,K-ATPase. *Int. Rev. Cytol. C* **137**, 139–168.

Nernst, W. (1888). On the kinetics of substances in solution. Translated from *Z. Phys. Chem.* **2**, 613–622, 634–637. In "Cell Membrane Permeability and Transport" (G.R. Kepner, Ed.), pp. 174–183. Dowden, Hutchinson & Ross, Stroudsburg, PA, 1979.

Nicoll, R. A. (1988). The coupling of neurotransmitter receptors to ion channels in the brain. *Science* **241**, 545–551.

Nicoll, R. A., Malenka, R. C., and Kauer, J. A. (1990). Functional comparison of neurotransmitter receptor subtypes in mammalian central nervous system. *Physiol. Rev.* **70**, 513–565.

Nowycky, M. C., Fox, A. P., and Tsien, R. W. (1985). Three types of neuronal calcium channel with different calcium agonist sensitivity. *Nature (London)* **316**, 440–443.

Overton, E. (1902). Beiträge zur allgemeinen Muskelund Nerven physiologie. II. Ueber die Urentbehrlichkeit von Natrium- (oder Lithium-) Ionen für den Contractsionact des Muskel. *Pfluegers Arch. Ges. Physiol. Menschen Tiere.* **92**, 346–386.

Pedersen, P. L., and Carafoli, E. (1987). Ion motive ATPases. I. Ubiquity, properties, and significance to cell function. *Trends Biochem. Sci.* **12**, 146–150.

Pennefather, P., Lancaster, B., Adams, P. R., and Nicoll, R. A. (1985). Two distinct Ca-dependent K currents in bullfrog sympathetic ganglion cells. *Proc. Natl. Acad. Sci. USA* **82**, 3040–3044.

Perez-Reyes, E., Cribbs, L. L., Daud, A., Lacerda, A. E., Barclay, J., Williamson, M. P., Fox, M., Rees, M., and Lee, J.-H. (1998). Molecular characterization of a neuronal low-voltage-activated T-type calcium channel. *Nature* **391**, 896–900.

Phillips, C. G. (1956). Intracellular records from betz cells in the cat. *Q. J. Exp. Physiol.* **41**, 58–69.

Prince, D. A., Lux, H. D., and Neher, E. (1973). Measurements of extracellular potassium activity in cat cortex. *Brain Res.* **50**, 489–495.

Regehr, W. G., and Tank, D. W. (1994). Dendritic calcium dynamics. *Curr. Opin. Neurobiol.* **4**, 373–382.

Reiner, P. B., and McGeer, E. G. (1987). Electrophysiological properties of cortically projecting histamine neurons of the rat hypothalamus. *Neurosci. Lett.* **73**, 43–47.

Reithmeier, R. A. F. (1994). Mammalian exchangers and cotransporters. *Curr. Opin. Cell Biol.* **6**, 583–594.

Ritchie, J. M., and Rogart, R. B. (1977). Density of sodium channels in mammalian myelinated nerve fibers and nature of the axonal membrane under the myelin sheath. *Proc. Natl. Acad. Sci. USA* **74**, 211–215.

Schwartzkroin, P. A., and Mathers, L. H. (1978). Physiological and morphological identification of a nonpyramidal hippocampal cell type. *Brain Res.* **157**, 1–10.

Shu, Y., Duque, A., Yu, Y., Haider, B., McCormick, D.A. (2007) Properties of action potential initiation in neocortical pyramidal cells: Evidence from whole cell axon recordings. *J. Neurophysiol.* **97**, 746-760.

Skou, J. C. (1957). The influence of some cations on an adenosine triphosphatase from peripheral nerves. *Biochim. Biophys. Acta* **23**, 394–401.

Skou, J. C. (1988). Overview: The Na,K pump. In "Methods in Enzymology" (S. Fleischer and B. Fleischer, Eds.), Vol. 156, pp. 1–25. Academic Press, Orlando, FL.

Stafstrom, C. E., Schwindt, P. C., and Crill, W. E. (1982). Negative slope conductance due to a persistent subthreshold sodium current in cat neocortical neurons *in vitro*. *Brain Res.* **236**, 221–226.

Stea, A., Soong, T. W., and Snutch, T. P. (1995). Voltage-gated calcium channels. In "Ligand and Voltage-Gated Ion Channels" (A. North, Ed.), pp. 113–152. CRC Press, Boca Raton, FL.

Steriade, M., and McCarley, R. W. (2005). "Brainstem Control of Wakefulness and Sleep." Springer, New York.

Steriade, M., McCormick, D. A., and Sejnowski, T. (1993). Thalamocortical oscillations in the sleep and aroused brain. *Science* **262**, 679–685.

Storm, J. F. (1990). Potassium currents in hippocampal pyramidal cells. *Prog. Brain Res.* **83**, 161–187.

Stuart, G., Spruston, N., Sakmann, B., Hausser, M. (1997). Action potential initiation and backpropagation in neurons of the mammalian CNS. *Trends Neurosci.* **10**, 125–131.

Stuhmer, W., Conti, F., Suzuki, H., Wang, X., Noda, M., Yahadi, N., Kobu, H., and Numa, S. (1989). Structural parts involved in activation and inactivation of the sodium channel. *Nature (London)* **339**, 597–603.

Tasaki, I., Polley, E. H., and Orrego, F. (1954). Action potentials from individual elements in cat geniculate and striate cortex. *J. Neurophysiol.* **17**, 454–474.

Thomas, R. C. (1972). Electrogenic sodium pump in nerve and muscle cells. *Physiol. Rev.* **52**, 563–594.

Thompson, S. M., Deisz, R. A., and Prince, D. A. (1988). Relative contributions of passive equilibrium and active transport to the distribution of chloride in mammalian cortical neurons. *J. Neurophysiol.* **60**, 105–124.

Tomita, T. (1965). Electrophysiological study of the mechanisms subserving color coding in the fish retina. *Cold Spring Harbor Symp. Quant. Biol.* **30**, 559–566.

Tsien, R. W., Ellinor, P. T., and Horne, W. A. (1991). Molecular diversity of voltage-dependent Ca^{2+} channels. *Trends Pharmacol. Sci.* **12**, 349–354.

Vandermaelen, C. P., and Aghajanian, G. K. (1983). Electrophysiological and pharmacological characterization of serotonergic dorsal raphe neurons recorded extracellularly and intracellularly in rat brain slices. *Brain Res.* **289**, 109–119.

Vassilev, P. M., Scheuer, T., and Catterall, W. A. (1988). Identification of an intracellular peptide segment involved in sodium channel inactivation. *Science* **241**, 1658–1661.

Vassilev, P., Scheuer, T., and Catterall, W. A. (1989). Inhibition of inactivation of single sodium channels by a site-directed antibody. *Proc. Natl. Acad. Sci. USA* **86**, 8147–8151.

von Krosigk, M., Bal, T., and McCormick, D. A. (1993). Cellular mechanisms of a synchronized oscillation in the thalamus. *Science* **261**, 361–364.

Wang, Z., and McCormick, D. A. (1993). Control of firing mode of corticotectal and corticopontine layer V burst-generating neurons by norepinephrine, acetylcholine, and 1*S*,3*R*-ACPD. *J. Neurosci.* **13**, 2199–2216.

Wheeler, D. B., Randall, A., and Tsien, R. W. (1994). Roles of N-type and Q-type Ca^{2+} channels in supporting hippocampal synaptic transmission. *Science* **264**, 107–111.

Williams, J. T., North, R. A., Shefner, S. A., Nishi, S., and Egan, T. M. (1984). Membrane properties of rat locus coeruleus neurones. *Neurosci.* **13**, 137–156.

Yellen, G. (2002). The voltage-gated potassium channels and their relatives. *Nature* **419**, 35–42.

Young, J. Z. (1936). The giant nerve fibers and epistellar body of cephalopods. *Q. J. Microsc. Sci.* **78**, 367.

Zagotta, W. N., Hoshi, T., and Aldrich, R. W. (1990). Restoration of inactivation in mutants of Shaker potassium channels by a peptide derived from ShB. *Science* **250**, 568–571.

Molecular Properties of Ion Channels

David Matthew Young, Yuh Nung Jan, and Lily Yeh Jan

As discussed in Chapter 5, ion channels are present in most if not all cells. The cell membrane is normally impermeable to all but small, uncharged molecules, but ion channels serve as highly selective passageways through which ions can pass into or out of the cell. In the nervous system, ion channels set the resting membrane potential of neurons and muscles and control the firing pattern and waveform of action potentials by altering the balance of ions across the membrane. Ion channels alter their activities in response to the actions of transmitters and the metabolic state of the cell so as to modulate neuronal excitability (Hille, 2001). To fulfill these physiological functions, each type of ion channel allows only certain ions to pass through. To maintain levels of channel activity appropriate for a range of physiological conditions, it is important to regulate these channels to have just the right number of channels at the right locations on the cell membrane.

How does an ion channel selectively allow certain ions to pass through and determine when to open and when to close? How does a cell control the number of channels on its surface? How are the different channels distributed on the dendrite, soma, axon, and nerve terminals of a neuron? These key questions to the understanding of ion channel functions have been the focus of recent molecular studies. Following a brief overview of how ion channels are grouped into different families based on their molecular properties, some of these key issues will be discussed in more detail.

The first question to be examined in this chapter concerns *channel gating*—the opening and closing of channels:

- How does voltage open channels?
- How does calcium affect channel opening?
- How do transmitters and second messengers affect the opening and closing of channels?

- How could the metabolic state of a cell influence channel opening?

The second question concerns ion selectivity:

- Why do potassium channels let the larger potassium ions rather than the smaller sodium ions go through?
- The narrowest part of a sodium channel pore is actually larger than that of a potassium channel pore (Hille, 2001). How does a sodium channel allow sodium rather than potassium ions to pass through?
- How does a calcium channel allow only calcium ions to pass through under physiological conditions where calcium ions are far outnumbered by sodium ions of nearly the same size?
- Typically millions of ions stream single file through a channel per second, so that there is less than one microsecond of interaction between the channel and the permeant ion. How does the channel manage to distinguish between different types of ions?

The last issue to be dealt with in this chapter concerns the distribution of different types of ion channels in a neuron and the physiological significance of controlling the ion channel type and number in various compartments of the cell membrane.

FAMILIES OF ION CHANNELS

Ion channels are grouped into several families. Across families, channels bear a number of similar structural features that allow them to pass ions through membranes. The most fundamental feature

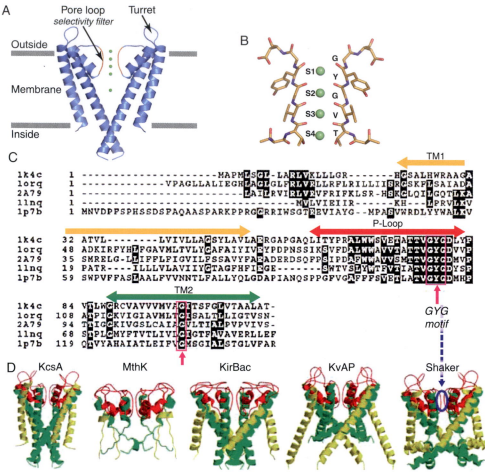

FIGURE 6.1 Examples of ion channel pores from various potassium channels. (A) A water-filled cavity is formed from four protein subunits, two of which are shown in the bacterial potassium channel, KcsA. The cavity creates a passageway through which ions can flow across the membrane, into or out of the cell. (B) The unique amino acid sequence of each family of channels allows it to selectively filter out particular ions. In the case of KcsA, K^+ but not Na^+ ions are allowed to pass through the selectivity filter, even though K^+ ions are bigger than Na^+ ions. S1-4 refers to the four K^+ ion-binding sites in the selectivity filter, each composed of eight oxygen atoms from the TVGYG signature motif. (C) Pore-region sequence alignments of five structurally known potassium channels are shown with the GYG signature motif boxed in magenta and other highly conserved regions labeled in black. (D) Structural comparison of the pore regions from the same five potassium channels. (A) and (B) adapted from Lockless *et al.* (PloS 2007, p. e121); (C) and (D) from Shrivastava and Bahar (Biophys J 2006, pp. 3929-3940).

of an ion channel is a *pore*, a tunnel that allows ions to stream across the membrane (see Fig. 6.1). The pore is made from transmembrane segments organized to form a gated hole through the membrane, whereas other transmembrane segments support this pore and influence its dynamics.

Unique amino acid sequences or assemblages of transmembrane segments or channel subunits distinguish channels from one another and confer unique ion selectivity or ion conductance properties. Channels in the same family typically share the same membrane topology for their pore-lining α subunits and display significant sequence similarity. Interestingly, some of these families bear weak though recognizable resemblance to one another, indicating that they are likely to be evolutionarily related.

The major families of ion channels are roughly grouped by what causes them to open and how many transmembrane segments they have. These families are as follows.

Voltage-Gated Ion Channels and Related Family Members (the 6-TM Family)

Voltage-gated channels (Fig. 6.2A) are opened by changes in membrane potential. The founding members of this family are voltage-gated sodium (Na_V), calcium (Ca_V) (Fig. 6.3A), and potassium (Kv) channels (Fig.6.3B) (Catterall, 1998, 2000; Jan and Jan, 1997; Plummer and Meisler, 1999), which each have unique pore structures that render them selectively permeable to their respective ions. The pore of

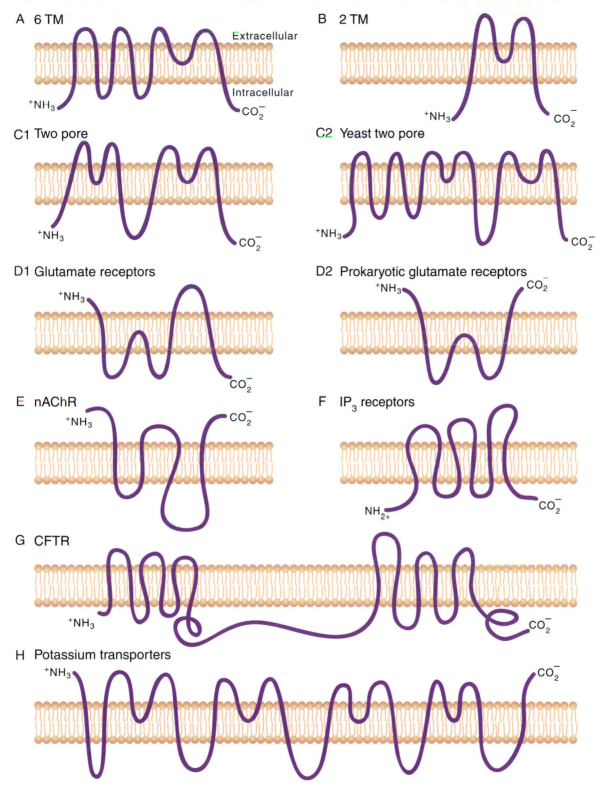

FIGURE 6.2 Several major families use similar designs. (A) The 6-TM family includes voltage-gated sodium, calcium, and potassium channels, hyperpolarization-activated cation channels, and cyclic nucleotide-gated cation channels. (B) The 2-TM family includes inwardly rectifying potassium channels and the bacterial potassium channel KcsA. The amiloride-sensitive epithelial sodium channel, the P2X ATP receptor, and a bacterial mechanosensitive cation channel also contain two transmembrane segments per subunit. (C) Two-pore potassium channels in the animal kingdom contain four transmembrane segments (left), whereas a yeast two-pore channel contains eight transmembrane segments (right). (D) Ionotropic glutamate receptors contain pore-lining domains that appear to be upside down relative to other channels in this figure. Glutamate receptors in the animal kingdom (left) are cation channels, whereas a prokaryotic glutamate receptor (right) is selective for potassium ions. (E) Nicotinic acetylcholine receptors (nAChR) are composed of five subunits, each with four transmembrane segments. The M2 subunit lines the pore and rotates to open the channel upon acetylcholine binding. (F) IP$_3$ is a major second messenger that leads to Ca^{2+} ion influx from internal stores by binding intracellular Ca channels, such as InsP$_3$R Ca^{2+} release channels. The NH$_2$-terminal region includes an IP$_3$ binding domain. (G) The pivotal role of the chloride channel Cystic Fibrosis Transmembrane Regulator (CFTR) becomes clear when mutations in the channel lead to a deficiency in movement of salt and water out of cells, blocking passageways and ultimately leading to increased morbidity and mortality in human patients. (H) Potassium transporters in bacteria and yeast contain four repeats of the 2-TM pore-lining structure.

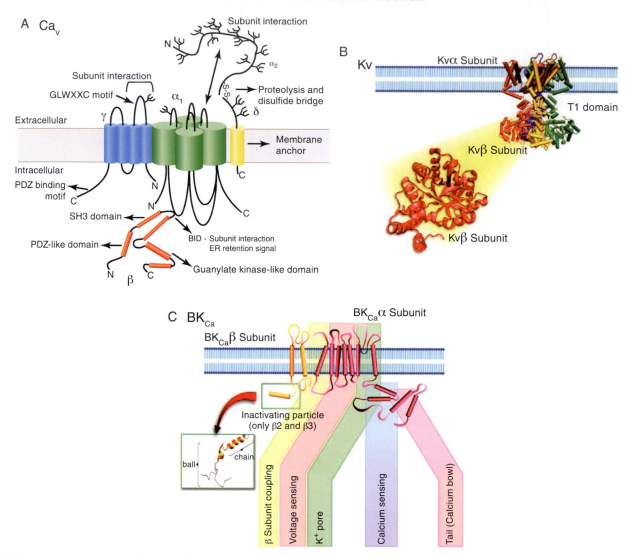

FIGURE 6.3 Voltage-gated Ca^{2+} and K^+ channels, key members of the voltage-gated ion channel family. (A) As with many other channels, Ca_V channels consist of many protein domains that allow the channel to be regulated by a variety of extra- and intracellular signals, in addition to voltage sensitivity through the α_1 subunit. (B) Voltage-gated K^+ channels consist of four α subunits that together form a pore for the passage of ions, as well as a cytoplasmic β subunit. (C) BK_{Ca} is an example of a K^+ channel that has an additional domain sensitive to Ca^{2+}. (A) from Arrikath and Campbell (Curr Op Neurobio 2003, pp. 298-307); (B) and (C) from Torres *et al.* (JBC 2007, pp. 24485-24489).

channels in this family is lined by four subunits, or in the case of voltage-gated sodium channels and calcium channels, four pseudosubunits linked together to form a large α subunit. Each subunit or pseudosubunit contains six transmembrane (TM) segments and an H5 region, or P ("pore") loop, in between the last two transmembrane segments. Although most voltage-gated channels are activated by depolarization, other family members include hyperpolarization-activated cation channels involved in rhythmic activities (Luthi and McCormick, 1998) and plant potassium channels activated by hyperpolarization (Gaymard *et al.*, 1996; Marten *et al.*, 1999; Schachtman *et al.*, 1992; Tang *et al.*, 2000).

In addition to voltage-gated family members, other members include calcium-activated potassium (Fig. 6.3B) (Meera *et al.*, 1997; Vergara *et al.*, 1998), cyclic nucleotide-gated cation, and transient receptor potential (TRP) channels that are important for sensory transduction (Caterina *et al.*, 2000; Davis *et al.*, 2000; Zufall *et al.*, 1997). These channels exhibit little voltage sensitivity. Like Ca_V channels, TRP channels are also Ca-permeable, 6-TM channels but activated not by voltage but by a variety of other stimuli, including second messengers in the phospholipase C (PLC) pathway or upon emptying internal calcium stores as in the case of TRPC channels, or heat and capsaicin as in the case of vanilloid (TRPV) receptors

(Venkatachalam and Montell, 2007). Whereas Ca_V channels contain four pseudosubunits in one α subunit, TRP channels are composed of subunits with a design similar to potassium channels (Montell, 2005).

Inwardly Rectifying Potassium Channels and Other Channels with Two Transmembrane Segments in Each Pore-Lining Subunit (the 2-TM Family)

Although potassium ions normally efflux out from the cell, inwardly rectifying potassium (Kir) channels are named for their ability to allow much larger potassium influx than efflux. Like Kv channels, Kir channels also have four α subunits lining the pore, but their transmembrane domains resemble the second half of the transmembrane domains of the Kv channel α subunit (Fig. 6.2B) (Jan and Jan, 1997). Other channels that also have two transmembrane segments per subunit, bearing little sequence similarity to Kir, include the amiloride-sensitive epithelial sodium channel (Sheng et al., 2000; Snyder et al., 1999), a bacterial mechanosensitive cation channel (Rees et al., 2000), the P_{2X} ATP receptor (Brake et al., 1994; Valera et al., 1994), and the FMRFamide-activated sodium channel (Coscoy et al., 1998). The last two are ligand-gated ion channels activated by the purinergic transmitter ATP and the peptide transmitter FMRFamide.

"Two-Pore" Potassium Channels (4-TM or 8-TM)

Each pore-lining α subunit of these channels appears to be a tandem dimer of two Kir-like, or one Kir-like and one Kv-like, α subunits (Fig. 6.2C) (Lesage and Lazdunski, 2000). These channels are thought to be "leak" potassium channels that are active at rest, contributing to the determination of the resting membrane potential. Some channels of this family may be modulated by volatile anesthetics (Patel et al., 1999).

Ionotropic Glutamate Receptors

This family includes the N-methyl D-aspartate (NMDA) and α-amino-3-hydroxy-5-methylisoxazole-4-proprionic acid (AMPA) receptors, known for their role in a learning and memory process called long-term potentiation (LTP) (see Chapters 11 and 19). The pore-lining domain of these ligand-gated ion channels is topologically equivalent to an "upside-down" Kir α subunit (Fig. 6.2D) (Wo and Oswald, 1995). A

prokaryotic potassium-selective glutamate receptor with this membrane topology (Chen et al., 1999) has been identified as a missing link between potassium channels and eukaryotic glutamate receptors, which are permeable to cations and contain one additional transmembrane segment at the C-terminus of the pore-lining domain.

Nicotinic Acetylcholine Receptors and Related Ionotropic Transmitter Receptors

Nicotinic acetylcholine receptors (nAChR) are well known for the role in receiving input at the neuromuscular junction (NMJ) that leads to muscular contraction, but nAChRs are found in both the CNS and PNS. This family of receptors contains five subunits, each with four transmembrane segments (Fig. 6.2E) (Miyazawa et al., 2003). The second (M2) transmembrane segment, the pore-lining helix, rotates upon acetylcholine binding so as to open the channel (Unwin, 1995). Like acetylcholine receptors, the 5-HT_3 serotonin receptor is permeable to cations and mediates fast excitatory synaptic transmission (Maricq et al., 1991). Other family members such as glycine receptors and $GABA_A$ receptors are permeable to anions and mediate fast inhibitory synaptic transmission (Galzi et al., 1992). These transmitter receptors differ in structure from ionotropic glutamate receptors (Sun et al., 2002).

Intracellular Calcium Channels

Calcium channels of likely six-transmembrane segments and a large cytoplasmic domain in each of their four subunits are responsible for releasing calcium from internal stores such as the endoplasmic reticulum (ER) (Foskett et al., 2007; George et al., 2004; Mikoshiba, 1997). Family members include the inositol 1,4,5-trisphosphate (IP_3) receptors that are activated by binding to the second messenger IP_3, and ryanodine receptors that can be activated by direct interaction with voltage-gated calcium channels on the cell membrane (Fig. 6.2F).

Chloride Channels

Chloride channels of 10–12 transmembrane segments per subunit of the CLC family are widely distributed, throughout the body as well as from prokaryotes to mammals (Fig. 6.2G) (Jentsch et al., 2005). The Cystic Fibrosis Transmembrane Regulator (CFTR) protein of the ATP-binding cassette (ABC) superfamily also forms chloride channels in the heart, airway epithelium, and exocrine tissue (Sheppard and Welsh, 1999).

Different Families of Ion Channels May Share Common Functional Modules

There is some resemblance between the pore-lining structures of different ion channels (e.g., voltage-gated ion channels of the 6-TM family, Kir channels of the 2-TM family, two-pore potassium channels, and ionotropic glutamate receptors). Likewise, there is a weak but discernible similarity between potassium transporters in bacteria and yeast and the pore-lining domain of potassium channels (Fig 6.2H) (Durell *et al.*, 1999; Jan and Jan, 1994); a transporter appears to contain four Kir-like pseudosubunits linked in tandem. Apparently, different ion channels and transporters may adopt the same fundamental structural motif for transporting ions across the membrane although the specific designs vary due to divergence of the physiological requirements.

Different Members of the Same Family May Have Divergent Functions

Whereas ion channels from the same family often have similar functional modules for ion permeation and/or channel gating, their functions can diverge to a remarkable degree. For example, voltage-gated ion channels are permeable to cations, but some members are selective to sodium, potassium, or calcium permeation, while others exhibit little selectivity among cations (Yellen, 2002). Also, not all members of voltage-gated ion channels are activated by depolarization; some are activated by hyperpolarization (Mannikko *et al.*, 2002), whereas still others show very weak, if any, voltage sensitivity. Not only could the functional modules have evolved to take on different functional characteristics, but divergent channel functions within the same family may also arise from differences in the temporal and spatial expression patterns of channels. Such divergence may be further encouraged following expression amplification by channel gene duplication.

In this chapter we will focus primarily on voltage-gated ion channels, which mediate action potentials and transmitter release, and inwardly rectifying potassium channels, which mediate slow synaptic potentials and possibly provide protection of neurons under metabolic stress. Chapter 11 provides a detailed description of ligand-gated ion channels that mediate synaptic transmission.

CHANNEL GATING

How Does Voltage Open Channels?

Controlling channel activity by voltage is key to neuronal excitability and signaling. As described in Chapter 5, voltage-gated sodium channels and potassium channels mediate the generation of action potentials, which allow signals to be propagated from one end of a neuron to the other (Hodgkin and Huxley, 1952; see also Chapter 7). Transmitters that bind to the receiving end of the neuron initiate an ion flux that changes the membrane potential locally. This voltage change alters the balance of open channels, leading to further changes in membrane potential farther along the neuron. Eventually, calcium entry due to activation of voltage-gated calcium channels in the nerve terminal triggers transmitter release onto a downstream neuron. The ability of these channels to be gated by voltage across the membrane is therefore fundamental to signaling in the nervous system. Intensive biophysical and molecular studies have yielded a framework for the mechanism of channel gating by voltage across the membrane.

Voltage-gated ion channels contain intrinsic voltage sensors. Voltage-gated ion channels typically are closed at the resting membrane potential but open upon membrane depolarization. These channels are intrinsically sensitive to membrane potential: built into the channel are voltage sensors that can detect changes in membrane potential and trigger conformational changes of the channel. This leads to movement of charges intrinsic to the channel protein; this *gating charge* and its resulting *gating current* have been measured in biophysical experiments (Keynes, 1994; Sigworth, 1994).

The S4 segment corresponds to the voltage sensor. Each of the four subunits or pseudosubunits of a voltage-gated ion channel contains an intrinsic voltage sensor, usually the fourth transmembrane segment S4, which bears basic, positively charged residues at every third position. Depolarization of the cell membrane causes the S4 segment to move outward relative to the side of the membrane that is inaccessible to water, as shown for the voltage-gated sodium channel and potassium channel (Fig. 6.4).

S4 moves in the voltage-gated sodium channel. Movement of the S4 segment in the fourth pseudosubunit of the voltage-gated sodium channel has been detected experimentally (Horn, 2000). By replacing an arginine of this S4 segment with a cysteine, which can react with thiol reagents such as MTSET, Horn and colleagues tested whether this cysteine is exposed to water. If the cysteine can access MTSET dissolved in water, the cysteine–thiol reaction sometimes results in an alteration in the kinetic properties of the channel, thereby providing an electrophysiological readout of the covalent modification. Remarkably, a cysteine at the position of the third arginine of this S4 segment is accessible from the cytoplasmic side of the membrane when the channel is closed but becomes accessible from the extracellular side when the channel

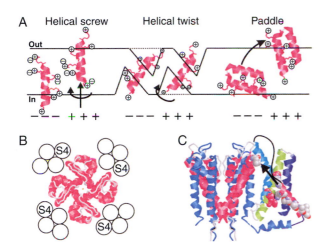

A Helical screw   Helical twist   Paddle

Out

In

− − − + + +   − − − + + +   − − − + + +

B

C

FIGURE 6.4 Various voltage-sensor movement models. (A) Three major movement models have been proposed for how the S4 voltage sensor moves outward during depolarization to open the channel. First proposed was the helical screw model, which involves both rotation and translational movement of the S4 segment. The helical twist model proposed rotation but not translation. A novel paddle model that suggests S4 movement as a hydrophobic cation through the lipid bilayer has stirred considerable controversy. (B) The position of the S4 segment has been another subject of considerable debate. The proposed position of S4 shown here with KcsA in the center places the S4 segment in the channel periphery, facing both lipids and proteins, based on the electrostatic effect of different charges introduced to the extracellular surface of the channel. (C) This position of S4 would allow large-scale translocation and is consistent with the helical screw model, shown here by the direction of the arrow. From Elinder *et al.* (Physio & Behav 2007, pp. 1-7).

is activated by depolarization, suggesting movement of the S4 segment (Yang *et al.*, 1996).

S4 moves in the voltage-gated potassium channel. In the homotetrameric Shaker potassium channel, each S4 segment contains seven basic residues, evenly spaced at every third position. As shown by Isacoff and colleagues, only the second arginine of the S4 segment is buried in the closed channel; a cysteine at that position cannot react with thiol reagents from either side of the membrane. As the channel opens upon membrane depolarization, the second arginine becomes exposed to the extracellular side of the membrane, whereas the third, fourth, and fifth basic residues move from the cytoplasmic side to sites buried in the membrane (Larsson *et al.*, 1996). A twist or rotation of the S4 segment probably takes place in this voltage-induced movement of the intrinsic voltage sensor (Cha *et al.*, 1999; Glauner *et al.*, 1999). The S4 movement can also be monitored by a rhodamine fluorophore that is covalently attached to the S4 segment and reports changes in its environment by changing its fluorescence. The time course of the fluorescence change parallels the time course of the gating current (Mannuzzu *et al.*, 1996),

allowing verification that the S4 movement reflects channel gating leading to channel opening by comparing the timing of fluorescence changes with changes in ion flux. Just where S4 is in the channel during hyperpolarization and depolarization is a hotly debated subject (Broomank *et al.*, 2003; Gandhi *et al.*, 2003; Jiang *et al.*, 2003, 2004; Laine *et al.*, 2003; Lee *et al.*, 2003).

The four S4 segments of four identical subunits undergo first independent and then concerted movements. The presence of four S4 segments that function as intrinsic voltage sensors of a voltage-gated ion channel accounts for the steep voltage dependence of channel activation. Upon depolarization, the S4 segments of the four identical subunits in a Shaker potassium channel move outward in two discernible steps, initially independently of one another and then cooperatively in a concerted step, leading to channel opening (Mannuzzu and Isacoff, 2000).

How Can a Voltage-Gated Ion Channel Not Stay Open Indefinitely upon Prolonged Depolarization?

Channels inactivate after they activate and open. Whereas some voltage-gated ion channels such as the M-type voltage-gated potassium channel stay open as long as the membrane potential is above the threshold for channel activation, most voltage-gated ion channels *inactivate* (see also Chapters 5 and 7). In other words, the channel stops conducting ions even though the membrane potential is maintained at a depolarized level. Channels can be inactivated in different ways. N-type ("fast") inactivation takes place via interaction of an "inactivation ball" at the N-terminus of the α subunit (or the β subunit) *near* the cytoplasmic end of the pore, whereas P-type and C-type ("slow") inactivation involves the extracellular end of the pore.

N-type inactivation takes place near the cytoplasmic end of the pore. N-type inactivation is also known as the "ball and chain" mechanism for inactivation (Armstrong and Bezanilla, 1977). The ball peptide at the N-terminus of the α or β subunit (Rettig *et al.*, 1994; Wallner *et al.*, 1999; Zagotta *et al.*, 1990), a part of the channel's cytoplasmic domain, appears to bind to a "receptor" within the channel that becomes accessible when the channel opens, resulting in blockade of ion permeation. The ball peptide resides at the N-terminus of the channel. Residues implicated in either electrostatic or physical interaction with the inactivation ball have been found in the sequences just preceding the first transmembrane segment S1 (Gulbis *et al.*, 2000), and the cytoplasmic loop connecting the S4 and the S5 segments, the S4-S5 loop (Isacoff *et al.*, 1991; Yellen, 1998). Whereas the inactivation ball at the N-terminus of the α or β subunit is thought to physically plug the pore at the cytoplasmic end, thereby blocking ion permeation, there is surprisingly little room that allows access of the inactivation ball to the pore. It is also conceivable that inactivation could result from conformational changes triggered by channel interaction with the ball, which functions even in isolation from the channel.

N-type inactivation may couple to voltage gating in different ways for the same channel, and the details of these interactions may vary for different channels. Some channels can inactivate only after some or all of their voltage sensors have moved in response to channel activation by membrane depolarization. Additionally, the voltage sensors may be "immobilized" in their activated configuration once the channel has entered the N-type inactivated state. These types of coupling would tend to simplify the state diagram of possible transitions among different states of the channel. In the extreme case, bringing the membrane potential from hyperpolarized to depolarized would cause a channel to shift from deactivated to activated state, and then to N-type inactivated state. Upon reversing the membrane potential to a hyperpolarized level, the channel would have to reverse course, going from N-type inactivated to activated state, and then to the deactivated state. This way the channel would open briefly after the membrane potential was brought back to a hyperpolarized level. Such reopening of voltage-gated calcium channels could cause the "delayed release" of transmitters observed following an action potential (Slesinger and Lansman, 1991).

P-type inactivation takes place near the extracellular end of the pore. P-type inactivation involves movements of pore-lining structures near the extracellular end of the pore (Loots and Isacoff, 1998; Olcese et al., 1997). After channel opening, movements of the P-region that forms the narrowest part of the pore halt ion permeation. The P-region in the bacterial KcsA potassium channel includes a pore loop (selectivity filter), with carbonyl groups surrounding the permeant ion (Doyle et al., 1998). Attached to the pore loop is a pore helix connected to a "turret," a short peptide loop projecting into the extracellular space. Movements of this turret on the extracellular side of the membrane accompany the P-type inactivation, as indicated by fluorescent changes of fluorophores attached to the turret and by state-dependent formation of disulfide bridge formation in this region (Gandhi et al., 2000; Loots and Isacoff, 1998).

C-type inactivation follows P-type inactivation and further prevents reactivation of the channel. Subsequent to P-type inactivation, more global movements of the channel take place and stabilize the S4 segment in the C-type inactivated state (Larsson and Elinder,

2000; Loots and Isacoff, 1998; Olcese et al., 1997; Yellen, 1998). Once the channel has entered the C-type inactivated state, a greater amount of hyperpolarization is necessary to revert the channel to the deactivated state so that the channel can once again be induced by depolarization to activate and open.

Channel inactivation may be caused by permeant ions. There are other forms of inactivation. For example, channels can be inactivated by the ions to which they are permeable, as in the case of calcium-induced calcium channel inactivation described later (Lee et al., 1999; Zühlke et al., 1999).

How Does Calcium Affect Channel Activity?

Neuronal activities may regulate channel activities via calcium. Besides changing the membrane potential, neuronal activities often cause increases of cytosolic calcium concentration, due to calcium entry through voltage-gated calcium channels or certain transmitter-gated ion channels, or due to release of calcium from internal stores by second messengers. Thus, calcium modulation of ion channels represents one way for channels to be regulated by neuronal activities. There are at least two different ways for calcium to modulate channel functions. Calcium may either directly interact with the channel protein or indirectly modulate channel activities via calcium-binding proteins such as calmodulin.

Direct calcium action is likely to underlie the calcium activation of the large conductance calcium-activated potassium (BK) channels. BK channels are sensitive to both voltage and calcium. Increasing intracellular calcium concentrations over six orders of magnitude causes the current-voltage dependence curve of channel activation to shift progressively to the left, meaning that increasing internal Ca^{2+} enables lower voltages to generate the same current through these channels (Barrett et al., 1982; Cui et al., 1997; Marty, 1981; Meera et al., 1997). How is this amazing feat accomplished? Recent molecular studies have provided valuable clues. The BK channels have a large cytoplasmic domain C-terminal to its 7-TM transmembrane domain. Preceding the six transmembrane segments that BK channels have in common with voltage-gated potassium channels is another transmembrane segment and an extracellular N-terminal domain (Meera et al., 1997). The large C-terminal domain contains multiple potential calcium-binding sites and an inhibitory region (Schreiber et al., 1999). This part of the C-terminal domain appears to stabilize the closed state(s), thereby inhibiting the channel in the absence of calcium. Calcium interaction with multiple calcium-binding sites in the C-terminal domain destabilizes the closed state(s), causing a shift of the voltage dependence curve of the channel to the left. Machinery for calcium gating must also be present outside the C-terminal domain, as BK channels truncated right after S6 are still gated by calcium (Piskorowski and Aldrich, 2002).

Calmodulin mediates modulation of many different ion channels. Calmodulin has four calcium-binding sites; calcium binding causes calmodulin to undergo conformation changes (Hoeflich and Ikura, 2002; Meador et al., 1993). Thus, channel activities may be modulated either by channels binding to calmodulin in a calcium-dependent manner or by sensing the conformation changes of calmodulin molecules that are bound to the channel constitutively, in the absence as well as in the presence of calcium. Such interactions underlie activation of the calcium-activated potassium channels of small and intermediate conductance (SK and IK) (Xia et al., 1998), calcium-induced calcium channel inactivation (Buddle et al., 2002), frequency-dependent facilitation of calcium channels (Lee et al., 1999; Zühlke et al., 1999), and calcium modulation of cyclic nucleotide-gated cation channels (Chen and Yau, 1994).

How Do Transmitters Cause Channels to Open or Close?

Metabotropic transmitter receptors are coupled to the trimeric G protein. Transmitter activation of G-protein-coupled receptors facilitates GTP exchange for GDP bound to the α subunit of the trimeric G protein, leading to the dissociation or rearrangement of the α-GTP subunit from the βγ subunit (Fig 6.5; see also Chapters 11 and 16) (Bunemann et al., 2003; Papin et al., 2005). Either or both of these G protein subunits may bind to channel proteins and modulate channel activities. Different isoforms of G-protein-coupled receptors couple to different G proteins, which may activate adenylyl cyclase (G_s), inhibit adenylyl cyclase (G_i), and activate phospholipase C (PLC) (G_q) or other signaling pathways. These G protein subunits may also modulate the activities of other effectors, thereby liberating second messengers such as calcium, cyclic nucleotides, phosphoinositides, and kinases. The various downstream second messengers in turn may alter channel activities.

Some channels are modulated directly by interaction with $G_{\beta\gamma}$ subunits. The βγ subunits of the G protein bind to G-protein-activated inwardly rectifying potassium channels (GIRK channels of the 2-TM Kir family) and cause channel activation. This allows inhibitory transmitters such as GABA to generate slow inhibitory postsynaptic potentials (IPSPs) in the brain, and the parasympathetic transmitter acetylcholine to slow the heart rate (Luscher et al., 1997; Wickman et al., 1998). (See Chapter 16 for a discussion of IPSPs in the CNS.) The βγ subunits also bind to calcium channels to cause channel inhibition (Herlitze et al., 1996; Ikeda, 1996), a likely mechanism for presynaptic inhibition given that calcium channels are found in a complex with synaptic proteins such as syntaxin and synaptotagmin necessary for vesicular binding and transmitter release (Catterall, 1998) (see also Chapter 8).

Multiple second messengers may converge on the same channel, resulting in integration of signaling processes. Second messenger systems are described in detail in Chapter 12. Here we briefly illustrate how they can exert major effects on ion channels. For example, protein kinase C (PKC) phosphorylation not only increases calcium channel activities but also prevents channel inhibition by the βγ subunits of the G protein (Zamponi et al., 1997). Likewise, the sensitivity of GIRK channels to the βγ subunits may be modulated by PIP_2 (Huang, et al., 1998; Gamper and Shapiro, 2007). In the heart, acetylcholine first activates GIRK channels via M2 muscarinic acetylcholine receptors due to direct action

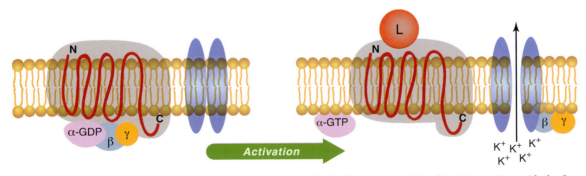

FIGURE 6.5 G-protein-activated inwardly rectifying potassium channels (Kir3) are activated by direct interaction with the βγ subunits of G protein. L represents the ligand for the G-protein-coupled receptor with seven transmembrane segments, e.g., the parasympathetic transmitter acetylcholine for slowing the heart rate or the inhibitory transmitter GABA for generating the slow inhibitory postsynaptic potential in the central nervous system.

of the $G_{\beta\gamma}$ subunits on the channel. The same transmitter may cause GIRK channel desensitization by activating M3 muscarinic acetylcholine receptors, which in turn activate PLC and reduce PIP_2 levels (Kobrinsky et al., 2000). PIP_2 also inhibits voltage-gated calcium channels by altering their voltage dependence for channel activation; this inhibition is alleviated by phosphorylation by protein kinase A (PKA) (Wu et al., 2002). In these examples multiple second messengers converge on the α subunit of the channel. In many other cases second messengers impinging on α subunits as well as the regulatory β subunits modulate channel activities.

How Could the Metabolic State of a Cell Influence Channel Activity?

One well-known example of metabolic regulation of channel activities is the ATP-sensitive potassium channel. Intracellular ATP, an indicator of a cell's energy status, may also alter the activity of channels. For example, intracellular ATP inhibits ATP-sensitive potassium channels. ATP-sensitive potassium channels open in response to increases in the blood sugar level and intracellular metabolic state of the pancreatic β cell to trigger insulin release. Similar or identical ATP-sensitive potassium channels are present in the brain, the heart, and the skeletal and smooth muscles (Ashcroft and Gribble,

1998; Ashcroft, 2000; Babenko et al., 1998; Nichols and Lopatin, 1997; Quayle et al., 1997). ATP-sensitive potassium channels in arterial smooth muscle may open during ischemia to cause dilation of blood vessels. Metabolic stress in central neurons may also activate these channels, thereby protecting the neurons from death. Besides metabolic regulation, transmitters and their G-protein-coupled receptors also modulate ATP-sensitive potassium channel activity. For example, vasodilators activate these channels in the smooth muscle via PKA, whereas vasoconstrictors inhibit these channels via PKC.

ATP and Mg-ADP mediate metabolic regulation of ATP-sensitive potassium channels. ATP-sensitive potassium channels are inhibited by ATP but stimulated by Mg-ADP. These channels contain not only four pore-lining subunits (Kir6.2) but also four regulatory subunits of the ATP-binding cassette (ABC) family (SUR1 or SUR2A/B) (Fig 6.6). ATP can act on the Kir6.2 subunits and cause channel inhibition, though the ATP sensitivity is greatly enhanced by interaction between Kir6.2 and the regulatory SUR subunits. In contrast, Mg-ADP acts on the SUR subunits to cause channel activation in a manner that requires the nucleotide-binding domains of the SUR, suggesting that ATP hydrolysis by the SUR subunits is necessary for channel regulation. The SUR subunits also mediate channel inhibition by sulphonylurea drugs such as

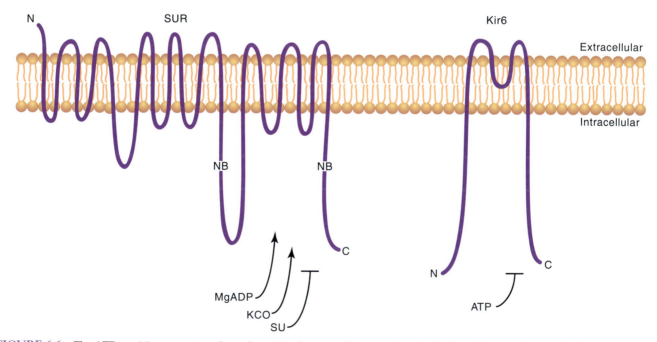

FIGURE 6.6 The ATP-sensitive potassium channels contain four pore-lining α subunits (Kir6) and four regulatory β subunits (SUR). SUR is a member of the ATP-binding cassette (ABC) family and contains two nucleotide-binding (NB) domains. ATP acts on Kir6 to inhibit the channel whereas Mg-ADP acts on SUR to activate the channel. Sulfonylurea (SU) drugs that inhibit the channel and KCO compounds that activate the channel also act on SUR.

glibenclamide and tolbutamide, used to treat type II diabetes. ATP-sensitive potassium channel openers (KCO) such as diazoxide act on the SUR subunits and stimulate ATP hydrolysis, a feature reminiscent of transporters in the ABC family. Whereas close relatives of SUR such as MRP transport hydrophobic substrates across the membrane, SUR interacts with the first transmembrane segment of Kir6.2 via its transmembrane domain. This interaction may potentially play a role in allowing the SUR subunit to regulate channel activity.

ION PERMEATION

Proper physiological functions of ion channels depend on their exquisite ion selectivity, a remarkable feat. How does a calcium channel allow only calcium ions to pass through even though sodium ions are of nearly the same size and much more abundant under physiological conditions? How does a potassium channel select for the larger potassium ions over the smaller sodium ions, given that millions of potassium ions stream through the channel in single file in a second so that there is less than 1 microsecond of interaction between the channel and a potassium ion? Given that the narrowest part of a sodium channel pore is actually larger than that of a potassium channel pore, how does a sodium channel allow sodium rather than potassium ions to pass through? These questions have attracted much scrutiny in biophysical and molecular studies, leading to the following model (Hille, 2001).

To achieve high selectivity without holding onto an ion for too long, ion channels have multiple binding sites for the permeant ion. The ion selectivity could be achieved if the channel contained a binding site with much higher affinity for calcium, or sodium, or potassium, than for other ions. But why then does the preferred ion not get stuck in the channel? One way to attain both exquisite ion selectivity and large ion flux through the channel would be to have more than one permeant ion in the channel pore, perhaps each interacting with a separate binding site. Electrostatic or other long-range interactions between these ions in the pore could facilitate their dissociation from their binding sites, thereby allowing rapid flow of permeant ions across the membrane.

Evidence for Multiple Ions Residing in a Channel Pore

Evidence for Ion-Binding Sites

A dependence of ion permeation on the composition and concentrations of ions implicates ion-binding sites in the channel. Early indications for binding sites arose from examining the amount of current flowing through a channel as a function of ion concentration; saturation of current level at high ion concentrations indicates the presence of at least one binding site. Another indication for ion-binding sites came from the "test of independence." If one assumes that permeant ions move through the channel, from one side of the membrane to the other, independently of one another, the permeability ratio for the two types of permeant ions being tested should vary with ion concentration in a certain way. This prediction does not fit the experimental data, indicating that there are ions in the channel pore, resulting in interdependence between ions as they permeate the channel (Hille, 2001).

Evidence for More Than One Binding Site

The presence of more than one ion in a channel pore is revealed by several different experiments. In the 1950s Hodgkin and Keynes (1955) showed that the ratio of potassium efflux and influx is 2.5, indicating that more than one of the monovalent potassium ions move in concert through the potassium channel (Fig. 6.7). Similarly, the voltage dependence of blocking an inwardly rectifying potassium channel by the monovalent cesium ion indicates that more than one cesium ion can reside in the channel pore at the same time (Hagiwara *et al.*, 1976; Hille, 2001).

Another indication is the so-called anomalous molefraction effect. First, let us suppose that a channel is permeable to two types of ions (Fig. 6.8). Compared to permeant ion B, permeant ion A binds to the channel's external binding site more tightly. Permeant ion A also moves through the channel at a slower rate. If the extracellular solution contains predominantly the B-type permeant ions mixed with a small amount of the A-type permeant ions, occupation of the channel's

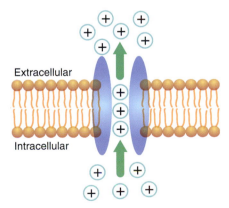

FIGURE 6.7 Movement of multiple (three in this example) ions in a channel pore in concert causes the ratio of potassium efflux to influx to be greater than one (2.5), even though each potassium ion carries one unit charge.

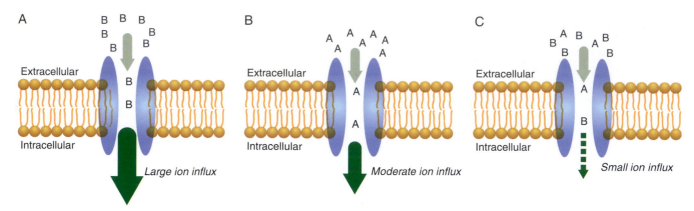

FIGURE 6.8   Anomalous mole-fraction effect. (A) In the presence of only permeant ion B the ion flux is large because permeant ion B moves through the channel at a fast rate. (B) In the presence of only permeant ion A the ion flux is moderate because permeant ion A binds more tightly to the channel's external binding site and moves through the channel more slowly. (C) In the presence of predominantly permeant ion B mixed with a small fraction of permeant ion A, the ion flux is smaller than in either of the preceding cases, because permeant ion A binds tightly to the external binding site in the channel and blocks passage of permeant ion B.

external binding site by permeant ion A would block the binding and passage of permeant B through the channel. Because of the difference in affinity, it would be more difficult for permeant ion B than for permeant ion A to displace permeant ion A from the external binding site to the next binding site. This way, the rate of ion flow may turn out to be smaller than when either permeant ion species alone is present. This apparently paradoxical, or anomalous, phenomenon arises when a channel contains more than one ion-binding site (Hagiwara *et al.*, 1977; Hille, 2001).

Crystallographic Evidence for Multiple Ions in a Pore

Direct evidence for the presence of multiple permeant ions in the channel pore is provided by visualization of the bacterial KcsA potassium channel in the crystal form (Fig. 6.1) (Doyle *et al.*, 1998). Two potassium ions are located within the upper half of the pore adjacent to the extracellular surface. These two potassium ions are coordinated by the backbone carbonyl groups of the pore loop, which is part of the H5 or P-region that connects the two transmembrane segments. The last transmembrane segment (S6) lines the inner half of the pore, which contains one discernible ion (Fig. 6.9).

The known crystal structure is for KcsA at pH 7.5 and may be a hybrid of an open conformation for the upper half and a closed conformation for the lower half near the cytoplasmic side of the membrane (Doyle *et al.*, 1998; Kuo *et al.*, 2003). KcsA channels are closed at pH 7 but can be activated at pH 4, due to conformational changes that include movements of the last transmembrane segment (Perozo *et al.*, 1999).

The exact form of the channel in its open conformation remains to be determined. Comparison between the structures of KcsA with another bacterial channel MthK indicates that the second transmembrane segment bends at a hinge point as the channel opens (Elinder *et al.*, 2007; Jiang *et al.*, 2002a,b, but see Webster *et al.*, 2004). For inward rectifier potassium (Kir) channels, the structure of KirBac1.1 in its closed conformation (Kuo *et al.*, 2003) is in close agreement with the model for Kir2.1 in its open conformation (Minor *et al.*, 1999), with respect to the pore-lining residues, lipid-facing residues, and residues buried within the channel protein. Two pairs of residues critical for holding Kir3.2 (GIRK2) channels in the closed conformation, identified from yeast screens of randomly mutagenized Kir3.2 channels (Yi *et al.*, 2001), reside in neighboring helices as the last transmembrane segment comes close to either the pore helix or the first transmembrane segment (Fig. 6.10) (Kuo *et al.*, 2003). Crystallization of a eukaryotic Kir3.1 (GIRK1)-prokaryotic Kir chimeric channel shows the cytoplasmic pore in two conformations, one dilated and the other constricted, through rigid-body displacements of channel subunits relative to one another (Nishida *et al.*, 2007).

The selectivity filter of the KcsA channel adopts different atomic structures depending on its surrounding ionic composition. Normally a lowered potassium concentration leads to an altered, nonconductive conformation in KcsA. Crystallography of a KcsA channel with a mutation in its selectivity filter showed the filter locked in a conductive conformation. Unlike the wild-type channel, the mutant conducted sodium ions in the absence of potassium, demonstrating that the channel's ability to adapt structurally to the presence

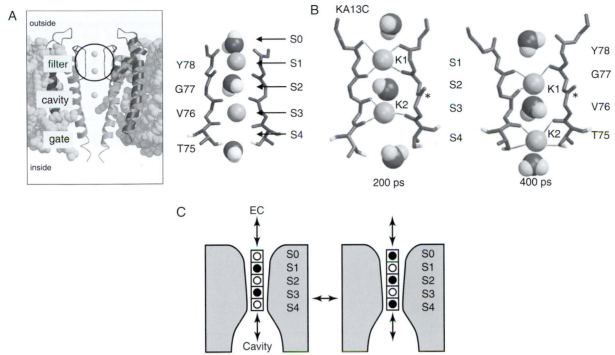

FIGURE 6.9 Multiple ion binding sites in the K$^+$ channel selectivity filter. (A) The KcsA channel pore region is composed of a gate, cavity, and selectivity filter. The filter contains multiple binding sites for K$^+$ ions as they pass through the pore. (B) Snapshots of the selectivity filter in a model of KcsA taken 200 ps apart show K$^+$ ions interacting with the filter polypeptide backbone while moving through the pore along with water molecules. (C) K$^+$ ions and water molecules alternate positions as they march through the pore. EC, extracellular. (A) and (B) from Shrivastava *et al.* (Biophys J 2002, pp. 633–645); (C) from Sansom *et al.* (BBA Biomem 2002, pp. 294–307).

of potassium or sodium ions is fundamental to ion selectivity (Fig. 6.11) (Valiyaveetil *et al.*, 2006). Recent crystallization of a Kv1.2-2.1 chimera at much higher resolution (2.4 Å resolution in the chimera versus 3.2 Å in KcsA) reveals a highly conserved, almost identical selectivity filter to that of KcsA (Doyle *et al.*, 1998; Long *et al.*, 2007).

The Ion Selectivity of Voltage-Gated Sodium Channels and Calcium Channels

Sodium Channels and Calcium Channels Have Wider Pores Than Those of Potassium Channels and Yet Have to Favor the Passage of Ions Smaller Than Potassium

It is of physiological importance to have voltage-gated sodium channels for the generation of action potentials in neurons, and to have voltage-gated calcium channels for triggering transmitter release and for allowing entry of calcium, which can function as a second messenger. Although sodium and calcium ions carry different amounts of charge, they are about the same size. How are channels designed to be selectively permeable to sodium, or to calcium?

Using organic ions of different sizes to gauge the pore dimension, Hille showed in the 1970s that the narrowest part of sodium channels and calcium channels are actually wider than that of potassium channels, though not quite large enough to allow permeation of fully hydrated ions (Hille, 2001). Thus, ion channels cannot simply discriminate different ions based on their size. Rather, the dehydration energy appears to be an important factor; the pore of a channel probably approximates the hydration shell of its permeant ion so as to ease the process of losing a significant fraction of the water molecules in the hydration shell as the ion moves through the channel. This consideration, combined with the features afforded by a multi-ion pore, may account for the different ion selectivity of these channels.

One Remarkable Feature of Voltage-Gated Calcium Channels Is the Dependence of Its Ion Selectivity on Ion Concentration

Under physiological conditions, with millimolar calcium ions in the extracellular medium, these channels are selectively permeable to calcium even though the extracellular sodium concentration is much higher,

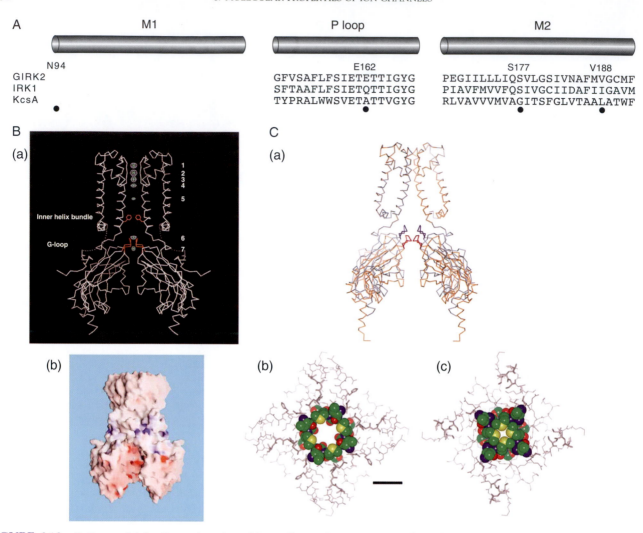

FIGURE 6.10 Gating model for G-protein-activated inwardly rectifying potassium channel Kir3.2 (GIRK2) deduced from yeast mutant screens. (A) Sequence alignment of the first transmembrane segment M1, the P loop and the second transmembrane segment M2 of GIRK2, IRK1 with an open probability close to one, and KcsA. Yeast screens of randomly mutagenized GIRK2 channels reveal that mutations of the four residues marked above the GIRK2 sequence cause the channel to be constitutively open. (B) (a) Side view of a chimeric Kir3.1-pro-karyotic Kir potassium channel showing the conduction pore in its dilated conformation. (b) Surface representation of the chimeric channel, with part of the channel removed to expose the interior of the cytoplasmic pore. (C) Models of the transmembrane helical arrangement in the open and closed forms. The pore region of the dilated (blue) and constricted (red) conformations have been superimposed in (a) to show the polypeptide movements from the open (b) to closed (c) positions of the channel. (A) Adapted from Yi *et al.* (Neuron 2001, pp. 657-667); (B) and (C) from Nishida *et al.* (EMBO 2007, pp. 4005-4015).

typically more than 100 mM. If the extracellular calcium concentration is reduced to submicromolar levels, calcium channels conduct sodium ions. At intermediate calcium concentrations, neither calcium ions nor sodium ions conduct currents through calcium channels (Fig. 6.12A) (Almers and McCleskey, 1984; Hess and Tsien, 1984).

This behavior of the calcium channel could be accounted for if the calcium channel had more than one binding site, with higher affinity for calcium than for sodium. In the absence of calcium ions, sodium ions occupy the binding sites and go through the channel.

As the calcium concentration is raised, one of the binding sites becomes occupied by calcium. The higher affinity of the binding site for calcium makes it difficult for a sodium ion to displace the calcium ion. Thus, the bound calcium ion in effect blocks sodium permeation through the channel. Having more than one calcium ion in the same channel would effectively reduce the affinity of these binding sites for calcium, due to the electrostatic repulsion between these doubly charged calcium ions. This explains why at low calcium concentrations it is much more likely to have calcium occupying only one of the binding sites. At sufficiently high

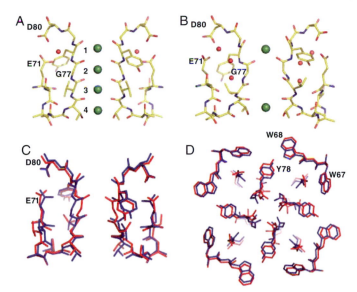

FIGURE 6.11 Conformational changes depending on K$^+$ ion concentration. The selectivity filter adopts different conformations in the presence of high K$^+$ ion concentration (A) and low-concentration (B), with lowered ion conductance in the low-concentration conformation. Water ions are depicted as red circles, while K$^+$ ions are green, with numbered K$^+$ ion-binding sites in the filter. Superimposed high (blue) and low (red) concentration conformations are shown from side view (C) as well as from a top view (D) extending 15 Å from the center of the filter. From Valiyaveetil et al. (Science 2006, pp. 1004-1007).

ION CHANNEL DISTRIBUTION

Neurons are highly polarized. Their dendrites receive synaptic inputs, thereby generating fast and slow synaptic potentials. Integration of these synaptic inputs sometimes leads to the generation of action potentials, which propagate from the soma along the axon to the nerve terminals. Transmitter release from the nerve terminals is triggered by the arrival of action potentials but may also be regulated by transmitters acting on receptors located at the nerve terminal. Proper function of these neuronal activities depends on adequate placement of ion channels of the appropriate number and type in each domain of the neuronal membrane. We will use a few examples to illustrate this point. Additional discussion of the role of voltage-dependent ion channels in dendrites can be found in Chapter 17.

Different ion channels are localized to different parts of the neuron. Whereas it has long been recognized that transmitter receptors are located near the sites of transmitter release, it is now evident that many other types of ion channels are also targeted to discrete regions of the neuronal membrane. Once the molecular entity of ion channels becomes known, specific probes can be generated to determine their expression patterns. These studies have revealed an intricate mosaic-like distribution of different ion channels. Presumably, ion channels are targeted to specific locations because their channel properties are most suited for the physiological functions at those sites. The possibilities of different channel isoforms to be regulated differently allow further dynamic modulation of channel properties by neuronal activities.

Different potassium channels are targeted to axons and dendrites. Myelinated axons have voltage-gated sodium channels confined to the node of Ranvier, gaps in the myelin sheath of nerve fibers (Salzer, 1997). These channels are flanked by certain isoforms of voltage-gated potassium channels (Kv1.1 and Kv1.2) forming two rings around the fiber in the juxtaparanodal regions, on opposite sides of the node (Rasband et al., 1998; Wang et al., 1993; Zhou et al., 1999), whereas KCNQ channels—voltage-gated potassium channels that activate at more hyperpolarized membrane potentials and show no inactivation—co-localize with voltage-gated sodium channels at the node (Devaux et al., 2004). Another member of the same family, Kv1.4, is found in patches along the axon and near the nerve terminals (Cooper et al., 1998), whereas members of a closely related family such as Kv4.2 are located on the dendrite and on the postsynaptic membrane (Alonso and Widmer, 1997; Sheng et al., 1993; Tkatch et al.,

calcium concentrations, under physiological conditions, it becomes more likely for a calcium ion in the external solution to displace the calcium ion at the external binding site and cause it to move to the next binding site, leading to calcium permeation.

Key molecular differences between voltage-gated calcium and sodium channels. In each of the four repeats of the voltage-gated calcium channel, there is a highly conserved glutamate in the H5 or P region. These negatively charged residues are crucial for the channel's affinity for calcium and probably form two binding sites for calcium (Fig. 6.12:B–D) (Yang et al., 1993). At the equivalent positions of the voltage-gated sodium channel, there are two glutamate, one alanine, and one lysine residues (Fig. 6.12E). Glutamate substitutions of these latter residues cause the mutant sodium channel to behave in a manner similar to calcium channels (Heinemann et al., 1992). In the absence of calcium, the mutant channels also exhibit altered selectivity among monovalent cations. Taken together these studies suggest that residues at the position of the highly conserved glutamate in the H5 or P region of voltage-gated calcium and sodium channels line the pore and interact with more than one permeant ion.

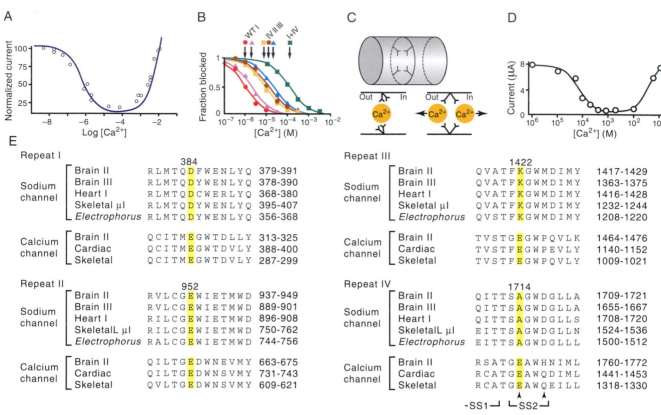

FIGURE 6.12 Dependence of voltage-gated calcium channel ion selectivity on calcium concentration. (A) In the presence of sodium ions and varying concentration of calcium ions, calcium channels are permeable to sodium ions at submicromolar calcium concentration. At submillimolar calcium concentration, a calcium ion occupies one binding site in the channel and blocks sodium permeation. At still higher calcium concentration, calcium may occupy multiple binding sites; the presence of multiple calcium ions in the same channel pore allows them to dissociate from the binding site more readily and pass through the channel. Adapted from Almers and McCleskey (1984). (B) The affinity of the calcium-binding site as indicated by the blocking action of calcium on lithium permeation is reduced by substituting a glutamate in the P loop with glutamine. WT, wild-type calcium channel. I, II, III, and IV indicate glutamine substitution in the first, second, third, and fourth repeats of the channel. I + IV indicates double mutations in the first and fourth repeats. (C) How the ring of four glutamate residues in the calcium channel pore might bind one or two calcium ions. (B) and (C) are adapted with permission from Macmillan Publishers Ltd. Yang, J., Ellinor, P.T., Sather, W.A., Zhang, J.F., and Tsien, R.W. (1993). Molecular determinants of Ca^2 selectivity and ion permeation in L-type Ca^2 channels [see comments]. *Nature* **366**, 158-161. (D) Glutamate substitution of lysine 1422 of the P loop in the third repeat of voltage-gated sodium channels causes the mutant channel to behave like a calcium channel. (E) Alignment of the P loop sequences for each of the four repeats of the voltage-gated sodium channels and calcium channels. (D) and (E) are adapted with permission from Macmillan Publishers Ltd. Heinemann *et al.* (Nature 1992, pp. 441-443).

2000). The large conductance (BK) calcium-activated potassium channels, by contrast, are present on the presynaptic membrane (Wanner *et al.*, 1999). Heteromeric channels formed by different α and β subunits may be localized to different domains of the neuronal membrane, further increasing the diversity of these channels (Rhodes *et al.*, 1995; Veh *et al.*, 1995).

Different calcium channels have different distributions. Pharmacological and electrophysiological studies have shown that N-type and P/Q-type voltage-gated calcium channels mediate calcium entry that triggers transmitter release from the nerve terminal, whereas L-type voltage-gated calcium channels are found in the soma. An even more refined picture has emerged from immunocytochemical studies (Catterall, 1998). α_{1B} subunits of N-type calcium channels and α_{1A} subunits of P/Q-type subunits are present in low densities in dendrites but at high density in presynaptic nerve terminals, where they co-localize with the SNARE proteins involved in synaptic vesicle docking and exocytosis. L-type calcium channels may contain α_{1C} or α_{1D} subunits, which are predominantly localized in cell bodies and proximal dendrites but have separate distribution patterns. α_{1D} subunits appear to be smoothly distributed on the cell membrane, whereas α_{1C} subunits form clusters extending far into dendrites and to postsynaptic membranes of glutamatergic synapses. In dendritic spines, R-type calcium channels primarily mediate calcium influx due to action potentials back-propagating from the soma into the dendrites (see later), but L-type calcium channels in these spines are responsible for inducing depression of calcium influx after brief trains of back-propagating action potentials (Yasuda *et al.*, 2003).

Action potentials may "back-propagate" from the soma to the dendrites or even be generated in the dendrite. Whereas it was thought earlier that action potentials are generated in the axon hillock region near the soma and propagate in one direction, down the axon, recent molecular and electrophysiological studies reveal many other possibilities (see also Chapter 17). First, voltage-gated potassium channels control whether action potentials will propagate past axonal branch points (Debanne *et al.,* 1997; Obaid and Salzberg, 1996). Second, not only can action potentials generated in or near the soma propagate "backward" into the dendrites via activation of dendritic voltage-gated sodium channels, but action potentials may also be initiated within the dendrites due primarily to activation of voltage-gated calcium channels (Schiller *et al.,* 1997; Spruston *et al.,* 1995). For example, back propagation of action potentials into dendrites has been observed not only in brain slice preparations but also in awake animals (Buzsaki and Kandel, 1998; Stuart *et al.,* 1997). And sensory stimulation of the whiskers sometimes generates calcium action potentials in pyramidal neurons in the somatosensory cortex of rodents (Helmchen *et al.,* 1999).

The extent of action potential back propagation varies with the types of neuron. In dopamine neurons in the midbrain and mitral cells in the olfactory bulb, neurons known to release transmitters from their dendrites, there is hardly any attenuation of action potentials as they propagate to dendrites. By contrast, the amplitude of dendritic action potentials decreases sharply within 0.1 mm from the soma of Purkinje cells in the cerebellum. In many other central neurons, including cortical neurons in layer 5, hippocampal pyramidal neurons in the CA1 region, and spinal neurons, the dendritic action potentials may be attenuated by ~50% over a distance of half a millimeter (Stuart *et al.,* 1997).

Action potential back propagation can be regulated. These observations indicate that prior neuronal activities may exert influence over subsequent synaptic inputs in dendrites via back-propagating action potentials. Repetitive firing results in a rapid decline of dendritic action potentials due to cumulative sodium channel inactivation, which is more prominent in hippocampal pyramidal apical dendrites (Colbert *et al.,* 1997; Mickus *et al.,* 1999; Spruston *et al.,* 1995). The lower conduction safety factor for action potential propagation is also evident from the tendency of back-propagating action potentials to fail at a branch point of dendrites (Mackenzie and Murphy, 1998; Stuart *et al.,* 1997). Moreover, in certain central neurons, the density of rapidly inactivating A-type voltage-gated potassium channels, which impede propagation,

is higher toward the distal ends of apical dendrites, and those channels at apical dendrites are more active due to a left shift in the voltage dependence of channel activation (Hoffman *et al.,* 1997). These spatial gradients present another factor that influences the extent of action potential back propagation. Both sodium channel inactivation and potassium channel activity can be further modulated by protein kinases and transmitters such as acetylcholine and dopamine, which stimulate metabotropic G-protein-coupled receptors (GPCRs) (Johnston *et al.,* 1999). Back-propagating action potentials and the extent to which they may alter synaptic potentials may therefore vary depending on the pattern of prior neuronal activities as well as the action of transmitters that stimulate metabotropic GPCRs.

Spatial gradients of ion channels allow synaptic potentials generated over large distances along dendrites to reach the cell soma with similar size and duration. As in the case of voltage-gated ion channels, the dendritic distribution is not uniform for channels that contribute to the neuron's input resistance and ionotropic transmitter receptors. These spatial gradients of channels play an important role in keeping the amplitude and waveform of the excitatory postsynaptic potential (EPSP) constant in some of the central neurons, even though they may be generated at synaptic sites that are up to 1 mm apart. If the channel distributions and membrane properties were uniform throughout the dendrite, the more distal on the dendrite an EPSP was generated, the smaller and longer lasting it would be when it reached the soma. This unevenness is prevented in certain central neurons by having a greater number of hyperpolarization-activated cation channels (I_h), which affect the input resistance and the membrane time constant, and a larger quantal size of transmitter response at the more distal dendrite (Magee, 1998; Magee and Cook, 2000; Williams and Stuart, 2000). Presumably, a greater number of glutamate receptors can be activated by glutamate released onto apical dendrites of these central neurons.

The distance a synaptic potential travels and the extent to which multiple synaptic potentials summate can be regulated. In addition to the spatial gradient of I_h channel density, there is a left shift for the voltage dependence of channel activation at more distal dendrites of certain central neurons (Magee, 1998), possibly due to a spatial gradient of second messengers. These channels are sensitive to modulation by transmitters and second messengers such as cyclic AMP (Luthi and McCormick, 1998). Thus, neuronal or synaptic activities that modulate channel activities through temporal and spatial patterns of transmitters and second messengers could profoundly alter the size and duration of subsequent synaptic potentials as well as the extent of temporal and spatial summation of synaptic inputs. Additional regulation may be imposed by interneurons that are tuned to different parameters of action potential firing patterns and make synaptic contacts with different parts of neurons (Pouille and Scaniziani, 2004).

SUMMARY

Signal processing in the nervous system is mediated by a wide variety of ion channels localized to different compartments of the highly polarized neuron. This mediation by channels can be better appreciated now that it is possible to follow the activity of single channels by calcium imaging and patch recording from dendrites as well as the soma and to examine channel distribution using molecular probes specific for individual channel types. Molecular analyses of the mechanisms for channel permeation and gating, as well as how these processes may be modulated by transmitters and second messengers mobilized by neuronal activities, provide further insight into the plasticity of these signaling processes.

References

Almers, W. & McCleskey, E.W. (1984). Non-selective conductance in calcium channels of frog muscle: calcium selectivity in a single-file pore. J Physiol (Lond) 353, pp. 585-608.

Alonso, G. & Widmer, H. (1997). Clustering of KV4.2 potassium channels in postsynaptic membrane of rat supraoptic neurons: an ultrastructural study. Neuroscience 77, pp. 617-621.

Armstrong, C.M. & Bezanilla, F. (1977). Inactivation of the sodium channel. II. Gating current experiments. J Gen Physiol 70, pp. 567-590.

Ashcroft, F.M. & Gribble, F.M. (1998). Correlating structure and function in ATP-sensitive K+ channels. Trends Neurosci 21, pp. 288-294.

Ashcroft, S.J.H. (2000). The beta-cell KATP channel. J Membrane Biol 176.

Babenko, A.P., Aguilar-Bryan, L. & Bryan, J. (1998). A view of sur/ KIR6.X, KATP channels. Annu Rev Physiol 60, pp. 667-687.

Barrett, J.N., Magleby, K.L. & Pallotta, B.S. (1982). Properties of single calcium-activated potassium channels in cultured rat muscle. J Physiol (Lond) 331, pp. 211-230.

Brake, A.J., Wagenbach, M.J. & Julius, D. (1994). New structural motif for ligand-gated ion channels defined by an ionotropic ATP receptor. Nature 371, pp. 519-523.

Broomank, A., Mannikko, R., Larsson, H.P. & Elinder F. (2003). Molecular movement of the voltage sensor in a K channel. J Gen Physiol 122, pp. 741-748.

Buddle, T., Meuth, S. & Pape, H.C. (2002). Calcium-dependent inactivation of neuronal calcium channels. Nature Reviews Neuroscience 3, pp. 673-684.

Bunemann, M., Frank, M. & Lohse, M.J. (2003). From the Cover: Gi Protein Activation in Intact Cells Involves Subunit Rearrangement Rather Than Dissociation. Proceedings of the National Academy of Sciences 100, pp. 16077-16082.

Buzsaki, G. & Kandel, A. (1998). Somadendritic backpropagation of action potentials in cortical pyramidal cells of the awake rat. J Neurophysiol 79, pp. 1587-1591.

Caterina, M.J., Leffler, A., Malmberg, A.B., Martin, W.J., Trafton, J., Petersen-Zeitz, K.R., Koltzenburg, M., Basbaum, A.I. & Julius, D. (2000). Impaired nociception and pain sensation in mice lacking the capsaicin receptor. Science 288, pp. 306-313.

Catterall, W.A. (1998). Structure and function of neuronal Ca^{2+} channels and their role in neurotransmitter release. Cell Calcium 24, pp. 307-323.

Catterall, W.A. (2000). From ionic currents to molecular mechanisms: the structure and function of voltage-gated sodium channels. Neuron 26, pp. 13-25.

Cha, A., Snyder, G.E., Selvin, P.R. & Bezanilla, F. (1999). Atomic scale movement of the voltage-sensing region in a potassium channel measured via spectroscopy. Nature 402, pp. 809-813.

Chen, G.-Q., Cui, C., Mayer, M.L. & Gouaux, E. (1999). Functional characterization of a potassium-selective prokaryotic glutamate receptor. Nature 402, pp. 817-821.

Chen, T.Y. & Yau, K.W. (1994). Direct modulation by Ca^{2+}-calmodulin of cyclic nucleotide-activated channel of rat olfactory receptor neurons. Nature 368, pp. 545-548.

Colbert, C.M., Magee, J.C., Hoffman, D.A. & Johnston, D. (1997). Slow recovery from inactivation of Na^+ channels underlies the activity-dependent attenuation of dendritic action potentials in hippocampal CA1 pyramidal neurons. J Neurosci 17, pp. 6512-6521.

Cooper, E.C., Milroy, A., Jan, Y.N., Jan, L.Y. & Lowenstein, D.H. (1998). Presynaptic localization of Kv1.4-containing A-type potassium channels near excitatory synapses in the hippocampus. J Neurosci 18, pp. 965-974.

Coscoy, S., Lingueglia, E., Lazdunski, M. & Barbry, P. (1998). The Phe-Met-Arg-Phe-amide-activated sodium channel is a tetramer. J Biol Chem 273, pp. 8317-8322.

Cui, J., Cox, D.H. & Aldrich, R.W. (1997). Intrinsic voltage dependence and Ca^{2+} regulation of mslo large conductance Ca-activated K^+ channels. J Gen Physiol 109, pp. 647-673.

Davis, J.B., Gray, J., Gunthorpe, M.J., Hatcher, J.P., Davey, P.T., Overend, P., Harries, M.H., Latcham, J., Clapham, C., Atkinson, K., Hughes, S.A., Rance, K., Grau, E., Harper, A.J., Pugh, P.L., Rogers, D.C., Bingham, S., Randall, A. & Sheardown, S.A. (2000). Vanilloid receptor-1 is essential for inflammatory thermal hyperalgesia. Nature 405, pp. 183-187.

Debanne, D., Guerineau, N.C., Gahwiler, B.H. & Thompson, S.M. (1997). Action-potential propagation gated by an axonal I(A)-like K^+ conductance in hippocampus [published erratum appears in Nature 1997 Dec 4;390(6659):536]. Nature 389, pp. 286-289.

Devaux, J.J., Kleopa, K.A., Cooper, E.C. & Scherer, S.S. (2004). KCNQ2 is a nodal K^+ channel. J. Neurosci 24, pp. 1236-1244.

Doyle, D.A., Morais Cabral, J.H., Pfuetzner, R.A., Kuo, A., Gulbis, J.M., Cohen, S.L., Chait, B.T. & MacKinnon, R. (1998). The structure of the potassium channel: molecular basis of K^+ conduction and selectivity. Science 280, pp. 69-77.

Durell, S.R., Hao, Y., Nakamura, T., Bakker, E.P. & Guy, H.R. (1999). Evolutionary relationship between K+ channels and symporters. Biophysical Journal 77, pp. 775-788.

Elinder, F., Nilsson, J. & Arhem, P. (2007). On the opening of voltage-gated ion channels. Physiology & Behavior 92, pp. 1-7.

Foskett, J.K., White, C., Cheung, K. & Mak, D.D. (2007). Inositol trisphosphate receptor Ca^{2+} release channels. Physiol Rev 87, pp. 593-658.

Galzi, J.L., Devillers-Thiery, A., Hussy, N., Bertrand, S., Changeux, J.P. & Bertrand, D. (1992). Mutations in the channel domain of a neuronal nicotinic receptor convert ion selectivity from cationic to anionic. Nature 359, pp. 500-505.

Gamper, N. & Shapiro, M.S. (2007). Regulation of ion transport proteins by membrane phosphoinositides. Nat Rev Neurosci 8, pp. 921-934.

Gandhi, C.S., Clark, E., Loots, E., Pralle, A. & Isacoff, E.Y. (2003). The orientation and molecular movement of a K$^+$ channel voltage-sensing domain. Neuron 40, pp. 515-525.

Gandhi, C.S., Loots, E. & Isacoff, E.Y. (2000). Reconstructing voltage sensor-pore interaction from a fluorescence scan of a voltage-gated K$^+$ channel. Neuron 27, pp. 585-595.

Gaymard, F., Cerutti, M., Horeau, C., Lemaillet, G., Urbach, S., Ravallec, M., Devauchelle, G., Sentenac, H. & Thibaud, J.B. (1996). The baculovirus/insect cell system as an alternative to Xenopus oocytes. First characterization of the AKT1 K$^+$ channel from Arabidopsis thaliana. J Biol Chem 271, pp. 22863-22870.

George, C.H., Jundi, H., Thomas, N.L., Scoote, M., Walters, N., Williams, A.J. & Lai, F.A. (2004). Ryanodine receptor regulation by intramolecular interaction between cytoplasmic and trans-membrane domains. Mol Biol Cell 15, pp. 2627-2638.

Glauner, K.S., Mannuzzu, L.M., Gandhl, C.S. & Isacoff, E.Y. (1999). Spectroscopic mapping of voltage sensor movement in the Shaker potassium channel. Nature 402, pp. 813-817.

Gulbis, J.M., Zhou, M., Mann, S. & MacKinnon, R. (2000). Structure of the cytoplasmic beta subunit-T1 assembly of voltage-dependent K$^+$ channels. Science 289, pp. 123-127.

Hagiwara, S., Miyazaki, S. & Rosenthal, N.P. (1976). Potassium current and the effect of cesium on this current during anomalous rectification of the egg cell membrane of a starfish. J Gen Physiol 67, pp. 621-638.

Hagiwara, S., Miyazaki, S., Krasne, S. & Ciani, S. (1977). Anomalous permeabilities of the egg cell membrane of a starfish in K$^+$-Tl$^+$ mixtures. J Gen Physiol 70, pp. 269-281.

Heinemann, S.H., Terlau, H., Stuhmer, W., Imoto, K. & Numa, S. (1992). Calcium channel characteristics conferred on the sodium channel by single mutations. Nature 356, pp. 441-443.

Helmchen, F., Svoboda, K., Denk, W. & Tank, D.W. (1999). In vivo dendritic calcium dynamics in deep-layer cortical pyramidal neurons. Nat Neurosci 2, pp. 989-996.

Herlitze, S., Garcia, D.E., Mackie, K., Hille, B., Scheuer, T. & Catterall, W.A. (1996). Modulation of Ca^{2+} channels by G-protein beta gamma subunits. Nature 380, pp. 258-262.

Hess, P. & Tsien, R.W. (1984). Mechanism of ion permeation through calcium channels. Nature 309, pp. 453-456.

Hille, B. (2001). Ionic Channels of Excitable Membranes. Sinauer Associates, Inc: Sunderland, MA.

Hodgkin, A.L. & Huxley, A.F. (1952). A quantitative description of membrane current and its application to conduction and excitation in nerve. J Physiol 117, pp. 500-544.

Hodgkin, A.L. & Keynes, R.D. (1955). The potassium permeability of a giant nerve fibre. J Physiol 128, pp. 61-88.

Hoeflich, K.P. & Ikura, M. (2002). Calmodulin in action: diversity in target recognition and activation mechanisms. Cell 108, pp. 738-742.

Hoffman, D.A., Magee, J.C., Colbert, C.M. & Johnston, D. (1997). K$^+$ channel regulation of signal propagation in dendrites of hippo-campal pyramidal neurons. Nature 387, pp. 869-875.

Horn, R. (2000). A new twist in the saga of charge movement in voltage-dependent ion channels. Neuron 25, pp. 511-514.

Huang, C.L., Feng, S. & Hilgemann, D.W. (1998). Direct activation of inward rectifier potassium channels by PIP2 and its stabilization by Gbetagamma. Nature 391, pp. 803-806.

Ikeda, S.R. (1996). Voltage-dependent modulation of N-type calcium channels by G-protein beta gamma subunits. Nature 380, pp. 255-258.

Isacoff, E.Y., Jan, Y.N. & Jan, L.Y. (1991). Putative receptor for the cytoplasmic inactivation gate in the Shaker K$^+$ channel. Nature 353, pp. 86-90.

Jan, L.Y. & Jan, Y.N. (1994). Potassium channels and their evolving gates. Nature 371, pp. 119-122.

Jan, L.Y. & Jan, Y.N. (1997). Cloned potassium channels from eu-karyotes and prokaryotes. Annu Rev Neurosci 20, pp. 91-123.

Jentsch, T.J., Neagoe, I. & Scheel, O. (2005). CLC chloride channels and transporters. Curr Opin Neurobiol 15, pp. 319-325.

Jiang, Q.X., Wang, D.N. & MacKinnon (2004). Electron microscopic analysis of KvAP voltage-dependent K$^+$ channels in an open conformation. Nature 430, pp. 806-810.

Jiang, Y., Lee, A., Chen, J., Cadene, M., Chalt, B.T. & MacKinnon, R. (2002a). Crystal structure and mechanism of a calcium-gated potassium channel. Nature 417, pp. 515-522.

Jiang, Y., Lee, A., Chen, J., Cadene, M., Chalt, B.T. & MacKinnon, R. (2002b). The open pore conformation of potassium channels. Nature 417, pp. 523-526.

Jiang, Y., Lee, A., Chen, J., Ruta, V., Cadene, M., Chalt, B.T. & MacKinnon, R. (2003). X-ray structure of a voltage-dependent K$^+$ channel. Nature 423, pp. 33-41.

Johnston, D., Hoffman, D.A., Colbert, C.M. & Magee, J.C. (1999). Regulation of back-propagating action potentials in hippocam-pal neurons. Curr Opin Neurobiol 9, pp. 288-292.

Keynes, R.D. (1994). The kinetics of voltage-gates ion channels. Q Rev Biophys 27, pp. 339-434.

Kobrinsky, E., Mirshahi, T., Zhang, H., Jin, T. & Logothetis, D.E. (2000). Receptor-mediated hydrolysis of plasma membrane mes-senger PIP2 leads to K$^+$-current desensitization. Nat Cell Biol 2, pp. 507-514.

Kuo, A., Gulbis, J.M., Antcliff, J.F., Rahman, T., Lowe, E.D., Zimmer, J., Cuthbertson, J., Ashcroft, F.M., Ezaki, T. & Doyle, D.A. (2003). Science 300, pp. 1922-1926.

Laine, M., Lin, M.C.A., Bannister, J.P.A., Silverman, W.R., Mock, A.F., Roux, B. & Papazian, D.M. (2003). Atomic proximity between S4 segment and pore domain in Shaker potassium channels. Neuron 39, pp. 467-481.

Larsson, H.P. & Elinder, F. (2000). A conserved glutamate is important for slow inactivation in K$^+$ channels. Neuron 27, pp. 573-583.

Larsson, H.P., Baker, O.S., Dhillon, D.S. & Isacoff, E.Y. (1996). Transmembrane movement of the Shaker K$^+$ channel S4. Neuron 16, pp. 387-397.

Lee, A., Wong, S.T., Gallagher, D., Li, B., Storm, D.R., Scheuer, T. & Catterall, W.A. (1999). Ca^{2+}/calmodulin binds to and modulates P/Q-type calcium channels. Nature 399, pp. 155-159.

Lee, H.C., Wang, J.M. & Swartz, K.J. (2003). Interaction between extracellular hanatoxin and the resting conformation of the voltage-sensor paddle in Kv channels. Neuron 40, pp. 527-536.

Lesage, F. & Lazdunski, M. (2000). Molecular and functional properties of two pore domain potassium channels. Am J Physiol 279, pp. F793-F801.

Lockless, S.W., Zhou, M., MacKinnon, R. (2007). Structural and thermodynamic properties of selective ion binding in a K$^+$ channel. PLoS Biol 5, pp. e121 EP.

Long, S.B., Tao, X., Campbell, E.B., MacKinnon, R. (2007). Atomic structure of a voltage-dependent K$^+$ channel in a lipid mem-brane-like environment. Nature 450, pp. 376-382.

Loots, E. & Isacoff, E.Y. (1998). Protein rearrangements underlying slow inactivation of the Shaker K$^+$ channel. J Gen Physiol 112, pp. 377-389.

Luscher, C., Jan, L.Y., Stoffel, M., Malenka, R.C. & Nicoll, R.A. (1997). G protein-coupled inwardly rectifying K$^+$ channels (GIRKs) mediate postsynaptic but not presynaptic transmitter actions in hippocampal neurons [published erratum appears in Neuron 1997 Oct;19(4):following 945]. Neuron 19, pp. 687-695.

Luthi, A. & McCormick, D.A. (1998). H-current: properties of a neuronal and network pacemaker. Neuron 21, pp. 9-12.

Mackenzie, P.J. & Murphy, T.H. (1998). High safety factor for action potential conduction along axons but not dendrites of cultured hippocampal and cortical neurons. J Neurophysiol 80, pp. 2089-2101.

Magee, J.C. (1998). Dendritic hyperpolarization-activated currents modify the integrative properties of hippocampal CA1 pyramidal neurons. J Neurosci 18, pp. 7613-7624.

Magee, J.C. & Cook, E.P. (2000). Somatic EPSP amplitude is independent of synapse location in hippocampal pyramidal neurons [see comments]. Nat Neurosci 3, pp. 895-903.

Mannikko, R., Elinder, F. & Larsson, H.P. (2002). Voltage-sensing mechanism is conserved among ion channels gated by opposite voltages. Nature 419, pp. 837-841.

Mannuzzu, L.M. & Isacoff, E.Y. (2000). Independence and cooperativity in rearrangements of a potassium channel voltage sensor revealed by single subunit fluorescence. J Gen Physiol 115, pp. 257-268.

Mannuzzu, L.M., Moronne, M.M. & Isacoff, E.Y. (1996). Direct physical measure of conformational rearrangement underlying potassium channel gating. Science 271, pp. 213-216.

Maricq, A.V., Peterson, A.S., Brake, A.J., Myers, R.M. & Julius, D. (1991). Primary structure and functional expression of the 5HT3 receptor, a serotonin-gated ion channel. Science 254, pp. 432-437.

Marten, I., Hoth, S., Deeken, R., Ache, P., Ketchum, K.A., Hoshi, T. & Hedrich, R. (1999). AKT3, a phloem-localized K^+ channel, is blocked by protons. Proc Natl Acad Sci USA 96, pp. 7581-7586.

Marty, A. (1981). Ca-dependent K channels with large unitary conductance in chromaffin cell membranes. Nature 291, pp. 497-500.

Meador, W.E., Means, A.R. & Quiocho, F.A. (1993). Modulation of calmodulin plasticity in molecular recognition on the basis of x-ray structures. Science 262, pp. 1718-1721.

Meera, P., Wallner, M., Song, M. & Toro, L. (1997). Large conductance voltage- and calcium-dependent K^+ channel, a distinct member of voltage-dependent ion channels with seven N-terminal transmembrane segments (S0-S6), an extracellular N terminus, and an intracellular (S9-S10) C terminus. Proc Natl Acad Sci USA 94, pp. 14066-14071.

Mickus, T., Jung, H. & Spruston, N. (1999). Properties of slow, cumulative sodium channel inactivation in rat hippocampal CA1 pyramidal neurons. Biophys J 76, pp. 846-860.

Mikoshiba, K. (1997). The InsP3 receptor and intracellular Ca^{2+} signaling. Curr Opin Neurobiol 7, pp. 339-345.

Minor, D.L., Jr., Masseling, S.J., Jan, Y.N. & Jan, L.Y. (1999). Transmembrane structure of an inwardly rectifying potassium channel. Cell 96, pp. 879-891.

Miyazawa, A., Fujiyoshi, Y. & Unwin, N. (2003). Structure and gating mechanism of the acetylcholine receptor pore. Nature 423, pp. 949-955.

Montell, C., 2005. The TRP superfamily of cation channels. Sci STKE 2005, re3.

Nichols, C.G. & Lopatin, A.N. (1997). Inward rectifier potassium channels. Annu Rev Physiol 59, pp. 171-191.

Nishida, M., Cadene, M., Chait, B.T., & MacKinnon, R. (2007). Crystal structure of a Kir3.1-prokaryotic Kir channel chimera. EMBO J 26, pp. 4005-4015.

Obaid, A.L. & Salzberg, B.M. (1996). Micromolar 4-aminopyridine enhances invasion of a vertebrate neurosecretory terminal arborization: optical recording of action potential propagation using an ultrafast photodiode-MOSFET camera and a photodiode array. J Gen Physiol 107, pp. 353-368.

Olcese, R., Latorre, R., Toro, L., Bezanilla, F. & Stefani, E. (1997). Correlation between charge movement and ionic current during slow inactivation in Shaker K^+ channels. J Gen Physiol 110, pp. 579-589.

Papin, J.A., Hunter, T., Palsson, B.O. & Subramaniam, S. (2005). Reconstruction of cellular signalling networks and analysis of their properties. Nat Rev Mol Cell Biol 6, pp. 99-111.

Patel, A.J., Honore, E., Lesage, F., Fink, M., Romey, G. & Lazdunski, M. (1999). Inhalational anesthetics activate two-pore-domain background K^+ channels. Nat Neurosci 2, pp. 422-426.

Perozo, E., Cortes, D.M. & Cuello, L.G. (1999). Structural rearrangements underlying K^+-channel activation gating. Science 285, pp. 73-78.

Piskorowski, R. & Aldrich, R.W. (2002). Calcium activation of BKCa potassium channels lacking the calcium bowl and RCK domains. Nature 420, pp. 499-502.

Plummer, N.W. & Meisler, M.H. (1999). Evolution and diversity of mammalian sodium channel genes. Genomics 57, pp. 323-331.

Pouille, F. & Scanziani, M. (2004). Routing of spike series by dynamic circuits in the hippocampus. Nature 429, pp. 717-723.

Quayle, J.M., Nelson, M.T. & Standen, N.B. (1997). ATP-sensitive and inwardly rectifying potassium channels in smooth muscle. Physiol Rev 77, pp. 1165-1232.

Rasband, M.N., Trimmer, J.S., Schwarz, T.L., Levinson, S.R., Ellisman, M.H., Schachner, M. & Shrager, P. (1998). Potassium channel distribution, clustering, and function in remyelinating rat axons. J Neurosci 18, pp. 36-47.

Rees, D.C., Chang, G. & Spencer, R.H. (2000). Crystallographic analyses of ion channels: lessons and challenges. J Biol Chem 275, pp. 713-716.

Rettig, J., Heinemann, S.H., Wunder, F., Lorra, C., Parcej, D.N., Dolly, J.O. & Pongs, O. (1994). Inactivation properties of voltage-gated K^+ channels altered by presence of beta-subunit. Nature 369, pp. 289-294.

Rhodes, K.J., Keilbaugh, S.A., Barrezueta, N.X., Lopez, K.L. & Trimmer, J.S. (1995). Association and colocalization of K^+ channel alpha- and beta-subunit polypeptides in rat brain. J Neurosci 15, pp. 5360-5371.

Salzer, J.L. (1997). Clustering sodium channels at the node of Ranvier: close encounters of the axon-glia kind. Neuron 18, pp. 843-846.

Sansom, M.S.P., Shrivastava, I.H., Bright, J.N., Tate, J., Capener, C.E. & Biggin, P.C. (2002). Potassium channels: structures, models, simulations. Biochimica et Biophysica Acta (BBA) - Biomembranes 1565, pp. 294-307.

Schachtman, D.P., Schroeder, J., Lucas, W.J., Anderson, J.A. & Gaber, R.F. (1992). Science 258, pp. 1654-1658.

Schiller, J., Schiller, Y., Stuart, G. & Sakmann, B. (1997). Calcium action potentials restricted to distal apical dendrites of rat neocortical pyramidal neurons. J Physiol (Lond) 505, pp. 605-616.

Schreiber, M., Yuan, A. & Salkoff, L. (1999). Transplantable sites confer calcium sensitivity to BK channels. Nature Neurosci 2, pp. 416-421.

Sheng, M., Liao, Y.J., Jan, Y.N. & Jan, L.Y. (1993). Presynaptic A-current based on heteromultimeric K^+ channels detected in vivo. Nature 365, pp. 72-75.

Sheng, S., Li, J., McNulty, K.A., Avery, D. & Kleyman, T.R. (2000). Characterization of the selectivity filter of the epithelial sodium channel. J Biol Chem 275, pp. 8572-8581.

Sheppard, D.N. & Welsh, M.J. (1999). Structure and function of the CFTR chloride channel. Physiol Rev 79, pp. S23-S45.

Shrivastava, I.H. & Bahar, I. (2006). Common mechanism of pore opening shared by five different potassium channels. Biophys J 90, pp. 3929-3940.

Shrivastava, I.H., Tieleman, D.P., Biggin, P.C., & Sansom, M.S.P. (2002). K+ versus Na+ ions in a K channel selectivity filter: a simulation study. Biophys J 83, pp. 633-645.

Sigworth, F.J. (1994). Voltage gating of ion channels. Q Rev Biophys 27, pp. 1-40.

Slesinger, P.A. & Lansman, J.B. (1991). Reopening of Ca2+ channels in mouse cerebellar neurons at resting membrane potentials during recovery from inactivation. Neuron 7, pp. 755-762.

Snyder, P.M., Olson, D.R. & Bucher, D.B. (1999). A pore segment in DEG/ENaC Na+ channels. J Biol Chem 274, pp. 28484-28490.

Spruston, N., Schiller, Y., Stuart, G. & Sakmann, B. (1995). Activity-dependent action potential invasion and calcium influx into hippocampal CA1 dendrites [see comments]. Science 268, pp. 297-300.

Stuart, G., Spruston, N., Sakmann, B. & Hausser, M. (1997). Action potential initiation and backpropagation in neurons of the mammalian CNS. TINS 20, pp. 125-131.

Sun, Y., Ofson, R., Horning, M., Armstrong, N., Mayer, M. & Gouaux, E. (2002). Mechanism of glutamate receptor desensitization. Nature 417, pp. 248-253.

Tang, X.D., Marten, I., Dietrich, P., Ivashikina, N., Hedrich, R. & Hoshi, T. (2000). Histidine(118) in the S2-S3 linker specifically controls activation of the KAT1 channel expressed in Xenopus oocytes. Biophys J 78, pp. 1255-1269.

Tkatch, T., Baranauskas, G. & Surmeier, D.J. (2000). Kv4.2 mRNA abundance and A-type K+ current amplitude are linearly related in basal ganglia and basal forebrain neurons. J Neurosci 20, pp. 579-588.

Torres, Y.P., Morera, F.J., Carvacho, I. & Latorre, R. (2007). A marriage of convenience: beta-subunits and voltage-dependent K+ channels. J Biol Chem 282, pp. 24485-24489.

Unwin, N. (1995). Acetylcholine receptor channel imaged in the open state. Nature 373, pp. 37-43.

Valera, S., Hussy, N., Evans, R.J., Adami, N., North, R.A., Surprenant, A. & Buell, G. (1994). A new class of ligand-gated ion channel defined by P2x receptor for extracellular ATP. Nature 371, pp. 516-519.

Valiyaveetil, F.I., Leonetti, M., Muir, T.W., & MacKinnon, R. (2006). Ion selectivity in a semisynthetic K+ channel locked in the conductive conformation. Science 314, pp. 1004-1007.

Veh, R.W., Lichtinghagen, R., Sewing, S., Wunder, F., Grumbach, I.M. & Pongs, O. (1995). Immunohistochemical localization of five members of the Kv1 channel subunits: contrasting subcellular locations and neuron-specific co-localizations in rat brain. European J Neurosci 7, pp. 2189-2205.

Venkatachalam, K. & Montell, C. (2007). TRP channels. Annu Rev Biochem 76, pp. 387-417.

Vergara, C., Latorre, R., Marrion, N.V. & Adelman, J.P. (1998). Calcium-activated potassium channels. Curr Opin Neurobiol 8, pp. 321-329.

Wallner, M., Meera, P. & Toro, L. (1999). Molecular basis of fast inactivation in voltage and Ca2+-activated K+ channels: a transmembrane beta-subunit homolog. Proc Natl Acad Sci USA 96, pp. 4137-4142.

Wang, H., Kunkel, D.D., Martin, T.M., Schwartzkroin, P.A. & Tempel, B.L. (1993). Heteromultimeric K+ channels in terminal and juxtaparanodal regions of neurons. Nature 365, pp. 75-79.

Wanner, S.G., Koch, R.O., Koschak, A., Trieb, M., Garcia, M.L., Kaczorowski, G.J. & Knaus, H.G. (1999). High-conductance calcium-activated potassium channels in rat brain: pharmacology, distribution, and subunit composition. Biochem 38, pp. 5392-5400.

Webster, S.M., del Camino, D., Dekker, J.P. & Yellen, G. (2004). Intracellular gate opening in Shaker K+ channels defined by high-affinity metal bridges. Nature 428, pp. 864-868.

Wickman, K., Nemec, J., Gendler, S.J. & Clapham, D.E. (1998). Abnormal heart rate regulation in GIRK4 knockout mice. Neuron 20, pp. 103-114.

Williams, S.R. & Stuart, G.J. (2000). Site independence of EPSP time course is mediated by dendritic I(h) in neocortical pyramidal neurons. J Neurophysiol 83, pp. 3177-3182.

Wo, Z.G. & Oswald, R.E. (1995). Unraveling the modular design of glutamate-gated ion channels. Trend Neurosci 18, pp. 161-168.

Wu, L., Bauer, C.S., Zhen, X.G., Xie, C. & Yang, J. (2002). Dual regulation of voltage-gated calcium channels by Ptdlns(4,5)P2. Nature 419, pp. 947-952.

Xia, X.M., Fakler, B., Rivard, A., Wayman, G., Johson-Pais, T., Keen, J.E., Ishii, T., Hirschberg, B., Bond, C.T., Lutsenko, S., Maylie, J. & Adelman, J.P. (1998). Mechanism of calcium gating in small-conductance calcium-activated potassium channels. Nature 395, pp. 503-507.

Yang, J., Ellinor, P.T., Sather, W.A., Zhang, J.F. & Tsien, R.W. (1993). Molecular determinants of Ca2+ selectivity and ion permeation in L-type Ca2+ channels [see comments]. Nature 366, pp. 158-161.

Yang, N., George, A.L.J. & Horn, R. (1996). Molecular basis of charge movement in voltage-gated sodium channels. Neuron 16, pp. 113-122.

Yasuda, R., Sabatini, B. & Svoboda, K. (2003). Plasticity of calcium channels in dendritic spines. Nature Neurosci 6, pp. 948-955.

Yellen, G. (1998). The moving parts of voltage-gated ion channels. Q Rev Biophys 31, pp. 239-295.

Yellen, G. (2002). The voltage-gated potassium channels and their relatives. Nature 419, pp. 35-42.

Yi, B.A., Lin, Y., Jan, Y.N. & Jan, L.Y. (2001). Yeast screen for constitutively active mutant G protein-activated potassium channels. Neuron 29, pp. 657-667.

Zagotta, W.N., Hoshi, T. & Aldrich, R.W. (1990). Restoration of inactivation in mutants of Shaker potassium channels by a peptide derived from ShB. Science 250, pp. 568-571.

Zamponi, G.W., Bourinet, E., Nelson, D., Nargeot, J. & Snutch, T.P. (1997). Crosstalk between G proteins and protein kinase C mediated by the calcium channel alpha1 subunit [see comments]. Nature 385, pp. 442-446.

Zhou, L., Messing, A. & Chiu, S.Y. (1999). Determinants of excitability at transition zones in Kv1.1-deficient myelinated nerves. J Neurosci 19, pp. 5768-5781.

Zufall, F., Shepherd, G.M. & Barnstable, C.J. (1997). Cyclic nucleotide gated channels as regulators of CNS development and plasticity. Curr Opin Neurobiol 7, pp. 404-412.

Zühlke, R.D., Pitt, G.S., Deisseroth, K., Tsien, R.W. & Reuter, H. (1999). Calmodulin supports both inactivation and facilitation of L-type calcium channels. Nature 399, pp. 159-162.

Dynamical Properties of Excitable Membranes

Douglas A. Baxter and John H. Byrne

The nervous system functions to encode, process, store, and transmit information. To perform these tasks, neurons have evolved sophisticated means of generating electrical and chemical signals. Chapters 12 and 14 describe several aspects of intracellular chemical signaling mechanisms (e.g., second-messenger cascades and genetic-regulatory networks) and Chapters 5, 8, and 16 describe several aspects of electrical signaling in neurons (e.g., the membrane potential, the action potential, and synaptic transmission). The goals of this chapter are to provide a more detailed mathematical description of neuronal excitability and introduce several mathematical tools based on the theory of nonlinear dynamical systems that can be used to analyze neuronal excitability. These mathematical tools (e.g., phase plane analysis, bifurcation theory) provide graphical or geometric representations of the system dynamics and can be used to understand, predict, and interpret biophysical features such as threshold phenomena and oscillatory and bursting behavior, as well as the mechanisms of bistability and hysteresis. Such analyses can provide novel insights into the capabilities of individual neurons to process and store information.

THE HODGKIN–HUXLEY MODEL

Since 1952, the understanding and methods for studying the biophysics of excitable membranes have been profoundly influenced by the pioneering work of Alan Hodgkin and Andrew Huxley. Using the techniques of voltage clamping, which had recently been developed by Cole (1949) and Marmont (1949) and improved upon by Hodgkin et al. (1952) (see also Box 5.3), Hodgkin and Huxley characterized the time- and voltage-dependency of the ionic conductances that underlie an action potential in the squid giant axon (Hodgkin et al. 1952; Hodgkin and Huxley 1952a–c). In addition, they developed a cogent mathematical model that accurately predicted the waveform of the action potential and several other physiological properties such as the refractory period, propagation of the action potential along an axon, anode break excitation, and accommodation (Hodgkin and Huxley 1952d). In biology, quantitatively predictive theories are rare, and this work stands out as one of the most successful combinations of experimental and computational approaches to understanding a fundamentally important issue in neuroscience. For their seminal work on neuronal excitability, Hodgkin and Huxley received the Nobel Prize in Physiology or Medicine in 1963. (To view biographical information about Hodgkin and Huxley and other Nobel laureates, you may wish to visit the Nobel Prize Web site at http://www.nobelprize.org. The lectures that Hodgkin (1964) and Huxley (1964) delivered when they received the Nobel Prize were published in 1964. In addition, Hodgkin (1976, 1977) and Huxley (2000, 2002) have published informal narratives describing events surrounding these studies, and others have reviewed the work of Hodgkin and Huxley from a historical perspective (Cole 1968; Hille 2001; Nelson and Rinzel 1998; Rinzel 1990).) Their empirical studies and quantitative model remain influential to the present day and provide an analytical framework for investigating and modeling a large class of diverse membrane phenomena. The first half of this chapter will review the steps taken by Hodgkin and Huxley to derive a mathematical description of neuronal excitability. The second half of this chapter will discuss applying techniques from the mathematical field of nonlinear dynamics to the analysis of neuronal excitability.

Perhaps the best way to begin a discussion of the Hodgkin–Huxley (HH) model is to analyze the *equivalent electrical circuit* employed by Hodgkin and Huxley to represent a patch of membrane (Fig. 7.1). In this approach, the membrane is considered to be an electrical circuit composed of a capacitive element (C_m) in parallel with conductances (g), which are in series with a battery. The capacitive element represents the dielectric properties of the lipid bilayer of biological membranes. The conductances represent channels in the membrane through which ions can pass (see Chapter 6), and the batteries represent the electrochemical potential gradient that is associated with a given species of ion (see Chapter 5). In the equivalent circuit, the current that flows across the membrane (I_m) has two major components, one associated with charging the membrane capacitance (I_{Cm}) and one associated with the movement of ions across the membrane (I_{ionic}). Thus:

$$I_m = I_{Cm} + I_{ionic} = C_m \frac{dV_m}{dt} + I_{ionic} \qquad (7.1)$$

where V_m is potential across the membrane, and t is time.

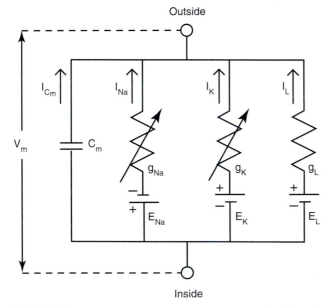

FIGURE 7.1 Equivalent electrical circuit proposed by Hodgkin and Huxley for a patch of squid giant axon. The electrical circuit has four parallel branches. The capacitive branch (C_m) represents the dielectric properties of the membrane. The variable resistors (resistors with arrows) represent voltage- and time-dependent conductances of Na^+ (g_{Na}) and K^+ (g_K), whereas the leakage conductance (g_L) is constant. Each conductance element is in series with a battery (E_{Na}, E_K, E_L) that represents its electromotive force. Adapted from Hodgkin and Huxley (1952d).

Separating the Membrane Current into Its Major Components

As a first step in their analysis, Hodgkin and Huxley separated the membrane current into its major components (i.e., the capacitive, I_{Cm}, and ionic, I_{ionic}, currents). It was the voltage-clamp technique that provided the method for this separation. By using a feedback circuit (see Chapter 5), the voltage-clamp technique maintains (i.e., "clamps") the membrane potential at a designated voltage (Hodgkin *et al.* 1952). In a typical experiment (Fig. 7.2), the membrane potential was held at voltage (i.e., the *holding voltage*, V_h), which was often near the resting potential of the neuron, and voltage-clamp commands (i.e., the *command voltage*, V_c) stepped the membrane potential from V_h to various depolarized or hyperpolarized levels for a few milliseconds and then back to V_h. The stepwise depolarizations (or hyperpolarizations) of the membrane have two advantages for measuring ionic current. First, except for the brief moment of transition between V_h and V_c, the membrane potential is constant (i.e., $dV_m/dt = 0$). Thus, the capacitive current is eliminated and the ionic current can be measured in isolation. Second, by keeping the voltage constant, the time dependency of the ionic currents can be measured.

The voltage-clamp technique provides a quantitative measure of the ionic currents that flow through an excitable membrane such as the squid giant axon. An example of the types of membrane currents that were recorded by Hodgkin and Huxley is presented in Figure 7.2A (see also Chapter 5). When the membrane potential was depolarized, a transient inward current (i.e., the flow of positive charge into the cell) was observed. This transient inward current (i.e., I_{early}) was followed by a sustained outward current (i.e., I_{late}), which represents the flow of positive charge out of the cell. The time course and magnitude of these ionic currents depended markedly on the magnitude of the depolarizing step (i.e., V_c). For example, the amplitude of I_{early} increased and then decreased as V_c became more positive, whereas I_{late} increased monotonically with V_c (Fig. 7.2B). Because of the voltage dependency of some membrane currents, the underlying channels, through which the ions flow, are often referred to as *voltage-gated channels*.

The second step in the analysis was to separate the complex ionic current into its components. By altering the ionic composition of the external solutions (e.g., substituting choline for Na^+) (Hodgkin and Huxley 1952a), the inward current (i.e., I_{early}) was eliminated (Fig. 7.3), leaving only outward current (i.e., I_{late}), which was assumed to be carried by K^+ (Huxley 1951). Thus, the ionic current was subdivided into two primary components: one carried by Na^+ (I_{Na}) and another carried by

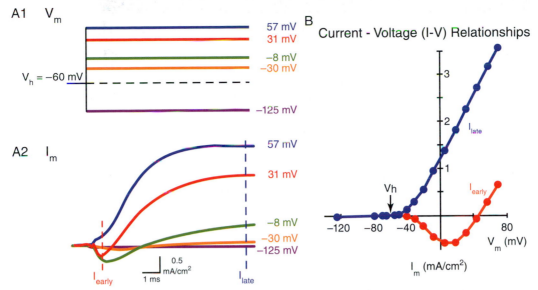

FIGURE 7.2 Membrane currents (I_m) recorded with voltage clamp of squid axon. Data are from an experiment by Hodgkin *et al.* 1952) (i.e., axon 41 at 3.8°C). (A1) In this simple voltage-clamp protocol, the membrane potential (V_m) was held near its rest potential, which is taken to be −60 mV (i.e., $V_h = V_m = −60$ mV). Voltage-clamp commands (V_c) were used to step the membrane potential to various hyperpolarized and depolarized potentials, which are indicated next to each trace. (Hodgkin and Huxley used a different convention for labeling membrane potentials and displaying currents; see Box 7.1 for an explanation. In this and all subsequent figures, the data have been modified to reflect modern conventions.) (A2) Responses elicited by a series of voltage-clamp steps. Successive current traces have been superimposed. Inward current is indicated by a downward deflection, and outward current is indicated by an upward deflection. (Only the response immediately prior to and during the initial phase of the voltage-clamp step is shown. The response at the end of the voltage-clamp step is not illustrated.) The time course, direction, and magnitude of I_m vary with V_m. (B) Current-voltage relation (i.e., *I-V plot*) from voltage-clamp experiment that is illustrated in panel A. (Additional data points are included in the I-V plot that were not illustrated in panel A.) The magnitude of the currents at 0.5 and 8 ms (I_{early} and I_{late}, respectively) are plotted as functions of V_m. Data from Hodgkin, A.L. and Huxley, A.F. (1952d). A quantitative description of membrane current and its application to conduction and excitation in nerve. *J. Physiol. (Lond.)* **117:** 500-544.

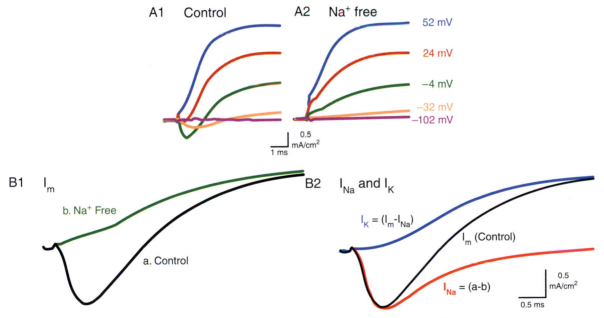

FIGURE 7.3 Separation of membrane current (I_m) into components carried by Na⁺ (I_{Na}) and K⁺ (I_K). Data are from an experiment by Hodgkin and Huxley (1952a) (axon 21 at 8.5°C). (A) Total membrane currents (I_m) were measured in control (A1) and Na⁺-free (A2) solutions. The membrane potential was held at –60 mV and was stepped to the various potentials indicated to the right of each trace. (Only the response immediately prior to and during the voltage-clamp step is illustrated.) In control solutions, command voltages that depolarized the membrane potential elicited an inward current followed by an outward current. After the Na⁺ in the solution was replaced with impermeant choline ions, the inward component of I_m was abolished and only the outward component remained. (B1) Enlarged view of the membrane currents elicited by the voltage-clamp step to –4 mV, in panel A. Membrane currents were first elicited in control solution (i.e., trace a) and again after removal of Na⁺ (trace b), which blocked the inward component of I_m. (B2) Subtracting the response in the Na⁺-free solution from the response in control solution (i.e., trace a – trace b) revealed the amplitude and time course of the Na⁺-dependent component of I_m (i.e., I_{Na}). Similarly, I_K could be isolated by subtracting I_{Na} from I_m. Data from Hodgkin, A.L. and Huxley, A.F. (1952d). A quantitative description of membrane current and its application to conduction and excitation in nerve. *J. Physiol. (Lond.)* **117:** 500-544.

K^+ (I_K). In addition, there is a small component that is referred to as the *leakage current* (I_l), which represents ions flowing through non-voltage-gate channels. Thus, the total ionic current was expressed as

$$I_{ionic} = I_{Na} + I_K + I_l \qquad (7.2)$$

In addition to separating the ionic current into its components, it was necessary to determine the

relationship between ionic current and membrane potential at a constant permeability. To examine this issue, Hodgkin and Huxley developed a voltage-clamp protocol that measured what they referred to as the *instantaneous current-voltage (I-V) relationship* (Hodgkin and Huxley 1952b). In this protocol, two voltage-clamp commands were applied to the axon (Fig. 7.4). The first voltage-clamp command (V_1) was

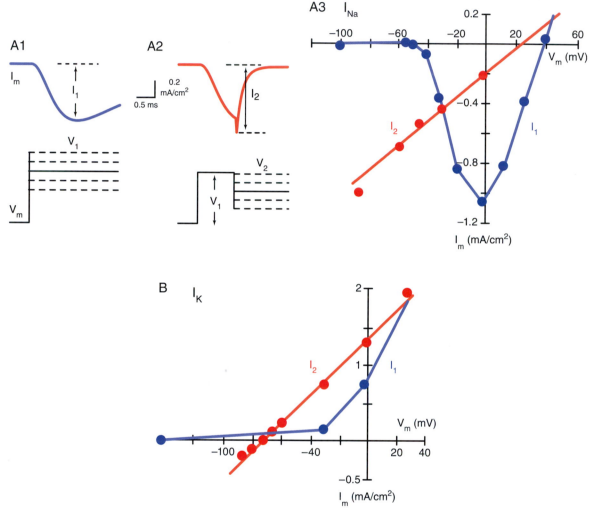

FIGURE 7.4 Instantaneous current-voltage (I-V) relations for the early and late components of I_m (i.e., I_{Na} and I_K, respectively). (A1) In a simple voltage-clamp protocol, the membrane potential is stepped from a holding potential to various depolarized (or hyperpolarized) command voltages (V_1), which activate ion channels and elicit the flow of membrane currents (I_1). (A2) To measure the instantaneous I-V relation, two command-voltage steps are used. The first voltage-clamp step (V_1) is to a fixed potential. In studies of Na$^+$ channels, $V_1 = -31$ mV, and in studies of K$^+$ channels, $V_1 = 24$ mV. V_1 was very brief and had a duration of ~1.5 ms in studies of Na$^+$ channels and was <1 ms in studies of K$^+$ channels. The second voltage-clamp step (V_2) varied between -100 mV and 50 mV. The instantaneous current (I_2) is measured immediately (i.e., within ~30 μs) of the second voltage-clamp step. (A3) Current-voltage relationships for the early inward component of I_m (i.e., I_{Na}). The current-voltage relation of the maximum inward current during a single voltage-clamp step (i.e., I_1 in panel A) is extremely nonlinear (blue curve). In contrast, the current-voltage relation of the instantaneous current at the beginning of the second voltage-clamp step (i.e., I_2 in panel B) is approximately linear (red curve). Data are from an experiment by Hodgkin and Huxley (1952b) (i.e., axon 31 at 4°C). (B) Current-voltage relationships for the late outward component of I_m (i.e., I_K), which were measured while the axon was bathed in Na$^+$-free saline. The current-voltage relationship of current measured ~0.6 ms after the beginning of the first voltage-clamp step (i.e., I_1) versus V_1 is nonlinear (blue trace), whereas the current-voltage relationship of the instantaneous current (i.e., I_2) is approximately linear (red trace). Data are from an experiment by Hodgkin and Huxley (1952b) (axon 26 at 20°C). The linear I-V relationships of the instantaneous currents (i.e., I_2) was in striking contrast to the extremely nonlinear I-V relationships that were obtained when the currents were measured at later intervals (i.e., I_1). The curvature of the I_1 I-V relationships (i.e., blue traces in panels A3 and B) reflects the voltage- and time-dependent opening of Na$^+$ and K$^+$ channels. In contrast, the linear nature of the instantaneous current-voltage relationships (i.e., red traces in panels A3 and B) indicate that the flow of ionic currents in open channels obey Ohm's law. Data from Hodgkin, A.L. and Huxley, A.F. (1952a). Currents carried by sodium and potassium ions through the membrane of the giant axon of Loligo. *J. Physiol. (Lond.)* **116**: 449-472.

brief and had a fixed amplitude, whereas the amplitude of the second voltage-clamp step (V_2) was varied. V_1 served to activate (i.e., open) the Na$^+$ channels (i.e., the channels that mediate I_{early} in Fig. 7.2). After the Na$^+$ channels were activated by V_1, the membrane potential was suddenly stepped to V_2 and the current (I_2) was measured within the first few microseconds of V_2 (i.e., at the instant the membrane potential changed from V_1 to V_2). (Because I_2 is measured at the "tail end" of V_1, such currents are also referred to as *tail currents*.) The purpose of this protocol was to measure the current without time-dependent influences. Immediately after the step from V_1 to V_2, the level of activation (and inactivation) is constant because it has not yet had time to change. Thus, only the driving force $\Delta V = (V_m - E_{ion})$, where $V_m = V_2$, differs for different values of V_2. The results indicated that the I-V relationship of I_2 was approximately linear. The instantaneous I-V relationship of K$^+$ channels was also studied, and a similar linear relationship was observed. These results indicated that under normal ionic conditions the flow of ionic current in open Na$^+$ and K$^+$ channels obeys *Ohm's law* (i.e., $\Delta V = I \times R$, where $G = 1/R$, and R stands for resistance and G for its inverse, conductance). It follows, therefore, that by using their empirical measurements of I_{Na} and I_K, Hodgkin and Huxley were able to determine g_{Na} and g_K (see later).

Ohm's law implies that the individual ionic currents in Eq. 7.2 were proportional to the conductance (i.e., g) times the *driving force* (i.e., the difference between the membrane potential, V_m, and the Nernst potential, E_{ion}, for a given ion species), resulting in a general equation for an ionic current of the form

$$I_{ion}(V_m, t) = g_{ion}(V_m, t)(V_m - E_{ion}) \qquad (7.3)$$

where $I_{ion}(V_m, t)$ is the current created by the movement of a given species of ion across the membrane, $g_{ion}(V_m, t)$ represents the voltage- and time-dependent conductance (i.e., permeability) of the membrane to that ionic species, and E_{ion} is the Nernst potential of the ion (see Chapter 5). Thus, I_{Na}, I_K, and I_L were described by

$$I_{Na}(V_m, t) = g_{Na}(V_m, t)(V_m - E_{Na}) \qquad (7.4)$$

$$I_K(V_m, t) = g_K(V_m, t)(V_m - E_K) \qquad (7.5)$$

$$I_l = g_l(V_m - E_l) \qquad (7.6)$$

By substituting Eq. 7.2 into Eq. 7.1, the description of membrane current becomes

$$I_m = C_m \frac{dV_m}{dt} + I_{Na} + I_K + I_l \qquad (7.7)$$

Finally, Eq. 7.7 can be expanded by including Eqs. 7.4, 7.5, and 7.6 and becomes

$$I_m = C_m \frac{dV_m}{dt} + g_{Na}(V_m, t)(V_m - E_{Na}) \\ + g_K(V_m, t)(V_m - E_K) + g_l(V_m - E_l) \qquad (7.8)$$

The final stage of the analysis was to characterize the active conductances (i.e., $g_{Na}(V_m, t)$ and $g_K(V_m, t)$).

Analyzing the Time and Voltage Dependency of Ionic Conductances

To analyze the ionic conductances, it was first necessary to devise a method for obtaining measures of g_{Na} and g_K. The linear nature of the instantaneous I-V relationship for I_{Na} and I_K (see earlier) indicated that these ionic currents and their underlying ionic conductances were related by Ohm's law. This provided Hodgkin and Huxley with the means of calculating g_{Na} and g_K. From Eqs. 7.4 and 7.5, it is possible to define ionic conductances as

$$g_{Na}(V_m, t) = \frac{I_{Na}(V_m, t)}{V_m - E_{Na}} \qquad (7.9)$$

$$g_K(V_m, t) = \frac{I_K(V_m, t)}{V_m - E_K} \qquad (7.10)$$

Thus, changes in the conductances g_{Na} and g_K during a voltage-clamp step could be calculated by applying Eqs. 7.9 and 7.10 to separated ionic currents. Figure 7.5 illustrates data obtained by Hodgkin and Huxley in which the magnitude and time course of g_K and g_{Na} were calculated for two different voltage-clamp steps. As discussed in Chapter 5, these conductances have several striking properties. First, the magnitude of the conductances increases with more positive values of V_m. Second, the rising phase of the conductances become more rapid with increasing V_m. Third, there is delay in the onset of the change in the conductances, particularly for g_K. Finally, the increase in g_{Na} is transient, whereas the increase in g_K is not.

To explain these experimental data, Hodgkin and Huxley suggested a model that could account for the voltage- and time-dependent properties of g_K and g_{Na}. This model is often referred to as the *gate model* (Fig. 7.6; see also Chapter 5). The model assumes that the macroscopic conductances as measured with the voltage-clamp procedure arise from the combined effects of many individual ion channels, each with a microscopic conductance to a specific species of ion (i.e., K$^+$ or Na$^+$). Each individual channel has one or more "gates" that regulate the flow of ions through the channel. Each gate can be in one of two states: open or closed. When all the gates for a particular channel are

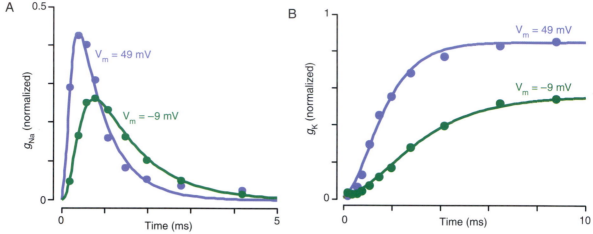

FIGURE 7.5 Experimental voltage-clamp data illustrating the voltage- and time-dependent properties of the g_{Na} (A) and g_K (B). By recording membrane currents in solutions of different ionic composition (e.g., normal seawater versus Na^+-free seawater), Hodgkin and Huxley (1952a) were able to isolate I_{Na} and I_K (see also Fig. 7.3 and Chapter 5). The conductances were calculated using Eqs. 7.9 and 7.10. Although Hodgkin and Huxley examined a wide range of membrane potentials, the results from only two voltage-clamp steps are illustrated. The blue traces are responses elicited by a command voltage (V_c) to 49 mV, and the green traces are responses elicited during a voltage-clamp step to −9 mV. The voltage-clamp steps are not illustrated, but they begin at $t = 0$ and extend beyond the end of the illustrations. In the filled circles are data from an experiment by Hodgkin and Huxley (axon 17), and the solid lines are best-fit curves for Eqs. 7.24 and 7.32. Data from Hodgkin, A.L. and Huxley, A.F. (1952b). The components of membrane conductance in the giant axon of Loligo. *J. Physiol. (Lond.)* **116**: 473-496.

open, ions can pass through the channel (i.e., the single channel conductance is >0). If any of the gates are closed, ions cannot pass through the channel (i.e., the single channel conductance is 0). The status of a gate (i.e., open versus closed) was assumed to be controlled by distribution of one or more charged "particles" within the membrane (i.e., *gating particles* or *gating charges*). These gating particles act as "molecular voltmeters," and as the electrical field across the membrane changes, the distribution of these gating particles is altered such that the gates transition between states. A gate is open only when all the particles that are associated with the gate are in a permissive state. The possible molecular structures responsible for gating particles are described in Chapters 5 and 6.

Mathematically, the voltage dependence of channel opening (and closing) can be derived using the *Boltzmann equation* of statistical mechanics, which describes the equilibrium distribution of independent particles in force fields. From the Boltzmann principle, the proportion of gating particles that are at a location associated with a *permissive state* (P_i) is related to the proportion of gating particles in *nonpermissive* locates (P_o) by the function

$$\frac{P_i}{P_o} = \exp\left(\frac{w + zeV}{kT}\right) \quad (7.11)$$

where w is the work required to move the gating particle from the nonpermissive state to the permissive state, z is the valence of the particle, e is the elementary charge of the particle, V is the potential difference

between the inside and outside of the membrane, k is Boltzmann's constant, and T is the absolute temperature. Since $P_i + P_0 = 1$, Eq. 7.11 can be rearranged to give the proportion of gating particles in the permissive state as a function of voltage:

$$p_i = \frac{1}{1 + \exp\left(\dfrac{-(w + zeV)}{kT}\right)} \quad (7.12)$$

Eq. 7.12 quantifies the voltage dependence of gating in the system and is sometimes referred to as the *activation function of a channel*. As V increases in Eq. 7.12 (i.e., the membrane potential is depolarized), the proportion of gating particles in the permissive state approaches unity (i.e., $P_i \rightarrow 1$). With an increasing proportion of gating particles in the permissive state, a greater number of channels is likely to be open, and thus, the macroscopic membrane conductance will increase. Figure 7.7 illustrates semilogarithmic plots of Eq. 7.12 that are superimposed on the maximum values of g_{Na} and g_K that were measured during a voltage-clamp experiment. The two curves are very similar in shape. One of the most striking properties of the curves is the extreme steepness of the relation between ionic conductance and membrane potential. At low depolarizations (i.e., near the resting potential), the curves approach straight lines. Since the ordinate is plotted on a logarithmic scale, this means that the peak conductances increase exponentially with membrane depolarization, until at high depolarizations, the curves reach a maximal

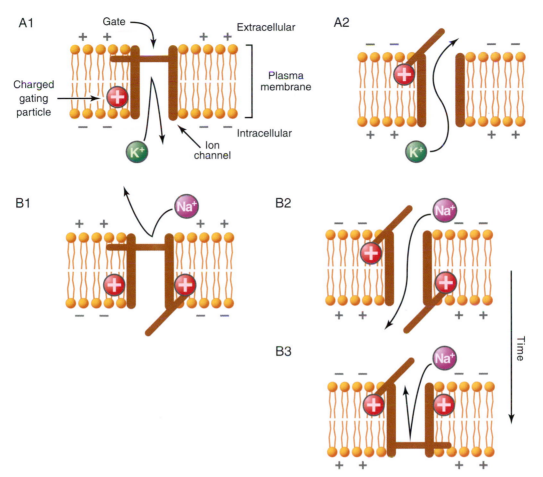

FIGURE 7.6 Schematic representation of the gate model that was proposed by Hodgkin and Huxley (1952d). The gate model assumes that many individual ion channels, each with a small ionic conductance, determine the behavior of the macroscopic membrane conductance. The ion channels have "gates" that are controlled by voltage-sensitive gating charges or particles. If the gating particles are in the permissive state, then the gates are open and ions can pass through the channel. Otherwise, the gates are closed and passage of ions through the channel is blocked. In this example, each gate is regulated by a single gating particle and the gating particles are assumed to be positively charged. Other scenarios are possible, however. (A) K$^+$ channels are regulated by a single activation gate, which is embodied in the HH model as n (see Eq. 7.33). At the resting potential (as indicated by the minus signs, "$-$", on the intracellular surface and the plus signs, "$+$", on the extracellular surface), the gating particles are primarily distributed at locations within the membrane that are nonpermissive, i.e., the gate is closed (A1). As the membrane potential is depolarized (A2; note translocation of "$-$" and "$+$" signs), the probability increases that a gating particle will be located in a position that is permissive, i.e., the gate is open (i.e., activation). Potassium ions (K$^+$) can flow through the open channel. (B) Na$^+$ channels are regulated by two gates: an activation gate and an inactivation gate. These two gates are embodied in the HH model as m and h, respectively (see Eq. 7.25). Unlike the activation gate, the inactivation gate is normally open at the resting potential (B1). Upon depolarization (B2), the probability increases that the activation gate will open, whereas the probability that the inactivation gate will remain open decreases. While both gates are open, Na$^+$ passes through the channel. As the depolarization continues, however (B3), the inactivation gate closes and ions can no longer pass through the channel (i.e., inactivation). Data from Hodgkin, A.L. and Huxley, A.F. (1952d). A quantitative description of membrane current and its application to conduction and excitation in nerve. *J. Physiol. (Lond.)* **117**: 500-544.

value and flatten (i.e., saturate). These maximal conductances are denoted $\bar{g}_{Na}$ and $\bar{g}_{K}$. To describe the time- and voltage-dependency of g_{Na} and g_K (see Fig. 7.5), $\bar{g}_{Na}$ and $\bar{g}_K$ must be multiplied by coefficients that represent the fraction of the maximum conductances expressed at any given time and at any given membrane potential. Thus, g_{Na} and g_K can be written in a general form as

$$g_{Na}(V_m, t) = y_{Na}(V_m, t)\bar{g}_{Na} \qquad (7.13)$$

$$g_K(V_m, t) = y_K(V_m, t)\bar{g}_K \qquad (7.14)$$

where y_{Na} and y_K are functions of one or more gating variables (y_i) that vary between zero and one.

The excellent agreement between the observed voltage dependency of g_{Na} and g_K and the Boltzmann equation (i.e., Fig. 7.7) lent support to Hodgkin and Huxley's proposition that changes in ionic permeability depended on the movement of some component of the membrane that behaved as though it were a charged particle. In addition to providing a possible mechanism for the voltage dependency (i.e., the nonlinear I-V relations in Fig. 7.2) of membrane

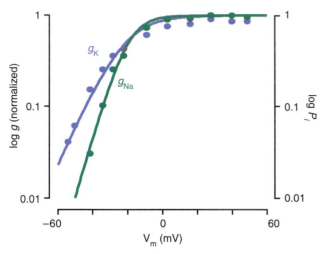

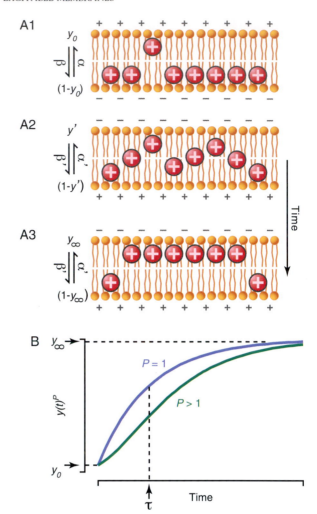

FIGURE 7.7 Voltage dependence of ionic conductances. The maximum g_{Na} and g_K were measured at several different membrane potentials under voltage clamp. The filled circles represent data from an experiment by Hodgkin and Huxley (axon 17) and the solid lines represent best-fit curves of the form given in Eq. 7.12. (The calculations assumed $w = 0$.) The most striking feature of the conductances is the extreme steepness of the relation between ionic conductance and the membrane potential. A depolarization of only ~4 mV can increase g_{Na} by e-fold (e ≈ 2.72), while the corresponding figure for g_K is ~5 mV. Adapted from Hodgkin and Huxley (1952a).

conductance, the gate model also offered a possible explanation for the time dependency of ionic conductances (Fig. 7.8). The gate model assumed that the rate of change in an ionic conductance following a step depolarization was governed by the rate of redistribution of the gating particles within the membrane and that the transitions between *permissive* (i.e., open or activated) and *nonpermissive* (i.e., closed or deactivated) *states* can be described by the following first-order kinetic model:

$$(1 - y) \underset{\beta_y(V_m)}{\overset{\alpha_y(V_m)}{\rightleftharpoons}} y \qquad (7.15)$$

where y is the probability of finding a single gating particle in the permissive state, $(1-y)$ is the probability of finding the particle in the nonpermissive state, and $\alpha_y(V_m)$ and $\beta_y(V_m)$ are voltage-dependent rate constants describing the rate at which a particle moves from nonpermissive to permissive states (α_y) and from permissive to nonpermissive states (β_y). If a gate is regulated by a single particle, then the probability that the gate will open over a short interval of time is proportional to the probability of finding the gate closed multiplied by the opening rate constant (i.e., $\alpha_y(V_m)(1 - y)$). Conversely, the probability that the gate will close over a short interval of time is proportional to the probability of finding the gate open multiplied by the closing rate constant

FIGURE 7.8 Kinetics of the increasing probability of channel activation during a voltage-clamp step. (A) Changes in the membrane potential alter the distribution of gating particles within the membrane. In this example, gating particles are assumed to have a positive charge and the permissive site for opening a gate is assumed to be at the outer surface of the membrane (see Fig. 7.6). (A1) At the resting potential, the probability of a gating particle being in a permissive state (y_0) is low, whereas the probability of a gating particle being in a nonpermissive state ($1 - y_0$) is high. (A2) Immediately following a depolarization (note the translocation of the "−" and "+" signs), the gating particles begin to redistribute. The rate of movement of the gating particles within the membrane is described by Eq. 7.16. The voltage-dependent rate constant α represents the rate at which particles move from the inner to outer surface (i.e., from nonpermissive to permissive states), and β is the rate of reverse movement. (A3) Eventually, the distribution reaches a steady state in which the probability of a gating particle being in a permissive state has increased (y_∞). (B) The kinetics of the redistribution of gating particles is described by Eq. 7.18. If a channel is controlled by a single gating particle (i.e., $P = 1$ in Eq. 7.20), then the solution is a simple exponential. If a channel is controlled by several identical and independent gating particles (i.e., $P > 1$), then a delay is noted in the change in conductance. Data from Hodgkin, A.L. and Huxley, A.F. (1952a). Currents carried by sodium and potassium ions through the membrane of the giant axon of Loligo. *J. Physiol. (Lond.)* **116**: 449–472.

(i.e., $\beta_y(V_m)y$). The rate at which the open probability for a single gating particle changes following a change in membrane potential is given as the difference of these two terms:

$$\frac{dy}{dt} = \alpha_y(V_m)(1-y) - \beta_y(V_m)y \qquad (7.16)$$

The first term in Eq. 7.16 (i.e., $\alpha_y(V_m)(1-y)$) describes the opening (i.e., *activation*) of the gate and the second term (i.e., $\beta_y(V_m)y$) describes the closing (i.e., *deactivation*) of the gate.

Although y is usually taken to represent the probability of finding a single gate in the open state, it can also be interpreted as the fraction of open gates in a large population of gates, and $(1-y)$ would be the fraction of gates in the closed state. If the membrane potential is voltage-clamped to some fixed value, then the fraction of gates in the open state will eventually reach a steady-state value (i.e., $dy/dt = 0$) as $t \to \infty$. Solving Eq. 7.16 for the steady-state value ($y_\infty(V_m)$) yields

$$y_\infty(V_m) = \frac{\alpha_y(V_m)}{\alpha_y(V_m) + \beta_y(V_m)} \qquad (7.17)$$

During a voltage-clamp step, the time course for approaching this steady state (i.e., the solution of a first-order kinetic expression like Eq. 7.16) is described by a simple exponential function

$$y(t) = y_\infty(V_c) - (y_\infty(V_c) - y_0)\exp^{-t/\tau_y(V_c)} \qquad (7.18)$$

where y_0 is the initial value of y (i.e., the value of y at the holding potential, V_h) and the time constant, $\tau_y(V_m)$, is given by

$$\tau_y(V_m) = \frac{1}{\alpha_y(V_m) + \beta(V_m)} \qquad (7.19)$$

If P independent and identical gating particles are involved in gating a channel, then the probability that all the particles will simultaneously be in the permissive state is the product of their individual probabilities (i.e., $y(t)^P$). Thus, by substituting Eq. 7.19 and including the possibility of more than one gating particle, the time course for approaching the steady state becomes

$$y(t)^P = \left(y_\infty(V_c) - (y_\infty(V_c) - y_0)\exp^{-(\alpha_y(V_c)+\beta_y(V_c))t}\right)^P \qquad (7.20)$$

As the number of gating particles increases (i.e., $P > 1$), a delay and a sigmoidal rising phase are introduced to the time course of $y(t)^P$ (Fig. 7.8B).

Instead of using the rate constants α_y and β_y, Eq. 7.16 can be written in terms of the steady-state value $y_\infty(V_m)$

(i.e., Eq. 7.17) and the voltage-dependent time constant (i.e., Eq. 7.19). Thus, Eq. 7.16 becomes

$$\frac{dy}{dt} = \frac{y_\infty(V_m) - y_0}{\tau_y(V_m)} \qquad (7.21)$$

Eq. 7.21 indicates that for a fixed voltage (V_m), the gating particle (y) approaches the steady-state value ($y_\infty(V_m)$) exponentially with the time constant $\tau_y(V_m)$. Although Eqs. 7.16 and 7.21 are equivalent, Eq. 7.21 is simpler to interpret and more conveniently fit to experimental data. In addition, from Eqs. 7.17 and 7.19, it is possible to calculate α_y and β_y from experimental data:

$$\alpha_y(V_m) = \frac{y_\infty(V_m)}{\tau_y(V_m)} \qquad (7.22)$$

$$\beta_y(V_m) = \frac{1 - y_\infty(V_m)}{\tau_y(V_m)} \qquad (7.23)$$

where y_∞ and τ_y are measured empirically (see later).

In the discussion provided previously, the descriptions of gating particles (i.e., Eqs. 7.13–7.21) have been presented using generalized notation that can be applied to a wide variety of conductances. The key remaining tasks in the development of the HH model are determining the number and type of gating particles that regulate g_{Na} and g_K and quantitatively describing the voltage dependency of the rate constants that govern these gating particles. In brief, this was done by measuring the $\tau_y(V_m)$ and $y_\infty(V_m)$ from the time records of g_{Na} and g_K (e.g., Fig. 7.5) and then calculating the rate constants using Eqs. 7.22 and 7.23. The values for the rate constants were plotted as functions of membrane potential, and the data were fit with empirically derived exponential functions (see later).

Characterizing the Na$^+$ Conductance

As illustrated in Fig. 7.5A, the dynamics of g_{Na} are complex. This led Hodgkin and Huxley to postulate that g_{Na} was regulated by two types of gates. One gate regulated the activation of g_{Na} and was termed m, and the other gate regulated inactivation and was termed h. To analyze the empirical voltage-clamp data and extract values for g_{Na}, τ_m, τ_h, and P (see Eqs. 7.13, 7.20, 7.22, and 7.23), an exponential function was fit to the time course of g_{Na} during voltage-clamp steps to various membrane potentials. Hodgkin and Huxley used an exponential function to fit the time course of g_{Na} during a voltage-clamp step (Fig. 7.9A):

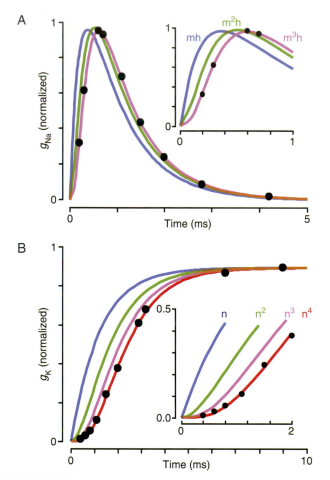

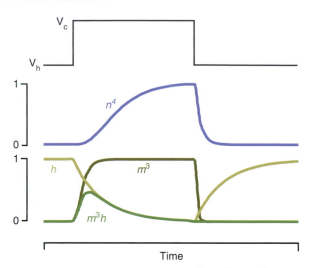

FIGURE 7.10 Temporal dynamics of n^4, h, m^3, and m^3h during a hypothetical voltage-clamp step. Initially, the membrane potential is voltage-clamped at a holding potential (V_h) near the resting potential (V_m). A command voltage (V_c) briefly steps the membrane potential to a depolarized level. During the voltage step, n and m are being activated, and thus, the curves for n^4 and m^3 followed the $(1 - \exp(-t/\tau))^P$ time course. Conversely, h is inactivated by depolarization, and thus, the curve for h follows the $\exp(-t/\tau)$ time course. In this schematic, the values for the time constants τ_m, τ_n, and τ_h were adjusted to a ratio of 1:4:5, and the duration of the voltage-clamp step was the equivalent of $20 \times \tau_m$. The curves illustrate how n^4 and m^3h closely imitate the time courses of g_{Na} and g_K that are observed empirically (compare with Fig. 7.9). Data from Hodgkin, A.L. and Huxley, A.F. (1952d). A quantitative description of membrane current and its application to conduction and excitation in nerve. *J. Physiol. (Lond.)* **117**: 500-544.

FIGURE 7.9 Estimating the number of gating particles that regulate each Na$^+$ and K$^+$ channel. The filled circles represent data from an experiment by Hodgkin and Huxley (axon 17). The inserts illustrate an enlargement of the first few milliseconds of the response. The ionic conductances, g_{Na} (A) and g_K (B), increase with a delay during a voltage-clamp step. This observation suggests that $P > 1$ in Eq. 7.20 (see also Fig. 7.8). To estimate appropriate values for P, empirical data are fit (solid lines) with equations similar to Eq. 7.20 (i.e., Eqs. 7.24 and 7.32), where the value for P is increased from 1 until an adequate fit of the data is achieved. The initial delay in g_{Na} is well fit with $P = 3$, while the corresponding value for g_K is $P = 4$. Adapted from Hodgkin and Huxley (1952d).

$$g_{Na}(t) = g'_{Na}(V_c)\left(1 - \exp^{-t/\tau_m}\right)^3 \exp^{-t/\tau_h} \quad (7.24)$$

where $g'_{Na}(V_c)$ is the value that g_{Na} would attain during the voltage-clamp step if h remained at its resting value (i.e., $h = 1$). This extrapolation is illustrated in Fig. 7.10. During the voltage-clamp step to V_c, m changes from its resting value, which in this example is 0, to a new steady-state level, which in this example is 1 (the curve labeled m^3). Conversely, the resting level of h in this example is 1, and during the voltage-clamp step to V_c, h approaches a new steady-state value, which in the example is 0 (curve labeled h). In the absence of h (i.e.,

the curve labeled m^3) the activation of g_{Na} attains a higher level and is maintained as compared to when h is included in the calculation (i.e., the curve labeled m^3h). The multiplicative interaction of the activation and inactivation gating variables (i.e., m and h) produces the transient response of g_{Na} during a voltage-clamp step and reduces the apparent magnitude of activation g_{Na}. The best fit of Eq. 7.24 to voltage-clamp data provided Hodgkin and Huxley with measurements of g_{Na}, τ_m, and τ_h at each value of V_c. In addition, the best fit of Eq. 7.24 to the voltage-clamp data also indicated that $P = 3$ (see Eq. 7.20), which suggested that three gating particles regulated the activation gate of g_{Na} (see Fig. 7.8) and which produced a sigmoidal time course of $m(t)$ (see insert in Fig. 7.9). Thus, g_{Na} was described by

$$g_{Na} = \bar{g}_{Na}m^3h \quad (7.25)$$

where

$$\frac{dm}{dt} = \alpha_m(V_m)(1 - m) - \beta_m(V_m)m$$

$$\frac{dh}{dt} = \alpha_h(V_m)(1 - h) - \beta_h(V_m)h \quad (7.26a, b)$$

or the equivalent expressions

$$\frac{dm}{dt} = \frac{m_\infty(V_m) - m}{\tau_m(V_m)}$$

$$\frac{dh}{dt} = \frac{h_\infty(V_m) - h}{\tau_h(V_m)}$$

(7.27a, b)

where

$$m_\infty(V_m) = \frac{\alpha_m(V_m)}{\alpha_m(V_m) + \beta_m(V_m)}$$

$$\tau_m(V_m) = \frac{1}{\alpha_m(V_m) + \beta_m(V_m)}$$

(7.28a, b)

$$h_\infty(V_m) = \frac{\alpha_h(V_m)}{\alpha_h(V_m) + \beta_h(V_m)}$$

$$\tau_h(V_m) = \frac{1}{\alpha_h(V_m) + \beta_h(V_m)}$$

(7.29a, b)

The time constants (i.e., $\tau_m(V_m)$ and $\tau_h(V_m)$) and the steady-state values of m and h (i.e., $m_\infty(V_m)$ and $h_\infty(V_m)$) were measured by fitting Eq. 7.24 to the voltage-clamp records of I_{Na} (Fig. 7.11A), and from these data the rate constants (i.e., $\alpha_m(V_m)$, $\beta_m(V_m)$, $\alpha_h(V_m)$, and $\beta_h(V_m)$) were calculated using Eqs. 7.22 and 7.23. The values for the rate constants were plotted as functions of voltage, and expressions for the voltage dependency of the rate constants were derived. Empirically, Hodgkin and Huxley derived the following equations for the rate constants:

$$\alpha_m(V_m) = \frac{0.1(V_r - V_m + 25)}{\exp\left(\dfrac{V_r - V_m + 25}{10}\right) - 1}$$

$$\beta_m(V_m) = 4\exp\left(\frac{V_r - V_m}{18}\right)$$

(7.30a, b)

$$\alpha_h(V_m) = 0.07\exp\left(\frac{V_r - V_m}{20}\right)$$

$$\beta_h(V_m) = \frac{1}{\exp\left(\dfrac{V_r - V_m + 30}{10}\right) + 1}$$

(7.31a, b)

where V_r is the resting potential (which is usually taken to be either −60 or −65 mV), V_m is the membrane potential in units of mV, and the rate constants (i.e., α_i and β_i) are expressed in units of ms^{-1}. (It should be noted that the conventions used by Hodgkin and Huxley to describe voltage were different from those in use today. Box 7.1 explains this difference.)

Although not indicated in Eqs. 7.26 and 7.27, the kinetics of the ionic conductances are influenced by temperature. Hodgkin and Huxley performed most of their voltage-clamp experiments with the preparations cooled to ~6°C. The cooler temperature slowed the kinetics of the membrane currents, which in turn

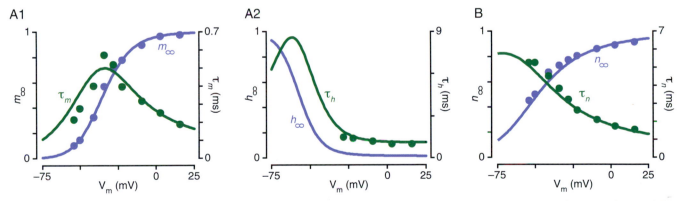

FIGURE 7.11 Voltage dependences of gating variables and their respective time constants. The time constants (τ_m, τ_h, and τ_n) and the steady-state activation (m_∞ and n_∞) and inactivation (h_∞) gating variables are plotted as functions of membrane potential (V_m) for the squid giant axon at 6.3°C. The filled circles represent data from an experiment by Hodgkin and Huxley (axon 17). The smooth curves were calculated using Eqs. 7.28 (panel A1), 7.29 (panel A2), and 7.36 (panel B). (A1) Activation gating variable (m_∞) and its time constant (τ_m) for Na$^+$ conductance. (A2) Inactivation gating variable (h_∞) and its time constant (τ_h) for Na$^+$ conductance. (B) Activation gating variable (n_∞) and its time constant (τ_n) for K$^+$ conductance. The steady-state inactivation h_∞ is a monotonically decreasing function of voltage, whereas the steady-state activation variables m_∞ and n_∞ increase with membrane depolarization. Note that in Eqs. 7.25 and 7.33, the activation gating variables for Na$^+$ (i.e., m) and K$^+$ (i.e., n) are raised to the third and fourth powers, respectively. Thus, the functional voltage-dependent activation of I_{Na} and I_K is much steeper than represented in this illustration. Adapted from Hodgkin and Huxley (1952d).

BOX 7.1

BOX 7.1

UPDATING THE PARAMETERS IN THE HODGKIN–HUXLEY EQUATIONS

Although the original papers of Hodgkin *et al.* 1952, and Hodgkin and Huxley 1952a–d were published over 50 years ago, this series of papers remains influential to the present day. The papers have been reprinted on several occasions (Cooke and Lipkin 1952; Hodgkin and Huxley 1990; Moore and Stuart 2000), and the detailed descriptions of what have became known as the HH equations (Cole *et al.* 1955) are often included in modern textbooks about cellular neurophysiology and computational neuroscience (Bower and Beeman 1998; Byrne and Schultz 1994; Cronin 1987; Dayan and Abbott 2001; DeSchutter 2001; Hille 2001; Johnston and Wu 1997; Koch 1999; MacGregor 1987; Tuckwell 1998; Ventriglia 1994; Weiss 1997). Although these modern descriptions of the HH equations are similar to those in the original publications, the present-day values for the parameters often do not appear to agree with those used by Hodgkin and Huxley in 1952.

This apparent discrepancy arises from the conventions used by Hodgkin and Huxley to represent voltage. In their original series of papers, Hodgkin *et al.* (1952) choose to regard the resting potential as a positive quantity and the action potential as a negative (i.e., downward) deflection. In addition, the variable V (voltage) in the HH equations denoted the *displacement* of the membrane potential from its resting value. Thus, Hodgkin and Huxley defined V as $V = E - E_r$, where E was the absolute value of the membrane potential and E_r was the absolute value of the resting potential. With their choice of conventions, depolarizations of the membrane potential were negative values (downward deflections) and inward membrane currents had a positive sign (upward deflections). For additional explanation of the conventions used by Hodgkin and Huxley, see Rinzel (1998).

This aspect of the papers by Hodgkin and Huxley conflicts with the current practices in which the intracellular electrode measures the membrane potential with respect to an external ground (i.e., the resting potential is a negative quantity), action potentials and depolarizations are positive (upward) deflections, inward membrane currents have a negative sign (downward deflections), and the voltage variable (V_m) in modern representations of the HH equations denotes the absolute membrane potential. The HH equations can be recast into modern conventions for polarity and using absolute membrane potential by simple subtraction and multiplication (Palti 1971a,b). In the present chapter, the HH equations for a spaced-clamped patch of membrane, as well as the data that are plotted in the figures, were expressed using the modern conventions where V_m is the absolute membrane potential and V_r is the absolute value for the resting potential, which was usually taken to be −60 mV. At a conceptual level, the choice of conventions for membrane currents and voltage is inconsequential. It does matter a great deal, however, when one wishes to implement and simulate the HH model.

---

made them easier to record and analyze. The standard parameters of the HH equations reflect a temperature of 6.3°C. The kinetics of the HH equations can be adjusted to reflect some other temperature (T) by multiplying the kinetic equations by a temperature coefficient $\Phi = Q_{10}^{\Delta T/10}$ (see later) (FitzHugh 1966, Hodgkin and Huxley 1952d).

Characterizing the K$^+$ Conductance

Unlike I_{Na}, I_K did not inactivate during voltage-clamp steps (e.g., Fig. 7.3B2). Thus, Hodgkin and Huxley postulated that g_K was governed by a single type of gate, which governed the activation of g_K and which they termed n. To determine values for g_K, τ_n, and p, the time course of g_K during voltage clamps to various membrane potential was best described by

$$g_k(t) = \left(g_\infty^{1/4}(V_c) - \left(g_\infty^{1/4}(V_c) - g_0^{1/4} \right) \exp^{-t/\tau_n(V_c)} \right)^4 \quad (7.32)$$

which indicated that four particles regulated the activation gate of g_K (Fig. 7.9B). The fourth power produces a sigmoidal time course of $n(t)$ (Fig. 7.10). Thus, g_K was described by

$$g_k = \bar{g}_K n^4 \quad (7.33)$$

where

$$\frac{dn}{dt} = \alpha_n(V_m)(1 - n) - \beta_n(V_m)n \quad (7.34)$$

or the equivalent expression

$$\frac{dn}{dt} = \frac{n_\infty(V_m) - n}{\tau_n(V_m)} \quad (7.35)$$

where

$$n_\infty(V_m) = \frac{\alpha_n(V_m)}{\alpha_n(V_m) + \beta_n(V_m)}$$

$$\tau_n(V_m) = \frac{1}{\alpha_n(V_m) + \beta_n(V_m)} \quad (7.36a, b)$$

The time constant (i.e., $\tau_n(V_m)$) and steady values of n (i.e., $n_\infty(V_m)$) were measured (Fig. 7.11B), and from these data the rate constants (i.e., $\alpha_n(V_m)$ and $\beta_n(V_m)$) were calculated (Fig. 7.12B). The empirically determined expressions for the voltage dependency of the rate constants are

$$\alpha_\infty(V_m) = \frac{0.01(V_r - V_m + 10)}{\exp\left(\dfrac{V_r - V_m + 10}{10}\right) - 1}$$

$$\beta_n(V_m) = 0.125 \exp\left(\frac{V_r - V_m}{80}\right) \quad (7.37a, b)$$

where V_r is the resting potential of the cell in units of mV, V_m is the membrane potential in units of mV, and $\alpha_n(V_m)$ and $\beta_n(V_m)$ are given in units of ms^{-1}.

Simulations of the Hodgkin–Huxley Equations

By incorporating Eqs. 7.25 and 7.33 into Eq. 7.8, it is possible to write a single equation that describes the total membrane current (I_m):

$$I_m = C_m \frac{dV}{dt} + \bar{g}_{Na} m^3 h (V_m - E_{Na}) + \bar{g}_k n^4 (V_m - E_K) + \bar{g}_l (V_m - E_l)$$

$$(7.38)$$

This nonlinear differential equation, in addition to the three linear differential equations that describe the temporal evolution of the rate constants (i.e., Eqs. 7.26a, b and 7.34) constitutes the four-dimensional Hodgkin and Huxley model for a space-clamped patch of membrane (i.e., Fig. 7.1). These equations and their associated algebraic functions and parameters were derived to mathematically describe the magnitude and time course of I_{Na} and I_K produced by a series of voltage-clamp-step depolarizations. As a first step toward validating their model, Hodgkin and Huxley tested the ability of the model to correctly calculate the total membrane current during a series of voltage-clamp steps. At a constant voltage, $dV/dt = 0$ and the steady-state values of the rate constants (i.e., $\alpha(V_m)$ and $\beta(V_m)$) are constant. The solution is then obtained directly in terms of the expressions given for $m(V_m,t)$, $h(V_m,t)$, $n(V_m,t)$ (i.e., Eqs. 7.26a, b, and 7.34). Using only a mechanical desk calculator (see Box 7.2), Andrew Huxley computed I_m for a number of different voltages and compared these computations to similar empirical data. This comparison is illustrated in Fig. 7.13. There is excellent agreement between the calculated membrane currents and empirical voltage-clamp records, which lent credence to the model.

The overriding goal of Hodgkin and Huxley's quantitative analysis of voltage-clamp currents, however, was to explain neuronal excitability, and the ultimate test of the model was to see if it could quantitatively describe the action potential. Thus, Hodgkin and Huxley concluded their studies with calculations of the membrane potential changes predicted by their equations. In the absence of the feedback amplifier (i.e., without the voltage clamp), dV/dt was no longer constant, and there is no explicit solution to Eq. 7.38.

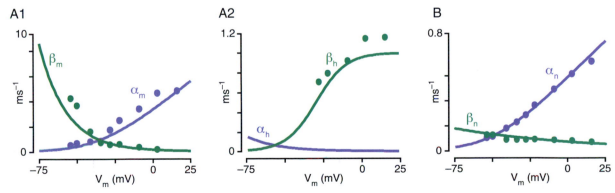

FIGURE 7.12 Voltage-dependences of the rate coefficients for the HH model of a squid giant axon at 6.3°C. The filled circles represent data from an experiment by Hodgkin and Huxley (axon 17). The smooth curves were calculated from Eqs. 7.30 (panel A1), 7.31 (panel A2), and 7.37 (panel B). (A1) Forward (α_m) and backward (β_m) rate coefficients for activation of the Na$^+$ conductance. (A2) Forward (α_h) and backward (β_h) rate coefficients for inactivation of the Na$^+$ conductance (B) Forward (α_n) and backward (β_n) rate coefficients for activation of the K$^+$ conductance. The forward rate coefficients for activation of Na$^+$ and K$^+$ (α_m and α_n, respectively) increase with membrane depolarization because m and n gating particles move into a permissive state in response to membrane depolarization (see Fig. 7.6). Conversely, the forward rate coefficient for inactivation of Na$^+$ (α_h) decreases with membrane depolarization because the h gating particle moves into a nonpermissive state in response to membrane depolarization. Data from Hodgkin, A.L. and Huxley, A.F. (1952d). A quantitative description of membrane current and its application to conduction and excitation in nerve. *J. Physiol. (Lond.)* **117**: 500-544.

BOX 7.2

COMPUTING SOLUTIONS TO THE HODGKIN–HUXLEY EQUATIONS

Hodgkin and Huxley conducted their voltage-clamp experiments on the squid giant axon at the Marine Biological Laboratory in Plymouth, England. Although the Plymouth laboratory was badly damaged during the great air raids of 1941, it was partially rebuilt by the time Hodgkin and Huxley arrived in July of 1949. With the help of Bernard Katz and their improved voltage-clamp apparatus, it took them only a month to obtain all of the voltage-clamp records that were used in the five papers published in 1952 (Hodgkin *et al.* 1952; Hodgkin and Huxley 1952a–d). Upon returning to the University of Cambridge, they spent the next 2 years analyzing the data and preparing the manuscripts. By March 1951, they had settled on a set of the equations and parameters that adequately described the time course, magnitude, and voltage dependency of the membrane currents that were observed during voltage-clamp steps. It was by no means a foregone conclusion, however, that these same equations would describe the behavior of the membrane under its normal operating conditions. Thus, the final stage of their analysis was to calculate the response of their mathematical representation of the nerve to the equivalent of an electrical stimulus. If the calculations produced an action potential that agreed favorably with experimental data, this would help validate their model.

Hodgkin and Huxley planned to solve their equations on the first electronic computer at the University of Cambridge, the EDSAC (electronic delay storage automatic calculator). Construction of the EDSAC was completed in 1949, and at that time, it was the state-of-the-art for electronic digital computing (Wheeler 1992a,b). EDSAC was approximate 12 feet by 12 feet in size, had a power consumption of 12 Kw, contained some 3000 vacuum tubes, and had about 2000 bytes of nonrandom access memory, which was constructed from a series of 5-foot-long tubes filled with mercury. EDSAC operated at 500 KHz and could perform 650 instructions per second. Division of two numbers took the EDSAC about 200 msec. Unfortunately for Hodgkin and Huxley, the EDSAC was undergoing major modifications in March of 1951 and would be unavailable for 6 months.

Hodgkin and Huxley overcame this setback by solving the differential equations numerically with a mechanical Brunsviga desk calculator. This was a laborious task. A space-clamped action potential (e.g., Fig. 7.14A) took a matter of days to compute, and a propagated action potential (e.g., Fig. 15 of Hodgkin and Huxley 1952d) took a matter of weeks. In addition to the space-clamped and propagated action potentials, Hodgkin and Huxley's computations included the impedance changes and the total movements of Na^+ and K^+ ions into and out of the axon during an action potential; recovery during the relative refractory period; anode break excitation; and the oscillatory response of the membrane to a rectangular pulse of current. These results were published in 1952 (Hodgkin and Huxley 1952d) and showed surprisingly good agreement with the available empirical data from the giant axon. This agreement suggested that the formulations developed by Hodgkin and Huxley were substantially correct.

The scope of Hodgkin and Huxley's computations was limited, however, by the fact that an automatic computer was unavailable. The manual methods of solving the HH equations were so laborious as to discourage more detailed and broader investigations of the ability of the HH equations to predict and interpret the well-established and fundamental characteristics of nerve behavior. For example, the all-or-none nature of initiating an action potential was considered to be a key feature of neuronal excitability, but it was unknown whether the HH equations manifested this key feature.

To address these issues, Kenneth Cole and his colleagues (1955) wrote the machine language program necessary to run simulations of the HH equations on the first fully operational stored-program electronic computer in the United States, the SEAC (standards eastern automatic computer). Construction of the SEAC was completed in 1950. Its design and capabilities were comparable to the EDSAC (Kirsch 1998). Although it was not their goal to cross-check the original calculations of Hodgkin and Huxley, the first computer simulation of the HH equations by Cole *et al.* in 1955 was of a space-clamped action potential (see Fig. 1 of Cole *et al.* 1955). The results from Huxley's hand calculations and from the computer simulation were indistinguishable. Moreover, the SEAC calculations provided evidence that the HH equations manifest an all-or-none response to current stimulation. Although some (including Huxley 1959) doubted the threshold phenomena observed by Cole *et al.*, the independent replication of the action potential calculation served to increase confidence in the HH equations.

BOX 7.2 (continued)

In theory, using a computer should increase the speed and accuracy of solving the HH equations. In practice, however, this was not always the case in the early days of computers. Although SEAC could calculate the space-clamped action potential in about 30 min, accessing the computer proved very slow (FitzHugh 1960). Solutions took a week or more, including the time for relaying instructions to the programming and operating technicians, scheduling time on the computer, and receiving the results. In addition, a flaw was detected in the machine language program for the first computer simulation of the HH equations (FitzHugh and Antosiewicz 1959). The program contained division-by-zero errors in the calculations of α_m and α_n. The major effect of these errors was to produce a spurious saddle point (see Box 7.4). It was an unfortunate accident that the spurious saddle point appeared near the membrane potential at which the threshold was believed to occur. In 1959, Fitz-Hugh and Antosiewicz reprogrammed the HH equations in FORTRAN, avoiding the division by zero errors, and using an IBM 704, re-examined the issue of all-or-none responses in the HH equations. On recalculation, the all-or-none threshold was lost, and the HH equations were found to manifest a "quasi-threshold" phenomenon (Fitz-Hugh and Antosiewicz 1959); i.e., over a sufficiently small range of stimulus intensities the amplitude of the action potential increases continuously from a "none" to an "all" state (see Fig. 2 of FitzHugh and Antosiewicz 1959). The sharpness of threshold phenomenon is determined by the steepness of the stimulus-response curve for the peak amplitude versus stimulus intensity. The HH equations manifest a very sharp threshold. An increase in the stimulus intensity of only one part in 10^8 was sufficient to distinguish between a very small graded response and a complete action potential. Thus, the lack of a mathematically correct threshold (i.e., saddle point) seemed entirely academic, and faith in the HH equations was not shaken. Indeed, the computational prediction of quasi-threshold behavior was subsequently confirmed with experimental studies (Cole *et al.* 1970), which further increased confidence in the HH equations.

At present, access to computers is no longer a limiting factor to individuals who wish to simulate the HH model. Commonly available personal computers are more than adequate to calculate solutions to the HH equations. In addition, simulating the HH model is no longer limited to individuals with skills necessary to develop the computer programs required to solve the HH equations. It is possible to solve the HH equations using commonly available spreadsheet programs (Brown 1999, 2000) and there are many freely available and user-friendly software packages that have been developed to simulate HH-type models of neurons. Software packages that are specifically designed to build and simulate models of HH-type neurons and neural circuits are commonly referred as *neurosimulators*. Some examples of neurosimulators include GENESIS (Bower and Beeman 1998), NEURON (Hines and Carnevale 1997), and SNNAP (Ziv *et al.* 1994). There are several reviews that describe the features and availability of neurosimulators (DeSchutter 1989, 1992; Hayes *et al.* 2002).

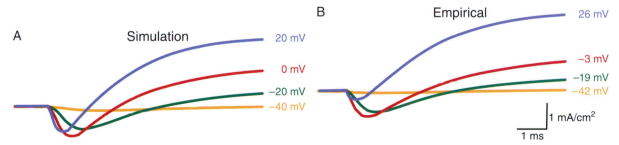

FIGURE 7.13 Comparison of simulated membrane current and empirical observations. (A) As a first test of their model, Hodgkin and Huxley used Eq. 7.38 to compute the membrane current that would be elicited by a series of voltage-clamp steps. Four of the calculated traces are illustrated. They represent the predicted total membrane current during voltage-clamp steps to −40, −20, 0, and 20 mV. (B) Empirical data from one of Hodgkin and Huxley's voltage-clamp experiments (axon 31 at 4°C). The empirical data were collected from a series of voltage-clamp steps to −42, −19, −3, and 26 mV. There was excellent agreement between the simulated and empirical data, which suggests that the HH equations captured the salient features of the experimental data. Data from Hodgkin, A.L. and Huxley, A.F. (1952d). A quantitative description of membrane current and its application to conduction and excitation in nerve. *J. Physiol. (Lond.)* **117**: 500-544.

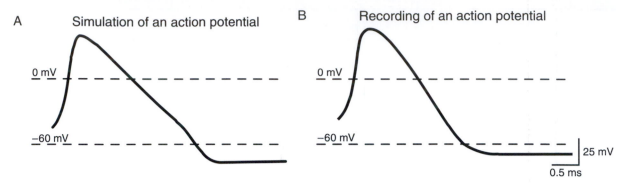

FIGURE 7.14 Comparison of simulated space-clamped action potential and empirical observations. If a giant axon is stimulated simultaneously over a substantial portion of its length (e.g., by applying a shock via a long internal electrode), all points within that length will undergo an action potential simultaneously (see Chapter 5). There will be no difference of potential along the axis of the nerve fiber, and therefore no longitudinal current. This type of action potential is referred to as a *space-clamped* (as opposed to a propagated) action potential. (A) Simulation of a space-clamped action potential. The solution of Eq. 7.38 describes the membrane potential of a space-clamped patch of membrane. If the equivalent of a suprathreshold electrical shock is incorporated in Eq. 7.38, an action potential is produced. This example was produced by Hodgkin and Huxley, who used a mechanical desk calculator to solve the set of differential equations that have come to be known as the HH equations (Eqs. 7.39–7.42). (B) An action potential recorded under experimental conditions that matched those simulated in panel A (i.e., spaced-clamped patch of membrane, temperature of 6°C). There is good agreement between the theoretical and empirical action potentials. This agreement suggests that the formulations developed by Hodgkin and Huxley accurately represented the processes that underlie the action potential. Adapted from Hodgkin, A.L. and Huxley, A.F. (1952d). A quantitative description of membrane current and its application to conduction and excitation in nerve. *J. Physiol. (Lond.)* **117**: 500–544.

Using numerical methods, Hodgkin and Huxley solved the equations and computed an action potential waveform that duplicated with remarkable accuracy the naturally occurring action potential (Fig. 7.14). In addition to describing the space-clamped action potential, Hodgkin and Huxley demonstrated the considerable power of their model to predict many other properties of neuronal excitability, including subthreshold responses, a sharp threshold for firing, membrane conductance changes during an action potential, the effects of temperature on the action potential waveform, propagated action potentials, ionic fluxes, absolute and relative refractory periods, anode break excitation, and accommodation. Although the model has some limitations (see Box 7.3), the remarkable success of the HH model to accurately describe such a wide array of phenomena remains to this day a triumph of classical biophysics in understanding a fundamental neuronal property (i.e., excitability).

One of the great advantages of having a mathematical model of a complex process (e.g., neuronal excitability) is that it provides an opportunity to examine the component processes in ways that may not be experimentally possible. For example, it is possible to calculate the time courses and magnitudes of the different ionic currents, conductances, and gating variables during an action potential (see Chapter 5). Similarly, it is possible to calculate the individual ionic currents during voltage-clamp steps (Fig. 7.15). Given that the tools (i.e., computers and software) necessary

to simulate the HH model are readily available (see Box 7.2), anyone who wishes to explore the rich dynamical properties of these equations can easily do so.

Summary

The HH model describes neuronal excitability in terms of four variables: the membrane potential, $V_m(t)$, and three gating variables, $m(V,t)$, $h(V,t)$, and $n(v,t)$, which describe the permeability (i.e., conductance) of the membrane to Na^+ and K^+ (i.e., g_{Na} and g_K). The magnitude of the activation gating variables (i.e., $m(V,t)$ and $n(V,t)$, increases with increasing depolarization, whereas the magnitude of the inactivation gating variable (i.e., $h(V,t)$) decreases with depolarization. The gating variables were described by first-order differential equations with two voltage-dependent terms: the steady-state activation or inactivation (i.e., $m(V_m)$, $h_\infty(V_m)$, $n_\infty(V_m)$) and the time constant ($\tau_m(V_m)$, $\tau_h(V_m)$, $\tau_n(V_m)$). Thus, the HH model (Hodgkin and Huxley 1952d) for a spaced-clamped patch of membrane (i.e., an isopotential compartment) is a system of four ordinary differential equations:

$$I_m = C_m \frac{dV}{dt} + \bar{g}_{Na} m^3 h (V_m - E_{Na}) + \bar{g}_K n^4 (V_m - E_K)$$
$$+ \bar{g}_l (V_m - E_l) \tag{7.39}$$

$$\frac{dm}{dt} = \Phi(T) \frac{m_\infty(V_m) - m}{\tau_m(V_m)}$$
$$= \Phi(T)[\alpha_m(V_m)(1 - m) - \beta_m(V_m)m] \tag{7.40}$$

BOX 7.3

LIMITATIONS OF THE HODGKIN–HUXLEY EQUATIONS

The HH model has been so widely accepted as a paradigm for excitable membranes that its appropriateness for the squid giant axon itself has generally not been questioned. The model fails, however, to provide a good description for some electrophysiological properties of the axon. For example, in response to a relatively long duration, suprathreshold current pulse, the axon generally produces a single action potential (i.e., accommodation). In contrast, the HH model predicts sustained spiking activity throughout the stimulus (see Fig. 3 of Clay 1998). This discrepancy is attributed to the assumption by Hodgkin and Huxley that the activation (m) and inactivation (h) of the Na$^+$ conductance were independent processes, whereas empirical evidence indicates that the two processes are coupled (Bezanilla and Armstrong 1977). A revised version of the HH model (Clay 1998; Vandenberg and Bezanilla 1991), which incorporates coupling between activation and inactivation, provides a better description of the squid axon.

Other assumptions inherent in the HH model have been examined and found to be only approximately valid. In the model, temperature is assumed to affect only the kinetics of the ionic conductances (see Eqs. 7.40–7.42). The conductance of an ionic channel is also altered by temperature, albeit to a relatively small degree (Hille 2001). If the HH equations are modified to incorporate an effect of temperature on the conductances, a better fit of empirical data is achieved (Fitz-Hugh 1966). Another assumption within the model is that the flow of ionic current through open channels obeys Ohm's law. Current data, however, suggest that the linearity is only approximate and holds neither under all ionic conditions nor in Na$^+$ and K$^+$ channels of all organisms (Hille 2001). For example, Na$^+$ channels are not ohmic in nodes of Ranvier (Dodge and Frankenhaeuser 1959).

Rather than detracting from the HH model, these experimental results highlight some of the advantages of formulating a detailed, quantitative model. First, the formulation of the model forces one to clearly and quantitatively state the assumptions that underlie the model and to evaluate the impact of these assumptions on the behavior of the model. This procedure, in turn, provides guidelines for future experimental studies that can directly test the enumerated assumptions. Second, the model provides a modifiable framework with which new data and concepts can be incorporated and evaluated. Hodgkin and Huxley were well aware that their model had limitations. In their discussion of the model (Hodgkin and Huxley 1952d), they acknowledged the shortcomings of the model and pointed out some discrepancies between the calculated and observed behavior of the squid giant axon. For example, the waveform of the calculated action point had sharper peak and small "hump" in the falling phase that was not present in the recorded action potential (closely compare the two action potentials in Fig. 7.14). As Huxley stated in 1964 *"I would not like to leave you with the impression that the particular equations we produced in 1952 are definitive....Hodgkin and I feel that these equations should be regarded as a first approximation which needs to be refined and extended in many ways in the search for the actual mechanisms of the permeability changes on the molecular scale."* Even if its details cannot be taken literally, the HH model continues to have important general properties with mechanistic implications that are helping to direct future studies, both empirical and computational.

$$\frac{dh}{dt} = \Phi(T)\frac{h_\infty(V_m) - h}{\tau_h(V_m)}$$

$$= \Phi(T)[\alpha_h(V_m)(1-h) - \beta_h(V_m)h] \qquad (7.41)$$

$$\frac{dn}{dt} = \Phi(T)\frac{n_\infty(V_m) - n}{\tau_n(V_m)}$$

$$= \Phi(T)[\alpha_n(V_m)(1 - n) - \beta_n(V_m)n] \qquad (7.42)$$

where the coefficient $\Phi(T)$ describes the effects of temperature on the three gating variables (see following), and the voltage- and time-dependency of the gating variables are given by

$$m_\infty(V_m) = \frac{\alpha_m(V_m)}{\alpha_m(V_m) + \beta_m(V_m)}$$

$$\tau_m(V_m) = \frac{1}{\alpha_m(V_m) + \beta_m(V_m)} \qquad (7.43a, b)$$

$$h_\infty(V_m) = \frac{\alpha_h(V_m)}{\alpha_h(V_m) + \beta_h(V_m)}$$

$$\tau_h(V) = \frac{1}{\alpha_h(V_m) + \beta_h(V_m)} \qquad (7.44a, b)$$

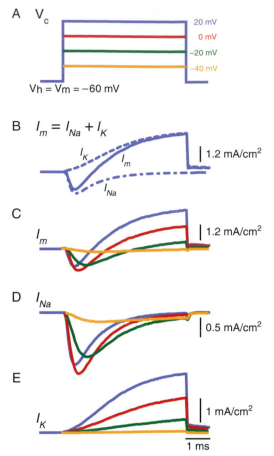

$$n_\infty(V_m) = \frac{\alpha_n(V_m)}{\alpha_n(V_m) + \beta_n(V_m)}$$

$$\tau_n(V_m) = \frac{1}{\alpha_n(V_m) + \beta_n(V_m)} \qquad (7.45a,b)$$

These equations relate the membrane potential (i.e., V_m) to the permeability of the membrane to Na^+, K^+, and the nonspecific leakage of ions. The equations contain several auxiliary parameters representing equilibrium potentials of the ions (i.e., E_{Na}, E_K, and E_l): maximum ionic conductances (i.e., $\bar{g}_{Na}$, $\bar{g}_K$, and $\bar{g}_l$), temperature coefficient (i.e., Φ), and the capacitance of the membrane (i.e., C_m). Values for these parameters were obtained from analyses of experimental data (Hodgkin *et al.* 1952; Hodgkin and Huxley 1952a–c). The values used by Hodgkin and Huxley (1952d) to describe the squid giant axon were

$$\begin{array}{ll}
E_{Na} = V_r + 115 \text{ mV} & \bar{g}_{Na} = 120 \text{ mS/cm}^2 \\
E_K = V_r - 12 \text{ mV} & \bar{g}_K = 36 \text{ mS/cm}^2 \\
E_l = V_r + 10.613 \text{ mV} & \bar{g}_l = 0.3 \text{ mS/cm}^2 \\
\Phi(T) = 3^{(T-6.3)/10} & C_m = 1\,\mu\text{F/cm}^2
\end{array}$$

where V_r is resting membrane potential of the cell, T is temperature in degrees centigrade. The coefficient Φ provides the three gating variables with a Q_{10} of 3 and equals 1 at Hodgkin and Huxley's standard temperature of 6.3°C. The rate constants (i.e., α_i and β_i) were estimated by fitting empirically derived exponential functions of voltage to the experimental data. For the squid giant axon at a temperature of 6.3°C, these functions were (Hodgkin and Huxley 1952d)

$$\alpha_m(V_m) = \frac{0.1(V_r - V_m + 25)}{\exp\left(\dfrac{V_r - V_m + 25}{10}\right) - 1}$$

$$\beta_m(V_m) = 4 \exp\left(\frac{V_r - V_m}{18}\right) \qquad (7.46a,b)$$

$$\alpha_h(V_m) = 0.07 \exp\left(\frac{V_r - V_m}{20}\right)$$

$$\beta_h(V_m) = \frac{1}{\exp\left(\dfrac{V_r - V_m + 30}{10}\right) + 1} \qquad (7.47a,b)$$

FIGURE 7.15 Simulation of a voltage-clamp experiment. Solving the HH equations is much easier today than it was for Hodgkin and Huxley in 1952 (see Box 7.2). Digital computers are widely available, and there are many software packages available that have been designed to simulate neuronal properties. Such simulations can provide a useful tool for gaining insights into the complex and nonlinear processes that underlie neuronal excitability. (A) In these simulations, which were produced using the simulation package SNNAP (Ziv *et al.* 1994), the holding potential (V_h) was equal to the resting membrane potential (V_m), which was taken to be –60 mV. The command voltage (V_c) briefly steps the membrane potential to various depolarized values. The values of V_c are indicated next to each trace. The colors of each voltage-clamp step in panel A correspond to the membrane currents in panels B–E. (B) The simulated membrane current (I_m) is composed primarily of I_{Na} and I_K. (C) The total membrane current is computed by solving Eq. 7.38. I_m has an early inward component that is followed by a large, sustained outward component. The overall appearance of these simulated membrane currents agrees well with empirical data (e.g., Fig. 13B) and with the original calculations of Hodgkin and Huxley (e.g., Fig. 13A). (D) I_{Na} in isolation. Note that the largest I_{Na} trace (i.e., the red trace) does not occur during the largest voltage-clamp step (i.e., the blue trace). As V_c approaches E_{Na}, the driving force for I_{Na} approaches zero and the magnitude of the current decreases (see Eq. 7.3). (E) I_K in isolation. Unlike I_{Na}, I_K increases monotonically as a function of increasing V_c. In addition, because the description of g_K (i.e., Eq. 7.33) has only an activation-gating variable, I_K is sustained throughout the duration of the voltage-clamp step. Data from Hodgkin, A.L. and Huxley, A.F. (1952d). A quantitative description of membrane current and its application to conduction and excitation in nerve. *J. Physiol. (Lond.)* **117**: 500-544.

$$\alpha_n(V_m) = \frac{0.01(V_r - V_m + 10)}{\exp\left(\dfrac{V_r - V_m + 10}{10}\right) - 1}$$

$$\beta_n(V_m) = 0.125 \exp\left(\frac{V_r - V_m}{80}\right)$$

$$(7.48a, b)$$

The work of Hodgkin and Huxley was a landmark in the field of biophysical research. Their protocol involved voltage-clamp analysis of membrane currents, separating the membrane current into its components, developing and fitting kinetic schemes for the time- and voltage-dependences of the ionic conductances, and finally, reconstructing the action potential. This work established a precedent for combining experimental and computational techniques to explore excitable membrane systems. This interdisciplinary approach is still commonly used, and Hodgkin–Huxley formalisms remain a cornerstone of quantitative models of neuronal excitability.

A GEOMETRIC ANALYSIS OF EXCITABILITY

The HH equations constitute a remarkably successful quantitative model. With reasonable accuracy, the model describes the membrane currents and the action potential of squid giant axon, as well as a number of other dynamical properties of neuronal excitability, such as anode break excitation, accommodation, and the refractory period (see Chapter 5). Although this quantitative, conductance-based model has been enormously fruitful in terms of providing a mathematical framework for modeling neuronal excitability, it has a serious drawback, i.e., its numerical complexity. The HH equations are highly nonlinear (i.e., m is raised to the third power and n is raised to the fourth power) and complex (i.e., the model consists of four coupled differential equations, a large number of algebraic equations, and a host of parameters). This complexity makes it difficult to intuitively understand the workings of the model. Indeed, Andrew Huxley may have said it best when he stated (Huxley 1964), "*Very often my expectations turned out to be wrong, and an important lesson I learned from these manual computations was the complete inadequacy of one's intuition in trying to deal with a system of this degree of complexity.*"

To gain intuitive insight into the dynamical properties of neurons, it is helpful to examine less complex models that manifest the salient features of neuronal excitability. By exploiting a low-dimensional (i.e., two or three differential equations) model of an excitable membrane and applying techniques from the mathematical field of nonlinear dynamics, many dynamical properties of neurons (e.g., threshold behavior, excitability, repetitive firing, autonomous bursting) can be understood, predicted, and interpreted. Others have used this approach and found that it is considerably easier—from a numerical and a conceptual point of view—to study the dynamical properties of neurons described by reduced models rather than simulating the behavior of biophysically complex neurons (Abbott 1994; Av-Ron *et al.* 1991; Alexander and Cai 1991; Bertram 1994; Bertram *et al.* 1995; Butera *et al.* 1996; Butera 1998; Canavier *et al.* 2002; Ermentrout 1996; FitzHugh 1960; Gall and Zhou 2000; Hoppensteadt and Izhikevich 2001; Izhikevich 2000; Kepler *et al.* 1992; Koch 1999; Krinskii and Kokoz 1973; Rinzel 1985; Rinzel and Ermentrout 1998) (Box 7.5) (see also Chapter 14). The second half of this chapter will illustrate how to develop reduced models of excitability and analyze their dynamical properties.

Two-Dimensional Reduction of the Hodgkin–Huxley Model

The HH model can be reduced to a two-variable model by identifying and combining variables with similar timescales and biophysical roles and allowing relatively fast variables to be instantaneous. For example, the time constant for m is an order of magnitude faster than those of h or n (Fig. 7.11). Thus, it is reasonable to approximate m by $m_\infty(V_m)$, which eliminates one of the differential equation (i.e., Eq. 7.26a). In addition, the variables n and h evolve on similar timescales and with an approximately constant relationship between their values (i.e., the sum of $n(V_m, t)$ and $-h$ (V_m, t) is approximately constant). Thus, it is reasonable to combine h and n into a single "recovery" variable, which has been termed w (FitzHugh 1961; Rinzel 1985). Using a single recovery variable replaces two of the differential equations (i.e., Eqs. 7.26b and 7.34) with a single expression for dw/dt (see later). By incorporating these approximations and some additional simplifications in the algebraic equations, it is possible to produce a tractable, two-variable model (i.e., a model with only dV_m/dt and dw/dt) that is versatile, manifests many of the salient features of neuronal excitability, and manifests a rich and diverse array of dynamical properties.

The reduced model (Av-Ron *et al.* 1991) includes a linear leakage current (I_l) and an externally applied stimulus current (I_s), as well as a time- and voltage-dependent inward current (I_{Na}) and outward current (I_K). The inward current is rapidly activated by depolarization and contributes to its own further activation by producing further depolarization (i.e., autocatalytic or positive feedback). This positive feedback process has an instantaneous dependence on membrane potential. There is also a slower, negative feedback process. With depolarization, w increases, leading to activation of I_K and inactivation of I_{Na}. The dynamics of this two-dimensional system are described by the following differential equations:

$$\frac{dV_m}{dt} = \frac{-(I_{Na} + I_K + I_l - I_s)}{C_m} \quad (7.49)$$

$$\frac{dw}{dt} = \Phi(T) \frac{w_\infty(V_m) - w}{\tau_w(V_m)} \quad (7.50)$$

where V_m is the absolute membrane potential, C_m is the membrane capacitance, and $\Phi(T)$ is similar to the temperature coefficient in the HH equations. The currents are described by

$$I_{Na} = \bar{g}_{Na} m_\infty^{mp}(V_m)(1 - w(V_m, t))(V_m - E_{Na}) \quad (7.51)$$

$$I_K = \bar{g}_K \left(\frac{w(V_m, i)}{s} \right)^{wp} (V_m - E_K) \quad (7.52)$$

$$I_l = \bar{g}_l(V_m - E_l) \quad (7.53)$$

where $\bar{g}_i$ represents the maximum conductances of a given ionic current, E_i represents the equilibrium potential for a given ionic current, mp and wp are parameters, and s is a scaling factor that allows the magnitude of recovery variable (i.e., w) to be different for I_{Na} and I_K. The following equations give the voltage dependencies of the steady state of the recovery variable (i.e., w_∞), the steady state of the Na^+ activation variable (i.e., m_∞), and the relaxation function of the recovery variable (i.e., τ_w):

$$w_\infty(V_m) = \frac{1}{1 + \exp\left(-2a^{(w)}\left(V_m - V_{1/2}^{(w)}\right)\right)} \quad (7.54)$$

$$m_\infty(V_m) = \frac{1}{1 + \exp\left(-2a^{(m)}\left(V_m - V_{1/2}^{(m)}\right)\right)} \quad (7.55)$$

$$\tau_w(V_m) = \frac{1}{\bar{\lambda} \exp\left(a^{(w)}\left(V_w - V_{1/2}^{(w)}\right)\right) + \lambda \exp\left(-a^{(w)}\left(V_m - V_{1/2}^{(w)}\right)\right)} \quad (7.56)$$

where Eqs. 7.54 and 7.55 are sigmoid curves, V is the voltage for the half maximal value, and parameter $a^{(i)}$

controls the slope of the curve at this midpoint (i.e., inflection point). Eq. 7.56 is a bimodal sigmoid and the parameter λ represents the effects of $\Phi(T)$ in Eq. 7.40. For additional details of this reduced model, see Av-Ron *et al.* (1991).

This reduced model (Eqs. 7.49 and 7.50) can exhibit two major types of qualitative behavior. For some parameter regimes, the model is normally quiescent, but excitable in that it can fire one or more action potentials in response to transient stimuli (Fig. 7.16A). For different parameter regimes, the model fires action potentials in a repetitive fashion (i.e., limit cycle oscillations) (Fig. 7.17B). In addition, for some parameter regimes, the model can manifest at least one type of bistability (Fig. 7.19B), which underlies some types of burst firing. The same nonlinear properties that allow the model to fire a single action potential also enable it to function as a nonlinear oscillator, generating sustained rhythmic activity. How can such distinct behaviors emerge from such a simple system, and how can the behavior of the system be predicted? With parameter variations, a series of examples can be generated in a way that explains how the nonlinear properties of this simple two-dimensional model endow the system with such rich dynamical behaviors. Several methods (e.g., phase planes, nullclines, stability, and bifurcations) also can be used to develop an intuitive understanding of the workings of this model (see Box 7.4). This approach closely follows that used by Rinzel and Ermentrout (1998) to analyze a different model (i.e., the Morris–Lecar model) and has been used by Av-Ron *et al.* (1991, 1993; Av-Ron 1994) and by Canavier *et al.* (2002) to analyze the two-variable model previously described (i.e., Eqs. 7.49 and 7.50).

The Action Potential Trajectory in the Phase Plane

Several methods can be used to visualize and analyze the dynamical behavior of the reduced model. For example, after numerically integrating the differential equations, the dependent variables (i.e., $V_m(t)$ and $w(t)$) can be plotted as a function of time (Fig. 7.16A1 and A2). Although this method of visualizing dynamical behavior may be familiar and is useful for comparing temporal relationship between the variables, it is not analytically powerful. A more valuable way to view the response of multiple variables is by *phase plane* profiles (i.e., curves of one dependent variable against another; Fig. 7.16A3). The solution path in the phase plane is referred to as a *trajectory*. The phase plane is completely filled with trajectories, since each point can serve as an initial

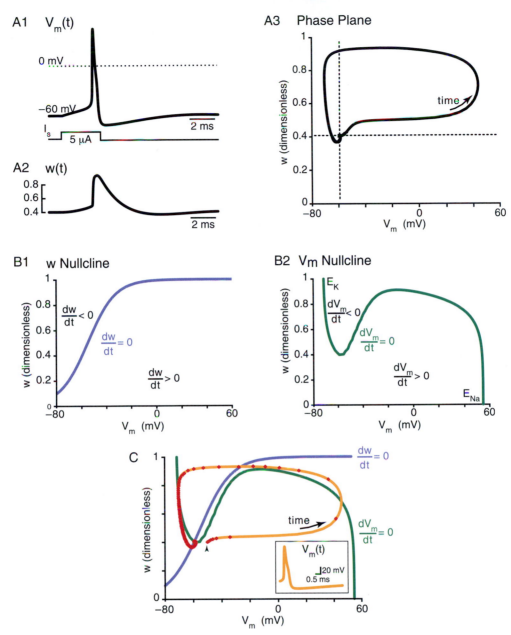

FIGURE 7.16 Two-dimensional model of neuronal excitability. The reduced model (Eqs. 7.49 and 7.50) is excitable and produces an action potential in response to a brief, suprathreshold stimulus. (A1) Time-series plot of the computed membrane potential, $V_m(t)$. The resting membrane potential is -60 mV, and in response to a brief depolarizing current (i.e., $I_s = 5$ μA cm^{-2} for 3 ms; lower trace), the model produces an action potential. The parameters of the model (see later) were adjusted to closely approximate the squid giant axon. (A2) Time-series plot of recovery variable, $w(t)$. At the resting potential, $w(t)$ has a value of $\sim$0.4. During the action potential, the magnitude of $w(t)$ increases, and this increase helps to restore the rest state. This variable acts much like the n and h gating variables of the Hodgkin and Huxley model (see Chapter 5). (A3) The phase plane of the reduced model has coordinates V_m and w. The evolution of the variables is a point moving through this phase plane (black line). The direction of the trajectory (i.e., time) is indicated by the arrow. The values of w and V_m at the resting potential are indicated by the dashed lines. In the absence of any external stimulus, the trajectory (i.e., solution path) remains at the resting values of V_m and w (i.e., the intersection of the dashed lines). The brief increase in I_s (see panel A1) displaces the trajectory away from the resting point. If the displacement crosses an apparent threshold along the V_m axis, the trajectory follows a large amplitude pseudo-orbit through the phase plane and back to the resting point. The large excursion of the trajectory represents the projection of the action potential in the phase plane. (B) Nullclines of the reduced model. Nullclines in the phase plane are curves along which derivatives of a given variable are constant and equal to zero (i.e., $dV_m/dt = 0$, $dw/dt = 0$). In addition, nullclines divide the phase plane into regions where the derivatives have a constant sign. (B1) w nullcline (blue line). The w nullcline specified by the steady-state function of w (Eq. 7.54). (B2) V_m nullcline (green line). The V_m nullcline represents the pairs of values of $V_m(t)$ and $w(t)$ for which the net current is equal to zero (see Eq. 7.49). E_K and E_{Na} indicate the equilibrium potentials of I_{Na} and I_K. (C) The trajectory of an action potential (yellow line) and the nullclines for V_m (green line) and w (blue line) are superimposed in the V_m-w plane.

Legend continued on next page

FIGURE 7.16 (*continued*) The intersection of the nullclines represents the resting point (i.e., quiescent state) in the phase plane. In response to an instantaneous displacement of the system (arrowhead) that crosses the threshold for generating an action potential, the system produces an action potential. Although time is not explicitly plotted in the phase plane, the red dots along the action potential trajectory are separated by 50 μs. Thus, it is possible to visualize the speed of the solution path through the phase plane. The action potential trajectory passes through four regions in the phase plane. The initial upstroke of the spike (i.e., V_m is increasing) occurs in the region where both dV_m/dt and dw/dt are greater than zero. The trajectory crosses the V_m nullcline (i.e., the peak of the action potential) and enters a region where $dV_m/dt < 0$ (i.e., the falling phase of the action potential) and $dw/dt > 0$. As the trajectory crosses the w nullcline, it enters a region where both dV_m/dt and dw/dt are less than zero (i.e., the absolute refractory period; see Chapter 5). Finally, the trajectory crosses the V_m nullcline a second time (i.e., the minimum of the afterhyperpolarization) and enters a region where $dV_m/dt > 0$ and $dw/dt < 0$ (i.e., the relative refractory period; see Chapter 5). (Insert) The time-series plot of $V_m(t)$ that is equivalent to the trajectory in the phase plane. For these simulations $C_m = 1$ μF/cm^2, $\bar{g}_{Na} = 120$ mS/cm^2, $\bar{g}_K = 36$ mS/cm^2, $\bar{g}_l = -0.3$ mS/cm^2, $E_{Na} = 55$ mV, $E_K = -72$ mV, $E_l = -49.4$ mV, $V_{1/2}^{(m)} = -33$ mV, $V_{1/2}^{(w)} = -55$ mV, $a^{(m)} = 0.055$, $a^{(w)} = 0.045$, $mp = 3$, $wp = 4$, $s = 1.3$, and $\bar{\lambda} = 0.2$. With these parameters, the reduced model simulates the properties of the squid giant axon. Adapted from Av-Ron et al. (1991).

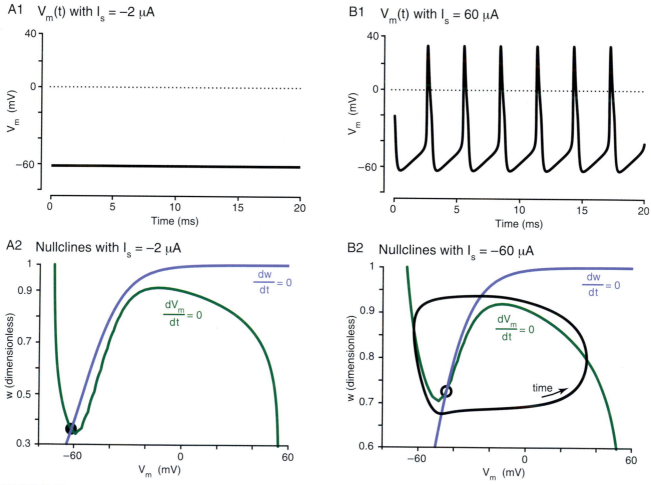

FIGURE 7.17 Oscillations emerge as fixed points lose stability. In a phase plane, the intersection of nullclines defines a fixed point (i.e., $dV_m/dt = dw/dt = 0$). Fixed points can be either stable or unstable. (A1) When the value of $I_s = -2$ μA cm^{-2}, the computed membrane potential is quiescent. Although sufficiently large perturbation will elicit an action potential (not shown), the system will ultimately return to the stable resting state. (A2) The fixed point is indicated with a filled black circle and occurs on a stable branch of the V_m nullcline (green line). Thus, the fixed point is stable. (B1) When the value of $I_s = 60$ μA cm^{-2}, the computed membrane potential is oscillating (i.e., repetitive spiking). Although perturbations may transiently alter this pattern of firing, the system will ultimately return to this exact pattern of repetitive spiking. (B2) The fixed point (open circle at the intersection of the nullclines) is now located on the unstable branch of the V_m nullcline, and thus, the resting state is not stable. The closed curve (black line) represents the trajectory of the system in the phase plane and corresponds to a stable limit cycle. With the exception of I_s, all values for parameters are as in Fig. 7.16.

BOX 7.4

FIXED POINTS AND BIFURCATIONS IN NONLINEAR DYNAMICAL SYSTEMS

Systems that change, that evolve in time, are referred to as *dynamical systems*. Mathematical representations of dynamical systems fall into two general categories: *difference equations* (or *iterative maps*) and *differential equations*. Iterative maps are used to represent systems where time is taken to be discrete, whereas differential equations are used to represent systems where time is taken to be continuous. Differential equations are widely used in science and engineering. The two most common types of differential equations are *ordinary differential equations* (*ODEs*) in which there is only one independent variable (e.g., time, *t*), and partial differential equations in which there are two or more independent variables, such as time (*t*) and space (*x*). Both types of differential equations were used by Hodgkin and Huxley to model action potentials (i.e., the spaced-clamped versus the propagated action potential). The solution and analysis of partial differential equations is complex, however. Thus, ODEs are more commonly used and there are many tools available to investigate and understand their dynamical behavior (Abraham and Shaw 1992; Ermentrout 1996, 2000; FitzHugh 1969; Izhikevich 2000; Jackson 1991; Pavlidis 1973; Strogatz 1994; Tufillaro *et al.* 1992).

Dynamical systems are described in terms of how many differential equations are included in the system (i.e., an *"nth-order"* or *"n-dimensional"* system) and whether they are *linear* or *nonlinear*. For example, consider the following general system:

$$\frac{dx_1}{dt} = k_1 x_1 + k_2 x_2 \tag{7.57}$$

$$\frac{dx_2}{dt} = k_3 x_1 + k_4 x_2 \tag{7.58}$$

where k_i are constants and x_i, are variables. This is a second-order or two-dimensional system, because there are two ODEs. In addition, these ODEs are referred to as *coupled* because x_1 is defined in terms of x_2, and vice versa. This is also a linear system, because all the x_i on the right-hand side appear to the first power only (see also Chapter 14). Otherwise, the system would be nonlinear. Typical nonlinear terms are products (e.g., $x_i x_j$), powers (e.g., x_i^3), and functions (e.g., $\sin x_i$).

Unlike linear systems (e.g., Eqs. 7.57 and 7.58), most nonlinear systems are impossible to solve analytically and must be studied by using techniques such as numerical integration of the equations, geometric methods such

as phase plane analysis, and stability theory. For example, consider a general form for a nonlinear, two-dimensional system:

$$\frac{1}{\mu}\frac{dx_1}{dt} = f_1(x_1, x_2) \tag{7.59}$$

$$\frac{dx_2}{dt} = f_2(x_1, x_2) \tag{7.60}$$

where the functions f_1 and f_2 are determined by the problem at hand and x_1 and x_2 are variables. For example, f_i may describe processes underlying neuronal excitability, and the variables x_1 and x_2 might represent membrane potential, gating variables for membrane conductances, or the intracellular concentration of a second messenger. The parameter μ can be thought of as a rate constant that scales the relative rates of the two functions. When $0 < \mu < 1$, then Eq. 7.59 is referred to as the *slow subsystem* and Eq. 7.60 is referred to as the *fast subsystem*. Numerical integration of Eqs. 7.59 and 7.60 will produce a set of ordered pairs of real numbers $x_1(t)$ and $x_2(t)$; where $x_1(0)$ and $x_2(0)$ represent the initial values (i.e., *initial conditions*) of the two variables (i.e., at time $t = 0$) and $x_1(t)$ and $x_2(t)$ represent the values of the two variables at time t (i.e., the *temporal evolution* of the variables). A common method for visualizing these solutions is to plot $x_1(t)$ and/or $x_2(t)$ verses time (e.g., Figs. 7.16A1 and A2). Such a plot is referred to as a *time series* plot of the variables. Alternatively, an abstract space with coordinates (x_1, x_2) can be constructed. In this space, the solution ($x_1(t)$, $x_2(t)$) corresponds to a point moving along a curve (e.g., Fig. 7.16A3). This curve (i.e., the solution to the system of differential equations) is referred to as a *trajectory*, and the direction of motion along a trajectory is often indicated by an arrowhead. The abstract space is called the *phase space* for the system, and the *phase portrait* of the system shows the overall picture of trajectories in phase space. Because Eqs. 7.59 and 7.60 constitute a two-dimensional system, the phase space of the system is a plane (i.e., a *phase plane*).

If a trajectory asymptotically approaches a constant, time-independent solution, then this point in the phase space is referred to as a *stable fixed point* (e.g., Fig. 7.17A2). (Note: Dynamical systems theory is rife with conflicting terminology, and different terms often are used for the same thing. For example, fixed points are also referred to as *points of equilibrium,* or *singularities,* or *critical points.* This lack of a standard terminology can be a source of great frustration and confusion for individuals not intimately

BOX 7.4 *(continued)*

involved in the field of dynamical systems theory.) In the phase plane, stable fixed points often are indicated by solid black dots. Alternatively, if a trajectory of a nonlinear system asymptotically approaches a time-dependent solution that precisely returns to itself in a time T (i.e., the period), then this periodic solution is referred to as a *stable limit cycle* or simply a *limit cycle* (e.g., Fig. 7.17B2). A limit cycle is represented as a closed curve on a phase plane. These two asymptotically stable trajectories (i.e., stable fixed point and limit cycle) are examples of *attractors*, because trajectories approach and coalesce on them. If a phase plane has only one attractor, then all trajectories ultimately lead to that solution, which is referred to as *globally attracting* (e.g., Fig. 7.16C). Alternatively, if a phase plane has more than one attractor, then the system can manifest more than one stable steady state and, thus, is referred to as *multistable* (e.g., Fig. 7.19B; see also Chapter 14).

Because of the special topological properties of the plane, phase plane analyses can provide fundamental insights into the dynamical properties of a two-dimensional system such as Eqs. 7.59 and 7.60. For example, the Jordan Curve Theorem implies that the only attractors in the phase plane are limit cycles and fixed points. The fixed points in a phase plane can be identified by plotting the *nullclines* (e.g., Fig. 7.16B). A nullcline is a curve in the phase plane along which the rate of change of a particular variable is zero (i.e., $dx_i/dt = 0$). Nullclines are useful because they break up the phase plane into regions in which the derivative of each variable has a constant sign and because any place the two nullclines intersect is a fixed point (i.e., $dx_1/dt = dx_2/dt = 0$).

Fixed points have several features that can be used to classify them. For example, a fixed point can be either *stable* (e.g., Fig. 7.17A2) or *unstable* (e.g., Fig. 7.17B2). A fixed point is stable if all sufficiently small perturbations away from it dampen out with time (i.e., the solution returns to the fixed point). Alternatively, if the perturbation grows with time, the fixed point is unstable. The stability of fixed points can be defined more rigorously in mathematical terms, and readers who want a more detailed discussion of this matter should consult one of the many excellent textbooks that deal with nonlinear systems (e.g., Strogatz 1994).

In addition to stability, fixed points can be classified on the basis of how trajectories behave in the neighborhood of the fixed point. Such behavior is often referred to as the *flow* or *motion*. For example, trajectories approach *stable nodes* and leave *unstable nodes*. Trajectories in the neighborhood of the *saddle point* are hyperbolic (i.e., they do not approach the saddle point but pass by it, looking somewhat like members of a family of hyperbolas near their common center). Saddle points organize boundaries (i.e., a *threshold separatrix*) between classes of trajectories with qualitatively different properties. For example, on one side of the separatrix may reside a stable node that represents a quiescence or rest state (e.g., the resting membrane potential of a neuron), and on the other side may reside a large amplitude trajectory that starts and ends near the equilibrium (e.g., a spike). Thus, small perturbations of the solution decay if they do not lead beyond the separatrix, whereas those crossing the separatrix grow away exponentially. A system with a saddle point has well-defined threshold and all-or-none behavior. Because the solution eventually returns to the stable node (i.e., rest state), however, the system is not oscillatory, but rather it is excitable. Thus, a saddle point can be used to describe threshold phenomenon mathematically.

Fixed points and closed orbits can be created, destroyed, or destabilized as parameters are varied. If the phase portrait changes its topological structure as a parameter is varied, this is termed a *bifurcation*. Examples include changes in the number or stability of fixed points, closed orbits, or saddle connections as a parameter is varied. Bifurcations are most clearly illustrated in what is termed a *bifurcation diagram* (e.g., Fig. 7.18B). A bifurcation diagram plots a system parameter (e.g., μ in Eq. 7.59), which is referred to as the *bifurcation parameter*, on the horizontal axis and a representation of an attractor (e.g., x_i in Eqs. 7.59 and 7.60) on the vertical axis (see also Chapter 14). As the value for the parameter is systematically varied the stability of the fixed points (and closed orbits) will change, which is usually indicated by a branch in the bifurcation diagram. Bifurcations are important because they provide insights into when transitions and instabilities may occur as some control parameter is varied.

For those who wish to explore the behavior of dynamical systems, dynamical systems software has recently become available for personal computers. In general, with these software packages all one has to do is type in the equations and parameters; the program solves the equations numerically and provides analytical tools. For example, the software package XPPAUT can plot variables in two- and three-dimensional phase space, calculate and plot nullclines, analyze the stability of fixed points, and perform bifurcation analyses (Ermentrout 2002). The features and availability of several of these software packages were recently reviewed (Ermentrout 2002; Hayes *et al.* 2002; Hubbard and West 1992; Kocak 1989).

condition for the model. For example, the simulation illustrated in Fig. 7.16A1–2 began with the initial conditions $V_m(0) = -59.407$ mV and $w(0) = 0.402$. This point is marked in the V_m-w plane by the dashed lines in Fig. 7.16A3. With these initial conditions the solution of the model does not change over time, that is, the system is quiescent. Since there is no tendency for the variables to change, this point is referred to as a *stable fixed point* or *steady state*. If the system is subjected to a sufficiently large perturbation (e.g., a brief depolarizing current is injected), then the solution path rapidly evolves along a curve in the phase plane (i.e., the black line in Fig. 7.16A3). This trajectory represents the evolution of an action potential in the V_m-w phase plane. Although time is not explicitly plotted in the phase plane, the direction of the solution is usually indicated by an arrow. The trajectory of the action potential begins and ends at the fixed point. Thus, this fixed point is an example of an *attractor*. Indeed, for the set of parameter values used in Fig. 7.16 and $I_s = 0$ μA, all trajectories will ultimately return to this fixed point, which is said to be *globally attracting*.

To gain a more complete understanding of the dynamical behavior of the reduced model, it is useful to combine the phase plane profiles with a nullcline analysis. A *nullcline* for a given variable is a curve in the phase plane along which one of the derivatives is constant and is equal to zero. In addition, nullclines divide the phase plane into regions where the derivatives have a constant sign (see later). Nullclines for the reduced model are illustrated in Fig. 7.16B. The nullcline associated with the slow variable (i.e., w) is specified by the steady-state w curve (i.e., Eq. 7.54). If the system is currently located on the w nullcline, then its imminent trajectory must be horizontal, because only V_m can change (i.e., $dw/dt = 0$). Horizontal movements to the right of the w nullcline represent depolarizations, which in turn would cause w to increase. Conversely, horizontal movements to the left of the w nullcline represent hyperpolarization, which in turn would cause w to decrease. Thus, w is decreasing (i.e., $dw/dt < 0$) in the region to the left of the w nullcline and w is increasing (i.e. $dw/dt > 0$) in regions to the right of the w nullcline. The nullcline associated with the fast variable (i.e., V_m) is a cubic function (Fig. 7.16B2) and is composed of pairs of values of $V_m(t)$ and $w(t)$ for which the net current is equal to zero (see Eq. 7.49). If the evolution of the system brings it onto the V_m nullcline, its imminent trajectory must be vertical, because only w can change (i.e., $dV_m/dt = 0$). Downward vertical movements represent a decrease in w, which in turn would cause V_m to increase (see Eq. 7.51).

Conversely, upward vertical movements represent an increase in w, which in turn would cause V_m to decrease. Thus, V_m is decreasing (i.e., $dV_m/dt < 0$) in regions above the V_m nullcline, and V_m is increasing (i.e., $dV_m/dt > 0$) in regions below the V_m nullcline. Given that the rate of change of V_m is fast compared to that of w, the nullclines define two very important features of the phase plane. First, the fixed points of the system can be predicted from the nullclines. At intersections of the nullclines (i.e., $dV_m/dt = dw/dt = 0$), there is no tendency for any variable to change, so these intersections represent fixed points or steady states. A fixed point does not guarantee a quiescent membrane potential, however. Fixed points may not be stable to small perturbations (see later). Second, the qualitative dynamics of the system can be predicted from the nullclines. The qualitative prediction is possible because the system will quickly relax to near the potential nullcline and then move in the direction dictated by whether w, the slow variable, is increasing or decreasing. Furthermore, the stability of a fixed point can be predicted based on which branch of the V_m nullcline contains it (under the assumption that voltage changes much more rapidly than w). In the middle branch, positive feedback dominates, so a fixed point in the branch is unstable with respect to perturbations. The other branches are stable. For the system to be excitable or oscillatory, w must be decreasing above the w nullcline (left stable branch) and increasing below it (right stable branch). This will cause a loop in the trajectory as shown in Fig. 7.16C.

The configuration of the nullclines can be changed by altering values of the parameters. For example (Fig. 7.17), the nullclines can be altered by varying the parameter I_s (i.e., the externally applied stimulus current). If the value of I_s is set to -2 μA cm^{-2}, the system is quiescent (Fig. 7.17A1). In the quiescent case, the single fixed point (filled circle in Fig. 7.17A2) falls on the left branch of the V_m nullcline and thus is stable. Although the system is quiescent, it remains excitable, and a sufficiently large perturbation will elicit an action potential (e.g., Fig. 7.16A1). If the value of I_s is set to 60 μA cm^{-2}, the dynamical behavior of the system is qualitatively different. The system continuously generates action potentials with a constant interspike interval (Fig. 7.17B1). This type of activity is often referred to as *beating* or *pacemaker activity*. In the pacemaker case, the fixed point is moved onto the unstable middle branch (open circle in Fig. 7.17B2). Thus, rather than returning to a single point in the phase plane, the solution path continually travels along a closed curve in the phase plane (black line in Fig. 7.17B2), which is referred to as a *limit cycle*.

This limit cycle is stable, and if the system is momentarily perturbed, the trajectory will only transiently leave the limit cycle. Regardless of the magnitude of the perturbation, the solution path will ultimately return to the limit cycle.

Bifurcations and Bistability

A nonlinear system can make a transition from a stable fixed point to a limit cycle in several ways (for review, see Izhikevich 2000). Such transitions are termed *bifurcations*. A bifurcation occurs any time the phase portrait (i.e., the overall picture of trajectories in the phase space) is changed to a topologically nonequivalent portrait by a change in the value of a control or *bifurcation parameter*. For example, increasing I_s from 0 μA to 60 μA causes the reduced model to transition (i.e., a bifurcation) from a resting state to an oscillatory state (Fig. 7.18A). In the phase portrait, this transition would represent the loss of a stable fixed point (e.g., Fig. 7.17A2) and the emergence of a stable limit cycle (e.g. Fig. 7.17B2). Conversely, if I_s is decreased from 60 μA cm^{-2} to 0 μA cm^{-2}, there is a bifurcation of the oscillatory state, which is represented in the phase portrait as a loss of the stable limit cycle and the emergence of a stable fixed point.

The bifurcation is most clearly illustrated in what is termed a *bifurcation diagram* (Fig. 7.18B; see also Chapter 14). A bifurcation diagram plots a system parameter (e.g., I_s) on the horizontal axis and a system variable (e.g., V_m) on the vertical axis. For example, in Fig. 7.18B, the value of I_s was systematically increased from 0 to 400 μA cm^{-2}. At each new value of I_s, the steady-state value of V_m (i.e., the value of V_m at the fixed point) was determined and the stability of the fixed point was determined. For values of I_s between 0 and 16.31 μA cm^{-2} the fixed point is stable, which is indicated by the solid black line. At $I_s = 16.31$ μA cm^{-2}, the fixed point loses stability and a periodic solution emerges (i.e., a bifurcation). The transition illustrated in Fig. 7.18 is a type of bifurcation known as a Hopf bifurcation (filled circle labeled HB). For values of I_s between 16.31 and 336.8 μA cm^{-2}, the fixed point remains unstable, as indicated by the dashed black line. At $I_s = 336.8$ μA cm^{-2}, the periodic solution loses stability and a stable fixed point emerges, again via a Hopf bifurcation. The fixed point remains stable for all further increases in I_s.

Figure 7.18B also illustrates the maximum and minimum values of V_m for the oscillatory response (the yellow lines). Just as a fixed point can be stable or unstable, a periodic solution can also be stable or unstable.

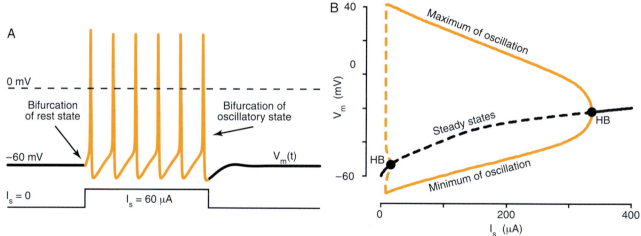

FIGURE 7.18   Bifurcations between quiescent and oscillatory states. As parameters are changed, nonlinear systems can make transitions between a stable steady state (i.e., a quiescent state) and a limit cycle (i.e., oscillations). These transitions are referred to as *bifurcations*. (A) When $I_s = 0$ μA cm^{-2}, the system is in a stable steady state. When $I_s = 60$ μA cm^{-2}, the resting state loses stability and stable oscillations emerge. If I_s is returned to 0 μA cm^{-2}, the periodic solution loses stability, and the system returns to a stable steady state. With the exception of I_s, all values for parameters are as in Fig. 7.16. (B) A bifurcation diagram plots the steady states (black line) of a system as the value of a parameter is systematically varied. Stable steady states are indicated by solid line and unstable steady states are indicated by dashed line. For values of I_s between 0 and 16.31 μA cm^{-2}, the steady state is stable and the system is quiescent. At $I_s = 16.31$ μA cm^{-2}, however, the steady state loses stability and oscillations emerge. The mathematical characteristics of this bifurcation classify it as a Hopf bifurcation (point HB). For a review of the mathematical criteria used to classify various types of bifurcations, see Strogatz (1994). The periodic solution loses stability as $I_s > 336.8$ μA cm^{-2}, and once again the system is quiescent (i.e., the steady state is stable). The second transition is also a Hopf bifurcation. In addition to the steady states, the bifurcation diagram plots the maximum and minimum values of V_m during periodic solutions (yellow line). Periodic solutions also can be either unstable (dashed line) or stable (solid line). With the exception of I_s, all values for parameter were as in Fig. 7.16. In this and subsequent figures, the bifurcation analyses were performed using the AUTO (Doedel 1981) algorithm, which is incorporated into the XPPAUT (Ermentrout 2002) software package.

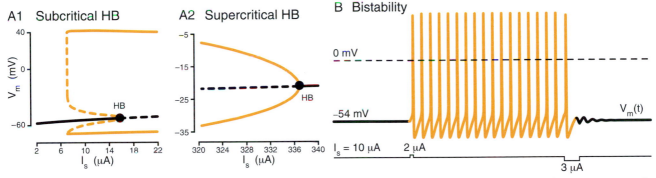

FIGURE 7.19 Bistability. In a nonlinear system, it is possible for more than one stable solution to coexist in a given parameter regime. Thus, the system is said to be bistable (i.e., two stable solutions) or multistable (more than two stable solutions). For an example of a multistable system see Canavier *et al.* (1993) and Butera (1998). (A) The regions of the two Hopf bifurcations in Fig. 7.18 are expanded and replotted. (A1) The first Hopf bifurcation (i.e., $I_s = 16.31$ μA cm^{-2}) is a subcritical Hopf bifurcation. The unstable branch of the periodic solution folds back over a region in which the steady state is stable. This results in a parameter region where the stable steady state and stable periodic solution overlap. Thus, the system can reside on either of the two stable attractors and perturbations can switch the system between the two stable states (see later). (A2) The second Hopf bifurcation is a supercritical Hopf bifurcation. The amplitude of the periodic solution decreases in amplitude through exponentially damped oscillations. (B) At the beginning of the simulation, the system resides in a stable resting state. A brief perturbation (see trace labeled I_s), however, induces a switch from the stable resting state to the stable oscillatory state. A second perturbation can switch the system back to the stable resting state. At a subcritical Hopf bifurcation, the trajectory can be induced to "jump" between distant attractors. Thus, large amplitude oscillations appear dramatically from the resting state. With the exception of I_s, all values for parameter are as in Figure 7.16. At the beginning of the simulation (i.e., at $t = 0$), the values for the two state variables (i.e., $V_m(t)$ and $w(t)$) were initially set to $V_m(0) = -54.314$ mV and $w(0) = 0.515$. The values of state variables at the beginning of a simulation are referred to as the *initial conditions*.

The unstable periodic solutions are indicated by the dashed yellow line, and the stable limit cycles are indicated by the solid yellow line. When a Hopf bifurcation leads to *unstable* periodic solutions (i.e., the Hopf bifurcation that occurs at $I_s = 16.31$ μA), then the bifurcation is termed *subcritical* (Fig. 7.19A1). When a Hopf bifurcation leads to only *stable* periodic solutions (i.e., the Hopf bifurcation that occurs at $I_s = 336.8$ μA cm^{-2}), it is termed *supercritical* (Fig. 7.19A2). The two Hopf bifurcations are illustrated in greater detail in Fig. 7.19. The supercritical Hopf bifurcation is the simpler type, and because of its appearance in a bifurcation diagram, it also is referred to as a *pitchfork bifurcation*. At the supercritical Hopf bifurcation, the amplitude of the periodic solution slowly decreases in amplitude until a stable steady state is reached. This lethargic decay is called *critical slowing down* in the physics literature (Strogatz 1994). In contrast to the supercritical Hopf bifurcation, the subcritical Hopf bifurcation has a relatively complex configuration (Fig. 7.19A1). The emergent branches of the unstable periodic solutions (dashed yellow lines) fold back into the parameter region where the steady state is stable. This backward folding has several important implications. First, because of this backward fold, there is a parameter region where the stable fixed points (i.e., solid black line) and the stable limit cycle (i.e., solid yellow lines) overlap. In this parameter region, the model is *bistable*, i.e., both a stable quiescent and a stable oscillatory solution coexist (see also Chapter 14). In the phase plane (not shown), this type of bistability would

be represented by a stable fixed point at the intersection of the nullclines, and surrounding the stable fixed point would first be an unstable period orbit and then a stable limit cycle. Note, bistability can result from other types of bifurcation schema, which in turn would have different properties and different phase plane portraits. For some examples of different mechanisms for achieving bistability, see Bertram *et al.* 1995, Canavier *et al.* 2002, Izhikevich 2000, or Rinzel and Ermentrout 1998. Second, the existence of different stable states allows for the possibility of "jumps" between the two stable states and of *hysteresis* (i.e., the lagging of an effect behind its cause). Finally, as the solution jumps from the stable steady state to the stable periodic solution, large amplitude oscillations appear dramatically from the resting state. In the physics literature, this dramatic jumping between states is referred to as *explosive instability* or a *blow-up* (Strogatz 1994).

Bistability can endow a neuron with some very interesting dynamical properties. For example, in the bistable regime, sufficiently large perturbations can induce a jump or switch from a resting state (i.e., stable steady solution) to a spiking state (i.e., stable oscillatory state) and vice versa. Figure 7.19B illustrates such a case. At the beginning of the simulation (i.e., at $t = 0$), the values for the two state variables (i.e., $V_m(t)$ and $w(t)$) were initially set to $V_m(0) = -54.314$ mV and $w(0) = 0.515$. The values of state variables at the beginning of a simulation are referred to as the *initial conditions*. With these initial conditions and parameter values, the system resides on the stable fixed point and is quiescent (black line in the $V_m(t)$ trace of Fig. 7.19A). In the absence of any large perturbations, the system will remain in this resting state indefinitely. However, a sufficiently large perturbation (e.g., a $2\ \mu A\ cm^{-2}$, 3 ms injection of current) can send the solution path off the stable fixed point and onto the stable limit cycle. In the absence of any large perturbations, the system will remain in the oscillatory state indefinitely. However, a sufficiently large perturbation (e.g., a $-3\ \mu A\ cm^{-2}$, 12 ms injection of current) can send the solution path off the stable limit cycle and back onto the stable fixed point. Such behavior has been observed empirically, for example, in the squid giant axon (Guttman *et al.* 1980) and in the bursting neuron R15 of *Aplysia* (Lechner *et al.* 1996). The switch from the resting state to the oscillatory state is often referred to as *hard excitation,* and the switch from the oscillatory state to the resting state is often referred to as *annihilation.* Moreover, this bistable behavior is critical for the occurrence of bursting when a slow conductance is added to the system (see later).

Dynamical Underpinnings of Bursting Activity

In contrast to pacemaker activity (i.e., continuous spiking activity with an approximately constant interspike interval), *bursting activity* is characterized by the clustering of spikes into groups that are separated by quiescent periods. Bursting cells can be classified as conditional or endogenous bursters. *Endogenous bursters* fire in a bursting pattern in the absence of any input, whereas *conditional bursters* can fire in bursts if they receive appropriate input from other neurons within a network. Endogenous bursting activity relies on bistability in the system.

In the example of bistability previously described (Fig. 7.19), the bifurcation parameter was an externally applied current. Other, more physiologically relevant, parameters can achieve similar results. For example, adjusting the magnitude of an ionic conductance such as $\bar{g}_K$ (i.e., the maximal conductance of I_K; see Eq. 7.52) can alter the dynamical properties of the system (Av-Ron *et al.* 1991). Figure 7.20 illustrates a bifurcation diagram in which $\bar{g}_K$ was used as the bifurcation parameter. As the value of $\bar{g}_K$ is decreased, the steady state loses stability and an oscillatory state emerges via a subcritical Hopf bifurcation. This oscillatory state represents the pacemaker activity, which is similar to that illustrated in Fig. 7.17B1. With further reductions in $\bar{g}_K$, the oscillatory state loses stability and a stable steady state emerges via a subcritical Hopf bifurcation. Thus,

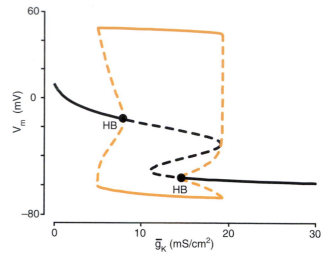

FIGURE 7.20 Bifurcation diagram as a function of $\bar{g}_K$. As in previous bifurcation diagrams, stable solutions are indicated by solid lines and unstable solutions are indicated by dashed lines. The steady states are indicated by the black line and the periodic solutions are indicated by the yellow line. Moving from right to left along the *x*-axis, the maximum conductance for the K+ current (i.e., $\bar{g}_K$) decreases. As $\bar{g}_K$ is decreased, the steady state loses stability and oscillations emerge via a subcritical Hopf bifurcation (point HB). As $\bar{g}_K$ continues to decrease, the periodic solution loses stability and a stable steady state emerges via a second subcritical Hopf bifurcation. With the exception of $\bar{g}_K$, all values for parameters are as in Fig. 7.16. Adapted from Av-Ron *et al.* (1991).

there are two parameter regions where bistability exists, and in principle, the system could be made to burst if the value of $\bar{g}_K$ were forced to traverse back and forth between parameter regions associated with a stable steady state and a stable limit cycle.

Generally, bursting cannot occur in a two-variable model (Rinzel and Ermentrout 1998). Bursting can be realized, however, by incorporating an additional process (i.e., a relatively slow negative feedback process). To illustrate the general principles underlying burst firing, the reduced model presented previously (i.e., Eqs. 7.49 and 7.50) is extended to include two more currents (I_{Ca} and $I_{K,Ca}$; see also Chapter 5) and an additional variable ($[Ca^{2+}]$), which describes the intracellular concentration of Ca^{2+}. Although several schema are possible, the two additional currents considered here are an inward Ca^{2+} current (I_{Ca}) that activates with increasing depolarization, and an outward K+ current that is activated by intracellular Ca^{2+} ($I_{K,Ca}$). The two new currents are described by

$$I_{K,Ca} = \bar{g}_{K,Ca}z([Ca^{2+}])(V_m - E_K) \tag{7.61}$$

$$I_{Ca} = \bar{g}_{Ca}m_\infty^{mp}(V_m)(1 - w(V_m,t))(V_m - E_{Ca}) \tag{7.62}$$

where $\bar{g}_i$ represents the maximal conductances of the ionic currents, E_i represents their respective equilibrium potentials, and z is a function that describes the Ca^{2+}-dependency of the $I_{K,Ca}$, which is given by

$$z([Ca^{2+}]) = \frac{[Ca^{2+}]}{[Ca^{2+}] + K_d} \quad (7.63)$$

where $[Ca^{2+}]$ represents the concentration of intracellular Ca^{2+} and K_d is the concentration at which $I_{K,Ca}$ is half activated. The dynamics of intracellular Ca^{2+} are described by

$$\frac{d[Ca^{2+}]}{dt} = K_p(-I_{Ca}) - R[Ca^{2+}] \quad (7.64)$$

where K_p is a conversion factor from current to concentration and R is the removal rate constant. (See Chapter 14 for more detailed models of Ca^{2+} regulation.) Note that by extending Eq. 7.49 to include $I_{K,Ca}$, the total K^+ conductance of the reduced model becomes

$$g_K^{(total)} = \bar{g}_K \left(\frac{w(V_m, t)}{S}\right)^{wp} + \bar{g}_{K,Ca} Z([Ca^{2+}]) \quad (7.65)$$

The conductances $\bar{g}_K$ and $\bar{g}_{K,Ca}$ are determined such that the system traverses a region of bistability (see Fig. 7.20). For additional details, see Av-Ron et al. (1991, 1993).

Figure 7.21 illustrates a bursting solution to the three-variable model (Eqs. 7.49, 7.50, and 7.64). During the quiescent phase, V_m (Fig. 7.21A) ramps up and fast subthreshold oscillations give rise to a burst of action potentials. During the burst of action potentials, $[Ca^{2+}]$ increases (Fig. 7.21B), which in turn increases the magnitude of $I_{K,Ca}$ (Fig. 7.21C). Once $g_K^{(total)}$ becomes sufficiently large, the oscillation is halted. Calcium is removed during the quiescent period, which in turn reduces $I_{K,Ca}$, and oscillations return. This pattern of activity does not require an external stimulus and will continue indefinitely. Thus, this type of bursting behavior is that of an *endogenous burster*. This type of bursting activity has been observed in mesencephalic trigeminal sensory neurons (Pedroarena et al. 1999).

Although phase plane analysis cannot provide a full description for higher-order systems such as the three-variable model (i.e., Eqs. 7.49, 7.50, and 7.64), judicious two-dimensional projections can yield useful insights into the dynamical behavior of higher-order systems. Canavier et al. (1993, 1994) provide an example of how a phase plane can be used to analyze an 11-order model of a bursting neuron. Figure 7.22A illustrates a bifurcation diagram for the three-variable model. To construct this diagram, the observation was made that the intracellular concentration of Ca^{2+} (i.e., Eq. 7.64) changed very slowly

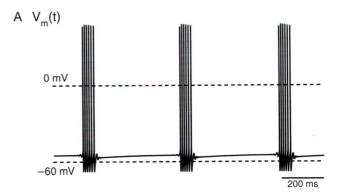

A $V_m(t)$

B $[Ca^{2+}]$ (t)

C $I_{K,ca}$ (t)

FIGURE 7.21 Elliptic bursting in a three-dimensional model. In the absence of any external stimulus, the three-variable model generates sustained bursting activity. (A) Time-series plot of the computed membrane potential ($V_m(t)$). Brief bursts of action potentials are separated by quiescent periods. Small subthreshold oscillations can be seen waxing and waning before and after each burst of spikes. (B) Time-series plot of the computed intracellular concentration of Ca^{2+} ($[Ca^{2+}](t)$). Calcium enters the cell during the spikes and slowly accumulates during the burst. During the intervening quiescent period, the levels of Ca^{2+} fall. Note the slow timescale with which levels of Ca^{2+} vary. (C) Time-series plot of $I_{K,Ca}(t)$ during the bursting activity. As Ca^{2+} accumulates during the burst (see panel B), the magnitude of $I_{K,Ca}$ increases. Once the total K^+ conductance (see Eq. 7.65) is sufficiently large, the spiking activity is terminated. As the level of Ca^{2+} slowly decreases (see panel B) during the quiescent period, $I_{K,Ca}$ begins to decrease. Once the total K^+ conductance has decreased sufficiently, spiking resumes. This cycle of events repeats indefinitely and does not require an external stimulus. Thus, this bursting activity is intrinsic (or endogenous) to the system. With the exception of $\bar{g}_{Ca} = 5$ mS/cm², $E_{Ca} = 124$ mV, $\bar{g}_K = 12$ mS/cm², $E_l = -50$ mV, $\bar{g}_{K,Ca} = 0.5$ mS/cm², $K_d = 0.5$ mM, $K_p = 0.00052$, and $R = 0.0045$, all values for parameter are as in Fig. 7.16. Adapted from Av-Ron et al. (1993).

relative to the other state variables (i.e., $V_m(t)$ and $w(t)$). Thus, the function z, which is directly related to the intracellular concentration of Ca^{2+} (see Eq. 7.63), can be treated as a *parameter* (i.e., a bifurcation parameter) rather than as a *function* when analyzing the dynamics of V_m and w. Because V_m and w are relatively fast as compared to $[Ca^{2+}]$ (and z),

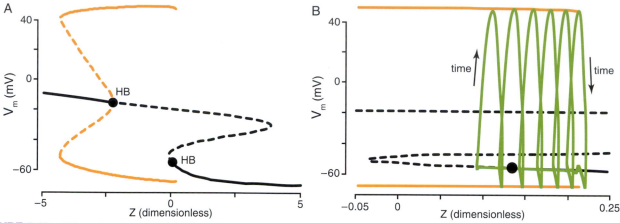

FIGURE 7.22   Bifurcation diagram of the three-dimensional system with z as the bifurcation parameter. Although z is a variable, it is sufficiently slow relative to changes in either $V_m(t)$ or $w(t)$, which constitute a fast subsystem, that it can be treated as a parameter for the purpose of generating this bifurcation analysis. (A) Bifurcation diagram using z as the control parameter. As before, solid lines indicate stable solutions and dashed lines indicated unstable solutions. The black line indicates the steady states and yellow line indicates periodic solutions. From right to left (i.e., as the magnitude of z is decreased), the steady state loses stability via a subcritical Hopf bifurcation (point HB at $z = 0.13$). The unstable oscillations that are born at the subcritical Hopf bifurcation (see Fig. 7.20) terminate in a saddle loop (not shown), and a stable periodic solution emerges beyond the saddle loop bifurcation. As the magnitude of z continues to decrease, the stable periodic solution destabilizes, and a stable steady state is re-established, again via a Hopf bifurcation (point HB at $z = -2.3$). (B) Projection of the bursting solution in the $z - V_m$ phase plane (green line) onto the bifurcations diagram for positive values of z. Note the resolution of the bifurcation has been substantially increased. The direction of the solution path (i.e., time) is indicated by the arrows. The trajectory tracks the fast subsystem (i.e. spiking in the periodic solution) until it reaches the unstable regime. At which point, the trajectory "falls" onto the stable branch of the steady state. As intracellular levels of Ca^{2+} decrease (see Fig. 7.21B) and z decreases, the trajectory is forced past the point where the stable steady state disappears (black, filled circle). Once in this unstable regime, the trajectory "jumps" onto the stable branch of the periodic solution. Thus, bursting emerges as the slow subsystem (i.e., $z(t)$) forces the trajectory to slowly drift through bifurcations in the fast subsystem. With the exception of z, all values for parameters are as in Fig. 7.21.

Eqs. 7.49 and 7.50 (i.e., the equations that describe V_m and w) are termed the *fast subsystem* and Eq. 7.64 (i.e., the equation that describes $[Ca^{2+}]$) is termed the *slow subsystem*. For each value of z, the dynamics of V_m and w can be analyzed. For some values of z, V_m and w may oscillate (i.e., spiking activity), whereas for other values of z, V_m and w may settle to a fixed point (i.e., quiescent membrane potential). As the magnitude of z is varied, the fast subsystem may undergo bifurcations where V_m and w are first at a fixed point and then begin to oscillate, or conversely, where V_m and w are oscillating and then become quiescent (see Fig. 7.22).

In Fig. 7.22A, the branch of the steady states (black line with stable steady states represented by a solid line and unstable steady states represented by a dashed line) forms the S-shaped curve and the oscillatory solutions are represented by the forked curve (yellow line with stable oscillatory states represented by a solid line and unstable oscillation represented by a dashed line). As the magnitude of z decreases (i.e., moving from right to left along the x-axis), the stable steady state destabilizes at a turning point (point HB at $z = 0.13$), which is subcritical Hopf bifurcation. The unstable oscillations born at this Hopf bifurcation, which are not illustrated, terminate in a saddle loop. A stable periodic solution

emerges from this saddle loop bifurcation. As the magnitude of z continues to decrease, the stable periodic solutions destabilize, again via a subcritical Hopf bifurcation (point HB at $z = -2.3$), and a stable steady state is re-established. Although negative values of z are physiologically irrelevant, the bifurcation diagram serves to illustrate that the stable steady states of the three-variable model destabilize via subcritical Hopf bifurcations, and that for physiologically relevant values of z (i.e., $z \geq 0$), the stable periodic solutions disappear abruptly when they reach the saddle node of periodics. For more detailed explanations of this bifurcation diagram, see Bertram *et al.* (1995), Canavier *et al.* (2002), Rinzel and Ermentrout (1998), and Izhikevich (2001).

The transitions between resting and oscillatory states can be visualized by projecting the burst solution in the z-V plane onto the bifurcation diagram (Fig. 7.22B). The resolution of the bifurcation diagram has been increased to focus attention on the ranges of values for z that are observed during the bursting activity (see Fig. 7.21). The trajectory (green line) tracks the stable periodic solution (i.e., the fast subsystem) until it loses stability (i.e., at the SNP). During spiking activity, the levels of Ca^{2+} increase, and hence the magnitude of z also increases (see Eq. 7.63 and Fig. 7.21). Thus, the general movement of the trajectory along the stable branch of the periodic solution is from left to right. At the termination of the stable branch of the periodic solution (i.e., at the SNP), the trajectory then "falls" onto the stable branch of the steady state (i.e., a quiescent membrane potential). During the quiescent phase, the level of Ca^{2+} decreases, and hence the magnitude of z also decreases (see Fig. 7.21). The general movement of the trajectory along the stable branch of the steady-state solution is from right to left. As z decreases (see Fig. 7.21B), the trajectory slowly moves through the Hopf bifurcation, and as the steady state loses stability, the trajectory "jumps" onto the stable branches of the periodic solution, and the general movement of the trajectory once again is from left to right. Thus, the oscillations in the intracellular levels of Ca^{2+}, which in turn induce oscillations in z and $I_{K, Ca}$, drive the trajectory back and forth through the bifurcation between spiking and nonspiking.

The envelope of $V_m(t)$ during the burst activity of Fig. 21A is "elliptical" as the subthreshold oscillations gradually increase (i.e., wax) in amplitude before a burst and gradually decrease (i.e., wane) at the end of the burst. Thus, this type of bursting activity often is referred to as *elliptic bursting*. Other qualitative types of bursting have been observed, such as *parabolic bursting* (i.e., the frequency of spiking activity gradually increases and then decreases during the burst) and *square-wave bursting* (i.e., the burst of spikes ride atop a plateau-like depolarization). Bursting activity can also be classified on the basis of the types of bifurcations that give rise to the oscillatory states. The history of formally classifying bursting began with Rinzel (1987), who contrasted the bifurcation mechanisms inherent in the elliptic, parabolic, and square-wave bursters. Bertram *et al.* (1995) suggested referring to the types of bursting using Roman numerals (i.e., elliptic bursting is Type I, parabolic bursting is Type II, and square-wave bursting is Type III). There are many ways that a nonlinear system can make a transition from a stable resting state to a stable periodic solution, and there are many possible combinations. Izhikevich (2000) has described all possible combinations and reviewed examples of each.

BOX 7.5

ALTERNATIVE MODELS OF EXCITABILITY

Over the past 50 years, the formalisms developed by Hodgkin and Huxley have been used as a starting point for developing conductance-based models. These models were usually developed to simulate specific systems, such as the endogenous bursting cell R15 in *Aplysia* (Canavier *et al.* 1991) or the thalamocortical relay neurons (McCormick and Huguenard 1992). These models often incorporate descriptions of 3 to 10 ionic conductances and may include additional variables that describe processes such as the dynamics of intracellular levels of Ca^{2+} or the dynamics of a second messenger such as cAMP. Thus, these models are high dimensional and difficult to analyze. Others have taken an alternative approach and developed low-dimensional models. Although these reduced models often lack biophysical detail, they still manifest salient features of neuronal excitability (e.g., threshold behavior and excitability) and have provided some fundamental insights into excitability. Three of these reduced models will be described briefly.

The FitzHugh–Nagumo Model

Perhaps the best-known and most widely used low-dimensional model of an excitable system is the FitzHugh–Nagumo model (FitzHugh 1961; Nagumo *et al.* 1962) (for reviews see Koch 1999; Murray 1993; Rinzel 1990). This model qualitatively describes the events occurring in an excitable membrane. The FitzHugh–Nagumo equations are often written, assuming dimensionless variables, in this general form:

$$\frac{dv}{dt} = f(v) - w + I_s \qquad (7.66)$$

$$\frac{dw}{dt} = \phi(v - \gamma w) \qquad (7.67)$$

where $f(v)$ is the cubic

$$f(v) = v(1 - v)(v - a) \qquad (7.68)$$

Here v corresponds to membrane potential, w is a recovery variable (i.e., combines the effects of h and n), which activates slowly (i.e., the rate constant, ϕ, is less than 1), and γ is a positive constant. The cubic expression represents a nonlinear I-V relationship, which follows from the assumption that inward current activates instantaneously; a is the voltage threshold for this current. The qualitative behavior of FitzHugh–Nagumo is very similar to that of the HH system in that for some parameter regimes the model has a globally stable resting state yet

---

BOX 7.5   *(continued)*

is excitable in that a sufficiently large perturbation will elicit a spike. The system does not have a saddle node, and thus, manifests a pseudo-threshold, which is similar to the HH system (see Box 7.2). In addition, with sustained current injection, a limit cycle appears and the system oscillates. The simplicity of this model has made it a model of choice for nonlinear dynamical analyses of excitability and for simulations of large-scale networks of excitable elements.

The Morris–Lecar Model

The Morris–Lecar model (Morris and Lecar 1981) was formulated and studied in the context of investigating electrical activity in barnacle muscle fibers. The model incorporates a voltage-gated Ca^{2+} current and a voltage-gated, delay-rectifier K^+ current; neither current inactivates. The Morris–Lecar equations are often written, assuming dimensionless variables, in this general form:

$$C\frac{dv}{dt} = I_s - \bar{g_{Ca}}m_\infty(v)(v - v_{Ca}) - \bar{g_K}w(v)(v - v_K)$$
$$- \bar{g_l}(v - v_l) \qquad (7.69)$$

$$\frac{dw}{dt} = \Phi\frac{w_\infty(v) - w}{\tau_w(v)} \qquad (7.70)$$

where

$$m_\infty(v) = 0.5\left(1 + \tanh\frac{v - v_1}{v_2}\right) \qquad (7.71)$$

$$w_\infty(v) = 0.5\left(1 + \tanh\frac{v - v_3}{v_4}\right) \qquad (7.72)$$

$$\tau_w(v) = \frac{1}{\cosh\dfrac{v - v_3}{v_4}} \qquad (7.73)$$

and where v_i are parameters, w is the fraction of K^+ channels open.

The qualitative behavior of the Morris–Lecar model is very similar to that of the HH system in that for some parameter regimes the model has a globally stable resting state yet is excitable. The system manifests a pseudo-threshold and oscillates with sustained current injection. The Morris–Lecar model has been studied extensively by Rinzel and Ermentrout (1998). They demonstrated an intriguing feature of this system. For some parameter regimes the oscillations emerge via a subcritical Hopf bifurcation. Thus, this two-dimensional system manifests bistability. Indeed, with different parameter regimes the two-variable system manifests

tristability (i.e., two distinct stable steady states and a stable limit cycle coexist for a given parameter regime). The greatest advantage of the model has come from an expanded version, which includes a third variable that describes the dynamics of intracellular Ca^{2+}, and by adding a Ca^{2+}-dependent K^+ current. This three-variable system can simulate a variety of bursting behaviors. The simplicity of this model and its rich dynamical repertoire make it an excellent choice for nonlinear dynamical analyses of complex neuronal firing patterns such as bursting.

The Hindmarsh–Rose Model

Although similar to the FitzHugh–Nagumo model, the Hindmarsh–Rose model (Hindmarsh and Rose 1982) was developed from first principles with the assumptions that the rate of change of membrane potential (dx/dt) depends linearly on z (an externally applied current) and y (an intrinsic current). The forms of the functions and values for the parameters were selected to fit data from a large neuron in the visceral ganglion of the pond snail *Lymnaea stagnalis*. The equations for this reduced model are

$$\frac{dx}{dt} = -a(f(x) - y - z) \qquad (7.74)$$

$$\frac{dy}{dt} = b(f(x) - q\exp(rx) + s - y) \qquad (7.75)$$

where $f(x)$ is given by the cubic

$$f(x) = cx^3 + dx^2 + ex + h \qquad (7.76)$$

where a-e, h, q, r, and s are constants. This two-variable system offers some advantages, such as a better fit of the frequency–current relationship and a better fit to aspects of the spike waveform. As with the Morris–Lecar model, however, the greatest advantages of the Hindmarsh–Rose model have come from extended versions. A three-dimensional version (Rose and Hindmarsh 1984, 1985, 1989a,b; Wang 1993) manifests bursting activity and multistability, including multirhythmicity (i.e., multiple forms of bursting activity coexisting for a single parameter regime). Additional extensions of the model (Rose and Hindmarsh 1989c) have been used to investigate the dynamical properties of thalamic neurons. Of the reduced models of excitability, the Hindmarsh–Rose model appears to exhibit the richest repertoire of dynamical properties.

Summary

Some neurons are normally quiescent, but excitable in that they can fire one or more action potentials in response to transient stimuli. Others are capable of firing action potentials in a repetitive fashion in the absence of external stimuli. The same nonlinear properties of the cell membrane that allow a neuron to fire a single action potential can enable it to function as a nonlinear oscillator, generating sustained rhythmic activity. The simplest type of repetitive action potential firing is pacemaker activity in which single, apparently identical, action potentials are generated at relatively fixed intervals. Another type of rhythmic firing, bursting, is characterized by action potentials clustered into bursts that are separated by quiescent periods. To generate either an action potential or a sustained oscillation of any kind, two opposing processes are required, one relatively rapid autocatalytic process and one somewhat slower, restorative process. For rhythmic activity to be sustained, the opposing processes must have the appropriate steady-state characteristics and kinetics to continue to alternately dominate the system dynamics. In addition to the two processes required for action potential generation, a bursting oscillation requires at least one additional, slower process to modulate action potential firing so as to group them in bursts.

Acknowledgments

We thank Dr. C.C. Canavier for her assistance in the initial implementation of the reduced model and her comments on a previous version of this manuscript. We also thank Drs. E. Av-Ron and P. Smolen for their comments on a previous version of the manuscript.

References

Abbott, L.F. (1994). Single neuron dynamics: an introduction. In: *Neural Modeling and Neural Networks* (F. Ventriglia, ed.), pp. 57-78. Pergamon Press, New York.

Abraham, R.H. and Shaw, C.D. (1992). *Dynamics: The Geometry of Behavior*. Addison-Wesley Publishing Company, Redwood City, CA.

Alexander, J.C. and Cai, D.-Y. (1991). On the dynamics of bursting systems. *J. Math. Biol.* **29**: 405-423.

Av-Ron, E. (1994). The role of a transient potassium current in a bursting neuron model. *J. Math. Biol.* **33**: 71-87.

Av-Ron, E., Parnas, H. and Segel, L. (1991). A minimal biophysical model for an excitable and oscillatory neuron. *Biol. Cybern.* **65**: 487-500.

Av-Ron, E., Parnas, H. and Segel, L. (1993). A basic biophysical model for bursting neurons. *Biol. Cybern.* **69**: 87-95.

Bertram, R. (1994). Reduced-system analysis of the effects of serotonin on a molluscan burster neuron. *Biol. Cybern.* **70**: 359-368.

Bertram, R., Butte, M.J., Kiemel, T. and Sherman, A. (1995). Topological and phenomenological classification of bursting oscillations. *Bull. Math. Biol.* **57**: 413-439.

Bezanilla, F. and Armstrong, C.M. (1977). Inactivation of the sodium channel. I. Sodium current experiments. *J. Gen. Physiol.* **70**: 549-566.

Bower, J.M. and Beeman, D. (1998). *The Book of GENESIS*. Springer-Verlag Publishers, Santa Clara, CA.

Brown, A.M. (1999). A methodology for simulating biological systems using Microsoft EXCEL. *Comput. Methods Programs Biomed.* **58**: 181-190.

Brown, A.M. (2000). Simulation of axonal excitability using a spreadsheet template created in Microsoft EXCEL. *Comput. Methods Programs Biomed.* **63**: 47-54.

Butera, R.J. (1998). Multirhythmic bursting. *Chaos* **8**: 274-284.

Butera, R.J., Clark, J.W. and Byrne, J.H. (1996). Dissection and reduction of a modeled bursting neuron. *J. Comput. Neurosci.* **3**: 199-223.

Byrne, J.H. and Schultz, S.G. (1994). *An Introduction to Membrane Transport and Bioelectricty*. Raven Press, New York.

Canavier, C.C., Baxter, D.A. and Byrne, J.H. (2002). Repetitive action potential firing. In: *Encyclopedia of Life Sciences* (N.P. Group, ed.), online (*http://www.els.net*). Grove's Dictionaries, New York.

Canavier, C.C., Baxter, D.A., Clark, J.W. and Byrne, J.H. (1993). Nonlinear dynamics in a model neuron provide a novel mechanism for transient synaptic inputs to produce long-term alterations of postsynaptic activity. *J. Neurophysiol.* **69**: 2252-2257.

Canavier, C.C., Baxter, D.A., Clark, J.W. and Byrne, J.H. (1994). Multiple modes of activity in a model neuron suggest a novel mechanism for the effects of neuromodulators. *J. Neurophysiol.* **72**: 872-882.

Canavier, C.C., Clark, J.W. and Byrne, J.H. (1991). Simulation of the bursting activity of neuron R15 in *Aplysia*: role of ionic currents, calcium balance, and modulatory transmitters. *J. Neurophysiol.* **66**: 2107-2124.

Clay, J.R. (1998). Excitability of squid giant axon revisited. *J. Neurophysiol.* **80**: 903-913.

Cole, K.S. (1949). Dynamic electrical characteristics of the squid axon membrane. *Arch. Sci. Physiol.* **22**: 253-258.

Cole, K.S. (1968). *Membranes, Ions, and Impulses*. University of California Press, Berkeley.

Cole, K.S., Antosiewicz, H.A. and Rabinowitz, P. (1955). Automatic computation of nerve excitation. *J. Soc. Indust. Appl. Math.* **3**: 153-172.

Cole, K.S., Guttman, R. and Bezanilla, F. (1970). Nerve excitation without threshold. *Proc. Natl. Acad. Sci. USA* **65**: 884-891.

Cooke, I. and Lipkin, M. (Eds.) (1972). *Cellular Neurophysiology: A Source Book*. Holt, Rinehart and Winston, Inc., New York.

Cronin, J. (1987). *Mathematical Aspects of Hodgkin-Huxley Neural Theory*. Cambridge University Press, New York.

Dayan, P. and Abbott, L.F. (2001). *Theoretical Neuroscience: Computational and Mathematical Modeling of Neural Systems*. The MIT Press, Cambridge.

DeSchutter, E. (1989). Computer software for development and simulation of compartmental models of neurons. *Comput. Biol. Med.* **19**: 71-81.

DeSchutter, E. (1992). A consumer guide to neuronal modeling software. *Trends Neurosci.* **15**: 462-464.

DeSchutter, E. (Ed.) (2001). *Computational Neuroscience: Realistic Modeling for Experimentalists*. CRC Press, New York.

Dodge, F. and Frankenhaeuser, B. (1959). Sodium currents in the myelinated nerve fibre of *Xenopu laevis* investigated with the voltage clamp technique. *J. Physiol. (Lond.)* **148**: 188-200.

Doedel, E.J. (1981). AUTO: a program for the automatic bifurcation and analysis of autonomous systems. *Cong. Num.* **30**: 265-284.

Ermentrout, G.B. (1996). Type I membranes, phase resetting curves, and synchrony. *Neural Comp.* **8**: 979-1001.

Ermentrout, G.B. (2002). *Simulating, Analyzing and Animating Dynamical Systems: A guide to XPPAUT for Researchers and Students.* SIAM, Philadelphia.

FitzHugh, R. (1960). Thresholds and plateaus in the Hodgkin-Huxley equations. *J. Gen. Physiol.* **43**: 867-896.

FitzHugh, R. (1961). Impulses and physiological states in theoretical models of nerve membrane. *Biophys. J.* **1**: 445-466.

FitzHugh, R. (1966). Theoretical effects of temperature on threshold in the Hodgkin-Huxley nerve model. *J. Gen. Physiol.* **49**: 989-1005.

FitzHugh, R. (1969). Mathematical models of excitation and propagation in nerve. In: *Biological Engineering* (H.P. Schwan, ed.), pp. 1-85. McGraw-Hill Book Company, New York.

FitzHugh, R. and Antosiewicz, H.A. (1959). Automatic computation of nerve excitation-detailed correction and addition. *J. Soc. Indust. Appl. Math.* **7**: 447-458.

Gall, W.G. and Zhou, Y. (2000). An organizing center for planar neural excitability. *Neurocomp.* **32-33**: 757-765.

Guttman, R., Lewis, S. and Rinzel, J. (1980). Control of repetitive firing in squid axon membrane as a model for a neuroneoscillator. *J. Physiol. (Lond.)* **305**: 377-395.

Hayes, R., Byrne, J.H. and Baxter, D.A. (2002). Neurosimulation: tools and resources. In: *The Handbook of Brain Theory and Neural Networks* (M.A. Arbib, ed.) (in press). MIT Press, Cambridge.

Hille, B. (2001). *Ionic Channels of Excitable Membranes.* Sinauer Associates, Inc., Sunderland, MA.

Hindmarsh, J.L. and Rose, R.M. (1982). A model of the nerve impulse using two first-order differential equations. *Sci.* **296**: 162-164.

Hindmarsh, J.L. and Rose, R.M. (1984). A model of neuronal bursting using three coupled first order differential equations. *Proc. R. Soc. Lond. B* **221**: 87-102.

Hines, M.L. and Carnevale, N.T. (1997). The NEURON simulation environment. *Neural Comput.* **9**: 1179-1209.

Hodgkin, A.L. (1964). The ionic basis of nervous conduction. *Science* **145**: 1148-1154.

Hodgkin, A.L. (1976). Chance and design in electrophysiology: an informal account of certain experiments on nerve carried out between 1934 and 1952. *J. Physiol. (Lond.)* **263**: 1-21.

Hodgkin, A.L. (1977). *The Pursuit of Nature: Informal Essays on the History of Physiology.* Cambridge University Press, Cambridge.

Hodgkin, A.L. and Huxley, A.F. (1952a). Currents carried by sodium and potassium ions through the membrane of the giant axon of *Loligo. J. Physiol. (Lond.)* **116**: 449-472.

Hodgkin, A.L. and Huxley, A.F. (1952b). The components of membrane conductance in the giant axon of *Loligo. J. Physiol. (Lond.)* **116**: 473-496.

Hodgkin, A.L. and Huxley, A.F. (1952c). The dual effect of membrane potential on sodium conductance in the giant axon of *Loligo. J. Physiol. (Lond.)* **116**: 497-506.

Hodgkin, A.L. and Huxley, A.F. (1952d). A quantitative description of membrane current and its application to conduction and excitation in nerve. *J. Physiol. (Lond.)* **117**: 500-544.

Hodgkin, A.L. and Huxley, A.F. (1990). A quantitative description of membrane current and its application to conduction and excitation in nerve. 1952. *Bull. Math. Biol.* **52**: 5-23.

Hodgkin, A.L., Huxley, A.F. and Katz, B. (1952). Measurements of current-voltage relations in the membrane of the giant axon of *Loligo. J. Physiol. (Lond.)* **116**: 424-448.

Hoppensteadt, F. and Izhikevich, E. (2001). Canonical neural models. In: *Brain Theory and Neural Networks* (M.A. Arbib, ed.), pp. 1-7. The MIT Press, Cambridge.

Hubbard, J.H. and West, B.H. (1992). *MacMath: A Dynamical Systems Software Package for the Macintosh.* Springer, New York.

Huxley, A.F. (1951). The ionic basis of electrical activity in nerve and muscle. *Biol. Rev.* **26**: 339-409.

Huxley, A.F. (1959). Can a nerve propagate a subthreshold disturbance? *J. Physiol. (Lond.)* **148**: 80-81P.

Huxley, A.F. (1964). Excitation and conduction in nerve: quantitative analysis. *Science* **145**: 1154-1159.

Huxley, A.F. (2000). Reminiscences: working with Alan, 1939-1952. *J. Physiol. (Lond.)* **527P**: 13S.

Huxley, A.F. (2002). Hodgkin and the action potential 1939-1952. *J. Physiol. (Lond.)* **538**: 2.

Izhikevich, E. (2001). Synchronization of elliptic bursters. *SIAM Review* **43**: 315-344.

Izhikevich, E.M. (2000). Neural excitability, spiking and bursting. *Int. J. Bifurcation Chaos* **10**: 1171-1266.

Jackson, E.A. (1991). *Perspectives of Nonlinear Dynamics.* Cambridge University Press, Cambridge.

Johnston, D. and Wu, S.M. (1997). *Foundations of Cellular Neurophysiology.* The MIT Press, Cambridge.

Kepler, T.B., Abbott, L.F. and Marder, E. (1992). Reduction of conductance-based neuron models. *Biol. Cybern.* **66**: 381-387.

Kirsch, R.A. (1998). SEAC and start of image processing at the National Bureau of Standards. *IEEE Ann. History Comput.* **20**: 7-13.

Kocak, H. (1989). *Differential and Difference Equations Through Computer Experiments*, 2nd ed. Springer, New York.

Koch, C. (1999). *Biophysics of Computation: Information Processing in Single Neurons.* Oxford University Press, New York.

Krinskii, V.I. and Kokoz, Y.M. (1973). Analysis of equations of excitable membranes-I. Reduction of the Hodgkin-Huxley equations to a second order system. *Biofizika* **18**: 506-511.

Lechner, H.A., Baxter, D.A., Clark, J.W. and Byrne, J.H. (1996). Bistability and its regulation by serotonin in the endogenously bursting neuron R15 in *Aplysia. J. Neurophysiol.* **75**: 957-962.

MacGregor, R.J. (1987). *Neural and Brain Modeling.* Academic Press, Inc., New York.

Marmont, G. (1949). Studies on the axon membrane. I. A new method. *J. Cell. Comp. Physiol.* **34**: 351-382.

McCormick, D.A. and Huguenard, J. (1992). A model of the electrophysiological properties of thalamocortical relay neurons. *J. Neurophysiol.* **68**: 1384-1440.

Moore, J.W. and Stuart, A.E. (2000). *Neurons in Action: Computer Simulations with NeuroLab.* Sinauer Associates, Inc., Publishers, Sunderland, MA.

Morris, C. and Lecar, H. (1981). Voltage oscillations in the barnacle giant muscle fiber. *Biophys. J.* **35**: 193-213.

Murray, J.D. (1993). *Mathematical Biology.* Springer, New York.

Nagumo, J.S., Arimato, S. and Yoshizawa, S. (1962). An active pulse transmission line simulating a nerve axon. *Proc. IRE* **50**: 2061-2070.

Nelson, M. and Rinzel, J. (1998). The Hodgkin-Huxley model. In: *The Book of Genesis: Exploring Realistic Neural Models with the GEeneral NEural SImulation System* (J.M. Bower and D. Beeman, eds.), pp. 29-49. Spring-Verlag, New York.

Palti, Y. (1971a). Description of axon membrane ionic conductances and currents. In: *Biophysics and Physiology of Excitable Membrane* (W.J. Adelman, ed.), pp. 168-182. Van Nostrand Reinhold Company, New York.

Palti, Y. (1971b). Digital computer solutions of membrane currents in the voltage clamped giant axon. In: *Biophysics and Physiology of Excitable Membranes* (W.J. Adelman, ed.), pp. 183-193. Van Nostrand Reinhold Company, New York.

Pavlidis, T. (1973). *Biological Oscillators: Their Mathematical Analysis.* Academic Press, New York.

Pedroarena, C.M., Pose, I.E., Yamuy, J., Chase, M.H. and Morales, F.R. (1999). Oscillatory membrane potential activity in the soma of a primary afferent neuron. *J. Neurophysiol.* **82**: 1465-1476.

Rinzel, J. (1985). Excitation dynamics: insights from simplified membrane models. *Fed. Proc.* **44**: 2944-2946.

Rinzel, J. (1990). Electrical excitability of cells, theory and experiment: review of the Hodgkin-Huxley foundation and an update. *Bull. Math. Biol.* **52**: 5-23.

Rinzel, J. (1998). The Hodgkin-Huxley Model. In: *The Book of Genesis: Exploring Realistic Neural Networks with the GEneral NEural SImulation System* (J.M. Bower and D. Beeman, ed.), Springer-Verlag, New York.

Rinzel, J. and Ermentrout, G.B. (1998). Analysis of neural excitability and oscillations. In: *Methods in Neuronal Modeling: From Ions to Networks*, 2nd ed. (C. Koch and I. Segev, ed.), pp. 251-292. The MIT Press, Cambridge.

Rinzel, J. and Lee, Y.S. (1987). Dissection of a model for neuronal parabolic bursting. *J. Math. Biol.* **25**: 653-675.

Rose, R.M. and Hindmarsh, J.L. (1985). A model of a thalamic neuron. *Proc. R. Soc. Lond. B* **225**: 161-193.

Rose, R.M. and Hindmarsh, J.L. (1989a). The assembly of ionic currents in a thalamic neuron I. The three-dimensional model. *Proc. R. Soc. Lond. B* **237**: 267-288.

Rose, R.M. and Hindmarsh, J.L. (1989b). The assembly of ionic currents in a thalamic neuron II. The stability and state diagrams. *Proc. R. Soc. Lond. B* **237**: 289-312.

Rose, R.M. and Hindmarsh, J.L. (1989c). The assembly of ionic current in a thalamic neuron III. The seven-dimensional model. *Proc. R. Soc. Lond. B* **237**: 313-334.

Strogatz, S.H. (1994). *Nonlinear Dynamics and Chaos: With Applications to Physics, Biology, Chemistry and Engineering.* Perseus Books, New York.

Tuckwell, H.C. (1988). *Introduction to Technical Neurobiology: Volume 2 Nonlinear and Stochastic Theories.* Cambridge University Press, Cambridge.

Tufillaro, N.B., Abbott, T. and Reilly, J. (1992). *An Experimental Approach to Nonlinear Dynamics and Chaos.* Addison-Wesley Publishing Company, Redwood City, CA.

Vandenberg, C.A. and Bezanilla, F. (1991). A sodium gating model based on single channel, macroscopic ionic, and gating currents in the squid giant axon. *Biophys. J.* **60**: 1511-1533.

Ventriglia, F. (Eds.) (1994). *Neural Modeling and Neural Networks.* Pergamon Press, New York.

Wang, X.-J. (1993). Genesis of bursting oscillations in the Hindmarsh-Rose model and homoclinicity to a chaotic saddle. *Physica D* **62**: 263-274.

Weiss, T.F. (1997). *Cellular Biophysics. Volume 2: Electrical Properties.* The MIT Press, Cambridge.

Wheeler, D.J. (1992a). The EDSAC programming systems. *IEEE Ann. History Comput.* **14**: 34-40.

Wheeler, J. (1992b). Applications of the EDSAC. *IEEE Ann. History Comput.* **14**: 27-33.

Ziv, I., Baxter, D.A. and Byrne, J.H. (1994). Simulator for neural networks and action potentials: description and application. *J. Neurophysiol.* **71**: 294-308.

Release of Neurotransmitters

Robert S. Zucker, Dimitri M. Kullmann,

and Thomas L. Schwarz

The synapse is the point of functional contact between one neuron and another. It is the primary place at which information is transmitted from neuron to neuron in the central nervous system or from neuron to target (gland or muscle) in the periphery. The simplest way for one cell to inform another of its activity is by direct electrical interaction, in which the current generated extracellularly from the action potential in the first cell passes through neighboring cells. Owing to the shunting of current by the highly conductive extracellular fluid, a 100 mV action potential may generate only 10–100 µV in a neighboring neuron. This coupling can be improved if neighboring cells are joined by a specialized conductive pathway through gap junctions (see Chapter 15); even then, a presynaptic spike is not likely to generate more than about 1 mV postsynaptically, unless the presynaptic process is nearly as large or larger than the postsynaptic process. This biophysical constraint limits the number of presynaptic cells that can converge on and influence a postsynaptic cell, and such electrical connections normally can be only excitatory and short lasting, are bidirectional in transmission, and show little plasticity or modifiability. They have limited potential for complex computation but can be useful when a postsynaptic neuron must be activated with high reliability and speed or when concurrent activity in a large number of presynaptic afferents must be signaled.

ORGANIZATION OF THE CHEMICAL SYNAPSE

Most interneuronal communication relies on the use of a chemical intermediary, or *transmitter*, secreted subsequent to action potentials by presynaptic cells to influence the activity of postsynaptic cells. In chemical transmission, a single action potential in a small presynaptic terminal can generate a large *postsynaptic potential* (PSP) (as large as tens of millivolts). This is accomplished by the release of thousands to hundreds of thousands of molecules of transmitter that can bind to postsynaptic *receptor molecules* and open (or close) as many ion channels in about 1 ms. There is room for many afferents (often thousands) to interact and influence a postsynaptic neuron, and the effect can be either excitatory or inhibitory, depending on the ions that permeate the channels operated by the receptor. The resulting responses are either *excitatory postsynaptic potentials* (EPSPs) or *inhibitory postsynaptic potentials* (IPSPs), depending on whether they drive the cell toward a point above or below its firing threshold. Different afferents can have different effects, with different strengths and kinetics, on each other as well as on postsynaptic cells. These differences depend on the identity of the transmitter(s) released and the receptors present (see Chapters 9, 11, and 16). Chemical synapses are often modified by prior activity in the presynaptic neuron.

Chemical synapses are also particularly subject to modulation of presynaptic ion channels by substances released by the postsynaptic or neighboring neurons. This flexibility is essential for the complex processing of information that neural circuits must accomplish, and it provides an important locus for modifiability of neural circuits underlying adaptive processes such as learning (Chapter 19).

Transmitter Release Is Quantal

One of the first applications of the microelectrode was the discovery that transmitter release is *quantal* in nature (Katz, 1969). Transmitter is released spontaneously in multimolecular packets called *quanta* in the absence of presynaptic electrical activity. Each packet generates a small postsynaptic signal—either a *miniature excitatory* or a *miniature inhibitory postsynaptic potential* (mEPSP or mIPSP, respectively, or just mPSP); under voltage clamp, a *miniature excitatory* or a *miniature inhibitory postsynaptic current* (mEPSC or mIPSC, respectively, or just mPSC) is generated. An action potential tremendously, but very briefly, accelerates the rate of secretion of quanta and synchronizes them to evoke a PSP. Vertebrate skeletal neuromuscular junctions are frequently used as model synapses, because both receptors and nerve terminals are relatively accessible for anatomical, electrophysiological, and biochemical studies. At the neuromuscular junction, the motor nerve forms a cluster of small unmyelinated processes that lie in shallow gutters in the muscle to form a structure called an *end plate*, and PSPs, PSCs, mPSPs, and mPSCs are called *end-plate potentials* (EPPs), *end-plate currents* (EPCs), *miniature end-plate potentials* (mEPPs), and *miniature end-plate currents* (mEPCs), respectively.

Why is transmission quantized? Neural circuits must process complex and quickly changing information fast enough to generate timely appropriate responses. This requires rapid transmission across synapses. Fast-acting chemical synapses accomplish this by concentrating transmitter in membrane-bound structures, ~50 nm in diameter, called *synaptic vesicles* and docking these vesicles at specialized sites called *active zones* along the presynaptic membrane (Fig. 8.1A). Vesicles not docked at the membrane are clustered behind it and associated with cytoskeletal elements (Heuser, 1977). Action potentials release transmitter by depolarizing the presynaptic membrane and opening Ca^{2+} channels that are strategically colocalized with the synaptic vesicles in the active zone (Robitaille *et al.*, 1990). The local intense rise in Ca^{2+} concentration triggers the fusion of docked vesicles with the plasma

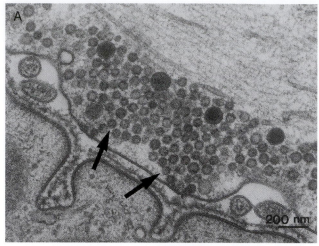

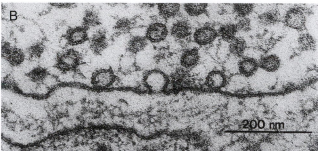

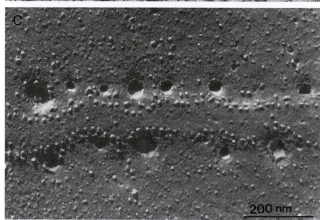

FIGURE 8.1 Ultrastructural images of synaptic vesicle exocytosis. Synapses from frog sartorius neuromuscular junctions were quick-frozen milliseconds after stimulation in 4-aminopyridine to broaden action potentials and enhance transmission. (A) A thin section from which water was replaced with organic solvents (freeze substitution) and fixed in osmium tetroxide, showing vesicles clustered in the active zone, some docked at the membrane (arrows). (B) Shortly (5 ms) after stimulation, vesicles were seen to fuse with the plasma membrane. (C) After freezing, presynaptic membranes were freeze-fractured and a platinum replica was made of the external face of the cytoplasmic membrane leaflet. Vesicles fuse about 50 nm from rows of intramembranous particles thought to include Ca^{2+} channels. (A) and (B) from Heuser (1977); (C) from Heuser and Reese (1981). © Heuser and Reece, 1981. Originally published in the Journal of Cell Biology, 88, 564–580.

membrane (called *exocytosis;* Figs. 8.1B and C) (Heuser and Reese, 1981) and the release of their contents into the narrow *synaptic cleft* (about 100 nm wide) separating the presynaptic terminal from high concentrations of postsynaptic receptors. The fusion of one vesicle releases about 5000 transmitter molecules within a millisecond (Fletcher and Forrester, 1975; Kuffler and Yoshikami, 1975; Whittaker, 1988) and generates the quantal response recorded postsynaptically. No membrane carrier can release so much transmitter this fast, nor can a pore or channel unless some mechanism exists to concentrate the transmitter behind the pore, which may be

regarded as the function of synaptic vesicles. Evidence that transmitter is released from vesicles and that one quantum is due to exocytosis of a vesicle is summarized in Box 8.1 (Betz and Bewick, 1993; Eccles, 1964; Edwards, 1992; Heuser and Reese, 1973; Heuser *et al.,* 1979; Hurlbut *et al.,* 1990; Large and Rang, 1978; Murthy and Stevens, 1998; Prior, 1994; Ryan *et al.,* 1997; Schiavo *et al.,* 1994; Searl *et al.,* 1991; Torri-Tarelli *et al.,* 1985, 1992; Van der Kloot, 1988, 1991; Van der Kloot and Molgó, 1994; von Wedel *et al.,* 1981).

At neuromuscular junctions, transmitter from one vesicle diffuses across the synaptic cleft in 2 μs and

---

BOX 8.1

EVIDENCE THAT A QUANTAM IS A VESICLE

Transmitter Is Released from Vesicles

1. All chemically transmitting synaptic terminals contain presynaptic vesicles (Eccles, 1964).
2. Synaptic vesicles concentrate and store transmitter (Edwards, 1992).
3. Rapid freezing of neuromuscular junctions during stimulation shows vesicle exocytosis occurring at the moment of transmitter release (Torri-Tarelli *et al.,* 1985).
4. Intravesicular proteins appear on the external terminal surface after secretion (von Wedel *et al.,* 1981; Torri-Tarelli *et al.,* 1992).
5. Retarding the filling of vesicles by using transport inhibitors (e.g., vesamicol for acetylcholine) or by reducing the transvesicular pH gradient generates a class of small mEPSPs that probably represent partially filled vesicles; drugs that enhance vesicle loading increase mEPSP size (Prior, 1994; Searl *et al.,* 1991; Van der Kloot, 1991).
6. Quantal size is independent of membrane potential or cytoplasmic acetylcholine concentration altered osmotically (Van der Kloot, 1988).
7. Synaptic vesicles formed by endocytosis load with extracellular electron-dense and fluorescent dyes (horseradish peroxidase and FM1–43, respectively) after nerve stimulation; the dye is released by subsequent stimulation (Betz and Bewick, 1993; Heuser and Reese, 1973).
8. False transmitters synthesized from choline derivatives load slowly into cholinergic vesicles; they are co-released with acetylcholine in proportion to their concentrations in vesicles (Large and Rang, 1978).

9. Clostridial toxins that interfere with the synaptic vesicle–plasma membrane interaction block neurosecretion (Schiavo *et al.,* 1994).

One Quantum Is One Vesicle

1. The number of acetylcholine molecules in isolated vesicles corresponds to the number of molecules released in a quantum (Fletcher and Forrester, 1975; Kuffler and Yoshikami, 1975; Whittaker, 1988).
2. When release is enhanced and the collapse of vesicle fusion images is prolonged by treatment with the potassium channel blocker 4-aminopyridine to broaden action potentials, the number of vesicle fusions observed corresponds to the number of quanta released by an action potential (Heuser *et al.,* 1979). In these special circumstances, several vesicles are released at each active zone (Fig. 8.1C).
3. The number of vesicles present in nerve terminals corresponds to the total store of releasable quanta. When endocytosis is blocked by the temperature-sensitive *Drosophila* mutant *shibire* (Van der Kloot and Molgó, 1994) or pharmacologically (Hurlbut *et al.,* 1990) and the motor nerve is stimulated to exhaustion, the number of quanta released corresponds to the original number of presynaptic vesicles.
4. The statistical variations in quantal release match the statistical variations in vesicle release and recovery measured with staining and destaining of lipophilic dyes (see Box 8.3 and section titled "Quantal Analysis") (Ryan *et al.,* 1997; Murthy and Stevens, 1988).

reaches a concentration of about 1 mM at the postsynaptic receptors (Matthews-Bellinger and Salpeter, 1973). These receptors bind transmitter rapidly, opening from 1000 to 2000 postsynaptic ion channels (Van der Kloot *et al.*, 1994) (two molecules of transmitter must bind simultaneously to receptors to open each channel; see Chapters 11 and 16). Each channel has a 25 pS conductance and remains open for about 1.5 ms, admitting a net inflow of 35,000 positive ions. A single action potential in a motor neuron can release 300 quanta within about 1.5 ms along a junction that contains about 1000 active zones. The resulting postsynaptic depolarization, which begins after a *synaptic delay* of about 0.5 ms and reaches a peak of tens of millivolts, is typically sufficient to generate an action potential in the muscle fiber.

At fast central synapses, postsynaptic cells make contact with presynaptic axon swellings called *varicosities* when they occur along fine axons and *boutons* when they are located at the tips of terminals. Each varicosity or bouton contains one active zone or a few of them. The postsynaptic process is often on a fine dendritic branch or tiny spine with a length of a few micrometers, having a very high input resistance and capable of generating active propagating responses (Chapter 17). At inhibitory GABAergic synapses and excitatory glutamatergic synapses (Edwards *et al.*, 1990; Jonas *et al.*, 1993), each action potential releases from 5 to 10 quanta, and each quantum released elevates the transmitter concentration (Clements *et al.*, 1992; Tang *et al.*, 1994; Tong and Jahr, 1994) in the cleft to about 1 mM and activates about 30 ion channels. At excitatory synapses, this release may be sufficient to generate EPSPs of 1 mV or less in amplitude, clearly subthreshold for generating action potentials. But central neurons often receive thousands of inputs, each of which has a "vote" on how the cell should respond (see Chapter 16). No input has absolute, or even majority, control over postsynaptic cell activity, but the matching of quantal size to input resistance ensures that inputs are reasonably effective. Consequently, at synapses onto larger central neurons with lower input resistances, quanta open between 100 and 1000 postsynaptic channels.

Synaptic Vesicles Are Recycled

A constant supply of vesicles filled with transmitter must be available for release from the nerve terminal at all times. Maintaining this supply requires the efficient recycling of synaptic vesicles. For this purpose, two partly overlapping cycles are used: one for the components of the synaptic vesicle membrane and another for the vesicle contents (transmitter substances). The cycles overlap from the time of transmitter packaging into vesicles until exocytosis. The cycles are distinct during the stages in which vesicle membrane and transmitter are recovered for reuse. The various steps of these cycles are common to all chemical synapses and are summarized in Fig. 8.2.

Vesicle Membrane Cycle

The components of the synaptic vesicle membrane are initially synthesized in the cell body before being transported to nerve terminals by fast axoplasmic transport (Bennett and Scheller, 1994; Jahn and Südhof, 1994) (see Chapter 2). Within the nerve terminal, the synaptic vesicles are loaded with transmitter and either anchored to each other and actin filaments (McGuiness *et al.*, 1989) or targeted to plasma membrane docking sites at active zones. These docking sites are also rich in clusters of high-voltage-activated Ca^{2+} channels (Haydon *et al.*, 1994; Robitaille *et al.*, 1990), mainly N- and P/Q-type Ca^{2+} channels, depending on the synapse (Dunlap *et al.*, 1995; Wheeler *et al.*, 1994; see Chapters 5 and 6). Depolarization of the plasma membrane by an invading action potential opens these voltage-dependent Ca^{2+} channels to admit Ca^{2+} ions in the neighborhood of docked vesicles. The local high concentration of Ca^{2+} resulting from the opening of multiple Ca^{2+} channels triggers exocytosis. After exocytosis, some vesicles may rapidly reclose, but most fuse fully with the plasma membrane (Granseth *et al.*, 2006). The latter are recovered by *endocytosis*, a budding off of the vesicular membrane to form a new "coated" vesicle covered by the protein *clathrin*. Endocytosis may also be regulated by presynaptic $[Ca^{2+}]$ (Schweizer and Ryan, 2006). Recovered vesicular membrane often fuses to form large membranous sacs, called *endosomes* or *cisternae*, from which new synaptic vesicles are formed. The molecular mechanisms of the vesicle cycle of exo- and endocytosis are discussed later in this chapter.

Transmitter Cycle

The steps of the transmitter cycle vary with the type of transmitter (for additional details see Chapters 9 and 10). Some transmitters are synthesized from precursors in the cytoplasm before transport into synaptic vesicles, whereas other transmitters are synthesized in synaptic vesicles from transported precursors. Peptide transmitters are synthesized exclusively in the cell body and not locally recycled. At most synapses, a transporter that harnesses the energy in the proton gradient across the vesicular membrane functions to concentrate transmitter (or

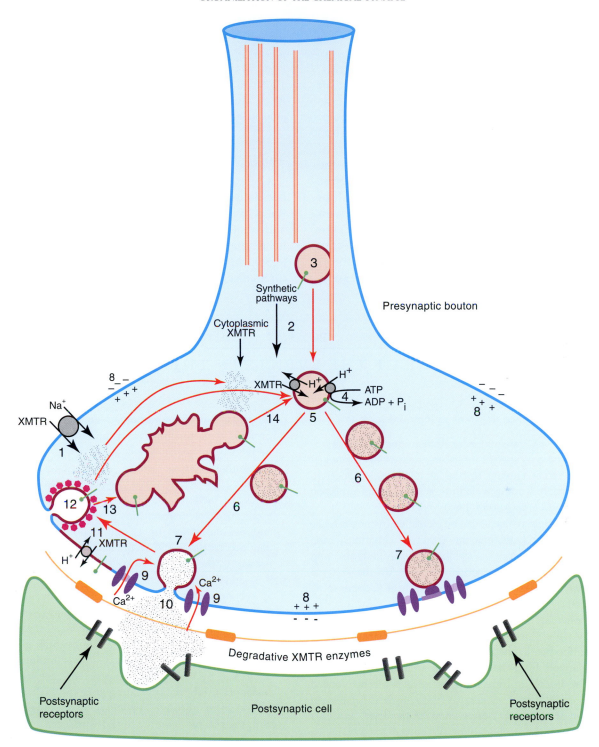

FIGURE 8.2 Steps in the life cycle of synaptic vesicles: (1) Na$^+$-dependent uptake of transmitter (XMTR) or XMTR precursors into the cytoplasm, (2) synthesis of XMTR, (3) delivery of vesicle membrane containing specialized transmembrane proteins by axoplasmic transport on microtubules, (4) production of transvesicular H$^+$ gradient by vacuolar ATPase, (5) concentration of XMTR in vesicles by H$^+$/XMTR antiporter, (6) synapsin I-dependent anchoring of vesicles to actin filaments near active zones, (7) releasable vesicles docked in active zones near Ca^{2+} channels, (8) depolarization of nerve terminal and presynaptic bouton by action potential, (9) opening of Ca^{2+} channels and formation of regions of local high [Ca^{2+}] ("Ca^{2+} microdomains") in active zones, (10) triggering of exocytosis of docked vesicles comprising quantal units of XMTR released by overlapping Ca^{2+} microdomains, (11) nonquantal leakage of XMTR through vesicle membrane fused with plasma membrane and exposure of vesicle proteins to synaptic cleft, (12) recovery of vesicle membrane by dynamin-dependent endocytosis of clathrin-coated vesicles, (13) fusion of coated vesicles with endosomal cisternae, (14) formation of synaptic vesicles from endosomes. Also shown are postsynaptic receptors with multiple XMTR binding sites and extracellular XMTR-degradative enzymes in synaptic cleft.

transmitter precursors) in vesicles (Edwards, 1992). The pH gradient arises from the action of a vacuolar proton ATPase that uses the energy of ATP hydrolysis to transport protons into vesicles. After exocytosis, released transmitter diffuses across the synaptic cleft and rapidly binds to receptors. As transmitter falls off receptors, it is typically recovered from the synaptic cleft by sodium-dependent uptake transporters (see Chapter 9). At cholinergic synapses, acetylcholine is hydrolyzed to acetate and choline by the enzyme acetylcholinesterase present in the synaptic cleft. This enzyme is saturated by the initial gush of transmitter following exocytosis but can keep up with its subsequent slower release from receptors. The choline so produced is recovered by a presynaptic choline transporter and made available for the synthesis of new transmitter. Much of the evidence for the steps outlined in Fig. 8.2 comes from ultrastructural and pharmacological experiments. Some of this evidence is outlined in Box 8.2 (Chan *et al.*, 1991; Cooper *et al.*, 1991; Edwards *et al.*, 1985; Elmgvist and Quastel, 1965; Llinás *et al.*, 1991; Parsons *et al.*, 1993; Van der Bliek and Meyerowitz, 1991) and Box 8.3 (Ryan *et al.*, 1993).

Summary

Chemical synapses are ideally suited to permit one neuron to rapidly and effectively excite or inhibit the activity of another cell. A diversity of transmitters and receptors guarantees a multiplicity of postsynaptic responses. The opportunity for presynaptic and postsynaptic interactions between inputs provides for marvelously complex computational capabilities.

BOX 8.2

EVIDENCE FOR SOME OF THE EVENTS IN THE LIFE HISTORY OF VESICLES

Numbers refer to the steps in Figure 8.2.

1. Uptake of transmitter or transmitter precursors is prevented by specific inhibitors, such as *hemicholinium-3* block of choline uptake at cholinergic synapses, ultimately leading to failure of synaptic transmission (Elmgvist and Quastel, 1965).
2. Cholinergic synapses can be identified by the presence of the synthetic enzyme choline acetyltransferase, GABAergic synapses by the enzyme glutamic acid decarboxylase, adrenergic synapses by the enzyme dopamine β-hydroxylase, and so forth (Cooper *et al.*, 1991).
3. Vesicular transport into nerve terminals is blocked by inhibitors of axoplasmic transport such as antibodies to the microtubule motor protein *kinesin* (Bennett and Scheller, 1994; Jahn and Südhof, 1994).
4. The storage of transmitter in vesicles can be blocked by inhibitors of vacuolar ATPase, such as *bafilomycin A_1*, or of an H^+-dependent transporter, such as *vesamicol* for acetylcholine (Parsons *et al.*, 1993).
5. Dephosphorylation of synapsin I inhibits vesicle movements and transmission, whereas its phosphorylation by Ca^{2+}-calmodulin-dependent kinase II protects against this inhibition (Llinás *et al.*, 1991; McGuiness *et al.*, 1989).
6. Toxins from *Clostridium* bacteria, which proteolyze the vesicular protein *synaptobrevin* or plasma membrane proteins *SNAP-25* and *syntaxin*, block exocytosis, whereas mutants deficient in the vesicle protein synaptotagmin and injection of peptides derived from *synaptotagmin* show defects in evoked transmitter release (more details later in this chapter).
7. Block of action potential propagation by local application of *tetrodotoxin* prevents transmission, and depolarization by elevating potassium in the bath accelerates mPSP frequency, as long as Ca^{2+} is present in the medium (Katz, 1969).
8. 9. N- and P/Q-type calcium channel antagonists, such as ω-conotoxin and ω-agatoxin IVA, prevent Ca^{2+} influx and block transmission at many synapses (Wheeler *et al.*, 1994; Dunlap *et al.*, 1995).
10. Cholinergic synapses show a nonquantal leak of acetylcholine that is enhanced after stimulation; it is blocked by vesamicol, the vesicular acetylcholine transport inhibitor, indicating that the leak is due to transport through vesicular membrane fused with the plasma membrane (Edwards *et al.*, 1985).
11. Endocytosis is blocked at high temperature in the *shibire* mutant of *Drosophila*, which affects the protein *dynamin* in endocytosis of coated vesicles (Chan *et al.*, 1991; Van der Bliek and Meyerowitz, 1991).

Morphological evidence for steps 13 and 14 in Figure 8.2 is given in Box 8.3.

BOX 8.3

HISTOLOGICAL TRACERS CAN BE USED TO FOLLOW VESICLE RECYCLING

An elegant picture of the life history of synaptic vesicles comes from studies using electron-dense or fluorescent markers of intracellular regions that have been in contact with the extracellular space. *Horseradish peroxidase* (HRP) is an enzyme that catalyzes the oxidation of diaminobenzidine, forming an electron-dense product that can easily be identified in tissues fixed with osmium tetroxide for electron microscopy; *FM1–43* is an amphipathic styryl dye that becomes highly fluorescent on partitioning into cell membranes. When frog muscles were soaked in HRP and the motor neurons were stimulated at 10 Hz for 1 min, the enzyme appeared in coated vesicles in nerve terminals in regions outside active zones. After more prolonged stimulation, most of the HRP collected in endosomal cisternae, owing to the fusion of endocytotic vesicles with these organelles. When the HRP was washed out and the neurons were rested for an hour before fixation, HRP appeared in small clear synaptic vesicles in active zones. When rested neurons were stimulated again before fixation, this time in the absence of HRP, the filled vesicles gradually disappeared owing to their release by exocytosis (Heuser and Reese, 1973).

Another study traced the uptake of FM1–43 into living motor nerve terminals with the use of confocal fluorescence microscopy. High-frequency stimulation for just 15 s in FM1–43 was marked by uptake of dye into nerve terminals. More prolonged stimulation followed by a period of rest without the dye in the bath resulted in the persistent staining of synaptic vesicles in active zones. Subsequent stimulation at 10 Hz gradually destained the terminals in minutes; destaining required the presence of Ca^{2+} in the medium and represented exocytosis of stained vesicles. After about 1 min, the rate of destaining decreased as the vesicle pool began to be diluted with unstained vesicles newly recovered by endocytosis (Betz and Bewick, 1993). Exposing dissociated hippocampal neurons to FM1–43 at various times after stimulation showed that endocytosis proceeded for about 1 min after exocytosis. Cells loaded with dye and then restimulated began to destain about 30 s after endocytosis, which is a measure of the time needed for recycling of recovered vesicles into the pool of releasable vesicles (Ryan et al., 1993). These experiments provide a dynamic view of the life cycle of synaptic vesicles.

The packaging of transmitter into vesicles and its release in quanta enable a single action potential to secrete hundreds of thousands of molecules of transmitter almost instantaneously at a synapse onto another cell. Neurochemical and ultrastructural studies have provided a rich picture of the life cycle of synaptic vesicles from their exocytosis at active zones to their recovery by endocytosis, their refilling with transmitter, and redocking at release sites.

EXCITATION–SECRETION COUPLING

Shortly after an action potential invades presynaptic terminals at fast synapses, the synchronous release of many quanta of transmitter generates the postsynaptic potential. Since the work of Locke in 1894, the presence of calcium in the external medium has been known to be a requirement for transmission. What is the central role of Ca^{2+} in triggering neurosecretion?

Calcium Triggers Release of Transmitters at Internal Sites

Calcium was originally believed to act at an external site to enable neurons to release transmitter. The pioneering work of Bernard Katz (1969) and his co-workers showed that Ca^{2+} acts intracellularly. This conclusion is based on many lines of evidence:

1. Calcium must be present only at the moment of invasion of the nerve terminal by an action potential for transmitter to be released.
2. Calcium entry is retarded by a large presynaptic depolarization, and transmitter release is delayed until the voltage gradient is reversed at the end of the pulse, whereupon Ca^{2+} enters and release occurs as an off-EPSP until Ca^{2+} channels close. Sodium influx is not necessary for secretion, and K^+ ions also play no role.
3. Elevation of intracellular $[Ca^{2+}]$ accelerates the spontaneous release of quanta of transmitter (Steinbach and Stevens, 1976; Rahamimoff et al.,

1980). Stimulation in a $[Ca^{2+}]$-free medium reduces intracellular $[Ca^{2+}]$ and mEPSP frequency.

4. The presence of Ca^{2+} channels in presynaptic terminals is shown by the ability to stimulate local action potentials that trigger release in a high-$[Ca^{2+}]$ medium when Na^+ action potentials are blocked with tetrodotoxin and K^+ channels are blocked with tetraethylammonium.

5. Divalent cations that permeate Ca^{2+} channels, such as Ba^{2+} and Sr^{2+}, support transmitter release, although only weakly. Cations that block Ca^{2+} channels, such as Co^{2+} and Mn^{2+}, block transmission (Augustine et al., 1987); Mg^{2+} reduces transmission, perhaps by screening fixed surface charge and effectively hyperpolarizing the nerve (Muller and Finkelstein, 1974).

6. Transmission depends nonlinearly on $[Ca^{2+}]$ in the bath, varying with the fourth power of $[Ca^{2+}]$, whereas Ca^{2+} influx remains a linear function of $[Ca^{2+}]$, indicating a high degree of Ca^{2+} cooperativity in triggering exocytosis (Llinas et al., 1981).

7. At giant synapses in the stellate ganglion of squid, voltage-clamp recording of the presynaptic Ca^{2+} current reveals a close correspondence between Ca^{2+} influx and transmitter release, including an association between the off-EPSP and a delay in Ca^{2+} current until the end of large pulses (called a *tail current*) (Llinás et al., 1981).

8. Action potentials trigger no phasic release of transmitter when Ca^{2+} influx is blocked, even when presynaptic Ca^{2+} is tonically elevated by photolysis of photosensitive Ca^{2+} chelators; however, the elevated presynaptic $[Ca^{2+}]$ accelerates the frequency of mEPSPs (Mulkey and Zucker, 1991).

Vesicles Are Released by Calcium Microdomains

Single action potentials generate a Ca^{2+} rise of about 10 nM, which lasts a few seconds (Charlton et al., 1982; Zucker et al., 1991). This increment in $[Ca^{2+}]$ is a small fraction of the typical resting $[Ca^{2+}]$ of 100 nM. How can such a tiny change in $[Ca^{2+}]$ trigger a massive synchronous release of quanta, and why is secretion so brief compared with the duration of the $[Ca^{2+}]$ change? As mentioned earlier, postsynaptic responses begin only 0.5 ms after an action potential invades nerve terminals. This synaptic delay includes the time taken for Ca^{2+} channels to begin to open after

the peak of the action potential (300 μs) (Llinás et al., 1981), leaving only about 200 μs after that for transmitter secretion and the start of a postsynaptic response. At this time, Ca^{2+} has barely begun to diffuse away from Ca^{2+} channel mouths. In an aqueous solution, $[Ca^{2+}]$ would be confined mainly to within 1 μm of channel mouths estimated roughly from the solution of the diffusion equation for a brief influx of M moles of Ca^{2+},

$$[Ca^{2+}] = \frac{M}{8(\pi Dt)^{3/2}} e^{-r^2/4Dt}$$

where t is time after the influx, r is distance from the channel mouth, and D is the diffusion constant for Ca^{2+}, $\sim 6 \times 10^{-6}$ cm^2 s^{-1}. In the cytoplasm, Ca^{2+} diffusion is retarded by intracellular organelles and the presence of millimolar concentrations of fast-acting protein-associated Ca^{2+} binding sites with an average dissociation constant of a few micromolar. Together, these effects restrict Ca^{2+} microdomains to about 50 nm around channel mouths.

Furthermore, the 200 μs preceding the postsynaptic response must include not only the time required for Ca^{2+} to reach its target but also the time required for Ca^{2+} to bind and initiate exocytosis and for transmitter to diffuse across the synaptic cleft, bind to receptors, and begin to open channels. Thus, the presynaptic Ca^{2+} targets must be located within a few tens of nanometers of Ca^{2+} channel mouths. Neuromuscular junctions that are fast-frozen during the act of secretion show vesicle fusion images in freeze-fracture planes of the presynaptic membrane about 50 nm from intramembranous particles thought to be Ca^{2+} channels (see Fig. 8.1C). Solution of the diffusion equation for a steady-point source of Ca^{2+} influx in the presence of a nearly immobile fast-binding Ca^{2+} buffer reveals that approximately 100 μs after a Ca^{2+} channel opens, $[Ca^{2+}]$ increases to more than 10 mM at 50 nm from its source and to more than 100 μM at a distance of 10 nm (Llinás et al., 1981).

This calculation considers only what happens in the neighborhood of a single open Ca^{2+} channel. However, when individual Ca^{2+} channels are labeled with biotinylated ω-conotoxin tagged with colloidal gold particles, more than 100 channels per active zone are seen in terminals of chick parasympathetic ganglia (Haydon et al., 1994). Any vesicle docked at such an active zone is likely to be surrounded by as many as 10 Ca^{2+} channels within a 50 nm distance. Even though not all these channels will open during each action potential, more than one channel is likely to open, so a vesicle will be influenced by Ca^{2+} entering through several nearby channels. At the squid giant synapse, more than 50 channels open in each ~ 0.6

µm² active zone, whereas 10 channels open within the more compact active zones of frog saccular hair cells (Roberts, 1994; Yamada and Zucker, 1992). The Ca²⁺ microdomains of these channels overlap at single vesicles, and they cooperate in triggering secretion of a vesicle. Calculations of diffusion of Ca²⁺ ions from arrays of Ca²⁺ channels in the presence of a saturable buffer indicate that the [Ca²⁺] at sites where neurotransmitter release is triggered may reach 10–200 µM or higher (Fig. 8.3).

Three indications that [Ca²⁺] in fact reaches very high levels in active zones during action potentials are

1. Highly localized [Ca²⁺] levels greater than 100 µM have been measured in presynaptic submembrane regions of squid giant synapses likely to be within active zones by using the low-affinity Ca²⁺-sensitive photoprotein n-aequorin-J (Llinás *et al.*, 1992).

FIGURE 8.3 Microdomains with high Ca²⁺ concentrations form in the cytosol near open Ca²⁺ channels and trigger the exocytosis of synaptic vesicles. (A) In this adaptation of a model of Ca²⁺ dynamics in the terminal, a set of Ca²⁺ channels is spaced along the *x* axis, as if in a cross section of a terminal. The channels have opened and, while they are open, the cytosolic Ca²⁺ concentration (*y*-axis) is spatially inhomogeneous. Near the mouth of the channel, the influx of Ca²⁺ drives the local concentration to as high as 800 µM, but within just 50 nm of the channel, the concentration drops off to about 25 µM. The channels are irregularly spaced but are often sufficiently close to one another that their clouds of Ca²⁺ can overlap and sum. (B) In the active zone (gray), an action potential has opened a fraction of the Ca²⁺ channels and microdomains of high cytosolic Ca²⁺ (pink) arise around these open channels as Ca²⁺ flows into the cell. In the rest of the cytoplasm, the Ca²⁺ concentration is at resting levels (0.10 µM), but within these microdomains, and particularly near the channel mouth, Ca²⁺ concentrations are much higher, as in (A). Synaptic vesicles docked and primed at the active zone may come under the influence of one or more of these microdomains and thereby be triggered to fuse with the membrane. (C) A few milliseconds after the action potential, the channels have closed and the microdomains have dispersed. The overall Ca²⁺ concentration in the terminal is now slightly higher (0.11 µM) than before the action potential. If no other action potentials occur, the cell will pump the extra Ca²⁺ out across the plasma membrane and restore the initial condition after several hundred milliseconds. (A) is adapted from Roberts, W.M. (1994); Localization of calcium signals by a mobile calcium buffer in frog saccular hair cells. *J. Neurosci.* **14,** 3246-3262. The numerical details are for hair cell synapses only.

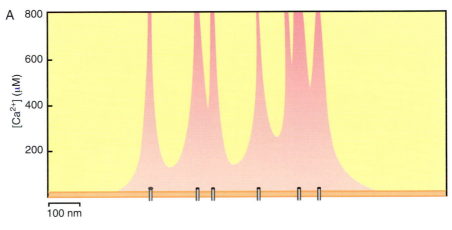

A

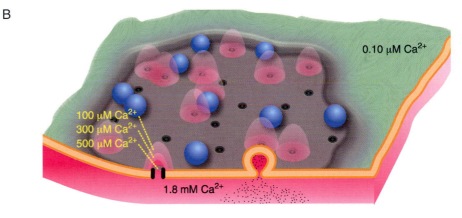

B

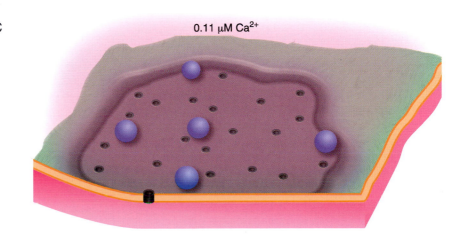

C

2. Estimates of [Ca^{2+}] based on the activity of Ca^{2+}-activated K$^+$ channels in active zones of mechanosensory hair cells are similar (Roberts, 1990).

3. At most rapidly transmitting synapses, transmitter release is blocked only by presynaptic injection of at least millimolar concentrations of fast high-affinity Ca^{2+} chelators, indicating that release is triggered locally by high concentrations of Ca^{2+} (Adler et al., 1991).

Vesicle Exocytosis Is Normally Triggered by Overlapping Ca^{2+} Channel Microdomains of High [Ca^{2+}]

Although release of a quantum of transmitter subsequent to the opening of a single presynaptic Ca^{2+} channel has been observed (Stanley, 1993), exocytosis may normally be due to Ca^{2+} entering through clusters of Ca^{2+} channels in active zones and contributing to local high [Ca^{2+}] at docked vesicles:

1. When transmitter release is increased under voltage clamp with pulses of increasing amplitude, a third-order power law relationship exists between presynaptic Ca^{2+} current and postsynaptic response (Augustine and Charlton, 1986). If each vesicle were released by Ca^{2+} entering through a single Ca^{2+} channel, then increasing depolarizations should recruit additional channel openings and proportionally more vesicle releases (Simon and Llinás, 1985). However, if Ca^{2+} channel microdomains from neighboring clustered Ca^{2+} channels overlap at docked vesicles, the [Ca^{2+}] at each vesicle will rise with increasing depolarization as more channels are recruited, and some cooperativity of Ca^{2+} action in triggering secretion will be expressed (Zucker and Fogelson, 1986).

2. In some neurons, more than one Ca^{2+} channel type contributes to secretion (Dunlap et al., 1995; Wheeler et al., 1994). When contributions of each channel type are isolated pharmacologically, their combined effects add nonlinearly, much as would be predicted by a fourth-order cooperativity, indicating that the Ca^{2+} microdomains of different channels overlap and summate at the Ca^{2+} sensors at vesicles docked within individual active zones.

3. At brain stem calyx of Held synapses (Borst and Sakmann, 1996), presynaptic injection of the slow-acting Ca^{2+} buffer ethylene glycol bis (b-aminoethyl ether)-N,N,N',N'-tetraacetic acid (EGTA) reduces transmission, indicating that the target for Ca^{2+} action is not particularly close to any one Ca^{2+} channel, but rather is affected by Ca^{2+} ions entering through many channels (Meinrenken et al., 2002).

4. When transmitter release is increased by prolonging presynaptic depolarizations (e.g., by broadening action potentials with K$^+$ channel blockers), more channels are not likely to be opened simultaneously. Rather, as channels that open early in the action potential close, others open; so the pattern of presynaptic Ca^{2+} microdomains is not so much intensified as prolonged, leading to a more nearly linear relationship between increases in Ca^{2+} influx and transmitter release (Zucker et al., 1991).

5. Large depolarizations admit little Ca^{2+} as they approach the Ca^{2+} equilibrium potential; they are therefore accompanied by a reduced Ca^{2+} current and reduced transmitter release during a pulse. However, large depolarizations can release more transmitter than can small depolarizations evoking a given macroscopic Ca^{2+} current (Augustine et al., 1985; Llinás et al., 1981). This apparent voltage dependence of transmitter release may be due to the different spatial profiles of [Ca^{2+}] in the active zone, with greater overlap of [Ca^{2+}] from the larger number of more closely apposed open Ca^{2+} channels during large depolarizations (Zucker and Fogelson, 1986).

The Exocytosis Trigger Must Have Fast, Low-Affinity, Cooperative Ca^{2+} Binding

The brevity of the synaptic delay implies not only that Ca^{2+} acts near Ca^{2+} channels to evoke exocytosis but also that Ca^{2+} must bind to its receptor extremely rapidly. This is confirmed by the finding that presynaptic injection of relatively slow Ca^{2+} buffers such as EGTA have almost no effect on transmitter release to single action potentials. Only millimolar concentrations of fast Ca^{2+} buffers such as 1,2-bis(2-amino-phenoxy)ethane-N,N,N',N'-tetraacetic acid (BAPTA), with on-rates of about 5×10^8 M^{-1} s^{-1}, can capture Ca^{2+} ions before they bind to the secretory trigger (Adler et al., 1991), indicating that the on-rate of Ca^{2+} binding to this trigger is similarly fast. At a rate of 5×10^8 M^{-1} s^{-1}, 10–100 μM [Ca^{2+}] reach equilibrium with its target in about 50–500 μs.

From the dependence of transmitter release on external [Ca^{2+}], it is known that at least four Ca^{2+}

ions cooperate in the release of a vesicle. The off-rate of Ca^{2+} dissociation from these sites also must be fast, at least 10^3 s^{-1}, to account for the rapid termination of transmitter release (0.25 ms time constant) after Ca^{2+} channels close and Ca^{2+} microdomains collapse. The high temperature sensitivity of the time course of transmitter release ($Q_{10} \approx 3$) indicates that exocytosis is rate limited by a step with a high-energy barrier (Yamada and Zucker, 1992). This step is likely to be the process of exocytosis itself. If Ca^{2+} binding is not rate limiting, its dissociation rate must be substantially faster than 10^3 s^{-1}. This means that the affinity of the secretory trigger for Ca^{2+} is low, with a dissociation constant (K_D) above 10 μM.

The Ca^{2+}-binding trigger is not saturated under normal conditions, because increasing $[Ca^{2+}]$ in the bath increases release. Furthermore, because of the speed with which Ca^{2+} binds to its sites, this reaction nearly equilibrates during the typical 0.5–1.0 ms that $[Ca^{2+}]$ remains high before Ca^{2+} channels close at the end of an action potential. If $[Ca^{2+}]$ reaches 10–100 μM or more in equilibrium with unsaturated release sites, the affinity of at least some of those sites binding Ca^{2+} must be ~10–100 μM or lower.

These predictions are consistent with experiments in which neurosecretion is triggered by photolysis of caged Ca^{2+} chelators such as DM-nitrophen. Partial photolysis of partially Ca^{2+}-loaded DM-nitrophen (Landò and Zucker, 1994) or addition of other Ca^{2+} buffers (Bollmann and Sakmann, 2005) generates a $[Ca^{2+}]$ "spike" of a duration similar to the lifetime of Ca^{2+} microdomains around Ca^{2+} channels opened by an action potential. This spike results in a postsynaptic response that closely resembles the normal EPSC at crayfish neuromuscular junctions, confirming that no presynaptic depolarization is necessary to obtain high levels of phasic transmitter release. Secretion depended on the fourth power of peak $[Ca^{2+}]$, and about 10–25 μM Ca^{2+} activated release at the same rate as an action potential.

In similar experiments on retinal bipolar neurons from fish (Heidelberger et al., 1994), fully loaded DM-nitrophen was photolyzed to produce a stepped increase in $[Ca^{2+}]$ while secretion was monitored as an increase in membrane capacitance, a measure of cell membrane area increased by fusion of vesicles. The Ca^{2+} concentration had to be raised by more than 20 μM before a fast phase of secretion developed. The sharp Ca^{2+} dependence of release and short synaptic delays were fitted by a model with a high degree of positive Ca^{2+} cooperativity, in which four successive Ca^{2+} ions bind with affinities increasing (or K_D decreasing) from 140 to 9 μM, followed by a Ca^{2+}-independent rate-limiting step. In contrast, at the calyx of Held

synapse, a $[Ca^{2+}]$ level of about 10 μM was sufficient to activate release at a rate similar to that in an EPSP (Schneggenburger and Neher, 2000), possibly without positive cooperativity (Bollmann et al., 2000). These experiments differentiate Ca^{2+} receptors triggering release at various synapses.

Calcium Ions Must Mobilize Vesicles to Docking Sites at Slowly Transmitting Synapses

Most peptidergic synapses and some synapses releasing biogenic amines display kinetics remarkably different from those of fast synapses. In these slower synapses, single action potentials often have no discernible postsynaptic effect. During repetitive stimulation, postsynaptic responses rise slowly, often with a delay of seconds from the beginning of stimulation, and persist just as long after stimulation ceases. Such slow responses are due to many factors: the postsynaptic receptors may have intrinsically sluggish second messengers or G proteins (Chapters 11 and 16); the postsynaptic receptors are often distant from release sites, so extracellular diffusion takes significant time; and release starts after the beginning of stimulation and continues after stimulation stops. Given these limitations, it is not surprising that single quanta are never discernible, either as spontaneous PSPs or as components of evoked responses.

The ultrastructural anatomy of presynaptic terminals of slowly transmitting synapses also is different from that of fast synapses. Transmitter is stored in large, dense core vesicles scattered randomly throughout the cytoplasm; vesicles do not tend to cluster at active zones or line up at the membrane, docked and ready for release (De Camilli and John, 1990; Leenders et al., 1999). Nevertheless, there is no doubt that transmitter is released from vesicles, because it is both stored and often synthesized in them, and they can be seen to undergo exocytosis during high-frequency stimulation causing high rates of release (Verhage et al., 1994).

A High-Affinity Calcium Binding Step Controls Secretion of Slow Transmitters

Calcium ions are required for excitation–secretion coupling in slow synapses, but the dependence of release on $[Ca^{2+}]$ is linear, in contrast with fast synapses (Sakaguchi et al., 1991). Furthermore, because few vesicles are predocked at active zones, most of those released by repetitive activity are not exposed to the local high $[Ca^{2+}]$ near Ca^{2+} channels. Thus, an important event triggered by Ca^{2+} influx in action potentials is likely to be the translocation of dense core vesicles to plasma membrane release sites, followed by exocytosis. This process has a very different dependence on $[Ca^{2+}]$ than does the release of docked vesicles. Measurements of $[Ca^{2+}]$ during stimulation indicate that release correlates well with $[Ca^{2+}]$ levels in the low micromolar range above a minimum, or threshold, level of a few hundred nanomolar (Lindau *et al.*, 1992; Peng and Zucker, 1993). Such a high affinity docking process may also be the rate-limiting step for secretion from some tonic ribbon synapses, such as photoreceptors (Thoreson *et al.*, 2004).

A striking difference between the release of fast transmitters, such as γ-aminobutyric acid (GABA) and glutamate, and peptide transmitters, such as cholecystokinin, was found in studies of *synaptosomes*, isolated nerve terminals prepared from homogenized brain tissue by differential centrifugation (Verhage *et al.*, 1991). When terminals were depolarized to admit Ca^{2+} through Ca^{2+} channels, the amino acid transmitters GABA and glutamate were released at much lower levels of bulk cytoplasmic $[Ca^{2+}]$ than when Ca^{2+} was admitted more uniformly and gradually across the membrane by use of the Ca^{2+}-transporting ionophore ionomycin. Peptides were released at the same low levels of $[Ca^{2+}]$ no matter which method was used to elevate $[Ca^{2+}]$. Thus, only amino acids were sensitive to the difference in $[Ca^{2+}]$ gradients imposed by the two methods and were preferentially released by local high submembrane $[Ca^{2+}]$ caused by depolarization. Apparently, peptides are released by a high-affinity rate-limiting step not especially sensitive to submembrane $[Ca^{2+}]$ levels.

Slow and Fast Transmitters May Be Co-Released from the Same Neuron Terminal

Some neurons have both small synaptic vesicles containing acetylcholine or glutamate and large, dense core vesicles containing neuropeptides (Lundberg and Hökfelt, 1986). Often, the two transmitters act on different targets. Single action potentials release only the fast transmitter, so different patterns of activity can have very different relative effects on the targets. For example, postganglionic parasympathetic nerves to the salivary gland release acetylcholine, which stimulates salivation, and vasoactive intestinal peptide, which stimulates vasodilation. Many examples of the co-release of multiple transmitters have been described.

Summary

Ca^{2+} acts as an intracellular messenger tying the electrical signal of presynaptic depolarization to the act of neurosecretion. At fast synapses, Ca^{2+} enters through clusters of channels near docked synaptic vesicles in active zones. It acts at extremely short distances (tens of nanometers) in remarkably little time (200 μs) and at very high local concentrations (10–100 μM), in calcium microdomains, by binding cooperatively to a low-affinity receptor with fast kinetics to trigger exocytosis. Some transmitters, such as peptides and some biogenic amines, are stored in larger, dense core vesicles not docked at the plasma membrane in active zones. Release of these transmitters, as well as their diffusion to postsynaptic targets and their postsynaptic actions, is much slower than that of transmitters such as acetylcholine and amino acids at fast synapses. Release of slow transmitters depends linearly on $[Ca^{2+}]$ and may be governed by a Ca^{2+}-sensitive rate-limiting step different from that triggering exocytosis of docked vesicles at fast synapses.

THE MOLECULAR MECHANISMS OF THE NERVE TERMINAL

To release neurotransmitter in response to an action potential, a synaptic vesicle must fuse with the plasma membrane with great rapidity and fidelity and thus the synapse requires an effective and well-regulated molecular machine. The mechanisms of the terminal must also include the means to load the vesicle with transmitter, to dock the vesicle near the membrane so that it can fuse with a short latency, to define a release site on the plasma membrane, and to restrict fusion to the active zone rather than other points on the surface of the terminal or axon. Additionally, a reserve of synaptic vesicles must be held near the active zone and those vesicles must be recruited to the plasma membrane as needed. The number of vesicles that are ready and waiting to fuse must be strictly determined, and the protein and lipid components of the vesicle must be recycled to form a new vesicle after fusion has occurred. For each of these processes, a molecular understanding remains incomplete, but rapid scientific progress in this field has made considerable headway.

A Cycle of Membrane Trafficking

Active neurons need to secrete transmitter in a constant, ongoing fashion. A bouton in the CNS, for example, may contain a store of 200 vesicles, but if it releases even one of these with each action potential and if the cell is firing at an unexceptional rate such as 5 Hz, the store of vesicles would be consumed within less than a minute. This calculation exemplifies the need for an efficient mechanism to recycle and reload vesicles within the terminal. Transport of newly synthesized vesicles from the cell body would be far too slow to support such a demand, and the axonal traffic needed to supply an entire arbor of nerve terminals would be staggering. As a consequence, only peptide neurotransmitters are supplied in this fashion, because they must be synthesized in the endoplasmic reticulum and then sent down the axon. Not surprisingly, therefore, vesicles with peptide transmitters are released at very low rates. For most neurotransmitters, however, the exocytosis of a synaptic vesicle is rapidly followed by its endocytosis and within approximately 30 s the vesicle is again available for release (Ryan and Smith, 1995). This pathway is sometimes referred to as the *exo–endocytic cycle* (Fig. 8.2). Moreover, each step in this cycle represents a potential control point for modulating the efficacy of the synapse. Modulation of the strength or fidelity of synaptic signaling, commonly known as *synaptic plasticity*, plays an important role both in the development of synaptic connections and in the functioning of the mature nervous system (see also Chapter 19). Indeed, regulation by Ca^{2+} and other second messenger systems is known to affect the docking, fusing, and recycling of vesicles at some synapses and is also likely to regulate the balance between reserve stores and those vesicles actively engaged in the exo–endocytic cycle (Beutner *et al.*, 2001; Dinkelacker *et al.*, 2000; Neher and Zucker, 1993; Smith *et al.*, 1998; Wierda *et al.*, 2007). Understanding the mechanisms of this modulation is an important goal and will certainly require a detailed understanding of the fundamental machinery itself. The question has been approached through the combined use of biochemical, genetic, and biophysical techniques.

Transmitter Release Is Rapid

As discussed earlier in this chapter, the delay between the arrival of an action potential at a terminal and the secretion of the transmitter can be less than 200 μs. This places some severe constraints on the fusion mechanism. Vesicles must already be present at the release sites, as there is no time to mobilize them from a distance. A catalytic cascade during fusion, such as that involved in phototransduction or in excitation–contraction coupling in smooth muscle, would also be far too slow for excitation–secretion coupling at the nerve terminal. Indeed, even a single

bimolecular catalytic step might be too slow for such short latencies. Models therefore favor the idea that a fusion-ready complex of the vesicle and plasma membrane is preassembled at release sites and that Ca^{2+} binding need only trigger a simple conformation change in this complex to open a pathway for the transmitter to exit the vesicle. Because the volume of the synaptic vesicle is small, the diffusion of transmitter from the vesicle proceeds almost instantaneously as soon as a pore has opened between the vesicle lumen and the extracellular space. This structure is referred to as the *fusion pore*, but its biochemical nature is unknown. Thus, the time-critical step comes between the influx of Ca^{2+} and the formation of the fusion pore. The complete merging of the vesicle and plasma membrane, if it occurs at all, can occur on a slower time course. The movement of the vesicle to the release site and any biochemical events that need to occur to reach the fusion-ready state can also be slower. These largely theoretical steps are sometimes referred to as "docking" and "priming." Docking and priming a vesicle cannot be too slow, however; a CNS synapse is estimated to have 2 to 20 vesicles in this fusion-ready state (Harris and Sultan, 1995; Stevens and Tsujimoto, 1995), and, therefore, if a synapse is to respond faithfully to a sustained train of action potentials, it must be able to replace the fusion-ready vesicles with a time course of seconds. The rate at which this occurs may determine some of the dynamic properties of the synapse.

The short latency of transmission would seem to preclude the involvement of ATP hydrolysis at the fusion step. This is born out by numerous physiological studies in which exocytosis persists after Mg^{2+}-ATP has been dialyzed from the cell and in which all the Mg^{2+} has been chelated (which would render any residual ATP inert to most enzymes) (Ahnert-Hilger *et al.*, 1985; Hay and Martin, 1992; Holz *et al.*, 1989; Parsons *et al*, 1995). Thus, if energy is needed to fuse the membranes, it should be stored in the fusion-ready state of the vesicle–membrane complex and released on addition of Ca^{2+}.

Transmitter Release Is a Cell Biological Question

Exocytosis and endocytosis are not unique to neurons; these processes go on in every eukaryotic cell. Moreover, exocytosis itself is only one representative of a general class of membrane-trafficking steps in which one membrane-bound compartment must fuse with another. Other examples would include the fusion of recycling vesicles with endosomes, transport from the endoplasmic reticulum (ER) to the Golgi, or transport from endosomal compartments to lysosomes. In each case, the same biophysical problem must be overcome, and the mechanisms for all these membrane fusion steps appear to have much in common.

For a vesicle to fuse with the plasma membrane, or any other target, there is a large energy barrier to surmount. To bring the lipid bilayers within a few nanometers of one another so that they can fuse, the hydration shell around the polar lipid head groups must be disrupted. Simply to split open each membrane so that the bilayer of one could be connected to the bilayer of the other would require exposing the hydrophobic core of each membrane to the

aqueous milieu of the cytoplasm, and this barrier is sufficiently great that it does not occur under normal conditions. Thus a specialized mechanism is required to bring the membranes close together and then drive fusion. At present, it appears that all the membrane-trafficking steps within the cell use a similar set of proteins to accomplish this task (Fig. 8.4). As is discussed later, homologues of proteins found at the synapse and known to be essential for exocytosis have also been shown to function in ER-to-Golgi transport in mammalian cells, in endosomal fusion, and in exocytosis in yeast. Indeed, it appears that representatives of this core set of proteins are present

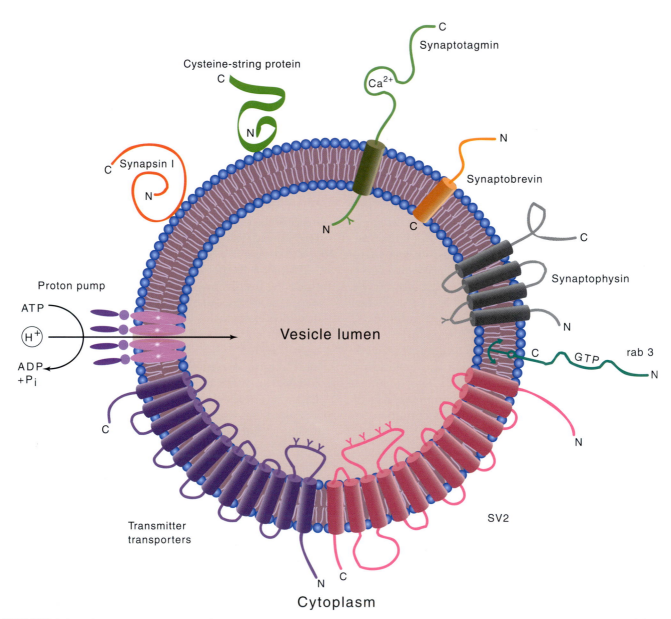

FIGURE 8.4 Schematic representation of the structure and topology of the major synaptic vesicle membrane proteins (see also Table 8.1). N, amino terminal; C, carboxy terminal.

on every trafficking vesicle or target membrane and required for every fusion step (Chen and Scheller, 2001). This discovery, which grew from the conjunction of independent studies of different model systems (Bennett and Scheller, 1993; Fischer von Mollard *et al.*, 1991), has led to the exciting hypothesis that all intracellular membrane fusions will be united by a single and universal mechanism. To the extent that this proves true, it will be a great boon to cell biology: experiments in one model system, for example, the highly developed genetic analysis of membrane fusion in yeast, can be absorbed into neuroscience. Similarly, the abundance of synaptic vesicles for biochemical analysis and the unparalleled precision of electrophysiological assays of single vesicle fusions can deepen the understanding of other cellular events. Will these disparate membrane fusion events be truly identical in their mechanisms? The jury is still out. Most likely, however, different membrane fusions will present variations on a common theme; the fundamental processes will be adapted to the specific requirements of each physiological step. Perhaps an instructive analogy may be drawn with muscle contraction (Leavis and Gergely, 1984): actin and myosin offer a fundamentally conserved mechanism for producing force and movement, and every muscle contains isoforms of actin, heavy and light myosin chains, troponins, and tropomyosins. But skeletal muscle and smooth muscle differ enormously in the details of how Ca^{2+} interacts with these proteins and regulates their activity. Regulation via troponins or via myosin light chain kinase, differences in the intrinsic ATPase rates of the myosins, and other critical features have adapted these related sets of proteins to their particular purpose in the individual type of muscle. Similarly, the need for extremely tight control of fusion in the nerve terminal—for rapid rates of fusion with very short latencies to occur in brief bursts at precise points on the plasma membrane—may cause some profound differences in how the synaptic isoforms of these proteins function compared with the isoforms involved in general cellular traffic.

Identifying the Synaptic Proteins: Vesicle Purification

The most important starting point in the elucidation of the machinery of the nerve terminal was the observation that synaptic vesicles can be purified. Synaptic vesicles are abundant in nervous tissue (Fig. 8.5) and, owing to their unique physical properties (uniform small diameter and low buoyant density), can be purified to homogeneity by simple subcellular fractionation techniques (Carlson *et al.*, 1978; Nagy *et al.*, 1976). As a result, at a biochemical level, synaptic vesicles are among the most thoroughly characterized organelles. One of the first sources for the purification of synaptic vesicles was the electric organ of marine elasmobranchs. This structure, a specialized adaptation of the neuromuscular junction, is highly enriched in synaptic vesicles. It has also proven simple to isolate synaptic vesicles from mammalian brain. The protein compositions of synaptic vesicles from these sources are remarkably similar, demonstrating the evolutionary conservation of synaptic vesicle function. This similarity also points to the fact that many of the proteins present on the synaptic vesicle membrane perform general functions that are not restricted to a single class of transmitter.

Our knowledge of many of the important proteins in vesicle fusion commenced with the purification and characterization of synaptic vesicles. These vesicles contained a discrete set of abundant proteins, and individual proteins could be isolated and subsequently cloned. This analysis has recently been elevated to a more comprehensive level by the application of mass spectrometry to purified vesicles (Takamori *et al.*, 2006). The major constituents of the synaptic vesicle are shown in Fig. 8.4. For some of these proteins there are well-established functions, but for others the functions remain uncertain (Table 8.1). The proton transporter, for example, is an ATPase that acidifies the lumen of the vesicle. The resulting proton gradient provides the energy by which transmitter is moved into the vesicle. This task is carried out by another vesicular protein, a vesicular transporter that allows protons to move down their electrochemical gradient (leaving the vesicle) in exchange for moving the transmitter into the lumen. Several vesicular transporters, all structurally related, are known, and by their substrate selectivity they help to specialize a terminal for release of the appropriate transmitter (see later). Two other large proteins with multiple transmembrane domains are abundant in vesicles: SV2 (synaptic vesicle protein 2) and synaptophysin. The functional significance of these proteins remains elusive despite extensive biochemical and genetic characterization (Bajjalieh *et al.*, 1992; Brose and Rosenmund, 1999; Buckley *et al.*, 1997; Crowder *et al.*, 1999; Custer *et al.*, 2006; Feany *et al.*, 1992; McMahon *et al.*, 1996; Südhof *et al.*, 1987). Other vesicular proteins are discussed later in this chapter.

Transmitter release depends on more than just vesicular proteins; proteins of the plasma membrane and cytoplasm are also important. In many cases, the identification of these additional components or an appreciation of their importance to the synapse

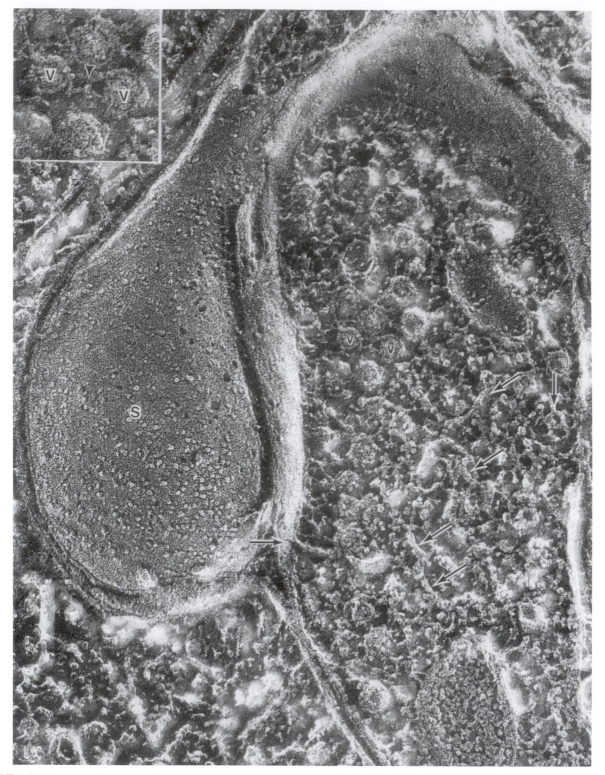

FIGURE 8.5 Structure of a synapse between a parallel fiber and a Purkinje cell spine (S) in the cerebellum. The sample was rapidly frozen and then freeze-fractured, shallow-etched, and rotary-shadowed to reveal the details of the synaptic architecture. Inset shows synaptic vesicles (V), and arrows indicate actin attachments. From Landis *et al.* (1988), *Neuron* 1: 201–209. Copyright © by Cell Press.

TABLE 8.1   Function of Synaptic Vesicle Proteins

Protein	Function
Proton pump	Generation of electrochemical gradient of protons
Vesicular transmitter transporter	Transmitter uptake into vesicle
VAMP/ synaptobrevin	Component of SNARE complex; acts in a late, essential step in vesicle fusion
Synaptotagmin	Ca^{2+} binding; possible trigger for fusion and component of vesicle docking at release sites via interactions with SNARE complex and lipid; promotes clathrin-mediated endocytosis by binding AP-2 complex
Rab3	Possible role in regulating vesicle targeting and availability
Synapsin	Likely to tether vesicle to actin cytoskeleton
Cysteine string protein	Promotes reliable coupling of action potential to exocytosis
SV2	Unknown
Synaptophysin	Unknown

derived from investigations of vesicular proteins. Two examples serve to illustrate this point. The first example begins with synaptotagmin, an integral membrane protein of the synaptic vesicle that was purified from synaptic vesicles and cloned (Bixby and Reichardt, 1985; Perin *et al.*, 1990). The portion of synaptotagmin that extends into the cytoplasm (the majority of the protein; see Fig. 8.4) was subsequently used for affinity column chromatography. In this manner, a protein called *syntaxin* was identified as a synaptotagmin-binding protein (Bennett *et al.*, 1992). This protein resides in the plasma membrane and is now appreciated as one of the critical players in vesicle fusion (see later). More recently, biochemical experiments have identified a new protein, called *complexin,* that binds to a syntaxin-containing complex. Complexin, as discussed later, appears to have a crucial function in regulating the fusion of vesicles (Giraudo *et al.*, 2006; McMahan *et al.*, 1995). Thus, subsequent to the isolation of an abundant vesicular protein, biochemical assays have led to a fuller picture. A second and similar example began with the realization that an abundant small GTP-binding protein, rab3, was present on the vesicle surface (Südhof *et al.*, 1987). This protein, discussed further later, may regulate vesicle availability or docking to release sites. Subsequently, rabphilin was identified on the basis of its affinity for the GTP-bound form of rab3 (Shirataki

et al., 1993). Rabphilin lacks a transmembrane domain but is recruited to the surface of the synaptic vesicle by binding to rab3. Rabphilin may have a role in modulating transmission, particularly in mossy fiber terminals of the hippocampus (Lonart and Südhof, 1998). Subsequently, many additional rab3-binding proteins have been identified, including RIM (rab3-interacting molecule), a component of the active zone (Dresbach *et al.*, 2001; Wang *et al.*, 1997). Rab3 can exist in either GTP- or GDP-bound states, and additional factors that regulate these states were also identified: a GDP dissociation inhibitor (GDI), GDP/GTP exchange protein (GEP), and GTPase-activating protein (GAP) are found in the synaptic cytosol (Südhof, 1997). Thus, from the identification of a synaptic vesicle protein, an array of additional factors have come to light, all of which are likely to figure in the exo–endocytic cycle.

Genetics Identifies Synaptic Proteins

Genetic screens have provided an independent method for identifying the machinery of transmitter release. One of the most fertile screens was carried out not in the nervous system *per se*, but rather in yeast. Because membrane trafficking in yeast is closely parallel to vesicle fusion at the terminal, mutations that alter the secretion of enzymes from yeast can be a springboard for the identification of synaptic proteins. In the early 1980s a series of such screens were carried out (Novick *et al.*, 1980, 1981) and a collection of more than 50 mutants was obtained. In many of these mutants, post-Golgi vesicles accumulated in the cytoplasm and thus the mutation appeared to block a late stage of transport, such as the targeting or fusion of these vesicles at the plasma membrane. Screens for suppressors and enhancers of these secretion mutations uncovered further components (Aalto *et al.*, 1993). Subsequently, excellent *in vitro* assays have been established in which to study the fusion of vesicles derived from yeast with their target organelles (Conradt *et al.*, 1994; Mayer *et al.*, 1996; Wickner and Haas, 2000). Among the secretion mutants and their interacting genes were homologues of some of the proteins discussed previously: sec4 encodes a small GTP-binding protein like rab3, and Sso1 and Sso2 encode plasma membrane proteins that are homologues of syntaxin (Aalto *et al.*, 1993; Salminen and Novick, 1987). The sec1 gene encodes a soluble protein with a very high affinity for Sso1, and the mammalian homologue of this protein, n-sec1, is tightly bound to syntaxin in nerve terminals and has an essential function in transmission (Aalto *et al.*, 1991; Pevsner *et al.*, 1994). Yeast is not the only organism in which a genetic screen uncovered an important protein for the synapse: the unc–13 mutation of *Caenorhabditis elegans*, for example, identified a component of the active zone membrane that may be important in priming vesicles for fusion and in the modulation of the synapse (Maruyama and Brenner, 1991; Richmond *et al.*, 1999; Wierda *et al.*, 2007).

Genetics has further contributed to the understanding of synaptic proteins by allowing tests of the significance of an identified protein for synaptic transmission. Such studies have been carried out in *C. elegans, Drosophila*, and mice and can reveal either an absolute requirement for the protein, as in the case of syntaxin mutants in *Drosophila* (Burgess *et al.*, 1997; Schulze *et al.*, 1995), or relatively subtle effects, as in the case of rab3 or SV2 mutations in mice (Custer *et al.*, 2006; Schluter *et al.*, 2004).

From biochemical purifications, *in vitro* assays, genetic screens, and fortuitous discoveries, an ever-growing list of nerve terminal proteins has been assembled (Table 8.1 and Table 8.2). The manner in which these proteins coordinate the release of transmitter as well as all the other cell biological functions of the exo–endocytic cycle remains uncertain, but a consensus has emerged in recent years that puts one set of proteins at the core of the vesicle fusion.

The Mechanism of Membrane Fusion: SNAREs and the Core Complex

Three synaptic proteins, VAMP (vesicle-associated membrane protein)/synaptobrevin, syntaxin, and SNAP-25 (synaptosome-associated protein of 25 kDa), are capable of forming an exceptionally tight complex that is generally referred to as either the *core complex* or *SNARE complex* (Söllner *et al.*, 1993; Sutton *et al.*, 1998). The interaction of these three is essential for synaptic transmission and likely to lie very close to or indeed at the final fusion step of exocytosis (Fig. 8.6 and Fig. 8.7). What are these proteins? VAMP (also called *synaptobrevin*) was among the first synaptic vesicle proteins to be cloned (Trimble *et al.*, 1988). It is anchored to the synaptic vesicle by a single transmembrane domain and has a cytoplasmic domain that contributes a coiled-coil strand to the core complex. Syntaxin has a very similar structure but is located primarily in the plasma membrane (though some is present on vesicles as well) (Bennett *et al.*, 1992). SNAP-25 is also a protein primarily of the plasma membrane but, unlike the others, lacks a transmembrane domain and is instead anchored in its central region by acylations (Chapman *et al.*, 1994). SNAP-25 contributes two strands to the SNARE complex. Thus the interactions of these proteins can be envisioned as

TABLE 8.2   Additional Proteins Implicated in Transmitter Release

Protein	Function
Syntaxin	SNARE protein present on plasma membrane (and on synaptic vesicles to a lesser extent); forms core complex with SNAP-25 and VAMP/synaptobrevin; essential for late step in fusion
SNAP-25	SNARE protein present on plasma membrane (and on synaptic vesicles to a lesser extent); forms core complex with syntaxin and VAMP/synaptobrevin; essential for late step in fusion
Nsec-1/munc-18	Syntaxin-binding protein required for all membrane traffic to the cell surface; likely bound to syntaxin before and after formation of SNARE complex
Complexin	SNARE-binding protein; inhibits vesicle fusion(?)
Tomosyn	SNARE-binding protein; inhibits vesicle fusion
Snapin	Binds SNAP-25; associated with synaptic vesicles; unknown function
NSF	ATPase that can disassemble the SNARE complex; likely to disrupt complexes after exocytosis
α-SNAP	Cofactor for NSF in SNARE complex disassembly
unc-13/munc-13	Active zone protein; vesicle priming for release; modulation of transmission
Rabphilin	C2 domain protein; Ca^{2+}-binding protein; binds rab3 and associates with synaptic vesicle; modulation of transmission
DOC2	C2 domain protein; Ca^{2+}-binding protein; binds munc-18; unknown function
RIM1 and related proteins	Active zone proteins; bind rab3; modulation of transmission
Piccolo	Likely scaffolding protein to tether vesicles near active zone
Bassoon	Likely scaffolding protein to tether vesicles near active zone
Exocyst (sec6/8 complex)	Marks plasma membrane sites of vesicle fusion in yeast; synaptic role uncertain

bringing closely together the two membranes that are meant to fuse. The proteins of this complex are archetypes of a class of membrane-trafficking proteins collectively called the SNAREs. The vesicle-associated proteins, such as VAMP, are referred to as *v-SNAREs*, and those of the target membrane, such as SNAP–25 and syntaxin, are referred to as *t-SNAREs*.

A SNARE complex is found at each membrane-trafficking step within a eukaryotic cell (Fig. 8.7).

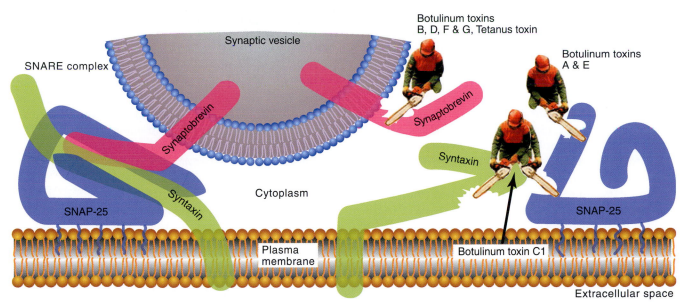

FIGURE 8.6 SNARE proteins and the action of the clostridial neurotoxins. The SNARE complex shown on the left brings the vesicle and plasma membranes into close proximity and likely represents one of the last steps in vesicle fusion. Vesicular VAMP, also called *synaptobrevin*, binds with the syntaxin and SNAP-25 that are anchored to the plasma membrane. Transmitter release can be blocked by tetanus toxin and the botulinum toxins—proteases that cleave specific SNARE proteins.

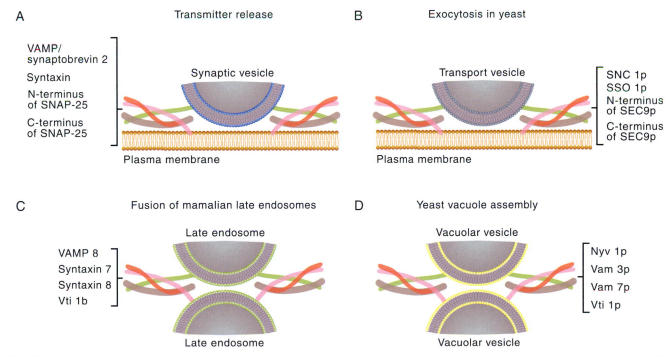

FIGURE 8.7 Neurotransmitter release probably shares a core mechanism with all membrane fusion events within eukaryotic cells. The fusion of synaptic vesicles (A) is driven by a particular complex of four coiled-coil domains contributed by three different proteins. Exocytosis in yeast (B), the fusion of late endosomes in mammalian cells (C), and the fusion of vacuolar vesicles in yeast (D) exemplify the closely related four-stranded coiled-coil complexes that are required to drive fusion in other membrane-trafficking steps.

Through a combination of SNARE proteins on the opposing membranes, a four-stranded coiled coil is formed. Alongside the example of the synapse in Fig. 8.7A, three analogous cases are shown from exocytosis in yeast (Fig. 8.7B), the fusion of late endosomes with one another (Fig. 8.7C) (Antonin et al., 2000), and the fusion of vesicles that form the yeast vacuole (Fig. 8.7D) (Sato et al., 2000; Wickner and Haas, 2000). Abundant genetic and biochemical data argue for an essential role for SNAREs, and the combination of yeast genetics, in vitro assays, synaptic biochemistry, and synaptic physiology has had a synergistic effect in advancing the field. It was the discovery that secretory mutations in yeast were homologues to the synaptic SNAREs that provided some of the first functional data on these proteins (Bennett and Scheller, 1993), while data on synapses first placed the proteins on vesicles and the plasma membrane (Bennett et al., 1992; Chapman et al., 1994; Trimble et al., 1988). Exploring the mechanism of the SNAREs similarly brought together neurons and yeast. The crystal structure, for example, was solved first for the synaptic proteins (Sutton et al., 1998). For crystallization, however, the transmembrane domains and, of course, the membranes themselves were absent. Thus it is not possible to conclude from the structure alone that the SNAREs formed a bridge between the membranes rather than complexes within the same membrane. In an in vitro assay with purified yeast vesicles, however, it was possible to establish this point (Nichols et al., 1997).

Some of the strongest evidence for an essential role for SNAREs at the synapse has come from the study of a potent set of eight neurotoxins produced by clostridial bacteria. These toxins (tetanus toxin and the family of related botulinum toxins) have long been known to block the release of neurotransmitter from the terminal. The discovery that they do so by proteolytically cleaving individual members of the SNARE complex (Link et al., 1992; Schiavo et al., 1992) provided neurobiologists with a set of tools with which to probe SNARE function (Fig. 8.6). Each of the toxins comprises a heavy chain and a light chain that are linked by disulfide bonds (Pellizzari et al., 1999). The heavy chain binds the toxin to surface receptors on neurons and thereby enables the toxin to be endocytosed. Once inside the cell, the disulfide bond is reduced and the free light chain enters the cytoplasm of the cell. This light chain is the active portion of the toxin and is a member of the Zn^{2+}-dependent family of proteases. The catalytic nature of the toxin accounts for its astonishing potency; a few tetanus toxin light chains, for example, at a synapse can suffice to proteolyze all the VAMP/synaptobrevin and thereby shut down transmitter release. The toxins are highly specific, recognizing unique sequences within an individual SNARE protein as summarized in Fig. 8.6. VAMP/synaptobrevin can be cleaved not only by tetanus toxin but by the botulinum toxin types B, D, F, and G, and each toxin cleaves at a different peptide bond within the structure (Schiavo et al., 1992; Yamasaki et al., 1994). SNAP-25 is cleaved by botulinum toxins A and E, but again at different sites (Blasi et al., 1993a; Schiavo et al., 1993). Botulinum toxin C1 cleaves both syntaxin (Blasi et al., 1993b) and, less efficiently, SNAP-25 (Foran et al., 1996).

What precisely is the function of the SNARE proteins in promoting transmitter release? How does the assembly of this complex relate to membrane fusion? Studies with botulinum toxins, yeast mutants, mutants of Drosophila and C. elegans, and permeabilized mammalian cells all place the SNAREs late in the process. Synapses that lack an individual SNARE, for example, synapses whose VAMP or syntaxin have been mutated or cleaved by toxin, have the expected population of synaptic vesicles, and these vesicles accumulate at active zones in the expected manner. There is, however, disagreement as to whether these vesicles are "docked" in a tight association with the plasma membrane in the absence of the SNARE complex (Hammarlund et al., 2007; Hunt et al., 1994). Despite the presence of vesicles, these synapses are incapable of secreting transmitter. Thus the SNAREs appear to be essential for a step concurrent either with or after the docking of the vesicle at the release site, but before the fusion pore opens and transmitter can diffuse into the cleft. The details of how the complex functions, however, are less certain. One attractive model (Chen and Scheller, 2001) is that the energy released by the formation of this very high affinity complex is used to drive together the two membranes. A loose complex of the SNARES would form and then "zipper up" and pull the membranes together. This may correspond to the actual fusion of the membranes or, alternatively, to a priming step that requires a subsequent rearrangement of the lipids to open the pore that will connect the vesicle lumen to the extracellular space. Evidence from the fusion of yeast vacuolar vesicles (Peters and Mayer, 1998; Peters et al., 2001) implicates a distinct downstream step, regulated by Ca^{2+}/calmodulin and involving subunits of the proton pump, but whether or not this is true of neurons as well remains unknown. Vesicles consisting of only lipid bilayers and SNAREs can fuse in vitro, suggesting that no other proteins will be essential for the fusion step (Nickel et al., 1999).

One additional function may reside with the SNARE proteins: the identification of an appropriate target membrane (McNew et al., 2000; Rothman and

Warren, 1994). Within a cell, there are myriad membrane compartments with which a transport vesicle might fuse: How then is specificity achieved? The great diversity of SNAREs (Fig. 8.7) may account for some of this specificity because not all combinations of v- and t-SNAREs form functional complexes. This potential mechanism, however, is likely to be only a part of the story. Particularly in the nerve terminal, it appears that synaptic vesicles can find the active zone even in the absence of the relevant SNAREs (Broadie *et al.*, 1995; Hunt *et al.*, 1994). In addition, the t-SNAREs syntaxin and SNAP-25 can be present along the entire axon and thus are inadequate in explaining the selective release of transmitter at synapses and active zones (Garcia *et al.*, 1995; Sesack and Snyder, 1995).

NSF: An ATPase for Membrane Trafficking

At some point as vesicles move through their exo–endocytic cycle, energy must be added to the system. *N*-Ethylmaleimide-sensitive factor (NSF), an ATPase involved in membrane trafficking, is one likely source. NSF was first identified as a required cytosolic factor in an *in vitro* trafficking assay (Block *et al.*, 1988). The importance of NSF was confirmed when it was found to correspond to the yeast sec18 gene, an essential gene for secretion (Eakle *et al.*, 1988). NSF hexamers bind a cofactor called a-SNAP (soluble NSF attachment protein) or sec17, and this complex, in turn, can bind to the SNARE complex. When Mg-ATP is hydrolyzed, the SNARE complex is disrupted into its component proteins (Söllner *et al.*, 1993). Originally it was speculated that this disassembly of the complex might correspond to fusion—that the action of NSF might catalytically wrench the SNAREs so that the membranes were brought together. As discussed previously, however, a late role for ATP is unlikely in transmitter release, and *in vitro* SNAREs can stimulate fusion without NSF present. More recent models put NSF action well before or after the fusion step (Banerjee *et al.*, 1996; Mayer *et al.*, 1996). If SNARE complexes form between VAMP, syntaxin, and SNAP-25 all in the same membrane, these futile complexes can be split apart by NSF so that productive complexes bridging the membrane compartments can be formed. After fusion, the tight SNARE complex needs to be disrupted so that the VAMP can be recycled to synaptic vesicles while the other SNAREs remain on the plasma membrane. If, indeed, the energy of forming a tight SNARE complex is part of the energy that drives fusion, NSF, by restoring the SNAREs to their dissociated, high-energy state, will be an important part of the energetics of membrane fusion.

Docking and Priming the Vesicles for Fusion

Although SNAREs are thought to act late in the fusion reaction and fusion itself is an extremely rapid state, many preparatory and regulatory steps may precede the action of the SNAREs. These steps must tether the vesicle at an appropriate release site in the active zone and hold the vesicle in a fusion-ready state. These mechanisms remain among the most obscure at the synapse, but the list of proteins that may participate is growing. One example is the protein n-sec1 (also called munc18), the neuronal homologue of the product of the yeast sec1 gene. Mutations of this protein prevent trafficking in both yeast and higher organisms. This protein binds to syntaxin with a very high affinity and, when so bound, may hold syntaxin in a closed state that prevents syntaxin from binding to SNAP-25 or VAMP (Garcia *et al.*, 1994; Pevsner *et al.*, 1994). However, n-sec1 also can bind to syntaxin and the assembled SNARE complex (Dulubova *et al.*, 2007). It appears likely that n-sec1 serves two functions: it may promote membrane fusion by priming syntaxin and interacting with the SNARE complex during fusion, but n-sec1 may also be a negative regulator, keeping syntaxin closed and inert until an appropriate vesicle or signal displaces n-sec1 and allows a SNARE complex to form. Another synaptic protein, tomosyn, also serves as a negative regulator of SNARE complex formation. Tomosyn, a soluble protein, can bind to complexes of syntaxin and SNAP-25 and prevent the incorporation of VAMP/synaptobrevin into the complex and thereby inhibit release (McEwen *et al.*, 2006).

One of the most interesting of the proteins that are thought to regulate docking and priming via the SNARE complex is unc-13, a protein first discovered in *C. elegans* by virtue of its neurological phenotype when mutated (Maruyama and Brenner, 1991). Unc-13 and its mammalian homologs (Munc-13s) appear necessary in the transformation of syntaxin from a closed state to an open state that permits SNARE complex assembly (Hammarlund *et al.*, 2007; Richmond *et al.*, 2001). Unc-13 contains a binding site for diacylglycerol and for Ca^{2+}/calmodulin and has a critical role in the modulation of transmitter release by those second messengers (Junge *et al.*, 2004; Rosenmund *et al.*, 2002; Wierda *et al.*, 2007). Thus unc-13 is not only an important piece of the release machinery but also a control point for synaptic plasticity.

Rab3 is another protein for which a priming or regulatory role is often invoked at the synapse. This small GTP-binding protein, mentioned previously, is the homologue of the yeast sec4 gene product. In yeast, and at other membrane-trafficking steps within mammalian cells, rab proteins have essential roles

(Gorvel *et al.*, 1991; Salminen and Novick, 1987). They appear to help a vesicle to recognize its appropriate target and begin the process of SNARE complex formation. At the synapse, however, the significance of rab3 is still uncertain. Though it is clearly associated with synaptic vesicles, genetic disruption of all four isoforms of rab3 in mice causes only subtle alterations in synaptic properties (Schluter *et al.*, 2004). Whether this is due to additional, redundant rab proteins or whether the rab family has been relegated to a more minor role at the synapse remains to be determined.

Because synaptic vesicles dock and fuse specifically at the active zone, this region of the nerve terminal membrane must have unique properties that promote docking and priming. The special nature of this domain is easily discernible in electron micrographs: a "fuzz" of electron-dense material can be observed opposite the postsynaptic density. In some synapses, such as photoreceptors, hair cells, and many insect synapses, the structures are more elaborate and include ribbons, dense bodies, and T-bars that extend into the cytoplasm and appear to have a special relationship with the nearby pool of vesicles. Recent advances in electron microscopy have allowed a more detailed look at the association of vesicles and plasma membrane at the active zone (Harlow *et al.*, 2001). At the neuromuscular junction of the frog (Fig. 8.1), it has been revealed that the electron-dense "fuzz" adjacent to the presynaptic plasma membrane is actually a highly ordered structure—a lattice of proteins that connect the vesicles to a cytoskeleton, to one another, and to the plasma membrane (Fig. 8.8). The molecules that correspond to these structures are not yet known. A few proteins, however, are known to be concentrated in the active zone or in the cloud of vesicles near the active zone. These proteins, piccolo, bassoon, RIM1, Cast/ELKS, and unc-13, may be a part of the machinery that defines the active zone as the appropriate target for synaptic vesicle fusion (Garner *et al.*, 2000; Ohtsuka *et al.*, 2002).

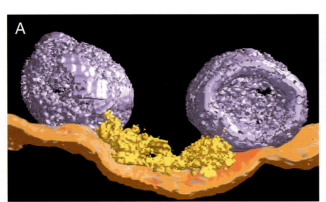

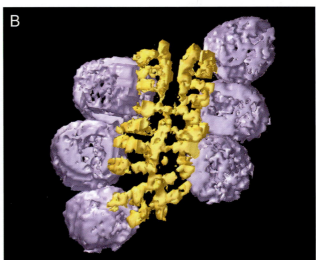

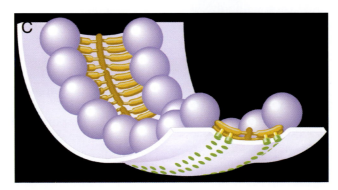

FIGURE 8.8 Fine structure of the active zone at a neuromuscular junction. (A) As revealed by electron tomography, synaptic vesicles (silver) are seen docked adjacent to the plasma membrane (tan) and associated with the proteins that make up the active zone (gold). (B) Viewed from the cytoplasmic side, the proteins are seen to extend from the vesicles and connect in the center. (C) Schematic rendering of an active zone based on the tomographic analysis. An ordered structure of ribs, pegs, and beams aligns the vesicles and connects them to the plasma membrane and to one another. Reproduced with permission from Macmillan Publishers Ltd. Harlow, M.L., Ress, D., Stoscheck, A., Marshall, R.M., and McMahan, U.J. (2001). The architecture of active zone material at the frog's neuromuscular junction. *Nature* **409**, 479–484.

In addition to the specializations of the active zone, additional machinery must be present to preserve a dense cluster of synaptic vesicles extending approximately 200 nm back from the active zone (Fig. 8.1). The vesicles in this domain are not likely to be releasable within microseconds of the arrival of an action potential but instead are likely to represent a reserve pool from which vesicles can be mobilized to release sites on the plasma membrane. The equilibrium between this pool and the vesicles actually at the membrane may be an important determinant of the number of vesicles released per impulse but remains poorly understood. One protein that is likely to play a role in the maintenance of the reserve pool is synapsin, a family of peripheral membrane proteins on synaptic vesicles (De Camilli et al., 1983; Südhof et al., 1989). Synapsins can also bind actin filaments and thus may provide a linker that tethers the vesicles in the cluster to the synaptic cytoskeleton (Fig. 8.5). Disruption of this link can cause the vesicle cluster to be diminished (Pieribone et al., 1995) and reduce the number of vesicles in the releasable pool of the terminal (Ryan et al., 1996). Synapsin has attracted considerable interest because it is the substrate for phosphorylation by both cAMP and Ca^{2+}-dependent protein kinases (Greengard et al., 1993). These phosphorylations may influence the availability of reserve vesicles for recruitment to release sites.

Coupling the Action Potential to Vesicle Fusion

The most striking difference between synaptic transmission and traffic between other cellular compartments is the rapid triggering of fusion by action potentials. As already discussed, the opening of Ca^{2+} channels and the focal rise of intracellular Ca^{2+} activate the fusion machinery. How does the terminal sense the rise in Ca^{2+}? What is the Ca^{2+} trigger and how does it open the fusion pore? Is it a single Ca^{2+}-binding protein or do several components respond to the altered Ca^{2+} concentration? Does Ca^{2+} remove a brake that normally prevents a docked, primed vesicle from fusing, or does Ca^{2+} induce a conformational change that is actively required to promote fusion? Whence does the steep, exponential relationship of release to intracellular Ca^{2+} arise? These questions are an active area of investigation and debate.

Some clues may come from other systems: Ca^{2+}-dependent membrane fusion is not unique to the synapse. Ca^{2+} can trigger both exocrine and endocrine secretion. Furthermore, Ca^{2+} released from intracellular stores now appears to be essential in trafficking steps in yeast that previously had been viewed as constitutive and unregulated. In yeast vacuolar fusion, for example, calmodulin senses a local increase in Ca^{2+} and triggers fusion (Peters and Mayer, 1998). Calmodulin, however, has not been the leading candidate for the synaptic trigger. The affinity of calmodulin for Ca^{2+} has generally been taken to be too high to explain the relatively high levels of Ca^{2+} (at least $10\,\mu M$) that are needed to evoke transmitter release. However, the apparent affinity of calmodulin for Ca^{2+} is very dependent on the proteins to which calmodulin binds, and the four Ca^{2+}-binding sites of free calmodulin probably have an average affinity of about $11\,\mu M$ (Jurado et al., 1999). Evidence for involvement of calmodulin in synaptic transmission continues to arise (Arredondo et al., 1998; Chamberlain et al., 1995; Chen and Scheller, 2001; Junge et al., 2004; Quetglas et al., 2000), suggesting that yeast vacuolar traffic and synaptic transmission may not be as divergent as one might think. At present, however, a modulatory role for calmodulin is favored over a requirement in the final triggering step at the synapse.

The leading candidate for synaptic Ca^{2+} sensor is synaptotagmin, an integral membrane protein of the synaptic vesicle (Fig. 8.4) (Brose et al., 1992; Geppert et al., 1994; Li et al., 1995; Perrin et al., 1991). Synaptotagmin has a large cytoplasmic portion that comprises two Ca^{2+}-binding C2 domains, called C2A and C2B. These domains can also interact with the SNARE complex proteins and with phospholipids in a Ca^{2+}-dependent manner. Consistent with the hypothesis that synaptotagmin is the trigger for vesicle fusion, mutations that remove the most abundant isoform of synaptotagmin, synaptotagmin 1, profoundly reduce synaptic transmission in flies, worms, and mice (Broadie et al., 1994; Di Antonio et al., 1993; DiAntonio and Schwarz, 1994; Geppert et al., 1994; Littleton et al., 1993; Nonet et al., 1993) while having little effect on or enhancing the rate of spontaneous release of transmitter. Mutations of the Ca^{2+}-binding residues of the C2B domain have a very potent effect in disrupting transmission (Mackler et al., 2002), more so than changes to the C2A domain (Fernandez-Chacon et al., 2002; Robinson et al., 2002), suggesting a particularly critical role for Ca^{2+}-binding at that site. However, the loss of synaptotagmin 1 does not completely abolish Ca^{2+}-evoked synaptic transmission (Broadie et al., 1994; Geppert et al., 1994; Nonet et al., 1993), indicating that either an unidentified additional synaptotagmin isoform drives this residual vesicle fusion or that an additional Ca^{2+} sensor is also involved. Synaptotagmin is likely to be involved in endocytosis and potentially in vesicle docking as well (Haucke and De Camilli, 1999; Jorgensen et al., 1995; Reist et al., 1998; Zhang et al., 1994), which has complicated the analysis. These processes, as mentioned previously, are also likely to be regulated by Ca^{2+}, and the Ca^{2+}-binding sites on synaptotagmin may be relevant for this regulation as well.

Fresh insight into how synaptotagmin promotes vesicle fusion has come from studies of complexin, a small cytoplasmic protein that can bind to SNARE complexes. Complexin appears to prevent membrane fusion when it is bound to the complex. In the presence of Ca^{2+}, however, and only in the presence of Ca^{2+}, synaptotagmin will displace complexin from the SNAREs and promote fusion. Thus the triggering event may involve the synaptotagmin-dependent removal of a brake (complexin) and subsequent interactions of synaptotagmin with both the SNAREs and membrane lipids that lead to the fusion of the membranes (Giraudo et al., 2006; McMahon et al., 1995).

Packaging Transmitter into the Vesicle

A central requirement of quantal synaptic transmission is the synchronous release of thousands of molecules of transmitter from the presynaptic nerve terminal. This requirement is partly met by the capacity of synaptic vesicles to accumulate and store high concentrations of transmitter. In cholinergic neurons, the concentration of acetylcholine within the synaptic vesicle can reach 0.6 M, more than 1000-fold greater than that in the cytoplasm (Carlson et al., 1978; Nagy et al., 1976). Two synaptic vesicle proteins mediate the uptake of transmitter: the vacuolar proton pump and a family of transmitter transporters. The vacuolar proton pump is a multisubunit ATPase that catalyzes the translocation of protons from the cytoplasm into the lumen of a variety of intracellular organelles, including synaptic vesicles (Nelson, 1992). The resulting transmembrane electrochemical proton gradient is used as the energy source for the active uptake of transmitter-by-transmitter transporters. Transmitter uptake has been characterized in isolated synaptic vesicle preparations in which at least four types of distinct transporters have been identified: one for acetylcholine, another for biogenic amines (catecholamines and serotonin), a third for the excitatory amino acid glutamate, and the fourth for inhibitory amino acids (GABA and glycine) (Edwards, 1992). At least one gene for each of these classes of transmitter transporter has now been cloned (Bellocchio et al., 2000; Reimer et al., 1998; Takamori et al., 2000). As expected, these distinct transporters are differentially expressed by neurons. The type of transporter expressed in a cell dictates the type of transmitter stored in the synaptic vesicles of a particular neuron, and when investigators drive the expression of a glutamate transporter, for example, in a GABA-releasing neuron, they can trick the cell into now releasing glutamate.

The vesicular transporters are integral membrane proteins with 12 membrane-spanning domains that display sequence similarity with bacterial drug resistance transporters. The synaptic vesicle transporters are clearly distinct from the plasma membrane transmitter transporters that remove transmitter from the synaptic cleft and thereby contribute to the termination of synaptic signaling (see Fig. 8.2 and Chapter 9). The distinguishing characteristics include their transport topology, energy source, pharmacology, and structure (Reimer et al., 1998).

In contrast to small chemical transmitters, proteinaceous signaling molecules, including neuropeptides and hormones, are typically stored in granules that are larger and have a higher electron density than synaptic vesicles. The contents of these granules are not recycled at the release sites; as a result, their replenishment requires new protein synthesis followed by packaging into secretory vesicles in the cell body. Because of the slow kinetics of their release, the slow responsiveness of their postsynaptic receptors, and their inability to be locally recycled, proteinaceous signaling molecules typically mediate regulatory functions.

Endocytosis Recovers Synaptic Vesicle Components

After exocytosis, the components of the synaptic vesicle membrane must be recovered from the presynaptic plasma membrane, as discussed previously. Vesicle recycling is accomplished by either of two mechanisms. The first is simply a reversal of the fusion process. In this case, a fusion pore opens to allow transmitter release and then rapidly closes to re-form a vesicle. Often nicknamed "kiss and run," it has the theoretical advantage that it would allow all the vesicular components to remain together on a single vesicle that would be immediately available for reloading with transmitter. The mechanism is potentially quick and energetically efficient. This mechanism is employed in some systems (Monck and Fernandez, 1992), but its relevance for the synapse is questionable and perhaps quite limited (Dickman et al., 2005; Granseth et al., 2006; Sankaranarayaran and Ryan, 2001).

The predominant pathway for synaptic vesicle recycling is more likely to be endocytosis. Endocytosis of synaptic vesicle components, like receptor-mediated endocytosis in other cell types, is mediated by vesicles coated with the protein clathrin (Granseth et al., 2006; Maycox et al., 1992). Accessory proteins select the cargo incorporated into these vesicles as they assemble. One class of accessory proteins known as AP–2 displays a high affinity for synaptotagmin (Zhang et al., 1994), which may be important, therefore, in recruiting the clathrin coat to the vesicle. The final pinching off of the clathrin-coated vesicle requires the protein dynamin, which can form a ringlike collar around the neck of an endocytosing vesicle (De Camilli *et al.*, 1995; Warnock and Schmid, 1996). A crucial role for dynamin in synaptic vesicle recycling is most clearly demonstrated in a temperature-sensitive *Drosophila* mutant known as *shibire*. The *shibire* gene encodes the *Drosophila* homologue of dynamin (Chen *et al.*, 1991; Van der Bliek and Meyerowitz, 1991). At the nonpermissive temperature, the *shibire* mutant flies rapidly become paralyzed owing to nearly complete depletion of synaptic vesicles from their nerve terminals. Dynamin is a GTPase whose activity is modulated by calcium-regulated phosphorylation and dephosphorylation (Robinson *et al.*, 1994). Thus, the endocytic recycling of synaptic vesicles provides another site at which calcium regulates the synaptic vesicle exo–endocytic cycle. When the components of the synaptic vesicle membrane are recovered in clathrin-coated vesicles, recycling is completed by vesicle uncoating and, perhaps, passage through an endosomal compartment in the nerve terminal (see Fig. 8.2). Other proteins, such as synaptojanin and endophilin, accelerate the process of retrieving vesicles by this classical pathway (Dickman *et al.*, 2005).

Summary

The life of the synaptic vesicle involves much more than just the Ca^{2+}-dependent fusion of a vesicle with the plasma membrane. It is a cyclical progression that must include endocytosis, transmitter loading, docking, and priming steps as well. In many regards, this cell biological process shares mechanistic similarities with membrane trafficking in other parts of the cell and with simpler organisms such as yeast. The interaction of the vesicular and plasma membrane proteins of the SNARE complex—VAMP/synaptobrevin, syntaxin, and SNAP–25—is an essential, late step in fusion. Many other proteins have been identified that are likely to precede the action of the SNAREs, regulate the SNAREs, and recycle them. Components of the synaptic vesicle membrane, the presynaptic plasma membrane, and the cytoplasm all contribute to the regulation of synaptic vesicle function. Together, these proteins build on the fundamental core apparatus to create an astonishingly accurate, fast, and reliable means of delivering transmitter to the synaptic cleft.

QUANTAL ANALYSIS

A quantitative description of the signal passing across a synapse is of utmost importance in understanding the function of the nervous system. This signal is the final output of all the integrative processes taking place in the presynaptic cell, and a complete statistical description should be able to capture the flow of information between neurons, as well as between neurons and effector cells. At many chemical synapses, a quantitative description is also a source of unique insight into the biophysics of transmission. The postsynaptic signal often fluctuates from trial to trial in a *quantal* manner; that is, it adopts preferred levels, which arise from the summation of various numbers of discrete events, thought to result from the release of individual vesicles of neurotransmitter (Katz, 1969, Martin, 1977). Examination of the trial-to-trial amplitude fluctuation of the synaptic signal allows the size of the *quantum*, as well as the average number of quanta released for a given presynaptic action potential, to be estimated. This approach is also a source of insight into the probabilistic processes underlying transmitter release from the presynaptic terminal and into the mechanisms by which transmission can be modified by physiological, pharmacological, and pathological phenomena.

Transmission at the Frog Neuromuscular Junction Is Quantized

The quantal nature of transmission was first demonstrated in the early 1950s by Bernard Katz and his colleagues, who studied the frog motor end plate. By recording from a muscle fiber immediately under a branch of the motor axon, they measured the postsynaptic potential both at rest and in response to stimulation of the axon. Spontaneous signals (mEPPs) were observed to occur at random intervals, measuring between 0.1 and 2 mV in amplitude (Fatt and Katz, 1952). Stimulating the presynaptic axon produced a postsynaptic signal up to 100 times larger. At any one site, the *spontaneous mEPPs* were of roughly the same amplitude, with a coefficient of variation of 30%, compatible with the intermittent release of multimolecular packets of transmitter from the presynaptic terminal. When the presynaptic axon was stimulated under conditions designed to depress transmitter release to very low levels, the evoked EPP fluctuated from trial to trial between preferred amplitudes, which coincided with integral multiples of the mEPP amplitude (Boyd and Martin, 1956; del Castillo and Katz, 1954; Liley, 1956) (see Fig. 8.9). This implied that the mEPP was a *quantal building block*, variable numbers of which were released to make up the evoked signal. The relative numbers of trials resulting in 0, 1, 2, . . ., quanta were well described by a Poisson distribution, a statistical distribution that arises in many instances where a random process operates (Box 8.4), implying that the process governing quantal release may also depend on a simple underlying mechanism.

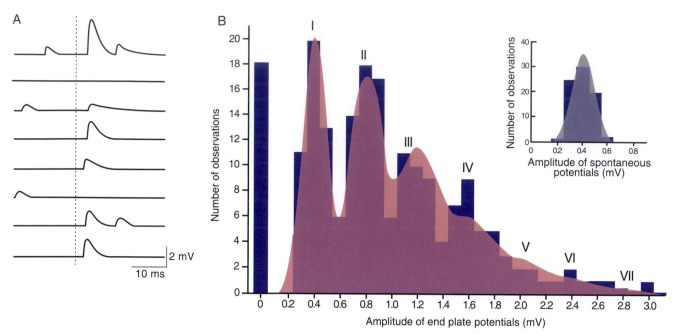

FIGURE 8.9 Quantal transmission at the neuromuscular junction. (A) Intracellular recordings from a rat muscle fiber in response to repeated presynaptic stimulation of the motor axon. Extracellular $[Ca^{2+}]$ and $[Mg^{2+}]$ were kept low and high, respectively, to depress transmission to a very low level. The size of the postsynaptic response, seen after the stimulus (not shown) that occurs at the dashed vertical line, fluctuated from trial to trial, with some trials giving failures of transmission. Spontaneous mEPPs, occurring in the background, had approximately the same amplitude as the smallest evoked EPPs, implying that they arose from the release of single quanta of acetylcholine. (B) The peak amplitudes of 200 evoked EPPs from a similar experiment, plotted as an amplitude histogram. Eighteen trials resulted in failures of transmission (indicated by the bar at 0 mV), and the rest gave EPPs whose amplitude tended to cluster at integral multiples of 0.4 mV. This coincides with the mean amplitude of the spontaneous mEPPs, whose amplitude distribution is shown in the inset together with a Gaussian fit. The shading through the EPP histogram is a fit obtained by assuming a Poisson model of quantal release (see Box 8.4). The parameters describing this model were the average number of quanta released, m, obtained by dividing the mean EPP amplitude by the mean mEPP amplitude, and the quantal amplitude, Q, and its variance, Var_Q, obtained from the mEPP amplitude distribution. There is a good agreement with the observed amplitude distribution. The Poisson model predicts 19 failures. Roman numerals indicate the number of quanta corresponding to each component in the distribution. (A) Adapted from Liley, A.W. (1956). The quantal components of the mammalian end-plate potential. *J. Physiol.* (B) from Boyd and Martin (1956).

On the basis of this evidence, Katz and colleagues proposed the following model of transmission, which has gained wide acceptance and is referred to as the *standard Katz model*.

- Arrival of an action potential at the presynaptic terminal briefly raises the probability of release of quanta of transmitter.
- Several quanta are available to be released, and every quantum gives roughly the same electrical signal in the postsynaptic cell. This is the quantal amplitude, Q, which sums linearly with all other quanta released.
- The average number of quanta released, m, is given by the product of n, the number of available quanta, and p, the average release probability: $m = np$. The relative probability of observing 0, 1, 2, ..., n quanta released is then given by a binomial distribution, with parameters n and p (see Box 8.4).

- Under conditions of depressed transmission, p is low, and the system approximates a Poisson process. This is the limiting case of the binomial distribution where p tends to 0 and n tends to ∞ and is determined by the unique parameter m (see Box 8.4).

The standard Katz model has been supported by similar experiments in other preparations, which have shown that evoked EPSPs, IPSPs, EPSCs, or IPSCs tend to cluster near preferred values corresponding to integral multiples of a unit. In many cases, this quantal amplitude corresponds closely to the amplitude of spontaneous miniature postsynaptic signals (potentials or currents) occurring in the absence of presynaptic action potentials (Fig. 8.10) (Paulsen and Heggelund, 1994).

A major impetus for accurate measurement of quantal parameters is that it may help determine the locus of modulatory effects on synaptic

BOX 8.4

BINOMIAL AND POISSON MODELS

According to the binomial description, n quanta can be released in response to a presynaptic action potential, each of which has a probability, p, of being discharged. For convenience, let us define q as the probability that a given quantum is not discharged in a given trial: $q = 1 - p$. In any given trial, the number of quanta observed is between 0 and n. Imagine that only three quanta are available ($n = 3$), each of which has a 40% chance of being discharged in response to the action potential ($p = 0.4$, $q = 0.6$). The average number of quanta released is $m = np = 1.2$.

The probability that no quanta are released is $q^3 = 0.216$.

The probability that only one quantum is released is the sum of the probabilities that only the first is released, only the second is released, and only the third is released: $3pq^2 = 0.432$.

The probability that two of three are released is, conversely, $3p^2q = 0.288$.

Finally, the probability that all three are released is $p^3 = 0.064$.

The relative probabilities of observing $0, 1, \ldots, n$ quanta are the coefficients of the expansion of $G = (p + qz)^n = 0.216 + 0.432z + 0.288z^2 + 0.064z^3$. This is known as the *generating function* for the binomial distribution, and z is simply a "dummy variable"; that is, it serves no function except to allow the binomial expansion. These coefficients can be obtained from the binomial distribution, which gives the probability of observing $0, 1, \ldots, n$ quanta released for any n and p. Writing P_x for the probability of observing x quanta, we obtain

$$P_x = \frac{n!}{(n-x)!x!}p^x q^{n-x}.$$

Poisson Model

As n becomes very large and p very small, the probability of observing $0, 1, \ldots,$ quanta is equally well described by a Poisson distribution. This is a limiting case of the binomial distribution when n tends to ∞ and p tends to 0, and, instead of two parameters (n and p), it is described by the sole parameter m. Again, m is the average value for P_x. The relative probability of observing $0, 1, \ldots,$ quanta is now given by

$$P_x = \frac{m^x e^{-m}}{x!}$$

For the same average quantal content m as in the binomial model, the relative probabilities of observing $0, 1, \ldots,$ quanta are now approximately

$$P_0 \cong 0.30,$$

$$P_1 \cong 0.36,$$

$$P_2 \cong 0.21,$$

$$P_3 \cong 0.08,$$

$$P_4 \cong 0.02,$$

.

.

.

Compound or Nonuniform Binomial Model

In the binomial model, what happens if different release sites have different, but still independent, probabilities? Let us take the following example: $p_1 = 0.1$, $p_2 = 0.2$, $p_3 = 0.9$ (again defining $q_1 = 1 - p_1$, $q_2 = 1 - p_2$, $q_3 = 1 - p_3$). The average quantal content is again $p_1 + p_2 + p_3 = 1.2$.

$$P_0 = q_1q_2q_3 = 0.07,$$
$$P_1 = p_1q_2q_3 + q_1p_2q_3 + q_1q_2p_3 = 0.67,$$
$$P_2 = q_1p_2p_3 + p_1q_2p_3 + p_1p_2q_3 = 0.23,$$
$$P_3 = p_1p_2p_3 = 0.01.$$

Generally, P_x are again obtained from the coefficients of the polynomial expansion of the generating function:

$$G = (p_1 + q_1z)(p_2 + q_2z)\ldots(p_n + q_nz)$$

$$= \Pi_k(p_k + q_kz), k = 1, \ldots, n.$$

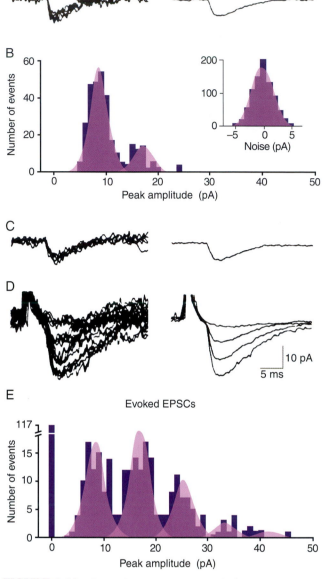

FIGURE 8.10 Quantal transmission in a thalamic neuron in a guinea pig brain slice. (A) Several spontaneous EPSCs superimposed (left), and the average time course (right). (B) An amplitude histogram showing that these events are clustered principally near 8.3 pA, with a smaller peak near 17 pA possibly representing the synchronous release of two quanta. (C) When presynaptic action potentials were abolished by tetrodotoxin, mean amplitudes of mEPSCs and spontaneous EPSCs were similar, implying that most were uniquantal. (D) EPSCs evoked by presynaptic stimulation in the optic tract. (E) Amplitude histogram of (D), showing clear clustering at integral multiples of approximately 8.3 pA. The superimposed Gaussian curves in (B) and (E) have approximately the same variance as the background noise, implying that quantal variability was negligible. From Paulsen and Heggelund (1994).

transmission. An increase or decrease in the probability of presynaptic transmitter release should be detected as a change in quantal content, *m*, whereas an alteration in the postsynaptic density or efficacy of receptors should be detected as a change in quantal amplitude, *Q*.

Biophysical Phenomena Underlie the Quantal Parameters

Considerable effort has been directed at establishing the physical correlates of the quantal parameters, *Q*, *n*, and *p*.

Quantal Parameter Q

A mEPP was thought to be too large to be accounted for by the release of an individual molecule of acetylcholine, because low concentrations of exogenous acetylcholine generated responses much smaller than an mEPP (Fatt and Katz, 1952). Vesicles were subsequently observed in electron micrographs of presynaptic terminals (see Chapter 1). Because these vesicles could act as packaging devices for transmitter, it was proposed that the quantal amplitude, *Q*, represents the discharge of one vesicle into the synaptic cleft. A large body of evidence (Box 8.1) has since led to almost universal agreement that *Q* is indeed the postsynaptic response to exocytosis of a single vesicle of transmitter.

Figure 8.11 (Bartol *et al.*, 1991) shows a computer simulation of the diffusion of molecules of acetylcholine and binding to receptors at the neuromuscular junction: acetylcholine spreads out in a disk in the synaptic cleft and binds most of the underlying postsynaptic receptors, although many more spare receptors beyond the edge of the disk remain unoccupied by transmitter (Matthews-Bellinger and Salpeter, 1973).

Quantal Parameter n

The physical correlate of *n* has been more elusive. In the original description of quantal transmission, *n* represented the number of releasable quanta. As the ultrastructure of the presynaptic terminal was elucidated, however, vesicles were found to be clustered near specializations, or active zones, in the presynaptic membrane. Each active zone is composed of a dense bar on the cytoplasmic face of the terminal membrane, bordered by rows of intramembranous particles, thought to be voltage-sensitive calcium channels (Heuser and Reese, 1973). As stated earlier, entry of calcium ions through these channels raises the intracellular calcium concentration in a small volume immediately adjacent to the channels, triggering the exocytosis. Can *n* then be the number of active zones, if they are release sites? At the frog neuromuscular junction, simultaneously evoked quanta seem to summate linearly; however, when acetylcholine is added to the bath the relationship between depolarization and acetylcholine concentration is nonlinear (Hartzell *et al.*, 1975). The discrepancy between the linear summation of quanta and the nonlinear dose dependency of the response to acetylcholine can be explained by proposing that the multiple quanta making up an EPP activate separate populations of

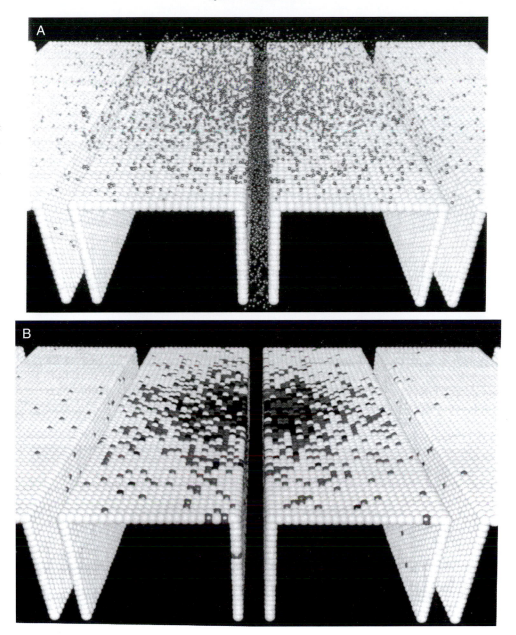

FIGURE 8.11 Monte Carlo simulation of quantal release of acetylcholine at the neuromuscular junction. The postsynaptic receptors are represented as a sheet of spheres—white if unliganded, gray if singly liganded, and black if doubly liganded. Presynaptic structures and the ends of the junctional folds are not shown, and the effect of acetylcholinesterase is not modeled. (A) Thirty microseconds after synchronous release of 9500 molecules of acetylcholine (small gray spheres) from a point source opposite the central fold. (B) Postsynaptic response, showing an effectively saturated area at the center, opposite the release site, surrounded by singly bound and unbound receptors. From Bartol *et al.* (1991).

receptors. This could occur if each active zone normally releases only one quantum and is consistent with the view that n is indeed the number of active zones.

Quantal Parameter p

Parameter p is the probability of exocytosis in response to a presynaptic action potential. Because this interval is of finite duration, it is more strictly a time integral of the probability during this transient event (Katz and Miledi, 1965). Moreover, because discharge of a quantum may leave a release site empty, p should be treated as a product of two probabilities: (1) that a release site is occupied by a quantum (p_1) and (2) that a presynaptic action potential evokes release (p_2) (Zucker, 1973).

With improved ultrastructural resolution of presynaptic terminals, it has become apparent that some vesicles are especially intimately related to the active zone. The number of such "docked" vesicles may represent a readily releasable pool (Schikorski and Stevens, 1997), which corresponds to the product of n and p_1.

The Standard Katz Model Does Not Always Apply

Before accepting the standard Katz model in all its details, we more closely examine some of its implications to see how they tally with our knowledge of the underlying molecular mechanisms.

Quantal Uniformity

At the vertebrate neuromuscular junction, the amount of acetylcholine released into the synaptic cleft determines the quantal size (Fletcher and Forrester, 1975; Kuffler and Yoshikami, 1975; Whittaker, 1988). Thus, for the quantal amplitude to be constant at different release sites, a uniform population of vesicles must be available to be released, and similar number of receptors with identical properties must exist opposite each release site. Electron microscopic images of vesicles in the presynaptic terminal indicate that their diameters are indeed remarkably uniform, although whether the neurotransmitter content of the vesicles is unvarying is not known. Similarly, although postsynaptic receptors are clustered opposite the active zone, their density and properties may not be uniform between different sites.

Uniform and Independent Release Probabilities

The release sites must be identical, with a uniform probability of exocytosis. If this condition were not satisfied, evoked signals would still cluster at integral multiples of the quantal amplitude, but the relative proportion of trials resulting in $0, 1, \ldots, n$ quanta would no longer be described by a simple binomial (or Poisson) distribution. As a limiting case, if p at some sites is effectively 0, then the meaning of n is questionable.

Rapid and Synchronous Transmitter Release

All-or-none exocytosis is clearly necessary for quantization and supported by freeze-fracture images of terminals taken during intense evoked release (Heuser *et al.*, 1979). However, all-or-none exocytosis may not be the only mode of transmitter release, because secretory vesicles in mast cells can release some of their contents through a fusion pore that opens reversibly without necessarily leading to full exocytosis (Alvarez de Toledo *et al.*, 1993). Whether this mode of release also occurs in synapses remains to be determined.

Quantization in the size of the evoked response can also be concealed by asynchrony of transmitter release from individual sites. (Isaacson and Walmsley, 1995).

Ion Channel Noise

Stochastic properties of postsynaptic ligand-gated ion channels must not add excessive variability to the size of the postsynaptic signal. Again, if this condition were not satisfied, it would be difficult to identify the quantal amplitude, and clustering of amplitudes at integral multiples of the quantum would be concealed. If we assume that individual ionophores act independently of one another, the variance of the quantal current arising from their stochastic opening is described by the binomial formula

$$Var = i^2 k p_o (1 - p_o),$$

where i, k, and p_o refer to the single-channel current, the number of channels, and their probability of opening in response to transmitter release, respectively. Because the average quantal current amplitude is ikp_o, the coefficient of variation of the quantal amplitude is

$$\sqrt{(1 - p_o)/kp_o}$$

A low quantal variability, which is required for quantal behavior to be detected, therefore implies either a large number of ionophores, k, or a high probability of opening, p_o, in response to transmitter release.

Postsynaptic Summation and Distortion of Signals

Postsynaptic currents or potentials arising from different release sites must sum linearly. If this is not satisfied, clustering of evoked postsynaptic signals may not occur at integral multiples of a quantal amplitude. If the postsynaptic membrane becomes appreciably depolarized as a result of the activation of many receptors, quanta may no longer summate linearly, either because the driving force for ion fluxes decreases or because voltage-gated channels open to cause regenerative currents to flow.

Stationarity

The state of the synapse must be relatively stable with time. A drift in the release probability with time could preclude a binomial or Poisson model, and changes in the quantal amplitude could prevent clear clustering in the distribution of evoked signals.

Thus, many of the requirements for the standard Katz model cannot realistically be expected to hold in all cases. In the presence of nonuniformity of release probability, the trial-to-trial amplitude fluctuation of the postsynaptic signal is unlikely to be described by a binomial or Poisson model. Indeed, the evidence that these simple probabilistic models are correct is far from compelling. On the other hand, the fact that evoked synaptic signals are often found clustered at integral multiples of an underlying unit strongly argues that vesicle filling and the postsynaptic phenomena determining the quantal amplitude are sufficiently uniform to ensure that a more general quantal description of transmission applies.

Central Nervous System Synapses Behave Differently from the Frog End Plate

A number of differences have emerged between quantal transmission at the neuromuscular junction and in central synapses in vertebrates.

One-Quantum Release

A correlation of histological and electrophysiological evidence obtained in the same preparation has led to the proposal that many individual terminals in the CNS have only one release site, which releases at most one vesicle at a time (Gulyas et al., 1993; Korn et al., 1981; Silver et al., 2003; Walmsley, 1991). There are some notable exceptions to this rule. Calyceal synapses in the mammalian brain stem auditory pathway, for instance, have multiple active zones, and glutamate released from one release site can interact with transmitter released from neighboring sites (Trussell et al., 1993). Other "giant" synapses in the CNS include those formed by climbing fibers on Purkinje cells, and by mossy fibers on pyramidal neurons in the CA3 region of the hippocampus. Evidence for multivesicular release has, however, also been obtained at small glutamatergic synapses, where the postsynaptic signal amplitude can, under certain conditions, vary with the release probability even when failures of release are excluded (Christie and Jahr, 2006; Oertner et al., 2002). Near-synchronous release of two vesicles of GABA has also been reported at small inhibitory synapses in the cerebellar cortex (Auger et al., 1998).

Nonuniform Release Probabilities

In the mammalian spinal cord, release probabilities may vary between individual sites supplied by an individual muscle afferent. Postsynaptic signals have been shown to fluctuate in a manner that cannot be described by a binomial model, unless the individual release probabilities are allowed to vary (Jack et al., 1981; Redman, 1990; Walmsley, 1991). Because the release sites are often segregated in different terminals, they may be subject to differing amounts of tonic presynaptic inhibition mediated by axo-axonic synapses.

Relatively Few Receptors Are Available to Detect Presynaptic Transmitter Release

A major difference between vertebrate CNS synapses and the neuromuscular junction is that the quantal amplitude is often determined not only by the vesicle contents but also by the number of available receptors (Edwards et al., 1990; Jonas et al., 1993; Korn and Faber, 1987, 1991; Redman, 1990; Walmsley, 1991). At some excitatory synapses, the glutamate content of a quantum appears to be sufficient to bind a large proportion of the available postsynaptic receptors (Clements et al., 1992; Tang et al., 1994; Tong and Jahr, 1994). Fewer than 100 receptors open, compared with 1000–2000 at the neuromuscular junction (Edwards et al., 1990; Jonas et al., 1993).

Different Receptors May Sample Different Quantal Contents

In contrast to the neuromuscular junction, CNS synapses frequently have several pharmacologically distinct postsynaptic receptors. Although little is known of quantal signaling via metabotropic receptors, glutamatergic synapses contain different combinations of AMPA, kainate, and N-methyl-D-aspartate (NMDA) receptors, all of which can open in response to glutamate release (see Chapter 11). The quantal content sampled by NMDA receptors is often larger than that sampled by AMPA receptors (Isaac et al., 1995; Kullmann, 1994; Liao et al., 1995). Because NMDA receptors are unable to open at hyperpolarized membrane potentials, such synapses are functionally silent in the absence of postsynaptic depolarization. This discrepancy in signaling by AMPA and NMDA receptors is partly explained by differential expression of receptors at synapses (Nusser et al., 1998), but, in addition, differences in the affinity of AMPA and NMDA receptors for glutamate may also play a role (Kullmann, 2003).

Spontaneous Miniature Postsynaptic Signals

In the central nervous system, spontaneous mEPSCs and mIPSCs vary widely in amplitude (Edwards et al., 1990; Jonas et al., 1993; Manabe et al., 1992) implying a quantal coefficient of variation considerably greater than that at the neuromuscular junction: between 40 and 80% instead of 30%. At first sight, this would preclude unambiguous peaks in histograms of evoked signals. However, quantal variability must be divided into variability from trial to trial at an individual release site (intrasite) and variability among sites (intersite). Spontaneous mEPSCs and mIPSCs arise from a large number of different sites, so their amplitude range includes both sources of quantal variability. If intrasite quantal variability were very large, then we would not expect to be able to detect quantal clustering in the amplitudes of evoked synaptic signals. The fact that such clustering is sometimes seen (see Fig. 8.10) implies that intrasite variability can be modest, and the wide range of amplitudes of miniature events principally indicates a large intersite variability. Thus, the mean amplitude of spontaneous miniature events cannot be used as a guide to the quantal amplitude underlying an evoked synaptic signal.

Quantal Parameters Can Be Estimated from Evoked and Spontaneous Signals

The goal is to establish, with a reasonable degree of precision, the quantal amplitude, Q, the average quantal content, m, and, if appropriate, the number of release sites, n, and the average release probability, p. The realization that release sites in the CNS may not always be uniform (Redman, 1990; Walmsley, 1991) makes a complete statistical description of transmission much more difficult to achieve. The parameters that need estimation must then include the release probability and the quantal amplitude and variability at each site. In principle, it should be possible to estimate the quantal parameters from the probability density of a statistic measured from the evoked postsynaptic signal. Most workers have measured the peak amplitude of the postsynaptic voltage or current on a large number of trials and displayed the results in the form of a histogram. Considerable information can also be obtained from the amplitude distribution of spontaneous miniature events. Two major obstacles, *sampling artifact* and *noise*, immediately arise.

Because the data sample is finite, the true probability density of a desired statistic is not known; only an approximation can be obtained from the recordings. This problem can be mitigated by obtaining a larger sample, but the sample is generally limited by nonstationarity in the recording and time constraints imposed by the experiment. Noise also conceals the true amplitude of spontaneous or evoked signals. If the noise amplitude is comparable to the quantal amplitude, then not only can spontaneous miniature events be missed, but features of the probability density of evoked signals also can be concealed. Relying entirely on visual inspection to determine whether the peaks and troughs in an amplitude histogram

are "genuine" (i.e., that they do not arise from sampling artifact and noise) is misleading. A number of different computational approaches has been developed to overcome this obstacle. These approaches differ in the degree to which they rely on assumptions about the underlying probabilistic process. Clearly, if the assumptions are incorrect, then nothing has been achieved, because the parameters will have been estimated incorrectly.

Spontaneous Miniature Signals

If Q can be obtained from the amplitude distribution of mEPSCs or mIPSCs, then the average quantal content, m, can be obtained by dividing the average evoked signal amplitude by Q. This can be done only when the amplitude distribution of spontaneous signals is narrow, and when there is no *a priori* reason that the quanta underlying the evoked signals should be different.

Spontaneous miniature signals can also be used to detect changes in quantal parameters caused by a conditioning treatment affecting a large number of synapses; if the average amplitude of mEPSCs or mIPSCs becomes larger after an experimental perturbation, the implication is that a widespread increase in quantal amplitude has been distributed among the synapses that give rise to the miniature currents. Changes in the frequency of spontaneous miniature events, on the other hand, usually imply an alteration in the average release probability (Van der Kloot, 1991; Van der Kloot and Molgó, 1994). An important difficulty with analysis of spontaneous miniature signals in CNS neurons is that their amplitude distribution is generally skewed, with a long tail toward larger values. At the other end of the distribution, small events often fall at the threshold for detection. To compare mEPSCs or mIPSCs obtained before and after a manipulation, cumulative distributions are generally easier to interpret than raw histograms (Van der Kloot, 1991). A genuine widespread change in amplitude is then seen as a shift in the position of the cumulative distribution, whereas a change in frequency should have no effect on the position of the line, other than that which can be accounted for by sampling artifact (Fig. 8.12). The Kolmogorov–Smirnov test can then be applied to test the hypothesis that any difference between the two curves arose by chance.

Multimodal amplitude distributions have occasionally been described for spontaneous miniature signals (Edwards *et al.*, 1990; Jonas *et al.*, 1993). When the modes are at equal intervals on the amplitude axis, multiquantal release, possibly arising from regenerative processes in the presynaptic terminal, is

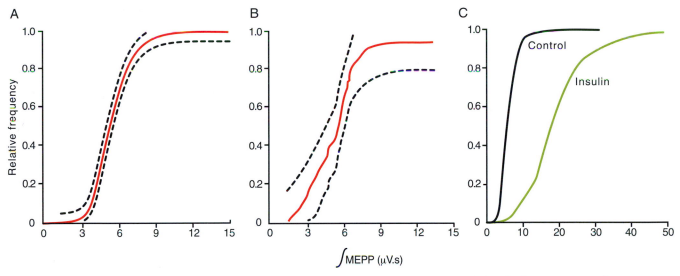

FIGURE 8.12   Cumulative distribution of mEPPs. (A) One thousand consecutive mEPPs recorded at the frog neuromuscular junction are plotted cumulatively. The dashed lines are the 95% confidence limits, calculated by applying Kolmogorov–Smirnov statistics. (B) The first 100 mEPPs are plotted in the same way, showing that, as the sample size is reduced, the confidence limits broaden. (C) Effect of insulin on the cumulative distribution. The curve is shifted to the right, indicating an increase in quantal amplitude. From Van der Kloot (1991).

implied, although sampling error as a source of spurious peaks must be ruled out.

The principal shortcoming of miniature spontaneous signal analysis is that, generally, the synapses giving rise to the spontaneous signals cannot be identified unambiguously, although localized application of hypertonic sucrose or barium can selectively increase the frequency of discharge in relatively restricted areas (Bekkers and Stevens, 1989; Fatt and Katz, 1952).

Quantal Amplitude Estimation

When Q cannot be estimated from the amplitude distribution of spontaneous mEPSCs or mIPSCs, it can often be obtained from the positions of the peaks in histograms of evoked signals. However, plotting data in the form of histograms can result in misleading estimates of Q in the presence of noise and finite sampling; spurious peaks and troughs can emerge, depending on the position and width of the bins. An approach for attacking this problem is to convolve the data with a *kernel*, generally a Gaussian function (Silverman, 1986). *Convolution* describes the mathematical equivalent of "smearing" one function across another. Gaussian kernel convolution has the effect of producing a smooth function with inflections that are no longer susceptible to *binning artifact*. The main pitfall of this approach is that the optimal width of the smoothing kernel cannot be known *a priori,* and, if it is too narrow, artifactual inflections, arising from noise and finite sampling, will still appear in the convolved function.

Whatever method is used to display the data, the question remains whether all the inflections in the resulting function could have arisen by chance because of finite sampling. A testable null hypothesis is that the underlying function is in fact continuous; that is, transmission is not quantal. This hypothesis can be modeled by choosing a unimodal distribution with the same overall shape as that of the data distribution, but without any peaks or troughs. If this hypothesis can be rejected at a given degree of confidence, the implication is that the clustering in the data sample does indeed indicate a genuine underlying quantized process. An easily implemented general method is to draw random samples repeatedly from the smooth function, with a sample size equal to that of the data. If the peaks and troughs in these random samples are never or only rarely as prominent as those in the data sample, then the null hypothesis can be rejected. This is an example of a *Monte Carlo test* (Horn, 1987).

Poisson Model

If the Poisson model holds, m can be estimated by counting the quanta released on each of a large number of trials (Isaacson and Walmsley, 1995). Alternatively, since the variance of a Poisson distribution is equal to its mean, m can be estimated from the trial-to-trial variability in the number of quanta released. However, it is rarely possible to count quanta unambiguously, principally because recordings are affected by noise and because the quantal amplitude has some intrinsic variability, and so indirect methods frequently have to be used to estimate m. The first of these is to count the proportion of trials that result in a failure of transmission (N_0 of N trials). The first term of the Poisson expansion (Box 8.4) gives the probability of observing zero quanta released and is equal to e^{-m}. Therefore, m is given by taking the natural logarithm of the inverse of this ratio: $m = \log_e(N/N_0)$. The second indirect approach to estimating m is to measure the coefficient of variation of the postsynaptic response, CV_R. CV_R must be corrected for two other sources of variability in the postsynaptic response: (1) variability in quantal size, expressed as the quantal coefficient of variation, CV_Q; and (2) background noise variance, Var_I, which can be measured separately by collecting data in the absence of evoked activity. If we assume that the variances arising from the Poisson process, quantal variability, and noise add linearly, m is then given by the formula

$$m = \frac{1 + CV_Q^2}{CV_R^2 - Var_I/\bar{R}^2},$$

where $\bar{R}$ is the average evoked response amplitude (McLachlan, 1978). The variance method of estimation is often used incorrectly when the necessary corrections for quantal size and noise fluctuations are ignored.

Agreement among the estimates of m obtained with all these methods constitutes circumstantial evidence in favor of the model.

Binomial Models

The binomial model has more parameters than the Poisson (n and p replace m). If the number of trials resulting in $0, 1, \ldots, n$ trials is known unambiguously, then n and p can be estimated as follows. From the binomial theorem, the variance of the number of quanta, Var_m is equal to $np(1 - p)$. It follows that $p = 1 - Var_m/m$ and $n = m/p$. If the number of quanta released cannot be estimated unambiguously, then p can be obtained from the proportion of failures of transmission. Because $N_0 = N(1 - p)^n$, it follows that $p = 1 - (N_0/N)^{1/n}$. The usefulness of this method is limited by the requirements that there be an appreciable proportion of failures and that n be known. Alternatively, the variance method may be used, again with an appropriate correction for the quantal variability and background noise (McLachlan, 1978):

$$p = 1 + CV_Q^2 - \frac{(\bar{R} \cdot CV_R^2 - Var_I/\bar{R})}{\bar{Q}}.$$

This method requires that estimates be made of both the average quantal size, $\bar{Q}$ and its coefficient of variation, CV_Q, severely limiting its usefulness.

If we relax the assumption of uniform release probabilities while continuing to assume that

different sites are independent of one another, then a nonuniform or compound binomial model must be applied (Jack et al., 1981). In this case, the desired parameters include the individual release probabilities: $p_1, p_2, \ldots, p_n$. If the sample were perfect, they could be obtained by treating the observed proportions of trials resulting in $0, 1, \ldots, n$ quanta as the polynomial expansion of $\Pi_k (p_k + q_K z)$ (see Box 8.4). Solving the polynomial would then yield $p_1, p_2, \ldots, p_n$ (Jack et al., 1981). In practice, however, the sample is incomplete; that is, some rare events may never have been observed, and others may be spuriously overrepresented. Root-finding algorithms generally yield complex roots in this situation. An alternative approach is to use a numerical optimization, that is, to find the release probabilities that give the best agreement with the data, taking into account the fact that the data sample is incomplete (Kullmann, 1989).

Noise Deconvolution

In many cases, the proportion of trials resulting in $0, 1, \ldots, n$ trials cannot be determined unambiguously because of excessive noise, and the assumption that Poisson or simple binomial statistics apply is untenable. How then can one resolve the underlying quantal process at the synapses under investigation? The method of noise deconvolution (Edwards et al., 1976) again relies on the assumption that noise adds linearly to the synaptic signal, which means that the sampled probability density function is a convolution of the underlying quantal density function with the noise density function. Because the noise can be measured independently, by recording the background signal in the absence of evoked synaptic activity, it should be possible to undo the convolution to reconstruct the probability density function that describes the underlying signal.

This operation is not trivial, because the evoked signal and noise samples are finite: their true probability density functions are not known, and only an approximation can be obtained from the measured signals. The underlying noise-free probability density function must therefore be estimated by applying an optimization method. The underlying function is generally assumed to comprise a number of discrete components representing different numbers of quanta released. The task is then formally equivalent to solving a *mixture problem*, in which the data are sampled from a mixture of overlapping distributions, or components, each having a membership (probability), mean amplitude, and variance that need to be estimated. Optimization algorithms work as follows: (1) the data distribution is compared with an initial solution reconvolved with the noise function; (2) the solution is then adjusted to improve the goodness of fit; and (3) the cycle is repeated until no further improvement is detectable. The best results are obtained by maximizing likelihood, and a robust and versatile algorithm to use for this purpose is known as the *expectation–maximization algorithm* (Kullmann, 1989; Stricker and Redman, 1994). A number of constraints can be imposed on the solution to accommodate physiological assumptions. As a rule, as more constraints are imposed, the quantal parameters are more accurately estimated, but only as long as the underlying assumptions are justified.

A major obstacle is that the number of components in the solution, which cannot generally be known *a priori*, is a critical parameter. As the number of parameters available to fit the data is increased, the maximum likelihood value increases, because finer details of the data distribution, many of which are due to sampling error and noise, can be accounted for. An alternative approach, which avoids the problem of overfitting, is to treat the underlying probability density function not as a mixture of discrete components but as a continuous function. The solution is biased toward the flattest, most featureless function that is just compatible with the data. This method, known as *maximum entropy noise deconvolution*, can give an estimate of quantal amplitude if there are periodic inflections in the solution (Kullmann and Nicoll, 1992).

Figure 8.13 shows the results of applying several deconvolution methods to an amplitude histogram.

Model Discrimination

An important goal of parameter estimation is to choose between different models of transmission. The simplest approach is to ask if a given model is able to fit the data, by applying a conventional goodness-of-fit test, such as the χ^2 (chi squared) test. If the fit is unsatisfactory, the model can tentatively be rejected with the corresponding degree of confidence. However, the model's being in good agreement with the data does not necessarily mean that the assumptions underlying the solution are correct, because many alternative models also may give adequate fits.

Confidence Intervals

Confidence intervals must be estimated for quantal parameters, as for any statistic. Such an estimation can be difficult for any but the simplest model because the parameter space has many dimensions, and even the number of dimensions is often unknown. Resampling methods that rely on repeating the optimization on a large number of random samples drawn from the original data set (Efron and

Tibshirani, 1993) must be used with caution, and it is important in all cases to be aware of the limitations and biases of optimization algorithms by testing them extensively with Monte Carlo simulations.

Quantal Analysis Can Shed Light on the Mechanisms of Modulation of Synaptic Transmission

As mentioned previously, a comparison of quantal parameters before and after a treatment that alters synaptic strength can potentially indicate how this alteration is expressed. Ideally, the mean release probability, number of release sites, and quantal amplitude could be estimated at different time points to determine how they change. In practice, it is often difficult to establish all these parameters unambiguously. This difficulty does not preclude separating changes in one parameter from changes in another parameter, because certain statistics that reflect the trial-to-trial variability of transmission change in characteristic ways. This approach is known as *variance analysis*.

Coefficient of Variation

With the use of the coefficient of variation of the evoked signal, the binomial and Poisson models allow inferences to be made about the site of modulation of transmission without the need to estimate the quantal parameters (McLachlan, 1978). To correct for the background noise variance, Var_I, the coefficient of variation of the underlying signal, CV_S, is given by

$$CV_S = \frac{\sqrt{Var_R - Var_I}}{\bar{R}}.$$

CV_S is determined by probabilistic quantal release as well as by the quantal variability and is a useful statistic because it is dimensionless. It can be used to distinguish between changes in quantal amplitude and changes in quantal content. Briefly, if CV_S changes with a conditioning treatment that alters the average amplitude of the postsynaptic signal, the implication is a change in quantal content. If, conversely, CV_S is unaffected, the implication is that the conditioning treatment altered quantal amplitude.

If a Poisson model is assumed, further information can be obtained by plotting the ratio of $1/CV_S^2$ before and after a manipulation against the corresponding ratio of mean amplitude $\bar{R}$ (Fig. 8.14) (Manabe *et al.*, 1993). Because the variance of a Poisson distribution is equal to its mean, a change in quantal content, m, should cause an excursion along the line of identity. A change in quantal amplitude, Q, on the other hand, should have no effect on $1/CV_S^2$, so the data points

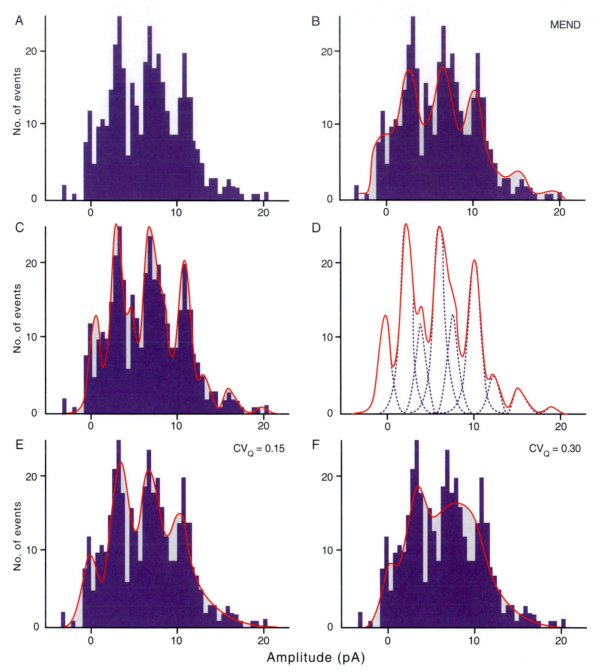

FIGURE 8.13 Noise deconvolution. (A) Amplitude histogram for 400 EPSCs recorded in a CA1 cell in a hippocampal slice in response to repeated stimulation of afferent fibers. EPSCs appear to cluster at integral multiples of approximately 3.6 pA. The continuous line in (B) is the maximum entropy noise deconvolution solution. This function, convolved with the noise, just fits the data at the 5% level of confidence (i.e., a curve any smoother and more featureless would have to be rejected at the 5% level). The periodic inflections seen in this function imply that clustering results not simply from noise, sampling, and binning artifact but also from an underlying quantal process. (C) The result of maximum likelihood deconvolution, with nine underlying components, each with the same variance as the background noise but with no constraint on their amplitudes and probabilities. The continuous line represents the solution reconvolved with the noise, showing a very good fit to the data. The underlying components (dashed lines) are plotted in (D), together with their sum (continuous line). Although the agreement of this solution with the data is excellent, some of its features may arise from sampling artifact and noise. (E, F) A quantal model has been fitted to the data; that is, the components have been constrained to occur at equal intervals, with the first component at 0, and with a quantal coefficient of variation (CV_Q) of 0.15 (E) or 0.3 (F). The maximum likelihood solution in (F), but not in (E), can be rejected at the 5% level, implying that $CV_Q < 0.3$.

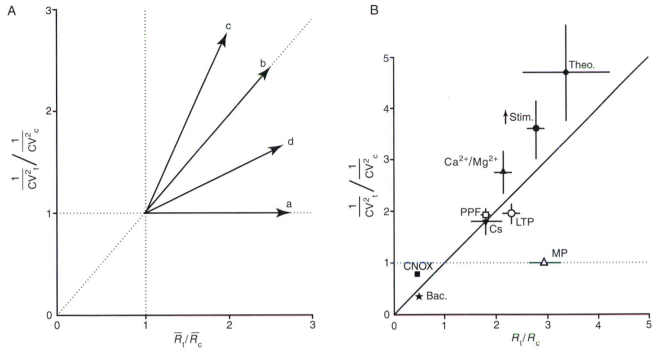

FIGURE 8.14 Coefficient of variation method for determining the site of modulation of synaptic transmission. (A) The possible excursions in the ratio of $1/CV_s^2$ plotted against the ratio of the mean response amplitude R for a manipulation that increases synaptic strength. Subscripts "c" and "t" refer to the control and test conditions, respectively. A horizontal excursion (a) implies an increase in quantal amplitude, Q, for instance, through a postsynaptic increase in the number of receptors. If a Poisson model applies, an excursion along the 45° line (b) implies an increase in m. If a simple binomial model holds, the ratio should fall on the 45° line for an increase in n and above it (c) for an increase in p. Ratios falling below the 45° line and above the horizontal (d) imply an increase in both quantal content and quantal amplitude. Conversely, modulations that decrease synaptic strength should cause the ratios to fall to the left of the point (1, 1) and below it. These rules apply only if a Poisson or simple binomial model holds. (B) Experimental results obtained by recording from CA1 hippocampal pyramidal cells with various modulations. Each point shows the mean of several cells (with standard errors). Increasing the driving force for the synaptic current by changing the postsynaptic membrane potential (MP) produces no change in $1/CV_s^2$ as expected from a purely postsynaptic modification. Extracellular theophylline or Cs^+, an increase in the extracellular $[Ca^{2+}]/[Mg^{2+}]$ ratio, and facilitation by a conditioning prepulse (PPF) cause the ratios to fall in region (c) of (A), as expected from an increase in p. Increasing the stimulus strength also causes the points to fall in region (c), although a simple binomial model predicts that an increase in n should cause the ratio to fall on the 45° line. Long-term potentiation of transmission causes the points to fall in region (d), implying an increase in both quantal content and quantal amplitude. Conversely, baclofen decreases transmitter release, and the glutamate receptor antagonist CNQX decreases quantal amplitude. (B) Reproduced with permission from Macmillan Publishers Ltd. Manabe, T., Renner, P., and Nicoll, R. (1992). Postsynaptic contribution to long-term potentiation revealed by the analysis of miniature synaptic currents. *Nature (London)* **355**, 50-55.

should fall on the horizontal line. If the points fall between the line of identity and the horizontal line, we can conclude that both quantal content and quantal amplitude changed. If a simple binomial model is assumed, then the results are slightly different, because the variance $np(1-p)$ is less than the mean np. A plot of the ratio of $1/CV_s^2$ against the ratio of $\bar{R}$ in this case falls on the line of identity for manipulations that increase n and above it for manipulations that increase p.

This method, although easy to use, depends heavily on the assumption that a simple binomial or Poisson model applies. As soon as this assumption is relaxed, a wide range of explanations can be put forward for virtually any outcome (Faber and Korn, 1992). The method is also sensitive to changes in the quantal coefficient of variation, so it must be assumed

that changes in Q are accompanied by proportional changes in $\sqrt{Var_Q}$. It is, however, often possible to test whether these assumptions are correct, by deliberately applying manipulations that are known to alter either n, or p, or Q. For instance, n can sometimes be altered by varying the number of presynaptic axons stimulated, while p can be manipulated in relative isolation by altering the extracellular $[Ca^{2+}]/[Mg^{2+}]$ ratio or by applying drugs known to act presynaptically. And Q can be scaled by applying low concentrations of postsynaptic receptor blockers or by manipulating the driving force for the synaptic current. This set of experiments establishes characteristic trajectories for a plot of ratios of $1/CV_s^2$ against ratios of $\bar{R}$, which may or may not coincide with those expected of binomial or Poisson models. These trajectories can then be compared with the effect of the new manipulation under investigation. If the observed ratio of $1/CV_S^2$ plotted against the ratio $\bar{R}$ runs along one of the trajectories established for manipulations of either n, p, or Q, then it can be inferred that an alteration in the corresponding parameter has occurred. Although this is a potentially powerful approach to verify the validity of variance analysis, it is still potentially flawed if the synaptic plasticity under investigation is not reproduced by any of the experimental alterations of n, p, or Q. A notable example is the controversy over the site of expression of long-term potentiation (LTP): although variance analysis consistently shows an increase in quantal content, implying a presynaptic alteration in n and/or p, an alternative explanation that has received substantial support from alternative methods is that postsynaptic receptor clusters are uncovered at previously silent sites (see Chapter 19). That is, a postsynaptic modification, with Q at some sites switching to nonzero values, mimics a presynaptic alteration.

Manipulating the Release Probability Experimentally Can Yield an Estimate of Quantal Parameters

The previous section described how experimental manipulation of the quantal parameters can be used to test the validity of the variance method. Although this method can shed light on how quantal parameters change with an alteration in synaptic strength, it is not designed to determine the absolute values of these parameters. Paradoxically, manipulating the release probability can actually allow an estimate to be obtained of the quantal parameters under conditions where they cannot be ascertained from the other methods previously described (Clements and Silver, 2000). This approach is analogous to nonstationary variance analysis, a powerful method used to estimate single-channel conductance from membrane currents. For a simple binomial model, the trial-to-trial variance of the postsynaptic response is given by: $Var_R = npQ^2 (1 + CV_Q^2) - np^2Q^2$. (The term $(1 + CV_Q^2)$ takes into account the contribution of quantal variability within release sites, assuming that the quantal coefficient of variation CV_Q is uniform across different sites.) This relationship can be rewritten as a function of the mean postsynaptic response amplitude $\bar{R}$: $Var_R = A\bar{R} - B\bar{R}^2$. The quantal parameter estimates are then given by $Q = A/(1 + CV_Q^2)$, $p = B \cdot \bar{R}(1 + CV_Q^2)/A$, and $n = 1/B$. In practice the method works as follows. The transmitter release probability is manipulated by varying the extracellular $[Ca^{2+}]/[Mg^{2+}]$ ratio or by applying drugs that act presynaptically, to construct a *variance–mean*

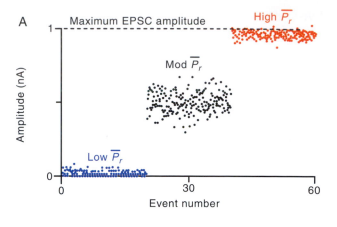

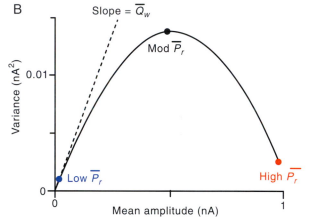

FIGURE 8.15 Variance–mean analysis. (A) Simulated synaptic response amplitudes collected under conditions of low, moderate, and high release probability. The variance is maximal with an intermediate release probability. (B) The variance–mean relationship for a binomial model is parabolic. The initial slope of the plot (dashed line) yields an estimate of the mean quantal amplitude. The number of release sites and the mean release probability can be estimated from the shape of the parabola. From Clements and Silver (2000).

plot (Fig. 8.15). If the binomial model holds, this has a parabolic shape. Fitting the formula $Var_R = A\bar{R} - B\bar{R}^2$ then yields estimates for Q, n, and p. If the simple binomial assumption is relaxed, the parabolic shape is often preserved or can become skewed. Under these conditions, the estimates for Q and p can still be informative, although they should be more correctly interpreted as weighted means of the underlying parameters, because release sites with low probabilities and/or low quantal amplitudes contribute less to the postsynaptic response variability. This method can be useful even if it is not possible to increase the mean release probability sufficiently to obtain a clear parabolic variance–mean relationship: the initial slope can be fitted with $Var_R = A\bar{R}$, yielding an estimate of Q. In this situation, however, it is not possible to estimate n and p.

Estimation of Transmitter Release Probability Does Not Rely Exclusively on Quantal Analysis of Evoked Postsynaptic Signals

With the development of fluorescent dyes that label presynaptic vesicles as they are recycled (Box 8.3), it has become possible to estimate the probability with which vesicles are released following presynaptic action potentials. This approach has given an elegant confirmation of the quantal hypothesis (Ryan et al., 1997).

An alternative method for estimating the release probability at glutamatergic synapses makes use of the pharmacological properties of N-methyl-D-aspartate (NMDA) receptors (Rosenmund et al., 1993). After application of MK-801, an irreversible open channel blocker, the size of the population synaptic signal mediated by NMDA receptors gradually decays with repeated stimulation of presynaptic fibers. Because the rate at which the postsynaptic signal decays is related to the probability that postsynaptic receptors are activated by released glutamate, this method gives an indirect estimate of the probability of transmitter release. Interestingly, the time course of the decay cannot be described by a single exponential, implying that the release probability must vary considerably across the population of synapses contributing to the signal.

Another indirect method for monitoring changes in transmitter release probability is to examine the short-term facilitation or depression when two or more presynaptic stimuli are delivered in rapid succession, and then to repeat this measurement after a manipulation that alters the strength of the synapse. In general, either increasing or reducing the probability of transmitter release has a relatively greater effect on the size of the response to the first pulse than on that to the second. Therefore, a change in the ratio of the responses to the two pulses implies that at least part of the effect of the manipulation is mediated presynaptically.

Summary

Quantal analysis has greatly improved our understanding of biophysical and pharmacological mechanisms of transmission. Numerical methods can be used to estimate the probability of transmitter release and the size of the postsynaptic effect of an individual quantum of neurotransmitter. Although these methods must be applied with caution, they yield a unique insight into the mechanisms of synaptic plasticity, both at the neuromuscular junction and in the central nervous system.

SHORT-TERM SYNAPTIC PLASTICITY

Chemical synapses are not static transmitters of information. Their effectiveness waxes and wanes, depending on frequency of stimulation and history of prior activity (Zucker and Regehr, 2002). At most synapses, repetitive high-frequency stimulation (called a *tetanus*) is initially dominated by a growth in successive PSP amplitudes, called *synaptic facilitation*. This process builds to a steady state within about 1 s and decays equally rapidly when stimulation stops. Decay is measured by single test stimuli given at various intervals after a conditioning train. Facilitation can often be divided into two exponential phases, called its first and second components, and may reach appreciable levels (e.g., a doubling of PSP size) after a single action potential. At most synapses, a slower phase of increase in efficacy, which has a characteristic time constant of several seconds and is called *augmentation*, succeeds facilitation. Finally, with prolonged stimulation, some synapses display a third phase of growth in PSP amplitude that lasts minutes and is called *potentiation*.

Often, a phase of decreasing transmission, called *synaptic depression*, is superimposed on these processes. Synaptic depression leads to a dip in transmission during repetitive stimulation, which often tends to overlap and obscure the augmentation and potentiation phases. When stimulation ceases, recovery from the various processes occurs in the same order as their development during the tetanus, with facilitation decaying first, then depression and augmentation, and finally potentiation (Fig. 8.16). Thus, potentiation is often visible in isolation only long after a tetanus and is thus called *post-tetanic potentiation* (PTP). At some synapses, even longer-lasting effects (persisting for hours), named *long-term potentiation* (LTP) (Baxter et al., 1985; Minota et al., 1991), have been observed. This long-term potentiation should not be confused with the form of synaptic plasticity bearing the same name and prominent at mammalian cortical synapses (Chapter 19). In almost all synapses in which a quantal analysis has been done, all these forms of synaptic plasticity (except some forms of cortical LTP) are due to changes in the number of quanta released by action potentials. When binomial parameters were estimated, correlated changes in p and n were usually observed. This result is expected if release sites with nonuniform p become more effective while "silent" sites are recruited during enhanced transmission, and vice versa during depression.

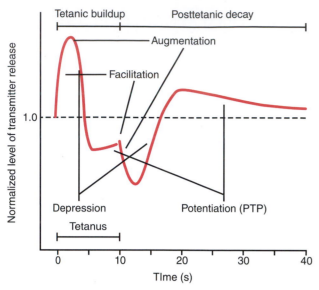

FIGURE 8.16　Accumulation of the effects of facilitation, augmentation, depression, and potentiation on transmitter release by each action potential in a tetanus, and the post-tetanic decay of these phases of synaptic plasticity measured by single stimuli after the tetanus.

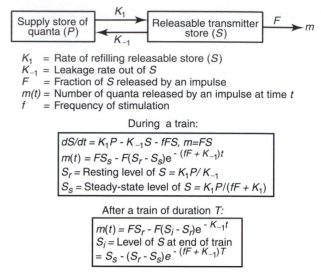

FIGURE 8.17　The classic depletion model of synaptic depression.

Depression May Arise from Depletion of Readily Releasable Transmitter or from Autoinhibition

In contrast with the growth phases in synaptic plasticity, the rate at which depression develops usually depends on stimulation frequency, whereas recovery from depression proceeds with a single time constant of seconds to minutes in different preparations. At many synapses, depression is relieved when transmission is reduced by lowering $[Ca^{2+}]$ or raising $[Mg^{2+}]$ in the medium. These characteristics are consistent with depression being due to the depletion of a readily releasable store of docked or nearly docked vesicles and recovery being due to their replenishment from a supply store (Fig. 8.17) (Zucker, 1989). In fact, styryl dyes appear to label distinct reserve and readily releasable pools of vesicles at *Drosophila* neuromuscular junctions (Kuromi and Kidokoro, 1998). The parameters of such a depletion model can be estimated from the rate of recovery from depression, which gives the rate of refilling the releasable store, and the fractional drop in PSPs given at short intervals, which gives the fraction of the releasable store liberated by each action potential.

Recent data suggest that the classic depletion model is too simple. As the readily releasable store is depleted, both the probability of its releasing a vesicle (Dobrunz and Stevens, 1997; Wu and Borst, 1999) and its maximum capacity (Dobrunz and Stevens, 1997) appear to drop. More realistic models take account of the finite capacity of the readily releasable vesicle pool (Stevens and Wesseling, 1999; Wu and Betz, 1998). At the frog neuromuscular junction depression also involves a reduction in the rates of endocytotic vesicle recovery and of their transport into the readily releasable store (Wu and Betz, 1998). At other synapses a Ca^{2+}/calmodulin-dependent target (Sakaba and Neher, 2001) speeds mobilization of vesicles into the readily releasable pool to partially compensate for its depletion (Dittman and Regehr, 1998; Stevens and Wesseling, 1998; Wang and Kaczmarek, 1998). This target may be Ca^{2+}/calmodulin-dependent phosphorylation of synapsin to mobilize vesicles by releasing them from tethering to actin filaments (Sun *et al.*, 2006).

It is now clear that vesicle depletion cannot account for all forms of synaptic depression. At some synapses, depression is due to an inhibitory action of released transmitter on presynaptic receptors called *autoreceptors*. For example, in rat hippocampal cortex, depression of GABA responses is blocked by antagonists of presynaptic $GABA_B$ receptors (Davies *et al.*, 1990). At synapses made by dorsal root ganglion neurons and by some brainstem neurons, depression appears to be due to inactivation of Ca^{2+} channels during repetitive activity (Jia and Nelson, 1986; Xu and Wu, 2005). This mechanism, however, has been specifically rejected at other synapses (Charlton *et al.*, 1982). Moreover, in other brain-stem neurons and at some central synapses in the sea slug *Aplysia*, depression is due in part to desensitization of postsynaptic cholinergic and AMPA-type glutamatergic receptors (Antzoulatos *et al*, 2003; Otis *et al.*, 1996; Wachtel and Kandel, 1971). Depression can also be caused by a reduction in sensitivity of NMDA-type glutamatergic receptors arising from postsynaptic Ca^{2+} accumulation (Mennerick and Zorumski, 1996).

Facilitation, Augmentation, and Potentiation Are Due to Effects of Residual Ca^{2+}

With few exceptions (Wojtowicz and Atwood, 1988), all the phases of increased short-term plasticity are Ca^{2+} dependent in the sense that little or no facilitation, augmentation, or potentiation is generated by stimulation in Ca^{2+}-free medium. Originally, these phases of increased transmission were thought to be due to the effect of residual Ca^{2+} remaining in active zones after presynaptic activity and summating with Ca^{2+} influx during subsequent action potentials to generate slightly higher peaks of $[Ca^{2+}]$ (Katz and Miledi, 1968; Zucker, 1989). Owing to the highly nonlinear dependence of transmitter release on $[Ca^{2+}]$, a small residual $[Ca^{2+}]$ could activate a substantial increase in phasic transmitter release during an action potential while simultaneously increasing MPSP frequency. Temporal correlations between increased spike-evoked transmission after single action potentials or tetani and increases in mPSP frequency supported this idea. But the tetanic accumulation of facilitation, augmentation, and potentiation did not accord quantitatively with predictions of a model of the accumulation of Ca^{2+} acting at one site (Magleby and Zengel, 1982). Simulations of expected levels of peak and residual $[Ca^{2+}]$ levels also were unable to account for the full magnitude of facilitation (Yamada and Zucker, 1992). Presynaptic $[Ca^{2+}]$ measurements using fura-2 confirmed the persistence of residual $[Ca^{2+}]$ after repetitive stimulation during facilitation, augmentation, PTP, and LTP but showed that it was too weak to explain augmentation or potentiation by simply summating with peak $[Ca^{2+}]$ from Ca^{2+} channels at release sites (Atluri and Regehr, 1996; Delaney et al., 1991; Delaney and Tank, 1994). These findings led to the proposal that, in addition to summating with peak $[Ca^{2+}]$ transients at release sites, Ca^{2+} acts to increase transmission at one or more targets distinct from the sites triggering exocytosis.

Two possibilities exist. Residual-free Ca^{2+} could act in equilibrium with such sites to increase transmission (Matveev et al., 2002), or Ca^{2+} could bind to these targets and activate processes that increase release after residual Ca^{2+} has dissipated (Yamada and Zucker, 1992). The latter idea is suggested by experiments in which facilitation, augmentation, or potentiation persists when residual Ca^{2+} should be absorbed by presynaptic introduction of exogenous chelators; however, results of this sort of experiment have not been consistent (Zucker, 1994). The former

idea is supported by experiments in which photolabile BAPTA derivatives were injected presynaptically at crayfish neuromuscular junctions (Kamiya and Zucker, 1994). Post-tetanic photolysis to produce enough chelator to suddenly remove residual Ca^{2+} after conditioning stimulation sharply reduced facilitation within a few milliseconds, and it reduced augmentation and potentiation within about 1 s. These results suggest that Ca^{2+} can "prime" subsequent phasic release by action potentials by acting at two additional targets distinct from the exocytosis trigger: a fast site responsible for facilitation and a slow one for augmentation and potentiation.

Several additional indications of separate sites of Ca^{2+} action in synaptic plasticity are as follows:

1. Facilitation grows while secretion decays after one action potential at very low temperature at frog neuromuscular junctions (Van der Kloot, 1994).
2. Sr^{2+} and Ba^{2+} selectively enhance facilitation and augmentation, respectively, of both evoked transmitter release and mEPP frequency at frog neuromuscular junctions (Zengel and Magleby, 1980, 1981).
3. *Drosophila* mutants defective in enzymes affecting cAMP-dependent phosphorylation show reduced facilitation and potentiation (Zhong and Wu, 1991).
4. Other transformed *Drosophila* carrying an inhibitor of Ca^{2+}-calmodulin-dependent protein kinase II have impaired facilitation, augmentation, and potentiation (Wang et al., 1994).
5. Mice lacking synapsin I show a specific defect in facilitation but not potentiation in hippocampal pyramidal cells (Rosahl et al., 1993).
6. Facilitation is approximately linearly related to residual $[Ca^{2+}]_i$, while secretion involves a high degree of Ca^{2+} cooperativity (Korogod et al., 2005; Vyshedskiy and Lin, 1997; Wright et al., 1996).
7. Some synapses have an extremely low basal probability of vesicular release to an action potential, which grows dramatically during repetitive activity. At such strongly facilitating synapses, vesicles that are releasable in a train (the readily releasable pool) are not all immediately releasable but first must undergo a final Ca^{2+}-dependent "priming" step (Millar et al., 2005).
8. Calmodulin-dependent binding of Ca^{2+} to munc13 isoforms that interact with RIM1 mediates at least the augmentation phase of post-tetanic enhancement of transmission (Junge et al., 2004, Rosenmund et al., 2002).

9. One mechanism by which Ca^{2+} can facilitate release is by binding to highly saturable endogenous buffers like calbindin-D28k, reducing its capability of rapidly capturing Ca^{2+} entering in successive action potentials (Matveev et al., 2004). This is the dominant mechanism for facilitation at some (Blatow et al., 2003), but not other (Rozov et al., 2001), vertebrate central synapses.

The duration of augmentation is set by the time course for removal of residual Ca^{2+} from boutons after a moderate tetanus. Potentiation lasts longer after a strong tetanus because residual Ca^{2+} is present longer, owing to overloading of the processes responsible for removing excess Ca^{2+} from neurons. These processes include Ca^{2+} extrusion pumps, such as the plasma membrane ATPase and Na^+–Ca^{2+} exchange, and Ca^{2+} uptake into organelles such as endoplasmic reticulum and mitochondria (David et al., 1998; Fossier et al., 1992; Tang and Zucker, 1997). The accumulation of Ca^{2+} in mitochondria during a tetanus and its gradual post-tetanic release into cytoplasm produce a prolonged elevation in residual $[Ca^{2+}]_i$ and lead to PTP (Tang and Zucker, 1997). In addition, Na^+ accumulation during tetanic activity prolongs residual Ca^{2+} by reducing its extrusion by Na^+–Ca^{2+} exchange (Zhong et al., 2001). PTP can also arise from a modulation of Ca^{2+} channels (Habets and Borst, 2006).

Potentiation has also been attributed to Ca^{2+}-calmodulin-dependent kinase II phosphorylation of synapsin I (Greengard et al., 1993). However, calmodulin inhibitors and kinase inhibitors and mutants failed to affect transmission, facilitation, augmentation, or potentiation at crayfish neuromuscular junctions and mammalian cortical synapses (Zucker, 1994). But short-term synaptic enhancement in *Drosophila* (see earlier) and LTP in bullfrog sympathetic ganglia may depend on this enzymatic pathway (Baxter et al., 1985; Minota et al., 1991).

Summary

Short-term synaptic plasticity allows synaptic strength to be modulated as a function of prior activity. Synapses may show a decline in transmission (depression) or an increase in synaptic efficacy, with time constants ranging from seconds (facilitation and augmentation) to minutes (potentiation or PTP) to hours (LTP); many synapses show a mixture of several of these phases. Depression may be due to depletion of a readily releasable supply of vesicles or to the inhibitory action of transmitter or enzymatically produced transmitter products on presynaptic autoreceptors. Depression makes synapses selectively responsive to brief stimuli or to changes in level of activity. Frequency-dependent increases in synaptic efficacy are due to the effects of residual presynaptic Ca^{2+} acting to modulate the release process. At least part of Ca^{2+} action is mediated by separate targets for facilitation and for augmentation and potentiation. PTP is prolonged after a long tetanus because residual Ca^{2+} remains in synaptic terminals for minutes after such stimulation. These frequency-dependent increases in synaptic efficacy allow synapses to distinguish significant signals from noise and respond to selected patterns of activity.

References

Aalto, M. K., Ronne, H., and Keranen, S. (1993). Yeast syntaxins Sso1p and Sso2p belong to a family of related membrane proteins that function in vesicular transport. *Embo J.* **12**, 4095–4104.

Aalto, M. K., Ruohonen, L., Hosono, K., and Keranen, S. (1991). Cloning and sequencing of the yeast *Saccharomyces cerevisiae* SEC1 gene localized on chromosome IV. *Yeast* **7**, 643–650.

Adler, E. M., Augustine, G. J., Duffy, S. N., and Charlton, M. P. (1991). Alien intracellular chelators attenuate neurotransmitter release at the squid giant synapse. *J. Neurosci.* **11**, 1496–1507.

Ahnert-Hilger, G., Bhakdi, S., and Gratzl, M. (1985). Minimal requirements for exocytosis: A study using PC 12 cells permeabilized with staphylococcal alpha-toxin. *J. Biol. Chem.* **260**, 12730–12734.

Alvarez de Toledo, G., Fernandez-Chacon, R., and Fernandez, J. M. (1993). Release of secretory products during transient vesicle fusion. *Nature (London)* **363**, 554–558.

Antonin, W., Holroyd, C., Fasshauer, D., Pabst, S., Von Mollard, G. F., and Jahn, R. (2000). A SNARE complex mediating fusion of late endosomes defines conserved properties of SNARE structure and function. *EMBO J.* **19**, 6453–6464.

Antzoulatos, E. G., Cleary, L. J., Eskin, A., Baxter, D. A., and Byrne, J. H. (2003). Desensitization of postsynaptic glutamate receptors contributes to high-frequency homosynaptic depression of *Aplysia* sensorimotor connections. *LearnMem.* in press.

Arredondo, L., Nelson, H. B., Beckingham, K., and Stern, M. (1998). Increased transmitter release and aberrant synapse morphology in a Drosophila *calmodulin* mutant. *Genetics* **150**, 265–274.

Atluri, P. P., and Regehr, W. G. (1996) Determinants of the time course of facilitation at the granule cell to Purkinje cell synapse. *J. Neurosci.* **16**, 5661–5671.

Auger, C., Kondo, S., and Marty, A. (1998). Multivesicular release at single functional synaptic sites in cerebellar stellate and basket cells. *J. Neurosci.* **18**, 4532–4547.

Augustine, G. J., and Charlton, M. P. (1986). Calcium-dependence of presynaptic calcium current and post-synaptic response at the squid giant synapse. *J. Physiol. (London)* **381**, 619–640.

Augustine, G. J., Charlton, M. P., and Smith, S. J. (1985). Calcium entry and transmitter release at voltage-clamped nerve terminals of squid. *J. Physiol. (London)* **367**, 163–181.

Augustine, G. J., Charlton, M. P., and Smith, S. J. (1987). Calcium action in synaptic transmitter release. *Annu. Rev. Neurosci.* **10**, 633–693.

Bajjalieh, S. M., Peterson, K., Shinghal, R., and Scheller, R. H. (1992). SV2, a brain synaptic vesicle protein homologous to bacterial transporters. *Science* **257**, 1271–1273.

Banerjee, A., Barry, V. A., DasGupta, B. R., and Martin, T. F. (1996). N-Ethylmaleimide-sensitive factor acts at a prefusion ATP-dependent step in Ca^{2+}-activated exocytosis. *J. Biol. Chem.* **271**, 20223–20226.

Bartol, T. M., Land, B. R., Salpeter, E. E., and Salpeter, M. M. (1991). Monte Carlo simulation of miniature endplate current generation in the vertebrate neuromuscular junction. *Biophys. J.* **59**, 1290–1307.

Baxter, D. A., Bittner, G. D., and Brown, T. H. (1985). Quantal mechanism of long-term synaptic potentiation. *Proc. Natl. Acad. Sci. USA* **82**, 5978–5982.

Bekkers, J. M., and Stevens, C. F. (1989). NMDA and non-NMDA receptors are co-localized at individual excitatory synapses in cultured rat hippocampus. *Nature (London)* **341**, 230–233.

Bellocchio, E. E., Reimer, R. J., Fremeau, R. T., Jr., and Edwards, R. H. (2000). Uptake of glutamate into synaptic vesicles by an inorganic phosphate transporter. *Science* **289**, 957–960.

Bennett, M. K., Calakos, N., and Scheller, R. H. (1992). Syntaxin: A synaptic protein implicated in docking of synaptic vesicles at presynaptic active zones. *Science* **257**, 255–259.

Bennett, M. K., and Scheller, R. H. (1993). The molecular machinery for secretion is conserved from yeast to neurons. *Proc. Natl. Acad. Sci. USA* **90**, 2559–2563.

Bennett, M. K., and Scheller, R. H. (1994). A molecular description of synaptic vesicle membrane trafficking. *Annu. Rev. Biochem.* **63**, 63–100.

Betz, W. J., and Bewick, G. S. (1993). Optical monitoring of transmitter release and synaptic vesicle recycling at the frog neuromuscular junction. *J. Physiol. (London)* **460**, 287–309.

Beutner, D., Voets, T., Neher, E., and Moser, T. (2001). Calcium dependence of exocytosis and endocytosis at the cochlear inner hair cell afferent synapse. *Neuron* **29**, 681–690.

Bixby, J. L., and Reichardt, L. F. (1985). The expression and localization of synaptic vesicle antigens at neuromuscular junctions *in vitro*. *J. Neurosci.* **5**, 3070–3080.

Blasi, J., Chapman, E. R., Link, E., Binz, T., Yamasaki, S., De Camilli, P., Südhof, T. C., Niemann, H., and Jahn, R. (1993a). Botulinum neurotoxin A selectively cleaves the synaptic protein SNAP-25. *Nature* **365**, 160–163.

Blasi, J., Chapman, E. R., Yamasaki, S., Binz, T., Niemann, H., and Jahn, R. (1993b). Botulinum neurotoxin C1 blocks neurotransmitter release by means of cleaving HPC-1/syntaxin. *EMBO J.* **12**, 4821–4828.

Blatow, M., Caputi, A., Burnashev, N., Monyer, H., and Rozov, A. (2003). Ca^{2+} buffer saturation underlies paired pulse facilitation in calbindin-D28k-containing terminals. *Neuron* **38**, 79–88.

Block, M. R., Glick, B. S., Wilcox, C. A., Wieland, F. T., and Rothman, J. E. (1988). Purification of an N-ethylmaleimide-sensitive protein catalyzing vesicular transport. *Proc. Natl. Acad. Sci. USA* **85**, 7852–7856.

Bollmann, J. H., and Sakmann, B. (2005). Control of synaptic strength and timing by the release-site Ca^{2+} signal. *Nat. Neurosci.* **8**, 426–434.

Bollmann, J. J., Sakmann, B., and Borst, J. G. G. (2000). Calcium sensitivity of glutamate release in a calyx-type terminal. *Science* **289**, 953–956.

Borst, J. G. G., and Sakmann, B. (1996). Calcium influx and transmitter release in a fast CNS synapse. *Nature* **383**, 431–434.

Boyd, I. A., and Martin, A. R. (1956). Spontaneous subthreshold activity at mammalian neuromuscular junctions. *J. Physiol. (London)* **132**, 74–91.

Broadie, K., Bellen, H. J., DiAntonio, A., Littleton, J. T., and Schwarz, T. L. (1994). Absence of synaptotagmin disrupts excitation–secretion coupling during synaptic transmission. *Proc. Natl. Acad. Sci. USA* **91**, 10727–10731.

Broadie, K., Prokop, A., Bellen, H. J., O'Kane, C. J., Schulze, K. L., and Sweeney, S. T. (1995). Syntaxin and synaptobrevin function downstream of vesicle docking in *Drosophila*. *Neuron* **15**, 663–673.

Brose, N., Petrenko, A. G., Südhof, T. C., and Jahn, R. (1992). Synaptotagmin: A calcium sensor on the synaptic vesicle surface. *Science* **256**, 1021–1025.

Brose, N., and Rosenmund, C. (1999). SV2: SVeeping up excess Ca^{2+} or tranSVorming presynaptic Ca^{2+} sensors? *Neuron* **24**, 766–768.

Buckley, K. M., Floor, E., and Kelly, R. B. (1987). Cloning and sequence analysis of cDNA encoding p38, a major synaptic vesicle protein. *J. Cell Biol.* **105**, 2447–2456.

Burgess, R. W., Deitcher, D. L., and Schwarz, T. L. (1997). The synaptic protein syntaxin1 is required for cellularization of *Drosophila* embryos. *J. Cell Biol.* **138**, 861–875.

Carlson, S. S., Wagner, J. A., and Kelly, R. B. (1978). Purification of synaptic vesicles from elasmobranch electric organ and the use of biophysical criteria to demonstrate purity. *Biochemistry* **17**, 1188–1199.

Chamberlain, L. H., Roth, D., Morgan, A., and Burgoyne, R. D. (1995). Distinct effects of alpha-SNAP, 14–3–3 proteins, and calmodulin on priming and triggering of regulated exocytosis. *J. Cell Biol.* **130**, 1063–1070.

Chan, M. S., Obar, R. A., Schroeder, C., Austin, T. W., Poodry, C. A., Wadsworth, S. A., and Vallee, R. B. (1991). Multiple forms of dynamin are encoded by *shibire*, a *Drosophila* gene involved in endocytosis. *Nature (London)* **351**, 583–586.

Chapman, E. R., An, S., Barton, N., and Jahn, R. (1994). SNAP-25, a t-SNARE which binds to both syntaxin and synaptobrevin via domains that may form coiled coils. *J. Biol. Chem.* **269**, 27427–27432.

Charlton, M. P., Smith, S. J., and Zucker, R. S. (1982). Role of presynaptic calcium ions and channels in synaptic facilitation and depression at the squid giant synapse. *J. Physiol. (London)* **323**, 173–193.

Chen, M. S., Obar, R. A., Schroeder, C. C., Austin, T. W., Poodry, C. A., Wadsworth, S. C., and Vallee, R. B. (1991). Multiple forms of dynamin are encoded by *shibire*, a *Drosophila* gene involved in endocytosis. *Nature* **351**, 583–586.

Chen, Y. A., and Scheller, R. H. (2001). SNARE-mediated membrane fusion. *Nat. Rev. Mol. Cell. Biol.* **2**, 98–106.

Christie, J.M., and Jahr, C.E. (2006). Multivesicular release at Schaffer collateral-CA1 hippocampal synapses. *J Neurosci* **26**, 210–216.

Clements, J. D., Lester, R. A. J., Tong, G., Jahr, C. E., and Westbrook, G. L. (1992). The time course of glutamate in the synaptic cleft. *Science* **258**, 1498–1501.

Clements, J. D., and Silver, R. A. (2000). Unveiling synaptic plasticity: A new graphical and analytical approach. *Trends Neurosci.* **23**, 105–113.

Conradt, B., Haas, A., and Wickner, W. (1994). Determination of four biochemically distinct, sequential stages during vacuole inheritance *in vitro*. *J. Cell Biol.* **126**, 99–110.

Cooper, J. R., Bloom, F. E., and Roth, R. H. (1991). "The Biochemical Basis of Neuropharmacology," 6th ed. Oxford Univ. Press, New York.

Crowder, K. M., Gunther, J. M., Jones, T. A., Hale, B. D., Zhang, H. Z., Peterson, M. R., Scheller, R. H., Chavkin, C., and Bajjalieh, S. M. (1999). Abnormal neurotransmission in mice lacking synaptic vesicle protein 2A (SV2A). *Proc. Natl. Acad. Sci. USA* **96**, 15268–15273.

Custer, K. L., Austin, N. S., Sullivan, J. M., and Bajjalieh, S. M. (2006). Synaptic Vesicle Protein 2 enhances release probability at quiescent synapses. *J. Neurosci.* **26**, 1303–1313.

David, G., Barrett, J. N., and Barrett, E. F. (1998). Evidence that mitochondria buffer physiological Ca^{2+} loads in lizard motor nerve terminals. *J. Physiol.* **509**, 59–65.

Davies, C. H., Davies, S. N., and Collingridge, G. L. (1990). Paired-pulse depression of monosynaptic GABA-mediated inhibitory postsynaptic responses in rat hippocampus. *J. Physiol. (London)* **424**, 513–531.

De Camilli, P., Harris, S. M., Jr., Huttner, W. B., and Greengard, P. (1983). Synapsin I (Protein I), a nerve terminal-specific phosphoprotein. II. Its specific association with synaptic vesicles demonstrated by immunocytochemistry in agarose-embedded synaptosomes. *J. Cell Biol.* **96**, 1355–1373.

De Camilli, P., and Jahn, R. (1990). Pathways to regulated exocytosis in neurons. *Annu. Rev. Physiol.* **52**, 625–645.

De Camilli, P., Takei, K., and McPherson, P. S. (1995). The function of dynamin in endocytosis. *Curr. Opin. Neurobiol.* **5**, 559–565.

del Castillo, J., and Katz, B. (1954). Quantal components of the end-plate potential. *J. Physiol. (London)* **124**, 560–573.

Delaney, K. R., and Tank, D. W. (1994). A quantitative measurement of the dependence of short-term synaptic enhancement on presynaptic residual calcium. *J. Neurosci.* **14**, 5885–5902.

Delaney, K. R., Zucker, R. S., and Tank, D. W. (1991). Presynaptic calcium in motor nerve terminals associated with posttetanic potentiation. *J. Neurosci.* **9**, 3558–3567.

DiAntonio, A., Parfitt, K. D., and Schwarz, T. L. (1993). Synaptic transmission persists in synaptotagmin mutants of *Drosophila*. *Cell* **73**, 1281–1290.

DiAntonio, A., and Schwarz, T. L. (1994). The effect on synaptic physiology of synaptotagmin mutations in *Drosophila*. *Neuron* **12**, 909–920.

Dickman, D. K., Horne, J. A., Meinertzhagen, I. M., and Schwarz, T. L. (2005). A slowed classical pathway rather than kiss-and-run mediates endocytosis at synapses lacking synaptojanin and endophilin. *Cell* **123**, 521–533.

Dinkelacker, V., Voets, T., Neher, E., and Moser, T. (2000). The readily releasable pool of vesicles in chromaffin cells is replenished in a temperature-dependent manner and transiently overfills at 37°C. *J. Neurosci.* **20**, 8377–8383.

Dittman, J. S., and Regehr, W. G. (1998). Calcium dependence and recovery kinetics of presynaptic depression at the climbing fiber to Purkinje cell synapse. *J. Neurosci.* **18**, 6147–6162.

Dobrunz, L. E., and Stevens, C. F. (1997). Heterogeneity of release probability, facilitation, and depletion at central synapses. *Neuron* **18**, 995–1008.

Dresbach, T., Qualmann, B., Kessels, M. M., Garner, C. C., and Gundelfinger, E. D. (2001). The presynaptic cytomatrix of brain synapses. *Cell. Mol. Life Sci.* **58**, 94–116.

Dulubova, I., Khvotchev, M., Liu, S., Huryeva, I., Sudhof, T. C., and Rizo, J. (2007). Munc18-1 binds directly to the assembled SNARE complex. *Proc. Natl. Acad. Sci. USA* **104**, 2697–2702.

Dunlap, K., Luebke, J. I., and Turner, T. J. (1995). Exocytotic Ca^{2+} channels in mammalian central neurons. *Trends Neurosci.* **18**, 89–98.

Eakle, K. A., Bernstein, M., and Emr, S. D. (1988). Characterization of a component of the yeast secretion machinery: Identification of the SEC18 gene product. *Mol. Cell. Biol.* **8**, 4098–4109.

Eccles, J. C. (1964). "The Physiology of Synapses." Springer-Verlag, Berlin. (A comprehensive summary of early work.)

Edwards, C., Doležal, V., Tuček, S., Zemková, H., and Vyskočil, F. (1985). Is an acetylcholine system transport system responsible for nonquantal release of acetylcholine at the rodent myoneural junction? *Proc. Natl. Acad. Sci. USA* **82**, 3514–3518.

Edwards, F. A., Konnerth, A., and Sakmann, B. (1990). Quantal analysis of inhibitory synaptic transmission in the dentate gyrus of rat hippocampal slices: A patch-clamp study. *J. Physiol. (London)* **430**, 213–249.

Edwards, F. R., Redman, S. J., and Walmsley, B. (1976). Statistical fluctuation in charge transfer at Ia synapses on spinal motoneurones. *J. Physiol. (London)* **259**, 665–688.

Edwards, R. H. (1992). The transport of neurotransmitters into synaptic vesicles. *Curr. Opin. Neurobiol.* **2**, 586–594.

Efron, B., and Tibshirani, R. (1993). "An Introduction to the Bootstrap." Chapman & Hall, New York.

Elmqvist, D., and Quastel, D. M. J. (1965). Presynaptic action of hemicholinium at the neuromuscular junction. *J. Physiol. (London)* **177**, 463–482.

Faber, D. S., and Korn, H. (1992). Application of the coefficient of variation method for analyzing synaptic plasticity. *Biophys. J.* **60**, 1288–1294.

Fatt, P., and Katz, B. (1952). Spontaneous sub-threshold activity at motor nerve endings. *J. Physiol. (London)* **119**, 109–128.

Feany, M. B., Lee, S., Edwards, R. H., and Buckley, K. M. (1992). The synaptic vesicle protein SV2 is a novel type of transmembrane transporter. *Cell* **70**, 861–867.

Fernandez-Chacon, R., Shin, O. H., Konigstorfer, A., Matos, M. F., Meyer, J. C., Garcia, J., Gerber, S. H., Rizo, J., Sudhof, T. C., and Rosenmund, C. (2002). Structure/function analysis of Ca^{2+} binding to the C2A domain of synaptotagmin. *J. Neurosci.* **22**, 8438–8446.

Fischer von Mollard, G., Südhof, T. C., and Jahn, R. (1991). A small GTP-binding protein dissociates from synaptic vesicles during exocytosis. *Nature* **349**, 79–81.

Fletcher, P., and Forrester, T. (1975). The effect of curare on the release of acetylcholine from mammalian motor nerve terminals and an estimate of quantum content. *J. Physiol. (London)* **251**, 131–144.

Foran, P., Lawrence, G. W., Shone, C. C., Foster, K. A., and Dolly, J. O. (1996). Botulinum neurotoxin C1 cleaves both syntaxin and SNAP-25 in intact and permeabilized chromaffin cells: Correlation with its blockade of catecholamine release. *Biochem.* **35**, 2630–2636.

Fossier, P., Baux, G., Trudeau, L. -E., and Tauc, L. (1992). Involvement of Ca^{2+} uptake by a reticulum-like store in the control of transmitter release. *Neurosci.* **50**, 427–434.

Garcia, E. P., Gatti, E., Butler, M., Burton, J., and De Camilli, P. (1994). A rat brain Sec1 homologue related to Rop and UNC18 interacts with syntaxin. *Proc. Natl. Acad. Sci. USA* **91**, 2003–2007.

Garcia, E. P., McPherson, P. S., Chilcote, T. J., Takei, K., and De Camilli, P. (1995). rbSec1A and B colocalize with syntaxin 1 and SNAP-25 throughout the axon, but are not in a stable complex with syntaxin. *J. Cell Biol.* **129**, 105–120.

Garner, C. C., Kindler, S., and Gundelfinger, E. D. (2000). Molecular determinants of presynaptic active zones. *Curr. Opin. Neurobiol.* **10**, 321–327.

Geppert, M., Goda, Y., Hammer, R. E., Li, C., Rosahl, T. W., Stevens, C. F., and Südhof, T. C. (1994). Synaptotagmin I: A major Ca^{2+} sensor for transmitter release at a central synapse. *Cell* **79**, 717–727.

Giraudo, C. G., Eng, W. S., Melia, T. J., and Rothman, J. E. (2006). A clamping mechanism involved in SNARE-dependent exocytosis. *Science* **313**, 676–680.

Gorvel, J. P., Chavrier, P., Zerial, M., and Gruenberg, J. (1991). rab5 controls early endosome fusion *in vitro*. *Cell* **64**, 915–925.

Granseth, B., Odermatt, B., Royle, S.J., and Lagnado, L. (2006). Clathrin-mediated endocytosis is the dominant mechanism of vesicle retrieval at hippocampal synapses. *Neuron* **51**, 773–786.

Greengard, P., Valtorta, F., Czernik, A. J., and Benfenati, F. (1993). Synaptic vesicle phosphoproteins and regulation of synaptic function. *Science* **259**, 780–785.

Gulyas, A.I., Miles, R., Sik, A., Toth, K., Tamamaki, N., and Freund, T.F. (1993). Hippocampal pyramidal cells excite inhibitory neurons through a single release site. *Nature* **366**, 683–687.

Habets, R.L., and Borst, J.G. (2006). An increase in calcium influx contributes to post-tetanic potentiation at the rat calyx of Held synapse. *J. Neurophysiol.* **96**, 2868–2876.

Hammarlund, M., Palfreyman, M. T., Watanabe, S., Olsen, S., and Jorgensen, E. M. (2007). Open syntaxin docks synaptic vesicles. *PLoS Biol.* **5**, e198.

Harlow, M. L., Ress, D., Stoschek, A., Marshall, R. M., and McMahan, U. J. (2001). The architecture of active zone material at the frog's neuromuscular junction. *Nature* **409**, 479–484.

Harris, K. M., and Sultan, P. (1995). Variation in the number, location and size of synaptic vesicles provides an anatomical basis for the nonuniform probability of release at hippocampal CA1 synapses. *Neuropharmacol.* **34**, 1387–1395.

Hartzell, H. C., Kuffler, S. W., and Yoshikami, D. (1975). Postsynaptic potentiation: Interaction between quanta of acetylcholine at the skeletal neuromuscular synapse. *J. Physiol. (London)* **251**, 427–463.

Haucke, V., and De Camilli, P. (1999). AP-2 recruitment to synaptotagmin stimulated by tyrosine-based endocytic motifs. *Science* **285**, 1268–1271.

Hay, J. C., and Martin, T. F. (1992). Resolution of regulated secretion into sequential MgATP-dependent and calcium-dependent stages mediated by distinct cytosolic proteins. *J. Cell. Biol.* **119**, 139–151.

Haydon, P. C., Henderson, E., and Stanley, E. F. (1994). Localization of individual calcium channels at the release face of a presynaptic nerve terminal. *Neuron* **13**, 1275–1280.

Heidelberger, R., Heinemann, C., Neher, E., and Matthews, G. (1994). Calcium dependence of the rate of exocytosis in a synaptic terminal. *Nature (London)* **371**, 513–515.

Heuser, J. E. (1977). Synaptic vesicle exocytosis revealed in quick-frozen frog neuromuscular junctions treated with 4-aminopyridine and given a single electrical shock. *Soc. Neurosci. Symp.* **2**, 215–239. (A fine anatomical analysis of vesicle recycling.)

Heuser, J. E., and Reese, T. S. (1973). Evidence for recycling of synaptic vesicle membrane during transmitter release at the frog neuromuscular junction. *J. Cell Biol.* **57**, 315–344.

Heuser, J. E., and Reese, T. S. (1981). Structural changes after transmitter release at the frog neuromuscular junction. *J. Cell Biol.* **88**, 564–580.

Heuser, J. E., Reese, T. S., Dennis, M. J., Jan, Y., Jan, L., and Evans, L. (1979). Synaptic vesicle exocytosis captured by quick freezing and correlated with quantal transmitter release. *J. Cell Biol.* **81**, 275–300.

Holz, R. W., Bittner, M. A., Peppers, S. C., Senter, R. A., and Eberhard, D. A. (1989). MgATP-independent and MgATP-dependent exocytosis: Evidence that MgATP primes adrenal chromaffin cells to undergo exocytosis. *J. Biol. Chem.* **264**, 5412–5419.

Horn, R. (1987). Statistical methods for model discrimination: Applications to gating kinetics and permeation of the acetylcholine receptor channel. *Biophys. J.* **51**, 255–263.

Hunt, J. M., Bommert, K., Charlton, M. P., Kistner, A., Habermann, E., Augustine, G. J., and Betz, H. (1994). A post-docking role for synaptobrevin in synaptic vesicle fusion. *Neuron* **12**, 1269–1279.

Hurlbut, W. P., Iezzi, N., Fesce, R., and Ceccarelli, B. (1990). Correlation between quantal secretion and vesicle loss at the frog neuromuscular junction. *J. Physiol. (London)* **425**, 501–526.

Isaac, J. T., Nicoll, R. A., and Malenka, R. C. (1995). Evidence for silent synapses: Implications for the expression of LTP. *Neuron* **15**, 427–434.

Isaacson, J. S., and Walmsley, B. (1995). Counting quanta: Direct measurements of transmitter release at a central synapse. *Neuron* **15**, 875–884.

Jack, J. J. B., Redman, S. J., and Wong, K. (1981). The components of synaptic potentials evoked in cat spinal motoneurones by impulses in single group Ia afferents. *J. Physiol. (London)* **321**, 65–96.

Jahn, R., and Südhof, T. C. (1994). Synaptic vesicles and exocytosis. *Annu. Rev. Neurosci.* **17**, 219–246.

Jia, M., and Nelson, P. G. (1986). Calcium currents and transmitter output in cultured spinal cord and dorsal root ganglion neurons. *J. Neurophysiol.* **56**, 1257–1267.

Jonas, P., Major, G., and Sakmann, B. (1993). Quantal components of unitary EPSCs at the mossy fibre synapse on CA3 pyramidal cells of rat hippocampus. *J. Physiol. (London)* **472**, 615–663.

Jorgensen, E. M., Hartwieg, E., Schuske, K., Nonet, M. L., Jin, Y., and Horvitz, H. R. (1995). Defective recycling of synaptic vesicles in synaptotagmin mutants of *Caenorhabditis elegans*. *Nature* **378**, 196–199.

Junge, H. J., Rhee, J. S., Jahn, O., Varoqueaux, F., Spiess, J., Waxham, M. N., Rosenmund, C., and Brose, N. (2004). Calmodulin and Munc13 form a Ca^{2+} sensor/effector complex that controls short-term synaptic plasticity. *Cell* **118**, 389–401.

Jurado, L. A., Chockalingam, P. S., and Jarrett, H. W. (1999). Apocalmodulin. *Physiol. Rev.* **79**, 661–682.

Kamiya, H., and Zucker, R. S. (1994). Residual Ca^{2+} and short-term synaptic plasticity. *Nature (London)* **371**, 603–606.

Katz, B. (1969). "The Release of Neural Transmitter Substances." Thomas, Springfield, IL. (Describes the classic experiments on transmitter release.)

Katz, B., and Miledi, R. (1965). The release of acetylcholine from nerve endings by graded electric pulses. *Proc. R. Soc. London Ser. B* **167**, 28–38.

Katz, B., and Miledi, R. (1968). The role of calcium in neuromuscular facilitation. *J. Physiol. (London)* **195**, 481–492.

Korogod, N., Lou, X., and Schneggenburger, R. (2005). Presynaptic Ca^{2+} requirements and developmental regulation of post-tetanic potentiation at the calyx of Held. *J. Neurosci* **25**, 5127–5137.

Korn, H., and Faber, D. S. (1987). Regulation and significance of probabilistic release mechanisms at central synapses. *In* "Synaptic Function" G. Edelman, W. E. Gall, and W. M. Cowan, Eds.), pp. 57–108. Wiley, New York.

Korn, H., and Faber, D. S. (1991). Quantal analysis and synaptic efficacy in the CNS. *Trends Neurosci.* **14**, 439–445.

Korn, H., Triller, A., Mallet, A., and Faber, D. S. (1981). Fluctuating responses at a central synapse: n of binomial fit predicts number of stained presynaptic boutons. *Science* **213**, 898–901.

Kuffler, S. W., and Yoshikami, D. (1975). The number of transmitter molecules in a quantum: An estimate from iontophoretic application of acetylcholine at the neuromuscular junction. *J. Physiol. (London)* **251**, 465–482.

Kullmann, D. M. (1989). Applications of the expectation–maximization algorithm to quantal analysis of postsynaptic potentials. *J. Neurosci. Methods* **30**, 231–245.

Kullmann, D. M. (1994). Amplitude fluctuations of dual-component EPSCs in hippocampal pyramidal cells: Implications for long-term potentiation. *Neuron* **12**, 1111–1120.

Kullmann, D. M. (2003). Silent synapses: What are they telling us about long-term potentiation? *Philos. Trans. R. Soc. Lond. B Biol. Sci.* **358**, 727–733.

Kullmann, D. M., and Nicoll, R. A. (1992). Long-term potentiation is associated with increases in quantal content and quantal amplitude. *Nature (London)* **357**, 240–244.

Kuromi, H., and Kidokoro, Y. (1998). Two distinct pools of synaptic vesicles in single presynaptic boutons in a temperature-sensitive *Drosophila* mutant, *shibire*. *Neuron* **20**, 917–925.

Landò, L., and Zucker, R. S. (1994). Ca^{2+} cooperativity in neurosecretion measured using photolabile Ca^{2+} chelators. *J. Neurophysiol.* **72**, 825–830.

Large, W. A., and Rang, H. P. (1978). Variability of transmitter quanta released during incorporation of a false transmitter into cholinergic nerve terminals. *J. Physiol. (London)* **285**, 25–34.

Leavis, P. C., and Gergely, J. (1984). Thin filament proteins and thin filament-linked regulation of vertebrate muscle contraction. *CRC Crit. Rev. Biochem.* **16**, 235–305.

Leenders, A. G., Scholten, G., Wiegant, V. M., Da Silva, F. H., and Ghijsen, W. E. (1999). Activity-dependent neurotransmitter release kinetics: Correlation with changes in morphological distributions of small and large vesicles in central nerve terminals. *Eur. J. Neurosci.* **11**, 4269–4277.

Li, C., Ullrich, B., Zhang, J. Z., Anderson, R. G., Brose, N., and Südhof, T. C. (1995). Ca^{2+}-dependent and -independent activities of neural and non-neural synaptotagmins. *Nature* **375**, 594–599.

Liao, D., Hessler, N.A., and Malinow, R. (1995). Activation of postsynaptically silent synapses during pairing-induced LTP in CA1 region of hippocampal slice. *Nature* **375**, 400–404.

Liley, A. W. (1956). The quantal components of the mammalian end-plate potential. *J. Physiol. (London)* **133**, 571–587.

Lindau, M., Stuenkel, E. L., and Nordmann, J. J. (1992). Depolarization, intracellular calcium and exocytosis in single vertebrate nerve endings. *Biophys. J.* **61**, 19–30.

Link, E., Edelmann, L., Chou, J. H., Binz, T., Yamasaki, S., Eisel, U., Baumert, M., Sudhof, T. C., Niemann, H., and Jahn, R. (1992). Tetanus toxin action: Inhibition of neurotransmitter release linked to synaptobrevin proteolysis. *Biochem. Biophys. Res. Commun.* **189**, 1017–1023.

Littleton, J. T., Stern, M., Schulze, K., Perin, M., and Bellen, H. J. (1993). Mutational analysis of *Drosophila* synaptotagmin demonstrates its essential role in Ca^{2+}-activated neurotransmitter release. *Cell* **74**, 1125–1134.

Llinás, R., Gruner, J. A., Sugimori, M., McGuiness, T. L., and Greengard, P. (1991). Regulation by synapsin I and Ca^{2+}-calmodulin-dependent protein kinase II of the transmitter release at the squid giant synapse. *J. Physiol. (London)* **436**, 257–282.

Llinás, R., Steinberg, I. Z., and Walton, K. (1981). Relationship between presynaptic calcium current and postsynaptic potential in squid giant synapse. *Biophys. J.* **33**, 323–351.

Llinás, R., Sugimori, M., and Silver, R. B. (1992). Microdomains of high calcium concentration in a presynaptic terminal. *Science* **256**, 677–679.

Locke, F. S. (1894). Notiz über den Einfluβ physiologischer Kochsalzlösung auf die elektrische Erregbarkeit von Muskel und Nerv. *Zentralbl. Physiol.* **8**, 166–167.

Lonart, G., and Südhof, T. C. (1998). Region-specific phosphorylation of rabphilin in mossy fiber nerve terminals of the hippocampus. *J. Neurosci.* **18**, 634–640.

Lundberg, J. M., and Hökfelt, T. (1986). Multiple co-existence of peptides and classical transmitters in peripheral autonomic and sensory neurons: Functional and pharmacological implications. *Prog. Brain Res.* **68**, 241–262.

Mackler, J. M., Drummond, J. A., Loewen, C. A., Robinson, I. M., and Reist, N. E. (2002). The C(2)B Ca(2+)-binding motif of synaptotagmin is required for synaptic transmission in vivo. *Nature* **418**, 340–344.

Magleby, K. L., and Zengel, J. E. (1982). A quantitative description of stimulation-induced changes in transmitter release at the frog neuromuscular junction. *J. Gen. Physiol.* **30**, 613–638.

Manabe, T., Renner, P., and Nicoll, R. (1992). Postsynaptic contribution to long-term potentiation revealed by the analysis of miniature synaptic currents. *Nature (London)* **355**, 50–55.

Manabe, T., Wyllie, D. J. A., Perkel, D. J., and Nicoll, R. A. (1993). Modulation of synaptic transmission and long-term potentiation: Effects on paired pulse facilitation and EPSC variance in the CA1 region of the hippocampus. *J. Neurophysiol.* **70**, 1451–1459.

Martin, A. R. (1977). Junctional transmission. II. Presynaptic mechanisms. *In* "Handbook of Physiology" (E. Kandel, Ed.), Sect. 1, pp. 329–355. Am. Physiol. Soc., Bethesda, MD.

Maruyama, I. N., and Brenner, S. (1991). A phorbol ester/diacylglycerol-binding protein encoded by the *unc-13* gene of *Caenorhabditis elegans*. *Proc. Natl. Acad. Sci. USA* **88**, 5729–5733.

Matthews-Bellinger, J., and Salpeter, M. M. (1973). Distribution of acetylcholine receptors at frog neuromuscular junctions with a discussion of some physiological implications. *J. Physiol. (London)* **279**, 197–213.

Matveev, V., Sherman, A., and Zucker, R. S. (2002). New and corrected simulations of synaptic facilitation. *Biophys. J.* **83**, 1368–1373.

Matveev, V., Zucker, R. S., and Sherman, A. (2004). Facilitation through buffer saturation: Constraints on endogenous buffering properties. *Biophys. J.* **86**, 2691–2709.

Maycox, P. R., Link, E., Reetz, A., Morris, S. A., and Jahn, R. (1992). Clathrin-coated vesicles in nervous tissue are involved primarily in synaptic vesicle recycling. *J. Cell Biol.* **118**, 1379-1388.

Mayer, A., Wickner, W., and Haas, A. (1996). Sec18p (NSF)-driven release of Sec17p (alpha-SNAP) can precede docking and fusion of yeast vacuoles. *Cell* **85**, 83–94.

McEwen, J. M., Madison, J. M., Dybbs, M., and Kaplan, J. M. (2006). Antagonistic regulation of synaptic vesicle priming by Tomosyn and UNC-13. *Neuron* **51**, 303–315.

McGuiness, T. L., Brady, S. T., Gruner, J. A., Sugimori, M., Llinás, R., and Greengard, P. (1989). Phosphorylation-dependent inhibition by synapsin I of organelle movement in squid axoplasm. *J. Neurosci.* **9**, 4138–4149.

McLachlan, E. M. (1978). The statistics of transmitter release at chemical synapses. *Int. Rev. Physiol.* **17**, 49–117.

McMahon, H. T., Bolshakov, V. Y., Janz, R., Hammer, R. E., Siegelbaum, S. A., and Südhof, T. C. (1996). Synaptophysin, a major synaptic vesicle protein, is not essential for neurotransmitter release. *Proc. Natl. Acad. Sci. USA* **93**, 4760–4764.

McMahon, H. T., Missler, M., Li, C., and Sudhof, T. C. (1995). Complexins: Cytosolic proteins that regulate SNAP receptor function. *Cell* **83**, 111–119.

McNew, J. A., Parlati, F., Fukuda, R., Johnston, R. J., Paz, K., Paumet, F., Söllner, T. H., and Rothman, J. E. (2000). Compartmental specificity of cellular membrane fusion encoded in SNARE proteins. *Nature* **407**, 153–159.

Meinrenken, C.J., Borst, J.G., and Sakmann, B. (2002). Calcium secretion coupling at calyx of held governed by nonuniform channel-vesicle topography. *J. Neurosci.* **22**, 1648–1667.

Mennerick, S., and Zorumski, C. F. (1996). Postsynaptic modulation of NMDA synaptic currents in rat hippocampal microcultures by paired-pulse stimulation. *J. Physiol.* **490**, 405–417.

Millar, A. G., Zucker, R. S., Ellis-Davies, G. C., Charlton, M. P., and Atwood, H. L. (2005). Calcium sensitivity of neurotransmitter release differs at phasic and tonic synapses. *J. Neurosci.* **25**, 3113–3125.

Minota, S., Kumamoto, E., Kitakoga, O., and Kuba, K. (1991). Long-term potentiation induced by a sustained rise in the intraterminal Ca^{2+} in bull-frog sympathetic ganglia. *J. Physiol. (London)* **435**, 421–438.

Monck, J. R. and Fernandez, J. M. (1992). The exocytotic fusion pore. *J. Cell Biol.* **119**, 1395–1404.

Mulkey, R. M., and Zucker, R. S. (1991). Action potentials must admit calcium to evoke transmitter release. *Nature (London)* **350**, 153–155.

Muller, R. U., and Finkelstein, A. (1974). The electrostatic basis of Mg^{2+} inhibition of transmitter release. *Proc. Natl. Acad. Sci. USA* **71**, 923–926.

Murthy, V. N., and Stevens, C. F. (1998). Synaptic vesicles retain their identity through the endocytic cycle. *Nature* **392**, 497–501.

Nagy, A., Baker, R. R., Morris, S. J., and Whittaker, V. P. (1976). The preparation and characterization of synaptic vesicles of high purity. *Brain Res.* **109**, 285–309.

Neher, E., and Zucker, R. S. (1993). Multiple calcium-dependent processes related to secretion in bovine chromaffin cells. *Neuron* **10**, 21–30.

Nelson, N. (1992). The vacuolar H^+-ATPase: One of the most fundamental ion pumps in nature. *J. Exp. Biol.* **172**, 19–27.

Nichols, B. J., Ungermann, C., Pelham, H. R., Wickner, W. T., and Haas, A. (1997). Homotypic vacuolar fusion mediated by t- and v-SNAREs. *Nature* **387**, 199–202.

Nickel, W., Weber, T., McNew, J. A., Parlati, F., Söllner, T. H., and Rothman, J. E. (1999). Content mixing and membrane integrity during membrane fusion driven by pairing of isolated v-SNAREs and t-SNAREs. *Proc. Natl. Acad. Sci. USA* **96**, 12571–12576.

Nonet, M. L., Grundahl, K., Meyer, B. J., and Rand, J. B. (1993). Synaptic function is impaired but not eliminated in C. elegans mutants lacking synaptotagmin. *Cell* **73**, 1291–1305.

Novick, P., Ferro, S., and Schekman, R. (1981). Order of events in the yeast secretory pathway. *Cell* **25**, 461–469.

Novick, P., Field, C., and Schekman, R. (1980). Identification of 23 complementation groups required for post-translational events in the yeast secretory pathway. *Cell* **21**, 205–215.

Nusser, Z., Lujan, R., Laube, G., Roberts, J. D., Molnar, E., and Somogyi, P. (1998) Cell type and pathway dependence of synaptic AMPA receptor number and variability in the hippocampus. *Neuron* **21**, 545–559.

Oertner, T. G., Sabatini, B. L., Nimchinsky, E. A., and Svoboda, K. (2002). Facilitation at single synapses probed with optical quantal analysis. *Nat. Neurosci.* **5**, 657–664.

Ohtsuka, T., Takao-Rikitsu, E., Inoue E., Inoue, M., Takeuchi, M., Matsubara, A., Deguchi-Twarada, M., Satoh, K., Morimoto, K., Nakanishi, H., and Takai, Y. (2002). Cast: A novel protein of the cytomatrix at the active zone of synapses that forms a ternary complex with RIM-1 and Munc13-1. *J. Cell Biol.* **158**, 577–590.

Otis, T., Zhang, S., and Trussell, L. O. (1996). Direct measurement of AMPA receptor desensitization induced by glutamatergic synaptic transmission. *J. Neurosci.* **16**, 7496–7504.

Parsons, S. M., Prior, C., and Marshall, I. G. (1993). Acetylcholine transport, storage, and release. *Int. Rev. Neurobiol.* **35**, 279-390.

Parsons, T. D., Coorssen, J. R., Horstmann, H., and Almers, W. (1995). Docked granules, the exocytic burst, and the need for ATP hydrolysis in endocrine cells. *Neuron* **15**, 1085–1096.

Paulsen, O., and Heggelund, P. (1994). The quantal size at retinogeniculate synapses determined from spontaneous and evoked EPSCs in guinea-pig thalamic slices. *J. Physiol. (London)* **480**, 505–511.

Pellizzari, R., Rossetto, O., Schiavo, G., and Montecucco, C. (1999). Tetanus and botulinum neurotoxins: Mechanism of action and therapeutic uses. *Philos. Trans. R. Soc. London Ser. B* **354**, 259–268.

Peng, Y. -Y., and Zucker, R. S. (1993). Release of LHRH is linearly related to the time integral of presynaptic Ca^{2+} elevation above a threshold level in bullfrog sympathetic ganglia. *Neuron* **10**, 465–473.

Perin, M. S., Brose, N., Jahn, R., and Südhof, T. C. (1991). Domain structure of synaptotagmin (p65). *J. Biol. Chem.* **266**, 623–629.

Perin, M. S., Fried, V. A., Mignery, G. A., Jahn, R., and Südhof, T. C. (1990). Phospholipid binding by a synaptic vesicle protein homologous to the regulatory region of protein kinase C. *Nature* **345**, 260–263.

Peters, C., and Mayer, A. (1998). Ca^{2+}/calmodulin signals the completion of docking and triggers a late step of vacuole fusion. *Nature* **396**, 575–580.

Peters, C., Bayer, M. J., Buhler, S., Andersen, J. S., Mann, M., and Mayer, A. (2001). Trans-complex formation by proteolipid channels in the terminal phase of membrane fusion. *Nature* **409**, 581–588.

Pevsner, J., Hsu, S. C., and Scheller, R. H. (1994). n-Sec1: A neural-specific syntaxin-binding protein. *Proc. Natl. Acad. Sci. USA* **91**, 1445–1449.

Pieribone, V. A., Shupliakov, O., Brodin, L., Hilfiker-Rothenfluh, S., Czernik, A. J., and Greengard, P. (1995). Distinct pools of synaptic vesicles in neurotransmitter release. *Nature* **375**, 493–497.

Prior, C. (1994). Factors governing the appearance of small-mode miniature endplate currents at the snake neuromuscular junction. *Brain Res.* **664**, 61–68.

Quetglas, S., Leveque, C., Miquelis, R., Sato, K., and Seagar, M. (2000). Ca^{2+}-dependent regulation of synaptic SNARE complex assembly via a calmodulin- and phospholipid-binding domain of synaptobrevin. *Proc. Natl. Acad. Sci. USA* **97**, 9695–9700.

Rahamimoff, R., Lev-Tov, A., and Meiri, H. (1980). Primary and secondary regulation of quantal transmitter release: Calcium and sodium. *J. Exp. Biol.* **89**, 5–18.

Redman, S. (1990). Quantal analysis of synaptic potentials in neurons of the central nervous system. *Physiol. Rev.* **70**, 165–198.

Reimer, R. J., Fon, E. A., and Edwards, R. H. (1998). Vesicular neurotransmitter transport and the presynaptic regulation of quantal size. *Curr. Opin. Neurobiol.* **8**, 405–412.

Reist, N. E., Buchanan, J., Li, J., DiAntonio, A., Buxton, E. M., and Schwarz, T. L. (1998). Morphologically docked synaptic vesicles are reduced in *synaptotagmin* mutants of *Drosophila*. *J. Neurosci.* **18**, 7662–7673.

Richmond, J. E., Davis, W. S., and Jorgensen, E. M. (1999). UNC-13 is required for synaptic vesicle fusion in C. elegans. *Nat. Neurosci.* **2**, 959–964.

Roberts, W. M. (1994). Localization of calcium signals by a mobile calcium buffer in frog saccular hair cells. *J. Neurosci.* **14**, 3246–3262.

Roberts, W. M., Jacobs, R. A., and Hudspeth, A. J. (1990). Colocalization of ion channels involved in frequency selectivity and synaptic transmission at presynaptic active zones of hair cells. *J. Neurosci.* **10**, 3664–3684.

Robinson, I. M., Ranjan, R., and Schwarz, T. L. (2002). Synaptotagmins I and IV promote transmitter release independently of Ca(2+) binding in the C(2)A domain. *Nature* **418**, 336–340.

Robinson, P. J., Liu, J. P., Powell, K. A., Fykse, E. M., and Südhof, T. C. (1994). Phosphorylation of dynamin I and synaptic-vesicle recycling. *Trends Neurosci.* **17**, 348–353.

Robitaille, R., Adler, E. M., and Charlton, M. P. (1990). Strategic location of calcium channels at transmitter release sites of frog neuromuscular synapses. *Neuron* **5**, 773–779.

Rosahl, T. W., Geppert, M., Spillane, D., Herz, J., Hammer, R. E., Malenka, R. C., and Südhof, T. C. (1993). Short-term synaptic plasticity is altered in mice lacking synapsin I. *Cell* **75**, 661–670.

Rosenmund, C., Clements, J. D., and Westbrook, G. L. (1993). Nonuniform probability of glutamate release at a hippocampal synapse. *Science* **262**, 754–757.

Rosenmund, C., Sigler, A., Augustin, I., Reim, K., Brose, N., and Rhee, J. S. (2002). Differential control of vesicle priming and short-term plasticity by Munc13 isoforms. *Neuron* **33**, 411–424.

Rothman, J. E., and Warren, G. (1994). Implications of the SNARE hypothesis for intracellular membrane topology and dynamics. *Curr. Biol.* **4**, 220–233.

Rozov, A., Burnashev, N., Sakmann, B., and Neher, E. (2001). Transmitter release modulation by intracellular Ca^{2+} buffers in facilitating and depressing nerve terminals of pyramidal cells in layer 2/3 of the rat neocortex indicates a target cell-specific difference in presynaptic calcium dynamics. *J. Physiol.* **531**, 807–826.

Ryan, T. A., and Smith, S. J. (1995). Vesicle pool mobilization during action potential firing at hippocampal synapses. *Neuron* **14**, 983–989.

Ryan, T. A., Li, L., Chin, L. S., Greengard, P., and Smith, S. J. (1996). Synaptic vesicle recycling in synapsin I knock-out mice. *J. Cell Biol.* **134**, 1219–1227.

Ryan, T. A., Reuter, H., Wendland, B., Schweizer, F. E., Tsien, R. W., and Smith, S. J. (1993). The kinetics of synaptic vesicle recycling measured at single presynaptic boutons. *Neuron* **11**, 713–724.

Ryan, T. A., Reuter, H., and Smith, S. J. (1997). Optical detection of a quantal presynaptic membrane turnover. *Nature* **388**, 478–482.

Sakaba, T., and Neher, E. (2001). Calmodulin mediates rapid recruitment of fast-releasing synaptic vesicles at a calyx-type synapse. *Neuron* **32**, 1119–1131.

Sakaguchi, M., Inaishi, Y., Kashihara, Y., and Kuno, M. (1991). Release of calcitonin gene-related peptide from nerve terminals in rat skeletal muscle. *J. Physiol. (London)* **434**, 257–270.

Salminen, A., and Novick, P. J. (1987). A ras-like protein is required for a post-Golgi event in yeast secretion. *Cell* **49**, 527–538.

Sankaranarayanan, S., and Ryan, T. A. (2001). Calcium accelerates endocytosis of vSNAREs at hippocampal synapses. *Nat. Neurosci.* **4**, 129–136.

Sato, T. K., Rehling, P., Peterson, M. R., and Emr, S. D. (2000). Class C Vps protein complex regulates vacuolar SNARE pairing and is required for vesicle docking/fusion. *Mol. Cell* **6**, 661–671.

Schiavo, G., Benfenati, F., Poulain, B., Rossetto, O., Polverino de Laureto, P., DasGupta, B. R., and Montecucco, C. (1992). Tetanus and botulinum-B neurotoxins block neurotransmitter release by proteolytic cleavage of synaptobrevin. *Nature* **359**, 832–835.

Schiavo, G., Rossetto, O., Catsicas, S., Polverino de Laureto, P., DasGupta, B. R., Benfenati, F., and Montecucco, C. (1993). Identification of the nerve terminal targets of botulinum neurotoxin serotypes A, D, and E. *J. Biol. Chem.* **268**, 23784–23787.

Schiavo, G., Rossetto, O., and Montecucco, C. (1994). Clostridial neurotoxins as tools to investigate the molecular events of transmitter release. *Semin. Cell Biol.* **5**, 221–229.

Schikorski, T., and Stevens, C. F. (1997) Quantitative ultrastructural analysis of hippocampal excitatory synapses. *J. Neurosci.* **17**, 5858–5867.

Schluter, O. M., Schmitz, F., Jahn, R., Rosenmund, C., and Sudhof, T. C. (2004). A complete genetic analysis of neuronal rab3 function. *J. Neurosci.* **24**, 6629–6637.

Schneggenburger, R., and Neher, E. (2000). Intracellular calcium dependence of transmitter release rates at a fast central synapse. *Nature* **406**, 889–893.

Schulze, K. L., Broadie, K., Perin, M. S., and Bellen, H. J. (1995). Genetic and electrophysiological studies of *Drosophila* syntaxin-1A demonstrate its role in nonneuronal secretion and neurotransmission. *Cell* **80**, 311–320.

Schweizer, F.E., and Ryan, T.A. (2006). The synaptic vesicle: Cycle of exocytosis and endocytosis. *Curr. Opin. Neurobiol.* **16**, 298–304.

Searl, T., Prior, C., and Marshall, I. G. (1991). Acetylcholine recycling and release at rat motor nerve terminals studied using (–)-vesamicol and troxpyrrolium. *J. Physiol. (London)* **444**, 99–116.

Sesack, S. R., and Snyder, C. L. (1995). Cellular and subcellular localization of syntaxin-like immunoreactivity in the rat striatum and cortex. *Neurosci.* **67**, 993–1007.

Shirataki, H., Kaibuchi, K., Sakoda, T., Kishida, S., Yamaguchi, T., Wada, K., Miyazaki, M., and Takai, Y. (1993). Rabphilin–3A, a putative target protein for smg p25A/rab3A p25 small GTP-binding protein related to synaptotagmin. *Mol. Cell. Biol.* **13**, 2061–2068.

Silver, R.A., Lubke, J., Sakmann, B., and Feldmeyer, D. (2003). High-probability uniquantal transmission at excitatory synapses in barrel cortex. *Science* **302**, 1981–1984.

Silverman, B. W. (1986). "Density Estimation for Statistics and Data Analysis." Chapman & Hall, London.

Simon, S. M., and Llinás, R. R. (1985). Compartmentalization of the submembrane calcium activity during calcium influx and its significance in transmitter release. *Biophys. J.* **48**, 485–498.

Smith, C., Moser, T., Xu, T., and Neher, E. (1998). Cytosolic Ca^{2+} acts by two separate pathways to modulate the supply of release-competent vesicles in chromaffin cells. *Neuron.* **20**, 1243–1253.

Söllner, T., Bennett, M. K., Whiteheart, S. W., Scheller, R. H., and Rothman, J. E. (1993). A protein assembly–disassembly pathway *in vitro* that may correspond to sequential steps of synaptic vesicle docking, activation, and fusion. *Cell* **75**, 409–418.

Stanley, E. F. (1993). Single Ca^{2+} channels and acetylcholine release at a presynaptic nerve terminal. *Neuron* **11**, 1007–1011.

Steinbach, J. H., and Stevens, C. F. (1976). Neuromuscular transmission. *In* "Frog Neurobiology" (R. Llinás and W. Precht, Eds.), pp. 33–92. Springer-Verlag, Berlin.

Stevens, C. F., and Tsujimoto, T. (1995). Estimates for the pool size of releasable quanta at a single central synapse and for the time required to refill the pool. *Proc. Natl. Acad. Sci. USA* **92**, 846–849.

Stevens, C. F., and Wesseling, J. F. (1998). Activity-dependent modulation of the rate at which synaptic vesicles become available to undergo exocytosis. *Neuron* **21**, 415–424.

Stevens, C. F., and Wesseling, J. F. (1999). Identification of a novel process limiting the rate of synaptic vesicle cycling at hippocampal synapses. *Neuron* **24**, 1017–1028.

Stricker, C., and Redman, S. (1994). Statistical models of synaptic transmission evaluated using the expectation–optimization algorithm. *Biophys. J.* **67**, 656–670.

Südhof, T. C., Lottspeich, F., Greengard, P., Mehl, E., and Jahn, R. (1987). A synaptic vesicle protein with a novel cytoplasmic domain and four transmembrane regions. *Science* **238**, 1142–1144.

Südhof, T. C., Czernik, A. J., Kao, H. -T., Takei, K., Johnston, P. A., Horiuchi, A., Kanazir, S. D., Wagner, M. A., Perin, M. S., de Camilli, P., and Greengard, P. (1989). Synapsins: Mosaics of shared and individual domains in a family of synaptic vesicle phosphoproteins. *Science* **245**, 1474–1480.

Südhof, T. C. (1997). Function of Rab3 GDP–GTP exchange. *Neuron* **18**, 519–522.

Sun, J., Bronk, P., Liu, X., Han, W., and Südhof, T. C. (2006). Synapsins regulate use-dependent synaptic plasticity in the calyx of Held by a Ca^{2+}/calmodulin-dependent pathway. *Proc. Natl. Acad. Sci. USA* **103**, 2880–2885.

Sutton, R. B., Fasshauer, D., Jahn, R., and Brunger, A. T. (1998). Crystal structure of a SNARE complex involved in synaptic exocytosis at 2.4 A resolution. *Nature* **395**, 347–353.

Takamori, S., Rhee, J. S., Rosenmund, C., and Jahn, R. (2000). Identification of a vesicular glutamate transporter that defines a glutamatergic phenotype in neurons. *Nature* **407**, 189–194.

Takamori, S., Holt, M., Stenius, K., Lemke, E. A., Grønborg, M., Riedel, D., Urlaub, H., Schenck, S., Brügger, B., Ringler, P.,

Müller, S. A., Rammner, B., Gräter, F., Hub, J. S., De Groot, B. L., Mieskes, G., Moriyama, Y., Klingauf, J., Grubmüller, H., Heuser, J., Wieland, F., and Jahn, R. (2006). Molecular anatomy of a trafficking organelle. *Cell* **127**, 831–846.

Tang, C.-M., Margulis, M., Shi, Q.-Y., and Fielding, A. (1994). Saturation of postsynaptic glutamate receptors after quantal release of transmitter. *Neuron* **13**, 1385–1393.

Tang, Y. -G., and Zucker, R. S. (1997). Mitochondrial involvement in post-tetanic potentiation of synaptic transmission. *Neuron* **18**, 483–491.

Thoreson, W.B., Rabl, K., Townes-Anderson, E., and Heidelberger, R. (2004). A highly Ca^{2+}-sensitive pool of vesicles contributes to linearity at the rod photoreceptor ribbon synapse. *Neuron* **42**, 595–605.

Tong, G., and Jahr, C. E. (1994). Multivesicular release from excitatory synapses of cultured hippocampal neurons. *Neuron* **12**, 51–59.

Torri Tarelli, F., Bossi, M., Fesce, R., Greengard, P., and Valtorta, F. (1992). Synapsin I partially dissociates from synaptic vesicles during exocytosis induced by electrical stimulation. *Neuron* **9**, 1143–1153.

Torri-Tarelli, F., Grohovaz, F., Fesce, R., and Ceccarelli, B. (1985). Temporal coincidence between synaptic vesicle fusion and quantal secretion of acetylcholine. *J. Cell Biol.* **101**, 1386–1399.

Trimble, W. S., Cowan, D. M., and Scheller, R. H. (1988). VAMP-1: A synaptic vesicle-associated integral membrane protein. *Proc. Natl. Acad. Sci. USA* **85**, 4538–4542.

Trussell, L. O., Zhang, S., and Raman, I. M. (1993). Desensitization of AMPA receptors upon multiquantal neurotransmitter release. *Neuron* **10**, 1185–1196.

Van der Bliek, A. M., and Meyerowitz, E. M. (1991). Dynamin-like protein encoded by the *Drosophila shibire* gene associated with vesicular traffic. *Nature* **351**, 411–414.

Van der Kloot, W. (1988). Acetylcholine quanta are released from vesicles by exocytosis (and why some think not). *Neurosci.* **24**, 1–7.

Van der Kloot, W. (1991). The regulation of quantal size. *Prog. Neurobiol.* **36**, 93–130.

Van der Kloot, W. (1994). Facilitation at the frog neuromuscular junction at 0°C is not maximal at time zero. *J. Neurosci.* **14**, 5722–5724.

Van der Kloot, W., and Molgó, J. (1994). Quantal acetylcholine release at the vertebrate neuromuscular junction. *Physiol. Rev.* **74**, 899–991. (An extremely comprehensive modern review.)

Van der Kloot, W., Balezina, O. P., Molgó, J., and Naves, L. A. (1994). The timing of channel opening during miniature endplate currents at the frog and mouse neuromuscular junctions: Effects of fasciculin-2, other anti-cholinesterases and vesamicol. *Pfluegers Arch.* **428**, 114–126.

Verhage, M., Ghijsen, W. E. J. M., and Lopes da Silva, F. H. (1994). Presynaptic plasticity: The regulation of Ca^{2+}-dependent transmitter release. *Prog. Neurobiol.* **42**, 539–574.

Verhage, M., McMahon, H. T., Ghijsen, W. E. J. M., Boomsma, F., Scholten, G., Wiegant, V. M., and Nicholls, D. G. (1991). Differential release of amino acids, neuropeptides, and catecholamines from isolated nerve terminals. *Neuron* **6**, 517–524.

von Wedel, R. J., Carlson, S. S., and Kelly, R. B. (1981). Transfer of synaptic vesicle antigens to the presynaptic plasma membrane during exocytosis. *Proc. Natl. Acad. Sci. USA* **78**, 1014–1018.

Vyshedskiy, A., and Lin, J. -W. (1997). Activation and detection of facilitation as studied by presynaptic voltage control at the inhibitor of the crayfish opener muscle. *J. Neurophysiol.* **77**, 2300–2315.

Wachtel, H., and Kandel, E. R. (1971). Conversion of synaptic excitation to inhibition at a dual chemical synapse. *J. Neurophysiol.* **34**, 56–68.

Walmsley, B. (1991). Central synaptic transmission: Studies at the connection between primary muscle afferents and dorsal spinocerebellar tract (DSCT) neurones in Clarke's column of the spinal cord. *Prog. Neurobiol.* **36**, 391–423.

Wang, J., Renger, J. J., Griffith, L. C., Greenspan, R. J., and Wu, C. -F. (1994). Concomitant alterations of physiological and developmental plasticity in Drosophila CaM kinase II-inhibited synapses. *Neuron* **13**, 1373–1384.

Wang, L.-Y., and Kaczmarek, L. K. (1998). High-frequency firing helps replenish the readily releasable pool of synaptic vesicles. *Nature* **394**, 384–388.

Wang, Y., Okamoto, M., Schmitz, F., Hofmann, K., and Südhof, T. C. (1997). Rim is a putative Rab3 effector in regulating synaptic–vesicle fusion. *Nature* **388**, 593–598.

Warnock, D. E., and Schmid, S. L. (1996). Dynamin GTPase, a force-generating molecular switch. *Bioessays* **18**, 885–893.

Wheeler, D. B., Randall, A., and Tsien, R. W. (1994). Roles of N-type and Q-type Ca^{2+} channels in supporting hippocampal synaptic transmission. *Science* **264**, 107–111.

Whittaker, V. P. (1988). Model cholinergic systems: An overview. *Handb. Exp. Pharmacol.* **86**, 3–22.

Wickner, W., and Haas, A. (2000). Yeast homotypic vacuole fusion: A window on organelle trafficking mechanisms. *Annu. Rev. Biochem.* **69**, 247–275.

Wierda, K. D., Toonen, R. F., de Wit, H., Brussaard, A. B., and Verhage, M. (2007). Interdependence of PKC-dependent and PKC-independent pathways for presynaptic plasticity. *Neuron* **54**, 275–290.

Wojtowicz, J. M., and Atwood, H. L. (1988). Presynaptic long-term facilitation at the crayfish neuromuscular junction: Voltage-dependent and ion-dependent phases. *J. Neurosci.* **8**, 4667–4674.

Wright, S. N., Brodwick, M. S., and Bittner, G. D. (1996). Calcium currents, transmitter release and facilitation of release at voltage-clamped crayfish nerve terminals. *J. Physiol.* **496**, 363–378.

Wu, L.-G., and Betz, W. J. (1998). Kinetics of synaptic depression and vesicle recycling after tetanic stimulation of frog motor nerve terminals. *Biophys. J.* **74**, 3003–3009.

Wu, L.-G., and Borst, J. G. G. (1999). The reduced release probability of releasable vesicles during recovery from short-term synaptic depression. *Neuron* **23**, 821–832.

Xu, J., and Wu, L. G. (2005). The decrease in the presynaptic calcium current is a major cause of short-term depression at a calyx-type synapse. *Neuron* **46**, 633–645.

Yamada, W. M., and Zucker, R. S. (1992). Time course of transmitter release calculated from simulations of a calcium diffusion model. *Biophys. J.* **61**, 671–682.

Yamasaki, S., Baumeister, A., Binz, T., Blasi, J., Link, E., Cornille, F., Roques, B., Fykse, E. M., Südhof, T. C., Jahn, R., and Niemann, H. (1994). Cleavage of members of the synaptobrevin/VAMP family by types D and F botulinal neurotoxins and tetanus toxin. *J. Biol. Chem.* **269**, 12764–12772.

Zengel, J. E., and Magleby, K. L. (1980). Differential effects of Ba^{2+}, and Sr^{2+}, and Ca^{2+} on stimulation-induced changes in transmitter release at the frog neuromuscular junction. *J. Gen. Physiol.* **76**, 175–211.

Zengel, J. E., and Magleby, K. L. (1981). Changes in miniature endplate potential frequency during repetitive nerve stimulation in the presence of Ca^{2+}, Ba^{2+}, and Sr^{2+} at the frog neuromuscular junction. *J. Gen. Physiol.* **77**, 503–529.

Zhang, J. Z., Davletov, B. A., Südhof, T. C., and Anderson, R. G. (1994). Synaptotagmin I is a high affinity receptor for clathrin AP-2: Implications for membrane recycling. *Cell* **78**, 751–760.

Zhong, N., Beaumont, V., and Zucker, R. S. (2001). Roles for mitochondrial and reverse mode Na^+/Ca^{2+} exchange and the plasmalemma Ca^{2+} ATPase in post-tetanic potentiation at crayfish neuromuscular junctions. *J. Neurosci.* **21**, 9598–9607.

Zhong, Y., and Wu, C.-F. (1991). Altered synaptic plasticity in *Drosophila* memory mutants with a defective cyclic AMP cascade. *Science* **251**, 198–201.

Zucker, R. S. (1973). Changes in the statistics of transmitter release during facilitation. *J. Physiol. (London)* **229**, 787–810.

Zucker, R. S. (1989). Short-term synaptic plasticity. *Annu. Rev. Neurosci.* **12**, 13–31.

Zucker, R. S. (1994). Calcium and short-term synaptic plasticity. *Neth. J. Zool.* **44**, 495–512.

Zucker, R. S., and Fogelson, A. L. (1986). Relationship between transmitter release and presynaptic calcium influx when calcium enters through discrete channels. *Proc. Natl. Acad. Sci. USA* **83**, 3032–3036.

Zucker, R. S., Delaney, K. R., Mulkey, R., and Tank, D. W. (1991). Presynaptic calcium in transmitter release and posttetanic potentiation. *Ann. N. Y. Acad. Sci.* **635**, 191–207.

Zucker, R. S., and Regehr, W. G. (2002). Short-term synaptic plasticity. *Annu. Rev. Physiol.* **64**, 355–405.

Pharmacology and Biochemistry of Synaptic Transmission: Classical Transmitters

Ariel Y. Deutch and Robert H. Roth

The study of the nervous system 100 years ago was a period of claim and counterclaim, confusion, and recrimination—not unlike politics today or, for that matter, science. The reason for this rowdy transition to the twentieth century was a body of accumulating data that argued against the prevailing wisdom, which was that the brain is a large continuous network, with each of its cells in physical contact. The remarkably sharp eye and interpretative powers of the Spaniard Santiago Ramón y Cajal revealed a very different picture, in which neurons, the units of the brain, are independent structures (see Shepherd, 1991, and Chapter 1). Although it would take another 50 years for electron microscopic data to be obtained that provided the final confirmation of Cajal's hypothesis, his view of neurons as the independent units of the nervous system had gained widespread acceptance much earlier. In turn, this acceptance brought about a new debate: how do neurons that are not in physical contact communicate? The answer is not static but is evolving continuously. In this chapter we discuss briefly several means through which cells communicate with each other and then discuss in considerable detail the major means of communication, chemical synaptic transmission. Another means of communication, electrical transmission through gap junctions, is discussed in Chapter 15.

As discussed in Chapter 1, neurons vary widely in form and function but share certain structural characteristics. From the cell body emanates processes (axons and dendrites) representing polarized compartments of the cell. Axons can be short or long and remain local or project to distant areas. In contrast, dendrites are local. The general concept arose that axons transmit information, which is conveyed to the dendrites or soma of follower cells. The critical gap between the transmitting element of the neurons (axon) and the recipient zone of the follower cell (e.g., the dendrite) is the area across which transmission of information occurs; this area was termed the *synapse* by Charles Sherrington (Shepherd, 1991). Thus, there were *presynaptic* and *post*synaptic neurons. This general conceptual framework remains in place today, although there are many exceptions, including dendrites that release transmitters and axons that receive inputs from other neurons. One other characteristic proposed by Sherrington is central to the concept of chemical communication between neurons: synaptic transmission does not follow all-or-none rules but is graded in strength and flexible (Sherrington, 1906).

DIVERSE MODES OF NEURONAL COMMUNICATION

The controversy surrounding the nature of neuronal communication—chemical or electrical—was in full force for the first half of the twentieth century, even though evidence was marshaled in support of the chemical mode of communication midway through the nineteenth century (see Valenstein,

2005). In 1849, Claude Bernard noted that curare, the active constituent of a poison applied to arrows in South America, blocks nerve-to-muscle neurotransmission (Bernard, 1849). This effect was subsequently shown to be due to binding of curare to postsynaptic receptors for the neurotransmitter acetylcholine (ACh), thus disrupting neuromuscular transmission. Half a century later Thomas Elliott (1905) observed that adrenaline caused contractions of a smooth muscle that had been deprived of its neural innervation, indicating that muscle contraction depended on the action of molecules liberated from nerves. In a seminal series of studies using the isolated frog heart, Otto Loewi (1921) demonstrated chemical transmission by showing that ACh was released on nerve stimulation and activated a target muscle.

Although it is now clear that the major means through which neurons convey information to other cells is through chemical communication, neurons also use several other processes for intercellular communication. These include electrical synaptic transmission (see Chapter 15), ephaptic interactions, and autocrine, paracrine, and long-range signaling, to which molecules produced by both neural and non-neural cells contribute. The nonsynaptic mode of intercellular communication with the longest range (distance) is humoral or hormonal signaling. For example, some peripheral hormones can enter the central nervous system to modulate neuronal activity. The neuronal targets of these hormones have specific receptors that respond to the humoral signal, which may evoke either a short-term response (acutely change neuronal activity) or long-lasting changes in gene expression by targeting nuclear hormone receptors.

Molecules produced by neurons can also be used in intercellular communication that does not require synaptic specializations. Factors that are secreted by neurons or diffuse from the cells in which they are generated include classic neurotransmitters, neuropeptides, and neuroactive steroids, as well as gases such as nitric oxide and carbon monoxide and substances that resemble the active consituents of marijuana. These factors may activate receptors on the same cell that releases them (see Fig. 9.1), or regulate through paracrine signaling nearby cells (Fig. 9.1), and the gaseous transmitters and some other compounds may act as retrograde neurotransmitters, with a "post"synaptic neuron releasing a substance to signal a "pre"synaptic element, i.e., working backward. Soluble factors can act on high- or low-affinity receptors and can act either locally or over some distance. The role of such molecules is thought to be primarily in modulating neural activity, although they may also provide trophic support ("guidance cues") for neurons that are elaborating their processes (axons) as they route to make connections. A more extended discussion of the "unconventional" transmitters as well as peptide transmitters and growth factors can be found in the next chapter.

Cells may also signal one another through structural components and adhesion molecules that are bound to cell surfaces. Adhesion molecules are typically transmembrane proteins that have an extracellular segment that mediates adhesion interactions across the intercellular space and a cytoplasmic domain that is linked to signal transduction pathways or the cytoskeleton. The intercellular domains of adhesion molecules may be "activated" by the binding of the extracellular segment to a cognate membrane-bound "ligand" protruding from another cell. Only in the context of diffusible transmitters is this a strange idea, with molecules reaching across the synapse to "touch and kiss" and thereby drive intracellular signaling cascades. In many cases cell-specific adhesion molecules and extracellular matrix components are critical signals during neuronal development and provide cues for cell differentiation.

The most intimate nonsynaptic mode of intercellular communication is the ephapse, in which electrical impulses in one cell or extracellular ion accumulation in the vicinity of one cell can directly affect the activity of an adjacent cell. In this case (see Fig. 9.1), the only morphological requirement is closely apposed membranes. Ephaptic interactions are either transient or sustained, and can be induced by various treatments.

The two categories of intercellular communication considered to be true synaptic transmission are electrical (see Chapter 15) and chemical synapses (Fig. 9.1). We focus on chemical synaptic transmission, which is thought to be the dominant mode of interneuronal communication.

CHEMICAL TRANSMISSION

Chemically mediated transmission is the major means by which a signal is communicated from one nerve cell to another and is the mode of neuronal communication on which this and the next chapter focus. The general acceptance of chemical communication between neurons as the dominant mode of signaling required that certain criteria be met for a compound to be accepted as a neurotransmitter. There is a relatively small number of compounds that meet all these criteria as "classic" transmitters, and

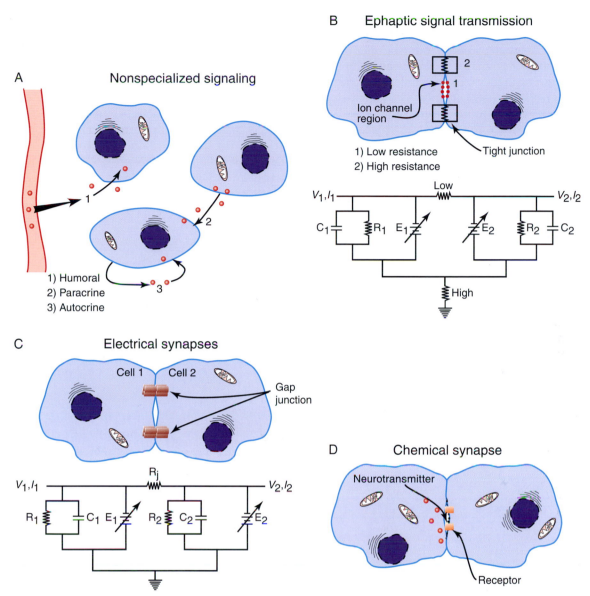

FIGURE 9.1 The multiple modes of intercellular signaling. (A) Substances (red) produced outside the nervous system (1) or by cells within the CNS (2, 3) can affect neuronal activity, acting through (1) humoral, (2) paracrine, and (3) autocrine mechanisms. (B) Ephaptic transmission. Two apposed cell membranes showing regions of low (1, red label indicating channels) and high (2, sawtooths, representing tight junctions) resistances. The electrical equivalent circuit is shown below. Communication between these cells is permitted through membrane regions of low resistance. The presence of tight junctions (2) between apposed cells favors an increase in current density by preventing current flow into the bulk extracellular space. Charges accumulate in the narrow intercellular space and affect capacitance and resistive components of the cell membranes. (C) Electrical synapses. Gap junction channels provide low-resistance pathways between adjacent cells, allowing direct communication between the cytoplasms of both cells. In contrast with the ephaptic mode of transmission, current flows directly from cell to cell and not through the extracellular space. The electrical equivalent circuit is shown below, differing from that of the ephapse in (B) primarily in the absence of a resistance to ground. (D) Chemical synapses. Neurotransmitters released from the presynaptic terminal (on the left) diffuse across the cleft to bind to postsynaptic receptors, thereby opening ion channels and increasing conductance of the postsynaptic membrane, producing currents that excite or inhibit the cell.

these were all identified by the mid-twentieth century. Since then it has become apparent that there is a much larger number of chemical messengers that broadly qualify as intercellular transmitters, although these often do not meet the classic criteria.

The Criteria for Definition as a Neurotransmitter

Neurotransmitters are endogenous substances that are released from neurons, act on receptors typically located on the membranes of postsynaptic cells, and produce a functional change in the properties of the target cell. Over the years there has been general agreement that several criteria should be met before a substance can be designated a neurotransmitter.

First, a neurotransmitter must be synthesized by and released from neurons. This usually means that the presynaptic neuron should contain a transmitter and the appropriate enzymes required for synthesis of that neurotransmitter. However, synthesis in the nerve terminal is not an absolute requirement. For example, peptide transmitters are synthesized in the cell body and transported to their release sites in axon terminals (see Chapter 10).

Second, the substance should be released from nerve terminals in a chemically or pharmacologically identifiable form, i.e., one should be able to isolate the transmitter and characterize its chemical structure.

Third, a neurotransmitter should reproduce at the postsynaptic cell the specific effects that are observed in response to stimulation of the presynaptic neuron. Moreover, the concentrations of applied transmitter should be similar to those measured after release of the neurotransmitter in response to nerve stimulation.

Fourth, the effects of a putative neurotransmitter should be blocked by known competitive antagonists of the transmitter in a dose-dependent manner. In addition, treatments that inhibit synthesis of the candidate transmitter should block the effects of presynaptic stimulation.

Fifth, there should be appropriate active mechanisms to terminate the action of the putative neurotransmitter. Such mechanisms can include enzymatic degradation and reuptake of the substance into the presynaptic neuron or glial cells through specific transporter molecules.

The Five Steps of Chemical Neurotransmission: Synthesis, Storage, Release, Receptor Binding, and Inactivation

The general mechanisms of chemical synaptic transmission are depicted in Fig. 9.2. Synaptic transmission consists of a number of steps, and each of these steps is a potential site of drug action.

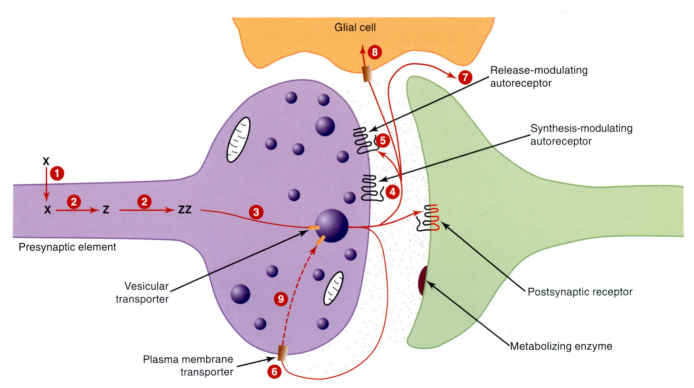

FIGURE 9.2 Schematic representation of the life cycle of a classic neurotransmitter. After accumulation of a precursor amino acid (X) into the neuron (step 1), the amino acid precursor is sequentially metabolized (step 2) to yield the mature transmitter (ZZ). The transmitter is then accumulated into vesicles by the vesicular transporter (step 3), where it is poised for release and protected from degradation. The released transmitter can interact with postsynaptic receptors (step 4) or autoreceptors (step 5) that regulate transmitter release, synthesis, or firing rate. Transmitter actions are terminated by means of a high-affinity membrane transporter (step 6) that is usually associated with the neuron that released the transmitter. Alternatively, the actions of the transmitter may be terminated by means of diffusion (step 7) or by accumulation into glia through a membrane transporter (step 8). When the transmitter is taken up by the neuron, it is subject to metabolic inactivation (step 9).

1. *Biosynthesis of the neurotransmitter in the presynaptic neuron.* For the transmitter to be synthesized, precursors should be present in appropriate sites in the neuron. The enzymes that catalyze the conversion of precursor(s) into transmitter should be present in an active form and localized to the appropriate compartment in the neuron, and any necessary cofactors for activity of the synthesizing enzyme should be present. The synthesis of neurotransmitters has long been an important site for clinically useful drugs. An example is α-methyl-*p*-tyrosine, a drug used in the treatment of an adrenal gland tumor (pheochromocytoma) that causes strikingly high blood pressure and increases the risk of strokes. This tumor releases massive amounts of norepinephrine, a neurotransmitter that acts to constrict blood vessels and increase cardiac output and thereby elevates blood pressure. α-Methyl-*p*-tyrosine blocks the synthesis of catecholamines such as norepinephrine and thus prevents the actions of these amine transmitters on target cells, lowering blood pressure.

2. *Storage of the neurotransmitter or its precursor or both in the presynaptic nerve terminal.* Classic and peptide transmitters are stored in synaptic vesicles, where they are sequestered and protected from enzymatic degradation and are readily available for release. In the case of so-called "classic" neurotransmitters (acetylcholine, biogenic amines, and amino acids), the synaptic vesicles are small (~50 nm in diameter). In contrast, neuropeptide transmitters in the brain are stored in large dense-core vesicles (~100 nm in diameter), which usually release their contents in response to high firing rates or burst firing of neurons. Because most neurotransmitters are synthesized in the cytosol of neurons, an active process must accumulate the transmitter into the vesicle; the proteins responsible for this action are the vesicular transporters.

3. *Release of the neurotransmitter into the synaptic cleft.* The transmitter-containing vesicle fuses with the cellular membrane, and release (exocytosis) of the transmitter occurs. Neurons use two pathways to secrete proteins. The release of most neurotransmitters occurs by a regulated pathway controlled by extracellular signals. The mechanisms involved in neurotransmitter release are discussed more fully in Chapter 8. There is a second (constitutive) pathway that is not triggered by extracellular stimulation and is used to secrete membrane components, viral proteins, and extracellular matrix molecules.

Some unconventional transmitters (e.g., growth factors) are in part synthesized and released in the constitutive pathway.

4. *Binding and recognition of the neurotransmitter by target receptors.* Transmitters that are released interact with receptors located on the target (postsynaptic) cell. These receptors fall into two broad classes. The first are metabotropic receptors, which are coupled to intracellular G proteins as effectors (see Chapters 11, 12, and 16). The other group of receptors, termed *ionotropic receptors,* form ion channels that are either ligand- or voltage-gated and through which ions such as sodium and calcium enter the cell. We usually think of neurotransmitter receptors as being localized to the postsynaptic neuron. However, receptors are also found on presynaptic neurons, which can respond to release of the transmitter from that cell (autoreceptors) or to a transmitter release from a different cell (heteroceptor). Autoreceptors can be thought of as a part of a homeostatic feedback system that modulates transmitter release or synthesis or the firing rate of neurons. A more thorough discussion of neurotransmitter receptors may be found in Chapter 11.

5. *Inactivation and termination of the action of the released transmitter.* Continuous activation of a neuron by a presynaptic input does not convey meaningful dynamic information to the recipient neuron and often has adverse effects; examples include tetanus in muscles or seizure discharges in neurons. There are multiple processes to terminate the action of neurotransmitters, both active and passive. The active mechanisms include reuptake of the transmitter through specific membrane transporter proteins (usually on the presynpatic axon) and enzymatic degradation of the released transmitter to an inactive substance. In addition, glial cells can accumulate certain released transmitters (see also Chapter 3). Diffusion of the transmitter away from the synapse is a passive mechanism that can inactivate neuronal signaling.

The five steps just described form a logical scaffold for understanding chemical neurotransmission. However, there are particular and peculiar intricacies for each of the steps and for each of the many neurotransmitters. Catecholamines are a structurally defined group of neurotransmitters that have been extensively studied, and which this chapter will focus on as illustrative examples. This discussion is followed by a brief examination of neurotransmission involving four other classic neurotransmitters. These include the indoleamine serotonin (5-hydroxytryptamine), acetylcholine, and the amino acids GABA (γ-aminobutyric acid) and glutamate. We also note the key differences between classic and other (nonclassic) neurotransmitters or chemical messengers, among which are peptide neurotransmitters and unconventional transmitters such as nitric oxide and growth factors. These are described in detail in the next chapter.

CLASSIC NEUROTRANSMITTERS

The term *classic* is used to distinguish acetylcholine, the catecholamines, serotonin, and the amino acid transmitters from other neurotransmitters. Although the designation is somewhat arbitrary, classic transmittters can be differentiated from others on several grounds. As discussed previously and in Chapter 8, storage vesicles, when present, are smaller for classic transmitters. In addition, classic transmitters or their metabolic products are subject to active reuptake by the presynaptic cell and can be viewed as homeostatically conserved, but there is no energy-dependent high-affinity reuptake process for nonclassic transmitters. Finally, most classic transmitters are enzymatically synthesized in the nerve terminal; in contrast, peptides and some unconventional transmitters are synthesized in the soma from a precursor protein and then transported to the nerve terminal.

The Catecholamine Neurotransmitters

The term *catecholamine* generally refers to organic compounds with a catechol nucleus (a benzene ring with two adjacent hydroxyl substitutions) and an amine group. In practice, however, the term is used to describe the endogenous compounds dopamine (dihydroxyphenylethylamine), norepinephrine, and epinephrine. These three neurotransmitters are formed by successive enzymatic steps requiring distinct enzymes (see Fig. 9.3). The localization of particular synthesizing enzymes to different cells results in distinct dopamine-, norepinephrine-, and epinephrine-containing neurons in the brain. The catecholamines also have transmitter roles in the peripheral nervous system and in the periphery also have certain hormonal functions.

In the peripheral nervous system, dopamine is present mainly as a precursor for norepinephrine but also has important biological activity in the kidney. Norepinephrine is the postganglionic sympathetic neurotransmitter in mammals; in contrast, epinephrine is the sympathetic transmitter in frogs. Despite this species difference in sympathetic nervous system characteristics, the biochemical aspects of neurotransmission as a general rule are remarkably constant across vertebrate species and, indeed, invertebrates.

Catecholamine Biosynthesis

The amino acids phenylalanine and tyrosine are precursors for catecholamines. These amino acids are present in high concentrations in both the plasma and brain. In mammals, tyrosine can be derived from dietary phenylalanine by an enzyme (phenylalanine hydroxylase) that is mainly found in the liver. Phenylketonuria, a disorder caused by insufficient amounts of phenylalanine hydroxylase, results in very high plasma and brain levels of phenylalanine. Unless dietary phenylalanine intake is restricted, this can result in intellectual impairment (see Box 9.1). Catecholamines are formed in the brain, adrenal chromaffin cells, and sympathetic nerves. The processes regulating catecholamine synthesis are generally the same in these different tissues.

Catecholamine synthesis is usually considered to begin with tyrosine, which represents a branch point for many important biosynthetic processes in animal tissues. The sequence of enzymatic steps in the synthesis of catecholamines from tyrosine was first postulated by Blaschko in 1939 and confirmed by Nagatsu and coworkers in 1964, when they demonstrated that the enzyme tyrosine hydroxylase (TH) converts the amino acid L-tyrosine into 3,4-dihydroxyphenylalanine (L-dopa). All the component enzymes in the catecholamine biosynthetic pathway have been purified to homogeneity, which has allowed detailed analyses of the kinetics, substrate specificity, and cofactor requirements of these enzymes and aided in the development of useful inhibitors of the enzymes. Moreover, the development of antibodies against the purified enzymes has permitted the precise localization of the enzymes by immunohistochemical techniques.

The hydroxylation of L-tyrosine by TH results in the formation of the dopamine precursor L-dopa, which is almost immediately metabolized to dopamine by L-aromatic amino acid decarboxylase (AADC). The decarboxylation step is so rapid that one cannot routinely measure L-dopa in brain without first inhibiting AADC. In dopamine-containing neurons of the brain the decarboxylation of L-dopa to dopamine is the final step in transmitter synthesis. However, in neurons using norepinephrine (also known as noradrenaline) or epinephrine (adrenaline) as transmitters, the enzyme dopamine beta-hydroxylase (DBH) is present; this enzyme oxidizes dopamine to yield norepinephrine. Finally, in neurons in which epinephrine is the transmitter, a third enzyme, phenylethanolamine N-methyltransferase (PNMT), converts norepinephrine

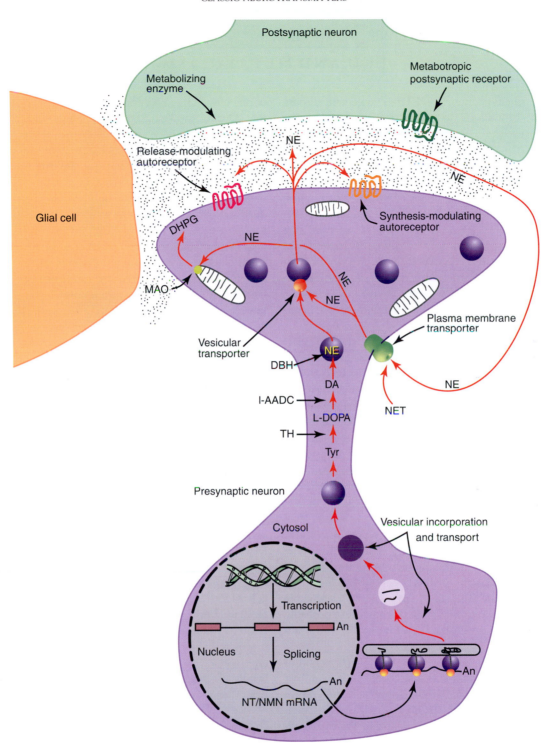

FIGURE 9.3 Characteristics of a norepinephrine (NE)-containing catecholamine neuron. On accumulation of tyrosine (Tyr) by the neuron, tyrosine is sequentially metabolized by tyrosine hydroxylase (TH) and L-aromatic amino acid decarboxylase (L-AADC) to dopamine (DA). The DA is then accumulated by the vesicular monoamine transporter. In dopaminergic neurons this is the final step. However, in this noradrenergic neuron, the DA is metabolized to NE by dopamine-b-hydroxylase (DBH), which is found in the vesicle. Once NE is released, it can interact with postsynaptic noradrenergic receptors or different types of presynaptic noradrenergic autoreceptors. The accumulation of NE by the high-affinity membrane norepinephrine transporter (NET) terminates the extracellular actions of NE. Once accumulated by the neuron, the NE can be metabolized to inactive species (for example, MHPG) by key degradative enzymes, such as monoamine oxidase (MAO), or taken back up by the vesicular transporter.

BOX 9.1

PKU AND METABOLISM

Classic phenylketonuria (PKU) is a genetic disease caused by mutations in the enzyme phenylalanine hydroxylase (PAH), resulting in loss of the enzyme's ability to hydroxylate phenylalanine (Phe) to tyrosine. PAH plays an integrated dual role in the metabolism of humans and other mammals. First, it provides an endogenous supply of tyrosine, thereby making consumption of this amino acid unnecessary for normal growth. Second, the reaction catalyzed by PAH is an essential step in the complete oxidation of Phe. When PAH levels are low or absent, as in PKU, blood levels of Phe are typically 20- to 50-fold higher than normal. A tiny fraction of the elevated Phe is converted to the phenylketone phenylpyruvic acid, which is excreted in the urine, hence the name of the disease.

The increased concentration of Phe seen in PKU spares the body, but spoils—indeed, devastates—the developing brain, and typically leads to severe mental retardation unless steps are taken to limit dietary Phe intake. The majority of untreated PKU patients suffers severe intellectual impairment (IQ <20). They also have a somewhat higher incidence of seizures and tend to have fair skin and hair. The latter effect is due to the inhibition of melanin formation by the excess Phe.

PKU is inherited as an autosomal recessive trait. The vast majority of PKU babies is conceived when both parents are heterozygotes, each one harboring one normal gene and one "PKU" gene. Thus, on average, one-fourth of the children born to such parents have PKU, one-fourth are normal, and half are heterozygotes. The incidence of heterozygosity for PKU is about 1 in 55.

The mechanism by which hyperphenylalaninemia damages the developing brain is unknown but probably involves competition by the high Phe levels with brain uptake of other essential amino acids.

Phenylalanine hydroxylase functions *in vivo* as part of a complex multicomponent system consisting of two other enzymes, dihydropteridine reductase and pterin 4α-carbinolamine dehydratase, and a nonprotein coenzyme, tetrahydrobiopterin (BH4). During the hydroxylation reaction, BH4 is stoichiometically oxidized; i.e., for every molecule of Phe converted to tyrosine by PAH, a molecule of BH4 is oxidized. The function of the two ancillary enzymes is to regenerate BH4.

With the realization that the hydroxylating system consists of four essential components, it was predicted that there might be variant forms of PKU caused by lack of one of the other components of the hydroxylating system. During the last 20 years, patients with these predicted variants have been described. Defects in the reductase or in one of the several enzymes essential for the synthesis of BH4 (but not the one due to defects in the dehydratase) were originally called "lethal" or "malignant" PKU. In all probability, these variants are deadly because BH4 and the reductase are also essential for the functions of tyrosine hydroxylase (TH) and tryptophan hydroxylase (TPH), thus leading to deficits in catecholamine and serotonin systems. These patients therefore suffer from three different metabolic lesions. Fortunately, these variants are extremely rare, accounting for between 1 and 2% of all PKU patients, and can be treated (with varying degrees of success) by feeding them the compounds beyond the metabolic blocks in TH and TPH (i.e., L-dopa and 5-hydroxytryptophan) and, when needed, large doses of BH4.

If one can dare say that there is anything fortunate about this dreadful disease, it is that it is extremely rare, with an average incidence of about 1/12,000. The frequency, however, varies widely among different ethnic groups, being only 1/200,000 in Japan but as high as 1/5000 in Ireland.

One auspicious feature of the disease is that the affected infants are essentially normal at birth, which raised the hope that some way might be found to prevent the intellectual deterioration of PKU. This hope was realized about 50 years ago with the introduction of a low-Phe diet (not a no-Phe diet!), which has proven to be largely if not totally effective in preventing brain damage, at least as reflected by the normal IQ of PKU patients who are started on the diet shortly after birth. The low-Phe diet is a heavy burden for both patients and their families, and it was once hoped that the diet could be discontinued after 6 or 7 years. It now appears that a longer period is beneficial. The goal of the diet is to keep blood Phe levels from rising no more than five- to six-fold normal levels.

Women with PKU who are contemplating having children raise dietary issues as well. If women with PKU went off the diet at some earlier time, they must resume it before they become pregnant, or the fetus risks *in utero* damage caused by the mother's high levels of Phe, a condition called *maternal PKU.*

Since the *sine qua non* of the successful dietary treatment of PKU is to start the diet as soon as possible after birth, its success was closely tied to the development of a cheap and into epinephrine. Thus, a neuron that uses epinephrine as its transmitter contains four enzymes (TH, AADC, DBH, PNMT), which sequentially metabolize tyrosine to epinephrine. Noradrenergic neurons express only the enzymes TH, AADC, and DBH, and thus norepinephrine cannot be further metabolized to epinephrine. Similarly, because dopamine neurons lack DBH and PNMT, the catecholamine end product in these neurons is the transmitter dopamine. The enzymes and cofactors taking part in the synthesis of the catecholamines are illustrated in Fig. 9.4.

Tyrosine hydroxylase. In human beings, a single tyrosine hydroxylase gene gives rise to four TH mRNA species through alternative splicing (Lewis *et al.*, 1993). In contrast, in most primates two TH isoforms are present. Still different is the rat, which possesses but a single form of TH. It has been speculated that the different forms of TH in human beings are associated with differences in activity of the enzyme, but conclusive data on the functional significance of the various isoforms are lacking.

Tyrosine hydroxylase function is determined by two factors: changes in enzyme activity (the rate at which the enzyme converts the precursor to its product) and changes in the amount of enzyme protein present. One determinant of TH activity is phosphorylation of the enzyme (Fig. 9.4), which takes place at four different serine sites at the N terminus of the protein (Haycock and Haycock, 1991). These four serine residues are differently phosphorylated by various kinases. A second means of regulating the activity of the enzyme is through end-product inhibition: catecholamines can inhibit the activity of TH through competition for a required pterin cofactor for the enzyme (see Cooper *et al.*, 2002).

An increased neuronal demand for catecholamine synthesis can be met by inducing new TH protein or by activating (by phosphorylation) the enzyme. The degree to which increases in catecholamine synthesis depend on *de novo* synthesis of new enzyme protein

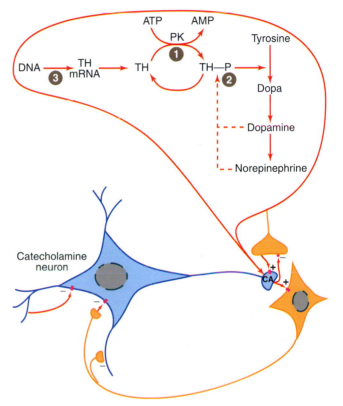

FIGURE 9.4 Schematic for the regulation of TH enzymatic activity. The numbered sites depict the three major types of regulation from TH phosphorylation (1), accomplished by the action of specific protein kinases (PK), to end-product inhibition (2) to changes in TH gene transcription (3) and the subsequent increase in protein.

or changes in enzymatic activity differs in various catecholamine neurons. For example, increased synthetic demand in brain-stem noradrenergic neurons is accomplished primarily by increasing TH gene expression, with the resultant increase in TH protein. In contrast, the same conditions and treatments that elicit increased TH gene expression in brain-stem noradrenergic neurons fail to increase TH mRNA levels in midbrain dopamine neurons. In these dopamine cells it appears that synthesis is regulated primarily by altering the activity of TH enzyme.

The synthesis of catecholamines starts with the entry of tyrosine into the brain. This process is an energy-dependent one at which tyrosine competes with large neutral amino acids as a substrate for a transporter. Because brain levels of tyrosine are high enough to saturate TH, under basal conditions synthesis of catecholamines is not increased by administration of tyrosine. An exception is synthesis in catecholamine neurons that have a high basal firing rate, such as dopamine neurons that innervate the prefrontal cortex, particularly under pathological conditions (Tam *et al.*, 1990).

Since TH is saturated by tyrosine under basal conditions, TH is the rate-limiting step for catecholamine synthesis under basal conditions. However, under conditions of neuronal activation, the enzyme responsible for norepinephrine synthesis, DBH, becomes rate limiting (Scatton *et al.*, 1984), and thus under certain conditions tyrosine availability may regulate catecholamine synthesis.

Tyrosine hydroxylase is a mixed-function oxidase that has moderate substrate specificity, hydroxylating phenylalanine as well as tyrosine. The major substrate in brain is tyrosine. TH function requires a biopterin cofactor and iron (Fe^{2+}). Tetrahydrobiopterin (BH4) is an essential cofactor for tyrosine hydroxylase as well as several other eyzymes (Thorny *et al.* 2000). Because the levels of the reduced BH4 cofactor are not saturated under basal conditions, endogenous levels of BH4 are important in regulating TH activity. The rate-limiting step in the synthesis of the BH4 cofactor is at GTP cyclohydrolase (Ichinose *et al.*, 1994). Thus, the activity of this enzyme is critical in regulating tyrosine hydroxylase and other pteridine-dependent enzymes. BH4 is also of clinical significance: a metation in the gene encoding GTP cyclohydrolase 1 causes a movement disorder called dopa-responsive dystonia.

L-Aromatic amino acid decarboxylase. The L-dopa formed by tyrosine hydroxylation is quickly decarboxylated by L-aromatic amino acid carboxylase (also known in the brain as "dopa decarboxylase") to the neurotransmitter dopamine. AADC has low substrate specificity and decarboxylates tryptophan as well as tyrosine. Because AADC is present in both catecholaminergic and serotonergic neurons, it plays an important role in the synthesis of both types of transmitters. In dopaminergic neurons, AADC is the final enzyme of the synthetic pathway.

Dopamine does not cross the blood–brain barrier. In contrast, L-dopa readily enters the brain. Accordingly, the dopamine precursor has achieved fame as

the primary approach for treating Parkinson disease, which is due to loss of dopamine in the striatum. L-dopa treatment of Parkinson disease patients increases brain dopamine concentrations and thus provides symptomatic relief. Although L-dopa readily enters the brain, there are decarboxylating enzymes in the liver and capillary endothelial cells that readily degrade the dopamine precursor, preventing appreciable amounts of L-dopa from reaching the brain. L-dopa is therefore administered in combination with a peripheral decarboxylase inhibitor that does not readily enter the brain. The administration of this peripheral decarboxylase inhibitor protects L-dopa from metabolism before it enters the brain, and thereby sharply increases central dopamine concentrations.

The AADC gene in mammalian organisms has multiple promotor sequences that lead to one AADC transcript being expressed in the brain and another in peripheral tissues. AADC mRNA is expressed in all catecholamine- and indoleamine-containing neurons in the CNS.

Levels of L-dopa are virtually unmeasurable in the CNS under basal conditions. This is because the activity of AADC is so high that L-dopa is converted into dopamine almost instantaneously. AADC requires pyridoxal 5-phosphate as a cofactor. The regulation of AADC has not been as intensively studied as that of TH, but converging data suggest that AADC is regulated primarily through induction of new protein rather than changes in activity. It is interesting to note that although L-dopa is used extensively in Parkinson disease, recent studies suggest that chronic administration of L-dopa or other dopamine agonists may decrease the activity of endogenous AADC. Thus, as is often the case with therapeutic administration of pharmacological agents, these drugs can have side effects or even be counter-therapeutic.

Dopamine β-hydroxylase. Noradrenergic and adrenergic neurons contain the enzyme dopamine-β-hydroxylase (DBH), a mixed-function oxidase that converts dopamine into norepinephrine. In noradrenergic neurons, this is the final step in catecholamine synthesis. Humans appear to possess two different DBH mRNAs that are generated from a single gene.

DBH is a copper-containing glycoprotein that requires ascorbic acid as the electron source during the hydroxylation of dopamine. Dicarboxylic acids such as fusaric acid are not absolute requirements but stimulate the enzymatic conversion of dopamine into norepinephrine. DBH does not have a high degree of substrate specificity and *in vitro* oxidizes almost any phenylethylamine to its corresponding phenylethanolamine. For example, in addition to forming norepinephrine from dopamine, DBH converts tyramine into octopamine and α-methyldopamine into α-methylnorepinephrine. Interestingly, many of the resultant structurally similar metabolites can replace norepinephrine at the noradrenergic nerve ending, acting as false neurotransmitters. More recently receptors for the trace amines such as octopamine have been cloned and characterized (Zucchi *et al.*, 2006). These receptors are expressed in both brain and gut, and may in part be responsible for the side effects of certain antidepressants and for therapeutic effects of other drugs.

The regulation of DBH is less completely understood than that of TH. It appears that under conditions that increase the activity of noradrenergic neurons not only does TH activity increase, but so does that of DBH, which actually becomes rate limiting at high firing rates of noradrenergic neurons. Because DBH is localized to the vesicle, when noradrenergic neurons are activated DBH becomes saturated and more dopamine accumulates in the vesicle; upon release of these vesicles there is substantial dopamine as well as norepinephrine release.

Phenylethanolamine **N-methyltransferase.** Phenylethanolamine *N*-methyltransferase (PNMT) is present at high levels in the adrenal medulla, where it methylates norepinephrine to form epinephrine, the major adrenal catecholamine. A single PNMT gene with three exons has been cloned. The transcript is present in the adrenal medulla and in the brain stem, and thus PNMT is also formed in two nuclei of the brain stem as well as the adrenal gland.

PNMT requires *S*-adenosylmethionine as the methyl donor for methylation of the amine nitrogen of norepinephrine. The enzyme has modest substrate specificity and will transfer methyl groups to the nitrogen atom on a variety of beta-hydroxylated amines. However, adrenal and brain PNMT are distinct from nonspecific *N*-methyltransferases found in the lung, which can methylate indoleamines such as serotonin.

The regulation of PNMT activity in the brain has not been extensively studied. In the adrenal gland, glucocorticoids regulate activity of the enzyme, and activity is increased in response to nerve growth factor.

Storage of Catecholamines and Their Enzymes

Vesicular storage. It has been known since the 1960s that much of the norepinephrine in sympathetic nerve endings and in adrenal chromaffin cells is present in highly specialized subcellular particles termed *granules*. Similarly, most of the norepinephrine and other catecholamines in the brain are stored in vesicles.

Vesicular storage of transmitters serves as a depot from which the transmitter can be released by appropriate physiological stimuli. Catecholamines are stored in small vesicles located near the synapse, where they are ready for fusion with the cellular membrane and subsequent exocytosis. In addition to being the substrate of release, the accumulation of catecholamines in the highly acidic vesicle prevents auto-oxidation of the amie and offers protection from metabolic inactivation by intraneuronal enzymes or attack by toxins that have gained entry into the neuron. Thus, the vesicle is a favorable environment for catecholamines offering strong protection against a variety of potentially harmful challenges.

The norepinephrine-synthesizing enzyme DBH differs from the other catecholamine-synthesizing enzymes by being present in the vesicles rather than the cytosol. This means that only after dopamine is accumulated into the vesicles by the vesicular monoamine transporter (VMAT) is dopamine metabolized to norepinephrine. As noted earlier, because DBH is a vesicular enzyme and the rate-limiting step in norepinephrine synthesis when noradrenergic neurons are strongly activated, noradrenergic axons can also release dopamine.

Vesicular monoamine transporters. The ability of vesicles to take up dopamine or other catecholamines depends on the presence of the VMAT (Weihe and Eiden, 2000). The VMAT is distinct from the neuronal membrane transporter (which is discussed in detail later) in terms of both substrate affinity and localization. Two vesicular monoamine transporter genes have been cloned. One is found in the adrenal cells that synthesize and release monoamines. The other is present in catecholamine and serotonin neurons. VMAT2, the isoform found in the brain, shows modest substrate specificity and transports catecholamines and serotonin, as well as histamine, into vesicles. VMATs are Mg^{2+}-dependent and inhibited by reserpine, a drug that disrupts vesicular storage of monoamines (Henry *et al.*, 1998).

Reserpine has been used in India for centuries as a folk medicine to treat hypertension and psychoses. The discovery that reserpine depletes vesicular stores of monoamines was critical to the understanding of the mechanisms through which reserpine alleviates psychotic symptoms, and sheds light on the means through which certain toxins can cause a Parkinson-disease-like syndrome. The use of reserpine in the treatment of hypertension and psychoses was reported in international journals in the early 1930s, but the therapeutic actions of reserpine were not widely appreciated in Western medicine until a generation later. At that time, Bernard Brodie and coworkers discovered that reserpine depleted brain stores of serotonin. Contemporaneously, it became known that the hallucinogen LSD (lysergic acid diethylamide) is structurally related to serotonin, which resulted in the proposal that the antipsychotic actions of reserpine were due to its ability to deplete brain stores of serotonin. However, it was soon realized that reserpine depletes both serotonin and catecholamines in the brain, and thus the antipsychotic effects of reserpine might be due to either serotonin or catecholamine depletion (or both).

To determine which transmitters were more important, in the early 1950s Arvid Carlsson and colleagues (see Carlsson, 1972) administered catecholamine and serotonin precursors to reserpine-treated rats to replenish monoamine levels and then examined locomotor activity, which is severely depressed by reserpine treatment. Motor function was restored by administration of the dopamine precursor L-dopa but not by the serotonin precursor 5-hydroxytryptophan.

Dopamine was considered to be only a precursor to norepinephrine at that time, and an independent role for dopamine was not recognized. Carlsson found that despite the ability of L-dopa to restore motor function in rats treated with reserpine to deplete catecholamines, L-dopa did not restore brain concentrations of norepinephrine and epinephrine. This led Carlsson to demonstrate that L-dopa treatment of reserpinized animals increased brain dopamine levels, suggesting that dopamine was a transmitter in its own right. Moreover, the ability of L-dopa to replenish dopamine was interpreted to suggest that the primary mechanism through which reserpine exerts antipsychotic effects is by disrupting dopamine transmission. These and subsequent studies led to the hypothesis that dysfunction of central dopamine systems underlies schizophrenia. This hypothesis soon became the dominant view guiding schizophrenia research. Interestingly, recent data suggest that to treat most effectively the full spectrum of psychotic symptoms in schizophrenia, drugs that are antagonists at both dopamine and serotonin receptors may be superior to drugs that are antagonists only at dopamine receptors.

A critical function of vesicles is to determine how much transmitter is contained in a single release, i.e., quantal release of transmitter. This is perhaps best shown in a study that examined the effects of adenovirus-mediated transfection of VMAT2 in cultures of ventral midbrain DA neurons. This study found that overexpression of the vesicular transporter in small synaptic vesicles increased both transmitter quantal size and frequency of release (Pothos *et al.*, 2000). These data are consistent with the recruitment of vesicles that do not normally release DA.

The cloning of the VMATs revealed a significant homology of these vesicular transporters to a group of bacterial antibiotic drug resistance transporters, suggesting a role for VMAT in detoxification. This indeed is the case. VMAT allows the vesicles to sequester toxins and thereby reduce toxicity. This is best exemplified by studies in VMAT2 knockout mice and heterozygous mice bearing one copy of VMAT2 (see Edwards, 1993). In such heterozygous mice there is an increase in the toxicity of DA neurons in response to administration of the parkinsonian toxin MPTP. The loss of VMAT means less sequestration of the toxin, which can then exert its toxic actions by targeting mitochondrial respiration.

One final function of VMAT appears to be the indirect role that the transporter plays in the generation of neuromelanin, black-pigmented deposits that accumulate in midbrain DA neurons as a function of age and are responsible for the term *substantia nigra* (black stuff). Neuromelanin can be induced in rat dopamine neurons of the substantia nigra treated with the dopamine precursor L-dopa. This effect of L-dopa is abolished in cells in which VMAT2 is overexpressed.

Release of Catecholamines

Catecholamine release typically occurs by the same Ca^{2+}-dependent process (exocytosis) that has been described for other transmitters (see Chapter 8 for a more extended discussion). This occurs at the synaptic cleft, where the presynaptic axon terminal is apposed to a specialized postsynaptic density. It has also been suggested that synapses are present at varicosities seen along the axon (resembling beads along an axon string).

Catecholamine release has also been observed to occur through at least two other mechanisms. First, catecholamines can be released by a reversal of the catecholamine (cell membrane) transporters. This occurs in response to certain drugs (e.g., amphetamine) and has been reported to occur following the application of excitatory amino acids as well. Second, dopamine (and perhaps other catecholamines) can be released from dendrites through a process that may not always involve conventional exocytosis, since some studies have found that dendritic release is not Ca^{2+}-dependent.

Regulation of catecholamine synthesis and release by autoreceptors. The enzymes that control the synthesis of catecholamines can be regulated at the transcriptional level and by post-translational modifications that alter enzymatic activity. In addition, the synthesis of catecholamines can be regulated by the interaction of the catecholamine released from the nerve terminal with specific autoreceptors that are located on the nerve terminal. Similarly, the release of catecholamines is regulated by autoreceptors, as is the firing rate of catecholaminergic neurons.

Dopamine autoreceptors are perhaps the best characterized of the catecholamine autoreceptors. Autoreceptors exist on most parts of the neuron, including the soma, the dendrites, and nerve terminals. They can be defined functionally in relation to the events that they regulate. Thus, synthesis-, release-, and impulse-modulating dopamine autoreceptors have been described (see Cooper *et al.*, 2002). All three types of dopamine autoreceptors belong to the D2 family of dopamine receptors, which includes three cloned receptors (D_2, D_3, and D_4). Although it is clear that there are D_2 autoreceptors, there has been considerable controversy surrounding the presence of D_3 autoreceptors. However, even assuming that there are D_3 as well as D_2 autoreceptors, there are three functionally different types of dopaminergic autoreceptors, raising the possibility that the same receptor protein may couple to functionally different autoreceptor roles through distinct transduction mechanisms.

Release-modulating autoreceptors appear to be a common regulatory feature on catecholamine neurons and other neurons that use classic transmitters. Dopamine that is released from the neuron interacts with an autoreceptor and dampens further release of the transmitter. This process can be thought of as a homeostatic feedback mechanism. Because intracellular levels of dopamine regulate tyrosine hydroxylase activity by binding the pterin cofactor, changes in the release of dopamine may also alter synthesis of transmitter.

Autoreceptors also directly regulate the synthesis of dopamine. Again, dopamine when released acts homeostatically at the synthesis-modulating autoreceptor to control synthesis: dopamine agonists decrease synthesis, whereas dopamine antagonists increase synthesis of the transmitter. Interestingly, synthesis-modulating autoreceptors are not found on all dopamine neurons. Some midbrain dopamine neurons that project to the prefrontal cortex appear to lack synthesis-modulating autoreceptors, as do the tuberoinfundibular dopamine neurons of the hypothalamus. Because release-modulating autoreceptors may indirectly regulate synthesis, the presence of synthesis-modulating autoreceptors may not be necessary in certain neurons.

Impulse-modulating autoreceptors are located on the soma and dendrites of dopamine neurons and regulate the firing rate of dopamine neurons. As noted earlier, because the release of dopamine can alter dopamine synthesis, impulse-modulating autoreceptors can also be expected to change dopamine synthesis. Thus, all three types of dopamine autoreceptors may regulate synthesis. This interdependence of regulatory processes over catecholamine neurons appears to be characteristic of monoamine neurons.

Although dopamine autoreceptors have been perhaps the most intensely studied autoreceptors, there are also norepinephrine autoreceptors that regulate release. However, the direct regulation of norepinephrine synthesis by synthesis-modulating autoreceptors on noradrenergic neurons is not well established. We do know, however, that there are two norepinephrine autoreceptors. One of these, an a_2 receptor, inhibits norepinephrine release, but a second norepinephrine receptor actually facilitates release. Little is known about the role of autoreceptors in regulating epinephrine release in the CNS.

Inactivation of Catecholamine Neurotransmission

Continuous stimulation of neuronal receptors is not a desirable condition on two grounds. The first is that the nonpathological activity of neurons is not continuous, but fluctuates, with neurons sometimes firing and sometimes not. Accordingly, continuous stimulation does not convey information concerning the activity of the presynaptic neuron accurately to the follower cell. This can be most easily understood with respect to receptors that form ion channels, in which continued action of a neurotransmitter would lead to inappropriate ion concentrations across the membrane and thus disrupt neurotransmission. The second reason that continuous stimulation is not desirable is that such continuous stimulation is typically pathological, resulting in damage to and toxic loss of postsynaptic neurons.

There are several different mechanisms for terminating the actions of a catecholamine. Perhaps the simplest is for the transmitter to diffuse away from the synaptic area, ultimately being diluted to concentrations too low to act on receptors. More critical are active modes of terminating transmitter action, including the uptake of catecholamines by neuronal-membrane-associated transporter proteins. In turn, this may be followed by uptake of the catecholamines into storage vesicles, from where they can be reused, or catabolism by monoamine oxidase (MAO). A third possibility is direct catabolism by catechol-*O*-methyltransferase (COMT).

Enzymatic inactivation of catecholamines. Enzymatic inactivation was originally thought to be the major means by which catecholamines are inactivated in the CNS, but now appears to play a secondary role in the termination of action of catecholamines and most other classic transmitters. Nonetheless, enzymatic inactivation remains important for two reasons.

The first is that certain drug treatments for neuropsychiatric disorders are based on manipulation of the key enzymes that degrade catecholamines (see Box 9.2). Second, enzymatic inactivation is the major mode of terminating the action of circulating catecholamines in the bloodstream.

Two major enzymes take part in catecholamine catabolism: MAO and COMT. Either enzyme can act independently or on the products of the other, leading to catecholamine metabolites that are deaminated, O-methylated, or both. COMT is a relatively nonspecific enzyme that transfers methyl groups from the

---

BOX 9.2

MAO AND COMT INHIBITORS IN THE TREATMENT OF NEUROPSYCHIATRIC DISORDERS

Depression

One hypothesis concerning the pathophysiology of depression posits a decrease in noradrenergic tone in the brain. MAO_A inhibitors such as tranylcypromine effectively increase norepinephrine levels (as well as dopamine and serotonin concentrations) and were once a mainstay of the treatment of depression. More recently the use of MAO inhibitors in depression has been largely supplanted by the introduction of drugs that increase extracellular norepinephrine levels by blocking reuptake of the transmitter (tricyclic antidepressants) and other agents to increase serotonin or dopamine levels by blocking SERT or DAT [*fluoxetine* (Prozac) and *bupropion* (Welbutrin), respectively].

The treatment of depression with MAO_A inhibitors, although still useful for certain patients who do not respond to other antidepressants, is marred by a large number of side effects. Among the most serious side effects is hypertensive crisis. Patients who are treated with MAO_A inhibitors and eat foods that contain large amounts of tyramine (such as aged cheeses) cannot metabolize the ingested tyramine. Because tyramine releases catecholamines from nerve endings and relatively small amounts of tyramine increase blood pressure significantly, a marked increase in blood pressure and a high risk for stroke may develop.

Parkinson Disease

Deprenyl, a specific inhibitor of MAO_B, has been used as an initial treatment for Parkinson disease (PD). The use of deprenyl in the treatment of PD and the rationale for its use were based on data from studies of a neurotoxin, 1-methyl-4-phenyl-1,2,3,6-tetrahydropyridine (MPTP). The systemic administration of MPTP to humans and other primates results in a relatively specific degeneration of midbrain dopamine neurons and a marked parkinsonian syndrome. MPTP toxicity was first noted in a group of opiate addicts. In an attempt to synthesize a designer

drug that was a meperidine (Demerol) derivative, the structurally related MPTP was inadvertently produced. Addicts who injected this drug developed a severe parkinsonian syndrome. Subsequent animal studies showed that MPTP itself is not toxic, but that the active metabolite of MPTP, MPP^+, is highly toxic. The formation of MPP^+ from MPTP is catalyzed by MAO_B, and animal studies soon revealed that treatment with MAO inhibition by deprenyl could prevent MPTP toxicity.

The realization that MPTP administration rather faithfully reproduces the cardinal signs and symptoms of Parkinson disease reawakened interest in environmental toxins as a cause of PD. This interest led to the idea that treatment with deprenyl might be useful in slowing the progression of PD, putatively caused by an environmental toxin. The first evaluation of clinical trials of newly diagnosed PD patients indicated that daily administration of deprenyl increased the amount of time required before patients needed other drugs for the relief of symptoms; however, when the MAO_B inhibitor was withdrawn, patients treated with deprenyl regressed and appeared no better than untreated subjects. It now appears that the actions of deprenyl are due at least in part to the symptomatic improvement that results from increasing dopamine levels by inhibiting degradation of the transmitter rather than to slowing of the progression of PD. Moreover, low levels of methamphetamine are generated by the metabolism of deprenyl; because methamphetamine potently releases dopamine from nerve terminals, this would result in a symptomatic improvement.

Catechol O-methyltransferase, which together with MAO is responsible for the enzymatic degradation of catecholamines, is also a target in the treatment of PD. Two inhibitors of COMT, one that is active peripherally and another that is active both peripherally and centrally, are used to prevent the enzymatic inactivation of L-dopa. By inhibiting COMT these drugs prolong the therapeutic action of L-dopa and smooth out the characteristic fluctuations in therapeutic response to dopa.

BOX 9.2 *(continued)*

Schizophrenia

Changes in catecholamine function have been a subject of intense scrutiny in schizophrenia, with much attention focusing on a loss of dopaminergic tone in the prefrontal cortex. One allelic variant of the COMT gene substitutes a single methionine for a valine, and results in a much reduced activity of the enzyme. Recent data have examined COMT alleles for full versus low COMT activity in normal subjects and

schizophrenics. Individuals bearing the allele that confers lower COMT activity have improved performance on cognitive tasks that involve the prefrontal cortex; the performance of schizophrenic persons on these tasks is impaired. There is a significant increase in transmission of the COMT allele conferring high enzyme activity to schizophrenic subjects, and it has been proposed that COMT activity may confer increased risk of developing schizophrenia.

---

donor S-adenosylmethionine to the m-hydroxy group of catechols. COMT is found in both peripheral tissues and central nervous system and is the major means of inactivating circulating catecholamines that are released from the adrenal gland.

Two forms of MAO have been identified on the basis of substrate specificities and selective enzyme inhibitors. MAO_A has a high affinity for norepinephrine and serotonin and is selectively inhibited by clorgyline. In contrast, MAO_B has a higher affinity for o-phenylethylamines and is selectively inhibited by the monoamine oxidase inhibitor (MAOI) deprenyl. Both MAO_A and MAO_B are associated with the outer mitochondrial membrane. The MAOs oxidatively deaminate catecholamines and their O-methylated derivatives to form inactive and unstable aldehyde derivatives. These aldehydes can be further catabolized by dehydrogenases and reductases to form corresponding acids and alcohols.

Neuronal catecholamine transporters. The reuptake of a released neurotransmitter is the major mode of terminating the synaptic actions of transmitters in the brain. In addition, the accumulation of the transmitter by these membrane transporters also allows intracellular enzymes that degrade the transmitter to act, thus bolstering the actions of extracellular enzymes.

Neuronal reuptake of catecholamines, and indeed of all transmitters for which a reuptake process has been identified, has several characteristics (Clark and Amara, 1993). The reuptake process is energy dependent and saturable, and depends on Na^+ cotransport as well as requiring extracellular Cl^-. Because reuptake depends on coupling to the Na^+ gradient across the neuronal membrane, toxins that inhibit Na^+, K^+-ATPase inhibit reuptake. However, under certain conditions, the coupling of transporter function to Na^+ flow may lead to local changes in

the membrane Na^+ gradient and thereby paradoxically extrude ("release") the transmitter.

The membrane catecholamine transporters are not Mg^{2+} dependent and are not inhibited by reserpine. These characteristics distinguish the neuronal membrane transporters from the monoamine transporters localized to neuronal vesicles. The catecholamine transporters are localized to neurons; although there appears to be a reuptake process that accumulates catecholamines in glial cells, the process is not a high-affinity one and the functional significance of the glial reuptake of catecholamines remains unknown.

Two distinct mammalian catecholamine transporter proteins, the dopamine transporter (DAT) and norepinephrine transporter (NET), have been cloned and characterized pharmacologically. The two transporters share significant sequence homology and are members of a class of transporter proteins (including serotonin and amino acid transmitter transporters) with 12 transmembrane domains. Neither transporter is very specific, with each accumulating both dopamine and norepinephrine. In fact, NET has a higher affinity for dopamine than for norepinephrine. A specific transporter for epinephrine-containing neurons has been identified in the frog but not in mammalian species.

The regional distribution of DAT and NET largely follows the expected localization to dopamine and norepinephrine neurons, respectively. However, DAT does not appear to be expressed in all dopamine cells. The tuberoinfundibular dopamine neurons, which are hypothalamic cells that release dopamine into the pituitary portal blood system, lack demonstrable DAT mRNA and protein. Because dopamine released from tuberoinfundibular neurons is carried away in the vasculature, the existence of a transporter protein on these dopamine neurons would be superfluous. However,

DAT is also not expressed or expressed at very low levels on some dopamine axons that innervate the prefrontal cortex.

Although studies defining the cellular localization of the two catecholamine transporters did not uncover many surprises, immunohistochemical studies of the subcellular localization of the transporters did yield an unexpected finding. The use of antibodies generated against DAT revealed that the transporter is typically expressed outside of the synapse, in the extrasynaptic region of the axon terminal. This finding suggests that the transporter may be used to inactivate (accumulate) dopamine that has escaped from the synaptic cleft and, thus, that diffusion is the initial process by which dopamine is removed from the synapse. This observation is consistent with recent studies indicating that perisynaptic concentrations of dopamine can reach 1.0 mM or more, a value roughly comparable to the affinity of the cloned DAT for DA. Receptors for dopamine and many other transmitters are also found extrasynaptically (indeed, along the length of axons); this observation, coupled with the presence of catecholamine transporters to extrasynaptic regions, suggests that extrasynaptic ("paracrine" or volume) neurotransmission (Zoli et al., 1999) may be of considerable importance for catecholaminergic signaling.

How neurotransmitter transporter proteins are regulated has been a major area of research in recent years. Chronic treatment with inhibitors of catecholamine reuptake alters the number of transporter sites, but the precise regulatory mechanisms remain unclear. There are phosphorylation sites on DAT and NET, and thus changes in neurotransmitter release may alter function through interaction with autoreceptors and subsequent activation of serine–threonine kinases.

The DAT knockout has been particularly useful in clarifying the role of DAT in dopaminergic neurons and, by extension, the role of other monoamine transmitter transporters. The constitutive loss of the DA transporter results in a remarkably wide array of deficits in dopaminergic function, ranging from an increase in extracellular DA levels and delayed clearance of released DA to a striking decrease in tissue concentrations of DA in the face of increased DA synthesis (Gainetdinov et al., 1998). In addition, there is a striking loss of autoreceptor-mediated tone, including deficits in release-, synthesis-, and impulse-modulating autoreceptor function (Jones et al., 1999). The alterations in DA knockout mice have been suggested to reflect a disinhibition of tyrosine hydroxylase due to a lack of intraneuronal DA to provide feedback inhibition of the enzyme and a markedly increased rate of DA turnover, such that synthesis and release of the neurons are accelerated. Interestingly, many of these biochemical deficits seen in the DAT knockout mice resemble the normal "physiological" functions of dopamine neurons that innervate the prefrontal cortex (see Roth and Elsworth, 1995).

Psychostimulants, such as cocaine and amphetamine, exert their effects on arousal by increasing extracellular levels of catecholamines; cocaine also increases extracellular serotonin levels. The mechanism through which psychostimulants increase catecholamine levels is by blocking DAT and NET. In particular, cocaine shows a very high affinity for the dopamine transporter; amphetamine is a less potent inhibitor of reuptake but also induces release of catecholamines from the cytoplasm. Studies in mice with targeted null mutations of the DAT have surprisingly revealed that these mice will still self-administer cocaine. However, mice with double knockouts of both the DA and serotonin transporters fail to self-administer cocaine (Sora et al., 2000). Cocaine administration results in sharp increases in both extracellular dopamine and serotonin, and the failure of DAT–serotonin transporter (SERT) knockout mice to sustain cocaine self-administration suggests that both dopamine and serotonin are critical for the effects of cocaine. Studies in animals with transient suppression of DAT and SERT expression are needed to ensure that the SERT is not compensating developmentally for the loss of DAT.

The NET is also a target of clinically important drugs. The tricyclic antidepressants potently inhibit norepinephrine reuptake, with significantly weaker effects on the dopamine and serotonin transporters. In addition, new agents that inhibit NET selectively without significantly decreasing serotonin or dopamine reuptake are now being used in the treatment of depression. Mice with targeted null mutations of NET act like antidepressant-treated wild-type mice; interestingly, NET knockout mice are hyper-responsive to psychostimulant-elicited locomotor stimulation (Xu et al., 2000).

Serotonin

Scientists were aware in the nineteenth century of a substance found in the blood that induces powerful contractions of smooth muscle organs. However, more than a century passed until Page and his collaborators in 1948 succeeded in isolating the compound (which they proposed to be a possible cause of high blood pressure) from platelets. At the same time, Italian researchers characterized a substance present in high concentrations in intestinal mucosa that caused contractions of gastrointestinal smooth muscle. The material isolated from blood platelets was given the name *serotonin,* and the substance isolated from the intestinal tract was called *enteramine.* Subsequently, both materials were purified and crystallized, and shown to be the identical substance, 5-hydroxytryptamine (5-HT), usually referred to as serotonin. The laboratory synthesis of serotonin soon allowed direct comparison of serotonin with the purified compound isolated from platelets, which conclusively demonstrated that serotonin possessed all the biological features of the natural substance.

Serotonin is found in neurons as well as several types of peripheral cells, including platelets, mast cells, and enterochromaffin cells. In fact, the brain accounts for only about 1% of body stores of serotonin.

Although the purification and identification of serotonin were based on studies of blood pressure regulation, the possible relation of serotonin to psychiatric disorders propelled research on the central effects of serotonin. The observation that the indole structure of serotonin was similar to that of the psychedelic agent LSD and a number of other psychotropic compounds soon led to theories linking abnormalities of serotonin function to various psychiatric disorders, including schizophrenia and depression. This linkage remains a major focus of research on central serotonergic systems.

The basic principles of the biochemical neuropharmacology of synaptic transmission as revealed by studies of catecholamines are also applicable to neurons that use serotonin as a transmitter. We therefore outline the nature of chemical transmission in serotonergic neurons, focusing on differences that are unique to serotonergic neurons.

Synthesis of Serotonin

The basic outline of serotonin biosynthesis is very similar to that of catecholamine transmitters: an amino acid gains entry to the central nervous system and is metabolized in specific neurons via a series of enzymatic steps to yield serotonin.

Once tryptophan enters the serotonergic neuron it is hydroxylated by tryptophan hydroxylase, the rate-limiting step in serotonin synthesis (see Fig. 9.5). The resultant serotonin precursor, 5-hydroxytryptophan (5-HTP), is subsequently decarboxylated by aromatic amino acid decarboxylase. Thus, only two critical enzymes (tryptophan hydroxylase and AADC) are involved in the synthesis of serotonin (Cooper *et al.,* 2002; Hensler, 2006).

Tryptophan hydroxylase. The rate-limiting step in serotonin synthesis is tryptophan hydroxylase. However, in contrast to the catecholamines, the availability of the precursor amino acid tryptophan plays an important role in regulating the synthesis of serotonin.

Because serotonin cannot cross the blood–brain barrier, brain cells must synthesize the amine. Tryptophan is present in high concentrations in the plasma, and changes in dietary sources of tryptophan can substantially alter brain levels of serotonin. An active uptake process facilitates entry of tryptophan into the brain. However, other large neutral aromatic amino acids compete for this transport process. Accordingly, brain levels of tryptophan are determined by plasma concentrations of competing neutral amino acids as well as the plasma levels of tryptophan itself.

There are two genes that encode for two different tryptophan hydroxylase proteins, TPH1 and TPH2. The latter is the major form found in brain. In contrast, TPH1 is mainly present in peripheral tissues but is found in lower levels in the brain.

Neurons must be able to adapt to short- or long-term demands on activity. In serotonergic neurons, the synthesis of serotonin from tryptophan is increased in a frequency-dependent manner in response to electrical stimulation of serotonergic cells. Tryptophan hydroxylase requires both molecular oxygen and a reduced pterin cofactor.

L-Aromatic Amino Acid Decarboxylase

Aromatic amino acid decarboxylase metabolizes the serotonin precursor 5-HTP to the transmitter serotonin. This is the same enzyme found in catecholaminergic neurons. Just as in catecholamine cells, in which the precursor L-dopa is almost instantaneously converted into dopamine by AADC, the precursor 5-HTP is so rapidly decarboxylated in serotonergic cells that central concentrations of 5-HTP under basal conditions are negligible. Thus, because AADC is not saturated with 5-HTP under physiological conditions, it is possible to increase the content of serotonin in brain not only by increasing the dietary intake of tryptophan but also by administering 5-HTP, which readily enters the brain.

Alternative tryptophan metabolic pathways. Although serotonin is generally thought of as the final product of tryptophan synthesis, in one part of the brain serotonin can be further metabolized. In the pineal gland, serotonin is metabolized to the hormone melatonin (5-methoxy-N-acetyltryptamine), which is thought to play important roles in both sexual behavior and sleep. The production of melatonin

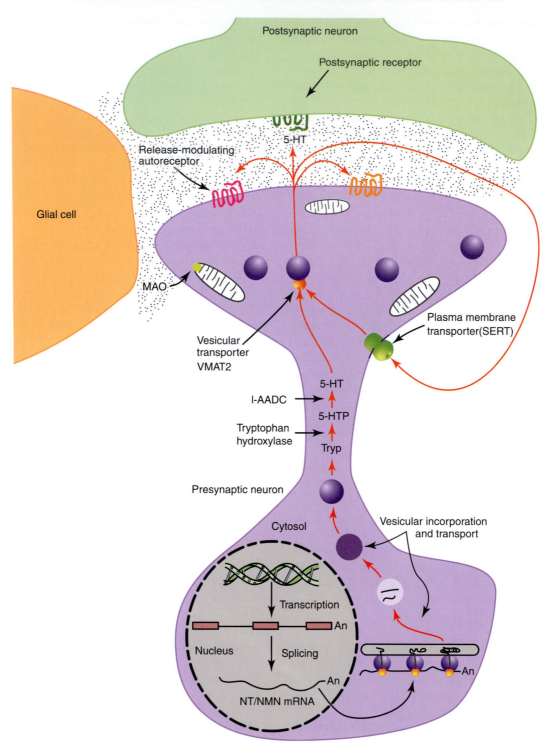

FIGURE 9.5 Depiction of a serotonergic neuron. Tryptophan (Tryp) in the neuron is sequentially metabolized by tryptophan hydroxylase and L-AADC to yield serotonin (5-HT). The serotonin is accumulated by the vesicular monoamine transporter. When released, the serotonin can interact with both postsynaptic receptors and presynaptic autoreceptors. 5-HT is taken up by the high-affinity serotonin transporter (SERT), and once inside the neuron it can be reaccumulated by vesicular transporter or metabolically inactivated by MAO and other enzymes.

from serotonin requires two enzymatic steps: N-acetylation of serotonin to form N-acetylserotonin, which is rapidly methylated by 5-hydroxyindole-O-methyltransferase to melatonin.

In peripheral tissues, most tryptophan is not metabolized to serotonin but is instead metabolized in a separate series of reactions called the *kynurenine pathway*. Recent data indicate that the kynurenine shunt is also present in the CNS and leads to the accumulation of neuroactive substances that may be of clinical importance in cases of trauma and stroke (Stone, 1993). The two major tryptophan metabolites that are generated by the kynurenine shunt are quinolinic acid and kynurenic acid. Quinolinic acid is a potent agonist at *N*-methyl-D-aspartate (NMDA) receptors and causes neurotoxicity and convulsions. In contrast, kynurenine is an antagonist at NMDA receptors. Considerable effort is being directed to determining the role that these tryptophan metabolites may play in neurological disorders (Schwarcz, 2004).

Storage and Release of Serotonin

Vesicular accumulation and storage of serotonin. Serotonin is stored primarily in vesicles and is released by an exocytotic mechanism. VMAT2, the same transporter that accumulates dopamine in catecholamine neurons, also transports serotonin into vesicles in serotonergic neurons. It has been suggested that serotonin-containing (but not catecholamine-containing) vesicles may express a specific high-affinity serotonin-binding protein.

Since catecholamines and serotonin share a common vesicular transporter, it is not surprising that reserpine, which depletes vesicular stores of catecholamines, also depletes serotonin from serotonin neurons.

Regulation of serotonin synthesis and release. There are some important regulatory differences between catecholaminergic and serotonergic neurons. Serotonin neurons are sensitive to changes in plasma levels of the precursor amino acid tryptophan, and thus dietary changes can regulate serotonin levels in the brain. In addition, it appears that increases in intracellular serotonin levels do not significantly alter serotonin synthesis *in vivo*; in contrast, in catecholaminergic neurons transmitter synthesis is influenced by end-product inhibition.

Short-term requirements for increases in serotonin synthesis appear to be accomplished by a Ca^{2+}-dependent phosphorylation of tryptophan hydroxylase, which changes its kinetic properties without necessitating the synthesis of more enzyme. In contrast, serotonergic neurons respond to the need for long-term increases in serotonin availability by the induction of (new) tryptophan hydroxylase protein.

Serotonin autoreceptors regulate serotonin release and synthesis. As is the case in catecholamine neurons, there are functionally dissociable somatodendritic and terminal autoreceptors on serotonin neurons. The impulse-modulating autoreceptor located on somatodendritic areas of serotonin neurons is a 5-HT_{1A} receptor; this receptor is also found as a heteroceptor on nonserotonergic neurons. There has been some difficulty in untangling which of two other serotonin receptors is the primary autoreceptor governing serotonin release at the axon terminal, the 5-HT_{1D} or 5-HT_{1B} receptor. Part of the problem is because of species differences in expression between these two 5-HT_1 isoforms. In addition, the lack of drugs that discriminate well between these different isoforms has made determination of which 5-HT_1 isoform is more important for regulating axonal release of serotonin difficult; available drugs show some degree of preference for one site over another *in vitro*, but when examined *in vivo* the difference in affinity of the various agents is smaller.

Inactivation of Released Serotonin

As in the case of the catecholamine neurotransmitters, reuptake serves as a major means of terminating the action of serotonin. Released serotonin is taken up by a plasma membrane carrier, the serotonin transporter. In addition, the same enzymatic inactivation that is operative in catecholamine neurons also is found in serotonin neurons.

Reuptake. Serotonin that is released into the synapse is inactivated primarily by the reuptake of the transmitter by a plasma membrane serotonin transporter. The serotonin transporter (SERT) has been cloned and sequenced and belongs to the same family of 12-transmembrane-domain transporters as the catecholamine transporters. SERT has in common with other transporter family members an absolute requirement for Na^+ co-transport.

SERT is also an important clinical target for therapeutic drugs. Just as the norepinephrine transporter is the target of tricyclic antidepressant drugs, SERT is the target of the class of antidepressant drugs termed *selective serotonin reuptake inhibitors* (SSRIs), which have become the initial choice of medications for treatment of major depression. There are also mixed SERT–NET inhibitors used to treat depression. The ability of antidepressant drugs to alter monoamine inactivation by disrupting serotonin and norepinephrine transporters or by disrupting enzymatic inactivation of the monoamines has led to the dominant theories of the pathogenesis of depression, which suggests a critical modulatory role for norepinephrine and serotonin (Heninger *et al.*, 1996).

As noted previously, cocaine and other psychostimulants block the dopamine transporter and thereby sharply increase extracellular dopamine levels. However, cocaine also increases extracellular serotonin levels. Interestingly, even though the dopamine transporter is a major target of cocaine, DAT knockout mice continue to self-administer cocaine, as do SERT knockout mice. However, in mice bearing double DAT–SERT knockouts, cocaine self-administration is reduced, suggesting that both transporters must be targeted for the rewarding effects of psychostimulants to be manifested.

Enzymatic degradation. The primary catabolic pathway for serotonin is oxidative deamination by the enzyme monoamine oxidase. The product of this reaction, 5-hydroxyindole acid aldehyde, is further oxidized to 5-hydroxyindoleacetic acid (5-HIAA) or can be reduced to 5-hydroxytryptophol. In the brain and cerebrospinal fluid 5-HIAA is the primary metabolite of serotonin. Monoamine oxidase inhibitors increase serotonin levels and have been used extensively as antidepressants. MAOIs are effective antidepressants but have the potential to produce serious side effects.

γ-Aminobutyric Acid: The Major Inhibitory Neurotransmitter

A number of amino acids fulfill most of the criteria for neurotransmitters. The three best studied of these are GABA, the major inhibitory transmitter in brain; glutamate, which is the major excitatory transmitter in brain; and glycine, another inhibitory amino acid (Olsen and Betz, 2006). The broad principles outlined in the discussion of catecholamine neurotransmitters are also applicable to the amino acid transmitters, although certain aspects of the synthesis of amino acid transmitters are less completely understood compared with the catecholamines.

A major difference between the biogenic amines transmitters and the amino acid transmitters is that the latter are derived from intermediary glucose metabolism. This dual role for the amino acid transmitters dictates that there must be mechanisms to segregate the transmitter and general metabolic pools of the amino acid transmitters. A second difference between amino acid and biogenic amine transmitters is that amino acid transmitters released from neurons are readily taken up by glial cells as well as neurons. We review GABA as a prototypic amino acid transmitter, focusing on differences between the catecholamine transmitters and GABA.

GABA was discovered in 1950 by Eugene Roberts, whose subsequent study (Roberts, 1986) revealed that GABA has a neurotransmitter role. GABA is ubiquitous in the CNS, as might be expected for a transmitter derived from the metabolism of glucose. Although the presence of GABA as a transmitter in neurons is widespread, it nonetheless has a distinct distribution. Although it was originally thought that (with a few exceptions) GABA was a neurotransmitter in local circuit interneurons but not in projection neurons, it has become apparent that there are many examples of GABAergic projection neurons.

GABA Biosynthesis

Several aspects of the synthesis of GABA differ from that of the monoamines (see Fig. 9.6). These differences are due to precursors of GABA being part of cellular intermediary metabolism rather than dedicated solely to a neurotransmitter synthetic pool.

The GABA shunt and GABA transaminase. GABA is ultimately derived from glucose metabolism. α-Ketoglutarate formed by the Krebs (tricarboxylic acid) cycle is transaminated to the amino acid glutamate by the enzyme GABA alpha-oxoglutarate transaminase (GABA-T). In those cells in which GABA is used as a transmitter, the presence of the enzyme glutamic acid decarboxylase (GAD) permits the formation of GABA from glutamate derived from α-ketoglutarate.

One unusual feature of the GABA synthetic pathway is that intraneuronal GABA is inactivated by the actions of GABA-T, which appears to be associated with mitochondria (Fig. 9.6). Thus, GABA-T is both a key synthetic enzyme and a degradative enzyme. GABA-T metabolizes GABA to succinic semialdehyde, but only if α-ketoglutarate is present to receive the amino group that is removed from GABA. This unusual GABA shunt serves to maintain supplies of GABA.

Glutamic acid decarboxylase. The critical biosynthetic enzyme for the neurotransmitter GABA is glutamic acid decarboxylase (GAD). GAD is localized exclusively in the central nervous system to neurons that use GABA as a transmitter.

There are two isoforms of GAD, which are encoded by two distinct genes (Erlander and Tobin, 1991). These two isoforms, designated GAD65 and GAD67 in accord with their molecular weights, exhibit somewhat different intracellular distributions, suggesting that the two GAD forms may be regulated in different ways (Soghomonian and Martin, 1988). GAD requires a pyridoxal phosphate cofactor for activity. GAD65 and GAD67 differ significantly in their affinity for this cofactor: GAD65 shows a relatively high affinity for the cofactor, whereas the larger GAD isoform does not. The affinity of GAD65 for the cofactor results in the ability of GAD65 enzyme

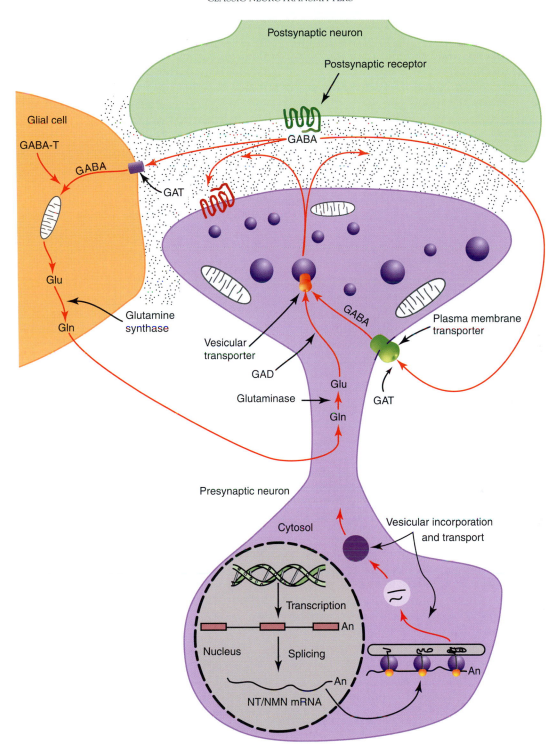

FIGURE 9.6 Schematic depiction of the life cycle of a GABAergic neuron. α-ketoglutarate formed in the Krebs cycle is transaminated to glutamate (Glu) by GABA transaminase (GABA-T). The transmitter GABA is formed from the Glu by glutamic acid decarboxylase (GAD). GABA that is released is taken by high-affinity GABA transporters (GAT) present on neurons and glia. Gln, glutamine.

activity to be efficiently and quickly regulated. In contrast, the activity of GAD67 is determined through induction of new enzyme protein rather than through post-translational mechanisms.

A major question concerning amino acid transmitters is how the transmitter pools are kept distinct from the general metabolic pools in which the amino acids serve. GAD is necessary for synthesis of the transmitter GABA, and the presence of the GAD mRNAs or proteins are markers of GABAergic neurons. GAD is a cytosolic enzyme, but GABA-T, which converts a-ketoglutarate into the GAD substrate glutamate, is present in mitochondria. Thus, the metabolic pool is present in the mitochondria, but glutamate destined for the transmitter pool must be exported from the mitochondria to the cytosolic compartment. This export process is poorly understood.

Glutamate is not only a precursor to the formation of GABA but also is the major excitatory neurotransmitter. GAD is not present in neurons in which glutamate functions as a transmitter, and thus glutamatergic neurons do not use GABA as a transmitter. What prevents GABA neurons from using the precursor glutamate as a transmitter was thought to mainly reflect two different biosynthetic enzymes for glutamate as a transmitter and as a metabolic intermediary, but it appears clear that vesicular transporters specific for GABA or glutamate are the major means of segregation (Takamori *et al.*, 2000). A specific form of glutaminase (a phosphate-activated glutaminase (PAG)) has been proposed to be responsible for the synthesis of the transmitter pool of glutamate. PAG is localized to certain vesicles. Because both GABA and glutamate cause very rapid changes in postsynaptic neurons, one depolarizing neurons and the other hyperpolarizing them, it is not surprising (and probably is fortunate) that the two amino acid transmitter pools are not generally colocalized. However, in the rat olfactory bulb and the chicken retina (which is a neural tissue), anatomical studies have suggested that there are a few isolated neurons in which GABA and glutamate are colocalized (Quaglino *et al.*, 1999). The functional significance of such an arrangement is not clear.

Storage and Release of GABA

Vesicular inhibitory amino acid transporter. A vesicular transporter in GABAergic cells accumulates GABA. The transporter was cloned on the basis of homology to unc-47 in Caenorhabditis elegans (McIntire *et al.*, 1997), a strategy of moving from the worm to mammalian species that has proven to be a very useful strategy for identifying a variety of mammalian transmitter-related genes. The vesicular

GABA transporter differs from the two VMATs by belonging to a different class, having 10 rather than 12 transmembrane domains. The vesicular GABA transporter shares with the VMATs, however, a lack of substrate specificity and will transport the inhibitory transmitter glycine as well as GABA. Consistent with this pharmacology, the vesicular GABA transporter has been found in glycine as well as GABA-containing neurons. Accordingly, it has been suggested that the transporter can be more accurately designated as a vesicular inhibitor amino acid transporter (Gasnier, 2004; Weihe and Eiden, 2000). Interestingly, there are some rare GABA neurons that lack the transporter, raising the specter of another (related) transporter in these neurons or, alternatively, some unique functional attribute of these cells.

Regulation of GABA release by autoreceptors. The major postsynaptic GABA receptor is the $GABA_A$ receptor, which contains the chloride ion channel (see Chapters 11 and 16). This multimeric receptor complex is formed by a number of different subunit proteins. Pharmacological studies indicate that autoreceptor-mediated regulation of GABA neurons takes place predominantly through $GABA_B$ receptors located on GABAergic nerve terminals. Immunohistochemical studies have revealed that both $GABA_B$ and $GABA_A$ receptors are present on postsynaptic non-GABAergic neurons. It is possible that these $GABA_A$ postsynaptic receptors respond to GABA released from a neuron that is presynaptic to another GABA neuron expressing the $GABA_A$ site. Because an anatomical arrangement of one GABA neuron terminating on another GABA cell would have the same functional consequence as an autoreceptor (decreasing subsequent transmitter release), it has been difficult to distinguish between true autoreceptors and heteroreceptors *in vivo*.

Inactivation of Released GABA

Uptake of several transmitters by glial cells as well as neurons has been reported. The dual glial–neuronal reuptake is common in neurons using amino acid transmitters, probably because amino acids can play dual roles as both transmitters and metabolic intermediaries. However, the ability of glia to avidly accumulate GABA and other amino acids distinguishes amino acid transmitters from other classic transmitters.

GABA transporter proteins. Reuptake is the primary mode of inactivation of GABA that is released from neurons. At least three specific GABA transporter (GAT) proteins are expressed in the CNS, providing a diverse means of regulating GABA neurons (see Cherubini and Conti, 2001). In addition, a betaine transporter that accumulates GABA has been cloned. Two types of GABA transporters were long known as being neuronal and glial and were defined on the basis of pharmacological specificity. However, the cloning of GABA transporters, which belong to the same family of transporter genes that includes the catecholamine transporters, revealed an unexpected finding. *In situ* hybridization and immunohistochemical studies revealed that one of the GATs found in brain, which on pharmacological grounds was defined as a "glial" transporter, is present in both neurons and glia. Moreover, the other GATs appear to be expressed in both neurons and glia.

The presence of multiple transporter proteins for the same transmitter, all localized to neurons, differs from the situation for catecholamine transmitters, in which a single membrane-associated transporter protein with relatively poor substrate specificity is found in a neuron. An obvious question arises: Why are there multiple transporters for GABA? GATs are expressed in both GABAergic neurons and non-GABAergic cells (presumably cells that receive a GABA innervation). However, it is not clear if multiple GATs are found in the same cell, and the precise intracellular localization of the transporter proteins is not yet known. It is possible that different transporters are targeted differently in the cell. For example, one might be present in dendrites and another expressed in axons, with corresponding different functional requirements. Another possibility is that the GATs that have been cloned may serve as cotransporters for other amino acids. For example, transporters for β-alanine and taurine have not been cloned, but these amino acids are accumulated by GATs. Finally, it is possible that one or more of these transporters frequently works in the outward direction, serving as a paradoxical mechanism for the release of GABA.

Enzymatic inactivation of GABA. GABA-T is both a synthetic and a degradative enzyme, with both enzymatic functions acting to conserve the transmitter pool of GABA. GABA-T is a particulate enzyme that is present in high concentration in GABAergic neurons. GABA-T is found in non-GABAergic as well as GABA-containing neurons and is present in a number of peripheral tissues. Electron microscope data suggest that GABA-T is associated with mitochondria. However, pharmacological studies of various subcellular fraction preparations suggest that the activity of GABA-T associated with synaptosomes that contain mitochondria is less than that seen in synaptosomal membrane fractions (without mitochondria), suggesting that GABA may be metabolized either extraneuronally or in postsynaptic neurons.

Glutamate and Aspartate: The Excitatory Amino Acid Transmitters

Excitatory amino acid transmitters account for most of the fast synaptic transmission that occurs in the mammalian CNS. Glutamate and aspartate are the major excitatory amino acid neurotransmitters, but several related amino acids, including *N*-acetylaspartylglutamate, also appear to have neurotransmitter roles. The excitatory amino acids, like the inhibitory amino acid transmitter GABA, participate in intermediary metabolism as well as cellular communication; the problem of dissociating neurotransmitter from metabolic roles therefore holds for excitatory amino acids. The intertwining of the transmitter roles of amino acids and intermediary metabolism makes it difficult to fulfill all the criteria that would give amino acids fully legitimate status as neurotransmitters. Despite these issues, it is now widely accepted that glutamate and aspartate function as excitatory transmitters in the CNS. We briefly consider glutamate biosynthesis and regulation, focusing on the differences between excitatory and inhibitory amino acid transmitters. Many of the general principles addressed in the section on GABA are applicable to glutamate and therefore not discussed in detail.

Biosynthesis of Glutamate

Although glutamic acid is present in very high concentrations in the CNS, brain glutamate and aspartate levels are derived solely by local synthesis from glucose, because neither amino acid crosses the blood–brain barrier. Two processes contribute to the synthesis of glutamate in the nerve terminal. As mentioned previously (in the section on GABA), glutamate is formed from glucose through the Krebs cycle and transamination of α-Ketoglutarate. In addition, glutamate can be formed directly from glutamine (see Fig. 9.7). Because glutamine is synthesized in glial cells, there is an unusual degree of interaction between glia and neurons in the determination of availability of the transmitter pool of glutamate. The glutamine that is formed in glia is transported into nerve terminals and then locally converted by glutaminase into glutamate (Hassel and Dingledine, 2006). Thus, the synthesis of glutamate depends critically on the enzyme glutaminase. A phosphate-activated glutaminase (PAG) has been suggested to be the specific form of the enzyme responsible for the synthesis of the transmitter pool of glutamate. However, PAG is also found in relatively high concentrations in peripheral tissues such as the liver (Conti and Minelli, 1994). PAG is localized to mitochondria; as discussed in the section on GABA, the process by which glutamate is exported to allow vesicular storage of the transmitter remains poorly understood.

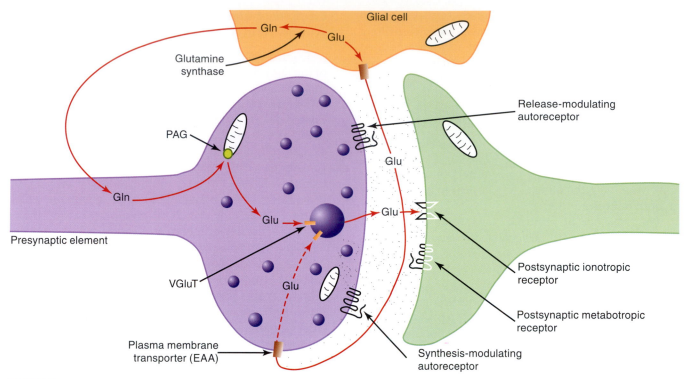

FIGURE 9.7 Depiction of an excitatory amino acid (glutamate) synapse. Glutamate, synthesized via metabolic pathways, is concentrated through a vesicular transporter into secretory granules. After release from the presynaptic terminal, glutamate can interact with postsynaptic and/or release-modulating receptors. Glutamate is then cleared from the synaptic region by the high-affinity plasma membrane transporters or by recycling through adjacent glia.

Storage and Release of Glutamate

Vesicular glutamate transporter. Glutamate is stored in synaptic vesicles from which the transmitter is released in a calcium-dependent manner on depolarization of the nerve terminal. Although the vesicular storage of glutamate was convincingly demonstrated quite some time ago and was well characterized biochemically, the cloning of multiple vesicular glutamate transporters lagged behind (Bellocchio *et al.*, 2000). This is in part due to the fact that the vesicular glutamate transporter is not related to other known transmitter transporters, although it shares significant sequence homology with EAT-4, a worm protein implicated in glutamatergic transmission. The first vesicular glutamate transporter was identified as a protein that was previously suggested to mediate the sodium-dependent transport of inorganic phosphate across the membrane. There are now three known vesicular glutamate transporters, all of which are densely expressed in axon terminals of glutamate neurons (Fremeau *et al.*, 2004). The mRNA encoding VGLUT1 is expressed by cortical neurons, while that encoding VGLUT2 is mainly expressed by subcortical glutamate neurons. The third vesicular glutamate transporter is much less abundant. There are now examples of colocalization of all three types of

vesicular glutamate transporters, although the functional significance of this arrangement is not clear.

Regulation of glutamate release. The release of glutamate from nerve terminals is regulated by a metabotropic autoreceptor (see Chapter 11 for discussion of metabotropic receptors). Indeed, eight different receptors (and various splice variants) that constitute three distinct classes of metabotropic glutamate receptors have been identified. The release-modulating autoreceptor is a member of one class of metabotropic receptors, the class II metabotropic glutamate receptors, that are negatively coupled to adenyl cyclase. Both members of the class II family (the mGluR$_2$ and mGluR$_3$ receptors) have been localized to presynaptic glutamatergic axon terminals, and a large number of studies has revealed that class II receptors function as release-modulating glutamate receptors. In addition, electrophysiological studies has suggested an impulse-modulating glutamate autoreceptor, which is thought to be either an mGluR$_1$ or mGluR$_5$ site. The mGluR$_4$ receptor has been localized to presynaptic nerve terminals but is present on nonexcitatory nerve terminals and thus is likely to be a heteroreceptor rather than an autoreceptor. With eight different primary metabotropic glutamatergic receptors, it is reasonable to ask the question: Why have so many? The answer may be that these receptors subserve an extremely broad array of functions and appear to be critically involved in regulating not only glutamatergic function but also the activity (including release) of a dizzying number of transmitters, ranging from classic transmitters such as dopamine to peptide transmitters such as substance P (Cartmell and Schoepp, 2000). There is currently considerable attention devoted to developing the use of drugs that target metabotropic glutamate receptors in neuropsychiatric disorders, including schizophrenia and Parkinson disease.

Inactivation of Glutamate

Glutamate inactivation occurs predominantly by reuptake of the amino acid by dicarboxylic acid plasma membrane transporters. In contrast to GABA and other classic transmitters, there does not appear to be a significant role for enzymatic inactivation of glutamate. The extent to which diffusion regulates synaptic and extracellular levels of glutamate is not clear.

Five glutamate transporters have been cloned, with some localized to glia and others to neurons. Electron microscopic studies suggest that certain glutamate transporters are heavily expressed in astrocytes but relatively weakly present in neurons. The glutamate transporters accumulate L-glutamate and D- and L-aspartate; although the affinities of the transporters are similar for glutamate they differ for other amino acids. The transporters have distinct brain distributions, and even the glial transporters exhibit regional and intracellular differences in expression (Chaudhry et al., 1996), underscoring the heterogeneity of glia as well as neurons.

The presence of certain glutamate transporters on glial cells is consistent with the intricate interplay of glial and neuronal elements in the synthesis of glutamate. Recent data have found that glutamate is not only taken up by glia cells, but also released from glia as a signaling molecule (Haydon and Carmignoto, 2006). Thus, because glutamate released from neurons is accumulated by glia and then metabolized to glutamine, there is ultimately a complex multicellular recycling of glutamate. The fate of glutamate accumulated by the neuronal glutamate transporter is unclear. It has not been established if glutamate released from a given neuron is taken up by a glutamate transporter on that particular neuron or, alternatively, by glutamate transporters on other neurons or glia.

Acetylcholine

Much of our basic understanding of chemical synaptic transmission is based on studies of acetylcholine (ACh), the first transmitter identified. First noted as the vagal stuff of Loewi (1921) and subsequently demonstrated to be responsible for transmission at the neuromuscular junction by Loewi and Navratil, it has been a century since ACh was first proposed as a transmitter.

A key reason ACh has assumed such a prominent role in guiding studies of neurotransmitters has been the relative ease with which ACh can be studied. Acetylcholine is the transmitter at the neuromuscular junction, and thus both the nerve terminal and its target can be readily accessed for experimental manipulations. Subsequent investigations also focused on another peripheral site, the superior cervical ganglion, which was also easy to isolate and study. Lessons learned from experiments conducted on these peripheral tissues have shaped our current approaches to defining the characteristics of neurotransmitters in the brain.

The ability to expose and maintain isolated preparations of the neuromuscular junction permitted electrophysiological and biochemical studies of synaptic transmission. Electrophysiological studies revealed fast excitatory responses of muscle fibers to stimulation of the nerve innervating the muscle. The presence of miniature end-plate potentials (MEPPs) in the muscle fiber was noted, and Fatt and Katz (1952) demonstrated that these MEPPs resulted from the slow "leakage" of ACh, with each MEPP representing the release of transmitter in one vesicle (termed a *quantum*) (see Chapter 8 for additional details on MEPPs). Overt depolarization generated an increase in the number of quanta released over a given period. In addition, studies of the neuromuscular junction allowed detailed analyses of the enzymatic inactivation of ACh, setting the reference for subsequent studies.

Over the past half-century many of the rules that govern ACh neurotransmission have been shown to be general principles that apply to other transmitters. For example, the concept of the quantal nature of neurotransmission is central to current ideas of transmitter release. Although the discovery of different neurotransmitters has expanded our knowledge, studies of ACh continue to provide a foundation for modern concepts of chemical neurotransmission.

Acetylcholine Synthesis

The synthesis of ACh is arguably the most simple transmitter synthesis, with but a single step: the acetyl group from acetyl-coenzyme A is transferred to choline by the enzyme choline acetyltransferase (ChAT). The requirements for ACh synthesis are correspondingly few: the substrate choline, the donor acetyl-coenzyme A, and the enzyme ChAT (see Fig. 9.8).

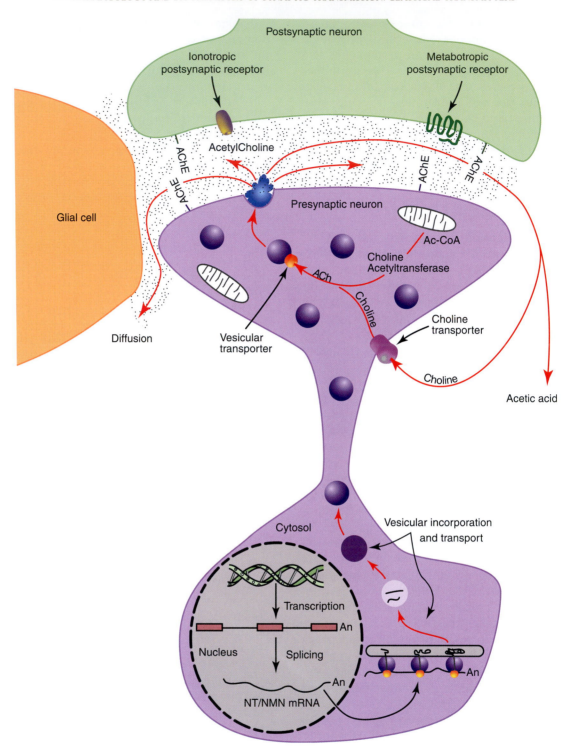

FIGURE 9.8 Acetylcholine (ACh) synthesis, release, and termination of action. A choline transporter accumulates choline. The enzyme choline acetyltransferase (ChAT) acetylates the choline using acetyl-CoA (Ac-CoA) to form the transmitter ACh, which is accumulated into vesicles by the vesicular transporter. The released ACh may interact with postsynaptic muscarinic or nicotinic cholinergic receptors, or can be taken up into the neuron by a choline transporter. Acetylcholine can be degraded after release by the enzyme acetylcholinesterase (AChE).

The acetyl-CoA that serves as the donor is derived from pyruvate generated by glucose metabolism. This obligatory dependence on a metabolic intermediary is similar to the situation present in GABA synthesis, where the immediate precursor glutamate is formed from α-Ketoglutarate. Acetyl-CoA is localized to mitochondria. Because the synthetic enzyme ChAT is cytoplasmic, acetyl-CoA must exit the mitochondria to gain access to ChAT; the specifics of this process are poorly understood.

Choline acetyltransferase. Choline acetyltransferase is the definitive marker for cholinergic neurons (Wu and Hersch, 1994). Multiple mRNAs encode ChAT, resulting from differential use of three promoters and alternative splicing of the 5′ non-coding region of the enzyme. In the rat the different transcripts encode the same protein but in humans give rise to multiple forms of the enzyme, including both active and inactive (truncated) forms. The functional significance of the different transcripts under normal conditions is a topic of considerable interest. Myasthenia gravis, a disease marked by decreased muscle activity, is linked to a variety of deficits in neuromuscular cholinergic function; in congenital forms of myasthenia mutations in both nicotinic receptors and AChE have been found. Recently, a particular ChAT mutation has been linked to a form of myasthenia that is characterized by often fatal episodes of apnea (Ohno *et al.*, 2001).

Although ChAT is the sole enzyme in ACh synthesis, ChAT is not the rate-limiting step in ACh synthesis. When ChAT activity is measured *in vitro* it is much greater than would be expected on the basis of ACh synthesis *in vivo*. The reason for this discrepancy has been suggested to be related to the need to transport acetyl-CoA from the mitochondria to the cytoplasm, which may be rate limiting in ACh synthesis. Alternatively, intracellular choline concentrations may ultimately determine the rate of ACh synthesis. This latter speculation has led to the use of choline precursors in attempts to enhance ACh synthesis in Alzheimer disease, in which there is a marked decrease in ACh in the cerebral cortex. Attempts have been made to treat Alzheimer disease with lecithin, a choline precursor; unfortunately, lecithin does not appear to diminish dementia, although it does markedly increase bad breath!

Acetylcholine Storage and Release

Vesicular cholinergic transporter. ACh is synthesized by ChAT and transported into vesicles by the vesicular cholinergic transporter (VAChT). This transporter is distinct from the membrane transporter that accumulates choline. VAChT was cloned on the basis of homology to a C. elegans gene (unc-17) that encodes a protein that is homologous with VMAT (Roghani *et al.*, 1994). VAChT is expressed in cholinergic neurons throughout the brain.

When the mammalian VAChT was initially cloned it was noted that the human VAChT is present in chromosome 10, near the gene for ChAT. It was subsequently demonstrated that the VAChT is unique in that its entire coding region is contained in the first intron of the ChAT gene (Usdin *et al.*, 1995). This suggests that both genes are coordinately regulated, a suspicion that has been confirmed (Bernard *et al.*, 1995).

Cholinergic autoreceptor function. Cholinergic release-modulating autoreceptors have been identified both in peripheral tissues and in the brain. This receptor is a muscarinic cholinergic receptor, rather than the nicotinic cholinergic receptor found at the neuromuscular junction. A cholinergic receptor regulating release of ACh from cholinergic axons in the cortex, hippocampus, and striatum has been identified, and converging data indicate that this receptor is an M2 cholinergic receptor, one of five muscarinic receptors; the M2 site is also found on noncholinergic (cholinoreceptive) neurons, where it serves as a conventional heteroceptor. There are few direct data that point to the presence of a synthesis-modulating cholinergic autoreceptor.

Inactivation of Acetylcholine

Acetylcholinesterase. The primary mode of inactivation of ACh appears to be enzymatic. Several enzymes that hydrolyze choline esters can degrade ACh, but the major esterase in the central nervous system is acetylcholinesterase (AChE; see Box 9.3).

The enzymatic inactivation of ACh is simply the hydrolysis of ACh to choline. Two groups of cholinesterases have been defined on the basis of substrate specificity, acetyl cholinesterases and butyrylcholinesterases (Taylor and Brown, 2006). The first are relatively specific for ACh and present in high concentration in the brain, where there are multiple AChE species (Fernandez *et al.*, 1996). Butyrylcholinesterases also efficiently hydrolyze choline esters but are found primarily in the liver, with lower levels being present in the adult brain.

AChE is present in high concentration in cholinergic neurons; however, AChE is also present in moderately high concentration in some noncholinergic neurons that receive cholinergic inputs (i.e., that are cholinoreceptive). This observation is consistent with the fact that AChE is a secreted enzyme that is associated with the cell membrane. Thus, ACh hydrolysis takes place extracellularly, and the choline generated is conserved by the high-affinity reuptake process.

BOX 9.3

ACETYLCHOLINESTERASE INHIBITORS, NERVE GASES, AND PHARMACOTHERAPY

The enzymatic inactivation of acetylcholine (ACh) has been fertile ground for the development of a large number of pharmaceutical agents. Anticholinesterases such as sarin are potent neurotoxins and have been used as nerve gases since World War I. Other anticholinesterases include organophosphates (such as parathion), which are widely used insecticides. Anticholinesterases, whether the target is a human or a tomato hornworm, function in the same way: instead of the released ACh leading to discrete single depolarizations of muscle fibers, the accumulation of acetylcholine at the neuromuscular junction leads to muscle fibrillation and ultimately depolarization inactivation of the muscle; i.e., the muscle is so excited it stops!

Anticholinesterases have some less aggressive uses as well. Competitive neuromuscular blocking agents such as succinylcholine are used as an adjunct to anesthetics during surgery to increase muscle relaxation; conversely, anticholinergics can be used to reverse the muscle paralysis caused by succinylcholine. Anticholinesterases are the mainstay of treatment of myasthenia gravis, a disorder of the neuromuscular junction that is usually marked by the presence of anti-nicotinic receptor antibodies. Attempts have also been made to treat Alzheimer disease, in which there is a sharp decrease in cortical ACh, by administering an anticholinesterase to inhibit breakdown of ACh. Unfortunately, this approach has not proven very effective.

In addition to its role in inactivating released acetylcholine, AChE has been proposed to function as a chemical messenger in the CNS (Greenfield, 1991). The release of AChE from neurons in the substantia nigra and cerebellum is calcium dependent, and cerebellar release is enhanced by electrical stimulation of cerebellar afferents. Electrophysiological studies have revealed that AChE elicits changes in the threshold for Ca^{2+} spikes, and local application of AChE enhances the responses of cerebellar neurons to glutamate and aspartate, the transmitters present in the climbing and mossy fiber innervations of the cerebellum (Appleyard and Jahnsen, 1992).

There are several AChE species, all of which are encoded by a single gene that is alternatively spliced (Schumacher et al., 1988), with tissue-specific expression of different transcripts (Seidman et al., 1995). Among the multiple mRNAs encoding AChE is one that represents the primary form of the enzyme expressed in brain and muscle.

High-affinity choline transporter. Choline is found in the plasma in high concentration. A low-affinity reuptake process for choline is widely distributed in the body; however, both low-affinity and high-affinity choline uptake processes are present in brain. Cholinergic neurons in the brain express a sodium-dependent transporter that is saturated at plasma levels of choline, consistent with the high-affinity component of choline uptake.

The choline transporter, in contrast to other plasma membrane transmitter transporters, is not directly involved in termination of the action of release transmitter. Since ACh is hydrolyzed by AChE, enzymatic inactivation is the major means of terminating cholinergic transmission. In the absence of direct evidence that choline binds with high affinity to cholinergic receptors, it appears that the function of the choline transporter is conservation of transmitter stores.

The choline transporter has long been known but was not cloned until recently (Okuda et al., 2000). Several years ago a putative high-affinity choline transporter was cloned. However, this transporter was present in peripheral tissues as well as brain and, when expressed in cell lines, displayed kinetics different from those of CNS tissue. Moreover, this transporter was not sensitive to hemicholinium 3, which blocks the high-affinity cholinergic reuptake process. It now appears that this gene cloned encoded a creatinine transporter rather than the choline transporter (Happe and Murrin, 1995). The high-affinity choline transporter appears to play important roles in a variety of homeostatic processes relating to cholinergic function, including cholinergic receptors (Bazalakova and Blakely, 2006; Bazalakova et al., 2007).

Paralleling the situation surrounding the identification of the vesicular glutamate transporter gene, which was cloned on the basis of a related worm gene, a cDNA encoding the rat high-affinity choline transporter was recently cloned based on similarities to *cho–1* in *C. elegans*. The vesicular cholinergic transporter is not homologous to other neurotransmitter transporters, being related instead to the sodium-dependent glucose transporter family.

Why Do Neurons Have So Many Transmitters?

We have discussed in varying amounts of detail a moderate number, but not all, classic transmitters. About a dozen classic transmitters and literally dozens of neuropeptides function as transmitters. Still more molecules serve as "unconventional" transmitters, including growth factors and gases such as nitric oxide. If the role of neurotransmitters is to serve as a chemical bridge that conveys information between two spatially distinct cells, why have so many chemical messengers?

Afferent Convergence on a Common Neuron

Perhaps the simplest explanation for multiple transmitters is that many nerve terminals abut on a single neuron. A neuron must be able to distinguish between the multiple inputs that bring information to the neuron. To some degree, this can be accomplished by the site on a neuron at which an afferent (input) terminates: at the cell body, axon, or dendritic shaft or spine. However, because many afferents terminate in close proximity, another means of distinguishing the inputs and their information is necessary. One way in which this can be accomplished is by chemically coding afferent neurons. The information conveyed by distinct transmitters is then distinguished by the different receptors present on the targeted neuron and their various transduction mechanisms.

Neurotransmitter Colocalization

A major conceptual change in the neurosciences over the past generation has been the realization that a cell can use more than one neurotransmitter. The idea that a neuron is limited to one transmitter can be traced to Henry Dale, or, more properly, to an informal restatement of what is termed *Dale's principle*. About 60 years ago, Dale posited that a metabolic process that takes place in the cell body can reach or influence events in all the processes of the neuron. John Eccles restated Dale's view to suggest that a neuron releases the same transmitter at all its processes. Illustrating the dangers of the scientific equivalent of sound bites, this principle was soon misinterpreted to indicate that only a single transmitter can be present in a given neuron. This is clearly not the case. Neurons can colocalize two or more transmitters. For example, a neuron can use both a classic transmitter such as dopamine and a peptide transmitter such as neurotensin. Indeed, it now appears that few if any neurons contain only one transmitter, and there are many cases in which three or even four transmitters are found in a single neuron.

The presence of multiple transmitters in a single neuron may indicate that different transmitters are used by a neuron to signal different functional states to its target cell. For example, the firing rates of neurons differ considerably, and thus it may be useful for a neuron to encode fast firing by one transmitter and slower firing by another transmitter. In addition, the firing pattern of cells is of significance. For example, a neuron might show an absolute firing rate of five impulses per second. This frequency could result from a neuron discharging every 200 ms or discharging five times in an initial 200 ms period followed by 800 ms of silence. Recent data indicate that in cases in which there is a colocalization of a peptide and a classic transmitter, peptide transmitters are often released at higher firing rates and particularly under burst-firing patterns.

In many ways, the different biosynthetic strategies used by peptides and classic transmitters may lead to differential release. Classic transmitters can be rapidly replaced because their synthesis occurs in nerve terminals. In contrast, peptide transmitters are synthesized in the cell body and transported to the terminal. It is therefore useful to conserve peptide transmitters for situations of high demand, because they would otherwise be rapidly depleted.

Transmitter Release from Different Processes

The restatement of Dale's principle by Eccles held that a transmitter or other protein is present in all processes of a neuron. However, it now appears that a transmitter can be specifically localized to different parts of a neuron (see Deutch and Bean, 1995). For example, in the marine mollusck *Aplysia*, different transmitters are targeted to different processes of a single neuron (Sossin *et al.*, 1990). If a transmitter is restricted to a particular part of a neuron, it follows that the neuron would need multiple transmitters to account for different release sites. Considerable evidence supports distinct spatial localizations of receptors (e.g., for ionotropic glutamate receptors) on a neuron, and even indicates that there is movement and clustering of receptors to maximize information transfer from pre- to postsynaptic neurons.

Synaptic Specializations Versus Nonjunctional Appositions between Neurons

In addition to the diversity in transmitters that may result from transmitters being targeted to different intraneuronal sites, the anatomical relationships between one cell and its follower may contribute to the need for different transmitters. We usually think of synaptic specializations (see Chapters 1 and 8) as the morphological substrate of communication between two neurons. However, there may also be nonsynaptic forms of communication between two neurons. These could occur across distances that are smaller (e.g., gap junctions) or much larger than the separation of pre- and postsynaptic neurons by the synaptic cleft. The requirements for transmitter action would differ from those discussed previously if the distance traversed by a transmitter is larger than that typically present at a synaptic apposition. Thus, transmitters that lack an efficient reuptake system, such as peptide transmitters, might be favored at nonsynaptic sites. Because a single neuron can form both synaptic and nonsynaptic specializations, a single neuron may require more than one neurotransmitter.

Fast Versus Slow Responses of Target Neurons to Neurotransmitters

We have seen that different firing rates or patterns may be accompanied by changes in the amount of transmitter being released from a neuron. As described in detail in Chapter 16, the postsynaptic response to a transmitter occurs over different timescales. For example, transmitter activation of ionotropic receptors (i.e., those that form ion channels) leads to very rapid changes, because the ionic gradients across the cell are almost instantaneously changed. In contrast, metabotropic receptors that respond to catecholamines and peptide transmitters are coupled to intracellular events via various transduction molecules, such as G proteins, and respond to neurotransmitter stimulation on a slower timescale than is seen when ionotropic receptors are activated. This difference in temporal response characteristics is useful, because it allows the receptive neuron to respond differently to a stimulus, depending on the antecedent activity in the cell. A transmitter can change the response characteristics of a particular cell to subsequent stimuli on the order of seconds or even minutes, and thus short-term changes can occur independent of changes in gene expression.

Nontransmitter Roles of Neurotransmitters

Over the past few years several of the key proteins involved in regulating chemical neurotransmission have been identified based on homologies to proteins found in invertebrate species, such as the worm *Caenorhabditis elegans* and the fly *Drosophila melanogaster*.

It now appears that some of the molecular players in neurotransmission are found even in plants! Plant homologues of glutamate receptors have been identified and shown to be important in regulating diverse functions, ranging from calcium utilization to morphogenesis (Brenner *et al.*, 2000; Kim *et al.*, 2001),

and geneological analysis has suggested that these glutamate receptors may predate the divergence of plants and animals (Chiu *et al.*, 1999). As nervous systems have become elaborate through evolution, many neurotransmitter-related proteins have roles that are not related to transmitter function or alternatively are involved in less discrete and more spatially elaborate signaling.

An example is acetylcholine. As discussed previously, ChAT is a cytosolic protein that drives the synthesis of ACh from choline and acetyl-CoA; however, one form of human ChAT is localized to the nucleus, where it seems unlikely to play a transmitter role (Resendes *et al.*, 1999). ChAT mRNA is found in the testes, where it is translated and the protein appears in spermatozoa (Ibanez *et al.*, 1991). Moreover, ChAT mRNAs have been reported to be present in lymphocytes, as have certain muscarinic cholinergic receptors (Kawashima *et al.*, 1998).

In addition, AChE mRNAs appear to be present in lymphocytes, where both acetylcholinesterase and butyrylcholinesterase enzyme activity has been reported as decreased in Alzheimer disease (Bartha *et al.*, 1987; Inestrosa *et al.*, 1994). AChE is present in high abundance in bone marrow cells and peripheral blood cells in certain types of leukemias (Lapidot-Lifson *et al.*, 1989). Recent data indicate that inhibition of AChE gene expression in bone marrow cultures suppresses apoptosis (programmed cell death) and leads to progenitor cell expansion (Soreq *et al.*, 1994), suggesting a role for AChE in the development of leukemias.

The presence of neurotransmitter-related proteins in peripheral tissues is not restricted to molecules related to ACh function. Three dopamine receptor mRNAs, encoding the D3, D4, and D5 receptors, are present in lymphocytes, and expression of the D3 receptor transcript is increased in schizophrenic subjects (Ilani *et al.*, 2001; Kwak *et al.*, 2001).

It is relatively easy to envision how transmitter receptors that are expressed on peripheral nonneural tissues can respond to transmitters present in the periphery, essentially functioning as hormonal signals. Thus, dopamine or other catecholamines that are circulating at low levels in the periphery may bind to DA receptors present on lymphocytes. However, another means of signaling is via axonal noradrenergic innervation of nonneural immune tissues in the periphery. Thus, sympathetic noradrenergic fibers innervate not only the vasculature but also primary (bone marrow) and secondary (spleen, lymph nodes) immune lymphoid structures (Felton *et al.*, 1985). This noradrenergic innervation may be the central regulator of peripheral immune and stress responses, since chemical lesions of the sympathetic nervous system markedly alter T- and B-cell proliferation and activity (Madden *et al.*, 2000). Not only are the transmitters of neurons communicating between neural and immune system cells the same, but there are similarities between the structural substrates of communication between different nervous system cells (neurons) and immune system cells, through junctional specializations called *synapses* (see Trautmann and Vivier, 2001).

SUMMARY

Classic neurotransmitters are small molecules that are derived from amino acids or intermediary metabolism and share several characteristics. The sequential actions of key enzymes result in the biosynthesis of these transmitters, usually in the general vicinity of where they will be released. The synthesized transmitter is stored in vesicles where it is poised for release and protected from degradation; the vesicular transporters also sequester xenobiotics and thus protect the neuron from certain toxins. Neurotransmitter release is elicited by depolarization and is calcium dependent. The action of the released neurotransmitter is terminated by a reuptake mechanism involving plasma membrane transporters and by enzymatic means.

The criteria for designation as a classic transmitter have been based on experiments conducted in sites that were easily accessible (such as the neuromuscular junction). Although many of the key principles of chemical synaptic transmission have been found to be the same in other areas that are less accessible to experimental manipulation (neurons in the brain), our ideas of the defining characteristics of transmitters have evolved to account for new knowledge and the emergence of many exceptions to the rules enunciated previously. The relatively high concentrations of classic transmitters permitted the easy measurement of these compounds, and thus the measurement of transmitter release became a key criterion for defining a neurotransmitter. Unfortunately, transmitter release has proven to be a difficult criterion to meet for many putative transmitters discovered over the past 30 years. Nevertheless, the increasing sensitivity of analytical techniques coupled with the ingenuity of neuroscientists led to the uncovering of a large number of peptides, growth factors, and even gases that function as transmitters. We explore in the next chapter the similarities and differences of the classic transmitters with these new kids on the block. These differences have often illuminated unknown fundamental processes of neurons and expanded our concept of information flow between neurons.

References

Appleyard, M., and Jahnsen, H. (1992). Actions of acetylcholinesterase in the guinea-pig cerebellar cortex *in vitro. Neurosci.* **47**, 291–301.

Bartha, E., Szelenyi, J., Szilagyi, K., Venter, V., Thu Ha, N. T., Paldi-Haris, P., and Hollan, S. (1987). Altered lymphocyte acetylcholinesterase activity in patients with senile dementia. *Neurosci. Lett.* **79**, 190–194.

Bazalakova, M. H., Wright, J., Schneble, E. J., McDonald, M. P., Heilman, C. J., Levey, A. I., and Blakely, R. D. (2007). Deficits in acetylcholine homeostasis, receptors and behaviors in choline transporter heterozygous mice. *Genes Brain Behav.* **6**, 411–424.

Bazalakova, M. H., and Blakely, R. D. (2006) The high-affinity choline transporter: A critical protein for sustaining cholinergic signaling as revealed in studies of genetically altered mice. *Handbook Exp. Pharmacol.* **175**, 525–544.

Bellocchio, E. E., Reimer, R. J., Fremeau, R. T., Jr., and Edwards, R. H. (2000). Uptake of glutamate into synaptic vesicles by an inorganic phosphate transporter. *Science* **289**, 957–960.

Bernard, C. (1849). Action physiologique des venins (curare). *C. R. Seances Soc. Biol. Ses Fil.* **1**, 90.

Bernard, S., Varoqui, H., Cervine, R., Israel, M., Mallet, J., and Diebler, M. F. (1995). Coregulation of two embedded gene products, choline acetyltransferase and the vesicular acetylcholine transporter. *J. Neurochem.* **65**, 939–942.

Brenner, E. D., Martinez-Barboza, N., Clark, A. P., Liang, Q. S., Stevenson, D. W., and Coruzzi, G. M. (2000). Arabidopsis mutants resistant to *S*(+)-beta-methyl-alpha, beta-diaminopropionic acid, a cycad-derived glutamate receptor agonist. *Plant Physiol.* **124**, 1615–1624.

Carlsson, A. (1972). Biochemical and pharmacological aspects of parkinsonism. *Acta Neurol. Scand. (Suppl.)* **51**, 11–42.

Cartmell, J., and Schoepp, D. D. (2000). Regulation of neurotransmitter release by metabotropic glutamate receptors. *J. Neurochem.* **75**, 889–907.

Chaudhry, F. A., Lehre, K. P., van Lookeren Campagne, M., Otterson, O. P., Danbolt, N. C., and Storm-Mathisen, J. (1996). Glutamate transporters in glial plasma membranes: Highly differentiated localizations revealed by quantitative ultrastructural immunocytochemistry. *Neuron* **15**, 711–720.

Cherubini, E., and Conti, F. (2001). Generating diversity at GABAergic synapses. *Trends Neurosci.* **24**, 155–162.

Chiu, J., DeSalle, R., Lam, H. M., Meisel, L., and Coruzzi, G. (1999). Molecular evolution of glutamate receptors: A primitive signaling mechanism that existed before plants and animals diverged. *Mol. Biol. Evol.* **16**, 826–838.

Clark, J. A., and Amara, S. G. (1993). Amino acid neurotransmitter transporters: Structure, function, and molecular diversity. *BioEssays* **15**, 323–332.

Conti, F., and Minelli, A. (1994). Glutamate immunoreactivity in rat cerebral cortex is reversibly abolished by 6-diazo-5-oxo-L-norleucine (DON), an inhibitor of phosphate-activated glutaminase. *J. Histochem. Cytochem.* **42**, 717–726.

Cooper, J. R., Bloom, F. E., and Roth, R. H. (2002). "The Biochemical Basis of Neuropharmacology," 8th ed. Oxford Univ. Press, New York.

Deutch, A. Y., and Bean, A. J. (1995). Colocalization in dopamine neurons. In "Psychopharmacology: The Fourth Generation of Progress" (F. E. Bloom and D. J. Kupfer, Eds.), Raven Press, New York, pp. 197–206.

Edwards, R. H. (1993) Neural degeneration and the transport of neurotransmitters. *Ann. Neurol.* **34**, 638–645.

Elliott, T. R. (1905). On the action of adrenaline. *J. Physiol.* **32**, 401.

Erlander, M. G., and Tobin, A. J. (1991) The structural and functional heterogeneity of glutamic acid decarboxylase: A review. *Neurochem. Res.* **16**, 215–226.

Fatt, P., and Katz, B. (1952). Spontaneous subthreshold activities at motor nerve endings. *J. Physiol.* **117**, 109–128.

Felton, D. L., Felton, S. Y., Carlson, S. L., Olschowka, J. A., and Livnat, S. (1985). Noradrenergic and peptidergic innervation of lymphoid tissue. *J. Neuroimmunol.* **135** (Suppl. 2), 755–765.

Fernandez, H. L., Moreno, R. D., and Inestrosa, N. C. (1996). Tetrametric (G4) acetylcholinesterase: Structure, localization, and physiological regulation. *J. Neurochem.* **66**, 1335–1346.

Fremeau, R. T., Voglmaier, S., Seal, R. P., and Edwards, R. H. (2004). VGLUTs define subsets of excitatory neurons and suggest novel roles for glutamate. *TINS* **27**, 98–102.

Gainetdinov, R. R., Jones, S. R., Fumagalli, F., Wightman, R. M., and Caron, M. G. (1998). Re-evaluation of the role of the dopamine transporter in dopamine system homeostasis. *Brain Res. Rev.* **26**, 148–153.

Gasnier, B. (2004) The SLC32 transporter, a key protein for the synaptic release of inhibitory amino acids. *Pflugers Arch.* **447**, 756–759.

Greenfield, S. A. (1991). A non-cholinergic role of ACHE in the substantia nigra: From neuronal secretion to the generation of movement. *Mol. Cell. Neurobiol.* **11**, 55–77.

Happe, H. K., and Murrin, L. C. (1995). In situ hybridization analysis of CHOTT, a creatine transporter, in the rat central nervous system. *J. Comp. Neurol.* **351**, 94–103.

Hassel, B. and Dingledine, R. (2006) Glutamate. In "Basic Neurochemistry" (G. J. Siegel, R. W. Albers, S. Brady, and D. L. Price, Eds.), 7th ed., Elsevier–Academic Press, San Diego, CA, pp. 267–290.

Haycock, J. W., Haycock, D. A. (1991). Tyrosine hydroxylase in rat brain dopaminergic nerve terminals: Multiple-site phosphorylation *in vivo* and in synaptosomes. *J. Biol. Chem.* **266**, 5650–5657.

Haydon, P.G., and Carmignoto, G. (2006). Astrocyte control of synaptic transmission and neurovascular coupling. *Physiol. Rev.* **86**, 1009–1031.

Heninger, G. R., Delgado, P. L., and Charney, D. S. (1996). The revised monoamine theory of depression: A modulatory role for monoamines, based on new findings from monoamine depletion experiments in humans. *Pharmacopsychiatry* **29**, 2–11.

Hensler, J. (2006). Serotonin. In "Basic Neurochemistry" (G. J. Siegel, R.W. Albers, S. Brady, and D. L. Price, Eds.), 7th ed., Elsevier–Academic Press, San Diego, CA, pp. 227–248.

Henry, J. P., Sagne, C., Bedet, C., and Gasnier, B. (1998). The vesicular monoamine transporter: From chromaffin granule to brain. *Neurochem. Int.* **32**, 227–246.

Ibáñez, C. F., Pelto-Huikko, M., Söder, O., Ritzèn, E. M., Hersh, L. B., Hökfelt, T., and Persson, H. (1991). Expression of choline acetyltransferase mRNA in spermatogenic cells results in an accumulation of the enzyme in the postacrosomal region of mature spermatozoa. *Proc. Natl. Acad. Sci. USA* **88**, 3676–3680.

Ichinose, H., Ohye, T., Takahashi, E., Seki, N., Hori, T., Segawa, M., Nomura, Y., Endo, K., Tanaka, K., Tanaka, H., and Tsuji, S. (1994). Hereditary progressive dystonia with marked diurnal fluctuation caused by mutations in the GTP cyclohydrolase I gene. *Nat. Genet.* **8**, 236–242.

Ilani, T., Ben-Shachar, D., Strous, R. D., Mazor, M., Sheinkman, A., Kotler, M., and Fuchs, S. (2001). A peripheral marker for schizophrenia: Increased levels of D3 dopamine receptor mRNA in blood lymphocytes. *Proc. Natl. Acad. Sci. USA* **98**, 625–628.

Inestrosa, N. C., Alarcon, R., Arriagada, J., Donoso, A., Alvarez, J., and Campos, E. O. (1994). Blood markers in Alzheimer disease: Subnormal acetylcholinesterase and butyrylcholinesterase in lymphocytes and erythrocytes. *J. Neurol. Sci.* **122**, 1–5.

Jones, S. R., Gainetdinov, R. R., Hu, X. T., Cooper, D. C., Wightman, R. M., White, F. J., and Caron, M. G. (1999). Loss of autoreceptor functions in mice lacking the dopamine transporter. *Nat. Neurosci.* **2**, 649–655.

Kawashima, K., Fujii, T., Watanabe, Y., and Misawa, H. (1998). Acetylcholine synthesis and muscarinic receptor subtype mRNA expression in T-lymphocytes. *Life Sci.* **62**, 1701–1705.

Kim, S. A., Kwak, J. M., Jae, S. K., Wang, M. H., and Nam, H. G. (2001). Overexpression of the AtGluR2 gene encoding an Arabidopsis homolog of mammalian glutamate receptors impairs calcium utilization and sensitivity to ionic stress in transgenic plants. *Plant Cell Physiol.* **42**, 74–84.

Kwak, Y. T., Koo, M. S., Choi, C. H., and Sunwoo, I. (2001). Change of dopamine receptor mRNA expression in lymphocyte of schizophrenic patients. *BMC Med. Genet.* **2**, 3.

Lapidot-Lifson, Y., Prody, C. A., Ginzberg, D., Meytes, D., Zakut, H., and Soreq, H. (1989). Coamplification of human acetylcholinesterase and butrylcholinesterase genes in blood cells: Correlation with various leukemias and abnormal megakaryacytopoiesis. *Proc. Natl. Acad. Sci. USA* **86**, 4715–4719.

Lewis, D. A., Melchitzky, D. S., and Haycock, J. W. (1993). Four isoforms of tyrosine hydroxylase are expressed in human brain. *Neurosci.* **54**, 477–492.

Loewi, O. (1921). Uber Humorale Ubertragbarkeit Herznervenwirkung. *Pfluegers Arch Ges. Physiol. Menschen Tiere* **189**, 239.

Madden, K. S., Stevens, S. Y., Felton, D. L., and Bellinger, D. L. (2000). Alterations in T lymphocyte activity following chemical sympathectomy in young and old Fisher 344 rats. *J. Neuroimmunol.* **103**, 131–145.

McIntire, S. L., Reimer, R. J., Schuske, K., Edwards, R. H., and Jorgensen, E. M. (1997). Identification and characterization of the vesicular GABA transporter. *Nature* **389**, 870–876.

Ohno, K., Tsujino, A., Brengman, J. M., Harper, C. M., Bajzer, Z., Udd, B., Beyring, R., Robb, S., Kirkham, F. J., and Engel, A. G. (2001). Choline acetyltransferase mutations cause myasthenic syndrome associated with episodic apnea in humans. *Proc. Natl. Acad. Sci. USA* **98**, 2017–2022.

Okuda, T., Haga, T., Kanai, Y., Endou, H., Ishihara, T., and Katsura, I. (2000). Identification and characterization of the high-affinity choline transporter. *Nat. Neurosci.* **3**, 120–125.

Olsen, R. W., and Betz, H. (2006). GABA and glycine. In "Basic Neurochemistry" (G. J. Siegel, R. W. Albers, S. Brady, and D. L. Price, Eds.), 7th ed., pp. 291–302, Elsevier–Academic Press, San Diego, CA.

Pothos, E. N., Larsen, K. E., Krantz, D. E., Liu, Y., Haycock, J. W., Setlik, W., Gershon, M. D., Edwards, R. H., and Sulzer, D. (2000). Synaptic vesicle transporter expression regulates vesicle phenotype and quantal size. *J. Neurosci.* **20**, 7297–7306.

Quaglino, E., Giustetto, M., Panzanelli, P., Cantino, D., Fasolo, A., and Sassoè-Pognetto, M. (1999). Immunocytochemical localization of glutamate and gamma-aminobutyric acid in the accessory olfactory bulb of the rat. *J. Comp. Neurol.* **408** , 61–72.

Resendes, M. C., Dobransky, T., Ferguson, S. S., Rylett, R. J. (1999). Nuclear localization of the 82 kDA form of human choline acetyltransferase. *J. Biol. Chem* **274**, 19417–19421.

Roberts, E. (1986). GABA: The road to neurotransmitter status. *In* "Benzodiazepine/GABA Receptors and Chloride Channels: Structural and Functional Properties" (R. W. Olsen and J. C. Venter, Eds.), Liss, New York, pp. 1–39.

Roghani, A., Feldman, J., Kohan, S. A., Shirzadi, A., Gundersen, C. B., Brecha, N., and Edwards, R. H. (1994). Molecular cloning of a putative vesicular transporter for acetylcholine. *Proc. Natl. Acad. Sci. USA* **91**, 10620–10624.

Roth, R. H., and Elsworth, J. D. (1995). Biochemical pharmacology of midbrain dopamine neurons. *In* "Psychopharmacology: The Fourth Generation of Progress" (F. E. Bloom and D. J. Kupfer, Eds.), Raven Press, New York, pp. 227–243.

Scatton, B., Dennis, T., and Curet, O. (1984). Increase in dopamine and DOPAC levels in noradrenergic nerve terminals after electrical stimulation of the ascending noradrenergic pathways. *Brain Res.* **298**, 193–196.

Schumacher, M., Maulet, Y., Camp, S., and Taylor, P. (1988). Multiple messenger RNA species give rise to the structural diversity of acetylcholinesterase. *J. Biol. Chem.* **263**, 18979–18987.

Schwarcz, R. (2004). The kynurenine pathway of tryptophan degradation as a drug target. *Curr. Opin. Pharmacol.* **4**, 12–17.

Seidman, S., Sternfeld, M., Ben Azziz-Aloya, R., Timberg, R., Kaufer-Nachum, D., and Soreq, H. (1995). Synaptic and epidermal accumulations of human acetylcholinesterase are encoded by alternative 3′-terminal exons. *Mol. Cell. Biol.* **15**, 2993–3002.

Shepherd, G. M. (1991). "Foundations of the Neuron Doctrine." Oxford Univ. Press, New York.

Sherrington, C. S. (1906). "The Integrative Action of the Nervous System." Scribner's, New York.

Soghomonian, J. J., and Martin, D. L. (1988). Two isoforms of glutamate decarboxylase: Why? *Trends Pharmacol. Sci.* **19**, 500–505.

Soreq, H., Pantinkin, D., Lev-Lehman, E., Grifman, M., Ginzberg, D., Eckstein, F., and Zakut, H. (1994). Antisense oligonucleotide inhibiton of acetylcholinesterase gene expression induces progenitor cell expansion and suppresses hematopoeietic apoptosis ex vivo. *Proc. Natl. Acad. Sci. USA.* **91**, 7907–7911.

Sossin, W. S., Sweet-Cordero, A., and Scheller, R. H. (1990). Dale's hypothesis revisited: Different neuropeptides derived from a common prohormone are targeted to different processes. *Proc. Natl. Acad. Sci. USA* **87**, 4845–4848.

Stone, T. W. (1993). Neuropharmacology of quinolinic and kynurenic acids. *Pharmacol. Rev.* **45**, 309–379.

Takamori, S., Rhee, J. S., Rosenmund, C., and Jahn, R. (2000). Identification of a vesicular glutamate transporter that defines a glutamatergic phenotype in neurons. *Nature* **407**, 189–194.

Tam, S. Y., Elsworth, J. D., Bradberry, C. W., and Roth, R. H. (1990). Mesocortical dopamine neurons: High basal firing frequency predicts tyrosine dependence of dopamine synthesis. *J. Neural Transm. (Gen. Sect.)* **81**, 97–110.

Taylor, P. and Brown, J. H. (2006). Acetylcholine. *In* "Basic Neurochemistry" (G. J. Siegel, R. W. Albers, S. Brady, and D. L. Price, Eds.), 7th ed., pp. 185–210, Elsevier–Academic Press, San Diego, CA.

Thorny, B., Auerbach, G., and Blau, N. (2000). Tetrahydrobiopterin biosynthesis, regeneration and functions. *Biochem. J.* **347**, 1–16.

Trautmann, A., and Vivier, E. (2001). Agrin: A bridge between the nervous and immune systems. *Science* **292**, 1667–1668.

Usdin, T. B., Eiden, L. E., Bonner, T. I., and Erickson, J. D. (1995). Molecular biology of the vesicular ACh transporter. *Trends Neurosci.* **18**, 218–224.

Valenstein, E. S. (2005). "The War of the Soups and the Sparks." Columbia Univ. Press, New York.

Weihe, E., and Eiden, L. E. (2000). Chemical neuroanatomy of the vesicular amine transporters. *FASEB J.* **14**, 2435–2449.

Wu, D., and Hersch, L. B. (1994). Choline acetyltransferase: Celebrating its fiftieth year. *J. Neurochem.* **62**, 1653–1663.

Xu, F., Gainetdinov, R. R., Wetsel, W. C., Jones, S. R., Bohn, L. M., Miller, G. W., Wang, Y. M., and Caron, M. G. (2000). Mice lacking the norepinephrine transporter are supersensitive to psychostimulants. *Nat. Neurosci.* **3**, 465–471.

Zoli, M., Jansson, A., Sykova, E., Agnati, L. F., and Fuxe, K. (1999). Volume transmission in the CNS and its relevance for neuropsychopharmacology. *Trends Pharmacol. Sci.* **20**, 142–150.

Zucchi, R., Chiellini, G., Scanlan, T. S., and Grandy, D. K. (2006). Trace amine-associated receptors and their ligands. *Br. J. Pharmacol.* **149**, 967–978.

Nonclassic Signaling in the Brain

Ariel Y. Deutch, Andrea Giuffrida, and James L. Roberts

As described in Chapter 9, chemically mediated transmission is the major means by which a signal is communicated from one nerve cell to another. The neuron doctrine, developed in the late nineteenth and early twentieth centuries, put forth that neurons are not part of a continuous physical network but are discrete, spatially distinct elements. This formulation led to the revolutionary idea that the release of some chemical substance from a nerve cell might influence another nerve cell or target. Today it is widely accepted that neuronal communication involves the release of specific neurotransmitters from one neuronal element to affect another element. However, there are a variety of forms of chemical communication that do not conform to the basic concept of synaptic signaling. These range from humoral influences, such as those of steroids and sugars, derived from peripheral sources to a cornucopia of different peptide–proteins, to readily diffusible gases produced within the brain. The goal of this chapter is to outline the different types of chemical communication, with an emphasis on describing how these signaling systems function in the brain.

In general, the nonclassic neurotransmitters share many of the same fundamental properties of the classic neurotransmitters with a few notable exceptions. They are not always locally synthesized, some being derived from other tissues in the body or regions of the brain. Nor are they always stored to await a specific release signal; in some cases the signaling mechanism for release is the same as the stimulus for synthesis, and thus these messengers are released as quickly as they can be synthesized.

PEPTIDE NEUROTRANSMITTERS

There are many more peptide transmitters than classic transmitters. There are some similarities between these two classes of transmitters, but also as many differences. Both classic and peptide transmitters are typically well conserved across species. In fact, many of the peptide transmitters, or closely related peptides, were initially isolated from amphibian species. Moreover, both classic and peptide transmitters are released in a calcium-dependent manner. However, the biosynthetic mechanisms and the modes of inactivation of peptide and classic transmitters are quite different. We discuss first the question of the significance of multiple neurotransmitters. We then describe the general principles of peptide transmitter biosynthesis and inactivation, which are illustrated by examining in detail one particular peptide transmitter, neurotensin.

Why Have So Many Transmitters?

Chemical neurotransmission, as we have loosely defined it, appears to be overwhelmingly redundant. There are about a dozen classic transmitters and literally dozens of neuropeptides that function as transmitters. If the role of neurotransmitters is to serve as a chemical bridge that conveys information between two spatially distinct cells, why have so many chemical messengers?

Several different factors, ranging from the intracellular localization of transmitters to the different firing rates and patterns of neurons, probably contribute to the need for multiple transmitters. The characteristics of neurons and neuronal communication that may require multiple transmitters are discussed as follows.

Afferent Convergence on a Common Neuron

Perhaps the simplest explanation for multiple transmitters is that many afferents terminate on a single neuron. It is apparent that a neuron must be able to distinguish between the multiple afferent inputs that bring information to the neuron. To some degree this can be accomplished by the site on a neuron at which an afferent terminates: at the axon, or soma, or dendritic spine, or shaft. However, since many afferents terminate in relatively close approximation, another means of distinguishing the inputs and their information is necessary. One way this can be accomplished is by chemically coding afferent neurons. The information conveyed by distinct transmitters is then distinguished by the different receptors present on a neuron and their various transduction mechanisms.

Colocalization of Neurotransmitters

A major conceptual revolution in the neurosciences over the past generation has been the realization that more than one neurotransmitter can be in the same cell. The idea that a neuron is limited to one transmitter can be traced to Sir Henry Dale or, more properly, to an informal restatement of what is termed *Dale's principle.* About 60 years ago Dale posited that a metabolic process that occurs in the cell body can reach or influence events in all the processes of the neuron. Sir John Eccles restated Dale's view to suggest that a neuron releases the same transmitter at all its processes. Illustrating the dangers of the scientific equivalent of sound bites, this principle was soon misinterpreted to indicate that a single transmitter is present in a given neuron. This is clearly not the case. Neurons can colocalize two or more transmitters. For example, a neuron can use both a classic transmitter such as dopamine and a peptide transmitter such as neurotensin. Indeed, it now appears that few if any neurons contain only one transmitter, and in several cases three or even four transmitters have been found in a single neuron.

The presence of multiple transmitters in a single neuron may suggest that the information that neurons transmit to follower cells is encoded by different transmitters for different functional states. For example, the firing rates of neurons differ considerably, and thus it may be useful for neurons to encode fast firing by one transmitter and a slower firing frequency by another transmitter. In addition, the firing pattern of cells is of significance. For example, a neuron might show an absolute rate of five spikes every second. This frequency could result from a neuron discharging every 200 ms, or by discharging five times in an initial 200 ms period followed by 800 ms of silence. Recent data indicate that in cases of colocalization of peptides and classic transmitters, peptide transmitters are often released at higher firing rates and particularly under burst firing patterns (Bean and Roth, 1992).

In many ways, the different biosynthetic strategies used by peptides and classic transmitters may lead to differential release. Classic transmitters can be rapidly synthesized because their synthesis takes place in nerve terminals, in vesicles in which the biosynthetic enzymes are stored. In contrast, peptide transmitters must be synthesized in the cell body and transported to the terminal. Thus, it might be useful to conserve peptide transmitters for situations of high demand, since they would be rapidly depleted otherwise.

Transmitter Release from Different Processes: Axonal Versus Dendritic Release

Neurons are polarized cells. This means that not only during development but also as mature cells they possess specialized regions for different functions, which may be thought of as heads and tails, or axons and dendrites. The prototypic site of release of transmitter is the region of the axon terminal that synapses onto a postsynaptic cell. However, transmitters are also released from dendrites. In addition, studies of peripheral nerves suggest that transmitters can be released from varicosities that are seen along the preterminal axons of some neurons. These different sites of transmitter release may be occupied by different transmitters.

As noted in Chapter 9, Eccles's restatement of Dale's principle states that a transmitter or other protein is present in all processes of a neuron. However, it appears that a transmitter can also be specifically localized to different parts of a neuron. For example, in the marine mollusk *Aplysia* different transmitters are targeted to different processes of a single neuron. Although differential targeting of transmitters in mammalian neurons has not been conclusively demonstrated, if a transmitter were restricted to a particular part of a neuron, it follows that the neuron would need multiple transmitters to account for different release sites. Consistent with this suggestion is the observation that a variety of neurotransmitter receptors are differentially distributed on mammalian neurons.

Synaptic Specializations Versus Nonjunctional Appositions between Neurons

In addition to the diversity in transmitters that may result from transmitters being targeted to different intraneuronal sites, the anatomical relationships between one cell and its follower may possibly contribute to the need for different transmitters. We typically think of synaptic specializations (see Chapter 2) as the morphological substrate of communication between two neurons. However, there may also be nonjunctional appositions between two neurons, and it appears likely that transmitters are released at these sites.

The requirements for transmitter action at a nonjunctional apposition and a synapse would differ, since the distance traversed by the transmitter molecule would be larger than at a synaptic apposition. Thus, transmitters that lack an efficient reuptake system, such as peptide transmitters, might be favored at nonjunctional synapses. Since a single neuron can form both synaptic specializations and nonjunctional appositions, a single neuron may require more than one neurotransmitter.

Fast Versus Slow Responses of Target Neurons to Neurotransmitters

We have seen that different firing rates or patterns are accompanied by changes in the transmitter being released from a neuron. The postsynaptic response to a transmitter can occur over different timescales. For example, receptors that form ion channels lead to very rapid changes on stimulation by a released transmitter, since the ionic gradients across the cell are almost instantaneously changed. In contrast, metabotropic receptors that respond to classical transmitters and peptide transmitters are coupled to intracellular events through specific transduction molecules, such as G proteins. Thus, metabotropic receptors respond to neurotransmitter stimulation on a slower timescale than do ionotropic receptors (see also Chapters 11 and 16). This difference in temporal response characteristics is useful, since it allows the receptive neuron to respond differently to a stimulus depending on the antecedent activity in the neuron. A transmitter can change the response characteristics of a particular cell to subsequent stimuli on the order of seconds or even minutes and, thus, can occur independent of changes in gene expression.

Comparison of Synthesis and Inactivation of Peptide and Classic Transmitters

There are two major differences between classic and peptide transmitters. Peptide transmitters are synthesized in the cell body, rather than at the terminal processes of neurons; this has significant functional consequences. In addition, peptide transmitters are inactivated by enzymatic actions and not by a reuptake process.

Synthesis and Storage of Peptide Transmitters

Classic transmitters are synthesized in the process (axon, dendrite) from which they are released, but peptide transmitters are not. In the majority of cases, peptide transmitter genes encode a prohormone, a larger precursor protein from which the peptide transmitter is subsequently cleaved. This synthesis from a larger precursor protein allows for an additional layer of synthetic strategy (see Box 10.1). The prohormone is incorporated into secretory granules after translation, where it can be acted on by peptidases, called *prohormone convertases* (Seidah and Chrétien, 1999), to form the functional neuropeptide. In contrast, classic transmitters are formed by successive small enzymatic transformations of a transmitter precursor, rather than from a larger precursor, and do not require transport to distal processes. Although the synthesis of a prohormone is the major strategy used for generation of peptide transmitters, certain small peptides can be enzymatically synthesized. An example is carnosine (N-β-alanyl-L-histidine), which is synthesized by the enzyme carnosine synthase.

Increases in the amount of a classic transmitter that is available for release occur by local synthesis. However, increasing the amount of a peptide transmitter requires an increase in gene expression of the prohormone mRNA, either by transcription of the gene or by stabilization of the mRNA (see Chapter 13); the subsequent delivery of the prohormone–peptide-containing granules to the terminal via axonal transport may take hours. Thus, classic transmitters can respond to increased demand for transmitter release quite rapidly, but peptide transmitters cannot. This difference in biosynthetic strategies contributed to the initial difficulties in localizing the sources of peptide-containing innervations of certain brain regions. Because peptides or their prohormones are transported from the soma immediately after translation from mRNA, the cell body region and proximal processes of these neurons typically contain very low concentrations of the peptide transmitters. Although immunohistochemical methods can easily localize the cell bodies of classic transmitters, to demonstrate the cell bodies in which peptide transmitters are formed it is usually necessary to disrupt microtubule-mediated axonal transport to allow the peptide to accumulate in the soma.

The methods of storage of peptide and classic transmitters also differ. Classic neurotransmitters

BOX 10.1

COORDINATE SYNTHESIS OF MULTIPLE PEPTIDES IN A SINGLE PRECURSOR

The biosynthetic pathway for a neuropeptide follows that of most secreted proteins in that there is a signal peptide at the N terminus that directs the polyribosome to the rough endoplasmic reticulum (RER) for co-translational vesicular discharge of the prohormone into the lumen of the RER. This also dictates a minimal size to the prohormone of approximately 60–70 amino acids for the signal peptide to emerge from the ribosome and be recognized by the signal recognition particle. As mature neuropeptides range in size from few to tens of amino acids, nature has evolved to use the extra "spacer" peptide material in many instances. Figure 10.1 shows many of the different strategies that have evolved, from coordinate synthesis of different peptides that work at different receptors to multiple copies of the same peptide. The presence of multiple neuropeptides within one precursor protein can also allow for differential processing of the same prohormone in different cell types. For example, the proopiomelanocortin (POMC) prohormone is processed to adrenocorticotrophic hormone (ACTH) in the anterior pituitary, but ACTH is further processed to α-melanocyte-stimulating hormone (α-MSH) in the hypothalamus. Another interesting strategy is reflected in the synthesis of two neuroendocrine peptides, vasopressin and oxytocin, which are made in the brain and released for action into the peripheral circulation. In this case, the prohormone also contains the binding protein, neurophysin, which aids in the transport of the neuropeptide in the bloodstream. All these different strategies give an additional level of complexity to synthesis of neuropeptides, adding to their uniqueness from the classic neurotransmitters.

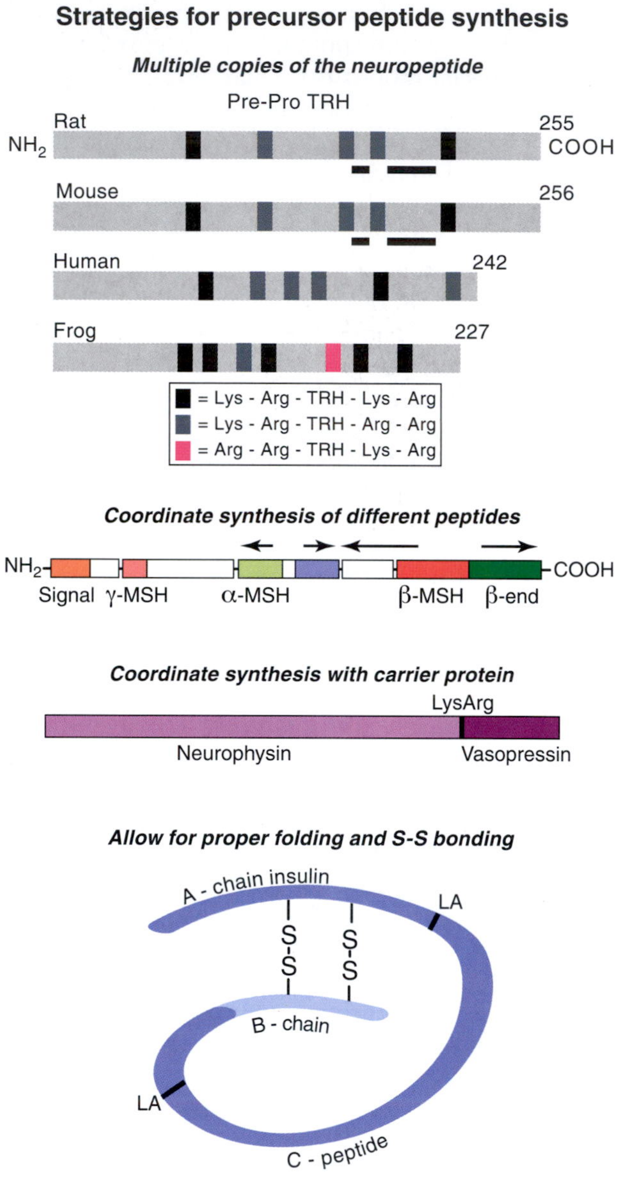

Strategies for precursor peptide synthesis

FIGURE 10.1   The synthesis of neuropeptides in a larger precursor form allows for multiple different synthetic strategies.

are stored in small (~50 nm) synaptic vesicles; neuropeptide transmitters are stored in large (~100 nm) dense-core vesicles. Since peptide transmitters are typically released at a high neuronal firing frequency or burst firing pattern, it is reasonable to assume that there are different mechanisms for the exocytosis and subsequent release of peptide and classic transmitter vesicles. Recent data suggest that there are distinct but related molecular mechanisms that subserve release of small and large dense-core vesicles (Bean *et al.*, 1995). The release of peptide transmitters, like the classic transmitters, is calcium dependent (see Chapter 8 for further discussion).

Inactivation of Peptide Transmitters

Different strategies are used to synthesize peptide and classic transmitters; these differences are paralleled by differences in inactivation of the released transmitter. Classic transmitters have high-affinity reuptake processes that remove the transmitter from the synaptic or extracellular space. In contrast, peptide transmitters appear to be inactivated enzymatically or by diffusion but lack a high-affinity reuptake process. The enzymatic inactivation of peptide transmitters also differs from that of classic transmitters. Since peptide transmitters are short chains of amino acids, the inactivating enzymes show specificity for certain types of amino acids at the cleavage site but are not specific to any single peptide (see Box 10.2). For example, the metallo-endopeptidase that inactivates enkephalins, which

---

BOX 10.2

NEUROPEPTIDE-METABOLIZING ENZYMES COME IN A VARIETY OF DIFFERENT FORMS

Neuropeptide metabolism takes place after the peptide has been released into the extracellular space. Unlike protein synthesis, the cleavage of a peptide bond requires no outside energy input, such as ATP, and hence can take place readily in the extracellular environment. The neuropeptidases that mediate peptide metabolism are a more diverse group than the subtilisin-like family of enzymes that constitute the majority of the prohormone convertases (Seidah and Chrétien, 1999). Essentially all classes of peptidases have been shown to be involved in neuropeptide metabolism, from the Zn-containing endopeptidases (which cleave internal to the peptides) enkephalinase and thimet oligopeptidase, to the aspartyl-peptidase, post-proline cleavage enzyme, to exopeptidases (which cleave from the outside in) like aminopeptidase N. Often, the N and C termini of neuropeptides are blocked by post-translational modifications such as N-acetylation and C-amidation, which protect them from exopeptidase degradation. Thus, the neuropeptides are protected from degradation until they are cleaved internally by endopeptidases. The peptidases are broadly distributed in the nervous system, reflecting the broad distribution of neuropeptides.

Several of these metabolizing peptidases, like enkephalinase, are membrane-anchored proteins, synthesized as classic membrane proteins with an N-terminal signal sequence and a membrane-spanning C-terminal sequence that locks the peptidase with its active site facing the lumen of the RER, which subsequently becomes an extracellular plasma membrane-facing enzyme. Others, however, are synthesized as soluble cytosolic enzymes, such as thimet oligopeptidase, that are released from neurons or glia via a yet uncharacterized mechanism. These peptidases either remain soluble, become associated with extracellular matrix, or become anchored on the extracellular face of the cell and function to metabolize neuropeptides, either degrading or converting them to a different biological activity (see Fig. 10.2). An exciting area of current research deals with the factors involved in regulating the location of these metabolizing peptidases at the cell surface and their proximity to the receptors that bind the peptides they metabolize. This becomes a crucial issue since peptides are not directly released into the synapse and generally must diffuse to their site of action, providing an opportunity for peptidase action.

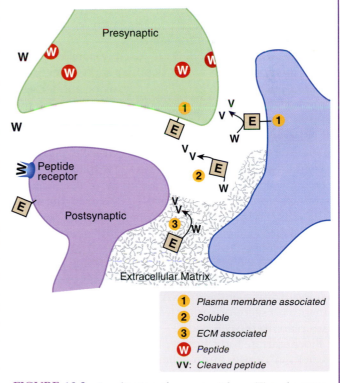

FIGURE 10.2   Localization of neuropeptidases (E) in the extracellular environment.

are small pentapeptide opioid-like transmitters, is frequently called *enkephalinase* but is also critically involved in the inactivation of other neuropeptides, such as neurotensin and somatostatin. One final difference in inactivation of peptide and classic transmitters is the product. In the case of classic transmitters that are catabolized, the product is inactive at the receptor site, hence the term *inactivation*. However, in the case of peptide transmitters, certain peptide fragments derived from the enzymatic "inactivation" of the peptide are biologically active. For example, the peptide angiotensin I is metabolized to yield angiotensin II and angiotensin III, which are successively more active than the parent peptide. Hence, it is difficult to distinguish between extracellular synthetic processing (of a prohormone) and inactivation. The peptide that is stored in vesicles and then released is therefore considered the transmitter, although the actions of certain peptidases may lead to other biologically active fragments. This is an exciting area of investigation in neuroscience as more becomes understood on how the peptide transmitters are metabolized *in vivo* (see Box 10.2).

Approaches to the Study of Neuropeptide Synthesis and Release

The unusual aspects of peptide transmitter biosynthesis and inactivation have resulted in different methods being emphasized in the study of peptide transmitters and classic transmitters.

Anatomical approaches are used extensively to define the neurons in which peptides serve as neurotransmitters. Immunohistochemistry, using antibodies generated against the prohormone, can be used to identify neurons in which peptide transmitters are synthesized. Similarly, antibodies against the peptide transmitter itself are often used to localize the peptide to cell bodies, particularly when axonal transport has been blocked by prior administration of colchicine. Recent anatomical studies have emphasized the use of *in situ* hybridization histochemistry, in which cRNA probes or oligonucleotides complementary to a defined sequence of an mRNA are used to identify the cells in which the gene encoding a peptide prohormone is transcribed. The advantage of this approach is that localization of the mRNA is unaffected by axonal transport of secretory granules since peptide synthesis takes place in the cell body.

Biochemical studies of peptide synthesis and inactivation also use approaches different from those undertaken in the study of classic transmitters. The synthesis of peptide transmitters has been extensively studied by following the rapid incorporation of radiolabeled amino acids into peptides, in so-called "pulse–chase" experiments. More recent studies have emphasized molecular approaches analyzing specific mRNAs, since a prohormone from which the peptide will subsequently be cleaved is directly transcribed from mRNA.

The release of peptides has been studied in slices of brain, much as has been the case with classic transmitters. *In vivo* studies of release have not been possible until relatively recently but can now be accomplished. The measurement of metabolites as an index of release is not a good method, since the peptide fragments generated may be similar to those of other peptides and since larger peptides are further degraded to smaller peptide fragments. This requires antisera with exquisite specificity for the parent peptide but not fragments; such antibodies are rare indeed. Nonetheless, if one knows the principal peptide fragments catabolically generated from the released peptide, one can use immunoassays coupled with chromatographic separation methods to measure levels of the parent peptide and certain peptide fragments to gain an appreciation of the release of the peptide. The introduction of *in vivo* microdialysis methods has been useful for the measurement of extracellular concentrations of neuropeptides, particularly small neuropeptides. Since the released peptide in relatively quick order diffuses across the dialysis membrane, while the peptidases are generally too large to do so, one can obtain measurable levels of peptides in the dialysate. The dialysis approach, however, requires very sensitive analytic methods because of the poor recovery of peptides and does not offer good temporal resolution. Another *in vivo* approach has been to insert electrodes coated with antibodies into a specific area of the brain for a short period. The removed probes are then exposed to a radiolabeled peptide, which will bind to receptors that are not already occupied by the peptide that was released endogenously. This method does not allow repeated measurements and requires very precise experiments to optimize the time the probe is left in the brain.

Pharmacological studies of neuropeptide transmitters have been hampered by the lack of specific compounds that interact with peptide receptors. Many specific peptide analogues of neurotransmitter peptides have been synthesized. However, while these peptides interact specifically with appropriate receptors, peptides and proteins do not readily enter the brain and, thus, have been of very limited utility in *in vivo* studies. Recently, several nonpeptide antagonists have been developed that enter the central nervous system with relative ease. In contrast, there are very few nonpeptide agonists that can be used to study central peptide transmitters; a notable exception is morphine and related opiates used to label central opioid receptors.

NEUROTENSIN AS AN EXAMPLE OF PEPTIDE NEUROTRANSMITTERS

Neurotensin (NT) is a peptide of 13 amino acids that is expressed in the central nervous system and in peripheral tissues, particularly the small intestine. In addition, another peptide termed *neuromedin N* (NMN) is also transcribed from the gene that encodes NT. NMN is a structurally related hexapeptide that occurs in mammals; a nearly identical peptide called LANT-6 is found in birds. Peptides often have central and peripheral roles, one involving chemical communication between neurons and the other involving varied peripheral functions. Indeed, neurotensin and many other mammalian brain peptides are structurally similar to peptides that are found in nonmammalian organisms, particularly amphibian species, and a large number of peptides were discovered in the skin of certain toads. An example is xenopsin, a neurotensin (NT)-like octapeptide, which was isolated from the skin of *Xenopus*.

Neurotensin Synthesis

A 170-amino-acid prohormone precursor of NT is encoded by a single gene that is transcribed to yield two mRNAs (see Fig. 10.3). The smaller transcript is the predominant form in the intestine, while both mRNA species are present in equal abundance in most brain areas (Kislauskis *et al.*, 1988). The 170-amino-acid precursor contains one copy each of NT and NMN. However, differential processing of the precursor can occur, leading to different molar ratios of the two peptides.

NT is present in very high concentrations in the small intestine and in lower amounts in the stomach and large intestine. NMN is also present in these tissues, but NT:NMN molar ratios differ across the different tissues, suggesting differential enzymatic processing of the prohormone or the generation of different transcripts. Because NT and NMN are contained in the same exon of the NT–NMN gene, differences in relative abundance of NT and NMN are due to differential processing of the precursor.

Storage and Release of Neurotensin

Neurotensin and NMN, and other peptide transmitters, are stored in large dense-core vesicles. Studies of the release of NT and NMN have been examined in the hypothalamus, an area of the brain enriched in NT and NMN. Superfusing hypothalamic tissue with high concentrations of K^1 to depolarize cells evokes release of both NT and NMN (Kitabgi *et al.*, 1992); the molar ratio of NT:NMN release is virtually identical to that in extracts of hypothalamic tissue. The release of NT and NMN is not seen in tissue perfused with a medium that lacks calcium. Thus, the release of the peptides is calcium dependent (reflecting dependence of vesicular docking with the membrane and exocytosis on calcium) and is evoked by stimulation. These criteria are considered necessary for the designation of a compound as a neurotransmitter.

As noted earlier, the release of peptides such as NT is impulse dependent. NT release is regulated by both the firing frequency and firing pattern of neurons: when neurons discharge rapidly in bursts, NT release is markedly enhanced (Bean and Roth, 1992).

Inactivation of Neurotensin

There is no high-affinity reuptake process for peptides. Accordingly, other than possible diffusion, the major means of inactivating peptide transmitters that are released from a neuron is enzymatic.

NT is enzymatically inactivated by a group of enzymes known as metallo-endopeptidases. In particular, three of these endopeptidases (known with great flair as EP24.11, EP24.15, and EP24.16) are involved in the catabolism of NT. In both brain and gut, NT is hydrolyzed by combinations of the three endopeptidases (Kitabgi *et al.*, 1992). Endopeptidase 24.11 (also known as *enkephalinase* for its well-characterized actions on enkephalins) cleaves NT at two sites, the Pro^{10}–Tyr^{11} and the Tyr^{11}–Ile^{12} residues. Endopeptidase 24.15 acts at the Arg^8–Arg^9 site, and endopeptidase 24.16 acts at the same Pro^{10}–Tyr^{11} site as enkephalinase. Thus, the enzymes involved in peptide degradation, including those that metabolize NT, do not exhibit a high degree of specificity (see Box 10.2).

We noted in the previous chapter that certain proteins that have important functions in neurotransmission (such as acetylcholinesterase) may have very different functions in other (nonneural) systems. Enkephalinase, which hydrolyzes NT, somatostatin, and enkephalins, appears to be similar to acetylcholinesterase by having a role in leukemia. The common acute lymphoblastic leukemia antigen (CALLA) is a protein expressed by most acute lymphoblastic leukemias and normal lymphoid progenitors. The human CALLA cDNA encodes a protein that appears nearly identical to that of enkephalinase (Shipp *et al.*, 1989).

As noted earlier, there is no high-affinity reuptake process for peptides. Nonetheless, certain peptides, including NT, have been shown to be accumulated by neurons. The high-affinity NT receptor is a G-protein-coupled receptor; such receptors, on binding of ligand,

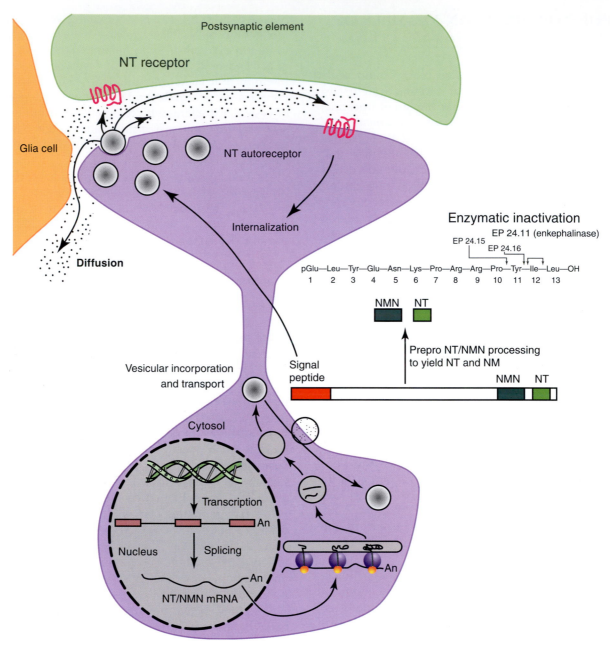

FIGURE 10.3 Schematic illustration of neurotensin (NT) and neuromedin N (NMN) biosynthesis and action, from gene transcription to translation and processing in the vesicular secretory pathway to diffusion from the synaptic cleft and inactivation by a group of endopeptidases (EP) extracellularly.

are internalized and thus accumulate the ligand into the neuron. A series of studies has revealed that NT, when injected into the striatum, is internalized by nerve terminals that express the high-affinity NT receptor; once internalized, the peptide dissociates from the receptor and is retrogradely transported to the cell bodies of origin, which are dopaminergic neurons in the midbrain.

The functional significance of internalization of NT and its subsequent retrograde transport is not clear. It could conceivably serve as a growth factor, regulating the targeted growth and survival of afferents to the striatum. An intriguing possibility is that NT may actually function over a relatively extended period to alter the transmitter function of the cell (Deutch and Zahm, 1992). Local injections of NT into the striatum result in retrograde labeling of midbrain dopamine neurons and increase the number of midbrain cells expressing tyrosine hydroxylase mRNA.

Coexistence of Neurotensin and Classic Transmitters

Neurotensin is found in many dopaminergic neurons in mammalian species, including certain hypothalamic and midbrain dopamine cells. In particular, certain midbrain DA neurons of rodents use NT as a transmitter. Since all of the neurotensin in the prefrontal cortex of the rat is thought to be contained in DA axons, the study of the rodent prefrontal cortex has afforded a unique opportunity to investigate interactions between colocalized transmitters *in vivo*. Such studies have revealed that neurotensin release in the prefrontal cortex is increased when neuronal firing is increased, or when DA neurons enter into a burstlike firing pattern (Bean and Roth, 1992). In addition, there appears to be a close association between the dopaminergic autoreceptor present on DA axons in the prefrontal cortex and NT release from these axons. In contrast to most forebrain areas innervated by DA axons, transmitter release but not synthesis is regulated by DA autoreceptors in the prefrontal cortex. Low autoreceptor selective doses of dopamine agonists, which decrease subsequent dopamine release, enhance neurotensin release from these axons; conversely, antagonists at the release-modulating autoreceptor decrease NT release but enhance DA release. Thus, NT and dopamine release are regulated in a reciprocal fashion by autoreceptors in neurons in which both transmitters are present.

Interestingly, although neurotensin-containing cells are found in many areas of the primate brain, very few midbrain dopamine cells express the NT–NMN gene in primate species, and the projection target of these neurons is not known.

UNCONVENTIONAL TRANSMITTERS

The designation of a substance as a neurotransmitter rests on the fulfillment of certain criteria, which were discussed in Chapter 9. However, these criteria were formulated early in the modern neuroscience era and are based mainly on studies of peripheral sites, particularly the neuromuscular junction and superior cervical ganglion. Over the past several decades there have been remarkable increases in technical sophistication that allow us to measure substances in the brain that are present in minute quantities or are very unstable. The ability to measure these substances has opened the possibility that several substances that previously did not meet the requirements for designation as a neurotransmitter may indeed be transmitters.

The advances in technical wizardry, however, have also led us to unanticipated mechanisms of intercellular signaling that have required a new definition of neurotransmitters. One simple approach would be to designate as chemical mediator a neurotransmitter if it allows information to flow from one neuron to another. This definition circumvents the issues of glial contribution to the ionic milieu of the neuron, which certainly imparts information concerning the function of the glia (and parenthetically points to the increasing awareness of the active roles that glia play in the CNS; see Box 10.3). On the other hand, such a definition closely approximates the definition for a hormone, in that the temporal characteristics of transmitter action are not defined, nor is the distance of the target cell to the transmitter substance specified. Finally, this definition does not accommodate unconventional roles for transmitters, such as the regulation of neuronal development and intracellular trafficking of proteins. Since growth factors can clearly influence neuritic outgrowth and cell survival, as well as influence neuronal polarity and other various functions thought to involve changes in intracellular trafficking, the definition of transmitter changes as one considers what we have termed "unconventional" transmitters.

The discovery of peptidergic neurotransmitters over a generation ago was accompanied by considerable debate over the definition of a transmitter. Yet peptide transmitters are quite similar to classic transmitters, sharing such features as conventional storage, calcium-dependent release, and the ability to influence the activity (e.g., the firing rate) of postsynaptic neurons over a relatively brief interval. Given these basic similarities, and the broad acceptance of peptide transmitters into the language of neurotransmission, it seems necessary to contrast peptide transmitters with other substances that influence target cells in unique ways. Accordingly, we have designated these novel "transmitters" as unconventional, at least for a few years. Among these are growth factors, neuroactive gases (nitric oxide and carbon monoxide), lipid derivatives (endocannabinids) and neurosteroids.

Growth Factors

Growth factors are an extremely heterogeneous group of proteins that regulate the growth and differentiation of various cell types, some of which specifically target neurons (see Russell, 1995; Thoenen, 1995). One similarity between the neurotrophic growth factors and nitric oxide (NO) is that both are capable of influencing presynaptic cells. Thus, NO has been shown to diffuse out of cells and alter transmitter release from presynaptic axons in the hippocampus,

BOX 10.3

ASTROCYTES: FROM SUPPORT PLAYER TO CENTER STAGE

For much of the past century neuroscience has focused on the integrative activity of neurons. Although neurons are certainly essential elements in central nervous system function, it is increasingly clear that glial cells, particularly astrocytes, are important: these cells are excitable and can send and receive signals using mechanisms similar to those employed by neurons. The diversity of signaling and contributions of astrocytes was not appreciated initially because recordings of the membrane potential of an astrocyte offered little to observe. The membrane potential is stable and maintained at about 280 mV, and regardless of the magnitude of depolarization elicited by current injection, an action potential is never elicited. Although this stable negative resting potential is critical to one function of the astrocyte, the electrogenic uptake of glutamate from the synaptic cleft, it does not reflect the diversity of regulated biochemical signaling that we are beginning to appreciate in these nonneuronal cells.

Technical developments over the past 15 years have allowed us to determine that astrocytes exhibit a form of nonelectrical excitability that depends on calcium. Neurotransmitters, which we commonly think of as the signals used at neuronal chemical synapses, also elicit calcium oscillations in astrocytes. These calcium signals arise from the release of calcium from internal stores, which is gated by metabotropic receptor-dependent activation of phospholipase C and the production of inositol trisphosphate. What, however, are the functional consequences of such calcium signals? One of the major surprises in the study of astrocytes has been the observation that even relatively small elevations of astrocytic intracellular calcium lead to the calcium-dependent release of the chemical transmitter glutamate! In addition, ATP, D-serine, and prostaglandin E_2 can also be released by astrocytes in response to metabotropic receptor activation. Thus, astrocytic calcium excitability results in the release of a variety of chemical transmitters that can regulate both local astrocytes and neurons.

The concept of tripartite synaptic transmission has emerged in recent years, with the astrocyte being considered an active participant in the control of communication between two neurons. Many synapses are associated with an astrocytic process. A series of complex studies has been performed to determine if active signaling is present between the neuronal and glial elements of the tripartite synapse. For instance, stimulation of presynaptic neuronal afferents not only evokes postsynaptic potentials but also stimulates calcium oscillations in synaptically associated astrocytes. Moreover, experimentally induced calcium increases in astrocytes can modulate neighboring synapses as a result of the release of "gliotransmsitters." These functional studies suggest that astrocytes, as part of the tripartite synapse, integrate neuronal inputs and provide feedback regulation of the synapse by way of the calcium-dependent release of chemical transmitters.

Calcium signaling within astrocytes is a complex process and does not exhibit simple all-or-none features. Synaptic activity can evoke calcium elevations within local portions of the processes of a single astrocyte; such local increases in Ca^{21} do not necessarily propagate throughout the entire cell. However, increased neuronal activity can switch the behavior from a local, synaptically associated process to a more global, cell-wide Ca^{21} elevation. Moreover, cellwide Ca^{21} elevations can propagate to neighboring astrocytes: recent studies suggest that astrocytes may be interconnected with one another in short-range circuits. Since each astrocyte contacts thousands of synapses, the distance over which a calcium signal spreads within one astrocyte and whether signals spread between interconnected astrocytes are likely to play a major role in modulating synaptic transmission.

These are early days in a new field of study. The full extent to which astrocytes are involved in the regulation of the synapse remains to be determined. Although most studies of astrocytic function have been performed using cell cultures, recent slice and *in vivo* studies have made it clear that synaptically associated glia serve a regulatory role in brain regions such as the hippocampus, as well as regions including the retina and neuromuscular junction. Whether biochemical integration in these nonneuronal cells plays roles in neuronal processes such as synaptic plasticity awaits a new level of experimental study. Regardless, the fact that astrocytes integrate neuronal inputs and exhibit calcium excitability that controls the release of chemical transmitters changes the way in which we view the roles of these cells in nervous system function.

Philip G. Haydon, PhD

and neurotrophic factors provide trophic support for developing axons innervating a region in which growth factors are expressed in cells (hence the term *trophic factors*). Growth factors are also unconventional in that the release of these proteins may be through both the constitutive and regulated pathways. In many regards cytokines, generally inflammatory proteins, are similar to growth factors in their synthesis and release, and tend to serve as a conduit between the immune system and the central nervous system.

There are several different classes of growth factors, each class comprising many different growth factors or cytokines (see Fig. 10.4). For example, the neurotrophins include nerve growth factor, brain-derived-neurotrophic factor, neurotrophin 3 (NT-3), NT-4, and NT-5 (Chao, 2003). In many cases, growth factors were originally identified from functions in peripheral tissues but have subsequently been found to have profound effects in brain, for example, fibroblast growth factor. Unfortunately, there are many gaps in our knowledge of the basic cellular biology of any single growth factor, even one known as long as nerve growth factor. Accordingly, we draw on examples from several different growth factors to illustrate what may be general characteristics of relevance.

Synthesis of Growth Factors

Growth factors have been identified historically by examining the effects of crude extracts of certain tissues or biological fluids on a bioassay of cell survival or growth. Following the determination that some serum or tissue factor is capable of sustaining cell survival or differentiation, further purification of the crude extract is performed, culminating in the isolation and purification of the protein. The protein is then sequenced. This approach is tedious and requires very large amounts of starting material, as well as a sensitive and reliable bioassay. This approach is essentially the same—and therefore confronted with the same difficulties—as the early methods used to discover and characterize pituitary and central peptides. Contemporary approaches to the identification of growth factors emphasize cloning of growth factors on the basis of sequence homology; the cloned genes are then expressed in various cell lines and tested in various bioassays. This approach has allowed us to gain some insight into potential regulatory features of growth factors and their cellular biology.

Molecular approaches have provided interesting and unexpected information on the synthesis of growth factors. For example, certain growth factors are translated from multiple mRNAs. Brain-derived neurotrophic factor (BDNF) is a neurotrophin that has a heterogeneous distribution in the CNS and several distinct functions, some of which are those classically associated with growth factors and others that are more "transmitter-like." The BDNF gene has five different exons that encode the mature BDNF protein. There is a separate promotor for each of four 5¢ exons, and alternative use of these promoters gives rise to eight different BDNF mRNAs. Despite this wonderful complexity in the molecular "synthesis" of BDNF, there is only one mature BDNF protein; it has been speculated that various BDNF transcripts may give rise to different amounts of protein and are differentially associated with pre- and postsynaptic elements.

Surprisingly little is known about the post-translational processing of growth factors. In the case of the neurotrophin nerve growth factor (NGF), the mature protein is cleaved from a prohormone, in a manner similar to the synthesis of peptide transmitters. The processing of the NGF prohormone is unique, however, in that three subunits are present. One of these, the g subunit, is a serine protease that is thought to cleave the prohormone to yield the mature protein. Other neurotrophins, including BDNF and NT-3, are also formed by cleavage of a prohormone, although the identity and degree of specificity of the responsible enzymes are only now being determined. One final aspect of growth factor biosynthesis (in its broad definition) is their unusual multimeric assembly. The neurotrophins form biologically active heteromers *in vitro* but *in vivo* have been thought to form exclusively homodimers. However, recent data suggest that heterodimers of NGF, BDNF, and NT-3 can be formed both *in vitro* and *in vivo* (Heymach and Shooter, 1995).

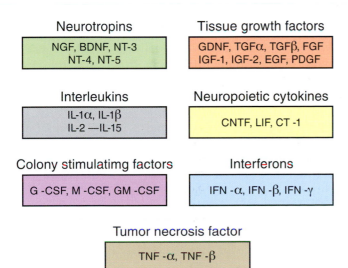

Neurotropins	Tissue growth factors
NGF, BDNF, NT-3 NT-4, NT-5	GDNF, TGFα, TGFβ, FGF IGF-1, IGF-2, EGF, PDGF

Interleukins	Neuropoietic cytokines
IL-1α, IL-1β IL-2 —IL-15	CNTF, LIF, CT -1

Colony stimulatimg factors	Interferons
G -CSF, M -CSF, GM -CSF	IFN -α, IFN -β, IFN -γ

Tumor necrosis factor

TNF -α, TNF -β

FIGURE 10.4   Outline of the different families of growth factors and cytokines in the brain.

Storage and Release of Growth Factors

Surprisingly little is known about the storage and release of growth factors. There have been very few studies of the storage of growth factors, particularly electron microscopic studies of intracellular localization. In part this reflects the difficulty in generating specific antibodies to growth factors, since individual growth factors (e.g., BDNF) within a family (neurotrophins) share a striking degree of homology with other members of the family (e.g., NT-3). However, the production of new specific antisera, in conjunction with studies of localization of growth factors in cells transfected with specific growth factor genes, has made it is possible to determine the intracellular localization of some growth factors.

Neurons secrete proteins by two distinct processes. The constitutive pathway for secretion is not triggered by extracellular stimulation, and is used to secrete membrane components, viral proteins, and extracellular matrix molecules. In this pathway there is continuous fusion of Golgi-derived vesicles with the plasma membrane. Growth factors have generally been considered to be secreted by the constitutive pathway. In contrast, the release of conventional neurotransmitters is controlled by extracellular signals and uses the so-called regulated pathway. For example, peptide protein precursors contain an N-terminal signal sequence, which targets the protein to the endoplasmic reticulum and then to the Golgi network ultimately to be packaged into vesicles.

Because some growth factors lack a signal sequence, the growth factors have generally been considered to be processed by the constitutive pathway, despite clear data indicating release of these proteins. Recent studies of neurotrophins suggest that BDNF and NGF are secreted through both the constitutive and regulated pathways. Using antibodies to examine the localization of BDNF in a transfected cell line, Goodman et al. (1996) showed that BDNF is present in chromogranin-containing secretory granules. This observation suggests that BDNF may be present in certain vesicles in neurons, but this remains to be shown.

Several published studies on BDNF and NGF suggest that under basal conditions there is constitutive release from the soma and proximal dendrites, but that depolarization induced by high potassium concentration or glutamate leads to regulated release from both the distal and proximal processes of the neuron (Goodman et al., 1996; Bloch and Thoenen, 1996). After seizures, BDNF can be seen to move into the axons of hippocampal pyramidal cells, and then into perikarya, suggesting that it is poised to be released; 2 to 3 h after seizures, BDNF immunoreactivity is observed in the neuropil surrounding pyramidal cells and in the area where mossy fibers terminate, suggesting that the growth factor has been released from somatodendritic sites (Wetmore et al., 1994). The activity-dependent release does not appear to depend on extracellular calcium, but is critically dependent on intracellular calcium stores (Bloch and Thoenen, 1996); this is distinct from release of conventional transmitters, which has an absolute dependence on extracellular calcium. Thus, the release of (at least certain) growth factors differs from that of conventional transmitters by occurring through both constitutive and regulated pathways and by not being dependent on extracellular calcium concentrations.

Functional Significance of Growth Factors as Neurotransmitters

Growth factors have a wide array of functions. As indicated by their name, growth factors support the development and differentiation of neurons. Thus, the survival and axonal growth of many neurons to their final target require certain factors expressed in the targeted cells. These functions are developmentally specific. Once the axon of a neuron reaches its target and survives, it usually does not require further support from growth factors under normal conditions. This ability of growth factors to determine survival, differentiation, and final target destination is certainly not part of the normal list of functions ascribed to neurotransmitters. It is not clear that the release of the growth factor in this context is a chemical signal that conveys information to neurons, as opposed to providing critical sustenance.

What information then supports the contention that growth factors may be unconventional transmitters, even under the broad definition that we have used? First, growth factors are stored and released from neurons. The release of growth factors is now thought to occur through the regulated as well as constitutive pathways, suggesting some specificity in release. Some data are consistent with the speculation that BDNF is stored in vesicles. The synthesis and release of growth factors are also under transynaptic control. For example, expression of NGF and BDNF is controlled by neuronal activity, with glutamate and acetylcholine increasing expression and GABA decreasing expression of these neurotrophic factors. Moreover, the induction of BDNF by various treatments is regionally, spatially, and temporally distinct. Thus, seizures result in patterns of increases of various BDNF exon-specific mRNAs that differ across hippocampal subregions, and the specific pattern of promotor activation depends on the stimulus (Kokaia *et al.*, 1994). These characteristics all suggest that in addition to the synthesis and storage of neurotrophins in neurons, the synthesis and release of these growth factors are regulated by neuronal activity, and thus growth factors may participate in both receiving information and conveying information across cells.

The second characteristic suggesting a transmitter role for growth factors is that they appear to regulate other neurons. The receptors for neurotrophins, the trk receptors, are expressed in neurons, are members of a class of transmembrane receptor tyrosine kinases, and have an intracellular catalytic domain that is activated on ligand binding. Among the functions subserved by growth factors are the regulation of growth, differentiation, migration, transcription, and protein synthesis in neurons. In addition, neurotrophins appear to change the functional activity of other cells. BDNF has been shown to stimulate phosphoinositide turnover in neurons (Widmer *et al.*, 1993), and thus shares with NO the ability to regulate a key intracellular transduction mechanism. BDNF has also been shown to increase NT-3 expression in the cerebellum and hippocampus (Leingartner *et al.*, 1994; Lindholm *et al.*, 1994), again suggesting that growth factors regulate the functional activity of target cells.

The previously described data suggest that growth factors can regulate the functional activity of neurons but do not address a physiological role for growth factors independent of their effects on growth and survival. However, several recent studies have suggested that BDNF may regulate hippocampal function during the induction of long-term potentiation (LTP). Mice carrying a deletion in the coding region of the BDNF gene show significantly reduced LTP (Korte *et al.*, 1995), while BDNF administration to hippocampal slices markedly enhances LTP (Patterson *et al.*, 1992).

Nitric Oxide and Carbon Monoxide: Gases as Unconventional Transmitters

Nitrates have been extensively used in the treatment of cardiovascular disorders to dilate blood vessels of the heart and thereby relieve the symptoms of angina, but the mechanisms of action of nitroglycerine and similar nitrates were not known until recently. In 1980, an endothelium-derived relaxing factor present in cells lining blood vessels was shown to potently and rapidly dilate blood vessels. Shortly thereafter it was demonstrated that this endothelial factor was a gas, nitric oxide. In addition, it was observed that glutamate, acting at NMDA receptors, releases a factor that causes vasodilation and increases cGMP levels in brain. It soon became apparent that the endothelium-derived relaxing factor and the glutamate-induced factor that dilate blood vessels were the same factor, and that nitric oxide (NO) was present in neurons as well as vasculature. These data, coupled with the identification of neuronal as well as vascular isoforms of the NO synthetic enzyme (nitric oxide synthase), led to the concept that NO is a molecule involved in intercellular communication, including neurotransmission.

Nitric oxide is a well-known air pollutant and, thus, an unlikely candidate for a neurotransmitter. The idea that an unstable toxic gas could be a transmitter led to several questions concerning the nature of neurotransmission, the most obvious being how a gas can be stored for release in an impulse-dependent manner. Because the simple answer is it can't, the classic definition of a neurotransmitter has become blurred or untenable, depending on one's perspective. Many theories can accommodate one exception. However, subsequent studies revealed that NO is not the only gaseous neurotransmitter and that carbon monoxide plays a similar transmitter-like role (Dawson and Snyder, 1994). These findings led to the realization that neurotransmitters as classically defined may be the exception rather than the norm. The exceptions posed by NO to the dogma of traditional neurotransmitters are substantial. For example, NO is not stored in cells, is not exocytotically released, lacks an active process that terminates its action, does not interact with specific membrane receptors on target cells, and regulates the function of synaptic terminals proximal to the neuron in which NO is synthesized. It is not difficult to understand the skepticism that met the hypothesis that gases such as NO could be neurotransmitters, or to have some sympathy for the view that NO is not a transmitter but some alien event, intent on making neuroscientists question their most cherished beliefs.

Synthesis of Nitric Oxide

The synthesis of NO is not complicated, consisting of one step: the conversion of L-arginine to NO and citrulline (see Fig. 10.5). The enzyme responsible for this step is nitric oxide synthase (NOS). Three distinct isoforms of NOS have been cloned. Macrophage-inducible NOS (iNOS) is present in microglia, while endothelial NOS (eNOS) is found in the endothelial cells lining blood vessels, and the localization of neuronal NOS (nNOS) is obvious. All three forms require tetrahydrobiopterin as a cofactor and NADPH as a coenzyme. In fact, NOS is identical to the previously

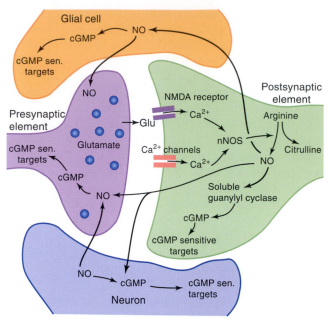

FIGURE 10.5 Schematic representation of a nitric oxide (NO)-containing neuron. NO is formed from arginine by the actions of different nitric oxide synthases (NOS). NO freely diffuses across cell membranes and can thereby influence both presynaptic neurons (such as the glutamatergic presynaptic neuron in the figure) or other cells that are not directly apposed to the NOS-containing neuron; these other cells can be neurons or glia.

described enzyme NADPH diaphorase, which is expressed in high concentrations in several types of neurons in the CNS.

Regulation of Nitric Oxide Levels

NOS is among the most regulated of enzymes, and NO levels in various tissues can be modified in several ways. Although the general use of the term *inducible NOS* for macrophage-inducible NOS may lead one to suspect that only iNOS is regulated, all NOS isoforms are regulated. Regulatory processes include phosphorylation (which decreases NOS activity) and hormonal control. In addition, levels of NOS can be modified by direct inhalation of NO.

Storage, Release, and Inactivation of Nitric Oxide

The criteria for classic transmitter status include intraneuronal synthesis of the substance and release from these cells by calcium-dependent exocytosis. These criteria imply an intermediary storage step. However, NO is not stored in vesicles, since NO is an uncharged molecule that diffuses freely across cell membranes. Nor is NO released by calcium-dependent exocytosis.

Although NO is synthesized in neurons, the ability of the gas to diffuse across membranes allows it to modify the activity of other cells. The lack of storage

conditions and exocytotic release suggests that NO simply modifies the activity of targets in a tonic fashion, but the tight regulation of NOS and the critical observation that neuronal stimulation elicits NO release indicate a more phasic (and, hence, more "transmitter-like") role.

NO also differs from conventional transmitters by lacking an active process to terminate its actions. Instead, inactivation of NO is essentially passive: NO decays spontaneously to nitrate with a half-life of less than 30 s. Another means of terminating the action of NO is by reaction of the gas with iron-containing compounds, including hemoglobin.

Actions of Nitric Oxide Synthase on Receptive Neurons and Functional Effects

Yet another unique aspect of NO as a neurotransmitter is that NO does not interact with specific receptor proteins on postsynaptic target cells, but instead interacts with second-messenger molecules in the target cell to initiate a cascade of intracellular processes. In particular, NO stimulates guanlylyl cyclase to increase cGMP levels.

A final aspect of NO that confounds the traditional boundaries of neurotransmitters is the cellular target of NO. Conventional neurotransmitters have effects on postsynaptic cells that are "downstream" of the neuron releasing the transmitter. In contrast, in many cases NO acts as a retrograde messenger, modifying the metabolism and release of transmitter from presynaptic terminals. For example, NO appears to be released from certain hippocampal neurons after NMDA receptor stimulation to enhance transmitter release from presynaptic elements; this role has been suggested to be of critical importance in long-term potentiation (O'Dell *et al.*, 1994). The release of NO on NMDA receptor stimulation may also underlie potential neurotoxic actions of NO, which is highly reactive. It has been suggested that an excess release of NO associated with NMDA receptor stimulation may be important in the toxic cell loss caused by strokes.

Carbon Monoxide

A large and still growing body of literature documents the function of NO in the brain and argues for NO as an unconventional neurotransmitter. Despite these data, the acceptance of NO as an unconventional transmitter would have been delayed considerably were it not for data indicating that a sibling of NO, carbon monoxide (CO), is also a neurotransmitter.

There are several striking parallels between NO and CO: both are gases that diffuse freely across cells, freely target guanlylyl cyclase rather than a membrane protein as a receptor, and lack a cellular storage compartment. Although CO has not yet received the attention devoted to NO, awareness of the role of CO as an unconventional transmitter is growing.

CO is formed by the enzyme heme oxygenase, which catalyzes the conversion of heme to biliverdin with the accompanying liberation of CO. There are two major isoforms of heme oxygenase (HO). HO-1 is an inducible enzyme that is present in glia and a few neurons, and is present in very high concentrations in the liver and spleen. In contrast, HO-2 is constitutively active rather than inducible, and is found in high concentration in central neurons, particularly in the cerebellum and hippocampus (Zakhary *et al.*, 1996). Cytochrome P450 reductase is a necessary electron donor for heme oxygenase and NOS.

Functional studies of CO are relatively recent, and it is difficult to gauge the degree to which CO may parallel NO. Recent data, however, do suggest that CO (like NO) may be an endothelium-derived relaxing factor (Zakhary *et al.*, 1996). Moreover, recent studies of eNOS knockout mice indicate that cerebral circulatory responses are relatively normal, despite the absence of eNOS, and suggest that CO may function cooperatively with NO to maintain vascular tone.

Endocannabinoids

The recreational and medicinal properties of cannabis-derived preparations have been known for centuries. Although the main psychoactive ingredient of marijuana, D^9-tetrahydrocannabinol (THC), was isolated in 1964, the pharmacological actions of cannabinoids have remained a mystery until the late 1980s, when experimental evidence for the existence of a "marijuana" receptor culminated with the cloning of the first cannabinoid receptor. As the discovery of opioid receptors in the 1970s led to the identification of morphin-like ligands in the brain—the enkephalins and the endorphins—the presence of cannabinoid receptors in the body spurred the search for their physiological ligands (endocannabinoids). Two arachidonic acid-derivatives were found to have cannabinoid-like activity and activate cannabinoid receptors: anandamide (from the Sanskrit word *ananda* = internal bliss) and 2-arachidonyl glycerol (2-AG). In the last decade, other signaling lipids, such as N-arachidonyl dopamine (NADA) and noladin ether, have been added to the endocannabinoid family, which most likely will continue to grow; however, the functional role of these lipid mediators remains largely unknown. Endocannabinoids are highly abundant in the central nervous system, as well as in

peripheral sites, including the digestive, circulatory, immune, reproductive, and endocrine systems, and bind with similar affinities to the two cannabinoid receptor subtypes identified so far, named CB_1 and CB_2. CB_1 receptors mediate the psychotropic effects of cannabimimetic drugs and are expressed predominantly in the central nervous system. The CB_1 receptor gene is intronless, and its structure is highly conserved across rodents and humans. On the other hand, CB_2 receptors are primarily localized in cells of lymphoid or myeloid origin involved in immune and inflammatory responses, and the encoding gene is divergent across different species.

Endocannabinoids like anandamide and NADA can also activate other receptors, such as the "transient receptor potential vanilloid type 1" (TRPV1) (Fig. 10.6A), the pharmacological target of the active ingredient of chilly pepper (Ross, 2003), and the

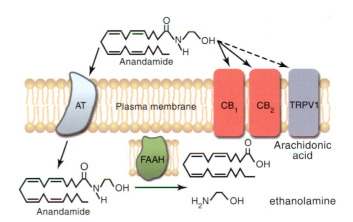

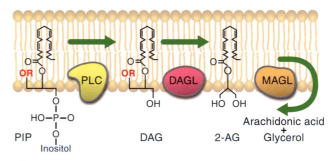

FIGURE 10.6 In neurons, membrane depolarization, increased intracellular calcium, and/or stimulation of G-protein-coupled receptors induce the synthesis of the endocannabinoids anandamide and 2-AG from phospholipid precursors located in the plasma membrane. (A) Endocannabinoids can bind CB_1 and CB_2 receptors. Anandamide can also bind, although with lower affinity, TRPV1 receptors. Endocannabinoids are cleared away from their targets by a putative anandamide transporter (AT). Inside the cell, anandamide is hydrolyzed into arachidonic acid and ethanolamine by the fatty acid amidohydrolase (FAAH). (B) A diacylglycerol lipase (DAGL) and a monoacylglycerol lipase promote the hydrolysis of 2-AG into arachidonic acid and glycerol.

peroxisome proliferators-activated receptors (PPAR), a super family of ligand-activated transcription factors that are involved in fatty acid metabolism, energy homeostasis, and regulation of inflammation and food intake (O'Sullivan, 2007). Although the low intrinsic efficacy of endocannabinoids at TRPV1 receptors have raised questions whether these lipids might serve as physiological ligands (endovanilloids), their efficacy and potency at these receptors can be enhanced by drugs that inhibit endocannabinoid inactivation.

Synthesis and Inactivation of the Endocannabinoids

As other lipid signaling molecules, endocannabinoids are released on demand via stimulus-dependent cleavage of membrane phospholipid precursors. This reaction appears to be initiated by activation of neurotransmitter receptors, as indicated by the enhanced outflow of anandamide in rat striatum following stimulation of dopamine D2-like receptors (Giuffrida *et al.,* 1999). Similarly, application of cholinergic and glutamatergic agonists increases 2-AG production *in vitro* and *in vivo*.

The putative biosynthetic pathway of anandamide involves the phospholipid precursor N-aracidonoyl phosphatidylethanolamine (NAPE), which is cleaved by a NAPE-specific phospholipase D (NAPE-PLD) to yield anandamide and phosphatidic acid. However, the observation that anandamide biosynthesis is unaffected in NAPE-PLD knockout mice suggests that other phospholipases might be involved.

The biological actions of anandamide are terminated by a two-step mechanism consisting of carrier-mediated transport into cells (Beltramo *et al.,* 1997) followed by enzymatic hydrolysis (Fig. 10.6A). In contrast with the transport systems for classical neurotransmitters, anandamide reuptake is neither dependent on external Na^+ ions nor affected by metabolic inhibitors, amino acid transmitters (such as glutamate or g-amino-butyrate), or biogenic amines (such as dopamine or norepinephrine), suggesting that it may be mediated via a process of carrier-facilitated diffusion. However, skepticism has been raised about the existence of an endocannabinoid transporter, and its molecular identity remains undetermined.

Once internalized, anandamide is hydrolyzed into arachidonic acid and ethanolamine by a fatty acid amido hydrolase (FAAH) (Fig. 10.6A), an intracellular membrane-bound enzyme that acts as a general hydrolytic enzyme with wide substrate preference (Cravatt *et al.,* 1996). Like other hydrolase enzymes, FAAH may act in reverse, catalyzing the synthesis of anandamide from free arachidonate and ethanolamine. However, the high K_M values reported for anandamide

synthase activity suggest that, under normal conditions, FAAH acts predominantly as a hydrolase. Anandamide also can be metabolized by COX-2, LOX, and cytochrome P450 into prostanoid-like compounds whose physiological roles have not been explored.

The biosynthesis of the second endocannabinoid, 2-AG, occurs via phospholipase C (PLC)-mediated hydrolysis of a phospholipids precursor to produce diacylglycerol (DAG), which is subsequently converted into 2-AG by a diacylglycerol lipase (DAGL) (Fig. 10.6B). As in the case of anandamide, other enzymes might produce 2-AG, such as phospholipase A1 (PLA1), which yields a lysophospholipid hydrolyzed to 2-AG by lyso-PLC activity. 2-AG is inactivated by reuptake through the same carrier system as anandamide and is subsequently hydrolyzed intracellularly into glycerol and arachidonic acid by a monoacylglycerol lipase (MAGL) (Dinh *et al.,* 2002). Enzymatic oxygenation of 2-AG by COX-2 into prostaglandin glycerol esters has also been reported.

Because of the rapid deactivation process, the endocannabinoids may primarily act near their sites of synthesis, by binding to and activating cannabinoid receptors on the surface of neighboring cells.

Functional Significance of the Endocannabinoids

In the last decade, the development of analytical methods for endocannabinoid detection and the availability of pharmacological tools and the knockout technology to modulate cannabinoid receptor function have provided insight into the pathophysiological roles played by the endocannabinoid system. There is overwhelming evidence that anandamide and 2-AG share some of the biological properties of plant-derived cannabinoids, although they display significant differences, which may involve activation of noncannabinoid receptors and/or unidentified targets.

Endocannabinoids contribute to the regulation of numerous physiological functions, such as pain processing, motor activity, blood pressure, cognition, cell growth, reproduction (e.g., embryo implantation), and synaptic plasticity. In particular, electrophysiological studies have shown that endocannabinoids are released postsynaptically and travel in a retrograde direction to inhibit or modulate presynaptic neurotransmitter release by activating CB_1 receptors (Wilson and Nicoll, 2001). As this form of short-term synaptic plasticity occurs at either glutamatergic or GABAergic synapses, endocannabinoids can reduce excitatory or inhibitory inputs to neurons depending on the localization of CB_1 receptors within the brain circuitries being stimulated. Endocannabinoids also influence long-term synaptic plasticity by inhibiting long-term potentiation (LTP) and depression (LTD), two *in vitro* models of learning and memory. Given this diversified neuromodulatory activity, the therapeutic potentials of drugs targeting the endocannabinoid system are remarkable. Besides the well-known antihemetic and appetite-stimulating properties of cannabinoid drugs, their possible applications for the treatment of neurological and psychiatric disorders and drug addiction are currently under investigation.

Steroid Hormones

Steroid hormones are classically understood as hormones derived from peripheral endocrine glands that communicate back to the brain the physiological status of the animal. It was believed that the steroids acted via their well-characterized nuclear receptors, suggesting that they would have a slow time course of action, many minutes to days. Studies in the last decade, however, have expanded and refined this concept to include local synthesis and/or modification of steroids in the brain and new sites of steroid hormone action.

It has long been understood that the brain is capable of metabolizing peripheral steroids to modify their function in the brain. For example, in the male, the primary circulating androgen, testosterone, is converted into estrogen by the enzyme aromatase, which then acts on estrogen receptors (see Fig. 10.7). Indeed, there are more estrogen receptors than androgen receptors in male brain tissue! The brain also produces some unique steroids as discussed in following text.

More recently it has become clear that the brain can actually synthesize steroids *de novo* (see Fig. 10.7). All the enzymes involved in the synthesis of steroids have been found in the brain. Cultures of brain cells have also been shown to convert labeled cholesterol into labeled mature steroids, showing that those enzymes indeed function. It is believed that much of the *de novo* synthesis takes place in glia as opposed to neurons, although this is still a controversial point.

There are several steroid metabolites (enriched in the brain and, hence, termed *neurosteroids*) that have a modulatory effect on GABA$_A$ receptor, sigma opiate receptor, or NMDA receptor function (Gibbs *et al.*, 1999; Monet *et al.*, 1995). The sulfated derivatives of 3β-hydroxy-Δ5-steroids, DHEA, or pregnenalone are inhibitory to GABA and hence are excitatory. On the other hand, 3α,5α-tetrahydroprogesterone enhances the effect of GABA and is inhibitory to neuronal depolarization. Thus, depending on which enzymes are present in a given brain region for steroid modification, there can be differential effects on neurotransmitter

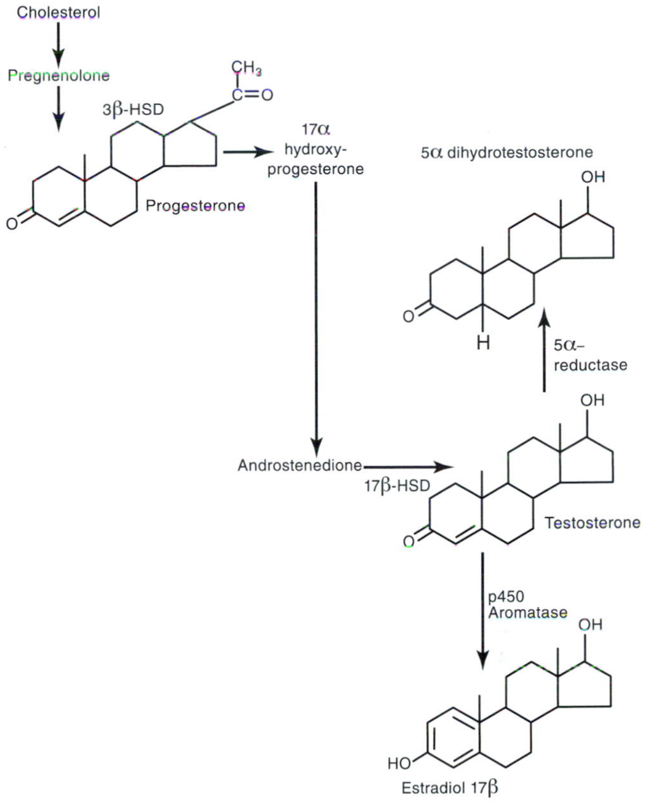

FIGURE 10.7 Neurosteroid synthesis and metabolism in the brain. While the brain can synthesize all these steroids *de novo* from circulating cholesterol, circulating steroids such as testosterone and progesterone can also enter the brain and be metabolized further to other steroids.

action. It is currently unclear whether the source of these steroids is local synthesis or modification of gonad- or adrenal-derived steroids brought to the brain by the bloodstream. In any event, it is clear that these freely diffusing substances can have a great effect on brain function.

Steroids are capable of functioning as neurotransmitters in a variety of circumstances, even though they are "unconventional" in their mode of synthesis and action. In addition to their action via nuclear receptors, they can directly modulate multiple neurotransmitter receptors and also affect signal transduction systems directly. The action of steroids via nuclear receptors and direct effects on transcription is well understood (see Chapter 13). There is now an emerging body of research that points to a cytosolic site of action of steroids, which results in a more rapid effect of steroids on neuronal function. Studies on the electrical activity of neurons showed that steroids could change the electrical properties within seconds to a few minutes, much too quickly to be explained by nuclear transcription events. At first this was passed off as nonspecific membrane effects of the steroids, e.g., the effect of the lipid-soluble steroid intercalating into the plasma membrane. There was, however, a specificity to the action; for example, different glucocorticoids would work, while sex steroids would not, even though their basic chemical structures were similar. There are now multiple reports of membrane-associated steroid receptors that can mediate steroid-dependent functions (reviewed in Lösel and Wehling, 2003; Watson and Gametchu, 1999). Some studies have shown that these "membrane" steroid-associated events are dependent on expression of the well-characterized "nuclear" steroid receptor, and suggest that the nuclear steroid receptor can also associate with membrane elements and mediate steroid signaling via cytosolic pathways (see Fig. 10.8) (Kim *et al.*, 1999; Razandi *et al.*, 2002). For example, it is becoming well established that estrogen can also elicit its action via the MAP kinase pathway, independent of direct estrogen receptor action in the nucleus (Edwards and Boonyaratanakornkit, 2003). Thus, steroid hormones can have profound effects at all levels of signaling in the brain, from electrical activity to synaptic transmission to long-term changes in gene expression.

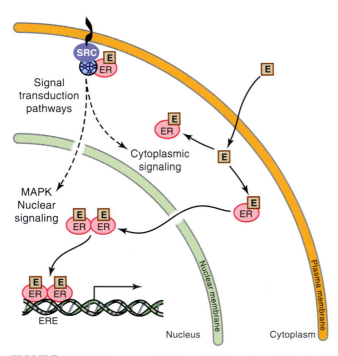

FIGURE 10.8 Estrogen acts in the cytoplasm to mediate signal transduction via the MAP kinase pathway. While the nuclear estrogen receptor appears to be capable of mediating the activation of MAP kinase, it still remains controversial as to whether a true plasma-membrane-associated receptor can also mediate the estrogenic effects.

SYNAPTIC TRANSMITTERS IN PERSPECTIVE

We have discussed, in what may seem to be overwhelming detail, the biochemical aspects of chemically coded interneuronal transmission. During this discussion several things have become apparent. One is that it is difficult to discuss the biochemical nature of synaptic transmission without reference to other critical information about the structure and function of neurons. Neuroscience is multidisciplinary, requiring an appreciation of several different aspects of cellular function to come to grips with basic principles of integrated neuronal function. In addition, synaptic transmission is a dynamic process that is constantly changing; befitting this situation, the study of synaptic transmission is also dynamic, requiring frequent reevaluation and revision.

This can be most clearly seen in the changing definitions of a neurotransmitter. Over the years there has been a shift from a rather strict definition based on criteria that evolved primarily from studies of peripheral sites. The status of classic transmitters as the sole occupants of the royal family of transmitters was challenged a generation ago with the emergence of peptide transmitters, the pretenders to the throne. Sufficient data amassed to establish that the peptides were at least cousins, if not siblings or offspring, of the classic transmitters. Soon after this trauma came the unexpected finding that more than one transmitter is present in a neuron. And now we are just beginning to come to grips with the concept that there may be transmitters that are gases!

The new developments in unconventional transmitters have been paralleled by a greater appreciation of the intricacies of classic transmitters. The use of the terms *conventional transmitters* and *unconventional transmitters* reflects our current unease with the expanding definition of transmitters. However, all aspects of neuroscience are expanding, which is part of the reason for the excitement of neuroscience.

References

Bean, A. I., and Roth, R. H. (1992). Dopamine-neurotensin interactions in mesocortical neurons: Evidence from microdialysis studies. *Ann. NY Acad. Sci.* **668**, 43–53.

Bean, A. J., Zhang, X., and Hokfelt, T. (1995). Peptide secretion: What do we know? *FASEB J.* **8**, 630–638.

Beltramo, M. et al. (1997). Functional role of high-affinity anandamide transport, as revealed by selective inhibition. *Science.* **277**, 1094–1097.

Bloch, A., and Thoenen, H. (1996). Localization of cellular storage compartments and sites of constitutive and activity-dependent release of nerve growth factor (NGF) in primary cultures of hippocampal neurons. *Mol. Cell Neurosci.* **7**, 173–190.

Chao, M. V. (2003). Neurotrophins and their receptors: A convergence point for many signaling pathways. *Nat. Rev.* **4**, 299–309.

Cravatt, B. F., Giang, D. K., Mayfield, S. P., Boger, D. L., Lerner, R. A., and Gilula, N. B. (1996). Molecular characterization of an enzyme that degrades neuromodulatory fatty-acid amides. *Nature* **384**, 83–87.

Dawson, T. M., and Snyder, S. H. (1994). Gases as biological messengers: Nitric oxide and carbon monoxide in the brain. *J. Neurosci.* **14**, 5147–5159.

Deutch, A. Y., and Zahm, D. S. (1992). The current status of neurotensin-dopamine interactions: Issues and speculations. *Ann. NY Acad. Sci.* **668**, 232–252.

Dinh, T. P., Carpenter, D., Leslie, F. M., Freund, T.F., Katona, I., Sensi, S. L., Kathuria, S., and Piomelli, D. (2002). Brain monoglyceride lipase participating in endocannabinoid inactivation. *Proc. Natl. Acad. Sci.* **99**, 10819–10824.

Edwards, D. P., and Boonyaratanakornkit, V. (2003). Rapid extranuclear signaling by the estrogen receptor (ER): MNAR couples ER and Src to the MAP kinase signaling pathway. *Mol. Interven.* **3**, 12–15.

Gibbs, T. T., Yagoubi, N., Weaver, C. E., Park-Chung, M., Russek, S. J., and Farb, D. H. (1999). Modulation of ionotropic glutamate receptors by neuroactive steroids. *In* "Neurosteroids: A New Regulatory Function in the Nervous System" (E. E. Baulieu, P. Robel, and M. Schumacher, Eds.), pp. 167–190, Humana Press, Totowa, NJ.

Giuffrida, A., Parsons, L. H., Kerr, T. M., Rodríguez de Fonseca, F., Navarro, M., and Piomelli, D. (1999). Dopamine activation of endogenous cannabinoid signaling in dorsal striatum. *Nature Neurosci.* **2**, 358–363.

Goodman, L. J., Valverde, J., Lim, F., Geschwind, M. D., Federoff, H. J., Geller, A. I., and Hefti, F. (1996). Regulated release and polarized localization of brain-derived neurotrophic factor in hippocampal neurons. *Mol. Cell Neurosci.* **7**, 222–238.

Heymach, J. V., Jr., and Shooter, E. M. (1995). The biosynthesis of neurotrophin heterodimers by transfected mammalian cells. *J. Biol. Chem.* **270**, 12297–12304.

Kim, H. P., Lee, J. Y., Jeong, J. K., Bae, S. W., Lee, H. K., and Jo, I. (1999). Nongenomic stimulation of nitric oxide release by estrogen is mediated by estrogen receptor α localized in caveolae. *Biochem. Biophys. Res. Commun.* **263**, 257–262.

Kislauskis, E., Bullock, B., McNeil, S., and Dobner, P. R. (1988). The rat gene encoding neurotensin and neuromedin N: Structure, tissue-specific expression, and evolution of exon sequences. *J. Biol. Chem.* **263**, 4963–4968.

Kitabgi, P., De Nadal, F., Rovere, C., and Bidard, J. N. (1992). Biosynthesis, maturation, release, and degradation of neurotensin and neuromedin N. *Ann. NY Acad. Sci.* **668**, 30–42.

Kokaia, Z., Metsis, M., Kokaia, M., Bengzon, J., Elmer, E., Smith, M. J., Timmusk, T., Siesjo, B. K., Persson, H., and Lindvall, O. (1994). Brain insults in rats induce increased expression of the BDNF gene through differential use of multiple promoters. *Eur. J. Neurosci.* **6**, 587–596.

Korte, M., Carroll, P., Wolf, E., Brem, G., Thoenen, H., and Bonhoeffer, T. (1995). Hippocampal long-term potentiation is impaired in mice lacking brain-derived neurotrophic factor. *Proc. Natl. Acad. Sci. USA* **92**, 8856–8860.

Leingartner, A., Heisenberg, C. P., Kolbeck, R., Thoenen, H., and Lindholm, D. (1994). Brain-derived neurotrophic factor increase neurotrophin-3 expression in cerebellar granule cells. *J. Biol. Chem.* **269**, 828–830.

Lindholm, D., da Penha Berzaghi, M., Cooper, J., Thoenen, H., and Castren, E. (1994). Brain-derived neurotrophic factor and neurotrophin-4 increase neurotrophin-3 expression in rat hippocampus. *Int. J. Dev. Neurosci.* **12**, 745–751.

Lösel, R., and Wehling, M. (2003). Nongenomic actions of steroid hormones. *Nat. Rev. Mol. Cell Biol.* **4**, 46–55.

Monet, P. P., Mahe, V., Robel, P., Baulieu, E. E. (1995). Neurosteroids via sigma receptors modulate the [^{3}H] norepinephrine release evoked by NMDA in the rat hippocampus *Proc. Natl. Acad. Sci. USA* **92**, 3774–3777.

O'Dell, T. J., Huang, P. L., Dawson, T. M., Dinnerman, J. L., Snyder, S. H., Kandel, E. R., and Fishman, M. C. (1994). Endothelial NOS and the blockade of LTP by NOS inhibitors lacking neuronal NOS. *Science* **265**, 542–546.

O'Sullivan, S. E. (2007) Cannabinoids go nuclear: Evidence for activation of peroxisome proliferator-activated receptors. *Br. J. Pharmacol.* **152**, 576–582.

Patterson, S.L., Grover, L.M., Schwartzkroin, P.A., and Bothwell, M. (1992). Neurotrophin Expression in rat hippocampal slices: A stimulus paradigm inducing LTP in CAI crokes incress in BDNF and NT-3 mRNAs. *Neuron* 9, 1081–1088.

Razandi, M., Oh, P., Pedram, A., Schnitzer, J., and Levin, E. R. (2002). ERs associate with and regulate the production of caveolin: Implications for signaling and cellular actions. *Mol. Endocrinol.* **16**, 100–115.

Ross, R. A., (2003). Anandamide and vanilloid TRPV1 receptors. *Br. J. Pharmacol.* **140**, 790–801.

Russell, D. S. (1995). Neurotrophins: Mechanisms of action. *Neuroscientist* 1, 3–6.

Seidah, N. G., and Chrétien, M. (1999). Proprotein and prohormone convertases: A family of subtilases generating diverse bioactive polypeptides. *Brain Res.* **848**, 45–62.

Shipp, M. A., Vijayaraghavan, J., Schmidt, E. V., Masteller, E. L., D'Adamio, L., Hersch, L. B., and Reinherz, E. L. (1989). Common acute lymphoblastic leukemia antigen (CALLA) is active neutral endopeptidase 24.11 ("enkaphalinase"): Direct evidence by cDNA transfection analysis. *Proc. Natl. Acad. Sci. USA* **86**, 297–301.

Thoenen, H. (1995). Neurotrophins and neuronal plasticity. *Science* **270**, 593–598.

Watson, C. S., and Gametchu, B. (1999). Membrane-initiated steroid actions and the proteins that mediate them. *Proc. Soc. Exp. Biol. Med.* **220**, 9–19.

Wetmore, C., Olson, L., and Bean, A. J. (1994). Regulation of brain-derived neurotrophic factor (BDNF) expression and release from hippocampal neurons is mediated by non-NMDA type glutamate receptors. *J. Neurosci.* **14**, 1688–1700.

Widmer, H. R., Ohsawa, F., Knusel, B., and Hefti, F. (1993). Down-regulation of phosphatidylinositol response to BDNF and NT-3 in cultures of cortical neurons. *Brain Res.* **614**, 325–334.

Wilson, R. I., and Nicoll, R. A. (2001). Endogenous cannabinoids mediate retrograde signalling at hippocampal synapses. *Nature* **410**, 588–592.

Zakhary, R., Gaine, S. P., Dinerman, J. L., Ruat, M., Flavahan, N. A., and Snyder, S. H. (1996). Heme oxygenase 2: Endothelial and neuronal localization and role in endothelial-dependent relaxation. *Proc. Natl. Acad. Sci. USA* **93**, 795–798.

Neurotransmitter Receptors

M. Neal Waxham

Chemical synaptic transmission plays a fundamental role in the process of neuron-to-neuron and neuron-to-muscle communication. The type of receptors present in the plasma membrane determines in large part the nature of the response of a neuron or muscle cell to a neurotransmitter. The nature of the response can be either through the direct opening of an ion channel (ionotropic receptors) or through alteration of the concentration of intracellular metabolites (metabotropic receptor) (see also Chapter 16). The response magnitude is determined by receptor number, the "state" of the receptors, and the amount of transmitter released. Finally, the sign of the response can be inhibitory or excitatory. The temporal and spatial summation of information conveyed by receptor activation determines whether the postsynaptic cell will fire an action potential or the muscle will contract. As one can see, there is remarkable flexibility and diversity in molding the response to neurotransmitter by constructing a synapse with the desired receptor types.

There exist two broad classifications for receptors. An ionotropic receptor is a relatively large, multisubunit complex typically composed of five individual proteins that combine to form an ion channel through the membrane (Fig. 11.1A). These ion channels exist in a closed state in the absence of neurotransmitter and are largely impermeable to ions. Neurotransmitter binding induces rapid conformational changes that open the channel, permitting ions to flow down their electrochemical gradients. Changes in membrane current resulting from ligand binding to ionotropic receptors are generally measured on a millisecond timescale. The ion flow ceases when transmitter dissociates from the receptor or when the receptor becomes desensitized, a process discussed in more detail later in this chapter.

In contrast, a metabotropic receptor is composed of a single polypeptide (Fig. 11.1B) and exerts its effects not through the direct opening of an ion channel but rather by binding to and activating GTP-binding proteins (G proteins). Transmitters that activate metabotropic receptors typically produce responses of slower onset and longer duration (from tenths of seconds to potentially hours) owing to the series of enzymatic steps necessary to produce a response. The metabotropic receptors have more recently been named *G-protein-coupled receptors*, or GPCRs for short, to more accurately capture their properties, and the latter nomenclature is adopted in this chapter.

We consider the structure of the ionotropic receptor family first and then turn to a description of the structure of GPCRs. In each section, information is presented to establish a general structural model of each receptor type. These models are then used to guide the description of other related ionotropic receptors or GPCRs. The order in which receptor types are presented is based predominantly on structural relatedness and should not be interpreted as representing their relative importance in the function of the nervous system.

IONOTROPIC RECEPTORS

All ionotropic receptors are membrane-bound protein complexes that form an ion-permeable pore in the membrane. By comparing the amino acid sequences of the cloned ionotropic receptors one can deduce that they are similar in overall structure, although two independent ancestral genes have given rise to two distinct families. One family includes

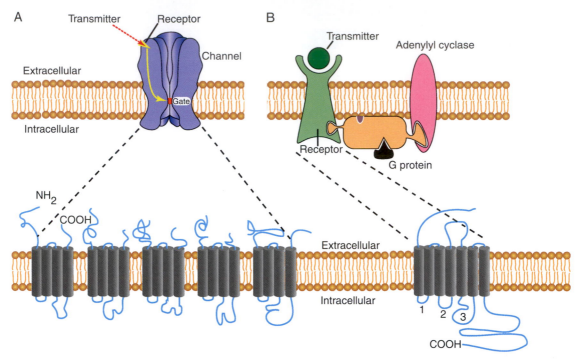

FIGURE 11.1 Structural comparison of ionotropic and metabotropic receptors. (A) Ionotropic receptors bind transmitter, and this binding directly translates into the opening of the ion channel through a series of conformational changes. Ionotropic receptors are composed of multiple subunits. Shown are the five subunits that together form the functional nAChR. Note that each nAChR subunit wraps back and forth through the membrane four times and that the mature receptor is composed of five subunits. (B) Metabotropic receptors bind transmitter and, through a series of conformational changes, bind to G proteins and activate them. G proteins then activate enzymes such as adenylyl cyclase to produce cAMP. Through the activation of cAMP-dependent protein kinase, ion channels become phosphorylated, which affects their gating properties. Metabotropic receptors are single subunits. They contain seven transmembrane-spanning segments, with the cytoplasmic loops formed between the segments providing the points of interactions for coupling to G proteins. Adapted from Kandel *et al.* (1991).

the nicotinic acetylcholine receptor (nAChR), the γ-aminobutyric acid A (GABA$_A$) receptor, the glycine receptor, and one subclass of serotonin receptors (Ortell and Lunt, 1995). The other family comprises the many types of ionotropic glutamate receptors (Hollmann and Heinemann, 1994).

The understanding of ionotropic receptor structure and function has expanded enormously in the past 25 years. Molecular approaches have provided elegant and extensive descriptions of gene families encoding different receptors, and systems for expressing cloned cDNAs have permitted detailed structure–function analysis of each receptor subtype. Expression of subunits independently and together has resulted in a detailed concept of the necessity and sufficiency of the multisubunit nature of the ionotropic receptor family. With the addition of biophysical and X-ray structural analysis, events associated with the opening of at least one ionotropic receptor, the nAChR, are available at nearly atomic resolution (Unwin, 1993a,b, 1995).

The nAChR Is a Model for the Structure of Ionotropic Receptors

The nAChR is so named because the plant alkaloid nicotine can bind to the ACh binding site and activate the receptor. Nicotine is therefore called an *agonist* of ACh because it binds to the receptor and opens it. In contrast, *antagonists* are molecules that bind to the receptor and inhibit its function. Agonists and antagonists are powerful tools that permit characterization of the structure and function of individual receptor subtypes.

More is known about the structure of the nAChR than any other ionotropic receptor, primarily because electric organs of certain species of fish, such as the *Torpedo* ray, contain nearly crystalline arrays of this molecule (Fig. 11.2). The electric organ is a specialized form of skeletal muscle that has the potential to generate large voltages (as much as 500 V in some cases) from the simultaneous opening of arrays of ion channels activated through the binding of ACh.

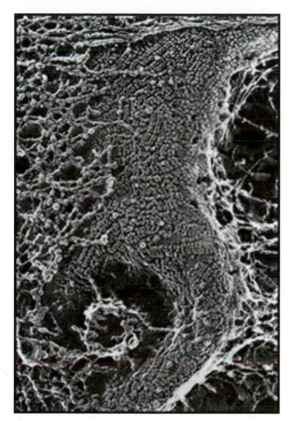

FIGURE 11.2 Panoramic view of the postsynaptic membrane of an electrocyte in the *Torpedo* electric organ, revealed by "deep-etch" electron microscopy. The vaselike structure in the center of the field is the external surface of the postsynaptic membrane, which is revealed by removal of the basal lamina. Clusters and linear arrays of 8 to 9 nm protrusions can be clearly seen. These represent the AChR oligomers. To the left of the vaselike structure, a lacelike basal lamina lies above the membrane, obscuring it from view. To the right of the vaselike structure, the postsynaptic membrane has been freeze-fractured away, thus revealing an underlying meshwork of cytoplasmic filaments that supports the postsynaptic membrane and its receptors × 175,000. Original courtesy of J. Heuser.

The majority of biochemical and structural analyses of nAChRs have been done on receptors isolated from the ray electric organ. Purification of nAChRs was aided significantly by utilization of a toxin from snake venom called α-*bungarotoxin*. Affinity columns constructed with α-bungarotoxin bind to nAChR with high affinity and specificity, providing a means of purifying nearly homogeneous nAChRs in a single chromatographic step.

The nAChR Is a Heteromeric Protein Complex with a Distinct Architecture

The structure of the nAChR is typical of ionotropic receptors. The nAChR purified as described from *Torpedo* is composed of five subunits (see Fig. 11.1)

and has a native molecular mass of approximately 290 kDa. The subunits are designated α, β, γ, and δ, and each receptor complex contains two copies of the α subunit. The subunits are homologous membrane-bound proteins that assemble in the bilayer to form a ring enclosing a central pore. Pioneering electron microscopic analyses by Nigel Unwin have provided the best image of the structural appearance of the nAChR (Fig. 11.3). In fact, the nAChR is the only membrane-bound neurotransmitter receptor for which high-resolution structural information is available. The extracellular domains of the subunits together form a funnel-shaped opening that extends approximately 100 Å outward from the outer leaflet of the plasma membrane. The funnel at the outer portion of the receptor has an inside diameter of 20–25 Å (Unwin, 1993a). The funnel shape is thought to concentrate and force ions and transmitter to interact with amino acids in the limited space of the pore without producing a major barrier to diffusion. This funnel narrows near the center of the lipid bilayer to form the domain of the receptor that determines the opened or closed state of the ion pore. The intracellular domain of the receptor forms short exits for ions traveling into the cell and an entrance for ions traveling out of the cell. The intracellular domain also establishes the association of the receptor with other intracellular proteins that determine the subcellular localization of the nAChR. The arrangement of the subunits in the receptor is somewhat debatable; however, most data support a model whereby the β subunit lies between the two α subunits (Unwin, 1993a,b, 1995).

Each nAChR Subunit Has Multiple Membrane-Spanning Segments

The primary structure of each nAChR subunit was obtained by the efforts of Shosaka Numa and his colleagues (Noda *et al.*, 1982, 1983). The deduced amino acid sequence from cloned mRNAs indicates that the nAChR subunits range in size from 40 to 65 kDa. A general domain structure for each subunit was derived from primary sequence data and toxin- and antibody-binding studies. Each subunit consists of four transmembrane-spanning segments referred to as TM1–TM4 (Fig. 11.4). Each segment is composed mainly of hydrophobic amino acids that stabilize the domain within the hydrophobic environment of the lipid membrane. The four transmembrane domains are arranged in an antiparallel fashion, wrapping back and forth through the membrane. The N terminus of each subunit extends into the extracellular space, as does the loop connecting TM2 and TM3 as

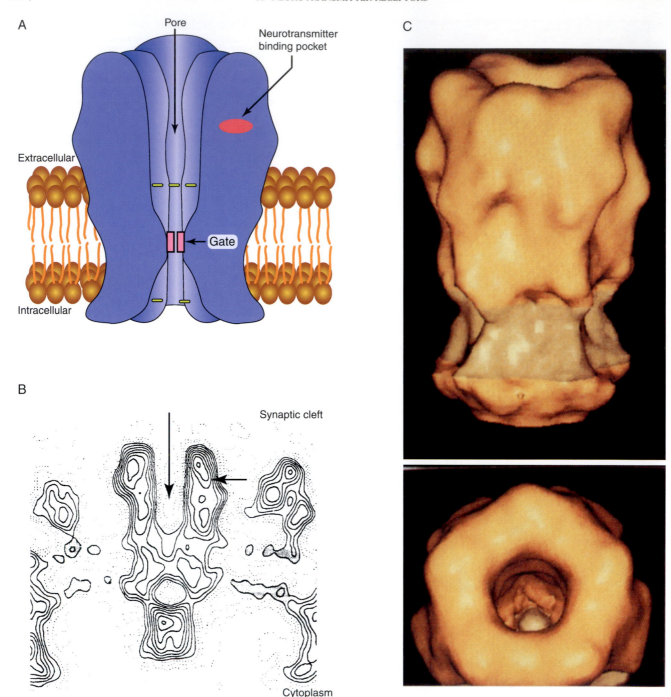

FIGURE 11.3   (A) Vertical section diagramming the structure of nAChR as it is believed to exist in the membrane. Note that the funnel-shaped structure narrows to a small central point referred to as the *gate*. Strategically placed rings of negatively charged amino acids on both sides of the gate form part of the selectivity filter for positively charged ions. The approximate position of the neurotransmitter binding site is shown in relation to the gate and the plasma membrane. (B) Protein density map derived from reconstructions of nAChR imaged by cryoelectron microscopy. The vertical arrow indicates the direction of ion flow from outside to inside within the funnel-shaped part of the receptor. The horizontal arrow indicates the predicted position of the neurotransmitter binding site that resides approximately 30 Å above the bilayer. The additional protein density attached to the bottom of the receptor is suggested to be a protein that anchors the nAChR to synapses. (C) Three-dimensional computer rendering of the nAChR. (Top) Side view of nAChR similar to that in (A). The darker shaded area near the bottom of the receptor delineates the approximate location where the receptor contacts the lipid bilayer. (Bottom) A view looking down into the funnel-shaped opening of the receptor. Note that the funnel narrows forming the gate.

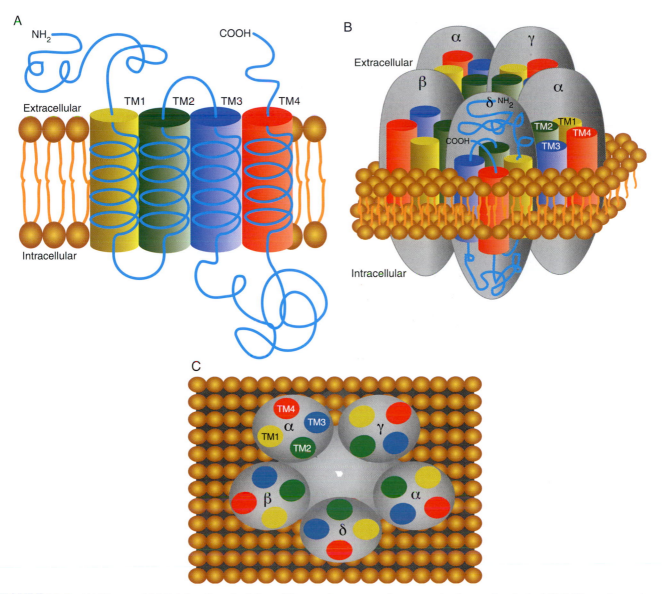

FIGURE 11.4 (A) Diagram highlighting the orientation of the membrane-spanning segments of one subunit of nAChR. The amino and carboxy termini extend in the extracellular space. The four membrane-spanning segments are designated TM1–TM4. Each forms an α helix as it traverses the membrane. (B) Side view of the five subunits in their approximate positions within the receptor complex. There are two α subunits present in each nAChR. (C) Top view of all five subunits highlighting the relative positions of their membrane-spanning segments, TM1–TM4, and the position of TM2 that lines the channel pore. Adapted from Kandel *et al.* (1991).

well as the C terminus. The amino acids linking TM1 and TM2 and those linking TM3 and TM4 form short loops that extend into the cytoplasm.

The Structure of the Channel Pore Determines Ion Selectivity and Current Flow

In the model shown in Fig. 11.4C, each subunit of the nAChR can be seen to contribute one cylindrical component (representing a membrane-spanning segment) that presents itself to a central cavity that forms the ion channel through the center of the complex. The membrane-spanning segments that line the pore are the five TM2 regions, one contributed by each subunit. The amino acids that compose the TM2 segment are arranged in such a way that three rings of negatively charged amino acids are oriented toward the central pore of the channel (Fig. 11.5). These rings of negative charge appear to provide much of the selectivity filter so that only cations can

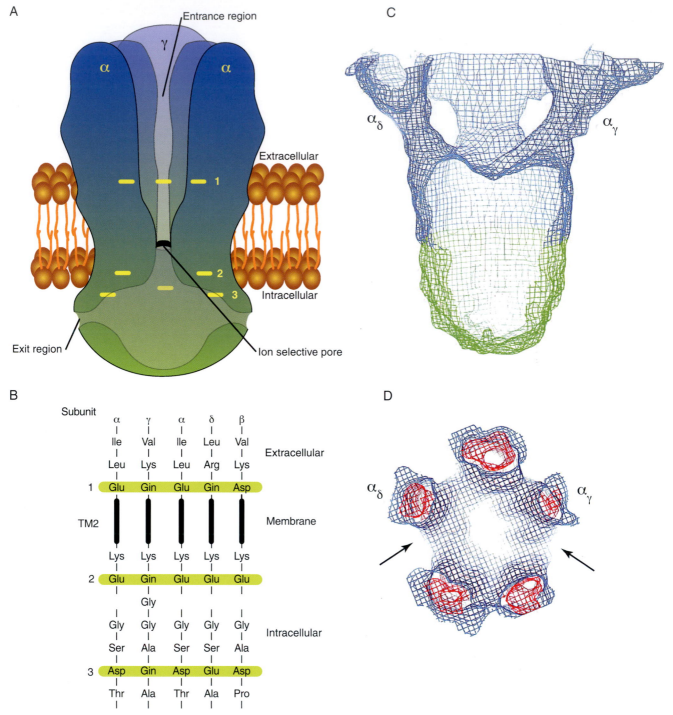

FIGURE 11.5 (A) Vertical section highlighting the relative positions of the three rings of negatively charged amino acids that help form the cation selectivity of nAChR. The regions where ions exit or enter from the intracellular side of the receptor are disposed laterally at the base of the receptor. (B) Amino acid sequence of each of the TM2 membrane-spanning segments of the five nAChR subunits. Numbers 1–3 correspond to the positions of the amino acids taking part in the formation for the three rings of negatively charged amino acids that determine the cation selectivity of the pore. Aspartate (Asp) and glutamate (Glu) are negatively charged amino acids. (C) Wireframe portrayal of the protein density distribution of the intracellular portion of nAChR. The front portion of the receptor was cut away to reveal the inverted cone-shaped cavity of the intracellular domain. The green wireframe represents protein density contributed by the anchoring protein rapsyn. (D) Wireframe portrayal of the protein density distribution of the intracellular domain of nAChR looking downward from within the receptor. The arrows indicate the major gaps in the lateral walls of the receptor where ions enter and exit.

pass through the central channel, whereas anions are largely excluded owing to charge repulsion (Imoto *et al.*, 1988; Karlin, 1993). The nAChR is permeable to most cations, such as Na$^+$, K$^+$, and Ca^{2+}, although monovalent cations are preferred. This mechanism for selectivity is poor in relation to the selectivity described for the family of voltage-gated ion channels (e.g., voltage-gated Ca^{2+} channels; see Chapters 5 and 6). From analysis of the permeation of various-sized cations, the dimensions of the pore forming the final barrier for ion permeation were estimated to be approximately 8.5 Å (Hille, 1992). This size is in excellent agreement with the measurements of 9–10 Å for the pore diameter from images of the open state of the receptor (Unwin, 1995). The restricted physical dimensions of the pore contribute greatly to the selectivity for particular ions. When the pore of the nAChR opens, positively charged ions move down their respective electrochemical gradients, resulting in an influx of Na$^+$ and Ca^{2+} and an efflux of K$^+$. A coarse filtering that also contributes to selectivity appears to be a shielding effect produced by other negatively charged amino acids surrounding the outer channel region of the receptor.

Ions do not directly enter or exit through the central pore of the cytoplasmic end of the nAChR. Two narrow openings are present on the lateral aspects of the cytoplasmic portion of the receptor through which ions must travel to exit or gain access to the central pore (Figs. 11.5A, C, and D) (Miyazawa *et al.*, 1999). α-Helical rods extending down from each subunit form an inverted pentagonal cone to produce these openings. Although too large (8 x 15 Å) to be a significant barrier to ion flow, these lateral pores could serve as an additional filtering step for the passage of certain ions.

There Are Two Binding Sites for ACh on the nAChR

Each receptor complex has two ACh binding sites that reside in the extracellular domain and lie approximately 30 Å from the outer leaflet of the membrane (see Fig. 11.3). The ACh binding site is formed for the most part by six amino acids in the α subunits; however, amino acids in both the γ and the δ subunits also contribute to binding (Karlin, 1993) (Fig. 11.6A). Mutations introduced at these critical amino acids in the α subunit significantly attenuate ligand binding. The two binding sites are not equivalent because of the receptor's asymmetry due to the different neighboring subunits (either γ or δ) adjacent to the two α subunits. Significant cooperativity also exists

within the receptor molecule, and so binding of the first molecule of ACh enhances binding of the second (Changeux *et al.*, 1984). A higher-resolution structure of the nAChR (Miyazawa *et al.*, 1999) revealed that access to the ACh binding sites appears to be through small channels that open into the interior mouth of the pore (Fig. 11.6B and C). Thus, ACh molecules must enter the pore and traverse these channels to gain access to their binding sites. It is speculated that similar attractive forces that bring positively charged ions into the pore also attract the positively charged ACh molecules favoring entrance into the channels that lead to the ACh binding sites. Two adjacent Cys residues (Cys-192 and Cys-193) in each α subunit form a disulfide bond that also appears to contribute to the stability of the ACh binding pocket (Fig. 11.6A). These Cys residues are highly conserved in most ionotropic receptors and must form an essential bond for stabilizing high-affinity neurotransmitter binding. α-bungarotoxin binds to the α subunit in close proximity to the two adjacent Cys residues (Karlin, 1993).

Opening of the nAChR Occurs Through Concerted Conformational Changes Induced by ACh Binding

When the nAChR binds two molecules of ACh, the channel opens almost instantaneously (time constants for opening are approximately 20 μs (Colquhoun and Ogden, 1988; Colquhoun and Sakmann, 1985)), thus permitting the passage of ions (see Fig. 16.4 of Chapter 16). A model developed from electron micrographic reconstructions of the nicotine-bound form of the nAChR indicates that the closed-to-open transition is associated with a rotation of the TM2 segments (Unwin, 1995) (Fig. 11.7). The TM2 segments are helical and exhibit a kink in their structure that forces a Leu residue from each segment into a tight ring that effectively blocks the flow of ions through the central pore of the receptor. When the TM2 segments rotate because of ACh binding, the kinks also rotate, relaxing the constriction formed by the Leu ring, and ions can then permeate the pore. The rotation also orients a series of Ser and Thr residues (amino acids with a polar character) into the central area of the pore, which facilitates the permeation of water-solvated cations. As the resolution of the structure increases, refinements in this model are likely to be forthcoming; however, the architecture of the nAChR is well established and provides a structural framework with which all other ionotropic receptors can be compared.

The main features of the nAChR transition from a closed to open state are summarized in Figure 11.8.

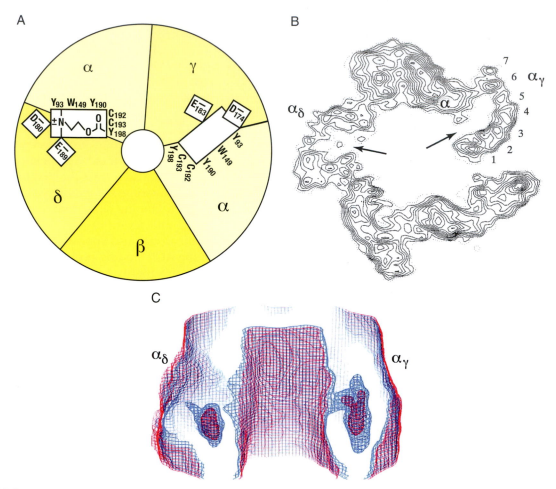

FIGURE 11.6 (A) Diagram of the relative positions of amino acids that form the Ach binding site in nAChR. The view is from above the receptor looking down into the pore. Each subunit is represented by a wedge. At the left, Ach is shown bound to its site at the interface between the a and d subunits. The length of the binding site is shown slightly contracted relative to the site (between the a and g subunits) without bound Ach. Critical amino acids for transmitter binding are indicated. Residues shown in boxes are amino acids predicted to make contact with the positively charged part of the Ach molecule. Note that many of the residues important for Ach binding are contributed by the a subunit. Cysteine (Cys) residues at positions 192 and 193 form a disulfide bond essential for stabilizing the Ach binding pocket. (B) A top-down view of the protein density map of nAChR sliced through the area where the Ach-binding areas reside. Note that Ach molecules must gain access to their binding sites through channels whose openings are on the inside of the funnel-shaped portion of the receptor. (C) Lateral view of a wireframe portrayal of nAChR at the level of the Ach binding sites. A portion of the receptor was cut away to highlight the fact that the cavities where Ach bind must be accessed from the pore of the receptor through short channels.

ACh gains access to its binding sites by entering the central pore of the receptor where it then enters small channels that provide access to the binding sites. Once both binding sites are occupied, the receptor rapidly opens to permit ion flow. It is the twisting of the TM2 segments induced by ACh binding that gates open the ion pore. In addition to negatively charged rings spaced within the central channel, charge screening also occurs in the lateral openings of the cytoplasmic domain of the receptor (Fig. 11.8). The ion selectivity of the nAChR for positively charged ions (Na^+ and K^+ mainly) comes from the charge screening at these different levels and the physical constriction of the gate of the pore.

The Muscle Form of the nAChR Is Very Similar to the nAChR from *Torpedo*

nAChRs at the neuromuscular junction are a concentrated collection of homogeneous receptors having a structure similar to that of the *Torpedo* electric organ. This similarity is not surprising, because the electric organ is a specialized form of muscle tissue. The adult form of the muscle receptor has the pentameric structure $\alpha_2\beta\varepsilon\delta$. An embryonic form of the receptor has an analogous structure, except that the ε subunit is replaced by a unique γ subunit. The embryonic and adult subunits of both mouse and bovine muscle receptors have been cloned and expressed in heterologous systems, such as the *Xenopus laevis* oocyte

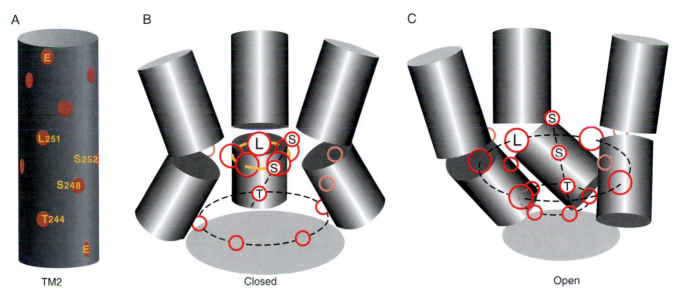

FIGURE 11.7 (A) Relative positions of amino acids in the TM2 segment of one of the nAChR a subunits modeled as an a helix. The glutamate residues (E) that form parts of the negatively charged rings for ion selectivity are shown at the top and bottom of the helix. (B) Arrangement of three of the TM2 segments of nAChR modeled with the receptor in the closed (Ach-free) configuration. In the closed configuration, leucine (L) residues form a right ring in the center of the pore that blocks ion permeation. (C) Arrangement of the three TM2 segments after Ach binds to the receptor. In the open configuration, the construction formed by the ring of leucine (L) residues opens as the helices twist about their axes. Note that the polar serine (S) and threonine (T) residues align when Ach binds, which apparently helps the water-solvated ions travel though the pore. Adapted with permission from Macmillan Publishers Ltd. Unwin, N. (1995). Acetylcholine receptor channel imaged in the open state. *Nature* **373,** 37–43.

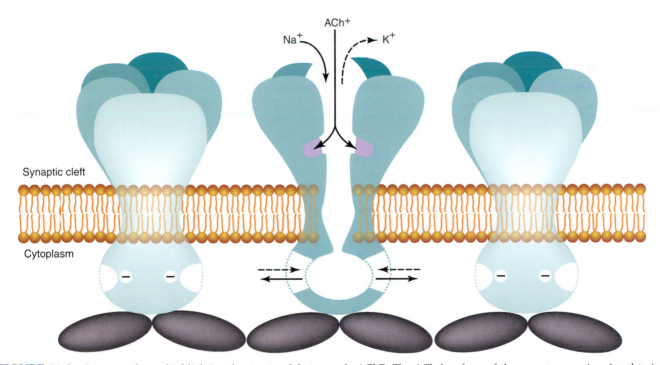

FIGURE 11.8 Summary figure highlighting the structural features of nAChR. The ACh-free form of the receptor remains closed to ion flow. ACh gains access to its binding site by entering the outer portion of the central pore of nAChR, which produces a relaxation of the central pore and an expansion of the holes in the lateral walls of the intracellular portion of the receptor. The protein rapsyn, represented by the gray ovals, anchors nAChR to synapses by interacting with the intracellular domain.

BOX 11.1

THE *XENOPUS* OOCYTE

The *Xenopus* oocyte has been used extensively to study the properties of cDNAs encoding receptor subunits and their mutated forms. In addition, the oocyte has been used to study how combinations of different subunits interact to produce receptors with different properties. The large size and efficient translational machinery of *Xenopus* oocytes make them ideal for electrophysiological analyses of cDNAs encoding prospective receptors and channels. For example, mRNAs produced by *in vitro* transcription of cDNAs encoding each of the individual nAChR subunits were introduced into oocytes by microinjection (A). Several days later, the oocytes were voltage-clamped to study the properties of the expressed channels (B). When ACh was applied through a separate pipet, a significant inward current was detected in the oocyte (C, panel 1). The response was specifically blocked by addition of an antagonist, tubocurarine (C, panel 2), and the block was reversed by a 15 min wash (C, panel 3). Details of this study indicate that all four of the nAChR subunits (α, β, γ, and δ) were required for ACh to produce an electrophysiological response (Mishina *et al.*, 1984). More recently, the patch-clamp technique has also been applied to oocyte expression of receptors to analyze the behavior of single channels.

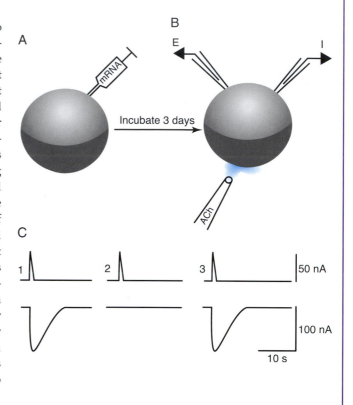

(Box 11.1) (Mishina *et al.*, 1984), and the receptors differ in both channel kinetics and channel conductance. These differences in channel properties appear to be necessary for the proper function of the nAChRs as they undergo the transition from developing to mature neuromuscular junction synapses.

The nAChR Matures as a Typical Membrane-Bound Protein and Has Well-Ordered Assembly

The pathway of the nAChR assembly in muscle is a tightly regulated process. For example, the five subunits of the nAChR have the potential to randomly assemble into 208 different combinations. Nevertheless, in vertebrate muscle, only one of these configurations ($\alpha_2\beta\epsilon\delta$) is typically found in mature tissue, indicating a very high degree of coordinated assembly and, ultimately, little structural variability (Green and Claudio, 1993; Paulson *et al.*, 1991). The

well-ordered assembly of specific intermediates is essential for this coordinated process, and the intermediates formed appear to start with a dimer between α and either ϵ (γ in mature muscle) or δ. The heterodimers then bind to β and to each other to form the final receptor (Gu *et al.*, 1991). An alternative pathway in which α, β, and γ first form a trimer has also been proposed (Green and Claudio, 1993). All this assembly takes place within the endoplasmic reticulum. During intracellular maturation, each subunit is glycosylated, and, if glycosylation is inhibited, the production of mature nAChRs decreases. Two highly conserved disulfide bonds in the N-terminal extracellular domain are essential for efficient assembly of the mature receptor. The first is between two adjacent Cys residues (Cys-192 and Cys-193) and, as noted, resides very close to the ACh binding site on the receptor (see Fig. 11.6). The second bond is between two Cys residues 15 amino acids apart, forming a loop in the extracellular domain.

nAChRs Are Concentrated and Anchored in the Postsynaptic Membrane

nAChRs are highly concentrated at the neuromuscular junction, which ensures rapid and reliable communication between the presynaptic motor neuron and postsynaptic muscle cell to induce contraction. The clustering of nAChRs begins during development and is mediated in part by the release of agrin from the motor neuron that binds to a tyrosine kinase in the postsynaptic membrane called *MuSK* (Sanes, 1997; Willman and Fuhrer, 2002). MuSK activation recruits members of the Src family of kinases that phosphorylates the nAChR leading to increased interactions with a multitude of postsynaptic signaling complexes ultimately leading to the recruitment of rapsyn. Rapsyn is a multidomain protein that binds directly to cytoplasmic tails of the nAChR subunits along with other proteins, including itself. Rapsyn thus forms the critical scaffolding function that clusters nAChRs at the neuromuscular junction. Rapsyn binds tightly to the nAChR and can be identified in reconstructed images of the receptor (Figs. 11.3C and 11.8). In addition to the nAChR, rapsyn also binds to other important muscle proteins including MuSK and the dystrophin–utrophin glycoprotein complex (dystrophin is the molecule that when mutated leads to certain types of muscular dystrophy; Sanes, 1997). As the junctions mature, a complex web of protein–protein interactions occur, including association with the underlying actin cytoskeleton (Fig. 11.9). Activity of the nAChR itself is also critical for its stabilization at the neuromuscular junction. The lifetime of the nAChR decreases from 14 d to less than a day following block of the receptor with α-bungarotoxin. In total, a complex interplay of signals coming from the presynaptic motor neuron are integrated by the postsynaptic receptors to concentrate and stabilize nAChRs at the neuromuscular junction.

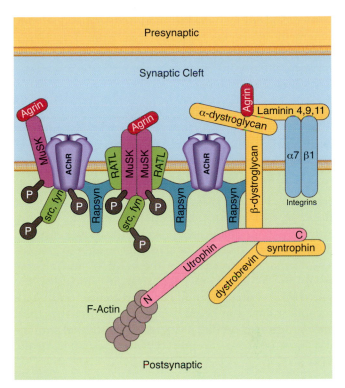

FIGURE 11.9 Diagram of nAChR clustering at the neuromuscular junction. Rapysn is a major anchoring protein at the neuromuscular junction that binds to itself and to the nAChR that concentrates and stabilizes nAChRs. The development and stabilization of the neuromuscular junction is mediated by a number of signaling cascades, only a few of which are shown. For example, agrin released from the presynaptic motor neuron binds to a number of proteins associated with the postsynaptic membrane including the tyrosine kinase MuSK (*muscle-specific kinase*). MuSK activation by agrin recruits and activates the soluble tyrosine kinases Src and Fyn, which further modify a number of proteins. RATL (*rapsyn-associated linker protein*) is a membrane-bound protein that binds to both MuSK and to rapsyn to anchor MuSK at the neuromuscular junction. Agrin also interacts with the dystroglycans that make up the dystrophin complex important for the maintenance of the neuromuscular junction. Rapsyn also binds to the utrophin complex that anchors the overlying protein complex to the actin cytoskeleton. Adapted from Willmann and Fuhrer (2002).

Phosphorylation Is a Common Post-Translational Modification of Receptors

Many ionotropic receptors, such as the nAChR, are phosphorylated, although the functional significance of the phosphorylation is not always evident. The nAChR is phosphorylated by at least three protein kinases: cAMP-dependent protein kinase (PKA; phosphorylates the γ and δ subunits), Ca^{2+}-phospholipid-dependent protein kinase (PKC; phosphorylates the δ subunit), and an unidentified tyrosine kinase that phosphorylates the β, γ, and δ subunits (Huganir and Greengard, 1990). The phosphorylation sites are all found in the intracellular loop between the TM3 and TM4 membrane-spanning segments. Phosphorylation by these three protein kinases appears to increase the rapid phase of desensitization of the receptor. Desensitization of receptors is a common observation, and this process limits the amount of ion flux through a receptor by producing transitions into a closed state (one that does not permit ion flow) in the continued presence of neurotransmitter. For the nAChR, the rate of desensitization has a time constant of approximately 50–100 ms. This rate appears to be too slow to have much significance in shaping the synaptic response at the neuromuscular junction, where the response typically lasts from 5 to 10 ms. This slow desensitization is not true of the brain forms of the nAChR and is discussed further in a later section of this chapter.

The Structures of Other Ionotropic Receptors Are Variations of the nAChR Structure

On the basis of similarity of structure, clear evolutionary relationships exist for the family of ionotropic receptors. Figure 11.10 shows an evolutionary tree for the family of ionotropic receptors related to nAChR. An early major subdivision separates those receptors permeable to anions from those permeable to cations. The former group includes the $GABA_A$ and glycine receptors, whereas the latter group includes the 5-hydroxytryptamine (5-HT_3, serotonin) and ACh receptors. One can begin to appreciate that structural similarities can predict a degree of functional similarity. Each of these receptor types is described in the next section.

Neuronal nAChRs Contain Two Types of Subunits

Structurally, neuronal nAChRs are similar to, yet distinct from, the *Torpedo* isoform of the receptor (Figs. 11.10 and Fig. 11.11). For example, the neuronal nAChR appears to have only two types of subunits, α and β, that combine to produce the functional receptor, and the majority of these receptors do not bind to α-bungarotoxin. At least nine different α subtypes (α1 being the muscle α subunit) have been identified, and some are species specific (α8 is found only in chicken, and α9 is found only in rat). Four different β subtypes (β1 being the muscle β subunit) have been identified. The neuronal β subunits are not closely related to the muscle β1 subunit and are sometimes referred to simply as *non-α subunits*. One structural feature that distinguishes neuronal α subunits from β subunits is the presence of particular Cys residues in the extracellular domain of the α subunit. Two of these Cys residues are adjacent to one another and form a disulfide bond. The β subunits do not have these adjacent Cys residues. Because these Cys residues are critical for ACh binding, the α subunits of neuronal AChRs, like muscle α subunits, contain the main contact points for ACh binding (Fig. 11.11). All the α and β genes encode proteins with four transmembrane-spanning segments (TM1–TM4). Although the physical structure of this receptor family has not been well characterized, it appears that each functional receptor is a pentameric assembly.

Structural Diversity of Neuronal nAChRs Produces Channels with Unique Properties

Neuronal nAChRs have diverse functions and are the receptors presumed to be responsible for the psychophysical effects of nicotine addiction. One major function of nAChRs in the brain is to modulate excitatory synaptic transmission through a presynaptic action (McGehee *et al.*, 1995). The diversity in function can be related to the heterogeneous structure contributed by the thousands of possible combinations between the different α and β subunits. Control mechanisms for receptor assembly in neurons do not appear to be as stringent as those of nAChR in *Torpedo* and muscle. Functional neuronal nAChRs can be assembled from a single subunit (as in α7, α8, and α9), and a single type of α subunit can be assembled with multiple types of β subunits (e.g., α3 with β2 or β4 or both) and vice versa (Fig. 11.11). These additional possibilities produce a staggering array of potential receptor molecules, each with distinct properties including differences in single-channel kinetics and rates of desensitization. This type of diversity is not unique to neuronal nAChRs. For most receptor classes studied in detail, diversity is the rule and not the exception. It is intriguing to speculate that subunit composition may also play roles in targeting the receptors to different intracellular locations.

Neuronal nAChRs exhibit a range of single-channel conductances between 5 and 50 pS, depending on the tissue analyzed or the specific subunits expressed. Most, but not all, are blocked by neuronal bungarotoxin, a snake venom distinct from α-bungarotoxin. All the neuronal nAChRs are cation-permeable channels that, in addition to permitting the influx of Na^+ and the efflux of K^+, permit an influx of Ca^{2+}. This Ca^{2+} permeability is greater than that for muscle nAChR (Vernino *et al.*, 1992) and is variable among the different neuronal receptor subtypes. Indeed, some receptors have very high Ca^{2+}–Na^+ permeability ratios; for example, α7 nAChRs exhibit a Ca^{2+}–Na^+ permeability ratio of nearly 20 (Seguela *et al.*, 1993). The Ca^{2+} permeability of the α7 nAChR can be eliminated by the mutation of a single amino acid residue in the second transmembrane domain (Glu-237 for Ala) without significantly affecting other aspects of the receptor (Bertrand *et al.*, 1993). This key Glu residue must lie within the pore of the receptor and presumably enhances the passage of Ca^{2+} ions through an interaction with its negatively charged side chain. Activation of α7 receptors through the binding of ACh could therefore produce a significant increase in the level of intracellular Ca^{2+} without the opening of voltage-gated Ca^{2+} channels. Subunits α7, α8, and

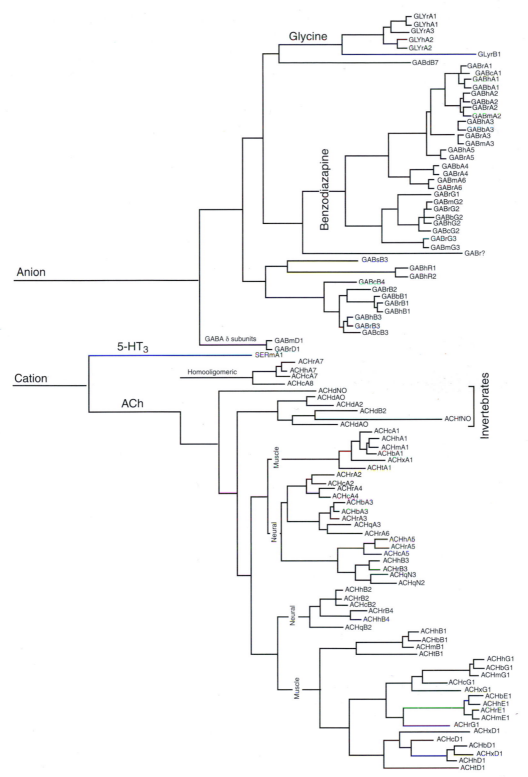

FIGURE 11.10 Evolutionary relationships of the family of cloned ionotropic receptor subunits. The nomenclature used to describe each receptor subunit is RRRsS#, where *RRR* represents the type of receptor, *s* the organism, *S* the subunit type, and # the subunit number. Type of receptor (RRR): Ach, acetylcholine; GAB, GABA; GLY, glycine; SER, serotonin. Organism (s): b, bovine; c, chicken; d, *Drosophila*; f, filaria; g, goldfish; h, human; l, locust; m, mouse; n, nematode; r, rat; s, snail; t, *Torpedo*; x, *Xenopus*. Subunit type (S): A, alpha; B, beta; G, gamma; D, delta; E, epsilon; R, rho; N, non-alpha; ?, undetermined. Adapted from Ortell and Lunt (1995).

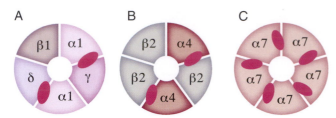

FIGURE 11.11 Diagrams of top-down views of AChR from muscle (A), one of the the neuronal AChRs composed of α2 and β4 subunits (B), and the homoligomeric form of neuronal AChR produced by assembly of α7 subunits (C). Purple ovals represent the ACh binding sites on each receptor complex. The ACh receptors from brain are diverse in both structure and properties due to the variety of different receptor complexes produced from subunit mixing.

α9 are also the α-bungarotoxin-binding subtypes of neuronal nAChRs. Other neuronal isoforms exhibit Ca^{2+}–Na^+ permeability ratios of about 1.0–1.5.

Neuronal nAChRs Desensitize Rapidly

For the nAChR from muscle, desensitization is minor and probably not of physiological significance in determining the shape of the synaptic response at the neuromuscular junction. However, for some neuronal nAChRs, desensitization likely plays a major role in determining the effects of the actions of ACh. Receptors composed of α7, α8, and certain α–β combinations exhibit desensitization time constants between 100 and 500 ms, whereas others exhibit desensitization constants between 2 and 20 s. Given the diverse functions of neuronal nAChRs, the variable rates of desensitization likely play important roles whereby this inherent property of the receptor shapes the physiological response generated from binding ACh.

One Serotonin Receptor Subtype, 5-HT₃, Is Ionotropic and Is a Close Relative of the nAChR

Serotonin (5-hydroxytryptamine, 5-HT) is historically thought of as a transmitter that binds to and activates only GPCRs (described in more detail later). The 5-HT₃ subclass is an exception forming an ionotropic receptor activated by binding serotonin. The 5-HT₃ receptor is permeable to Na^+ and K^+ ions and similar in many ways to nAChR in that both desensitize rapidly and are blocked by tubocurarine. From expression studies of the cloned cDNA (Maricq et al., 1991), it appears that the 5-HT₃ receptor is a homomeric complex composed of five copies of the same subunit. The deduced amino acid sequence of the cDNA indicates that the protein is 487 amino

acids long (56 kDa) and has a structure most analogous to the α7 subtype of neuronal nAChRs, which also forms a homo-oligomeric receptor.

The 5-HT₃ receptor is mostly impermeable to divalent cations. For example, Ca^{2+} is largely excluded from permeation and, in fact, effectively blocks current flow through the pore, even though the predicted pore size of the channel (7.6 Å) is approximately the same as that for the nAChR (8.4 Å). Apparently, other physical or electrochemical barriers limit the capacity of divalent ions to permeate the 5-HT₃ pore. Dose–response studies indicate that at least two ligand binding sites must be occupied for the channel to open; however, the binding of agonist and/or opening of the channel appears to be approximately 10 times slower than for most other ligand-gated ion channels. The functional significance or physical explanation of this slow opening is not known. The native 5-HT₃ receptor also exhibits desensitization (time constant 1–5 s), although the rate varies widely, depending on the methodology used for analysis and the source of receptor. Interestingly, this desensitization can be significantly slowed or enhanced by single amino acid substitutions at a Leu residue in the TM2 transmembrane-spanning segment of the subunit (Yakel et al., 1993).

The 5-HT₃ receptors are sparsely distributed on primary sensory nerve endings in the periphery and widely distributed at low concentrations in the mammalian CNS. The 5-HT₃ receptor is clinically significant because antagonists of 5-HT₃ receptors have important applications as antiemetics, anxiolytics, and antipsychotics.

GABA_A Receptors Are Related in Structure to the nAChRs but Exhibit an Inhibitory Function

Synaptic inhibition in the mammalian brain is mediated principally by GABA receptors. The most widespread ionotropic receptor activated by GABA is designated GABA_A. The subunits composing the GABA_A receptor have sequence homology with the nAChR subunit family, and the two families have presumably diverged from a common ancestral gene. In fact, the general structures of the two receptors appear to be quite similar. The GABA_A receptor is composed of multiple subunits, probably forming a heteropentameric complex of approximately 275 kDa. Five different types of subunits are associated with GABA_A receptors and are designated α, β, γ, δ, and ε. An additional subunit, ρ, is found predominantly in the retina, whereas the other subunits are widely distributed in the brain. Each subunit group also has different subtypes; for example, six different α, four β, four γ, and two ρ subunits have been identified. The predicted amino acid sequences indicate that each of these subunits has a molecular mass ranging between 48 and 64 kDa. Like neuronal nAChR, these subunits mix in a heterogeneous fashion to produce a wide array of GABA$_A$ receptors with different pharmacological and electrophysiological properties. The predominant GABA$_A$ receptor in brain and spinal cord is α1, β2, and γ2, with a likely stoichiometry of two α1s, two β2s, and one γ2. Expression of subunit cDNAs in oocytes indicates that the α subunit is essential for producing a functional channel. The α subunit also appears to contain the high-affinity binding site for GABA (Seighart, 1992).

The ion channel associated with the GABA$_A$ receptor is selective for anions (in particular, Cl$^-$), and the selectivity is provided by strategically placed positively charged amino acids near the ends of the ion channel (Barnard et al., 1987). When GABA binds to and activates this receptor, Cl$^-$ flows into the cell, producing a hyperpolarization by moving the membrane potential away from the threshold for firing an action potential (see also Chapter 16). The neuronal GABA$_A$ receptor exhibits multiple conductance levels, with the predominant conductance being 27–30 pS. Measurements and modeling of single-channel kinetics suggest that two sequential binding sites exist for anions within the pore (Bormann, 1988).

The GABA$_A$ Receptor Binds Several Compounds That Affect Its Properties

The GABA$_A$ receptor is an allosteric protein, its properties being modulated by the binding of a number of compounds. Two well-studied examples are barbiturates and benzodiazepines, both of which bind to the GABA$_A$ receptor and potentiate GABA binding. The net result is that in the presence of barbiturates or benzodiazepines or both, the same concentration of GABA will increase inhibition (see Fig. 16.8 of Chapter 16 for example). Benzodiazepine binding is conferred on the receptor by the γ subunit (Pritchett et al., 1989), but the presence of the α and β subunits is necessary for the qualitative and quantitative aspects of benzodiazepine binding. The benzodiazepine binding site appears to lie along the interface between the α and γ subunits, and only certain subtypes are sensitive to benzodiazepines. Benzodiazepine binding to GABA$_A$ receptors requires α1, α2, or α5 and γ2 or γ3; other subunit combinations are insensitive to benzodiazepines (Rudolph et al., 1999).

Picrotoxin, a potent convulsant compound, appears to bind within the channel pore of the GABA$_A$ receptor

and prevent ion flow (Seighart, 1992). Single-channel experiments indicate that picrotoxin either slowly blocks an open channel or prevents the GABA receptor from undergoing a transition into a long-duration open state. Apparently, barbiturates produce similar changes in channel properties, but they potentiate rather than inhibit GABA$_A$ receptor function. Bicuculline, another potent convulsant, appears to inhibit GABA$_A$ receptor channel activity by decreasing the binding of GABA to the receptor. Steroid metabolites of progesterone, corticosterone, and testosterone also appear to have potentiating effects on GABA currents that are similar in many ways to the action of barbiturates; however, the binding sites for these steroids and the barbiturates are distinct. Finally, penicillin directly inhibits GABA receptor function, apparently by binding within the pore and thus being designated an open channel blocker.

The physiological effects of compounds such as picrotoxin, bicuculline, and penicillin are striking. Each of these compounds at a sufficiently high concentration can produce widespread and sustained seizure activity. Conversely, many, but not all, of the sedative properties associated with barbiturates and benzodiazepines can be attributed to their ability to augment inhibition in the brain by enhancing GABA's inhibitory potency.

Interestingly, ρ-subunit-containing GABA receptors, found in abundance in the retina, are pharmacologically unique. They are resistant to bicuculline's inhibitory action, although they remain sensitive to blockage by picrotoxin. In addition, these retinal receptors are not sensitive to modulation by barbiturates or benzodiazepines. Thus, ρ-subunit-containing receptors are distinct from GABA$_A$ receptors and similar to receptors earlier designated GABA$_C$ (Bormann and Fiegenspan, 1995).

Several studies indicate that phosphorylation of the GABA$_A$ receptor likely modifies its functions; however, whether the receptor itself is phosphorylated in vivo and whether phosphorylation increases or decreases the current flowing through the channel remain debated. GABA$_A$ receptors are modulated by protein kinase A, protein kinase C, Ca^{2+}-calmodulin-dependent protein kinase, and an undefined protein kinase.

Glycine Receptor Structure Is Closely Related to GABA$_A$ Receptor Structure

Glycine receptors are the major inhibitory receptors in the spinal cord (Betz, 1991) and within the CNS, particularly in the brain stem; glycine receptors provide similar inhibitory functions. Glycine receptors are similar to GABA$_A$ receptors in that both are ion channels selectively permeable to the anion Cl$^-$ (see Fig. 11.9).

The structure of the glycine receptor is indicative of this similarity in properties. The native complex is approximately 250 kDa and composed of two main subunits, α (48 kDa) and β (58 kDa). The receptor appears to be pentameric, most likely composed of three α and two β subunits. Apparently, three molecules of glycine must bind to the receptor to open it to ion flow (Young and Snyder, 1974), suggesting that the α subunit may contain the glycine binding site. The glycine receptor has an open-channel conductance of approximately 35–50 pS, similar to that of the GABA$_A$ receptor. A potent antagonist of the glycine receptor is the compound strychnine.

Four distinct α subunits and one β subunit of the glycine receptor have been cloned. Each exhibits the typical predicted four transmembrane segments, and they are approximately 50% identical with one another at the amino acid level. Expression of a single α subunit in oocytes is sufficient to produce functional glycine receptors, indicating that the α subunit is the pore-forming unit of the native receptor. The β subunits play exclusively modulatory roles, affecting, for example, sensitivity to the inhibitory actions of picrotoxin. They are widespread in the brain, and their distribution does not specifically colocalize with glycine receptor α-subunit mRNA. The β subunits may serve other functions independent of their association with glycine receptor.

Clustering of GABA$_A$ and Glycine Receptors

During maturation of inhibitory synapses, gephyrin clusters beneath the postsynaptic membrane and recruits and localizes GABA$_A$ or glycine receptors. Gephyrin appears to serve an analogous function for the stabilization of inhibitory receptors that rapsyn plays for nAChRs. Gephyrin is a multidomain protein that interacts with the cytoplasmic domains of the GABA$_A$ or glycine receptor subunits (Moss and Smart, 2001; Salyed et al., 2007). Like rapsyn, gephyrin has the capacity to interact with itself and other proteins in addition to the inhibitory receptors and these interactions serve to restrict the lateral mobility of GABA$_A$ and glycine receptors in the plasma membrane. Ultimately these interactions promote the formation of postsynaptic inhibitory specializations.

Certain Purinergic Receptors Are Also Ionotropic

Purinergic chemical transmission is distributed throughout the body, and the receptor subtypes and myriad effects are considered in greater detail in the later section on GPCRs. Purinergic receptors bind to ATP (or other nucleotide analogs) or its breakdown product adenosine. ATP is released from certain synaptic terminals in a quantal manner and often packaged within synaptic vesicles containing another neurotransmitter, the best described being acetylcholine and the catecholamines.

Although not included in Fig. 11.10, a few purinergic receptor subtypes are related to the family of ionotropic receptors. Two subtypes of ATP-binding purinergic receptors (P2x and P2z) were discovered to be ionotropic receptors, but data on their functions and properties are sparse. P2x receptors appear to mediate a fast depolarizing response in neurons and muscle cells to ATP by the direct opening of a nonselective cation channel. cDNAs encoding the P2x receptor indicate that its structure comprises only two transmembrane domains, with some homology in its pore-forming region with K$^+$ channels (Brake et al., 1994; Valera et al., 1994). The P2z receptor also is a ligand-gated channel that permits permeation of either anions or cations and even molecules as large as 900 Da. Its primary structure has not yet been defined.

Glutamate Receptors Are Derived from a Different Ancestral Gene and Are Structurally Distinct from Other Ionotropic Receptors

Glutamate receptors are widespread in the nervous system, where they are responsible for mediating the vast majority of excitatory synaptic transmission in the brain and spinal cord. Early studies suggested that the glutamate receptor family was composed of several distinct subtypes. In the 1970s, Jeffrey Watkins and his colleagues significantly advanced this field by developing agonists that could pharmacologically distinguish between different glutamate receptor subtypes. Four of these agonists—N-methyl-D-aspartate (NMDA), amino-3-hydroxy-5-methylisoxazoleproprionic acid (AMPA), kainate, and quisqualate—are distinct in the type of receptors to which they bind and have been used extensively to characterize the glutamate receptor family (Hollmann and Heinemann, 1994; Watkins et al., 1990). A convenient distinction for describing the ionotropic glutamate receptors has been to classify them as either NMDA or non-NMDA subtypes, depending on whether they bind the agonist NMDA. Non-NMDA receptors also bind the agonist kainate or AMPA. Both NMDA and non-NMDA receptors are ionotropic. Quisqualate is unique within this group in having the capacity to activate both ionotropic and GPCR glutamate receptor subtypes (Hollmann and Heinemann, 1994). A family tree highlighting the evolutionary relationship of the glutamate receptors is shown in Figure 11.12.

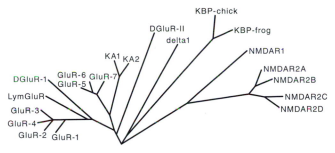

FIGURE 11.12 Evolutionary relationships of the ionotropic glutamate receptor family. Adapted from Hollmann and Heinemann (1994).

Non-NMDA Receptors Are a Diverse Family

In 1989, Stephen Heinemann and his colleagues reported the isolation of a cDNA that produced a functional glutamate-activated channel when expressed in *Xenopus* oocytes (Hollmann *et al.*, 1989). The initial glutamate receptor was termed GluR-K1, and the cDNA encoded a protein with an estimated molecular mass of 99.8 kDa. Not long after this original report, Heinemann's group (Boulter *et al.*, 1990), Peter Seeburg's group (Keinanen *et al.*, 1990), and Richard Axel's group (Nakanishi *et al.*, 1990) independently reported the isolation of families of glutamate receptor subunits, termed $GluR_1$–$GluR_4$ by Heinemann's group and GluRA–GluRD by Seeburg's group. Each GluR subunit consists of approximately 900 amino acids and has four predicted membrane-spanning segments (TM1–TM4). However, there is an important distinction in the TM2 domain making the GluRs distinct from the nAChR family. The native form of GluR subunits appears to be a tetrameric complex with an approximate molecular mass of 600 kDa (Blackstone *et al.*, 1992; Wenthold *et al.*, 1992). Thus, the size of the glutamate receptor is almost twice that of nAChR, mostly because of the large extracellular domain where glutamate binds to the receptor.

Unique Properties of Non-NMDA Receptors Are Determined by Assembly of Different Subunits

When the cDNAs encoding these receptors were expressed in either oocytes or HeK-293 cells, application of the non-NMDA receptor agonist AMPA produced substantial inward currents. In these same experiments, the agonist kainate was demonstrated to produce larger currents, mainly because of rapid and significant desensitization of the receptor when AMPA was used as the agonist. A striking observation was made from these expression studies. Specifically, when the $GluR_2$ subunit alone was expressed in the oocytes, little current was obtained when the preparation was exposed to agonist, unlike the large currents found when either $GluR_1$ or $GluR_3$ was expressed (Boulter *et al.*, 1990; Nakanishi *et al.*, 1990; Verdoorn *et al.*, 1991). $GluR_2$ subunits by themselves appear to form poorly conducting receptors. However, when $GluR_2$ is expressed with either $GluR_1$ or $GluR_3$, the behavior of the heteromeric receptor is distinctly different. Examination of the current–voltage relationships (i.e., I/V plots; see also Chapter 16) indicates that when $GluR_1$ and $GluR_3$ are expressed alone or together, they produce channels with strong inward rectification. Coexpression of $GluR_2$ with either $GluR_1$ or $GluR_3$ produces a channel with little rectification and a near-linear I/V plot. Further analyses (Hollmann and Heinemann, 1994) indicated that $GluR_1$ and $GluR_3$, either independently or when coexpressed, exhibited channels permeable to Ca^{2+}. In contrast, any combination of receptor that included the $GluR_2$ subunit produced channels impermeable to Ca^{2+}. Earlier single-channel analyses of glutamate receptors expressed in embryonic hippocampal neurons indicated that two distinct receptors are present: one relatively impermeable to Ca^{2+}, exhibiting a linear I/V plot, and another significantly permeable to Ca^{2+}, exhibiting a rectifying I/V plot (Iino *et al.*, 1990). Clearly, the properties of glutamate receptors can be quite different and initiate unique intracellular responses, depending on the subunit composition expressed in a particular neuron. In a series of elegant studies, the replacement of a single amino acid (Arg for Gln) in the second transmembrane region of the $GluR_2$ subunit (see Fig. 11.13B for identification of this amino acid) was shown to switch its behavior from a non-Ca^{2+}-permeable to a Ca^{2+}-permeable channel (Burnashev *et al.*, 1992; Hume *et al.*, 1991). Apparently, an Arg at this position blocks Ca^{2+} from traversing the pore formed in the center of the GluR channel.

Functional Diversity in GluRs Is Produced by mRNA Splicing and RNA Editing

Analysis of the mRNAs encoding GluR subunits indicated that each could be expressed in one of two splice variants, termed *flip* and *flop* (Sommer *et al.*, 1990). These flip and flop modules are small (38 amino acid) segments just preceding the TM4 transmembrane domain in all four GluR subunits. The receptor channel expressed from these splice variants has distinct properties, depending on which of the two modules is present. Specifically, the flop-containing receptors exhibit significantly greater magnitudes of desensitization during glutamate application. Therefore, GluRs with flop modules express smaller

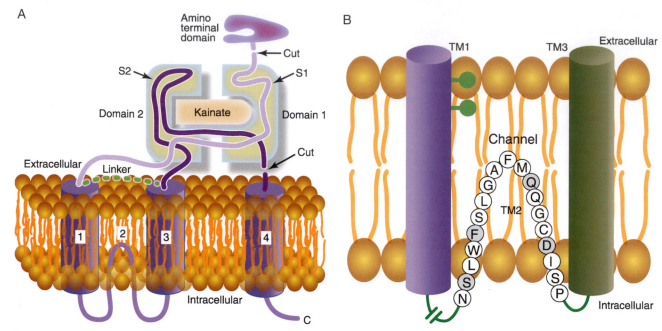

FIGURE 11.13   (A) Model of one of the subunits of the ionotropic glutamate receptor. The ionotropic glutamate receptors have four membrane-associated segments; however, unlike nAChR, only three of them completely traverse the lipid bilayer. TM2 forms a loop and re-exits into the cytoplasm. Thus, the large N-terminal region extends into the extracellular space, while the C terminus extends into the cytoplasm. Two domains in the extracellular segments associate with each other to form the binding site for transmitter, in this example kainate, a naturally occurring agonist of glutamate. (B) Enlarged area of the predicted structure and amino acid sequence of the TM2 region of the glutamate receptor, $GluR_3$. TM1 and TM3 are drawn as cylinders in the membrane flanking TM2. The residue that determines the Ca^{2+} permeability of the non-NMDA receptor is the glutamine residue (Q) highlighted in gray. In NMDA receptors, an asparagine residue at this same position is the proposed site of interaction with Mg^{2+} ions that produce the voltage-dependent channel block. The serine (S) and phenylalanine (F) also shaded in gray are highly conserved in the non-NMDA receptor family. The aspartate (D) residue is also conserved and thought to form part of the internal cation binding site. The break in the loop between TM1 and TM2 indicates a domain that varies in length among ionotropic glutamate receptors. Adapted from Wo and Oswald (1995).

steady-state currents than GluRs with flip modules. Both flip- and flop-containing GluRs are widely expressed in the brain with a few exceptions. One unique cell type appears to be pyramidal CA3 cells in the rat hippocampus, where the GluRs are deficient in flop modules. In neighboring CA1 pyramidal cells and dentate granule cells, flop-containing GluRs appear to dominate. The significance of these splice variations for information processing in the brain is not known, but the physiological prediction would be that CA3 neurons exhibit larger steady-state glutamate-activated currents owing to decreased desensitization from the absence of flop modules.

Typically, one believes that there is absolute fidelity in the process of transcribing DNA into mRNA and then into protein (i.e., the nucleotides present in the DNA are accurate predictors of the ultimate amino acid sequence of the protein). However, Peter Seeburg and his colleagues discovered a novel mechanism in the neuronal nucleus that edits mRNAs post-transcriptionally, and at least three of the four GluR subunits are subjected to this editing mechanism (Sommer *et al.*, 1991). In fact, one of the sites edited is the critical Arg residue regulating Ca^{2+} permeability in the $GluR_2$ subunit. At another edited site, Gly replaces Arg-764 in the $GluR_2$ subunit, and this editing also takes place in $GluR_3$ and $GluR_4$. The Arg-to-Gly conversion at amino acid 764 produces receptors that exhibit significantly faster rates of recovery from the desensitized state (Lomeli *et al.*, 1994). The extent to which other receptors or other protein molecules undergo this form of editing is an area rich for investigation. At a minimum, this editing mechanism produces dramatic differences in the function of GluRs.

Glutamate Receptors do not Conform to the Typical Four Transmembrane-Spanning Segment Structures Described for the nAChR

Although the field of glutamate receptors is advancing at a rapid pace, few structural data are available on the native molecule or on the topology of any single GluR subunit as it exists in the membrane. The receptor has a large extracellular domain that serves as the binding site for glutamate (Fig. 11.13A). Through the use of sophisticated genetic engineering, a crystal structure has been obtained of the glutamate binding site for GluR in the presence of the agonist kainate. Intricate interactions between the extracellular loops of the GluR subunits form the kainate binding sites (Fig. 11.13A).

Superficially, the remainder of the receptor was originally thought to resemble nAChR in having four TM segments that wrap back and forth through the membrane in an antiparallel fashion. However, the original model has now been proven incorrect by a number of elegant molecular and biochemical studies. The most recent information indicates that the TM2 membrane-spanning segment does not completely traverse the membrane (Fig. 11.13). Instead, it forms a kink within the membrane and enters back into the cytoplasm, similar in some ways to the pore-forming domain (P segment) of voltage-activated K^+ channels (Wo and Oswald, 1995) (see Chapter 6). An enlargement of this P segment (Fig. 11.12B) highlights the amino acids conserved in all the GluRs and further identifies the critical Gln (Q) residues responsible for Ca^{2+} permeability of the receptor. It also appears that glutamate receptors do not conform to the five-subunit structure of nAChR. There is both biochemical (Armstrong and Gouaux, 2000; Mano and Teichberg, 1998) and electrophysiological (Rosenmund et al., 1998) evidence that functional glutamate receptors are composed of four, not five, subunits. Thus, it appears glutamate receptors are rather highly divergent from the nAChR receptor family. In fact, their structure conforms more closely to the family of K^+ channels in that both appear tetrameric and have a unique P segment that forms the selectivity filter.

Other Non-NMDA GluRs Have Poorly Characterized Functions

Three other members, $GluR_5–GluR_7$, now form a second non-NMDA receptor subfamily, whose contribution to producing functionally distinct receptors is less well understood. Their overall structure is similar to that of $GluR_1–GluR_4$, and they exhibit about 40% sequence homology. However, their agonist-binding profile and their electrophysiological properties are distinct. They are expressed at lower levels in the brain than the $GluR_1–GluR_4$ family (Hollmann and Heinemann, 1994).

Two members of the glutamate receptor family, KA-1 and KA-2, are the high-affinity kainate-binding receptors found in brain. Clearly distinct from the glutamate receptors discussed so far, KA-1 and KA-2 are more similar to the $GluR_5–GluR_7$ subfamily than to the $GluR_1–GluR_4$ subfamily. Neither KA-1 nor KA-2 produces a functional channel when expressed in cells or oocytes, even though high-affinity kainate binding sites were detected. KA-1 does not appear to form functional receptors or channels with any of the other GluR subunits, and its physiological relevance remains obscure. It is expressed at high concentrations in only two cell types, hippocampal CA3 and dentate granule cells. KA-2 exhibits interesting properties when combined with other GluR subunits. For example, coexpression of $GluR_6$ and KA-2 produces functional receptors that respond to AMPA, although neither subunit itself responds to this agonist (Herb et al., 1992). This information indicates that agonist binding sites are at least partly formed at the interfaces between subunits.

Although other kainate-binding proteins and glutamate receptors have been described, their functions and biological significance are not currently understood. These receptors include two kainate-binding proteins, one from chicken and the other from frog, several invertebrate glutamate receptors, and two "orphan" receptors termed $\alpha 1$ and $\delta 2$ (Hollmann and Heinemann, 1994).

The NMDA Receptors Are a Family of Ligand-Gated Ion Channels That Are Also Voltage Dependent

NMDA receptors appear to be at least partly responsible for aspects of development, learning, and memory and neuronal damage due to brain injury. The particular significance of this receptor to neuronal function comes from two of its unique properties. First, the receptor exhibits associativity (see Chapter 19 for a more detailed discussion of associativity and the role of the NMDA receptor in memory mechanisms). For the channel to be open the receptor must bind glutamate and the membrane must be depolarized. This behavior is due to a Mg^{2+}-dependent block of the receptor at normal membrane resting potentials (Ascher and Nowak, 1988; Mayer and Westbrook, 1987) and gives rise to the dramatic voltage dependence of the channel (see Fig. 16.9). Second, the receptor permits a significant influx of Ca^{2+}, and increases in intracellular Ca^{2+} activate a variety of processes that alter the properties of the neuron. Excess Ca^{2+} is also toxic to neurons, and the hyperactivation of NMDA receptors is thought to contribute to a variety of neurodegenerative disorders.

Many pharmacological compounds produce their effects through interactions with the NMDA receptor. For example, certain hallucinogenic compounds, such as phencyclidine (PCP) and dizocilpine (MK-801), are effective blockers of the ion channel associated with the NMDA receptor (Fig. 11.14). These potent antagonists require the receptor channel to be open to gain access to their binding sites and are therefore referred to as *open-channel blockers.* They also become trapped when the channel closes and are therefore difficult to wash out of the NMDA receptor's channel. Antagonists for the glutamate binding site also have been developed, and some of the most well known are AP-5 and AP-7. These and other antagonists specific for the glutamate binding site also produce hallucinogenic effects in both animal models and humans. NMDA remains a specific agonist for this receptor; however, it is about one order of magnitude less potent than L-glutamate for receptor activation. L-glutamate is the predominant neurotransmitter that activates the NMDA receptor; however, L-aspartate can also activate the receptor, as can an endogenous dipeptide in the brain, N-acetylaspartylglutamate (Hollmann and Heinemann, 1994).

NMDA Receptor Subunits Show Similarity to Non-NMDA Receptor Subunits

The primary structure of the NMDA receptor was revealed in 1990 when Nakanishi and his colleagues isolated the first cDNA encoding a subunit of the NMDA receptor (Moriyoshi *et al.*, 1991). The first cloned subunit was aptly named *NMDAR$_1$*, and the deduced amino acid sequence indicated a protein of approximately 97 kDa, similar to other members of the GluR family. Four potential transmembrane domains were identified, and the current assumption is that four individual subunits compose the macromolecular NMDA receptor complex. However, recall that the transmembrane organization of GluR subunits indicates that TM2 does not fully traverse the membrane. It seems likely that the NMDA receptor subunits will also follow this recent modification of the model. The TM2 segment of each subunit clearly lines the pore of the NMDA receptor channel, as does the TM2 segment of the GluR subunits. In fact, a single Asn residue, analogous to that in the GluR$_2$ subunit, regulates the Ca^{2+} permeability of the NMDA receptor (Burnashev *et al.*, 1992; Mori *et al.*, 1992).

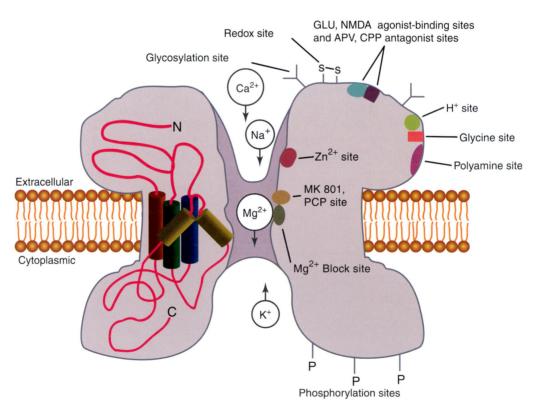

FIGURE 11.14   Diagram of an NMDA receptor highlighting binding sites for numerous agonists, antagonists, and other regulatory molecules. The location of these sites is a crude approximation for the purpose of discussion. Adapted from Hollmann and Heinemann (1994).

Mutation of this Asn residue markedly reduces Ca^{2+} permeability.

Three of the best-characterized facets of the NMDA receptor were found when the $NMDAR_1$ subunit was initially expressed by itself in oocytes, although currents were relatively small. These characteristics are (1) an Mg^{2+}-dependent voltage-sensitive ion channel block, (2) a glycine requirement for effective channel opening, and (3) Ca^{2+} permeability (Moriyoshi et al., 1991). As described later, other NMDAR subunits contribute to assembly of the receptors thought to exist in the nervous system.

Functional Diversity of NMDA Receptors Occurs Through RNA Splicing

At least eight splice variants have now been identified for the $NMDAR_1$ subunit and these variants produce differences, ranging from subtle to significant, in the properties of the expressed receptor (Hollmann and Heinemann, 1994). For example, $NMDAR_1$ receptors lacking a particular N-terminal insert owing to alternative splicing exhibit enhanced blockade by protons and exhibit responses that are potentiated by Zn^{2+} in micromolar concentrations. Zn^{2+} has classically been described as an NMDA receptor antagonist that significantly blocks its activation. Clearly, the particular splice variant incorporated into the receptor complex affects the types of physiological response generated. Spermine, a polyamine found in neurons and in the extracellular space, also slightly increases the amplitude of NMDA responses, and this modulatory effect also appears to be associated with a particular splice variant. The physiological role of spermine in regulating NMDA receptors remains unclear.

Multiple NMDA Receptor Subunit Genes Also Contribute to Functional Diversity

Four other members of the NMDA receptor family have been cloned ($NMDAR_2A$–$NMDAR_2D$), and their deduced primary structures are highly related. These four NMDA receptor subunits do not form channels when expressed singly or in combination unless they are coexpressed with $NMDAR_1$ (Kutsuwada et al., 1992; Meguro et al., 1992; Monyer et al., 1992). Apparently, $NMDAR_1$ serves an essential function for the formation of a functional pore by which activation of NMDA receptors permits the flow of ions. $NMDAR_2A$–$NMDAR_2D$ play important roles in modulating the receptor activity when mixed as heteromeric forms with $NMDAR_1$. Coexpression of $NMDAR_1$ with any of the other subunits produces much larger currents (from 5- to 60-fold greater) than when $NMDAR_1$ is expressed in isolation, and NMDA receptors expressed in neurons are likely to be hetero-oligomers of $NMDAR_1$ and $NMDAR_2$ subunits. The C-terminal domains of $NMDAR_2A$–$NMDAR_2D$ are quite large relative to the $NMDAR_1$C terminus, and they appear to play roles in altering channel properties and in affecting the subcellular localization of the receptors. All the NMDAR subunits have an Asn residue at the critical point in the TM2 domain essential for producing Ca^{2+} permeability. This Asn residue also appears to form at least part of the binding site for Mg^{2+}, which suggests that the sites for Mg^{2+} binding and Ca^{2+} permeation overlap (Burnashev et al., 1992; Mori et al., 1992).

The distribution of $NMDAR_2$ subunits is generally more restricted than the homogeneous distribution of $NMDAR_1$, with the exception of $NMDAR_2A$, which is expressed throughout the nervous system. $NMDAR_2C$ is restricted mostly to cerebellar granule cells, whereas 2B and 2D exhibit broader distributions. As noted, the large size of the C terminus of the $NMDAR_2$ subunit suggests a potential role in association with other proteins, possibly to target or restrict specific NMDA receptor types to areas of the neuron. Mechanisms related to receptor targeting are now becoming understood and will clearly play major roles in determining the efficacy of synaptic transmission (Ehlers et al., 1995; Komau et al., 1995).

NMDA Receptors Exhibit Complex Channel Properties

The biophysical properties of the NMDA receptor are complex (Ascher and Nowak, 1988; Mayer and Westbrook, 1987). The single-channel conductance has a main level of 50 pS; however, subconductances are evident, and different subunit combinations produce channels with distinct single-channel properties. A binding site for the Ca^{2+}-binding protein calmodulin has also been identified on the $NMDAR_1$ subunit (Ehlers et al., 1996). Binding of Ca^{2+}-calmodulin to NMDA receptors produces a fourfold decrease in open-channel probability. Ca^{2+} influx through the NMDA receptor could induce calmodulin binding and lead to an immediate short-term feedback inhibition, decreasing ion flow through the receptor.

Glutamate Receptors Cluster at Synapses

Glutamate receptors are concentrated at excitatory synapses through interactions with underlying scaffolding molecules (Fig. 11.15). These scaffolding molecules and a host of other structural and signaling

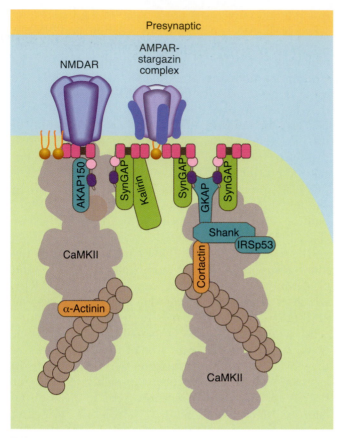

FIGURE 11.15 Diagram of glutamate receptor clustering at an excitatory synapse. The NMDA receptor interacts directly with PSD-95 through binding to one of PSD-95's three PDZ domains (the PDZ domains of PSD-95 are shown as pink squares). The AMPAR is associated with a protein called *stargazin* and stargazin interacts with one of the PDZ domains of PSD-95. Only a few of the many other signaling and scaffolding proteins at excitatory synapses are shown. AKAP150 is an A-kinase-anchoring protein of 150 kDa that binds to protein kinase A and other proteins; SynGAP is an abundant *syn*aptic-associated Ras *G*TP-ase-*a*ctivating protein that interacts with PSD-95; GKAP is a guanylate kinase-associated protein that interacts with PSD-95; and CaMKII is an abundant Ca^{2+}–calmodulin-activated protein kinase that interacts directly with the NMDAR. CaMKII also interacts with itself and with α-actinin, which is an actin-binding protein. This web of protein–protein interactions forms the electron-dense structures called the *postsynaptic densities* visible in electon micrographs of excitatory synapses. Adapted from Sheng and Hoogenraad (2007).

proteins form an electron-dense structure at excitatory synapses called the *postsynaptic density* (PSD; see also Chapter 8). One of the best known of these scaffolds is PSD-95, a multidomain protein enriched at excitatory synapses that helps organize receptors and signaling molecules. Through one of its multiple PDZ domains, PSD-95 binds directly to the cytoplasmic domain of $NMDAR_2$ subunits and anchors them tightly to the PSD (Sheng and Hoogenraad, 2007).

In fact, biochemically isolated PSDs stripped of the overlying membrane retain NMDARs, indicating the interaction is mediated by tight binding. PSD-95 and NMDARs are considered to be "core" PSD components, as their concentration is relatively consistent between PSDs. A number of other PSD-enriched proteins have been identified to bind to the C-terminal domain of NMDA receptor subunits, including the Ca^{2+}–calmodulin-activated protein kinase-CaMKII (Fig. 11.15).

Non-NMDA receptors do not interact directly with PSD-95, and their relative concentration in the PSD is dynamic. The up-and-down regulation of non-NMDA receptor number in the postsynaptic membrane alters the efficiency of synaptic transmission and provides a mechanism for the expression of plasticity at excitatory synapses. As such, identifying factors that influence the trafficking of non-NMDA receptors is an intensely studied area. One such factor is the protein stargazin that was recently discovered as a structural component of isolated non-NMDA receptors (Nicoll *et al.*, 2006; Ziff, 2007). Stargazin is one member of a family of proteins called *TARPs*, for *t*ransmembrane *A*MPA receptor *r*egulatory *p*roteins, that are important for the maturation and targeting of non-NMDA receptors to excitatory synapses (Fig. 11.15). TARPs appear to bind stoichiometrically to non-NMDA receptors and can be identified as additional protein density in electron microscopic reconstructions of isolated non-NMDA receptors (Nakagawa *et al.*, 2005). This situation is analogous to that described for rapsyn binding to the nAChR. However, TARPS are themselves membrane-spanning proteins, having four predicted transmembrane domains. TARPs also have a domain that interacts with PSD-95 (or other PDZ-containing proteins), and it is this interaction that leads to the association of non-NMDA receptors with synaptic specializations. TARPs also affect the maturation and channel properties of non-NMDA receptors in addition to their role in synaptic targeting (Nicoll *et al.*, 2006). Interestingly, TARP binding decreases when glutamate binds to the receptor, suggesting interplay exists between receptor use and TARP-mediated localization. There are other well-documented proteins that interact with the C-terminal domain of non-NMDA receptor subunits, some in an isoform-specific manner. This list includes GRIP/ABP, PICK-1, NSF, and SAP-97, and these proteins have been implicated in the maturation and trafficking of non-NMDA receptors. It is also well documented that specific residues in the C-terminal domain of non-NMDA receptors are phosphorylated, and phosphorylation regulates channel properties and surface expression. These post-translational modifications and interactions with other proteins must work in concert for the proper localization, stability, and function of non-NMDA receptors.

Summary

A general model for ionotropic receptors has emerged mainly from analyses of nAChR. Ionotropic receptors are large membrane-bound complexes generally composed of five subunits. The subunits each have four transmembrane domains, and the amino acids in the transmembrane segment TM2 form the lining of the pore. Transmitter binding induces rapid conformational changes that are translated into an increase in the diameter of the pore, permitting ion influx. Cation or anion selectivity is obtained through the coordination of specific negatively or positively charged amino acids at strategic locations in the receptor pore. How well the details of structural information obtained for nAChR will generalize to other ionotropic receptors awaits structural analyses of these other members. However, it is already clear that this model does not adequately describe the orientation of the transmembrane domains or the subunit number of the glutamate receptor family. The TM2 domain of glutamate receptors forms a hairpin instead of traversing the membrane completely, causing the remainder of the receptor to adopt an architecture different from that described for the nAChR family. Glutamate receptors are also composed of four, not five, subunits. These differences are perhaps not surprising given that the nAChR family and the glutamate receptor family appear to have arisen from two different ancestral genes.

G-PROTEIN-COUPLED RECEPTORS

The number of members in the G-protein-coupled receptor family is enormous, with more than 1000 identified and the number growing. The historical term *metabotropic* was used to describe the fact that intracellular metabolites are produced when these receptors bind ligand. However, there are now clearly documented cases where activation of metabotropic receptors does not produce alterations in metabolites but instead produces their effects by interacting with G proteins that alter the behavior of ion channels. Thus, these receptors are now referred to as *G-protein-coupled receptors*, or GPCRs.

When a GPCR is activated it couples to a G protein initiating the exchange of GDP for GTP, activating the G protein. Activated G proteins then couple to many downstream effectors, and most alter the activity of other intracellular enzymes or ion channels. Many of the G-protein target enzymes produce diffusible second messengers (metabolites) that stimulate further downstream biochemical processes, including the activation of protein kinases (see Chapter 12). Time is required for each of these coupling events, and the effects of GPCR activation are typically slower in onset than those observed following activation of ionotropic receptors. Because there is a lifetime associated with each intermediate, the effects produced by activation of GPCRs are also typically longer in duration than those produced by activation of ionotropic receptors. Most small neurotransmitters, such as ACh, glutamate, serotonin, and GABA, can bind to and activate both ionotropic and GPCRs. Thus, each of these transmitters can induce both fast responses (milliseconds), such as typical excitatory or inhibitory postsynaptic potentials, and slow-onset and longer-duration responses (from tenths of seconds to, potentially, hours). Other transmitters, like neuropeptides, produce their effects largely by binding only to GPCRs. These effects across multiple time domains provide the nervous system with a rich source for temporal information processing that is subject to constant modification. Currently, the GPCR family can be divided into three subfamilies on the basis of their structures: (1) the rhodopsin–adrenergic receptor subfamily, (2) the secretin–vasoactive intestinal peptide receptor subfamily, and (3) the metabotropic glutamate receptor subfamily (Strader *et al.*, 1995).

GPCR Structure Conforms to a General Model

A GPCR consists of a single polypeptide with a generally conserved structure. The receptor contains seven membrane-spanning helical segments that wrap back and forth through the membrane (Fig. 11.16). G-protein-coupled receptors are homologous to rhodopsin from both mammalian and bacterial sources, and detailed structural information on rhodopsin has been used to provide a framework for developing a general model for GPCR structures (Henderson *et al.*, 1990; Palczewski *et al.*, 2000). Aside from rhodopsin, two of the best structurally characterized GPCRs are the β-adrenergic receptor (βAR) and the muscarinic acetylcholine receptor (mAChR), and biochemical analyses to date support the use of rhodopsin as a structural framework for the family of GPCRs (Mizobe *et al.*, 1996).

The most conserved feature of GPCRs is the seven membrane-spanning segments; however, other

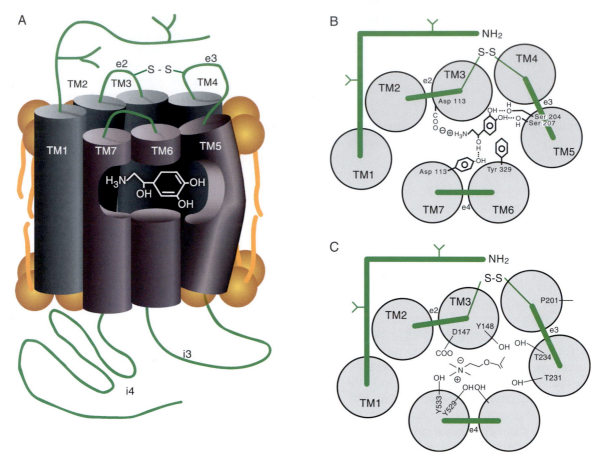

FIGURE 11.16 (A) Diagram showing the approximate position of the catecholamine binding site in βAR. The transmitter binding site is formed by amino acids whose side chains extend into the center of the ring produced by the seven transmembrane domains (TM1–TM7). Note that the binding site exists at a position that places it within the plane of the lipid bilayer. (B) A view looking down on a model of βAR identifying residues important for ligand binding. The seven transmembrane domains are represented as gray circles labeled TM1 though TM7. Amino acids composing the extracellular domains are represented as green bars labeled e1 through e4. The disulfide bond (–S–S–) that links e2 to e3 also is shown. Each of the specific residues indicated makes stabilizing contact with the transmitter. (C) A view looking down on a model of mAChR identifying residues important for ligand binding. Stabilizing contacts, mainly through hydroxyl groups (–OH), are made with the transmitter on four of the seven transmembrane domains. The chemical nature of the transmitter (i.e., epinephrine versus ACh) determines the type of amino acids necessary to produce stable interactions in the receptor binding site (compare B and C). Adapted from Strosberg (1990).

generalities can be made about their structure. The N terminus of the receptor extends into the extracellular space, whereas the C terminus resides within the cytoplasm (Fig. 11.16). Each of the seven transmembrane domains between the N and C termini consists of approximately 24 mostly hydrophobic amino acids. These seven TM domains associate together to form an oblong ring within the plasma membrane (Fig. 11.16B). Between each transmembrane domain is a loop of amino acids of various sizes. The loops connecting TM1 and TM2, TM3 and TM4, and TM5 and TM6 are intracellular and labeled i1, i2, and i3, respectively, whereas those between TM2 and TM3, TM4 and TM5, and TM6 and TM7 are extracellular and labeled e1, e2, and e3, respectively (see Fig. 11.16A for examples).

The Neurotransmitter Binding Site Is Buried in the Core of the Receptor

The neurotransmitter binding site for many GPCRs (excluding the metabotropic glutamate, GABA$_B$, and neuropeptide receptors) resides within a pocket formed in the center of the seven membrane-spanning segments (Fig. 11.16A). In the βAR, this pocket resides ~11 Å into the hydrophobic core of the receptor, placing the ligand binding site within the plasma membrane lipid bilayer (Kobilka, 1992; Mizobe *et al.*, 1996). Strategically positioned charged and polar residues in the membrane-spanning segments point inward into a central pocket that forms the binding site for the ligand. For example, Asn residues in the second and third segments, two Ser residues in the fifth segment, and a Phe residue in the sixth segment provide major contact points in the βAR binding site for the transmitter (Kobilka, 1992) (Fig. 11.16B). Replacing the Asp in TM3 with a Glu reduced transmitter binding by more than 100-fold, and replacement with a less conserved amino acid, such as Ser, reduces binding by more than 10,000-fold. Two Ser residues in TM5 are also essential for efficient transmitter binding and receptor activation, as are an Asp residue in TM2 and a Phe residue in TM6. In total, the two Asp, two Ser, and one Phe residues are highly conserved in all receptors that bind catecholamines. Variations in the amino acids at these five positions appear to provide the specificity between binding of different transmitters to the individual GPCRs.

The neurotransmitter binding site of the mAChRs, like that of β2AR, has been investigated in great detail (Fig. 11.16C). The Asp residue in TM3 is also critical for ACh binding to the mAChRs. Mutagenesis studies indicate important roles for Tyr and Thr residues in TM3, TM5, TM6, and TM7 in contributing to the ligand binding site for ACh. Interestingly, many of these mutations do not affect antagonist binding, indicating that distinct sets of amino acids participate in binding agonists and antagonists. When the TM domains are examined from a side view (Fig. 11.17), all the key amino acids implicated in agonist binding

lie at about the same level within the core of the receptor structure, buried approximately 10–15 Å from the surface of the plasma membrane (yellow boxed amino acids). An additional amino acid identified as essential for agonist binding of the mAChR is a Pro residue in TM4. This residue is also highly conserved among the GPCRs, and structural predictions suggest that it affects ligand binding not by interacting with agonist directly but by stabilizing a conformation essential for high-affinity binding. Structural predictions also place this Pro residue in the same plane as the Asp, Tyr, and Thr residues that form the ligand binding site of mAChR (Fig. 11.17).

Transmitter Binding Causes a Conformational Change in the Receptor and Activation of G Proteins

Proposed models for GPCR activation assume that the receptor can spontaneously isomerize between the inactive and active states (Perez et al., 1996; Premont et al., 1995). Only the active state interacts with G proteins in a productive fashion. This isomerization is analogous to the spontaneous isomerization proposed for ion channels as they oscillate between open and closed states. At equilibrium, in the absence of

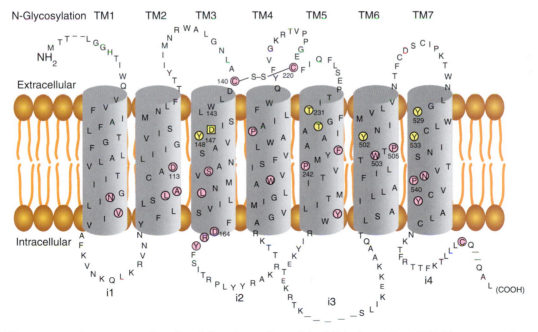

FIGURE 11.17 Amino acid sequence and predicted domain topology of the M3 isoform of mAChR. The transmembrane domains are TM1–TM7. The NH₂ terminus of the protein is at the left and extends into the extracellular space. The COOH terminus is intracellular and is at the right; i1 to i4 are the four intracellular domains. The conserved disulfide bond (–S–S–) connects extracellular loop 2 to loop 3. The dashes in the amino acid sequence represent inserts of various lengths that are not shown. Conserved amino acids for all members of the G-protein-coupled receptor family of receptors are marked in purple. The amino acids taking part in ACh binding to the receptor are highlighted in yellow. Note that all amino acids associated with ligand binding lie in approximately the same horizontal plane across the receptor. Adapted from Wess, J., *Trends Pharmacol. Sci.*, Vol. 14, 1993.

agonist, the inactive state of GPCRs is favored, and little G-protein activation occurs. Agonist binding stabilizes the active conformation and shifts the equilibrium toward the active form, and G-protein activation ensues. Conversely, receptor antagonists block G-protein activation through two proposed mechanisms: (1) negative antagonism in which antagonists bind to the inactive state of the receptor, thus favoring an equilibrium with the inactive form; and (2) neutral antagonism in which antagonists bind to both the active and inactive forms, thus stabilizing both and preventing a complete transition into the active form. This kinetic model indicates that agonist binding is not necessary for the receptor to undergo a transition into the active state; instead, it stabilizes the activated state of the receptor. This proposed model is supported by observations of both spontaneously arising and engineered mutants of βARs and αARs. Specific amino acid replacements produce receptors that exhibit constitutive activity in the absence of agonists (Perez *et al.*, 1996; Premont *et al.*, 1995). The amino acid changes apparently stabilize the active conformation of the molecule in a state more similar to the agonist-bound form of the receptor, leading to productive interactions with G proteins in the agonist-free state.

The Third Intracellular Loop Forms a Major Determinant for G-Protein Coupling

Extensive studies using site-directed mutagenesis and the production of chimeric molecules have revealed the domains and amino acids essential for G-protein coupling to GPCRs. Receptor domains within the second (i2) and third (i3) intracellular loops (Fig. 11.17) appear largely responsible for determining the specificity and efficiency of coupling for adrenergic and muscarinic cholinergic receptors and are the likely sites for G-protein coupling of the entire GPCR family. In particular, the 12 amino acids of the N-terminal region of the third intracellular loop significantly affect the specificity of G-protein coupling. Other regions in the C terminus of the third intracellular loop and the N-terminal region of the C-terminal tail appear to be more important for determining the efficiency of G-protein coupling than for determining its specificity (Kobilka, 1992). The third intracellular loop varies enormously in size among the different G-protein-coupled receptors, ranging from 29 amino acids in the substance P (a neuropeptide) receptor to 242 amino acids in mAChR (Strader *et al.*, 1994). The intracellular loop connecting TM5 and TM6 is the main point of receptor coupling to G proteins, and ligand binding to amino acids in

TM5 and TM6 may be responsible for triggering G protein–receptor interaction by transmitting a conformational change to the third intracellular loop (i3).

Specific Amino Acids Are Involved in Transducing Transmitter Binding into G-Protein Coupling

Residues associated with transmitting the conformational change induced by ligand binding to the activation of G proteins have been investigated with the use of the mAChRs. These studies revealed that an Asp residue in TM2 is important for receptor activation of G proteins, and altering the Asp by site-directed mutagenesis has a major negative effect on G protein–receptor activation (Fraser *et al.*, 1988, 1989). A Thr residue in TM5 and a Tyr residue in TM6 also are essential. Because these residues are connected by i3, they are assumed to play fundamental roles in transmitting the conformational change induced by ligand binding to the area of i3 essential for G-protein coupling and activation. When mutated, a Pro residue on TM7 produces a major impairment in the ability of the TM3 segment to induce activation of phospholipase C through a G protein and, presumably, is another key element in propagating the conformational changes necessary for efficient coupling to G proteins. As informative as mutagenesis studies can be, a true molecular understanding of the conformational changes induced by agonist binding will likely require a structural approach similar to that applied by Nigel Unwin to the nAChR.

As mentioned earlier, GPCRs are single polypeptides; however, they are clearly separable into distinct functional domains. For example, the β2AR can be physically split, with the use of molecular techniques, into two fragments, one fragment containing TM1–TM5 and the other containing TM6 and TM7. In isolation, neither of these fragments can produce a functional receptor; however, when coexpressed in the same cell, functional β2ARs that can bind ligand and activate G proteins are produced (Fig. 11.18). This remarkable experiment indicates that physical contiguity in the primary sequence is not essential for producing functional β2ARs, but it does emphasize the contribution of domains in the separate fragments (TM1–TM5 and TM6, TM7) to both ligand binding and G-protein coupling. Like the β2AR, the m2 and m3 members of the mAChR family can form functional receptors even if split into two separate domains. A fragment containing the first five TM domains, when expressed with a fragment containing TM6 and TM7, forms a functional receptor (Strosberg, 1990) (Fig. 11.18).

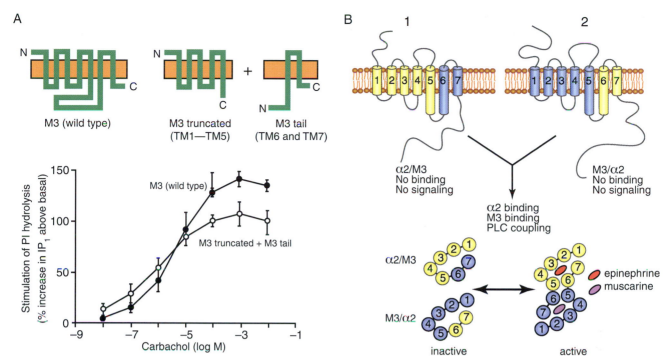

FIGURE 11.18 (A) mAChR can be split into two physically separated domains that, when added back together, retain the ability to bind transmitter and activate G proteins. (Upper left) Model of full-length mAChR; (upper right) two engineered pieces of the receptor. The graph indicates that, when coexpressed in the same cells, the two fragments can produce a functional mAChR that responds to the agonist carbachol producing activation of G protein and subsequent activation of an enzyme that hydrolyzes phosphatidylinositol (PI). Adapted from Wess, J., *Trends Pharmacol. Sci.*, Vol. 14, 1993. (B) Some GPCRs can function as dimers. In this example, chimeric receptors were produced between α2AR (a2) and mAChR(M3) by swapping certain TM domains through genetic engineering. When α2/M3 or M3/α2 are expressed separately, they are not active. However, if both chimeric molecules are expressed in the same cells, they form receptors that can be activated by either epinephrine or muscarine. The bottom panel shows a top-down view of how this domain swapping might occur when two molecules dimerize to produce receptors that can respond to both transmitters.

GPCRs Also Exist as Homo-Oligomers or Hetero-Oligomers

The observation that GPCRs can be physically split through genetic engineering and when recombined produce functional channels provided the first hint that full-length GPCRs might also oligomerize with each other into functional molecules. A test of such a hypothesis was accomplished by making chimeric receptors composed of TM domains 1–5 of α2-AR and TM domains 6 and 7 of the m3 muscarinic receptors and vice versa (Fig. 11.18B) (Maggio *et al.*, 1993). When either of these chimeric molecules was expressed in isolation, neither formed a functional receptor. However, when coexpressed, receptors were formed that bind both muscarinic and adrenergic ligands, and ligand binding led to functional activation of downstream effectors. Through domain swapping the ligand binding sites for both receptor ligands were reconstituted by oligomerization of the two chimeric receptors into one bifunctional chimeric

dimer (Fig. 11.18B). Oligomerization of GPCRs is also supported by crosslinking and immunoprecipitation experiments and with experiments examining the direct biophysical association of the receptors in living cells (Lee *et al.*, 2000; Maggio *et al.*, 1993; Overton and Blumer, 2000; Salahpour *et al.*, 2000). While some debate remains, the evidence now seems overwhelming that oligomerization of GPCRs is adding a new layer of complexity and diversity to the study of these receptors. The functional impacts of GPCR oligomerization are just beginning to be appreciated. Important functional consequences could relate to alterations in (1) ligand binding, (2) efficiency and specificity of coupling to downstream effectors, (3) subcellular localization, and (4) receptor desensitization. The evolving and apparently widespread nature of direct receptor–receptor interactions leads one to believe that our current understanding of neurotransmitter receptors and their biological impact will be undergoing continual modifications for many years to come.

G-Protein Coupling Increases the Affinity of the Receptor for Neurotransmitter

The affinity of GPCR for agonist increases when the receptor is coupled to the G protein. This positive feedback effectively increases the lifetime of the agonist-bound form of the receptor by decreasing the dissociation rate of the agonist. An excellent demonstration of this effect comes from studies using engineered βARs that are constitutively active in their ability to couple to G proteins. These mutant receptors show significantly increased affinity for agonists (Perez *et al.*, 1996). When G protein dissociates, the agonist-binding affinity of the receptor returns to its original state. The changes induced by ligand binding apparently stabilize the receptor in a conformation with both higher affinity for ligand and higher affinity for coupling to G proteins.

The Specificity and Potency of G-Protein Activation Are Determined by Several Factors

GPCRs associate with G proteins to transduce ligand binding into intracellular effects. This coupling step can lead to diverse responses, depending on the type of G protein and the type of effector enzyme present. In addition, ligand binding to a single subtype of GPCR can activate multiple G-protein-coupled pathways. Activated α2ARs have been shown to couple to as many as four different G proteins in the same cell (Strader *et al.*, 1994). Some of the specificity for G-protein activation can be determined by the specific conformations assumed by the receptor, and a single receptor can assume multiple conformations. For example, α2ARs can isomerize into at least two states. One state interacts with a G protein that couples to phospholipase C, and a second state interacts with G proteins that couple to both phospholipase C and phospholipase A2 (Perez *et al.*, 1996). Thus, a single GPCR can produce a diversity of responses, making it difficult to assign specific biological effects to individual receptor subtypes in all settings.

Activated GPCRs are free to couple to many G-protein molecules, permitting a significant amplification of the initial transmitter-binding event (Cassel and Selinger, 1977). This catalytic mechanism is referred to as *collision coupling* (Tolkovsky *et al.*, 1982), whereby a transient association between the activated receptor and the G protein is sufficient to produce the exchange of GDP for GTP, activating the G protein. Because enzymes such as adenylyl cyclase appear to be tightly coupled to the G protein, the rate-limiting step in the production of cAMP is the number of productive collisions between the receptor and the G protein. A constant GTPase activity hydrolyzes GTP, bringing the G protein and therefore the adenylyl cyclase back to the basal state. Transmitter concentration clearly plays a role in the number of activated receptors present at any given time, and GPCRs exhibit saturable dose–response curves. This apparent maximal rate is achieved when all of the G protein–cyclase complexes have become activated (more accurately, when the rate of formation is maximal with respect to the rate of GTP hydrolysis). A less intuitive consequence that evolves from these models is that receptor number can significantly affect the concentration of transmitter that produces a half-maximal response of cAMP accumulation. Thus, the larger the receptor number, the greater the probability that a productive collision will occur between an agonist-bound receptor and the G protein. Experimental evidence for this prediction was obtained for βAR expressed at various levels in eukaryotic cells. Increasing concentrations of βAR produced a decrease in the concentration of agonist required to produce half-maximal production of cAMP. Apparently, the cell can adjust the magnitude of its response by adjusting the number of receptors available for transmitter interaction. In addition, the important process of receptor desensitization can also regulate the number of receptors capable of productive G-protein interactions.

Receptor Desensitization Is a Built-In Mechanism for Decreasing the Cellular Response to Transmitter

Desensitization is a very important process whereby cells can decrease their sensitivity to a particular stimulus to prevent saturation of the system. Desensitization involves a complex series of events (Kobilka, 1992; Clark *et al.*, 1999). For GPCRs, desensitization is defined as an increase in the concentration of agonist required to produce half-maximal stimulation of, for example, adenylyl cyclase. In practical terms, desensitization of receptors produces less response for a constant amount of transmitter.

There are two known mechanisms for desensitization. One mechanism is a decrease in response brought about by the covalent modifications produced by receptor phosphorylation and is quite rapid (seconds to minutes). The other mechanism is the physical removal of receptors from the plasma membrane (likely through a mechanism of receptor-mediated endocytosis) and tends to require greater periods (minutes to hours). The latter process can be either reversible (sequestration) or irreversible (down-regulation).

The Rapid Phase of GPCR Desensitization Is Mediated by Receptor Phosphorylation

Desensitization of βARs appears to involve at least three protein kinases: PKA, PKC, and β-adrenergic receptor kinase (βARK; also referred to as G-protein receptor kinase (GRK)). Phosphorylation of ARs by PKA does not require that agonist be bound to the receptor and appears to be a general mechanism by which the cell can reduce the effectiveness of all receptors, independent of whether they are in the agonist-bound or unbound state (Fig. 11.19). This process is also referred to as *heterologous desensitization* because the receptor does not require bound agonist (for simplicity PKA is shown phosphorylating only the agonist-bound form of the receptor in Fig. 11.19). PKA and PKC phosphorylate sites on the third intracellular loop and possibly the C-terminal cytoplasmic domain. Phosphorylation of these sites functionally interferes with the receptor's ability to couple to G proteins, thus producing the desensitization (Fig. 11.19). Whether the same sites on βAR are phosphorylated by both PKA and PKC is controversial. Some researchers conclude that the effects of phosphorylation by either kinase on decreasing coupling of the receptor to G proteins are similar (suggesting that the sites phosphorylated are similar) (Huganir and Greengard, 1990). Others find that the effects are additive (Yuan *et al.*, 1994). Although the details of the role played by each of these kinases are ambiguous, phosphorylation by either enzyme desensitizes the receptor.

G-protein receptor kinases (GRKs) can also phosphorylate GPCRs and lead to receptor desensitization

(Inglese *et al.*, 1993; Sterne-Marr and Benovic, 1995). Six members of the GRK family of kinases have been identified: rhodopsin kinase (GRK1), βARK (GRK2), and GRK3 through GRK6 (Premont *et al.*, 1995; Sterne-Marr and Benovic, 1995). GRK2 (originally called *β-adrenergic receptor kinase*, or βARK) is a Ser- and Thr-specific protein kinase initially identified by its capacity to phosphorylate βAR. GRK2 phosphorylates only the agonist-bound form of the receptor, usually when agonist concentrations reach the micromolar level, as typically found in the synaptic cleft. This process is referred to as *homologous desensitization* because the regulation is specific for those receptor molecules that are in the agonist-bound state. Phosphorylation of βAR by GRK2 does not substantially interfere with coupling to G proteins. Instead, an additional protein, arrestin, binds the GRK2-phosphorylated form of the receptor, thus blocking receptor–G-protein coupling (Fig. 11.19). This process is analogous to the desensitization of the light-sensitive receptor molecule rhodopsin produced by GRK1 phosphorylation and the binding of arrestin. The phosphorylation sites on βAR for GRK2 reside on the C-terminal cytoplasmic domain and are distinct from those phosphorylated by PKA.

The cycle of homologous desensitization starts with the activation of a GPCR, which induces activation of G proteins and dissociation of the βγ subunit complex from α subunits. At least one role for the βγ complex appears to be to bind to GRKs, which leads to their recruitment to the membrane in the area of the locally activated G-protein–receptor complex. The recruited GRK is then activated, leading to phosphorylation of the agonist-bound receptor and subsequent binding of arrestin. Arrestin binds to the same domains on the receptor necessary for coupling to G proteins, thus terminating the actions of the activated receptor (Fig. 11.20). The ensuing process of sequestration follows GPCR phosphorylation and arrestin binding.

Desensitization Can Also Be Produced by Loss of Receptors from the Cell Surface

Desensitization of GPCRs is also produced by removal of the receptor from the cell surface. This process can be either reversible (sequestration or internalization) or irreversible (down-regulation). Sequestration is the term used to describe the rapid (within minutes) but reversible endocytosis of receptors from the cell surface after agonist application (Fig. 11.20). Neither G-protein coupling nor receptor phosphorylation appears to be absolutely essential for this process, but phosphorylation by GRKs clearly

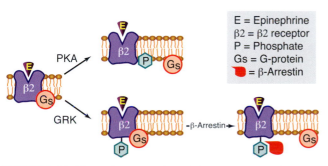

FIGURE 11.19 Different modes of desensitization of GPCRs. This diagram indicates that the epinephrine (E)-bound form of β2AR normally couples to the G protein Gₛ. PKA can phosphorylate the receptor, leading to an inhibition of binding to Gₛ. G-protein receptor kinase (GRK) also can phosphorylate the receptor; however, this phosphorylation does not directly interfere with binding to Gₛ. GRK phosphorylation is needed for the binding of another protein, β-arrestin, which, by its association with the receptor, prevents Gₛ from binding. Adapted from Ehlers *et al.* (1996).

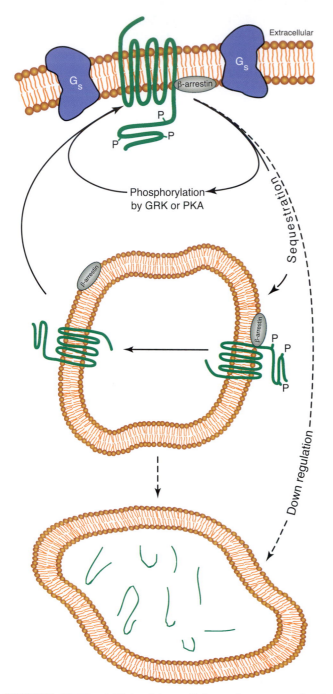

FIGURE 11.20 Additional intracellular pathways associated with desensitization of GPCRs. GPCRs are phosphorylated (noted with P) on their intracellular domains by PKA, GRK, and other protein kinases. The phosphorylated form of the receptor can be removed from the cell surface by a process called *sequestration* with the help of the adapter protein b-arrestin; thus fewer binding sites remain on the cell surface for transmitter interactions. In intracellular compartments, the receptor can be dephosphorylated and returned to the plasma membrane in its basal state. Alternatively, the phosphorylated receptors can be degraded (down-regulated) by targeting to a lysosomal organelle. Degradation requires replenishment of the receptor pool through new protein synthesis. Adapted from Kobilka (1992).

enhances the rate of sequestration (Ferguson *et al.*, 1995). The binding of arrestins to the phosphorylated receptor also enhances sequestration (Ferguson *et al.*, 1996). Thus, arrestin binding appears to promote not only rapid desensitization by disrupting the receptor–G-protein interaction but also receptor sequestration. Because the receptor can be functionally uncoupled from the G protein through the rapid phosphorylation-dependent phase of desensitization, the physiological role(s) of sequestration remains an open issue, although decreasing the number of receptor molecules on the cell surface would contribute to the overall process of desensitization to agonist. Receptor cycling through intracellular organelles is a trafficking mechanism that leads to an enhanced rate of dephosphorylation of the phosphorylated receptor, returning it to the cell surface in its basal state (Fig. 11.20) (Barak *et al.*, 1994).

Down-regulation occurs more slowly than sequestration and is irreversible (Fig. 11.20). The early phase (within 4h) may involve both a PKA-dependent and a PKA-independent process (Bouvier *et al.*, 1989; Proll *et al.*, 1992). This early phase of down-regulation is apparently due to receptor degradation after endocytotic removal from the plasma membrane. The later phases (>14h) of down-regulation appear to be further mediated by a reduction in receptor biosynthesis through a decrease in the stability of the receptor mRNA (Bouvier *et al.*, 1989) and a decreased transcription rate.

Other Post-Translational Modifications Are Required for Efficient Metabotropic Receptor Function

Like many proteins expressed on the cell surface, GPCRs are glycosylated, and the N-terminal extracellular domain is the site of carbohydrate attachment. Relatively little is known about the effect of glycosylation on the function of GPCRs. Glycosylation does not appear to be essential to the production of a functional ligand-binding pocket (Strader *et al.*, 1994), although prevention of glycosylation may decrease membrane insertion and alter intracellular trafficking of β2AR.

Another important structural feature of most GPCRs is the disulfide bond formed between two Cys residues present on the extracellular loops (e2 and e3; Figs. 11.16 and 11.17). Apparently, the disulfide bond stabilizes a restricted conformation of the mature receptor by covalently linking the two extracellular domains, and this conformation favors ligand binding. Disruption of this disulfide bond significantly decreases agonist binding (Kobilka, 1992).

A third Cys residue, in the C-terminal domain of GPCRs (Fig. 11.17, pink-circled C in i4), appears to serve as a point for covalent attachment of a fatty acid (often palmitate). Presumably, fatty acid attachment stabilizes an interaction between the C-terminal domain of a GPCR and the membrane (Casey, 1995). The full consequences of this post-translational modification are not understood, because replacing the normally palmitoylated Cys with an amino acid that cannot be acylated appears to have little effect on receptor binding; however, G-protein coupling may not be as efficient (O'Dowd *et al.*, 1989).

GPCRs Can Physically Associate with Ionotropic Receptors

There is now good evidence that GPCRs and ionotropic receptors can interact directly with each other (Liu *et al.*, 2000). $GABA_A$ receptors (ionotropic) were shown to couple to dopamine (D_5) receptors (GPCR) through the second intracellular loop of the γ subunit of the $GABA_A$ receptor and the C-terminal domain of the D_5, but not the D_1, receptor. Dopamine binding to D_5 receptors produced down-regulation of $GABA_A$ currents, and pharmacologically blocking the $GABA_A$ receptor produced decreases in cAMP production when cells were stimulated with dopamine. It further appeared that ligand binding to both receptors was necessary for their stable interaction. Whether this form of receptor regulation is unique to this pair of partners or is a widespread phenomenon remains an open question ripe for further investigation.

GPCRs All Exhibit Similar Structures

The family of GPCRs exhibits structural similarities that permit the construction of "trees" describing the degree to which they are evolutionarily related (Fig. 11.21). Some remarkable relations become evident in such an analysis. For example, the D_1 and D_5 subtypes of dopamine receptors are more closely related to α2AR than to the D_2, D_3, and D_4 dopamine receptors. The similarities and differences among GPCR families are highlighted in the remainder of this chapter.

Muscarinic ACh Receptors

Muscarine is a naturally occurring plant alkaloid that binds to muscarinic subtypes of AChRs and activates them. mAChRs play a dominant role in mediating the actions of ACh in the brain, indirectly producing both excitation and inhibition through binding to a family of unique receptor subtypes. mAChRs are found both presynaptically and postsynaptically,

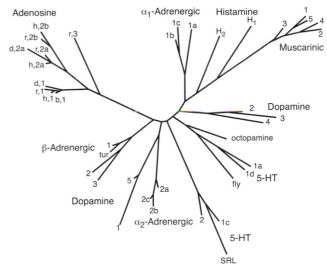

FIGURE 11.21 Evolutionary relationship of the metabotropic receptor family. To assemble this tree, sequence homologies in the transmembrane domains were compared for each receptor. Distance determines the degree of relatedness. r, rat; d, dog; h, human; tur, turkey; SRL, a putative serotonin receptor; 5-HT, 5-hydroxytryptamine (serotonin). Adapted from Linden, Chapter 21, Figure 2, "Basic Neurochemistry," 1994. Original tree construction was by William Pearson and Kevin Lynch, University of Virginia.

and ultimately, their main neuronal effects appear to be mediated through alterations in the properties of ion channels. Presynaptic mAChRs take part in important feedback loops that regulate neurotransmitter release. ACh released from the presynaptic terminal can bind to mAChRs on the same nerve ending, thus activating enzymatic processes that modulate subsequent neurotransmitter release. This modulation is typically an inhibition; however, activation of the m5 AChR produces an enhancement in subsequent release. These autoreceptors are an important regulatory mechanism for short-term (milliseconds to seconds) modulation of neurotransmitter release.

The family of mAChRs now includes five members (m1–m5), ranging from 55 to 70 kDa, and each of the five subtypes exhibits the typical architecture of seven TM domains. Much of the diversity in this family of receptors resides in the third intracellular loop (i3) responsible for the specificity of coupling to G proteins. The m1, m3, and m5 mAChRs couple predominantly to G proteins that activate the enzyme phospholipase C. The m2 and m4 receptors couple to G proteins that inhibit adenylyl cyclase, as well as to G proteins that directly regulate K^+ and Ca^{2+} channels. As is the case for other GPCRs, the domain near the N terminus of the third intracellular loop is important for the specificity of G-protein coupling. This domain is conserved in m1, m3, and m5 AChRs but is unique in m2 and m4. Several other important residues also have been identified for G-protein coupling. A particular Asp residue near the N terminus of the second intracellular loop (i2) is important for G-protein coupling, as are residues residing in the C-terminal region of the i3 loop.

The major mAChRs found in the brain are m1, m3, and m4, and each is diffusely distributed. The m2 subtype is the heart isoform and is not highly expressed in other organs. The genes for m4 and m5 lack introns, whereas those encoding m1, m2, and m3 contain introns, although little is known concerning alternatively spliced products of these receptors. Atropine is the most widely used antagonist for mAChR and binds to most subtypes, as does N-methylscopolamine. The antagonist pirenzipine appears to be relatively specific for the m1 mAChR, and other antagonists such as AF-DX116 and hexahydrosiladifenidol appear to be more selective for the m2 and m3 subtypes.

Adrenergic Receptors

The catecholamines epinephrine (adrenaline) and norepinephrine (noradrenaline) produce their effects by binding to and activating adrenergic receptors. Interestingly, epinephrine and norepinephrine can both bind to the same adrenergic receptor. Adrenergic receptors are currently separated into three families, $\alpha 1$, $\alpha 2$, and β. The $\alpha 1$ and $\alpha 2$ families are further subdivided into three subclasses each. Similarly, the β family also contains three subclasses ($\beta 1$, $\beta 2$, and $\beta 3$). The main adrenergic receptors in the brain are the $\alpha 2$ and $\beta 1$ subtypes. $\alpha 2$ARs have diverse roles, but the function that is best characterized (in both central and peripheral nervous tissue) is their role as autoreceptors (i.e., presynaptic receptors that bind transmitter and alter the release apparatus so that subsequent release is modulated, usually, in an inhibitory fashion). Different adrenergic receptor subtypes bind to G proteins that can alter the activity of phospholipase C, Ca^{2+} channels, and, probably best studied, adenylyl cyclase. For example, activation of $\alpha 2$ARs produces inhibition of adenylyl cyclase, whereas all βARs activate the cyclase.

Only a few agonists or antagonists cleanly distinguish the adrenergic receptor subtypes. One of them, isoproterenol, is an agonist that appears to be highly specific for βARs. Propranolol is the best-known antagonist for β receptors, and phentolamine is a good antagonist for α receptors but weakly binds at β receptors. The genomic organization of the different AR subtypes is unusual. Like many GPCRs, $\beta 1$ARs and $\beta 2$ARs are encoded by genes lacking introns. $\beta 3$ARs, which apparently have a role in lipolysis and are poorly characterized, are encoded by an intron-containing gene, as are αARs, providing an opportunity for alternative splicing as a means of introducing functional heterogeneity into the receptor.

Dopamine Receptors

Some 80% of the dopamine in the brain is localized to the corpus striatum, which receives major input from the substantia nigra and takes part in coordinating motor movements. Dopamine is also found diffusely throughout the cortex, where its specific functions remain largely undefined. However, many neuroleptic drugs appear to exert their effects by blocking dopamine binding, and imbalances in the dopaminergic system have long been associated with neuropsychiatric disorders.

Dopamine receptors are found both pre- and postsynaptically, and their structure is homologous to that of the receptors for other catecholamines (Bunzow et al., 1988; Civelli et al., 1993). Five subtypes of dopamine receptors can be grouped into two main classes, D_1-like and D_2-like receptors. D_1-like receptors include D_1 and D_5, and D_2-like receptors include D_2, D_3, and D_4 (see Fig. 11.21). The main distinction between these two classes is that D_1-like receptors activate adenylyl cyclase through interactions with G_s, whereas D_2-like receptors inhibit adenylyl cyclase and other effector molecules by interacting with G_i/G_o. D_1-like receptors are also slightly larger in molecular mass than D_2-like receptors. An additional point of interest, as noted previously, is that D_5 receptors selectively bind to $GABA_A$ receptors, impacting their function and vice versa (Liu et al., 2000). The deduced amino acid sequence for the entire family ranges from 387 amino acids (D_4) to 477 amino acids (D_5). The main structural differences between D_1-like and D_2-like receptors are that the intracellular loop between the sixth and seventh TM segments is larger in the D_2-like receptors, and the D_2-like receptors have smaller C-terminal intracellular segments. Two isoforms of the D_2 receptor have been isolated and are called D_2 long and D_2 short; alternative splicing generates these isoforms. D_2 long contains a 29-amino-acid insert in the large intracellular loop between the fifth and sixth membrane-spanning segments. Functional or anatomical differences have not yet been fully resolved for the short and long forms of D_2.

D_1-like receptors, like βARs, are transcribed from intronless genes (Dohlman et al., 1987). Conversely, all D_2-like receptors contain introns, thus providing for possibilities of alternatively spliced products. Post-translational modifications include glycosylation at one or more sites, disulfide bonding of the two Cys residues in e2 and e3, and acylation of the Cys residue in the C-terminal tail (analogous to the $\beta 2$AR).

The dopamine binding site includes two Ser residues in TM5 and an Asp residue in TM3, analogous to βAR.

Because of dopamine's presumed role in neuropsychiatric disorders, enormous effort has been focused on developing pharmacological tools for manipulating this system. Dopamine receptors bind amphetamines, bromocriptine, lisuride, clozapine, melperone, fluperlapine, and haloperidol. Because these drugs do not show great specificity for receptor subtypes, their usefulness for dissecting effects specifically related to binding to one or another dopamine receptor subtype is limited. However, their role in the treatment of human neuropsychiatric disorders is enormous.

Purinergic Receptors

Purinergic receptors bind to ATP (or other nucleotide analogs) or its breakdown product adenosine. Although ATP is a common constituent found within synaptic vesicles, adenosine is not and is therefore not considered a "classic" neurotransmitter. However, the multitude of receptors that bind and are activated by adenosine indicates that this molecule has important modulatory effects on the nervous system. Situations of high metabolic activity that consume ATP and situations of insufficient ATP-regenerating capacity can lead to the accumulation of adenosine. Because adenosine is permeable to membranes and can diffuse into and out of cells, a feedback loop is established in which adenosine can serve as a local diffusible signal that communicates the metabolic status of the neuron to surrounding cells and vice versa (Linden, 1994).

The original nomenclature describing purinergic receptors defined adenosine as binding to P1 receptors and ATP as binding to P2 receptors. Families of both P1 and P2 receptors have since been described, and adenosine receptors are now identified as A-type purinergic receptors, consisting of A_1, A_{2a}, A_{2b}, and A_3. ATP receptors are designated as P type and consist of P2x, P2y, P2z, P2t, and P2u. Recall that P2x and P2z subtypes are ionotropic receptors.

A-type receptors exhibit the classic arrangement of seven transmembrane-spanning segments but are typically shorter than most GPCRs, ranging in size between 35 and 46 kDa. The ligand binding site of A-type receptors is unique in that the ligand, adenosine, has no inherent charged moieties at physiological pH. A-type receptors appear to use His residues as their points of contact with adenosine, and, in particular, a His residue in TM7 is essential, because its mutation eliminates agonist binding. Other His residues in TM6 and TM7 are conserved in all A-type

receptors and may serve as other points of contact with agonists. Work with chimeric A-type receptors has further substantiated the importance of residues in TM5, TM6, and TM7 for ligand binding. A_1 receptors are highly expressed in the brain, and their activation down-regulates adenylyl cyclase and increases phospholipase C activity. A_{2a} and A_{2b} receptors are not as highly expressed in nervous tissue and are associated with the stimulation of adenylyl cyclase and phospholipase C, respectively. The A_3 subtype exhibits a unique pharmacological profile in that binding of xanthine derivatives, which blocks adenosine's action competitively, is absent. Very low levels of the A_3 receptor are found in brain and peripheral nervous tissue. The A_3 receptor appears to be coupled to the activation of phospholipase C.

The human A_1 receptor has a unique mode of receptor expression (Olah and Stiles, 1995). Introns in the 5' untranslated sequence of the mRNA, spliced in a tissue-specific manner, are capable of affecting the translational efficiency of the mRNA. Two extra start codons upstream from the start codon that initiates translation of the A_1 receptor exert a negative effect on translation. Mutating these two extra start codons can relieve the translational repression. This process is an effective way of controlling the level of receptor expression and may serve as a more general model for translational regulation for other mRNAs.

The P-type receptors, P2y, P2t, and P2u, are typical G-protein-linked GPCRs, mostly localized to the periphery. However, direct effects of ATP have been detected in neurons, and often the response is biphasic; an early excitatory effect followed, with its breakdown to adenosine, by a secondary inhibitory effect. Interestingly, P-type receptors exhibit a higher degree of homology to peptide-binding receptors than they do to the A-type purinergic receptors. As in A-type receptors, P-type receptors have a His residue in the third transmembrane domain; however, other sites for ligand binding have not been specifically identified.

Serotonin Receptors

Serotonin-containing cell bodies are found in the raphe nucleus in the brain stem and in nerve endings distributed diffusely throughout the brain (Julius, 1991). Serotonin has been implicated in sleep, modulation of circadian rhythms, eating, and arousal. Serotonin also has hormone-like effects when released in the bloodstream, regulating smooth muscle contraction and affecting the platelet-aggregating and immune systems.

Serotonin receptors are classified into four subtypes: 5-HT_1 to 5-HT_4, with a further subdivision of the 5-HT$_1$ subtypes. Recall that the 5-HT$_3$ receptor is ionotropic. The other 5-HT receptors exhibit the typical seven transmembrane-spanning segments, and all couple to G proteins to exert their effects. For example, 5-HT$_1$a, 5-HT$_1$b, 5-HT$_1$d, and 5-HT$_4$ either activate or inhibit adenylyl cyclase. 5-HT$_1$c and 5-HT$_2$ receptors preferentially stimulate activation of phospholipase C to produce increased intracellular levels of diacylglycerol and inositol 1,4,5-trisphosphate.

Serotonin receptors can also be grossly distributed into two groups on the basis of their gene structures. Both 5-HT$_1$c and 5-HT$_2$ are derived from genes that contain multiple introns. In contrast, similar to the βAR family, 5-HT$_1$ is coded by a gene lacking introns. Interestingly, 5-HT$_1$a is more closely related ancestrally to the βAR family than it is to other membranes of the serotonin receptor family and was originally isolated by using the cDNA for β2AR as a molecular probe (Kobilka *et al.*, 1987). This observation helps explain some pharmacological data suggesting that both 5-HT$_1$a and 5-HT$_1$b can bind certain adrenergic antagonists.

Glutamate GPCRs

The GPCRs that bind glutamate (metabotropic glutamate receptors, or mGluRs) are similar in general structure in having seven transmembrane-spanning segments to other GPCRs; however, they are divergent enough to have originated from a separate evolutionary-derived receptor family (Hollmann and Heinemann, 1994; Nakanishi, 1994). In fact, sequence homology between the mGluR family and the other GPCRs is minimal except for the GABA$_B$ receptor. The mGluR family is heterogeneous in size, ranging from 854 to 1179 amino acids. Both the N-terminal and C-terminal domains are unusually large for G-protein-coupled receptors. One great difference in the structures of mGluRs is that the binding site for glutamate resides in the large N-terminal extracellular domain and is homologous to a bacterial amino-acid-binding protein (Armstrong and Gouaux, 2000; O'Hara *et al.*, 1993). In most of the other families of GPCRs, the ligand-binding pocket is formed by the transmembrane segments partly buried in the membrane. In addition, mGluRs exist as functional dimers in the membrane in contrast to the single-subunit forms of most GPCRs (Kunishima *et al.*, 2000). These significant structural distinctions support the idea that mGluRs evolved separately from other GPCRs. The third intracellular loop, thought to be the major determinant responsible for G-protein coupling of mGluRs is relatively small, whereas the C-terminal domain is quite large. The coupling between mGluRs and their respective G proteins may be through

unique determinants that exist in the large C-terminal domain.

Currently, eight different mGluRs can be subdivided into three groups on the basis of sequence homologies and their capacity to couple to specific enzyme systems. Both mGluR1 and mGluR5 activate a G protein coupled to phospholipase C. mGluR1 activation can also lead to the production of cAMP and arachidonic acid, by coupling to G proteins that activate adenylyl cyclase and phospholipase A2 (Aramori and Nakanishi, 1992). mGluR5 seems more specific, activating predominantly the G-protein-activated phospholipase C.

The other six mGluR subtypes are distinct from one another in favoring either *trans*-1-aminocyclopentane-1,3-dicarboxylate (mGluR$_2$, mGluR$_3$, and mGluR$_8$) or 1,2-amino-4-phosphonobutyrate (mGluR$_4$, mGluR$_6$, and mGluR$_7$) as agonists for activation. mGluR$_2$ and mGluR$_4$ can be further distinguished pharmacologically by using the agonist 2-(carboxycyclopropyl)glycine, which is more potent at activating mGluR$_2$ receptors (Hayashi *et al.*, 1992). Less is known about the mechanisms by which these receptors produce intracellular responses; however, one effect is to inhibit the production of cAMP by activating an inhibitory G protein.

mGluRs are widespread in the nervous system and are found both pre- and postsynaptically. Presynaptically, they serve as autoreceptors and appear to participate in the inhibition of neurotransmitter release. Their postsynaptic roles appear to be quite varied and depend on the specific G protein to which they are coupled. mGluR$_1$ activation has been implicated in long-term synaptic plasticity at many sites in the brain, including long-term potentiation in the hippocampus and long-term depression in the cerebellum (see Chapter 19).

GABA$_B$ Receptors

GABA$_B$ receptors are found throughout the nervous system, where they are sometimes colocalized with ionotropic GABA$_A$ receptors. GABA$_B$ receptors are present both pre- and postsynaptically. Presynaptically, they appear to mediate inhibition of neurotransmitter release through an autoreceptor-like mechanism by activating K$^+$ conductances and diminishing Ca^{2+} conductances. In addition, GABA$_B$ receptors may affect K$^+$ channels through a direct physical coupling to the K$^+$ channel, not mediated through a G-protein intermediate. Postsynaptically, GABA$_B$ receptor activation produces a characteristic slow hyperpolarization (termed the *slow inhibitory postsynaptic potential*) through the activation of a K$^+$ conductance. This effect appears to be through a pertussis-toxin-sensitive G protein that inhibits adenylyl cyclase.

The cloning of the GABA$_B$ receptor (GABA$_B$R$_1$) revealed that it has high sequence homology to the family of glutamate GPCRs but shows little similarity to other G-protein-coupled receptors. The large N-terminal extracellular domain of the GABA$_B$ receptor is the presumed site of GABA binding. With the exception of this large extracellular domain, the GABA$_B$ receptor structure is typical of the GPCR family, exhibiting seven TM domains. The initial cloning of the GABA$_B$ receptor was made possible by the development of a high-affinity, high-specificity antagonist termed *CGP64213*. This antagonist is several orders of magnitude more potent at inhibiting GABA$_B$ receptor function than the more widely known antagonist saclofen. Baclofen, an analog of saclofen, remains the best agonist for activating GABA$_B$ receptors.

Functional GABA$_B$ receptors appear to exist primarily as dimers in the membrane (Jones *et al.*, 1998; Kaupmann *et al.*, 1998; White *et al.*, 1998). Expression of the cloned GABA$_B$R$_1$ isoform does not produce significant functional receptors. However, when coexpressed with the GABA$_B$R$_2$ isoform, receptors that are indistinguishable functionally and pharmacologically from those in brain were produced. In addition, GABA$_B$ dimers exist in neuronal membranes, and all data point to the conclusion that GABA$_B$ receptors dimerize and that the dimer is the functionally important form of the receptor. As noted earlier in this chapter (Fig. 11.18B), GPCRs can interact with themselves and other receptors. It is good to keep in mind that these types of direct receptor interactions may be more widespread than currently appreciated.

Peptide Receptors

Neuropeptide receptors are an immense family. Because of their diversity, they cannot be covered in detail in this chapter. Despite their diversity, none of the receptors that bind peptides appear to be coupled directly to the opening of ion channels. Neuropeptide receptors exert their effects either through the typical pathway of activation of G proteins or through a pathway related to activation of an intrinsic tyrosine kinase activity associated with the receptor.

The peptide-binding domain of neuropeptide receptors includes residues in both the large N-terminal extracellular domain and the transmembrane domain (Strader *et al.*, 1995). These additional stabilizing contacts presumably provide the receptors with their remarkably high affinity for neuropeptides (in the nanomolar concentration range). For example, residues in the first and second extracellular domains, as well as those in at least four of the TM domains of the NK1 neurokinin receptor, interact with substance P to form stabilizing contacts. Many small-molecule antagonists are known to inhibit activation of the NK1 neurokinin receptor, and these antagonists bind to some, but not all, of the same amino acids in the TM segments as does substance P. The possible mechanisms for inhibition of the peptide receptors range from complete structural overlap between agonist and antagonist binding to complete allosteric exclusion (Strader *et al.*, 1995). Knowledge of the activated structure of the neuropeptide receptors provides remarkable opportunities for future drug design.

Summary

GPCRs are single polypeptides composed of seven transmembrane-spanning segments. In general, the binding site for neurotransmitter is located within the core of the circular structure formed by these segments. Transmitter binding produces conformational changes in the receptor that expose parts of the i3 region, among others, for binding to G proteins. G-protein binding increases the affinity of the receptor for transmitter. Desensitization is common among GPCRs and leads to decreased response of the receptor to neurotransmitter by several distinct mechanisms. mGluRs are structurally distinct from other GPCRs; mGluRs have large N-terminal extracellular domains that form the binding site for glutamate. Otherwise, the basic structure of mGluRs appears to be similar to that of the rest of the GPCR family.

References

Aramori, I., and Nakanishi, S. (1992). Signal transduction and pharmacological characteristics of a metabotropic glutamate receptor, mGluR1, in transfected CHO cells. *Neuron* **8**, 757–765.

Armstrong, N., and Gouaux, E. (2000). Mechanisms for activation and antagonism of an AMPA-sensitive glutamate receptor: Crystal structures of the GluR2 ligand binding core. *Neuron* **28**, 165–181.

Ascher, P., and Nowak, L. (1988). The role of divalent cations in the N-methyl-D-aspartate responses of mouse central neurones in culture. *J. Physiol.* **399**, 247–266.

Barak, L. S., Tiberi, M., Freedman, N. J., Kwatra, M. M., Lefkowitz, R. J., and Caron, M. G. (1994). A highly conserved tyrosine residue in G protein-coupled receptors is required for agonist-mediated beta 2-adrenergic receptor sequestration. *J. Biol. Chem.* **269**, 2790–2795.

Barnard, E. A., Darlison, M. G., and Seeburg, P. (1987). Molecular biology of the GABAA receptor: The receptor/channel superfamily. *Trends Neurosci.* **10**, 502–509.

Bertrand, D., Galzi, J. L., Devillers-Thiery, A., Bertrand, S., and Changeux, J. P. (1993). Mutations at two distinct sites within the channel domain M2 alter calcium permeability of neuronal alpha 7 nicotinic receptor. *Proc. Natl. Acad. Sci. USA* **90**, 6971–6975.

Betz, H. (1991). Glycine receptors: Heterogeneous and widespread in the mammalian brain. *Trends Neurosci.* **14**, 458–461.

Blackstone, C. D., Moss, S. J., Martin, L. J., Levey, A. I., Price, D. L., and Huganir, R. L. (1992). Biochemical characterization and localization of a non-N-methyl-D-aspartate glutamate receptor in rat brain. *J. Neurochem.* **58**, 1118–1126.

Bormann, J. (1988). Electrophysiology of GABA_A and GABA_B receptor subtypes. *Trends Neurosci.* **11**, 112–116.

Bormann, J., and Fiegenspan, A. (1995). GABA_C receptors. *Trends Neurosci.* **18**, 515–519.

Boulter, J., Hollmann, M., O'Shea-Greenfield, A., Hartley, M., Deneris, E., Maron, C., and Heinemann, S. (1990). Molecular cloning and functional expression of glutamate receptor subunit genes. *Science* **249**, 1033–1037.

Bouvier, M., Collins, S., O'Dowd, B. F., Campbell, P. T., de Blasi, A., Kobilka, B. K., MacGregor, C., Irons, G. P., Caron, M. G., and Lefkowitz, R. J. (1989). Two distinct pathways for cAMP-mediated down-regulation of the beta 2-adrenergic receptor: Phosphorylation of the receptor and regulation of its mRNA level. *J. Biol. Chem.* **264**, 16786–16792.

Brake, A. J., Wagenbach, M. J., and Julius, D. (1994). New structural motif for ligand-gated ion channels defined by an ionotropic ATP receptor. *Nature* **371**, 519–523.

Bunzow, J. R., Van Tol, H. H., Grandy, D. K., Albert, P., Salon, J., Christie, M., Machida, C. A., Neve, K. A., and Civelli, O. (1988). Cloning and expression of a rat D2 dopamine receptor cDNA [see comments]. *Nature* **336**, 783–787.

Burnashev, N., Schoepfer, R., Monyer, H., Ruppersberg, J. P., Gunther, W., Seeburg, P. H., and Sakmann, B. (1992). Control by asparagine residues of calcium permeability and magnesium blockade in the NMDA receptor. *Science* **257**, 1415–1419.

Casey, P. J. (1995). Protein lipidation in cell signaling. *Science* **268**, 221–225.

Cassel, D., and Selinger, Z. (1977). Mechanism of adenylate cyclase activation by cholera toxin: Inhibition of GTP hydrolysis at the regulatory site. *Proc. Natl. Acad. Sci. USA* **74**, 3307–3311.

Changeux, J. P., Devillers-Thiery, A., and Chemouilli, P. (1984). Acetylcholine receptor: An allosteric protein. *Science* **225**, 1335–1345.

Civelli, O., Bunzow, J. R., and Grandy, D. K. (1993). Molecular diversity of the dopamine receptors. *Annu. Rev. Pharmacol. Toxicol.* **33**, 281–307.

Clark, R. B., Knoll, B. J., and Barber, R. (1999). Partial agonists and G protein-coupled receptor desensitization. *Trends Pharmacol. Sci.* **20**, 279–286.

Colquhoun, D., and Sakmann, B. (1985). Fast events in single-channel currents activated by acetylcholine and its analogues at the frog muscle end-plate. *J. Physiol. (London)* **369**, 501–557.

Colquhoun, D., and Ogden, D. C. (1988). Activation of ion channels in the frog end-plate by high concentrations of acetylcholine. *J. Physiol.* **395**, 131–159.

Dohlman, H. G., Caron, M. G., and Lefkowitz, R. J. (1987). A family of receptors coupled to guanine nucleotide regulatory proteins. *Biochemistry* **19/26**, 2657–2664.

Ehlers, M. D., Whittemore, G. T., and Huganir, R. L. (1995). Regulated subcellular distribution of the NR1 subunit of the NMDA receptor. *Science* **269**, 1734–1737.

Ehlers, M. D., Zhang, S., Bernhardt, J. P., and Huganir, R. L. (1996). Inactivation of NMDA receptors by direct interaction of calmodulin with the NR1 subunit. *Cell* **84**, 745–755.

Ferguson, S. S. G., Menard, L., Barak, L. S., Koch, W. J., Colapietro, A.-M., and Caron, M. G. (1995). Role of phosphorylation in agonist-promoted gb2-adrenergic receptor sequestration. *J. Biol. Chem.* **270**, 24782–24789.

Ferguson, S. S. G., Downey, W. E., Colapietro, A.-M., Barak, L. S., Menard, L., and Caron, M. G. (1996). Role of arrestin in mediating agonist-promoted G-protein coupled receptor internalization. *Science* **271**, 363–366.

Fraser, C. M., Chung, F. Z., Wang, C. D., and Venter, J. C. (1988). Site-directed mutagenesis of human beta-adrenergic receptors: Substitution of aspartic acid-130 by asparagine produces a receptor with high-affinity agonist binding that is uncoupled from adenylate cyclase. *Proc. Natl. Acad. Sci. USA* **85**, 5478–5482.

Fraser, C. M., Wang, C. D., Robinson, D. A., Gocayne, J. D., and Venter, J. C. (1989). Site-directed mutagenesis of m1 muscarinic acetylcholine receptors: Conserved aspartic acids play important roles in receptor function. *Mol. Pharmacol.* **36**, 840–847.

Green, W. N., and Claudio, T. (1993). Acetylcholine receptor assembly: Subunit folding and oligomerization occur sequentially. *Cell (Cambridge, Mass.)* **74**, 57–69.

Gu, Y., Forsayeth, J. R., Verrall, S., Yu, X. M., and Hall, Z. W. (1991). Assembly of the mammalian muscle acetylcholine receptor in transfected COS cells. *J. Cell Biol.* **114**, 799–807.

Hayashi, Y., Tanabe, Y., Aramori, I., Masu, M., Shimamoto, K., Ohfune, Y., and Nakanishi, S. (1992). Agonist analysis of 2-(carboxycyclopropyl)glycine isomers for cloned metabotropic glutamate receptor subtypes expressed in Chinese hamster ovary cells. *Br. J. Pharmacol.* **107**, 539–543.

Henderson, R., Baldwin, J. M., Ceska, T. A., Zemlin, F., Beckmann, E., and Downing, K. H. (1990). Model for the structure of bacteriorhodopsin based on high-resolution electron cryo-microscopy. *J. Mol. Biol.* **213**, 899–929.

Herb, A., Burnashev, N., Werner, P., Sakmann, B., Wisden, W., and Seeburg, P. H. (1992). The KA-2 subunit of excitatory amino acid receptors shows widespread. *Neuron* **8**, 775–785.

Hille, B. (1992). "Ionic Channels of Excitable Membranes." Sinauer, Sunderland, MA.

Hollmann, M., and Heinemann, S. (1994). Cloned glutamate receptors. *Annu. Rev. Neurosci.* **17**, 31–108.

Hollmann, M., O'Shea-Greenfield, A., Rogers, S. W., and Heinemann, S. (1989). Cloning by functional expression of a member of the glutamate receptor family. *Nature* **342**, 643–648.

Huganir, R. L., and Greengard, P. (1990). Regulation of neurotransmitter receptor desensitization by protein phosphorylation. *Neuron* **5**, 555–567.

Hume, R. I., Dingledine, R., and Heinemann, S. F. (1991). Identification of a site in glutamate receptor subunits that controls calcium permeability. *Science* **253**, 1028–1031.

Iino, M., Ozawa, S., and Tsuzuki, K. (1990). Permeation of calcium through excitatory amino acid receptor channels in cultured hippocampal neurones. *J. Physiol.* **424**, 151–165.

Imoto, K., Busch, C., Sakmann, B., Mishina, M., Konno, T., Nakai, J., Bujo, H., Mori, Y., Fukuda, K., and Numa, S. (1988). Rings of negatively charged amino acids determine the acetylcholine receptor channel conductance. *Nature* **335**, 645–648.

Inglese, J., Freedman, N. J., Koch, W. J., and Lefkowitz, R. J. (1993). Structure and mechanism of the G protein-coupled receptor kinases. *J. Biol. Chem.* **268**, 23735–23738.

Jones, K. A., Borowsky, B., Tamm, J. A., Craig, D. A., Durkin, M. M., Dai, M., Yao, W. J., Johnson, M., Gunwaldsen, C., Huang, L. Y., Tang, C., Shen, Q., Salon, J. A., Morse, K., Laz, T., Smith, K. E., Nagarathnam, D., Noble, S. A., Branchek, T. A., and Gerald, C. (1998). GAGA_B receptors function as a heteromeric assembly of the subunits GABA_BR1 and GABA_BR2. *Nature* **396**, 674–679.

Julius, D. (1991). Molecular biology of serotonin receptors. *Annu. Rev. Neurosci.* **14**, 335–360.

Kandel, E. R., Schwartz, J. H., and Jessell, T. M. (1991). "Principles of Neural Science," 3rd ed. Elsevier, New York/Amsterdam.

Karlin, A. (1993). Structure of nicotinic acetylcholine receptors. *Curr. Opin. Neurobiol.* **3**, 299–309.

Kaupmann, K., Malitschek, B., Schuler, V., Heid, J., Froestl, W., Beck, P., Mosbacher, J., Bischoff, S., Kulik, A., Shigemoto, R., Karschin, A., and Bettler, B. (1998). GABA$_B$-receptor subtypes assemble into functional heteromeric complexes. *Nature* **396**, 683–687.

Keinanen, K., Wisden, W., Sommer, B., Werner, P., Herb, A., Verdoorn, T. A., Sakmann, B., and Seeburg, P. H. (1990). A family of AMPA-selective glutamate receptors. *Science* **249**, 556–560.

Kobilka, B. (1992). Adrenergic receptors as models for G protein-coupled receptors. *Annu. Rev. Neurosci.* **15**, 87–114.

Kobilka, B. K., Frielle, T., Collins, S., Yang-Feng, T., Kobilka, T. S., Francke, U., Lefkowitz, R. J., and Caron, M. G. (1987). An intronless gene encoding a potential member of the family of receptors coupled to guanine nucleotide regulatory proteins. *Nature* **329**, 75–79.

Komau, H.-C., Schenker, L. T., Kennedy, M. B., and Seeburg, P. H. (1995). Domain interaction between NMDA receptor subunits and the postsynaptic density protein PSD-95. *Science* **269**, 1737–1740.

Kunishima, N., Shimada, Y., Tsuji, Y., Sato, T., Yamamoto, M., Kumasaka, T., Nakanishi, S., Jingami, H., and Morikawa, K. (2000). Structural basis of glutamate recognition by a dimeric metabotropic glutamate receptor. *Nature* **407**, 971–977.

Kutsuwada, T., Kashiwabuchi, N., Mori, H., Sakimura, K., Kushiya, E., Araki, K., Meguro, H., Masaki, H., Kumanishi, T., Arakawa, M., and Mishina, M. (1992). Molecular diversity of the NMDA receptor channel [see comments]. *Nature* **358**, 36–41.

Lee, S. P., O'Dowd, B. F., Ng, G. Y. K., Varghese, G., Akil, H., Mansour, A., Nguyen, T., and George, S. R. (2000). Inhibition of cell surface expression by mutant receptors demonstrates that D2 dopamine receptors exist as oligomers in the cell. *Mol. Pharmacol.* **58**, 120–128.

Linden, J. (1994). *In* "Basic Neurochemistry" (G. J. Siegel, B. W. Agranoff, R. W. Albers, and P. B. Molinoff, Eds.), pp. 401–416, Raven Press, New York.

Liu, F., Wan, Q., Pristupa, Z. B., Yu, X. -M., Want, Y. T., and Niznik, H. B. (2000). Direct protein–protein coupling enables cross-talk between dopamine D5 and g-aminobutyric acid A receptors. *Nature* **403**, 274–278.

Lomeli, H., Mosbacher, J., Melcher, T., Hoger, T., Geiger, J. R., Kuner, T., Monyer, H., Higuchi, M., Bach, A., and Seeburg, P. H. (1994). Control of kinetic properties of AMPA receptor channels by nuclear RNA editing. *Science* **266**, 1709–1713.

Maggio, R., Vogel, Z., and Wess, J. (1993). Coexpression studies with mutant muscarinic/adrenergic receptors provide evidence for intermolecular "cross-talk" between G-protein coupled receptors. *Proc. Natl. Acad. Sci. USA* **90**, 3103–3107.

Mano, I., and Teichberg, V.I. (1998). A tetrameric subunit stoichiometry for a glutamate receptor–channel complex. *NeuroReport* **26**, 327–331.

Maricq, A. V., Peterson, A. S., Brake, A. J., Myers, R. M., and Julius, D. (1991). Primary structure and functional expression of the 5HT3 receptor, a serotonin-gated ion channel. *Science* **254**, 432–437.

Mayer, M. L., and Westbrook, G. L. (1987). Permeation and block of N-methyl-D-aspartic acid receptor channels by divalent cations in mouse cultured central neurones. *J. Physiol.* **394**, 501–527.

McGehee, D. S., Heath, M. J., Gelber, S., Devay, P. and Role, L. W. (1995). Nicotine enhancement of fast excitatory synaptic transmission in CNS by presynaptic receptors [see comments]. *Science* **269**, 1692–1696.

Meguro, H., Mori, H., Araki, K., Kushiya, E., Kutsuwada, T., Yamazaki, M., Kumanishi, T., Arakawa, M., Sakimura, K., and Mishina, M. (1992). Functional characterization of a heteromeric NMDA receptor channel expressed from cloned cDNAs. *Nature* **357**, 70–74.

Mishina, M., Kurosaki, T., Tobimatsu, T., Morimoto, Y., Noda, M., Yamamoto, T., Terao, M., Lindstrom, J., Takahashi, T., Kuno, M., and Numa, S. (1984). Expression of functional acetylcholine receptor from cloned cDNAs. *Nature* **307**, 604–608.

Miyazawa, A., Fujiyoshi, Y., Stowell, M., and Unwin, N. (1999). Nicotinic acetylcholine receptor at 4.6A resolution: Transverse tunnels in the channel wall. *J. Mol. Biol.* **288**, 765–786.

Mizobe, T., Maze, M., Lam, V., Suryanarayana, S., and Kobilka, B. K. (1996). Arrangement of transmembrane domains in adrenergic receptors: Similarity to bacteriorhodopsin. *J. Biol. Chem.* **271**, 2387–2389.

Monyer, H., Sprengel, R., Schoepfer, R., Herb, A., Higuchi, M., Lomeli, H., Burnashev, N., Sakmann, B., and Seeburg, P. H. (1992). Heteromeric NMDA receptors: Molecular and functional distinction of subtypes. *Science* **256**, 1217–1221.

Mori, H., Masaki, H., Yamakura, T., and Mishina, M. (1992). Identification by mutagenesis of a Mg(2+)-block site of the NMDA receptor channel. *Nature* **358**, 673–675.

Moriyoshi, K., Masu, M., Ishii, T., Shigemoto, R., Mizuno, N., and Nakanishi, S. (1991). Molecular cloning and characterization of the rat NMDA receptor. *Nature* **354**, 31–37.

Moss, S.J., and Smart, T.G. (2001). Constructing inhibitory synapses. *Nature Rev. Neurosci.* **2**, 240–250.

Nakagawa, N., Cheng, Y., Ramm, E., Sheng, M., and Walz, T. (2005). Structure and different conformational states of native AMPA receptor complexes. *Nature* **433**, 545–549.

Nakanishi, N., Shneider, N. A., and Axel, R. (1990). A family of glutamate receptor genes: Evidence for the formation of heteromultimeric receptors with distinct channel properties. *Neuron* **5**, 569–581.

Nakanishi, S. (1994). Metabotropic glutamate receptors: Synaptic transmission, modulation, and plasticity. *Neuron* **13**, 1031–1037.

Nicoll, R.A., Tomita, S., and Bredt, D.S. (2006). Auxiliary subunits assist AMPA-type glutamate receptors. *Science* **311**, 1253–1256.

Noda, M., Takahashi, H., Tanabe, T., Toyosato, M., Furutani, Y., Hirose, T., Asai, M., Inayama, S., Miyata, T., and Numa, S. (1982). Primary structure of alpha-subunit precursor of *Torpedo californica* acetylcholine receptor deduced from cDNA sequence. *Nature* **299**, 793–797.

Noda, M., Takahashi, H., Tanabe, T., Toyosato, M., Kikyotani, S., Furtani, Y., Hirose, T., Takashima, H., Inayama, S., Miyata, T., and Numa, S. (1983). Structural homology of *Torpedo californica* acetylcholine receptor subunits. *Nature* **302**, 528–532.

O'Dowd, B. F., Hnatowich, M., Caron, M. G., Lefkowitz, R. J., and Bouvier, M. (1989). Palmitoylation of the human beta 2-adrenergic receptor: Mutation of Cys341 in the carboxyl tail leads to an uncoupled nonpalmitoylated form of the receptor. *J. Biol. Chem.* **264**, 7564–7569.

O'Hara, P. J., Sheppard, P. O., Thogersen, H., Venezia, D., Haldeman, B. A., McGrane, V., Houamed, K. M., Thomsen, C., Gilbert, T. L., and Mulvihill, E. R. (1993). The ligand-binding domain in metabotropic glutamate receptors is related to bacterial periplasmic binding proteins. *Neuron* **11**, 41–52.

Olah, M. E., and Stiles, G. L. (1995). Adenosine receptor subtypes: Characterization and therapeutic regulation. *Annu. Rev. Pharmacol. Toxicol.* **35**, 581–606.

Ortell, M. O., and Lunt, G. G. (1995). Evolutionary history of the ligand-gated ion-channel superfamily of receptors. *Trends Neurosci.* **18**, 121–128.

Overton, M. C., and Blumer, K. J. (2000). G-protein-coupled receptors function as oligomers *in vivo*. *Curr. Biol.* **10**, 341–344.

Palczewski, K., Kumasaka, T., Hori, T., Behnke, C. A., Motoshima, H., Fox, B. A., Le Trong, I. L., Teller, D. C., Okada, T., Stenkamp, R. E., Yamamoto, M., and Miyano, M. (2000). Crystal structure of rhodopsin: A G protein-coupled receptor. *Science* **289**, 739–745.

Paulson, H. L., Ross, A. F., Green, W. N., and Claudio, T. (1991). Analysis of early events in acetylcholine receptor assembly. *J. Cell Biol.* **113**, 1371–1384.

Perez, D. M., Hwa, J., Gaivin, R., Manjula, M., Brown, F., and Graham, R. M. (1996). Constitutive activation of a single effector pathway: Evidence for multiple activation states of a G protein-coupled receptor. *Mol. Pharmacol.* **49**, 112–122.

Premont, R. T., Inglese, J., and Lefkowitz, R. J. (1995). Protein kinases that phosphorylate activated G protein-coupled receptors. *FASEB J.* **9**, 175–182.

Pritchett, D. B., Sontheimer, H., Shivers, B. D., Ymer, S., Kettenmann, H., Schofield, P. R., and Seeburg, P. H. (1989). Importance of a novel GABAA receptor subunit for benzodiazepine pharmacology. *Nature* **338**, 582–585.

Proll, M. A., Clark, R. B., Goka, T. J., Barber, R., and Butcher, R. W. (1992). Adrenergic receptor levels and function after growth of S49 lymphoma cells in low concentrations of epinephrine. *Mol. Pharmacol.* **42**, 116–122.

Rosenmund, C., Stern-Bach, Y., and Stevens, C.F. (1998). The tetrameric structure of a glutamate receptor channel. *Science* **280**, 1596–1599.

Rudolph, U., Crestani, F., Benke, D., Brunig, I., Benson, J. A., Fritschy, J.-M., Martin, J. R., Bluethmann, H., and Mohler, H. (1999). Benzodiazepine actions mediated by specific g-aminobutyric acid$_A$ receptor subtypes. *Nature* **401**, 796.

Salahpour, A., Angers, S., and Bouvier, M. (2000). Functional significance of oligomerization of G-protein-coupled receptors. *Trends Endocrinol. Metab.* **11**, 163–168.

Salyed, T., Paarmann, I., Schmitt, B., Haeger, S., Sola, M., Schmalzing, G.M., Weissenhorn, W., and Betz, H. (2007). Molecular basis of gephyrin clustering at inhibitory synapses. *J. Biol. Chem.* **282**, 5625–5632.

Sanes, J.R. (1997). Genetic analysis of postsynaptic differentiation at the vertebrate neuromuscular junction. *Curr. Op. Neurobiol.* **7**, 93–100.

Seguela, P., Wadiche, J., Dineley-Miller, K., Dani, J. A., and Patrick, J. W. (1993). Molecular cloning, functional properties, and distribution of rat brain alpha 7: A nicotinic cation channel highly permeable to calcium. *J. Neurosci.* **13**, 596–604.

Seighart, W. (1992). GABAA receptors: Ligand-gated Cl- ion channels modulated by multiple drug-binding sites. *Trends Pharmacol. Sci.* **13**, 446–450.

Sheng, M., and Hoogenraad, C. C. (2007). The postsynaptic architecture of excitatory synapses: A more quantitative view. *Ann. Rev. Biochem.* **76**, 823–847.

Sommer, B., Keinanen, K., Verdoorn, T. A., Wisden, W., Burnashev, N., Herb, A., Kohler, M., Takagi, T., Sakmann, B., and Seeburg, P. H. (1990). Flip and flop: A cell-specific functional switch in glutamate-operated channels of the CNS. *Science* **249**, 1580–1585.

Sommer, B., Kohler, M., Sprengel, R., and Seeburg, P. H. (1991). RNA editing in brain controls a determinant of ion flow in glutamate-gated channels. *Cell* **67**, 11–19.

Sterne-Marr, R., and Benovic, J. L. (1995). Regulation of G protein-coupled receptors by receptor kinases and arrestins. *Vitam. Horm. (N.Y.)* **51**, 193–234.

Strader, C. D., Fong, T. M., Tota, M. R., Underwood, D., and Dixon, R. A. (1994). Structure and function of G protein-coupled receptors. *Annu. Rev. Biochem.* **63**, 101–132.

Strader, C. D., Fong, T. M., Graziano, M. P., and Tota, M. R. (1995). The family of G-protein-coupled receptors. *FASEB J.* **9**, 745–754.

Strosberg, A. D. (1990). Biotechnology of beta-adrenergic receptors. *Mol. Neurobiol.* **4**, 211–250.

Tolkovsky, A. M., Braun, S., and Levitzki, A. (1982). Kinetics of interaction between receptors, GTP protein, and the catalytic unit of turkey erythrocyte adenylate cyclase. *Proc. Natl. Acad. Sci. USA* **79**, 213–217.

Unwin, N. (1993a). Neurotransmitter action: Opening of ligand-gated ion channels. *Cell* 7(Suppl.), 31–41.

Unwin, N. (1993b). Nicotinic acetylcholine receptor at 9 A resolution. *J. Mol. Biol.* **229**, 1101–1124.

Unwin, N. (1995). Acetylcholine receptor channel imaged in the open state. *Nature* **373**, 37–43.

Valera, S., Hussy, N., Evans, R. J., Adami, N., North, R. A., Surprenant, A., and Buell, G. (1994). A new class of ligand-gated ion channel defined by P2x receptor for extracellular ATP [see comments]. *Nature* **371**, 516–519.

Verdoorn, T. A., Burnashev, N., Monyer, H., Seeburg, P. H., and Sakmann, B. (1991). Structural determinants of ion flow through recombinant glutamate receptor channels. *Science* **252**, 1715–1718.

Vernino, S., Amador, M., Luetje, C. W., Patrick, J., and Dani, J. A. (1992). Calcium modulation and high calcium permeability of neuronal nicotinic acetylcholine receptors. *Neuron* **8**, 127–134.

Watkins, J. C., Krogsgaard-Larsen, P., and Honore, T. (1990). Structure–activity relationships in the development of excitatory amino acid receptor agonists and competitive antagonists. *Trends Pharmacol. Sci.* **11**, 25–33.

Wenthold, R. J., Yokotani, N., Doi, K., and Wada, K. (1992). Immunochemical characterization of the non-NMDA glutamate receptor using subunit-specific antibodies: Evidence for a hetero-oligomeric structure in rat brain. *J. Biol. Chem.* **267**, 501–507.

White, J. H., Wise, A., Main, M. J., Green, A., Fraser, N. J., Disney, G. H., Barnes, A. A., Emson, P., Foord, S. M., and Marshall, F. H. (1998). Heterodimerization is required for the formation of a functional GABA$_B$ receptor. *Nature* **396**, 679–682.

Willman, R. and Fuhrer, C. (2002). Neuromuscular synaptogenesis: Clustering of acetylcholine receptors revisited. *Cell. Mol. Life Sci.* **59**, 1296–1316.

Wo, Z. G., and Oswald, R. E. (1995). Unraveling the modular design of glutamate-gated ion channels. *Trends Neurosci.* **18**, 161–168.

Yakel, J. L., Lagrutta, A., Adelman, J. P., and North, R. A. (1993). Single amino acid substitution affects desensitization of the 5-hydroxytryptamine type 3 receptor expressed in *Xenopus* oocytes. *Proc. Natl. Acad. Sci. USA* **90**, 5030–5033.

Young, A. B., and Snyder, S. H. (1974). The glycine synaptic receptor: Evidence that strychnine binding is associated with the ionic conductance mechanism. *Proc. Natl. Acad. Sci. USA* **71**, 4002–4005.

Yuan, N., Friedman, J., Whaley, B. S., and Clark, R. B. (1994). cAMP-dependent protein kinase and protein kinase C consensus site mutations of the adrenergic receptor. *J. Biol. Chem.* **269**, 23,032–23,038.

Ziff, E. B. (2007). TARPs and the AMPA receptor trafficking paradox. *Neuron* **53**, 627–633.

Intracellular Signaling

Howard Schulman

Almost all aspects of neuronal function, from its maturation during development, to its growth and survival, cytoskeletal organization, gene expression, neurotransmission, and use-dependent modulation, are dependent on intracellular signaling initiated at the cell surface. The response of neurons and glia to neurotransmitters, growth factors, and other signaling molecules is determined by their complement of expressed receptors and pathways that transduce and transmit these signals to intracellular compartments, and by the enzymes, ion channels, and cytoskeletal proteins that ultimately mediate the effects of the neurotransmitters. The molecules involved in signal transmission and transduction are highly represented in mammalian and invertebrate genomes. Individual neuronal responses are further determined by the concentration and localization of signal transduction components and modified by the prior history of neuronal activity. Several primary classes of signaling systems, operating at different time courses, provide great flexibility for intercellular communication. One class comprises ligand-gated ion channels, such as the nicotinic receptor considered in Chapter 11. This class of signaling provides fast transmission (see Chapter 16) that is activated and deactivated within 10 ms. It forms the underlying "hard wiring" of the nervous system that makes rapid multisynaptic computations possible. A second class consists of receptor tyrosine kinases, which typically respond to growth factors and to trophic factors and produce major changes in the growth, differentiation, or survival, of neurons (Chapter 10). A third and largest class utilizes G-protein-linked signals in a multistep process that slows the response from 100–300 ms to many minutes (see Chapter 16). The relatively slow speed is offset, however, by a richness in the diversity

of its modulation and its inherent capacity for amplification and plasticity. The initial steps in this signaling system typically generate a second messenger inside the cell, and this second messenger then activates a number of proteins, including protein kinases that modify cellular processes. Signal transduction also modulates the level of transcription of genes, which determine the differentiated and functional state of cells.

SIGNALING THROUGH G-PROTEIN-LINKED RECEPTORS

Signal transduction through G-protein-linked receptors requires three membrane-bound components: (1) a cell surface receptor that determines to which signal the cell can respond; (2) a G protein on the intracellular side of the membrane that is stimulated by the activated receptor; and (3) either an effector enzyme that changes the level of a second messenger or an effector channel that changes ionic fluxes in the cell in response to the activated G protein. The human genome encodes for more than 800 receptors for catecholamines, odorants, neuropeptides, and light that couple to one or more of the 16 identified G proteins. These, in turn, regulate one or more of more than two dozen different effector channels and enzymes. The key feature of this information flow is the ability of G proteins to detect the presence of activated receptors and amplify the signal by altering the activity of appropriate effector enzymes and channels.

G proteins are GTP-binding proteins that couple the activation of seven-helix receptors by neurotransmitters at the cell surface to changes in the activity of effector enzymes and effector channels. A common effector enzyme is adenylyl cyclase, which synthesizes cyclic AMP (cAMP)—an intracellular surrogate for the neurotransmitter, the first messenger. Phospholipase C (PLC), another effector enzyme, generates diacylglycerol (DAG) and inositol 1,4,5-trisphosphate (IP_3), the latter of which releases intracellular stores of Ca^{2+}. Information from an activated receptor flows to the second messengers that typically activate protein kinases, which modify a host of cellular functions. Cyclic AMP, Ca^{2+}, and DAG have in common the ability to activate protein kinases with broad substrate specificities. They phosphorylate key intracellular proteins, ion channels, enzymes, and transcription factors taking part in diverse cellular biological processes. The activities of protein kinases and phosphatases are in balance, constituting a highly regulated process, as revealed by the phosphorylation state of these targets of the signal transduction process. In addition to regulating protein kinases, second messengers such as cAMP, cyclic GMP (cGMP), Ca^{2+}, and arachidonic acid can directly gate, or modulate, ion channels. G proteins can also couple directly to ion channels without the interception of second messengers or protein kinases. In these diverse ways, a neurotransmitter outside the cell can modulate essentially every aspect of cell physiology and encode the history of cell stimuli in the form of altered activity and expression of its cellular constituents. An overview of G-protein signaling to protein kinases is presented in Figure 12.1.

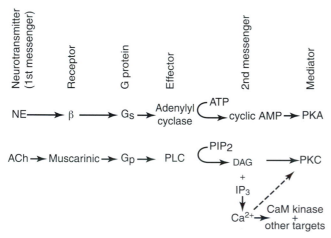

FIGURE 12.1 Overview of G-protein signaling to protein kinases. Norepinephrine (NE) and acetylcholine (ACh) can stimulate certain receptors that couple through distinct G proteins to different effectors, which results in increased synthesis of second messengers and activation of protein kinases (PKA and PKC). PLC, phospholipase C; PIP$_2$, phosphatidylinositol bisphosphate; DAG, diacylglycerol; CaM, Ca^{2+}-calmodulin-dependent; IP$_3$, inositol 1,4,5-trisphosphate.

G-Protein Signaling Operates on Common Principles

The many types of G proteins and the many types of effector enzymes have certain features in common (Hille, 1992). First, each receptor can couple to one or only a few G proteins, thus specifying the stimulus response. Second, the second messengers are typically synthesized in one or two steps from a precursor (e.g., ATP) that is readily available in those cells at high concentration but is itself inactive as a signaling molecule. G proteins can also stimulate enzymes that eliminate second messengers. Third, the initial signal can be greatly amplified: each receptor can activate many G protein molecules; each adenylyl cyclase can synthesize many cAMP molecules; and each protein kinase can phosphorylate many copies of each of its substrates. Fourth, the process is slower in onset and persists longer than ligand-gated ion channel signaling.

A fifth feature of G-protein signaling is the ability to orchestrate a variety of effects through the same second messenger. For example, the cAMP-dependent protein kinase (PKA), stimulated by serotonin exposure to sensory neurons in the mollusk *Aplysia*, phosphorylates and inhibits a K^+ channel (see Chapter 19). This inhibition results in a prolonged action potential and a greater influx of Ca^{2+} with each stimulation of the sensory neuron and in a sensitization of the withdrawal response regulated by this circuit. On a slower time course, cAMP and PKA can also modify carbohydrate metabolism to keep up with activity and the transcription of genes and translation of proteins that ultimately modify the number of synaptic contacts made by the sensory neuron after prolonged periods of activity or inactivity.

A nervous system with information flow by fast transmission alone would be capable of stereotyped or reflex responses. The large diversity of signaling molecules and their intracellular targets offer nearly unlimited flexibility of response over a broad timescale and with high amplification. This signaling is a key feature of neuronal plasticity, regulating every step of the way from neurotransmitter receptors and ion channels, signal transduction pathways, neurotransmitter synthesis, and release to the expression of genes in the nucleus that underlie synaptic changes linked to learning and memory (see Chapter 19 for extended discussion).

Receptors Catalyze the Conversion of G Proteins into the Active GTP-Bound State

G proteins undergo a molecular switch between two interconvertible states that are used to "turn on" or "turn off" downstream signaling. G proteins taking part in signal transduction utilize a regulatory motif that is seen in other GTPases engaged in protein synthesis and in intracellular vesicular traffic. G proteins are switched on by stimulated receptors, and they switch themselves off after a time delay. The G proteins are inactive when GDP is bound and are active when GTP is bound. The sole function of seven-helix receptors in activating G proteins is to catalyze an exchange of GTP for GDP. This is a temporary switch because G proteins are designed with a GTPase activity that hydrolyzes the bound GTP and converts the G protein back into the GDP-bound, or inactive, state. Thus, a G protein must continuously sample the state of activation of the receptor, and it transmits downstream information only while the neuron is exposed to neurotransmitter. A fast GTPase means that the signal transduction pathway is very responsive to the presence of neurotransmitter outside and can respond to individual stimuli even at high frequency. A slow GTPase provides greater amplification but cannot produce distinct responses to each stimulus presented at high frequency. The GTPase activity of G proteins serves as both a timer and an amplifier (Fig. 12.2).

The G-Protein Cycle

G proteins are trimeric structures composed of two functional units: (1) an α subunit (39–52 kDa) that catalyzes the GTPase activity and (2) a βγ dimer (35 and 8 kDa, respectively) that tightly interacts with the α subunit when bound to GDP (Birnbaumer, 2007; Neer, 1995; Stryer and Bourne, 1986). The role of the three subunits in the G-protein cycle is depicted in Figure 12.3. In the basal state, GDP is bound tightly to the α subunit, which is associated with the βγ pair

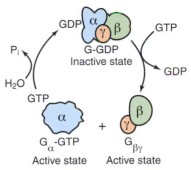

FIGURE 12.3 Interconversion, catalyzed by excited receptors, of G-protein subunits between inactive and active states. Displacement of GDP with GTP dissociates the inactive heterotrimeric G protein, generating α-GTP and βγ, both of which can interact with their respective effectors and activate them. The system converts into the inactive state after GTP has been hydrolyzed and the subunits have reassociated. From "Biochemistry," 4th ed. by Stryer. © 1995 by Lubert Stryer. Used with permission of W. H. Freeman and Company.

to form an inactive G protein. In addition to blocking interaction of the α subunit with its effector, the βγ pair increases the affinity of the α subunit for activated receptors. Binding of the neurotransmitter to the receptor produces a conformational change that positions previously buried residues that promote increased affinity of the receptors for the inactive G protein (Palczewski *et al.*, 2000). A given receptor can interact with only one or a limited number of G proteins, and the α subunit produces most of this specificity. Coupling with the activated receptor reduces the affinity of the α subunit for GDP, facilitating its dissociation and thus leaving the nucleotide binding site empty. Either GDP or GTP can bind to the vacant site; however, because the level of GTP in the cell is much greater than the level of GDP, the dissociation of GDP is usually followed by the binding of GTP. Thus, the receptor effectively catalyzes an exchange of GTP for GDP (Cassel and Selinger, 1978). Binding of GTP has two consequences: (1) it dissociates the G protein into α-GTP and βγ and (2) the α-GTP subunit has a reduced affinity for the stimulated receptor, leading to dissociation of the complex (Sprang *et al.*, 2007).

The GTP–GDP exchange is inherently very slow and ensures that very little of the G protein is in the on state under basal conditions. The level of G protein in the on state can increase from 1% to more than 50% of all G protein. The direction of the cycle is determined by the GTPase reaction, which uses the energy of GTP to make the reaction irreversible and maintain a low level of α-GTP in the basal state (Stryer and Bourne, 1986).

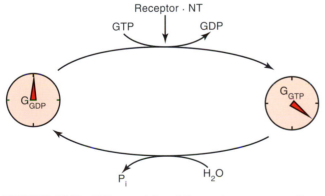

FIGURE 12.2 GTPase activity of G proteins serves as a timer and amplifier. Receptors activated by neurotransmitters (NT) initiate the GTPase timing mechanism of G proteins by displacement of GDP by GTP. Neurotransmitters thus convert G_{GDP} ("turned-off state") to G_{GTP} (time-limited "turned-on" state).

Information Flow

One of the more tense and public debates in signal transduction has been the question of whether the α subunit alone conveys information that specifies which effector is activated or whether the βγ pair has a role. One of the contestants even paid for a vanity license plate proclaiming "α not β." The α subunit was thought to be responsible for specifying which effector enzyme was activated by a G protein (Cassel and Selinger, 1978). This notion was eventually changed because of the finding that βγ can directly activate certain K^+ channels (Huang *et al.*, 1995; Logothetis *et al.*, 1987). The historic association of G-protein function with α has persisted for the purpose of nomenclature, with G_s and $α_s$ referring to the G protein and its corresponding α subunit, which stimulates adenylyl cyclase. These names have been retained even though it is now apparent that α and βγ subunits can both modify effector enzymes and channels and that a given α subunit may combine with a number of βγ pairs. The α subunits may act either independently or in concert with βγ (Clapham and Neer, 1993). Furthermore, β and γ subunits in a βγ pair can combine in many different ways. Other legacy terms include G_i, G_p, and G_o, used for G-protein activities that inhibited adenylyl cyclase, stimulated phospholipase, or were presumed to have other effects, respectively. At this stage, cloning and genomic sequencing have outpaced functional studies. The inherent affinities of βγ pairs for a particular α and spatial segregation of G proteins probably greatly limit the number of combinations of subunits. See Box 12.1.

Function of α Subunits

Determination of the crystal structures of different conformational states of some α subunits and βγ subunits has been a source of insight into their functional domains and the critical GTPase activity of all α subunits (Sondek *et al.*, 1996; Wall *et al.*, 1996). The α subunits are compact molecules that must accommodate a number of protein–protein interactions in addition to their GTP-hydrolyzing activity. The β subunit consists of repeating motifs arranged like blades of a propeller or wedges of a pie with a central tunnel (see next subsection). In the trimeric GDP-bound state, the α subunit is docked to the β subunit (Wall *et al.*, 1996). Receptors interact with the carboxy terminus of the α subunit; this interaction, along with other contacts, specifies which receptors can activate which α subunits (Conklin *et al.*, 1993). The guanine nucleotide binding site is highly conserved and altered by interaction with ligand-bound receptors to facilitate dissociation of GDP and exchange for GTP, thus producing an activated α-GTP conformation. The γ phosphate present in GTP, but not in GDP, interacts with a region of α that participates in the docking of α to βγ. The resulting conformational changes reduce the affinity between this surface and βγ, resulting in dissociation of the trimer to produce α and βγ capable of interacting with their respective effectors. Dissociation also releases the receptor, which can then catalyze activation of other G proteins. Thus, each functional half of the G protein inhibits the action of the other. Dissociation of these subunits exposes the residues needed for interactions with effectors. GTP hydrolysis reverses the conformational changes and enables a region of α that can interact with a binding pocket formed by the β subunit. The known effector functions of α include both stimulation and inhibition of adenylate cyclase that is sensitive to cholera toxin and pertussis toxin, respectively. In addition, it modulates activation of cGMP phosphodiesterase, PLC, and regulation of Na^+–K^+ exchange, PI3K, RhoGEF, and rasGAP (Birnbaumer, 2007; Clapham and Neer, 1993).

Function of βγ Subunits

The crystal structure of the βγ heterodimer both in its inactive form bound to the α subunit and in its free active form has been determined (Sondek *et al.*, 1996; Wall *et al.*, 1996). The C-terminal half of the β subunit has a seven-fold repeat of a structural motif termed the *WD repeat*, which interacts with γ and is the likely site of interaction between β and its effector enzymes and channels. The WD repeat is a sequence of 25–40 amino acids and ends with the amino acids tryptophan (W) and aspartic acid (D). Each repeat forms a wedge of a circular disk with a central tunnel in both the free and bound states. The γ subunit is in an extended conformation, circling and making contact with several WD domains of the β subunit. Prenylation of the C terminus of γ anchors the βγ dimer to membranes.

The effector functions of βγ dimers include inhibition of many adenylate cyclases and stimulation of adenylate cyclase types II and IV (with α). In addition, they regulate stimulation of phospholipase Cβ, K^+ and Ca^{2+} channels, phospholipase A_2, phosphatidylinositol-3-kinase, PKD, and dynamin in vesicle budding. The identified effector targets of βγ are several adenylyl cyclases, the β isoform of phospholipase C, ion channels (for K^+ and Ca^{2+}), phospholipase A2, and phosphatidylinositol 3-kinase (Birnbaumer, 2007; Clapham and Neer, 1993). The other functions of βγ are as follows. First, they keep G proteins

BOX 12.1

WHY ARE G-PROTEIN-REGULATED SYSTEMS SO COMPLEX?

Transmembrane signaling systems all contain two fundamental elements: one that recognizes an extracellular signal (a receptor) and another that generates an intracellular signal. These elements can be easily incorporated into a single molecule, for example, in receptor tyrosine kinases and guanylyl cyclases. Why then are G-protein-regulated systems so complex, minimally containing five gene products in the basic module (receptor, heterotrimeric G protein, and effector)? The design of these systems permits both integration and branching at its two interfaces: between receptor and G protein and between G protein and effector. Each component of the system can thus be regulated independently—transcriptionally, post-translationally, or by interactions with other regulatory proteins. Furthermore, hundreds of genes encode receptors for hormones and neurotransmitters, dozens of genes encode G-protein subunits (α, β, and γ), and dozens more genes encode G-protein-regulated effectors. Each cell in an organism thus has the opportunity to sample the genome and construct a highly customized and sophisticated switchboard in its plasma membrane, permitting the organism to make an extraordinary variety of responses to complex situations. The choices that each cell makes include much more than just the components of the basic modules, extending as well to regulators of G-protein-mediated pathways such as receptor kinases and GTPase-activating proteins. And, of course, the identity and concentrations of the components of the switchboard can be sculpted within minutes or hours to permit adaptation to developmental needs or environmental stresses.

The classic stress response of mammals to the hormone epinephrine provides an elegant example of the power of modular signal transduction systems. A single hormone is used to initiate responses of opposite polarity in very similar cells. Thus, vascular smooth muscle in skin contracts to minimize bleeding (if there is a wound) and maintain blood pressure, whereas vascular smooth muscle in skeletal muscle relaxes to provide increased blood flow during heightened physical activity. Smooth muscle in the intestine relaxes, whereas cardiac muscle is powerfully stimulated. In addition, hepatocytes hydrolyze glycogen to glucose, adipocytes hydrolyze triglycerides and release free fatty acids for fuel, and certain endocrine and exocrine secretions are stimulated or inhibited. Several distinct receptors for epinephrine are selectively expressed in various cell types to achieve this beneficial orchestration of stimulatory and inhibitory events. These receptors vary in their capacities to interact with G proteins from three different subfamilies (G_s, G_i, G_q) and thus to activate or inhibit several effects to achieve the desired responses. Each cell's choices of particular receptors, G proteins, and effectors permit additional choices of regulators of each of these components to adjust the magnitude and/or the kinetics of the response.

The past two decades of research in this area have witnessed identification and characterization of the molecular players involved in G-protein-mediated signaling and appreciation of the basic mechanisms that underlie the protein–protein interactions that drive these systems. Current research is expanding this basic core of knowledge: on the one hand, toward greater understanding of how the individual modules contribute to the integrated networks of intact cells and, on the other, toward elucidation of the physical and structural bases of these complex cellular reactions. We can begin to construct a movie of G-protein-mediated signaling at atomic resolution, and many of its most important frames are shown in Figure 12.4.

Alfred G. Gilman
Stephen R. Sprang

inactive in the basal state by complexing with α and reducing its intrinsic GTP–GDP exchange rate. Second, they help to target the α subunits to the membrane and increase the affinity of α-GDP for ligand-bound receptors. Third, both β and γ help to specify which receptors couple to the G protein. Fourth, the WD repeats of β serve as anchors for a protein kinase, called βARK, that terminates signaling by ligand-bound β-adrenergic receptors.

Examination and Manipulation of G-Protein-Coupled Signals

Neurotransmitters can produce their cellular effects by a variety of signal transduction pathways. A number of experimental tools and approaches must be used to delineate the pathway used in any system of interest. Table 12.1 summarizes some of these experimental tools.

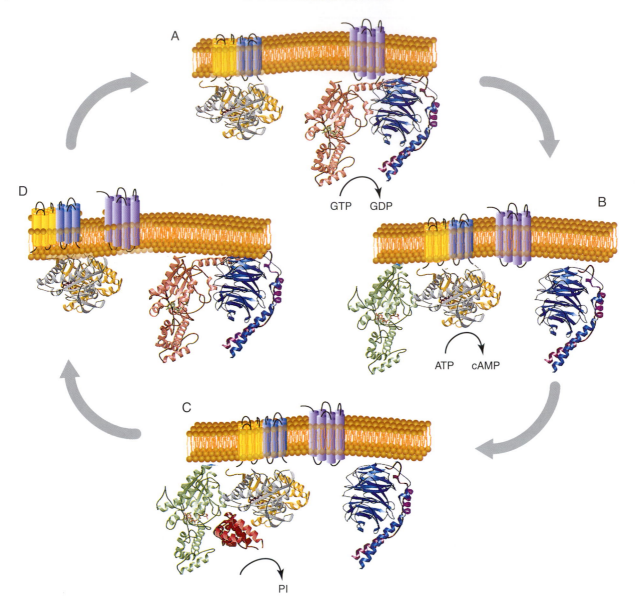

FIGURE 12.4   (A) G proteins are held in an inactive state because of very high affinity binding of GDP to their α subunits. When activated by an agonist, membrane-bound seven-helical receptors (right, glowing magenta) interact with heterotrimeric G proteins (α, amber; β, teal; γ, burgundy) and stimulate dissociation of GDP. This permits GTP to bind to and activate α, which then dissociates from the high-affinity dimer of β and γ subunits. (B) Both activated (GTP-bound) α (lime) and βγ are capable of interacting with downstream effectors. Also illustrated is the interaction of α$_s$-GTP with adenylyl cyclase (catalytic domains are mustard and ash). Adenylyl cyclase then catalyzes the synthesis of the second messenger cyclic AMP (cAMP) from ATP. (C) Signaling is terminated when α hydrolyzes its bound GTP to GDP. In some signaling systems, GTP hydrolysis is stimulated by GTPase-activating proteins or GAPs (cranberry) that bind to α and stablize the transition state for GTP hydrolysis. (D) Hydrolysis of GTP permits α-GDP to dissociate from its effector and associate again with βγ. The heterotrimeric G protein is then ready for another signaling cycle if an activated receptor is present. The figure is based on the original work of Mark Wall and John Tesmer.

Toxins

Differential sensitivities to cholera toxin and pertussis toxin can be used to implicate a G-protein-mediated pathway. Both G_s and transducin are sensitive to cholera toxin, which selectively ADP-ribosylates α$_s$ and α$_t$. The α subunits of G_i, G_o, and transducin, but not of G_s, are ADP-ribosylated by pertussis toxin. These toxins transfer the ADP-ribose moiety of NAD^+ to an arginine (cholera toxin) or cysteine (pertussis toxin) on the appropriate α subunit. The toxins act at distinct steps in the GTPase cycle to lock it in either the on or the off position (see Fig. 12.3). Cholera toxin inhibits the GTPase activity, thereby generating a persistently on state.

TABLE 12.1   Experimental Tools for and Approaches to Testing the Role of G Proteins

Cholera toxin and pertussis toxin

GTPγS or GTPβS, NaF

Antibody to G-protein subunits

Antisense oligonucleotides or RNA

Knockouts or other forms of gene disruption

Reconstitution from purified components

For example, cholera toxin treatment of G_s stimulates robust and continuous production of cAMP by adenylyl cyclase. In contrast, pertussis toxin acts on the inactive G protein, blocking its interaction with receptors so that it cannot be activated and thus remains in the GDP-bound, or off, state. Toxin sensitivity can be used to narrow the choice of possible G proteins taking part in a process or to block a known pathway.

Guanine Nucleotides and NaF

The GTPase cycle of all G proteins can also be modified for experimental purposes by GTPγS or GDPβS. GTPγS is a nonhydrolyzable analog of GTP with high affinity for the α subunit. Activation of any of the G proteins in the presence of GTPγS leads to the exchange of GTPγS for GDP to produce α-GTPγS. The G protein is thereby activated for prolonged periods because the GTPase cycle is blocked and remains so until GDP exchanges with GTPγS. A similar response is obtained by addition of aluminum fluoride. Fluoride forms a complex with trace amounts of Al^{3+} and binds to α-GDP. The aluminum fluoride moiety simulates the γ phosphate of GTP so that, like GTPγS, $\alpha\text{-GDP-AlF}_3$ persistently activates the G protein. (Coleman *et al.*, 1994; Sondek *et al.*, 1994). In contrast, GDPβS can exchange with GDP, leaving the G protein in the inactive state. GDPβS has a higher affinity than GDP for the G protein and, as a result, the GTPase cycle is slowed and spends more of its time with an inactive G protein.

Effector Enzymes, Channels, and Transporters Decode Receptor-Mediated Cell Stimulation in the Cell Interior

The function of the trimeric G proteins is to decode information about the concentration of neurotransmitters bound to appropriate receptors on the cell surface and convert this information into a change in the activity of enzymes and channels that mediate the effects of the neurotransmitter. The effector can be an enzyme that synthesizes or degrades a diffusible second messenger or it can be an ion channel. The number of identified effectors of G-protein signaling has increased markedly in the past few years and now also includes membrane transport proteins.

Response Specificity in G-Protein Signaling

The modular design of G-protein signaling may appear to be incapable of providing specificity. Receptors can stimulate one or more G proteins, G proteins can couple to one or more effector enzymes or channels, and the resulting second messengers will affect many cellular processes. Signals originating from activated receptors can either converge or diverge, depending on the receptor and on the complement of G proteins and effectors in a given neuron (Fig. 12.5) (Ross, 1989).

How can a neurotransmitter produce a specific response if G-protein coupling has the potential for such a diversity of effectors? A given neuron has only a subset of receptors, G proteins, and effectors, thereby limiting possible signaling pathways. Transducin, for example, is confined to the visual system, where the predominant effector is the cGMP phosphodiesterase and not adenylyl cyclase. A number of other factors combine to increase signal specificity (Hille, 1992; Neer, 1995; Sondek *et al.*, 1994).

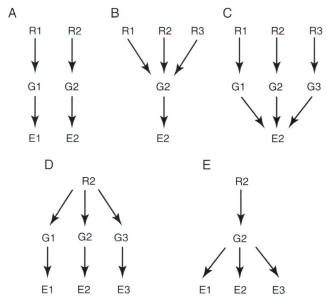

FIGURE 12.5   Signals can converge or diverge on the basis of interactions between receptors (R) and G proteins (G) and between G proteins and effectors (E). The complement of receptors, G proteins, and effectors in a given neuron determines the degree of integration of signals, as well as whether cell stimulation will produce a focused response to a neurotransmitter or a coordination of divergent responses. Adapted from Ross (1989).

Specificity and choice. Receptors and G proteins have higher intrinsic affinities and efficacies for modulating the activity of the "correct" G protein(s) and effector(s), respectively.

Spatial compartmentalization. Second-messenger systems can be compartmentalized, thus adding specificity and localized control of signaling. The same receptor may regulate a Ca^{2+} channel through one G protein at a nerve terminal and regulate PLCβ at a distal dendrite through another G protein. Although the addition of the neurotransmitter in culture would produce both effects, synaptic inputs would be able to elicit specific effects at nerve terminals or at dendrites.

GTPase activity. The degree of amplification by the G protein (based on GTPase) and by the effector (based on its specific activity or conductance) can determine which of the possible pathways is more prominent. Furthermore, some effectors appear to act as GTPase-activating proteins (GAPs), which modify the intrinsic GTPase activity of the G protein. Such an effector terminates signaling faster when stimulated by one G protein relative to another and fine-tunes the flow of information through the various forks of the signaling system. The signaling strength (i.e., the speed and efficiency) of any branch of the pathway can be modulated. Inherent affinities, level of expression of the various components, compartmentalization, GTPase rates, and GAP activity combine to produce either a well-focused response by a single pathway or a richer and more diffuse response through several pathways.

Fine-Tuning of cAMP by Adenylyl Cyclases

The level of cAMP is highly regulated owing to a balance between synthesis by adenylyl cyclases and degradation by cAMP phosphodiesterases (PDEs). Each of these enzymes can be independently regulated and manipulated. Adenylyl cyclase was the first G-protein effector to be identified, and now a group of related adenylyl cyclases is known to be differentially regulated by both α and βγ subunits. (Taussig and Gilman, 1995). G proteins can both activate and inhibit adenylyl cyclases either synergistically or antagonistically.

Adenylyl cyclases are large proteins of approximately 120 kDa consisting of a tandem repeat of the same structural motif: a short cytoplasmic region followed by six putative transmembrane segments and then a highly conserved catalytic domain of approximately 35 kDa on the cytoplasmic side (Krupinski *et al.*, 1989). The catalytic domains resemble each other as well as the catalytic domain of guanylyl cyclase. It is therefore likely that these two domains

interact with G_s, bind ATP, and catalyze its conversion into cAMP. Some isoforms are activated by calmodulin and have one calmodulin-binding domain in the link between the first catalytic domain and the second set of transmembrane sequences.

Differential regulation of adenylyl cyclase isoforms. All adenylyl cyclase isoforms are stimulated by G_s through its $α_s$ subunit. Mammals have at least nine adenylyl cyclase isoforms, designated I–IX (not including alternative splicing), that differ in their regulatory properties and tissue distribution. (Hanoune and Defer, 2001; Tang and Gilman, 1991; Taussig and Gilman, 1995). They can be minimally divided into at least three groups on the basis of additional regulatory properties (Fig. 12.6). Group A (types I, III, and VIII) possesses a calmodulin-binding domain and is activated by Ca^{2+}-calmodulin. Group B (types II and IV) is weakly responsive to direct interaction with $α_s$ or βγ but is highly activated when both are present (Krupinski *et al.*, 1989). As described later, this synergistic effect enables this cyclase to function as a coincidence detector (see also Chapter 19). Group C is typified by types V and VI (and IX), which differ from group A cyclases in their inhibitory regulation.

Inhibition of adenylyl cyclases. Adenylyl cyclases are also subject to several forms of inhibitory control. First, activation of all adenylyl cyclases can be antagonized to some extent by βγ released from abundant G proteins, such as G_i, G_o, and G_z, which complex with $α_s$-GTP and shift the equilibrium toward an inactive trimer by mass action. Second, either α or βγ subunits derived from G_i, G_o, or G_z can directly inhibit group A cyclases, and the α subunit from G_i or G_z can inhibit group C cyclases (Tang and Gilman, 1991). Many G proteins can generate βγ subunits

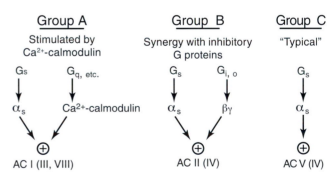

FIGURE 12.6 Isoforms of adenylyl cyclase (AC). All isoforms are stimulated by $α_s$ but differ in the degree of interaction with Ca^{2+}-calmodulin and with βγ derived from inhibitory G proteins. Not shown is the ability of excess βγ to complex with and $α_s$ inhibit group A and group C adenylyl cyclases. Adapted from Taussig and Gilman (1995).

capable of directly inhibiting group A and activating group B adenylyl cyclases. However, not all G proteins are sufficiently abundant to produce enough $\beta\gamma$ to bring about these effects. The level of G_s in particular is low; thus, the α_s derived from G_s is sufficient to activate adenylyl cyclases, but the $\beta\gamma$ derived from it is insufficient to directly inhibit or activate adenylyl cyclases. Therefore, the sources of $\beta\gamma$ for modulation of type I and II adenylyl cyclases are likely the abundant G proteins, such as G_i and G_o. This explains the apparent paradox that receptors that couple to G_s produce effects only through α_s, whereas receptors that couple to G_i produce effects through both α_i and $\beta\gamma$ even though they can share the same $\beta\gamma$.

Receptor coupling to adenylyl cyclase. Dozens of neurotransmitters and neuropeptides work through cAMP as a second messenger and do so by activation of either G_s to stimulate adenylyl cyclase or G_i or G_o to inhibit adenylyl cyclase. Among the neurotransmitters that increase cAMP are the amines norepinephrine, epinephrine, dopamine, serotonin, and histamine and the neuropeptides vasointestinal peptide (VIP) and somatostatin. In the olfactory system, a special form of G-protein α subunit, termed α_{olf}, serves the same function as α_s. Odorants are detected by several hundred seven-helix receptors that activate G_{olf}, which in turn activates the type III adenylyl cyclase in the neuroepithelium. Many of the same neurotransmitters activate distinct receptors that couple to G_i or G_o. They include acetylcholine, dopamine, serotonin, norepinephrine, and opiate peptides.

Adenylyl cyclases as coincidence detectors. The properties of adenylyl cyclases described in Figure 12.6 suggest an integrative capacity for adenylyl cyclase. Type I and type II adenylyl cyclases appear to be specifically designed to detect concurrent stimulation of neurons by two or more neurotransmitters (Bourne and Nicoll, 1993).

Type I adenylyl cyclase is stimulated by neurotransmitters that couple to G_s and by neurotransmitters that elevate intracellular Ca^{2+}. This adenylyl cyclase can convert the depolarization of neurons into an increase in cAMP (Wayman *et al.*, 1994), a likely basis for its role in several associative forms of learning (Levin *et al.*, 1992; Wu *et al.*, 1995).

Stimulation of type II adenylyl cyclase by α_s is conditional on the presence of $\beta\gamma$ derived from a G protein other than G_s, thus enabling the cyclase to serve as a coincidence detector (Federman *et al.*, 1992; Tang and Gilman, 1991). As indicated earlier, $\beta\gamma$ derived from G_s is not sufficient to produce synergistic activation of this enzyme. Thus, activation of a second receptor, presumably coupled to the abundant G_i and G_o, is needed to provide the $\beta\gamma$. In cortex and hippocampus, which contain the type II adenylyl cyclase, neurotransmitters coupling to G_i can potentiate increases in cAMP resulting from concurrent stimulation by neurotransmitters coupled to G_s.

Sources of Second Messengers: Phospholipids

Two phospholipids, phosphatidylinositol 4,5-bisphosphate (PIP_2) and phosphatidylcholine (PC), are primary precursors for a G-protein-based second-messenger system. Three second messengers—diacylglycerol, arachidonic acid and its metabolites, and elevated Ca^{2+}—are ultimately produced. A single step converts the inert phospholipid precursors into the lipid messengers (Divecha and Irvine, 1995). Ca^{2+} becomes functionally active by its regulated entry into the lumen of the cytosol or nucleus from a sequestered pool in the endoplasmic reticulum or extracellular space. DAG action is mediated by protein kinase C (Takai *et al.*, 1979; Tanaka and Nishizuka, 1994). Ca^{2+} has many cellular targets but mediates most of its effects through calmodulin, a Ca^{2+}-binding protein that activates many enzymes after it binds Ca^{2+}. One class of calmodulin-dependent enzymes is a family of protein kinases that enable Ca^{2+} signals to modulate a large number of cellular process by phosphorylation (Braun and Schulman, 1995).

Generation of DAG and IP_3 from G_q and G_i coupled to PLCβ. The phosphatidyl inositide-signaling pathway is just as prominent in neuronal signaling as the cAMP pathway and similar to it in overall design. Stimulation of a large number of neurotransmitters and hormones (including acetylcholine (M1, M3), serotonin ($5HT_2$, $5HT_{1C}$), norepinephrine (α_{1A}, α_{1B}), glutamate (metabotropic), neurotensin, neuropeptide Y, and substance P) is coupled to the activation of a phosphatidylinositide-specific PLC. In recent years, it has become clear that in addition to their function as membrane phospholipids, the phosphoinositides are precursors for second messengers (Berridge, 1993; Clapham, 1995; Di Paolo and De Camilli, 2006).

Phosphatidylinositol (PI) is composed of a diacylglycerol backbone with *myo*inositol attached to the *sn*–3 hydroxyl by a phosphodiester bond (Fig. 12.7). The six positions of the inositol are not equivalent: the 1-position is attached by a phosphate to the DAG moiety. PI is phosphorylated by PI kinases at the 4-position and then at the 5-position to form PIP_2. In response to the appropriate G-protein coupling, PLC hydrolyzes the bond between the *sn*–3 hydroxyl of the DAG backbone and the phosphoinositol to produce two second messengers: DAG, a hydrophobic molecule, and inositol 1,4,5-trisphosphate (IP_3), which is water soluble (Divecha and Irvine, 1995; Hokin and Hokin, 1955) (Fig. 12.8). Three

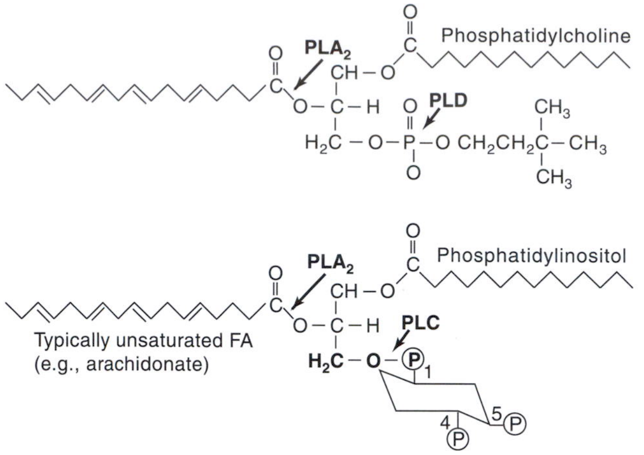

PLCβ is coupled to neurotransmitters by G_q and G_i (Clapham, 1995). A pertussis-toxin-insensitive pathway is mediated by a number of isoforms originally termed G_p, with the subscript "p" for PLC activation but more commonly referred to as G_q and α_q. G_q couples to PLCβ through its α subunit. There are several PLCβ isoforms, and they show distinct regulation by G proteins. One isoform is most sensitive to α_q; another is more sensitive to βγ (e.g., from G_i in a pertussis-toxin-sensitive pathway), and a third shows little activation by either G protein.

DAG is derived from activation of phospholipase D. A slower but larger increase in DAG can be generated by activation phospholipase D (PLD), which cleaves phosphatidylcholine to produce phosphatidic acid and choline. Dephosphorylation of phosphatidic acid produces DAG. The PLD pathway may be used by some mitogens and growth factors and likely contains a variety of activation schemes that may include G proteins (Divecha and Irvine, 1995; Nishizuka, 1995).

Regulation of PLCγ by receptor tyrosine kinases DAG and IP_3 is also produced when certain receptor tyrosine kinases are activated and is independent of G proteins. Stimulation of receptor tyrosine kinases such as epidermal growth factor (EGF) leads to their autophosphorylation on tyrosine residues and activation.

FIGURE 12.7 Structures of phosphatidylinositol and phosphatidylcholine. The sites of hydrolytic cleavage by PLC, PLD, and PLA_2 are indicated by arrows. FA, fatty acid.

classes of PLC that hydrolyze PIP_2 with some selectivity have been cloned and characterized. PLCβ, PLCγ, and PLCδ are soluble enzymes that have in common a catalytic domain structure but differ in their regulatory properties. G proteins couple to several variants of PLCβ. PLCγ is regulated by growth factor tyrosine kinases.

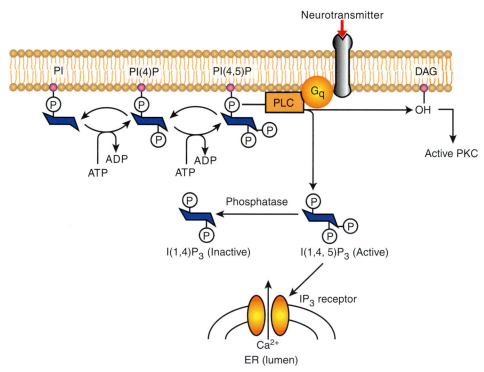

FIGURE 12.8 Schematic pathway of IP_3 and DAG synthesis and action. Stimulation of receptors coupled to G_q activates PLCβ, which leads to release of DAG and IP_3. DAG activates PKC, whereas IP_3 stimulates the IP_3 receptor in the endoplasmic reticulum (ER), leading to mobilization of intracellular Ca^{2+} stores. Adapted with permission from Macmillan Publishers Ltd. Berridge, M.J. (1993). Inositol triphosphate and calcium signalling. *Nature* **361**, 315–325.

Specific phosphotyrosine moieties on the receptor then recruit effector enzymes, such as PLCγ, that possess Src homology 2 (SH2) domains. These structural elements specifically recognize certain protein sequences with phosphotyrosine and lead to the translocation of effectors such as PLCγ to the receptor at the membrane. After binding to the receptor, PLCγ is activated by phosphorylation on tyrosine and hydrolyzes PIP_2 to DAG and IP_3.

Additional lipid messengers. DAG is itself a source of another lipid messenger, by the action of phospholipase A_2 (PLA_2), which releases the fatty acid, typically arachidonic acid, at the *sn*–2 position of the DAG backbone (Fig. 12.7). PLA_2 can be activated by βγ and α subunits of G proteins. A cytosolic PLA_2 may mediate growth factor signaling via mitogen-activated protein (MAP) kinases. Phosphorylation of this PLA_2 leads to its translocation to the membrane, where it can act on membrane phospholipids. Arachidonic acid has biological activity of its own in addition to serving as a precursor for prostaglandins and leukotrienes. Arachidonic acid and other cis-unsaturated fatty acids can modulate K^+ channels, PLCγ, and some forms of PKC. Other lipids also may generate signaling molecules.

A subfamily of lipid kinases that are specific for addition of a phosphate moiety on the 3-position, phosphoinositide 3-kinases (PI–3 kinases) also play a regulatory role (Rameh and Cantley, 1999). Depending on their preferred lipid substrate, they can produce PI-3-P, PI–3,4-P_2, PI–3,5-P_2, and PI–3,4,5-P_3. A number of signals, including growth factors, activate PI–3 kinases to generate these lipid messengers. In turn, these lipids then bind directly to a number of proteins and enzymes to modify vesicular traffic and activity of protein kinases involved in survival and cell death. There is also evidence that another lipid, sphingomyelin, is a precursor for intracellular signals as well.

IP_3, a potent second messenger that produces its effects by mobilizing intracellular Ca^{2+}. The main function of IP_3 is to stimulate the release of Ca^{2+} from intracellular stores (Streb *et al.*, 1983). The concentration of cytosolic-free Ca^{2+} is approximately 100 nM in unstimulated neurons, whereas its concentration in the extracellular space is 1.5–2.0 mM. This provides a tremendous driving force for movement down its concentration gradient; its reversal potential is more than 100 mV. Ca^{2+} is the most common second messenger in neurons, yet it can be neurotoxic. Neurons have therefore developed several mechanisms for maintaining a low interstimulus level of free Ca^{2+}. A Ca^{2+}-ATPase and a Na^+–Ca^{2+} exchanger in the plasma membrane catalyze the active transport of Ca^{2+} to the extracellular space, and a different Ca^{2+}-ATPase in the ER membrane sequesters Ca^{2+} in the ER network. Much of the Ca^{2+} in the ER is complexed with low-affinity binding proteins that enable the ER to concentrate Ca^{2+} yet enable Ca^{2+} to readily flow down its concentration gradient into the cell lumen on opening of Ca^{2+} channels in the ER. The ER is the major IP_3-sensitive Ca^{2+} store in cells (Fig. 12.8).

The IP_3 receptor is a macromolecular complex that functions as an IP_3 sensor and a Ca^{2+} release channel. It has a broad tissue distribution but is highly concentrated in the cerebellum. The IP_3 receptor is a tetramer of 313 kDa subunits with a single IP_3 binding site at its N terminal of each subunit, facing the cytoplasm. Ca^{2+} release by IP_3 is highly cooperative so that a small change in IP_3 has a large effect on Ca^{2+} release from the ER. The IP_3 receptor has low activity at either high or low levels of cytoplasmic Ca^{2+}, with peak release requiring 200–300 nM Ca^{2+}, a property that may be used in the generation of some Ca^{2+} waves (Berridge, 1993; Bootman *et al.*, 1997; Tsien and Tsien, 1990). The mouse mutants *pcd* and *nervous* have deficient levels of the IP_3 receptor and exhibit defective Ca^{2+} signaling, and a genetic knockout of the IP_3 receptor leads to motor and other deficits.

Termination of the IP_3 signal. IP_3 is a transient signal terminated by dephosphorylation to inositol. Inactivation is initiated either by dephosphorylation to inositol 1,4-bisphosphate (Fig. 12.8) or by an initial phosphorylation to a tetrakisphosphate form that is dephosphorylated by a different pathway. Both pathways have in common an enzyme that cleaves the phosphate on the 1-position. Complete dephosphorylation yields inositol, which is recycled in the biosynthetic pathway. Recycling is important because most tissues do not contain *de novo* biosynthetic pathways for making inositol. Thus, the phosphatases not only terminate the signal but also serve as a salvage step that may be particularly important when cells are actively undergoing PI turnover. It is intriguing that the simple salt Li^+ selectively inhibits the salvage of inositol by inhibiting the enzyme that dephosphorylates the 1-position and is common to the two pathways. This simple salt is the drug used to treat manic–depressive disorders. At therapeutic doses of Li^+, the reduced salvage of inositol in cells with high phosphoinositide signaling may lead to depletion of PIP_2 and a selective inhibition of this signaling pathway in active cells.

Calcium Ion

Calcium has a dual role as a carrier of electrical current and as a second messenger. Its effects are more diverse than those of other second messengers such as cAMP and DAG because its actions are mediated by a much larger array of proteins, including protein kinases (Carafoli and Klee, 1999). Furthermore, many signaling pathways directly or indirectly increase cytosolic Ca^{2+} concentration from 100 nM to 0.5–1.0 mM. The source of elevated Ca^{2+} can be either the ER or the extracellular space (Fig. 12.9). As indicated earlier, mobilization of ER Ca^{2+} is mediated by IP_3 derived from PLCβ activation through G proteins and from PLCγ activation by receptor tyrosine kinases acting on the IP_3 receptor. In addition, Ca^{2+} can activate its own mobilization through the ryanodine receptor on the ER. Mechanisms for Ca^{2+} influx from outside the cell include several voltage-sensitive Ca^{2+} channels and ligand-gated cation channels that are permeable to Ca^{2+} (e.g., nicotinic receptor and N-methyl-D-aspartate (NMDA) receptor). In the *Drosophila* visual system and in nonexcitable mammalian cells, depletion of Ca^{2+} from the cytosol and ER initiates an unknown signal that stimulates a low-conductance influx current called I_{CRAC} (Ca^{2+}-release-activated current) that replenishes ER stores.

Dynamics of Ca^{2+} signaling revealed by fluorescent Ca^{2+} indicators. We know a great deal about the spatial and temporal regulation of Ca^{2+} signals because of the development of fluorescent Ca^{2+} indicators. A variety of fluorescent compounds selectively bind Ca^{2+} at physiological concentration ranges and rapidly change their fluorescent properties upon binding Ca^{2+} to a fairly accurate measurement of ionized Ca^{2+} (Minta and Tsien, 1989). Digital fluorescence imaging can be used to detect Ca^{2+} in subcellular compartments such as dendrites and spines, the nucleus, and the cytosol and has demonstrated localized changes in free Ca^{2+} (see Chapter 19, Fig. 19.9).

Lack of uniformity in Ca^{2+} levels. The concentration of Ca^{2+} entering the cytosol through voltage-sensitive Ca^{2+} channels in the plasma membrane or through the IP_3 receptor in the ER is extremely high because of the large concentration gradient across these membranes. Relatively low-affinity Ca^{2+}-dependent processes can produce effects of Ca^{2+} near the membrane, such as synaptic release and modulation of Ca^{2+} channels. However, by the time Ca^{2+} diffuses a few membrane diameters away, it is rapidly buffered by many Ca^{2+}-binding proteins, and its concentration drops from 100 to 1 μM or less. The diffusion of Ca^{2+} is greatly slowed in biological fluid because of the high concentration of binding proteins (0.2–0.3 mM). Ca^{2+} diffuses a distance of 0.1–0.5 μm, and diffusion lasts approximately 30 ms before Ca^{2+} is bound. Ca^{2+} is therefore a second messenger that acts locally, a feature that makes Ca^{2+} subdomains possible where Ca^{2+} signaling is spatially segregated. In contrast, IP_3 is a global intermediate with an effective range that can span a typical soma before being terminated by dephosphorylation (Allbritton *et al.*, 1992).

Calmodulin-Mediated Effects of Ca^{2+}

Ca^{2+} acts as a second messenger to modulate the activity of many mediators. The predominant mediator of Ca^{2+} action is calmodulin, an abundant and ubiquitous 17 kDa calcium-binding protein. Ca^{2+} binds to calmodulin in the physiological range and converts it into an activator of many cellular targets (Cohen and Klee, 1988). Binding of Ca^{2+} to calmodulin produces a conformational change that greatly increases its affinity for target enzymes. Ca^{2+}-calmodulin binds and activates more than two dozen eukaryotic enzymes, including cyclic nucleotide PDEs, adenylyl cyclase, nitric oxide synthase, Ca^{2+}-ATPase, calcineurin (a phosphoprotein phosphatase), and several protein kinases (Fig. 12.9). This activation of calmodulin allows neurotransmitters that change the concentration of Ca^{2+} to affect dozens of cellular proteins, presumably in an orchestrated fashion. Ca^{2+} modulates some proteins and enzymes independently of calmodulin (Fig. 12.9).

Calmodulin interacts with its targets in several ways. The "conventional" interaction with enzyme targets requires a stimulated rise in Ca^{2+}. A second mode involves binding at basal Ca^{2+}, to proteins such as GAP-43 (neuromodulin), neurogranin, and unconventional myosins and may serve to localize calmodulin near other targets or reduce the level of free calmodulin (Persechini and Cronk, 1999). A third group of enzymes, which includes the inducible form of nitric oxide synthase, binds calmodulin in a

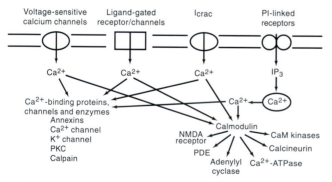

FIGURE 12.9 Multiple sources of Ca^{2+} converge on calmodulin and other Ca^{2+}-binding proteins. Cellular levels of Ca^{2+} can rise either by influx (e.g., calcium-release-activated calcium current (I_{CRAC}) channels) or by redistribution from intracellular stores triggered by IP_3. Calcium modulates dozens of cellular processes by the action of the Ca^{2+}-calmodulin complex on many enzymes, and calcium has some direct effects on enzymes such as PKC and calpain. CaM kinase, Ca^{2+}-calmodulin-dependent kinase.

manner that makes it sensitive to basal Ca^{2+}, and the enzymes are therefore active at basal Ca^{2+}.

Actions of enzymes and proteins modulated by calmodulin. Calmodulin has four Ca^{2+} binding sites, or binding folds, described as EF hands, a recurring Ca^{2+} binding structural motif. Calmodulin is composed of a number of helical segments designated by capital letters and separated by loops. Orientation of the helix–loop–helix EF segment is similar to that of the thumb and index finger, which positions amino acids in the loop for coordination with Ca^{2+}, hence the name EF hand. The ability of Ca^{2+} to be accommodated in an asymmetric coordination shell with multiple and distant amino acids, including uncharged oxygens, enables it to compete with Mg^{2+} and produce large conformation changes that are the basis for interconversion between inactive and active states of proteins.

Calmodulin recognizes a short segment of the enzymes that it regulates; however, there is no strict consensus sequence for calmodulin binding. X-ray crystallography and NMR show calmodulin to be in an extended structure composed of two globular regions, each containing a set of calmodulin folds separated by a long α-helical tether. Binding of Ca^{2+} allows movement of the globular regions around the calmodulin binding site, "gripping" it as would hands around a rope (Ikura *et al.*, 1992). The two lobes of this compact structure surround the target residues, making dozens of hydrophobic contacts as well as ionic interactions between Arg and Lys typically found in target sequences and Glu residues in calmodulin. This binding likely produces the necessary displacement of the calmodulin-binding domain for activation of the target enzymes.

Regulation of guanylyl cyclase by nitric oxide. An important target of Ca^{2+}-calmodulin is the enzyme nitric oxide synthase (NOS) (see Chapter 10). This enzyme synthesizes one of the simplest known messengers, the gas NO (Baranano *et al.*, 2001; Schmidt and Walter, 1994). Nitric oxide was first recognized as a signaling molecule that mediates the action of acetylcholine on smooth muscle relaxation (Furchgott and Zawadzki, 1980). In the pathway that led to its discovery, acetylcholine stimulates the PI signaling pathway in the endothelium to increase intracellular Ca^{2+}, which activates NOS so that more NO is made. NO then diffuses radially from the endothelial cells across two cell membranes to the smooth muscle cell, where it activates guanylate cyclase to make cGMP. This in turn activates a cGMP-dependent protein kinase that phosphorylates proteins, leading to a relaxation of muscle. In 1998, Robert F. Furchgott, Louis J. Ignarro, and Ferid Murad received the Nobel Prize for their discoveries concerning nitric oxide as a signaling molecule and therapeutic mediator in the cardiovascular system.

Let us now turn to the details of the NO pathway. We see that other pathways can activate NOS, mediate the actions of NO, stimulate guanylyl cyclases, and mediate the actions of cGMP.

Nitric oxide is derived from L-arginine in a reaction catalyzed by NOS, a complex enzyme that converts L-arginine and O_2 into NO and L-citrulline. NOS likely produces the neutral free radical NO as the active agent. NO lasts only a few seconds in biological fluids, and thus, no specialized processes are needed to inactivate this particular signaling molecule. As a gas, NO is soluble in both aqueous and lipid media and can diffuse readily from its site of synthesis across the cytosol or cell membrane and affect targets in the same cell or in nearby neurons, glia, and vasculature (Baranano *et al.*, 2001). NO produces a variety of effects, including relaxation of smooth muscle of the vasculature, relaxation of smooth muscle of the gut in peristalsis, and killing of foreign cells by macrophages (Schmidt and Walter, 1994). It was first recognized as a neuronal messenger that couples glutamate receptor stimulation to increases in cGMP. Analogs of L-arginine, such as nitroarginine and monomethyl arginine, block NOS unless there is an excess of L-arginine. Such inhibitors of NO synthase have been used to implicate NO in long-term potentiation and long-term depression in the hippocampus and cerebellum, respectively (Schuman and Madison, 1994) (see also Chapter 19).

Three classes of NO synthase have been characterized. Two of them are constitutively expressed and activated by Ca^{2+}-calmodulin whereas the third is induced by transcription and protein synthesis in response to cell stimulation, for example, in macrophages stimulated by cytokines. After translation, the inducible form is active at basal Ca^{2+} levels because it has a tightly bound calmodulin in a conformation that enhances its Ca^{2+} sensitivity greatly. The constitutive neuronal isoform is concentrated in cerebellar granule cells and likely provides the NO that activates guanylate cyclase in nearby Purkinje cells during the induction of long-term depression in the cerebellum (see Chapter 19).

Activation of guanylyl cyclases. Two types of guanylyl cyclase, a soluble one regulated by NO and a membrane-bound enzyme directly regulated by neuropeptides (Garbers and Lowe, 1994) synthesize cGMP from GTP in a reaction similar to the synthesis of cAMP from ATP. The soluble enzyme is a heterodimer, with catalytic sites resembling those of adenylyl cyclase and a heme group. NO activates the soluble enzyme by binding to the iron atom of the heme moiety. This is the basic mechanism for regulation of soluble guanylyl cyclases. Stimulation of guanylyl cyclase is the major, but not only, effect of NO in the brain and other tissues. A number of therapeutic muscle relaxants, such as nitroglycerin and nitroprusside, are NO donors that produce their effects by stimulating cGMP synthesis.

The membrane-bound guanylyl cyclases are transmembrane proteins with a binding site for neuroendocrine peptides on the extracellular side of the plasma membrane and a catalytic domain on the cytosolic side. Several isoforms of membrane-bound guanylyl cyclase, each with a binding site for a distinct neuropeptide such as atrial natriuretic peptide and brain natriuretic peptide, have been characterized. In the periphery, these peptides regulate sodium excretion and blood pressure; in the brain, their functions are less clear.

Cyclic GMP Phosphodiesterase, an Effector Enzyme in Vertebrate Vision

The versatility of G-protein signaling is illustrated in vertebrate phototransduction, in which a specialized G protein called *transducin* (G_t) is activated by light rather than by a hormone or neurotransmitter. Without transducin, we would not be able to see. Transducin stimulates cGMP phosphodiesterase, an effector enzyme that hydrolyzes cGMP and ultimately turns off the dark current. Nature has devised an elegant mechanism for using photons of light to modify a hormone-like molecule, retinal, that activates a seven-helix receptor called *rhodopsin* (Baylor, 1996). This receptor has a built-in prehormone that is converted into the active form by light. Light photoisomerizes the inactive 11-*cis*-retinal to the active all-*trans*-retinal, which functions as a neurotransmitter to activate its receptor. Activated rhodopsin triggers the GTP–GDP exchange of transducin, leading to dissociation of its α_t and $\beta\gamma$ subunits. The active species in transducin is the α subunit. It activates a soluble cGMP phosphodiesterase by binding to and displacing an inhibitory subunit of the enzyme. In the dark, retinal rods contain high levels of cGMP, which maintains a cGMP-gated channel permeable to Na^+ and Ca^{2+} in the open state and thus provides a depolarizing dark current. As the levels of cGMP drop, the channel closes to hyperpolarize the cell.

Rods can detect a single photon of light because the signal-to-noise ratio of the system is very low owing to a very low spontaneous conversion of the 11-*cis*-retinal into the all-*trans*-retinal. Furthermore, the amplification factor is quite high; one rhodopsin molecule stimulated by a single photon can activate 500 transducins. Transducin remains in the "on" state long enough to activate 500 PDEs. PDE is designed for speed and can hydrolyze 10^5 cGMP molecules in the second before it is deactivated by GTP hydrolysis and dissociated from transducin (Stryer, 1991). Cyclic GMP in rods regulates a cGMP-gated cation channel, leading to additional amplification of the signal.

Modulation of Ion Channels by G Protein

Each type of neuron has a repertoire of ion channels that give it a distinct response signature, and it is not surprising that several types of mechanisms regulate these channels. Channel modulation occurs via G proteins, second messengers and their cognate protein kinases that phosphorylate ion channels (see also Chapter 16) as well as by direct effects of G proteins.

The first ion channel demonstrated to undergo regulation by G proteins was the cardiac K^+ channel that mediates slowing of the heart by acetylcholine released from the vagus nerve. When this I_{KACh} channel is examined in a membrane patch delimited by the seal of a cell-attached electrode, addition of acetylcholine within the electrode increases the frequency of channel opening, whereas addition of acetylcholine to the cell surface outside the seal does not. Although acetylcholine stimulates muscarinic M2 receptors when added either inside or outside the seal, the receptors outside do not have access to the channels being recorded in the sealed patch because this signaling pathway does not include diffusible second messengers that can affect distant channels. The process is therefore described as membrane delimited, which is explained most simply by a direct interaction between the G protein and the channel. Subsequent studies have shown that the pathway is pertussis toxin sensitive and that purified G_i activated by GTPγS added to the underside of the patch will activate the channel.

Whether the active component of G_i is the α or the βγ subunit was controversial. Dogmas do not die easily and, for many years, βγ was not considered a direct activator or inhibitor of effector enzymes and channels. The I_{KACh} channel can be activated either by α_i or by βγ (Huang *et al.*, 1995; Wickman *et al.*, 1994).

Of the ion channels other than the K^+ channel, evidence is most compelling for the stimulation or inhibition of Ca^{2+} channel subtypes by G proteins. The central role played by Ca^{2+} in muscle contraction, in synaptic release, and in gene expression makes the modulation of Ca^{2+} influx a common target for regulation by neurotransmitters. In the heart, where L-type Ca^{2+} channels are critical for regulation of contractile strength, the Ca^{2+} current is enhanced by α_s formed by β-adrenergic stimulation of G_s. In contrast, N-type Ca^{2+} channels, which modulate synaptic release in nerve terminals, are often inhibited by muscarinic and α-adrenergic agents and by opiates acting at receptors coupled to G_i and G_o. In sympathetic ganglia, norepinephrine reduces synaptic release by inhibiting Ca^{2+} influx through the N channel by favoring the time spent in a low-open probability mode (Delcour and Tsien, 1993).

G-Protein Signaling Gives Special Advantages in Neural Transmission

The G-protein-based signaling system provides several advantages over fast transmission (Birnbaumer, 2007; Hille, 1992). These advantages include amplification of the signal, modulation of cell function over a broad temporal range, diffusion of the signal to a large cellular volume, cross talk, and coordination of diverse cell functions.

Amplification

Several thousand-fold amplification can be initiated by a single neurotransmitter–receptor complex that activates numerous G proteins, each of which activates many effector enzymes and channels. Each enzyme can generate many second-messenger molecules, and each channel allows the flux of many ions. As we see in the next section, second messengers often activate protein kinases that phosphorylate many substrates before deactivation.

Temporal Range

The sacrifice in speed relative to signaling by ligand-gated ion channels is compensated by a broad range of signaling that facilitates integration of signals by the G-protein system. Transmission through membrane-delimited coupling of ion channels to G proteins is relatively fast, with only some sacrifice in speed. Signaling that includes second messengers is much slower. It can be as fast as 100–300 ms, as in olfactory signaling in which cAMP and IP_3 take part, or it can take from seconds to minutes.

Spatial Range

A slower time frame means that cellular processes that are quite distant from the receptor can be modulated. Diffusion of second messengers such as IP_3, Ca^{2+}, and DAG can extend neurotransmission through the cell body and to the nucleus to alter gene expression.

Cross Talk

Both the signal transduction machinery and the ultimate mediators of their responses, such as the protein kinases, are capable of cross talk. This is seen in coincident detection of signals from two receptors converging on type I and type II adenylyl cyclase.

Coordinated Modulation

Neurotransmitters acting through G proteins can elicit a coordinated response of the cell that can modulate synaptic release, resynthesis of neurotransmitter, membrane excitability, the cytoskeleton, metabolism, and gene expression.

Summary

A major class of signaling using G-protein-linked signals affords the nervous system a rich diversity of modulation, amplification, and plasticity. Signals are mediated through second messengers activating proteins that modify cellular processes and gene transcription. A key feature is the ability of G proteins to detect the presence of activated receptors and amplify the signal through effector enzymes and channels. Phosphorylation of key intracellular proteins, ion channels, and enzymes activates diverse, highly regulated cellular processes. Specificity of response is ensured through receptors reacting with only a limited number of G proteins. The response of the system is determined by the speed of activation of GTPase. The function of G-protein subunits is now being elucidated. In addition to speed of response, the spatial compartmentalization of the system enables specificity and localized control of signaling. Phospholipids and phosphoinositols provide substrates for second-messenger signaling for G proteins. Stimulation of release of intracellular calcium is often the mediator of the signal. Calcium itself has a dual role as a carrier of electrical current and as a second messenger. Calmodulin is a key regulator that provides complexity and enhances specificity of the signaling system. Sensitivity of the system is imparted by an extremely robust amplification system, as seen in the visual system, which can detect single photons of light.

MODULATION OF NEURONAL FUNCTION BY PROTEIN KINASES AND PHOSPHATASES

Protein phosphorylation and dephosphorylation are key processes that regulate cellular function. They play a fundamental role in mediating signal transduction initiated by neurotransmitters, neuropeptides, growth factors, hormones, and other signaling molecules. The primary determinants of morphology and function of a cell are the protein constituents expressed in that cell. However, the functional state of many of these proteins is modified by phosphorylation–dephosphorylation, the most ubiquitous post-translational modification in eukaryotes. A fifth of all proteins may serve as targets for kinases and phosphatases. Phosphorylation or dephosphorylation can rapidly modify the function of enzymes, structural and regulatory proteins, receptors, and ion channels taking part in diverse processes without a need to change the level of their expression. As is described in greater detail in the following chapter, phosphorylation and dephosphorylation can also produce long-term alterations in cellular properties by modulating transcription and translation and changing the complement of proteins expressed by cells.

Protein kinases catalyze the transfer of the terminal, or γ, phosphate of ATP to the hydroxyl moieties of Ser, Thr, or Tyr residues at specific sites on target proteins. Most protein kinases are either Ser–Thr kinases or Tyr kinases, with only a few designed to phosphorylate both categories of acceptor amino acids. Protein phosphatases catalyze the hydrolysis of the phosphoryl groups from phosphoserine–phosphothreonine, phosphotyrosine, or both types of phosphorylated amino acids on phosphoproteins.

Regulation of the phosphorylation state of proteins is bidirectional, involving both kinases and phosphatases (Fig. 12.10). The phosphorylation state can be dynamically altered either upward or downward from the steady state, depending on the cell's inputs and its complement of kinases and phosphatases. The phosphorylation state of proteins *in vivo* ranges

widely, from minimal to almost fully phosphorylated, even in the absence of cell stimulation (Greengard *et al.*, 1998; Rosenmund *et al.*, 1994).

The activity of protein kinases and protein phosphatases is typically regulated either by a second messenger (e.g., cAMP or Ca^{2+}) or by an extracellular ligand (e.g., nerve growth factor). In general, the second-messenger-regulated kinases modify Ser and Thr, whereas the receptor-linked kinases modify Tyr. Among the hundreds of protein kinases and protein phosphatases in neurons, a relatively small number serve as master regulators to orchestrate neuronal function.

The cAMP-dependent protein kinase (PKA) is a prototype for the known regulated Ser–Thr kinases; they are similar in overall structure and regulatory design. PKA is emphasized here because the experimental strategies currently being used in the study of kinases have come from the investigation of PKA-mediated processes. As its name implies, cAMP-dependent protein kinase carries out the post-translational modification of numerous protein targets in response to signal transduction processes that act through G proteins and alter the level of cAMP in cells. PKA is the predominant mediator for signaling through cAMP, the only others being a cAMP-liganded ion channel in olfaction and an exchange protein directly activated by cAMP (Epac), which modulate GDP–GTP exchange for the small GTPases Rap1 and Rap2. Epacs are activated by cAMP independent of PKA action and regulate cell adhesion, cell-junction formation, and exocytosis (Bos, 2006). In a similar fashion, the related cGMP-dependent protein kinase (PKG) mediates most of the actions of cGMP. Ca^{2+}-calmodulin-dependent protein kinase II and several other kinases mediate many of the actions of stimuli that elevate intracellular Ca^{2+}. Finally, the PI signaling system increases both DAG and Ca^{2+}, which activate any of a family of protein kinases collectively called *protein kinase C*. Each of these kinases has a broad substrate specificity and is therefore able to phosphorylate diverse substrates throughout the cell. Phosphorylation and dephosphorylation are reversible processes, and the net activity of the two processes determines the phosphorylation state of each substrate. Among the many phosphoprotein phosphatases, a relatively small number exemplified by protein phosphatase 1 (PP-1), protein phosphatase 2A (PP-2A), and protein phosphatase 2B (PP-2B, or calcineurin) are responsible for most of the dephosphorylation at Ser and Thr residues on phosphoproteins that are under the regulation of the aforementioned kinases. The Nobel Prize for Physiology or Medicine was awarded to Edwin G. Krebs and

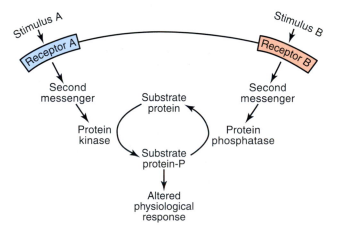

FIGURE 12.10 Regulation by protein kinases and protein phosphatases. Enzymes and other proteins serve as substrates for protein kinases and phosphoprotein phosphatases, which modify their activity and control them in a dynamic fashion. Multiple signals can be integrated at this level of protein modification. Adapted from Greengard *et al.* (1998).

Edmund H. Fischer in 1992 for their pioneering work on regulation of cell function by protein kinases and phosphatases.

Certain Principles Are Common in Protein Phosphorylation and Dephosphorylation

Protein kinases and protein phosphatases are described either as multifunctional, if they have a broad specificity and therefore modify many protein targets, or as dedicated, if they have a very narrow substrate specificity and may modify only a single protein target. The Ser–Thr kinases and phosphatases described here are multifunctional, but how is response specificity achieved with kinases and phosphatases that are designed to recognize many substrates? These enzymes are by no means promiscuous; their substrates conform either to a consensus sequence along the primary protein sequence (for the kinases) or to general features of the three-dimensional structure of the phosphoprotein (for the phosphatases). Furthermore, spatial positioning of kinases and their substrates in the cell either increases or decreases the likelihood of phosphorylation–dephosphorylation of a given substrate.

The amplification of signal transduction described earlier is continued during the transmission of the signal by protein kinases and protein phosphatases. In some cases, the kinases are themselves subject to activation by phosphorylation in a cascade in which one activated kinase phosphorylates and activates a second, and so on, to provide amplification and a switchlike response termed *ultrasensitivity* (Ferrell and Machleder, 1998). (See Chapter 14 for additional discussion of ultrasensitivity.)

Kinases and phosphatases integrate cellular stimuli and encode the stimuli as the steady-state level of phosphorylation of a large complement of proteins in the cell (Hunter, 1995). The phosphorylation state depends on the degree of activation or inactivation of the protein kinase or protein phosphatase, the affinity of the protein target for these enzymes, and the concentration and access of the kinase, phosphatase, and target protein. Some proteins are phosphorylated largely in the basal state and subject primarily to regulation of phosphatases. Distinct signal transduction pathways can converge on the same or different target substrates. In some cases, these substrates can be phosphorylated by several kinases at distinct sites.

Phosphorylation produces specific changes in the function of a target protein, such as increasing or decreasing the catalytic activity of an enzyme, the affinity of a protein with DNA, phospholipids, or other cellular constituents, and desensitization or localization of receptors. Any of several characteristics of ion channels can be altered by phosphorylation, including voltage dependence, probability of being opened, open and close time kinetics, and conductance. The number of possible effects is almost limitless and enables the fine-tuning of numerous cellular processes over broad timescales, from milliseconds to hours. Kinases and phosphatases do this fine-tuning by regulating the presence of a highly charged and bulky phosphoryl moiety on Ser, Thr, or Tyr at a precise location on the substrate protein. The phosphate may introduce a steric constraint at the surface of the protein in interactions with other cellular constituents, or the negative charge of the phosphoryl moiety may elicit a conformational change because of attractive or repulsive ionic interactions between the phosphorylated segment and other charged amino acids on the protein.

Finally, each of the three kinases described here is capable of functioning as a cognitive kinase, that is, a kinase capable of a molecular memory. Although each is activated by its respective second messenger, it can undergo additional modification that reduces its requirement for the second messenger. As is described in greater detail in Chapter 19, this molecular memory potentiates the activity of these kinases and may enable them to participate in aspects of neuronal plasticity.

cAMP-Dependent Protein Kinase Was the First Well-Characterized Kinase

Neurotransmitters that stimulate the synthesis of cAMP exert their intracellular effects primarily by activating PKA (Nairn *et al.*, 1985). The functions (and substrates) regulated by PKA include gene expression (cAMP response element-binding protein, or CREB), catecholamine synthesis (tyrosine hydroxylase), carbohydrate metabolism (phosphorylase kinase), cell morphology (microtubule-associated protein 2, or MAP-2), postsynaptic sensitivity (AMPA receptor), and membrane conductance (K^+ channel). Paul Greengard and Eric Kandel received the Nobel Prize for Physiology or Medicine in 2000 (along with Arvid Carlsson) for their discoveries concerning signal transduction via PKA and phosphoprotein phosphatases in the nervous system.

PKA is a tetrameric protein composed of two types of subunits: (1) a dimer of regulatory (R) subunits (either two RI subunits for type I PKA or two RII subunits for type II PKA) and (2) two catalytic subunits (C subunit) (Scott, 1991). Two or more isoforms of the RI, RII, and C subunits have distinct tissue and developmental patterns of expression but appear to function similarly. The C subunits are 40 kDa proteins that contain the binding sites for protein substrates and ATP. The R subunits are 49 to 51 kDa proteins that contain two cAMP binding sites. In addition, the R subunit dimer contains a region that interacts with cellular-anchoring proteins that serve to localize PKA appropriately within the cell.

The binding of second messengers by PKA and the other second-messenger-regulated kinases relieves an inhibitory constraint and thus activates the enzymes (Fig. 12.11). The C subunit has intrinsic protein kinase activity that remains inhibited as long as the C subunit is complexed with the R subunits in the tetrameric holoenzyme. Cyclic AMP activates the C subunit by facilitating dissociation of its inhibitory R subunits. The steady-state level of cAMP determines the fraction of PKA that is in the dissociated or active form. In this way PKA decodes cAMP signals into the phosphorylation of proteins and the resultant change in various cellular processes.

PKA is a member of a large family of protein kinases that have in common a significant degree of homology in their catalytic domains and are likely derived from an ancestral gene (Hanks and Hunter, 1995) (Fig. 12.12). This homology extends to the three-dimensional crystal structure, based on X-ray crystallography of PKA and a few other kinases. The catalytic domain comprises approximately 280 amino acids that may be in a subunit distinct from the regulatory domain, as in PKA, or in the same subunit, as in PKC and Ca^{2+}-calmodulin-dependent (CaM) kinases. The crystal structure of the C subunit complexed to a segment of protein kinase inhibitor (PKI), a selective high-affinity inhibitor of PKA, reveals that the C subunit is composed of two lobes

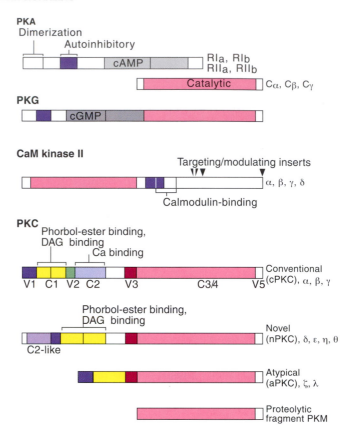

FIGURE 12.12   Domain structure of protein kinases. Protein kinases are encoded by proteins with recognizable structural sequences that encode specialized functional domains. Each of the kinases (PKA, PKG, CaM (Ca^{2+}-calmodulin-dependent) kinase II, and PKC) have homologous catalytic domains that are kept inactive by the presence of an autoinhibitory segment (blue segments). The regulatory domains contain sites for binding second messengers such as cAMP, cGMP, Ca^{2+}-calmodulin, DAG, and Ca^{2+}-phosphatidylserine. Alternative splicing creates additional diversity.

(Knighton *et al.*, 1991). A small N-terminal lobe contains a highly conserved region that binds Mg^{2+}-ATP in a cleft between the two lobes. A larger C-terminal lobe contains the protein–substrate recognition sites and the appropriate amino acids for catalyzing the transfer of the phosphoryl moiety from ATP to the polypeptide chain of the substrate. Inhibition by PKI is diagnostic of PKA involvement; PKI contains an autoinhibitory sequence resembling PKA substrates and is positioned in the catalytic site like a substrate, thus blocking access for substrates.

How can protein kinases utilize a homologous structure yet exhibit phosphorylation target specificity? Although the C-terminal lobes of all kinases may use a similar scaffold for their peptide binding and catalytic sites, distinct amino acids are positioned on this scaffold to produce specificity in peptide binding. Some substrates also have a distinct docking site

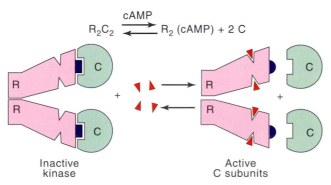

$$R_2C_2 \underset{\longleftarrow}{\overset{cAMP}{\longrightarrow}} R_2 (cAMP) + 2\ C$$

FIGURE 12.11   Activation of PKA by cAMP. An autoinhibitory segment (blue) of the regulatory subunit (R) dimer interacts with the substrate-binding domain of the catalytic (C) subunits of PKA, blocking access of substrates to their binding site. Binding of four molecules of cAMP reduces the affinity of R for C, resulting in dissociation of constitutively active C subunits.

for the kinase that localizes the kinase and serves to further increase the selectivity of phosphorylation (Smith *et al.*, 2000).

PKA phosphorylates Ser or Thr at specific sites in dozens of proteins. The sequences of amino acids at the phosphorylation sites are not identical, but a consensus sequence can be deduced from a comparison of these sequences (Kennelly and Krebs, 1991). PKA phosphorylates at sites with the consensus sequence Arg–Arg–X–Ser/Thr–Y, in which X can be one of many different amino acids and Y is a hydrophobic amino acid. Each kinase has a characteristic consensus sequence that forms the basis for distinct substrate specificities although many substrates are phosphorylated at "anomalous" (Walsh and Patten, 1994). The consensus sites of PKA, CaM kinase II, and PKC all include a basic residue on the substrate, and these kinases do share some target substrates.

A regulatory theme common to PKA, CaM kinase II, and PKC is that their second messengers activate them by displacing an autoinhibitory domain from the active site; that is, they activate by relieving an inhibitory constraint (Kemp *et al.*, 1994). Some of the contacts between the C and R subunits of PKA resemble those between the C subunit and its protein substrates. The R subunit blocks access of substrates by positioning a pseudosubstrate or autoinhibitory domain in the catalytic site. This segment of R resembles a substrate and binds to C as would a substrate or PKI (protein kinase inhibitor). Binding of cAMP to the R subunit near this autoinhibitory domain must disrupt its binding to the C subunit, thus leading to dissociation of an active C subunit (Su *et al.*, 1995). CaM kinase II and PKC likewise have autoinhibitory segments that are near the second-messenger binding sites and may be activated similarly (Kemp *et al.*, 1994) (see Fig. 12.12).

Functional differences between type I and type II PKA (which have C subunits in common but have different R subunits) may arise from differential targeting in cells and from differences in regulation by autophosphorylation. RII, but not RI, is autophosphorylated by its C subunit when it is in the holoenzyme form. This potentiates cAMP action by reducing the rate of reassociation of RII and C after a stimulus. Only anchoring proteins for RII have been characterized thus far (Beene and Scott, 2007).

Multifunctional CaM Kinase II Decodes Diverse Signals That Elevate Intracellular Ca^{2+}

Most of the effects of Ca^{2+} in neurons and other cell types are mediated by calmodulin, and many of the effects of Ca^{2+}-calmodulin are mediated by protein phosphorylation–dephosphorylation. In contrast with the cAMP system, both dedicated and multifunctional kinases are found in the Ca^{2+} signaling system (Schulman and Braun, 1999). Two kinases, MLCK and phosphorylase kinase, are each dedicated to the phosphorylation of a single substrate—myosin light chains and phosphorylase, respectively. The Ca^{2+} signaling system also contains a family of Ca^{2+}-calmodulin-dependent protein kinases with broad substrate specificity, including CaM kinases I, II, and IV; of these, CaM kinase II is the best characterized. CaM kinase II phosphorylates tyrosine-hydroxylase, MAP-2, synapsin I, calcium channels, Ca^{2+}-ATPase, transcription factors, and glutamate receptors and thereby regulates synthesis of catecholamines, cytoskeletal function, synaptic release in response to high-frequency stimuli, calcium currents, calcium homeostasis, gene expression, and synaptic plasticity, respectively. The enzyme is activated by Ca^{2+} influx or release from intracellular stores. It is found in every tissue but is particularly enriched in neurons, where it may account for as much as 2% of all hippocampal protein. It is found in the cytosol, in the nucleus, in association with cytoskeletal elements, and in postsynaptic thickening, termed the *postsynaptic density*, which is found in asymmetric synapses. CaM kinase II is a large multimeric enzyme, consisting of 12 subunits derived from four homologous genes (α, β, γ, and δ) that encode different isoforms of the kinase that range from 54 to 72 kDa per subunit. Multimers and heteromultimers of α- and β-CaM kinase II isoforms are found predominantly in brain, whereas the γ- and δ-CaM kinases are found throughout the body, including the brain.

The catalytic, regulatory, and targeting–association domains of CaM kinase II are all contained within a single polypeptide (Fig. 12.12.). The N-terminal half of each isoform contains the catalytic domain, which is highly homologous to the catalytic subunit of PKA and other Ser–Thr kinases. The middle region constitutes the regulatory domain, which contains an autoinhibitory domain with an overlapping calmodulin-binding sequence. The C-terminal end contains an association domain that allows 12 subunits (two rings of 6 subunits) to assemble into a multimer (Rosenberg *et al.*, 2006), as well as targets sequences that direct the kinase to distinct intracellular sites.

Regulation of the kinase by autophosphorylation is a critical feature of CaM kinase II. The kinase is inactive in the basal state because an autoinhibitory segment distorts the active site and sterically blocks access to its substrates (Rosenberg *et al.*, 2005). Binding of Ca^{2+}-calmodulin to the calmodulin-binding domain displaces the autoinhibitory domain from the catalytic site thus activating the kinase by enabling protein substrates and ATP to bind. The activated kinase autophosphorylates Thr-286 (in α-CaM kinase II and comparable Thr on other isoforms). Phosphorylation of this site disables the autoinhibitory segment by preventing it from reblocking the active site after calmodulin dissociates and thereby locks the kinase in a partially active state that is independent, or autonomous, of Ca^{2+}-calmodulin (Hanson et al., 1989; Miller and Kennedy, 1986; Saitoh and Schwartz, 1985) and can anchor to additional targets. Displacement of this domain also exposes a binding site for anchoring proteins that the activated kinase can bind (Bayer et al., 2001). Interestingly, such binding to both the NMDA receptor and eag K^+ channel leads to a persistently active kinase without the need for autophosphorylation and may do so by similarly keeping the autoinhibitory domain from reblocking the active site (Bayer et al., 2001; Sun et al., 2004).

An additional dramatic effect of autophosphorylation is that it enhances the affinity of the bound calmodulin by 400-fold, which it achieves by reducing the rate of dissociation of calmodulin from the kinase after Ca^{2+} levels are reduced below threshold. In essence, autophosphorylation traps bound calmodulin for several seconds and keeps the kinase active for a while after Ca^{2+} levels decline to baseline. The consequence of calmodulin trapping and disruption of the autoinhibitory domain is to prolong the active state of the kinase, a potentiation that led to its description as a cognitive kinase (Lisman, 1994; Schulman and Braun, 1999).

CaM kinase II is targeted to distinct cellular compartments. Differences between the four genes encoding CaM kinase II and between the two or more isoforms that are encoded by each gene by apparent alternative splicing reside primarily in a variable region at the start of the association domain (see Fig. 12.12). In some isoforms, this region contains an additional sequence of 11 amino acids that targets those isoforms to the nucleus. The major neuronal isoform, α-CaM kinase II, is largely cytosolic but is also found attached to postsynaptic densities and to synaptic vesicles and may therefore have several targeting sequences. Targeting to the NMDA-type glutamate receptor and to the eag K^+ channel occurs only after calmodulin activates the kinase and exposes a binding site (Bayer et al., 2001; Sun et al., 2004).

Protein Kinase C Is the Principal Target of the PI Signaling System

Protein kinase C (PKC) is a collective name for members of a relatively diverse family of protein kinases most closely associated with the PI signaling system. PKC is a multifunctional Ser–Thr kinase capable of modulating many cellular processes, including exocytosis and endocytosis of neurotransmitter vesicles, neuronal plasticity, gene expression, regulation of cell growth and cell cycle, ion channels, and receptors. The role of DAG generated during PI signaling was unclear until its link to PKC was established (Takai et al., 1979; Tanaka and Nishizuka, 1994). Many PKC isoforms also require an acidic phospholipid such as phosphatidylserine for appropriate activation. The kinase is also of interest because it is the target of a class of tumor promoters called phorbol esters. They activate PKC by simulating the action of DAG, bypassing the normal receptor-based pathway and inappropriately stimulating cell growth.

The PKC family of kinases is diverse in structure and regulatory properties (Tanaka and Nishizuka, 1994). Unlike PKA, PKC is a monomeric enzyme (78–90 kDa) with catalytic, regulatory, and targeting domains all on one polypeptide. Each isoform has a regulatory domain, with several subdomains, in its N-terminal half and a catalytic domain at the C terminal (Newton, 1995) (see Fig. 12.12). Only the first PKC isoforms to be characterized, now termed the conventional isoforms (or cPKC), have all the domains. The domains are referred to as (1) V1, which contains the autoinhibitory or pseudosubstrate sequence present in all isoforms; (2) C1, a cysteine-rich domain that binds DAG and phorbol esters; (3) C2, a region necessary for Ca^{2+} sensitivity and for binding to phosphatidylserine and to anchoring proteins; (4) V3, a protease-sensitive hinge; (5) C3/4, the catalytic domain; and (6) V5, which may also mediate anchoring. Another class of isoforms, termed novel PKCs (nPKC), lacks a true C2 domain and is therefore not Ca^{2+} sensitive. Another class is considered atypical (aPKC) because it lacks C2 and the first of two cysteine-rich domains that are necessary for DAG (or phorbol ester) sensitivity. This class is neither Ca^{2+} nor DAG sensitive. Not included is a DAG-interacting kinase originally designated as PKCμ and now termed PKD because its catalytic domain is different from the other PKC isoforms.

Activation of PKC is best understood for the conventional isoforms. Generation of DAG (specifically its sn-1,2-diacylglycerol isomer) resulting from stimulation of the PI signaling pathway increases the affinity of cPKC isoforms for Ca^{2+} and phosphatidylserine. Cell stimulation results in the translocation of cPKC from a variety of sites to the membrane or cytoskeletal elements where it interacts with PS-Ca^{2+}-DAG at the membrane (Kraft and Anderson, 1983; Newton, 1995; Ron et al., 1994; Zhang et al., 1995). Binding of the second messengers to the regulatory domain disrupts the nearby autoinhibitory domain, leading to a reversible activation of PKC by deinhibition, as is found for PKA and CaM kinase II (Muramatsu *et al.*, 1989).

Translocation is not restricted to the plasma membrane. Upon activation some PKC isoforms translocate reversibly to intracellular sites enriched with certain anchoring proteins for the activated form of PKC, termed RACK (receptors for activated C kinase) (Mochly-Rosen, 1995; Ron *et al.*, 1994). Activation may consist of both displacement of the autoinhibitory segment to unblock the catalytic site and displacement of an "auto-anchor" site to unblock the RACK binding site.

Prolonged activation of PKC can be produced by the addition of phorbol esters, which simulate activation by DAG but remain in the cell until they are washed out. In a matter of hours to days, such persistent activation by phorbol esters leads to a degradation of PKC. This phenomenon is sometimes used experimentally to produce a PKC-depleted cell (at least for phorbol-ester-binding isoforms) and thereafter to test for a loss of putative PKC functions.

Spatial Localization Regulates Protein Kinases and Phosphatases

Protein kinases and protein phosphatases are often localized near their substrates, or they translocate to their substrates upon activation to improve speed and specificity in response to neurotransmitter stimulation. PKA is targeted to intracellular sites on the cytoskeleton, membrane, and Golgi through interactions between the RII subunit and specific anchoring proteins (Beene and Scott, 2007). One of its anchoring proteins, A Kinase Anchoring Protein 79 (AKAP79) is an anchor for PKA, for PKC, and for calcineurin, the Ca^{2+}-calmodulin-dependent phosphatase (Klauck *et al.*, 1996). Better coordination of the phosphorylation–dephosphorylation of the same or different substrates may be achieved by placing calcineurin, PKC, and PKA in the same compartment through AKAPs. Another example of a signaling complex is the protein termed *yotiao*, which binds to the NMDA-type glutamate receptor and serves as an anchor for both PKA and a phosphatase (PP-1) (Westphal *et al.*, 1999).

The use of anchoring proteins has several consequences. First, rate of phosphorylation and specificity are enhanced when kinases or phosphatases are concentrated near intended substrates. Second, it increases the signal-to-noise ratio for substrates that are not near anchoring proteins by reducing basal-state phosphorylation. For example, PKA is anchored on the Golgi away from the nucleus so that phosphorylation in the basal state or even after a brief stimulus produces little phosphorylation of nuclear proteins. Prolonged stimuli, however, enable some C subunits to passively diffuse through nuclear pores and into nuclei, where they can participate in regulation of gene expression (Bacskai *et al.*, 1993). Termination of the nuclear action of C subunits is aided by PKI, which inhibit and export it back out of the nucleus (Wen *et al.*, 1995). Third, concentration of kinases via anchoring enables significant phosphorylation of nearby substrates even at low level of kinase activation, such as in the basal state. In such a situation the regulation of the substrate would occur by pathways that modulate its phosphatase or that modify the exposure of the phosphorylation site on the substrate to the nearby kinase.

PKA, CaM Kinase II, and PKC Are Cognitive Kinases

The ability of three major Ser–Thr kinases (PKA, CaM kinase II, and PKC) in brain to initiate or maintain synaptic changes that underlie learning and memory may require that they themselves undergo some form of persistent change in activity. As mentioned earlier, they have been described as cognitive kinases because they are capable of sustaining their activated states after their second messengers return to basal level and because their target substrates modulate synaptic plasticity (Schwartz, 1993).

cAMP-Dependent Protein Kinase

As is discussed in greater detail in Chapter 19, a role for PKA as a cognitive kinase can be seen in long-term facilitation of the withdrawal reflexes in *Aplysia* and in long-term potentiation in the vertebrate hippocampus. In *Aplysia*, stimulation of the body with a strong stimulus such as an electric shock facilitates the withdrawal response to a light touch delivered to another part of the animal. This is an example of a simple form of learning called *sensitization*. A single shock produces a short-lasting memory, but repeated shocks (training) produce a memory that can last several days. The shock stimulates the release of serotonin, which increases cAMP and PKA activity in the sensory neurons. Features of the behavioral training can be simulated in a system in which a single sensory neuron is co-cultured with a single motor neuron and serotonin is applied to the bath to mimic the effects of sensitizing stimuli. A single exposure to serotonin (or cAMP) produces short-term facilitation and a short-term increase in the phosphorylation of more than a dozen PKA substrates in these cells. However, repeated or prolonged exposure to these agents leads to long-term facilitation and an enhanced

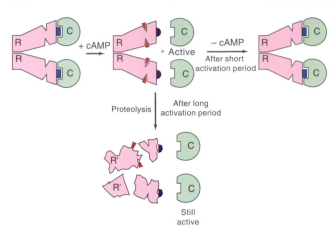

FIGURE 12.13 Long-term stimulation can convert PKA into a constitutively active enzyme. Dissociation of PKA R and C subunits is reversible with short-term elevation of cAMP. More prolonged activation results in loss of R subunits to proteolysis, resulting in an insufficient amount of R subunits to associate with and inhibit all C subunits after the cAMP stimulus terminates.

state of phosphorylation of the same set of proteins. This phenomenon is due to a PKA that is persistently active despite the fact that cAMP is no longer elevated (Chain *et al.*, 1999). A possible scheme for this phenomenon is shown in Figure 12.13 (see also Fig. 19.24). Phospho-RII and C subunits dissociate on elevation of cAMP, and the rate of reassociation following the decrease in cAMP levels is greatly reduced by the presence of phosphate on the R subunit, thus prolonging phosphorylation of various target proteins by the C subunit. The R subunits are more susceptible than the C subunits to proteolytic degradation in their dissociated state, and thus prolonged or repetitive stimulation leads to a preferential decrease in the inhibitory R subunits. At the end of the stimulation there is a slight excess of C subunits that remain persistently active because of insufficient R subunits. The various targets of PKA can then be phosphorylated by this active C subunit long after cAMP levels return to basal or prestimulus levels. Prolonged activation of PKA enables the C subunit to enter the nucleus and induce gene expression, and one of these genes facilitates further proteolysis of R. Although short-term facilitation does not require transcription of new genes or protein synthesis, the long-term effects on phosphorylation and on facilitation do require transcription of new genes and protein synthesis (Kaang *et al.*, 1993). In this interesting process, a molecular memory of appropriate stimulation by serotonin is encoded by a persistence of PKA activity that is regenerative.

Ca²⁺-Calmodulin-Dependent Protein Kinase

CaM kinase II has features of a cognitive kinase because it has a molecular memory of its activation that is based on autophosphorylation and it phosphorylates proteins that modulate synaptic plasticity (Lisman *et al.*, 2002). The biochemical properties of CaM kinase II suggest mechanisms by which appropriate stimulus frequencies can generate an autonomous enzyme (Fig. 12.14; see also Fig. 14.3 in Chapter 14). Autophosphorylation takes place within each holoenzyme but requires the phosphorylation of one subunit by a proximate neighbor. Individual stimuli may be too brief and available calmodulin may be limited so a single stimulus would activate only a few subunits per holoenzyme. At low stimulus frequency, the time between stimuli is sufficient for calmodulin to dissociate and the kinase to be dephosphorylated, and the same submaximal activation will occur with each stimulus. Brief stimuli at low frequency produce minimal autophosphorylation because autophosphorylation occurs only when two proximate neighboring subunits are simultaneously activated, apparently because Ca²⁺-calmodulin binding exposes the phosphorylation site on the "substrate" subunit (Hanson *et al.*, 1994). However, at higher frequencies, some subunits will remain autophosphorylated and bound to calmodulin so successive stimuli will result in more calmodulin bound

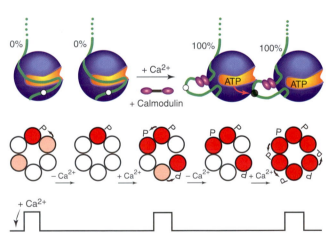

FIGURE 12.14 Frequency-dependent activation of CaM kinase II. Autophosphorylation occurs when both of two neighboring subunits in a holoenzyme are bound to calmodulin. At high frequency of stimulation (rapid Ca²⁺ spikes), the interspike interval is too short to allow significant dephosphorylation or dissociation of calmodulin, thereby increasing the probability of autophosphorylation with each successive spike. In a simplified CaM kinase with only six subunits, calmodulin-bound subunits are shown in pink, and autophosphorylated subunits with trapped calmodulin are shown in red. Adapted from Hanson and Schulman (1992).

per holoenzyme, which will make autophosphorylation more probable. The enzyme is therefore able to decode the frequency of cellular stimulation and translate this into a prolonged activated state. Low-frequency stimulation leads to submaximal activation of the kinase at each stimulus, whereas higher frequencies exceed a threshold beyond which stimulation leads to recruitment of additional calmodulin and a higher level of activation and autonomy with each spike (De Koninck and Schulman, 1998). See Chapter 14 for additional discussion of the dynamics and effects of CaM kinase II autophosphorylation.

CaM kinase II phosphorylates a number of substrates that affect synaptic strength. For example, phosphorylation of synapsin I, a synaptic vesicle protein (see Chapter 8), by CaM kinase II reduces the attachment of synaptic vesicles to actin at nerve terminals. Inhibition of CaM kinase II in hippocampal slices or just elimination of its autophosphorylation by an α-CaM kinase II mouse knock-in in which the critical Thr was replaced by Ala blocks autonomy and the induction of long-term potentiation (Giese et al., 1998). These mice are deficient in learning spatial navigational cues, one of the functions of the rodent hippocampus. The enzyme can therefore be appropriately described as a cognitive kinase with respect to its own molecular memory as well as its functional role in mediating aspects of synaptic plasticity.

Protein Kinase C

PKC also can be converted into a form that is independent, or autonomous, of its second messenger and can be described as a cognitive kinase. Before Ca^{2+} and DAG were known to have roles in the reversible activation of PKC, the PKC was identified as an inactive precursor that was activated *in vitro* by Ca^{2+}-dependent proteolysis to a constitutively active fragment termed *protein kinase M* (PKM). Physiological activation of PKC can lead to proteolyic removal of its inhibitory domain, thus converting it to a constitutively active kinase often termed *PKM*. For example, during the persistent phase of long-term potentiation, some of the PKC is converted to PKM (Serrano et al., 2005). PKC (and PKM) substrates associated with long-term potentiation include NMDA and AMPA receptors.

Protein Tyrosine Kinases Take Part in Cell Growth and Differentiation

Protein kinases that phosphorylate tyrosine residues on key proteins participate in numerous cellular processes and are usually associated with regulation of cell growth and differentiation. Signal transduction by protein tyrosine kinases often includes a cascade of kinases phosphorylating other kinases, eventually activating Ser–Thr kinases, which carry out the intended modification of a cellular process. There are two classes of protein tyrosine kinases. The first is a family of receptor tyrosine kinases that are activated by the binding of extracellular growth factors such as nerve growth factor, epidermal growth factor, insulin, and platelet-derived growth factor. The second family of protein tyrosine kinases such as c-Src are soluble kinases that also participate in regulation of cell growth, as well as in neuronal plasticity, but are indirectly activated by extracellular ligands.

Why have two sets of amino acids been chosen as targets for phosphorylation? First, the consequences of leaky, or "promiscuous," phosphorylation by a protein Ser–Thr kinase of an unintended target may affect metabolic activity or synaptic function but do not typically initiate irreversible and global functions such as cell growth and differentiation. The consequence of such inappropriate stimulation is seen in the effect of a variety of oncogenes that use altered forms of receptor tyrosine kinases or intermediates in their cascades to subvert normal cell growth. The cellular concentrations of protein Ser–Thr kinases and their targets are much higher than those of protein tyrosine kinases and their substrates. Inadvertent phosphorylation of targets that play critical roles in cell growth is less likely if these targets are regulated at tyrosine residues, which are not well recognized by the numerous protein Ser–Thr kinases. Second, introduction of a phosphotyrosine structure into a protein has a greater regulatory potential than does introduction of a phosphoserine or phosphothreonine. The three phosphorylated amino acids have in common an ability to produce conformational changes due to the extra charge or bulk of the phosphate. For example, in the activation of receptor tyrosine kinases, autophosphorylation displaces an inhibitory domain. In addition, however, the phosphotyrosine and nearby amino acid sequences can be recognized by various signal transduction effectors, such as PLCγ, that contain structural domains that bind to the tyrosine-phosphorylated kinase. The receptor tyrosine kinase thus becomes a platform for concentrating various signaling molecules at specific phosphotyrosine sites in its sequence. These signaling molecules either are activated directly by binding or are activated after having been phosphorylated by the receptor tyrosine kinase. It is easier to bind with the necessary strict specificity to segments of protein around phosphotyrosines, because of the aromatic side chain in tyrosine, and this may be an additional reason for use of Tyr as phosphotransferase targets.

Protein Phosphatases Undo What Kinases Create

Protein phosphatases in neuronal signaling are categorized as either phosphoserine–phosphothreonine phosphatases (PSPs) or phosphotyrosine phosphatases (PTPs) (Hunter, 1995; Mansuy and Shenolikar, 2006). The enzymes catalyze the hydrolysis of the ester bond of the phosphorylated amino acids to release inorganic phosphate and the unphosphorylated protein. Phosphatases control all the cellular processes of protein kinases, including neurotransmission, neuronal excitability, gene expression, protein synthesis, neuronal plasticity, and cell growth. A limited number of multifunctional PSPs account for most of such phosphatase activity in cells (Hunter, 1995; Mansuy and Shenolikar, 2006). They are categorized into six groups (1, 4, 5, 2A, 2B, and 2C) on the basis of their substrates, inhibitors, and divalent cation requirements (Table 12.2). Of these PSPs, only protein phosphatase 2B (PP-2B, or calcineurin) directly responds to a second messenger; it responds to increases in cellular Ca^{2+}. PP-1, -4, -5, -2A, and calcineurin are structurally related and differ from PP-2C. Little is known about the basis of substrate specificity of these phosphatases; examination of the primary sequences of their dephosphorylation sites reveals no obvious consensus. The specificity of PP-1 and PP-2A is particularly broad, and each can remove phosphates that were transferred by any of the protein kinases discussed herein as well as many other kinases. The phosphotyrosine phosphatases constitute a distinct and larger class of phosphatases, including PTPs with dual specificity for both phosphotyrosines and phosphoserines–phosphothreonines. PTPs are either soluble enzymes or membrane proteins with variable extracellular domains that enable regulation by extracellular binding of either soluble or membrane-bound signals.

Structure and Regulation of PP-1 and Calcineurin

PP-1 and calcineurin are the best-characterized phosphatases with regard to both structure and regulation. The domain structures of the catalytic subunits of PP-1 and calcineurin are depicted in Figure 12.15. PP-1 is a protein of 35–38 kDa; most of its sequence forms the catalytic domain, and its C terminal is the site of regulatory phosphorylation. The catalytic domains of PP-1, PP-2A, and calcineurin are highly homologous (Price and Mumby, 1999).

PP-1 and PP-2A are normally complexed in cells with specific anchoring or targeting subunits (Hubbard and Cohen, 1993; Price and Mumby, 1999). For example, PP-1 is attached to glycogen particles in liver, myofibrils in muscle, and unidentified targeting subunits in brain. Phosphorylation of the PP-1 targeting subunit in liver releases the catalytic subunit and results in reduced dephosphorylation of substrates near the targeting subunit because diffusion reduces the local concentration of PP-1. Targeting of PP-1 also modulates its regulation by natural inhibitors. As PP-1 dissociates from targeting subunits, it becomes susceptible to inhibition by inhibitor 2.

Inhibition of PP-1 by two other inhibitors, inhibitor 1 and its homolog DARPP-32 (dopamine and cAMP-regulated phosphoprotein, M_r 32,000), is conditional on the phosphorylation state of these inhibitors (Hemmings et al., 1984). Inhibitor 1 is broadly distributed in brain, whereas DARPP-32 is largely

---

TABLE 12.2 Categories of Protein Phosphatases

Phosphatase	Characteristic	Other inhibitors
PP-1	Sensitive to phospho-inhibitor 1, phospho-DARPP-32, and inhibitor 2; has targeting subunits	Weakly sensitive to okadaic acid
PP-4	Nuclear	Highly sensitive to okadaic acid
PP-5	Nuclear	Mildly sensitive to okadaic acid
PP-2A	Regulatory subunits	Highly sensitive to okadaic acid
	Does not require divalent cation	
PP-2B (calcineurin)	Ca^{2+}/calmodulin-dependent CnB regulatory subunit	FK506, cyclosporin
PP-2C	Requires Mg^{2+}	EDTA
Receptor PTPs[a]	Plasma membrane	Vanadate, tyrphostin, erbstatin
Nonreceptor PTPs	Various cellular compartments	Vanadate, tyrphostin
Dual-specificity PTPs	Nuclear (e.g., cdc25A/B/C and VH family)	Vanadate

[a]Protein phosphotyrosine phosphatases.
Adapted from Hunter (1995).

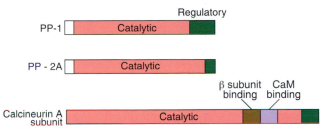

FIGURE 12.15 Domain structure of the catalytic subunits of some Ser–Thr phosphatases. The three major phosphoprotein phosphatases, PP-1, PP-2A, and calcineurin, have homologous catalytic domains but differ in their regulatory properties.

found in the medium spiny neurons in the neostriatum and in their terminals in the globus pallidus and substantia nigra. Both proteins inhibit only after they are phosphorylated by PKA or PKG. PKA also increases the susceptibility of PP-1 to inhibition by stimulating its release from targeting subunits. Because the substrates for PKA and PP-1 overlap somewhat, the rate and extent of phosphorylation of such substrates are enhanced by the ability of PKA to both catalyze their phosphorylation and block their dephosphorylation. Inhibitor 1, DARPP-32, and inhibitor 2 are all selective for PP-1. Highly selective inhibitors capable of penetrating the cell membrane are available for these phosphatases. Okadaic acid, a natural product of marine dinoflagellates, is a tumor promoter, but, unlike phorbol esters, it acts on PP-2A and PP-1, rather than on PKC.

Protein phosphatase 1. The X-ray structure of the catalytic subunit of PP-1 bound to the toxin microcystin, a cyclic peptide inhibitor, reveals PP-1 to be a compact ellipsoid with hydrophobic and acidic surfaces forming a cleft for binding substrates (Goldberg *et al.*, 1995). PP-1 is a metalloenzyme requiring two metals in the active site that likely take part in electrostatic interactions with the phosphate on substrates that aid in catalyzing the hydrolytic reaction. The phosphate would be positioned at the intersection of two grooves on the surface of the enzyme where binding to amino acid residues on the substrate would occur. Such binding would be blocked when phosphoinhibitor 1 or microcrystin LR binds to this surface. The same general structure of the catalytic domain is seen in calcineurin.

Calcineurin (PP-2B). Calcineurin is a Ca^{2+}-calmodulin-dependent phosphatase that is highly enriched in the brain. It is a heterodimer with a 60 kDa A subunit (CnA) that contains an N-terminal catalytic domain and a C-terminal regulatory domain that includes an autoinhibitory segment, a calmodulin-binding domain, and a binding site for the 19 kDa regulatory B subunit (CnB) (Rusnak and Mertz,

2000). CnB is a calmodulin-like Ca^{2+}-binding protein that binds to a hinge region of CnA. Regulation of calcineurin takes place in this region because it controls access of phosphoproteins to the catalytic site. Some activation of calcineurin is attained by binding of Ca^{2+} to CnB. Stronger activation is obtained by the binding of Ca^{2+}-calmodulin.

The Ca^{2+}-calmodulin sensitivity of calcineurin and CaM kinase II are quite different. Weak or low-frequency stimuli may selectively activate calcineurin, whereas strong or high-frequency stimuli activate CaM kinase II and calcineurin. This difference may play a role in bidirectional control of synaptic strength (depression versus potentiation) by low- and high-frequency stimulation (Lisman, 1994). (See also discussion of LTP and LTD in Chapter 19).

Additional regulation may be accorded by interaction of this hinge region with cyclophilin and FK506-binding protein (FKBP), proteins that bind the immunosuppressive agents cyclosporin and FK506, respectively. FKBP, is highly abundant in the brain, and its distribution resembles that of calcineurin. Both FK506 and cyclosporin A are membrane permeant and are highly potent and selective inhibitors of calcineurin. They are referred to as *immunophilins* because their ability to block the essential role of calcineurin in lymphocyte activation makes them effective immunosuppressants. The X-ray structure of calcineurin complexed with FK506 reveals a ternary complex in which FK506 is bound at the interface between FKBP and the regulatory domain of CnA. (Griffith *et al.*, 1995). CnB binds to one surface of an extended regulatory domain of CnA and FKBP–FK506 binds to the opposite surface. The FK506–FKBP complex is wedged between the regulatory domain and the catalytic site and likely inhibits calcineurin by making it difficult for phosphoproteins to have access to the catalytic site. It is unclear whether FKBP and calcineurin interact physiologically via a natural ligand that functions like FK506 to facilitate their interaction.

Protein Kinases, Protein Phosphatases, and Their Substrates Are Integrated Networks

Cross talk between protein kinases and protein phosphatases is key to their ability to integrate inputs into neurons (Cohen, 1992). Such cross talk is exemplified by the interaction of cAMP and Ca^{2+} signals through PKA and calcineurin, respectively. The medium spiny neurons in the neostriatum receive cortical inputs from glutamatergic neurons that are excitatory and nigral inputs by dopaminergic neurons that inhibit them. A possible signal transduction scheme for this regulation is shown in Figure 12.16.

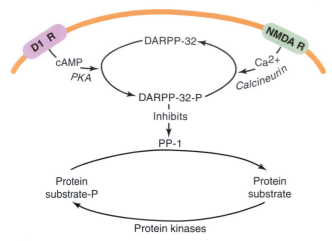

FIGURE 12.16 Cross talk between kinases and phosphatases. The state of phosphorylation of protein substrates is regulated dynamically by protein kinases and phosphatases. In the striatum, for example, dopamine stimulates PKA, which converts DARPP-32 into an effective inhibitor of PP-1. This increases the steady-state level of phosphorylation of a hypothetical substrate subject to phosphorylation by a variety of protein kinases. This action can be countered by NMDA receptor stimulation by another stimulus that increases intracellular Ca^{2+} and activates calcineurin. PP-1 is deinhibited and dephosphorylates the phosphorylated substrate when calcineurin deactivates DARPP-32-P. Adapted from Greengard *et al.* (1998).

A more complex scheme is shown in Figure 12.17 and described in Box 12.2. The key to the regulation is the bidirectional control of DARPP-32 phosphorylation (Greengard *et al.*, 1998; Svenningsson *et al.* 2004). Glutamate activates calcineurin by increasing intracellular Ca^{2+}, leading to the dephosphorylation and inactivation of phospho-DARPP-32. This releases inhibition of PP-1, which can then dephosphorylate a variety of substrates, including the Na^+, K^+-ATPase, and lead to membrane depolarization. This is countered by dopamine, which stimulates cAMP formation and activation of PKA, which then converts DARPP-32 into its phosphorylated (i.e., PP-1 inhibitory) state. Although PKA and calcineurin are acting in an antagonistic manner, they are not doing it by phosphorylating and dephosphorylating the ATPase. By their actions upstream, at the level of DARPP-32, the regulation of numerous target enzymes (e.g., Ca^{2+} channels and Na^+ channels) in addition to the ATPase can be coordinated.

Studying Cellular Processes Controlled by Phosphorylation–Dephosphorylation Requires a Set of Criteria

Major goals of signal transduction research are to delineate pathways by which signals such as neurotransmitters transduce their signals to modify cellular processes. We do not understand all signaling pathways, their cross talk with other pathways, or the physiological and pathophysiological roles that they subserve. Such investigation is often the start of a process to identify targets for therapeutic intervention in disease. Cellular and biochemical assays can often identify the entire signaling pathway, from stimulation of receptor, to generation of a second-messenger activation of a kinase or phosphatase, change in the phosphorylation state of the substrate, and an ultimate change in its functional state. Signal transduction research utilizes a variety of pharmacological inhibitors or activators of the signaling molecules complemented by genetic approaches that utilize transfection of activated forms of the kinases or phosphatases in question, siRNAs, transgenic animals, and mice with individual signaling components knocked out.

Summary

A morphology of a cell is determined by protein constituents. Its function is regulated by the phosphorylation or dephosphorylation of the proteins. Phosphorylation modifies the function of regulatory proteins subsequent to their genetic expression. The activities of the protein kinases and protein phosphatases are typically regulated by second messengers and extracellular ligands. Kinases and phosphatases integrate and encode stimulation of a large group of cellular receptors. The number of possible effects is almost limitless and enables the tuning of cellular processes over a broad timescale. Potentiation of kinase and phosphatase activity may be key elements in molecular memory and neuronal plasticity. Most of the effects of Ca^{2+} in cells are mediated by calmodulin, which in turn mediates changes in protein phosphorylation–dephosphorylation. The phosphoinositol signaling system is mediated through PKC, which modulates many cellular processes from exocytosis to gene expression. All three classes of enzymes discussed have been described as cognitive kinases because they are capable of sustaining their activated states after their second messenger stimuli have returned to basal levels. PKA has been implicated in learning and memory in *Aplysia* and in hippocampus, where it is involved in long-term potentiation. Protein phosphatases play an equally important role in neuronal signaling by dephosphorylating proteins. Cross talk between protein kinases and protein phosphatases is key to their ability to integrate inputs into neurons. A major effort of signal transduction research is to delineate the pathways through which the neurotransmitters' signals across the plasma membrane are transmitted to the ultimate cellular components to be modified.

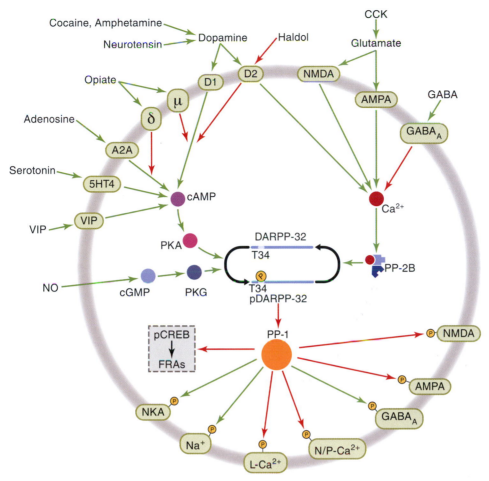

FIGURE 12.17 Signaling pathways in the neostriatum. Activation by dopamine of the D₁ subclass of dopamine receptors stimulates the phosphorylation of DARPP-32 at Thr-34. This is achieved through a pathway involving the activation of adenylyl cyclase, the formation of cAMP, and the activation of PKA. Activation by dopamine of the D₂ subclass of dopamine receptors causes the dephosphorylation of DARPP-32 through two synergistic mechanisms: D₂ receptor activation (i) prevents the D₁ receptor-induced increase in cAMP formation and (ii) raises intracellular calcium, which activates a calcium-dependent protein phosphatase, namely, calcineurin, calcium–calmodulin-dependent protein phosphatase. Activated calcineurin dephosphorylates DARPP-32 at Thr-34. Glutamate acts as both a fast-acting and a slow-acting neurotransmitter. Activation by glutamate of AMPA receptors causes a rapid response through the influx of sodium ions, depolarization of the membrane, and firing of an action potential. Slow synaptic transmission, in response to glutamate, results in part from activation of the AMPA and NMDA subclasses of the glutamate receptor, which increases intracellular calcium and the activity of calcineurin, and causes the dephosphorylation of DARPP-32 on Thr-34. All other neurotransmitters that have been shown to act directly to alter the physiology of dopaminoceptive neurons also alter the phosphorylation state of DARPP-32 on Thr-34 through the indicated pathways. Neurotransmitters that act indirectly to affect the physiology of these dopaminoceptive neurons also regulate DARPP-32 phosphorylation; for example, neurotensin, stimulating the release of dopamine, increases DARPP-32 phosphorylation; conversely, cholecystokinin (CCK), by stimulating the release of glutamate, decreases DARPP-32 phosphorylation. Antischizophrenic drugs and drugs of abuse, all of which affect the physiology of these neurons, also regulate the state of phosphorylation of DARPP-32 on Thr-34. For example, the antischizophrenic drug haloperidol (Haldol), which blocks the activation by dopamine of the D₂ subclass of dopamine receptors, increases DARPP-32 phosphorylation. Agonists for the m and d subclasses of opiate receptors block D₁ and A₂ₐ receptor-mediated increases in cAMP, respectively, and block the resultant increases in DARPP-32 phosphorylation. Cocaine and amphetamine, by increasing extracellular dopamine levels, increase DARPP-32 phosphorylation. Marijuana, nicotine, alcohol, and LSD, all of which affect the physiology of the dopaminoceptive neurons, also regulate DARPP-32 phosphorylation. Finally, all drugs of abuse have greatly reduced biological effects in animals with targeted deletion of the DARPP-32 gene. 5HT4, 5-hydroxytryptophan (serotonin) receptor 4; NKA, Na⁺, K⁺-ATPase; VIP, vasoactive intestinal peptide; L- and N/P-Ca²⁺, L-type and N/P-type calcium channels. Green arrows designate activation and red arrows inhibition. From Greengard *et al.* (1999).

INTERACTIONS OF SIGNAL TRANSDUCTION PATHWAYS IN THE BRAIN

An understanding of the signal transduction mechanisms by which neurotransmitters produce their effects on their target neurons and of the mechanisms by which coordination of various signal transduction pathways is achieved represents a major area of research in cellular neurobiology. The dopaminoceptive medium-sized spiny neurons, located in the neostriatum, have been studied in great detail with respect to these mechanisms. Figure 12.17 illustrates a portion of what is now known about interactions of signaling mechanisms in these neurons. Activation by dopamine of D_1 receptors increases cAMP, causing activation of PKA (cAMP-dependent protein kinase) and phosphorylation of DARPP-32 (*dopamine* + cAMP-regulated *phospho*protein, M_r32,000) on threonine-34. Conversely, glutamate, acting on NMDA receptors, increases $[Ca^{2+}]_i$, leading to the activation of calcineurin (protein phosphatase 2B, PP-2B) and dephosphorylation of phosphothreonine-34–DARPP-32. Neurotensin, VIP, NO (nitric oxide), and some other neurotransmitters increase the phosphorylation of DARPP-32 through a variety of signaling mechanisms. Dopamine (acting on D_2 receptors), CCK, GABA, and some other neurotransmitters decrease the state of phosphorylation of DARPP-32 through a variety of other signaling mechanisms. CK1 (casein kinase I) and CK2 (casein kinase II) phosphorylate DARPP-32 on residues other than threonine-34, causing it to undergo conformational changes. These changes result in phosphothreonine-34–DARPP-32 becoming a poorer substrate for calcineurin (in the case of CKI) or a better substrate for PKA (in the case of CKII). Antipsychotic drugs such as haloperidol (Haldol) increase the state of phosphorylation of DARPP-32 by blocking the dopamine-induced D_2 receptor-mediated activation of calcineurin.

The physiological consequences of phosphorylation of DARPP-32 on threonine-34 are profound. Thus, DARPP-32 in its threonine-34 phosphorylated, but not dephosphorylated, form acts as a potent inhibitor of PP-1 (protein phosphatase 1). PP-1 is a major serine–threonine protein phosphatase, which controls the state of phosphorylation of a variety of phosphoprotein substrates in the brain. These substrates include Na^+ channels; L-, N-, and P-type Ca^{2+} channels; the electrogenic ion pump Na^+, K^+-ATPase; the NR-1 subclass of glutamate receptors; and probably many more.

In summary, the DARPP-32–PP-1 cascade provides a mechanism by which a large number of neurotransmitters act in a complex, but coordinated, fashion to regulate the state of phosphorylation and activity of a variety of ion channels, ion pumps, and neurotransmitter receptors.

Adapted from Greengard *et al.* (1998).

References

Allbritton, N. L., Meyer, T., and Stryer, L. (1992). Range of messenger action of calcium ion and inositol 1,4,5-trisphosphate. *Science* **258**, 1812–1815.

Bacskai, B. J., Hochner, B., Mahaut-Smith, M., Adams, S. R., Kaang, B. K., Kandel, E. R., and Tsien, R. Y. (1993). Spatially resolved dynamics of cAMP and protein kinase A subunits in *Aplysia* sensory neurons. *Science* **260**, 222–226.

Baranano, D. E., Ferris, C. D., and Snyder, S. H. (2001). Atypical neural messengers. *Trends Neurosci.* **24**, 99–106.

Bayer, K.-U., De Koninck, P., Leonard, A. S., Hell, J. W., and Schulman, H. (2001). Interaction with the NMDA receptor locks CaM-KII in an active conformation. *Nature* **411**, 801–805.

Baylor, D. (1996). How photons start vision. *Proc. Natl. Acad. Sci. USA* **93**, 560–565.

Beene, D. L. and Scott, J. D. (2007). A-kinase anchoring proteins take shape. *Curr. Opin. Cell Biol.* **19**, 192–198.

Berridge, M. J. (1993). Inositol trisphosphate and calcium signalling. *Nature* **361**, 315–325.

Birnbaumer, L. (2007). Expansion of signal transduction by G proteins. The second 15 years or so: From 3 to 16 α subunits plus βγ dimers. *Biochim. Biophys. Acta.* **1768**, 772–793.

Bootman, M. D., Berridge, M. J., and Lipp, P. (1997). Cooking with calcium: The recipes for composing global signals from elementary events. *Cell* **91**, 367–373.

Bos, J. L. (2006). Epac proteins: Multi-purpose cAMP targets. *Trends Biochem. Sci.* **31**, 680–686.

Bourne, H. R., and Nicoll, R. (1993). Molecular machines integrate coincident synaptic signals. *Cell* **72**, 65–75.

Braun, A. P., and Schulman, H. (1995). The multifunctional calcium/calmodulin-dependent protein kinase: From form to function. *Annu. Rev. Physiol.* **57**, 417–445.

Carafoli, E., and Klee, C. (1999). "Calcium as a Cellular Regulator." Oxford Univ. Press, New York.

Cassel, D., and Selinger, Z. (1978). Mechanism of adenylyl cyclase activation through the β-adrenergic receptor: Catecholamine-induced displacement of bound GDP by GTP. *Proc. Natl. Acad. Sci. USA* **75**, 4155–4159.

Chain, D. G., Casadio, A., Schacher, S., Hegde, A. N., Valbrun, M., Yamamoto, N., Goldberg, A. L., Bartsch, D., Kandel, E. R., and Schwartz, J. H. (1999). Mechanisms for generating the autonomous cAMP-dependent protein kinase required for long-term facilitation in *Aplysia*. *Neuron* **22**, 147–156.

Clapham, D. E. (1995). Calcium signaling. *Cell* **80**, 259–268.

Clapham, D. E., and Neer, E. J. (1993). New roles for G-protein βγ-dimers in transmembrane signalling. *Nature* **365**, 403–406.

Cohen, P. (1992). Signal integration at the level of protein kinases, protein phosphatases and their substrates. *Trends Biochem. Sci.* **17**, 408–413.

Cohen, P., and Klee, C. B. (1988). Calmodulin. *In* "Molecular Aspects of Cellular Regulation," Vol. 5, pp. 396. Elsevier, Amsterdam.

Coleman, D. E., Berghuis, A. M., Lee, E., Linder, M. E., Gilman, A. G., and Sprang, S. R. (1994). Structures of active conformations of $G_{i\alpha 1}$ and the mechanism of GTP hydrolysis. *Science* **265**, 1405–1412.

Conklin, R. B., Farfel, Z., Lustig, K. D., Julius, D., and Bourne, H. R. (1993). Substitution of three amino acids switches receptor specificity of $G_{q\alpha}$ to that of $G_{i\alpha}$. *Nature* **369**, 274–276.

De Koninck, P., and Schulman, H. (1998). Sensitivity of Ca^{2+}/calmodulin-dependent protein kinase II to the frequency of Ca^{2+} oscillations. *Science* **279**, 227–230.

Delcour, A. H., and Tsien, R. W. (1993). Altered prevalence of gating modes in neurotransmitter inhibition of N-type calcium channels. *Science* **259**, 980–984.

Di Paolo, G. and De Camilli, P. (2006). Phosphoinositides in cell regulation and membrane dynamics. *Nature* **443**, 651–657.

Divecha, N., and Irvine, R. F. (1995). Phospholipid signaling. *Cell* **80**, 269–278.

Federman, A. D., Conklin, B. R., Schrader, K. A., Reed, R. R., and Bourne, H. R. (1992). Hormonal stimulation of adenylyl cyclase through G_i-protein βγ subunits. *Nature* **356**, 159–161.

Ferrell, J. E., Jr., and Machleder, G. M. (1998). The biochemical basis of an all-or-none switch in *Xenopus* oocytes. *Science* **280**, 895–898.

Furchgott, R. F., and Zawadzki, J. V. (1980). The obligatory role of the endothelial cells in the relaxation of arterial smooth muscle by acetylcholine. *Nature* **288**, 373–376.

Garbers, D. L., and Lowe, D. G. (1994). Guanylyl cyclase receptors. *J. Biol. Chem.* **269**, 30741–30744.

Giese, K. P., Fedorov, N. B., Filipkowski, R. K., and Silva, A. J. (1998). Autophosphorylation at Thr[286] of the α calcium-calmodulin kinase II in LTP and learning. *Science* **279**, 870–873.

Goldberg, J., Huang, H., Kwon, Y., Greengard, P., Nairn, A. C., and Kuriyan, J. (1995). Three-dimensional structure of the catalytic subunit of protein serine/threonine phosphatase-1. *Nature* **376**, 745–753.

Greengard, P., Allen, P. B., and Nairn, A. C. (1999). Beyond the dopamine receptor: the DARPP-32/protein phosphatase-1 cascade. *Neuron* **23**, 435–447.

Greengard, P., Nairn, A. C., Girault, J.-A., Quimet, C. C., Snyder, G. L., Fisone, G., Allen, P. B., Fienberg, A., and Nishi, A. (1998). The DARPP-32/protein phosphatase-1 cascade: A model for signal integration. *Brain Res. Rev.* **26**, 274–284.

Griffith, J. P., Kim, J. L., Kim, E. E., Sintchak, M. D., Thomson, J. A., Fitzgibbon, M. J., Fleming, M. A., Caron, P. R., Hsiao, K., and Navia, M. A. (1995). X-ray structure of calcineurin inhibited by the immunophilin-immunosuppressant FKBP12-FK506 complex. *Cell* **82**, 507–522.

Hanks, S. K., and Hunter, T. (1995). Protein kinases 6. The eukaryotic protein kinase superfamily: Kinase (catalytic) domain structure and classification. *FASEB J.* **9**, 576–596.

Hanoune, J., and Defer, N. (2001). Regulation and role of adenylyl cyclase isoforms. *Annu. Rev. Pharmacol. Toxicol.* **41**, 145–174.

Hanson, P. I., and Schulman, H. (1992). Neuronal Ca^{2+}/calmodulin-dependent protein kinases. *Annu. Rev. Biochem.* **61**, 559–601.

Hanson, P. I., Kapiloff, M. S., Lou, L. L., Rosenfeld, M. G., and Schulman, H. (1989). Expression of a multifunctional Ca^{2+}/calmodulin-dependent protein kinase and mutational analysis of its autoregulation. *Neuron* **3**, 59–70.

Hanson, P. I., Meyer, T., Stryer, L., and Schulman, H. (1994). Dual role of calmodulin in autophosphorylation of multifunctional CaM kinase may underlie decoding of calcium signals. *Neuron* **12**, 943–956.

Hemmings, H. C., Jr., Greengard, P., Tung, H. Y., and Cohen, P. (1984). DARPP-32, a dopamine-regulated neuronal phosphoprotein, is a potent inhibitor of protein phosphatase-1. *Nature* **310**, 503–505.

Hille, B. (1992). G protein-coupled mechanisms and nervous signaling. *Neuron* **9**, 187–195.

Hokin, L. E., and Hokin, M. R. (1955). Effects of acetylcholine on the turnover of phosphoryl units in individual phospholipids of pancreas slices and brain cortex slices. *Biochim. Biophys. Acta.* **18**, 102–110.

Huang, C.-L., Slesinger, P. A., Casey, P. J., Jan, Y. N., and Jan, L. Y. (1995). Evidence that direct binding of $G_{\beta\gamma}$ to the GIRK1G protein-gated inwardly rectifying K^+ channel is important for channel activation. *Neuron* **15**, 1133–1143.

Hubbard, M. J., and Cohen, P. (1993). On target with a new mechanism for the regulation of protein phosphorylation. *Trends Biochem. Sci.* **18**, 172–177.

Hunter, T. (1995). Protein kinases and phosphatases: The yin and yang of protein phosphorylation and signaling. *Cell* **80**, 225–236.

Ikura, M., Clore, G. M., Fronenborn, A. M., Zhu, G., Klee, C. B., and Bax, A. (1992). Solution structure of a calmodulin—target peptide complex by multidimensional NMR. *Science* **256**, 632–638.

Kaang, B. K., Kandel, E. R., and Grant, S. G. (1993). Activation of cAMP-responsive genes by stimuli that produce long-term facilitation in *Aplysia* sensory neurons. *Neuron* **10**, 427–435.

Kemp, B. E., Faux, M. C., Means, A. R., House, C., Tiganis, T., Hu, S.-H., and Mitchellhill, K. I. (1994). Structural aspects: Pseudosubstrate and substrate interactions. *In* "Protein Kinases" (J. R. Woodgett, Ed.), pp. 30–67, New York: Oxford Univ. Press.

Kennelly, P. J., and Krebs, E. G. (1991). Consensus sequences as substrate specificity determinants for protein kinases and protein phosphatases. *J. Biol. Chem.* **266**, 15555–15558.

Klauck, T. M., Faux, M. C., Labudda, K., Langeberg, L. K., Jaken, S., and Scott, J. D. (1996). Coordination of three signaling enzymes by AKAP79, a mammalian scaffold protein. *Science* **271**, 1589–1592.

Knighton, D. R., Zheng, J., Eyck, L. F. T., Xuong, N., Taylor, S. S., and Sowadski, J. M. (1991). Structure of a peptide inhibitor bound to the catalytic subunit of cyclic adenosine monophosphate-dependent protein kinase. *Science* **253**, 414–420.

Kraft, A. S., and Anderson, W. B. (1983). Phorbol esters increase the amount of Ca^{2+}, phospholipid-dependent protein kinase associated with plasma membrane. *Nature* **301**, 621–623.

Krupinski, J., Coussen, F., Bakalyar, H. A., Tang, W.-J., Feinstein, P. G., Orth, K., Slaughter, C., Reed, R. R., and Gilman, A. G. (1989). Adenylyl cyclase amino acid sequence: Possible channel- or transporter-like structure. *Science* **244**, 1558–1564.

Levin, L. R., Han, P.-L., Hwang, P. M., Feinstein, P. G., Davis, R. L., and Reed, R. R. (1992). The Drosophila learning and memory gene rutabaga encodes a Ca^{2+}/calmodulin-responsive adenylyl cyclase. *Cell* **68**, 479–489.

Lisman, J. (1994). The CaM kinase II hypothesis for the storage of synaptic memory. *Trends Neurosci.* **17**, 406–412.

Lisman, J., Schulman, H., and Clive, H. (2002). The molecular basis of CaMKII function in synaptic and behavioural memory. *Nat. Rev. Neurosci.* **71**, 175–190.

Logothetis, D. E., Kurachi, Y., Galper, J., Neer, E. J., and Clapham, D. E. (1987). The βγ subunits of GTP-binding proteins activate the muscarinic K^+ channel in heart. *Nature* **325**, 321–326.

Mansuy, I. M., and Shenolikar, S. (2006). Protein serine/threonine phosphatases in neuronal plasticity and disorders of learning and memory. *Trends Neurosci.* **29**, 679–686.

Miller, S. G., and Kennedy, M. B. (1986). Regulation of brain type II Ca^{2+}/calmodulin-dependent protein kinase by autophosphorylation: A Ca^{2+}-triggered switch. *Cell* **44**, 861–870.

Minta, A., and Tsien, R. Y. (1989). Fluorescent indicators for cytosolic calcium based on rhodamine and fluorescein chromophores. *J. Biol. Chem.* **264**, 8171–8178.

Mochly-Rosen, D. (1995). Localization of protein kinases by anchoring proteins: A theme in signal transduction. *Science* **268**, 247–251.

Muramatsu, M., Kaibuchi, K., and Arai, K. (1989). A protein kinase C cDNA without the regulatory domain is active after transfection *in vivo* in the absence of phorbol ester. *Mol. Cell Biol.* **9**, 831–836.

Nairn, A. C., Hemmings, H. C., Jr., and Greengard, P. (1985). Protein kinases in the brain. *Annu. Rev. Biochem.* **54**, 931–976.

Neer, E. J. (1995). Heterotrimeric G proteins: Organizers of transmembrane signals. *Cell* **80**, 249–257.

Newton, A. C. (1995). Protein kinase C: Structure, function, and regulation. *J. Biol. Chem.* **270**, 28495–28498.

Nishizuka, Y. (1995). Protein kinase C and lipid signaling for sustained cellular responses. *FASEB J.* **9**, 484–496.

Palczewski, K., Kumasaka, T., Hori, T., Behnke, C. A., Motoshima, H., Fox, B. A., Trong, I. L., Teller, D. C., Okada, T., Stenkamp, R. E., Yamamoto, M., and Miyano, M. (2000). Crystal structure of rhodopsin: A G protein-coupled receptor. *Science* **289**, 739–745.

Persechini, A., and Cronk, B. (1999). The relationship between the free concentrations of Ca^{2+} and Ca^{2+}-calmodulin in intact cells. *J. Biol. Chem.* **274**, 6827–6830.

Price, N. E., and Mumby, M. C. (1999). Brain protein serine/threonine phosphatases. *Curr. Opin. Neurobiol.* **9**, 336–342.

Rameh, L. E., and Cantley, L. C. (1999). The role of phosphoinositide 3-kinase lipid products in cell function. *J. Biol. Chem.* **274**, 8347–8350.

Ron, D., Chen, C. H., Caldwell, J., Jamieson, L., Orr, E., and Mochly-Rosen, D. (1994). Cloning of an intracellular receptor for protein kinase C: A homolog of the beta subunit of G proteins. *Proc. Natl. Acad. Sci. USA* **91**, 839–843.

Rosenberg, O.S., Deindl, S., Comolli, L.R., Hoelz, A., Downing, K. H., Nairn, A. C., and Kuriyan, J. (2006). Oligomerization states of the association domain and the holoenzyme of Ca^{2+}/CaM kinase II. *FASEB J.* **273**, 682–694.

Rosenberg, O.S., Deindl, S., Sung, R.J., Nairn, A. C., and Kuriyan, J. (2005). Structure of the autoinhibited kinase domain of CaMKII and SAXS analysis of the holoenzyme. *Cell* **123**, 849–860.

Rosenmund, C., Carr, D. W., Bergeson, S. E., Nilaver, G., Scott, J. D., and Westbrook, G. L. (1994). Anchoring of protein kinase A is required for modulation of AMPA/kainate receptors on hippocampal neurons. *Nature* **368**, 853–856.

Ross, E. M. (1989). Signal sorting and amplification through G protein-coupled receptors. *Neuron* **3**, 141–152.

Rusnak, F., and Mertz, P. (2000). Calcineurin: Form and function. *Physiol. Rev.* **80**, 1483–1521.

Saitoh, T., and Schwartz, J. H. (1985). Phosphorylation-dependent subcellular translocation of a Ca^{2+}/calmodulin-dependent protein kinase produces an autonomous enzyme in *Aplysia* neurons. *J. Cell Biol.* **100**, 835–842.

Schmidt, H. H., and Walter, U. (1994). NO at work. *Cell* **78**, 919–925.

Schulman, H., and Braun, A. (1999). Ca^{2+}/calmodulin-dependent protein kinases. *In* "Calcium as a Cellular Regulator" (E. Carafoli and C. Klee, Eds.), pp. 311–343, Oxford Univ. Press, New York.

Schuman, E. M., and Madison, D. V. (1994). Locally distributed synaptic potentiation in the hippocampus. *Science* **263**, 532–536.

Schwartz, J. H. (1993). Cognitive kinases. *Proc. Natl. Acad. Sci. USA.* **90**, 8310–8313.

Scott, J. D. (1991). Cyclic nucleotide-dependent protein kinases. *Pharmacol. Ther.* **50**, 123–145.

Serrano, P., Yao, Y., and Sacktor, T. C. (2005). Persistent phosphorylation by protein kinase Mζ maintains late-phase long-term potentiation. *J. Neurosci.* **25**, 1979–1984.

Smith, J. A., Poteet-Smith, C. E., Lannigans, D. A., Freed, T. A., Zoltosk, A. J., and Sturgill, T. W. (2000). Creation of a stress-activated p90 ribosomal S6 kinase: The carboxyl-terminal tail of the MAPK-activated protein kinases dictates the signal transduction pathway in which they function. *J. Biol. Chem.* **275**, 31588–31593.

Sondek, J., Bohm, A., Lambright, D. G., Hamm, H. E., and Sigler, P. B. (1996). Crystal structure of a G protein βγ dimer at 2.1 Å resolution. *Nature* **379**, 369–374.

Sondek, J., Lambright, D. G., Noel, J. P., Hamm, H. E., and Sigler, P. B. (1994). GTPase mechanism of G proteins from the 1.7 Å crystal structure of transducin α-GDP-AlF$_4^-$. *Nature* **372**, 276–279.

Sprang, S.R., Chen, Z., and Du, X. (2007). Structural basis of effector regulation and signal termination in heterotrimeric Gα proteins. *Adv. Protein Chem.* **74**, 1–65.

Streb, H., Irvine, R. F., Berridge, M. J., and Schulz, I. (1983). Release of Ca^{2+} from a nonmitochondrial intracellular store in pancreatic acinar cells by inositol-1,4,5-trisphosphate. *Nature* **306**, 67–69.

Stryer, L. (1991). Visual excitation and recovery. *J. Biol. Chem.* **266**, 10711–10714.

Stryer, L., and Bourne, H. R. (1986). G proteins: A family of signal transducers. *Ann. Rev. Cell Biol.* **2**, 391–419.

Su, Y., Dostmann, W. R., Herberg, F. W., Durick, K., Xuong, N. H., Ten Eyck, L., Taylor, S. S., and Varughese, K. I. (1995). Regulatory subunit of protein kinase A: Structure of deletion mutant with cAMP binding domains. *Science* **269**, 807–813.

Sun, X. X., Hodge, J. J., Zhou, Y., Nguyen, M. and Griffith, L. C. (2004). The eag potassium channel binds and locally activates calcium/calmodulin-dependent protein kinase II. *J. Biol. Chem.* **279**, 10206–10214.

Svenningsson, P., Nishi, A., Fisone, G., Girault, J.-A., Nairn, A. C., and Greengard, P. (2004). DARPP-32: An integrator of neurotransmission. *Annu. Rev. Pharmacol. Toxicol.* **44**, 269–296.

Takai, Y., Kishimoto, A., Kikkawa, U., Mori, T., and Nishizuka, Y. (1979). Unsaturated diacylglycerol as a possible messenger for the activation of calcium-activated, phospholipid-dependent protein kinase system. *Biochem. Biophys. Res. Commun.* **91**, 1218–1224.

Tanaka, C., and Nishizuka, Y. (1994). The protein kinase C family for neuronal signaling. *Annu. Rev. Neurosci.* **17**, 551–567.

Tang, W.-J., and Gilman, A. G. (1991). Type-specific regulation of adenylyl cyclase by G protein βγ subunits. *Science* **254**, 1500–1503.

Taussig, R., and Gilman, A. G. (1995). Mammalian membrane-bound adenylyl cyclases. *J. Biol. Chem.* **270**, 1–4.

Tsien, R. W., and Tsien, R. Y. (1990). Calcium channels, stores, and oscillations. *Annu. Rev. Cell. Biol.* **6**, 715–760.

Wall, M. A., Coleman, D. E., Lee, E., Iniguez-Lluhi, J. A., Posner, B. A., Gilman, A. G., and Sprang, S. R. (1996). The structure of the G protein heterotrimer $G_{i\alpha1}\beta_1\gamma_2$. *Cell* **83**, 1047–1058.

Walsh, D. A., and Patten, S. M. V. (1994). Multiple pathway signal transduction by the cAMP-dependent protein kinase. *FASEB J.* **8**, 1227–1236.

Wayman, G. A., Impey, S., Wu, Z., Kindsvogel, W., Prichard, L., and Storm, D. R. (1994). Synergistic activation of the type I adenylyl cyclase by Ca^{2+} and G_s-coupled receptors *in vivo*. *J. Biol. Chem.* **269**, 25400–25405.

Wen, W., Meinkoth, J. L., Tsien, R. Y., and Taylor, S. S. (1995). Identification of a signal for rapid export of proteins from the nucleus. *Cell* **82**, 463–473.

Westphal, R. S., Tavalin, S. J., Lin, J. W., Alto, N. M., Fraser, I. D., Langeberg, L. K., Sheng, M., and Scott, J. D. (1999). Regulation of NMDA receptors by an associated phosphatase–kinase signaling complex. *Science* **285**, 93–96.

Wickman, K. D., Iñiguez-Lluhi, J. A., Davenport, P. A., Taussig, R., Krapivinsky, G. B., Linder, M. E., Gilman, A. G., and Clapham, D. E. (1994). Recombinant G-protein βγ-subunits activate the muscarinic-gated atrial potassium channel. *Nature* **368**, 255–257.

Wu, Z. L., Thomas, S. A., Villacres, E. C., Xia, Z., Simmons, M. L., Chavkin, C., Palmiter, R. D., and Storm, D. R. (1995). Altered behavior and long-term potentiation in type I adenylyl cyclase mutant mice. *Proc. Natl. Acad. Sci. USA* **92**, 220–224.

Zhang, G., Kazanietz, M. G., Blumberg, P. M., and Hurley, J. H. (1995). Crystal structure of the cys2 activator-binding domain of protein kinase Cδ in complex with phorbol ester. *Cell* **81**, 917–924.

Regulation of Neuronal Gene Expression and Protein Synthesis

James L. Roberts and James R. Lundblad

INTRACELLULAR SIGNALING AFFECTS NUCLEAR GENE EXPRESSION

The previous chapter describes how signaling systems regulate the function of cellular proteins already expressed; another critical level of control exerted by these systems is their ability to regulate the synthesis of cellular proteins by regulating the expression of specific genes. For all living organisms, regulation of gene expression by intracellular signals is a fundamental mechanism of development, homeostasis, and adaptation to the environment. The regulation of gene expression by intracellular signals remains as the most important mechanism underlying the remarkable degree of functional plasticity exhibited by neurons. Alterations in gene expression underlie many forms of long-term changes in neural functioning, with a time course that ranges from hours to many years. Indeed, as discussed earlier and in Chapter 19, evidence now suggests that formation of long-term memories in many neural systems requires changes in gene expression and new protein synthesis. In addition to regulation of the levels of synthesis of new mRNAs from specific genes, changes in messenger RNA turnover and/or changes in the efficiency of synthesis of new proteins from a specific messenger RNA, both processes that result in altered levels of expression of a specific gene product, are observed.

The Information Content of the Genome

The double helix of DNA is an ideal molecule for the storage of information. DNA is a stable, linear polymer, and because of the properties of complementarity

between the individual strands of the double helix, can be readily replicated or serve as the template for the synthesis of RNA. The information content of DNA is expressed through other molecules: RNA and proteins, encoded by genes. What defines a gene? The human genome is estimated to contain 20,000 to 40,000 genes that encode protein-coding messenger RNAs (mRNAs) (International Human Genome Sequencing Consortium, 2001). In addition to mRNAs, gene-encoded RNA molecules include classes of RNAs with structural, functional, and other regulatory functions, as discussed later. Identification of DNA sequences that might encode a gene depends on comparison of genomic sequences with those detected in RNA transcripts and expressed sequence tags (ESTs), but also employs informatic analyses and the recognition of conserved DNA sequence features common to transcribed genomic sequences (promoter elements, splice acceptors and donors, and termination sequences). In addition, the presence of similar or homologous known genes in other species (i.e., orthologues or genes that code for corresponding proteins in different organisms) may provide clues as to the identity of a gene. An analysis of the proteins expressed may also be used to identify new genes by deducing the coding DNA sequence from the peptide sequence.

The total number of proteins expressed by an organism (the *proteome*), however, vastly exceeds the total gene number. Genes are sometimes duplicated, and duplicate copies may encode the same or a similar protein. As discussed in Chapter 10, post-translational processing also generates diversity in peptide products from a single protein precursor, and similarly some genes produce more than one mRNA and therefore more than one protein product.

The interrupted nature of genes leads to the generation of diversity in expression from a single gene via alternative RNA processing of a single primary transcript. In this mechanism, a single gene can generate multiple mRNAs that encode unique and distinct protein products. Newly synthesized, unprocessed RNA is termed the *primary transcript* (Fig. 13.1). The primary transcript for an mRNA contains sequences encompassing one gene, although the sequences encoding the polypeptide chain may not be contiguous. Noncoding tracts of DNA that interrupt the *exons*, or sequences ultimately found in mRNA, are called *introns*. Splicing removes the introns from the primary transcript to form a continuous sequence or mRNA that includes signals for translation initiation and termination. In addition, the 3′ ends of mRNA precursors are cleaved and undergo polyadenylation with 80 to 250 adenylate residues concomitantly with transcription. The use of alternative polyadenylation signals and splice donors and acceptors permits elaboration of more than one mRNA from a single primary transcript. It is estimated that more than 60% of human genes employ alternative splicing, often generating gene products with different functions.

Transcription of the calcitonin–calcitonin gene-related peptide (CGRP) gene (Lou and Gagel, 1998) illustrates how alternative RNA processing may generate diversity in the proteome from the relatively smaller gene complement of the genome. The calcitonin–CGRP gene is transcribed in both neuronal cells and the parafollicular or C cells of the thyroid but generates different mRNAs and protein products by both differential splicing and polyadenylation site usage (Fig. 13.2A). In thyroid parafollicular (or C) cells that produce the hormone calcitonin, splicing of the calcitonin–CGRP gene primary transcript includes exon 4, which contains the calcitonin peptide precursor coding sequence, and transcription termination and polyadenylation signals. In neuronal cells, splicing skips exon 4, instead incorporating alternative exons 5 and 6 (with an associated polyadenylation signal downstream of exon 6) into an mRNA-encoding CGRP.

As another level of post-transcriptional processing, RNA may undergo insertion, deletion, or substitution of nucleotides prior to translation by the process of *RNA editing*, generating an RNA sequence not represented in genomic DNA (Bass, 2002). The most common mode of RNA editing in mammalian cells is hydrolytic deamination of adenosine to inosine, read as a guanine base by the ribosome at affected codons, in a reaction that requires short stretches of double-stranded RNA. RNA editing of specific nucleotide residues may alter signals for RNA splicing, translation initiation, transport, or degradation, in addition to the translation or coding sequence. Analysis of sequence data indicates that, to a large extent, transcribed pre-mRNAs are edited but largely in

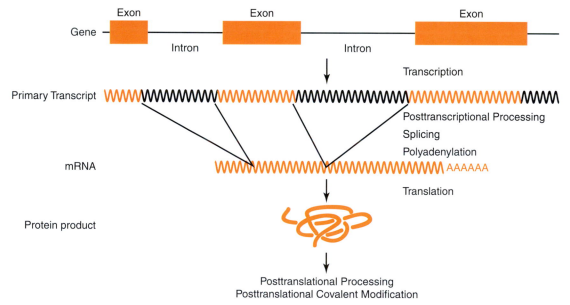

FIGURE 13.1   Expression of protein-encoding genes. Transcription of protein-encoding genes is mediated by RNA polymerase II. The primary RNA transcript undergoes post-transcriptional processing, including splicing of intron sequences and 3′ polyadenylation, to generate the mature messenger RNA (mRNA). Splicing, mRNA stability, and turnover are also important processes targeted by regulatory mechanisms. The function of the protein product of a gene may also be regulated by post-translational proteolytic processing, or covalent modifications.

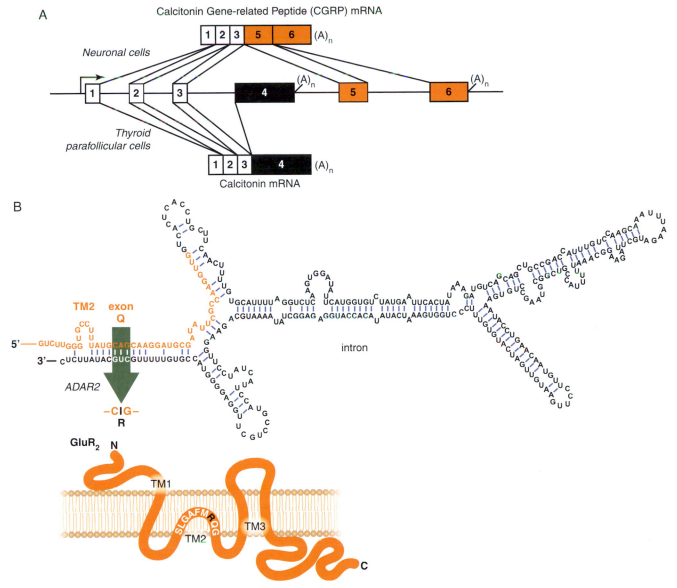

FIGURE 13.2 Alternative RNA processing. (A) The calcitonin–CGRP gene is a classic illustration of alternate RNA processing in the brain where a single gene transcript (hnRNA) can be processed to different mRNAs (modified from Lou and Gagel, 1998). (B) RNA editing occurs during the RNA splicing process and changes a nucleotide by enzymatic action, resulting in a different codon usage and hence an altered protein with functional consequences for the protein (modified from Seeburg *et al.*, 1998). Both processes yield additional diversity of gene products from the limited pool of gene present in the DNA and give additional complexity to brain function.

untranslated or noncoding regions, thus altering translation by RNA editing is rare. The enzymes responsible for this conversion, adenosine deaminases (ADARs), are relatively abundant in the central nervous system. Two well-characterized examples illustrate the consequences of RNA editing in modulating activities of receptors important in neuronal function, including the glutamate receptor (GluR) subtypes, as well as the serotonin (5-HT$_{2c}$) receptor (Seeburg *et al.*, 1998). For the glutamate receptor (GluR2), RNA editing changes the codon for a key

glutamine residue in the calcium conductance channel, to a codon encoding an arginine. This reduces the calcium conductance determined by this receptor channel. In these examples, the double-stranded structure is provided by a complementary intronic sequence looped back to the exonic target codon (Fig. 13.2B), indicating RNA editing occurs before splicing of the primary RNA transcript.

Post-transcriptional RNA editing changes codons in mammalian transcripts for subunits of glutamate receptor channels, leading to alterations in calcium ion flux (Figure 13.2B). In this example and others, the duplex RNA necessary for editing is formed by a section of intronic RNA that forms an imperfect duplex with the target sequence. Thus, RNA editing of these transcripts must occur prior to splicing events. Why does the GluR use this mechanism? The reason is not fully understood but may allow attenuation of GluR-dependent calcium permeability in certain cell types, permitting a single gene to encode functionally distinct receptors.

Noncoding RNAs in Decoding the Genome

RNA molecules serve additional functional roles in the transmission of the information content of the genome, and recent studies have identified a growing class of small RNA molecules with regulatory functions. It has been long known that noncoding RNAs play essential roles in information transfer. For example, small nuclear RNAs perform essential roles as components of the spliceosome nucleoprotein complex, and transfer RNAs (tRNAs) and ribosomal RNAs serve functional and structural roles in protein translation. For many of these RNAs they confer a structural role or contribute specificity in the activity of a nucleoprotein complex in the regulation of nucleic acid transactions by specific base pairing, thus directing protein components to specific targets.

Noncoding RNAs may participate in the phenomenon of RNA interference, a very important mechanism in regulating expression of the genome. RNA interference mechanisms have the potential for therapeutic use in human disease in silencing disease-causing genes. Small interfering RNAs (siRNAs) and micro RNAs (miRNAs) produced by transcription of the genome mediate sequence-specific gene silencing. Pathways that employ siRNA-mediated silencing direct sequence-specific cleavage and degradation with short, perfectly complementary sequences corresponding to specific mRNA targets. In another mechanism, micro RNAs are thought to interfere with translation by imperfect base pairing with targets.

These RNAs are produced from intergenic regions or introns and are of the class of small noncoding RNAs (ncRNAs) produced from the genome. The RNAs are processed both in the nucleus and cytoplasm from longer double-stranded species. In the cytoplasm, an RNase termed *Dicer* processes precursor RNAs into short duplex siRNA and miRNAs. The duplex RNA sequences are activated by a multiprotein complex (called RISC or *RNA*-induced *silencing* complex, in addition to a protein called *Argonaute*) that generates only the antisense strand (called the *guide strand*), which then targets the specific mRNA

sequence. If the sequence is a perfect match (i.e., a siRNA), the complex then cleaves the specific mRNA, which then leads to rapid degradation of the mRNA molecule. An imperfect match by hybridization of a miRNA results in suppression of translation, but sometimes also mRNA degradation in the cytoplasm.

Interactions of Specific DNA Sequences with Regulatory Proteins Control Both Basal and Signal-Regulated Transcription

The strict regulation of the expression of the genes composing the genome in a precise temporal and spatial pattern is essential for proper development, differentiation, and function of the nervous system. Although any of the steps from the transcription of a gene, to translation of the mRNA to a protein product, may be regulated (and examples of regulation at each have been catalogued), arguably the most important controls are exerted at the first step transcription initiation. For the DNA composing a gene, a fundamental distinction can be made between DNA sequences that are transcribed to RNA, and DNA sequences that exert control functions. Control sequences determine the beginnings and ends of segments of DNA that can be transcribed into RNA. These linked DNA sequences determine whether a potentially transcribed segment is actually transcribed in a particular cell and, if so, in what circumstances. These sequence elements determine the expression of genes (either enhancing or silencing transcription) by altering the chromatin structure encompassing the associated genomic region.

In contrast to prokaryotic organisms, the genomic DNA of eukaryotes is packaged with histones into a nucleoprotein complex to form chromatin (Workman and Kingston, 1998). As a consequence of condensing the genome into the confines of the nucleus, chromatin plays an essential role in regulation of the dynamic range of gene expression, in part by limiting the accessibility of DNA to the transcriptional machinery. Regulated gene expression conferred by the nucleotide sequence of the DNA itself is called *cis*-regulation because the control regions are physically linked on the DNA to the regions that can potentially be transcribed. The control sequences of genes influence the structure of chromatin by serving as high-affinity binding sites for regulatory proteins called *transcription factors* (or *transacting factors*, because they may be encoded anywhere in the genome rather than on the same stretch of DNA that they regulate). These DNA-bound transcription factors then contact basal transcription factors either directly or indirectly through coregulatory factors, or recruit enzyme complexes that modify the accessibility of the DNA in chromatin to the transcriptional machinery.

A multiprotein enzyme complex called *RNA polymerase II* carries out the transcription of protein-encoding genes into mRNA. This process is often divided into three steps: initiation of RNA synthesis, RNA chain elongation, and chain termination (Fig. 13.1) (Lee and Young, 2000). Although biologically significant regulation may occur at any of these steps, it is at the step of transcription initiation at which extracellular signals, such as neurotransmitters, hormones, drugs, and growth factors, exert their most significant control over the processes that gate the flow of information out of the genome.

The basic machinery for the synthesis of mRNA is largely common to all RNA polymerase II–transcribed genes; thus, additional layers of control permit the specificity that ensures appropriate temporal and conditional regulation of transcription. Controlling this process at the step of transcription initiation accomplishes this degree of specificity by (1) positioning RNA polymerase II at the correct start site of the gene to be transcribed and (2) controlling the efficiency of the initiation of RNA synthesis to produce the appropriate transcriptional rate for the circumstances of the cell (Tjian and Maniatis, 1994). The *cis*-regulatory elements that set the transcription start sites of genes are called *basal promoter elements* (Smale, 1994). Other *cis*-regulatory elements tether additional activator and repressor proteins to the DNA to regulate the overall transcriptional rate (Ptashne and Gann, 1997).

The Basal Promoter

The promoter regions of genes transcribed by RNA polymerase II contain distinct DNA sequences that function as basal promoter elements on which the basal transcription complex is assembled (Butler and Kadonaga, 2002; Smale, 1994). This complex is composed of a set of proteins, some of which recognize the specific DNA structural elements and some of which bind and position the RNA polymerase II. In contrast to other *cis*-regulatory sequences discussed in this chapter that confer tissue-specific, neurotransmitter, or hormonal regulation, the basal promoter determines the site of transcription initiation and the direction of RNA synthesis (Smale, 2001). Although specific promoter architecture varies considerably, the core elements of the basal promoter are largely conserved among genes (Fig. 13.3) The basal promoter of most of these genes contains a sequence rich in the nucleotides adenine (A) and thymine (T) located

between 25 and 30 bases upstream of the transcription start site, called the *TATA box* (Breathnach and Chambon, 1981). This sequence binds the transcription factor TATA-binding protein (TBP), a component of the TFIID multiprotein complex (Burley and Roeder, 1997). Another sequence located just upstream of the TATA box binds TFIIB in the complex with TBP and the TATA element (Lagrange *et al.*, 1998). Certain RNA polymerase II–transcribed promoters, most commonly those controlling "housekeeping genes," lack a TATA box; in such cases, cytosine–guanosine (CpG)-rich nucleotide sequences may stand in for the TATA box (Smale, 1997; Smale and Baltimore, 1989). For many promoters, the pyrimidine-rich initiator sequence (Inr), overlapping the transcription initiation site (23 to 15), serves as a recognition site for the TFIID associated factors $TAF_{II}250$ and $TAF_{II}150$ (Chalkley and Verrijzer, 1999) and RNA polymerase II (Weis and Reinberg, 1997). Other sequences sometimes present include the CCAAT box, located 70 to 80 bases upstream of the start site (Graves *et al.*, 1986; McKnight and Kingsbury, 1982), and a downstream promoter element (DPE) contacted by components of TFIID (Burke *et al.*, 1998; Butler and Kadonaga, 2002). The presence and combination of these elements determine the relative strength of the basal promoter. The DNA elements of the promoter sequentially recruit the basal transcription factors, beginning with the binding of TFIID to the TATA element (Burley and Roeder, 1997). These factors initiate events required for positioning the transcription initiation site, promoter melting, DNA unwinding, and the early events in the transition between transcription initiation and promoter clearance and elongation by the RNA polymerase complex (Orphanides *et al.*, 1996).

Sequence-Specific Transcription Factors

The basal transcription apparatus is not adequate to confer specificity of expression from a particular promoter or initiate more than low levels of transcription. To achieve the extremely broad dynamic range of regulated expression observed *in vivo*, this multiprotein assembly requires help from transcriptional activators and repressors, proteins that recognize and bind *cis*-regulatory elements found elsewhere within the gene. Because they are tethered to DNA by specific *cis*-regulatory recognition sequences, such proteins have been described as sequence-specific transcription factors (Pabo and Sauer, 1992).

Functional *cis*-regulatory elements are generally found within several hundred base pairs of the start site of the gene to which they are linked, but they can occasionally be found many thousands of

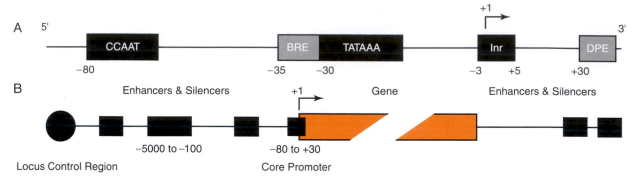

FIGURE 13.3   Promoter architecture. (A) Core promoter elements. Promoters may contain virtually any combination of these elements, although some promoters lack even the TATA sequence, the initiator (Inr), and downstream promoter elements (DPE). (B) Structure of the transcriptional regulatory region of a gene. Enhancer and silencer elements may be located at a distance flanking the structural gene sequence itself, or even within the transcribed portion of a gene. Other higher levels of regulation over larger domains of the genome are exerted by genetic control elements such as locus control regions.

nucleotides away, upstream or downstream of the start site, or even within the transcribed sequence (Guarente, 1988; McKnight and Tjian, 1986). They are generally composed of small modular DNA sequences, typically 7–12 bp in length and structured as inverted or direct repeats, each of which is a specific binding site for one or more transcription factors. Each gene has a particular combination of *cis*-regulatory elements, the nature, number, and spatial arrangement of which determine the gene's unique pattern of expression, including the cell types in which it is expressed, the times during development in which it is expressed, and the level at which it is expressed in adults both basally and in response to physiological signals. *Cis*-regulatory elements that activate expression of their target genes when appropriate transcription factors are bound are commonly termed *enhancers* (Fig. 13.3B). *Cis*-regulatory elements that repress expression of target genes are commonly termed *silencers*.

Sequence-specific transcription factors commonly comprise several physically distinct functional domains. In particular, such transcription factors frequently contain (1) a domain that recognizes and binds a specific nucleotide sequence (i.e., a *cis*-regulatory element), (2) transcription activation or repression domains that interact with either general transcription factors or transcriptional coregulatory proteins, and (3) cooperativity domains that permit interactions with other transcription factors (Johnson and McKnight, 1989; Triezenberg, 1995).

Many transcription factors are active only as dimers or higher-order complexes with other proteins. Transcription factors frequently interact with DNA as dimers, using intrinsic dimerization domains or forming dimers in a cooperative manner on binding to direct or inverted DNA repeats. For some of

these transcriptional regulatory proteins, oligomerization and DNA binding may be regulated by post-translational modifications (e.g., phosphorylation) or by small-molecule binding (e.g., the nuclear hormone receptors) (Beckett, 2001). In some cases heterodimerization may generate novel DNA recognition motifs from the contributions of the DNA-binding determinants of each partner. Within active transcription factor dimers, whether homodimers or heterodimers, both partners commonly contribute jointly to both the DNA binding domain and the activation domain. Dimerization can be a mechanism of either positive or negative control of transcription since a partner may negatively influence the activation domain within a dimeric transcription factor or even prevent DNA binding of the complex. As is discussed in the next section, the second messenger responsive CREB, CREM, and ATF family of bZIP proteins may heterodimerize through a common dimerization domain, a coiled-coil motif termed the *leucine zipper*, potentially combining proteins with distinct activities (Vinson *et al.*, 2002). In this way, the ability of transcription factors to form heterodimers and other multimers increases the diversity of transcription factor complexes that can form in cells and, as a result, increases the types of specific regulatory information that can be exerted on gene expression at a particular *cis*-acting DNA element.

How do sequence-specific transcription factors binding at distant sites influence transcription from a promoter? Transcriptional activators and repressors are hypothesized to alter transcription at a distance by direct protein–protein interactions with the promoter-bound basal transcriptional machinery. DNA-bound transcriptional activator proteins may directly contact one or more proteins within the basal transcription complex such as TFIID, TFIIB, or the RNA polymerase itself. Frequently, however, they do not interact with the basal transcription apparatus directly but through the mediation of adapter proteins termed *coactivators* (Naar *et al.*, 2001). Repressors may inhibit transcription from a promoter by direct interference with activator function. For example, some repressors inhibit DNA binding of an activator by competing for a DNA binding site or by inactivating DNA binding through heterodimerization; for example, the repressor domain of CREMα may interfere with the CREB activation domain through heterodimerization (Loriaux *et al.*, 1994). Others may interfere with activator function directly by physically masking a transactivation domain; for example, MDM2 masks the activity of the tumor suppressor and transcriptional activator p53 (Chene, 2003). In a manner analogous to activators, however, sequence-specific repressors may interfere with the basal transcriptional machinery directly or indirectly through recruitment of corepressor proteins (Burke and Baniahmad, 2000; Glass and Rosenfeld, 2000). In any of these models, transcription factors may communicate with the basal transcription apparatus, looping out intervening DNA and bringing linearly distant control regions into close proximity, providing a biochemical explanation for the distance- and orientation-independence

of *cis*-acting elements (Fig. 13.4) (Blackwood and Kadonaga, 1998; Ptashne and Gann, 1997).

Coactivators, Corepressors, and Chromatin Modification

The concept of transcriptional coactivators came from the observation that some activators contacted the TATA-binding protein (TBP)-associated components of the TFIID complex (Pugh and Tjian, 1990, 1992; Tanese *et al.*, 1991). Subsequently other factors have been found to mediate the effects of DNA-bound activators. For the most part, these proteins reside in large multifunctional complexes proposed to bridge activators to basal factors. For example, a variety of biochemically defined complexes functionally similar to the yeast Mediator complex have been found in metazoans; these complexes interact with the carboxyl-terminal domain of the largest subunit of RNA polymerase II to allow activator-dependent regulation from promoters *in vitro* (Malik and Roeder, 2000; Rachez and Freedman, 2001). The functions of the individual components of these large complexes remain unclear, but they may participate in preinitiation

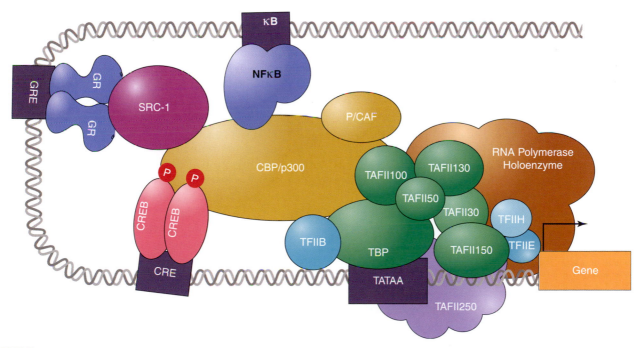

FIGURE 13.4 Transcriptional enhancer sequences act at a distance. DNA-bound activators (and repressors) may influence the activity of a distant promoter either directly by contacting the components of the basal transcription complex (such as TFIIB, TFIID, TFIIA, and even RNA polymerase) or indirectly through coregulatory proteins (coactivators or corepressors). The general coactivators CBP and p300 interact with a large number of DNA-bound transcriptional regulatory proteins including the phosphorylation-dependent transcriptional activator CREB, Fos–Jun, NF-κB, and the nuclear hormone receptors (e.g., the glucocorticoid receptor or GR). CBP and p300 also form complexes with other coactivators, including the nuclear hormone receptor coactivators (e.g., SRC-1) and PCAF, as well as components of the basal transcription complex, including TFIIB and TFIID, and the RNA polymerase holoenzyme.

complex formation or the transition from initiation to transcription elongation (Naar *et al.*, 2001).

Chromatin modification has emerged as a key mechanism of transcriptional regulation. Other factors identified as coactivators for specific transcription factors include the general factors CBP–p300 and P–CAF and the nuclear hormone receptor coactivators SRC-1 and ACTR (Glass and Rosenfeld, 2000; Goodman and Smolik, 2000). These proteins harbor histone acetylase enzymatic activity (Roth *et al.*, 2001) linking DNA-bound transcriptional activators to a modification long known to be associated with transcriptionally active chromatin.

Chromatin exerts an important level of regulation on expression of genes; DNA assembled into chromatin is largely transcriptionally inactive in the absence of the binding of gene-specific activators (Workman and Kingston, 1998). The nucleosome, the basic repeating unit of chromatin, consists of DNA wrapped around an octameric complex (see Fig. 13.5A) composed of a dimer of a tetramer of core histone proteins, H2A, H2B, H3, and H4 (Kornberg and Lorch, 1999). In the core nucleosome, 146 bp of DNA is wrapped 1.65 times around the octamer core. The core nucleosome with the linker histones H1 and H5 forms the 10 nm nucleosomal fiber or the "beads on a string" structure visible by electron microscopy. This chromatin fiber then assembles into higher-order structures to provide the degree of packaging required to condense nearly 2 meters of mammalian chromosomal DNA into the nucleus of a cell.

The overall effect of this high degree of DNA condensation is to limit the accessibility of DNA to both DNA-binding regulatory proteins and the transcriptional apparatus itself. As part of the mechanism of increasing transcription at a promoter, activators increase accessibility of DNA to the transcription apparatus by recruiting coactivator complexes that alter the structure or remodel the chromatin encompassing the target gene (Naar *et al.*, 2001; Narlikar *et al.*, 2002) (Fig. 13.5). The activities of chromatin-directed coactivators may be divided into two broad categories: (1) protein complexes with ATP-dependent chromatin remodeling activity that may "slide" or reposition nucleosomes on DNA (e.g., the SWI–SNF, ISWI, and Mi-2 families of complexes) and (2) proteins with histone acetyltransferase activity (e.g., the general coactivators CBP–p300 and P–CAF, the nuclear hormone receptor coactivators SRC-1 and ACTR, and the basal factor $TAF_{II}250$). Although the details of how these complexes work together remain unclear, both decondensation and histone acetylation are associated with activation of gene transcription. Countering this activity, corepressor complexes contain proteins with deacetylase activity (Knoepfler and Eisenman, 1999).

Histone acetylation has long been known to be associated with active chromatin. Lysine residues within the N-terminal tails of the histones (H3, H4, and, to lesser extent, H2A and H2B) in particular are substrates of the acetyltransferase activity of coactivators *in vitro* and *in vivo* (Roth *et al.*, 2001). The biochemical consequences of acetylation of these lysines and other post-translational modifications of the N-terminal histone tails, including phosphorylation and methylation, on the structure of the nucleosome are unclear. The crystal structure of the nucleosome suggests that neutralization of the basic charge of lysines within the N-terminal tails by acetylation may decrease binding to DNA or alter intranucleosomal protein–protein interactions promoting chromatin decondensation (Luger *et al.*, 1997). An alternative mechanism is that modification of histone tails may serve to mark these specific regions of chromatin for decondensation by chromatin-remodeling enzyme complexes by serving as docking sites for modification-specific protein interaction domains (Jenuwein and Allis, 2001; Turner, 2002). The bromodomain, found in many proteins implicated in transcriptional activation, including the histone acetylases CBP, p300, $TAF_{II}250$, P–CAF, and GCN5, is thought to target these proteins to acetylated histones in transcriptionally active chromatin, or euchromatin (see Fig. 13.5). The acetylated histones then attract bromodomain-containing components of the ATP-dependent SWI–SNF chromatin-remodeling complex (Turner, 2002). In contrast, a class of nuclear transcriptional repressor proteins containing a module called the *chromodomain* are recruited to lysine 9-methylated histone H3 (MeK9 H3) in transcriptionally silenced or heterochromatin (Richards and Elgin, 2002). Repression of transcription by cytosine methylation at CpG sites also involves histone modification. Proteins containing 5-methylcytosine binding domains (MBD1, MBD2, MBD3, MeCP2) participate in methylation-dependent repression by recruiting histone deacetylases (Bird, 2002). Recruitment to this type of "marked" chromatin (either histones or DNA) is thought to stabilize either the active or inactive chromatin state. Thus, DNA-bound activators and repressors may influence chromatin remodeling by recruiting coregulatory enzyme and the processes of covalent histone modification.

Restriction of Expression of Neurally Expressed Genes to the Nervous System

Covalent modification of chromatin (DNA methylation, and histone acetylation and methylation) not only exerts important controls on the short-term regulation of gene expression within a cell or tissue type,

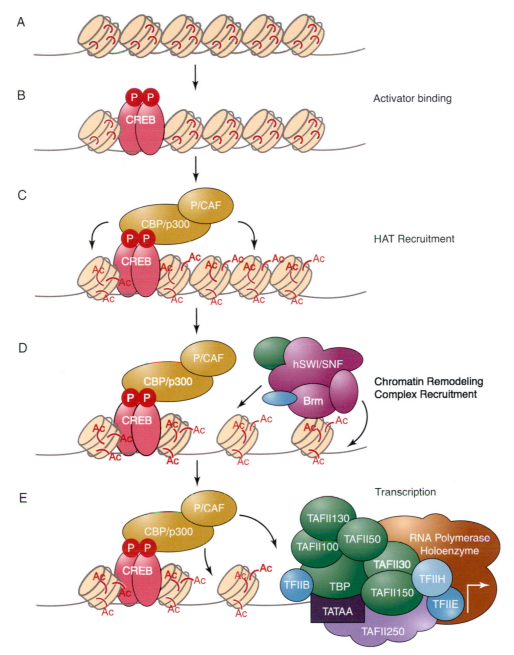

FIGURE 13.5 Transcriptional activators recruit coactivators with histone-modifying and chromatin-remodeling enzymatic activities. Histone acetylation is associated with activation of transcription from a promoter. (A) In the basal state, promoter-associated histones are relatively hypoacetylated. (B) A DNA-bound activator (e.g., protein kinase A-phosphorylated CREB) recruits coactivator complexes that harbor histone acetyltransferase activity (CBP–p300, PCAF). This leads to (C) acetylation of promoter-associated histones within N-terminal tails and (D) the recruitment of ATP-dependent chromatin-remodeling complexes (e.g., hSWI–SNF). (E) Relief of chromatin repression permits access by components of the basal transcriptional machinery, which are also contacted by the general coactivators CBP–p300.

but also is essential for establishing the stable patterns of gene expression that are heritable through cell division. The covalent modification of both histones and DNA underlies mitotically stable changes in gene function that are not explained by alterations in the sequence of DNA itself (Francis and Kingston, 2001). These *epigenetic* mechanisms determine the developmental patterns of

genes that are either expressed or silenced within specific cell lineages, the process of genomic imprinting, and X-chromosome inactivation.

The processes of cellular memory maintain the programs of gene expression in the nervous system. Although the entire complement of regulatory mechanisms defining the patterns of neuronal gene expression remain elusive, a number of genes restricted to neuronal tissue including the type II sodium channel and SCG10 contain a regulatory element called the *neuron restrictive silencer element* (NRSE) (Mori *et al.*, 1990), also known as *repressor element 1* (RE-1)(Maue *et al.*, 1990). NRSE–RE-1 functions to silence target genes specifically in nonneuronal cell types. NRSE–RE-1 recruits a DNA-binding protein called *repressor element 1 silencing transcription* (REST) factor (Chong *et al.*, 1995), also known as *neuron restrictive silencer factor* (NRSF) (Mori *et al.*, 1992; Schoenherr and Anderson, 1995). The REST–NRSF DNA binding domain consists of eight GL1-Krüpple zinc fingers and at least two repression domains. The carboxyl-terminal repression domain recruits the transcriptional corepressor CoREST (Andres *et al.*, 1999), whereas the amino-terminal repression domain associates with the mSin3–histone deacetylase complex (Holdener *et al.*, 2000) and the nuclear receptor corepressor N-CoR (Jepsen *et al.*, 2000). CoREST recruitment to DNA-bound REST–NRSF leads to silencing over a broad chromosomal region encompassing even promoters not directly targeted by REST–NRSF, through propagation of both DNA methylation and histone H3 lysine 9 methylation and recruitment of heterochromatic protein 1 (HP1), leading to stable chromatin condensation and transcriptional silencing in nonneuronal tissue (Lunyak *et al.*, 2002) (Fig. 13.6).

A Significant Consequence of Intracellular Signaling Is the Regulation of Transcription

Intracellular signals play a major role in the regulation of gene expression. Many activator proteins can participate in the assembly of the mature transcription apparatus only after a signal-directed change in subcellular localization (e.g., from the cytoplasm to the nucleus) or a post-translational modification, most commonly phosphorylation. Such alterations in location or conformation permit information obtained by the cell from its different signaling systems to regulate gene expression appropriate to the cells' status.

All diploid cells within an organism, starting with the fertilized one-cell embryo, contain a complete copy of the organism's genome. Differential expression of this common genome is required for the formation of distinct cell types during development, of crucial importance in the differentiation of thousands of distinct types of neurons found in the brain (see Chapter 1). The mechanisms by which these differentiated cells form are highly dependent on intercellular signaling. Much work in this area has been done in *Drosophila* and *Xenopus*, organisms in which viable embryos can be well studied in isolation.

In certain cases, restriction of the expression of a gene to specific cell types depends on the presence of critical transcription factors only in those cell types. For

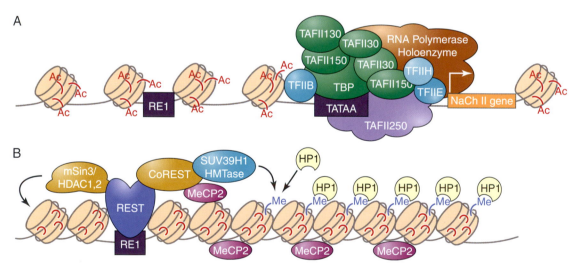

FIGURE 13.6 The transcriptional repressor REST–NRSF restricts the expression of neuronal genes to neural tissue. Neural tissue-specific expression of a number of genes is maintained by the transcriptional repressor REST (or NRSF). (A) In neuronal tissue, genes such as the voltage-gated sodium channel II gene are transcriptionally active. (B) In nonneuronal tissue, the transcriptional repressor REST–NRSF silences neuronal genes by recruitment of corepressor complexes, including CoREST and Sin3-associated histone deacetylase complexes (HDACs 1 and 2). The corepressor CoREST may also interact with other repressors, including MeCP2 (a 5-methylcytosine-directed DNA-binding protein) and the histone methyltransferase (HMTase) SUV39H1. Histone N-terminal tail methylation by HMTases at lysine 9 of histone H3 is associated with transcriptionally silenced chromatin or heterochromatin, recruiting heterochromatin-associated protein 1 (HP1). For some of these gene promoters, methylation at CpG sites is essential for CoREST recruitment and transcriptional repression. Through these mechanisms, DNA binding of REST–NRSF with CoREST recruitment may propagate a silenced state of chromatin to encompass adjacent clustered neuron-specific genes in nonneuronal tissue (modified from Lunyak *et al.*, 2002).

example, the pituitary hormones—growth hormone and prolactin—are expressed only in pituitary lactotrophs and somatotrophs because their required activator transcription factor, Pit 1, is expressed only in those two cell types in the mature organism (Nelson *et al.*, 1988). In other cases as discussed previously, genes restricted to the nervous system contain *cis*-regulatory elements that bind transcriptional repressor proteins in nonneuronal tissues; the presence of repressor proteins locks expression of those genes in that cell type.

The sequential expression, during development, of hierarchies of activator and repressor proteins depends initially on the asymmetric distribution of critical signaling molecules within the embryo, leading to differential gene expression within embryonic cells. As cells gain individual identities during development, cell–cell interactions mediated by contact or by the elaboration of intercellular autocrine, paracrine, or longer-range signaling continue the process of specifying the complement of genes expressed in target cells. Genes that are silent during particular phases of development may become unavailable for subsequent activation because they become wrapped in inactive chromatin structures.

Transcriptional Regulation by Intracellular Signals

As discussed earlier, all protein-encoding genes contain *cis*-regulatory elements that permit the genes to which they are linked to be activated or repressed by physiological signals. Intracellular signals can activate transcription factors through a variety of different general mechanisms, but each requires a translocation step by which the signal is transmitted through the cytoplasm to the nucleus. Some transcription factors are themselves translocated to the nucleus. For example, the transcription factor NF-kB is retained in the cytoplasm by its binding protein IkB; this interaction masks the NF-kB nuclear localization signal. Signal-regulated phosphorylation of IkB by protein kinase C and other protein kinases leads to dissociation of NF-kB, permitting it to enter the nucleus. Other transcription factors must be directly phosphorylated or dephosphorylated to bind DNA. For example, in many cytokine-signaling pathways, plasma membrane receptor tyrosine phosphorylation of transcription factors known as *signal transducers and activators of transcription* (STATs) permits their multimerization, which in turn permits both nuclear translocation and construction of an effective DNA binding site within the multimer. Yet other transcription factors are already bound to their cognate *cis*-regulatory elements within the nucleus under basal conditions and become able to activate transcription after phosphorylation. The transcription factor CREB, for example, is constitutively bound to cAMP response elements (CREs) found within many genes. The critical nuclear translocation step in CREB activation involves not the transcription factor itself, but the catalytic subunit of protein kinase A, which, on entering the nucleus, can phosphorylate CREB. Phosphorylation of CREB converts it into its active state by permitting it to interact with the adapter protein CBP, which can then contact the basal transcription apparatus and acetylate histones leading to chromatin remodeling (see Fig. 13.5).

ROLE OF cAMP AND Ca²¹ IN THE ACTIVATION PATHWAYS OF TRANSCRIPTION

As described in Chapter 12, the cAMP second-messenger pathway is among the best-characterized intracellular signaling pathways; a major feature of signaling by this pathway is the regulation of a large number of genes. Cyclic AMP response elements with the consensus nucleotide sequence of TGACGTCA have been identified in many genes expressed in the nervous system.

The consensus CRE sequence illustrates a common feature of many transcription-factor binding sites; it is a palindrome or inverted repeat. Examination of the sequence TGACGTCA readily reveals that the sequences on the two complementary strands, which run in opposite directions, are identical. Many *cis*-regulatory elements are perfect or approximate palindromes because many transcription factors bind DNA as dimers, in which each member of the dimer recognizes one of the "half-sites." CREB binds to CREs as a homodimer, with a higher affinity for perfectly palindromic than for asymmetric CREs.

When bound to a CRE, CREB activates transcription when it is phosphorylated at Ser-133 (Mayr and Montminy, 2001). It does so, as described earlier, because phosphorylated CREB, but not unphosphorylated CREB, recruits the adapter protein CBP into the transcription complex. CBP, in turn, interacts with the basal transcription complex and modifies histones to enhance the efficiency of transcription (Fig. 13.5).

The regulation of CREB activation by phosphorylation illustrates several general principles, including the requirement for nuclear translocation of protein kinases in cases where transcription factors are already found in the nucleus under basal conditions and the role of phosphorylation in regulating protein–protein interactions. An additional important principle illustrated by CREB is the convergence of signaling pathways. CREB is phosphorylated on Ser-133 by the free catalytic subunit of the cAMP-dependent protein kinase. However, CREB Ser-133 can also be phosphorylated by Ca^{++}-calmodulin-dependent protein kinases types II and IV (Sheng et al., 1990) and by RSK2, a kinase activated in growth factor pathways including Ras and MAP kinase (Xing et al., 1996). When each individual signal is relatively weak, convergence may be a critical mechanism resulting in specificity of gene regulation, with some genes being activated only when multiple pathways are stimulated. Some genes that contain CREs are known to be induced in more than an additive fashion by the interaction of cAMP and Ca^{++}, but how convergent phosphorylation on the same serine might produce synergy is not yet clear. Synergy is more readily understood in cases in which a particular protein is modified at two different sites, causing interacting conformational changes. In addition to Ser-133, CREB contains sites for phosphorylation by a variety of protein kinases, including glycogen synthase kinase 3 (GSK3), but the biological effects of phosphorylating these additional serines are not fully understood. These additional phosphorylation events may fine-tune the regulation of CREB-mediated transcription.

CREB illustrates yet another important principle of transcriptional regulation: CREB is a member of a family of related proteins. Many transcription factors are members of families; this permits complex forms of positive and negative regulation as discussed previously. CREB is closely related to other proteins called *activating transcription factors* (ATFs) and *CRE modulators* (CREMs), which result from alternative splicing of a single CREM gene. All these proteins bind CREs as dimers; many can dimerize with CREB itself. ATF-1 appears to be very similar to CREB in that it can be activated by both cAMP and Ca^{++} pathways. Many of the other ATF proteins and CREM isoforms can activate transcription; however, certain CREMs may act to repress it. These CREM isoforms lack the glutamine-rich transcriptional activation domain of CREB–ATF family members that are activators of transcription. Thus CREB–CREM homodimers may bind DNA but fail to activate transcription. Like CREB, many of the ATF proteins are constitutively synthesized, but ATF-3 and certain CREM isoforms are inducible in response to environmental stimuli. The new synthesis of transcription factors is yet another mechanism of gene regulation.

The dimerization domain used by the CREB–ATF proteins and several other families of transcription factors is called a *leucine zipper*. This domain was first identified in transcription factor C–EBP and is also used by the AP-1 family of transcription factors. The so-called leucine zipper actually forms a coiled coil. The dimerization motif is an α helix in which every seventh residue is a leucine; based on the periodicity of α helices, the leucines line up along one face of the helix two turns apart. The aligned leucines of the two dimerization partners interact hydrophobically and stabilize the dimer. In CREB, C–EBP, and the AP-1 family of proteins, the leucine zipper is at the carboxy terminus of the protein. Just upstream of the leucine zipper is a region of highly basic amino acid residues that form the DNA binding domain. Dimerization by means of the leucine zipper domain juxtaposes the adjacent basic regions of each of the partners; these juxtaposed basic regions then undergo a random coil-to-helix transition when they bind DNA in the major groove of DNA. This combination of motifs is why this superfamily of proteins is referred to as the *basic leucine zipper proteins* (bZIPs).

AP-1 Transcription Factors Are Derived Mainly from Cellular "Immediate-Early" Genes

Activator protein 1 (AP-1) is another family of bZIP transcription factors that play a central role in the regulation of neural gene expression by extracellular signals. The AP-1 family comprises multiple proteins that bind as heterodimers (and a few as homodimers) to the DNA sequence TGACTCA, the consensus AP-1 element that forms a palindrome flanking a central C or G. Although the AP-1 sequence differs from the CRE sequence by only a single base, this one-base difference strongly biases protein binding away from the CREB family of proteins. AP-1 sequences confer responsiveness to the protein kinase C pathway.

The AP-1 proteins generally bind DNA as heterodimers composed of one member each of two different families of related bZIP proteins, the Fos family and the Jun family. The known members of the Fos family are c-Fos, Fra-1 (Fos-related antigen-1), Fra-2, and FosB; there is also evidence for post-translationally modified forms of FosB. The known members of the Jun family are c-Jun, JunB, and JunD. Heterodimers form between proteins of the Fos family and proteins of the Jun family by means of the leucine zipper. Unlike the Fos proteins, c-Jun and JunD, but not JunB, can form homodimers that bind to AP-1 sites, albeit with far lower affinity than Fos–Jun heterodimers. The potential complexity of transcriptional regulation is greater still because some AP-1 proteins can heterodimerize through the leucine zipper with members of the CREB–ATF family, for example, ATF2 with c-Jun. AP-1 proteins can also form higher-order complexes with unrelated families of transcription factors.

In addition, AP-1 proteins can complex with and thus apparently inhibit the transcriptional activity of certain nuclear hormone receptors (discussed later).

Among the known Fos and Jun proteins, only JunD is expressed constitutively at high levels in many cell types. The other AP-1 proteins tend to be expressed at low or even undetectable levels under basal conditions but, with stimulation, may be induced to high levels of expression. Thus, unlike genes that are regulated by constitutively expressed transcription factors such as CREB, genes that are regulated by c-Fos–c-Jun heterodimers require new transcription and translation of their required regulatory factors.

Genes that are transcriptionally activated by synaptic activity, drugs, and growth factors have often been classified roughly into two groups. Genes, such as the *c-fos* gene itself, that are activated rapidly (within minutes), transiently, and without requiring new protein synthesis are often described as cellular *immediate-early genes* (IEGs). Genes that are induced or repressed more slowly (within hours) and dependent on new protein synthesis have been described as *late-response genes*. The term *IEG* was initially applied to describe viral genes that are activated "immediately" on infection of eukaryotic cells by utilization of pre-existing host cell transcription factors. Viral immediate-early genes generally encode transcription factors needed to activate viral "late" gene expression. This terminology has been extended to cellular (i.e., nonviral) genes with varying success. The terminology is problematic because many cellular genes are induced independent of protein synthesis, but with a time course intermediate between those of "classic" IEGs and late-response genes. In fact, some genes may be regulated with different time courses or requirements for protein synthesis in response to different intracellular signals. Moreover, many cellular genes regulated as IEGs encode proteins that are not transcription factors. Despite these caveats, the concept of IEG-encoded transcription factors in the nervous system is useful. Because of their rapid induction from low basal levels in response to neuronal depolarization (the critical signal being Ca^{++} entry) and second-messenger and growth factor pathways, several IEGs have been used as cellular markers of neural activation, permitting novel approaches to functional neuroanatomy.

The protein products of those cellular IEGs that function as transcription factors bind to *cis*-regulatory elements contained within a subset of late-response genes to activate or repress them. IEGs such as *c-fos* have therefore been termed third messengers in signal transduction cascades, with neurotransmitters designated as intercellular first messengers, and small intracellular molecules, such as cAMP and Ca^{++}, as second messengers. However, IEGs are not always a necessary step in signal-regulated expression of genes having roles in the differentiated function of neurons. In fact, many such genes, including many genes encoding neuropeptides such as proenkephalin (Konradi *et al.*, 1993) and prodynorphin (Cole *et al.*, 1995) and some genes encoding neurotrophic factors, are activated in response to neuronal depolarization or cAMP by phosphorylation of the constitutively expressed transcription factor CREB rather than by IEG third messengers. In sum, neural genes that are regulated by extracellular signals are activated or repressed with varying time courses by reversible phosphorylation of constitutively synthesized transcription factors and by newly synthesized transcription factors, some of which are regulated as IEGs.

Activation of the c-fos Gene

The *c-fos* gene is activated rapidly by neurotransmitters or drugs that stimulate the cAMP pathway or Ca^{++} elevation. Both pathways produce phosphorylation of transcription factor CREB (Sheng *et al.*, 1990). The *c-fos* gene contains three binding sites for CREB. The *c-fos* gene can also be induced by the Ras–MAP kinase pathway, which is activated by a number of growth factors. For example, neurotrophins, such as nerve growth factor (NGF), bind a family of receptor tyrosine kinases (Trks); NGF interacts with Trk A, which activates Ras. Ras then acts through a cascade of protein kinases including Raf and the cytoplasmic MAP kinase kinases (MAPKKs) MEK1–MEK2, which phosphorylate the MAP kinases ERK1–ERK2. These kinases translocate into the nucleus, where they can activate RSK2 to phosphorylate CREB, but they can also apparently directly phosphorylate other transcription factors such as the ternary complex factor Elk-1. Elk-1 binds along with the serum response factor (SRF) to the serum response element (SRE) within the *c-fos* gene and many other growth factor–inducible genes. Cross talk between neurotransmitter and growth factor signaling pathways has been documented with increasing frequency and likely plays an important role in the precise tuning of neural plasticity to diverse environmental stimuli.

Regulation of c-Jun

Expression of most of the proteins of the Fos and Jun families that constitute transcription factor AP-1 and the binding of AP-1 to DNA is regulated by extracellular signals. However, both Fos and Jun family members are phosphoproteins themselves, and AP-1-mediated transcription requires not only the new synthesis of AP-1 but also the phosphorylation of proteins within AP-1 complexes. Phosphorylation of c-Jun within its N-terminal activation domain has been shown to markedly enhance its ability to activate transcription without affecting its ability to form dimers or bind DNA. Other phosphorylation sites within c-Jun regulate its ability to bind DNA.

Phosphorylation and activation of c-Jun can result from the action of Jun N-terminal kinase (JNK). JNK is a member of the mitogen-activated protein kinase (MAPK) family of protein kinases whose mammalian members include the ERKs, p38, and JNK. In addition to cellular stressors, the inflammatory cytokines interleukin-1β (IL-1β) and tumor necrosis factor α (TNFα) have been shown to activate both JNK and p38. JNK has also been shown to be activated by neurotransmitters, including glutamate. Thus, AP-1-mediated transcription within the nervous system requires multiple steps beginning with the activation of genes encoding AP-1 proteins.

Growth Factors and Cytokines as Modulators of Gene Expression in the Nervous System

With respect to function, the boundary between trophic, or growth, factors and cytokines in the nervous system has become increasingly arbitrary. However, cell signaling mechanisms offer a useful means of distinction. Growth factors, such as the neurotrophins (e.g., nerve growth factor, brain-derived neurotrophic factor, and neurotrophin 3), epidermal growth factor (EGF), and fibroblast growth factor (FGF), act through receptor protein tyrosine kinases, whereas the cytokines, such as leukemia inhibitory factor (LIF), ciliary neurotrophic factor (CNTF), and interleukin-6 (IL-6), act through nonreceptor protein tyrosine kinases (see Chapter 10).

LIF, CNTF, and IL-6 subserve a wide array of overlapping functions inside and outside the nervous system, including hematopoietic and immunological functions outside the nervous system and regulation of neuronal survival, differentiation, and, in certain circumstances, plasticity within the nervous system. These peptides have marked homologies of their tertiary structures rather than their primary sequences, which presumably permit them to interact with related receptor complexes that contain a common signal-transducing subunit, gp130.

The receptors for these cytokines consist of a signal-transducing β component, which includes gp130 and, in some cases, additional subunits. As is typical of cytokine receptors, the cytoplasmic tails of the signal-transducing β components of the IL-6, LIF, and CNTF receptors lack kinase domains. Rather, the cytoplasmic domains interact with nonreceptor protein tyrosine

kinases (PTKs) of the Janus kinase (Jak) family, which include Jak1, Jak2, and Tyk2 (Fig. 13.7). Some cytokine receptors, such as the prolactin receptor, which interacts exclusively with Jak2, can interact only with a single Jak PTK. In contrast, the IL-6, LIF, and CNTF receptors can interact with multiple Jak PTKs, including Jak1, Jak2, and Tyk2. Presumably, the dimerization of receptors on ligand binding permits Jak-family PTKs to cross-phosphorylate each other.

Signal transduction to the nucleus includes tyrosine phosphorylation by the Jak PTKs of one or more of the STAT proteins mentioned earlier. The first STAT family members were identified as proteins binding to interferon-regulated genes but have subsequently been found to take part in the activity of multiple cytokines. On phosphorylation, STAT proteins form dimers through the association of SH2 domains, an important type of protein interaction domain described earlier. Dimerization is thought to trigger translocation to the nucleus, where STATS bind their cognate cytokine response elements. Different STATs become activated by different cytokine receptors, not because of differential use of Jak PTKs but because of specific coupling of certain STATs to certain receptors. Thus, for example, the IL-6 receptor preferentially activates STAT1 and STAT3; the CNTF receptor preferentially activates STAT3.

Signaling through the Jak–STAT systems is ultimately turned off by feedback from a STAT-induced gene product, suppressor of cytokine signaling (SOCS). SOCS binds to the activated Jak PTK and prevents it from phosphorylating the β component of the cytokine receptor and, thus, blocks subsequent STAT binding (see Fig. 13.7). Thus, like many other signaling cascades, this system has an inherent feedback mechanism that terminates the signal.

Many of these target genes integrate signals from a variety of signaling pathways. The *c-fos* gene, for example, contains an element called the SIS-inducible element (SIE), which binds STAT proteins; thus, *c-fos* gene expression can also be induced by cytokines. Cytokine response elements have now been identified within many neural genes, including vasoactive intestinal polypeptide and several other neuropeptide-encoding genes.

Nuclear Receptors Regulate Transcription Directly

The differentiation of many cell types in the brain is established by exposure to steroids and other small-molecule hormones. For example, exposure to estrogen or testosterone during critical developmental periods results in sexually dimorphic development of certain

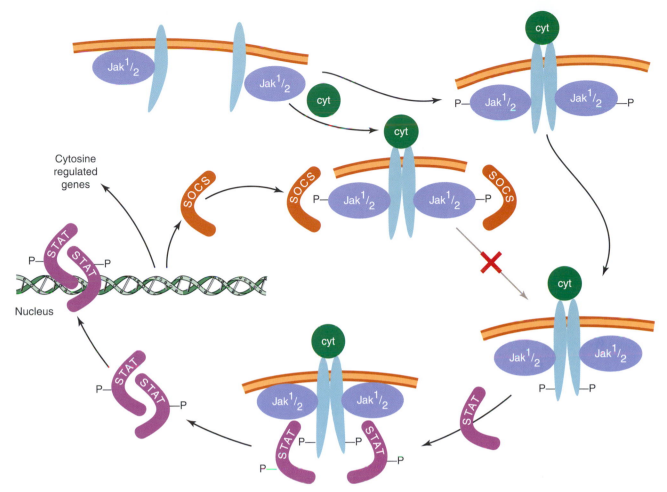

FIGURE 13.7 Cytokine regulation of gene transcription. Cytokines (Cyt) bind, causing dimerization and activation of Jak PTK (Jak 1–2) by phosphorylation, which subsequently phosphorylates the receptor. Phosphoreceptor recruits STAT proteins, which become phosphorylated by Jak PTK, which allows for dimerization and nuclear translocation. Phosphorylated STATs activate transcription of multiple genes, including SOCS, which feeds back to block cytokine receptor phosphorylation and turns off the signal.

brain nuclei. The small-molecule hormones, including the glucocorticoids, sex steroids, mineralocorticoids, retinoids, thyroid hormone, and vitamin D, are small lipid-soluble ligands that can diffuse across cell membranes. Unlike the other types of intercellular signals described previously, these small-molecule hormones bind their receptor inside the cell and subsequently affect specific gene transcription in the nucleus. In some cases, such as the estrogen receptor or retinoic acid receptor, the unliganded receptor is resident in the nucleus, awaiting the diffusing molecule. In other systems, such as with the glucocorticoid receptor, the receptor is in the cytoplasm and, on hormone binding, undergoes conformational change and dissociation of associated proteins, which unmasks a nuclear localization signal and DNA binding domain, similar to the activation of NFκB. A significant exception to this deals with the more recently accepted ability of small-molecule hormone receptors to affect signaling in the

cytoplasm without cycling through the nucleus. This is described in detail in Chapter 10.

Like the other transcription factors described in this chapter, the steroid hormone receptors are modular in nature (see Fig. 13.8A). Each has a transcriptional activation domain at its amino terminus, a DNA binding domain, and a ligand binding domain at its carboxy terminus. The DNA binding domains recognize specific palindromic or direct repeat DNA sequences, hormone response elements, within the regulatory regions of specific genes (Fig. 13.8B), binding in dimers. In the palidromic regulatory elements, the "spacer" DNA puts the same sequence (the complement from the opposite strand) one helical turn away and thus on the same side of the DNA surface, so the nuclear receptor homodimer can bind to an identical site in a head-to-head configuration.

As with the other transcription factors, once bound to the DNA, the nuclear hormone receptors interact

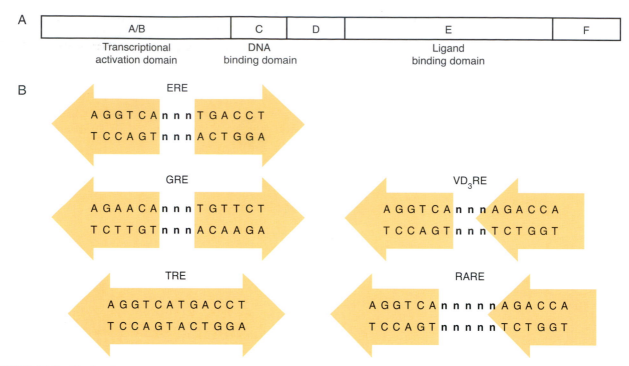

FIGURE 13.8 Nuclear hormone receptor structure and DNA binding sites. (A) Domain structure of the nuclear small-molecule hormone receptor. Domain A–B contains the transcriptional activation structures for interacting with other proteins, C is the highly homologous DNA binding domain, D represents the structure involved in stabilizing the dimer form of the receptor, and E contains the ligand (hormone) binding domain. (B) The estrogen receptor regulatory element (ERE), glucocorticoid receptor element (GRE), and thyroid hormone receptor element (TRE) are constructed of palindromic sequences of six nucleotides separated by three or zero nucleotides. The vitamin D_3 receptor regulatory element (VD$_3$RE) and retinoic acid receptor regulatory element (RARE) contain imperfect repeats separated by three to five nucleotides. These represent the consensus sequences, and there may be small variations in sequence that alter the affinity of the receptor for the DNA.

with transcriptional coactivators and/or repressors to modulate levels of transcription appropriate for the physiological state of the animal. The same molecular mechanisms described earlier are used to modulate transcription levels.

Nuclear hormone receptors also elicit transcriptional actions via interactions with other nuclear transcriptional machinery. This became apparent in earlier studies on specific gene promoters that contained no hormone receptor regulatory element in the DNA, yet were still transcriptionally sensitive to hormone action. For example, for some promoters, glucocorticoids modulated transcription but required an AP-1 element in the promoter to mediate the action (Jonat et al., 1990; Schüle et al., 1990). At the same time, similar observations were made for estrogen (Gaub et al., 1990). Subsequent studies have shown that the small-molecule hormone receptor can affect the recruitment of coactivators–corepressors without having to interact with the DNA (Göttlicher et al., 1998; Kushner et al., 2000). This type of action of nuclear hormone receptors has now been seen for other transcription factor systems such as NFκB and CREB, suggesting even greater complexity in transcriptional regulation. It provides the basis for understanding the broad changes in gene expression that occur within the brain in association with major changes in hormone levels associated with chronic stress or developmental events such as puberty.

Gene Expression Is Also Regulated at the Post-Transcriptional Level

The steady-state level of a specific mRNA is determined by its rate of synthesis (transcription) as well as its rate of degradation (mRNA turnover). Although modulation of the transcriptional rate remains the predominant mode for regulating gene expression in the nervous system, evidence continues to accumulate showing that regulation of mRNA turnover also plays an important role. Indeed, the coordination of transcription and mRNA turnover allows for exquisitely tight control over the level of an mRNA, allowing for rapid on/off signaling in some cases.

mRNA has several post-transcriptional modifications that help prevent it from endoribonuclease degradation: 7-Me-G capping at the 5' end and polyadenylation at the 3' end. During the lifetime of the mRNA, the poly(A) tail gradually shortens and finally the 5' 7-Me-G cap is removed and the mRNA becomes subject to rapid nuclease degradation (see Fig. 13.8). Early biochemical experiments classified mRNAs as either short-lived, with a $t_{1/2}$ of about 30–60 min, or long-lived, with a $t_{1/2}$ of 24–36 h, but now we know there is a broad range of half-lives ranging from 5 min to many days. Further studies have identified multiple sequence elements within the mRNA that mediate the relative stability of the different mRNAs.

The first characterized RNA structure mediating mRNA stability was an AU-rich element (ARE),

which is present in the 3′ untranslated region of the mRNA (see Fig. 13.9). These have been categorized into three classes, ARE I–III, based on sequence (Chen and Shyu, 1995), and all bind a variety of cytoplasmic proteins, which confer either stability or instability. Although the exact mechanism is still not fully understood, it appears that the proteins binding to the AREs in many cases cause the 3′ poly(A) tail to either shorten or stabilize, resulting in mRNA stability

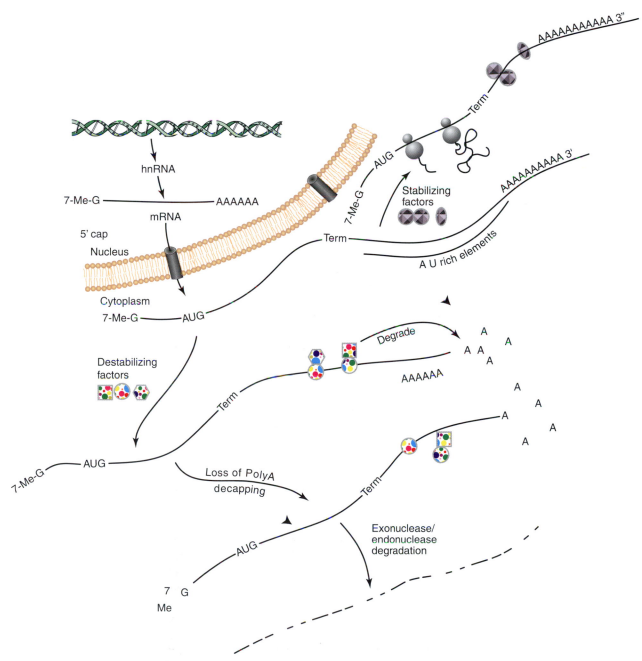

FIGURE 13.9 Schematic of regulation of mRNA turnover. After synthesis in the nucleus with post-transcriptional modification of both the 5′ and 3′ ends of the mRNA it is transported to the cytoplasm. There it can interact with destabilizing factors that lead to a loss of the poly (A) tail at the 3′ end, subsequent 7-Me-G decapping, and rapid degradation in the cytoplasm.

change. This mechanism is common with many cytokine, transcription factor, cell cycle regulator, or cell surface receptor mRNAs, all belonging to classes of neural genes that the organism would like to have under tight control. For example, the transcription factor c-Fos-encoding gene is rapidly induced on depolarization of a neuron, producing a rapid rise in c-fos mRNA. At the same time, however, there is also an induction of proteins that interact with the c-Fos mRNA ARE, resulting in instability and rapid degradation of the newly made c-fos mRNA. The net result is a transcriptional signal that lasts for only an hour or so after a neuron depolarizes.

Although the exact mechanisms by which signaling in a neuron or glia is translated into alteration of stability of a specific mRNA are still being elucidated, we are beginning to see a pathway that has much in common with the regulation of transcription. The family of signal transduction activators of the RNA (STAR) family of proteins is an excellent example (Lasko, 2003). These proteins bind mRNA through specific domains and can be subject to phosphorylation by activated tyrosine kinases. This modification changes how the STAR protein interacts with other proteins leading to alterations in mRNA stability and/or where the mRNA is located within a cell. For example, myelin basic protein (MBP) mRNA contains a STAR protein-binding sequence in its 3′ untranslated region that enhances the stability of MBP mRNA. In the quaking mouse, where myelination is sharply reduced, the mutation was traced to the QK1 gene (Hardy et al., 1996), a STAR protein that interacts with this MBP mRNA element, resulting in destabilization and mislocation of MBP mRNA in the oligodendrocyte (Li et al., 2000).

The issue of where a specific mRNA is located and translated in a neuron is also important to neuronal function. Specific mRNAs can be found asymmetrically distributed in subcellular compartments where the proteins they encode are used. For example, several synaptic protein-encoding mRNAs are located at the neuromuscular junction (Chakkalakal and Jasmin, 2002). There are also extensive examples of dendritic localization of specific mRNAs, and the mechanism of their translation is better understood (Kindler et al., 1997; Steward and Schuman, 2001). The presumption has been that there is local synthesis of proteins encoded by these dendritic mRNAs, allowing for rapid, local changes in proteins. Indeed, the various components of the translational machinery have been localized to the dendrites (Tiedge and Brosius, 1996), and protein synthesis has been shown to occur in isolated dendrites (Torre and Steward, 1992). The identification of localized synthesis of proteins is crucial to our understanding of how a neuron or glia can rapidly respond with changes in protein expression often at places far removed from the cell body.

Finally, the actual act of translation of an mRNA into a protein can also be regulated by modulation of proteins that bind the mRNA. In some cases, particularly in development, specific mRNAs are kept in an inactive state, awaiting the appropriate signals to begin translation. Again, proteins bind to specific mRNA sequences, effectively shutting down the mRNA, and cytoplasmic changes in polyadenylation of the mRNA appear to play an important role (Richter, 1999). Translation silencing is also important in avoiding translation until an mRNA has been transported to its subcellular site of translation (Richter and Lorenz, 2002; Wang et al., 2002). Other mechanisms involve turning up or down the level of translation of particular mRNAs. Again, in a manner analogous to regulation of transcription, specific proteins, which bind to the mRNA, interact with translation initiation factors altering the level of translation. Thus, there are multiple mechanisms, in addition to transcriptional regulation, by which different neurons and glia can adjust the level of synthesis of specific proteins to respond to the different developmental and physiological states.

SUMMARY

The formation of long-term memories requires changes in gene expression and new protein synthesis (see also Chapter 19). Control sequences on DNA determine which segments of DNA can be transcribed into RNA. It is at transcription initiation that extracellular signals such as neurotransmitters, hormones, drugs, and growth factors exert their most significant control. The transcription itself is carried out by RNA polymerases. The transcription is modulated by transcription factors that recruit the polymerases to the DNA. For example, the critical nuclear translocation step in the activation of transcription factor CREB involves the catalytic subunit of PKA, which can phosphorylate CREB on entering the nucleus.

The genes that encode the transcription factors themselves may respond quickly or slowly. These genes have been coined "third messengers" in signal transduction cascades. Cross talk between neurotransmitter and growth factor signaling pathways is likely to play an important role in the precise tuning of neuronal plasticity to diverse environmental stimuli.

The active, mature transcription complex is a remarkable architectural assembly of proteins assembled at the basal promoter—in most cases, at a sequence called the TATA box. In addition to RNA polymerase II, this complex includes a large number of associated general transcription factors, sequence-specific transcription factors, and intervening adapters. A wide variety of transcription factors bound to *cis*-regulatory elements elsewhere in the gene, but permitted to interact by the looping of DNA, join in the formation of the active transcription complex. This remarkable mechanism permits cells to exert exquisite control of the genes being transcribed in a variety of situations, for example, to govern appropriate entry or exit from the cell cycle, to maintain appropriate cellular identity, and to respond appropriately to extracellular signals.

Transcription can be regulated by many different extracellular signals modulated by a large array of signaling pathways (many including reversible phosphorylation) and a complex array of transcription factors. Most of these factors are members of families and regulate transcription only as multimers. Given this complexity, the potential for very precise regulation is clear, but the mechanisms by which such precision is achieved are not fully understood. In this chapter, regulation has been illustrated by only a few of the families of transcription factors. Those chosen appear to play important roles in the nervous system and illustrate many of the basic principles of gene regulation.

Gene expression is also regulated at the post-transcriptional level. The steady-state level of an mRNA is determined as much by its degradation as it is by its synthesis. As with transcriptional regulation, there are proteins that bind to specific sequences in the mRNA and vary the stability of the mRNA, its subcellular location within the cell, or the rate at which it is translated into protein.

References

Andres, M. E., Burger, C., Peral-Rubio, M. J., Battaglioli, E., Anderson, M. E., Grimes, J., Dallman, J., Ballas, N., and Mandel, G. (1999). CoREST: A functional corepressor required for regulation of neural-specific gene expression. *Proc. Natl. Acad. Sci. USA* **96**, 9873–9878.

Bass, B. L. (2002). RNA editing by adenosine deaminases that act on RNA. *Ann. Rev. Biochem.* **71**, 817–846.

Beckett, D. (2001). Regulated assembly of transcription factors and control of transcription initiation. *J. Mol. Biol.* **314**(3), 335–352.

Bird, A. (2002). DNA methylation patterns and epigenetic memory. *Genes Dev.* **16**, 6–21.

Blackwood, E. M., and Kadonaga, J. T. (1998). Going the distance: A current view of enhancer action. *Science* **281**, 61–63.

Breathnach, R., and Chambon, P. (1981). Organization and expression of eucaryotic split genes coding for proteins. *Annu. Rev. Biochem.* **50**, 349–383.

Burke, L. J., and Baniahmad, A. (2000). Co-repressors 2000. *FASEB J.* **14**, 1876–1888.

Burke, T. W., Willy, P. J., Kutach, A. K., Butler, J. E., and Kadonaga, J. T. (1998). The DPE, a conserved downstream core promoter element that is functionally analogous to the TATA box. *Cold Spring Harbor Symp. Quant. Biol.* **63**, 75–82.

Burley, S. K., and Roeder, R. G. (1997). Biochemistry and structural biology of transcription factor IID (TFIID). *Annu. Rev. Biochem.* **65**, 769–799.

Butler, J. E., and Kadonaga, J. T. (2002). The RNA polymerase II core promoter: A key component in the regulation of gene expression. *Genes Dev.* **16**, 2583–2592.

Chakkalakal, J. V., Jasmin, B. J. (2002). Localizing synaptic mRNAs at the neuromuscular junction: It takes more than transcription. *BioEssays* **25**, 25–31.

Chalkley, G. E., and Verrijzer, C. P. (1999). DNA binding site selection by RNA polymerase II TAFs: A TAF(II)250–TAF(II)150 complex recognizes the initiator. *EMBO J.* **18**, 4835–4845.

Chen, C. Y., and Shyu, A. B. (1995). AU-rich elements: Characterization and importance in mRNA decay. *Trends Biochem. Sci.* **20**, 465–470.

Chene, P. (2003). Inhibiting the p53–MDM2 interaction: An important target for cancer therapy. *Nat. Rev. Cancer* **3**, 102–109.

Chong, J. A., Tapia-Ramirez, J., Kim, S., Toledo-Aral, J. J., Zheng, Y., Boutros, M. C., Altshuller, Y. M., Frohman, M. A., Kraner, S. D., and Mandel, G. (1995). REST: A mammalian silencer protein that restricts sodium channel gene expression to neurons. *Cell* **80**, 949–957.

Cole, R. L., Konradi, C., Douglass, J., and Hyman, S. E. (1995). Neuronal adaptation to amphetamine and dopamine: Molecular mechanisms of prodynorphin gene regulation in rat striatum. *Neuron* **14**, 813–823.

Francis, N. J., and Kingston, R. E. (2001). Mechanisms of transcriptional memory. *Nat. Rev. Mol. Cell. Biol.* **2**, 409–421.

Gaub, M. P., Bellard, M., Scheuer, I., Chambon, P., and Sassone, C. P. (1990). Activation of the ovalbumin gene by the estrogen receptor involves the fos–jun complex. *Cell* **63**, 1267–1276.

Glass, C. K., and Rosenfeld, M. G. (2000). The coregulator exchange in transcriptional functions of nuclear receptors. *Genes Dev.* **14**, 121–141.

Goodman, R. H., and Smolik, S. (2000). CBP/p300 in cell growth, transformation, and development. *Genes Dev.* **14**, 1553–1577.

Göttlicher, M., Heck, S., and Herrlich, P. (1998). Transcriptional cross-talk, the second mode of steroid hormone receptor action. *J. Mol. Med.* **76**, 480–489.

Graves, B. J., Johnson, P. F., and McKnight, S. L. (1986). Homologous recognition of a promoter domain common to the MSV LTR and the HSV tk gene. *Cell* **44**, 565–576.

Guarente, L. (1988). UASs and enhancers: common mechanism of transcriptional activation in yeast and mammals. *Cell* **52**, 303–305.

Hardy, R. J., Loushin, C. L., Friedrich, V. L., Chen, Q., Ebersole, T. A., Lazzarini, R. A., and Artzt, K. (1996). Neural cell type specific expression of QKI proteins is altered in the quaking viable mutant mouse. *J. Neurosci.* **16**, 7941–7949.

Holdener, B. C., Mandel, G., and Kouzarides, T. (2000). The co-repressor mSin3A is a functional component of the REST–CoREST repressor complex. *J. Biol. Chem.* **275**, 9461–9467.

International Human Genome Sequencing Consortium (2001). Initial sequencing and analysis of the human genome. *Nature* **409**, 860–921.

Jenuwein, T., and Allis, C. D. (2001). Translating the histone code. *Science* **293**, 1074–1080.

Jepsen, K., Hermanson, O., Onami, T. M., Gleiberman, A. S., Lunyak, V., McEvilly, R. J., Kurokawa, R., Kumar, V., Liu, F.,

Seto, E., Hedrick, S. M., Mandel, G., Glass, C. K., Rose, D. W., and Rosenfeld, M. G. (2000). Combinatorial roles of the nuclear receptor corepressor in transcription and development. *Cell* **102**, 753–763.

Johnson, P. F., and McKnight, S. L. (1989). Eukaryotic transcriptional regulatory proteins. *Annu. Rev. Biochem.* **58**, 799–839.

Jonat, C., Rahmsdorf, H. J., Park, K. K., Cato, A. C., Gebel, S., Ponta, H., and Herrlich, P. (1990). Antitumor promotion and antiinflammation: Down-modulation of AP-1 (Fos/Jun) activity by glucocorticoid hormone. *Cell* **62**, 1189–1204.

Kindler, S., Mohr, E., and Richter, D. (1997). Quo vadis: Extrasomatic targeting of neuronal mRNAs in mammals. *Mol. Cell. Endocrin.* **128**, 7–10.

Knoepfler, P. S., and Eisenman, R. N. (1999). Sin meets NuRD and other tails of repression. *Cell* **99**, 447–450.

Konradi, C., Kobierski, L. A., Nguyen, T. V., Heckers, S. H., and Hyman, S. E. (1993). The cAMP-response-element-binding protein interacts, but Fos protein does not interact, with the proenkephalin enhancer in rat striatum. *Proc. Natl. Acad. Sci. USA* **90**, 7005–7009.

Kornberg, R. D., and Lorch, Y. (1999). Twenty-five years of the nucleosome, fundamental particle of the eukaryote chromosome. *Cell* **98**, 285–294.

Kushner, P. J., Agard, D. A., Greene, G. L., Scanlan, T. S., Shiau, A. K., Uht, R. M., Webb, P. (2000). Estrogen receptor pathways to AP-1. *J. Steroid Biochem. Mol. Bio.* **74**, 311–317.

Lagrange, T., Kapanidis, A. N., Tang, H., Reinberg, D., and Ebright, R. H. (1998). New core promoter element in RNA polymerase II-dependent transcription: Sequence-specific DNA binding by transcription factor IIB. *Genes Dev.* **12**, 34–44.

Lasko, P. (2003). Gene regulation at the RNA layer: RNA binding proteins in intercellular signaling networks. *Science's Stke* **179**, 1–8.

Lee, T. I., and Young, R. A. (2000). Transcription of eukaryotic protein-coding genes. *Annu. Rev. Genet.* **34**, 77–137.

Li, Z., Zhang, Y., Li, D., and Feng, Y. (2000). Destabilization and mislocalization of MBP mRNAs in quaking dysmyelination lacking the QKI RNA binding proteins. *J. Neurosci.* **20**, 4944–4953.

Loriaux, M. M., Brennan, R. G., and Goodman, R. H. (1994). Modulatory function of CREB.CREMα heterodimers depends upon CREMα phosphorylation. *J. Biol. Chem.* **269**, 28839–28843.

Lou, H., and Gagel, R.F. (1998). Alternative RNA processing: Its role in regulating expression of calcitonin/calcitonin gene-related peptide. *J. Endocrinol.* **156**, 401–405.

Luger, K., Mader, A. W., Richmond, R. K., Sargent, D. F., and Richmond, T. J. (1997). Crystal structure of the nucleosome core particle at 2.8 A resolution. *Nature* **389**, 251–260.

Lunyak, V. V., Burgess, R., Prefontaine, G. G., Nelson, C., Sze, S. H., Chenoweth, J., Schwartz, P., Pevzner, P. A., Glass, C., Mandel, G., and Rosenfeld, M. G. (2002). Corepressor-dependent silencing of chromosomal regions encoding neuronal genes. *Science* **298**, 1747–1752.

Malik, S., and Roeder, R. G. (2000). Transcriptional regulation through Mediator-like coactivators in yeast and metazoan cells. *Trends Biochem. Sci.* **25**, 277–283.

Maue, R. A., Kraner, S. D., Goodman, R. H., and Mandel, G. (1990). Neuron-specific expression of the rat brain type II sodium channel gene is directed by upstream regulatory elements. *Neuron* **4**, 223–231.

Mayr, B., and Montminy, M. (2001). Transcriptional regulation by the phosphorylation-dependent factor CREB. *Nat. Rev. Mol. Cell Biol.* **2**, 599–609.

McKnight, S., and Tjian, R. (1986). Transcriptional selectivity of viral genes in mammalian cells. *Cell* **46**, 795–805.

McKnight, S. L., Kingsbury, R. (1982). Transcriptional control signals of a eukaryotic protein-coding gene. *Science* **217**, 316–324.

Mori, N., Schoenherr, C., Vandenbergh, D. J., and Anderson, D. J. (1992). A common silencer element in the SCG10 and type II Na1 channel genes binds a factor present in nonneuronal cells but not in neuronal cells. *Neuron* **9**, 45–54.

Mori, N., Stein, R., Sigmund, O., Anderson, D. J. (1990). A cell type-preferred silencer element that controls the neural-specific expression of the SCG10 gene. *Neuron* **4**, 583–594.

Naar, A. M., Lemon, B. D., and Tjian, R. (2001). Transcriptional coactivator complexes. *Annu. Rev. Biochem.* **70**, 475–501.

Narlikar, G. J., Fan, H. Y., and Kingston, R. E. (2002). Cooperation between complexes that regulate chromatin structure and transcription. *Cell* **108**, 475–487.

Nelson, C., Albert, V. R., Elsholtz, H. P., Lu, L. I. W., Rosenfeld, M. G. (1988). Activation of cell-specific expression of rat growth hormone and prolactin genes by a common transcription factor. *Science* **239**, 1400–1405.

Orphanides, G., Lagrange, T., and Reinberg, D. (1996). The general transcription factors of RNA polymerase II. *Genes Dev.* **10**, 2657–2683.

Pabo, C. O., and Sauer, R. T. (1992). Transcription factors: Structural families and principles of DNA recognition. *Annu. Rev. Biochem.* **61**, 1053–1095.

Ptashne, M., and Gann, A. (1997). Transcriptional activation by recruitment. *Nature* **386**, 569–577.

Pugh, B. F., and Tjian, R. (1990). Mechanism of transcriptional activation by Sp1: Evidence for coactivators. *Cell* **61**, 1187–1197.

Pugh, B. F., and Tjian, R. (1992). Diverse transcriptional functions of the multisubunit eukaryotic TFIID complex. *J. Biol. Chem.* **267**, 679–682.

Rachez, C., and Freedman, L. P. (2001). Mediator complexes and transcription. *Curr. Opin. Cell. Biol.* **13**, 274–280.

Richards, E. J., and Elgin, S. C. (2002). Epigenetic codes for heterochromatin formation and silencing: Rounding up the usual suspects. *Cell* **108**, 489–500.

Richter, J. D. (1999). Cytoplasmic polyadenylation in development and beyond. *Microbio. Mol. Biol. Rev.* **63**, 446–456.

Richter, J. D., and Lorenz, L. J. (2002). Selective translation of mRNAs at synapses. *Curr. Opini. Neurobiol.* **12**, 300–304.

Roth, S. Y., Denu, J. M., and Allis, C. D. (2001). Histone acetyltransferases. *Annu. Rev. Biochem.* **70**, 81–120.

Schoenherr, C. J., and Anderson, D. J. (1995). The neuron-restrictive silencer factor (NRSF): A coordinate repressor of multiple neuron-specific genes. *Science* **267**, 1360–1363.

Schüle, R., Rangarajan, P., Kliewer, S., Ransone, L. J., Bolado, J., Yang, N., Verma, I. M., and Evans, R. M. (1990). Functional antagonism between oncoprotein c-Jun and the glucocorticoid receptor. *Cell* **62**, 1217–1226.

Seeburg, P.H., Higushi, M., and Sprengel, R. (1998). RNA editing of brain glutamate receptor channels: Mechanism and physiology. *Brain Res. Rev.* **26**, 217–229.

Sheng, M., McFadden, G., and Greenberg, M. E. (1990). Membrane depolarization and calcium induce c-fos transcription via phosphorylation of transcription factor CREB. *Neuron* **4**, 571–582.

Smale, S. T. (1994). Core promoter architecture for eukaryotic protein-coding genes. *In* "Transcription: Mechanisms and Regulation" (R. C. Conaway and J. W. Conaway, Eds.), Raven Press, New York, pp. 63–81.

Smale, S. T. (1997). Transcription initiation from TATA-less promoters within eukaryotic protein-coding genes. *Biochim. Biophys. Acta* **1351**, 73–88.

Smale, S. T. (2001). Core promoters: Active contributors to combinatorial gene regulation. *Genes Dev.* **15**, 2503–2508.

Smale, S. T., and Baltimore, D. (1989). The "initiator" as a transcription control element. *Cell* **57**, 103–113.

Steward, O., and Schuman, E. M. (2001). Protein synthesis at synaptic sites on dendrites. *Annu. Rev. Neurosci.* **24**, 299–325.

Tanese, N., Pugh, B. F., and Tjian, R. (1991). Coactivators for a pro-line-rich activator purified from the multisubunit human TFIID complex. *Genes Dev.* **5**, 2212–2224.

Tiedge, H., and Brosius, J. (1996). Translational machinery in dendrites of hippocampal neurons in culture. *J. Neurosci.* **16**, 7171–7181.

Tjian, R., and Maniatis, T. (1994). Transcription activation: A complex puzzle with few easy pieces. *Cell* **77**, 5–8.

Torre, E. R., and Steward, O. (1992). Demonstration of local protein synthesis within dendrites using a new cell culture system which permits the isolation of living axons and dendrites from their cell bodies. *J. Neurosci.* **12**, 762–772.

Triezenberg, S. J. (1995). Structure and function of transcriptional activation domains. *Curr. Opin. Genet. Dev.* **5**, 190–196.

Turner, B. M. (2002). Cellular memory and the histone code. *Cell* **111**, 285–291.

Vinson, C., Myakishev, M., Acharya, A., Mir, A. A., Moll, J. R., and Bonovich, M. (2002). Classification of human B-ZIP proteins based on dimerization properties. *Mol. Cell. Biol.* **22**, 6321–6335.

Wang, H., Iacoangeli, A., Popp, S., Muslimov, I. A., Imataka, H., Sonenberg, N., Lomakin, I. B., and Tiedge, H. (2002). Dendritic BC1 RNA: Functional role in regulation of translation initiation. *J. Neurosci.* **22**(23), 10232–10241.

Weis, L., and Reinberg, D. (1997). Accurate positioning of RNA polymerase II on a natural TATA-less promoter is independent of TATA-binding-protein-associated factors and initiator-binding proteins. *Mol. Cell. Biol.* **17**, 2973–2984.

Workman, J. L., and Kingston, R. E. (1998). Alteration of nucleosome structure as a mechanism of transcriptional regulation. *Annu. Rev. Biochem.* **67**, 545–579.

Xing, J., Ginty, D. D., and Greenberg, M. E. (1996). Coupling of the RAS–MAPK pathway to gene activation by RSK2, a growth factor-regulated CRED kinase. *Science* **273**, 959–963.

Modeling and Analysis of Intracellular Signaling Pathways

Paul. D. Smolen, Douglas A. Baxter, and John H. Byrne

As discussed in Chapter 12, sequences of biochemical reactions termed *intracellular signaling pathways* transmit signals from the extracellular medium to cytoplasmic or nuclear targets. Ca^{2+} influx can activate such pathways, as does binding of hormones, neurotransmitters, or growth factors to receptors. Signaling pathways transduce these stimuli into effects on the rates of specific cellular processes, such as the transcription of particular genes. In this way, extracellular signals can modulate long-term processes, such as neuronal growth or the formation of long-term memory. Indeed, biochemical signals may be regarded as a conduit of information within the nervous system, supplementing electrical activity (Katz and Clemens, 2001). This chapter provides an overview of concepts and techniques necessary to describe and analyze the operation of intracellular signaling pathways and metabolic pathways, with concise mathematical language (see also General References).

Information on the operation of biochemical signaling pathways is rapidly accumulating. For example, the expression of large groups of genes subsequent to activation of signaling pathways by hormones, neurotransmitters, or drugs can be followed with microarrays ("DNA chips") of specific DNA sequences. Modeling can provide a conceptual framework to assemble these complex data into concise pictures of the structure and operation of signaling pathways and metabolic pathways. Intuition fails to adequately deal with these complex biochemical systems, and equations are needed to make "word models" rigorous and check them for consistency (Marder, 2000). Once model equations are developed, computer simulations or algebraic calculations can display and

predict the behavior of biochemical systems more completely than can intuition. The necessity for computer simulations is particularly evident if the biochemical reactions are nonlinear. An enzyme-catalyzed biochemical reaction is *linear* if its rate is proportional to the product of reactant concentrations. If the reaction rate varies in a more complex manner, the reaction is *nonlinear*. Nonlinear reactions commonly display unexpected dynamics (see also Chapter 7, Box 7.4).

Models also help in understanding the effects of multiple feedback interactions within pathways, such as alteration of enzyme activities due to allosteric binding of downstream reaction products. Time delays, such as are required for macromolecular synthesis and intracellular transport, also add complexity to the behavior of biochemical systems. Time delays can be critical for sustaining oscillations in reaction rates and in metabolite concentrations. Such effects can be appreciated only with a mathematical model.

A model is particularly valuable if it succeeds in reproducing, or predicting, behaviors in addition to those it was originally constructed to replicate. For example, the Hodgkin–Huxley model of voltage-gated currents was constructed to reproduce the properties of those currents as observed in experiments with each current isolated (Hodgkin and Huxley, 1952). That the model would, in addition, succeed in simulating an action potential was not evident *a priori* and was of great significance. The simulation helped to confirm that the action potential resulted from the voltage- and time-dependence of the known voltage-gated currents in the squid axon. The success of this model also helped establish the utility of mathematical descriptions of neuronal properties. Chapters 5 and 7 discuss the Hodgkin–Huxley model in more detail. Also, a useful model must predict the results of future experiments. If the predictions bear out, the model gains credibility as a valid physiological description.

Models are also valuable in establishing principles governing the behavior of biochemical systems. Modeling of signaling pathways and gene regulation has established a unifying principle that feedback interactions commonly underlie complex behaviors. Negative feedback, in which a biochemical species acts to inhibit its own production, often underlies oscillations in biochemical concentrations or gene expression rates.

Biochemical models can be constructed at several levels. The most detailed is the *stochastic level*. Here, one keeps track of the numbers of each type of molecule over time. Such models can be required if the average numbers of key macromolecules are low. In a cell, only tens or hundreds of molecules of specific transcription factors or mRNAs may be present. Then, stochastic fluctuations in molecule numbers due to randomly timed creation and destruction of molecules can have considerable consequences. Irreducible individual variability in laboratory animals may be due to random fluctuations in the numbers of transcription factors and mRNAs during development (Gartner, 1990).

Lack of experimental data for molecule numbers and their fluctuations limits the applicability of stochastic models. More commonly, the biochemical concentrations are approximated as continuous variables. This *continuous level* uses differential equations for the rates of change of continuous concentration variables. Familiar quantities from biochemical kinetics—rate constants, Michaelis constants, enzyme activities—govern changes in concentrations. Most models discussed in this chapter use the continuous approximation. Computer simulations numerically integrate the differential equations to follow concentrations over time, and simulations are compared with experimental data. Values of model parameters, such as binding constants or rate constants, are adjusted to improve agreement between simulation and experiment. Finally, if the experimental data is quite sparse, a *logical or Boolean level* of modeling is sometimes used. Here, genes or enzymes are simply regarded as switches that are either on or off.

The choice of which level of modeling to use is mainly determined by the hypothesis to be tested. For example, if a specific enzyme is believed to be rate limiting for the progress of a biochemical pathway, it is appropriate to use the continuous level of modeling. The activity of the enzyme will be a parameter determining the rate of the reaction. Then it is possible to explore, in a computer simulation, how changing the activity affects the concentrations of both the reaction product and of all chemical species "downstream" in the pathway. It is commonly desirable to construct the simplest model capable of capturing the features of the system most relevant to the hypothesis being tested. Such a model highlights the essential features without excessive details. Separation of rapid and slow biochemical reactions is often used to simplify models. The number of equations can be reduced by assuming that the fastest biochemical reactions are always near equilibrium.

Another distinction is between *quantitative models* that describe the structure and behavior of specific biochemical pathways, and *qualitative models,* which generally have fewer equations and are constructed to embody biochemical elements common to multiple pathways. Such elements include feedback interactions, random fluctuations in molecule numbers, or formation of multiprotein complexes containing enzymes or transcription factors. This chapter will rely primarily on qualitative models. Such models illustrate how typical biochemical elements can give rise to specific types of pathway dynamics—that is, pathway behavior over time. These behaviors include oscillations in reaction rates, or switching between steady states of reaction rates and molecular concentrations. As one example, qualitative models of glycolysis incorporate feedback activation of the key glycolytic enzyme phosphofructokinase by its reaction product ADP (Goldbeter, 1996). These models have established that the feedback plays a role in sustaining oscillations in the rate of glycolysis (Dano *et al.,* 1999).

The discussion of modeling begins with a basic process common to all pathways: the transport of signaling molecules within the cell. The mechanism of transport can strongly affect the behavior of signaling pathways. Then, techniques for modeling sequences of enzyme-catalyzed biochemical reactions will be illustrated with examples relevant to neuronal function. The unifying principle that feedback interactions commonly underlie complex behaviors of biochemical systems will be developed. Following these topics, techniques for estimating model parameter values from data will be discussed, and the limitations of these techniques for determining *in vivo* values assessed. Also, the importance and methodology of assessing the sensitivity of model behavior to parameter values will be discussed. The method for simplifying models based on separation of rapid and slow biochemical reactions will be discussed.

Following these general topics, specific topics important for many models of biochemical pathways will be addressed. A method will be discussed for determining which parameters are key control points for the flux of metabolite through a pathway. Modeling of stochastic fluctuations in molecule numbers will be discussed. Finally, methods for modeling the regulation of gene expression will be discussed. Models of transcriptional regulation are of particular importance because regulation of gene expression is an essential part of signaling pathways that mediate long-term changes in cell properties. An example is long-term strengthening of synapses, commonly denoted *long-term potentiation* (LTP). LTP appears to be an essential process in the formation of long-term memory (Bliss *et al.*, 2006; Martin *et al.*, 2002) (see also Chapter 19).

For further study, a selection of textbooks is given in the General References list. These provide a more comprehensive treatment of the ways in which models can be constructed and analyzed, and of the mathematics required.

INTRACELLULAR TRANSPORT OF SIGNALING MOLECULES CAN BE MODELED AT SEVERAL LEVELS OF DETAIL

Intracellular messages can be transmitted by ions such as Ca^{2+}, by small molecules such as cAMP, or by movement of macromolecules such as enzymes or transcription factors.

Passive Diffusion Dominates for Ions and Small Molecules

Intracellular transport of ions and small molecules is generally diffusive. Modeling of *diffusive transport* due to random thermal collisions with other molecules will be discussed in some detail. Passive diffusion of macromolecules can be modeled by the same equations as for small molecules, but with much smaller diffusion coefficients.

The discussion will focus on techniques for modeling the diffusion of a specific ion, Ca^{2+}. Both free Ca^{2+} and Ca^{2+} bound to proteins like calmodulin can activate enzymes within signaling pathways. The intracellular movement of Ca^{2+} has been extensively studied experimentally and modeled in a variety of ways. The same equations and the same techniques for numerical simulations can be applied

to diffusion of other small molecules and ions. The diffusion coefficient must be corrected for differences in molecular weight. For free diffusion in aqueous solution, the diffusion coefficient is inversely proportional to the cube root of molecular weight. However, diffusion *in vivo* is not free, because *buffering*, or binding of small molecules or ions to sites on macromolecules, is common throughout the cytoplasm. Buffering will decrease diffusion coefficients below values in aqueous solution.

Concise Models of Ca^{2+} Exchange Between Compartments Represent Ca^{2+} Release and Uptake Mechanisms

Two basic approaches model Ca^{2+} transport. The simplest approach considers only Ca^{2+} exchange across membranes that separate intracellular compartments. The exchange can be driven by passive diffusion or by "pump" proteins that utilize ATP. Compartments are each assigned a *pool* of Ca^{2+}. Cytoplasmic Ca^{2+}, Ca^{2+} in the endoplasmic reticulum, mitochondrial Ca^{2+}, and Ca^{2+} in the nucleus might each be represented as a pool. Cytoplasmic Ca^{2+} could communicate with all the other pools. The time course of the amount of Ca^{2+} in each pool is followed during computer simulations. In the second approach, the spaces within intracellular compartments are subdivided into clusters of small volume elements, and the Ca^{2+} concentration in each small element is followed separately. Diffusion of Ca^{2+} within or between compartments is modeled as fluxes between volume elements. This approach is required if it is necessary to model concentration gradients within the cytoplasm or other compartments.

Suppose a single pool describes cytoplasmic Ca^{2+}. A single *ordinary differential equation* then describes the rate of change of cytoplasmic Ca^{2+} concentration, $[Ca^{2+}_{cyt}]$. The rate of change depends linearly on Ca^{2+} influx across the plasma membrane and on Ca^{2+} exchange between the cytoplasm and other compartments such as the endoplasmic reticulum,

$$\frac{d[Ca^{2+}_{cyt}]}{dt} = \lambda I_{Ca} + k_{exch}([Ca^{2+}_{ER}] - [Ca^{2+}_{cyt}]) - k_{Ca}[Ca^{2+}_{cyt}] \tag{14.1}$$

In Eq. 14.1, the first term on the right-hand side describes Ca^{2+} influx across the plasma membrane through an ion channel or set of channels. A scaling constant denoted by λ converts Ca^{2+} current to a rate of change of Ca^{2+} concentration. As discussed in Chapter 7, detailed expressions can be used to describe the voltage- and time-dependence of such ionic currents. The second term on the right-hand side of Eq. 14.1 describes exchange of Ca^{2+} between the endoplasmic reticulum (ER) and the cytoplasm. For illustrative purposes this term is very simple—proportional to the concentration difference between the pools. The last term in Eq. 14.1 describes Ca^{2+} extrusion across the membrane as a simple first-order process.

In models of specific cell types, more detailed kinetic expressions might be needed to describe Ca^{2+} release from the ER. For example, special expressions are necessary to describe stimulation of ER Ca^{2+} release by the second-messenger molecule inositol 1,4,5-trisphosphate (IP_3). Models describe Ca^{2+} release through the IP_3 receptor on the ER as increasing steeply with the third or fourth power of the cytoplasmic IP_3 concentration (Sneyd *et al.*, 2004; Wagner *et al.*, 1998). Ca^{2+} release through the IP_3 receptor also increases steeply with the cytoplasmic Ca^{2+} concentration. This is *positive feedback*, because an initial increase in cytoplasmic Ca^{2+} stimulates further release of Ca^{2+} from the ER and a further increase in cytoplasmic Ca^{2+}. Such positive feedback can help to generate Ca^{2+} waves and oscillations in a variety of cell types (Nakamura *et al.*, 1999; Sneyd *et al.*, 2004; Wagner *et al.*, 1998).

A more detailed description of Ca^{2+} extrusion from the cell can also be used. For example, the term

$$- k_{Ca} \frac{[Ca^{2+}_{cyt}]}{[Ca^{2+}_{cyt}] + K} \qquad (14.2)$$

describes Ca^{2+} extrusion across the plasma membrane by a saturable ion pump. The parameter K gives the value of $[Ca^{2+}]$ at which the pump rate is half-maximal. Saturable pumps driven by ATP hydrolysis are present in many cell types. A more complicated expression

$$- k_{Ca}([Ca^{2+}_{cyt}]^2 [Na^+_{external}]^3$$
$$- [Ca^{2+}_{external}]^2 [Na^+_{cyt}]^3) \qquad (14.3)$$

is sometimes used to describe removal of Ca^{2+} by a $Na^+ - Ca^{2+}$ exchanger present in neurons and other excitable cells (De Schutter and Smolen, 1998). This exchanger brings Na^+ into the cytoplasm while concurrently removing Ca^{2+}. In Eq. 14.3 there are two terms. The first term describes Ca^{2+} removal and the second term is necessary because the exchanger can also operate in reverse, bringing in Ca^{2+}. The powers of Na^+ and Ca^{2+} concentrations reflect the cyclic mechanism of the exchanger protein, which brings in approximately three Na^+ ions while extruding two Ca^{2+} ions.

To simulate changes in Ca^{2+} concentration described by Eq. 14.1, *Euler's method* is often used. A reasonable initial value for $[Ca^{2+}_{cyt}]$ is chosen (such as 50 nM, which is in the range expected if a neuron is hyperpolarized and there is little Ca^{2+} influx through ion channels). Then, at each small step in time, one calculates the rate of change of $[Ca^{2+}_{cyt}]$ from the right-hand side of Eq. 14.1. The rate of change is multiplied by the size of the time step, and the result is added to $[Ca^{2+}_{cyt}]$. This procedure is repeated until a time length sufficient to display the behavior of $[Ca^{2+}_{cyt}]$ during neuronal activity has been simulated. To determine whether the time step chosen is small enough, some simulations are typically repeated with a time step ~1/3 as long, and if the behavior of $[Ca^{2+}_{cyt}]$ does not vary significantly, the larger time step is considered adequate. This simple method of numerical integration can be applied to any differential equation discussed in this chapter. More complex integration methods also exist and these may, for many models, provide for much faster simulations. Textbooks on numerical analysis (e.g., Burden and Faires, 1993) can be consulted for details about those methods.

To Model Diffusion in Realistic Detail, Standard but Complex Equations Can Be Used

To more accurately model the intracellular diffusion of Ca^{2+}, other ions, or small molecules, the intracellular space is *discretized*, or divided into small volume elements. The evolution of the Ca^{2+} concentration within each element is simulated. This is the standard method for simulating the partial differential equation that describes diffusion within an extended region. A common implementation uses cubic volume elements. The cytoplasm or other region of interest is approximated by a large set of small boxes in *x, y, z* space. The overall set is chosen to have a shape that is a reasonable representation of the region of interest. For example, an approximate sphere with a "boxy" boundary may be used to represent the nucleus. The flux of Ca^{2+} between two cubic boxes depends on the cube size and the Ca^{2+} diffusion coefficient, D_{Ca}. The dependence is as follows:

$$Flux_{1->2} = D_{Ca} \frac{A}{L} ([Ca^{2+}]_1 - [Ca^{2+}]_2) \qquad (14.4)$$

In Eq. 14.4, L is the length of a cube edge, and A ($= L^2$) is the area of a cube face. $[Ca^{2+}]_1$ refers to the Ca^{2+} concentration in cube 1. Although Eq. 14.4 describes only the diffusion of free Ca^{2+}, the same type of equation can be used to describe the diffusion of Ca^{2+} bound to calmodulin or other species. Simulations using Eq. 14.4 can use Euler's integration method, as for Eq. 14.1. At each small time step, the flux of Ca^{2+} given by Eq. 14.4 is multiplied by the size of the time step, and the resulting quantity is added to the preceding Ca^{2+} concentration in volume element 2 while also being subtracted from the preceding Ca^{2+} concentration in volume element 1. At each time step, this calculation must be repeated for every pair of adjacent volume elements in the system.

In one application (Simon and Llinas, 1985, see also Fig. 8.3), Eq. 14.4 was used to simulate dynamics of localized Ca^{2+} peaks near clusters of open Ca^{2+} channels. It was found that near the cytoplasmic mouth of a channel, Ca^{2+} concentrations could reach tens of micromolar or higher. Such *Ca^{2+} microdomains* form and dissipate within microseconds of channel openings and closings (Shahrezaei and Delaney, 2005). Enzymes or other proteins positioned near clusters of channels can serve as sensors for such elevations of concentration, and intracellular signaling pathways could thereby be activated. The importance of Ca^{2+} microdomains for the release of neurotransmitter is discussed in Chapter 8.

Modeling diffusion with rectangular volume elements has a disadvantage of requiring very large amounts of computer time and memory because of the large number of elements required to discretize space in three dimensions. For example, if each axis is to be divided into 100 parts, a total of 1 million volume elements are needed. A great reduction in time and memory requirements can be achieved if the biological system can be approximated by a model where, because of *symmetry*, diffusion in one or two dimensions can be neglected. Then only two or one dimensions, respectively, need to be divided into small volume elements. For example, suppose diffusion within a spherically symmetric compartment, such as an ideal soma, is being modeled. If Ca^{2+} is assumed to enter and leave at equal rates over all of the outer surface, then Ca^{2+} concentration is independent of both angle coordinates within the sphere. Only the radial dimension needs to be discretized because diffusion is only along the radial direction from the surface to the center. The compartment can be divided radially into thin, concentric spherical shells (Fig. 14.1A). The concentration changes in each shell due to fluxes of Ca^{2+} between shells are simulated. An equation analogous to Eq. 14.4 is used. The algorithm is described in detail in Blumenfeld *et al.* (1992).

As a second example consider a portion of a dendrite or axon, which can be modeled as a cylinder. Suppose Ca^{2+} is assumed to enter one end and leave at the other end, without significant influx from the sides.

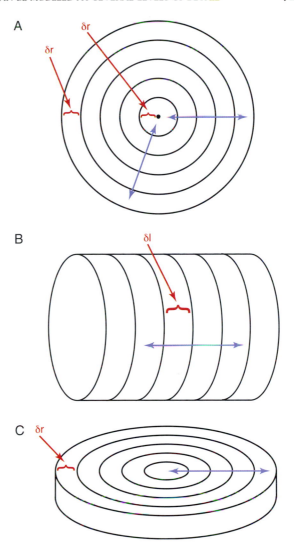

FIGURE 14.1 Symmetry can allow advantageous division of spatial regions into volume elements to model diffusion. (A) Spherically symmetric diffusion. A spherical region is divided into thin concentric cells of width δr. Diffusion of molecules between shells occurs in the radial direction (blue arrows). (B) Longitudinal diffusion in a cylinder (blue arrow). The cylinder is divided into thin slabs of width δl. (C) Radially symmetric diffusion in a disk-shaped region that approximates a synaptic cleft. The disk is divided into concentric annuli of width δr.

Then Ca^{2+} gradients across the width of the cylinder can be neglected, and Ca^{2+} diffusion occurs only along the length of the cylinder. Thus, only the dimension of length needs to be divided into volume elements. The cylinder can be divided into thin slabs, and the evolution of the concentration in each slab simulated (Fig. 14.1B). As a further example, diffusion of neurotransmitter within a synaptic cleft has been modeled by describing the cleft as a thin disk-shaped region and neglecting the gradient of neurotransmitter across the cleft. If neurotransmitter is released at the center, only diffusion in the radial direction needs to be considered. The disk can be discretized into concentric annular rings (Fig. 14.1C) and diffusion between the rings simulated. However, if neurotransmitter is released anywhere besides the disk center, diffusion in the angular direction cannot be neglected, and simulation requires discretization in both the radial and the angular dimensions.

These methods for modeling diffusion all assume that ionic movement is driven by concentration gradients to a much greater extent than by electrical potential gradients. This may not hold true within small structures such as dendritic spines. Electrical activity might cause large potential differences between the spine head and the dendritic shaft. The *electrodiffusion* modeling approach is more accurate for such situations. This approach uses more general and complex equations, with terms to account for ion movement due to diffusion and due to potential-driven drift. As an example, the Nernst–Planck equation, and discretization into small volume elements, has been used to model diffusion of several ionic species within a spine and the adjacent dendrite (Qian and Sejnowski, 1990). When constructing electrodiffusion models, it is necessary to consider which charged ions or molecules are likely to play the largest roles in carrying charges and dynamically altering potential gradients. Both the concentrations and mobilities of common charged species need to be considered when making these assessments.

Buffering Must Be Considered in Models

Binding of ions or small molecules to macromolecules can greatly slow diffusion. For example, a Ca^{2+} ion spends much of its time bound, so that the diffusion of Ca^{2+} ions is much slower in the cell than in water. Other ions, and small molecules, tend to have fewer binding sites than does Ca^{2+}, so that their diffusion is not slowed so much. For example, there are relatively few intracellular binding sites for the second-messenger IP_3. Therefore, diffusion of IP_3 can be considerably more rapid than diffusion of Ca^{2+}, even though the molecular weight of a Ca^{2+} ion is much lower (Jafri and Keizer, 1995).

For a given class of Ca^{2+} binding sites, denoted S and present at a total concentration S_{tot}, consider equilibration between free and bound Ca^{2+}. Using standard chemical kinetic notation, the dissociation and association rate constants for formation of S-Ca^{2+}

are respectively denoted k_b and k_f. Equilibration is described by the following differential equation:

$$\frac{d[Ca^{2+}]_{free}}{dt} = k_b[S\text{-}Ca^{2+}]$$
$$- k_f[S_{tot} - S\text{-}Ca^{2+}][Ca^{2+}]_{free} \quad (14.5)$$

In Eq. 14.5, [S-Ca^{2+}] denotes the concentration of S-Ca^{2+} complex, $[Ca^{2+}]_{free}$ denotes the concentration of unbound Ca^{2+}, and $[S_{tot} - S\text{-}Ca^{2+}]$ gives the concentration of free S sites. The total Ca^{2+} concentration C_{tot} is equal to [S-Ca^{2+}] + $[Ca^{2+}]_{free}$.

Euler's integration method, discussed previously, can be used with Eq. 14.5 to determine the time courses of the concentrations of free and bound Ca^{2+}. To model Ca^{2+} dynamics in neurons, terms must also be added to the right-hand side to describe the influx of Ca^{2+}_{free} to the cytoplasm through channels and its removal by pumps. The form of these terms is discussed earlier (Eqs. 14.1–14.3).

An equilibrium dissociation constant, K, is defined as k_b/k_f. At equilibrium, the system is unchanging, so that

$$\frac{d[Ca^{2+}]_{free}}{dt} = 0$$

Therefore, at equilibrium, the right-hand side of Eq. 14.5 can be set equal to zero. Then a rearrangement gives

$$K = k_b/k_f = \frac{[S][Ca^{2+}]_{free}}{[S\text{-}Ca^{2+}]} \quad (14.6)$$

When $[Ca^{2+}]_{free}$ is equal to K, half of the sites S are occupied by Ca^{2+}. This can be seen as follows. Eq. 14.6 simplifies to

$$\frac{[S]}{[S\text{-}Ca^{2+}]} = 1$$

that is, 50 % of the sites are occupied by Ca^{2+}.

When multiple classes of buffer sites (S_1, S_2...) are present, each class will have its own values of the parameters k_f, k_b, and S_{tot}. An analog of Eq. 14.5 can be written. The right-hand side of the analog contains separate terms for each value of k_f and k_b for each class.

Experimental data suggests that, to a reasonable approximation, total intracellular Ca^{2+} buffering can often be thought of as due to one or two "lumped" buffering species, or classes of buffer sites (Zhou and Neher, 1993). The effect of each species is described by different values of the equilibrium constant K and the total site concentration S_{tot}. A useful experimental quantity that helps determine values of K and S_{tot} is the *buffering capacity*. This is the ratio of the change in total (free + bound) Ca^{2+} to the change in free Ca^{2+} when a small amount of Ca^{2+} is added to the cytoplasm. An analysis of Ca^{2+} buffering in adrenal chromaffin cells found that buffering could be described by two "lumped" species, one immobile with a buffering capacity of ~30, and the other slowly diffusing and having a buffering capacity of ~10 (Zhou and Neher, 1993). In neurons, recent data suggests very high buffering capacity, over 100 (Muller *et al.*, 2005). Calculations using these estimates suggest that ~99% of cytosolic Ca^{2+} is bound in neurons and other excitable cells.

Molecular probes have been developed that monitor the intracellular movement of Ca^{2+} and other ions or small molecules, including Mg^{2+}, Zn^{2+}, and cAMP. For example, dyes such as Fura-2 have a fluorescence emission spectrum that is altered by Ca^{2+} binding. Experimental data characterizing Ca^{2+} gradients and movement is obtained by monitoring changes in the fluorescence of such dyes. Data obtained with such probes is vital to refining mathematical models that include transport. Models to simulate such data need to consider the diffusion of all important species including dye with and without bound Ca^{2+}. Buffering due to introduced probes can have complex effects on the diffusion coefficients of ions and small molecule (Gabso *et al.*, 1997). Introduction of a small probe molecule that binds Ca^{2+} can speed up the net rate of Ca^{2+} diffusion, even though the free concentration of Ca^{2+} is decreased. This occurs because, after the probe is added, a significant fraction of total Ca^{2+} is bound to highly mobile probe molecules instead of poorly mobile macromolecules.

Details of buffer and Ca^{2+} diffusion can be disregarded if Ca^{2+} dynamics are being modeled using separate intracellular "pools" (Eq. 14.1) with Ca^{2+} exchange across membranes separating pools. In this situation, it is common to regard the ratio of free Ca^{2+} to total Ca^{2+} as fixed within each pool. For example, if the overall buffering capacity of the cytoplasm is in the range of 100, a fixed ratio of 0.01 would be appropriate. In this case, only 1/100 of the Ca^{2+} entering or leaving the pool is reflected as a change in the concentration of free Ca^{2+}. The modification of Eq. 14.1 required is

$$\frac{d[Ca^{2+}{}_{cyt}]_{free}}{dt} = f\left(\lambda I_{Ca} + k_{exch}([Ca^{2+}{}_{ER}]_{free} - [Ca^{2+}{}_{cyt}]_{free}) - k_{Ca}[Ca^{2+}{}_{cyt}]_{free}\right)$$

$$(14.7)$$

In Eq. 14.7, f denotes the fixed ratio of free Ca^{2+} to total Ca^{2+}, and only concentrations of free Ca^{2+} are used to determine total Ca^{2+} fluxes between compartments or out of the cell because bound Ca^{2+} is generally not transported across membranes.

From Eq. 14.7, it follows that the effect of a small ratio f is to greatly slow down the rate of change of Ca^{2+} concentration. Slow oscillations in $[Ca^{2+}{}_{cyt}]_{free}$ (period of seconds) can result. Such oscillations are often coupled to oscillations of membrane potential, and sometimes bursts of action potentials are superimposed on the membrane potential oscillations (Amini *et al.*, 1999; Bertram *et al.*, 2004; Imtiaz *et al.*, 2006).

For modeling the diffusion of macromolecules, it is also desirable to include buffering. However, available data usually do not suffice to determine the number and distribution of binding sites for specific macromolecules. With this limitation, a reasonable approach is to estimate a plausible value for the macromolecular diffusion coefficient and then simulate experiments that measure macromolecular movement rates. A common experimental technique is *fluorescence recovery after photobleaching* (FRAP) (Sprague and McNally, 2005). This technique uses strong brief illumination of an intracellular region with a laser. The laser eliminates emission from a fluorescently tagged macromolecule within the region by photobleaching the fluorophore. Afterward, the time course of fluorescence recovery is monitored as macromolecule diffuses in from other regions. Fitting this time course with computer simulations provides an estimate for the diffusion coefficient.

Active Transport May Dominate Over Diffusion for Macromolecules

For macromolecules, intracellular movement is often via *active transport* requiring ATP. Active transport is especially important in cells with extended morphologies, such as neurons. Calculations indicate that passive diffusion of mRNAs and proteins from the soma into narrow dendrites would be much slower than observed rates of movement (Sabry *et al.*, 1995). Active transport consists of directed movement along cytoskeletal elements of specific macromolecules or vesicles containing macromolecules. The movement is mediated by motor proteins such as kinesin or dynein (see Chapter 2). For constraining models of active macromolecular transport, data specific to neurons needs to be used because there is great variability in transport rates between cell types. A reasonable approximation is to assume a constant drift of macromolecule, at a velocity that may vary between different regions of a neuron. Diffusion might be simply added to the motion due to active transport, providing an additional random motion. Diffusion might be ignored if data suggests active transport is much more important for particular macromolecules.

STANDARD EQUATIONS SIMPLIFY MODELING OF ENZYMATIC REACTIONS, FEEDBACK LOOPS, AND ALLOSTERIC INTERACTIONS

In cells, most biochemical reactions of interest are catalyzed by enzymes. Many enzymatic reactions have complex kinetic mechanisms, and specialized equations are needed to describe their rates in detail (Cornish-Bowden, 2004). However, when a series or group of reactions is being modeled, it is more common to use simplified equations for the individual reaction rates. A few such forms have become standard.

To review these equations, it is first helpful to review the definition of *reaction order*. A *zero-order reaction* converts substrate into product at a fixed rate independent of substrate concentration. A *first-order reaction* converts substrate into product at a rate proportional to substrate concentration. A *second-order reaction* creates product at a rate proportional to either the product of two substrate concentrations, or the square of a single substrate concentration. Higher-order reactions are similarly defined. In many cases, an enzyme will transform a single substrate, S, into a single product, P. The simplest assumption is that the rate of production of P is first-order with respect to S. This assumption can be useful when many reactions are being modeled, in order to reduce the complexity of the model and the computational time required for simulations (e.g., Bhalla and Iyengar, 1999). The next level of detail considers *saturation* of the reaction rate at high concentrations of S. Then the reaction no longer has a definite order, and a nonlinear equation must be used for the rate. The simplest such equation for the rate of production of P, d[P]/dt, is the standard Michaelis–Menten equation:

$$\frac{d[P]}{dt} = \frac{V_{max}[S]}{[S] + K_m} \qquad (14.8)$$

This equation has been found *in vitro* to accurately describe the rates of many enzymatic reactions. V_{max} represents the maximal enzyme velocity when it is fully saturated with the substrate S, and when $[S] = K_m$ the velocity is half-maximal. If the enzyme is saturated with substrate (high [S]), Eq. 14.8 reduces to a zero-order reaction. If two substrates S_1 and S_2 are converted into a single product P, then the simplest assumption is that the rate of production of P is second order, proportional to the product $[S_1]$ $[S_2]$. One can add more realism by using a product of Michaelis–Menten expressions. Suppose there also is first-order degradation of P. Then a differential equation for the *in vivo* rate of change of the concentration of P is

$$\frac{d[P]}{dt} = k_{max}\left(\frac{[S_1]}{[S_1] + K_1}\frac{[S_2]}{[S_2] + K_2}\right) - k_{deg}[P] \qquad (14.9)$$

Here, the first term on the right-hand side is a product of Michaelis–Menten expressions for $[S_1]$ and $[S_2]$. The maximal velocities V_{max} have been combined into the parameter k_{max}. The second term represents first-order degradation. K_1, K_2, k_{max}, and k_{deg} are parameters to be estimated from experimental data.

A model of a biochemical reaction pathway is composed of a set of differential equations such as Eq. 14.9. There is an equation for the rate of change of the concentration of each reaction substrate and product. Given initial values for the concentrations of all biochemical species, numerical integration of the differential equations is done as described previously for Eq. 14.1.

Allosteric Interactions Between Enzymes and Small Molecules Can Alter Enzyme Activities and Mediate Feedback

Binding of small molecules can alter an enzyme's conformation and alter the rate of the reaction catalyzed by the enzyme. Often, such allosteric interactions are with effector molecules not involved in the reaction. But for some enzymes (e.g., pyruvate kinase, phosphofructokinase) substrates or products can serve as allosteric effectors. Allosteric interactions can therefore mediate feedback and feed-forward interactions within a biochemical pathway, as well as cross talk between pathways.

In a *feedback interaction*, a product of an enzymatic reaction affects the activity of another enzyme earlier in the pathway, whereas in a *feed-forward interaction*, the affected enzyme is later in the pathway. With *cross talk*, a metabolite from one pathway affects an enzyme in another pathway. Two types of feedback can be distinguished. If the product of a later reaction acts to speed up an earlier reaction, *positive feedback* results; if the effect is to slow down the earlier reaction, *negative feedback* results. Positive feedback tends to drive a biochemical pathway to a state of maximal activity determined by the saturated rate of the slowest reaction, whereas negative feedback tends to drive the pathway to low activity. Graphically, such an interaction is often represented as in Figure 14.2A, where the "−" sign denotes negative feedback.

In models, allosteric interactions are commonly represented by *Hill functions* (Cornish-Bowden, 2004). These are saturable functions of the concentration of the effector molecule, denoted here by [L]. If L activates an enzyme, the enzyme activity is taken as proportional to a function of the nth power of [L]:

$$\frac{[L]^n}{[L]^n + K_H^n} \tag{14.10a}$$

When the Hill function of Eq. 14.10a is plotted versus effector concentration [L], the graph has a sigmoid shape. In Eq. 14.10a, the parameter K_H has units of concentration. When $[L] = K_H$, the activity has a value of 0.5. Over a range of [L] centered about K_H, the activity increases rather steeply to near its maximal value of 1. The parameter n is called the *Hill coefficient*. Greater values of n correspond to steeper sigmoids—a narrowing of the range of [L] over which the enzyme activity is significantly above 0 and below 1.

If L inhibits an enzyme, the enzyme activity is taken as proportional to a decreasing function of [L]

$$\frac{K_H^n}{[L]^n + K_H^n} \tag{14.10b}$$

As an example of how these functions are used in rate equations, inclusion of an increasing Hill function in Eq. 14.8 gives

$$\frac{d[P]}{dt} = \left(\frac{[L]^n}{[L]^n + K_H^n}\right)\frac{V_{max}[S]}{[S] + K_m} \tag{14.11}$$

An enzyme often has multiple binding sites for an allosteric effector. Greater values of the Hill coefficient often correspond to a larger number of binding sites for a given allosteric effector. Experimentally determined Hill coefficients are often taken as a rough indication of the number of binding sites. Instead of Hill functions, more complex expressions based on *Monod–Wyman–Changeux allosteric theory*, or

similar theories, are sometimes used (Changeux and Edelstein, 2005).

Figure 14.2B illustrates a relatively simple model (Cooper *et al.*, 1995) that incorporates Hill functions as well as first-order reactions. This model describes the relationship between cyclic AMP (cAMP)

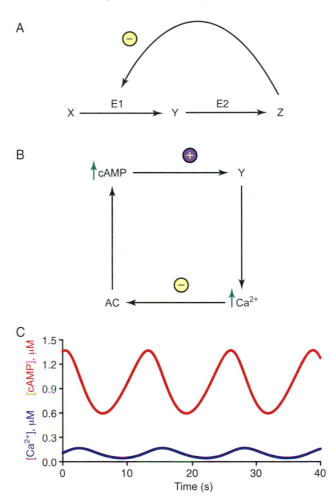

FIGURE 14.2 Feedback within two model signaling pathways. (A) Negative feedback within a simple, linear reaction scheme. Metabolite X is transformed into Y, which is transformed into end product Z. Z inhibits the enzyme E1 that catalyzes the X → Y reaction. (B) A model that captures some relationships between cAMP production and Ca^{2+} influx. cAMP activates Ca^{2+} channels (denoted Y). Ca^{2+} influx is thereby increased. Ca^{2+} influx inhibits adenylyl cyclase (AC). The overall effect is a negative-feedback loop in which cAMP inhibits its own production. (C) Oscillations in cAMP and Ca^{2+} are sustained by the negative-feedback loop. For this simulation, the values of the parameters in Eqs. 14.12–14.15 are

$$k_1 = 1.0\,sec^{-1}, \quad k_2 = 0.5\,sec^{-1}, \quad k_3 = 1.0\,sec^{-1},$$
$$k_4 = 0.5\,sec^{-1}, \quad k_5 = 0.45\,\mu M\,sec^{-1}, k_6 = 1.0\,sec^{-1},$$
$$k_7 = 0.11\,sec^{-1}, \quad k_8 = 2.0\,\mu M\,sec^{-1},$$
$$K_1 = 2.0\,\mu M, \quad K_2 = 0.2\,\mu M.$$

Initial values for the model variables at $t=0$ are cAMP=1.0 μM, AC=0.5 μM, Y=0.05 μM, Ca^{2+}=0.01 μM.

production and the rate of Ca^{2+} influx into a cell. The model can simulate persistent oscillations in [cAMP] if the parameter values fall within the proper range (see legend to Fig. 14.2).

The model contains four biochemical species: cAMP, Ca^{2+}, active adenylyl cyclase enzyme (denoted by AC), and active Ca^{2+} channels in the plasma membrane (denoted by Y). The rate of cAMP production is assumed to be proportional to the amount of active adenylyl cyclase enzyme. cAMP is assumed to be degraded by a first-order process. Therefore, the differential equation describing the rate of change of cAMP concentration is

$$\frac{d[cAMP]}{dt} = k_1[AC] - k_2[cAMP] \qquad (14.12)$$

The rate of Ca^{2+} influx is assumed to be proportional to the number of active Ca^{2+} channels in the plasma membrane. Ca^{2+} is extruded from the cell by a first-order process. Therefore, for the rate of change of Ca^{2+} concentration,

$$\frac{d[Ca^{2+}]}{dt} = k_3Y - k_4[Ca^{2+}] \qquad (14.13)$$

Ca^{2+} channels are assumed to be activated by the binding of cAMP. A Hill function characterizes activation by cAMP. Channels deactivate in a first-order fashion in the absence of cAMP. The differential equation for the number of active channels, denoted Y, is therefore written as

$$\frac{dY}{dt} = \frac{k_5[cAMP]^3}{[cAMP]^3 + K_1{}^3} - k_6Y \qquad (14.14)$$

Ca^{2+} is assumed to inhibit adenylyl cyclase. This inhibition can be represented by an equation in which Ca^{2+} increases the rate at which the variable AC decreases. This effect of Ca^{2+} can be described by a Hill function. The resulting differential equation for AC is

$$\frac{dAC}{dt} = k_7 - AC\frac{k_8[Ca^{2+}]^3}{[Ca^{2+}]^3 + K_2{}^3} \qquad (14.15)$$

Actual regulation of adenylyl cyclase is more complex than Eq. 14.15. Some neuronal isoforms of adenylyl cyclase are stimulated by elevations of Ca^{2+}, but others are inhibited (Chapters 12 and 19).

If Hill coefficients greater than 2 are chosen in Eqs. 14.14 and 14.15, the model can produce oscillations in [cAMP] and the other variables (Fig. 14.2C). Generation of oscillations requires a negative

feedback loop. This feedback loop operates as follows. cAMP activates Ca^{2+} channels, allowing the influx of Ca^{2+}. Ca^{2+}, in turn, inhibits adenylyl cyclase. The inhibition acts to decrease the concentration of cAMP, closing the loop.

POSITIVE AND NEGATIVE FEEDBACK CAN SUPPORT COMPLEX DYNAMICS OF BIOCHEMICAL PATHWAYS

Feedback Can Sustain Oscillations or Multistability of Concentrations and Reaction Rates

One interesting behavior in biochemical pathways is *persistent oscillations* in the rates of enzymatic reactions and in the concentrations of substrates and products. For example, oscillations have been observed in the metabolic flux through glycolysis and also in the rates of secretion of hormones such as insulin. Oscillations in reaction rates and concentrations commonly rely upon negative feedback to sustain oscillations (Goldbeter, 1996). Models of biochemical systems that rely upon positive feedback tend to display another type of behavior, termed *multistability*. This is the existence of multiple steady states for the concentrations of the chemical species. Each steady state corresponds to a different constant metabolic flux through the pathway. Each state is stable in that the system, if originally in one state, will return to that state following a small disturbance. However, large disturbances can switch the system between states.

Bistability is a specific type of multistability with two stable states. Bistability was discussed in Chapter 7 (Fig. 7.19) in the context of a model neuron that could exhibit a fixed, stable membrane potential but, following a brief current injection, would switch to a state of continuous electrical spiking. *Ca^{2+}-calmodulin-activated protein kinase II* (CaMKII) activity may provide an example of biochemical bistability relevant to learning and memory. As discussed in Chapter 12, CaMKII exists *in vivo* as a holoenzyme of multiple subunits (~12). CaMKII is activated by binding of Ca^{2+}-calmodulin complexes to the subunits. Each subunit of CaMKII can be phosphorylated on Thr-286. The enzymatic activity of a phosphorylated subunit is increased, and this activity persists when Ca^{2+}-calmodulin is not bound. A holoenzyme can undergo *autophosphorylation*, in which active subunits phosphorylate other subunits that have Ca^{2+}-calmodulin complexes bound (Zhabotinsky, 2000). Elevations of Ca^{2+} increase the frequency of CaMKII autophosphorylation reactions.

Modeling has suggested that CaMKII might act as a *bistable switch* (Holmes, 2000; Lisman and Zhabotinsky, 2001). For a given CaMKII holoenzyme, if a brief electrical stimulus caused Ca^{2+} influx leading to phosphorylation of a critical number of subunits on Thr-286, then these subunits might phosphorylate the remaining subunits. Then the holoenzyme would remain active after Ca^{2+} influx ceased. As mentioned in Chapters 12 and 19, prolonged activation of CaMKII may be important for long-term synaptic potentiation in response to correlated pre- and postsynaptic electrical activity.

The most recent models (Lisman and Zhabotinsky, 2001; Miller *et al.*, 2005) suggest that CaMKII could remain highly active and autophosphorylated only in the postsynaptic regions of dendrites. In other regions of the cytosol, the total concentration of CaMKII is much lower relative to phosphatases that counter the autophosphorylation reaction. Thus, in the postsynaptic region, CaMKII might act as a bistable switch in which a brief influx of Ca^{2+} would lead to long-term activation of CaMKII *via* sustained autophosphorylation. Figure 14.3A schematizes the autophosphorylation process, and Figure 14.3B illustrates how bistable activity of the CaMKII holoenzyme can be represented as a plot of enzyme activity versus $[Ca^{2+}]$.

Chapter 7 introduced the concept of a *bifurcation diagram*, in which the stable or unstable steady states of a variable in a model are plotted. The plot shows how the steady states change as a function of a model parameter. For example, in a model of neuronal excitability, a stable solution for membrane potential will change its value and perhaps lose stability if a conductance parameter or a background stimulus current is varied (see Figs. 7.18 and 7.20). Figure 14.3B is a bifurcation diagram showing how the activity of CaMKII varies as $[Ca^{2+}]$ is varied. The solid (upper and lower) portions of the black curve give the values of CaMKII activity that are stable at any given $[Ca^{2+}]$. If $[Ca^{2+}]$ lies within the shaded region, there are two possible values of CaMKII activity. The higher activity corresponds to a higher degree of CaMKII autophosphorylation. The dashed portion of the black curve corresponds to an intermediate activity that is unstable: a small change in $[Ca^{2+}]$ will cause CaMKII activity to move up or down to one of the two stable states. Prior to a stimulus that leads to an increase in Ca^{2+} influx, the point representing the state of CaMKII might lie in the lower steady state of the graph—low activity, at a low (resting) level of $[Ca^{2+}]$. This state is labeled 1. Electrical activity and consequent Ca^{2+} influx can move the Ca^{2+} concentration well to the right, from point 1 on the low steady

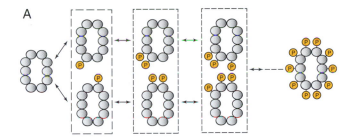

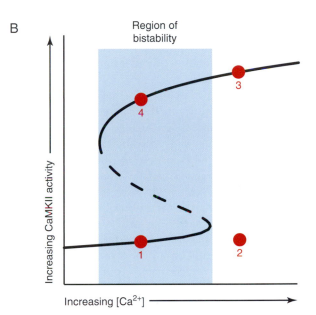

FIGURE 14.3 Autophosphorylation of CaMKII may lead to bistability in enzyme activity. (A) Schematic illustrating multiple phosphorylation states. Assuming 10 subunits per holoenzyme, there are a large number of possible states, but if CaMKII activity depends only on the number of phosphorylated subunits and not their location within the holoenzyme, there are only 11 states of differing enzyme activity. Five of these states are shown, corresponding to 0, 1, 2, 3, and 10 phosphorylated subunits. Phosphate groups are orange. A calmodulin molecule with bound Ca^{2+} (not shown) must be bound to a subunit for phosphorylation to be possible. (B) Schematic illustrating bistability. The *x*-axis represents intracellular Ca^{2+} concentration. For a range of Ca^{2+} concentrations, there is both a low and a high stable steady state for CaMKII activity (solid portions of curve, shaded region of graph). The two stable states are separated by an unstable steady state (dashed portion of curve).

state to point 2, where the lower steady state of CaMKII activity no longer exists. At point 2 only the state of high CaMKII activity is stable. If $[Ca^{2+}]$ stays above the shaded region for some critical time, Ca^{2+}-calmodulin-induced autophosphorylation will move the activity of CaMKII up to a high value, to point 3. Then, when Ca^{2+} concentration returns to a lower level, the state of CaMKII moves back to the left, to point 4. But the CaMKII activity still remains on the upper portion of the curve: high activity, with sustained autophosphorylation.

The significance of CaMKII bistability is somewhat controversial. Long-lasting changes in CaMKII activity have not yet been shown to be essential for long-term synaptic potentiation (LTP). Autophosphorylation of CaMKII does occur after electrical stimuli that produce LTP (Yamagata and Obata, 1998). When autophosphorylation is prevented by mutating Thr-286, LTP *induction* is severely impaired (Giese *et al.*, 1998). However, CaMKII autophosphorylation may not be required for LTP *maintenance*. Inhibitors of CaMKII applied after LTP induction appear not to affect maintenance (Chen *et al.*, 2001). A recent study suggests CaMKII activation is short-lived after LTP induction (Lengyel *et al.*, 2004).

Feedback due to allosteric regulation of enzymes can generate oscillations in the rate of metabolic flux through a biochemical pathway. The best-known example is oscillations in glycolysis. The enzyme phosphofructokinase (PFK) is largely responsible for glycolytic oscillations in experimental preparations (Dano *et al.*, 1999; Goldbeter, 1996). Glycolytic oscillations are likely to modulate the secretion of insulin (Bertram *et al.*, 2004; Prentki *et al.*, 1997). PFK is activated by one of its reaction products, adenosine diphosphate (ADP). The activation depends on a power of ADP concentration ([ADP]). Positive feedback occurs, with production of ADP progressively activating PFK until PFK activity and [ADP] plateau. Usually, positive feedback alone cannot support oscillations. However, in glycolysis, the substrate of PFK (fructose 6-phosphate) is supplied at a limited rate. Therefore, cycles of substrate depletion can occur. When PFK activity is high, the substrate is used up. A fall in the substrate concentration leads to a loss of PFK activity. PFK activity and [ADP] remain low while substrate accumulates again. Eventually, substrate accumulation raises the rate of production of ADP, and enough ADP is formed to reach the threshold for positive feedback to become effective. Then the cycle can begin anew. Several mathematical models have been developed that describe these oscillations (Goldbeter, 1996). With these models, bistability in PFK activity can also appear when different values of kinetic parameters are assumed. This is expected, since bistability is typical for models with positive feedback.

Several Mechanisms, Including Positive Feedback, Can Yield High Sensitivity of Response Magnitude to Signal Strength

Sometimes a modest change in the concentration of substrate, or a modest change in the activity of an enzyme, will greatly change the *net* rate of substrate–product conversion. This phenomenon is termed *ultrasensitivity*. Such switchlike behavior can be generated by several mechanisms. One mechanism relies on allosteric binding of a substrate to multiple sites on an enzyme in a cooperative fashion, such as that described by Eq. 14.10a. If the Hill coefficient of cooperativity is high (i.e., high n in Eq. 14.10a), then a small change in substrate level can yield a large change in enzyme activity. A second mechanism is *zero-order ultrasensitivity* (Goldbeter and Koshland, 1984). Suppose the enzyme catalyzing a given reaction is almost fully saturated with substrate, such that the reaction is almost zero-order. Then suppose there is a small increase in the enzyme concentration and thus in V_{max}. This gives a small reaction rate increase. To restore the rate to its original value, the substrate concentration must drop. However, since the enzyme is nearly saturated, a *large* drop in the concentration of substrate is required to give the small rate decrease. Alternatively, suppose other enzyme activities change, altering the metabolic steady state of the system, thus causing a decrease in the reaction rate under consideration. A large drop in substrate concentration will again result. Thus with zero-order ultrasensitivity, small changes in enzyme activity or reaction rate result in large changes in the ratio of substrate to product.

A well-known pathway in which zero-order ultrasensitivity may be important begins with hormonal or neurotransmitter receptors and ultimately activates mitogen-activated protein kinase (MAPK) (Fig. 14.4A). In this pathway, MAPK is itself a substrate for a kinase termed *MAP kinase kinase* (MAPKK, or MEK). Some isoforms of MAPK may be present in high enough concentrations to nearly saturate MAPKK. It has therefore been suggested that a small increase in the activity of MAPKK could cause, via zero-order ultrasensitivity, a large decrease in its substrate (inactive MAPK) and a large increase in phosphorylated and active MAPK. Experiment and modeling have suggested this occurs physiologically (Huang and Ferrell, 1996; Xiong and Ferrell, 2003).

Figure 14.4B illustrates stimulus-response curves for the doubly phosphorylated (PP) fractions of MAPK and MAPKK versus the strength of an input stimulus. These curves were calculated from a model (Huang and Ferrell, 1996) that uses the kinetic scheme of Figure 14.4A. The stimulus corresponds to a fixed activation level of MAPKKK in Figure 14.4A. The stimulus was assumed to be applied for long enough that all reactions in Figure 14.4A attained equilibrium. The rates of change of the amounts of active enzymes can therefore be set equal to zero. The resulting

A

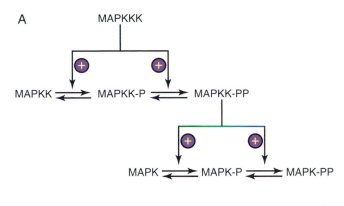

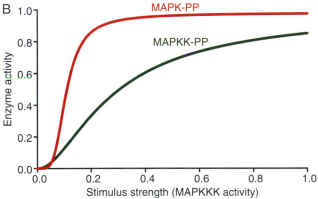

FIGURE 14.4 A mechanism for generating ultrasensitivity in the stimulus-response curve of a MAP kinase signaling cascade. (A) The kinetic scheme illustrating double phosphorylations of both MAPKK and MAPK. Phosphorylations are reversible. (B) Stimulus-response curves for the activities of MAPK (top, red curve) and MAPKK (green curve) based on the kinetic scheme of (A). For both enzymes, the activity is assumed proportional to the fraction that is doubly phosphorylated. The stimulus is a constant level of MAPKKK activation.

system of equations can be solved numerically to yield levels of doubly phosphorylated MAPKK and MAPK as a function of input stimulus strength. The steepness of the MAPK phosphorylation curve is due to zero-order ultrasensitivity.

Cross Talk Between Signaling Pathways May Shape Stimulus Responses

Electrical activity or exposure to neurotransmitters or growth factors can simultaneously activate a variety of signaling pathways in neurons. Cross talk between pathways can be mediated by a chemical species that is produced or consumed by reactions in two pathways. Cross talk can also be mediated by a species produced in one pathway that interacts allosterically with an enzyme in another pathway. Modeling of pathways can help identify points of possible cross talk and ranges of parameter values

that would make this cross talk significant. Experiments could then examine whether parameter values fall in these ranges and whether cross talk causes observable effects on system behavior.

A recent model (Bhalla, 2002; Bhalla and Iyengar, 1999) represents activation by glutamate of four signaling pathways at dendritic synapses of hippocampal neurons. These pathways activate protein kinase A (PKA), protein kinase C (PKC), MAPK, and CaMKII. Figure 14.5A illustrates points of cross talk between these pathways.

Experimental values for model parameters were sometimes not available for neurons, so values from other cell types had to be used in this model. Despite this disadvantage, the model is useful for suggesting points of cross talk. One possibility is a positive-feedback loop that increases both MAPK and PKC activity. MAPK can phosphorylate and activate the enzyme phospholipase A_2, whose activity generates the second messenger arachidonic acid (AA). AA activates PKC. In turn, PKC can phosphorylate and activate a guanine-nucleotide exchange factor, which then activates Ras GTPase. Ras can then further activate the MAPK signaling cascade, closing the positive-feedback loop. In simulations with the model, a brief glutamate exposure activated the positive-feedback loop. This exposure caused the model to switch from a stable state with a low, basal activity of MAPK and PKC to a stable state with a high activity of both enzymes. In neurons, induction by glutamate of a prolonged activation of MAPK and PKC might correspond to induction of a large change in synaptic strength, dependent on a long-lasting phosphorylation by MAPK or PKC of transcription factors such as cAMP – responsive element binding protein (CREB).

A second point of possible cross talk is a *gate* involving cAMP and CaMKII. As discussed previously, autophosphorylation of CaMKII after a brief influx of Ca^{2+} might convert CaMKII to a long-lasting, active state independent of Ca^{2+}. Such a state might be essential to maintain LTP. However, Ca^{2+} influx also elevates the concentration of Ca^{2+}-calmodulin, which in turn may activate a phosphatase, protein phosphatase I (PP1). PP1 can dephosphorylate CaMKII. If cAMP levels are also elevated by the same stimulus that causes Ca^{2+} influx, the autophosphorylation of CaMKII might have a greater chance of becoming self-sustaining. This is because cAMP activates PKA, which in turn leads to inhibition of PP1, thus "opening a gate" to allow a high level of CaMKII phosphorylation. This high level could be self-sustaining after cAMP and PP1 activity return to basal levels, because once a critical number of subunits in a CaMKII holoenzyme have been phosphorylated, autophosphorylation of the remaining subunits may

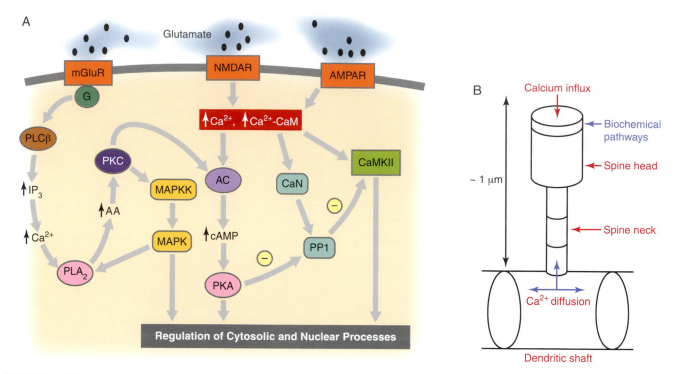

FIGURE 14.5 Aspects of a model that relates glutamate exposure at hippocampal synapses to long-term synaptic strengthening. (A) Biochemical pathways. Glutamate can act through metabotropic glutamate receptors (mGluR) to activate G proteins (G). Glutamate also acts through NMDA and AMPA receptors to increase levels of free Ca^{2+} and Ca^{2+} bound to calmodulin. These events lead to activation of phospholipase Cβ (PLCβ), CaMKII, calcineurin (CaN), adenylyl cyclase (AC), and PKA. Two forms of cross talk between these signaling pathways are illustrated. PKA activation leads to the inhibition of PP1. This inhibition relieves dephosphorylation of CaMKII by PP1, helping sustain CaMKII autophosphorylation. Also, MAPK activates phospholipase A_2 (PLA$_2$), and the resulting increase in arachidonic acid activates PKC. PKC in turn activates MAPKK, which further activates MAPK. MAPK, PKA, and CaMKII regulate gene expression as well as cytosolic components such as the cytoskeleton, which are essential for synaptic strengthening. (B) Part of a compartmental model of a hippocampal neuron. A spine and part of the dendritic shaft are shown. The spine head, the spine neck, and the dendritic shaft are each modeled as a series of cylindrical compartments. The compartments are necessary to model Ca^{2+} diffusion. Stimuli are modeled as pulses of Ca^{2+} influx into the spine. The components of the biochemical model of (A) are all assumed to exist in the terminal (uppermost) spine compartment. Ca^{2+} influx affects this model as discussed above.

be rapid enough to overcome the effects of PP1. Then the holoenzyme could remain phosphorylated and active until eventual degradation.

A model (Bhalla and Iyengar, 1999) suggested that this "gate" could be important. In this model, exposure led to a long-lasting increase in CaMKII activity only if PP1 was inhibited by PKA. In that case, only stimuli that increased both Ca^{2+} and cAMP past threshold levels were predicted to give a prolonged activation of CaMKII. Experiments (Brown *et al.*, 2000) have supported the importance of such a dual threshold of Ca^{2+} and cAMP levels in determining the degree of synaptic modification in hippocampal neurons. Addition of a cAMP analog together with electrical stimulation was observed to yield strong activation of CaMKII. Neither stimulus alone could produce strong CaMKII activation. Furthermore, only the combination of electrical stimulation and cAMP

analog could induce LTP. As noted previously, CaMKII activity is necessary for LTP induction, but its role in LTP maintenance is not yet empirically established.

The model of Figure 14.5A also helps to order elements of signaling pathways in a set of cause–effect relationships. For example, Figure 14.5A suggests that during the biochemical events following glutamate exposure, PKA activation lies "downstream" of both PKC and MAPK activation, but "upstream" of CaMKII activation. The model has also been used to discriminate between stimulus patterns such as those known to produce long-term potentiation (LTP) or long-term depression (LTD) of synapses (Bhalla, 2002; see also Chapter 19). For simulations of LTP and LTD, the biochemical model of Fig 14.5A was embedded within a *compartmental model* of a neuron, part of which is schematized in Figure 14.5B. The biochemical model (Fig. 14.5A) was assumed to operate in the head of the dendritic spine (top compartment in Fig. 14.5B). Stimuli were modeled as pulses of Ca^{2+} influx into the spine head. These pulses activate enzymes such as MAPK (Fig. 14.5A).

A compartmental model provides a method for simulating Ca^{2+} diffusion between different portions of the spine (different compartments) and between the spine and the dendritic shaft. Within these small structures, Ca^{2+} diffusion plays an essential role in shaping Ca^{2+} concentration changes following stimulation. Therefore, Ca^{2+} diffusion must be included to model the activation of neuronal biochemical pathways. Diffusion between compartments is modeled with a variant of Eq. 14.4. A compartmental model can also simulate the spread of voltage changes between portions of a neuron (e.g., action potentials).

Cross talk between pathways might act to degrade the specificity of stimulus-response relationships. With excessive cross talk, a given biochemical response could be activated by spurious stimuli. Therefore, Bardwell et al. (2007) suggest that signaling pathways are likely to have evolved to minimize nonspecific cross talk. These authors give concise, intuitive mathematical definitions of pathway specificity and fidelity, both of which are maximized in the absence of cross talk. They analyze several mechanisms for minimizing cross talk. Scaffolding proteins can physically separate enzymes and other components of one pathway from those of another. Cross-pathway inhibition, in which activation of pathway A represses flux through pathway B, can minimize nonspecific activation of pathway B. Combinatorial signaling, in which the target of a pathway requires a further unrelated input for full activation, is a generic way to increase fidelity of pathways.

The model of Figure 14.4 describes the pathways involved in the early induction of LTP and its persistence for ~1-2 h. LTP also exists in a more enduring form called late LTP. Late LTP (L-LTP) involves changes in gene expression whereas early LTP (E-LTP) does not. L-LTP, but not E-LTP, is blocked by protein synthesis inhibitors. A recent qualitative model (Smolen et al., 2006) describes activation of some of the essential convergent signaling pathways leading to the induction of L-LTP. In the model, at stimulated synapses, activation of MAPK, CaMKII, and PKA leads to phosphorylation of a synaptic "tag", which allows those synapses to "capture" newly synthesized proteins and thereby increase their strength. Experiments have verified the existence of such a synaptic tag (Reymann and Frey, 2007; Frey and Morris, 1997). Activation of MAPK and of

nuclear CaM kinase IV induces gene expression. In the model, this gene expression is represented simply and qualitatively as synthesis of a single generic protein, GPROD. The rate of increase of a synaptic weight is set equal to the product of [GPROD] and the synaptic tag variable. Figure 14.6A illustrates the signaling pathways of the model, and Figures 14.6B–C illustrate simulated L-LTP induction. L-LTP was induced by three tetanic stimuli, each 1 sec long, spaced 5 min apart. In Figure 14.6B, CaMKII deactivates within ~5 min after each tetanus, whereas MAPK and the synaptic tag both remain elevated for a few hours after tetani. These durations agree qualitatively with published data (Smolen et al., 2006). In Figure 14.6C, the percent elevation of the synaptic weight W is also similar to data. Subsequent return of W to baseline is very slow (timescale of months), in agreement with data demonstrating that L-LTP in vivo can last for months (Abraham et al., 2002).

This model has helped to clarify how convergence of multiple signaling pathways onto a single target—the synaptic weight—amplifies the stimulus response, allowing a brief (~1 sec) stimulus to be transduced into a long-lasting change in synaptic strength. Further stimulus amplification is provided by the requirement that multiple Ca^{2+} ions bind to CaM to activate CaMKII. If these modes of nonlinear stimulus amplification were not present, no significant change in synaptic weight could occur. Because of the difference in timescales, a linear stimulus-response relationship would transduce a brief synaptic stimulus into a completely negligible change in the much slower synaptic weight variable.

Another recent modeling study (Smolen et al., 2008) suggests that bistability of MAPK activity may play a role in maintaining late LTP. This bistability would rely on a positive feedback loop, with active MAPK indirectly promoting its own phosphorylation. Persistently active MAPK, in turn, would upregulate protein synthesis and synaptic strength. Further experiments are required to test this hypothesis.

MODEL DYNAMICS SHOULD USUALLY BE ROBUST TO PARAMETER VARIATION

Genetic and biochemical systems are commonly observed to be robust to large changes in parameters, such as genetic mutations that alter enzyme activities (Little et al., 1999; Wagner, 2000). The

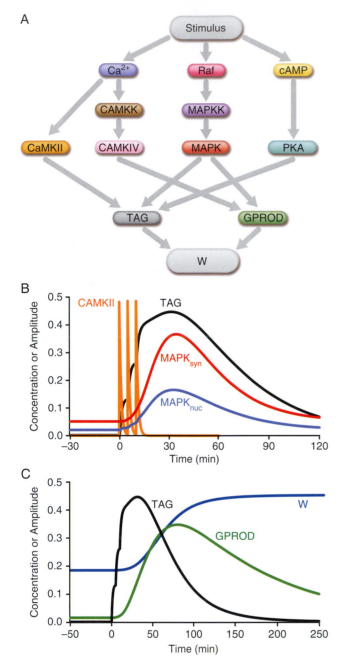

FIGURE 14.6 Dynamics of a model of the late phase of LTP. (A) Schematic of the model. Synaptic stimulation elevates Ca^{2+} and cAMP and activates Raf kinase, the initial enzyme in the MAPK cascade. Ca^{2+} activates CaMKII and CaM kinase kinase (CaMKK). CaMKK activates CaMKIV. cAMP activates PKA. Activated MAPK, PKA, and CaMKII all contribute to the setting of a synaptic tag (variable TAG) required for L-LTP. MAPK and CaMKIV induce expression of a generic gene whose protein product is denoted GPROD. L-LTP induction is modeled as an increase in a synaptic weight W. The rate of increase is proportional to the value of the synaptic tag and to the amount of gene product. (B) Time courses of the levels of active CaMKII, active synaptic MAPK, active nuclear MAPK, and the synaptic tag variable during and after three simulated tetanic stimuli. (C) Time courses of the gene product assumed necessary for L-LTP, the synaptic tag, and the synaptic weight variable W.

qualitative behavior of a gene network or biochemical pathway is often preserved due to evolved compensatory mechanisms. Therefore, viable models will usually *not* exhibit large changes in dynamics following small parameter changes. A standard method of assessing robustness to parameter changes is to define a set of sensitivities S_i, with the index i ranging over all model parameters p_i (Beck and Arnold, 1977; Frank, 1978). Let R denote the amplitude of the response of a signaling pathway to a fixed stimulus. For each p_i, a small change is made, and the resulting change in R is determined. The *relative sensitivity* S_i is then defined as the *relative*, or *fractional*, change in R divided by the *relative* change in p_i,

$$S_i = \frac{\Delta R / R}{\Delta p_i / p_i} \qquad (14.16)$$

The relative sensitivity has the advantage that it is independent of the magnitudes and units of R and the p_i.

R in Eq. 14.16 could also represent the time delay between a simulated stimulus and response, or the period of a simulated oscillation in gene transcription rate. In general, R is some quantity that has been measured and whose simulation is a major goal of model development.

The S_i in Eq. 14.16 are determined for small changes (10% or less) in the values of each individual parameter. For a model to be considered *robust* (i.e., not overly sensitive) the S_i should not be too large. They should generally not exceed ~10, unless there is experimental evidence that a response is particularly sensitive to a specific parameter. In practice, determinations of S_i are complemented by qualitative assessments of whether particular behaviors (e.g., oscillations in reaction rates) are preserved during modest parameter changes.

More detailed investigations vary multiple parameters simultaneously. The goal is to determine model behavior in regions of the multidimensional *parameter space* whose coordinate axes are the parameter values. For example, a region might support multiple stable solutions for metabolite concentrations and reaction rates, as does the blue region in Figure 14.3B. Characterizing the location and size of such a region could predict that particular *in vivo* conditions would support multistability. *Bifurcation analysis* (Ermentrout, 2001; Izhikevich, 2007; Wiggins, 1990) can help in determining and visualizing model behavior in parameter space. This technique allows determination of the curves or surfaces in parameter space at which model behavior undergoes significant changes. For example, with parameter values on one side of such a surface, the model variables may maintain a stable equilibrium. For parameter values on the other side of the surface, the equilibrium may disappear and persistent oscillations of model variables may appear. Software packages have been developed to determine and plot these curves or surfaces (e.g., AUTO, http://sourceforge.net/projects/auto2000/).

PARAMETER UNCERTAINTIES IMPLY THE MAJORITY OF MODELS ARE QUALITATIVE, NOT QUANTITATIVE, DESCRIPTIONS

Models of biochemical pathways usually contain parameters whose values are not well constrained by experiment. It is obligatory for investigators to state which parameters in their models are poorly constrained, because parameter values might later be found to differ considerably from those used in a given model. Such differences could falsify the model because simulation of experimental results might no longer be possible.

Standard experiments can be done to estimate some parameter values. For example, enzyme Michaelis constants (K_1 and K_2 in Eq. 14.9) are often estimated with preparations containing small amounts of enzyme. Other parameters are more difficult to estimate. For example, it is difficult to estimate the amount of active enzyme per cell. Therefore V_{max}'s of enzymes *in vivo* are often poorly constrained.

Many parameters such as reaction rate constants, enzyme activities, and Michaelis constants depend on the activity coefficients of enzymes and reactants. Activity coefficients are likely to be considerably different *in vivo* than *in vitro*. In cells, a high concentration of macromolecules creates a *macromolecular crowding* effect. Experiments with crowding by inert substances such as polyethylene glycol demonstrate that macromolecular crowding raises activity coefficients of all macromolecular species (Minton, 2001). This increase in activities preferentially increases association rates and consequently increases levels of aggregates at the expense of monomers. For example, consider the association and dissociation rates, R_f and R_b respectively, that

govern the formation of an AB dimer from monomers of A and B:

$$R_f = k_f \lambda_A [A]\ \lambda_B [B] \qquad (14.17a)$$

$$R_b = k_b \lambda_{AB} [AB] \qquad (14.17b)$$

In Eqs. 14.17a–b, k_f and k_b are standard rate constants; λ_A, λ_B, and λ_{AB} are activity coefficients for A and B monomers and for the AB dimer. It is seen that the concentrations of A, B, and AB are each multiplied by the corresponding activity coefficients. The association rate R_f contains the product of two activity coefficients, so R_f will tend to be enhanced more than R_b when activity coefficients are raised by macromolecular crowding. Therefore, formation of AB dimers will be favored.

Generally, crowding and increased association rates can help stabilize signaling complexes containing multiple enzymes. This stabilization can enhance the efficacy of metabolite transfer between enzymes and thereby enhance the flux through a metabolic pathway (Rohwer *et al.*, 1998).

Because of uncertainties in parameter values and activity coefficients, models of biochemical pathways should most commonly be thought of as qualitative descriptions, rather than quantitative. Simulations with such models are, however, valuable for characterizing the types of dynamics likely to be supported by the biochemical architecture of a pathway. Simulations that adjust parameters to reproduce experimental results can help estimate the qualitative importance of interactions, such as feedback loops. Parameter adjustment could, for example, suggest that a dissociation constant for an allosteric effector is quite large and that as a result, a feedback interaction mediated by that effector is weak. Further experiments are necessary to test such predictions.

Adjustment of parameter values is commonly done by trial and error. However, it is often preferable to use a computer program that repeats the simulation with many different sets of parameter values. In this case, after each repetition, a measure of the "distance" between the simulated and experimental time courses is computed. A commonly used measure sums the squares of the differences between simulated and experimental metabolite concentrations. The sum is taken over all concentrations and also over all experimental time points. Some type of optimization routine (Press, 1994) is included in such computer programs to adjust parameters in a direction suggested by the "distance" calculations for the previous few simulations. The goal is to find parameter values that minimize the distance. Parameter adjustment continues until one or more value sets are found for which the shapes and amplitudes of the simulated metabolite time courses approximate the experimental time courses. If this approximation proves impossible, a new model based upon new equations is needed. Following successful simulation, *key control parameters* can be defined as those parameters that are tightly constrained by the need to reproduce experimental data. Relatively small variations in these parameters cause significant changes in model behavior. Key control parameters are often candidate points for biological regulation of the signaling pathway. Experiments that determine values of these parameters constitute tests that can falsify a model.

Although full empirical characterization of *in vivo* parameters and processes cannot be achieved, models deficient in some details have often proven very useful in making predictions and interpreting experiments. Therefore, for modeling of incompletely understood *in vivo* systems to be useful, one should focus less on building a complete and precise model, than on building incomplete but falsifiable models that make clear, testable experimental predictions (Gutenkunst *et al.*, 2007).

SEPARATION OF FAST AND SLOW PROCESSES TO SIMPLIFY MODELS

Experimental data generally determines the timescale of interest for a model. For example, experimental data concerning gene expression will have a timescale of minutes or hours, whereas data concerning enzyme reaction rates may have a timescale of seconds or minutes. Models that simulate experimental data with a long timescale can be simplified, and the number of differential equations reduced, when processes or reactions with a fast timescale are assumed to be at equilibrium. Conversely, to simulate data with a fast timescale, the dynamics of slow processes can be neglected. A variable associated with a slow process—such as the concentration of a gene product—can simply be treated as a parameter. This separation into fast and slow timescales, and correspondingly fast and slow variables, is widely used for simplification of models. Chapter 7 has illustrated such a separation in a model of neuronal electrical activity. In that model, $[Ca^{2+}]$ was the slow variable, and the fast variables were membrane potential and a channel gating variable w (see Figs. 7.21 and 7.22). For further details on this very important technique, and for discussion of separation by spatial as well as time scales, see Chapter 7 of Murray (2003), Chapter 9 of Izhikevich (2007), and Ermentrout (2001).

MODELS HELP TO ANALYZE METABOLIC FLUX REGULATION

One goal of modeling biochemical pathways is to predict which parameters control most strongly the rate at which substance flows from initial substrate to final pathway product (e.g., the rate at which glucose is converted to pyruvate in glycolysis). This rate is termed the *metabolic flux* through the pathway. *Flux regulation* is necessary to keep metabolite concentrations within ranges appropriate to the state of the cell. Flux regulation of pathways is also essential for balancing the production and use of ATP. Commonly one asks, to which specific enzyme activities and amounts is the pathway flux most sensitive? *Metabolic control analysis* (MCA, also termed *metabolic control theory*) quantifies the importance of enzymes in regulating metabolic flux. The methodology of MCA will be briefly discussed. For a fuller treatment, Stephanopoulos *et al.* (1998) and Hornberg *et al.* (2007) should be consulted. In MCA, the sensitivities of the metabolic flux to changes in the activities of enzymes are expressed as *flux control coefficients* (FCCs). Denote the steady-state metabolic flux through a pathway by J and the activity of the ith enzyme in the pathway by E_i. An FCC, F^i, is defined as follows. An infinitesimal change in E_i, dE_i will yield an infinitesimal change in J, dJ. The *relative*, or *fractional*, change in J is then defined as dJ/J. Divide this relative change by the *relative* change in E_i, obtaining

$$F^i = \frac{dJ/_J}{dE_i/_{E_i}} = \frac{d \ln J}{d \ln E_i} \qquad (14.18)$$

To understand FCCs, it helps to examine the relationship between pathway flux and enzyme activity. The pathway flux J will saturate as the activity or concentration of any given enzyme is increased.

To obtain the FCC from a graph of J versus enzyme activity, the slope of the tangent line (dJ/dE_i) is multiplied by the enzyme activity and then divided by J. Thus, infinitesimal changes in the enzyme activity are used to define FCCs. For a substantial, empirically measured change in enzyme activity, the inferred slope of the flux versus activity relation (the fraction $\Delta J/\Delta E_i$) can be considerably different from the actual slope (the derivative dJ/dE_i). Thus, experiments can only approximate FCCs. MCA has been modified to use these approximate FCCs from responses to large changes (Small and Kacser, 1993). A *summation theorem* states that all the FCCs—one for each enzyme in the pathway—must add up to 1. A consequence is that most FCCs in long pathways must be small, so that the sensitivity of pathway flux to alterations in the activity of any one enzyme is usually low. By contrast, in short pathways several FCCs are often large, and the corresponding enzymes exert strong flux control.

One result of MCA has been to modify concepts of metabolic control guessed via intuition. For example, consider enzymes that are subject to feedback inhibition by their reaction products. These enzymes serve as control points of the metabolic flux through the pathway, because the feedback prevents the concentrations of their reaction products and of pathway end products from rising too high. Intuitively, one might expect that these enzymes are strongly rate limiting, in the sense that small changes in their activity would have large effects on the pathway flux. However, MCA has demonstrated that this expectation is commonly false (for details see Stephanopoulos et al., 1998). MCA may also be useful in drug design. By determining rate-limiting enzymes in metabolic pathways necessary for growth of pathogens or development of neuronal diseases such as Alzheimer or Parkinson, MCA may guide development of enzyme inhibitors as drugs (Hornberg et al., 2007).

Some applications of MCA have helped increase understanding of neuronal metabolism. For example, in several regions of rat brain, the flux control coefficient of the enzyme nitric oxide synthase (NOS) for the in vivo synthesis of nitric oxide (NO) has been determined (Salter, 1996). Administering varying dosages of an NOS inhibitor and measuring effects on the NO synthesis rate determined the FCC of NOS. In all regions, this FCC was found to be close to 1. Because the sum of all FCCs in a biochemical pathway must equal 1 by the summation theorem, other processes essential for NO synthesis must have low FCCs, and NOS must be rate limiting for NO production. In other biochemical pathways, this method might be adapted to quantify the flux control of any enzyme for which a specific inhibitor is available.

MCA has been extended to begin considering concentration gradients within cells. FCCs have quantified the control exerted by macromolecule diffusion coefficients on the flux through a simple model biochemical pathway (Kholodenko et al., 2000). However, empirical determination of diffusion FCCs has not yet been accomplished.

SPECIAL MODELING TECHNIQUES ARE REQUIRED IF ENZYMES ARE ORGANIZED IN MACROMOLECULAR COMPLEXES

To maximize the efficiency and specificity of reactions, many biochemical pathways are organized so that successive enzymes are positioned close to each other. This organization is mediated by multienzyme complexes, also termed *signaling complexes*. These contain groups of enzymes in combination with organizing "anchoring" proteins, and possibly with receptors for hormones or neurotransmitters. The anchoring proteins may bind to cytoskeletal elements. For example, complexes of glycolytic enzymes can be organized by binding to actin filaments (Fokina et al., 1997; Kurganov, 1986). In neurons, signaling pathways using isoforms of MAPK appear to be organized into complexes. Scaffold proteins termed *JIP proteins* bind to particular isoforms of MAPK and MAPKK (Davis, 2000) and interact with kinases that activate MAPKK. PKA, calcineurin, and protein kinase C (PKC) bind to A-kinase anchoring proteins (AKAPS), which bind cytoskeletal elements (Edwards and Scott, 2000). The NMDA receptor is in a large complex with more than 50 signaling proteins, kinases, and phosphatases (Husi et al., 2000).

Advantages of organizing signaling proteins into complexes are (1) to favor a rapid passage of signals through an enzymatic cascade, and (2) to prevent unwanted cross talk between signaling pathways. For example, if specific isoforms of MAPKK and MAPK are colocalized in complexes, then a given isoform of MAPKK will not be able to activate "improper" isoforms of MAPK. Also, a metabolite produced by an enzyme can be more efficiently passed on to the next enzyme if both are colocalized.

Modeling signaling pathways organized into complexes may require equations specific to the system under study. However, some general comments can be made. Intermediate metabolites might not equilibrate with the cytoplasm via diffusion, but instead *metabolic channeling* may dominate. Metabolic channeling is movement of a metabolite directly from one enzyme to the next within a complex. In this situation, the rate of use of the intermediate metabolite would not be a function of its bulk concentration. Rather, it might be a direct function of the rate of production of the metabolite. Denoting the rate of metabolite production as $R_{production}$, this situation could be modeled with a Michaelis–Menten expression for the rate of usage of the metabolite, R_{usage}, like so:

$$R_{usage} = \frac{V_{max} R_{production}}{R_{production} + K_S} \qquad (14.19)$$

Here, V_{max} and K_s correspond to the maximal velocity and Michaelis constant of the second enzyme in a standard Michaelis–Menten equation (Eq. 14.8). However, the units and values of these parameters are different in Eq. 14.19 than in Eq. 14.8. Metabolic control analysis can assess flux regulation within a pathway containing multienzyme complexes (Stephanopoulos *et al.*, 1998).

A recent model (Levchenko *et al.*, 2000) illustrates possible kinetic effects of a scaffold protein that is assumed to bind both MAPKK and MAPK, allowing interaction of the two enzymes. As shown in Figure 14.7, this model assumes reversible binding of MAPKK and MAPK to the scaffold protein. Both MAPKK and MAPK must be doubly phosphorylated to be active. In solution, these phosphorylations occur sequentially. However, the model assumes that when either enzyme is bound to the scaffold protein, both phosphorylations occur simultaneously—hence the absence of singly phosphorylated MAPKK and MAPK in Figure 14.7. Another assumption is that only inactive MAPK and MAPKK can bind to the scaffold. Thus, if MAPK or MAPKK is activated while bound to the scaffold, and then unbinds, it cannot rebind until the phosphates are removed. These assumptions have not been confirmed experimentally. However, they are necessary to support the results of simulations (Levchenko *et al.*, 2000). One result is that the presence of scaffold protein reduces the steepness of the dose-response curve for MAPK activation as a function of stimulus strength, with stimulus represented by MAPKKK activation. A second result is that for fixed stimulus strength, as the concentration of scaffold protein is increased, the percentage of activated MAPK at first increases, but then decreases. The increase is due to the assumption of simultaneous dual phosphorylations when MAPKK and MAPK are bound

to the scaffold, because simultaneous phosphorylations are more efficient than the sequential phosphorylations that occur in solution. The decrease of MAPK activation when the level of scaffold protein is made very high occurs because complexes of scaffold protein with only one inactive enzyme (MAPKK or MAPK) account for an increasing proportion of total enzyme as scaffold concentration is increased.

STOCHASTIC FLUCTUATIONS IN MOLECULE NUMBERS INFLUENCE THE DYNAMICS OF BIOCHEMICAL REACTIONS

As discussed in the introduction, in order to accurately simulate biochemical reactions with molecules that are present in low numbers in a cell, it is necessary to model random fluctuations in molecule copy numbers. Rather than using chemical concentrations as variables, this stochastic level of modeling uses the actual numbers of each molecular species. In this way, one can include *random variability* in the times of synthesis and degradation of individual macromolecules or metabolites. For some macromolecules, such as specific mRNAs or transcription factors, average molecule numbers per cell can be ≤ 100. In this case, random variability in single-molecule synthesis or degradation will drive *stochastic fluctuations* in macromolecule numbers (McAdams and Arkin, 1998, 1999). Fluctuations of the numbers of enzymes or metabolites in signaling pathways can cause large variability in the response of individual members of a cell population to a stimulus (McAdams and Arkin, 1999). Only a fraction of cells might give an observable response, such as differentiation following application of a hormone.

Modeling suggests that fluctuations will most commonly act to decrease the steepness of a nonlinear stimulus-response relationship, such as that in Figure 14.4B for MAPK activation (Berg *et al.*, 2000). However, in specific circumstances steepness can increase (Paulsson *et al.*, 2000). Also, via a phenomenon known as *stochastic resonance*, fluctuations can paradoxically increase the reliability or magnitude of responses to small signals (Stacey *et al.*, 2001). These results are qualitative, and the effects of fluctuations within specific signaling pathways in neurons have not been well quantified. It is difficult to estimate absolute molecule numbers for cells with a complex morphology, such as neurons. Such estimates would, however, be essential for more accurate modeling of fluctuations.

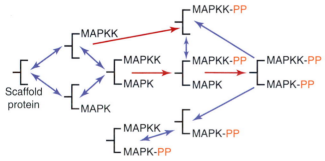

FIGURE 14.7 A model of a scaffold protein able to reversibly bind both MAPK and MAPKK. Nine biochemical species are possible: for example, free scaffold, scaffold with unphosphorylated MAPK and/or MAPKK bound, and scaffold with doubly phosphorylated MAPK and/or MAPKK bound. Transitions between species can be due to kinase association or dissocation from the scaffold (blue arrows) or to phosphorylation of kinases (red arrows). Reverse transitions due to dephosphorylation are not modeled. Dissociation of phosphorylated kinases is irreversible (single-direction blue arrows). For clarity, only some possible dissociations are shown.

To carry out simulations with fluctuations, the most common algorithm is that devised by Gillespie (Gillespie, 1977). This method takes variable time steps, from one reaction event to the next. The length of the time step and the type of the next reaction are both stochastic variables. The method is limited in that diffusion or spatial localization of molecules cannot be simulated. This limitation can be overcome by using an algorithm that represents each molecule as an individual object. At each time step this object is able to diffuse and also has probabilities of undergoing reactions. One software package that embodies this approach is StochSim, developed by C. Firth, D. Bray, and colleagues (Firth and Bray, 2001). Such a package is particularly useful in simulating the dynamics of a signaling complex of associated proteins, containing enzymes and receptors specific to one or more intracellular signaling pathways. Conformational changes, covalent modifications, and ligand binding are assumed to switch the complex between a series of possible states, and the state trajectory can be followed (Morton-Firth *et al.*, 1999). Such simulations will be needed to help understand how the NMDA receptor complex, or other neuronal complexes, transmits signals.

Consider a pathway for which a model based on continuous concentration variables predicts oscillations or multiple stable steady states of metabolite concentrations. Large fluctuations could disrupt the oscillations or destabilize the steady states (Barkai and Liebler, 2000; Smolen *et al.*, 1999). Therefore, predictions of behavior that neglect fluctuations may be false. Nevertheless, for many biochemical pathways, continuous models consisting of sets of differential equations (e.g., Eqs. 14.12–14.15) will remain essential because of insufficient data to justify a stochastic model. Generally, data used to construct a continuous model relies on large and reproducible changes in pathway fluxes and concentrations following strong stimuli. Because those responses are reproducible, a continuous model may be expected to reliably predict responses to new stimuli of similar strength to those used in model construction.

GENES CAN BE ORGANIZED INTO NETWORKS THAT ARE ACTIVATED BY SIGNALING PATHWAYS

Gene regulation is a common end point of biochemical signaling pathways. As discussed in Chapter 13, signaling pathways often activate proteins termed *transcription factors* (TFs) (Kewley *et al.*, 2004). Activation is often via phosphorylation of critical amino acid residues. Activated TFs regulate the transcription of genes by binding to nearby short segments of DNA. In Chapter 13, these segments were referred to as *cis*-regulatory elements. Another term is *response elements*. If these elements activate transcription, they are commonly termed *enhancers*; if they repress transcription, they are commonly termed *silencers*. Many genes are regulated by multiple TFs. Genes coding for TFs can be repressed or activated by TFs, including their own products. Large clusters of genes are often regulated in concert by biochemical signaling pathways that activate specific TFs. For example, activation of MAPK can lead to activation of hundreds of genes and repression of many others (Roberts *et al.*, 2000). *Gene networks* may be defined as gene clusters in which the expression of some members is regulated by the protein products of other members, or by a common input such as a hormone or neurotransmitter stimulus. The expression of network genes varies in a coordinated manner. Gene networks mediate long-term processes such as development and memory formation. Understanding these processes will therefore require understanding the dynamics of gene networks.

One gene network often implicated in the control of synaptic plasticity and memory formation is the network based on transcriptional regulation by the family of TFs that includes CREB (Bito *et al.*, 2003; De Cesare and Sasson-Corsi, 2000; Lonze and Ginty, 2002; Mayford and Kandel, 1999; Waltereit and Weller, 2003). Figure 14.8 illustrates aspects of this network, which are also discussed in Chapter 13. Neurotransmitters, such as serotonin, bind to receptors and activate G proteins. Production of intracellular second messengers such as cAMP is enhanced. Kinases such as PKA or MAPK are thereby activated. These phosphorylate CREB and related TFs. A positive-feedback loop appears to exist in this network (Fig. 14.8). In this loop, phosphorylated CREB, when bound to CREs in the vicinity of the *creb* promoter, activates *creb* transcription (Mohamed *et al.*, 2005; De Cesare and Sassone-Corsi, 2000; Walker *et al.*, 1995). There is also a negative feedback loop. CREB induces the gene for another TF of the CREM–CREB family (Chapter 13) termed *inducible Ca^{2+}/cAMP-responsive early repressor* (ICER) (Mioduszewska *et al.*, 2003). ICER is a powerful transcriptional repressor. Upon binding to CREs, ICER represses *creb* transcription, closing the loop.

Data essential for understanding the dynamics of gene networks is obtained by several methods. Sets of time courses of the mRNA levels expressed by

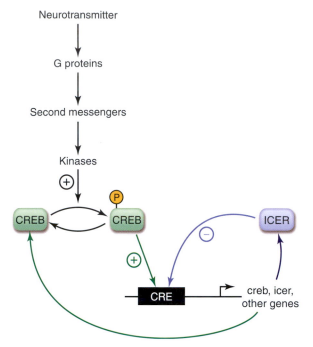

FIGURE 14.8 Signaling pathway involving transcriptional regulation by CREB. Neurotransmitters such as serotonin bind to receptors and act through G proteins to elevate levels of intracellular messengers (e.g., cAMP, Ca^{2+}). Kinases such as PKA are activated, resulting in phosphorylation of CREB and related TFs. Phosphorylated CREB stimulates the transcription of genes when bound to enhancer sequences termed $Ca^{2+}/cAMP$-response elements (CREs). Possible feedback interactions among the genes coding for CREB and related TFs are shown. A positive-feedback loop (green arrows) may regulate CREB synthesis. In this loop, CREB binds to CREs near *creb* and activates *creb* transcription. The repressor ICER is an element of a negative feedback loop (blue arrows). Transcription of *icer* is increased when the level of CREB increases, and ICER in turn can bind to CREs near the *creb* gene, repressing *creb* transcription.

large numbers of genes allow characterization of the response of cells or tissues to stimuli (Roberts *et al.*, 2000) or characterization of genetic disease states (Mirnics *et al.*, 2000). Current technology allows the expression time courses of ~10,000 genes to be simultaneously followed via quantification of their mRNAs. Cross talk between signaling pathways can take the form of convergent activation of groups of genes by several pathways and be identified from mRNA time courses. Understanding neural plasticity will require determining expression time courses for genes involved in synaptic modification. Further complexity exists because protein levels are often not well correlated with mRNA levels. Therefore, time courses of protein levels also need to be characterized (Ryu and Nam, 2000). Understanding genetic regulation further requires a framework for expressing the biochemical architecture of genetic systems and

for extracting genetic regulatory relationships from mRNA and protein expression time course. Modeling provides such a framework (Hasty *et al.*, 2001; Smolen *et al.*, 2000). Models are also important for predicting the response of gene networks to novel pharmacological or chemical agents.

Models of gene networks have often been qualitative, with modest numbers of equations and parameters. These models assess the behaviors of networks with common biochemical elements (Smolen *et al.*, 2000). Such elements include formation of TF dimers or oligomers, and positive and negative feedback loops in which TFs activate or repress transcription of their own or each other's genes. These models give insight into the behaviors more complex gene networks will exhibit, and will be used later to illustrate the dynamics generated by typical regulatory schemes. Relatively few models have considered regulation of mRNA processing or translation (Cao and Parker, 2001). Therefore, the following discussion focuses on transcriptional regulation.

In gene expression experiments, mRNAs are extracted at a series of time points following exposure to a hormone or other stimulus, and then hybridized to *DNA microarrays* to determine the relative amounts of different mRNAs at these time points (Gerhold *et al.*, 1999; Wen *et al.*, 1998). A technique termed *cluster analysis* is commonly applied to the time courses of the mRNAs (Ramoni *et al.*, 2002). This technique groups genes with expression time courses of similar shape. Such clusters are predicted to be regulated by a common factor, such as a specific TF. Several methods use clustered gene expression data to construct models of regulation within gene networks (e.g., Toh and Horimoto, 2002). Model parameter values are adjusted until the time courses for mRNA concentrations approximate the experimental responses. Limitations of some cluster analyses are (1) gene network regulation is assumed hierarchical, and (2) genes performing similar functions are assumed to exhibit similar expression patterns. Methods not subject to these limitations are being developed (Ihmels *et al.*, 2002; Imoto *et al.*, 2006; Yeung *et al.*, 2002). In one recent study, microarray data generated in gene disruption experiments was used to estimate a gene network and further used to identify or predict sets of genes or regulatory pathways affected by drugs (Imoto *et al.*, 2006). This methodology is expected to be important for drug development. Another recent study presents a method for combining and analyzing multiple time courses of expression of the genes in a network, with each time course generated by a different stimulus, in order to infer the network regulatory structure (Wang *et al.*, 2006).

METHODS EXIST TO MODEL GENE NETWORKS AT VERY DIFFERENT LEVELS OF DETAIL

In the *Boolean*, or *logical-network, method*, modeling is rather crude, but simulations are very rapid. The rate of transcription of each gene is assumed to be a Boolean logical variable—either on (a fixed transcription rate) or off. Control of gene expression is modeled by simple rules about which genes are activated or repressed by others. Such rules might state the following: Gene 1 is on only if gene 2 was on at the previous time step, and Gene 2 is on only if Gene 1 was off at the previous time step and Gene 3 was off. Boolean models can be efficiently constructed from large sets of gene expression data (Toh and Horimoto, 2002; Wen *et al.*, 1998).

With the continuous approach, differential equations are written to describe the rates of change of important mRNA and protein concentrations. These equations contain terms to represent processes such as macromolecular synthesis, degradation, and association of monomers into oligomers. Regulation of transcription is included in terms describing the synthesis of mRNA. As an example of a simple model that produces interesting behavior, consider the case of a single gene, *tf-a*, whose protein product is denoted TF-A. TF-A activates *tf-a* transcription, thereby forming a positive-feedback loop. It is assumed that two TF-A molecules must bind together, forming a dimer, which then binds to an enhancer sequence near the *tf-a* promoter (Fig. 14.9A). The corresponding differential equation for the rate of change of *tf-a* mRNA concentration is

$$\frac{d[tf\text{-}a\ \text{mRNA}]}{dt} = \frac{k_{max}[\text{TF-A}]^2}{[\text{TF-A}]^2 + K_d} - k_{degR}[tf\text{-}a\ \text{mRNA}] + R_{bas} \quad (14.20)$$

In Eq. 14.20, the first term on the right-hand side gives the transcription rate of *tf-a* as a saturable function of TF-A concentration. The second term accounts for first-order degradation of *tf-a* mRNA. The third term accounts for a low rate of transcription, R_{bas}, in the absence of activation of the *tf-a* gene by TF-A. At high [TF-A] the transcription rate approaches a maximal value k_{max}. TF-A dimers regulate transcription, and in Eq. 14.20, the square of TF-A concentration is used to roughly approximate the concentration of TF-A dimers. Justifying this approximation relies on the kinetic relationship that the rate of formation of TF-A dimers is proportional to the square of TF-A monomer concentration (for further detail, see Smolen

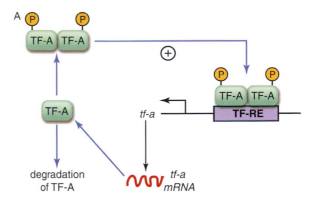

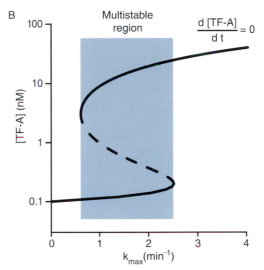

FIGURE 14.9 Positive autoregulation of a single gene can generate bistability. (A) Schematic of the model. Phosphorylated dimers of TF-A activate *tf-a* transcription when bound to a response element, termed a *TF-RE*, near the TF-A promoter. Degradation of TF-A protein is also indicated. (B) Bistability in the model of (A). For $0.6\ \text{min}^{-1} < k_{max} < 2.5\ \text{min}^{-1}$ (shaded region in figure), two stable steady-state solutions of [TF-A] exist (lower and upper portions of the steady-state curve along which $\frac{d[\text{TF-A}]}{dt} = 0$). There is an unstable steady solution between (dashed) the two stable steady states. Outside the brown region, there is only a single stable solution.

et al., 2000). As discussed previously (Eq. 14.17b), activity coefficients will also affect the dimer formation rate.

Along with Eq. 14.20, a second differential equation is needed for TF-A protein concentration. If TF-A protein was immediately translated from mRNA, a simple equation might be assumed, with two linear terms for TF-A synthesis and degradation:

$$\frac{d[\text{TF-A}]}{dt} = k_{2,f}[tf\text{-}a\ \text{mRNA}] - k_{2,d}[\text{TF-A}] \quad (14.21)$$

The relatively simple model of Eqs. 14.20–14.21 can generate complex behavior (Smolen *et al.*, 1998).

Figure 14.9B is a bifurcation diagram displaying how the dynamics of [TF-A] change as the parameter k_{max} is varied. With appropriate parameter values, the model is bistable. The curve traces the stable solutions of [TF-A] and of the *tf-a* transcription rate. Along this curve, the derivatives in Eqs. 14.20 and 14.21 are zero. At low values of k_{max} there is only one stable solution. Stimulus-induced phosphorylation of TF-A could make TF-A more effective at activating transcription, thereby increasing k_{max}. When k_{max} is raised into the shaded region in Fig. 14.9B, two stable steady states exist (lower and upper portions of curve) with an unstable solution between (middle, dashed portion of curve) the two stable steady states. For even higher values of k_{max} there is again a single stable solution.

In the bistable region, in the lower steady state, there is virtually no activation of the positive-feedback loop in which *tf-a* activates its own transcription, because [TF-A] is very low. A small degradation rate, proportional to [TF-A], is balanced by a small basal rate of *tf-a* transcription. If [TF-A] is increased, positive feedback occurs, and TF-A dimers activate *tf-a* transcription in a regenerative process. The rate of transcription and TF-A protein synthesis is then faster than TF-A degradation. At a high concentration of TF-A protein, the positive feedback saturates as the TF-RE enhancer becomes fully occupied with TF-A dimers. A maximal transcription rate is approached. But the TF-A degradation rate still increases with [TF-A] because it is directly proportional to [TF-A] in Eq. 14.21. Therefore, at a high value of [TF-A], the degradation rate of TF-A must "catch up" with the synthesis rate. The second steady state is thereby created. If the model is initially at either stable solution, it will return to it after a small induced change in [TF-A]. However, a large change in [TF-A] can switch the system between steady states. An example might be a stimulus-induced phosphorylation of TF-A that causes a large but temporary change in k_{max}. After k_{max} returns to its original value, [TF-A] would remain at the new value.

In comparing methods of modeling gene networks, the continuous method using differential equations is often preferred over the Boolean method. Greater physical accuracy is inherent in the use of continuous, rather than on–off, gene expression rates. Furthermore, oscillations in gene expression seen with a Boolean model may disappear when a more accurate, differential equation-based model is used (Bagley and Glass, 1996). Steady states may also disappear (Glass and Kaufman, 1973). Finally, bifurcation analysis (Izhikevich, 2007; Song *et al.*, 2007; Wiggins, 1990) can be applied to differential equation models, allowing visualization of how dynamics vary as a function of parameter values. However, the continuous method is more computationally intensive than the Boolean method, because the continuous method requires small time steps in simulations. For this reason, and because of the impracticality of thoroughly characterizing the expression kinetics of many genes, the Boolean method is sometimes regarded as reasonable for large gene networks. Recently, the Boolean method was used to reinforce and generalize the postulate that positive and negative biochemical feedback loops lead, respectively, to multistable or oscillatory dynamics (Kwon and Choh, 2007).

In the Boolean and continuous methods, intracellular transport of mRNA or protein can be modeled only in an *ad hoc* fashion. Time delays can be used to represent the time between synthesis of a macromolecule in one intracellular compartment and its arrival and function in another compartment. For active transport of macromolecules, this may be a reasonable approach (Smolen *et al.*, 1999). For diffusive transport, more accurate modeling requires discretizing the intracellular space into small volume elements, as discussed earlier for Ca^{2+} diffusion. It may be important to model transport, particularly in extended cells such as neurons. In neurons, delays of hours can be associated with transport of protein or mRNA from the soma to synapses.

Simulations of the effects of macromolecular transport on gene expression regulation have found large differences in gene expression dynamics depending upon whether transport from nucleus to cytoplasm and vice versa is assumed to be active or diffusive (Busenberg and Mahaffy, 1985; Smolen *et al.*, 1999). If active transport is modeled by a fixed time delay for transit, then longer delays tend to destabilize steady states of gene expression rates and favor oscillations. By contrast, in diffusion-dominated systems, slow diffusion tends to damp oscillations and yield steady states. These results suggest that dynamics of neuronal gene expression and synaptic modification may depend critically on the mode of macromolecular transport.

GENE NETWORK MODELS SUGGEST THAT FEEDBACK LOOPS AND PROTEIN DIMERIZATION CAN GENERATE COMPLEX DYNAMICS

Transcription factors commonly bind to their target sequences as dimers. Dimerization steepens the relationship between the level of TF and the strength of the regulatory effect. Steepening this relationship favors complex dynamics, such as multiple stable gene expression rates or oscillations in rates. Complex dynamics are also favored if the regulated gene lies within a positive or negative feedback loop in which a gene activates or represses its own expression. Keller (1995) analyzes four models that include feedback loops and dimerization of TFs. These models exhibit multiple stable gene expression rates. In each model, if dimerization does not occur, only a single steady state is obtained. However, dimerization of TFs is not an absolute requirement for multistability. Keller (1995) also gives conditions for multistability in two models with monomers. In one model, two TFs each repress the other's transcription, and in the second, a TF activates its transcription by binding to multiple enhancers.

Positive feedback is usually essential for multistability of gene expression rates (Thomas *et al.*, 1995). For example, both positive feedback and TF oligomerization are essential for bistability in a model of neural tissue development in the *Drosophila* embryo (Kerszberg and Changeux, 1998). This model simulates the development of a group of undifferentiated cells into a structure of neurons that resembles an embryonic neural tube. Bistability in gene expression rates is essential for simulating the development of well-defined, smooth boundaries between neural and nonneural cells. In the simulation proneural genes are on within a spatially restricted region, and off outside that region. A key element of this model is a switch in which bistability of gene expression rates is created by competition between homodimers of two TFs. Each homodimer activates expression of its own gene while repressing the gene for the other TF. Figure 14.10

illustrates this competition. This scheme constitutes a positive-feedback loop. An increase in the level of one TF, by repressing the formation of the second TF, favors a further increase in the first TF. In each stable solution, the gene for one TF is strongly activated by its own product, which simultaneously represses the other gene. If TF-1 levels are high in a specific cell, that cell will differentiate into a neuron. If TF-2 levels are high, then *tf-1* expression is repressed and the cell will not become a neuron. Specific *Drosophila* TFs were suggested for TF-1 and TF-2 (Kerszberg and Changeux, 1998).

In the complete model, protein–protein interactions between receptors on the membranes of adjacent cells were also included so that the development of each cell would be influenced by the state of neighboring cells. For each cell, the state of the neighboring cells influences the transcription of the TFs in Figure 14.10. Therefore, for each cell, the choice of whether to differentiate into a neuron or not depends on the state of the neighboring cell. These intercellular interactions are critical for forming boundaries between regions of neural and nonneural tissue (Kerszberg and Changeux, 1998).

Steady states of gene expression rates could constitute information that could be preserved through cycles of cell division (Keller, 1994). Multistability in gene networks has been hypothesized to be essential for cell differentiation (Laurent and Kellershohn, 1999), although one recent study argues against this (Chang *et al.*, 2006). Because multistability is a manifestation of a positive-feedback loop, it has been suggested that genes involved in such loops may regulate differentiation (Thomas *et al.*, 1995). A recent study suggests that in gene networks, the number of steady states is mainly determined by the number of self-regulatory genes, rather than by the total number of genes or interactions between genes (Mochizuki, 2005).

In model gene networks, negative feedback is generally necessary to sustain oscillations in gene expression rates (Smolen *et al.*, 2000). Many models describing oscillations also rely upon dimerization of TFs (Goldbeter, 1996; Smolen *et al.*, 2000). A simple model with negative feedback can be constructed by adding a second TF to the model of Figure 14.9A. As diagrammed in Figure 14.11A, this second TF, termed *TF-R*, represses transcription of both its own gene and that for TF-A. In turn, TF-A activates the transcription of both *tf-a* and *tf-r*. The differential equations corresponding to this model consist of (1) Eqs. 14.20 and 14.21 modified to include repression of *tf-a* transcription by TF-R, and (2) two differential equations that describe the rates of change of

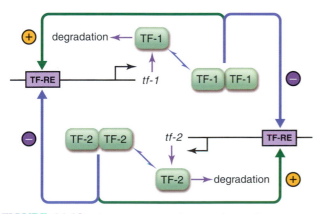

FIGURE 14.10 A two-gene regulatory scheme that generates bistability. The transcription factor TF-A activates expression of its own gene and represses the TF-B gene, whereas TF-B activates its own gene and represses the TF-A gene. Both TFs compete as homodimers for binding to response elements.

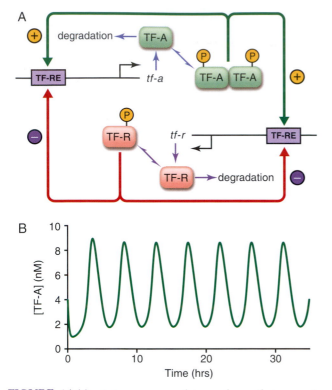

FIGURE 14.11 A two-gene regulatory scheme that generates oscillations. (A) Schematic of the model. A second transcription factor, TF-R, is added to the model of Figure 14.9A. TF-A activates transcription by binding to TF-RE response elements, which are present near the promoter regions of both *tf-a* and *tf-r*. TF-R represses transcription by competing with the TF-A dimer for binding to both TF-REs. (B) Sustained oscillations of [TF-A] produced by the model of (A).

[*tf-r* mRNA] and [TF-R] (Smolen *et al.*, 1999). The modified version of Eq. 14.20 is

$$\frac{d[tf\text{-}a\ \text{mRNA}]}{dt} = \left(\frac{k_{max}[\text{TF-A}]^2}{[\text{TF-A}]^2 + K_d}\right)\left(\frac{K_R}{[\text{TF-R}]^2 + K_R}\right)$$
$$-\ k_{degR}[tf\text{-}a\ \text{mRNA}] + R_{bas}$$

$$(14.22)$$

The model of Figure 14.11A readily generates stable oscillations in mRNA and protein levels (Fig. 14.11B).

An important synergistic application of modeling and experiment is construction of simple genetic systems that generate bistability or oscillations, combined with modeling based on equations similar to Eqs. 14.20–14.22. In one example a bistable switch was constructed based on repression by two genes of each other's transcription (Gardner *et al.*, 2000). Modeling predicted perturbations that would flip the switch

between states. In a second example a network of three genes was constructed (Elowitz and Leibler, 2000). Gene 1 repressed Gene 2, which repressed Gene 3, which repressed Gene 1. This network constitutes a negative-feedback loop because expression of Gene 1 ultimately leads to repression of Gene 1. This system generated oscillations in transcription rates. Modeling predicted conditions capable of sustaining oscillations.

Simple, designed genetic systems similar to the bistable switch or the three-gene network might prove useful for gene therapy. For example, transfection with several genes might confer novel drug responsiveness on target cells. Correction of a polygenetic disorder, in which several genes contribute to the observed phenotype, might also require transfection with multiple genes. Several promising methods for gene delivery to the brain are being developed, including viral transfection and cerebral injection of neural stem cells (Hsich *et al.*, 2002). Models are likely to be essential for predicting and understanding the responses of multiple transfected genes, or even a single gene, *in vivo*. Models have already helped to design gene vectors with maximal transfection efficiency (Varga *et al.*, 2001).

Positive and negative feedback loops within the gene network involving CREB and related TFs (Fig. 14.8) might support complex dynamics such as oscillations in transcription rates (Song *et al.*, 2007). Oscillations of CREB mRNA have been studied in endocrine cells (Walker *et al.*, 1995), and these feedback loops were suggested to be responsible.

Gene networks with negative feedback appear essential to drive *circadian rhythms*, which have evolved in most organisms exposed to daylight (Dunlap *et al.*, 1999; Goldbeter, 1996; Reppert and Weaver, 2001). These behavioral rhythms are self-sustaining in constant darkness. In circadian oscillators, the products of one or a few core genes repress their own transcription in a cyclic fashion. Following the onset of repression, there is a delay of several hours, after which repressing proteins are degraded, relieving repression (Dunlap *et al.*, 1999). For example, in *Drosophila* neurons that drive circadian behavior, a transcription factor termed *PER* represses its own gene. Over many hours, PER becomes multiply phosphorylated and then degrades. Degradation relieves repression of *per* by PER, and a surge of *per* transcription initiates the next cycle. The multiple phosphorylations of PER prior to degradation introduce a time delay, which appears to be important for generating the ~24-hour oscillation period. Positive-feedback regulation of *per* and other circadian genes also exists and shapes the oscillations

(Bae *et al.*, 2000). In mammals, several isoforms of PER exist, and PER proteins in concert with others mediate negative and positive regulation of core gene transcription (Shearman *et al.*, 2000). Multiple phosphorylations of mammalian PER occur before degradation and may be important for the 24-hour period (Lee *et al.*, 2001).

Models based on differential equations have been developed to represent circadian rhythms and their responses to stimuli. The organism most commonly modeled is *Drosophila* (Leloup and Goldbeter, 1998; Smolen *et al.*, 2004). Models for the generation of circadian rhythms in particular species are complex because of multiple genes and time delays. A simpler model is presented here, capturing essential processes common to many species. A TF is assumed to repress its own gene. Newly synthesized TF protein must be multiply phosphorylated prior to degradation. Thus there is a period of approximately 1 day during which TF protein is phosphorylated and degraded. Degradation allows initiation of another burst of *tf* transcription, beginning another oscillation. A time delay τ is included between transcription of *tf* mRNA and subsequent translation of protein. The model is schematized in Figure 14.12A.

This model simulates circadian oscillations as illustrated in Figure 14.12B, which displays time courses of mRNA and of total protein concentration. Oscillations are seen only if multiple (~8 or more) TF phosphorylations occur prior to degradation. This observation is consistent with experiments in *Drosophila* and other organisms (Dunlap *et al.*, 1999). The necessity for multiple phosphorylations follows from a general result that increasing the number of kinetic steps in a negative-feedback loop promotes oscillations (Griffith, 1968). Inclusion of the delay τ (~1 hr), is not necessary for oscillations but tends to increase their amplitude and stability.

In the model equations, the symbols P_0, P_1, ..., P_N denote a TF with 0, 1, ..., N phosphorylations (N = 8 − 10). Only P_N is degraded. k_{ph} is a first-order rate constant for the rate of phosphorylation. $[P_{tot}]$ is the total concentration of protein, $[P_{tot}] = [P_0] + [P_1] + ... + [P_N]$. The differential equation for TF mRNA has a synthesis term containing repression by TF protein ($[P_{tot}]$ in the denominator) with a Hill coefficient of n; n is typically 2–4. First-order degradation of mRNA is assumed

$$\frac{d[mRNA]}{dt} = v_R \frac{K_R^n}{[P_{tot}]^n + K_R^n} - k_d[mRNA]$$

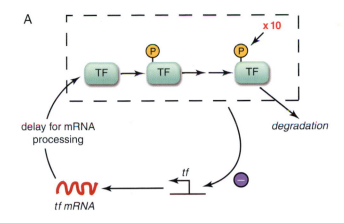

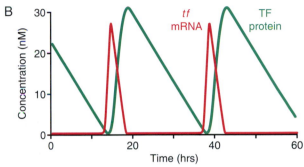

FIGURE 14.12 Generation of circadian rhythmicity by negative transcriptional feedback. (A) The kinetic scheme. A transcription factor (TF) can undergo multiple phosphorylation steps. Ten sequential phosphorylations are assumed. As indicated by the dashed box, all forms of TF protein are assumed capable of repressing *tf* transcription. A time delay is included between the appearance of *tf* mRNA and TF protein. Only fully phosphorylated TF can degrade, but it does so relatively rapidly, so that over the space of a day, virtually all TF protein becomes fully phosphorylated and then degrades. This relieves *tf* repression so that another "burst" of *tf* transcription can occur. (B) Circadian oscillations in the levels of *tf* mRNA and TF protein simulated by the model of (A).

Nonphosphorylated TF, P_0 is synthesized from mRNA after a delay τ (typically 1–2 hrs). P_0 disappears via phosphorylation to P_1

$$\frac{d[P_0]}{dt} = k_P[mRNA](t - \tau) - k_{ph}[P_0]$$

In simulations using this *delay differential equation*, the value of [mRNA] is stored at time t and used τ minutes later to compute the derivative of $[P_0]$.

A series of phosphorylations of P ensues until P_N is created. The reactions are assumed sequential so that phosphorylation i cannot occur until phosphorylation i − 1 has occurred:

$$\frac{d[P_i]}{dt} = -k_{ph}[P_i] + k_{ph}[P_{i-1}], \text{ for } i = 1, ... N-1$$

P_N is synthesized by phosphorylation and degraded in a Michaelis–Menten enzymatic reaction:

$$\frac{d[P_N]}{dt} = k_{ph}[P_{N-1}] - \frac{v_P[P_N]}{K_P + [P_N]}$$

RANDOM FLUCTUATIONS IN MOLECULE NUMBERS CAN STRONGLY INFLUENCE GENETIC REGULATION

As discussed previously, random variability in the times of individual macromolecular synthesis and degradation events is expected to generate stochastic fluctuations in the numbers of individual mRNAs and proteins. Recent observations indicate that the timing of transcription of individual mRNAs is indeed random (Elowitz et al., 2001; Zlokarnik et al., 1998). Also, for important mRNA or protein species, the average copy number present in the nucleus may be < 100, so that random fluctuations in copy number may be relatively large and could strongly influence the dynamics of gene expression. In neurons, late long-term potentiation of synapses lasting 24 hrs or more is dependent on transcription and translation. Therefore, stochastic fluctuations in mRNA and protein numbers may induce considerable synapse-to-synapse variability in the amount of potentiation. Such fluctuations have an intrinsic component (generated by the reactions in the network) and an extrinsic component (generated by fluctuations in other cellular processes that impinge on the network), and modeling can quantify the contributions of these components (Pedraza and van Oudenaarden, 2005).

An example of how fluctuations can destabilize steady states is shown in Figure 14.13. The simulations for this illustration used a set of four differential equations, similar to Eqs. 14.21 and 14.22, describing reciprocal regulation of two genes. The regulatory interactions were those of Figure 14.10. Figure 14.13A shows that without fluctuations, the model exhibits multistability. Initially, the concentrations of both tf-1 and tf-2 mRNA and protein are in a low, stable steady state. A large temporary increase in the basal transcription rate of tf-1 is needed to drive the illustrated transition to an upper stable steady state. Here, the concentrations of tf-1 mRNA and protein are high, whereas those of tf-2 mRNA and protein (not shown) are low.

Stochastic fluctuations in molecule numbers were then simulated with the Gillespie method. Now, when the numbers of each molecular species were initialized at values corresponding to the lower steady state of Figure 14.13A, the state was no longer

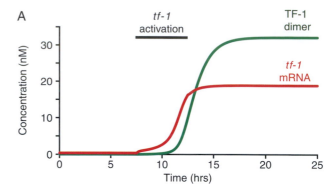

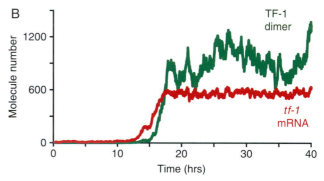

FIGURE 14.13 Stochastic fluctuations in molecule numbers may destabilize steady states of genetic regulatory systems. (A) Without fluctuations, a set of equations based on Figure 14.10A exhibits bistability. Initial levels of tf-1 (and tf-2) mRNA and protein are low (< 2 nM) and steady. At $t = 7.5$ hr the rate of tf-1 transcription is increased by a constant amount for 10 hrs (horizontal bar). The increase causes a transition to a stable state with high tf-1 mRNA and protein levels. The high levels persist after the imposed increase in transcription rate ceases. (B) The initial state of (A) is spontaneously destabilized within ~ 20 hrs when stochastic fluctuations are incorporated. No induced increase in tf-1 transcription is included in this simulation, which is otherwise the same as in (A) except for a scaling factor converting the units of the variables from concentrations to molecule numbers.

stable. After several hours, random fluctuations in molecule numbers would cross a threshold, initiating a positive-feedback loop in which either tf-1 or tf-2 would activate its own transcription. Figure 14.13B illustrates that the system would leave the initial state and settle in a new steady state, without an outside perturbation. In the new state, tf-1 transcription was high and tf-2 transcription low. Generally, stochastic fluctuations act to prevent gene networks from settling in states that are weakly stable.

Fluctuations commonly tend to destabilize periodic oscillations of gene expression (Barkai and Liebler, 2000). However, fluctuations can also create or enhance oscillations (Song et al., 2007; Vilar et al., 2002). This phenomenon is termed *stochastic resonance*.

SUMMARY

Mathematical models are essential for synthesizing large amounts of data concerning intracellular signaling pathway dynamics and regulation of gene expression. Standard modeling techniques have been developed to represent common biochemical elements such as allosteric interactions, feedback interactions in biochemical pathways and gene networks, and intracellular transport of molecules. Modeling of biochemical signaling pathways and of gene networks has suggested a unifying principle: that positive and negative feedback interactions commonly underlie complex dynamics of signaling pathways and gene expression. Such dynamics can include multiple stable states, or oscillations, of concentrations and reaction rates. Models are also necessary to assess the strength and significance of cross talk between signaling pathways.

For extended cells such as neurons, it may be important that models of intracellular signaling include transport of macromolecules or metabolites. Modeling random fluctuations in the numbers of macromolecules is important if the molecule number per cell is low (several hundred or less). Understanding variability in stimulus responses may require consideration of fluctuations in critical mRNAs and proteins. However, sufficient data to constrain models of fluctuations is not available for most signaling pathways or gene networks.

A combination of theory (metabolic control analysis) and experiment has helped determine principles of metabolic flux control in biochemical pathways. For example, the metabolic flux through a pathway is least likely to be sensitive to changes in the activity of enzymes subject to feedback regulation. Use of metabolic control analysis to interpret experimental data can help determine the rate-limiting reactions for the production of metabolites.

Regulation of gene expression can be modeled at a coarse level, in which genes are on–off switches, or at a finer level, in which gene expression rates are continuous variables. The finer level, using differential equations, is more accurate, but due to limited data for some gene networks, the on–off representation may sometimes be appropriate. Feedback gene regulation can generate complex dynamics such as multistability or oscillations. For further accuracy in models of small gene networks, random fluctuations in macromolecule numbers can be included. However, sufficient data to constrain such models is commonly not available.

For all models, it is essential to determine the sensitivity of model behavior to parameters and to assess which parameters are not well constrained by data. Experiments determining values of key control parameters for model behavior constitute stringent tests of a model. Uncertainties in parameters imply that biochemical models are most often qualitative rather than quantitative descriptions of *in vivo* systems. Reduction of the number of model equations to the minimum necessary to simulate experimental data is desirable in order to help determine the biochemical elements most important for system behavior. Separation of variables according to whether they vary on a fast or slow timescale can help with this simplification.

With the biochemical and gene expression data now becoming available, efficient characterization of coupled genetic and biochemical systems requires collaboration between biochemists, molecular biologists, and mathematical biologists. These collaborations will fit data into a mathematical framework and design experiments to test model predictions. This methodology will be necessary to predict and analyze the responses of neurons and organisms to physiological stimuli, environmental contaminants, and novel pharmaceutical agents.

References

General

Cornish-Bowden, A. (2004). *Fundamentals of Enzyme Kinetics*. Portland Press Limited, Portland.

Ellner, S. P., and Guckenheimer, J. (2006). *Dynamic Models in Biology*. Princeton University Press, Princeton.

Goldbeter, A. (1996). *Biochemical Oscillations and Cellular Rhythms*. Cambridge University Press, Cambridge.

Koch, C., and Segev, I. (1998). *Methods in Neuronal Modeling: From Synapses to Networks (2nd edition)*. MIT Press, Cambridge.

Murray, J. D. (2003). *Mathematical Biology (vols. I, II)*. Springer-Verlag, New York.

Smolen, P., Baxter, D. A., and Byrne, J. H. (2000). Mathematical modeling of gene networks. *Neuron* **26**: 567-580.

Specific

Abraham, W. C., Logan, B., Greenwood, J. M., and Dragunow, M. (2002). Induction and experience-dependent consolidation of stable long-term potentiation lasting months in the hippocampus. *J. Neurosci.* **22**: 9626-9634.

Amini, B., Clark, J. W., and Canavier, C. C. (1999). Calcium dynamics underlying pacemaker-like and burst firing oscillations in midbrain dopaminergic neurons, a computational study. *J. Neurophysiol.* **82**: 2249-2261.

Bae, K., Lee, C., Hardin, P. E., and Edery, I. (2000). dCLOCK is present in limiting amounts and likely mediates daily interactions between the dCLOCK-CYC transcription factor and the PER-TIM complex. *J. Neurosci.* **20**: 1746-1753.

Bagley, R. J., and Glass, L. (1996). Counting and classifying attractors in high dimensional dynamical systems. *J. Theor. Biol.* **183**: 269-284.

Bardwell, L., Zou, X., Nie, Q., and Komarova, N. L. (2007). Mathematical models of specificity in cell signaling. *Biophys. J.* **92**: 3425-3441.

Barkai, N., and Leibler, S. (2000). Circadian clocks limited by noise. *Nature* **403**: 267-268.

Beck, J. V., and Arnold, K. J. (1977). *Parameter Estimation in Engineering and Science*, pp. 17-24, 481-487. John Wiley, New York.

Berg, O. G., Paulsson, J., and Ehrenberg, M. (2000). Fluctuations and quality of control in biological cells: Zero-order ultrasensitivity reinvestigated. *Biophys. J.* **79**: 1228-1236.

Bertram, R., Satin, L., Zhang, M., Smolen, P., and Sherman, A. (2004). Calcium and glycolysis mediate multiple bursting modes in pancreatic islets. *Biophys. J.* **87**: 3074-3087.

Bhalla, U. S. (2002). Biochemical signaling networks decode temporal patterns of synaptic input. *J. Comp. Neurosci.* **13**: 49-62.

Bhalla, U. S., and Iyengar, R. (1999). Emergent properties of networks of biological signaling pathways. *Science* **283**: 381-386.

Bito, H., and Takemoto-Kimura, S. (2003). Ca^{2+}/CREB/CBP-dependent gene regulation: A shared mechanism critical in long-term synaptic plasticity and neuronal survival. *Cell Calcium* **34**: 425-430.

Bliss, T. V., Collingridge, G. L., Laroche, S. (2006). ZAP and ZIP, a story to forget. *Science* **313**: 1058-1059.

Blumenfeld, H., Zablow, L., and Sabatini, B. (1992). Evaluation of cellular mechanisms for modulation of Ca^{2+} transients using a mathematical model of fura-2 Ca^{2+} imaging in *Aplysia* sensory neurons. *Biophys. J.* **63**: 1146-1164.

Brown, G. P., Blitzer, R. D., Connor, J. H., Wong, T., Shenolikar, S., Iyengar, R., and Landau, E. M. (2000). Long-term potentiation induced by θ frequency stimulation is regulated by a protein phosphatase-1-operated gate. *J. Neurosci.* **20**: 7880-7887.

Burden, R. L., and Faires, J. D. (1993). *Numerical Analysis*. PWS-Kent Pub. Co., Boston.

Busenberg, S., and Mahaffy, J. M. (1985). Interaction of spatial diffusion and delays in models of genetic control by repression. *J. Math. Biology* **22**: 313-333.

Cao, D., and Parker, R. (2001). Computational modeling of eukaryotic mRNA turnover. *RNA* **7**: 1192-1212.

Chang, H. H., Oh, P. Y., Ingber, D. E., and Huang, S. (2006). Multistable and multistep dynamics in neutrophil differentiation. *BMC Cell Biol.* **7**: 11.

Changeux, J. P., and Edelstein, S. J. (2005). Allosteric mechanisms of signal transduction. *Science* **308**: 1424-1428.

Chen, H. X., Otmakhov, N., Strack, S., Colbran, R. J., and Lisman, J. E. (2001). Is persistent activity of calcium/calmodulin-dependent kinase required for the maintenance of LTP? *J. Neurophysiol.* **85**: 1368-1376.

Cooper, D., Mons, N., and Karpen, J. (1995). Adenylyl cyclases and the interaction between calcium and cAMP signaling. *Nature* **374**: 421-424.

Cornish-Bowden, A. (2004). *Fundamentals of Enzyme Kinetics*. Portland Press Limited, Portland.

Dano, S., Sorensen, P. G., and Hynne, F. (1999). Sustained oscillations in living cells. *Nature* **402**: 320-322.

Davis, R. J. (2000). Signal transduction by the JNK group of MAP kinases. *Cell* **103**: 239-252.

De Cesare, D., and Sassone-Corsi, P. (2000). Transcriptional regulation by cyclic AMP-responsive factors. *Prog. Nucleic Acid Res. Mol. Biol.* **64**: 343-369.

De Schutter, E., and Smolen, P. (1998). Calcium dynamics in large neuronal models. In: *Methods in Neuronal Modeling: from Synapses to Networks (2nd edition)*. Edited by Koch, C., and Segev, I. MIT Press, Cambridge.

Dunlap, J. C., Loros, J. J., Liu, Y., and Crosthwaite, S. K. (1999). Eukaryotic circadian systems: Cycles in common. *Genes Cells* **4**: 1-10.

Edwards, A. S., and Scott, J. D. (2000). A-kinase anchoring proteins: Protein kinase A and beyond. *Curr. Opin. Cell Biol.* **12**: 217-221.

Ellner, S. P., and Guckenheimer, J. (2006). *Dynamic Models in Biology*. Princeton University Press, Princeton.

Elowitz, M. B., and Leibler, S. (2000). A synthetic oscillatory network of transcriptional regulators. *Nature* **403**: 335-338.

Elowitz, M. B., Levine, A. J., Siggia, E. D., and Swain, P. S. (2001). Stochastic gene expression in a single cell. *Science* **297**: 1183-1186.

Ermentrout, B. (2001). Simplifying and reducing complex models. In: *Computational Modeling of Genetic and Biochemical Networks*. Edited by Bower, J., and Bolouri, H. MIT Press, Cambridge.

Firth, C. A., and Bray, D. (2001). Stochastic simulation of cell signaling pathways. In: *Computational Modeling of Genetic and Biochemical Networks*. Edited by Bower, J., and Bolouri, H. MIT Press, Cambridge.

Fokina, K. V., Dainyak, M. B., Nagradova, N. K., and Muronetz, V. I. (1997). A study on the complexes between human erythrocyte enzymes participating in the conversions of 1,3-diphosphoglycerate. *Arch. Biochem. Biophys.* **345**: 185-192.

Frank, P. M. (1978). *Introduction to System Sensitivity Theory*, pp. 9-10. Academic Press, New York.

Frey, U., and Morris, R. G. (1997). Synaptic tagging and long-term potentiation. *Nature* **385**: 533-536.

Gabso, M. G., Neher, E., and Spira, M. E. (1997). Low mobility of the Ca^{2+} buffers in axons of cultured *Aplysia* neurons. *Neuron* **18**: 473-481.

Gardner, T. S., Cantor, C. R., and Collins, J. J. (2000). Construction of a genetic toggle switch in *Escherichia coli*. *Nature* **403**: 339-342.

Gartner, K. (1990). A third component causing variability beside environment and genotype. A reason for the limited success of a 30-year long effort to standardize laboratory animals? *Lab. Anim.* **24**: 71-77.

Gerhold, D., Rushmore, T., and Caskey, C. T. (1999). DNA chips: Promising toys have become powerful tools. *Trends Bioch. Sci.* **24**: 168-173.

Giese, K. P., Fedorov, N. B., Filipkowski, R. K., and Silva, A. J. (1998). Autophosphorylation at Thr(286) of the alpha calcium-calmodulin kinase II in learning and memory. *Science* **279**: 870-873.

Gillespie, D. T. (1977). Exact stochastic simulation of coupled chemical reactions. *J. Phys. Chem.* **61**: 2340-2361.

Glass, L., and Kauffman, S. A. (1973). The logical analysis of continuous, non-linear biochemical control networks. *J. Theor. Biol.* **39**: 103-129.

Goldbeter, A. (1996). *Biochemical Oscillations and Cellular Rhythms*. Cambridge University Press, Cambridge.

Goldbeter, A., and Koshland, D. E. (1984). Ultrasensitivity in biochemical systems controlled by covalent modification: Interplay between zero-order and multistep effects. *J. Biol. Chem.* **259**: 14441-14447.

Griffith, J. S. (1968). Mathematics of cellular control processes. I. Negative feedback to one gene. *J. Theor. Biol.* **20**: 202-208.

Guntenkunst, R. N., Waterfall, J. J., Casey, F. P., Brown, K. S., Myers, C. R., and Sethna, J. P. (2007). Universally sloppy parameter sensitivities in systems biology models. *PLoS Comput. Biol.* **3**: e189.

Hasty, J., McMillen, D., Isaacs, F., and Collins, J. J. (2001). Computational studies of gene regulatory networks: In numero molecular biology. *Nat. Rev. Genet.* **2**: 268-279.

Hodgkin, A. L., and Huxley, A. F. (1952). A quantitative description of membrane current and its application to conduction and excitation in nerve. *J. Physiol. (Lond.)* **117**: 500-544.

Holmes, W. R. (2000). Models of calmodulin trapping and CaM kinase II activation in a dendritic spine. *J. Comp. Neurosci.* **8**: 65-85.

Hornberg, J. J., Bruggeman, F. J., Bakker, B. M., and Westerhoff, H. V. (2007). Metabolic control analysis to identify optimal drug targets. *Prog. Drug Res.* **64**: 173-189.

Hsich, G., Sena-Esteves, M., and Breakefield, X. O. (2002). Critical issues in gene therapy for neurologic disease. *Human Gene Ther.* **13**: 579-604.

Huang, C. F., and Ferrell, J. E. (1996). Ultrasensitivity in the mitogen-activated protein kinase cascade. *Proc. Natl. Acad. Sci. USA* **93**: 10078-10083.

Husi, H., Ward, M. A., Choudhary, J. S., Blackstock, W. P., and Grant, S. G. (2000). Proteomic analysis of NMDA receptor-adhesion protein signaling complexes. *Nat. Neurosci.* **3**: 661-669.

Ihmels, J., Friedlander, G., Bergmann, S., Sarig, O., Ziv, Y., and Barkai, N. (2002). Revealing modular organization in the yeast transcriptional network. *Nature Genet.* **31**: 370-377.

Imoto, S., Tamada, Y., Savoie, C. J., and Miyano, S. (2006). Analysis of gene networks for drug target discovery and validation. *Methods Mol. Biol.* **360**: 33-56.

Imtiaz, M. S., Katnik, C. P., Smith, D. W., and van Helden, D. F. (2006). Role of voltage-dependent modulation of store Ca^{2+} release in synchronization of Ca^{2+} oscillations. *Biophys. J.* **90**: 1-23.

Izhikevich, E. M. (2007). *Dynamical Systems in Neuroscience: The Geometry of Excitability and Bursting*. MIT Press, Cambridge.

Jafri, M. S., and Keizer, J. (1995). On the roles of Ca^{2+} diffusion, Ca^{2+} buffers, and the endoplasmic reticulum in IP3-induced Ca^{2+} waves. *Biophys. J.* **69**: 2139-2153.

Katz, P. S., and Clemens, S. (2001). Biochemical networks in nervous systems: Expanding neuronal information capacity beyond voltage signals. *Trends Neurosci.* **24**: 18-25.

Keller, A. (1994). Specifying epigenetic states with autoregulatory transcription factors. *J. Theor. Biol.* **170**: 175-181.

Keller, A. (1995). Model genetic circuits encoding autoregulatory transcription factors. *J. Theor. Biol.* **172**: 169-185.

Kerszberg, M., and Changeux, J. (1998). A simple molecular model of neurulation. *Bioessays* **20**: 758-770.

Kewley, R. J., Whitelaw, M. L., and Chapman-Smith, A. (2004). The mammalian basic helix-loop-helix family of transcriptional regulators. *Int. J. Biochem. Cell. Biol.* **36**: 189-204.

Kholodenko, B. N., Brown, G. C., and Hoek, J. B. (2000). Diffusion control of protein phosphorylation in signal transduction pathways. *Biochem. J.* **350**: 901-907.

Kurganov, B. I. (1986). The role of multienzyme complexes in integration of intracellular metabolism. *J. Theor. Biol.* **119**: 445-455.

Kwon, Y., and Cho, K. (2007). Boolean dynamics of biological networks with multiple coupled feedback loops. *Biophys. J.*, **92**: 2975-2981.

Laurent, M., and Kellershohn, N. (1999). Multistability: A major means of differentiation and evolution in biological systems. *Trends Bioch. Sci.* **24**: 418-422.

Lee, C., Etchegaray, J., Cagampang, F., Loudon, A. S., and Reppert, S. M. (2001). Posttranslational mechanisms regulate the mammalian circadian clock. *Cell* **107**: 855-867.

Leloup, J. C., and Goldbeter, A. (1998). A model for circadian rhythms in *Drosophila* incorporating the formation of a complex between the PER and TIM proteins. *J. Biol. Rhythms* **13**: 70-87.

Lengyel, I., Voss, K., Cammarota, M., Bradshaw, K., Brent, V., Murphy, K. P., Giese, K. P., Rostas, J. P., and Bliss, T. V. P. (2004). Autonomous activity of CaMKII is only transiently increased following the induction of long-term potentation in the rat hippocampus. *Eur. J. Neurosci.* **20**: 3063-3072.

Levchenko, A., Bruck, J., and Sternberg, P. W. (2000). Scaffold proteins may biphasically affect the levels of mitogen-activated protein kinase signaling and reduce its threshold properties. *Proc. Natl. Acad. Sci. USA* **97**: 5818-5823.

Lisman, J. E., and Zhabotinsky, A. M. (2001). A model of synaptic memory: A CaMKII / PP1 switch that potentiates transmission by organizing an AMPA receptor anchoring assembly. *Neuron* **31**: 191-201.

Little, J. W., Shepley, D. P., and Wert, D. W. (1999). Robustness of a gene regulatory circuit. *EMBO J.* **18**: 4299-4307.

Lonze, B. E., and Ginty, D. D. (2002). Function and regulation of CREB family transcription factors in the nervous system. *Neuron.* **35**: 605-623.

Marder, E. (2000). Models identify hidden assumptions. *Nature Neurosci. Suppl.* **3**: 1198.

Martin, S. J., and Morris, R. G. (2002). New life in an old idea: The synaptic plasticity and memory hypothesis revisited. *Hippocampus* **12**: 609-636.

Mayford, M., and Kandel, E. R. (1999). Genetic approaches to memory storage. *Trends Genet.* **15**: 463-470.

McAdams, H., and Arkin, A. (1998). Simulation of prokaryotic genetic circuits. *Ann. Rev. Biophys. Biom. Struct.* **27**: 199-224.

McAdams, H., and Arkin, A. (1999). It's a noisy business! Genetic regulation at the nanomolar scale. *Trends Genet.* **15**: 65-69.

Miller, P., Zhabotinsky, A. M., Lisman, J. E., and Wang, X. J. (2005). The stability of a stochastic CaMKII switch: Dependence on the number of enzyme molecules and protein turnover. *PLoS Biol.* **3**: e107.

Minton, A. P. (2001). The influence of macromolecular crowding and macromolecular confinement on biochemical reactions in physiological media. *J. Biol. Chem.* **276**: 10577-10580.

Mioduszewska, B., Jaworski, J., and Kaczmarek, L. (2003). Inducible cAMP early repressor (ICER) in the nervous system: A transcriptional regulator of neuronal plasticity and programmed cell death. *J. Neurochem.* **87**: 1313-1320.

Mirnics, K., Middleton, F. A., Marquez, A., Lewis, D. A., and Levitt, P. (2000). Molecular characterization of schizophrenia viewed by microarray analysis of gene expression in prefrontal cortex. *Neuron* **28**: 53-67.

Mochizuki, A. (2005). An analytical study of the number of steady states in gene regulatory networks. *J. Theor. Biol.* **236**: 291-310.

Mohamed, H. A., Yao, W., Fioravante, D., Smolen, P., and Byrne, J. H. (2005). cAMP response elements in *Aplysia creb1, creb2*, and *Ap-uch* promoters: implications for feedback loops modulating long-term memory. *J. Biol. Chem.* **280**: 27035-27043.

Morton-Firth, C. J., Shimizu, T. S., and Bray, D. (1999). A free-energy-based stochastic simulation of the Tar receptor complex. *J. Mol. Biol.* **286**: 1059-1074.

Muller, A., Kukley, M., Stausberg, P., Beck, H., Muller, W., and Dietrich, D. (2005). Endogenous Ca^{2+} buffer concentrations and Ca^{2+} microdomains in hippocampal neurons. *J. Neurosci.* **25**: 558-565.

Nakamura, T., Barbara, J., Nakamura, K., and Ross, W. N. (1999). Synergistic release of Ca^{2+} from IP$_3$-sensitive stores evoked by synaptic activation of mGluRs paired with backpropagating action potentials. *Neuron* **24**: 727-737.

Paulsson, J., Berg, O. G., and Ehrenberg, M. (2000). Stochastic focusing: Fluctuation-enhanced sensitivity of intracellular regulation. *Proc. Natl. Acad. Sci. USA* **97**: 7148-7153.

Pedraza, J. M., and van Oudenaarden, A. (2005). Noise propagation in gene networks. *Science* **307**: 1965-1969.

Prentki, M., Tornheim, K., and Corkey, B. E. (1997). Signal transduction mechanisms in nutrient-induced insulin secretion. *Diabetologia* **40** Suppl 2: S32-S34.

Press, W. H. (1994). *Numerical Recipes in C: The Art of Scientific Computing*. Cambridge University Press, Cambridge.

Qian, N., and Sejnowski, T. (1990). When is an inhibitory synapse effective? *Proc. Natl. Acad. Sci. USA* **87**: 8145-8149.

Ramoni, M. F., Sebastiani, P., and Kohane, I. S. (2002). Cluster analysis of gene expression dynamics. *Proc. Natl. Acad. Sci. USA* **99**: 9121-9126.

Reppert, S. M., and Weaver, D. R. (2001). Molecular analysis of mammalian circadian rhythms. *Ann. Rev. Physiol.* **63**: 647-676.

Reymann, K. G., and Frey, J. U. (2007). The late maintenance of hippocampal LTP: Requirements, phases, "synaptic tagging," "late-associativity," and implications. *Neuropharmacology* **52**: 24-40.

Roberts, C. J., Nelson, B., Marton, M. J., Stoughton, R., Meyer, M. R., Bennett, H. A., He, Y. D., Dai, H., Walker, W. L., Hughes, T. R., Tyers, M., Boone, C., and Friend, S. H. (2000). Signaling and circuitry of multiple MAPK pathways revealed by a matrix of global gene expression profiles. *Science* **287**: 873-880.

Rohwer, J. M., Postma, P. W., Kholodenko, B. N., and Westerhoff, H. V. (1998). Implications of macromolecular crowding for signal transduction and metabolite channeling. *Proc. Natl. Acad. Sci. USA* **95**: 10547-10552.

Ryu, D. D., and Nam, D. H. (2000). Recent progress in biomolecular engineering. *Biotechnol. Prog.* **16**: 2-16.

Sabry, J., O'Connor, T., and Kirschner, M. W. (1995). Axonal transport of tubulin in Ti1 Pioneer neurons in situ. *Neuron* **14**: 1247-1256.

Salter, M. (1996). Determination of the flux control coefficient of nitric oxide synthase for nitric oxide synthesis in discrete brain regions *in vivo*. *J. Theor. Biol.* **182**: 449-452.

Shahrezaei, V., and Delaney, K. R. (2005). Brevity of the Ca^{2+} microdomain and active zone geometry prevent Ca^{2+}-sensor saturation for neurotransmitter release. *J. Neurophysiol.* **94**: 1912-1919.

Shearman, L. P., Sriram, S., Weaver, D. R., Maywood, E. S., Chaves, I., Zheng, B., Kume, K., Lee, C. C., van der Horst, G., Hastings, M. H., and Reppert, S. M. (2000). Interacting molecular loops in the mammalian circadian clock. *Science* **288**: 1013-1019.

Simon, S., and Llinas, R. (1985). Compartmentalization of the submembrane calcium activity during calcium influx and its significance in transmitter release. *Biophys. J.* **48**: 485-498.

Small, J. R., and Kacser, H. (1993). Responses of metabolic systems to large changes in enzyme activities and effectors. *Eur. J. Bioch.* **213**: 613-640.

Smolen, P., Baxter, D. A., and Byrne, J. H. (1998). Frequency selectivity, multistability, and oscillations emerge from models of genetic regulatory systems. *Am. J. Physiol.* **274**: C531-C542.

Smolen, P., Baxter, D. A., and Byrne, J. H. (1999). Effects of macromolecular transport and stochastic fluctuations on the dynamics of genetic regulatory systems. *Am. J. Physiol.* **277**: C777-C790.

Smolen, P., Baxter, D. A., and Byrne, J. H. (2000). Mathematical modeling of gene networks. *Neuron* **26**: 567-580.

Smolen, P., Baxter, D. A., and Byrne, J. H. (2006). A model of the roles of essential kinases in the induction and expression of late long-term potentiation. *Biophys. J.* **90**: 2760-2775.

Smolen, P., Baxter, D. A., and Byrne, J. H. (2008). Bistable MAP kinase activity: a plausible mechanism contributing to maintenance of late long-term potentiation. *Am J. Physiol. Cell Physiol* **294**: C503-C515.

Smolen, P., Hardin, P. E., Lo, B. S., Baxter, D. A., and Byrne, J. H. (2004). Simulation of *Drosophila* circadian oscillations, mutations, and light responses by a model with VRI, PDP-1, and CLK. *Biophys. J.* **86**: 2786-2802.

Sneyd, J., Falcke, M., Dufour, J. F., and Fox, C. (2004). A comparison of three models of the inositol trisphosphate receptor. *Prog. Biophys. Mol. Biol.* **85**: 121-140.

Song, H., Smolen, P., Av-Ron, E., Baxter, D. A., and Byrne, J. H. (2007). Dynamics of a minimal model of interlocked positive and negative feedback loops of transcriptional regulation by CREB proteins. *Biophys. J.* **92**: 3407-3424.

Sprague, B. L., and McNally, J. G. (2005). FRAP analysis of binding: Proper and fitting. *Trends Cell Biol.* **15**: 84-91.

Stacey, W. C., and Durand, D. M. (2001). Synaptic noise improves detection of subthreshold signals in hippocampal CA1 neurons. *J. Neurophysiol.* **86**: 1104-1112.

Stephanopoulos, G. N., Aristidou, A., and Nielsen, J. (1998). *Metabolic Engineering, Principles and Methodologies*. Academic Press, San Diego.

Thomas, R., Thieffry, D., and Kaufman, M. (1995). Dynamical behaviour of biological regulatory networks—I. Biological role of feedback loops and practical use of the concept of the loop-characteristic state. *Bull. Math. Biol.* **57**: 247-276.

Toh, H., and Horimoto, K. (2002). Inference of a genetic network by a combined approach of cluster analysis and graphical Gaussian modeling. *Bioinformatics* **18**: 287-297.

Varga, C. M., Hong, K., and Lauffenburger, D. A. (2001). Quantitative analysis of synthetic gene delivery vector design properties. *Mol. Ther.* **4**: 438-446.

Vilar, J. M., Kueh, H. Y., Barkai, N., and Leibler, S. (2002). Mechanisms of noise resistance in genetic oscillators. *Proc. Natl. Acad. Sci. USA* **99**: 5988-5992.

Wagner, J., Li, Y. X., Pearson, J., and Keizer, J. (1998). Simulation of the fertilization Ca^{2+} wave in *Xenopus laevis* eggs. *Biophys. J.* **75**: 2088-2097.

Wagner, A. (2000). Robustness against mutations in genetic networks of yeast. *Nat. Genet.* **24**: 355-361.

Walker, W., Fucci, L., and Habener, J. (1995). Expression of the gene encoding transcription factor CREB: Regulation by follicle-stimulating hormone-induced cAMP signaling in primary rat sertoli cells. *Endocrinology* **136**: 3534-3545.

Waltereit, R., and Weller, M. (2003). Signaling from cAMP/PKA to MAPK and synaptic plasticity. *Mol. Neurobiol.* **27**: 99-106.

Wang, Y., Joshi, T., Zhang, X. S., Xu, D., and Chen, L. (2006). Inferring gene regulatory networks from multiple microarray datasets. *Bioinformatics* **22**: 2413-2420.

Wen, X. L., Fuhrman, S., Michaels, G., Carr, D., Smith, S., Barker, J., and Somogyi, R. (1998). Large-scale temporal gene expression mapping of central nervous system development. *Proc. Natl. Acad. Sci. USA* **95**: 334-339.

Wiggins, S. (1990). *Introduction to Applied Nonlinear Dynamical Systems and Chaos*. Springer-Verlag, Heidelberg.

Xiong, W., and Ferrell, J. E. (2003). A positive-feedback-based bistable "memory module" that governs a cell fate decision. *Nature* **426**: 460-465.

Yamagata, Y., and Obata, K. (1998). Dynamic regulation of the activated, autophosphorylated state of Ca^{2+}/calmodulin-dependent protein kinase II by acute neuronal excitation *in vivo*. *J. Neurochem.* **71**: 427-439.

Yeung, M. K. S., Tegner, J., and Collins, J. J. (2002). Reverse engineering gene networks using singular value decomposition and robust regression. *Proc. Natl. Acad. Sci. USA* **99**: 6163-6168.

Zhabotinsky, A. M. (2000). Bistability in the Ca^{2+}/calmodulin-dependent protein kinase-phosphatase system. *Biophys. J.* **79**: 2211-2221.

Zhou, Z., and Neher, E. (1993). Mobile and immobile calcium buffers in bovine adrenal chromaffin cells. *J. Physiol. (Lond.)* **469**: 245-273.

Zlokarnik, G., Negulescu, P., Knapp, T. E., Mere, L., Burres, N., Feng, L., Whitney, M., Roemer, K., and Tsien, R. Y. (1998). Quantitation of transcription and clonal selection of single living cells with β-lactamase as a reporter. *Science* **279**: 84-88.

Connexin- and Pannexin-Based Channels in the Nervous System: Gap Junctions and More

Juan C. Sáez and Bruce J. Nicholson

CELL INTERACTIONS IN THE NERVOUS SYSTEM: THE LARGER PICTURE

The coordination of cell functions is a challenge for all organ systems of the body. This is achieved in part through a variety of exocrine and endocrine mechanisms that involve the complex release of signals from cells, their detection by receptors in other cells, and the ultimate generation of an intracellular response in the target cell. However, there is also an ancient mechanism that evolved with the earliest of muticellular organisms whereby signals, nutrients, and other metabolites under about 1000 in MW are exchanged directly between adjacent cells without dilution through the extracellular environment. This occurs through structures called *gap junctions*. The nervous system is no exception to this; gap junctions are present at homocellular and heterocellular contacts between most cells in the CNS and PNS, including astrocytes, oligodendrocytes, microglia, endothelial and ependymal cells, and some neurons. Between neurons, gap junctions have a unique role in that they form electrical synapses, which contrast with and complement the role of the more extensively studied chemical synapses that predominate in the nervous systems of most animals above the Coelenterates. This chapter will discuss the functional significance of electrical synapses in the nervous system, as well as other roles of gap junctions associated with the intercellular transport of metabolites and signals between not only neurons but also many cell types within the CNS. The significance of hemichannels, which form away from sites of cell–cell contact and allow exchanges of these same molecules with the extracellular environment, will also be considered, along with roles of the gap junction as a nexus for signaling and an adhesive structure between cells of the nervous system. Misregulated or anomalous hemichannels and/or gap junction channels contribute to the manifestation of acquired and genetic pathologies, and their roles in specific disease states and knock-out phenotypes will be discussed.

GENERAL PROPERTIES AND STRUCTURE OF GAP JUNCTION CHANNELS AND HEMICHANNELS

Gap junctions are specialized membrane structures between closely apposed cells that permit limited cytoplasmic continuity between cells. A gap junctional plaque contains up to thousands of gap junction channels that span the plasma membranes of the two cells and the 2 nm wide extracellular "gap" that separates them, and gives them their name (Revel and Karnovsky, 1967). Each channel is formed by two hemichannels, or connexons, found at the appositional membrane of adjacent cells (Fig. 15.1). The hemichannel is composed of six protein subunits called *connexins* in vertebrates, or *innexins* in invertebrates. While these two groups of proteins have similar topologies and form analogous structures, they are unrelated in

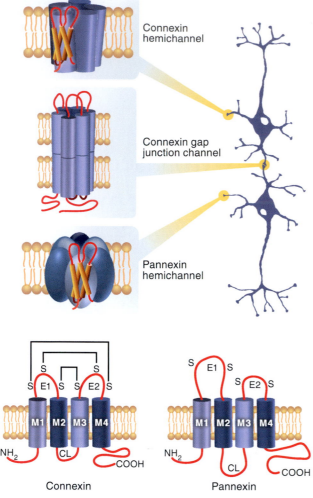

FIGURE 15.1 Schematic diagram of the distribution of gap junctions and hemichannels on a neuron. Gap junctions, representing large parallel arrays of intercellular channels composed of connexin proteins (two hexameric hemichannels docked head to head, top left insert), form electrical synapses between neurons. Hemichannels can also form on unapposed cell surfaces of neurons and most other cells in the CNS and are composed of either connexin (top right insert) or pannexin proteins (middle right insert). The topology of pannexins and connexins (bottom) is very similar, and while each have conserved cystines forming intramolecular disulfides in the extracellular loops (Foote *et al.*, 1998), there are three per loop in connexins and two per loop in innexins. There is no homology in their primary sequence, as pannexins evolved from the invertebrate gap junction protein family of innexins.

sequence. In fact, distant relatives of the innexins, called *pannexins*, are found in the vertebrates, including man, but as we will discuss later, their function as gap junctions has been replaced by connexins. Connexins form a highly conserved protein family encoded by 21 different genes in the human and 20 in the mouse genome (Willecke *et al.*, 2002).

Connexins

Connexins are widely expressed in mammalian tissues, and each connexin is commonly referred to according to its predicted molecular weight (e.g., connexin 43 [Cx43] weight ~43 kDa). For species other than human, a prefix is included (e.g. mCx43 for mouse, XeCx38 for Xenopus, etc.). A different nomenclature has been adopted for the genes, using the prefix GJ and grouping the connexins into four classes (A, B, C, and E) based on sequence homology. Within each class, individual genes are assigned an Arabic numeral that largely reflects their order of discovery (i.e., Cx43 is GJA1). A list of connexins found in the nervous system, along with both protein and gene names, and the cell types in which they are expressed, is provided in Table 15.1.

Pannexins and Innexins

By contrast, only three pannexin genes have been identified in mammals (designated Px1–Px3) (Litvin *et al.*, 2006). However, in invertebrates, innexins display a diversity comparable to the connexins, with 24 genes in *Caenorhabditis elegans* and 8 in *Drosophila melanogaster*. While an innexin-naming system has been adopted for the genes, the proteins are often referred to by complex names that relate to the mutant phenotype with which they are associated (e.g., in *C. elegans*, the Unc proteins associated with uncoordinated movement, and the Eat proteins associated with digestive tract problems, and in *Drosphila*, the Ogre protein associated with head and eye deformities and the ShakB protein associated with muscle tremors).

Channel Structure

Connexins and innexins are both tetra-membrane spanning proteins with N and C termini located in the cytoplasm. While both families have highly conserved cysteines in their extracellular loops, there are only two per loop in innexins and three in connexins. These have been shown to form intramolecular disulfides in connexins that are essential for docking of the hemichannels to form gap junctions (Foote *et al.*, 1998). Higher resolution structures of the pore, based on electron diffraction and cryoelectron microscopy, implicate two of the four transmembrane helices of each subunit as contributing to the pore (Fig. 15.2B), and based on a systematic screening of cysteine mutations for exposure to the pore, it has been suggested these may be the second and third transmembrane segments (Skerrett *et al.*, 2002). However, application of a similar strategy to hemichannels

TABLE 15.1 Connexin and Pannexin Expression in the Nervous System

Cell Type	Protein	Cell Type–Stage	Gene Name
Neurons	**Cx36**	broadly expressed	Gjd2
	Cx30.2	broadly expressed	Gjc1
	Cx45	retina–olfactory bulb	Gjd3
	Cx57	horizontal cells	Gja10
	Cx26	horizontal cells–hemichannels	Gjb2
	Px1	broadly epressed	Px1
	Px2	broadly expressed	Px2
Neurons (early development)	Cx43	stomach and intestine	Gja1
	Cx26	radial glia	Gjb2
	Cx32		Gjb1
Astrocytes	**Cx43**		Gja1
	Cx30		Gjb6
	Cx26		Gjb2
Oligodendrocytes	**Cx32**	also in Schwann cells	Gjb1
	Cx47		Gjc2
	Cx29		Gjc3
Microglia	Cx43	when activated	Gja1
	Cx32	when activated	Gjb1
	Px1		Px1
Vasculature (BBB)	Cx43	endothelium and myo-endothelium	Gja1
	Cx40	endothelium	Gja5
	Cx37	myo-endothelium	Gja4
Leptomeninges	**Cx26**		Gjb2
	Cx43		Gja1
	Cx30		Gjb6

Note: **Bold proteins** indicate the most abundant or widely expressed in that cell type.

has suggested that the first transmembrane domain may contribute to the pore in this form of the channel (Zhou et al., 1997; Kronengold et al., 2003). The most variable domains between members of the family are those located in the cytoplasm, with the central loop and C terminus varying significantly in size and sequence within the family. The C-terminal domain, in particular, contains many regulatory and binding sites, which have been well characterized in the case of Cx43 (Fig. 15.2C). It has not only been shown to modulate junctional activity through gating, degradation, or changes in biosynthesis, but it also has been implicated in mediating some of the effects of connexins through binding of signaling molecules and proteins

that mediate links to the cytoskeleton or other junctional complexes (Duffy et al., 2002; Giepmans, 2004). There have even been reports that the C terminus can be cleaved, move to the nucleus, and potentially affect transcription of other genes (Jiang and Gu, 2005).

Life Cycle of a Gap Junction

Depending on the connexin type, newly synthesized connexins are assembled in the ER (e.g., in the case of Cx26 and 32) or Golgi apparatus (e.g., in the case of Cx43) to form hemichannels (Martin et al., 2001; Sarma et al., 2002). The hemichannels are then transported in vesicles to the cell surface via a microtubules-dependent system (George et al., 1999; Martin et al., 2001; Shaw et al., 2007). Once inserted into the membrane at sites of cell–cell apposition, hemichannels diffuse laterally towards the center of a plaque to find and dock in series with hemichannels from an adjacent cell, thus forming gap junction channels (Fig. 15.2A). In time, new channels are formed at the periphery of a gap junction plaque while existing channels are removed from the central region by internalization into one of the two contacting cells. During the internalization process, cytoplasm from the adjacent cell is captured (Gaietta et al., 2002), presumably resulting in a small amount of intercellular trafficking of macromolecules. Internalized junctions are then degraded via proteosomes and/or lysosomes (Laing et al., 1997; Jordan et al., 2001). Hemichannels forming cell–cell gap junction channels do not seem to be reused. The half-life of several rodent connexins has been measured as 2 to 5 h (Laird, 2006).

Channel Properties of Gap Junctions

The rapid turnover of gap junctions indicates that intercellular coupling could be modulated by changes in synthesis and/or degradation rate of connexins, at least within reasonable time frames of development. More rapid regulation can be achieved by gating of the gap junction channels, which are usually open under resting conditions. These channels can be induced to close by changes in transjunctional (Vj) and sometimes transmembrane voltage (Vm), elevated cytoplasmic H^+, Ca^{2+} levels, certain lipophilic agents, and protein phosphorylation (Harris, 2001).

Gap junctions allow coordination of cellular responses by permitting the exchange of metabolites (e.g., ATP, ADP, glucose, glutamate, and glutathione) and second messengers [e.g., cAMP, Ca^{2+}, and inositol 1,4,5-triphosphate (IP_3)] (Sáez et al., 2003). They also coordinate electrical activity within cell communities through the passive spread of electronic potentials.

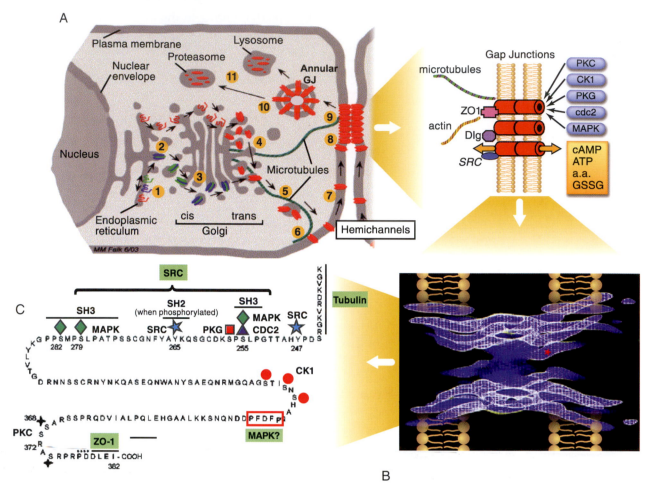

FIGURE 15.2 (A) Schematic diagram of the life cycle of connexins. Different connexin subunits (red, e.g., Cx43; blue, e.g., Cx32; and green, e.g., Cx26) are inserted into the ER, where some oligomerize (Cx32 and 26), while others do so only in the Golgi (Cx43). Hexamers are transported to the surface on microtubules to sites of close cell apposition. These hemichannels then diffuse to points of cell contact where they dock with a hemichannel from the apposing cell and assemble laterally into a gap junction. Gap junctions are removed by invagination of the whole structure into one cell to form an "annular gap junction" in the cell that is then targeted for proteosomal or lysosomal degradation. This whole process has a half life of 2–5 hours. The insert at the right shows how gap junctions, which serve as conduits for exchange of ions and other metabolites (yellow box) provide anchoring points for both cytoskeletal and signaling elements within the cell and are targets of regulation by several kinases (modified from Yeager *et al.*, 1998). (B) Detailed structure of a gap junction channel in profile derived from electron diffraction studies of isolated Cx43 gap junctions at 7.5 Å resolution (from Unger *et al.*, 1999). (C) Diagram of the C-terminal cytoplasmic domain of Cx43 that illustrates the number of phosphorylation sites (PKC, CKI, ERK, PKG, cdc2, and v-src) and binding sites for cytoskeletal (tubulin and ZO1 for actin binding) and signaling (src, MAPK) molecules that can potentially regulate these channels.

These functions depend, in part, on channel conductance and permeability properties. Some gap junction channels are slightly more permeable to anions, whereas most show a mild preference for cations or exhibit little charge selectivity (Sáez *et al.*, 2003). However, their selectivity for larger molecules is more marked. Some connexins form highly restrictive pores with size cut-offs below 400 Da, while others allow the passage of molecules well above 1000 Da (Harris, 2001). In addition, there is growing evidence that gap junctions composed of different connexins can show distinct selectivities for endogenous metabolites and signaling molecules (Goldberg *et.al.*, 1999; Bedner *et al.*, 2006) Therefore, the physiological role of gap

junctions is determined by their total expression levels and regulatory properties, as well as the specific signals and/or metabolites that each connexin isotype may allow to be transmitted.

Hemichannels: More Than Half of a Gap Junction

Recently, the presence of hemichannels on the cell surface (Figs. 15.1 and 15.2A) has been demonstrated using morphological, biochemical, electrophysiological, and functional criteria (Bennett *et al.*, 2003). Several connexins (e.g., Cx46, 50, 43, 32, 26, and XeCx38) expressed in exogenous systems have been shown to generate relatively nonselective currents in the plasma membrane that have been attributed to hemichannel opening (Retamal *et al.*, 2007a). These hemichannels are permeable to fluorescent permeability tracers (carboxyfluorescein, cascade blue, Lucifer Yellow, and calcein) as well as large ions like ethidium bromide. Physiological concentrations of extracellular Ca^{2+} (1–2 mM) and membrane depolarization maintain these hemichannels closed to prevent leak of metabolites from cells. Hemichannels have, however, been induced to open under normal physiological Ca^{2+} concentrations, as a response to metabolic inhibition or hypoxia reoxygenation, treatment with proinflammatory cytokines (Retamal *et al.*, 2007b), or mechanical stress on the membrane (Bao *et al.*, 2004a, b). In particular, Cx26 hemichannels seem less sensitive to voltage and have been shown to open in normal Ca^{2+} conditions without compromising the cell viability (González *et al.*, 2006).

It was mentioned previously that pannexins may have lost the ability of their ancestors, the innexins, to form gap junction channels, although Pnx1 has been reported to do so when expressed in Xenopus oocytes (Bruzzone *et al.*, 2003). However, there is ample evidence that Px1, at least, can form hemichannels. As both Cx43 and Px1 are widely expressed, they are coexpressed in many cell types (Barbe *et al.*, 2006; Laird, 2006; Boassa *et al.*, 2007; Shestopalov and Panchin, 2008), so careful dissection of their roles in different physiological responses is often required. Like connexins, Px1 hemichannels are activated by membrane depolarization and mechanical stimulation. However, they are also activated by enhanced intracellular $[Ca^{2+}]$ and ATP-mediated purinergic receptor (i.e., P2X) transactivation (Bao *et al.*, 2004b; Locovei *et al.*, 2006a; Locovei *et al.*, 2006b; Pelegrin and Surprenant, 2006; Locovei *et al.*, 2007) and are not sensitive to extracellular Ca^{2+} concentrations. While they can be blocked by some of the lipophilic agents that close connexin channels, they are insensitive to La^{3+} and heptanol (Retamal *et al.*, 2007a). Px1 also forms hemichannels of higher conductance (>500 pS) than most connexins (Retamal *et al.*, 2007a), with the possible exception of Cx37.

CONNEXINS IN CNS ONTOGENY

The distribution of connexins varies according to the developmental stage, cell type, and brain region (Dermietzel *et al.*, 1989; Batter *et al.*, 1992; Nadarajah *et al.*, 1998; Bittman *et al.*, 2002; Leung *et al.*, 2002; Maxeiner *et al.*, 2003; Nagy *et al.*, 2004; Söhl *et al.*, 2004; Van Bockstaele, 2004; Cina *et al.*, 2007; Orthmann-Murphy

et al. 2007). It is believed that gap junctions mediate the intercellular transfer of signals (e.g., second messengers and/or morphogenic agents) generating intracellular gradients that control ontogeny (Warner, 1992). Accordingly, neuroblasts, progenitor cells of neurons and macroglia (astrocytes and oligodendrocytes), are well coupled through gap junctions, primarily composed of Cx43. Coordinated rises in intracellular free Ca^{2+} concentration also occur early in neurogenesis in the ventricular zone of the embryonic neocortex, and it has also been proposed that connexins may help in the propagation of Ca^{2+} waves (Lo Turco and Kriegstein, 1991; Nadarajah *et al.*, 1998). The specific role of connexins in Ca^{2+} waves will be discussed later in considering the role of gap junctions in astrocytes.

During advanced neuronal differentiation, gap junctional communication is progressively downregulated (Naradajah *et al.*, 1998; Rozental *et al.*, 1998; Leung *et al.*, 2002), suggesting that decreased intercellular communication favors neuronal proliferation and differentiation. These findings are replicated in primary cultures of neuronal and glial progenitor cells. While neuronal phenotypic changes are associated with loss of Cx43, coupling reduction, and selective expression of Cx36, and possibly Cx45, in some neuronal populations, cells acquiring the astrocytic lineage remain well coupled through primarily Cx43 channels (Rozental *et al.*, 2000), although Cx30, and in some cases Cx26, expression has also been reported (Theis *et al.*, 2005).

The functional significance of these changes in connexin expression in different cell types has been demonstrated through several *in vivo* studies. The deletion of Cx43 in GFAP-positive cells led to disorganization in the development of several brain regions, including the cortex, hippocampus, and cerebellum (Wiencken-Barger *et al.*, 2007). These results are consistent with the effect of less specific pharmacological inhibition of gap junctions during neuronal and astroglial differentiation (Pleasure *et al.*, 1992; Bani-yaghoub *et al.*, 1999).

However, caution needs to be used in interpreting all these developmental effects as reflective of losses in intercellular exchanges of ions or signals. In a recent elegant study on the embryonic development of the mouse cortex, the ablation of Cx26 or 43 in radial glia, the neuronal stem cells of the cerebral cortex, led to a lack of migration of neurons to the outer layers of the cortex. The expression of selected point mutants of Cx26 and 43 demonstrated, however, that it was not the channel function of these proteins that was important, but their adhesive properties that were needed to promote neuronal migration along the radial glial tracks (Elias *et al.*, 2007). A functional role of hemichannels in neuronal differentiation has also been documented. The NGF-induced differentiation of PC12 cells overexpressing Cx43 or Cx32 is accompanied by process outgrowth due to enhanced ATP released through hemichannels in the absence of significant changes in gap junctional communication between cells (Belliveau *et al.*, 2006).

CONNEXINS IN NEURONS OF THE ADULT CNS

Electrical Synapses Versus Chemical Synapses

During the last half century, chemical synapses in all animal phyla have received particular attention. Numerous properties of the pre- as well post-synaptic organelles, their molecular components, physiological events, and functional roles of chemical synapses are described in Chapters 8–11, 16, and 19. Moreover, the involvement of chemical synapses in pathological conditions is well documented and frequently applied in therapies for humans, as discussed in Chapter 20.

Chemical synapses were thought of as the only pathway for functional cell–cell propagation of electrical action potentials around the late 1950s, until the work of Furshpan and Potter (1959) demonstrated direct passive electrical transmission in segments of giant fibers in crustaceans. All early studies of electrical synapses were done in lower-order vertebrates and invertebrates. However, after the discovery of their molecular constituents in mammals 22 years ago (Kumar and Gilula, 1986; Paul, 1986), the tools became available to demonstrate that the components of electrical synapses were expressed in most cellular elements of the nervous system in higher vertebrates as well. The presence of functional coupling of cells was first demonstrated by electrophysiological and dye transfer recordings, but later this was correlated with the structures we now recognize as gap junctions in electron microscopy of thin sections and freeze-fracture replicas (for review see Bennett and Zukin, 2004).

Several functional properties of electrical synapses distinguish them from chemical synapses, and allow them to play a complementary role in both central and peripheral nervous system function. Firstly, due to their complex structure, chemical synapses are typically unidirectional, the one exception being in the nerve net of the jellyfish *Cyanea* (Anderson, 1985; Anderson and Grünert, 1988), while electrical synapses are generally bidirectional, as they allow passive ion flux in both directions. The exceptions to

the latter are the few cases, all reported in invertebrates, where the Vm sensitivity of the connexins produces a rectifying electrical conductor (e.g., the crayfish giant axon; Furshpan and Potter, 1959). This bidirectional property is well suited to ensuring synchronous firing of neurons as seen in various regions of the mammalian brain. In general, electrical synapses play roles in propagation of excitatory impulses. However, if the hyperpolarizing chemical PSPs, as well as the hyperpolarizing phase of the action potential, can be electrotonically spread through gap junctions (e.g; in Mauthner neuron; Furukawa and Furshpan, 1963).

Secondly, due to the simple conductance pathway that does not require any vesicular release mechanisms, diffusion of neurotransmitter, and activation of a receptor, electrical synapses propagate potentials rapidly without the "pause" seen in chemical transmission. They effectively serve as a low-pass filter between neurons (Bennett and Spray, 1987; Galarreta and Hestrin, 1999; Gibson *et al.*, 1999). For this reason, electrical synapses feature heavily in escape responses, including the jump response of *Drosophila* (see review by Phelan, 2005), and the escape response of crayfish (see earlier) and teleost fishes (i.e., Mauthner neuron, Fig. 15.3).

Thirdly, the complexity of chemical synapses allows for multiple levels of regulation that lead to synaptic plasticity, a critical property of even the most primitive nervous systems to allow adaptation to the environment and ultimately "learning" (Chapter 19). The simple structure of gap junctions suggests that they are passive conductors that do not show modulation. However, we now know this view to be too simplistic. The efficacy of electrical synapses can be modulated over time through the rapid turnover of the proteins that allow for transcriptional driven changes in channel number within hours. Much more rapid modulation can also be achieved when the conductance changes in the surrounding membrane shunt the conductance and decrease the effectiveness of electrical coupling. This happens frequently in mixed chemical and electrical synapses (Fig. 15.3A), which are common in vertebrates, where the two types of synapse are located adjacent to one another (Llinás, 1988). Changes in the duration of the presynaptic impulse can also influence the effectiveness of electrical transmission if it leads to a closer match with the time constant of the postsynaptic membrane (Pereda and Faber, 1996). However, most importantly, in the mixed synapse club endings of the eigth nerve inputs to the Mauthner neuron of goldfish, the efficacy of electrical transmission was increased substantially, and over a prolonged period, following

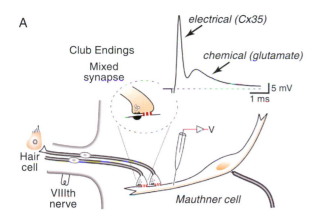

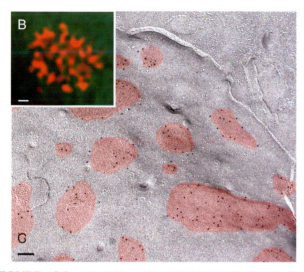

FIGURE 15.3 (A) Club endings exhibit mixed synaptic transmission. Typical experimental arrangement showing VIIIth nerve auditory primary afferents (which contact saccular hair cells; "hair cell") terminating as club endings on the ipsilateral M-cell lateral dendrite. Inset represents a club ending, at which both mechanisms of synaptic transmission, electrical (gap junction), and chemical, coexist. VIIIth nerve stimulation evokes a mixed electrical and chemical synaptic potential. The trace represents the average of 20 individual responses. (B–C) Cx35 is found at club ending gap junctions. (B) Laser scanning confocal immunofluorescence (average of three z sections) showing Cx35 at a club ending. There are multiple puncta immunoreactive for Cx35. Calibration: 1 μm. (C) FRIL image from a club ending showing Cx35 localization at the gap junctions. All 14 gap junctions in this image are labeled with 10 nm gold beads. Calibration: 0.1 μm. Figure was kindly provided by Dr. Alberto Pereda.

train of tetanic stimuli (Yang *et al.*, 1990). This post-tetanic potentiation (PTP), thought to be unique to chemical synapses, has been linked to release of endocanabinoids and activation of cAMP-dependent protein kinase (PKA) through dopamine receptors, and subsequent phosphorylation of Cx35, the connexin in these gap junctions (Fig. 15.3) (Cachope *et al.*, 2007).

Connexin Expression Patterns

As the only known component of electrical synapses between vertebrate neurons, connexins are broadly expressed in the CNS (Table 15.1). While Cx26, 29, 31.1, 32, 37, 43, 40, 47, and 57 transcript expression has been detected in diverse neuronal types at different developmental stages (Nadarajah *et al.*, 1998; Bittman *et al.*, 2002; Leung *et al.*, 2002; Maxeiner *et al.*, 2003; Nagy, 2004; Söhl *et al.*, 2004; Van Bockstaele, 2004; Cina *et al.*, 2007), only Cx36, 45, 57, and most recently Cx30.2 have been reproducibly identified at the ultrastructural level in neuronal gap junction structures of the adult rat brain (Rash *et al.*, 2000; Rash, 2004; Rash, 2005; Fukuda, 2006; Rash *et al.*, 2007). Cx36 is the most widely expressed neuronal gap junction protein, and immunogold labeling of freeze-fracture replicas has demonstrated its localization between neurons in the inferior olive, spinal cord, retina, olfactory bulbs, visual cortex, suprachiasmatic nucleus, and locus coeruleus. Cx30.2 is frequently coexpressed with Cx36 in interneurons (Kreusberg *et al.*, 2008). Cx45 neuronal gap junctions have been located in the retina and olfactory bulb (Rash *et al.*, 2000; Rash, 2004; Rash, 2005; Kamasawa *et al.*, 2005; Fukuda, 2006; Rash *et al.*, 2007), although expression of the Cx45 gene has also been detected in cerebral cortical, hippocampal, and thalamic neurons as well as basket and stellate cells of cerebellum (Maxeiner *et al.*, 2003). Cx57 expression appears restricted only to horizontal cells in the retina. Cx26 and Cx32 are expressed in pre-Bötzinger neurons of neonatal and adult rats, but their localization of gap junction structures has not been confirmed (Solomon *et al.*, 2001). In addition to connexins, it should be noted that Px1 and Px2 both have been found to be expressed in the CNS, particularly in the hippocampus (Bruzzone *et al.*, 2003).

Cells of CNS glands also express connexins. Rat pinealocytes express both Cx26 and Cx36 (Sáez *et al.*, 1991; Belluardo *et al.*, 2000), and anterior pituitary cells express Cx36 (Belluardo *et al.*, 2000), while a small percentage of cells containing luteinizing hormone present Cx43 immunoreactivity (Yamamoto *et al.*, 1993), as do folliculo-stellate cells (Shirasawa *et al.*, 2004).

Numerous cells of the peripheral nervous system express connexins and form functional gap junctions that are highly regulated by physiological conditions. For example, the expression of Cxs 32, 36, and 43 is developmentally regulated in the trigeminal motor nucleus, while Cx26 expression remains high throughout postnatal development. In the mesencephalic trigeminal nucleus, Cx26, 32, and 43 expression is intense throughout development, with only Cx36 showing a developmental regulation (Honma *et al.*, 2004). Connexins and gap junctions are also found in neurons of the enteric nervous system. In the stomach and intestines, Cx43 gap junctions are found between interstitial cells of Cajal (ICC) as well as between ICC and adjacent muscle layers of each tissue (Daniel and Wang, 1999; Seki and Komuro, 2006).

Studies on gap junctions in neuronal primary cultures have been limited by the loss of connexin expression, most likely due to the lack of regulatory factors in cultured cells that lead to a level of de-differentiation. One regulator of connexin expression in primary cultures is basic FGF. It increases gap junctional communication and Cx43 levels in cortical progenitor cells (Nadarajah et al., 1998). Moreover, basic FGF enhances cell–cell communication via gap junctions and levels of Cx43 in rat embryonic day 14 midbrain cultures (Siu et al., 2001).

The physiological significance of the specialized expression of connexins in different regions remains obscure. We do know that some connexins will form heterotypic channels with other connexins expressed in neighboring cells, but this is a selective process, the rules of which we still do not fully understand, although generally connexin within the same genetic groupings (A–E) prefer to interact. Regulatory properties of connexins vary significantly, including their sensitivity to voltage gating. However, the time course of gating in response to voltage differences between cells is slow (hundreds of milliseconds) compared to the time course of action potentials in neurons (tens of milliseconds). Permeability properties of different connexin channels also vary, as noted previously, but the understanding of this is in its infancy and provides little physiological insight at this time. It is notable that the major connexin between neurons in the CNS is Cx36, which has the smallest single-channel conductance and size cut-off for larger molecules of any connexin studied, indicating that it is largely specialized for electrical conductance rather than metabolite transfer.

Functional Roles

The roles of gap junction coupling of neurons in the CNS are likely to be primarily at the level of fine-tuning of oscillatory networks. This is both suggested by extensive modeling of electrical networks within the brain, and more recently, tested empirically through Cx36 genetic knock-outs in mice (Buhl et al., 2003). Cx36 is primarily expressed between interneurons at dendro–dendritic connections. Modeling studies have suggested that axo–axonal electrical coupling may also play a critical role in synchronization of neuronal firing patterns; it is likely that these gap junctions may be composed of Cx45. As yet no CNS-specific knock-out

has been developed, and the whole animal knock-out is lethal embryonically.

The major effects of loss of Cx36 expression in the CNS is a loss of synchronicity of selected oscillatory patterns mediated by interneurons in the inferior olive of the cerebellum (2–10 Hz subthreshold oscillations), the hippocampus (12–90 Hz gamma oscillations, but not the slower theta or the fast high-frequency oscillations), and the cortex (supra- and subthreshold oscillations, particularly among GABA interneurons). Typically the basic oscillatory behaviors are retained in the absence of Cx36, but their power and synchrony are reduced or lost. It has been proposed that in the cerebellum this would result in the loss of 10–20 ms of precision in the coordination of neuronal activity. Hence, it is not surprising that these mice show no major loss of motor skills or coordination yet do show more subtle defects in object recognition and memory and other behavioral impairments, as well as disrupted circadian rhythms (Hormuzdi et al., 2004; Frisch et al., 2005; Long et al., 2005). The role of coexpression of Cx30.2 in regulating these oscillations is unclear, as no defects in Cx30.2 −/− mice have been detected (Kreusberg et.al., 2008).

The Retina: A Case Study of Diversity

A particularly intriguing model for the study of diverse roles of connexins in regulating neuronal function is the retina (see Fig. 15.4 for overall circuitry). Cx36 is the most widely expressed, being found in cone cells, OFF cone bipolar cells, AII amacrine cells, and ganglion cells. Cx45 is expressed by ON cone bipolar cells and a small subset of amacrine cells and ganglion cells, while Cx57 is found exclusively in horizontal cells, which in the fish, at least, also express Cx26 but probably only in the hemichannel form (summarized in Söhl et al., 2004). Initially, the coupling of the photoreceptor cells (cones are coupled not only to one another by Cx36 but also to rod cells through an unknown connexin) seems counterintuitive, as it might be expected to reduce the ability to distinguish between closely spaced stimuli. However, it appears that the largest problem in retinal detection is the background noise generated by spontaneous firings of receptors in the absence of specific stimuli, and the coupling of cells greatly mutes this effect, thereby increasing signal-to-noise ratio.

The coupling between horizontal cells and AII amacrine cells has always made more intuitive sense, as these cells have been implicated in the regulation of the receptive field size in the retina, through center-surround inhibition effects (where responses around the central area of stimulation are suppressed) and

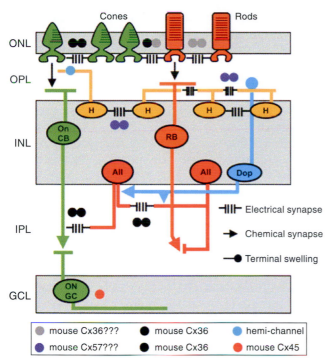

Cones   Rods

ONL

OPL

INL

IPL

GCL

| mouse Cx36??? | mouse Cx36 | hemi-channel |
| mouse Cx57??? | mouse Cx36 | mouse Cx45 |

—|||— Electrical synapse

→ Chemical synapse

—● Terminal swelling

FIGURE 15.4 A simplified diagram of the circuitry of the retina illustrates the dependence of rod signaling on Cx36 gap junctions. Cones and rods are coupled to themselves (by Cx36) and one another (through unknown heterotypic channels). Cones signal through chemical synapses to ON ganglion cells (GC) via ON cone bipolar cells (CB) (green pathway). Rod cells, however, chemically innervate rod bipolar cells (RB), which innervate AII amacrine cells (AC) (red pathway). These are electrically coupled to one another, and to ON CB cells. Thus, rods can signal to ganglion cells through either the RB–AII system, or via cones, but both circuits have at least one connection that is fully reliant on electrical coupling. Both AII and horizontal cells (H) are coupled to one another electrically, forming a network parallel to the surface of the retina that is important in adjusting receptive field size. The coupling of both these networks is regulated by light-sensitive dopaminergic neurons (Dop) (blue). From Hormuzdi et al., 2004.

integration of signals from multiple receptors, respectively. These effects are clearly important for balancing maximal acuity in bright light (photopic) conditions (minimized receptive fields) with maximal sensitivity in low light (mesopic and scotopic) conditions (integrating signals over a wider area). Thus, there might be expected a diurnal–nocturnal regulation of coupling to modulate receptive field size. Indeed, this is seen in both horizontal and AII amacrine cells, where light induces a dopaminergic response that elevates cAMP levels, which reduces coupling between these cells (Bloomfield et al., 1997). In the case of Cx36, this has been associated with PKA-sensitive sites on the cytoplasmic loop (Mitropulou and Bruzzone, 2003).

Horizontal cell connexins are also involved in a rather unusual means of influencing the electrical responsiveness of cone cells through a mechanism

called *ephaptic transmission* (summarized by Kamermans and Fahrenfort, 2004). In turtle retinas, Cx26 has been identified as being expressed in the complex "ribbon" synapse of cones, where the presynaptic cone synapse with the ON bipolar cells also involves postsynaptic contacts with horizontal cells. Cx26 is expressed by the horizontal cells at this site, but no gap junctions can be detected, indicating that they are likely to exist exclusively in hemichannel form. Cones are typically depolarized in the dark, resulting in release of glutamate and depolarization of horizontal cells. Light causes hyperpolarization of cone cells, and the loss of glutamate release results in a shift of the horizontal cell membrane potential to more negative values. The problem the retina faces is how to adjust the gain in this response to allow responses to small changes in light, but over many orders of magnitude differences in the ambient light conditions. This requires some kind of positive feedback loop that can reset the threshold for glutamate release at different light levels. This appears to occur through a change in the response levels of the Ca^{2+} channels that trigger glutamate release. The proposed mechanism relates to extracellular currents that are generated in the very narrow synaptic cleft of the cone–horizontal cell interaction. As the horizontal cells become more hyperpolarized in high light conditions, there is a stronger driving force for current across the Cx26 hemichannels clustered in this postsynaptic region (these channels are insensitive to Vm and remain open). This current flow results in a drop in the potential of the extracellular space, which is sensed by the cone cell membrane as a depolarization (i.e., the potential difference between intracellular and extracellular environments decreases). This results in activation of Ca^{2+} channels and release of glutamate, despite the fact that the absolute internal potential of the cone cells did not change. Connexin hemichannels are uniquely adapted to this role, although as of yet, Cx26 has not been detected in mammalian cone–horizontal cell synapses, which may mean other channels can also mediate this novel method for adjusting the gain of a synapse.

Perhaps the most surprising role of Cx36 coupling in the retina was revealed when it was ablated genetically in mice (Deans et al., 2002). This resulted in a loss of all rod signaling to ganglion cells under low light (scotopic) conditions. The reason for this is that rods do not make direct contact with rod-specific ganglion cells but rather connect to rod bipolar cells that innervate AII amacrine cells, which in turn synapse with ON cone bipolar cells, leading to activation of ganglion cells (Fig. 15.4). While much of this pathway is mediated by standard chemical synapses, the AII amacrine to ON cone bipolar cell connection is exclusively mediated by electrical synapses composed of Cx36.

An alternative route for rod activation of ganglion cells does exist, through their coupling to cone cells directly, which also activates ON cone bipolar cells; however, this coupling also requires Cx36 expression in the cone cells (Fig. 15.4). Hence, in the absence of Cx36, all connectivity between rod cells and ganglion cells is lost. An interesting side effect of the loss of AII amacrine cell–ON bipolar cell coupling was also observed that serves to remind us that a gap junction can serve roles beyond that of an electrical synapse. Cone cells and amacrine cells are glycinergic transmitters, yet the cone cells do not express a glycine transporter for recapturing the neurotransmitter. In the Cx36 −/− mice, glycine levels in the bipolar cells are significantly reduced, supporting a previous proposal (Cohen and Sterlin, 1986; Vaney et al., 1998) that they receive their glycine through gap junctions from AII amacrine cells that do express the glycine transporter.

ASTROGLIAL CONNEXINS

Connexin Expression Patterns

Cortical astrocytes present immunoreactivity for Cxs 26, 30, 40, 43, and 45 (Nagy et al., 1997, 1999; Dermietzel et al., 2000; Rash et al., 2000) but only Cxs 26, 30, and 43 show colocalization at ultrastructurally defined astroglial gap junctions (Rash et al., 2000; Nagy et al., 2001). Although Cx43 and Cx30 are both present in astrocytes of the visual cortex (Rochefort et al., 2005), characterization of unitary current events of gap junctions in cultured astrocytes are consistent with only homotypic Cx43 gap junction channels (Dermietzel et al., 1991; Giaume et al., 1991; Moreno et al., 1994; Kwak et al., 1995; Bukauskas et al., 2001). Consistent with this, cultured astrocytes from Cx43-deficient mice do not form functional gap junctions (Naus et al., 1997). Cultured cortical astrocytes also express Cx26, but it has been difficult to detect its expression at either cell–cell contacts (Martínez and Sáez, 1999) or unapposed cell membranes, where only Cx43 hemichannels are detected (Retamal et al., 2007a). Thus, Cx43 is thought to be the main functional connexin in cultured astrocytes. However, astrocytes cocultured with neurons with active chemical synapses also express Cx30 (Rouach et al., 2000).

Gap junctions between astrocytes are dynamically regulated and play diverse functions. Levels of astrocytic Cx43 are upregulated at about the time that the rat pineal gland becomes innervated (Berthoud and Sáez, 1993), suggesting that Cx43 expression is also regulated by neuronal activity. Because Cx43 presents multiple phosphorylation sites, astrocytic gap junctional communication can be rapidly increased or decreased upon activation of protein phosphatases or kinases. Accordingly, glutamate or high extracellular K^+ concentration enhance astrocyte gap junctional communication (Enkvist and McCarthy, 1994) through a mechanism linked to increases in Cx43 phosphorylation mediated by calmodulin kinase (De Pina-Benabou et al., 2001).

Heterocellular gap junctions between astrocytes and neurons have been described, with both dye transfer and electrical coupling of astrocytes and neurons being demonstrated in the locus coeruleus (Alvarez-Maubecin et al., 2000; Van Bockstaele et al., 2004) and astrocyte–neuron cocultures (Froés and de Carvalho, 1998; Froés et al., 1999; Rozental et al., 2001). Nevertheless, in vivo ultrastructurally defined neuron–astroglial gap junctions have not been observed (Rash et al., 2007), and it is likely that these would have to be restricted in order to prevent undue shunting of current from active neurons into the large population of glia. Electrical coupling mediated by Cx36 between neurons and microglia have been observed in cell cocultures (Dobrenis et al., 2005), but the physiological significance of this coupling is as yet undefined. However, there is growing evidence of gap junction formation between oligodendrocytes and astrocytes that would create a pan-glial syncytium (reviewed in Theis et al., 2005). The composition of the junctions that would mediate these connections, given the Connexin expression patterns of oligodendrocytes reviewed later, has yet to be defined, but based on what is known of heterotypic compatibility between connexins, it seems likely that this would be mediated either by Cx43/47 or Cx30/32 pairings of connexins (listed as astrocyte–oligodendrocyte side).

Functional Roles

Cortical astrocytes have been considered as "nursing cells" because they maintain the extracellular homeostasis and provide metabolic support to neurons for their normal functioning.

The "spatial buffering" of K^+ and small molecules like neurotransmitters (e.g., glutamate) is a very important function of glial networks, which form extensive syncytia through gap junction coupling (Fig. 15.5). K^+ and glutamate, which accumulate in the extracellular milieu surrounding foci of high neuronal activity, are taken up, either passively through K^+ channels, or actively via transporters, by astrocytes located in the vicinity. The potassium ions are diluted intracellularly to regions distal to the area of high neuronal activity while the glutamate is converted to glutamine, which

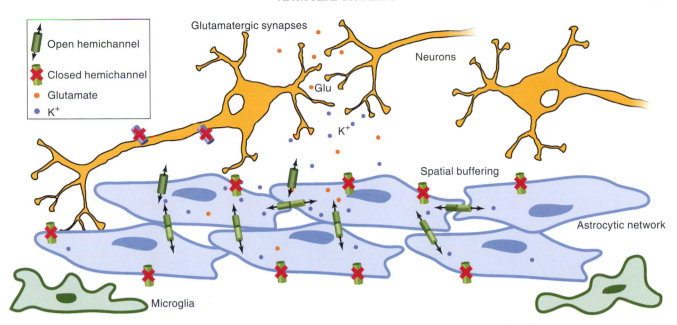

FIGURE 15.5   Scheme showing functional roles of gap junction channels between astrocytes. Regions of highly active neurons release into the extracellular fluid that can induce hyperexcitability and neuronal apoptosis. Surrounding astroglia take up both the K⁺ (blue dots) and glutamate (orange dots) either through open hemichannels or specific K⁺ channels or glutamate transporters, and distribute them through gap junctions throughout the astrocytic syncytium before releasing them at a remote site. It has been proposed that in cases of myelinated neurons (not shown), the oligodendrocytes that are in closest proximity to the axons may be the first site for taking up K⁺, which is then passed to the astrocytic population through heterotypoic Cx32/30 or Cx47/43 channels (see text).

recycles to neurons where it can be modified to glutamate and used as a neurochemical transmitter (Fig. 15.5). This function may be facilitated to the recently reported Cx43-induced regulation of astrocytic glutamate transporters (Figiel *et al.*, 2007).

The importance of this spatial buffering of K⁺ and glutamate stems from the fact that exposure of neurons to high levels of these components in the extracellular fluid leads to hyperexcitablility of the neurons, which is ultimately toxic. A phenomenon known as *spreading depression* (SD) is likely a product of this, where propagating waves of depolarization of neurons is followed by neuronal inactivation. It had been proposed, based on the application of pharmacological inhibitors of gap junctions, that astrocytic coupling may have exacerbated this process (Nakase and Naus, 2004). However, studies on Cx43 −/− mice have shown that loss of Cx43 leads to an enhanced rate of SD propagation (Theis *et al.*, 2003). The discrepancy in the results could either stem from the relatively nonspecific nature of gap junction pharmacological agents that could be toxic in their own right, or the involvement of other connexins than Cx43 in aspects of SD. In concert with its role in inhibiting SD, Cx43 has also been shown to serve a protective role in focal brain ischemia by reducing stroke volume (Nakase *et al.*, 2003).

Related to their role in spatial buffering, gap junctional coupling of astrocytes is also likely to be important in the distribution of energy resources throughout the brain. Under normal conditions, the main energy substrate of the brain is glucose metabolized via glycolysis in astrocytes and by oxidative phosphorylation in neurons (Kasischke *et al.*, 2004). The latter is favored by a tight metabolic balance between neurons and astrocytes, termed *neurometabolic coupling*. The glucose in this balance is provided by glycogen stored in astrocytes that can be mobilized over a period of seconds during periods of high neuronal activity. Immediately following the initial burst of glycogen-derived glucose, blood-borne glucose is shuttled over the blood–brain barrier through a mechanism termed *neurobarrier coupling* (Leybaert, 2005). These metabolic responses occur with a close spatial and temporal correlation between high neuronal activity and hyperemia (Roy and Sherrington, 1890).

Astroglial cells have been proposed as the mediators of changes in brain microvascular tone in response to neuronal activity, causing rapid and localized changes in cerebral blood flow upon demand (Fig. 15.6) (Ye *et al.*, 2003, Metea and Newman, 2006; Takano *et al.*, 2006). *In situ* studies, using chemical cell-type-specific ablations, demonstrated that arterial dilation in response to elevated neuronal activity (in

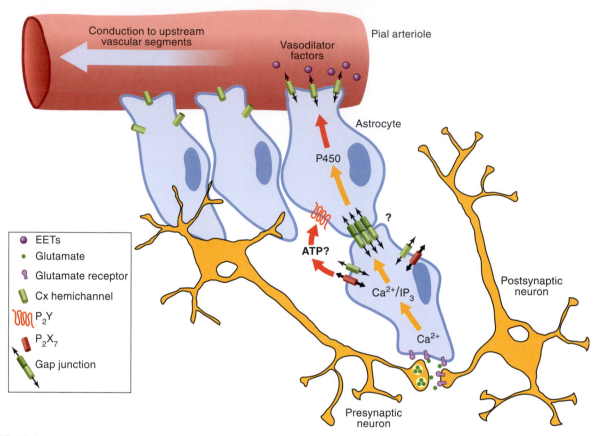

FIGURE 15.6 Scheme showing the proposed astrocyte signaling function in the neurovascular unit. Astrocytes have been shown to be essential for mediating vascular responses to high neuronal activity. One model is that a subset of astrocytes known to express glutamate receptors (instead of transporters) responds in the vicinity of the elevated neuronal activity and releases glutamate by taking up Ca^{2+}. This is then propagated between astrocytes in a Ca^{2+} wave, either directly through Cx43 gap junctions, or via extracellular release of ATP through Cx43 hemichannels and activation of P2Y purine receptors on adjacent cells, which initiate the Ca^{2+} response all over again. Ultimately, astrocytic end feet must release vasoactive factors to the arteriole, a step that could also be mediated by Cx43 hemichannels.

either pathogenic or normal range) required astrocytes, but not endothelial cells (Xu *et al.*, 2007). This supports the contention that astrocytes act as a conduit of forward signals during neurovascular coupling. That this also required longer distance propagation of the response through gap junctions was demonstrated by employing connexin-mimetic peptides specific to the extracellular domains of Cx43 that selectively prevent formation of Cx43 gap junctions through competition, and also induce closure of Cx43 hemichannels (Evans and Leybaert, 2007).

The propagation of signals through astrocytes that mediate changes in microvascular tone have been linked to intracellular Ca^{2+} concentration, and likely involve propagation of Ca^{2+} waves through the astrocytes (Ye *et al.*, 2003). This is a well-established behavior in astrocytes and many other cells in culture and can be mediated via two routes: the passage of IP_3 between cells through gap junctions, thus propagating

a regenerating wave of Ca^{2+} release from intracellular stores; or via an extracellular pathway involving ATP release (summarized in Iadecola and Needergaard, 2007). The ATP release in the latter mechanism may also be connexin dependent and has been proposed to occur through Cx43 hemichannels (Stout *et al.*, 2002). ATP then diffuses to adjacent cells and binds to the purinergic P2Y receptors, leading to activation of PLC, production of IP_3, and subsequent Ca^{2+} release from internal stores. In brain slices, extracellular ATP mediates the propagation of Ca^{2+} waves (Schipke *et al.*, 2002) that seem to coordinate neurovascular responses (Zonta *et al.*, 2003). Cx43 hemichannels may also play a role in the release of vasoactive signals at the glial end feet, as they remain separated from the arterioles by ~20 nm (Nagelhus *et al.*, 2004), and the signaling can be ablated through extracellular perfusion (Xu *et al.*, 2007). However, this connection has yet to be established.

CONNEXINS IN OLIGODENDROCYTES

Connexin Expression Patterns

Cx29, 32, and 47 have been detected in oligodendrocytes (Dermietzel *et al.,* 1997; Söhl *et al.,* 2001; Li *et al.,* 2004; Kamasawa *et al.,* 2005), and gap junction communication has been demonstrated between oligodendrocytes (Kettenmann *et al.,* 1983; Kettenmann and Ransom, 1988) and with their astrocytic neighbors (Orthmann-Murphy *et al.,* 2007). Electrophysiological recordings and immunolabeling experiments demonstrated that oligodendrocyte–astrocyte gap junctions are most likely composed of Cx47/Cx43 heterotypic channels at the stroma of the oligodendrocytes and Cx32/Cx30 heterotypic channels in the area of the myelin sheath (Orthmann-Murphy *et al.,* 2007).

Functional Roles

Gap junctions between oligodendrocytes and with astrocytes, linking into a larger pan-glial syncytium, contribute to spatial buffering of potassium released during neuronal activity, as this will be particularly concentrated near nodes of Ranvier where the myelin sheath is the first site for uptake of excess K^+ (Menichella *et al.,* 2006). In addition to their proposed neuroprotective role in spatial buffering, there is direct genetic evidence for a requirement for connexins in the maintenance and/or development of myelination of both peripheral and central neurons. This began with the first link of connexins to an inheritable disease by Fischbeck and colleagues who found that mutations of the Cx32 gene caused the X-linked version of Charcot-Marie-Tooth disease (CMTX), in which myelination of peripheral nerves fails in early adulthood, leading to partial leg paralysis (Bergoffen *et al.,* 1993). This is discussed in detail later, but is thought to arise from loss of gap junctions that form between the myelin wrappings to shorten diffusion of signals and nutrients between the soma and the innermost myelin wrapping of Schwann cells (Oh *et al.,* 1997; Abrams *et al.,* 2000). In oligodendrocytes, these "reflexive" gap junctions have not been confirmed, and only occasional CNS involvement is noted in CMTX patients (Taylor *et al.,* 2003). However, Cx47 −/− mice show vacuolation of central nerve fibers, particularly in the optic nerve, consistent with failure of myelination (Odermatt *et al.,* 2003), and Cx47 defects have been linked to abnormal CNS myelin in Pelizaeus–Merzbacher–like disease (Uhlenberg *et al.,* 2004). Most dramatically, mice with double knock-outs of Cx32 and 47 die at 6 weeks and appear to never develop appropriate myelination, suffer oligodendrocye death and axonal loss (Menichella *et al.,* 2003). Thus, while the role of connexins in forming a reduced diffusion pathway through myelin that is required for maintenance of peripheral myelin may not be so critical in the CNS, connexins do seem to be required for normal oligodendrocyte development and myelin formation.

CONNEXINS IN MICROGLIA

In the adult rat brain microglia are sparse and a few express low levels of Cx43 (Eugenin *et al.,* 2001). Under CNS-threatening conditions, microglia migrate to the affected area, forming clusters where they then express Cx43. In primary cultures, microglia are partially activated and express low levels of Cx36, Cx43, and Cx45, but they do not form functional gap junctions (Eugenín *et al.,* 2001; Dobrenis *et al.,* 2005). After treatment with proinflammatory compounds (e.g., peptidoglycans or IFN-γ with bacterial endotoxin or TNF-α), microglia express increased levels of Cx32 and Cx43 (Eugenín *et al.,* 2001; Garg *et al.,* 2005; Takeuchi *et al.,* 2006), resulting in functionally detectable coupling of the microglia (Eugenín *et al.,* 2001; Garg *et al.,* 2005). This effect may be linked to increases in the intracellular free Ca^{2+} concentration (Martínez *et al.,* 2002).

CONNEXINS IN THE BLOOD BRAIN BARRIER (BBB)

The BBB is formed by endothelial cells that line all cerebral microvessels. The permeability of this endothelium is unique in the vasculature and critical for regulating the access of compounds, including intravenously delivered drugs, to the brain. *In vivo,* these endothelial cells express Cx43 and Cx40 (Little *et al.,* 1995). Cell lines derived from the BBB are well coupled, and IP_3 induces Ca^{2+} waves that depend on extracellular ATP released through Cx hemichannels (Braet *et al.,* 2003). Other functional interactions of Cx40 and Cx43 gap junctions have been observed in primary culture of porcine BBB endothelial cells, which interact via tight junctions to establish the barrier function of this endothelium (Nagasawa *et al.,* 2006). Gap junction blockers oleamide and glycyrrhetinic acid inhibit the establishment of these tight junctions, as determined by measurement of paracellular flux of manitol, insulin, and ions (Nagasawa *et al.,* 2006), indicating an interaction between gap and tight junction structures. This is consistent with reported associations between Cx43 and cytosolic components of tight junctions like ZO-1 and Discs Large (Fig. 15.2; reviewed in Giepmans, 2004, and Laird, 2006). Cx37 and Cx43 are also expressed by myoendothelial cells (Haddock *et al.*, 2006), which allows for possible electrical coupling between the muscle and endothelial layers of the blood vessels that could facilitate the synchronization of vasomotor activity along the vessel length through Ca^{2+} waves (Haddock *et al.*, 2006). Although there is no *in vivo* evidence of astrocytes–endothelial cell coupling, which remain separated by a basal lamina, *in vitro* studies have documented weak electrical coupling (but no dye spread) between astrocytes and blood–brain barrier endothelial cells associated with the spread of both gap junction–dependent and –independent calcium waves (Braet *et al.*, 2001).

CONNEXINS IN EPENDIMAL CELLS AND LEPTOMENINGEAL CELLS

Beyond the major cell types, connexin expression has also been reported in the meninges lining the brain and its ventricles. The ependymocytes, glial cells that line the ventricles of the brain and the central canal of the spinal cord, are highly coupled through gap junctions, identified at the ultrastructural level (Rash *et al.*, 1997), composed of Cx26 and Cx43 (Dermietzel *et al.*, 1989; Yamamoto *et al.*, 1990; Yamamoto *et al.*, 1992), and possibly Cx30 (Kunzelmann, 1999; Nagy *et al.*, 1999). These junctional channels allow the synchronization of rhythmic ciliary beating. Ependymocytes also form Cx43-based gap junctions with astrocytes *in situ* (Rash *et al.*, 1997), although the functional significance of this remains unclear.

There are three layers of meninges around the brain and spinal cord: the outer layer (dura mater), the middle layer (arachnoid), and an inner layer (pia mater), connected to the arachnoid by numerous threadlike strands. The space under the arachnoid, the subarachnoid space, is filled with cerebrospinal fluid and contains blood vessels. Gap junction communication is widely extended in the developing and adult meninges, and strong coupling is seen in cultured leptomeningeal cells. (Dermietzel and Krause, 1991; Spray *et al.*, 1991). In cultured cells, coupling probably occurs through Cx26, Cx30, and Cx43, which are expressed at high levels in these cells (Nagy *et al.*, 1999; Mercier and Hatton, 2001).

PATTERN OF PANNEXIN LOCALIZATION IN BRAIN CELLS

In addition to the connexin family, three distant relatives of the invertebrate gap junction proteins (innexins), called *pannexins* (Px), have been cloned in mammals (Bruzzone *et al.*, 2003; Baranova *et al.*, 2004). While pannexins and innexins share a similar membrane topology to connexins (Fig. 15.1), their sequences are unrelated (Panchin, 2005). Px1 is also glycosylated during targeting to the plasma membrane (Boassa *et al.*, 2007), unlike connexins.

Expression Patterns of Pannexins in Mammalian CNS

Px1 is ubiquitously expressed in mammalian tissues, including the brain (Baranova *et al.*, 2004). It can be detected in several regions of the CNS including cortex, striatum, olfactory bulb, hippocampus, thalamus, inferior olive, inferior colliculus, amygdala, spinal cord, retina, and cerebellum (Bruzzone *et al.*, 2003; Ray *et al.*, 2005; Zappalà *et al.*, 2006). At the cellular level, Px1 is localized in different neuronal types, including olfactory bulb mitral cells, Purkinje cells, dopaminergic neurons, cholinergic neurons, and glutamatergic neurons (Bruzzone *et al.*, 2003; Ray *et al.*, 2005; Zappalà *et al.*, 2006). In the cerebral cortex and hippocampus, Px1 is localized in the postsynaptic cells (Zoidl *et al.*, 2007). Although Px1 has been detected in Bergmann glia in the cerebellum, it has not been found in other astroglial cells *in situ* (Ray *et al.*, 2005; Vogt *et al.*, 2005; Zappalà *et al.*, 2006), although its expression has been reported in cultured astrocytes, immature oligodendrocytes, and neurons under resting conditions. Px1 is also present in microglia (Shestopalov and Panchin, 2008).

Px2 is expressed exclusively in the brain (Panchin, 2005), including olfactory bulb, hippocampus, amygdala, superior colliculus, substantia nigra, cerebellum, hypothalamus, and spinal cord (Bruzzone *et al.*, 2003; Zappalà *et al.*, 2007). Under normal conditions hippocampal astrocytes do not express Px2. However, a transient expression of Px2 occurs in hippocampal astrocytes several hours after ischemia–reperfusion (Zappalà *et al.*, 2007), the functional significance of which is unknown.

Functional Roles

In exogenous expression systems (e.g., Xenopus oocytes, LNCaP, or C6 cells) the overexpression of Px1, but not Px2 or Px3, leads to gap junction formation

(Bruzzone *et al.*, 2003; Vanden Abeele *et al.*, 2006; Lai, 2007). However, there is no evidence that endogenously expressed pannexins form functional gap junctions in any system *in situ*. At unapposed plasma membranes, functional Px1, Px1/Px2, and Px3 hemichannels have been demonstrated in several exogenous expression systems (Bruzzone *et al.*, 2003; Bao *et al.*, 2004b; Bruzzone *et al.*, 2005; Barbe *et al.*, 2006; Locovei *et al.*, 2006a; Pelegrin and Surprenant, 2006; Peñuela *et al.*, 2007). Moreover, Px1 hemichannels have been recorded in hippocampal neurons exposed to oxygen–glucose deprivation (Thompson *et al.*, 2006), although not in cortical astrocytes under either resting conditions or after treatment with proinflammatory cytokines, or FGFs (Huang *et al.*, 2007; Retamal *et al.*, 2007b).

Px1 hemichannels are permeable to Ca^{2+} and small molecules such as ATP and ethidium (Locovei *et al.*, 2006b; Pelegrin and Surprenant, 2006; Pelegrin and Surprenant, 2007), calcein (Thompson *et al.*, 2006), sulforhodamine (Thompson *et al.*, 2006; Peñuela *et al.*, 2007), and carboxyfluorescein (Locovei *et al.*, 2006a). The activity of Px hemichannels is inhibited by low intracellular pH and some connexin blockers such as carbenoxolone as well as, to a lesser extent, 18-β-glycerrhetinic acid and flufenamic acid. However, they are insensitive to other known blockers of connexin hemichannels like 1-heptanol, 1-octanol, La^{3+}, and Gd^{3+} (Pelegrin and Surprenant, 2006; Retamal *et al.*, 2007b). This, and their large single-channel conductances (550 pS, Bao *et al.*, 2004b) allows their function to be distinguished from connexin-based hemichannels. However, in most cases the specific contribution of connexins and pannexins to hemichannel-related activity in neurons has not been resolved.

One unique behavior of pannexins has been observed in macrophages and may be relevant to a role for their expression in microglia. Activation of $P2X_7$ receptors by ATP has long been associated with the opening of large pores. It has been proposed that this may occur through a conformational change of the $P2X_7$ channel that is normally small-ion permeant, to allow movement of larger molecules (North, 2002). However, strong evidence has suggested recently that the release of larger molecules, like interleukin-1β from macrophages, is actually mediated by Px1 hemichannels that are activated through $P2X_7$ (Pelegrin and Surprenant, 2006) by a mechanism that remains to be elucidated but may involve a heteromeric protein complex. ATP-driven cryopyrin-mediated caspase1 activation in response to bacterial stimuli of macrophages has also been shown to be dependent on functional Px1 hemichannels (Kanneganti *et al.*, 2007).

GAP JUNCTION CHANNELS AND HEMICHANNELS IN ACQUIRED AND GENETIC PATHOLOGIES OF THE CNS

Changes in neuronal as well glial gap junction channels and hemichannels have been described in diverse pathological conditions, including epilepsy, schizophrenia, and drug addiction (McCracken *et al.*, 2005; Nilsen *et al.*, 2006; Aleksic *et al.*, 2007), but herein we describe those in which molecular mechanisms involving gap junction channels and/or hemichannels are better understood.

Ischemia Reperfusion, Trauma, and Inflammatory Response

The inflammatory response is a common pathophysiological process of most if not all acute and chronic diseases. Under threatening conditions, microglia and astrocytes are activated and show phenotypic changes related to the intensity, duration, and quality of the threat. At moderate threat levels, ATP is released from glia via Cx43 hemichannels, leading to an increase in extracellular levels of adenosine that mediates a preconditioning response (Lin *et al.*, 2008), which confers tissue resistance to successive deleterious insults within a limited period of time. At higher threat levels, both microglia and astrocytes manifest rapid phenotypical changes *ad hoc* to either neutralize the threatening agents or condition and/or quickly adapt to the new microenvironment (e.g., high extracellular concentrations of K^+ and proinflammatory molecules). In contrast, neurons cannot adapt quickly and show a progressive increase in electrical activity leading to excitotoxicity. As more cell debris is generated, more microglia become activated (Beyer *et al.*, 2000) and more proinflammatory molecules are released, generating a positive feedback mechanism. A similar mechanism can be triggered by pathogens or foreign molecules detected by microglia. Free radicals and proinflammatoy cytokines generated at injured loci enhance the hemichannel activity and reduce gap junctional communication in astroglial cells (Contreras *et al.*, 2002; Même *et al.*, 2006; Retamal *et al.*, 2007b). These changes deprive neurons from numerous glial protective functions, and they become more susceptible to changes in their environment and die.

In chronic diseases, this inflammatory mechanism can be activated by anomalous molecules such as Aβ, α-synuclein, and huntingtin (e.g. in Down's syndrome, Alzheimer's, Parkinson's, and Huntington's diseases), by long-lasting infections (e.g. in multiple sclerosis and HIV), or by oxidizing environments [e.g. in ALS (Dangond *et al.*, 2004) and diabetes (Münch *et al.*, 2003; Raza and John, 2004)]. The inflam-mation-induced edema also reduces tissue perfusion, leading to an ischemic event that worsens with the inflammatory response. Edema causes pressure that might reduce astroglial expression of Cx43 (Malone *et al.*, 2007).

The role of connexins in ischemic insult is likely to be closely linked to the phenomenon of spreading depression (SD), discussed earlier. While pharmaco-logical blockers of connexin function had been reported to exacerbate damage caused by spreading depression or ischemia, genetic ablation of Cx43 in mice indicated that Cx43 serves a protective role. Of the two results, the genetic strategy is less prone to side effects but is also specific to only one connexin subtype, so the overall role of connexins and gap junctions in this process still has to be resolved.

Genetic Pathologies

Mutations (i.e., mis-sense) of different connexins have been directly linked to a diverse array of human diseases, some of which have been associated with the nervous system, including peripheral nerve paral-ysis and the most common genetically inherited form of nonsyndromic deafness. Expression of mutated connexins in cells and mice have confirmed the caus-ative role of many of these connexin defects in producing specific phenotypes associated with the disease (White and Paul, 1999). Exogenous expression of the mutated connexins has revealed a variety of problems, from failures in biosynthesis, to changes in permeability or gating (e.g., failure to open, or shift in the gating profile to voltage, etc.). When the effects are specific to one function of the channels (e.g., abil-ity to form gap junctions, or hemichannels, or perme-ability to specific molecules), these mutations can provide valuable insights into the underlying cause of the disease. Coexpression of mutant and wild-type connexins has also been shown, in some cases, to reproduce the dominant or recessive nature of the disease (Skerrett *et. al.*, 2004). However, when consid-ering the function of these connexins *in situ*, it is important to remember that many tissues express more than one connexin type, so that only cells with-out the ability to compensate for the specific connexin defect might be affected by the mutation.

Cx32 and X-Linked Charcot-Marie-Tooth Disease (CMTX)

Bergoffen *et al.* (1993) described the first definitive link of connexins to human disease by tracing Cx32

mutations to being the causative factor behind the X-linked form of Charcot-Marie-Tooth (CMTX) dis-ease, a demyelinating neuropathy predominantly affecting the peripheral nervous system (PNS), with occasional CNS involvement (Taylor *et al.*, 2003). Sub-sequently, Cx47 mutations have been associated with abnormal CNS myelin in Pelizaeus–Merzbacher–like disease (Uhlenberg *et al.*, 2004). CMT occurs with a fre-quency of 1:3000 and is the most common type of inher-ited nerve disorder in children. With a frequency of about 10%, the X-chromosome-linked form of CMT is the second most common inherited neuropathy and is genetically defined by over 260 distinct mutations in the GJB1 gene encoding Cx32 (Nave *et al.*, 2007; Shy *et al.,* 2007). While many of the mutations in Cx32 cause accumulation of the protein within the cell (Deschenes *et al.,* 1997), many form functional channels with gating abnormalities (Oh *et al.*, 1997; Ressot *et al.*, 1998; Abrams *et al.*, 2001, 2002) or permeability changes (Oh *et al.*, 1997; Ressot *et al.*, 1998; Bicego *et al.,* 2006). Over-all, the evidence supports that the channel function of Cx32 is important for normal myelin structure.

Cx32 is located at the paranodes and Schmidt–Lanter-mann incisures of myelinating Schwann cells in periph-eral nerves (Scherer *et al.*, 1995), where "reflexive" gap junctions (connecting cytoplasmic domains of the same cell) between myelin layers are believed to reduce the dif-fusion distance between the Schwann cell nucleus and the myelin wrap closest to the axon (Oh *et al.*, 1997; Abrams *et al.*, 2000) that may be critical to the normal function of the axon–Schwann cell unit. While a role for these reflexive gap junctions is certainly consistent with most or all of the disease-linked mutants studied to date, it should be noted that several CMTX-linked mutations of Cx32 (S85C, D178Y, and F235C) have been associated to altered hemichannel properties, typically resulting in much higher open probabilities under physiological con-ditions (Castro *et al.*, 1999; Abrams *et al.*, 2002; Gómez-Hernández *et al.*, 2003; Liang *et al.*, 2005). Such an increase in Cx32 hemichannel activity in myelinating Schwann cells may induce damage through loss of ionic gradients and small metabolites and increased Ca²⁺ influx, thus providing a mechanism by which some Cx32 mutants may damage cells in which they are expressed.

Cx43 and Oculodentodigital Dysplasia (ODDD)

ODDD is a rare autosomal-dominant disorder characterized by craniofacial anomalies involving the teeth and skull, as well as fusion of the digits, as the name would suggest. Patients also manifest neu-rological symptoms, such as mental retardation, ataxia, neurogenic bladder, seizures, spasticity, and hearing impairment (Loddenkemper *et al.*, 2002; Kjaer *et al.*, 2004; Vitiello *et al.*, 2005). GJA1, encoding Cx43, is the only gene in which mutations have been found in patients or families affected with ODDD. So far, over 35 distinct mutations causing ODDD have been identified, involving most domains of Cx43, all with a negative influence on gap junction channel function (Paznekas *et al.*, 2003). The correlation of disease phenotype with altered hemichannel function is less clear, as in the cases that have been tested, some mutants cause loss of both gap junction and hemichannel activity (Lai *et al.*, 2006), while in others an increase in hemichannel activity was noted, correlated with an extended half life of the mutant, in the absence of any detectable gap junction function (Dobrowolski *et al.*, 2007). The latter study concluded that the presence of more open hemichannels than normal could aggravate the disease.

Patients with ODDD also present hearing defects, yet Cx43 is not the predominant connexin in the adult ear. This apparent contradiction might be explained if Cx43 is required for normal ear ontogeny. Since Cx43 is not expressed by most neurons, this suggests that the presence of central nervous system symptoms in ODDD patients is related to the altered function of other cell types, most likely astrocytes and microglia, or in neural progenitor cells that still express Cx43.

Cx26 (also Cx30 and 31) and Nonsyndromic Deafness

Mutations of GJB2 gene, encoding Cx26, are the most common cause of hereditary deafness, and more than 90 different mutations are associated with recessive forms of nonsyndromic hearing loss (Oguchi *et al.*, 2005). In fact GJB1 is one of the most highly mutated genes in the human genome, for largely the same reason why CFTR is heavily mutated: there is a string of Gs at the beginning of the gene that is prone to errors during replication. Among other tissues, the cochlea expresses Cx26 (Sohl and Willecke, 2004) in the nonsensory epithelial cells in the organ of Corti, stria vascularis, and type II fibrocytes of the connective tissue beneath the epithelial layer (Kikuchi *et al.*, 1995). Connexin 30 (Cx30) is also present in this tissue and colocalizes with Cx26, presumably forming heteromeric hemichannels (Forge *et al.*, 1999, 2003).

In the inner ear, Cx26 is thought to contribute in maintaining cochlear homeostasis by offering an intercellular pathway between the supporting cells of the organ of Corti for recycling the endolymphatic K^+ that is expelled from the sensory cells during auditory transduction (Kikuchi *et al.*, 1995; Zhao *et al.*, 2006). This is likely to be important for "spatial buffering" of K^+ as

described for astrocytes in the cortex, and consistent with this, mice with conditional knock-out of Cx26 in the ear show significantly enhanced death of the hair cells, consistent with hyperexcitability-induced apoptosis (Aarts and Tymianski, 2004). However, there is also a specialized need for K^+ recycling in the ear, as K^+ must diffuse intercellularly from its release from the hair cells through the supporting cells, to the spiral ligament of stria vascularis, until it is ultimately released into the endolymph by Na^+Cl^-/K^+ co-transporters or Na^+/K^+ antiporters. Gap junctions comprise an important part of this "circuit." This is required for normal hearing, as the endolymph maintains a high K^+ concentration, and hence potential ($\sim+80$ mV), that provides an enhanced driving potential across the hair cell membrane (Vm of ~160 mV), resulting in an enhanced signal upon activation of the hair cells (translating into an amplification of almost 90 decibels in effective hearing).

Although many deafness mutants are recessive, some do display dominant behaviors (Kelsell *et al.*, 1997), and this has been correlated with dominant-negative effects when coexpressed with wild type Cx26 in exogenous systems like the Xenopus oocyte (Skerrett *et al.*, 2004; Laird, 2008). As with the previous two diseases discussed, while disease-causing mutants are dispersed throughout the protein, there is a strong correlation of disease phenotype with channel function. Some effects of these mutants on hemichannels have been reported (Stong *et al.*, 2006), but in at least one case the dominant nature of the disease correlated better with the effects of the mutant in gap junctions than in hemichannels, where it showed recessive characteristics (Chen *et al.*, 2005). One particularly instructive deafness mutant (V85L) studied by Mammano and colleagues showed normal gap junction characteristics in terms of gating, conductance, and even passage of dyes between cells. Only when it was specifically tested for IP_3 permeability was its defect apparent (Beltramello *et al.*, 2005), suggesting that recycling of K^+ may occur not directly but through a regenerative mechanism analogous to Ca^{2+} waves. Of course, interpretation of all these findings is complicated by the coexpression of at least two other connexins in these tissues, each of which, when mutated, can lead to deafness.

SUMMARY AND PERSPECTIVE

In this chapter we have concentrated on the roles of connexins in vertebrates, yet invertebrate systems use electrical systems to control activity to an even greater extent. Composed of innexins, electrical synapses in invertebrates control coordination of muscle contractions as required in *C. elegans* for locomotion (unc7 and 9) or rhythmic contractions in the digestive system (eat5 and inx3 and 6), and escape responses such as the jump reflex in *D. melanogaster* (shakB lethal) and the tail flip of crustaceans, or transmission at the giant synapse of the squid. All these responses utilize the reduced synaptic time constant at electrical compared to chemical synapses to achieve rapid response times. Electrical synapses in the adult nervous systems of vertebrates are much more restricted in scope but similarly tend to be involved in rapid and bidirectional transmission as seen in escape reflexes (Mauthner neuron in teleost fish) or coordination of oscillatory activity in various brain regions like the hippocampus, cerebellum, and cortex. These oscillations are likely to serve in the fine-tuning of various behaviors, and possibly memories, but their roles are still ill-defined at this time. The role of neuronal coupling is better understood in the retina, which serves as a model system to illustrate multiple connexin roles. In vertebrates, the frequent coexistence of chemical and electrical synapses at the same cell contact allows for optimal flexibility in neuronal responses, including activity-dependent potentiation of electrical synapses that has usually been considered to occur only with chemical transmission.

While neuronal electrical synapses may be less common in vertebrate nervous systems, gap junctions form between many other cell types and mediate a variety of functions related to intercellular transfer of not only ions but also larger metabolites. Primary among these is the supporting role of astrocytes and oligodendrocytes in maintenance of neuronal health at sites of high neuronal activity by spatial buffering of K^+ and glutamate. Oligodendrocytes and their PNS cousins, Schwann cells, also require connexin expression for the formation and maintenance of myelination. Several diseases of the PNS and CNS associated with defects in Cx32, 47, 26, and 30 illustrate several of these functions. The chapter also emphasizes the functions of connexins beyond gap junctions, including their roles as hemichannels, composed of either connexins or the descendents of innexins in vertebrates, the Px1, and possibly 2. A role of connexins in adhesion during neuronal migration in early development is also described.

Compared to other ion channel classes and receptors, the study of connexins and particularly pannexins in the nervous system is in its infancy. We have already learned that they play very diverse roles in regulating CNS and PNS function, but a true understanding of the mechanisms underlying these roles lies ahead as we begin to understand the permeability and regulatory features of each connexin, which functions are redundant and unique, and how to more selectively probe their physiological relevance through more targeted genetic ablations. The next decade offers great prospects for a deeper understanding of the integrative effects of gap junctions on the function of the nervous system.

References

Aarts MM, Tymianski M. Molecular mechanisms underlying specificity of excitotoxic signaling in neurons. Curr. Mol. Med. 4: 137-147, 2004.

Abrams CK, Freidin MM, Verselis VK, Bennett MV, Bargiello TA. Functional alterations in gap junction channels formed by mutant forms of connexin 32: evidence for loss of function as a pathogenic mechanism in the X-linked form of Charcot-Marie-Tooth disease. Brain Res. 900: 9-25, 2001.

Abrams CK, Bennett MV, Verselis VK, Bargiello TA. Voltage opens unopposed gap junction hemichannels formed by a connexin 32 mutant associated with X-linked Charcot-Marie-Tooth disease. Proc. Natl. Acad. Sci. USA. 99: 3980-3984, 2002.

Abrams CK, Freidin M, Bukauskas F, Dobrenis K, Bargiello TA, Verselis VK, Bennett MV, Chen L, Sahenk Z. Pathogenesis of X-linked charcot-marie-tooth disease: differential effects of two mutations in connexin 32. J. Neurosci. 23: 10548-10558, 2003.

Aleksic B, Ishihara R, Takahashi N, Maeno N, Ji X, Saito S, Inada T, Ozaki N. Gap junction coding genes and schizophrenia: a genetic association study. J. Hum. Genet. 52: 498-501, 2007.

Alvarez-Maubecin V, Garcia-Hernandez F, Williams JT, Van Bockstaele EJ. Functional coupling between neurons and glia. J. Neurosci. 20: 4091-4098, 2000.

Anderson, PA. Physiology of a bidirectional, excitatory, chemical synapse. J. Neurophysiol. 53: 821-835, 1985.

Anderson PA, Grunert U. Three-dimensional structure of bidirectional, excitatory chemical synapses in the jellyfish Cyanea capillata. Synapse 2: 606-613, 1988.

Bani-Yaghoub M, Underhill TM, Naus CC. Gap junction blockage interferes with neuronal and astroglial differentiation of mouse P19 embryonal carcinoma cells. Dev Genet. 24: 69-81, 1999.

Bao L, Sachs F, Dahl G. Connexins are mechanosensitive. Am. J. Physiol. 287: C1389-C1395, 2004a.

Bao L, Locovei S, Dahl G. Pannexin membrane channels are mechanosensitive conduits for ATP. FEBS Lett. 572: 65-68, 2004b.

Baranova A, Ivanov D, Petrash N, Pestova A, Skoblov M, Kelmanson I, Shagin D, Nazarenko S, Geraymovych E, Litvin O, Tiunova A, Born TL, Usman N, Staroverov D, Lukyanov S, Panchin Y. The mammalian pannexin family is homologous to the invertebrate innexin gap junction proteins. Genomics. 83: 706-716, 2004.

Barbe MT, Monyer H, Bruzzone R. Cell-cell communication beyond connexins: the pannexin channels. Physiology (Bethesda) 21: 103-114, 2006.

Batter DK, Corpina RA, Roy C, Spray DC, Hertzberg EL, Kessler JA. Heterogeneity in gap junction expression in astrocytes cultured from different brain regions. Glia. 6: 213-221, 1992.

Bedner P, Niessen H, Odermatt B, Kretz M, Willecke K, Harz H. Selective permeability of different connexin channels to the second messenger cyclic AMP. J. Biol. Chem. 281: 6673-6681, 2006.

Belliveau DJ, Bani-Yaghoub M, McGirr B, Naus CC, Rushlow WJ. Enhanced neurite outgrowth in PC12 cells mediated by connexin hemichannels and ATP. J Biol Chem. 2006 Jul 28;281 (30): 20920-31.

Belluardo N, Mudò G, Trovato-Salinaro A, Le Gurun S, Charollais A, Serre-Beinier V, Amato G, Haefliger JA, Meda P, Condorelli DF. Expression of connexin36 in the adult and developing rat brain. Brain Res. 865:121-138, 2000.

Beltramello M, Piazza V, Bukauskas FF, Pozzan T, Mammano F. Impaired permeability to Ins(1,4,5)P3 in a mutant connexin underlies recessive hereditary deafness. Nat. Cell Biol. 7: 63-69, 2005.

Bennett MV, Contreras JE, Bukauskas FF, Sáez JC. New roles for astrocytes: gap junction hemichannels have something to communicate. Trends Neurosci. 26: 610-617, 2003.

Bennett MV, Zukin RS. Electrical coupling and neuronal synchronization in the mammalian brain. Neuron 41: 495-511, 2004.

Bennett MVL and Spray DC. Intercellular communication mediated by gap junctions can be controlled in many ways. In: G. Gall, WE. Gall, WH. Cowan (Eds.) Synaptic function pp. 109-134, 1987.

Bergoffen J, Scherer SS, Wang S, Scott MO, Bone LJ, Paul DL, Chen K, Lensch MW, Chance PF, Fischbeck KH. Connexin mutations in X-linked Charcot-Marie-Tooth disease. Science. 262: 2039-2042, 1993.

Berthoud VM, Sáez JC. Changes in connexin43, the gap junction protein of astrocytes, during development of the rat pineal gland. J. Pineal Res. 14(2): 67-72, 1993.

Beyer M, Gimsa U, Eyüpoglu IY, Hailer NP, Nitsch R. Phagocytosis of neuronal or glial debris by microglial cells: upregulation of MHC class II expression and multinuclear giant cell formation in vitro. Glia. 31: 262-266, 2000.

Bicego M, Morassutto S, Hernandez VH, Morgutti M, Mammano F, D'Andrea P, Bruzzone R. Selective defects in channel permeability associated with Cx32 mutations causing X-linked Charcot-Marie-Tooth disease. Neurobiol Dis. 21: 607-617, 2006.

Bloomfield SA, Xin D, Osborne T. Light-induced modulation of coupling between AII amacrine cells in the rabbit retina. Vis. Neurosci. 14: 565-576, 1997.

Boassa D, Ambrosi C, Qiu F, Dahl G, Gaietta G, Sosinsky G. Pannexin1 channels contain a glycosylation site that targets the hexamer to the plasma membrane. J. Biol. Chem. 282: 31733-31743, 2007.

Braet K, Paemeleire K, D'Herde K, Sanderson MJ, Leybaert L. Astrocyte-endothelial cell calcium signals conveyed by two signalling pathways. Eur. J. Neurosci. 13: 79-91, 2001.

Braet K, Aspeslagh S, Vandamme W, Willecke K, Martin PE, Evans WH, Leybaert L. Pharmacological sensitivity of ATP release triggered by photoliberation of inositol-1,4,5-trisphosphate and zero extracellular calcium in brain endothelial cells. J. Cell Physiol. 197: 205-213, 2003.

Bruzzone R, Hormuzdi SG, Barbe MT, Herb A, Monyer H. Pannexins, a family of gap junction proteins expressed in brain. Proc. Natl. Acad. Sci. USA. 100: 13644-13649, 2003.

Bruzzone R, Barbe MT, Jakob NJ, Monyer H. Pharmacological properties of homomeric and heteromeric pannexin hemichannels expressed in Xenopus oocytes. J. Neurochem. 92: 1033-1043, 2005.

Bukauskas FF, Bukauskiene A, Bennett MV, Verselis VK. Gating properties of gap junction channels assembled from connexin43 and connexin43 fused with green fluorescent protein. Biophys. J. 81:137-152, 2001.

Cachope R, Mackie K, Triller A, O'Brien J, Pereda AE. Potentiation of electrical and chemical synaptic transmission mediated by endocannabinoids. Neuron 56: 1034-1047, 2007.

Castro C, Gómez-Hernández JM, Silander K, Barrio LC. Altered formation of hemichannels and gap junction channels caused by C-terminal connexin-32 mutations. J. Neurosci. 19: 3752-3760, 1999.

Chen Y, Deng Y, Bao X, Reuss L, Altenberg GA. Mechanism of the defect in gap-junctional communication by expression of a connexin 26 mutant associated with dominant deafness. FASEB J. 19: 1516-1518, 2005.

Cina C, Bechberger JF, Ozog MA, Naus CC. Expression of connexins in embryonic mouse neocortical development. J. Comp Neurol. 504: 298-313, 2007.

Cohen E, Sterling P. Accumulation of (³H)glycine by cone bipolar neurons in the cat retina. J. Comp. Neurol. 250: 1-7, 1986.

Contreras JE, Sánchez HA, Eugenin EA, Speidel D, Theis M, Willecke K, Bukauskas FF, Bennett MV, Sáez JC. Metabolic inhibition induces opening of unapposed connexin 43 gap junction hemichannels and reduces gap junctional communication in cortical astrocytes in culture. Proc. Natl. Acad. Sci. USA 99: 495-500, 2002.

Contreras JE, Sánchez H, Véliz L, Bukauskas FF, Bennett MVL, Sáez JC. Role of connexin-based gap junction channels and hemichannels in ischemia-induced cell death in nervous tissue. Brain Res. Rev. 47: 290-303, 2004.

Dangond F, Hwang D, Camelo S, Pasinelli P, Frosch MP, Stephanopoulos G, Stephanopoulos G, Brown RH Jr, Gullans SR. Molecular signature of late-stage human ALS revealed by expression profiling of postmortem spinal cord gray matter. Physiol. Genomics 16: 229-239, 2004.

Daniel EE, Wang YF. Gap junctions in intestinal smooth muscle and interstitial cells of Cajal. Microsc. Res. Tech. 47: 309-320, 1999.

Deans MR, Volgyi B, Goodenough DA, Bloomfield SA, Paul DL. Connexin36 is essential for transmission of rod-mediated visual signals in the mammalian retina. Neuron 36: 703-712, 2002.

De Pina-Benabou MH, Srinivas M, Spray DC, Scemes E. Calmodulin kinase pathway mediates the K+-induced increase in gap junctional communication between mouse spinal cord astrocytes. J. Neurosci. 21: 6635-6643, 2001.

Dermietzel R, Traub O, Hwang TK, Beyer E, Bennett MV, Spray DC, Willecke K. Differential expression of three gap junction proteins in developing and mature brain tissues. Proc. Natl. Acad. Sci. USA 86: 10148-10152, 1989.

Dermietzel R, Hertberg EL, Kessler JA, Spray DC. Gap junctions between cultured astrocytes: immunocytochemical, molecular, and electrophysiological analysis. J. Neurosci. 11: 1421-1432, 1991.

Dermietzel R, Krause D. Molecular anatomy of the blood-brain barrier as defined by immunocytochemistry. Int. Rev. Cytol. 127: 57-109, 1991.

Dermietzel R, Farooq M, Kessler JA, Althaus H, Hertzberg EL, Spray DC. Oligodendrocytes express gap junction proteins connexin32 and connexin45. Glia. 20: 101-114, 1997.

Dermietzel R, Kremer M, Paputsoglu G, Stang A, Skerrett IM, Gomes D, Srinivas M, Janssen-Bienhold U, Weiler R, Nicholson BJ, Bruzzone R, Spray DC. Molecular and functional diversity of neural connexins in the retina. J. Neurosci. 20: 8331-8343, 2000.

Deschênes SM, Walcott JL, Wexler TL, Scherer SS, Fischbeck KH. Altered trafficking of mutant connexin32. J. Neurosci. 17: 9077-9084, 1997.

Dobrenis K, Chang HY, Pina-Benabou MH, Woodroffe A, Lee SC, Rozental R, Spray DC, Scemes E. Human and mouse microglia express connexin36, and functional gap junctions are formed between rodent microglia and neurons. J. Neurosci. Res. 82: 306-315, 2005.

Dobrowolski R, Sommershof A, Willecke K. Some oculodentodigital dysplasia-associated cx43 mutations cause increased hemichannel activity in addition to deficient gap junction channels. J. Membr. Biol. 219: 9-17, 2007.

Duffy HS, Delmar M, Spray DC. Formation of the gap junction nexus: binding partners for connexins. J Physiol (Paris) 96: 243-249, 2002.

Ebihara L, Steiner E. Properties of a nonjunctional current expressed from a rat connexin46 cDNA in Xenopus oocytes. J. Gen. Physiol. 102: 59-74, 1993.

Elias LA, Wang DD, Kriegstein AR. Gap junction adhesion is necessary for radial migration in the neocortex. Nature 448: 901-907, 2007.

Enkvist MO, McCarthy KD. Astroglial gap junction communication is increased by treatment with either glutamate or high K+ concentration. J. Neurochem. 62: 489-495, 1994.

Eugenín EA, Eckardt D, Theis M, Willecke K, Bennett MV, Sáez JC. Microglia at brain stab wounds express connexin 43 and in vitro form functional gap junctions after treatment with interferon-gamma and tumor necrosis factor-alpha. Proc. Natl. Acad. Sci. USA 98: 4190-4195, 2001.

Evans WH, Leybaert L. Mimetic peptides as blockers of connexin channel-facilitated intercellular communication. Cell Commun. Adhes. 14: 265-273, 2007.

Figiel M, Allritz C, Lehmann C, Engele J. Transgenic mice for conditional gene manipulation in astroglial cells. Glia. 55: 1565-1576, 2007.

Foote CI, Zhou L, Zhu X, Nicholson B J. The pattern of disulfide linkages in the extracellular loop regions of connexin32 suggests a model for the docking interface of gap junctions. J. Cell Biol. 140: 1187-1197, 1998.

Forge A, Becker D, Casalotti S, Edwards J, Evans WH, Lench N, Souter M. Gap junctions and connexin expression in the inner ear. Novartis Found. Symp. 219: 134-50; discussion 151-156, 1999.

Forge A, Marziano NK, Casalotti SO, Becker DL, Jagger D. The inner ear contains heteromeric channels composed of cx26 and cx30 and deafness-related mutations in cx26 have a dominant negative effect on cx30. Cell Commun. Adhes. 10: 341-346, 2003.

Frisch C, De Souza-Silva MA, Söhl G, Güldenagel M, Willecke K, Huston JP, Dere E. Stimulus complexity dependent memory impairment and changes in motor performance after deletion of the neuronal gap junction protein connexin36 in mice. Behav. Brain Res. 157: 177-185, 2005.

Fróes MM, de Carvalho AC. Gap junction-mediated loops of neuronal-glial interactions. Glia. 24: 97-107, 1998.

Fróes MM, Correia AH, Garcia-Abreu J, Spray DC, Campos de Carvalho AC, Neto MV. Gap-junctional coupling between neurons and astrocytes in primary central nervous system cultures. Proc. Natl. Acad. Sci. USA 96: 7541-7546, 1999.

Fukuda T, Kosaka T, Singer W, Galuske RA. Gap junctions among dendrites of cortical GABAergic neurons establish a dense and widespread intercolumnar network. J Neurosci. 26: 3434-3443, 2006.

Furshpan EJ, Potter, DD. Transmission at the giant motor synapses of the crayfish. J. Physiol. (Lond.) 145: 289-325, 1959.

Furukawa T, Furshpan EJ. Two inhibitory mechanisms in the Mauthner neurons of goldfish. J Neurophysiol. 26: 140-176, 1963.

Gaietta G, Deerinck TJ, Adams SR, Bouwer J, Tour O, Laird DW, Sosinsky GE, Tsien RY, Ellisman MH. Multicolor and electron microscopic imaging of connexin trafficking. Science 296: 503-507, 2002.

Galarreta M, Hestrin S. A network of fast-spiking cells in the neocortex connected by electrical synapses. Nature. 402: 72-75, 1999.

Garg S, Md Syed M, Kielian T. Staphylococcus aureus-derived peptidoglycan induces Cx43 expression and functional gap junction intercellular communication in microglia. J. Neurochem. 95: 475-483, 2005.

George CH, Kendall JM, Evans WH. Intracellular trafficking pathways in the assembly of connexins into gap junctions. J. Biol. Chem. 274: 8678-8685, 1999.

Gerido DA, White TW. Connexin disorders of the ear, skin, and lens. Biochim. Biophys. Acta. 1662: 159-170, 2004.

Giaume C, Fromaget C, el Aoumari A, Cordier J, Glowinski J, Gros D. Gap junctions in cultured astrocytes: single-channel currents and characterization of channel-forming protein. Neuron 6: 133-143, 1991.

Gibson JR, Beierlein M, Connors BW. Two networks of electrically coupled inhibitory neurons in neocortex. Nature. 402: 75-79, 1999.

Giepmans BNG. Gap junctions and connexin-interacting proteins. Cardiovasc. Res. 62: 233-245, 2004.

Goldberg GS, Lampe PD, Nicholson, BJ. Selective transfer of endogenous metabolites through gap junctions composed of different connexins. Nature Cell Biol. 1: 457-459, 1999.

Gómez-Hernández JM, de Miguel M, Larrosa B, González D, Barrio LC. Molecular basis of calcium regulation in connexin-32 hemichannels. Proc. Natl. Acad. Sci. USA 100: 16030-16035, 2003.

González D, Gómez-Hernández JM, Barrio LC. Species specificity of mammalian connexin-26 to form open voltage-gated hemichannels. FASEB J. 20: 2329-2338, 2006.

Haddock RE, Grayson TH, Brackenbury TD, Meaney KR, Neylon CB, Sandow SL, Hill CE. Endothelial coordination of cerebral vasomotion via myoendothelial gap junctions containing connexins 37 and 40. Am. J. Physiol. Heart Circ. Physiol. 291: H2047-H2056, 2006.

Harris, AL. Emerging issues of connexin channels: biophysics fills the gap. Q. Rev. Biophys. 34: 325-472, 2001.

Honma S, De S, Li D, Shuler CF, Turman JE Jr. Developmental regulation of connexins 26, 32, 36, and 43 in trigeminal neurons. Synapse 52: 258-271, 2004.

Hormuzdi SG, Filippov MA, Mitropoulou G, Monyer H, Bruzzone R. Electrical synapses: a dynamic signaling system that shapes the activity of neuronal networks. Biochim. Biophys. Acta. 1662: 113-137, 2004.

Huang Y, Grinspan JB, Abrams CK, Scherer SS. Pannexin1 is expressed by neurons and glia but does not form functional gap junctions. Glia. 55: 46-56, 2007.

Iadecola C, Nedergaard M. Glial regulation of the cerebral microvasculature. Nat. Neurosci. 10: 1369-1376, 2007.

Jiang, JX, Gu S. Gap junction- and hemichannel-independent actions of connexins. Biochim. Biophys. Acta. 1711: 208-214, 2005.

Jordan K, Chodock R, Hand AR, Laird DW. The origin of annular junctions: a mechanism of gap junction internalization. J. Cell Sci. 114(Pt 4): 763-773, 2001.

Kamasawa N, Sik A, Morita M, Yasumura T, Davidson KG, Nagy JI, Rash JE. Connexin-47 and connexin-32 in gap junctions of oligodendrocyte somata, myelin sheaths, paranodal loops and Schmidt-Lanterman incisures: implications for ionic homeostasis and potassium siphoning. Neuroscience 136: 65-86, 2005.

Kamermans M, Fahrenfort I. Ephaptic interactions within a chemical synapse: hemichannel-mediated ephaptic inhibition in the retina. Curr. Opin. Neurobiol. 14: 531-541, 2004.

Kanneganti TD, Lamkanfi M, Kim YG, Chen G, Park JH, Franchi L, Vandenabeele P, Núñez G. Pannexin-1-mediated recognition of bacterial molecules activates the cryopyrin inflammasome independent of Toll-like receptor signaling. Immunity 26: 433-443, 2007.

Kasischke KA, Vishwasrao HD, Fisher PJ, Zipfel WR, Webb WW. Neural activity triggers neuronal oxidative metabolism followed by astrocytic glycolysis. Science 305: 99-103, 2004.

Kelsell DP, Dunlop J, Stevens HP, Lench NJ, Liang JN, Parry G, Mueller RF, Leigh IM. Connexin 26 mutations in hereditary non-syndromic sensorineural deafness. Nature 387: 80-83, 1997.

Kettenmann H, Orkand RK, Schachner M. Coupling among identified cells in mammalian nervous system cultures. J. Neurosci. 3: 506-516, 1983.

Kettenmann H, Ransom BR. Electrical coupling between astrocytes and between oligodendrocytes studied in mammalian cell cultures. Glia. 1: 64-73, 1988.

Kikuchi T, Kimura RS, Paul DL, Adams JC. Gap junctions in the rat cochlea: immunohistochemical and ultrastructural analysis. Anat. Embryol. (Berl). 191: 101-118, 1995.

Kjaer KW, Hansen L, Eiberg H, Leicht P, Opitz JM, Tommerup N. Novel connexin 43 (GJA1) mutation causes oculo-dento-digital dysplasia with curly hair. Am. J. Med. Genet. A. 127: 152-157, 2004.

Kreuzberg MM, Deuchars J, Weiss E, Schober A, Sonntag S, Wellershaus K, Draguhn A, Willecke K. Expression of connexin30.2 in interneurons of the central nervous system in the mouse. Mol. Cell. Neurosci. 37: 119-134, 2008.

Kronengold J, Trexler EB, Bukauskas FF, Bargiello TA, Verselis VK. Single-channel SCAM identifies pore-lining residues in the first extracellular loop and first transmembrane domains of Cx46 hemichannels. J Gen Physiol. 122: 389-405, 2003.

Kumar NM, Gilula NB. Cloning and characterization of human and rat liver cDNAs coding for a gap junction protein. J.Cell Biol. 103: 767-776, 1986.

Kunzelmann P, Schröder W, Traub O, Steinhäuser C, Dermietzel R, Willecke K. Late onset and increasing expression of the gap junction protein connexin30 in adult murine brain and long-term cultured astrocytes. Glia. 25: 111-119, 1999.

Kwak BR, Sáez JC, Wilders R, Chanson M, Fishman GI, Hertzberg EL, Spray DC, Jongsma HJ. Effects of cGMP-dependent phosphorylation on rat and human connexin43 gap junction channels. Pflugers Arch. 430: 770-778, 1995.

Lai A, Le DN, Paznekas WA, Gifford WD, Jabs EW, Charles AC. Oculodentodigital dysplasia connexin43 mutations result in non-functional connexin hemichannels and gap junctions in C6 glioma cells. J. Cell Sci. 119: 532-541, 2006.

Lai CP, Bechberger JF, Thompson RJ, MacVicar BA, Bruzzone R, Naus CC. Tumor-suppressive effects of pannexin 1 in C6 glioma cells. Cancer Res. 67: 1545-1554, 2007.

Laing JG, Tadros PN, Westphale EM, Beyer EC. Degradation of connexin43 gap junctions involves both the proteasome and the lysosome. Exp. Cell Res. 236(2): 482-492, 1997.

Laird DW. Life cycle of connexins in health and disease. Biochem. J. 394: 527-543, 2006.

Laird DW. Closing the gap on autosomal dominant connexin-26 and connexin-43 mutants linked to human disease. J. Biol. Chem. 283: 2997-3001, 2008.

Leung DS, Unsicker K, Reuss B. Expression and developmental regulation of gap junction connexins cx26, cx32, cx43 and cx45 in the rat midbrain-floor. Int. J. Dev. Neurosci. 20: 63-75, 2002.

Leybaert L. Neurobarrier coupling in the brain: a partner of neurovascular and neurometabolic coupling? J. Cereb. Blood Flow Metab. 25: 2-16, 2005.

Li X, Ionescu AV, Lynn BD, Lu S, Kamasawa N, Morita M, Davidson KG, Yasumura T, Rash JE, Nagy JI. Connexin47, connexin29 and connexin32 co-expression in oligodendrocytes and Cx47 association with zonula occludens-1 (ZO-1) in mouse brain. Neuroscience 126: 611-630, 2004.

Liang GS, de Miguel M, Gómez-Hernández JM, Glass JD, Scherer SS, Mintz M, Barrio LC, Fischbeck KH. Severe neuropathy with leaky connexin32 hemichannels. Ann. Neurol. 57: 749-754, 2005.

Lin JH, Lou N, Kang N, Takano T, Hu F, Han X, Xu Q, Lovatt D, Torres A, Willecke K, Yang J, Kang J, Nedergaard M. A central role of connexin 43 in hypoxic preconditioning. J. Neurosci. 28: 681-695, 2008.

Little TL, Beyer EC, Duling BR. Connexin 43 and connexin 40 gap junctional proteins are present in arteriolar smooth muscle and endothelium in vivo. Am. J. Physiol. 268: H729-H739, 1995.

Litvin O, Tiunova A, Connell-Alberts Y, Panchin Y, Baranova A. What is hidden in the pannexin treasure trove: the sneak peek and the guesswork. J. Cell Mol. Med. 10: 613-634, 2006.

Llinás RR. The intrinsic electrophysiological properties of mammalian neurons: insights into central nervous system function. Science 242: 1654-1664, 1988.

Locovei S, Wang J, Dahl G. Activation of pannexin 1 channels by ATP through P2Y receptors and by cytoplasmic calcium. FEBS Lett. 580: 239-244, 2006a.

Locovei S, Bao L, Dahl G. Pannexin 1 in erythrocytes: function without a gap. Proc. Natl. Acad. Sci. USA 103: 7655-7659, 2006b.

Locovei S, Scemes E, Qiu F, Spray DC, Dahl G. Pannexin1 is part of the pore forming unit of the P2X(7) receptor death complex. FEBS Lett. 581: 483-488, 2007.

Loddenkemper T, Grote K, Evers S, Oelerich M, Stogbauer F. Neurological manifestations of the oculodentodigital dysplasia syndrome. J. Neurol. 249: 584-595, 2002.

Magistretti PJ, Pellerin L. Cellular mechanisms of brain energy metabolism and their relevance to functional brain imaging. Philos. Trans. R. Soc. Lond. B. Biol. Sci. 354: 1155-1163, 1999.

Malone P, Miao H, Parker A, Juarez S, Hernandez MR. Pressure induces loss of gap junction communication and redistribution of connexin 43 in astrocytes. Glia. 55: 1085-1098, 2007.

Martin PE, Errington RJ, Evans WH. Gap junction assembly: multiple connexin fluorophores identify complex trafficking pathways. Cell Commun. Adhes. 8: 243-248, 2001.

Martínez AD, Sáez JC. Arachidonic acid-induced dye uncoupling in rat cortical astrocytes is mediated by arachidonic acid byproducts. Brain Res. 816: 411-423, 1999.

Martínez AD, Eugenín EA, Brañes MC, Bennett MV, Sáez JC. Identification of second messengers that induce expression of functional gap junctions in microglia cultured from newborn rats. Brain Res. 943: 191-201, 2002.

Maxeiner S, Krüger O, Schilling K, Traub O, Urschel S, Willecke K. Spatiotemporal transcription of connexin45 during brain development results in neuronal expression in adult mice. Neuroscience. 119: 689-700, 2003.

McCracken CB, Hamby SM, Patel KM, Morgan D, Vrana KE, Roberts DC. Extended cocaine self-administration and deprivation produces region-specific and time-dependent changes in connexin36 expression in rat brain. Synapse 8: 141-150, 2005.

Même W, Calvo CF, Froger N, Ezan P, Amigou E, Koulakoff A, Giaume C. Proinflammatory cytokines released from microglia inhibit gap junctions in astrocytes: potentiation by beta-amyloid. FASEB J. 20: 494-496, 2006.

Menichella DM, Majdan M, Awatramani R, Goodenough DA, Sirkowski E, Scherer SS, Paul DL. Genetic and physiological evidence that oligodendrocyte gap junctions contribute to spatial buffering of potassium released during neuronal activity. J. Neurosci. 26: 10984-10991, 2006.

Mercier F, Hatton GI. Connexin 26 and basic fibroblast growth factor are expressed primarily in the subpial and subependymal layers in adult brain parenchyma: roles in stem cell proliferation and morphological plasticity? J. Comp. Neurol. 431: 88-104, 2001.

Metea MR, Newman EA. Glial cells dilate and constrict blood vessels: a mechanism of neurovascular coupling. J. Neurosci. 26: 2862-2870, 2006.

Mitropoulou G, Bruzzone R. Modulation of perch connexin35 hemi-channels by cyclic AMP requires a protein kinase A phosphory-lation site. J. Neurosci. Res. 72: 147-157, 2003.

Moreno AP, Rook MB, Fishman GI, Spray DC. Gap junction chan-nels: distinct voltage-sensitive and -insensitive conductance states. Biophys. J. 67: 113-119, 1994.

Münch G, Gasic-Milenkovic J, Dukic-Stefanovic S, Kuhla B, Hein-rich K, Riederer P, Huttunen HJ, Founds H, Sajithlal G. Micro-glial activation induces cell death, inhibits neurite outgrowth and causes neurite retraction of differentiated neuroblastoma cells. Exp. Brain Res. 150: 1-8, 2003.

Nadarajah B, Makarenkova H, Becker DL, Evans WH, Parnavelas JG. Basic FGF increases communication between cells of the developing neocortex. J. Neurosci. 18: 7881-7890, 1998.

Nagasawa K, Chiba H, Fujita H, Kojima T, Saito T, Endo T, Sawada N. Possible involvement of gap junctions in the barrier function of tight junctions of brain and lung endothelial cells. J. Cell Physiol. 208: 123-132, 2006.

Nagelhus EA, Mathiisen TM, Ottersen OP. Aquaporin-4 in the cen-tral nervous system: cellular and subcellular distribution and coexpression with KIR4.1. Neurosci. 129: 905-913, 2004.

Nagy JI, Ochalski PA, Li J, Hertzberg EL. Evidence for the co-local-ization of another connexin with connexin-43 at astrocytic gap junctions in rat brain. Neurosci. 78: 533-548, 1997.

Nagy JI, Patel D, Ochalski PA, Stelmack GL. Connexin30 in rodent, cat and human brain: selective expression in gray matter astro-cytes, co-localization with connexin43 at gap junctions and late developmental appearance. Neurosci. 88: 447-468, 1999.

Nagy JI, Li X, Rempel J, Stelmack G, Patel D, Staines WA, Yasu-mura T, Rash JE. Connexin26 in adult rodent central nervous system: demonstration at astrocytic gap junctions and colocali-zation with connexin30 and connexin43. J. Comp. Neurol. 441: 302-323, 2001.

Nakase T, Naus CC. Gap junctions and neurological disorders of the central nervous system. Biochim. Biophys. Acta. 1662: 149-158, 2004.

Nakase T, Fushiki S, Naus CCG. Astrocytic gap junctions composed of connexin 43 reduce apoptotic neuronal damage in cerebral ischemia. Stroke 34: 1987-1993, 2003.

Naus CC, Bechberger JF, Zhang Y, Venance L, Yamasaki H, Juneja SC, Kidder GM, Giaume C. Altered gap junctional communica-tion, intercellular signaling, and growth in cultured astrocytes deficient in connexin43. J. Neurosci. Res. 49: 528-540, 1997.

Nave KA, Sereda MW, Ehrenreich H. Mechanisms of disease: inherited demyelinating neuropathies—from basic to clinical research. Nat. Clin. Pract. Neurol. 3: 453-464, 2007.

Nilsen KE, Kelso AR, Cock HR. Antiepileptic effect of gap-junction blockers in a rat model of refractory focal cortical epilepsy. Epi-lepsia. 47: 1169-1175, 2006.

North RA. Molecular physiology of P2X receptors. Physiol. Rev. 82: 1013-1067, 2002.

Odermatt B, Wellershaus K, Wallraff A, Seifert G, Degen J, Euwens C, Fuss B, Büssow H, Schilling K, Steinhäuser C, Willecke K. Connexin 47 (Cx47)-deficient mice with enhanced green fluores-cent protein reporter gene reveal predominant oligodendrocytic expression of Cx47 and display vacuolized myelin in the CNS. J. Neurosci. 23: 4549-4559, 2003.

Oguchi T, Ohtsuka A, Hashimoto S, Oshima A, Abe S, Kobayashi Y, Nagai K, Matsunaga T, Iwasaki S, Nakagawa T, Usami S. Clini-cal features of patients with GJB2 (connexin 26) mutations: severity of hearing loss is correlated with genotypes and protein expression patterns. J. Hum. Genet. 50: 76-83, 2005.

Oh S, Ri Y, Bennett MV, Trexler EB, Verselis VK, Bargiello TA. Changes in permeability caused by connexin 32 mutations

underlie X-linked Charcot-Marie-Tooth disease. Neuron 19: 927-938, 1997.

Orthmann-Murphy JL, Freidin M, Fischer E, Scherer SS, Abrams CK. Two distinct heterotypic channels mediate gap junction coupling between astrocyte and oligodendrocyte connexins. J. Neurosci. 27: 13949-13957, 2007.

Panchin YV. Evolution of gap junction proteins—the pannexin alternative. J. Exp. Biol. 208: 1415-1419, 2005.

Paul, DL. Molecular cloning of cDNA for rat liver gap junction pro-tein. J. Cell Biol. 103: 123-134, 1986.

Paznekas WA, Boyadjiev SA, Shapiro RE, Daniels O, Wollnik B, Keegan CE, Innis JW, Dinulos MB, Christian C, Hannibal MC, Jabs EW. Connexin 43 (GJA1) mutations cause the pleiotropic phenotype of oculodentodigital dysplasia. Am. J. Hum. Genet. 72: 408-418, 2003.

Pelegrin P, Surprenant A. Pannexin-1 mediates large pore forma-tion and interleukin-1beta release by the ATP-gated P2X7 recep-tor. EMBO. J. 25: 5071-5082, 2006.

Pelegrin P, Surprenant A. Pannexin-1 couples to maitotoxin- and nigericin-induced interleukin-1beta release through a dye uptake-independent pathway. J. Biol. Chem. 282: 2386-2394, 2007.

Penuela S, Bhalla R, Gong XQ, Cowan KN, Celetti SJ, Cowan BJ, Bai D, Shao Q, Laird DW. Pannexin 1 and pannexin 3 are glycoproteins that exhibit many distinct characteristics from the connexin family of gap junction proteins. J. Cell Sci. 120: 3772-3783, 2007.

Pereda AE, Faber DS. Activity-dependent short-term enhancement of intercellular coupling. J Neurosci. 16: 983-992, 1996.

Phelan P. Innexins: members of an evolutionarily conserved family of gap-junction proteins. Biochim. Biophys. Acta. 1711: 225-245, 2005.

Pleasure SJ, Page C, Lee VM. Pure, postmitotic, polarized human neurons derived from NTera 2 cells provide a system for express-ing exogenous proteins in terminally differentiated neurons. J Neurosci. 12: 1802-1815, 1992.

Rash JE, Davidson KG, Kamasawa N, Yasumura T, Kamasawa M, Zhang C, Michaels R, Restrepo D, Ottersen OP, Olson CO, Nagy JI. Ultrastructural localization of connexins (Cx36, Cx43, Cx45), glutamate receptors and aquaporin-4 in rodent olfactory mucosa, olfactory nerve and olfactory bulb. J Neurocytol. 34: 307-341, 2005.

Rash JE, Duffy HS, Dudek FE, Bilhartz BL, Whalen LR, Yasumura T. Grid-mapped freeze-fracture analysis of gap junctions in gray and white matter of adult rat central nervous system, with evi-dence for a "panglial syncytium" that is not coupled to neurons. J. Comp. Neurol. 388: 265-292, 1997.

Rash JE, Olson CO, Davidson KG, Yasumura T, Kamasawa N, Nagy JI. Identification of connexin36 in gap junctions between neurons in rodent locus coeruleus. Neurosci. 147: 938-956, 2007.

Rash JE, Staines WA, Yasumura T, Patel D, Furman CS, Stelmack GL, Nagy JI. Immunogold evidence that neuronal gap junctions in adult rat brain and spinal cord contain connexin-36 but not connexin-32 or connexin-43. Proc. Natl. Acad. Sci. USA 97: 7573-7578, 2000.

Ray A, Zoidl G, Weickert S, Wahle P, Dermietzel R. Site-specific and developmental expression of pannexin1 in the mouse ner-vous system. Eur. J. Neurosci. 21: 3277-3290, 2005.

Raza H, John A. Glutathione metabolism and oxidative stress in neonatal rat tissues from streptozotocin-induced diabetic mothers. Diabetes Metab. Res. Rev. 20: 72-78, 2004.

Ressot C, Gomès D, Dautigny A, Pham-Dinh D, Bruzzone R. Connexin32 mutations associated with X-linked Charcot-Marie-Tooth disease show two distinct behaviors: loss of func-tion and altered gating properties. J. Neurosci. 18: 4063-4075, 1998.

Retamal MA, Cortés CJ, Reuss L, Bennett MVL, Sáez JC. S-nitrosylation and permeation through connexin 43 hemichannels in astrocytes: induction by oxidant stress and reversal by reducing agents. Proc. Natl. Acad. Sci. USA 103: 4475-4480, 2006.

Retamal MA, Schalper KA, Shoji KF, Orellana JA, Bennett MV, Sáez JC. Possible involvement of different connexin43 domains in plasma membrane permeabilization induced by ischemia-reperfusion. J. Membr. Biol. 218: 49-63, 2007a.

Retamal MA, Froger N, Palacios-Prado N, Ezan P, Sáez PJ, Sáez JC, Giaume C. Cx43 hemichannels and gap junction channels in astrocytes are regulated oppositely by proinflammatory cytokines released from activated microglia. J. Neurosci. 27: 13781-13792, 2007b.

Revel J-P, Karnovsky M. Hexagonal array of subunits in intercellular junctions of the mouse heart and liver. J. Cell Biol. 33: C7-C12, 1967.

Rochefort N, Quenech'du N, Ezan P, Giaume C, Milleret C. Postnatal development of GFAP, connexin43 and connexin30 in cat visual cortex. Brain Res. Dev. Brain Res. 160: 252-264, 2005.

Rouach N, Glowinski J, Giaume C. Activity-dependent neuronal control of gap-junctional communication in astrocytes. J. Cell Biol. 149: 1513-1526, 2000.

Roy CS, Sherrington C. On the regulation of the blood supply of the brain. J. Physiol. 11: 85-108, 1890.

Rozental R, Andrade-Rozental AF, Zheng X, Urban M, Spray DC, Chiu FC. Gap junction-mediated bidirectional signaling between human fetal hippocampal neurons and astrocytes. Dev. Neurosci. 23: 420-431, 2001.

Sáez JC, Berthoud VM, Brañes MC, Martínez AD, Beyer EC. Plasma membrane channels formed by connexins: their regulation and functions. Physiol. Rev. 83(4): 1359-1400, 2003.

Sáez JC, Berthoud VM, Kadle R, Traub O, Nicholson BJ, Bennett MV, Dermietzel R. Pinealocytes in rats: connexin identification and increase in coupling caused by norepinephrine. Brain Res. 568: 265-275, 1991.

Sarma JD, Wang F, Koval, M. Targeted gap junction protein constructs reveal connexin-specific differences in oligomerization. J. Biol. Chem. 277: 20911-20918, 2002.

Scherer SS, Deschenes SM, Xu YT, Grinspan JB, Fischbeck KH, Paul DL. Connexin32 is a myelin-related protein in the PNS and CNS. J. Neurosci. 15: 8281-8294, 1995.

Schipke CG, Boucsein C, Ohlemeyer C, Kirchhoff F, Kettenmann H. Astrocyte Ca^{2+} waves trigger responses in microglial cells in brain slices. FASEB J. 16: 255-257, 2002.

Seki K, Komuro T. Distribution of interstitial cells of Cajal and gap junction protein, Cx 43 in the stomach of wild-type and W/Wv mutant mice. Anat. Embryol. (Berl.) 206: 57-65, 2002.

Shaw RM, Fay AJ, Puthenveedu MA, von Zastrow M, Jan YN, Jan LY. Microtubule plus-end-tracking proteins target gap junctions directly from the cell interior to adherens junctions. Cell 128: 547-560, 2007.

Shestopalov VI, Panchin Y. Pannexins and gap junction protein diversity. Cell Mol. Life Sci. 65: 376-394, 2008.

Shirasawa N, Mabuchi Y, Sakuma E, Horiuchi O, Yashiro T, Kikuchi M, Hashimoto Y, Tsuruo Y, Herbert DC, Soji T. Intercellular communication within the rat anterior pituitary gland: X. Immunohistocytochemistry of S-100 and connexin 43 of folliculo-stellate cells in the rat anterior pituitary gland. Anat. Rec. A. Discov. Mol. Cell Evol. Biol. 278: 462-473, 2004.

Shy ME, Siskind C, Swan ER, Krajewski KM, Doherty T, Fuerst DR, Ainsworth PJ, Lewis RA, Scherer SS, Hahn AF. CMT1X phenotypes represent loss of GJB1 gene function. Neurology 68: 849-855, 2007.

SiuYi Leung D, Unsicker K, Reuss B. Gap junctions modulate survival-promoting effects of fibroblast growth factor-2 on cultured midbrain dopaminergic neurons. Mol. Cell Neurosci. 18: 44-55, 2001.

Skerrett IM, Aronowitz JA, Shin JH, Kasparek E, Cymes G, Cao FL, Nicholson BJ. Identification of amino acid residues lining the pore of the gap junction channel. J. Cell Biol. 159: 349-359, 2002.

Skerrett IM, Di W-L, Kasperek EM, Kelsell DP, Nicholson BJ. Aberrant gating, but a normal expression pattern, underlies the recessive phenotype of the deafness mutant Cx26M34T FASEB J. 18: 860-862, 2004.

Söhl G, Eiberger J, Jung YT, Kozak CA, Willecke K. The mouse gap junction gene connexin29 is highly expressed in sciatic nerve and regulated during brain development. Biol. Chem. 382: 973-978, 2001.

Söhl G, Maxeiner S, Willecke K. Expression and functions of neuronal gap junctions. Nature Rev. Neurosci. 6: 191-200, 2005.

Sohl G, Willecke K. Gap junctions and the connexin protein family. Cardiovasc. Res. 62: 228-232, 2004.

Solomon IC, Halat TJ, El-Maghrabi R, O'Neal MH 3rd. Differential expression of connexin26 and connexin32 in the pre-Bötzinger complex of neonatal and adult rat. J. Comp. Neurol. 440: 12-19, 2001.

Spray DC, Moreno AP, Kessler JA, Dermietzel R. Characterization of gap junctions between cultured leptomeningeal cells. Brain Res. 568: 1-14, 1991.

Stout CE, Costantin JL, Naus CC, Charles AC. Intercellular calcium signaling in astrocytes via ATP release through connexin hemichannels. J. Biol. Chem. 277: 10482-10488, 2002.

Takano T, Tian GF, Peng W, Lou N, Libionka W, Han X, Nedergaard M. Astrocyte-mediated control of cerebral blood flow. Nat. Neurosci. 9: 260-267, 2006.

Takeuchi H, Jin S, Wang J, Zhang G, Kawanokuchi J, Kuno R, Sonobe Y, Mizuno T, Suzumura A. Tumor necrosis factor-alpha induces neurotoxicity via glutamate release from hemichannels of activated microglia in an autocrine manner. J. Biol. Chem. 281: 21362-21368, 2006.

Taylor RA, Simon EM, Marks HG, Scherer SS. The CNS phenotype of X-linked Charcot-Marie-Tooth disease: more than a peripheral problem. Neurology 61: 1475-1478, 2003.

Theis M, Jauch R, Zhuo L, Speidel D, Wallraff A, Döring B, Frisch C, Söhl G, Teubner B, Euwens C, Huston J, Steinhäuser C, Messing A, Heinemann U, Willecke K. Accelerated hippocampal spreading depression and enhanced locomotory activity in mice with astrocyte-directed inactivation of connexin43. J. Neurosci. 23: 766-776, 2003.

Theis M, Söhl G, Eiberger J, Willecke K. Emerging complexities in identity and function of glial connexins. Trends in Neurosci. 28: 188-195, 2005.

Thompson RJ, Zhou N, MacVicar BA. Ischemia opens neuronal gap junction hemichannels. Science 312: 924-927, 2006.

Trexler EB, Bennett MV, Bargiello TA, Verselis VK. Voltage gating and permeation in a gap junction hemichannel. Proc. Natl. Acad. Sci. USA 93: 5836-5841, 1996.

Uhlenberg B, Schuelke M, Rüschendorf F, Ruf N, Kaindl AM, Henneke M, Thiele H, Stoltenburg-Didinger G, Aksu F, Topaloğlu H, Nürnberg P, Hübner C, Weschke B, Gärtner J. Mutations in the gene encoding gap junction protein alpha 12 (connexin 46.6) cause Pelizaeus-Merzbacher-like disease. Am. J. Hum. Genet. 75: 251-260, 2004.

Unger VM, Kumar NM, Gilula NB, Yeager M. Three-dimensional structure of a recombinant gap junction membrane channel. Science 283: 1176-1180, 1999.

Van Bockstaele EJ, Garcia-Hernandez F, Fox K, Alvarez VA, Williams JT. Expression of connexins during development and following manipulation of afferent input in the rat locus coeruleus. Neurochem. Int. 45: 421-428, 2004.

Vanden Abeele F, Bidaux G, Gordienko D, Beck B, Panchin YV, Baranova AV, Ivanov DV, Skryma R, Prevarskaya N. Functional implications of calcium permeability of the channel formed by pannexin 1. J. Cell Biol. 174: 535-546, 2006.

Vaney DI, Nelson JC, Pow DV. Neurotransmitter coupling through gap junctions in the retina. J. Neurosci. 18: 10594-10602, 1998.

Vitiello C, D'Adamo P, Gentile F, Vingolo EM, Gasparini P, Banfi S. A novel GJA1 mutation causes oculodentodigital dysplasia without syndactyly. Am. J. Med. Genet. A. 133: 58-60, 2005.

Vogt A, Hormuzdi SG, Monyer H. Pannexin1 and Pannexin2 expression in the developing and mature rat brain. Brain Res. Mol. Brain Res. 141: 113-120, 2005.

White TW, Paul DL. Genetic diseases and gene knockouts reveal diverse connexin functions. Annu. Rev. Physiol. 61: 283-310, 1999.

Wiencken-Barger AE, Djukic B, Casper KB, McCarthy KD. A role for Connexin43 during neurodevelopment. Glia. 55(7): 675-686, 2007.

Willecke K, Eiberger J, Degen J, Eckardt D, Romualdi A, Güldenagel M, Deutsch U, Söhl G. Structural and functional diversity of connexin genes in the mouse and human genome. Biol. Chem. 383: 725-737, 2002.

Xu HL, Mao L, Ye S, Paisansathan C, Vetri F, Pelligrino DA. Astrocytes are a key conduit for upstream signaling of vasodilation during cerebral cortical neuronal activation in vivo. Am. J. Physiol. 294: H622-H632, 2008.

Yamamoto T, Ochalski A, Hertzberg EL, Nagy JI. LM and EM immunolocalization of the gap junctional protein connexin 43 in rat brain. Brain Res. 508: 313-319, 1990.

Yamamoto T, Vukelic J, Hertzberg EL, Nagy JI. Differential anatomical and cellular patterns of connexin43 expression during postnatal development of rat brain. Dev. Brain Res. 66: 165-180, 1992.

Yamamoto T, Hossain MZ, Hertzberg EL, Uemura H, Murphy LJ, Nagy JI. Connexin43 in rat pituitary: localization at pituicyte and stellate cell gap junctions and within gonadotrophs. Histochem. 100: 53-64, 1993.

Yang XD, Korn H, Faber DS. Long-term potentiation of electrotonic coupling at mixed synapses. Nature 348: 542-545, 1990.

Ye ZC, Wyeth MS, Baltan-Tekkok S, Ransom BR. Functional hemichannels in astrocytes: a novel mechanism of glutamate release. J. Neurosci. 23: 3588-3596, 2003.

Yeager M, Unger VM, Falk MM. Synthesis, assembly and structure of gap junction intercellular channels. Curr. Opin. Struct. Biol. 8: 517-524, 1998.

Zappalà A, Cicero D, Serapide MF, Paz C, Catania MV, Falchi M, Parenti R, Pantò MR, La Delia F, Cicirata F. Expression of pannexin1 in the CNS of adult mouse: cellular localization and effect of 4-aminopyridine-induced seizures. Neurosci. 141: 167-178, 2006.

Zhang, Y, Tang W, Ahmad S, Sipp JA, Chen P, Lin X. Gap junction-mediated intercellular biochemical coupling in cochlear supporting cells is required for normal cochlear functions. Proc. Natl. Acad. Sci. USA 102: 15201-15206, 2005.

Zhao HB, Kikuchi T, Ngezahayo A, White TW. Gap junctions and cochlear homeostasis. J. Membr. Biol. 209: 177-186, 2006.

Zhou X-W, Pfahnl A, Werner R, Hudder A, Lianes A, Luebke A, Dahl, G. Identification of a pore lining segment in gap junction hemichannels. Biophys. J. 72: 1946-1953, 1997.

Zonta M, Angulo MC, Gobbo S, Rosengarten B, Hossmann KA, Pozzan T, Carmignoto G. Neuron-to-astrocyte signaling is central to the dynamic control of brain microcirculation. Nat. Neurosci. 6: 43-50, 2003.

Postsynaptic Potentials and Synaptic Integration

John H. Byrne

The study of synaptic transmission in the central nervous system provides an opportunity to learn more about the diversity and richness of mechanisms underlying this process and to learn the ways in which some of the fundamental signaling properties of the nervous system, such as action potentials and synaptic potentials, work together to process information and generate behavior.

Postsynaptic potentials (PSPs) in the CNS can be divided into two broad classes on the basis of mechanisms and, generally, duration, of these potentials. One class is based on the *direct* binding of a transmitter molecule(s) with a receptor–channel complex; these receptors are *ionotropic*. The structure of these receptors is discussed in detail in Chapter 11. The resulting PSPs are generally short lasting and hence are sometimes called *fast PSPs*; they have also been referred to as "classic" because they were the first synaptic potentials to be recorded in the CNS (Eccles, 1964; Spencer, 1977). The duration of a typical fast PSP is about 20 ms.

The other class of PSPs is based on the *indirect* effect of transmitter molecule(s) binding with a receptor. The receptors that produce these PSPs are *metabotropic*. As discussed in Chapter 11, the receptors activate G proteins and are therefore also called *G-protein-coupled receptors* (GPCR). They affect the channel either directly or through additional steps in which the level of a second messenger is altered. The changes in membrane potential produced by GPCRs can be long lasting and are therefore called *slow PSPs*. The mechanisms for fast PSPs mediated by ionotropic receptors are considered first.

IONOTROPIC RECEPTORS: MEDIATORS OF FAST EXCITATORY AND INHIBITORY SYNAPTIC POTENTIALS

The Stretch Reflex Is Useful to Examine the Properties and Functional Consequences of Ionotropic PSPs

The stretch reflex, one of the simpler behaviors mediated by the central nervous system, is a useful example with which to examine the properties and functional consequences of ionotropic PSPs. The tap of a neurologist's hammer to a ligament elicits a reflex extension of the leg, as illustrated in Figure 16.1. The brief stretch of the ligament is transmitted to the extensor muscle and detected by specific receptors in the muscle and ligament. Action potentials initiated in the stretch receptors are propagated to the spinal cord by afferent fibers. The receptors are specialized regions of sensory neurons with somata located in the dorsal root ganglia just outside the spinal column. The axons of the afferents enter the spinal cord and make excitatory synaptic connections with at least two types of postsynaptic neurons. First, a synaptic connection is made to the extensor motor neuron. As the result of its synaptic activation, the motor neuron fires action potentials that propagate out of the spinal cord and ultimately invade the terminal regions of the motor axon at neuromuscular junctions. There, acetylcholine (ACh) is released, nicotinic ACh receptors are activated, an end-plate potential (EPP) is produced, an action potential is initiated in the muscle

469

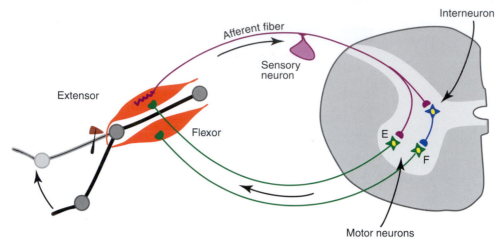

FIGURE 16.1 Features of the vertebrate stretch reflex. Stretch of an extensor muscle leads to the initiation of action potentials in the afferent terminals of specialized stretch receptors. The action potentials propagate to the spinal cord through afferent fibers (sensory neurons). The afferents make excitatory connections with extensor motor neurons (E). Action potentials initiated in the extensor motor neuron propagate to the periphery and lead to the activation and subsequent contraction of the extensor muscle. The afferent fibers also activate interneurons that inhibit the flexor motor neurons (F).

cell, and the muscle cell is contracted, producing the reflex extension of the leg. Second, a synaptic connection is made to another group of neurons called *interneurons* (nerve cells interposed between one type of neuron and another). The particular interneurons activated by the afferents are inhibitory interneurons, because activation of these interneurons leads to the release of a chemical transmitter substance that inhibits the flexor motor neuron. This inhibition tends to prevent an uncoordinated (improper) movement (i.e., flexion) from occurring. The reflex system illustrated in Figure 16.1 is also known as the *monosynaptic stretch reflex* because this reflex is mediated by a single ("mono") excitatory synapse in the central nervous system.

Figure 16.2 illustrates procedures that can be used to experimentally examine some of the components of synaptic transmission in the reflex pathway for the stretch reflex. Intracellular recordings are made from one of the sensory neurons, the extensor and flexor motor neurons, and an inhibitory interneuron. Normally, the sensory neuron is activated by stretch to the muscle, but this step can be bypassed by simply injecting a pulse of depolarizing current of sufficient magnitude into the sensory neuron to elicit an action potential. The action potential in the sensory neuron leads to a potential change in the motor neuron known as an *excitatory postsynaptic potential* (EPSP) (Fig. 16.2).

Mechanisms responsible for fast EPSPs mediated by ionotropic receptors in the CNS are fairly well known. Moreover, the ionic mechanisms for EPSPs in the CNS are essentially identical to the ionic mechanisms at the skeletal neuromuscular junction. Specifically, the transmitter substance released from the presynaptic terminal (Chapters 8–10) diffuses across the synaptic cleft, binds to specific receptor sites on the postsynaptic membrane (Chapter 11), and leads to a simultaneous increase in permeability to Na^+ and K^+, which makes the membrane potential move *toward* a value of about 0 mV. However, the processes of synaptic transmission at the sensory neuron–motor neuron synapse and the motor neuron–skeletal muscle synapse differ in two fundamental ways: (1) in the transmitter used and (2) in the amplitude of the PSP. The transmitter substance at the neuromuscular junction is ACh, whereas that released by the sensory neurons is an amino acid, probably glutamate (see Chapter 9). Indeed, glutamate is the most common transmitter that mediates excitatory actions in the CNS. The amplitude of the postsynaptic potential at the neuromuscular junction is about 50 mV; consequently, each PSP depolarizes the postsynaptic cell beyond threshold, so there is a one-to-one relationship between an action potential in the spinal motor neuron and an action potential in the skeletal muscle cell. Indeed, the EPP must depolarize the muscle cell by only about 30 mV to initiate an action potential, allowing a safety factor of about 20 mV. In contrast, the EPSP in a spinal motor neuron produced by an action potential in an afferent fiber has an amplitude of only about 1 mV. The mechanisms by which these small PSPs can trigger an action potential in the postsynaptic neuron are discussed in a later section of this chapter and in Chapter 17.

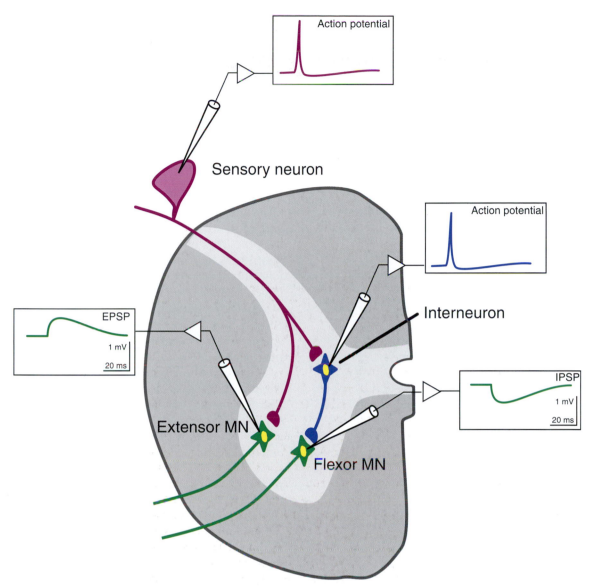

FIGURE 16.2 Excitatory (EPSP) and inhibitory (IPSP) postsynaptic potentials in spinal motor neurons. Idealized intracellular recordings from a sensory neuron, interneuron, and extensor and flexor motor neurons (MNs). An action potential in the sensory neuron produces a depolarizing response (an EPSP) in the extensor motor neuron. An action potential in the interneuron produces a hyperpolarizing response (an IPSP) in the flexor motor neuron.

Macroscopic Properties of PSPs Are Determined by the Nature of the Gating and Ion Permeation Properties of Single Channels

Patch-Clamp Techniques

Patch-clamp techniques (Hamill *et al.*, 1981) with which current flowing through single isolated receptors can be measured directly can be sources of insight into both the ionic mechanisms and the molecular properties of PSPs mediated by ionotropic receptors. This approach was pioneered by Erwin Neher and Bert Sakmann in the 1970s and led to their being awarded the Nobel Prize in Physiology or Medicine in 1991.

Figure 16.3 illustrates an idealized experimental arrangement of an "outside-out" patch recording of a single ionotropic receptor. The patch pipette contains a solution with an ionic composition similar to that of the cytoplasm, whereas the solution exposed to the outer surface of the membrane has a composition similar to that of normal extracellular fluid. The electrical potential across the patch, and hence the transmembrane potential (V_m), is controlled by the patch-clamp amplifier. The extracellular (outside) fluid is

A

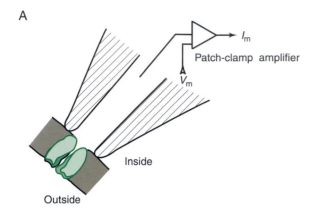

Inside

Outside

Single-channel current

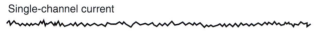

B

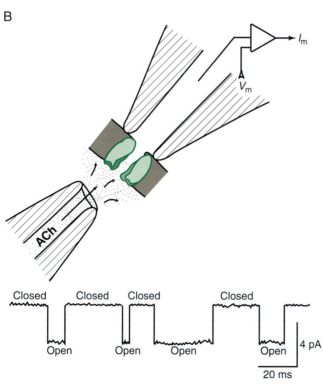

Closed	Closed	Closed	Closed

Open   Open   Open   Open   4 pA

20 ms

FIGURE 16.3 Single-channel recording of ionotropic receptors and their properties. (A) Experimental arrangement for studying properties of ionotropic receptors. (B) Idealized single-channel currents in response to application of ACh.

considered "ground." Transmitter can be delivered by applying pressure to a miniature pipette filled with an agonist (in this case, ACh), and the current (I_m) flowing across the patch of membrane is measured by the patch-clamp amplifier (Fig. 16.3). Pressure in the pipette that contains ACh can be continuous, allowing a constant stream of ACh to contact the membrane, or can be applied as a short pulse to allow a precisely

timed and discrete amount of ACh to contact the membrane. The types of recordings obtained from such an experiment are illustrated in the traces in Figure 16.3. In the absence of ACh, no current flows through the channel (Fig. 16.3A). When ACh is continuously applied, current flows across the membrane (through the channel), but the current does not flow continuously; instead, small steplike changes in current are observed (Fig. 16.3B). These changes represent the probabilistic (random) opening and closing of the channel.

Channel Openings and Closings

As a result of the type of patch-recording techniques heretofore described, three general conclusions about the properties of ligand-gated channels can be drawn. First, ACh, as well as other transmitters that activate ionotropic receptors, causes the opening of individual ionic channels (for a channel to open, usually two molecules of transmitter must bind to the receptor). Second, when a ligand-gated channel opens, it does so in an all-or-none fashion. Increasing the concentration of transmitter in the ejection microelectrode does not increase the permeability (conductance) of the channel; it increases its probability (P) of being open. Third, the ionic current flowing through a single channel in its open state is extremely small (e.g., 10^{-12} A); as a result, the current flowing through any single channel makes only a small contribution to the normal postsynaptic potential. Physiologically, when a larger region of the postsynaptic membrane, and thus more than one channel, is exposed to released transmitter, the net conductance of the membrane increases owing to the increased probability that a larger population of channels will be open at the same time. The normal PSP, measured with standard intracellular recording techniques (e.g., Fig. 16.2), is then proportional to the sum of the currents that flow through these many individual open channels. The properties of voltage-sensitive channels (see Chapters 5 and 6) are similar in that they, too, open in all-or-none fashion, and, as a result, the net effect on the cell is due to the summation of the currents flowing through many individual open ion channels. The two types of channels differ, however, in that one is opened by a chemical agent, whereas the other is opened by changes in membrane potential.

Statistical Analysis of Channel Gating and the Kinetics of the PSP

The experiment illustrated in Figure 16.3B was performed with continuous exposure to ACh. Under such conditions, the channels open and close repeatedly. When ACh is applied by a brief pressure pulse to more accurately mimic the transient release from the presynaptic terminal, the transmitter diffuses away before it can cause a second opening of the channel. A set of data similar to that shown in Figure 16.4A would be obtained if an ensemble of these openings were collected and aligned with the start of each opening. Each individual trace represents the response to each successive "puff" of ACh. Note that, among the responses, the duration of the opening of the channel varies considerably—from very short (less than 1 ms) to more than 5 ms. Moreover, channel openings are independent events. The duration of any one channel opening does not have any relationship to the duration of a previous opening. Figure 16.4B illustrates a plot that is obtained by adding 1000 of these individual responses. Such an

addition roughly simulates the conditions under which transmitter released from a presynaptic terminal leads to the near-simultaneous activation of many single channels in the postsynaptic membrane. (Note that the addition of 1000 channels would produce a synaptic current equal to about 4 nA.) This simulation is valid given the assumption that the statistical properties of a single channel over time are the same as the statistical properties of the ensemble at one instant of time (i.e., an ergotic process). The ensemble average can be fit with an exponential function with a decay time constant of 2.7 ms. An additional observation (explored in the next section) is that the value of the time constant is equal to the mean duration of the channel openings. The curve in Figure 16.4B is an indication of the probability that a channel will remain open for various times, with a high probability for short times and a low probability for long times.

The ensemble average of the single-channel currents (Fig. 16.4B) roughly accounts for the time course of the EPSP. However, note that the time course of the aggregate synaptic *current* can be somewhat faster than that of the excitatory postsynaptic *potential* in Figure 16.2. This difference is due to the charging of the membrane capacitance by a rapidly changing synaptic current. Because the single-channel currents were recorded with the membrane voltage clamped, the capacitive current $[I_c = C_m \cdot (dV/dt)]$ is zero. In contrast, for the recording of the postsynaptic potential in Figure 16.2, the membrane was not voltage clamped, and therefore as the voltage changes (i.e., dV/dt), some of the synaptic current charges the membrane capacitance (see Eq. 17).

Analytical expressions that describe the shape of the ensemble average of the open lifetimes and the mean open lifetime can be derived by considering that single-channel opening and closing is a stochastic process (Colquhoun and Hawkes, 1977, 1981, 1982; Colquhoun and Sakmann, 1981; Johnston and Wu, 1995; Sakmann, 1992). Relations are formalized to describe the likelihood (probability) of a channel being in a certain state. Consider the following two-state reaction scheme:

$$C \underset{\beta}{\overset{\alpha}{\rightleftharpoons}} O$$

In this scheme, α represents the rate constant for channel closing, and β is the rate constant for channel opening. The scheme can be simplified further if we consider a case in which the channel has been opened

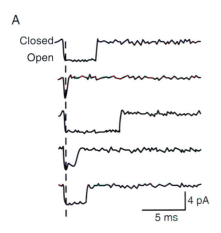

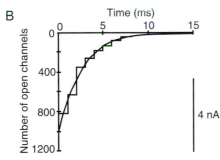

FIGURE 16.4 Determination of the shape of the postsynaptic response from the single-channel currents. (A) Each trace represents the response of a single channel to a repetitively applied puff of transmitter. The traces are aligned with the beginning of the channel opening (dashed line). (B) Addition of 1000 of the individual responses. If a current equal to 4 pA were generated by the opening of a single channel, then a 4 nA current would be generated by 1000 channels opening at the same time. The data are fitted with an exponential function having a time constant equal to $1/\alpha$ (see Eq. 16.9). Reprinted with permission from Sakmann (1992). Copyright 1992 American Association for the Advancement of Science.

by the agonist and the agonist is removed instantaneously. A channel so opened (at time *0*) will then close after a certain random time (Fig. 16.4). We first formulate an analytical expression that describes the probability that the channel is open (o) at some time (i.e., time *t*), given that it was open at time *0*. This expression is referred to as $P_{o/o}(t)$. To formulate an analytical expression for $P_{o/o}(t)$, first consider the probability that a channel will be *closed* (c) at time $t + \Delta t$, given that it was open at time *t*, in the limit that Δt is so small that we can ignore multiple events such as an opening followed by a closing. This term, which is referred to as $P_{c/o}(\Delta t)$, will equal $\alpha \Delta t$ (the product of the reverse rate constant and the time interval). Therefore, the probability $[P_{o/o}(\Delta t)]$ that a channel will be *open* at time $t + \Delta t$, given that it was open at time *t*, will equal $1 - \alpha \Delta t$ (i.e., 1 minus the probability that it will be closed at $t + \Delta t$). Finally, the probability that the channel will be open at time *t and* will be open at time $t + \Delta t$ can be described by

$$P_{o/o}(t + \Delta t) = P_{o/o}(t)\, P_{o/o}(\Delta t). \quad (16.1)$$

By substituting and factoring, we obtain

$$P_{o/o}(t + \Delta t) = P_{o/o}(t)(1 - \alpha \Delta t), \quad (16.2)$$

$$P_{o/o}(t + \Delta t) = P_{o/o}(t) - \alpha \Delta t P_{o/o}(t), \quad (16.3)$$

$$(P_{o/o}(t + \Delta t) - P_{o/o}(t))/\Delta t = -\alpha P_{o/o}(t). \quad (16.4)$$

Note that as $\Delta t \to 0$, the left-hand term of Eq. 16.4 defines the derivative. Thus,

$$\frac{dP_{o/o}(t)}{dt} = -\alpha P_{o/o}(t). \quad (16.5)$$

This differential equation is satisfied by an exponential function. Consequently,

$$P_{o/o}(t) = e^{-\alpha t}. \quad (16.6)$$

We can now determine the probability $[P_{c/o}(t)]$ that the channel is closed at time *t*, given that it was open at time *0*. This will simply be $1 - P_{o/o}(t)$. Therefore,

$$P_{c/o}(t) = 1 - e^{-\alpha t}. \quad (16.7)$$

The function $P_{c/o}(t)$ represents the cumulative distribution function (or simply the distribution function) for the channel (i.e., the probability that a channel will be closed by time *t*). This quantity is called the *cumulative distribution* because it is equal to the sum, or integral, over the probabilities that the channel closes at each of the preceding times. Distribution functions satisfy the relationship

$$0 \leq P(t) \leq 1. \quad (16.8)$$

Note that, for Eq. 16.7, at *t = 0* the probability of a channel being closed is 0 and, at $t = \infty$, the probability of a channel being closed is 1. To obtain an equation for the probability that a channel closing occurs in exactly some period $t + \Delta t$ as Δt approaches 0, we need to determine the *probability density function* $[p(t)]$, which is defined as the first derivative of the cumulative distribution function (Papoulis, 1965). Thus, the probability density function is

$$P(t) = \alpha e^{-\alpha t}. \quad (16.9)$$

Note that the distribution of open lifetimes illustrated in Figure 16.4B corresponds well to that predicted by Eq. 16.9. With an analytical expression for the probability density function in hand, we can now determine another important property of channels—the mean open lifetime. The mean open lifetime can be obtained by taking the average of the probability density function (i.e., the expected value). Operationally, we multiply *t* and *p(t)* and integrate between time 0 and ∞. Thus,

$$\text{mean open time} = \int_0^\infty t\, \alpha\, e^{-\alpha t}\, dt = \frac{1}{\alpha}. \quad (16.10)$$

Note that the mean open time is the time constant of the cumulative distribution function (Eq. 16.7) and the probability density function (Eq. 16.9) of the channel.

Gating Properties of Ligand-Gated Channels

Although statistical analysis can be a valuable source of insight into the statistical nature of the gating process and the molecular determinants of the macroscopic postsynaptic potential, the description in the preceding section is a simplification of the actual processes. Specifically, a more complete description must include the kinetics of receptor binding and unbinding and the determinants of the channel opening, as well as the fact that channels display rapid transitions between open and closed states during a single agonist receptor occupancy. Thus, the open states illustrated in Figures 16.3B and 16.4A represent the period of a burst of extremely rapid openings and closings. If the bursts of rapid channel openings and closings are thought of, and behave functionally as a single continuous channel closure, the formalism developed in the preceding section is a reasonable approximation for many ligand-gated channels. Nevertheless, a more complex reaction scheme is necessary to quantitatively explain the available data. Such a scheme would include the following states:

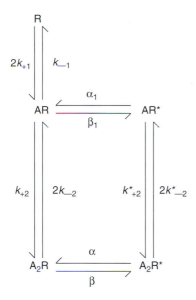

where R represents the receptor, A the agonist, and the α's, β's, and k's the forward and reverse rate constants for the various reactions. A_2R^* represents a channel opened as a result of the binding of two agonist molecules. The asterisk indicates an open channel (Colquhoun and Sakmann, 1981; Sakmann, 1992). Note that the lower part of the reaction scheme is equivalent to that developed earlier, that is,

$$C \underset{\beta}{\overset{\alpha}{\rightleftharpoons}} O$$

With the use of probability theory, equations describing the transitions between the states can be determined. The approach is identical to that used in the simplified two-state scheme. However, the mathematics and analytical expressions are more complex because of the interactions among transitions and the multiple dimensionality of the variables (Colquhoun and Hawkes, 1977, 1981, 1982). For some receptors, additional states must be represented. For example, as described in Chapter 11, some ligand-gated channels exhibit a process of desensitization in which continued exposure to a ligand results in channel closure.

The Null (Reversal) Potential and Slope of I–V Relationships

What ions are responsible for the synaptic current that produces the EPSP? Early studies of the ionic mechanisms underlying the EPSP at the skeletal neuromuscular junction yielded important information. Specifically, voltage-clamp and ion substitution experiments indicated that the binding of transmitter

to receptors on the postsynaptic membrane led to a simultaneous increase in Na^+ and K^+ permeability that depolarized the cell toward a value of about 0 mV (Fatt and Katz, 1951; Takeuchi and Takeuchi, 1960). These findings are applicable to the EPSP in a spinal motor neuron produced by an action potential in an afferent fiber and have been confirmed and extended at the single-channel level.

Figure 16.5 illustrates the type of experiment in which the analysis of single-channel currents can be a source of insight into the ionic mechanisms of EPSPs. Transmitter is delivered to the patch while the membrane potential is systematically varied (Fig. 16.5A). In the upper trace, the patch potential is –40 mV. The ejection of transmitter produces a sequence of channel openings and closings, the amplitudes of which are constant for each opening (i.e., about 4 pA). Now consider the case in which the transmitter is applied when the potential across the patch is –20 mV. The frequency

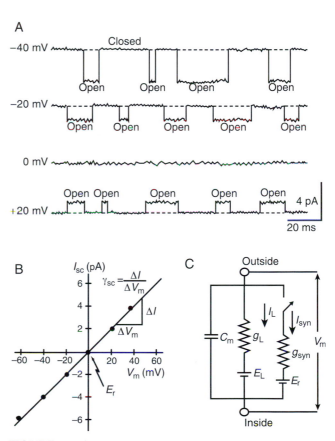

FIGURE 16.5 Voltage dependence of the current flowing through single channels. (A) Idealized recording of an ionotropic receptor in the continuous presence of agonist. (B) I–V relationship of the channel in (A). (C) Equivalent electrical circuit of a membrane containing that channel. Abbreviations: γ_{sc}, single-channel conductance; I_L, leakage current; I_{sc}, single-channel current; g_L, leakage conductance; g_{syn}, macroscopic synaptic conductance; E_L, leakage battery; E_r, reversal potential.

of the responses, as well as the mean open lifetimes, is about the same as when the potential was at –40 mV, but now the amplitude of the single-channel currents is decreased uniformly. Even more interestingly, when the patch is artificially depolarized to a value of about 0 mV, an identical puff of transmitter produces no current in the patch. If the patch potential is depolarized to a value of about +20 mV and the puff again delivered, openings are again observed, but the flow of current through the channel is reversed in sign; a series of upward deflections indicates outward single-channel currents. In summary, there are downward deflections (inward currents) when the membrane potential is at –40 mV, no deflections (currents) when the membrane is at 0 mV, and upward deflections (outward currents) when the membrane potential is moved to +20 mV.

The simple explanation for these results is that no matter what the membrane potential, the effect of the transmitter binding with receptors is to produce a permeability change that tends to move the membrane potential toward 0 mV. If the membrane potential is more negative than 0 mV, an inward current is recorded. If the membrane potential is more positive than 0 mV, an outward current is recorded. If the membrane potential is at 0 mV, there is no deflection because the membrane potential is already at 0 mV. At 0 mV, the channels are opening and closing as they always do in response to the agonist, but there is no net movement of ions through them. This 0 mV level is known as the *synaptic null potential* or *reversal potential*, because it is the potential at which the sign of the synaptic current reverses. The fact that the experimentally determined reversal potential equals the calculated value obtained by using the Goldman–Hodgkin–Katz (GHK) equation (Chapter 5) provides strong support for the theory that the EPSP is due to the opening of channels that have equal permeabilities to Na^+ and K^+. Ion substitution experiments also confirm this theory. Thus, when the concentration of Na^+ or K^+ in the extracellular fluid is altered, the value of the reversal potential shifts in a way predicted by the GHK equation. (Some other cations, such as Ca^{2+}, also permeate these channels, but their permeability is low compared with that of Na^+ and K^+.)

Different families of ionotropic receptors have different reversal potentials because each has unique ion selectivity. In addition, it should now be clear that the sign of the synaptic action (excitatory or inhibitory) depends on the value of the reversal potential relative to the resting potential. If the reversal potential of an ionotropic receptor channel is more positive than the resting potential, opening of that channel will lead to a depolarization (i.e., an EPSP). In contrast,

if the reversal potential of an ionotropic receptor channel is more negative than the resting potential, opening of that channel will lead to a hyperpolarization, that is, an inhibitory postsynaptic potential (IPSP), which is the topic of a later section in this chapter.

Plotting the average peak value of the single-channel currents (I_{sc}) versus the membrane potential (transpatch potential) at which they are recorded (Fig. 16.5B) can be a source of quantitative insight into the properties of the ionotropic receptor channel. Note that the current–voltage (I–V) relationship is linear; it has a slope, the value of which is the single-channel conductance, and an intercept at 0 mV. This linear relationship can be put in the form of Ohm's law ($I = G \cdot \Delta V$). Thus,

$$I_{sc} = \gamma_{sc} \cdot (V_m - E_r), \tag{16.11}$$

where γ_{sc} is the single-channel conductance and E_r is the reversal potential (here, 0 mV).

Summation of Single-Channel Currents

We now know that the sign of a synaptic action can be predicted by knowledge of the relationship between the resting potential (V_m) and the reversal potential (E_r), but how can the precise amplitude be determined? The answer to this question lies in understanding the relationship between the synaptic conductance and the extrasynaptic conductances. These interactions can be rather complex (see Chapter 17), but some initial understanding can be obtained by analyzing an electrical equivalent circuit for these two major conductance branches. We first need to move from a consideration of single-channel conductances and currents to that of macroscopic conductances and currents. The postsynaptic membrane contains thousands of any one type of ionotropic receptor, and each of these receptors could be activated by transmitter released by a single action potential in a presynaptic neuron. Because conductances in parallel add, the total conductance change produced by their simultaneous activation would be

$$g_{syn} = \gamma_{sc} \cdot P \cdot N, \tag{16.12}$$

where γ_{sc}, as before, is the single-channel conductance, P is the probability of opening of a single channel (controlled by the ligand), and N is the total number of ligand-gated channels in the postsynaptic membrane. The macroscopic postsynaptic current produced by the transmitter released by a single presynaptic action potential can then be described by

$$I_{syn} = g_{syn} \cdot (V_m - E_r). \tag{16.13}$$

Equation 16.13 can be represented physically by a voltage (V_m) measured across a circuit consisting of a resistor (g_{syn}) in series with a battery (E_r). An equivalent circuit of a membrane containing such a conductance is illustrated in Figure 16.5C. Also included in this circuit is a membrane capacitance (C_m), a resistor representing the leakage conductance (g_L), and a battery (E_L) representing the leakage potential. (Voltage-dependent Na^+, Ca^{2+}, and K^+ channels that contribute to the generation of the action potential have been omitted for simplification.)

The simple circuit allows the simulation and further analysis of the genesis of the PSP. Closure of the switch simulates the opening of the channels by transmitter released from some presynaptic neuron (i.e., a change in P of Eq. 16.12 from 0 to 1). When the switch is open (i.e., no agonist is present and the ligand-gated channels are closed), the membrane potential (V_m) is equal to the value of the leakage battery (E_L). Closure of the switch (i.e., the agonist opens the channels) tends to polarize the membrane potential toward the value of the battery (E_r) in series with the synaptic conductance. Although the effect of the channel openings is to depolarize the postsynaptic cell *toward* E_r (0 mV), this value is never achieved, because the ligand-gated receptors are only a small fraction of the ion channels in the membrane. Other channels (such as the leakage channels, which are not affected by the transmitters) tend to hold the membrane potential at E_L and prevent the membrane potential from reaching the 0 mV level. In terms of the equivalent electrical circuit (Fig. 16.5C), g_L is much greater than g_{syn}.

An analytical expression that can be a source of insight into the production of an EPSP by the engagement of a synaptic conductance can be derived by examining the current flowing in each of the two conductance branches of the circuit in Figure 16.5C. As previously shown (Eq. 16.13), the current flowing in the branch representing the synaptic conductance is equal to

$$I_{syn} = g_{syn} \cdot (V_m - E_r).$$

Similarly, the current flowing through the leakage conductance is equal to

$$I_L = g_L \cdot (V_m - E_L). \tag{16.14}$$

By conservation of current, the two currents must be equal and opposite. Therefore,

$$g_{syn} \cdot (V_m - E_r) = -g_L \cdot (V_m - E_L).$$

Rearranging and solving for V_m, we obtain

$$V_m = \frac{g_{syn}E_r + g_L E_L}{g_{syn} + g_L} \tag{16.15}$$

Note that when the synaptic channels are closed (i.e., switch open), g_{syn} is 0 and

$$V_m = E_L.$$

Now consider the case of the ligand-gated channels being opened by release of transmitter from a presynaptic neuron (i.e., switch closed) and a neuron with $g_L = 10$ nS, $E_L = -60$ mV, $g_{syn} = 0.2$ nS, and $E_r = 0$ mV. Then

$$V_m = \frac{(0.2 \times 10^{-9} \cdot 0) + (10 \times 10^{-9} \cdot -60)}{10.2 \times 10^{-9}}$$

$$= -59 \text{ mV}$$

Thus, as a result of the closure of the switch, the membrane potential has changed from its initial value of -60 mV to a new value of -59 mV; that is, an EPSP of 1 mV has been generated.

The preceding analysis ignored the membrane capacitance (C_m), the charging of which makes the synaptic potential slower than the synaptic current. Thus, a more complete analytical description of the postsynaptic factors underlying the generation of a PSP must account for the fact that some of the synaptic current will flow into the capacitive branch of the circuit. Again, by conservation of current, the sum of the currents in the three branches must equal 0. Therefore,

$$0 = C_m \frac{dV_m}{dt} + I_L + I_{syn}, \tag{16.16}$$

$$0 = C_m \frac{dV_m}{dt} + g_L \cdot (V_m - E_L) + g_{syn}(t) \cdot (V_m - E_r), \tag{16.17}$$

where $C_m (dV_m/dt)$ is the capacitative current.

By solving for V_m and integrating the differential equation, we can determine the magnitude and time course of a PSP. An accurate description of the kinetics of the PSP requires that the simple switch closure (all-or-none engagement of the synaptic conductance) be replaced with an expression [$g_{syn}(t)$] that describes the dynamics of the change in synaptic conductance with time. Equation 16.9, which describes the dynamics of channel closure, could be used as an approximation of these effects, but a more accurate simulation requires an expression that also describes the kinetics of channel opening (which, in Eq. 16.9, is assumed to be instantaneous) (Magleby and Stevens, 1972).

Nonlinear I–V Relationships of Some Ionotropic Receptors

For many PSPs mediated by ionotropic receptors, the current–voltage relationship of the synaptic current is linear or approximately linear (Fig. 16.5B). Such ohmic relations are typical of nicotinic ACh channels and AMPA (alpha-amino-3-hydroxyl-5-methyl-4-isoxazole-propionate) glutamate channels (as well as many receptors mediating IPSPs, for example, Fig. 16.10). The linear I–V relationship is indicative of a channel whose conductance is not affected by the potential across the membrane. Such linearity should be contrasted with the steep voltage dependency of the conductance of channels underlying the initiation and repolarization of action potentials (Chapters 5 and 7).

NMDA (N-methyl-D-aspartate) glutamate channels are a class of ionotropic receptors that have nonlinear current–voltage relationships. At negative potentials, the channel conductance is low even when glutamate is bound to the receptor. As the membrane is depolarized, the conductance increases and the current flowing through the channel increases, resulting in the type of I–V relationship illustrated in Figure 16.6A. This nonlinearity is represented by an arrow through the resistor representing this synaptic conductance in the equivalent circuit of Figure 16.6B. The nonlinear I–V relationship of the NMDA receptor can be explained by a voltage-dependent block of the channel by Mg^{2+} (Fig. 16.7). At normal values of the resting potential, the pore of the channel is blocked by Mg^{2+}. Thus, even when glutamate binds to the receptor (Fig. 16.7B), the blocked channel prevents ionic flow (and an EPSP). The block can be relieved by depolarization, which presumably displaces the Mg^{2+} from the pore (Fig. 16.7B). When the pore is unblocked, cations (i.e., Na^+, K^+, and Ca^{2+}) can readily flow through the channel, and this flux is manifested in the linear part of the I–V relationship (Fig. 16.6A). AMPA channels (Fig. 16.7A) are not blocked by Mg^{2+} and have linear I–V relationships (Fig. 16.5B).

Inhibitory Postsynaptic Potentials Decrease the Probability of Cell Firing

Some synaptic events *decrease* the probability of generating action potentials in the postsynaptic cell. Potentials associated with these actions are called *inhibitory postsynaptic potentials*. Consider the inhibitory interneuron illustrated in Figure 16.2. Normally, this interneuron is activated by summating EPSPs from converging afferent fibers. These EPSPs summate in space and time such that the membrane potential of the interneuron reaches threshold and fires an action potential. This step can be bypassed by artificially depolarizing the interneuron to initiate an action potential. The consequences of that action potential from the point of view of the flexor motor neuron are illustrated in Figure 16.2. The action potential in the interneuron produces a transient increase in the membrane potential of the motor neuron. This transient hyperpolarization (the IPSP) looks very much like the EPSP, but it is reversed in sign.

What are the ionic mechanisms for these fast IPSPs, and what is the transmitter substance? Because the membrane potential of the flexor motor neuron is about –65 mV, one might expect an increase in the conductance to some ion (or ions) with an equilibrium potential (reversal potential) more negative than –65 mV. One possibility is K^+. Indeed, the K^+ equilibrium potential in spinal motor neurons is about –80 mV; thus, a transmitter substance that produced a selective increase in K^+ conductance would lead to an IPSP. The K^+ conductance increase would move the membrane potential from –65 mV toward the K^+ equilibrium potential of –80 mV. Although an increase in K^+ conductance mediates IPSPs at some inhibitory synapses (see following text and Fig. 16.10), it does not at the synapse between the inhibitory interneuron and the spinal motor neuron. At this particular synapse, the IPSP seems to be due to a selective increase in Cl^- conductance. The equilibrium potential for Cl^- in spinal motor neurons is about –70 mV. Thus, the transmitter substance released by the inhibitory neuron diffuses across the cleft and interacts with receptor sites on the postsynaptic membrane. These receptors are normally closed, but when opened they become selectively permeable to Cl^-. As a result of the increase in Cl^- conductance, the membrane potential moves from a resting value of –65 mV toward the Cl^- equilibrium potential of –70 mV.

As in the sensory neuron–spinal motor neuron synapse, the transmitter substance released by the inhibitory interneuron in the spinal cord is an amino acid, but in this case the transmitter is glycine. The toxin strychnine is a potent antagonist of glycine

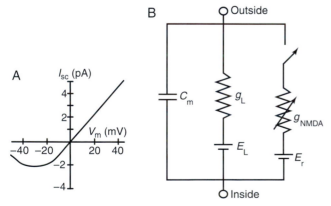

FIGURE 16.6 (A) I–V relationship of the NMDA receptor. (B) Equivalent electrical circuit of a membrane containing NMDA receptors.

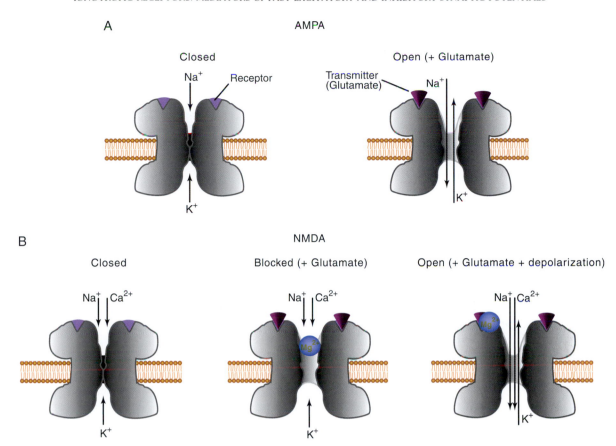

FIGURE 16.7 Features of AMPA and NMDA glutamate receptors. (A) AMPA receptors: (left) in the absence of agonist, the channel is closed; (right) glutamate binding leads to channel opening and an increase in Na^+ and K^+ permeability. AMPA receptors that contain the GluR2 subunit are impermeable to Ca^{2+}. (B) NMDA receptors: (left) in the absence of agonist, the channel is closed; (middle) the presence of agonist leads to a conformational change and channel opening, but no ionic flux occurs, because the pore of the channel is blocked by Mg^{2+}; (right) in the presence of depolarization, the Mg^{2+} block is removed and the agonist-induced opening of the channel leads to changes in ion flux (including Ca^{2+} influx into the cell).

receptors. Although glycine was originally thought to be localized to the spinal cord, it is also found in other regions of the nervous system. The most common transmitter associated with inhibitory actions in many areas of the brain is γ-aminobutyric acid (GABA) (see Chapter 9).

GABA receptors are divided into three major classes: $GABA_A$, $GABA_B$, and $GABA_C$ (Billington et al., 2001; Bormann, 1988; Bormann and Feigenspan, 1995; Bowery, 1993; Cherubini and Conti, 2001; Gage, 2001; Moss and Smart, 2001). As discussed in Chapter 11, $GABA_A$ receptors are ionotropic receptors, and, like glycine receptors, binding of transmitter leads to an increased conductance to Cl^-, which produces an IPSP. $GABA_A$ receptors are blocked by bicuculline and picrotoxin. A particularly striking aspect of $GABA_A$ receptors is their modulation by anxiolytic benzodiazepines. Figure 16.8 illustrates the response of a neuron to GABA before and after treatment with diazepam (Bormann, 1988). In the presence of diazepam, the response is greatly potentiated. In contrast

to $GABA_A$ receptors that are pore-forming channels, $GABA_B$ receptors are G-protein-coupled (see also Chapter 11). $GABA_B$ receptors can be coupled to a variety of different effector mechanisms in different neurons. These mechanisms include decreases in Ca^{2+} conductance, increases in K^+ conductance, and modulation of voltage-dependent A-type K^+ current. In hippocampal pyramidal neurons, the $GABA_B$-mediated IPSP is due to an increase in K^+ conductance (see Fig. 16.10). Baclofen is a potent agonist of $GABA_B$ receptors, whereas phaclofen is a selective antagonist. $GABA_C$ receptors are pharmacologically distinct from $GABA_A$ and $GABA_B$ receptors and are found predominantly in the vertebrate retina. $GABA_C$ receptors, like $GABA_A$ receptors, are Cl^--selective pores.

Ionotropic receptors that lead to the generation of IPSPs and ionotropic receptors that lead to the generation of EPSPs have biophysical features in common. Indeed, the analyses of the preceding section are generally applicable. A quantitative understanding of the

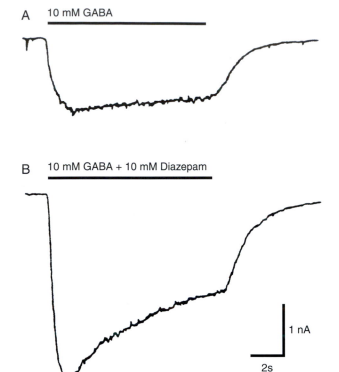

FIGURE 16.8 Potentiation of GABA responses by benzodiazepine ligands. (A) Brief application (bar) of GABA leads to an inward Cl⁻ current in a voltage-clamped spinal neuron. (B) In the presence of diazepam the response is significantly enhanced. From Bormann (1988).

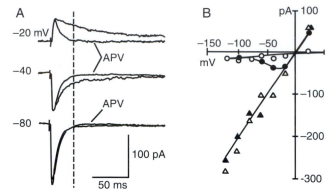

FIGURE 16.9 Dual-component glutamatergic EPSP. (A) The excitatory postsynaptic current was recorded before and during the application of APV at the indicated membrane potentials. (B) Peak current–voltage relationships are shown before (solid triangles) and during (open triangles) the application of APV. The current–voltage relationships measured 25 ms after the peak of the EPSC (dotted line in A) before (solid circles) and during (open circles) application of APV are also shown. Reprinted with permission from Hestrin *et al.* (1990).

effects of the opening of glycine or GABA receptors can be obtained by using the electrical equivalent circuit of Figure 16.5C and Eq. 16.15, with the values of g_{syn} and E_r appropriate for the respective ionotropic receptor. Interactions between excitatory and inhibitory conductances can be modeled by adding additional branches to the equivalent circuit (see Fig. 16.15D and Chapter 17).

Some PSPs Have More Than One Component

The transmitter released from a presynaptic terminal diffuses across the synaptic cleft, where it binds to ionotropic receptors. In many cases, the postsynaptic receptors are homogeneous. In other cases, the same transmitter activates more than one type of receptor. A major example of this type of heterogeneous postsynaptic action is the simultaneous activation by glutamate of NMDA and AMPA receptors on the same postsynaptic cell. Figure 16.9 illustrates such a dual-component glutamatergic EPSP in the CA1 region of the hippocampus. The cell is voltage clamped at various fixed holding potentials, and the macroscopic synaptic currents produced by activation

of the presynaptic neurons are recorded. The experiment is performed in the presence and absence of the agent 2-amino-5-phosphonovalerate (APV), which is a specific blocker of NMDA receptors. When the cell is held at a potential of +20 or −40 mV, APV leads to a dramatic reduction of the late, but not the early, phase of the excitatory postsynaptic current (EPSC). In contrast, when the potential is held at −80 mV, the EPSC is unaffected by APV. These results indicate that the PSP consists of two components: (1) an early AMPA-mediated component and (2) a late NMDA-mediated component. In addition, the results indicate that the conductance of the non-NMDA component is linear, whereas the conductance of the NMDA component is nonlinear. The I–V relationships of the early (peak) and late (at approximately 25 ms) components of the EPSC are plotted in Figure 16.9 (Hestrin *et al.*, 1990).

Dual-component IPSPs are also observed in the CNS. Stimulation of afferent pathways to the hippocampus results in an IPSP in a pyramidal neuron, which has a fast initial inhibitory phase followed by a slower inhibitory phase (Fig. 16.10). Application of the GABA_A antagonists blocks the early inhibitory phase, whereas the GABA_B receptor antagonist phaclofen blocks the late inhibitory phase (not shown). The early and late IPSPs can also be distinguished based on their ionic mechanisms. Hyperpolarizing the membrane potential to −78 mV nulls the early response, but at this value of membrane potential the late response is still hyperpolarizing (Figs. 16.10A and B). Hyperpolarizing the membrane potential to values more negative than −78 mV reverses the sign

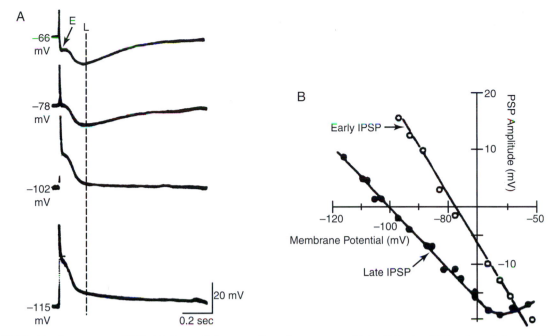

FIGURE 16.10 Dual-component IPSP. (A) Intracellular recordings from a pyramidal cell in the CA3 region of the rat hippocampus in response to activation of mossy fiber afferents. With the membrane potential of the cell at the resting potential, afferent stimulation produces an early (E) and late (L) IPSP. With increased hyperpolarizing produced by injecting constant current into the cell, the early component reverses first. At more negative levels of the membrane potential, the late component also reverses. This result indicates the ionic conductance underlying the two phases is distinct. (B) Plots of the change in amplitude of the early (measured at 25 ms) and late (measured at 200 ms, dashed line) response as a function of the membrane potential. The reversal potentials of the early and late components are consistent with a GABAA-mediated chloride conductance and a GABAB-mediated potassium conductance, respectively. From Thalmann (1988).

of the early response, but the slow response does not reverse until the membrane is made more negative than about −100 mV (Thalmann, 1988). The reversal potentials are consistent with a fast Cl⁻-mediated IPSP, produced by fast opening of GABA$_A$ receptors, and a slower K⁺-dependent IPSP, produced by GABA$_B$ receptors, which activate G-protein-activated inwardly rectifying K⁺ channels (GIRKs). In mutant mice that lack a specific GIRK (i.e., GIRK2), the GABA$_B$ response in hippocampal pyramidal neurons is reduced or absent (Luscher et al., 1997).

Dual-component PSPs need not be strictly inhibitory or excitatory. For example, a presynaptic cholinergic neuron in the mollusk Aplysia produces a diphasic excitatory–inhibitory (E–I) response in its postsynaptic follower cell. The response can be simulated by local discrete application of ACh to the postsynaptic cell (Fig. 16.11) (Blankenship et al., 1971). The ionic mechanisms underlying this synaptic action were investigated in ion substitution experiments, which revealed that the dual response is due to an early Na⁺-dependent component followed by a slower Cl⁻-dependent component. Molecular mechanisms underlying such slow synaptic potentials are discussed next.

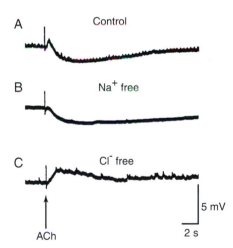

FIGURE 16.11 Dual-component cholinergic excitatory–inhibitory response. (A) Control in normal saline. Ejection of ACh produces a rapid depolarization followed by a slower hyperpolarization. (B) In Na⁺-free saline, ACh produces a purely hyperpolarizing response, indicating that the depolarizing component in normal saline includes an increase in g_{Na}. (C) In Cl⁻-free saline, ACh produces a purely depolarizing response, indicating that the hyperpolarizing component in normal saline includes an increase in g_{Cl}. Reproduced with permission from Blankenship, J.E., Wachtel, H., and Kandel, E.R. (1971). Ionic mechanisms of excitatory, inhibitory and dual synaptic actions mediated by an identified interneuron in abdominal ganglion of Aplysia. J. Neurophysiol. **34**, 76–92.

Summary

Synaptic potentials mediated by ionotropic receptors are the fundamental means by which information is rapidly transmitted between neurons. Transmitters cause channels to open in an all-or-none fashion, and the currents through these individual channels summate to produce the macroscopic postsynaptic potential. The sign of the postsynaptic potential is determined by the relationship between the membrane potential of the postsynaptic neuron and the ion selectivity of the ionotropic receptor.

METABOTROPIC RECEPTORS: MEDIATORS OF SLOW SYNAPTIC POTENTIALS

A common feature of the types of synaptic actions heretofore described is the direct binding of the transmitter with the receptor–channel complex. An entirely separate class of synaptic actions has as its basis the indirect coupling of the receptor with the channel. Two major types of coupling mechanisms have been identified: (1) coupling of the receptor and channel through an intermediate regulatory protein, such as a G protein (see previous discussion of $GABA_B$ responses); and (2) coupling through a diffusible second-messenger system. The coupling through a diffusible second-messenger system is the focus of this section.

A comparison of the features of direct, fast ionotropic-mediated and indirect, slow metabotropic-mediated synaptic potentials is shown in Figure 16.12. Slow synaptic potentials are not observed at every postsynaptic neuron, but Figure 16.12A illustrates an idealized case in which a postsynaptic neuron receives two inputs, one of which produces a conventional fast EPSP and the other of which produces a slow EPSP. An action potential in neuron 1 leads to an EPSP in the postsynaptic cell with a duration of about 30 ms (Fig. 16.12B). This type of potential might be produced in a spinal motor neuron by an action potential in an afferent fiber. Neuron 2 also produces a postsynaptic potential (Fig. 16.12C), but its duration (note the calibration bar) is more than three orders of magnitude greater than that of the EPSP produced by neuron 1.

How can a change in the postsynaptic potential of a neuron persist for many minutes as a result of a single action potential in the presynaptic neuron? Possibilities include a prolonged presence of the transmitter due to continuous release, slow degradation, or slow

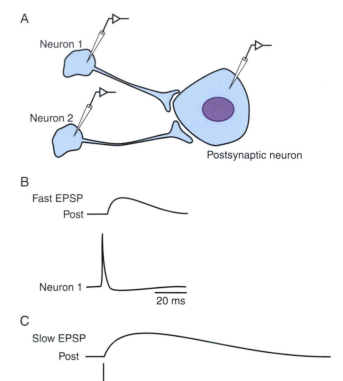

FIGURE 16.12 Fast and slow synaptic potentials. (A) Idealized experiment in which two neurons (1 and 2) make synaptic connections with a common postsynaptic follower cell (Post). (B) An action potential in neuron 1 leads to a conventional fast EPSP with a duration of about 30 ms. (C) An action potential in neuron 2 also produces an EPSP in the postsynaptic cell, but the duration of this slow EPSP is more than three orders of magnitude greater than that of the EPSP produced by neuron 1. Note the change in the calibration bar.

reuptake of the transmitter, but the mechanism here involves a transmitter-induced change in the metabolism of the postsynaptic cell. Figure 16.13 compares the general mechanisms for fast and slow synaptic potentials. Fast synaptic potentials are produced when a transmitter substance binds to a channel and produces a conformational change in the channel, causing it to become permeable to one or more ions (both Na^+ and K^+ in Fig. 16.13A). The increase in permeability leads to a depolarization associated with the EPSP. The duration of the synaptic event critically depends on the amount of time during which the transmitter substance remains bound to the receptors. Acetylcholine, glutamate, and glycine remain bound only for a very short period. These transmitters are removed by diffusion, enzymatic breakdown, or reuptake into the presynaptic cell. Therefore, the duration of the synaptic potential is directly related to the lifetimes of the

A

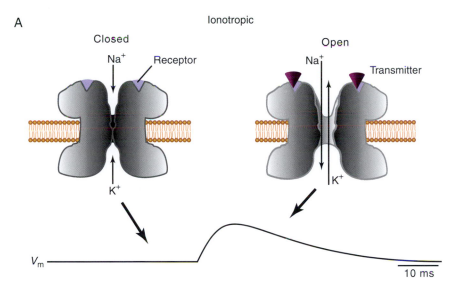

B

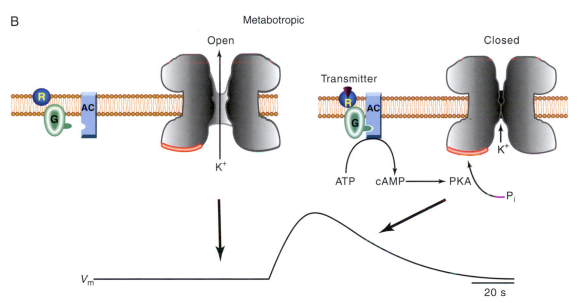

FIGURE 16.13 Ionotropic and metabotropic receptors and mechanisms of fast and slow EPSPs. (A, left) Fast EPSPs are produced by the binding of transmitter to specialized receptors that are directly associated with an ion channel (i.e., a ligand-gated channel). When the receptors are unbound, the channel is closed. (A, right) Binding of the transmitter to the receptor produces a conformational change in the channel protein such that the channel opens. In this example, the channel opening is associated with a selective increase in the permeability to Na$^+$ and K$^+$. The increase in permeability results in the EPSP shown in the trace. (B, left) Unlike fast EPSPs that are due to the binding of a transmitter with a receptor–channel complex, slow EPSPs are due to the activation of receptors (metabotropic) that are not directly coupled to the channel. Rather, the coupling takes place through the activation of one of several second-messenger cascades, in this example, the cAMP cascade. A channel that has a selective permeability to K$^+$ is normally open. (B, right) Binding of the transmitter to the receptor (R) leads to the activation of a G protein (G) and adenylyl cyclase (AC). The synthesis of cAMP is increased, cAMP-dependent protein kinase (protein kinase A, PKA) is activated, and a channel protein is phosphorylated. The phosphorylation leads to closing of the channel and the subsequent depolarization associated with the slow EPSP shown in the trace. The response decays owing to both the breakdown of cAMP by cAMP-dependent phosphodiesterase and the removal of phosphate from channel proteins by protein phosphatases (not shown).

opened channels, and these lifetimes are relatively short (see Fig. 16.4B).

One mechanism for a slow synaptic potential is shown in Figure 16.13B. In contrast with the fast PSP for which the receptors are actually part of the ion channel complex, the channels that produce the slow synaptic potentials are not directly coupled to the transmitter receptors. Rather, the receptors are physically separated and exert their actions indirectly through changes in metabolism of specific second-messenger systems. Figure 16.13B illustrates one type of response in *Aplysia*, for which the cAMP–protein kinase A system is the mediator, but other slow PSPs use other second-messenger–kinase systems (e.g., the protein kinase C system). In the cAMP-dependent slow synaptic responses in *Aplysia*, transmitter binding to membrane receptors activates G proteins and stimulates an increase in the synthesis of cAMP. Cyclic AMP then leads to the activation of cAMP-dependent protein kinase (protein kinase A, PKA), which phosphorylates a channel protein or protein associated with the channel (Siegelbaum *et al.*, 1982). A conformational change in the channel is produced, leading to a change in ionic conductance. Thus, in contrast with a direct conformational change produced by the binding of a transmitter to the receptor–channel complex, in this case, a conformational change is produced by protein phosphorylation. Indeed, phosphorylation-dependent channel regulation is a fairly general feature of slow PSPs. However, channel regulation by second messengers is not exclusively produced by phosphorylation. In one family of ion channels, the channels are gated or regulated directly by cyclic nucleotides. These cyclic nucleotide-gated channels require cAMP or cGMP to open but have other features in common with members of the superfamily of voltage-gated ion channels (Kaupp, 1995; Zimmermann, 1995).

Another interesting feature of slow synaptic responses is that they are sometimes associated with decreases rather than increases in membrane conductance. For example, the particular channel illustrated in Figure 16.13B is selectively permeable to K^+ and is normally open. As a result of the activation of the second messenger, the channel closes and becomes less permeable to K^+. The resultant depolarization may seem paradoxical, but recall that the membrane potential is due to a balance between the resting K^+ and Na^+ permeability. The K^+ permeability tends to move the membrane potential toward the K^+ equilibrium potential (−80 mV), whereas the Na^+ permeability tends to move the membrane potential toward the Na^+ equilibrium potential (+55 mV). Normally, the K^+ permeability predominates, and the resting membrane potential is close to, but not equal to, the K^+ equilibrium potential. If K^+ permeability is decreased because some of the channels close, the membrane potential will be biased toward the Na^+ equilibrium potential and the cell will depolarize.

At least one reason for the long duration of slow PSPs is that second-messenger systems are slow (from seconds to minutes). Take the cAMP cascade as an example. Cyclic AMP takes some time to be synthesized, but, more importantly, after synthesis, cAMP levels can remain elevated for a relatively long period (minutes). The duration of the elevation of cAMP depends on the actions of cAMP phosphodiesterase, which

breaks down cAMP. However, duration of an effect could outlast the duration of the change in the second messenger because of persistent phosphorylation of the substrate protein(s). Phosphate groups are removed from the substrate proteins by protein phosphatases. Thus, the net duration of a response initiated by a metabotropic receptor depends on the actions of not only the synthetic and phosphorylation processes but also the degradative and dephosphorylation processes.

The activation of a second messenger by a transmitter can have a localized effect on the membrane potential through phosphorylation of membrane channels near the site of a metabotropic receptor. The effects can be more widespread and even longer lasting than depicted in Figure 16.13B. For example, second messengers and protein kinases can diffuse and affect more distant membrane channels. Moreover, a long-term effect can be induced in the cell by altering gene expression. For example, protein kinase A can diffuse to the nucleus, where it can activate proteins that regulate gene expression. Detailed descriptions of second messengers and their actions are given in Chapters 12, 13, 14, and 19.

Summary

In contrast to the rapid responses mediated by ionotropic receptors, responses mediated by metabotropic receptors (i.e., GPCRs) are generally relatively slow to develop and persistent. These properties arise because metabotropic responses can involve the activation of second-messenger systems. By producing slow changes in the resting potential, metabotropic receptors provide long-term modulation of the effectiveness of responses generated by ionotropic receptors. Moreover, these receptors, through the engagement of second-messenger systems, provide a vehicle by which a presynaptic cell can not only alter the membrane potential but also produce widespread changes in the biochemical state of a postsynaptic cell.

INTEGRATION OF SYNAPTIC POTENTIALS

The small amplitude of the EPSP in spinal motor neurons (and other cells in the CNS) poses an interesting question. Specifically, how can an EPSP with an amplitude of only 1 mV drive the membrane potential of the motor neuron (i.e., the postsynaptic neuron) to threshold and fire the spike in the motor neuron that is necessary to produce the contraction of the muscle? The answer to this question lies in the principles of temporal and spatial summation.

When the ligament is stretched (Fig. 16.1), many stretch receptors are activated. Indeed, the greater the stretch, the greater the probability of activating a larger number of the stretch receptors; this process is referred to as *recruitment*. However, recruitment is not the complete story. The principle of frequency coding in the nervous system specifies that the greater the intensity of a stimulus, the greater the number of action potentials per unit time (frequency) elicited in a sensory neuron. This principle applies to stretch receptors as well. Thus, the greater the stretch, the greater the number of action potentials elicited in the stretch receptor in a given interval and therefore the greater the number of EPSPs produced in the motor neuron from that train of action potentials in the sensory cell. Consequently, the effects of activating multiple stretch receptors add together (spatial summation), as do the effects of multiple EPSPs elicited by activation of a single stretch receptor (temporal summation). Both of these processes act in concert to depolarize the motor neuron sufficiently to elicit one or more action potentials, which then propagate to the periphery and produce the reflex.

Temporal Summation Allows Integration of Successive PSPs

Temporal summation can be illustrated by firing action potentials in a presynaptic neuron and monitoring the resultant EPSPs. For example, in Figures 16.14A and 16.14B, a single action potential in sensory neuron 1 produces a 1 mV EPSP in the motor neuron. Two action potentials in quick succession produce two EPSPs, but note that the second EPSP occurs during the falling phase of the first, and the depolarization associated with the second EPSP adds to the depolarization produced by the first. Thus, two action potentials produce a summated potential that is about 2 mV in amplitude. Three action potentials in quick succession would produce a summated potential of about 3 mV. In principle, 30 action potentials in quick succession would produce a potential of about 30 mV and easily drive the cell to threshold. This summation is strictly a passive property of the cell. No special ionic conductance mechanisms are necessary. Specifically, the postsynaptic conductance change (g_{syn} in Eq. 16.13) produced by the second of two successive action potentials adds to that produced by the first. In addition, the postsynaptic membrane has a capacitance and can store charge. Thus, the membrane temporarily stores the charge of the first EPSP, and the charge from the second EPSP is added to that of the first. However, the "time window" for this

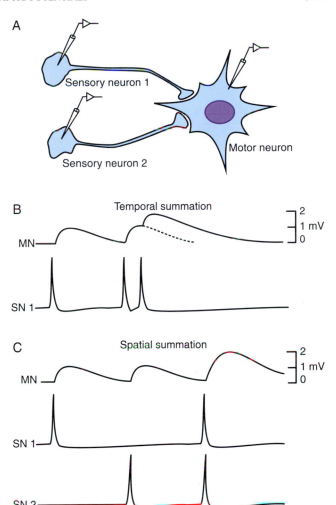

FIGURE 16.14 Temporal and spatial summation. (A) Intracellular recordings are made from two idealized sensory neurons (SN1 and SN2) and a motor neuron (MN). (B) Temporal summation: A single action potential in SN1 produces a 1 mV EPSP in the MN. Two action potentials in quick succession produce a dual-component EPSP, the amplitude of which is approximately 2 mV. (C) Spatial summation: Alternative firing of single action potentials in SN1 and SN2 produce 1 mV EPSPs in the MN. Simultaneous action potentials in SN1 and SN2 produce a summated EPSP, the amplitude of which is about 2 mV.

process of temporal summation very much depends on the duration of the postsynaptic potential, and temporal summation is possible only if the presynaptic action potentials (and hence postsynaptic potentials) are close in time to each other. The time frame depends on the duration of changes in the synaptic conductance and the time constant (Chapter 4). Temporal summation, however, is rarely observed to be linear as in the preceding examples, even when the postsynaptic conductance change (g_{syn} in Eq. 16.13) produced by the second of two successive action potentials is identical to that produced

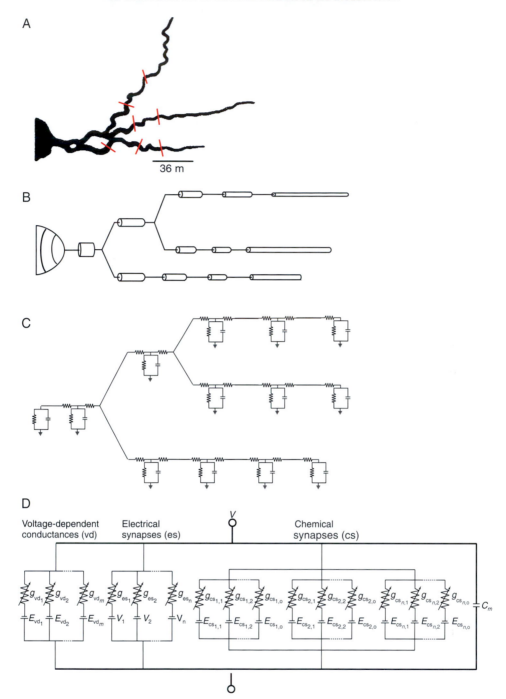

FIGURE 16.15 Modeling the integrative properties of a neuron. (A) Partial geometry of a neuron in the CNS revealing the cell body and pattern of dendritic branching. (B) The neuron modeled as a sphere connected to a series of cylinders, each of which represents the specific electrical properties of a dendritic segment. (C) Segments linked with resistors representing the intracellular resistance between segments, with each segment represented by the parallel combination of the membrane capacitance and the total membrane conductance. Reprinted with permission from Koch and Segev (1989). Copyright 1989 MIT Press. (D) Electrical circuit equivalent of the membrane of a segment of a neuron. In D, the segment has a membrane potential V and a membrane capacitance C_m. Currents arise from three sources: (1) m voltage-dependent conductances (g_{vd_1}–g_{vd_m}), (2) n conductances due to electrical synapses (g_{es_1}–g_{es_n}), and (3) n times o time-dependent conductances due to chemical synapses with each of the n presynaptic neurons ($g_{cs_{1,1}}$–$g_{cs_{n,o}}$). E_{vd} and E_{cs} are constants and represent the values of the equilibrium potential for currents due to voltage-dependent conductances and chemical synapses, respectively. V_1–V_n represent the value of the membrane potential of the coupled cells. Reproduced with permission from Ziv, I., Baxter, D.A., and Byrne, J.H. (1994). Simulator for neural networks and action potentials: description and application. *J. Neurophysiol.* **71**, 294–308.

by the first (i.e., no presynaptic facilitation or depression), the synaptic current is slightly less because the first PSP reduces the driving force ($V_m - E_r$) for the second. Interested readers should try some numerical examples.

Spatial Summation Allows Integration of PSPs from Different Parts of a Neuron

Spatial summation (Fig. 16.14C) requires a consideration of more than one input to a postsynaptic neuron. An action potential in sensory neuron 1 produces a 1 mV EPSP, just as it did in Figure 16.14B. Similarly, an action potential in a second sensory neuron by itself also produces a 1 mV EPSP. Now, consider the consequences of action potentials elicited simultaneously in sensory neurons 1 and 2. The net EPSP is equal to the summation of the amplitudes of the individual EPSPs. Here, the EPSP from sensory neuron 1 is 1 mV, the EPSP from sensory neuron 2 is 1 mV, and the summated EPSP is approximately 2 mV (Fig. 16.14C). Thus, spatial summation is a mechanism by which synaptic potentials generated at different sites can summate. Spatial summation in nerve cells is influenced by the space constant—the ability of a potential change produced in one region of a cell to spread passively to other regions of a cell (see Chapter 4).

Summary

Whether a neuron fires in response to synaptic input depends, at least in part, on how many action potentials are produced in any one presynaptic excitatory pathway and on how many individual convergent excitatory input pathways are activated. The summation of EPSPs in time and space is only part of the process, however. The final behavior of the cell is also due to the summation of inhibitory synaptic inputs in time and space, as well as to the properties of the voltage-dependent currents (Fig. 16.15) in the soma and along the dendrites (Koch and Segev, 1989; Ziv et al., 1994). For example, voltage-dependent conductances such as the A-type K^+ conductance have a low threshold for activation and can thus oppose the effectiveness of an EPSP to trigger a spike. Low-threshold Na^+ and Ca^{2+} channels can boost an EPSP. Finally, we need to consider that the spatial distribution of the various voltage-dependent channels, ligand-gated receptors, and metabotropic receptors is not uniform (see also Chapter 6). Thus, each segment of the neuronal membrane can perform selective integrative functions. Clearly, this system has an enormous capacity for the local processing of information and for performing logical operations. The flow of information in dendrites and the local processing of neuronal signals are discussed in Chapter 17. Several software packages are available for the development and simulation of realistic models of single neurons and neural networks (Hayes et al., 2003). One, Simulator for Neural Networks and Action Potentials (SNNAP) (Baxter and Byrne, 2007; http://snnap.uth.tmc.edu/), provides mathematical descriptions of ion currents, intracellular second messengers, and ion pools, and allows simulation of current flow in multicompartment models of neurons.

References

General

Burke, R. E., and Rudomin, P. (1977). Spatial neurons and synapses. In "Handbook of Physiology" (E. R. Kandel, Ed.), Sect. 1, **Vol. 1**, Part 2, pp. 877–944. American Physiological Society, Bethesda, MD.

Byrne, J. H., and Schultz, S. G. (1994). "An Introduction to Membrane Transport and Bioelectricity," 2nd ed. Raven Press, New York.

Cowan, W. M., Sudhof, T. C., and Stevens, C. F. (Eds.) (2001). "Synapses." Johns Hopkins Univ. Press, Bethesda, MD.

Hille, B. (Ed.) (2001). "Ion Channels of Excitable Membranes," 3rd ed. Sinauer Associates, Sunderland, MA.

Shepherd, G. M. (Ed.) (2004). "The Synaptic Organization of the Brain," 5th ed. Oxford Univ. Press, New York.

Cited

Baxter, D. A., and Byrne, J. H. (2007). Simulator for neural networks and action potentials (SNNAP): Description and application. In "Methods in Molecular Biology: Neuroinformatics" (C. Crasto, Ed.), pp. 127–154. Humana Press Inc., Totowa.

Billington, A., Ige, A. O., Bolam, J. P., White, J. H., Marshall, F. H., and Emson, P. C. (2001). Advances in the molecular understanding of GABA$_B$ receptors. *Trends Neurosci.* **24**, 277–282.

Blankenship, J. E., Wachtel, H., and Kandel, E. R. (1971). Ionic mechanisms of excitatory, inhibitory and dual synaptic actions mediated by an identified interneuron in abdominal ganglion of Aplysia. *J. Neurophysiol.* **34**, 76–92.

Bormann, J. (1988). Electrophysiology of GABA$_A$ and GABA$_B$ receptor subtypes. *Trends Neurosci.* **11**, 112–116.

Bormann, J., and Feigenspan, A. (1995). GABA$_C$ receptors. *Trends Neurosci.* **18**, 515–519.

Bowery, N. G. (1993). GABA$_B$ receptor pharmacology. *Annu. Rev. Pharmacol. Toxicol.* **33**, 109–147.

Cherubini, E., and Conti, F. (2001). Generating diversity at GABAergic synapses. *Trends Neurosci.* **24**, 155–162.

Colquhoun, D., and Hawkes, A. G. (1977). Relaxation and fluctuations of membrane currents that flow through drug-operated channels. *Proc. R. Soc. London Ser. B* **199**, 231–262.

Colquhoun, D., and Hawkes, A. G. (1981). On the stochastic properties of single ion channels. *Proc. R. Soc. London Ser. B* **211**, 205–235.

Colquhoun, D., and Hawkes, A. G. (1982). On the stochastic properties of bursts of single ion channel openings and of clusters of bursts. *Proc. R. Soc. London Ser. B* **300**, 1–59.

Colquhoun, D., and Sakmann, B. (1981). Fluctuations in the microsecond time range of the current through single acetylcholine receptor ion channels. *Nature (London)* **294**, 464–466.

Eccles, J. C. (1964). "The Physiology of Synapses." Springer-Verlag, New York.

Fatt, P., and Katz, B. (1951). An analysis of the end-plate potential recorded with an intra-cellular electrode. *J. Physiol. (London)* **115**, 320–370.

Gage, P. W. (2001). Activation and modulation of neuronal K^+ channels by GABA. *Trends Neurosci.* **15**, 46–51.

Hamill, O. P., Marty, A., Neher, E., Sakmann, B., and Sigworth, J. (1981). Improved patch-clamp techniques for high-resolution current recording from cells and cell-free membrane patches. *Pflügers Arch.* **391**, 85–100.

Hayes, R. D., Byrne, J. H., and Baxter, D. A. (2003). Neurosimulation: Tools and resources. *In* "The Handbook of Brain Theory and Neural Networks," 2nd ed. (M. A. Arbib, Ed.), pp. 776–780. MIT Press, Cambridge, MA.

Hestrin, S., Nicoll, R. A., Perkel, D. J., and Sah, P. (1990). Analysis of excitatory synaptic action in pyramidal cells using whole-cell recording from rat hippocampal slices. *J. Physiol. (London)* **422**, 203–225.

Johnston, D., and Wu, S. M.-S. (1995). "Foundations of Cellular Neurophysiology." MIT Press, Cambridge, MA.

Kaupp, U. B. (1995). Family of cyclic nucleotide gated ion channels. *Curr. Opin. Neurobiol.* **5**, 434–442.

Koch, C., and Segev, I. (1989). "Methods in Neuronal Modeling." MIT Press, Cambridge, MA.

Luscher, C., Yan, L. Y., Stoffel, M., Malenka, R. C., and Nicoll, R. A. (1997). G protein-coupled inwardly rectifying K^+ channels (GIRKs) mediate postsynaptic but not presynaptic transmitter actions in hippocampal neurons. *Neuron* **19**, 687–695.

Magleby, K. L., and Stevens, C. F. (1972). A quantitative description of end-plate currents. *J. Physiol. (London)* **223**, 173–197.

Moss, S. J., and Smart, T. G. (2001). Constructing inhibitory synapses. *Nat. Rev. Neurosci.* **2**, 240–250.

Papoulis, A. (1965). "Probability, Random Variables, and Stochastic Processes." McGraw-Hill, New York.

Sakmann, B. (1992). Elementary steps in synaptic transmission revealed by currents through single ion channels. *Science* **256**, 503–512.

Siegelbaum, S. A., Camardo, J. S., and Kandel, E. R. (1982). Serotonin and cyclic AMP close single K^+ channels in *Aplysia* sensory neurones. *Nature (London)* **299**, 413–417.

Spencer, W. A. (1977). The physiology of supraspinal neurons in mammals. *In* "Handbook of Physiology" (E. R. Kandel, Ed.), Sect. 1, **Vol. 1**, Part 2, pp. 969–1022. American Physiological Society, Bethesda, MD.

Takeuchi, A., and Takeuchi, N. (1960). On the permeability of end-plate membrane during the action of transmitter. *J. Physiol. (London)* **154**, 52–67.

Thalmann, R. H. (1988). Evidence that guanosine triphosphate (GTP)-binding proteins control a synaptic response in brain: Effect of pertussis toxin and GTPgS on the late inhibitory postsynaptic potential of hippocampal CA3 neurons. *J. Neurosci.* **8**, 4589–4602.

Zimmermann, A. L. (1995). Cyclic nucleotide gated channels. *Curr. Opin. Neurobiol.* **5**, 296–303.

Ziv, I., Baxter, D. A., and Byrne, J. H. (1994). Simulator for neural networks and action potentials: Description and application. *J. Neurophysiol.* **71**, 294–308.

Complex Information Processing in Dendrites

John. H. Byrne and Gordon M. Shepherd

A hallmark of neurons is the variety of their dendrites. The branching patterns are dazzling and the size range astounding, from the large trees of cortical pyramidal neurons to the tiny size of a retinal bipolar cell, which would fit comfortably within the cell body of a pyramidal neuron (see Fig. 17.1)! A main challenge in modern neuroscience is understanding the molecular and functional properties of these structures, and their significance for information processing by neurons.

The first step in meeting this challenge is to understand how information processing occurs by passive spread of electrical current. This was the subject of Chapter 4. In this chapter we build on this foundation to understand the more complex types of information processing that can occur by interactions between active properties within branching trees. We ask the fundamental questions: (1) what are the principles of information processing in these dendritic trees, and (2) how are these dendrites with these properties adapted for the operational tasks of a specific neuron type within the microcircuits characteristic of that region? For further orientation, the student is referred to Segev *et al.* (1995), Koch (1999), Shepherd (2004), and Stuart *et al.* (2008).

STRATEGIES FOR STUDYING COMPLEX DENDRITES

Strategies for answering these two questions may be illustrated by the synaptic organization of two cell types, the mitral and granule cells of the olfactory bulb, shown in the lower left-hand corner of Figure 17.1. A first step in the modern approach is to break down a complex dendritic tree into functional compartments so that the integrative actions within the tree can be identified at successive levels of functional organization. As illustrated in Figure 17.2, these levels start with the *individual synapse*, which may be on a dendritic branch, as in the mitral cell, or on a dendritic spine, as in the granule cell. The next level is in terms of *local patterns of synaptic connections*. Successive levels involve not only *larger extents of dendritic branches*, until one reaches the level of *distinct dendritic compartments*—in the case of the mitral cell, distal tuft, primary dendrite, and secondary dendrites—but at each level a dendritic compartment also includes the cells interacting synaptically with that compartment. At the highest level is the global summation at the axon hillock and the global output through the axon. A similar analysis applies to the granule cell, except that it lacks an axon.

This approach allows one to identify the synaptic interactions within a dendritic tree as constituting a hierarchy, within which the specific pattern of interactions at a given level forms the fundamental integrative unit for the next level in the hierarchy. These integrative units are sometimes referred to as *microcircuits*, defined as a specific pattern of interactions performing a specific functional operation (Shepherd, 1977). In a computer microcircuit, a particular circuit configuration can be useful in many different contexts; similarly, in the brain, this gives the hypothesis that a particular microcircuit may be useful in different cells in different contexts at the equivalent level of organization. By this means the principles of information

489

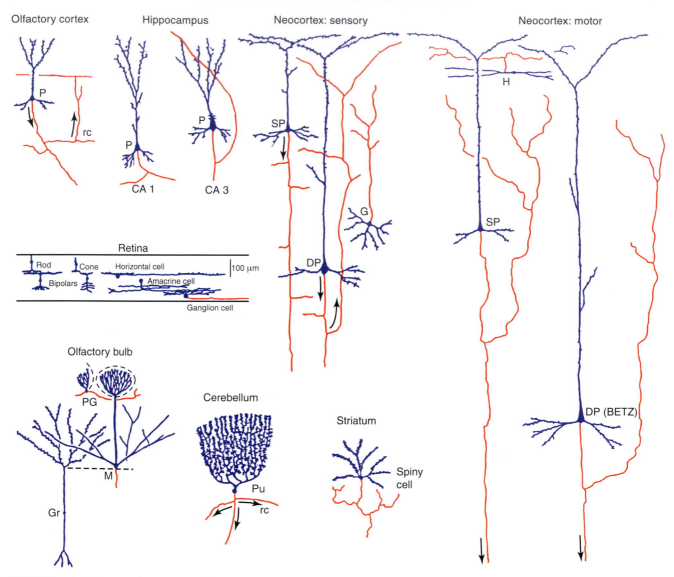

FIGURE 17.1   Varieties of neurons and dendritic trees. P, pyramidal neuron; rc, recurrent collateral; SP, small pyramidal neuron; DP, deep pyramidal neuron; G, granule cell; Gr, granule cell (olfactory); M, mitral cell; PG, periglomerular cell; Pu, Purkinje cell. Modified from Shepherd (1992).

processing across different cell types in different regions and phyla can be identified. We will use this approach to parse the organization of the dendrites of other representative cell types, including those in Figure 17.1.

BUILDING PRINCIPLES STEP BY STEP

As discussed in Chapter 4, the neuron processes information through five basic types of activity: intrinsic, reception, integration, encoding, and output. We saw that understanding how these activities are integrated within the neuron starts with the rules of passive current spread. Many of the principles were first worked out in the dendrites of neurons that lack axons or the ability to generate action potentials. There are many examples in invertebrate ganglia. In vertebrates, they include the retinal amacrine cell and the olfactory granule cell. These studies have shown that a dendritic tree by itself is capable of performing many basic functions required for information processing, such as the generation of intrinsic activity, input–output functions for feature extraction, parallel processing, signal-to-noise enhancement, and oscillatory activity. These cells demonstrate that *there is no one thing that dendrites do; they do whatever is required to process information within their particular neuron or neuronal circuit,* with or without an axon.

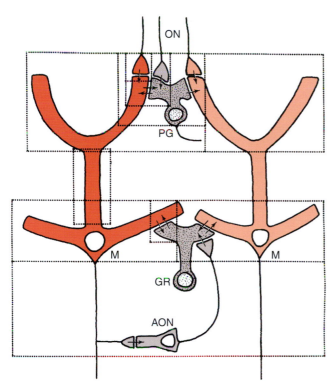

FIGURE 17.2 Compartmentalization (dashed lines) of olfactory bulb neurons to identify the functional subunits of dendrites and their relation to different levels of synaptic organization. ON, olfactory nerve; PG, periglomerular cell; M, mitral cell; GR, granule cell; AON, anterior olfactory nucleus. Reproduced with permission from Shepherd, G.M. (1977). The olfactory bulb: a simple system in the mammalian brain. In "Handbook of Physiology", Sect. I, The Nervous System; Part I, Cellular Biology of Neurons (ed. E.R. Kandel), pp. 945–968. Bethesda, MD, American Physiological Society.

We also need to recognize that *information in dendrites can take many forms.* There are actions of neuropeptides on membrane receptors and internal cytoplasmic or nuclear receptors; actions of second and third messengers within the neuron; movement of substances within the dendrites by diffusion or by active transport; and changes occurring during development. All these types of cellular traffic and information flow in dendrites are coming under direct study (Stuart *et al.,* 2007). The student should review these subjects in earlier chapters. This chapter focuses on information processing in dendrites involving electrical signaling mechanisms by synapses and voltage-gated channels.

We will focus on how this information processing takes place in neurons with axons. Neurons with axons may be classified into two groups, as suggested originally by Camillo Golgi in 1873: those with long axons and those with short axons. Long axon (output) cells tend to be larger than short axon (local) cells and have therefore been more accessible to experimental analysis. Indeed, virtually everything known about the functional relations between dendrites and axons has been

obtained from studies of long axon cells. Consequently, much of what we *think* we understand about those relations in short axon cells is only by inference.

As noted in the analysis of the passive properties of neurons in Chapter 4, there are a number of sites on the Web that support the analysis of complex neurons and their active dendrites. For orientation to the molecular properties of dendritic compartments of different neurons discussed in this chapter, consult senselab.med.yale.edu/neurondb; for computational models based on those properties, consult senselab.med.yale.edu/modeldb. For the structures of dendrites, see synapse-web.org/lab/lab.stm, Cell Centered Database (ccdb.ucsd.edu/CCDBWebSite/index.html), and neuromorpho.org/neuroMorpho/index.jsp.

AN AXON PLACES CONSTRAINTS ON DENDRITIC PROCESSING

As we saw in Chapter 4, Figure 4.1, the neuron has five essential functions related to signal processing: generation, reception, integration, encoding, and output. The presence of an axon places critical constraints on the dendritic processing that leads to axonal output.

The first principle is *if a neuron has an axon, it has only one.* This near universal "single axon rule" is remarkable and still little understood. It results from developmental mechanisms that provide for differentiation of a single axon from among early undifferentiated processes. These mechanisms are being analyzed especially in neuronal cultures (Craig and Banker, 1994). The principle means that for dendritic integration to lead to output from the neuron to distant targets, all the activity within the dendrites must eventually be funneled into the origin of the axon in the single axon hillock. Therefore, in these cells the flow of information in dendrites has an overall orientation. Ramón y Cajal and the classical anatomists called this the *law of dynamic polarization of the neuron* (Ramón y Cajal, 1911; summarized in Shepherd, 1991). We thus have a *principle of global output:*

In order to transfer information between regions, the information distributed at different sites within a dendritic tree of an output neuron must be encoded, for global output at a single site at the origin of the axon.

A related principle is that the main function of the axon in long axon cells is to support the generation of action potentials in the axon hillock–initial segment region. By definition, action potentials there have thresholds for generation; thus, the *principle of frequency encoding of global output* in an axonal neuron is as follows:

The results of dendritic integration affect the output through the axon by initiating or modulating action potential generation in the axon hillock–initial segment. Global output from dendritic integration is therefore encoded in impulse frequency in a single axon.

Classically it has been known that the axons of most output neurons are so long that the only significant signals reaching their axon terminals are the digital all-or-nothing action potentials carrying a frequency code. However, biology always produces exceptions. Recent research has shown that the synaptic potentials within the soma–dendrites may spread sufficiently in some axons to modulate the membrane potentials of the axon terminals (Shu *et al.,* 2006). This effect would presumably be most significant in short axon cells, where, as noted previously, the understanding of signal processing is most limited. In these cells, it appears that the axon may carry the outcome of soma–dendritic integration mainly in a digital (impulse frequency) form but with a contribution from analog (synaptic potential amplitude) signals.

A further consequence of the spatial separation of dendrites and axon is that some of the activity within a dendritic tree will be below threshold for activating an axonal action potential; we thus have the *principle of subthreshold dendritic activity*:

A considerable amount of subthreshold activity, including local active potentials, can affect the integrative states of the dendrites and any local outputs but not necessarily directly or immediately affect the global output of the neuron.

We turn now to the functional properties that allow dendritic trees to process information within these constraints.

DENDRODENDRITIC INTERACTIONS BETWEEN AXONAL CELLS

We first recognize, from the example in Figure 17.2, that axonal cells, as well as anaxonal cells, can have outputs through their dendrites (Rall and Shepherd, 1966). This is against the common assumption that if a neuron has an axon, all the output goes through the axon. There are many examples in invertebrates.

Neurite–Neurite Synapses in Lobster Stomatogastric Ganglion

One of the first examples in invertebrates was in the stomatogastric ganglion of the lobster (Selverston *et al.,* 1976). Neurons were recorded intracellularly and stained with Procion yellow. Serial electron micrographic reconstructions showed the synaptic relations between stained varicosities in the processes and their neighbors (the processes are equivalent to dendrites, but are often referred to as *neurites* in the invertebrate literature). In many cases, a varicosity could be seen to be not only presynaptic to a neighboring varicosity but also postsynaptic to that same process. It was concluded that synaptic inputs and outputs are distributed over the entire neuritic arborization. Polarization was not from one part of the tree to another. "Bifunctional" varicosities appeared to act as local input–output units, similar to the manner in which granule cell spines appear to operate (see later). Similar organization has been found in other types of stomatogastric neurons (Fig. 17.3A).

Sets of these local input–output units, distributed throughout the neuritic tree, participate in the generation and coordination of oscillatory activity involved in controlling the rhythmic movements of the stomach. In a current model of this oscillatory circuit, these interactions are mutually inhibitory (Fig. 17.3B).

In summary, *a cell with an axon can have local outputs through its dendrites as well as distant outputs through its axon,* which may be involved in specific local functions such as generating oscillatory circuits.

PASSIVE DENDRITIC TREES CAN PERFORM COMPLEX COMPUTATIONS

Another principle that carries over from axonless cells is the ability of the dendrites of axonal cells to carry out complex computations with mostly passive properties. This is exemplified by neurons that are motion detectors.

Motion detection is a fundamental operation carried out by the nervous systems of most species; it is essential for detecting prey and predator alike. In invertebrates, motion detection has been studied especially in the brain of the blowfly. In the lobula plate of the third optic neuropil are tangential cells (LPTCs) that respond to preferential direction (PD) of motion with increased depolarization due to sequential responses across their dendritic fields. This response has been modeled by Reichardt and colleagues by a series of elementary motion detectors

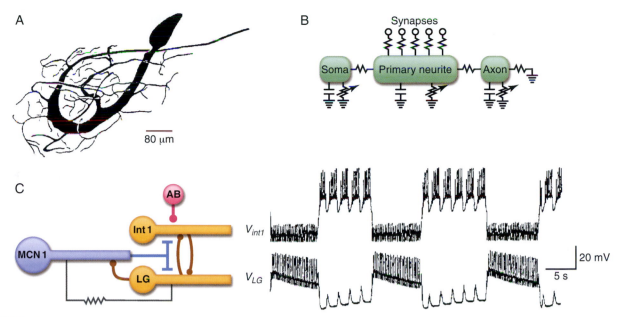

FIGURE 17.3 Local synaptic input–output sites are widely found within the neuropil of invertebrate ganglia. (A) Output neuron with many neurite branches in the gastric mill ganglion of the lobster, (B) compartmental representation of stomatogastric neuron, (C) model of rhythm generating circuit of the gastric mill of the lobster, involving neurite–neurite interactions. (A) and (B) reproduced with permission from Golowasch, J., and Marder, E. (1992). Ionic currents of the lateral pyloric neuron of the stomatogastric ganglion of the crab. *J. Neurophysiol.* **67, 2,** 318–331. (C) from Manor *et al.* (1999).

(EMDs) in the dendrites. A compartmental model (Single and Borst, 1998) reproduces the experimental results and theoretical predictions by showing how local modulations at each EMD are smoothed by integration in the dendritic tree to give a smoothed high-fidelity global output at the axon (see Fig. 17.4A). In the model, spatial integration is largely independent of specific electrotonic properties but depends critically on the geometry and orientation of the dendritic tree.

In vertebrates, motion detection is built into the visual pathway at various stages in different species: the retina, midbrain (optic tectum), and cerebral cortex. Studies in the optic tectum have revealed cells with splayed uniplanar dendritic trees and specialized distal appendages that appear highly homologous across reptiles, birds, and mammals (Fig. 17.4B) (Luksch *et al.*, 1998). Physiological studies are needed to test the hypothesis that these cells perform operations through their dendritic fields similar to those of LPTC cells in the insect. To the extent that this is borne out, it will support a principle of *motion detection through spatially distributed dendritic computations* that is conserved across vertebrates and invertebrates.

Directional selectivity of dendritic processing was predicted by Rall (1964) from his studies of dendritic electrotonus (see Chapter 4). In an electrotonic cable, summation of EPSPs moving away from a recording site produces a plateau of ever-decreasing potentials, whereas summation of EPSPs moving toward a recording site produces an accumulating peak of potential. This is one of several possible mechanisms that are under current investigation in different cell types.

SEPARATION OF DENDRITIC FIELDS ENHANCES COMPLEX INFORMATION PROCESSING

An important feature of many types of neuron is a separation of their dendritic fields, which has important functional consequences. We saw this principle in the compartmental organization of the mitral cell (Figs. 17.1 and 17.2), where the primary dendritic tuft receives the olfactory nerve input whereas the secondary dendritic branches are specialized for a completely different function: self and lateral inhibition, as we explain later.

Pyramidal cells in the cerebral cortex also show a clear separation into apical and basal dendrites (Figs. 17.1 and 17.2). The apical dendrite extends across different layers, allowing fibers within those layers from different cells to modulate the transfer of activity from the distal tuft toward the cell body. This kind of modulation is absent in the mitral cell but key in the pyramidal neuron.

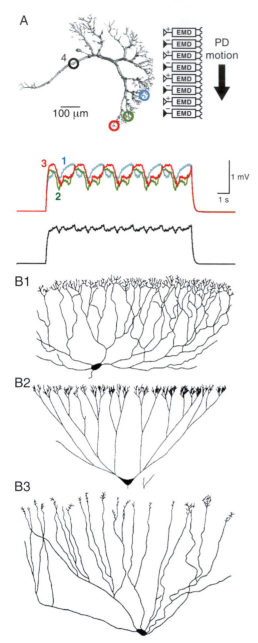

FIGURE 17.4 Dendritic systems as motion detectors. (A) A computational model of a motion detector neuron in the visual system of the fly, consisting of elementary motion detector (EMD) units in its dendritic tree activated by the preferential direction (PD) of motion. Local modulations of the individual EMDs are integrated in the dendritic tree to give smooth global output in the axon (lower black trace) (Single and Borst, 1998). (B) Dendritic trees of neurons in the optic tectum of lizard (B1), chick (B2), and gray squirrel (B3). The architecture of the dendritic branching patterns and distal specialization for the reception of retinal inputs is highly homologous (references in Luksch *et al.*, 1998).

Within the basal dendrites, the placement of inputs is critical. An example has been shown in experiments in which excitatory and inhibitory inputs can be independently targeted to the same or different dendritic branches. As illustrated in Figure 17.5,

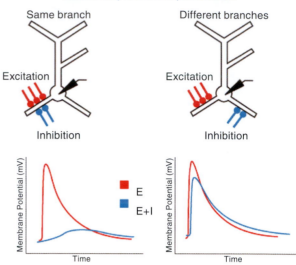

FIGURE 17.5 Importance of the locations of interacting synaptic responses within a dendritic field. Left, excitation and inhibition converging on the same dendritic branches produces sharp reduction of the excitatory response recorded at the soma (see electrode). Right, inhibition on a different set of branches has little effect in reducing the excitatory response recorded at the soma. This illustrates a practical application of the principle illustrated in Figure 4.13 for electrotonic relations between synaptic conductances in individual branches. From Mel and Schiller (2004).

synaptic inhibition has little effect on synaptic excitation when the two are targeted to different branches but has a profound effect when on the same branches. This is a clear example of the interaction of synaptic conductances illustrated in Figure 4.13 in Chapter 4.

As a final example, cells with separate dendritic fields are critical for selectively summing their synaptic inputs to mediate directional selectivity in the auditory system (Sorensen and Rubel, 2006).

DISTAL DENDRITES CAN BE CLOSELY LINKED TO AXONAL OUTPUT

An obvious problem for a neuron with an axon is that the distal branches of dendritic trees are a long distance from the site of axon origin at or near the cell body. The common perception is that these distal dendrites are too distant from the site of axonal origin and impulse generation to have more than a slow and weak background modulation of impulse output, and that the only synapses that can bring about rapid signal processing by the neuron are those located on the soma or proximal dendrites.

This perception is so ingrained in our visual impression of what is near and far that it is difficult to accept that it is wrong. It is, however, disproved by many kinds of neurons in which specific inputs are located preferentially on their distal dendrites. An example is the mitral (and tufted) cells in the olfactory bulb, which we met in Figures 17.1 and 17.2. In this cell the input from the olfactory nerves ends on the most distal dendritic branches in the glomeruli; in rat mitral cells, this may be 400–500 μm or more from the cell body, in turtle, 600–700 μm. The same applies to their targets, the pyramidal neurons of the olfactory cortex, where the input terminates on the spines of the most distal dendrites in layer I. In many other neurons, a given type of input terminates over much or all of the dendritic tree; such is the case, for example, for climbing fiber and parallel fiber inputs to the cerebellar Purkinje cells. All these neuron types are shown in Figure 17.1.

How do distal dendrites in these neurons effectively control axonal output? Some of the important properties underlying this ability are summarized in Table 17.1. We consider several examples next, as well as in later sections.

Large-Diameter Dendrites

The simplest way to enhance spread of a signal through dendrites is by a large diameter. It already has been illustrated for the spread of current in a branching cable in Chapter 4 (Fig. 4.4). However, space is at a premium in the central nervous system, so there is a trade-off between diameter and length. This is why other membrane properties become important in overcoming distance.

High Specific Membrane Resistance

A key property is the specific membrane resistance (R_m) of the dendritic membrane. The functional significance of R_m is discussed in Chapter 4 (see Fig. 4.3).

TABLE 17.1 Properties That Increase the Effectiveness of Distal Synapses in Effecting Axonal Output

---

Higher membrane resistance

Larger distal synaptic conductances

Voltage-gated channels, which do the following:

  increase EPSP amplitude

  generate large-amplitude slow action potentials give rise to forward-propagating full action potentials, local "hot spots" that set up fast prepotentials

  function as coincidence detectors to summate responses

  mediate "pseudosaltatory conduction" toward the soma through individual active sites or clusters

---

Traditionally, the argument was that if R_m is relatively low, the characteristic length (length constant) of the dendrites will be relatively short, the electrotonic length will be correspondingly long, and synaptic potentials will therefore decrement sharply in spreading toward the axon hillock. However, as discussed in Chapter 4, intracellular recordings indicated that R_m is sufficiently high that the electrotonic lengths of most dendrites are relatively short, in the range of 1–2 (Johnston and Wu, 1995), and patch recordings suggest much higher R_m values, indicating electrotonic lengths less than 1. Thus, a relatively high R_m seems adequate for close electrotonic linkage between distal dendrites and somas, at least in the steady state.

Low K Conductances

An important factor controlling effective membrane resistance is K conductances. Chapter 4 discusses how a K channel, I_h, can affect the summation of EPSPs in striatal spiny cells. There is increasing evidence that dendritic input conductance is controlled by different types of K currents (Magee, 1999; Midtgaard et al., 1993). When dendritic K conductances are turned off, R_m increases and dendritic coupling to the soma is enhanced. These conductances also control back-propagating action potentials, as discussed in Chapter 4 and below.

Large Synaptic Conductances

A potentially important property is the amplitude of the conductance generated by the synapse itself. Early studies showed that in motor neurons, distal excitatory synaptic potentials were many times the amplitude of proximal synapses (Redman and Walmsley, 1983). This observation would account for the fact that the unitary synaptic response recorded at the soma slows with increasing distance in the dendrites but maintains a constant amplitude of approximately 100 μV. A corresponding increase in synaptic conductance has been shown in the distal dendrites of cortical pyramidal neurons (Magee, 2000). Patch recordings show that, whereas inhibitory postsynaptic currents (IPSCs) are similar in amplitude whether recorded from the distal dendrites or the soma, excitatory postsynaptic currents (EPSCs) are larger when recorded from distal dendrites than from the soma (Fig. 17.6; Andrásfalvy and Mody, 2006). It is hypothesized that this observation reflects receptor channels composed of different subunits. Research is needed to determine which synaptic protein subunits are involved to give these differences in conductance in specific cells.

Postsynaptic currents at
different distances in the dentritic tree

A. Excitatory Evoked mEPSCs

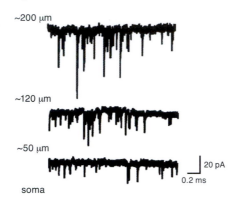

B. Inhibitory Evoked mIPSCs

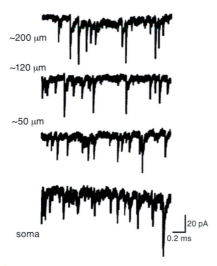

FIGURE 17.6 Larger excitatory synaptic currents may be present in the apical dendrite of a cortical pyramidal neuron. (A) Excitatory. Evoked miniature excitatory postsynaptic currents (mEPSCs) in patch recordings at three distances from the soma. Note the increase in amplitude with increasing distance. (B) Inhibitory. Same for inhibitory mIPSCs. Note relatively constant amplitudes with distance. From Andrásfalvy and Mody (2006).

Voltage-Gated Depolarizing Conductances

For transient responses, the electrotonic linkage becomes weaker because of the filtering effect of the capacitance of the membrane, and it is made worse by a higher R_m, which increases the membrane time constant, thereby slowing the spread of a passive potential (see Chapter 4). This disadvantage can be overcome by depolarizing voltage-gated conductances: Na^+, Ca^{2+}, or both. These add a wide variety of signal processing mechanisms to dendrites. As indicated in

Table 17.1, they include boosting EPSP amplitudes, generating large-amplitude slow pacemaker potentials underlying the spontaneous activity of a neuron, supporting full back-propagating or forward-propagating action potentials in the dendrites, forming "hot spots" that set up fast prepotentials at branch points in the distal dendrites, and functioning as coincidence detectors. Interactions between active sites, such as spines with voltage-gated conductances, can give rise to a sequence of activation of those sites, resulting in "pseudosaltatory conduction" through the dendritic tree between active sites or active clusters of sites.

Some of these active properties also contribute to complex information processing capabilities, including logic operations, as discussed further later.

Summary

These examples illustrate an important principle of distal dendritic processing:

Distal dendrites can mediate relatively rapid, specific information processing, even at the weakest levels of detection, in addition to slower modulation of overall neuronal activity. The spread of potentials to the site of global output from the axon is enhanced by multiple passive and active mechanisms.

DEPOLARIZING AND HYPERPOLARIZING DENDRITIC CONDUCTANCES INTERACT DYNAMICALLY

We see that depolarizing conductances increase the excitability of distal dendrites and the effectiveness of distal synapses, whereas K^+ conductances reduce the excitability and control the temporal characteristics of the dendritic activity. This balance is thus crucial to the functions of dendrites. Figure 17.7 summarizes data showing how these conductances vary along the extents of the dendrites of olfactory mitral cells and hippocampal and neocortical pyramidal neurons.

The significance of a particular density of channel needs to be judged in relation to the electrotonic properties discussed in Chapter 4. For instance, a given conductance has more effect on membrane potential in smaller distal branches because of the higher input resistance (see Fig. 4.14). Dendritic conductances are crucial in setting the intrinsic excitability state of the neuron. In the motor neuron, for example, the neuron can alternate between bistable states dependent on the

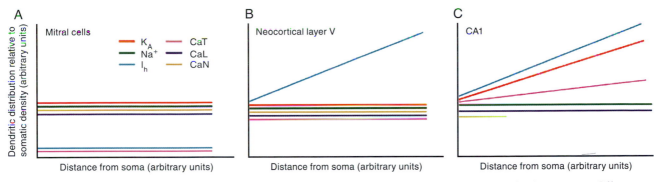

FIGURE 17.7   Graphs of the distribution of different types of intrinsic membrane conductances along the dendritic trees in different types of neurons. Reproduced with permission from Macmillan Publishers Ltd. Migliore, M. and Shepherd, G.M. (2002). Emerging rules for the distributions of active dendritic conductances. *Nature Neurosci. Revs.* **3**, 362–370.

activation of dendritic metabotropic glutamate receptors (Svirskie *et al.,* 2001). The significance of these and other conductance interactions for the firing properties of different cell types is discussed in following text.

These combinations of ionic conductances occur within the larger framework of the morphological types of dendritic trees, particularly whether they arise from thick or thin trunks. This has given rise to a classification of dendritic types on integrative principles that cuts across the traditional classification of neuron types (Migliore and Shepherd, 2002, 2005). This is a first step toward a deeper insight into canonical types of input–output operations that are carried out by dendritic trees.

As noted previously, the combinations of properties within different dendritic compartments of a given neuron type can be searched in an online database (senselab.med.yale.edu/neurondb) and models based on these compartmental representations of many types of neurons can be accessed and run at senselab.med.yale.edu/modeldb.

Summary

The combination of conductances at different levels of the dendritic tree involves a delicate balance between depolarizing and hyperpolarizing actions over different time periods. These combinations vary in different morphological types of neurons. They also contribute to a new classification of dendrites according to a *principle of multiple criteria for dendritic classification*:

Dendritic trees can be categorized functionally on the basis of a combination of branch morphology, functional ionic current type, and genetic channel subunit type. These categories appear to define canonical integrative properties that extend across classical morphological categories.

THE AXON HILLOCK–INITIAL SEGMENT ENCODES GLOBAL OUTPUT

In cells with long axons, activity in the dendrites eventually leads to activation and modulation of action potential output in the axon. A key question is the precise site of origin of this action potential. This question was one of the first to be addressed in the rise of modern neuroscience; the historical background is summarized in Box 17.1. These studies established the classical model: the lowest threshold site for action potential generation is in the axonal initial segment.

Definitive analysis was achieved by Stuart and Sakmann (1994) using dual-patch recordings from cortical pyramidal neurons under infrared differential contrast microscopy. This approach has provided the breakthrough for subsequent analyses of dendritic properties and their coupling to the axon (see later). As shown in Figure 17.10, with depolarization of the distal dendrites by injected current or excitatory synaptic inputs, a large amplitude depolarization is produced in the dendrites, which spreads to the soma. Despite its lower amplitude, soma depolarization is the first to initiate the action potential. Subsequent studies with triple-patch electrodes have shown that the action potential actually arises first in the initial segment and first node.

MULTIPLE IMPULSE INITIATION SITES ARE UNDER DYNAMIC CONTROL

In addition to the evidence for action potential initiation in the axon hillock, another line of work has provided evidence for shifting of the site under

BOX 17.1

CLASSICAL STUDIES OF THE ACTION POTENTIAL INITIATION SITE

Fuortes and colleagues (1957) were the first to deduce that an EPSP spreads from the dendrites through the soma to initiate the action potential in the region of the axon hillock and the initial axon segment. They suggested that the action potential has two components: (1) an A component that is normally associated with the axon hillock and initial segment and (2) a B component that is normally associated with retrograde invasion of the cell body. Because the site of action potential initiation can shift under different membrane potentials, they preferred the noncommittal terms "A" and "B" for the two components as recorded from the cell body. In contrast, Eccles (1957) referred to the initial component as the *initial segment* (IS) component and to the second component as the *somadendritic* (SD) component (Fig. 17.8).

Apart from the motor neuron, the best early model for intracellular analysis of neuronal mechanisms was the crayfish stretch receptor, described by Eyzaguirre and Kuffler (1955). Intracellular recordings from the cell body showed that stretch causes a depolarizing receptor potential equivalent to an EPSP, which spreads through the cell to initiate an action potential. It was first assumed that this action potential arose at or near the cell body. Edwards and Ottoson (1958), working in Kuffler's laboratory, tested this postulate by recording the local extracellular current in order to locate precisely the site of inward current associated with action potential initiation. Surprisingly, this site turned out to be far out on the axon, some 200 μm from the cell body (Fig. 17.9). This result showed that potentials generated in the distal dendrites can spread all the way through the dendrites and soma well out into the initial segment of the axon to initiate impulses. It further showed that the action potential recorded at the cell body is the backward-spreading impulse from the initiation site. Edwards and Ottoson's study was important in establishing the basic model of impulse initiation in the axonal initial segment.

Gordon M. Shepherd

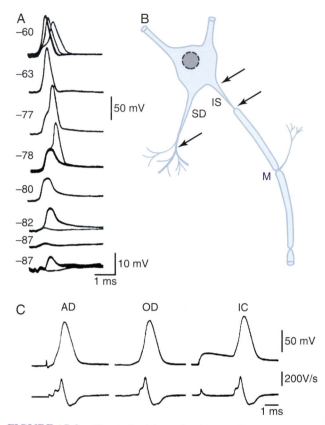

FIGURE 17.8 Classical evidence for the site of action potential initiation. Intracellular recordings were from the cell body of the motor neuron of an anesthetized cat. (A) Differential blockade of an antidromic impulse by adjusting the membrane potential by holding currents. Recordings reveal the sequence of impulse invasion in the myelinated axon (recordings at –87 mV, two amplifications), the initial segment of the axon (first component of the impulse beginning at –82 mV), and the somadendritic region (large component beginning at –78 mV). (B) Sites of the three regions of impulse generation (M, myelinated axon; IS, initial segment; SD, soma and dendrites); arrows show probable sites of impulse blockade in A. (C) Comparison of intracellular recordings of impulses generated antidromically (AD), synaptically (orthodromically, OD), and by direct current injection (IC). Lower traces indicate electrical differentiation of these recordings showing the separation of the impulse into the same two components and indicating that the sequence of impulse generation from the initial segment into the somadendritic region is the same in all cases. From Eccles (1957).

BOX 17.1 *(continued)*

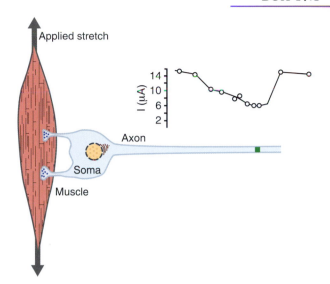

FIGURE 17.9 Classical demonstration of the site of impulse initiation in the stretch receptor cell of the crayfish. Moderate stretch of the receptor muscle generated a receptor potential that spread from the dendrites across the cell body into the axon. The excitability curve (shown above the axon) was obtained by passing current between the electrodes and finding the current (I) intensity needed to evoke an impulse response. This experiment showed that the trigger zone was several hundred micrometers out on the axon (green region in axon). Modified with permission from Ringham, G.L. (1971). Origin of nerve impulse in slowly adapting stretch receptor of crayfish. *J. Neurophysiol.* **33,** 773–786.

References

Eccles, J. C. (1957). "The Physiology of Nerve Cells." Johns Hopkins Univ. Press, Baltimore.

Edwards, C., and Ottoson, D. (1958). The site of impulse initiation in a nerve cell of a crustacean stretch receptor. J. Physiol. (Lond.) **143,** 138–148.

Eyzaguirre, C., and Kuffler, S. W. (1955). Processes of excitation in the dendrites and in the soma of single isolated sensory nerve cells of the lobster and crayfish. J. Gen. Physiol. **39,** 87–119.

Fuortes, M. G. E., Frank, K., and Banker, M. C. (1957). Steps in the production of motor neuron spikes. J. Gen. Physiol. **40,** 735–752.

dynamic conditions. This line began with extracellular recordings of a "population spike" that appears to propagate along the apical dendrites toward the cell body in hippocampal pyramidal cells (Anderson, 1960). This finding was supported by the recording in these cells of "fast prepotentials" at dendritic "hot spots" (Spencer and Kandel, 1961), and by current source density calculations in cortical pyramidal neurons (Herreras, 1990). In recordings from dendrites in tissue slices in CA1 hippocampal pyramidal neurons, weak synaptic potentials elicited action potentials near the cell body (Richardson *et al.,* 1987), but this site shifted to proximal dendrites with stronger synaptic excitation (Turner *et al.,* 1991). This finding confirmed the suggestion of M. G. F. Fuortes and K. Frank that the site can shift under different stimulus conditions, and was consistent with the stretch receptor, where larger receptor potentials shift the initiation site closer to the cell body (see Box 17.1).

The olfactory mitral cell is a favorable model for studying this question, because it is unusual in that all its excitatory inputs are restricted to its distal dendritic tuft. As illustrated in Figure 4.15, at weak levels of electrical shocks to the olfactory nerves, the site of action potential initiation is at or near the soma, as in the classical model (see Fig. 4.15).

As the level of distal excitatory input is increased, dual-patch recordings show clearly that the action potential initiation site shifts gradually from the soma to the distal dendrite (see Fig. 4.15). Thus the site of impulse initiation is not fixed in the mitral cell but varies with the intensity of distal excitatory input. The action potential is due to Na^+ channels distributed along the extent of the primary dendrite. The mechanism by which passive electrotonic potential spread along the dendrite controls the site of action potential initiation in these experiments is discussed in Chapter 4 (Fig. 4.15). The reader will also recognize that the site can also be shifted to distal dendrites by synaptic inhibition applied to the soma through dendrodendritic synapses.

Summary

The low threshold of the initial axonal segment favors its being the site of action potential output for a wide range of dendritic activity, but the site can shift with strongly depolarizing dendritic input. This introduces the *principle of the dynamic control of action potential initiation*:

The site of global output through action potential initiation from a neuron can shift between first

A   *Somatic current pulse*

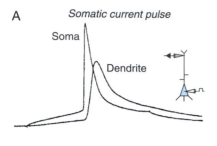

B   *Dendritic current pulse*

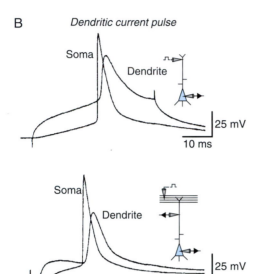

FIGURE 17.10 Direct demonstration of the impulse-initiation zone and back propagation into dendrites using dual-patch recordings from soma and dendrites of a layer V pyramidal neuron in a slice preparation of the rat neocortex. (A) Depolarizing current injection in either the soma or the dendrite elicits an impulse first in the soma. (B) The same result is obtained with synaptic activation of layer I input to distal dendrites. Note the close similarity of these results to the earlier findings in the motor neuron (Fig. 17.8). Reproduced with permission from Macmillan Publishers Ltd. Stuart, G.J., and Sakmann, B. (1994). Active propagation of somatic action potentials into neocortical pyramidal cell dendrites. *Nature (Lond.)* **367,** 6–72.

axon node, initial segment, axon hillock, soma, proximal dendrites, and distal dendrites, depending on the dynamic state of dendritic excitability.

RETROGRADE IMPULSE SPREAD INTO DENDRITES CAN HAVE MANY FUNCTIONS

In addition to identifying the preferential site for action potential initiation in the axonal initial segment, the experiments of Stuart and Sackmann (1994) showed clearly that the action potential does not merely spread passively back into the dendrites but actively back-propagates. Note that we distinguish between passive electrotonic "spread" and active "propagation" of the action potential (see Box 4.2 in Chapter 4).

What is the function of the dendritic action potential? Experimental evidence shows that it can have a variety of functions.

Dendrodendritic Inhibition

A specific function for an action potential propagating from the soma into the dendrites was first suggested for the olfactory mitral cell, where mitral-to-granule dendrodendritic synapses are triggered by the action potential spreading from the soma into the secondary dendrites (Fig. 17.11A, B). Because of the delay in activating the reciprocal inhibitory synapses from the granule cells, self-inhibition of the mitral cell occurs in the wake of the passing impulse; the two do not collide. The mechanism operates similarly with both active back propagation and passive electrotonic spread into the dendrites, as tested in computer simulations. Functions of dendrodendritic inhibition include center-surround antagonism mediating the abstraction of molecular determinants underlying the discrimination of different odor molecules, storing of olfactory memories at the reciprocal synapses, and generation of oscillating activity in mitral and granule cell populations (Egger and Urban, 2006; Schaefer and Margrie, 2007; Shepherd *et al.*, 2004).

Intercolumnar Connectivity

Recent research has given new insight into the function of the action potential in the mitral cell lateral dendrite. Because the action potential can propagate away from the cell body throughout the length of the dendrite (Fig. 17.11B; Xiong and Chen, 2002), it enables activation of granule cells independent of distance. Connectivity of mitral cells to distant groups of granule cells, arranged in columns in relation to glomeruli, has been demonstrated by pseudorabies viral tracing (Willhite *et al.*, 2006), and activation of distant granule cells by means of such connectivity has been shown in realistic computational studies (Migliore and Shepherd, 2007). This has led to the hypothesis that the lateral dendrite can function to activate ensembles of granule cell columns processing similar aspects of an odor map, with the added flexibility that the dendrite can be modulated by granule cell inhibition throughout its length. The diagram in Fig. 17.11A thus provides an updated representation of the functional subunits and microcircuits formed by the mitral and granule cells shown in Figures 17.1 and 17.2.

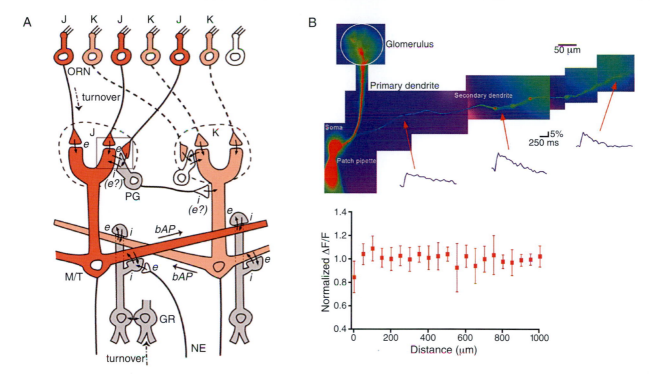

FIGURE 17.11 A. Microcircuit organization of the olfactory bulb. An action potential in the mitral cell body sets up a back-spreading/back-propagating impulse (bAP) into the secondary dendrites, activating both feedback and lateral inhibition of the mitral cells by columns of granule cells (GR) acting through the dendrodendritic pathway (e, i). Abbreviations: J, K, subsets of olfactory receptor neurons (ORN); PG, periglomerular cells; M/T, mitral or tufted cells; NE, centrifugal norepinephrine fiber. From Shepherd *et al.* (2007). (C) Ability of an action potential to invade the length of a mitral cell secondary dendrite, as shown by Ca^{2+} fluorescence. Fluorescence measurements are plotted in the graph below, showing full propagation up to 1000 microns. From Xiong and Chen (2002).

Boosting Synaptic Responses

In several types of pyramidal neurons, active dendritic properties appear to boost action potential invasion so that summation with EPSPs occurs that makes the EPSPs more effective in spreading to the soma.

Resetting Membrane Potential

A possible function of a back-propagating action potential is that the Na^+ and K^+ conductance increases associated with active propagation wipe out the existing membrane potential, resetting the membrane potential for new inputs.

Synaptic Plasticity

The action potential in the dendritic branches presumably depolarizes the spines (because of the favorable impedance matching, as discussed in Chapter 4), which means that the impulse depolarization would summate with the synaptic depolarization of the spines. This process would enable the spines to function as coincidence detectors and implement changes in synaptic plasticity (see discussion on Hebbian synaptic mechanisms in Chapter 19). This postulate has been tested by electrophysiological recordings (Spruston *et al.*, 1995) and Ca^{2+} imaging (Yuste *et al.*, 1994). Activity-dependent changes of dendritic synaptic potency are not seen with passive retrograde depolarization but appear to require actively propagating retrograde impulses (Spruston *et al.*, 1995).

Frequency Dependence

Trains of action potentials generated at the soma–axon hillock can invade the dendrites to varying extents. Proximal dendrites appear to be invaded throughout a high-frequency burst, whereas distal dendrites appear to be invaded mainly by the early action potentials (Callaway and Ross, 1995; Regehr *et al.*, 1989; Spruston *et al.*, 1995; Yuste *et al.*, 1994). Activation of Ca^{2+}-activated K^+ conductances and subsequent hyperpolarizatons by early impulses may effectively switch off the distal dendritic compartment.

Nonuniform propagation of spikes due to spike after hyperpolarizations (AHP) may contribute to the function of sensory processing and synaptic plasticity in invertebrate neurons with complex neuritic processes. For example, the touch sensory cells

(T cells) in the leech have an extensive neuritic arborization that spans three segmental ganglia and that innervates three receptive fields (a major receptive field and anterior and posterior minor receptive fields) (Fig. 17.12A). In addition, T cells manifest an AHP following a sustained discharge of action potentials (Baylor and Nicholls, 1969; Nicholls and Baylor, 1968a,b). It has been suggested that the functional role of the AHP may be to increase the threshold of the T cell and reduce the excitability of the neuron, and thereby, block the propagation of action potentials from the minor receptive fields to the soma (Mar and Drapeau 1996; Van Essen, 1973; Yau, 1976). A potential consequence of conduction block would be to silence portions of the synaptic contacts of the

T cell and thus regulate the flow of information from the T cell to its postsynaptic targets (Gu, 1991; Gu et al., 1989; Macagno et al., 1987). Scuri et al. (2002) found that the amplitude of the AHP in T cells is modulated by spiking activity. Moreover, the activity-dependent increases in the AHP are correlated with decreasing EPSPs in a postsynaptic neuron. These results suggest that the activity-dependent modulation of the AHP and the subsequent induction of conduction block may play a role in regulating synaptic efficacy. Similar processes have been suggested to underlie presynaptic inhibition in other systems (e.g., Hatt and Smith, 1976; Luscher, 1998; Luscher and Shiner, 1990a,b; Rudomin et al., 1998; Wall, 1995).

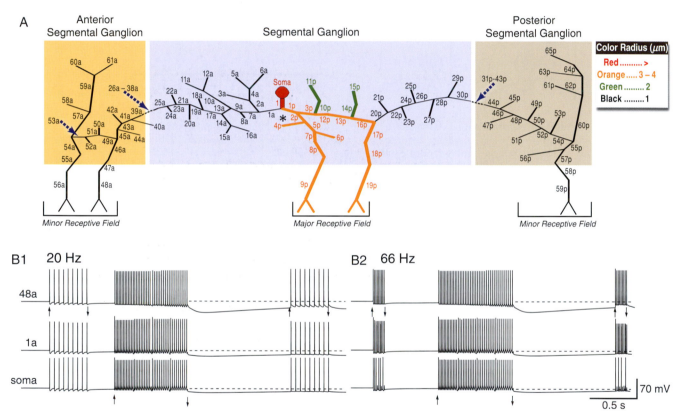

FIGURE 17.12 Contribution of the afterhyperpolarization (AHP) to frequency-dependent conduction failure in touch sensory cells (T cells) of the leech. (A) Diagram of the morphological features of a model of the T cell. Each compartment represents a Hodgkin–Huxley-type electrical circuit, which contained a capacitance and passive and four active conductances and a Na$^+$/K$^+$ pump in parallel. The compartments are connected by axial resistances. The designation "a" indicates compartments anterior to the soma, whereas the designation "p" indicates compartments posterior to the soma. Brackets below indicate the receptive fields for the neurites (dendrites): major (9p, 19p), anterior minor (56a, 48a), and posterior minor (59p), extending throughout three ganglia. (B) Activity- and frequency-dependent conduction failure of action potentials transmitted from the anterior minor receptive field to the soma before and after the induction of an AHP and at two different frequencies of activity. The AHP was induced by a brief, high-frequency burst of activity in the soma. (B1) Spikes were elicited in the compartment 48a at 20 Hz. At this relatively low frequency of spike activity, the conduction of spikes from the anterior minor receptive field to the soma was 1:1 (first burst of spikes) without reduction, even in presence of an AHP (final burst of spikes). Thus, the AHP does not affect spike transmission at low frequencies of activity. (B2) Spikes were elicited in the compartment 48a at 66 Hz. Due to the impedance mismatch at the 1a–1 branching point (see asterisk), the mean frequency of spikes that were transmitted to the soma was reduced to ~41 Hz with further reduction by the AHP to ~25 Hz. Thus, the AHP contributes to frequency-dependent conduction failure at higher frequencies. Arrows up and down indicate the onset and offset of stimuli, respectively, and the compartment into which the stimulus was injected. Modified from Cataldo et al. (2005).

To investigate the role that the activity-dependent modulation of the AHP may play in T cells, Cataldo *et al.* (2005) used the neurosimulator SNNAP to develop a multicompartmental model of the T cell (Fig. 17.12A). The model incorporated empirical data that describe the geometry of the cell and activity-dependent changes of the AHP. The simulations indicated that at some branching points, activity-dependent increases of the AHP reduced the number of spikes transmitted from the minor receptive fields to the soma and beyond (Fig. 17.12B). In addition, simulations suggested that AHP could modulate, under some circumstances, transmission from the soma to the synaptic terminals, suggesting that the AHP can regulate spike conduction within the presynaptic arborizations of the cell and could in principle contribute to the synaptic depression that is correlated with increases in the AHP.

Retrograde Actions at Synapses

The retrograde action potential can contribute to the activation of neurotransmitter release from the dendrites. The clearest example of this is the olfactory mitral cell as already described. Dynorphin released by synaptically stimulated dentate granule cells can affect the presynaptic terminals (Simmons *et al.*, 1995). In the cerebral cortex there is evidence that GABAergic interneuronal dendrites act back on axonal terminals of pyramidal cells and that glutamatergic pyramidal cell dendrites act back on axonal terminals of the interneurons (Zilberter, 2000). The combined effects of the axonal and dendritic compartments of both neuronal types regulate the normal excitability of pyramidal neurons and may be a factor in the development of cortical hyperexcitability and epilepsy.

Conditional Axonal Output

Because of the long distance between distal dendrites and initial axonal segment, we may hypothesize that the coupling between the two is not automatic. Indeed, conditional coupling dependent on synaptic inputs and intrinsic activity states at intervening dendritic sites appears to be fundamental to the relation between local dendritic inputs and global axonal output (Spruston, 2000).

Summary

The action potential arising at the initial axonal segment has two functions: propagating into the axon to carry the global output to the axon terminals, and

propagating retrogradely through the soma into the dendrites. In the dendrites the action potential can carry out many distinct functions, as described previously.

When the retrograde action potential has been activated by EPSPs spreading from the dendrites, we call the action potential *back propagating,* that is, back toward the site of the initial input. When the retrograde action potential propagates through the soma into previously unactivated dendrites, we can still call it back propagating in the sense of backward with regard to the law of dynamic polarization, which applied to the overall flow of activity is from distal dendrites to soma and axon, or we can consider it as propagating retrogradely, to distinguish it from back propagating toward a distal input site.

EXAMPLES OF HOW VOLTAGE-GATED CHANNELS ENHANCE DENDRITIC INFORMATION PROCESSING

It is commonly believed that active dendrites are a modern concept, but in fact this idea is as old as Ramón y Cajal; he assumed that dendrites conduct impulses like axons do. However, with the first intracellular recordings in the 1950s it appeared that dendritic membranes were mostly passive. We have noted that studies since then have increasingly documented the widespread distribution and numerous functions of voltage-gated channels in dendritic membranes.

These channels are the principal means for enhancing the information processing capabilities of complex dendrites. Detailed analysis of active dendritic properties began with computational studies of olfactory mitral cells and experimental studies of cerebellar Purkinje cells. Since then, studies of active dendritic properties have proliferated, particularly since the introduction of the patch recording method. Several types of neurons have provided important models for the possible functional roles of active dendritic properties.

Purkinje Cells

The cerebellar Purkinje cell has the most elaborate dendritic tree in the nervous system, with more than 100,000 dendritic spines receiving synaptic inputs from parallel fibers and mossy fibers. The basic distribution of active properties in the Purkinje cell was indicated by the pioneering experiments of Llinás

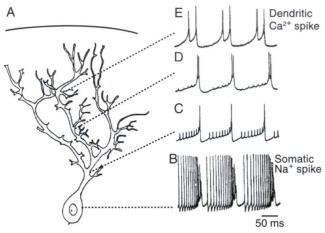

FIGURE 17.13 Classical demonstration of the difference between soma and dendritic action potentials. (A) Drawing of a Purkinje cell in the cerebellar slice. (B) Intracellular recordings from the soma showing fast Na⁺ spikes. (C–E) Intracellular recordings from progressively more distant dendritic sites; fast soma spikes become small due to electrotonic decrement and are replaced by large-amplitude dendritic Ca²⁺ spikes. Spread of these spikes to the soma causes an inactivating burst that interrupts the soma discharge. Adapted from Llinás and Sugimori (1980).

and Sugimori (1980) in tissue slices (Fig. 17.13). The action potential in the cell body and axon hillock is due mainly to fast Na⁺ and delayed K⁺ channels; there is also a Ca²⁺ component. The action potential correspondingly has a large amplitude in the cell body and decreases by electrotonic decay in the dendrites. In contrast, recordings in the dendrites are dominated by slower "spike" potentials that are Ca²⁺ dependent due to a P-type Ca²⁺ conductance (see Fig. 17.13). These spikes are generated from a plateau potential due to a persistent Na⁺ current (I_{NaP}).

There are two distinct operating modes of the Purkinje cell in relation to its distinctive inputs. Climbing fibers mediate strong depolarizing EPSPs throughout most of the dendrites that appear to give rise to synchronous Ca²⁺ dendritic action potentials throughout the dendritic tree, which then spread to the soma to elicit the bursting "complex spike" in the axon hillock. In contrast, parallel fibers are active in small groups, giving rise to smaller populations of individual EPSPs possibly targeted to particular dendritic regions (compartments) (see also Chapters 18 and 19).

The Purkinje cell thus illustrates several of the principles we have discussed. Subthreshold amplification through active dendritic properties may enhance the effect of a particular set of input fibers in controlling or modulating the frequency of Purkinje cell action potential output in the axon hillock. The Purkinje cell is subjected to local inhibitory control by stellate cell synapses targeted to specific dendritic compartments,

and to global inhibitory control of axonal output by basket cell synapses on the axonal initial segment.

Medium Spiny Cell

A different instructive example of the role of active dendritic properties is found in the medium spiny cell of the neostriatum (Figs. 17.14A, B). The passive electrotonic properties of this cell are described in Chapter 4 (Fig. 4.12). Inputs to a given neuron from the cortex are widely distributed, meaning that a given neuron must summate a significant number of

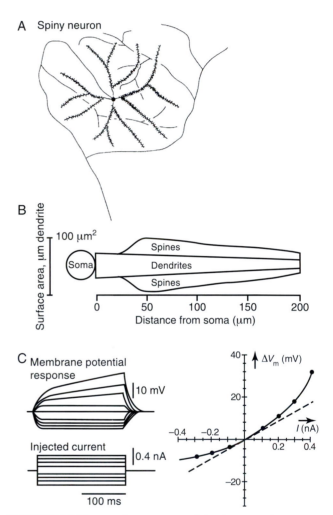

FIGURE 17.14 Dendritic spines and dendritic membrane properties interact to control neuronal excitability. (A) Diagram of a medium spiny neuron in the caudate nucleus; (B) plot of surface areas of different compartments showing a large increase in surface area due to spines; and (C) intracellular patch-clamp analysis of medium spiny neuron showing inward rectification of the membrane that controls the response of the dendrites to excitatory synaptic inputs (cf. Chapter 4, Fig. 4.12). From Wilson, C. (1998). Basal ganglia. In "The Synaptic Organization of the Brain", 4th Ed. (G. Shepherd, ed.), pp. 329–375. Oxford Univ. Press, New York. Reproduced with permission from Oxford University Press.

synaptic inputs before generating an impulse response. The responsiveness of the cell is controlled by its cable properties; individual responses in the spines are filtered out by the large capacitance of the many dendritic spines so that individual EPSPs recorded at the soma are small.

With synchronous specific inputs, larger summated EPSPs depolarize the dendritic membrane strongly. The dendritic membrane contains inwardly rectifying channels (I_h) (Fig. 17.14C), which reduce their conductance upon depolarization and thereby increase the effective membrane resistance and shorten the electrotonic length of the dendritic tree. Large depolarization also activates high-threshold (HT) Ca^{2+} channels, which contribute to large-amplitude, slow depolarizations. These combined effects change the neuron from a state in which it is insensitive to small noisy inputs into a state in which it gives a large response to a specific input and is maximally sensitive to additional inputs. Through this voltage-gated mechanism, a neuron can enhance the effectiveness of distal dendritic inputs, not by boosting inward Na^+ and K^+ currents, but by reducing outward shunting K^+ currents. This exemplifies the principle of dynamic control over dendritic properties through interactions involving K^+ conductances mentioned earlier.

Pyramidal Neurons

Active properties of the apical dendrite of hippocampal pyramidal neurons have been amply documented by patch recordings (Magee and Johnston, 1995). In contrast to the Purkinje cell, both fast Na^+ and Ca^{2+} conductances have been shown throughout the dendritic tree of the pyramidal neuron by electrophysiological and dye-imaging methods (see Fig. 17.6). Activation of low-threshold Na^+ channels is believed to play an important role in triggering the higher-threshold Ca^{2+} channels. Similar results have been obtained in studies of pyramidal neurons of the cerebral cortex.

At the simplest level, the output pattern of a neuron depends on its dendritic properties and their interaction with the soma. This is exemplified by the generation of a burst response in a pyramidal neuron. EPSPs spread through the dendrite, activating fast Na^+ and then HT Ca^{2+} channels that give a subthreshold boost to the EPSP. The enhanced EPSP spreads to the soma–axon hillock, triggering a Na^+ action potential. This propagates into the axon and also back-propagates into the dendrites, eliciting a slower all-or-nothing Ca^{2+} action potential. This large-amplitude, slow depolarization then spreads through the dendrites and back to the soma, triggering a train of action potentials that form a burst response.

This sequence of events is simulated most accurately by a realistic multicompartmental model of the dendritic tree. However, the essence can be contained in a two-compartment model representing the soma and dendritic compartments (Fig. 17.15). The model sequence emphasizes not only the importance of the interplay between the different types of channels but also the critical role of the compartmentalization of the neuron into dendritic and somatic compartments so that they can interact in controlling the intensity and time course of the impulse output.

This simpler model would argue that the specific form of the input–output transformation does not depend on a specific distribution of active channels in the dendritic tree. Na^+ and Ca^{2+} channels are in fact distributed widely in pyramidal neuron dendrites. In computational simulations, grouping channels in different distributions may have little effect on the input–output functions of a neuron (Mainen and Sejnowski, 1995). However, there is evidence that subthreshold amplification by voltage-gated channels may tend to occur in the more proximal dendrites of some neurons (Yuste and Denk, 1995). In addition, the dendritic trees of some neurons are clearly divided into different anatomical and functional subdivisions, as discussed in the next section.

Summary

These are only a few examples of the range of operations carried out by complex dendrites. These dendritic operations are embedded in the circuits that control behavior. Many further examples could be mentioned; for instance, the way motor neuron intrinsic properties are involved in the activation patterns

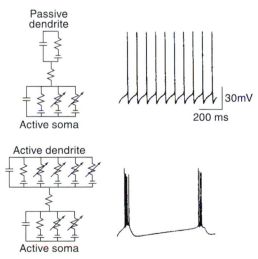

FIGURE 17.15 Generation of a burst response by interactions between soma and dendrites. From Pinsky and Rinzel (1994).

of motor units controlling the limbs (Gorassini *et al.*, 1999). Thus, for each neuron, the dendritic tree constitutes an expanded field of local units essential to the circuits underlying behavior.

DENDRITIC SPINES ARE MULTIFUNCTIONAL MICROINTEGRATIVE UNITS

Much of the complex processing that takes place in dendrites involves inputs through dendritic spines, the tiny outcroppings from the dendritic surface. Their electrotonic properties were described in relation to Figure 4.14. The very small size of dendritic spines has made it difficult to study them directly. However, examples have already been given of spines with complex information processing capacities, such as granule cell spines in the olfactory bulb and spines of medium spiny neurons in the striatum. In cortical neurons, spines have been implicated in cognitive functions from observations of dramatic changes in spine morphology in relation to different types of mental retardation and different hormonal exposures.

One of the most fertile hypotheses, by Rall (1974), is that changes in the dimensions of the spine stem control the effectiveness of coupling of the synaptic response in the spine head to the rest of the dendritic branch, and could therefore provide a mechanism for learning and memory (see also Harris and Kater, 1994; Shepherd, 1996; Yuste and Denk, 1995). For example, an activity-dependent decrease in stem diameter could increase the input resistance of the spine head, increasing an EPSP amplitude, which could have local effects on subsequent responses and also decrease the coupling to the parent dendrite. In addition to these electrotonic effects, a decrease in stem diameter could also increase the biochemical compartmentalization of the spine head (see Fig. 4.14).

Computational models have been very useful in testing these hypotheses, as well as suggesting other possible functions, such as the dynamic changes of electrotonic structure in medium spiny cells of the basal ganglia (see earlier discussion). With the development of more powerful light microscopical methods, such as two-photon laser confocal microscopy, it has become possible to test these hypotheses directly by imaging Ca^{2+} fluxes in individual spines in relation to synaptic inputs and neuronal activity (Fig. 17.16).

Evidence for active properties of dendrites has suggested that the spines may also have active properties. Thus, spines may be devices for nonlinear thresholding

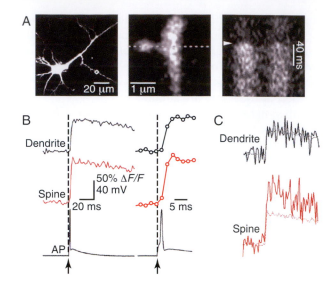

FIGURE 17.16 Calcium transients can be imaged in single dendritic spines in a rat hippocampal slice. (A) Fluo-4, a calcium-sensitive dye, injected into a neuron enables an individual spine to be imaged under two-photon microscopy. (B) An action potential (AP) induces an increase in Ca^{2+} in the dendrite and a larger increase in the spine (averaged responses). (C) Fluctuation analysis indicated that spines likely contain up to 20 voltage-sensitive Ca^{2+} channels; single channel openings could be detected, which had a high (0.5) probability of opening following a single action potential. From Sabatini and Svoboda (2000).

operations, either through voltage-gated ion channels (see Fig. 17.7) or through voltage-dependent synaptic properties such as *N*-methyl-D-aspartate (NMDA) receptors. This could powerfully enhance the information processing capabilities of spiny dendrites.

Computational simulations have shown that logic operations are inherent in coincidence detection by synapses (Koch *et al.*, 1983), and these operations are enhanced by active dendritic sites such as spines (Rall and Segev, 1987). An example is an AND operation performed by two dendritic spines with Hodgkin–Huxley-type active kinetics, with intervening passive dendritic membrane. As illustrated in Figure 17.17, simultaneous synaptic input of 1 nS conductance to spines 1 and 2 gives rise to action potentials within both spines, which spread passively to activate action potentials in spines 3 and 4. Sequential coincidence detection by active spines can thus bring boosted synaptic responses close to the soma.

Further computational experiments have shown that spines can function as OR gates or as AND–NOT gates, which together with AND gates, provide the basic operations for a digital computer. This does not necessarily mean that dendrites, or the brain, operate like digital computers by digital logic; rather, it shows that simple logic operations are inherent in dendrites, a starting point for investigating the actual

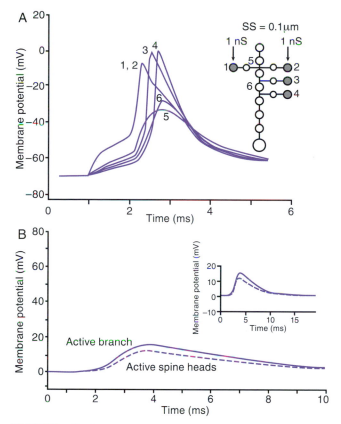

FIGURE 17.17 Logic operations are inherent in coincidence detection by active dendritic sites. The example is an AND operation performed by two dendritic spines with Hodgkin–Huxley-type active kinetics, with intervening passive dendritic membrane. (A) Simultaneous synaptic input of 1 nS conductance to spines 1 and 2 gives rise to action potentials within both spines, which spread passively to activate action potentials in spines 3 and 4. Sequential coincidence detection by active spines can thus bring boosted synaptic responses close to the soma. From Shepherd and Brayton (1987). (B) Recording of boosted spine responses at the soma shows their similarity to the slow time course of classical EPSPs due to the electrotonic properties of the intervening dendritic membrane (see text). Inset shows B on slower time sweep to indicate resemblance to commonly recorded EPSPs. SS, spine stem diameter. From Shepherd et al. (1989).

kinds of information processing that the brain uses. It is also apparent that activity-dependent changes in the coupling between the voltage-dependent spine response and the parent dendrite provide a mechanism for adjustments of synaptic weights that could underlie learning and memory as postulated by Rall (1974). These local adjustments would reflect not properties of the synapses themselves, as Hebb suggested, but nonlinear properties in the spines and dendrites. Thus was introduced the concept of a hierarchy of local active dendritic units, from individual spines to sets of spines to dendritic branches (Koch et al., 1983; Rall and Segev, 1987; Shepherd and Brayton, 1987), which carry out local information

processing and information storage. A recent study has reported direct support for this prediction, providing evidence using two-photon microscopy and glutamate application that "local spikes selectively respond only to appropriately correlated input" and concluding that "compartmentalized changes in branch excitability could store multiple complex features of synaptic input" (Losonczy et al., 2008).

In addition to these functions underlying normal functioning of dendrites, the morphological characteristics of a spine may be used to isolate functional properties that result from pathological processes. One such suggestion is that spines may function as compartments to isolate changes at the synapse, such as influx of excess Ca^{2+} that occurs in ischemia due to stroke, that lead to degenerative changes that are harmful to the rest of the neuron (Volfovsky et al., 1999).

The range of functions that have been hypothesized for spines is partly a reflection of how little direct evidence we have of specific properties of spines. It also indicates that the answer to the question "What is the function of the dendritic spine?" is unlikely to be only one function but rather a range of functions that in a given neuron is tuned into the specific operations of that neuron. The concept of the spine as a microcompartment that integrates a range of functions is now well established (Harris and Kater, 1994; Shepherd, 1996; Yuste and Denk, 1995). A spiny dendritic tree is thus covered with a large population of microintegrative units. We thus have an important principle:

The effect of any given spine on the action potential output of the neuron should be assessed with regard not only to the distant cell body and axon hillock, but rather first with regard to its effect on its neighboring microintegrative units.

SUMMARY: THE DENDRITIC TREE AS A COMPLEX INFORMATION PROCESSING SYSTEM

Dendrites are the primary information processing substrate of the neuron. They allow the neuron wide flexibility in carrying out the operations needed for processing information in the spatial and temporal domains within nervous centers. The main constraints on these operations are the rules of passive electrotonic spread (Chapter 4) and the rules of nonlinear thresholding at multiple sites within the complex geometry of dendritic trees. Many specific types of information processing can be demonstrated in dendrites, such as logic operations, motion detection, oscillatory activity, lateral inhibition, and network control of sensory processing and motor control. These types are possible for both cells without axons and cells with axons, the latter operating in addition within constraints that govern local versus global outputs and sub- versus suprathreshold activities. Several of these operations, as reviewed earlier in the chapter, are summarized in the diagram of Figure 17.18.

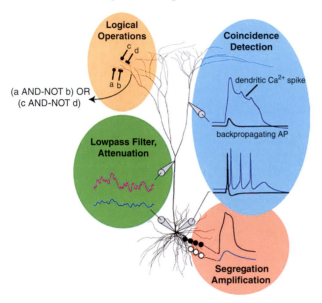

FIGURE 17.18 Summary of some of the functions of dendritic tree of cortical pyramidal neurons that have been demonstrated experimentally and computationally and discussed in this chapter. From Mel and Schiller (2004).

Spines add a dimension of local computation to dendritic function that is especially relevant to mechanisms for learning and memory (see also Chapter 19). Although spines seem to distance synaptic responses from directly affecting axonal output, many cells demonstrate that distal spine inputs carry specific information.

The key to understanding how all parts of the dendritic tree, including its distal branches and spines, can participate in mediating specific types of information processing is to recognize the tree as a complex system of active nodes. From this perspective, if a spine can affect its neighbor, and that spine its neighbor, a dendritic tree becomes a cascade of decision points, with multiple cascades operating over multiple overlapping timescales (see Fig. 17.19A, B). Far from being a single node, as in the classical concept of McCulloch and Pitts (1943) (Fig. 17.19C) and classical neural network models, the complex neuron is a system of nodes in itself, within which *the dendrites constitute a kind of neural microchip for complex computations.* The global output becomes the summation of all the logic operations taking place in the dendrites (Shepherd and Brayton, 1987). The neuron as a single node, so feeble in its information processing capacities, is replaced by the neuron as a powerful complex multinodal system.

The range of operations of which this complex system is capable continues to expand. A formal representation of the dendrite as a multinodal system has been applied to the ensemble of thin oblique dendrites that are emitted by the apical dendrite of a CA1

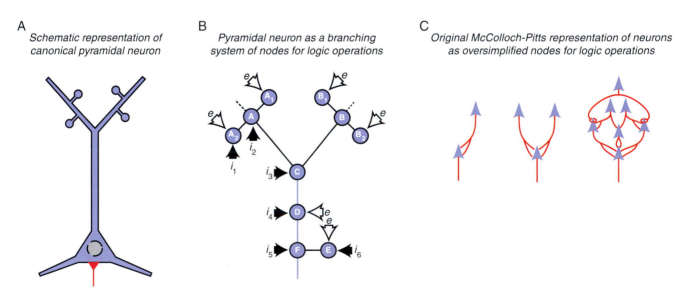

FIGURE 17.19 The dendritic tree as a complex system of logic nodes. (A) A simplified representation of a cortical pyramidal cell. (B) Conversion to a representation in terms of logic nodes and interconnections. (C) Comparison with the concept introduced by McCulloch and Pitts (1943) of the neuron as a functional node for carrying out logic operations, but in which the dendritic tree is ignored and the entire neuron is reduced to a single computational node. e, excitatory synapse; i, inhibitory synapse. From Shepherd, G.M. (1994). "Neurobiology," (3rd Ed.). Oxford Univ. Press, New York. Reproduced with permission from Oxford University Press.

Mapping a multinode dendrite neuron into a two-layer neural network

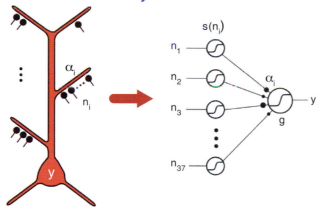

FIGURE 17.20 How can the kinds of complex operations illustrated in Figures 17.18 and 17.19 be incorporated into network models? One way is illustrated here, in which the input–output operations of the thin oblique dendritic branches of the apical dendrite of a CA1 pyramidal neuron are represented, together with the summing node at the soma, as a two-layer neural network. From Poirazi *et al.* (2003).

pyramidal neuron. As shown in Figure 17.20, these thin dendrites, as with spines in Figure 17.19, can be represented by individual summing nodes. The cell can then be mapped onto a two-layer "neural network," in which the first layer consists of the synaptic inputs to the oblique nodes, whose outputs are then summed at the cell body for final thresholding and global output (Poirazi *et al.*, 2003). Most of the input to the oblique dendrites is believed to be involved in the generation of long-term potentiation, a candidate model for learning and memory (Chapter 19). The two-layer conceptual approach thus may be a bridge between realistic multicompartmental models and single-node neural networks in the study of brain mechanisms in learning and memory. Exploring the information processing capacities of the brain at the level of real dendritic systems, by both experimental and theoretical methods, thus presents one of the most exciting challenges for neuroscientists at present and into the future.

References

Anderson, P. (1960). Interhippocampal impulses. II. Apical dendritic activation of CA1 neurons. Acta Physiol. Scand. **48**, 178–208.

Andrásfalvy, B. K., and Mody, I. (2006). Differences between the scaling of miniature IPSCs and EPSCs recorded in the dendrites of CA1 mouse pyramidal neurons. J. Physiol. (Lond.) **576**, 191–196.

Baylor, D. A., and Nicholls, J. G. (1969). After-effects of nerve impulses on signaling in the central nervous system of the leech. J. Physiol. (Lond.) **203**, 571–589.

Cajal, S. R. (1911). *Histologie du systeme nerveux* (Vol. 2). Paris: Maline.

Callaway, J. C., and Ross, W. N. (1995). Frequency-dependent propagation of sodium action potentials in dendrites of hippocampal CA1 pyramidal neurons. J. Neurophysiol. **74**, 1395–1403.

Cataldo, E., Brunelli, M., Byrne, J. H., Av-Ron, E., Cai, Y., and Baxter, D. A. (2005). Computational model of touch mechanoafferent (T cell) of the leech: Role of afterhyperpolarization (AHP) in activity-dependent conduction failure. J. Comput. Neurosci. **18**, 5–24.

Craig, A. M., and Banker, G. (1994). Neuronal polarity. Annu. Rev. Neurosci. **17**, 267–310.

Egger, V., and Urbann, N. N. (2006). Dynamic connectivity in the mitral cell-granule cell microcircuit. Semin. Cell Dev. Biol. **17**, 424–432.

Golowasch, J., and Marder, E. (1992). Ionic currents of the lateral pyloric neuron of the stomatogastric ganglion of the crab. J. Neurophysiol. **67**, 2, 318–331.

Gorassini, M., Bennett, D. J., Kiehn, O., Eken, T., and Hultborn, H. (1999). Activation patterns of hindlimb motor units in the awake rat and their relation to motoneuron intrinsic properties. J. Neurophysiol. **82**, 709–717.

Gu, X., Macagno, E., and Muller, K. J. (1989). Laser microbeam axotomy and conduction block show that electrical transmission at a central synapse is distributed at multiple contacts. J. Neurobiol. **20**, 422–434.

Gu, X. (1991). Effect of conduction block at axon bifurcations on synaptic transmission to different postsynaptic neurones in the leech. J. Physiol. (Lond.) **441**, 755–778.

Harris, K. M., and Kater, S. B. (1994). Dendritic spines: Cellular specializations imparting both stability and flexibility to synaptic function. Annu. Rev. Neurosci. **17**, 341–371.

Hatt, H., and Smith, D. O. (1976) Synaptic depression related to presynaptic axon conduction block. J. Physiol. (Lond.) **259**, 367–393.

Herreras, O. (1990). Propagating dendritic action potential mediates synaptic transmission in CA1 pyramidal cells in situ. J. Neurophysiol. **64**, 1429–1441.

Johnston, D. A., and Wu, S. M.-S. (1995). "Foundations of Cellular Neurophysiology." MIT Press, Cambridge, MA.

Koch, C. (1999). "Biophysics of Computation: Information Processing in Single Neurons." Oxford Univ. Press, New York.

Koch, C., Poggio, T., and Torre, V. 1983. Nonlinear interactions in a dendritic tree: Localization, timing, and role of information processing. Proc. Nat. Acad. Sci. USA **80**: 2799–17802.

Llinas, R., and Sugimori, M. (1980). Electrophysiological properties of in vitro Purkinje cell dendrites in mammalian cerebellar slices. J. Physiol. (Lond.) **305**, 197–213.

Losonczy, A., Makara, J. K., and Magee, J. C. (2008). Compartmentalized dendritic plasticity and input feature storage in neurons. Nature **452**, 436–441.

Luksch, H., Cox, K., and Karten, H. J. (1998). Bottlebrush dendritic endings and large dendritic fields: Motion-detecting neurons in the tectofugal pathway. J. Comp. Neurol. **396**, 399–414.

Luscher, H. R., and Shiner, J. S. (1990a). Computation of action potential propagation and presynaptic bouton activation in terminal arborizations of different geometries. Biophys. J. **58**, 1377–1388.

Luscher, H. R., and Shiner, J. S. (1990b). Simulation of action potential propagation in complex terminal arborizations. Biophys. J. **58**, 1389–1399.

Luscher, H. R. (1998). Control of action potential invasion into terminal arborizations. In: "Presynaptic Inhibition and Neural Control" (P. Rudomin, R. Romo, and L. Mendell, Eds.), pp. 126–137. Oxford Univ. Press, New York.

Macagno E., Muller, J. M., and Pitman, R. M. (1987). Conduction block silences parts of a chemical synapse in the leech central nervous system. J. Physiol. (Lond.) 387, 649–664.

Magee, J. C. (1999). Voltage-gated ion channels in dendrites. In: "Dendrites" (G. Stuart, N. Spruston, and M. Hausser, Eds.), pp. 139–160. Oxford Univ. Press, New York.

Magee, J. C. (2000). Dendritic integration of excitatory synaptic input. Nature Neurosci. 1, 181–190.

Magee, J. C., and Johnston, D. (1995). Characterization of single voltage-gated Na^+ and Ca^{2+} channels in apical dendrites of rat CA1 pyramidal neurons. J. Physiol. (Lond.) 487, 67–90.

Mainen, Z. E., and Sejnowski, T. J. (1995). Influence of dendritic structure on firing pattern in model neocortical neurons. Nature 382, 363–365.

Manor, Y., Nadim, E., Epstein, S., Ritt, J., Marder, E., and Kopell, N. (1999). Network oscillations generated by balancing graded asymmetric reciprocal inhibition in passive neurons. J. Neurosci. 19, 2765–2779.

Mar, A., and Drapeau, P. (1996) Modulation of conduction block in leech mechanosensory neurons. J. Neurosci. 16, 4335–4343.

Matus, A., and Shepherd, G. M. (2000). The millennium of the dendrite? Neuron 27, 431–434.

McCulloch, W. S., and Pitts, W. H. (1943). A logical calculus of the ideas immanent in nervous activity. Bull. Math. Biophys. 5, 115–133.

Mel, B. W., and Schiller, J. (2004). On the fight between excitation and inhibition: Location is everything. Sci. STKE Sep 7 (250), E44.

Midtgaard, J., Lasser-Ross, N., and Ross, W. N. (1993). Spatial distribution of Ca^{2+} influx in turtle Purkinje cell dendrites in vitro: Role of a transient outward current. J. Neurophysiol. 70, 2455–2469.

Migliore, M., and Shepherd, G. M. (2002). Emerging rules for the distributions of active dendritic conductances. Nature Neurosci. Revs. 3, 362–370.

Migliore, M., and Shepherd, G. M. (2005). Opinion: An integrated approach to classifying neuronal phenotypes. Nat. Rev. Neurosci. 10, 810–818.

Migliore, M., and Shepherd, G. M. (2007). Dendritic action potentials connect distributed dendrodendritic circuits. J. Comput. Neurosci. Aug 3, 2007 [E-pub. ahead of print in 2008].

Nicholls, J. G., and Baylor, D. A. (1968a). Specific modalities and receptive fields of sensory neurons in CNS of the leech. J. Neurophysiol. 31, 740–756.

Nicholls, J. G., and Baylor, D. A. (1968b). Long-lasting hyperpolarization after activity of neurons in leech central nervous system. Science 162, 279–281.

Pinsky, P. E., and Rinzel, J. (1994). Intrinsic and network rhythmogenesis in a reduced Traub model for CAS neurons. J. Comput. Neurosci. 1, 39–60.

Poirazi, P., Brannon, T., and Mel, B. W. (2003). Pyramidal neuron as two-layer neural network. Neuron 37, 989–999.

Poirazi, P., and Mel, B. W. (2001). Impact of active dendrites and structural plasticity on the memory capacity of neural tissue. Neuron 29, 779–796.

Rall, W. (1964). Theoretical significance of dendritic trees for neuronal input-output relations. In: "Neural Theory and Modelling" (R. E. Reiss, Ed.), pp. 73–97. Stanford, CA, Stanford Univ. Press.

Rall, W., and Shepherd, G. M. (1968). Theoretical reconstruction of field potentials and dendrodendritic synaptic interactions in olfactory bulb. J. Neurophysiol. 31, 884–915.

Rall, W. (1974). Dendritic spines, synaptic potency and neural plasticity. In "Cellular Mechanisms Subserving Changes in

Neuronal Activity" (C. D. Woody, K. A. Brown, T. J. Crow, and J. D. Knipsel, Eds.). Brain Information Service Research Report No. e., UCLA, Los Angeles, pp 13–21). Reproduced in: Segev et al (1995).

Rall, W., and Segev, I. (1987). Functional possibilities for synapses on dendrites and on dendritic spines. In: "Synaptic Function" (G. M. Edelman, W. F. Gall, and W. M. Cowan, Eds.), pp. 605–636. John Wiley & Sons, New York.

Redman, S. J., and Walmsley, B. (1983). Amplitude fluctuations in synaptic potentials evoked in cat spinal motoneurons at identified group in synapses. J. Physiol. (Lond.) 343, 135–145.

Regehr, W. G., Connor, J. A., and Tank, D. W. (1989). Optical imaging of calcium accumulation in hippocampal pyramidal cells during synaptic activation. Nature (Lond.) 341, 533–536.

Richardson, T. L., Turner, R. W., and Miller, J. J. (1987). Action-potential discharge in hippocampal CA1 pyramidal neurons. J. Neurophysiol. 58, 981–996.

Ringham, G. L. (1971). Origin of nerve impulse in slowly adapting stretch receptor of crayfish. J. Neurophysiol. 33, 773–786.

Rudomin P., Romo R., and Mendell L. (1998). "Presynaptic Inhibition and Neural Control." Oxford Univ. Press, New York.

Sabatini, B. L., and Svoboda, K. (2000). Analysis of calcium channels in single spines using optical fluctuation analysis. Nature 408, 589–593.

Schaefer, A. T., and Margrie, T. W. (2007). Spatiotemporal representations in the olfactory system. Trends Neurosci. 30, 92–100.

Scuri, R., Mozzachiodi, R., and Brunelli, M. (2002). Activity-dependent increase of the AHP amplitude in T sensory neurons of the leech. J. Neurophysiol. 88, 2490–2500.

Segev, I., Rinzel, J., and Shepherd, G. M. (Eds.). (1995). "The Theoretical Foundation of Dendritic Function: Selected Papers of Wilfrid Rail." MIT Press, Cambridge, MA.

Selverston., A. L., Russell, D. E., and Miller, J. P. (1976). The stomato-gastric nervous system: Structure and function of a small neural network. Prog. Neurobiol. 37, 215–289.

Shepherd, G. M. (1977). The olfactory bulb: A simple system in the mammalian brain. In: "Handbook of Physiology," Sect. 1, The Nervous System; Part 1, Cellular Biology of Neurons (E. R. Kandel, Ed.), pp. 945–968. American Physiological Society, Bethesda, MD.

Shepherd, G. M. (1991). "Foundations of the Neuron Doctrine." Oxford Univ. Press, New York.

Shepherd, G. M. (1992). Canonical neurons and their computational organization. In: "Single Neuron Computation" (T. McKenna, J. Davis, and S. E. Zornetzer, Eds.), pp. 27–59. MIT Press, Cambridge, MA.

Shepherd, G. M. (1994). "Neurobiology," 3rd ed. Oxford Univ. Press, New York.

Shepherd, G. M. (1996). The dendritic spine: A multifunctional integrative unit. J. Neurophysiol. 75, 2197–2210.

Shepherd, G. M. (Ed.) (2004). "The Synaptic Organization of the Brain. Fifth Edition." Oxford Univ. Press, New York.

Shepherd, G. M., and Brayton, R. K. (1987). Logic operations are properties of computer-simulated interactions between excitable dendritic spines. Neurosci. 21, 151–166.

Shepherd, G. M., Chen, W. R. and Greer, C. A. (2004). Olfactory bulb. In "The Synaptic Organization of the Brain" (G. M. Shephred, Ed.), pp. 165–216. New York, Oxford University Press.

Shepherd, G. M., Chen, W. R., Willhite, D., Migliore, M., and Greer, C. A. (2007). The olfactory granule cell: From classical enigma to central role in olfactory processing. Brain Res. Rev. 55, 373–382.

Shepherd, G. M., Woolf, T. B., and Cernevals, N. T. (1989). Comparisons between active properties of distal dendritic branches and spines: Implications for neuronal computations. J. Cogn. Neurosci. **1**, 273–286.

Shu, Y., Hasenstaub, A., Duque A., Yu, Y., and McCormick, D.A. (2006). Modulation of intracortical sunaptic potentials by presynaptic somatic membrane potential. Nature. **441**, 761–765.

Simmons, M. L., Terman, G. W., Gibbs, S. M., and Chavkin, C. (1995). L-type calcium channels mediate dynorphin neuropeptide release from dendrites but not axons of hippocampal granule cells. Neuron **14**, 1265–1272.

Single, S., and Borst, A. (1998). Dendritic integration and its role in computing image velocity. Science **281**, 1848–1850.

Sorensen, S. A., and Rubel, E. W. (2006). The level and integrity of synaptic input regulates dendrite structure. J. Neurosci. **26**, 1539–1550.

Spruston, N. (2000). Distant synapses raise their voices. Nature Neurosci. **3**, 849–851.

Spruston, N., Schiller, Y., Stuart, G., and Sakmann, B. (1995). Activity-dependent action potential invasion and calcium influx into hippocampal CA1 dendrites. Science **268**, 297–300.

Stern, P., and Marx, J. (2000). Beautiful, complex and diverse specialists. Science **290**, 735.

Stuart, G. J., and Sakmann, B. (1994). Active propagation of somatic action potentials into neocortical pyramidal cell dendrites. Nature (Lond.) **367**, 6–72.

Stuart, G., Spruston, N., and Hausser, M., eds. (2007). "Dendrites." Second edition. Oxford Univ. Press, New York.

Svirskie, G., Gutman, A., and Hounsgaard, J. (2001). Electrotonic structure of motoneurons in the spinal cord of the turtle: Inferences for the mechanisms of bistability. J. Neurophysiol. **85**, 391–399.

Turner, R. W., Meyers, E. R., Richardson, D. L., and Barker, J. L. (1991). The site for initiation of action potential discharge over the somatosensory axis of rat hippocampal CA1 pyramidal neurons. J. Neurosci. **11**, 2270–2280.

Van Essen, D. C. (1973). The contribution of membrane hyperpolarization to adaptation and conduction block in sensory neurones of the leech. J. Physiol. (Lond.) **230**, 509–534.

Volfovsky, N., Parnas, H., Segal M., and Korkotian, E. (1999). Geometry of dendritic spines affects calcium dynamics in hippocampal neurons: Theory and experiments. J. Neurophysiol. **82**, 450–462.

Wall, P. (1995) Do nerve impulses penetrate terminal arborizations? A pre-synaptic control mechanism. Trends Neurosci. **18**, 99–103.

Willhite, D. C., Nguyen, K. T., Masurkar A. V., Greer, C. A., Shepherd, G. M., and Chen, W. R. (2006). Viral tracing identifies distributed columnar organization in the olfactory bulb. Proc Natl Acad Sci U S A **103**, 12592–12597.

Wilson, C. (1998). Basal ganglia. In: "The Synaptic Organization of the Brain," 4th ed. (G. Shepherd, Ed.), pp. 329–375. Oxford Univ. Press, New York.

Xiong, W., and Chen, W. R. (2002). Dynamic gating of spike propagation in the mitral cell lateral dendrites. Neuron **34**, 115–126.

Yau, K. W. (1976). Receptive field, geometry and conduction block of sensory neurones in the central nervous system of the leech. J. Physiol. (Lond.) **263**, 513–538.

Yuste, R., and Denk, W. (1995). Dendritic spines as basic functional units of neuronal integration in dendrites. Nature (Lond.) **375**, 682–684.

Yuste, R., Gutnick, M. J., Saar, D., Delaney, K. D., and Tank, D. W. (1994). Calcium accumulations in dendrites from neocortical neurons: An apical band and evidence for functional compartments. Neuron **13**, 23–43.

Zilberter, Y. (2000). Dendritic release of glutamate suppresses synaptic inhibition of pyramidal neurons in rat neocortex. J. Physiol. **528**, 489–496.

Information Processing in Neural Networks

James J. Knierim

INFORMATION PROCESSING

The essential functions of the brain and nervous system are to collect information about the external world and the internal state of the body, interpret that information, determine how that information conforms with the needs and goals of the organism, and formulate an appropriate behavioral response (if necessary) to accomplish those goals. Understanding how neural circuits support these functions is a primary goal of systems, behavioral, and cognitive neuroscience. This task requires knowledge of the following interrelated questions: (1) What information is *represented* in each part of the circuit? (2) How is this information *encoded* in neural firing patterns? (3) How is the information *decoded* by downstream neurons? (4) What are the *computations* that describe the transformation of input representations to output representations?

What Information Is Represented?

In order for information to be processed by a system, the information must be explicitly represented in that system. Thus, a first goal toward understanding how a neural circuit processes information is to determine precisely what information is being represented. Often we have an understanding of this question at only a rudimentary level. For example, based on its anatomical connectivity with the eyes, we can deduce that the lateral geniculate nucleus (LGN) of the thalamus represents visual information (Fig. 18.1A). Although this level is a starting point, one needs to know more precisely what aspect of visual information is being represented (Fig. 18.1B). Is LGN activity a representation of the point-by-point distribution of luminance intensity at each location in the retina? Is it a more complex representation, perhaps of the local contrast in luminance between each point and a surrounding region? Answers to these questions are necessary to understand the functions of neural circuits.

How Is the Information Encoded?

In addition to understanding *what* is represented in a neural circuit, one needs to understand *how* that information is represented. What is the neural code that the brain uses to convey information? Let us use the number 21 to describe the difference between a representation and a code. There are numerous codes that can be used to represent this number. Three examples are 21, 10101, and XXI. In the first two cases, Arabic numerals are used to represent the number, but the numerals are used in two different coding schemes (decimal and binary) to represent the same number. In the last case, roman letters and a third coding scheme are used to represent the number. Similarly, in the nervous system, there are different ways in which neurons can represent information. Information can be represented by the firing rate of an individual neuron, by the average rate of an entire population of neurons, by the precise timing of firing among two or more neurons, or by other coding schemes. Understanding what information is represented in a circuit and how it is encoded are intertwined questions. It is difficult to answer one precisely without knowing the answer to the other—one reason why it is so difficult to decipher complex neural circuitry.

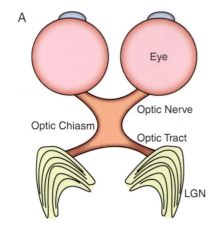

FIGURE 18.1 (A) The lateral geniculate nucleus (LGN) of the thalamus receives input from the eyes. (B) Two possible representations of visual information that may be encoded in a visual region: local luminance across the image (top) or local contrast in luminance (bottom). A key step in understanding the processing of a neural circuit is to determine explicitly what information is represented in the circuit.

How Is the Information Decoded?

Information that is encoded into a representation is useful only if it can be decoded by a downstream receiver, in our case a neuron or circuit. To interpret the meaning of the symbol "10101," one needs to know the original coding scheme in order to determine if it represents the number 21 or the number 10,101. In the nervous system, because of the stochastic nature of neural signals, decoding is not necessarily the simple inverse of encoding. Encoding and decoding each follow their own statistical rules, as will be described later. Thus, the effectiveness of a neural coding scheme depends on the ability of downstream neurons to decode the signal. No matter how cleverly or efficiently a neural coding scheme may represent information, the representation is useless if the neural circuits that receive that information are unable to decode it. Thus, the evolution of neural encoding schemes is constrained by the ability of neurons and neural circuits to decode the information.

What Are the Computations?

None of the concepts discussed to this point have involved actual information processing. That is, if the only operations performed on a piece of information are to encode it into a representation and then subsequently decode it, then no information processing has occurred. Rather, information processing requires that a representation be transformed in a meaningful way. This transformation may be as simple as changing a representation from one coding scheme to another. For example, it is often beneficial to transform a decimal number into a binary or hexadecimal representation for ease of computation in a computer. More typically, however, the transformation involves the calculation of a new parameter or value that is then represented in the output of the circuit. For example, an input in a visual system circuit may be a representation of the level of brightness at each location in the visual field, and the output of that circuit may be a representation of local *contrast* in brightness at each location in the visual field (Fig. 18.1B). Such a representation (which is generally thought to occur in the retina, with the retinal ganglion cells providing the output signal of local contrast; Kuffler, 1953) would be useful in creating stable visual perceptions over a wide range of illumination levels (e.g., a bright object looks bright in both well-lit and dark rooms, even though the amount of light actually being reflected off the object can differ by orders of magnitude; Gregory, 1977). Understanding the computations that occur in a neural circuit—the rules that determine how input representations are transformed to output representations—is the ultimate goal in understanding the detailed functions of the network and the mechanisms used to perform those functions.

NEURAL REPRESENTATION

Representation in a Simple Circuit

Many neural circuits are fairly simple and relatively well understood. These circuits may be little more than a sensory neuron synapsing with a motor neuron, with perhaps an interneuron in the middle. In this case, the processing is described in simple terms, such as "a stimulus causes a sensory neuron to fire, which causes a motor neuron to fire, which causes a muscle to twitch." Although this is a perfectly adequate description of a simple circuit, it is also useful to consider the circuit in terms of the concepts of representation, coding, and decoding. These concepts are essential for describing more complex neural circuits that are formed from networks of thousands or more neurons.

One of the simplest neural circuits is the stretch reflex (also called the *deep tendon reflex*) (Nolte, 2002; Pearson and Gordon, 2000). This is the reflex that your doctor tests when she taps your knee to elicit a knee-jerk response (Fig. 18.2). In this circuit, the hammer tap to the tendon causes the muscle to stretch and activates stretch receptors in the muscle spindle. Action potentials of the innervating sensory neurons travel to the spinal cord and elicit transmitter release at the synapse between the sensory neuron and a motor neuron. The motor neuron becomes depolarized, and action potentials travel down its axon to the same muscle, causing a contraction that makes the lower leg jerk forward. In this example, the activity of the sensory neuron is a representation of the change in length of the muscle. Under normal conditions, the stretch reflex is involved in such functions as the automatic maintenance of posture. If the muscle begins to stretch, the stretch reflex can counteract that stretch automatically by contracting the muscle, thus maintaining the body in a stable position. The rate at which the sensory neuron fires is the code that conveys information about how fast the muscle is stretching—the faster the firing, the faster the stretch (Hunt, 1990). The motor neuron activity is a representation of the force needed to contract the muscle and counteract the stretch. It also uses a rate code—the faster the firing, the more force will be generated by the muscle when it contracts (Monster and Chan, 1977). Note that the motor neuron must decode the signals that it receives from the sensory neuron. In this case, the task is fairly simple. As the sensory neuron fires spikes at a higher rate, the motor neuron is depolarized more strongly, and the biophysics of action potential generation cause the motor neuron to fire more spikes. Note, however, that the rate of action potential generation in the sensory neuron does not necessarily equal the rate of action potential generation in the motor neuron. A stretch that produces firing at 10 Hz in the sensory neuron may require activity of 20 Hz in the motor neuron to contract the muscle and counteract the stretch. This computation is the information processing that is performed in the circuit, as the representation of muscle stretch is transformed into a representation of appropriate muscle force, each using a rate code appropriate for its function.

Although an individual stretch receptor represents the length and the dynamic change in length of a muscle fiber, these neurons can be part of a larger system that represents something fundamentally different. That is, the system of stretch receptors in the entire body forms an implicit representation of the position of the body and limbs with respect to each other (called the sense of *proprioception*). Thus, at higher levels of organization, simple representations can be combined into larger representations that convey information about more holistic properties. A brain area that receives input from stretch receptors (and other proprioceptors) transforms the information about muscle length and joint position into a representation of body position.

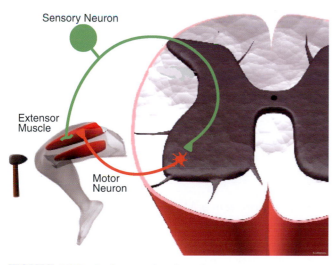

FIGURE 18.2 In the stretch reflex, the sensory neuron uses a rate code to represent the change in length of the muscle. The motor neuron represents the amount of force needed to contract the muscle to counteract the stretch. The motor neuron also uses a rate code, and must compute its appropriate firing rate based on the firing rate of the sensory neuron.

Brain Maps

As discussed earlier, information can be represented in the firing of single neurons. In more complicated neural systems, information can also be represented in the form of topographic maps of a particular stimulus.

These maps are composed of individual neurons that may represent a particular value of a stimulus dimension, and the neurons are organized anatomically such that the stimulus dimension is mapped onto the brain structure. Thus, information about the value of that stimulus may be encoded by the anatomical location of the neuron within the representation.

Homunculus

The most famous brain map is the "homunculus" in the motor cortex described by the neurosurgeon Wilder Penfield in his classic studies (Fig. 18.3) (Penfield and Erickson, 1941). By electrically stimulating different locations on the surface of the motor cortex, Penfield discovered that he could elicit predictable body movements depending on the site of the stimulation. Nearby cortical sites cause movements of nearby body parts. Furthermore, more cortical space is devoted to body parts that perform complex movements (e.g., the hand) than to body parts that perform simple movements (e.g., the leg). Thus, the motor cortex contains a map of the body on the cortical sheet, forming a two-dimensional representation of body movement.

A similar set of maps is found in the neighboring somatosensory cortex (Kaas *et al.*, 1979). Primary somatosensory cortex contains four different sensory maps of the body. Each map receives input from a different set of sensory receptors, from either the skin or deep tissue. Thus, each map represents a different type of sensory stimulation across the body surface, and the location of an active neuron on the cortex encodes what body part is being stimulated and by what type of stimulation. This location coding, in which information is represented by the physical location of a neuron in a structure, is called a *place code*.

Visual Field

The visual cortex similarly contains a multitude of areas that represent topographic maps, in this case maps of the visual field (Felleman and Van Essen, 1991). For example, individual neurons in the primary visual cortex (called *striate cortex*, or V1) respond to visual stimulation from a restricted location in the visual field, and nearby neurons respond to stimulation of nearby visual regions (Fig. 18.4; Van Essen *et al.*, 1984). Moreover, the overall visuotopic representation across the surface of V1 is subdivided into representations of other stimulus dimensions (not shown). Thus, information about whether the information comes from the right or left eye is segregated into interdigitating stripes (resembling those of a zebra) called *ocular dominance columns* (LeVay *et al.*, 1975, 1985). Information about the orientation of a stimulus is segregated into pinwheel-like structures (Blasdel and Salama, 1986). Information about the color of a stimulus is segregated into polka-dot-like "blobs" in the superficial cortical layers (Carroll and Wong-Riley, 1984; Livingstone and Hubel, 1984). Overall, V1 contains multiple representations of the visual image, woven together into a holistic representation of various dimensions of a visual stimulus. This segregated information is then routed to mid-level visual areas, each with its own visuotopic map (Livingstone and Hubel, 1988). These visual areas contain neurons that are selective for the different visual attributes. One area (the middle temporal area, or MT) is selective for moving stimuli, and MT is thought to represent motion across the visual field. Another area (visual area 4, or V4) contains neurons that are selective for form and color, and V4 is thought to represent various aspects of the form and identity of visual objects. The segregation of visual information into these brain areas presumably allows the efficient processing of each attribute by specialized, dedicated circuitry in each area.

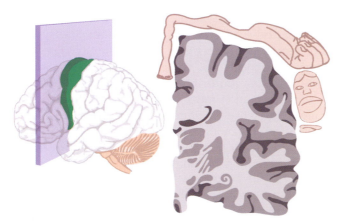

FIGURE 18.3 The motor cortex represents body movements using a topographic map of the body surface along the cortical surface. This figure shows a cross section through the primary motor cortex. Electrical stimulation of the medial wall of the cortex produces movements of the foot or leg, whereas stimulation of the lateral cortex produces movement of the face or hand. Notice that more cortex is devoted to body parts that make fine-scale movements, such as the hand, compared to body parts that make only coarse movements, such as the leg.

Auditory Frequency Map

As a final example, auditory cortex contains a tonotopic map of frequency. Different neurons are selectively responsive to different frequencies of auditory stimulation, and these neurons are arranged such that nearby neurons respond to similar frequencies. Thus, the audible frequency spectrum is represented across the surface of auditory cortex (Kilgard and Merzenich, 1998; Sally and Kelly, 1988).

Visual Field

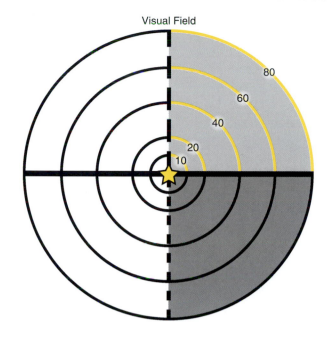

Primary Visual Cortex of Left Hemisphere

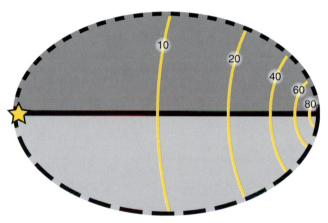

FIGURE 18.4 The primary visual cortex represents the visual field (top) as a distorted topographic map (bottom). The right visual field (shaded gray) is represented in the left hemisphere. The upper visual field (light gray) is represented ventral to the lower visual field (dark gray). Like the motor cortex (Fig. 18.3), more cortical space is devoted to regions that require fine-scale processing. Thus, the region of high-acuity vision (the fovea, denoted by a star, and the surrounding 10°) takes up ~50% of the primary visual cortex, whereas the far periphery takes up only a small fraction of cortical space (Van Essen *et al.*, 1984).

Nontopographic Organization

It is a matter of active debate whether the place code embodied by topographic maps plays an important role in sensorimotor processing, or whether it is an epiphenomenon of development or reflects merely an efficient anatomical wiring scheme. Indeed, not all brain systems represent information in topographic maps. The further removed from the sensory periphery that

a brain area is located, the less topographically organized it appears to be (Van Essen *et al.*, 1990). Primary sensory regions have the most precise topography, whereas mid-level regions become less precise. For example, visual receptive fields of V1 neurons are very small, allowing V1 to represent visual space in a very precise, fine-scale map. In V4 and MT, by contrast, the receptive fields tend to be larger. Thus, as more information about the visual stimulus itself becomes abstracted and represented in this area, precise information about location in the visual field is diminished. Finally, in the highest-level areas of visual cortex in the temporal lobe, visual receptive fields can cover the entire visual field. At these levels, location in visual space is no longer the major dimension that is mapped on the cortical surface. Presumably, some other, high-order dimension about the identity of visual objects is mapped at these levels. Alternatively, there may be no dimension that is explicitly represented as a place code at these levels, either because the nature of high-order processing is not conducive to a place code or because the anatomical arrangements of inputs to higher-order cortical areas preclude the formation of such a code.

ENCODING AND DECODING

As described in the section "Neural Representation," neurons and brain areas form various types of representations of a particular attribute or set of attributes. These representations must be formed by some type of coding scheme that allows the value of the attribute to be represented in the language of neuronal communication, that is, action potentials and synaptic potentials. There are various types of encoding schemes that the nervous system can utilize.

Rate Codes

By far the most commonly studied coding scheme is the rate code, which we already discussed in relation to the stretch reflex. In a rate code, the value of the attribute being represented by a neuron is encoded in the firing rate of the neuron, measured over some discrete time interval. Most sensory neuron tuning curves are constructed with the assumption that the neurons utilize a rate code. For example, an orientation tuning curve for a neuron in V4 is shown in Figure 18.5A (McAdams and Maunsell, 1999). The experimenters presented a bar of light to the receptive field of the neuron at different orientations and measured the number of spikes that was elicited for each orientation. Dividing the number of spikes by the amount of time that

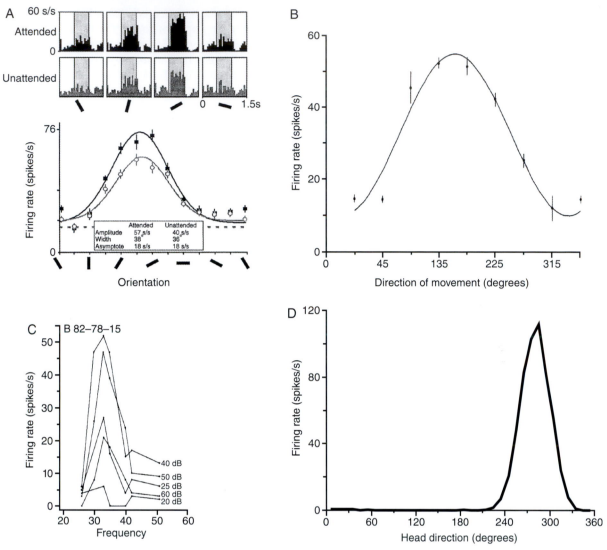

FIGURE 18.5 Neural tuning curves based on rate coding. (A) Orientation tuning curve from visual cortex. The firing rate of the neuron changes as a function of the orientation of the bar of light presented in the cell's receptive field. The top shows peristimulus time histograms of the firing rate of the neuron as a function of time; the shaded regions represent the stimulus presentation time. The bottom shows a tuning curve based on the spike firing histograms. Note that the shape of the tuning curve can vary based on cognitive factors such as attention, as black indicates conditions when the monkey paid attention to the stimulus and white indicates conditions when the stimulus was unattended. Modified with permission from McAdams and Maunsell (1999). (B) A tuning curve for direction of motion from the primary motor cortex. Reproduced with permission from Georgopoulos *et al.* (1982). (C) A tuning curve for sound frequency from auditory cortex. Note that the tuning curves vary as a function of sound intensity (in decibels) as well as frequency. Modified with permission from Phillips, D.P., Orman, S.S. (1984). *J. Neurophysiol.* **51**, 147–163. (D) A tuning curve for head direction from the anterior dorsal thalamus of an awake, behaving rat. From D. Yoganarasimha and J. J. Knierim (unpublished data).

the bar was presented at each orientation produces the firing rate of the neuron in response to each orientation. Based on the firing rate, many visual cortex neurons are shown to be selective, or tuned, to a particular orientation, in that the firing rate to a narrow range of orientations is larger than the firing rate to the other orientations. Thus, the orientation of the bar of light is encoded by the firing rate of the neuron. As shown by Figure 18.5A, the tuning curve itself can be modulated by other factors, such as attention.

In other brain systems and circuits, rate codes are used to represent other attributes (Fig. 18.5B–D). For example, motor cortex neurons change their firing rate as a function of the direction and force necessary to produce a desired movement (Evarts, 1968; Georgopoulos *et al.*, 1982). Auditory cortex neurons have tuning curves in which firing rate changes as a function of both the amplitude and the frequency of the sound (Phillips and Orman, 1984). Head direction cells of the limbic system change their firing rate as a function of the direction that the head is pointing in an allocentric (i.e., world-centered) coordinate frame (Taube, 2007). Notice the differences in the shape and width of the different tuning curves. For example, head direction cells are very sharply tuned, whereas motor cortex neurons are more broadly tuned.

Temporal Codes

Rate codes integrate the number of spikes over a particular time interval, ignoring the precise timing of when individual spikes occurred. A temporal code, in contrast, uses the timing of neural activity to encode information about a stimulus. For example, significant information about visual stimuli is encoded in the timing of action potentials in visual cortex. Figure 18.6 shows the responses of a neuron in a monkey's inferotemporal cortex when it was presented with two checkerboard-like visual stimuli (Richmond *et al.*, 1987). The raster plots at the bottom show the individual spikes elicited on each of the individual trials, and the spike density functions at the top estimate the firing probability of the neuron as a function of time. When averaged over the entire stimulus presentation time, the firing rate of the neuron does not distinguish the two patterns clearly, as it fires at similar rates to both. However, when the shapes of the spike density functions are considered, there are clear and consistent differences in the timing of the neuronal firing to each pattern. The neuron responded to the stimulus in Figure 18.6A with a brief burst, followed by a quiet period, and then a period of sustained firing. In contrast, the response to the stimulus in Figure 18.6B was delayed by approximately 100 msec and consisted of a single-peaked response profile. Thus, even though the overall firing rate of the neuron does not distinguish these two patterns, the temporal firing profile carries information that could potentially allow a downstream neuron to decode which pattern was presented. Moreover, the response to other patterns showed both rate and temporal differences, showing that a neuron can use both rate codes and temporal codes simultaneously.

Other temporal codes make use of the global rhythms that can be generated by oscillatory activity in neural networks. One example is the firing of projection neurons (PN) of the antennal lobe of the locust olfactory system (Fig. 18.7; Wehr and Laurent, 1996). As in the visual system example earlier, a PSTH analysis reveals that the response of a projection neuron to different odors varies not only in firing probability but also in time. At a fine temporal resolution, the firing of these neurons is phase-locked to the oscillating

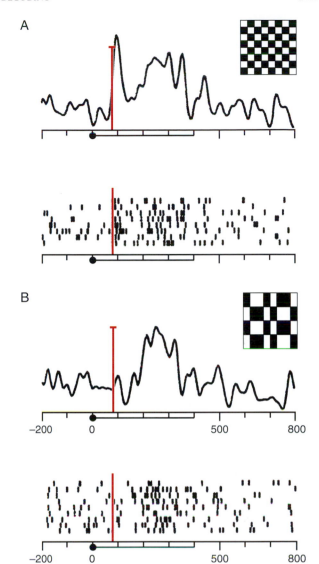

FIGURE 18.6 Temporal coding of checkerboard-like stimuli called *Walsh patterns* by inferotemporal cortex neurons. (A) The top graph shows a spike density function, and the bottom graph shows raster plots of individual spikes on each presentation. The horizontal line under each graph represents the stimulus duration. The cell fired an initial transient burst, was quieter for ~100 msec, and then fired again. (B) The same cell fired at a similar overall rate to a different pattern, but the temporal pattern of activity was different from that of A. For pattern B, the cell did not produce the initial transient burst but instead fired only during the second interval after a delay of ~100 msec relative to pattern A. Reproduced with permission from Richmond, B.J., Optican, L.M., Podell, M., and Spitzer, H. (1987). *J. Neurophysiol.* **57**, 132–146.

local field potential (LFP) of the system; that is, each spike tends to occur at the same phase of the sinusoidal LFP. (The LFP is the electrical field generated by the overall activity in a restricted region of the nervous system, reflecting the electrical activity of all the cell bodies, fibers, and synaptic terminals in the region.) Information about odor identity is encoded

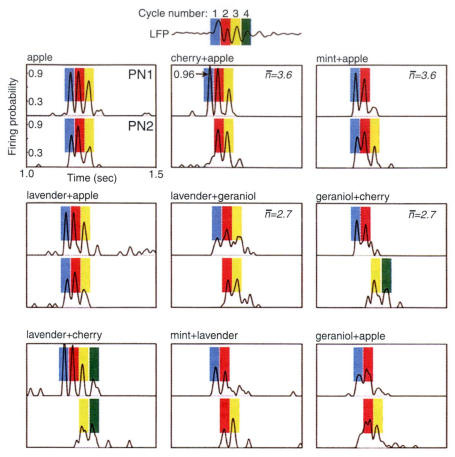

FIGURE 18.7   Temporal coding in the locust olfactory system. (Top) Four oscillations are apparent in the local field potential (LFP). (Bottom) Peristimulus histograms of firing probability as a function of time for two neurons (PN1 and PN2) presented with nine different odor combinations. Each odor stimulus produces a different combination of firing rates (probabilities) and temporal patterns relative to the oscillations of the LFP in the two projection neurons. Reproduced with permission from Macmillan Publishers Ltd. Wehr, M., and Laurent, G. (1996). *Nature* **384**, 162–166.

in the particular pattern of LFP cycles in which the cell fires. For example, PN1 and PN2 fire at three of the four LFP cycles when presented with the odor of apple, but PN2 fires only during cycles 2 and 3 when presented with the combination of cherry + apple. Thus, a downstream neuron can determine whether the odor was apple or cherry + apple by decoding whether PN2 fired on all three cycles of the LFP or on only the second and third cycles. As shown in Figure 18.7, other odors elicited distinct patterns of firing between the two neurons in terms of both firing probability and time.

Another example of a temporal code that uses spike timing relative to the oscillation of an LFP is the theta-phase precession of hippocampus place cells (Fig. 18.8; Huxter *et al.*, 2003; O'Keefe and Recce, 1993; Skaggs *et al.*, 1996). A place cell is a neuron that fires in a restricted location (its "place field") as an animal explores an environment. During exploration, a prominent, 8–12 Hz, sinusoidal rhythm (the theta rhythm)

dominates the LFP, and the place cell fires in bursts that are time-locked to the rhythm. As an animal moves through the place field, however, the bursts advance to earlier and earlier phases of the cycle (i.e., the firing "precesses" through the theta cycle). This precession can be seen clearly by comparing the relative timing of the spike bursts (red tick marks in Fig. 18.8C) with the constant reference phase of the theta oscillation (blue tick marks). As the rat moves from left to right through the place field, the timing of the spike burst shifts to increasingly earlier phases of the theta cycle. There are two coding implications of this precession. First, the exact phase of theta at which the cell fires a burst encodes information about precisely where the animal is located within the cell's place field, providing a more precise measure of location. Second, the temporal order of firing of partially overlapping place fields is preserved within each theta cycle, encoding the order of locations that the animal visits (i.e., its trajectory; Fig. 18.8D).

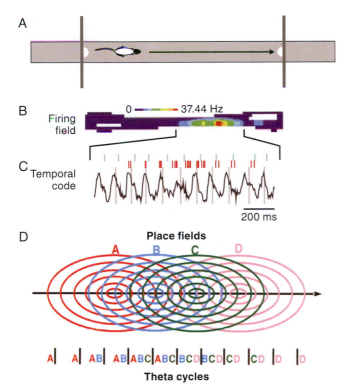

A

B Firing field

0 ——— 37.44 Hz

C Temporal code

200 ms

D **Place fields**

A   B   C   D

A | A | AB | AB | ABC | ABC | BCD | BCD | CD | CD | D | D

Theta cycles

FIGURE 18.8 Temporal coding in the rat hippocampus place cell system. (A) A rat runs back and forth on a linear track for food reward at each end. (B) A firing rate map for one hippocampal cell shows a firing field of elevated activity at a single location on the track. The color code denotes mean firing frequency at each location. (C) The local field potential (black trace) shows the characteristic theta rhythm. Each burst of spikes from the place cell (red ticks) occurs at successively earlier phases of the theta rhythm (compare relative locations of red and blue ticks). (D) Because of the theta "phase-precession" effect shown in C, place fields that occur in a particular order on the track (ABCD) preserve their firing order in a temporally compressed manner within each theta cycle, which may be important for the storage of spatiotemporal sequences of activity in the hippocampus. Reproduced with permission from Macmillan Publishers Ltd. (A), (B), and (C) from Huxter, J., Burgess, N., and O'Keefe, J. (2003). *Nature* **425**, 828–832; (D) from Skaggs, W.E., McNaughton, B.L., Wilson, M.A., and Barnes, C.A. (1996). *Hippocampus* **6**, 149–172.

Multiplexed Codes

A neuron is not limited to using only a single code. We have already seen that individual neurons can convey information by both firing rate (or probability) and temporal codes. It is possible that different types of information are encoded by different coding schemes in a way that allows the neuron to convey different types of information simultaneously in its firing patterns (called *multiplexing*). For example, the cell may encode one variable in its firing rate and another variable in its timing relative to an LFP or relative to the firing of another neuron (Huxter *et al.*, 2003). The downstream neurons would then be

required to decode the signal in such a way that distinguishes the information encoded by rate and the information encoded by timing.

Population Codes: Coarse Coding

Many neurons are broadly tuned to a particular dimension. For example, motor cortex neurons that are tuned for direction of motion can respond to directions far away from their preferred directions (e.g., Fig. 18.5B). In other words, the range of directions that cause the cell to fire is ~270°. If the tuning curve is shaped as a Gaussian profile, then the rate code presents an inherent ambiguity. If the cell is firing at half its maximum rate, does this firing rate signal that the movement is in a direction to the left or right of the preferred direction? If an individual neuron is so broadly tuned, how can we make the precisely controlled movements required by most of our daily activities? A partial answer to these questions is that the brain uses a population code to control the precise direction of movement. Rather than relying on the firing of a single, direction-tuned neuron in isolation, the brain integrates the firing of that neuron with the firing of many other neurons, each tuned to different preferred directions. The concept is illustrated in Figure 18.9. Each cell has a broad directional tuning curve. If cells 1 and 2 are active but cell 3 is silent, however, a downstream structure receiving information from all three cells can narrow down the desired direction to the overlapping region of cells 1 and 2. With input from many such broadly tuned neurons, movement direction can be encoded

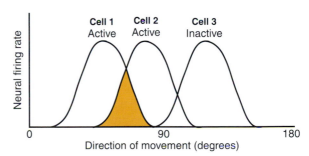

FIGURE 18.9 Population coding. Three hypothetical motor cortex cells are demonstrated with coarse tuning curves. The activity of each cell is ambiguous as to the precise direction of motion of a limb. For example, if cell 1 is active at half its maximal firing rate, it is ambiguous as to whether the direction of movement is to the right or left of its preferred firing direction. However, knowing the activity of a population of neurons allows a more precise decoding of movement direction. If both cells 1 and 2 are firing at their half-maximal rate, and cell 3 is inactive, then the possible direction of movement that can cause this activity pattern is limited to a smaller range (shaded region). The population activity of many such cells can encode the movement direction with high precision.

with great precision. This is called a *coarse coding scheme*, in which the individual units are broadly (coarsely) tuned for a particular attribute, but the activity of a large population of such cells can encode the attribute with high precision.

A well-known example of such population encoding is the activity of neurons in the primary motor cortex. These neurons are coarsely tuned for the direction of motion of a part of the body (e.g., a forelimb; Fig. 18.5B). Knowing the firing of a single neuron will allow an observer to predict the general motion of the arm, but with little precision. However, knowing the firing activity of a population of such neurons, each with its own preferred direction of motion, allows an observer to calculate the precise direction of motion by calculating an average firing rate of each neuron weighted by the preferred firing direction of that neuron (Fig. 18.10A). Such a population vector can reconstruct with high precision the precise motion direction of the limb (Georgopoulos *et al.*, 1988). When applied to place cells of the hippocampus, similarly calculated population vectors reproduce the precise trajectory of a rat as it moves around an environment (Fig. 18.10B; Wilson and McNaughton, 1993).

What is the advantage of using a population code of many coarsely tuned neurons? If each motor cortex neuron fired specifically for a very small range of movement, then one could determine the motion direction directly from the firing of that single neuron with high precision, without the necessity of resorting to a population code. The answer appears to be that population coding is a more efficient method to encode many variables at high precision with fewer neurons. As an example, consider the cone system for color vision. With only three types of cones, each coarsely tuned to a broad range of light wave frequency, the brain can encode countless variations in color by the relative firing rates of these types of cones. Similarly, many fewer neurons are needed to encode motion direction or object orientation if the brain uses a population code of coarsely tuned neurons. Each neuron can take part in many different representations, thereby increasing greatly the information capacity of the system without increasing the number of neurons. Such a system is also robust to damage. Since the information is encoded in the relative firing of many neurons, no single neuron is essential, and the system is less prone to catastrophic failure if a subset of the neurons is damaged.

Sparse Codes

Even though the coarse coding scheme can be very efficient, there are brain systems that use a sparse coding system instead. In a sparse code, each unit responds to a

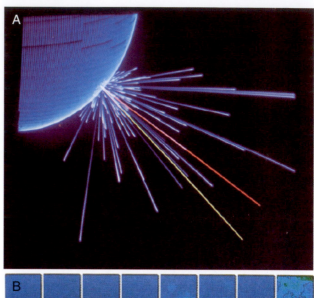

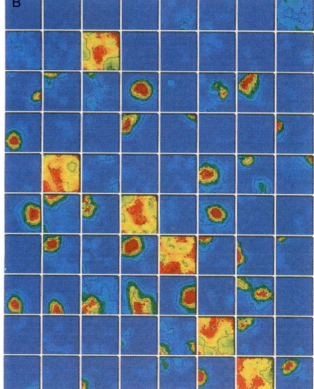

Trajectory reconstruction

FIGURE 18.10   (A) Population vectors from motor cortex. Each blue line indicates the preferred firing direction of a motor cortex neuron. The length of the line indicates the neuron's mean firing rate for a particular movement. Calculating the average of all blue vectors produces the red population vector, which is very close to

Legend continued on next page

very specific, narrow range of stimuli and subsequently takes part in a small number of representations. One example is the song circuit of songbirds. In parts of this system, the cells fire very specifically in a narrow temporal window corresponding to a particular syllable of the song (Hahnloser *et al.*, 2002). Thus, an observer can reliably tell what portion of the song is being sung just by knowing whether that particular cell fired. Another example comes from recordings of human subjects (Quiroga *et al.*, 2005). These patients had electrodes implanted in their temporal lobes in order to determine the origin of epileptic seizures, and they volunteered to have individual neurons recorded in visual perception experiments. It was found that the firing of the neurons was highly selective. One neuron was tested across 87 stimuli and was found to respond strongly and exclusively to pictures of the actress Jennifer Aniston. No other face or object elicited a response from the cell. A final example is the firing of granule cells of the dentate gyrus of the hippocampus. These cells have spatially selective firing fields in an environment, similar to the place cells of the hippocampus proper (Jung and McNaughton, 1993). However, it is estimated that only ~2% of granule cells are active in any environment (Chawla *et al.*, 2005), compared to ~40% of CA1 place cells (Guzowski *et al.*, 1999; Wilson and McNaughton, 1993). Unlike a coarse population code, in which a cell may be active in many environments and thus be a component of many representations, a sparse code requires that the cell be part of only a limited number of representations, and the resolution of that representation depends more strongly on a narrow range of firing for each individual neuron (Barlow, 1972).

It was stated earlier that coarse coding is an efficient method for encoding many stimuli at a level of resolution much higher than the numbers of neurons themselves. A sparse coding scheme, on the other hand, seems rather inefficient. If each neuron responds to a single, narrow range of a stimulus dimension and can take part in only a few such representations, the system must require many more neurons to encode numerous representations at high resolution. What is the advantage of this seemingly inefficient coding scheme? The answer appears to depend on the memory-storage properties of the network. If the function of the network is to store information in memory, then the coarse coding scheme presents a problem called *interference*. The more memories that need to be stored in the network, the more likely it is that two memories will interfere with each other if many neurons take part in both memories. More memories can be stored in a network of sparsely tuned neurons with little interference and recall errors than in a coarsely tuned network (Marr, 1969; McNaughton and Morris, 1987; Rolls and Treves, 1998). Thus, the nature of the coding scheme depends to a large extent on the specific function of the network. A population code of many overlapping, coarsely tuned neurons is an efficient code for such structures as primary sensory and motor areas. However, for networks that are involved in storing representations in memory, a sparse coding scheme is more likely to allow many memories to be stored with less interference.

Encoding Versus Decoding

In a discussion of neural codes, it is important to distinguish between neural encoding and neural decoding. We have come across both processes so far in this chapter. Neurons have to take the information provided by their upstream neurons and encode that information into a particular representation. Downstream neurons that receive action potentials from these neurons have to decode that signal in order to perform some processing and create a new representation, which must be encoded into the firing of that neuron. For example, movement direction is encoded in the firing rate of individual, direction-tuned neurons. As we have seen, because of the coarse nature of the tuning curves, the information about movement direction cannot be decoded at the single-cell level. Rather, decoding neurons (or systems of neurons) must have access to a population of direction-tuned neurons in order to determine precisely the direction of intended movement. Thus, a neural code is only as good as the ability of the downstream neurons to decode it. For example, a temporal code in which stimulus identity is encoded in the timing relationships between action potentials and oscillatory local field potentials is effective only if the downstream neurons (1) have access (at some level) to both the LFP activity and the spiking activity and (2) have mechanisms to compare the two types of signals and determine the relationship.

The distinction between neural encoding and decoding has been formalized using the methods of Bayesian statistics (Rieke *et al.*, 1997). Neural encoding addresses how a neuron responds to a particular stimulus. This can be represented by the entity P(response | stimulus): the probability of a particular response given the presentation of a particular stimulus. Neural decoding, on the other hand, addresses how a neuron interprets the responses of its afferents in order to determine what original stimulus generated those responses. This is represented by the entity P(stimulus | response): the probability that a given stimulus was presented, given the response of the neuron. By Bayesian statistics, these two entities can be related by the equation

$$P(stimulus \mid response) = P(response \mid stimulus) \times P(stimulus)/P(response).$$

As seen by this equation, the problems of encoding and decoding are not simple inverses of each other. Rather, decoding a stimulus [calculating P(stimulus | response)] entails not only knowing the encoding scheme [P(response | stimulus)] but also knowing (1) the probability that a particular stimulus occurred out of all possible stimuli [P(stimulus)] and (2) the general probability that the particular response occurred, regardless of what stimulus was presented [P(response)]. Thus, neural encoding and decoding are inherently statistical processes. Neurons are stochastic devices that respond variably to the same stimulus and have inherent ambiguities in their responses. The job of a neural decoder is to determine the probability that a given pattern of input corresponds to a particular stimulus or particular state of a neural representation. To do this, it needs to know not only the encoding scheme but also information about the statistical probabilities of those states and the statistical probabilities of the individual neural responses.

To illustrate these concepts, imagine you are looking at a figure in your living room and need to determine the identity of the figure by decoding the population activity of your inferotemporal cortex. By knowing the coding scheme, you eliminate all possibilities except two: The figure is either your aunt or an elephant, but the neural code is so similar for the two that the identity is ambiguous. However, the probability that there is an elephant in your living room is much less than the probability that your aunt is in the living room, and thus a neural system that has prior knowledge about the probabilities of such inputs (either from prior experience or hard-wired through evolution) will successfully decode the image as that of your aunt. If it really was an elephant, the system would have to come up with a very unique code to represent this improbable stimulus. That is, if the probability of the stimulus being an elephant is very low, then the probability of the particular response profile for "elephant in my living room" must also be very low (i.e., very unique) in order for

the system to decode it properly rather than attributing the response to a more likely stimulus. Thus, the decoder system may recognize that it is highly unlikely that an elephant is in the living room, but if the code for elephant is highly unique, then there is no other way to interpret the signal.

ICONIC NEURAL CIRCUITS

The first sections of this chapter introduced the general concepts of representation and encoding that are essential for understanding neural processing. In the following sections, we explore different types of iconic neural circuits that are used to carry out the functions of neural encoding, representation, and computation. These iconic circuits are the building blocks of complex neural networks.

Excitation: Feedforward, Feedback, and Lateral

The simplest circuit in the body is a monosynaptic reflex, such as the stretch reflex (Fig. 18.2). This circuit is an example of the most straightforward connection pattern in the nervous system, the feedforward connection. In an excitatory *feedforward connection* (Fig. 18.11A), a neuron makes a synapse with another neuron, such as the sensory neuron synapsing with the motor neuron in the stretch reflex. Feedforward connections imply a certain degree of hierarchy in a circuit, in that information is being transmitted from a lower level to a higher level (e.g., a neuron in the primary visual cortex sending an axon to a higher-order visual area). In contrast, a *feedback connection* (Fig. 18.11B) refers to a connection that goes from a higher-order neuron to a lower-order neuron (or, in some cases, to itself). Feedback connections allow the output of a circuit to influence the input neurons in subsequent cycles of processing. Finally, a third type of excitatory connection is a *lateral connection* (Fig. 18.11C). In circuits that contain many neurons at each processing level, the neurons within a level can communicate through both local and long-range lateral connections. These connections allow sophisticated computation to occur within a particular level of a circuit before the output is transmitted to the next level via feedforward connections.

Inhibition: Feedforward, Feedback, and Lateral

Of equal importance to the functioning of neural circuitry are inhibitory connections, in which the firing of a neuron inhibits the firing of a downstream

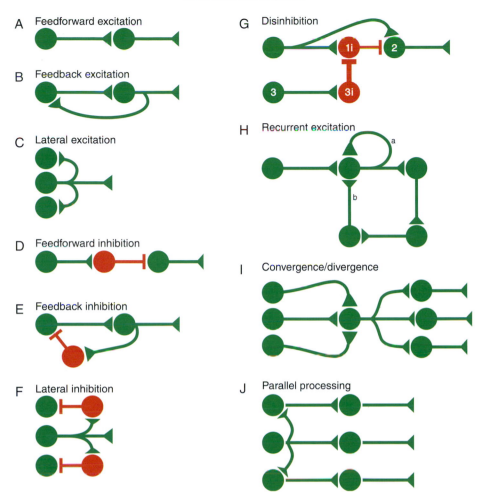

FIGURE 18.11 Iconic neural circuits. Green neurons are excitatory and red neurons are inhibitory.

neuron to which it is connected. Inhibitory circuits can be feedforward, feedback, or lateral, just as excitatory circuits. Often, an inhibitory circuit is a combination of a long-range excitatory projection neuron that synapses onto a local inhibitory interneuron, which then synapses onto another projection neuron. Feedforward inhibition is used when the output of one level of a circuit requires the activity of the next level to be decreased (Fig. 18.11D). A simple example is the autogenic inhibition reflex, in which activity of the Golgi tendon organ of the muscle (which signals the amount of force being generated by the muscle) causes the inhibition of the motor neuron that innervates the same muscle. Similarly, feedback inhibition is useful when the output of a higher-order level of a circuit is needed to shut down the input activity to that circuit (Fig. 18.11E). Finally, lateral inhibition among neurons at the same level of processing is a common feature of many circuits, in which the activity of one set of neurons can shut down the activity of

other neurons at the same level (Fig. 18.11F). As we shall see later, feedforward, feedback, and lateral inhibition can serve other useful functions when they are incorporated into more complex neural circuitry.

Disinhibition and Rebound Excitation

In an apparent paradox, inhibitory neurons can also be used to indirectly cause net excitation in a circuit. A neuron may receive both excitatory and inhibitory inputs but may be silent because the inhibitory input is stronger than the excitatory input. However, if the inhibitory neuron is itself inhibited, the excitatory input then causes the output neuron to fire. This process is called *disinhibition*. For example, Figure 18.11G illustrates a circuit in which neuron 2 receives excitatory input from neuron 1 and inhibitory input from inhibitory interneuron 1i. If the inhibition is equal to the excitation, neuron 2 will not fire many spikes. However, if inhibitory interneuron 3i is excited by neuron 3, it will inhibit interneuron 1i, shutting it down and allowing neuron 1 to drive neuron 2. Thus, the activity of inhibitory interneuron 3i results in a net excitatory output of the circuit by disinhibiting output neuron 2.

In many cases, a neuron that is tonically inhibited will fire action potentials when the inhibition is released, even in the absence of external excitation (i.e., neuron 2 will fire even if neuron 1 is silent). In such cases, biophysical mechanisms within the output neuron itself cause the cell to fire when it becomes disinhibited. This process is called *rebound excitation*. Thus, disinhibition and rebound excitation are two mechanisms that can use inhibitory neurons in a circuit to produce net excitatory output. A prime example is the output of the cerebellum, where Purkinje cells fire tonically to inhibit the neurons in the cerebellar output nuclei (see later). When the Purkinje cells reduce their firing rate, the output nucleus cells are disinhibited and released to relay their signals to their efferent structures.

Recurrent Connections and Processing Loops

Some brain circuits make use of recurrent circuitry, which is a hybrid between feedforward, feedback, and lateral circuitry (Fig. 18.11 H). In recurrent connections, the output neurons of a circuit make connections with themselves (Fig. 18.11 Ha). Thus, the neurons can be thought to be sending feedforward connections to themselves, but these connections also have the properties of feedback connections that can alter their own output. Recurrent connectivity may also occur in a more indirect fashion, in which chains of feedforward connections loop backward and eventually connect back onto the initial processing stages, forming a processing loop (Fig. 18.11 Hb). In this case, the computations reflected in the output of the loop can alter the processing that occurs within the loop. These forms of recurrent circuitry reflect a powerful neural architecture that allows neural circuits to engage in complex, nonlinear processing that is thought to underlie much of the sophisticated dynamics of neural information processing.

Convergence and Divergence

In complex circuits composed of large numbers of neurons, such as in neocortex, an important principle is the degree of convergence and divergence in the connectivity. *Convergence* refers to the degree to which a neuron receives input from large numbers of other neurons. *Divergence* refers to the degree to which a neuron projects to large numbers of target neurons (Fig. 18.11I). To illustrate these principles, consider two

brain areas that are connected to each other. If a neuron in area A projects to hundreds or thousands of neurons in area B (one-to-many connection), then the output of that neuron can influence the activity of a large fraction of the neurons in the downstream structure. If the neuron projects to only one or a few neurons in area B (one-to-one connection), then it can affect only a small proportion of neurons. The former case is considered to be a divergent connection pattern, whereas the latter case is a restricted connection pattern. Conversely, each target neuron in area B can receive input from a single neuron in area A, or it can receive input from thousands of neurons in area A. The latter case is considered to display a highly convergent pattern. Such a neuron is able to integrate input from many different upstream neurons and have its activity affected by countless patterns of activity from the input neurons. At this level of organization, each neuron may take part in a mixture of divergent and convergent connectivity schemes. That is, area A may send highly divergent connections to area B, but area B neurons receive input from only a very few neurons. Similarly, area A may send restricted projections to area B, but each area B neuron receives thousands of such restricted connections. Finally, both the area A and area B connections may be highly localized or highly divergent and convergent. The degree of divergence and convergence reflects the requirements of different parts of the circuit to broadcast widely the outputs of their computations and integrate information from many different sources in order to perform the computations.

Parallel Processing

An important principle in neural circuitry is parallel processing. Unlike a strictly serial computer, in which the execution of each line of code has to be completed before the next line of code can be executed, the brain operates more like a parallel processing computer, in which many lines of code are executed simultaneously (Fig. 18.11 J). For example, in the visual system, there are parallel streams of processing that start in the retina and remain somewhat segregated through many stages of processing (i.e., through the LGN, primary visual cortex, and higher cortical areas; Livingstone and Hubel, 1988; Ungerleider and Mishkin, 1982). One stream of processing starts out with retinal cells that respond to low-contrast stimuli (the M pathway, named for the magnocellular layer of the LGN) and is the start of the stream that eventually forms the dorsal pathway (the "where" pathway) in the parietal cortex. Another stream starts out with retinal cells that respond to color stimuli (the P pathway, named for the parvocellular layer of the LGN) and is part of the stream that eventually forms the ventral pathway (the "what" pathway) in the inferotemporal cortex. Thus, the brain processes in parallel information about the identity of visual objects as well as information about the location of visual objects (although there is a large amount of cross talk between these parallel streams; DeYoe and Van Essen, 1988). Each processing stream, in turn, is composed of substreams that mix a combination of serial and parallel processing. In ways that are not well understood, these parallel streams are somehow integrated to form a unitary perception. Such parallel processing occurs in all sensory and motor systems of the brain, and presumably allows different parts of a task to be divided and processed with specialized circuitry appropriate for the task, in a faster and more efficient manner than would be possible with a strictly serial processing architecture.

Emergent Properties of Complex Networks

The iconic circuits of Figures 18.11A–J are the building blocks used to create vastly complicated networks required for the enormously sophisticated processing performed by the nervous system. Even the simple stretch reflex of Figure 18.2 has numerous elaborations (not shown) that are required to make the reflex work appropriately under countless behavioral conditions (e.g., collaterals that inhibit the antagonist muscle of the limb). One principle that has been derived from the study of complex networks of thousands of neurons is that these networks display emergent properties that are not present in the simple components. An example of an emergent property in the realm of thermodynamics is temperature. The temperature of a gas is a reflection of the average energy of motion of its component molecules. Each molecule itself does not have a temperature; rather, temperature is an emergent feature of the collective activity of all the molecules in the gas. Similarly, neural networks can display emergent phenomena that arise only from the collective activity of many neurons. For example, memories can be stored in a distributed fashion among many synapses in a network. No single neuron stores the memory. Instead, the memory can be recalled only by the collective activity of the entire network.

PLASTICITY

Neural circuits are so varied and complex, typically encompassing hundreds or thousands of neurons and connections, that they usually cannot be genetically hard-wired. Rather, neural circuits are remarkably plastic, as they continually evolve over the lifespan of the organism. Thus, neural plasticity, which was once thought to be the exclusive realm of development and learning phenomena, is now considered to be important for the normal functioning of almost any neural circuit. As the organism changes during the lifespan, by growing bigger, fatter, older, less flexible, and so on, brain circuits must adapt in order to continue to meet the needs of the organism. Thus, if eyesight deteriorates and glasses must be worn, the neural circuits of the vestibulo-ocular reflex must adapt to maintain steady gaze in the face of changing optics. If muscle elasticity or bone mass changes, circuits in the spinal cord and motor system must adapt to produce appropriately timed and forceful movements. Neural plasticity continually adjusts the circuit to calibrate its performance in the face of continual environmental and body changes. This plasticity fine-tunes the response properties of neurons in order to optimize them for the particular statistics of their inputs. Thus, neural representations of particularly important auditory frequencies or motor outputs, for example, will become more finely tuned and occupy larger areas of the brain as the result of experience (Buonomano and Merzenich, 1998).

LTP, LTD, and STDP

Long-term potentiation (LTP) is the most well-characterized form of synaptic plasticity. Originally described in the hippocampus and long associated with the learning and memory functions of that structure (Bliss and Lomo, 1973), various forms of LTP have since been identified in numerous brain regions, including the neocortex (Malenka and Bear, 2004). As described more thoroughly in Chapter 19, "classic" LTP requires activation of the NMDA receptor (NMDAR) on the postsynaptic membrane, which opens a pore to allow Ca^{++} ions to enter the cell. The NMDA receptor acts as a coincidence detector of presynaptic and postsynaptic cell activity, as the channel opening requires not only the binding of glutamate (released by the presynaptic cell) but also the removal of a magnesium block by the depolarization of the postsynaptic cell. This property allows NMDAR-dependent LTP to fulfill Hebb's postulate, which stipulated that the connection strength between two neurons should increase if the presynaptic neuron continually takes part in causing the postsynaptic neuron to fire (Hebb, 1949).

The flip side of long-term potentiation is long-term depression (LTD) (Dudek and Bear, 1992). Neurons also need a mechanism to decrease the strength of their synaptic connections. In some cases, this weakening occurs because two cells fire in an uncorrelated manner. Working together, LTP and LTD allow cells that fire in a correlated manner to increase their connection strength with each other, while decreasing the strength of connectivity with neurons that do not take part in the current activity pattern. From such a mechanism, the Hebbian notion of a cell assembly forms: populations of neurons that excite each other and form a neural representation of an external stimulus, body movement, or internal cognitive state (Hebb, 1949).

Hebb's postulate of synaptic strengthening implied a temporal order rule for spike firing. The synapse between neuron A and neuron B should strengthen only if neuron A fired before neuron B. If neuron A fired after neuron B, then it could not have been involved in causing neuron B's activity, and therefore its connection to neuron B should not be strengthened. Early work on LTP demonstrated that this temporal asymmetry was indeed a property of LTP in the hippocampus (Levy and Steward, 1983). More recent work showed that LTP can be elicited at some synapses if a single presynaptic spike precedes a single postsynaptic spike (within a ~20 msec time window), whereas LTD can be elicited if the presynaptic spike follows the postsynaptic spike (Markram *et al.*, 1997). This phenomenon, called *spike timing-dependent plasticity* (STDP), may allow circuits to organize into what Hebb called a *phase sequence* (Hebb, 1949). That is, if one cell assembly reliably causes a second cell assembly to fire, the neurons in the first assembly strengthen their connections to the neurons in the second assembly. Future activation of the first assembly could then automatically activate the second assembly, which would activate a third assembly, and so on. Such mechanisms might allow neural circuits to store sequences of activation patterns in the network, which could endow the circuits with the ability to predict future brain states on the basis of current sensory input or brain activity. This ability would be useful in learning stimulus–stimulus associations; learning stimulus–response associations; adapting sensory systems to tune their response profiles to the dynamic, statistical regularities of their inputs; and storing sequences of experience in a way that promotes subsequent episodic memory retrieval.

Homeostasis and Synaptic Scaling

Theoretical work on neural networks has demonstrated that a major problem with neural plasticity is that highly interconnected networks are susceptible to positive feedback loops that cause runaway excitation. That is, if two neurons that fire together increase

their synaptic strength, this would cause them to fire even more strongly together, further increasing their strength of connection and so on. Eventually, as the firing of the cells becomes greater, they can cause other cells to reach threshold and fire, resulting in the incorporation of the new cells into the population assembly. Eventually, the network would become saturated, as all cells become active. Similar problems, in the reverse direction, can occur with runaway depression, leading to a complete lack of activity in the network. To prevent these states, neural circuits have mechanisms that keep their overall firing rates within certain ranges that are presumably optimal for the circuit's processing (Turrigiano, 1999). We have already encountered one such mechanism, feedback inhibition (Fig. 18.11E). Another mechanism, synaptic scaling, causes the nonspecific depression or augmentation of large numbers of synapses in a circuit based on the overall activity level of the circuit, in order to keep the network activity within a prescribed range (Turrigiano *et al.*, 1998). Such homeostatic mechanisms are increasingly recognized as crucial components in the functioning of neural circuits in both invertebrate and vertebrate nervous systems.

Excitability

Since the discovery of LTP in the 1970s, most work on plasticity in neural circuits has focused on synaptic plasticity. However, other cellular mechanisms also play an important role. A key mechanism is the alteration of neuronal excitability. A neuron can fire more spikes in response to a given set of inputs if its membrane properties have changed in a way to make it hyperexcitable. Recent work on *Aplysia, Hermissenda,* and CA1 pyramidal cells has demonstrated that changes in membrane excitability can play a large role in addition to synaptic plasticity in tuning the function of a neural circuit (Brembs *et al.*, 2002; Crow, 2004; Disterhoft and Oh, 2006).

EXAMPLE CIRCUITS

We end this chapter by describing a number of real neural circuits that illustrate some of the concepts elucidated earlier. The circuits range from relatively simple invertebrate preparations, in which individual cell types and connectivity patterns can be identified, to more complex mammalian circuits, in which large numbers of neurons form complex neural networks whose connectivity patterns can be described only in statistical terms.

Aplysia Gill Withdrawal Reflex

One of the classic circuits studied in neuroscience is the siphon–gill withdrawal circuit of the marine snail *Aplysia* (Fig. 18.12). This circuit mediates a sensorimotor reflex similar to the stretch reflex of mammals mentioned earlier. The circuit contains ~24 sensory receptors that innervate the siphon, a spoutlike organ used to expel seawater and waste (Kandel, 2000). The sensory receptors synapse onto motor neurons that innervate the gill. They also innervate both excitatory and inhibitory interneurons. Thus, the circuit contains both feedforward excitation and inhibition. When the siphon is stimulated with a tactile stimulus, the sensory neurons fire and cause the motor neurons to fire, resulting in the defensive gill-withdrawal reflex. This circuit shows adaptive plasticity. After repeated stimulation of the siphon, the gill withdrawal becomes weaker, as the result of less transmitter release at the synapses onto the motor neuron. This form of learning is called *habituation*. A further aspect to the circuit involves input from the tail region of the animal. A strong shock to the tail causes subsequently larger gill withdrawal responses to the siphon stimulation, a form of learning called *sensitization*. This learning results from changes in the strength of synaptic connections between sensory neurons and their follower interneurons and motor neurons. (See Chapter 19 for a more detailed treatment of the cellular and biochemical mechanisms underlying learning in this circuit.)

Lobster Stomatogastric Ganglion

The stomatogastric ganglion (STG) of lobsters and crabs has been extensively studied as a model central pattern generator (CPG). CPGs are neural circuits that produce stereotyped, rhythmic patterns of activity in the absence of sensory feedback. The STG is composed of approximately 30 identified neurons, which are connected primarily by inhibitory synapses (Selverston *et al.*, 1998). There are two primary rhythms associated with the STG, the pyloric rhythm and the gastric mill rhythm. The pyloric rhythm is produced by three groups of neurons that fire rhythmic bursts of action potentials in a specific temporal order relative to each other. This circuit is associated with the pyloric muscle, and the pyloric rhythm is used to constrict and dilate the pylorus in order to filter food particles as they pass to the midgut. The gastric mill rhythm is a more complex rhythm used to grind the internal "teeth" of the gastric mill to pulverize the food.

The essential principles of the pyloric rhythm circuit can be illustrated in a simplified wiring diagram (Fig. 18.13). The anterior burster–pyloric dilator (AB–PD)

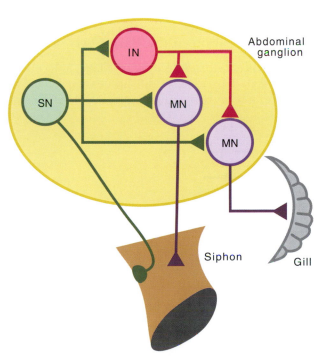

FIGURE 18.12 Circuitry of the *Aplysia* siphon-gill withdrawal circuit. A stimulus to the siphon causes the gill and siphon to withdraw in a reflex action. After repeated stimulation, the behavioral response decreases because less transmitter is released from the sensory neurons. This simple form of learning is called *habituation*.

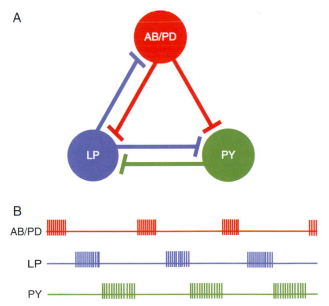

FIGURE 18.13 Pyloric rhythm circuit of the lobster stomatogastric ganglion. (A) Schematic diagram of the three cell types that produce the rhythm. Note that the synapses are inhibitory. (B) The cells fire in a stereotyped sequence of activity, with the electrotonically coupled AB/PD cells firing first, followed by the LP cells, and finally the PY cells.

component represents an electrotonically coupled set of neurons that acts as a pacemaker for the circuit. These neurons are connected to the lateral pyloric (LP) and pyloric (PY) neurons through inhibitory synapses. Intrinsic bursting properties, coupled with the inhibitory synapses, combine to produce the stereotyped, triphasic, oscillatory output of the circuit shown in Figure 18.13. The prevalence of mutually inhibitory circuitry illustrates the use of rebound excitation in the performance of neural circuits. The AB–PD neurons appear to start the process, firing with an intrinsic rhythm and inhibiting the other neurons in the circuit. When the AB–PD neurons stop firing (due to intrinsic cellular mechanisms), the LP and PY neurons are released from inhibition. A combination of intrinsic and circuit parameters in these cells determines that the LP neuron will fire first, followed by the PY neuron, before the AB–PD pair begins to fire again and the next cycle of the rhythm begins. How the different cellular and network parameters combine to produce this rhythm is an area of intense investigation at cellular, physiological, and computational levels of analysis (Prinz *et al.*, 2004; Soto-Trevino *et al.*, 2001).

Locust Antennal Lobe–Mushroom Body

The antennal lobe of the insect brain is the analog of the olfactory bulb in vertebrates (Laurent, 2002). Its projection neurons (PNs) synapse onto Kenyon cells (KCs) in the mushroom body, which is involved in olfactory memory (Fig. 18.14). The PNs encode odors using a temporal, coarse coding scheme, in which each neuron responds to many different odors with a characteristic pattern of temporally modulated

responses (Fig. 18.7; Perez-Orive *et al.*, 2002). The KCs, on the other hand, employ a sparse coding scheme, in which each cell selectively fires a brief burst of spikes to only a small number of odors. As discussed earlier, such sparse coding is an advantageous strategy for a memory system that needs to store a number of different patterns with minimum interference. Thus, the PN–KC synapse embodies a transformation from a coarse code to a sparse code. This transformation is apparently accomplished with the aid of a feedforward inhibitory circuit. In addition to their projections to the mushroom bodies, the PNs send collaterals to an area called the *lateral horn* (LH). The LH cells send inhibitory projections to the mushroom bodies. LH cells are very broadly tuned to odors, responding to many odors with little discrimination. Because KCs receive inputs from only a small fraction of PNs, and because the PN responses are temporally modulated, the excitatory drive onto KCs rarely reaches firing threshold. The feedforward inhibition from the LH neurons magnifies this effect. The combination of these properties ensures that the KCs will fire only rarely and very selectively, which presumably allows the mushroom body to store large numbers of olfactory memories in a distributed code with little interference among the many different stored memories (Laurent, 1999; Perez-Orive *et al.*, 2002).

Sound Localization in the Barn Owl

Barn owls are extraordinarily proficient at localizing prey in the dark by using auditory cues to calculate the location of a sound source (Konishi, 2003). Sounds are

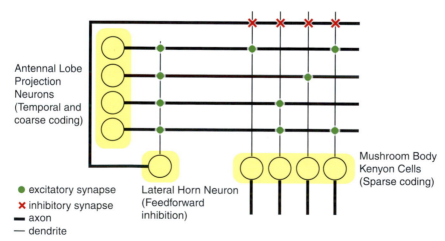

FIGURE 18.14 Circuit of the locust antennal lobe–mushroom body system. Antennal lobe projection neurons make sparse connections with Kenyon cells of the mushroom body and strong connections with feedforward inhibitory neurons of the lateral horn. The combination of temporally specific firing, sparse connectivity, and strong feedforward inhibition turns the coarse-coded antennal lobe representation into a sparse-coded mushroom body representation.

localized in the vertical plane (elevation) by differences in intensity between the right and left ear (due to an anatomical asymmetry between the owl's ears). Sounds are localized in the horizontal place (azimuth) by differences in the arrival of the sound waves between the two ears. That is, a sound originating on the animal's left will arrive at the left ear before it arrives at the right ear. This interaural time difference is the basis for calculating the azimuthal coordinate of the sound source (Fig. 18.15; Carr and Konishi, 1988, 1990).

Neurons in the cochlear nucleus magnocellularis (NM) project bilaterally to the brain-stem nucleus laminaris (NL). NM neurons are tuned to the frequency of the auditory stimulus (Fig. 18.5C), and a tonotopic map of frequency is present in the nucleus. Neurons project bilaterally to the right and left NL. However, the ipsilateral projection enters the dorsal part of NL and the axons run roughly parallel within the nucleus. The contralateral projection, in contrast, enters the ventral part of NL and the axons run parallel with the ipsilateral projections, but interdigitated

with them and running in the opposite direction. The result of the anatomical organization of this circuit is that the axons act as delay lines that allow the NL neurons to become selective for precise interaural time differences. As a result, each NL contains a map of contralateral and frontal space (frontal space is mapped on both sides). That is, if a sound source is located directly in front of the animal, both ears will be activated simultaneously, and only a neuron near the ventral border of each NL will fire, as the action potentials from the ipsilateral and contralateral projections reach it simultaneously (i.e., it acts as a "coincidence detector" of right and left ear activity). If the sound is to the right of the animal, causing the right NM cell to fire before the left NM cell, then the left NL neuron tuned to no delay will not fire because, by the time the left action potentials have arrived, the right action potentials have already passed by. Note, however, that a neuron in a more dorsal part of the left NL will be perfectly situated to receive the simultaneous activity from both ears. This neuron will thus become selective for interaural

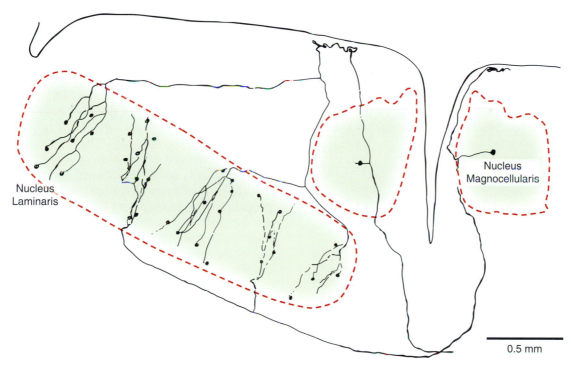

0.5 mm

FIGURE 18.15   Delay lines and coincidence detection in the barn owl sound localization system. Axons from the ipsilateral nucleus magnocellularis innervate the nucleus laminaris from the dorsal side. Axons from the contralateral nucleus magocellularis innervate the nucleus laminaris from the ventral side and send interdigitating projections parallel to the contralateral projections. Individual neurons in the nucleus laminaris receive input from both projections. Because the contralateral and ipsilateral projections convey sound information from different ears, the cells in nucleus laminaris are excited when they receive coincident input from both ears. Cells at the dorsal part of the nucleus will receive coincident input when the contralateral ear receives sound before the ipsilateral ear, as it takes longer for the signal to be transmitted over the longer axons. Cells in the ventral part of nucleus laminaris will receive coincident input when the contralateral and ipsilateral ears receive sound simultaneously. Thus, the nucleus laminaris contains an auditory map of contralateral and frontal space, computed from interaural time differences. The owl uses this representation to detect the location of the sound source in the azimuthal (horizontal) plane. Adapted with permission from Carr and Konishi (1988).

time differences in which the right ear is activated first. Thus, the NL–NM circuit produces a transformation of a representation of auditory stimulus frequency to a representation of interaural time differences, which are smoothly mapped along the dorsal–ventral axis of the NM (tone is mapped along the orthogonal axis). This mapping produces a place code, in which the dorsal–ventral location of the active neuron signifies the computed time difference. Downstream structures combine this time-difference code with the intensity-difference code to create a map of sound location in two dimensions, which ultimately allows the barn owl to zero in on its prey in total darkness.

Orientation Tuning in V1

One of the most extensively studied transformations of neural representation in the mammalian brain is the generation of orientation selectivity in the primary visual cortex (Fig. 18.16). Neurons in the lateral geniculate nucleus (LGN) of the thalamus are not selective for stimulus orientation, as they respond to bars or gratings of all orientations. They have circular receptive fields with a center-surround organization, such that the cell is excited by the onset of spots of light in the center of the receptive field and inhibited by spots of light in the surrounding annulus. If a small spot of light is presented only in the center, the cell responds vigorously. If the spot of light is made larger, such that it encroaches on both the center and surround regions, the excitatory center and the inhibitory surround cancel each other, and the cell responds weakly (if at all). This center-surround organization suggests that the LGN cells encode the local contrast of light intensity, rather than the absolute value of light intensity, which is a property necessary for the visual system to maintain constant visual perceptions in the face of orders of magnitude of variation of overall background illumination (e.g., the difference between a darkly lit room and a bright, sunny day).

If the LGN receptive field is stimulated by a bar of light, the cell will respond to bars of all orientations, provided that the bar of light encroaches on its excitatory center. However, in cats, the cells that receive input from the LGN, layer IV of the primary visual cortex, demonstrate highly selective responses for particular orientations. What is the circuit that transforms the orientation-nonselective representation of the LGN to the orientation-selective representation of the cortex? David Hubel and Torsten Wiesel, in their pioneering work on the response properties of visual cortex that earned them a Nobel Prize, proposed that the orientation selectivity could arise from a particular circuit-wiring diagram (Hubel and Wiesel, 1962). If a cortical cell received input from particular LGN cells that had their receptive fields aligned in a row, the cortical cell would fire most strongly to an oriented bar of light that fell along that line (Fig. 18.16). If the bar of light encroached on all of the LGN excitatory centers, this would provide maximum excitatory drive to the cortical cell. If the bar of light was rotated away from the preferred orientation, it would fall onto fewer of the excitatory center regions and encroach on more of the inhibitory surround regions of the LGN cells, thus providing weaker excitatory drive to the cortical cell. If the bar of light were orthogonal to the preferred orientation, then the excitatory and inhibitory regions of the LGN inputs would be roughly equal (or the inhibition would exceed the excitation), and the cell would not fire at all. According to this model, the geometric pattern of connectivity of specific LGN cells onto specific cortical cells causes the transformation of a representation of local contrast in luminance to a representation of the orientation of a small patch of the visual field, which is a step toward the construction of a representation of the figures and background of a visual scene.

Because of the complexity of the cortex, it has been difficult to determine whether the Hubel and Wiesel model is the actual mechanism by which the brain generates orientation selectivity. However, over the years a number of experiments have provided strong indirect evidence in support of the model. For example, recordings from the LGN fibers that innervate a

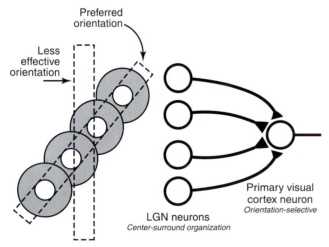

FIGURE 18.16 Model of orientation tuning proposed by Hubel and Wiesel (1962). The center-surround receptive fields of LGN neurons are shown on the left. If a primary visual cortex neuron received input only from LGN cells that were aligned in a particular orientation, then the cortical neuron would be maximally excited when a bar of light of the same orientation was presented. Bars of light at other orientations would excite fewer of the LGN afferents, causing the cortical cell to fire at a lower rate.

particular patch of cortex have receptive fields that align along the preferred orientation of the cortical cells in that patch, which is a strong prediction of the Hubel and Wiesel model (Chapman *et al.*, 1991). Further enhancements of the model have been proposed that incorporate feedforward inhibition, recurrent excitation, and both lateral and feedback connectivity to account for other properties displayed by orientation-selective neurons. For example, the width of cortical tuning curves remains constant with increasing contrast of the visual image, even though the responses of LGN cells increase and the Hubel–Wiesel model predicts that the tuning curve should become broader. A particular form of feedforward inhibition, however, in which LGN cells with opposite polarity (i.e., inhibitory centers and excitatory surrounds) make inhibitory connections onto the same cortical cells that receive excitatory connections from their complementary LGN cells, has been proposed as a mechanism that allows the cortical cells to maintain narrowly tuned responses over a large range of luminance contrast (Ferster and Miller, 2000). Thus, the feedforward circuitry may provide the initial basis for orientation selectivity, which is then fine-tuned by other circuit enhancements to produce the contrast-invariant responses that are necessary for the extraction of orientation over wide ranges of stimulus contrast.

Cerebellum

The cerebellum is a remarkable structure that encompasses only 10% of the volume of the human brain but contains more than half the brain's neurons. The circuit diagram of the cerebellum is relatively simple and nearly identical throughout the entire structure (Fig. 18.17). The inputs to the cerebellum, in contrast, are quite variable, including both motor systems and multiple sensory systems. The combination of a wide variety of input systems and a stereotyped, repeating pattern of internal connectivity suggests that the entire cerebellum is involved in one particular type of computation that is performed on input from multiple brain systems and then relayed back to those systems (Eccles *et al.*, 1967).

The cerebellum receives two basic types of inputs (Eccles *et al.*, 1967). Mossy fibers originate mainly from the pontine nuclei and the spinal cord, and they terminate on the cerebellar deep nuclei (the sole output of the cerebellum) and on granule cells of the cerebellar cortex. Each mossy fiber innervates hundreds of granule cells. The granule cells, in turn, send axons

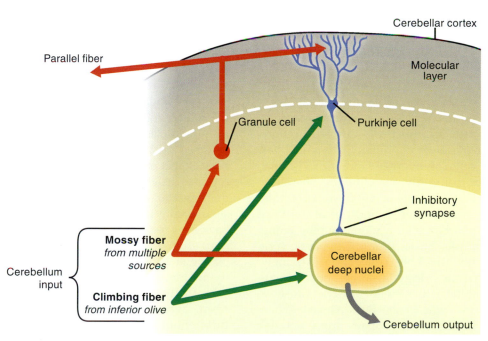

FIGURE 18.17 Schematic diagram of some main features of cerebellum circuitry. The cerebellum receives input from mossy fibers and climbing fibers. The mossy fibers synapse onto cerebellar deep nuclei cells and onto granule cells of the cerebellar cortex, which send parallel fibers that synapse onto the inhibitory output neurons of the cortex, the Purkinje cells. Climbing fibers synapse onto deep nuclei cells and also make powerful synapses directly onto Purkinje cells and act as an error signal to drive cerebellar learning. (See Chapter 19 for a more detailed treatment of the circuitry of the cerebellum and its role in learning.)

to the surface of the cerebellar cortex that bifurcate and extend along the surface of the cortex in one dimension, parallel with the axons of all the other granule cells (hence, they are called *parallel fibers*). The parallel fibers make *en passant* synapses onto the elaborately branched Purkinje cells, which have two-dimensional dendritic arbors that are oriented perpendicularly to the parallel fibers. Each parallel fiber contacts hundreds of Purkinje cells. Thus, because of the enormous divergence of connectivity in the mossy fiber–granule cell synapse and the parallel fiber–Purkinje cell synapse, and the enormous convergence of input from thousands of parallel fibers onto a single Purkinje cell, the activity of the Purkinje cell can potentially be influenced by hundreds of thousands of mossy fiber inputs.

Because of the large number of inputs, Purkinje cells fire at tonically high rates (Eccles *et al.*, 1967; Gilbert and Thach, 1977). The synapse from Purkinje cells to cerebellar nucleus cells is inhibitory, so the Purkinje cells tonically inhibit the output of the cerebellum. In order for the nucleus cells to fire, the inhibition must be reduced by an appropriately timed pause in Purkinje cell firing. Plasticity at the parallel fiber–Purkinje cell synapse appears to tune each Purkinje cell to select specific combinations of mossy fiber–parallel fiber inputs, from the countless numbers of possible combinations (Marr, 1969). If each mossy fiber represents the value of a single parameter about the state of the body or about external stimuli, the Purkinje cell can tune itself to represent a specific combination of these parameters that has particular adaptive significance to the organism, and to reduce its firing rate when this specific combination of inputs occurs.

How does this selection of inputs occur? The other major type of input to the cerebellum, the climbing fibers that originate in the inferior olivary nucleus of the brain stem, plays a critical role in this selection. Unlike mossy fibers, each climbing fiber innervates only 10 Purkinje cells, on average, and each Purkinje cell receives input from only one climbing fiber. The climbing fiber makes multiple synaptic contacts onto the Purkinje cell as it weaves in and out of the Purkinje cell dendritic arbor, like a vine climbing along the branches of a tree. These multiple synaptic contacts make the climbing fiber connection extremely powerful, such that the firing of a single action potential in the climbing fiber will always cause a specialized, calcium spike in the Purkinje cell. When this calcium spike occurs, synapses from parallel fibers that are active at the same time become weaker, through a cerebellum-specialized LTD mechanism. The climbing fibers are thought to convey an "error signal" to the cerebellum, in the engineering sense of the term. That is, whenever a system is not producing

appropriate behavior, the mismatch between the desired output and the actual output (the error) is reported to the cerebellum by the climbing fibers (Gilbert and Thach, 1977). The mossy fibers convey information about the intended movement, the state of the environment, and the state of the body at that time, and the synapses from active parallel fibers either weaken (through LTD) if there is an increase in climbing fiber spikes or strengthen (through LTP) if there is a decrease in climbing fiber spikes. Thus, whenever that particular combination of mossy fibers is activated in the future, the Purkinje cell activation will be altered to increase the probability that the cerebellum nucleus output produces a movement that is closer to the intended movement (Medina *et al.*, 2002; Ohyama *et al.*, 2003; Raymond *et al.*, 1996).

Hippocampus Recurrent Collateral Circuitry

The hippocampus is a brain structure that is critical for spatial learning, contextual learning, and episodic memory (Nadel *et al.*, 1985; O'Keefe and Nadel, 1978; Vargha-Khadem *et al.*, 1997). It appears to receive a spatial signal from the medial entorhinal cortex (Hafting *et al.*, 2005) and a nonspatial signal from the lateral entorhinal cortex (Hargreaves *et al.*, 2005), and combines these signals into a conjunctive representation of objects in space (or events in context) critical for episodic memory (the conscious recollection of events from one's past). The hippocampus contains numerous parallel pathways. In addition to the well-known "trisynaptic loop" pathway from the entorhinal cortex → dentate gyrus → CA3 → CA1, there are other direct pathways from EC → CA3 and EC → CA1 (Witter and Amaral, 2004). The functions of these parallel pathways are unknown. However, a prominent feature of hippocampal circuitry that has received great attention from theoretical neuroscientists is the recurrent collateral circuitry of the CA3 region. CA3 pyramidal cells send axons to CA1 in the Schaffer collateral system, but these axons also send recurrent collaterals that make excitatory feedback synapses onto the same CA3 cells. This circuitry may endow CA3 with the properties of an autoassociative network (Fig. 18.18). That is, patterns of activity within the CA3 network can become associated with themselves through a Hebbian learning mechanism, which could cause activity to reverberate throughout the network and maintain a stable state of sustained activity in the network (a so-called "attractor" state) in the absence of external excitation into the system. This property has generated considerable theoretical interest because it appears to have many properties that are thought to represent

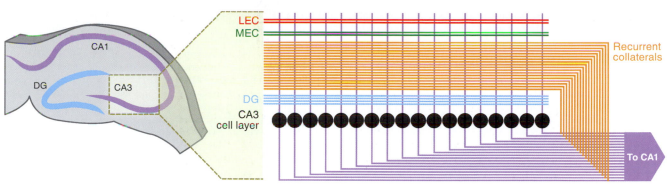

FIGURE 18.18 Schematic diagram of the recurrent collateral system of the CA3 field of the hippocampus. CA3 neurons receive direct input from the medial entorhinal cortex (MEC) and the lateral entorhinal cortex (LEC), as well as a strong projection from the dentate gyrus (DG). CA3 sends output to CA1 (the Schaffer collaterals) but also sends recurrent collaterals that innervate other CA3 cells. This recurrent collateral system has been hypothesized to endow the CA3 network with the properties of an autoassociative memory system, including the ability to reconstruct complete output patterns when presented with only partial input.

hallmarks of memory processing (Marr, 1971; McNaughton and Morris, 1987; Rolls and Treves, 1998).

One such property is called *pattern completion*. Consider a situation in which information about salient landmarks in an environment enters the hippocampus (perhaps through the lateral entorhinal pathway). This information is combined with spatial input from the medial entorhinal pathway, and through a Hebbian learning process, these inputs drive a pattern of CA3 cells that forms a unique representation of that particular environment. An important consequence of the recurrent collateral circuitry is that the CA3 representation also becomes associated with itself, again through Hebbian learning. This learning makes the CA3 representation robust to degradation in the quality of the input patterns. In the future, if the animal enters the same environment under circumstances in which some of the landmarks are occluded or degraded (e.g., in darkness), the entorhinal input will be degraded and will cause only a fraction of the CA3 representation to be activated. However, the active CA3 cells will be able to activate the remaining CA3 cells to complete the full neural pattern. (See Chapter 19, Figure 19.42 for further discussion of CA3-inspired artificial neural networks.) Although these ideas have generated considerable theoretical interest (Marr, 1971; McNaughton and Morris, 1987; Rolls and Treves, 1998), it is only in recent years that they have undergone rigorous experimental tests in behaving animals (Lee *et al.*, 2004; Leutgeb *et al.*, 2004, 2005, 2007; McHugh *et al.*, 2007; Nakazawa *et al.*, 2002; Wills *et al.*, 2005). The evidence is supportive, although inconclusive, and further experimental tests will be required to determine the validity of these models.

SUMMARY

Understanding how a brain system processes information requires knowing what information is represented in the system, how the information is encoded and decoded, and what the computations are that transform the information throughout different stages of processing. Neural systems can represent information in a variety of ways, including topographic maps of stimulus dimensions as well as nontopographic representations. Neurons also use a variety of coding schemes, including rate codes, temporal codes, and population codes. Any encoding scheme is constrained by the ability of downstream neurons to decode the information. Different neural circuits have evolved to accomplish the information processing tasks of the nervous system. The building blocks of these circuits include feedforward, feedback, and lateral connections; excitatory and inhibitory connections; and recurrent collaterals. Circuits show varying degrees of divergence, convergence, and parallel processing, depending on the information processing task at hand. Because the systems need to adapt to countless behavioral and perceptual conditions, plasticity is ubiquitous throughout most brain systems. These principles are exemplified in a number of well-studied neural circuits, from both vertebrate and invertebrate preparations.

References

Barlow, H. B. (1972). Single units and sensation: a neuron doctrine for psychology? *Perception* **1**, 371-394.

Blasdel, G. G., and Salama, G. (1986). Voltage-sensitive dyes reveal a modular organization in monkey striate cortex. *Nature* **321**, 579-585.

Bliss, T. V., and Lomo, T. (1973). Long-lasting potentiation of synaptic transmission in the dentate area of the anaesthetized rabbit following stimulation of the perforant path. *J. Physiol.* **232**, 331-356.

Brembs, B., Lorenzetti, F. D., Reyes, F. D., Baxter, D. A., and Byrne, J. H. (2002). Operant reward learning in *Aplysia*: neuronal correlates and mechanisms. *Science* **296**, 1706-1709.

Buonomano, D. V., and Merzenich, M. M. (1998). Cortical plasticity: from synapses to maps. *Annu. Rev. Neurosci.* **21**, 149-186.

Carr, C. E., and Konishi, M. (1988). Axonal delay lines for time measurement in the owl's brainstem. *Proc. Natl. Acad. Sci. USA* **85**, 8311-8315.

Carr, C. E., and Konishi, M. (1990). A circuit for detection of interaural time differences in the brain stem of the barn owl. *J. Neurosci.* **10**, 3227-3246.

Carroll, E. W., and Wong-Riley, M. T. (1984). Quantitative light and electron microscopic analysis of cytochrome oxidase-rich zones in the striate cortex of the squirrel monkey. *J. Comp. Neurol.* **222**, 1-17.

Chapman, B., Zahs, K. R., and Stryker, M. P. (1991). Relation of cortical cell orientation selectivity to alignment of receptive fields of the geniculocortical afferents that arborize within a single orientation column in ferret visual cortex. *J. Neurosci.* **11**, 1347-1358.

Chawla, M. K., Guzowski, J. F., Ramirez-Amaya, V., Lipa, P., Hoffman, K. L., Marriott, L. K., Worley, P. F., McNaughton, B. L., and Barnes, C. A. (2005). Sparse, environmentally selective expression of Arc RNA in the upper blade of the rodent fascia dentata by brief spatial experience. *Hippocampus* **15**, 579-586.

Crow, T. (2004). Pavlovian conditioning of *Hermissenda*: current cellular, molecular, and circuit perspectives. *Learn. Mem.* **11**, 229-238.

DeYoe, E. A., and Van Essen, D. C. (1988). Concurrent processing streams in monkey visual cortex. *Trends Neurosci.* **11**, 219-226.

Disterhoft, J. F., and Oh, M. M. (2006). Learning, aging and intrinsic neuronal plasticity. *Trends Neurosci.* **29**, 587-599.

Dudek, S. M., and Bear, M. F. (1992). Homosynaptic long-term depression in area CA1 of hippocampus and effects of N-methyl-D-aspartate receptor blockade. *Proc. Natl. Acad. Sci. USA* **89**, 4363-4367.

Eccles, J. C., Ito, M., and Szentagothai, J. (1967). "The Cerebellum as a Neuronal Machine." Springer, New York.

Evarts, E. V. (1968). Relation of pyramidal tract activity to force exerted during voluntary movement. *J. Neurophysiol.* **31**, 14-27.

Felleman, D. J., and Van Essen, D. C. (1991). Distributed hierarchical processing in the primate cerebral cortex. *Cereb. Cortex.* **1**, 1-47.

Ferster, D., and Miller, K. D. (2000). Neural mechanisms of orientation selectivity in the visual cortex. *Annu. Rev. Neurosci.* **23**, 441-471.

Georgopoulos, A. P., Kalaska, J. F., Caminiti, R., and Massey, J. T. (1982). On the relations between the direction of two-dimensional arm movements and cell discharge in primate motor cortex. *J. Neurosci.* **2**, 1527-1537.

Georgopoulos, A. P., Kettner, R. E., and Schwartz, A. B. (1988). Primate motor cortex and free arm movements to visual targets in three-dimensional space. II. Coding of the direction of movement by a neuronal population. *J. Neurosci.* **8**, 2928-2937.

Gilbert, P. F., and Thach, W. T. (1977). Purkinje cell activity during motor learning. *Brain Res.* **128**, 309-328.

Gregory, R. L. (1977). "Eye and Brain: The Psychology of Seeing." McGraw-Hill, New York.

Guzowski, J. F., McNaughton, B. L., Barnes, C. A., and Worley, P. F. (1999). Environment-specific expression of the immediate-early gene Arc in hippocampal neuronal ensembles. *Nat. Neurosci.* **2**, 1120-1124.

Hafting, T., Fyhn, M., Molden, S., Moser, M. B., and Moser, E. I. (2005). Microstructure of a spatial map in the entorhinal cortex. *Nature* **436**, 801-806.

Hahnloser, R. H., Kozhevnikov, A. A., and Fee, M. S. (2002). An ultra-sparse code underlies the generation of neural sequences in a songbird. *Nature* **419**, 65-70.

Hargreaves, E. L., Rao, G., Lee, I., and Knierim, J. J. (2005). Major dissociation between medial and lateral entorhinal input to dorsal hippocampus. *Science* **308**, 1792-1794.

Hebb, D. O. (1949). "The Organization of Behavior." Wiley, New York.

Hubel, D. H., and Wiesel, T. N. (1962). Receptive fields, binocular interaction and functional architecture in the cat's visual cortex. *J. Physiol.* **160**, 106-154.

Hunt, C. C. (1990). Mammalian muscle spindle: peripheral mechanisms. *Physiol. Rev.* **70**, 643-663.

Huxter, J., Burgess, N., and O'Keefe, J. (2003). Independent rate and temporal coding in hippocampal pyramidal cells. *Nature* **425**, 828-832.

Jung, M. W., and McNaughton, B. L. (1993). Spatial selectivity of unit activity in the hippocampal granular layer. *Hippocampus* **3**, 165-182.

Kaas, J. H., Nelson, R. J., Sur, M., Lin, C. S., and Merzenich, M. M. (1979). Multiple representations of the body within the primary somatosensory cortex of primates. *Science* **204**, 521-523.

Kandel, E. R. (2000). Cellular mechanisms of learning and the biological basis of individuality. In "Principles of Neural Science" (E. R. Kandel, J. H. Schwartz, and T. M. Jessell, Eds.), pp. 1247-1279. McGraw-Hill, New York.

Kilgard, M. P., and Merzenich, M. M. (1998). Cortical map reorganization enabled by nucleus basalis activity. *Science* **279**, 1714-1718.

Konishi, M. (2003). Coding of auditory space. *Annu. Rev. Neurosci.* **26**, 31-55.

Kuffler, S. W. (1953). Discharge patterns and functional organization of mammalian retina. *J. Neurophysiol.* **16**, 37-68.

Laurent, G. (1999). A systems perspective on early olfactory coding. *Science* **286**, 723-728.

Laurent, G. (2002). Olfactory network dynamics and the coding of multidimensional signals. *Nat. Rev. Neurosci.* **3**, 884-895.

Lee, I., Yoganarasimha, D., Rao, G., and Knierim, J. J. (2004). Comparison of population coherence of place cells in hippocampal subfields CA1 and CA3. *Nature* **430**, 456-459.

Leutgeb, J. K., Leutgeb, S., Moser, M. B., and Moser, E. I. (2007). Pattern separation in the dentate gyrus and CA3 of the hippocampus. *Science* **315**, 961-966.

Leutgeb, J. K., Leutgeb, S., Treves, A., Meyer, R., Barnes, C. A., McNaughton, B. L., Moser, M. B., and Moser, E. I. (2005). Progressive transformation of hippocampal neuronal representations in "morphed" environments. *Neuron* **48**, 345-358.

Leutgeb, S., Leutgeb, J. K., Treves, A., Moser, M. B., and Moser, E. I. (2004). Distinct ensemble codes in hippocampal areas CA3 and CA1. *Science* **305**, 1295-1298.

LeVay, S., Connolly, M., Houde, J., and Van Essen, D. C. (1985). The complete pattern of ocular dominance stripes in the striate cortex and visual field of the macaque monkey. *J. Neurosci.* **5**, 486-501.

LeVay, S., Hubel, D. H., and Wiesel, T. N. (1975). The pattern of ocular dominance columns in macaque visual cortex revealed by a reduced silver stain. *J. Comp. Neurol.* **159**, 559-576.

Levy, W. B., and Steward, O. (1983). Temporal contiguity requirements for long-term associative potentiation/depression in the hippocampus. *Neuroscience* **8**, 791-797.

Livingstone, M., and Hubel, D. (1988). Segregation of form, color, movement, and depth: anatomy, physiology, and perception. *Science* **240**, 740-749.

Livingstone, M. S., and Hubel, D. H. (1984). Anatomy and physiology of a color system in the primate visual cortex. *J. Neurosci.* **4**, 309-356.

Malenka, R. C., and Bear, M. F. (2004). LTP and LTD: an embarrassment of riches. *Neuron* **44**, 5-21.

Markram, H., Lubke, J., Frotscher, M., and Sakmann, B. (1997). Regulation of synaptic efficacy by coincidence of postsynaptic APs and EPSPs. *Science* **275**, 213-215.

Marr, D. (1969). A theory of cerebellar cortex. *J. Physiol.* **202**, 437-470.

Marr, D. (1971). Simple memory: a theory for archicortex. *Philos. Trans. R. Soc. Lond. B. Biol. Sci.* **262**, 23-81.

McAdams, C. J., and Maunsell, J. H. (1999). Effects of attention on orientation-tuning functions of single neurons in macaque cortical area V4. *J. Neurosci.* **19**, 431-441.

McHugh, T. J., Jones, M. W., Quinn, J. J., Balthasar, N., Coppari, R., Elmquist, J. K., Lowell, B. B., Fanselow, M. S., Wilson, M. A., and Tonegawa, S. (2007). Dentate gyrus NMDA receptors mediate rapid pattern separation in the hippocampal network. *Science* **317**, 94-99.

McNaughton, B. L., and Morris, R. G. M. (1987). Hippocampal synaptic enhancement and information storage within a distributed memory system. *Trends Neurosci.* **10**, 408-415.

Medina, J. F., Nores, W. L., and Mauk, M. D. (2002). Inhibition of climbing fibres is a signal for the extinction of conditioned eyelid responses. *Nature* **416**, 330-333.

Monster, A. W., and Chan, H. (1977). Isometric force production by motor units of extensor digitorum communis muscle in man. *J. Neurophysiol.* **40**, 1432-1443.

Nadel, L., Wilner, J., and Kurtz, E. M. (1985). Cognitive maps and environmental context. In "Context and Learning" (P. D. Balsam and A. Tomie, Eds.), pp. 385-406. Earlbaum, Hillsdale, NJ.

Nakazawa, K., Quirk, M. C., Chitwood, R. A., Watanabe, M., Yeckel, M. F., Sun, L. D., Kato, A., Carr, C. A., Johnston, D., Wilson, M. A., and Tonegawa, S. (2002). Requirement for hippocampal CA3 NMDA receptors in associative memory recall. *Science* **297**, 211-218.

Nolte, J. (2002). "The Human Brain: Introduction to Its Functional Anatomy." Mosby, St. Louis.

O'Keefe, J., and Nadel, L. (1978). "The Hippocampus as a Cognitive Map." Clarendon Press, Oxford.

O'Keefe, J., and Recce, M. L. (1993). Phase relationship between hippocampal place units and the EEG theta rhythm. *Hippocampus* **3**, 317-330.

Ohyama, T., Nores, W. L., Murphy, M., and Mauk, M. D. (2003). What the cerebellum computes. *Trends Neurosci.* **26**, 222-227.

Pearson, K., and Gordon, J. (2000). Spinal reflexes. In "Principles of Neural Science" (E. R. Kandel, J. H. Schwartz, and T. M. Jessell, Eds.), pp. 713-736. McGraw-Hill, New York.

Penfield, W., and Erickson, T. C. (1941). "Epilepsy and Cerebral Localization." Charles C. Thomas, Springfield, IL.

Perez-Orive, J., Mazor, O., Turner, G. C., Cassenaer, S., Wilson, R. I., and Laurent, G. (2002). Oscillations and sparsening of odor representations in the mushroom body. *Science* **297**, 359-365.

Phillips, D. P., and Orman, S. S. (1984). Responses of single neurons in posterior field of cat auditory cortex to tonal stimulation. *J. Neurophysiol.* **51**, 147-163.

Prinz, A. A., Bucher, D., and Marder, E. (2004). Similar network activity from disparate circuit parameters. *Nat. Neurosci.* **7**, 1345-1352.

Quiroga, R. Q., Reddy, L., Kreiman, G., Koch, C., and Fried, I. (2005). Invariant visual representation by single neurons in the human brain. *Nature* **435**, 1102-1107.

Raymond, J. L., Lisberger, S. G., and Mauk, M. D. (1996). The cerebellum: a neuronal learning machine? *Science* **272**, 1126-1131.

Richmond, B. J., Optican, L. M., Podell, M., and Spitzer, H. (1987). Temporal encoding of two-dimensional patterns by single units in primate inferior temporal cortex. I. Response characteristics. *J. Neurophysiol.* **57**, 132-146.

Rieke, F., Warland, D., de Ruyter van Steveninck, R., and Bialek, W. (1997). "Spikes: Exploring the Neural Code." MIT Press, Cambridge, MA.

Rolls, E. T., and Treves, A. (1998). "Neural Networks and Brain Function." Oxford University Press, Oxford.

Sally, S. L., and Kelly, J. B. (1988). Organization of auditory cortex in the albino rat: sound frequency. *J. Neurophysiol.* **59**, 1627-1638.

Selverston, A., Elson, R., Rabinovich, M., Huerta, R., and Abarbanel, H. (1998). Basic principles for generating motor output in the stomatogastric ganglion. *Ann. NY Acad. Sci.* **860**, 35-50.

Skaggs, W. E., McNaughton, B. L., Wilson, M. A., and Barnes, C. A. (1996). Theta phase precession in hippocampal neuronal populations and the compression of temporal sequences. *Hippocampus* **6**, 149-172.

Soto-Trevino, C., Thoroughman, K. A., Marder, E., and Abbott, L. F. (2001). Activity-dependent modification of inhibitory synapses in models of rhythmic neural networks. *Nat. Neurosci.* **4**, 297-303.

Taube, J. S. (2007). The head direction signal: origins and sensorymotor integration. *Annu. Rev. Neurosci.* **30**, 181-207.

Turrigiano, G. G. (1999). Homeostatic plasticity in neuronal networks: the more things change, the more they stay the same. *Trends Neurosci.* **22**, 221-227.

Turrigiano, G. G., Leslie, K. R., Desai, N. S., Rutherford, L. C., and Nelson, S. B. (1998). Activity-dependent scaling of quantal amplitude in neocortical neurons. *Nature* **391**, 892-896.

Ungerleider, L. G., and Mishkin, M. (1982). Two cortical visual systems. In "Analysis of visual behavior" (D. J. Ingle, M. A. Goodale, and R. J. W. Mansfield, Eds.), pp. 549-586. MIT Press, Cambridge, MA.

Van Essen, D. C., Felleman, D. J., DeYoe, E. A., Olavarria, J., and Knierim, J. (1990). Modular and hierarchical organization of extrastriate visual cortex in the macaque monkey. *Cold Spring Harbor Symp. Quant. Biol.* **55**, 679-696.

Van Essen, D. C., Newsome, W. T., and Maunsell, J. H. (1984). The visual field representation in striate cortex of the macaque monkey: asymmetries, anisotropies, and individual variability. *Vision Res.* **24**, 429-448.

Vargha-Khadem, F., Gadian, D. G., Watkins, K. E., Connelly, A., Van Paesschen, W., and Mishkin, M. (1997). Differential effects of early hippocampal pathology on episodic and semantic memory. *Science* **277**, 376-380.

Wehr, M., and Laurent, G. (1996). Odour encoding by temporal sequences of firing in oscillating neural assemblies. *Nature* **384**, 162-166.

Wills, T. J., Lever, C., Cacucci, F., Burgess, N., and O'Keefe, J. (2005). Attractor dynamics in the hippocampal representation of the local environment. *Science* **308**, 873-876.

Wilson, M. A., and McNaughton, B. L. (1993). Dynamics of the hippocampal ensemble code for space. *Science* **261**, 1055-1058.

Witter, M. P., and Amaral, D. G. (2004). Hippocampal formation. In "The Rat Nervous System," 3rd ed. (G. Paxinos, Ed.), pp. 635-704. Elsevier, Amsterdam.

Learning and Memory: Basic Mechanisms

John H. Byrne, Kevin S. LaBar, Joseph E. LeDoux,

Glenn E. Schafe, J. David Sweatt, and Richard F. Thompson

Previous chapters in this book described the various components of nerve cells and their biophysical and biochemical properties as well as the ways in which neurons are connected to each other to process information and generate behavior. This chapter describes the ways in which these components and properties of the nervous system are used to mediate two of its most important functions: learning and memory. Possible subcellular modifications that underlie learning and memory are discussed in the first half of this chapter. The latter half of the chapter is an extension of a discussion from Chapters 16 and 18 on ways in which specific neural circuits can generate behavior and ways in which learning can change these behaviors and circuits.

Neuroscientists are beginning to have a reasonably detailed cellular and molecular theory of simple forms or aspects of learning and memory. The field is experiencing great synergism from the fusion of two research traditions. The "top-down" approach starts with the behavioral facts and laws, identifies the critical circuits, and then localizes the neuronal mechanisms responsible for changes in the modified circuits. The "bottom-up" approach begins by exploring neuronal modifications that seem to be promising candidate mechanisms for supporting plasticity in circuits that control the behavior(s) of interest.

LONG-TERM SYNAPTIC POTENTIATION AND DEPRESSION

The "bottom-up" approach is exemplified by studies on the cellular neurobiology of long-term synaptic potentiation (LTP) and depression (LTD). These activity-dependent neuronal changes have been traced for hours, days, and even months. Long-term potentiation of synaptic transmission is currently one of the most widely studied and popular candidate mechanisms for a cellular phenomenon underlying long-term storage of information in the nervous system. This section will briefly describe the overall phenomenon and basic attributes of LTP and its functional opposite, LTD.

The particular circuits and neuronal connections that underlie most forms of mammalian learning and memory are mysterious at present, especially for hippocampus-dependent forms of learning. There is little understanding of the means by which complex spatial memories are stored and recalled at the neural circuit level—a very important avenue of research that will be touched upon later in this chapter. Thus in many ways the study of LTP serves as a surrogate for directly studying hippocampus-dependent memory. LTP can be viewed only as a surrogate at present, because there are very few studies available directly implicating LTP (especially hippocampal

539

LTP, which has been most widely studied) in defined memory behaviors. Even considering the vast numbers of published studies of LTP, the causal link between LTP and memory *per se* has not been established conclusively. This is in contrast to many of the other types of memory that will be described in the following sections, such as sensitization and classical conditioning in *Aplysia*, eye-blink conditioning, and amygdala-dependent fear conditioning, where the underlying circuitry is much better understood.

Thus, the first part of this chapter will focus on LTP and LTD as *candidate* cellular and synaptic mechanisms for complex hippocampus- and cortex-dependent forms of memory. LTP and LTD will be emphasized for three main reasons. First, they have been extensively studied and are the forms of synaptic plasticity best understood at the molecular level. Second, they are robust forms of synaptic plasticity and worthy of description in their own right. Finally, they are specific candidate cellular mechanisms for mediating certain forms of associative learning, spatial learning, and adaptive change in the mammalian CNS.

Hebb's Postulate

Despite the various caveats concerning the specific role of LTP in hippocampus-dependent memory formation, one general hypothesis for memory storage is available and broadly accepted. This hypothesis is that *memories are stored as alterations in the strength of synaptic connections between neurons in the CNS*. The significance of this general hypothesis should be emphasized—this is one of the few areas of contemporary cognitive research for which there is a unifying hypothesis. However, it does not imply that changes in synaptic strength are the only mechanism for memory storage (see Box 19.2).

This general hypothesis has a solid underlying rationale. Learning and memory manifest themselves as a change in an animal's behavior, and scientists capitalize upon this in order to study these phenomena by observing and measuring changes in an animal's behavior in the wild or in experimental situations. However, all the behavior exhibited by an animal is a result of activity in the animal's nervous system. The nervous system comprises many kinds of cells, but the primary functional units of the nervous system are neurons. As neurons are cells, all of an animal's behavioral repertoire is a manifestation of an underlying cellular phenomenon. By extension, changes in an animal's behavior such as occurs with learning must also be subserved by an underlying cellular change.

The vast majority of the communication between neurons in the nervous system occurs at synapses. As synapses mediate the neuron–neuron communication that underlies an animal's behavior, changes in behavior are ultimately subserved by alterations in the nature, strength, or number of interneuronal synaptic contacts in the animal's nervous system. The capacity for alterations of synaptic connections between neurons is referred to as *synaptic plasticity*, and as described previously, one of the great unifying theories to emerge from neuroscience research in the last century was that synaptic plasticity subserves learning and memory. LTP (of some sort at least) is the specific form of synaptic plasticity that is the leading candidate as a mechanism subserving behavior-modifying changes in synaptic strength that mediate higher-order learning and memory in mammals.

One of the pioneers in advancing the idea that changes in neuronal connectivity are a mechanism for memory was the Canadian psychologist Donald Hebb, who published his seminal formulation as what is now generally known as *Hebb's postulate*:

> When an axon of cell A … excites cell B and repeatedly or persistently takes part in firing it, some growth process or metabolic change takes place in one or both cells so that A's efficiency as one of the cells firing B is increased.
> D. O. Hebb, *The Organization of Behavior*, 1949.

A BREAKTHROUGH DISCOVERY: LTP IN THE HIPPOCAMPUS

As a young postdoctoral researcher, Tim Bliss set out to find an example of "Hebbian" changes in synaptic function, that is, long-lasting changes that were triggered by persistent or repeated firing. By teaming up with Terje Lomo in Per Anderson's laboratory in Oslo, he did just that. The seminal report by Bliss and Lomo in 1973, describing a phenomenon they termed *long-term potentiation* of synaptic transmission, set the stage for what is now over three decades of progress in understanding the basics of long-term synaptic alteration in the CNS.

In their experiments Bliss and Lomo recorded synaptic responses in the dentate gyrus, stimulating the perforant path inputs from the entorhinal cortex (Bliss and Lomo, 1973). They used extracellular stimulating and recording electrodes implanted into the animal, and the basic experiment was begun by recording baseline synaptic transmission in this pathway. They discovered that a brief period of high-frequency (100 Hz "tetanic") stimulation led to a robust increase in the strength of synaptic connections between the perforant path inputs from the entorhinal cortex onto the dentate granule neurons in the dentate gyrus. They also observed an increased likelihood of the cells firing action potentials in response to a constant synaptic input, a phenomenon they termed *E–S* (EPSP-to-spike) *potentiation*. These two phenomena together were termed *LTP*. LTP lasted many hours in this intact rabbit preparation. The appeal of LTP as an analog of memory was immediately apparent: it is a long-lasting change in neuronal function that is produced by a brief period of unique stimulus, exactly the sort of mechanism that had long been postulated to be involved in memory formation. This pioneering work of Bliss and Lomo set in motion a several-decades-long pursuit by numerous investigators geared toward understanding the attributes and mechanisms of LTP. Much of the progress in this area is described in the following sections.

THE HIPPOCAMPAL CIRCUIT AND MEASURING SYNAPTIC TRANSMISSION IN THE HIPPOCAMPAL SLICE

Bliss and Lomo performed their experiment using stimulating and recording in an anesthetized animal using implanted electrodes. Since then, this preparation has been largely supplanted by the use of recordings from hippocampal slices maintained *in vitro* (Fig. 19.1). As most of the LTP experiments that will be described in the remainder of this section use this type of preparation, the next few paragraphs will describe the hippocampal neuronal and synaptic circuit and give an overview of extracellular recording in a typical LTP experiment in a hippocampal slice.

The main information processing circuit in the hippocampus is the relatively simple trisynaptic pathway, and much is preserved in transverse slices across the long axis of the hippocampus (Fig. 19.1). Various types of LTP can be induced at all three of these synaptic sites, and we will discuss later some mechanistic differences among the various types of LTP. Most experiments on the basic attributes and mechanisms of LTP have been studies of the synaptic connections between axons from area CA3 pyramidal neurons that extend into area CA1. These are the synapses onto CA1 pyramidal neurons that are known as the *Schaffer collateral* inputs.

The main excitatory (i.e., glutamatergic) synaptic circuitry in the hippocampus consists of three modules (Johnston and Amaral, 1998; Naber and Witter, 1998; van Groen and Wyss, 1990). Information enters

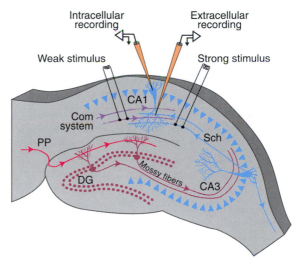

FIGURE 19.1   Schematic of a transverse hippocampal brain slice preparation from the rat. Two extracellular stimulating electrodes are used to activate two nonoverlapping inputs to pyramidal neurons of the CA1 region of the hippocampus. Both inputs consisted of axons of the Schaffer collateral/commissural (Sch/com) system. By suitably adjusting the current intensity delivered to the stimulating electrodes, different numbers of Sch/com axons can be activated. In this way, one stimulating electrode was made to produce a weak postsynaptic response and the other to produce a strong postsynaptic response. Sometimes three or more stimulating electrodes are used. Also illustrated is an extracellular recording electrode placed in the stratum radiatum (the projection zone of the Sch/com inputs) and an intracellular recording electrode in the stratum pyramidale (the cell body layer). Also indicated is the mossy fiber projection from the granule cells of the dentate gyrus (DG) to the pyramidal neurons of the CA3 region. Adapted from Barrionuevo and Brown (1983).

the dentate gyrus of the hippocampal formation from cortical and subcortical structures via the perforant path (PP) inputs from the entorhinal cortex (Fig. 19.1). These inputs make synaptic connections with the dentate granule (DG) cells of the dentate gyrus. After synapsing in the dentate gyrus, information is moved to area CA3 via the mossy fiber pathway, which consists of the axonal outputs of the dentate granule cells and their connections with pyramidal neurons in area CA3. After synapsing in area CA3, information is moved to area CA1 via the Schaffer collateral path, which consists largely of the axons of area CA3 pyramidal neurons along with other projections from area CA3 of the contralateral hippocampus as well. After synapsing in CA1, information exits the hippocampus via projections from CA1 pyramidal neurons and returns to subcortical and cortical structures.

The connections in this synaptic circuit are retained in a fairly impressive manner in transverse slices of the hippocampus, as the inputs, "trisynaptic circuit," and outputs are laid out in a generally laminar fashion along the long axis of the hippocampal formation. This configuration is a great advantage for *in vitro* electrophysiological experiments. It is important to emphasize that the trisynaptic circuit outlined earlier is a great oversimplification, as there are many additional synaptic components of the hippocampus. For example, inhibitory GABAergic interneurons make synaptic connections with all the principal excitatory neurons outlined previously. These GABAergic inputs serve in both a feedforward and feedback fashion to control excitability (see Chapter 18). Many recurrent and collateral excitatory connections are present between the excitatory pyramidal neurons as well, particularly in the CA3 region. A direct projection from the entorhinal cortex to the distal regions of CA1 pyramidal neuron dendrites is a pathway known as the *stratum lacunosum moleculare.*

In addition, many modulatory pathways projection into the hippocampus and make synaptic connections with the principal neurons. These inputs are via long projection fibers from various nuclei in the brain stem region, and they are by and large not directly excitatory or inhibitory but rather serve to modulate synaptic connectivity in a fairly subtle way. There are four predominant extrinsic modulatory projections into the hippocampus: norepinephrine (NE)-containing fibers that project from the locus ceruleus; dopamine (DA)-containing fibers that arise from the substantia nigra; acetylcholine (ACh) input from the medial septal nucleus; and 5-hydroxytryptamine (5HT, serotonin) from the raphe nuclei.

LTP of Synaptic Responses

In a popular variation of the basic LTP experiment, extracellular field potential recordings in the dendritic regions of area CA1 are utilized to monitor synaptic transmission at Schaffer collateral synapses (Fig. 19.2). A bipolar stimulating electrode is placed in the stratum radiatum subfield of area CA1 and stimuli (typically constant current pulses ranging from 1–30 µA) are delivered. Stimuli delivered in this fashion stimulate the output axons of CA3 neurons that pass nearby, causing action potentials to propagate down these axons. Cellular responses to this stimulation are recorded using extracellular or intracellular electrophysiologic recording techniques.

The typical waveform in an extracellular recording consists of a *fiber volley*, which is an indication of the presynaptic action potential arriving at the recording site, and the excitatory postsynaptic potential (EPSP) itself. The EPSP responses are a manifestation of synaptic activation (depolarization) in the CA1 pyramidal neurons. For measuring "field"

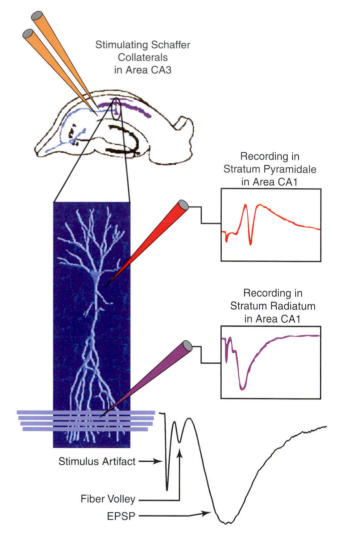

FIGURE 19.2 Recording configuration and typical physiologic responses in a hippocampal slice recording experiment. Electrode placements and responses from stratum pyramidale (cell body layer) and stratum oriens (dendritic regions) are shown. In addition, the typical waveform of a population EPSP is illustrated, showing the stimulus artifact, fiber volley, and population EPSP. Illustration and data by Joel Selcher.

(i.e., extracellularly recorded) EPSPs, the parameter typically measured is the initial slope of the EPSP waveform. Absolute peak amplitude of EPSPs can also be measured, but the initial slope is the preferred index because the initial slope is less subject to contamination from other sources of current flow in the slice. For example, currents are generated by feedforward inhibition due to GABAergic neuron activation. Also, if the cells fire action potentials, the later stages of the EPSP can be contaminated, even when recording from the dendritic region.

Because extracellular field recordings measure responses from a population of neurons, EPSPs recorded in this fashion are referred to as *population EPSPs* (pEPSPs). Note that pEPSPs are downward deflecting for stratum radiatum recordings (Fig. 19.2). When recording from the cell body layer (stratum pyramidale) the EPSP is an upward deflection, and if the cells fire action potentials, the EPSP has superimposed upon it a downward-deflecting "spike," the population spike.

As a prelude to starting an LTP experiment, input–output (I/O) functions for stimulus intensity versus EPSP magnitude are recorded in response to increasing intensities of stimulation (Fig. 19.3A). For the remainder of the experiment, the test stimulus intensity is set to elicit an EPSP that is approximately 35–50% of the maximum response recorded during the I/O measurements. Baseline synaptic transmission at this constant test stimulus intensity is usually monitored for a period of 15–20 min to assure a stable response. Once the health of the hippocampal slice is confirmed as indicated by a stable baseline synaptic response, LTP can be induced using any one of a wide variety of different LTP induction protocols. Many popular variations include a single or repeated period of 1 s, 100 Hz stimulation (with delivery of the 100 Hz trains separated by 20 s or more) where stimulus intensity is at a level necessary

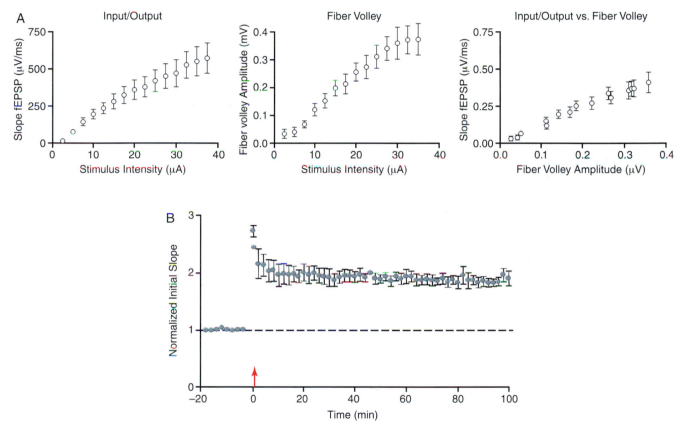

FIGURE 19.3   An input–output curve and typical LTP experiment. The top panel shows the relationship of EPSP magnitude versus stimulus intensity (in microAmps), and the same data converted to an input–output relationship for EPSP versus fiber volley magnitude in order to allow an evaluation of postsynaptic response versus presynaptic response in the same hippocampal slice. The lower panel illustrates a typical high-frequency stimulation-induced potentiation of synaptic transmission in area CA1 of a rat hippocampal slice *in vitro*. The arrow indicates the delivery of 100 Hz (100 pulses/second) synaptic stimulation. Data courtesy of Ed Weeber and Coleen Atkins. In most pharmacologic experiments using physiologic recordings in hippocampal slice preparations, effects of drug application on baseline synaptic transmission can be evaluated by simply monitoring EPSPs before and after drug application, using a constant stimulus intensity. A more elaborate alternative is to produce input–output curves for EPSP initial slope (or magnitude) versus stimulus intensity for the presynaptic stimulus. These types of within-slice experiments are very straightforward, but in some experimental comparisons this type of within-preparation design is not possible. For example, if one is comparing a wild-type with a knockout animal there of necessity must be a comparison across preparations. How does one evaluate if there is a difference in basal synaptic transmission in this situation? The principal confound is that while one has control over magnitude of the stimulus one delivers to the presynaptic fibers, differences in electrode placement, slice thickness, etc. from preparation to preparation cause variability in the magnitude of the synaptic response elicited by a constant stimulus amplitude. One commonly used approach in order to compare from one preparation (or animal strain) to the next is to quantitate the EPSP relative to the amplitude of the fiber volley in that same hippocampal slice. The rationale is that the fiber volley, which represents the action potentials firing in the presynaptic fibers, is a presynaptic physiologic response from within the same slice and that one can at least normalize the EPSP to a within-slice parameter. The underlying assumption is that the magnitude of the fiber volley is representative of the number of axons firing an action potential. Although not a perfect control, evaluating input–output relationships for fiber volley magnitude versus EPSP is a great improvement when making comparisons between different types of animals. If differences are observed, an increase in the fiber volley amplitude–EPSP slope relationship suggests an augmentation of synaptic transmission.

for approximately half-maximal stimulation (Fig. 19.3, bottom panel). A variation is a "strong" induction protocol, where LTP is induced with three pairs of 100 Hz, 1 s stimuli and where stimulus intensity is near that necessary for a maximal EPSP. This latter protocol gives robust LTP that lasts for essentially as long as one can keep the hippocampal slice alive.

A final major variation in LTP induction protocol is high-frequency stimulation patterned after the endogenous hippocampal theta rhythm. These protocols are based on the natural occurrence of an increased rate of hippocampal pyramidal neuron firing while a rat or mouse is exploring and learning about a new environment. Under these circumstances hippocampal pyramidal neurons fire bursts of action potentials at about 5 bursts/s (i.e., 5 Hz). This is the hippocampal "theta" rhythm that has been described in the literature. One variation of LTP-inducing stimulation that mimics this pattern of firing is referred to as *theta-frequency stimulation* (TFS), which consists of 30 s of single stimuli delivered at 5 Hz. Another variation, *theta burst stimulation* (TBS), consists of three trains of stimuli delivered at 20 s intervals, each train composed of 10 stimulus bursts delivered at 5 Hz, with each burst consisting of four pulses at 100 Hz

(Fig. 19.4A). It is worth noting that these patterns of stimulation, which are based on naturally occurring firing patterns *in vivo*, lead to LTP in hippocampal slice preparations just as the less natural 1 s, 100 Hz stimulation protocols do.

Short-Term Plasticity: PTP and PPF

Two types of short-term plasticity are exhibited at hippocampal Schaffer collateral synapses and elsewhere that are activity-dependent just as is LTP. These are paired-pulse facilitation (PPF) and posttetanic potentiation (PTP). PPF is a form of short-term synaptic plasticity that is commonly held to be due to residual calcium augmenting neurotransmitter release presynaptically. When two single stimulus pulses are applied with interpulse intervals ranging from 20–300 msec, the second EPSP produced is larger than the first (Fig. 19.5). This effect is referred to as PPF. The role of this type of synaptic plasticity in the behaving animal is unknown.

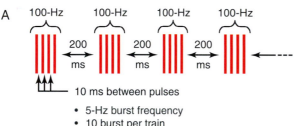

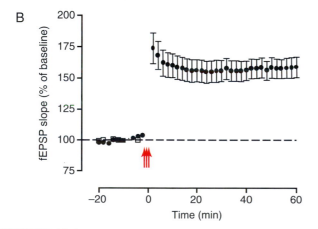

FIGURE 19.4　LTP triggered by theta burst stimulation in the mouse hippocampus. (A) Schematic depicting theta burst stimulation. This LTP induction paradigm consists of three trains of 10 high-frequency bursts delivered at 5 Hz. (B) LTP induced with theta burst stimulation (TBS–LTP) in hippocampal area CA1. The three red arrows represent the three TBS trains.

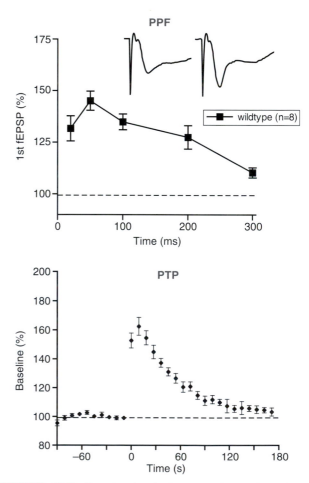

FIGURE 19.5　Paired-pulse facilitation (left panel) and posttetanic potentiation (right panel). See text for details. Data for both panels courtesy of Michael Levy.

However, it clearly is a robust form of temporal integration of synaptic transmission and could be used in information processing behaviorally. The second form of short-term plasticity, PTP, is a large enhancement of synaptic efficacy observed after brief periods of high-frequency synaptic activity. For example, in experiments where LTP is induced with one or two 1 s, 100 Hz tetani, a large and transient increase in synaptic efficacy is produced immediately after high-frequency tetanus (Fig. 19.5). The mechanisms for PTP and PPF are not fully understood but likely involve changes in presynaptic calcium (see also Chapter 8).

NMDA Receptor–Dependence of LTP

In 1983 Graham Collingridge made the breakthrough discovery that induction of tetanus-induced forms of LTP are blocked by blockade of a specific subtype of glutamate receptor, the NMDA receptor (Collingridge *et al.*, 1983). Collingridge's fascinating discovery was that the glutamate analog amino-phosphono-valeric acid (APV), an agent that selectively blocks the NMDA subtype of glutamate receptor, could block LTP induction while leaving baseline synaptic transmission entirely intact. This was the first experiment to give a specific molecular insight into the mechanisms of LTP induction. The associate properties of the NMDA receptor that allow it to function in this unique role of triggering LTP are important and were described in Chapters 11 and 16; they will be reviewed again briefly in the following section.

Pairing LTP

Early studies of LTP used mostly high-frequency (100 Hz) stimulation, in repeated 1 s long trains, as the LTP-inducing stimulation protocol. Although these protocols are still widely used to good effect, intracellular recording and patch-clamp techniques that measure electrophysiologic responses in single neurons have also been used more recently in studies of LTP. With these recording configurations, synaptic potentiation can be induced using tetanic stimulation or theta-pattern stimulation and measure LTP as an increase in postsynaptic currents through glutamate-gated ion channels, or as an increase in postsynaptic depolarization when monitoring the membrane potential. Moreover, using techniques such as patch clamping that allow control of the membrane potential of the postsynaptic neuron also allows for some sophisticated variations of the LTP induction paradigm. One particularly important series of experiments discovered that LTP can be induced by pairing repeated single presynaptic stimuli with postsynaptic membrane

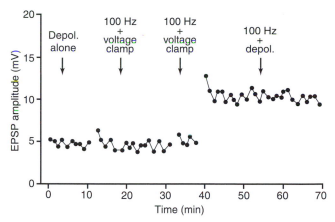

FIGURE 19.6 Pairing LTP. Intracellular recordings were made from a pyramidal cell in area CA1 of the hippocampus, with stimulation applied to the Sch/com inputs. Excitatory postsynaptic potential (EPSP) amplitudes are plotted as a function of time of occurrence (arrows) of three manipulations: an outward current step alone (depol. alone) or synaptic stimulation trains delivered while applying either a voltage clamp (100 Hz + voltage clamp) or an outward current step (100 Hz + depol.). Only presynaptic stimulation combined with postsynaptic depolarization resulted in lasting potentiation. Each point is the average of five consecutive EPSP amplitudes. Adapted from Kelso *et al.* (1986).

depolarization, so-called pairing LTP (Kelso *et al.*, 1986; Wigstrom and Gustafsson, 1986) (Fig. 19.6).

The basis for pairing LTP comes from one of the fundamental properties of the NMDA receptor (Fig. 19.7). The NMDA receptor is both a glutamate-gated channel and a voltage-dependent one. The simultaneous presence of glutamate and a depolarized membrane is necessary and sufficient (when the co-agonist glycine is present) to gate the channel. Pairing synaptic stimulation with membrane depolarization provided via the recording electrode (plus the low levels of glycine always normally present) opens the NMDA receptor channel and leads to the induction of LTP.

How is it that the NMDA receptor triggers LTP? The NMDA receptor is a calcium channel, and its gating leads to elevated intracellular calcium in the postsynaptic neuron. It is this calcium influx that triggers LTP, and indeed many subsequent chapters in this volume deal with the various processes this calcium influx triggers. It is important to remember that it is not necessarily the case that every calcium molecule involved in LTP induction actually comes through the NMDA receptor. Calcium influx through voltage-gated calcium channels (VGCC) in the membrane and calcium released from intracellular stores may also be involved (Fig. 19.8). The gating of the NMDA receptor–channel involves a voltage-dependent Mg^{2+} block of the channel pore. Depolarization of the

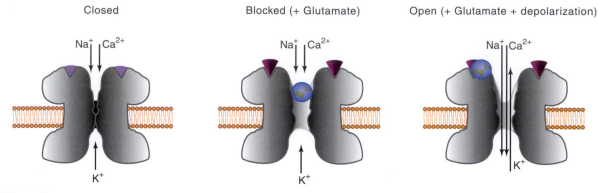

FIGURE 19.7   Coincidence detection by the NMDA receptor. The simultaneous presence of glutamate and membrane depolarization is necessary for relieving Mg^{++} blockade and allowing calcium influx.

membrane in which the NMDA receptor resides is necessary to drive the divalent Mg^{2+} cation out of the pore, which then allows influx of calcium ions. Thus, the simultaneous occurrence of both glutamate in the synapse and a depolarized postsynaptic

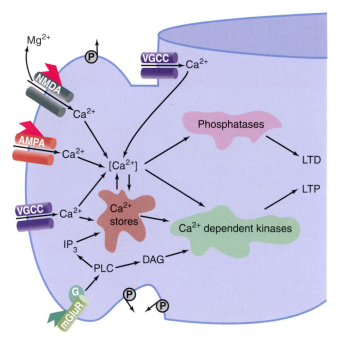

FIGURE 19.8   Events leading to LTP or LTD. The schematic depicts a postsynaptic spine with various sources of Ca^{2+}. The NMDA receptor–channel complex admits Ca^{2+} only after depolarization removes the Mg^{2+} block. Ca^{2+} may also enter through ligand-gated AMPA receptor channel or voltage-gated Ca^{2+} channels (VGCC), which may be located on the spine head or dendritic shaft. Also, certain subtypes of metabotropic glutamate receptors (mGluRs) are coupled positively to phospholipase C (PLC), which cleaves membrane phospholipids into inositol trisphosphate (IP$_3$) and diacylglycerol (DAG). Increased levels of IP$_3$ lead to the release of intracellular Ca^{2+} stores, whereas increases in DAG activate Ca^{2+}-dependent enzymes. Ca^{2+} pumps (P), located on the spine head, neck, and dendritic shaft, are hypothesized to help isolate [Ca^{2+}]$_i$ changes in the spine head from those in the dendritic shaft.

membrane is necessary to open the channel and allow LTP-triggering calcium into the postsynaptic cell.

These properties, glutamate dependence *and* voltage dependence, of the NMDA receptor allow it to function as a coincidence detector (Fig. 19.7). This is a critical aspect of NMDA receptor regulation and allows for a unique contribution of the NMDA receptor to information processing at the molecular level. Using the NMDA receptor the neuron can trigger a unique event, calcium influx, specifically when a particular synapse is both active presynaptically (glutamate is present in the synapse) and postsynaptically (when the membrane is depolarized).

This feature of the NMDA receptor confers a computational property of associativity on the synapse. This attribute is nicely illustrated by "pairing" LTP, as described previously, where low-frequency synaptic activity paired with postsynaptic depolarization can lead to LTP (Fig. 19.6). The associative property of the NMDA receptor allows for many other types of sophisticated information processing as well, however. For example, activation of a weak input to a neuron can induce potentiation, provided a strong input to the same neuron is activated at the same time (Barrionuevo and Brown, 1983) (Fig. 19.9). These particular features of LTP induction have stimulated a great deal of interest, as they are reminiscent of classical conditioning with depolarization and synaptic input roughly corresponding to unconditioned and conditioned stimuli, respectively.

The associative nature of NMDA receptor activation allows for synapse specificity of LTP induction as well, which has been shown to occur experimentally. If postsynaptic depolarization is paired with activity at one set of synaptic inputs to a cell, while leaving a second input silent or active only during periods at which the postsynaptic membrane is near the resting potential, then selective potentiation of the paired input pathway occurs (Fig. 19.9).

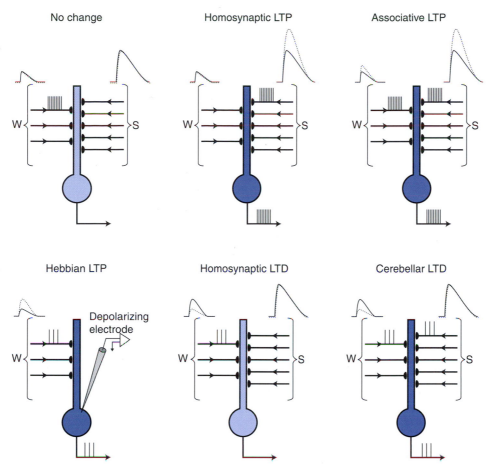

FIGURE 19.9   Various types of synaptic activity–modification relationships. Schematic neurons are shown receiving nonoverlapping inputs that are either weak (W, few axons) or strong (S, many axons). Action potentials on axons are denoted by short vertical lines that are closely spaced (high frequency) or widely spaced (low frequency). Activity in synapses or cells is indicated by black shading. Postsynaptic potentials before inducing stimulation are given by solid lines; those after are given by dotted lines. No change is produced when a single weak input is stimulated at high frequency. Homosynaptic LTP is induced in the synapses of a strong input when it receives high-frequency stimulation, but the unstimulated synapses of the weak input are unchanged. Associative LTP occurs when a weak input (that would not potentiate if stimulated by itself) is stimulated concurrently with a strong input, resulting in potentiation of both inputs. Hebbian LTP occurs when a weak input (that would not potentiate if stimulated by itself) is stimulated at the same time depolarization is induced in the postsynaptic cell through injection of an intracellular current. Homosynaptic ("telencephalic") LTD is induced only in synapses receiving low-frequency (1–5 Hz) stimulation; unstimulated synapses are unchanged. Cerebellar LTD occurs when parallel fibers (represented here by the W input) are stimulated simultaneously with climbing fibers (represented here by the S input). Parallel fibers undergo LTD, but climbing fibers do not. Adapted from Brown *et al.* (1990).

Similarly, in field stimulation experiments LTP is restricted to tetanized pathways—even inputs convergent on the same dendritic region of the postsynaptic neuron are not potentiated if they receive only baseline synaptic transmission in the absence of synaptic activity sufficient to adequately depolarize the postsynaptic neuron (Andersen *et al.*, 1977). This last point illustrates the basis for LTP "cooperativity." LTP induction in extracellular stimulation experiments requires cooperative interaction of afferent fibers, which in essence means there is an intensity threshold for triggering LTP induction. Sufficient total synaptic activation by the input fibers must be achieved such that the postsynaptic membrane is adequately depolarized to allow opening of the NMDA receptor (McNaughton *et al.*, 1978).

Dendritic Action Potentials

In the context of the functioning hippocampal neuron *in vivo* the associative nature of NMDA receptor activation means that a given neuron must reach a critical level of depolarization in order for LTP to occur at any of its synapses. Specifically, in the physiologic context the hippocampal pyramidal neuron generally must reach the threshold for firing an action potential, although there are some exceptions to this generalization. Although action potentials are triggered in the active zone of the cell body, hippocampal pyramidal neurons along with many other types of CNS neurons can actively propagate action potentials into the dendritic regions: the so-called *back-propagating* action potential, or "BPAP" (Magee and Johnston, 1997) (Fig. 19.10). These dendritic action potentials are just like action potentials propagated down axons in that they are mediated predominantly by voltage-dependent ion channels such as sodium channels. The penetration of the back-propagating action potential into the dendritic region provides a wave of membrane depolarization that allows for the opening of the voltage-dependent

NMDA receptor–ion channels. *Active* propagation of the action potential is necessary because the biophysical properties of the dendritic membrane dampen the passive propagation of membrane depolarization, thus an active process such as action potential propagation is required (see Chapters 4 and 17). As a generalization, in many instances in the intact cell back-propagating action potentials are what allow sufficient depolarization to reach hippocampal pyramidal neuron synapses in order to open NMDA receptors. In an ironic twist, this has brought us back to a more literal reading of Hebb's postulate, where as we discussed in the introduction to this chapter, Hebb actually specified *firing* of the postsynaptic neuron as being necessary for the strengthening of its connections.

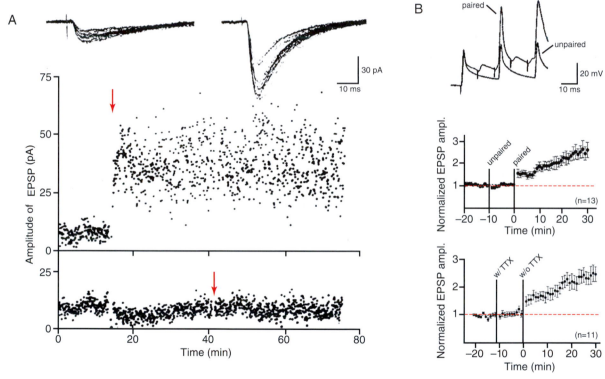

FIGURE 19.10 Pairing LTP produced by postsynaptic depolarization and back-propagating action potentials. (A) LTP of synaptic transmission induced by pairing postsynaptic depolarization with synaptic activity. The traces in the upper panels illustrate postsynaptic currents recorded directly from the postsynaptic neuron using voltage-clamp techniques. The two graphs are the result of a pairing LTP experiment (upper), and control, nonpaired pathway (lower). In the pairing LTP experiment, hippocampal CA1 pyramidal neurons were depolarized from −70 mV to 0 mV while the paired pathway was stimulated at 2 Hz 40 times. Control received no stimulation during depolarization. From Malinow and Tsien (1990). (B) Pairing small EPSPs with back-propagating dendritic action potentials induces LTP. Upper traces show that subthreshold EPSPs paired with back-propagating action potentials increase dendritic action potential amplitude. Recording is at approximately 240 μm from soma, that is, in the dendritic tree of the neuron. Action potentials were evoked by 2 ms current injections through a somatic whole-cell electrode at 20 ms intervals. Alone, action potential amplitude was small (unpaired). Paired with EPSPs (five stimuli at 100 Hz), the action potential amplitude increased greatly (paired). Graph below illustrates grouped data showing normalized EPSP amplitude after unpaired and paired stimulation. The pairing protocol shown in A was repeated five times at 5 Hz at 15 s intervals for a total of two times. In lower graph, a similar pairing protocol was given with and without applying the sodium channel blocker tetrodotoxin (TTX, to block action potential propagation) to the proximal apical dendrites to prevent back-propagating action potentials from reaching the synaptic input sites. LTP was induced only when action potentials fully back-propagated into the dendrites. Reproduced with permission from Magee and Johnston (1997).

In fact, the timing of the arrival of a dendritic action potential with synaptic glutamate input appears to play an important part in precise, timing-dependent triggering of synaptic plasticity in the hippocampus (Magee and Johnston, 1997) (Fig. 19.11). A critical timing window is involved vis-à-vis back-propagating action potentials: glutamate arrival in the synaptic cleft must slightly precede the back-propagating action potential in order for the NMDA receptor to be effectively opened. This timing dependence arises in part due to the time required for glutamate to bind to and open the NMDA receptor. The duration of an action potential is quite short, so in essence the glutamate must be there first and already bound to the receptor in order for full activation to occur. (Additional factors are also involved; for a discussion, see Bi and Poo, 1998; Johnston *et al.*, 2000; Kamondi *et al.*, 1998; and Linden, 1999.)

This order-of-pairing specificity allows for a precision of information processing: not only must the membrane be depolarized, but also, as a practical matter, the cell must fire an action potential. Moreover, the timing of the back-propagating action potential arriving at a synapse must be appropriate. It is easy to imagine how the nervous system could capitalize on these properties to allow for forming precise timing-dependent associations between two events.

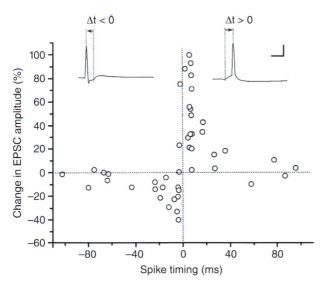

FIGURE 19.11 The timing of back-propagating action potentials with synaptic activity determines whether synaptic strength is altered, and in which direction. Precise timing of the arrival of a back-propagating action potential (a "spike") with synaptic glutamate determines the effect of paired depolarization and synaptic activity. A narrow window when the arrival of the synaptic EPSP immediately precedes or follows the arrival of the back-propagating action potential determines whether synaptic strength is increased, decreased, or remains the same. See text for additional discussion. Figure adapted from Bi and Poo (1998).

One twist to the order-of-pairing specificity is that if the order is reversed and the action potential arrives before the EPSP, then synaptic depression is produced. The mechanisms for this attribute are under investigation at present: one hypothesis is that the backward pairing by various potential mechanisms leads to a lower level of calcium influx, which produces synaptic depression (see later).

NMDA RECEPTOR–INDEPENDENT LTP

Although the vast majority of studies of LTP and its molecular mechanisms have investigated NMDA receptor–dependent processes, there also are several types of NMDA receptor–independent LTP. The next section will briefly describe a few different types of NMDA receptor–independent LTP as background material, and highlight them as important areas of investigation.

200 Hz LTP

NMDA receptor–independent LTP can be induced at the Schaffer collateral synapses in area CA1, the same synapses discussed thus far. This allows for somewhat of a compare-and-contrast of two different types of LTP at the same synapse. A protocol that elicits NMDA receptor–independent LTP in area CA1 is the use of four 1/2 s, 200 Hz stimuli separated by 5 s (Grover and Teyler, 1990). LTP induced with this stimulation protocol is partially insensitive to NMDA receptor–selective antagonists such as APV (Fig. 19.12). It is interesting that simply doubling the rate of tetanic stimulation from 100 Hz to 200 Hz appears to shift activity-dependent mechanisms for synaptic potentiation into NMDA receptor independence. At the simplest level of thinking this indicates that there is some unique type of temporal integration occurring at the higher-frequency stimulation that allows for superseding the necessity for NMDA receptor activation. What might the 200 Hz stimulation be uniquely stimulating? One appealing hypothesis arises from the observation that 200 Hz LTP is blocked by blockers of voltage-gated calcium channels (VGCC) (Fig. 19.12). Thus, the current working model is that 200 Hz stimulation elicits sufficiently large and prolonged membrane depolarization, resulting in the opening of voltage-dependent calcium channels, to trigger elevation of postsynaptic calcium sufficient to trigger LTP synaptic potentiation. One observation consistent with this hypothesis is that

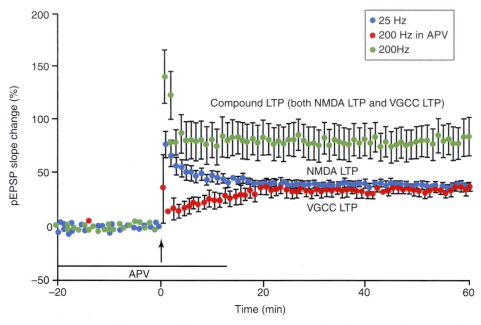

FIGURE 19.12 Composite graph of multiple forms of LTP. Superimposition of 25 Hz LTP (blue circles) and 200 Hz in APV LTP (red circles) reveal that the two forms achieve stable potentiation at similar magnitude (60 min post tetanus). A 200 Hz tetanus in standard medium (green circles) results in a compound LTP twice as large as the NMDA- and VGCC-mediated components. pEPSP, population EPSP. Modified with permission from Cavus, I., and Teyler, T.J. (1996). Two forms of long-term potentiation in area CA1 activate different signal transduction pathways. *J. Neurophysiol.* **76**, 3038–3047.

injection of postsynaptic calcium chelators blocks 200 Hz stimulation–induced LTP.

Mossy Fiber LTP in Area CA3

The predominant model system for studying NMDA receptor–independent LTP is not the Schaffer collateral synapses, but rather the mossy fiber inputs into area CA3 pyramidal neurons (Fig. 19.13). Considerable excitement accompanied the discovery of NMDA receptor–independent LTP at these synapses by Harris and Cotman (Harris and Cotman, 1986). The mossy fiber synapses are unique, large synapses with unusual presynaptic specializations, and there has been much interest in comparing the attributes and mechanisms of induction of mossy fiber LTP (MF-LTP) with those of NMDA receptor–dependent LTP in area CA1. However, subsequent progress in investigating the mechanistic differences between these two types of LTP has been relatively slow for several reasons. First, the experiments are technically difficult physiologically—typically area CA3 is the first part of the hippocampal slice preparation to die *in vitro*. The local circuitry in area CA3 is complex, with many recurrent excitatory connections between neurons: synapses that also are plastic and exhibit NMDA receptor–dependent LTP. Most problematic has been an ongoing controversy about the necessity

of postsynaptic events, especially elevations of postsynaptic calcium, for the induction of mossy fiber LTP. There basically are two schools of thought on mossy fiber LTP. One line of thinking is that MF-LTP is entirely presynaptic in its induction and

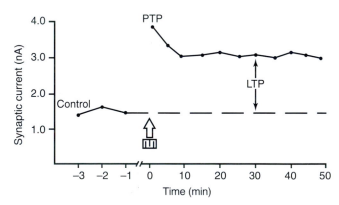

FIGURE 19.13 Mossy fiber EPSC amplitudes plotted over time, before and after the induction of LTP. Brief tetanic stimulation was applied at the time indicated (striped bar and arrow). Note the change in the timescale at the time of stimulation. Each data point is the average of five EPSCs obtained from a holding potential of –90 mV. The tetanic stimulation induced post-tetanic potentiation (PTP) and long-term potentiation (LTP). Modified with permission from Barrionuevo, G., Kelso, S., Johnston, D., and Brown, T.H. (1986). Conductance mechanism responsible for long-term potentiation in monosynaptic and isoloated excitatory synaptic inputs to the hippocampus. *J. Neurophysiol.* **55**, 540–550.

expression (Zalutsky and Nicoll, 1990). A second line of thinking is that MF-LTP has a requirement for postsynaptic signal transduction events for its induction (see, e.g., Kapur *et al.*, 1998; Yeckel *et al.*, 1999).

A ROLE FOR CALCIUM INFLUX IN NMDA RECEPTOR–DEPENDENT LTP

In contrast to the analyses of mossy fiber LTP, the analyses of NMDA receptor–dependent LTP at Schaffer collateral synapses has achieved a broad consensus of a necessity for elevations of postsynaptic calcium for triggering LTP (Lynch *et al.*, 1983). In fact, this is one of the few areas of LTP research where there is almost universal agreement. The case for a role for elevated postsynaptic calcium in triggering LTP is quite clear-cut and solid. It is well established and has been reviewed adequately a sufficient number of times (Chittajallu *et al.*, 1998; Johnston *et al.*, 1992; Nicoll and Malenka, 1995), so this material will be presented only in overview here. A principal line of evidence is that injection of calcium chelators postsynaptically blocks the induction of LTP. Also, inhibitors of a variety of calcium-activated enzymes also block LTP induction, including when they are specifically introduced into the postsynaptic neuron. Fluorescent imaging experiments using calcium-sensitive indicators have clearly demonstrated that postsynaptic calcium is elevated with LTP-inducing stimulation. Finally, elevating postsynaptic calcium is sufficient to cause synaptic potentiation (although there has been some controversy on this point). Thus, the hypothesis of a role for postsynaptic calcium elevation in triggering LTP is on a solid experimental footing.

Presynaptic Versus Postsynaptic Mechanisms

One of the most intensely studied and least satisfactorily resolved aspects of LTP concerns the locus of LTP maintenance and expression. One component of LTP is an increase in the EPSP, which could arise from increasing glutamate concentrations in the synapse or by increasing the responsiveness to glutamate by the postsynaptic cell (Fig. 19.14). The "pre" versus "post" debate is whether the relevant changes reside presynaptically and manifest as an increase in neurotransmitter release or similar phenomenon, or whether they reside postsynaptically as a change in glutamate receptor responsiveness and so on. Over the last 15 years or so there have been numerous experiments performed to try to address this question, and as of yet there is no clear consensus answer.

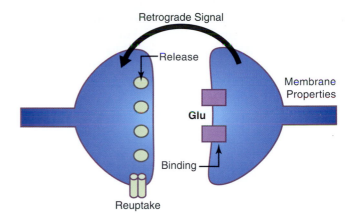

Presynaptic = Altered
- Neurotransmitter amount in vesicles
- Number of vesicles released
- Kinetics of release
- Glutamate reuptake
- Probability of vesicle fusion

Postsynaptic = Altered
- Number of AMPA receptors
- Insertion of AMPA receptors
- Ion flow through AMPA channels
- Membrane electrical properties

Additional possibilities include changes in number of total synaptic connections between two cells

FIGURE 19.14 Potential sites of synaptic modification in LTP. A wide variety of different specific molecular sites are candidates for subserving increased synaptic transmission in LTP. It is important to note that potentially all could be occurring simultaneously at a given cell or synapse.

In some of the earliest studies to begin to analyze LTP mechanistically it became clear that infusing compounds into the postsynaptic cell led to a block of LTP. A few of these studies involving calcium chelators were described in the last section. If compounds that are limited in their distribution to the postsynaptic compartment block LTP, the most parsimonious hypothesis is that LTP resides postsynaptically. However, shortly thereafter evidence began to accumulate suggesting that there were presynaptic changes involved in LTP expression as well. For example, various types of "quantal" analyses (Chapter 8) that had been successfully applied at the neuromuscular junction to dissect presynaptic changes from postsynaptic changes suggested that LTP is associated with changes presynaptically. In a series of investigations several laboratories used whole-cell recordings of synaptic transmission in hippocampal slices and found an increase in the probability of release, a strong indicator of presynaptic changes in classic quantal analysis (Bekkers and Stevens, 1990; Dolphin *et al.*, 1982; Malgaroli *et al.*, 1995; Malinow and Tsien, 1990; Malinow, 1991;

Zakharenko *et al.*, 2001). These findings fit nicely with earlier studies from Tim Bliss's laboratory suggesting an increase in glutamate release in LTP as well (Dolphin *et al.*, 1982). Given findings supporting postsynaptic locus on the one hand and presynaptic locus on the other, why not just hypothesize that there are changes both presynaptically and postsynaptically? The rub came in that some of the quantal analysis results seemed to exclude postsynaptic changes as occurring.

These findings in the early 1990s ushered in an exciting phase of LTP research that was important independent of the pre-versus-post debate *per se*. If there are changes presynaptically but these changes are triggered by events originating in the postsynaptic cell, as the earlier inhibitor-perfusion experiments had indicated, then the existence of a *retrograde messenger* is implied. A retrograde messenger is a compound generated in the postsynaptic compartment that diffuses back to and signals changes in the presynaptic compartment—the opposite (retrograde) direction from normal synaptic transmission. Moreover, if the compound is generated intracellularly in the postsynaptic neuron, then the compound must be able to traverse the postsynaptic membrane somehow. The data supporting presynaptic changes in LTP implied the existence of such a signaling system, and this hypothesis launched a number of interesting and important experiments to determine what types of molecules might serve such a role; some of these are highlighted in Figure 19.15.

However, in the mid-1990s the pre–post pendulum began to swing back in the opposite direction, toward the postsynaptic side. Several groups found evidence for postsynaptic changes that could account for the apparently presynaptic changes identified by quantal analysis studies. Specifically, evidence was generated for what are termed *silent synapses* (Fig. 19.16). These are synapses that contain NMDA receptors but no AMPA receptors—they are capable of synaptic plasticity mediated by NMDA receptor activation but are physiologically silent in terms of baseline synaptic transmission. Silent synapses are rendered active by NMDA receptor–triggered activation of latent AMPA receptors postsynaptically. Such uncovering of silent AMPA receptors could involve membrane insertion or post-translational activation of already-inserted receptors. Activation of silent synapses is a postsynaptic mechanism that could explain the effects (decreased failure rate, e.g.) in quantal analysis experiments that implied presynaptic changes. Thus there is now an argument that all LTP physiology and biochemistry at Schaffer collateral synapses could be postsynaptic.

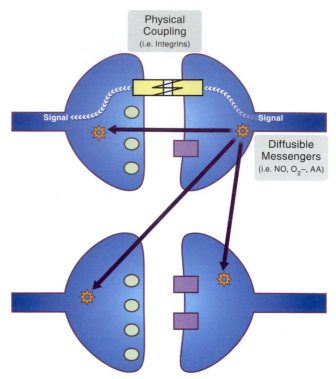

FIGURE 19.15 Potential mechanisms for retrograde signaling. A number of different specific retrograde signaling mechanisms have been identified, including utilization of membrane-permeant species such as nitric oxide (NO), superoxide (O_2 anion), or arachidonic acid (AA), as trans-synaptic protein species such as integrins.

This model for conversion of silent synapses into active synapses by AMPA receptor insertion at Schaffer collateral synapses is an entirely postsynaptic phenomenon. However, there has been a variation of this idea proposed that has been referred to as a "whispering" synapse. A whispering synapse has both AMPA and NMDA receptors in it, but because of a number of hypothetical factors such as glutamate affinity differences between NMDA and AMPA receptors, kinetics of glutamate elevation in the synapse, or spatial localization of the receptors, the AMPA receptors are silent. A whispering synapse is converted to being fully active by a presynaptic mechanism. An increase in glutamate release presynaptically, resulting in an elevation of glutamate levels in the synapse, then allows the effective activation of preexisting AMPA receptors with baseline synaptic transmission. By this mechanism a synapse that was previously silent with respect to baseline synaptic transmission is rendered detectably active. However, this alternative mechanism requires no change in the postsynaptic compartment whatsoever.

Like the retrograde messenger hypothesis, the silent synapses hypothesis has also led to a number of important and interesting experiments that warrant

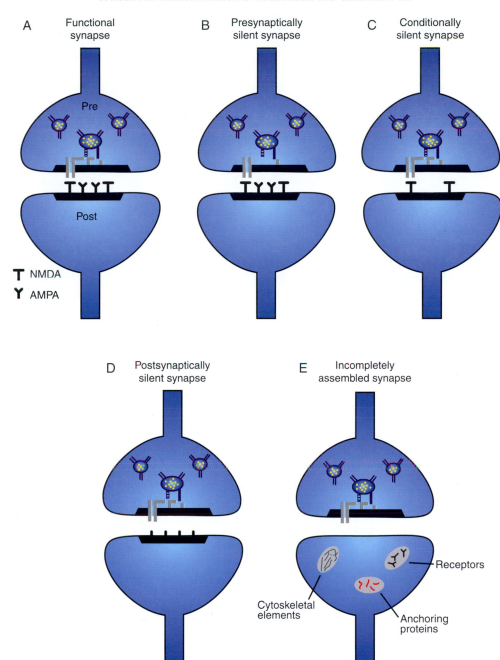

FIGURE 19.16 Possible types of silent synapse. (A) Complete synapse, possessing all the necessary pre- and postsynaptic components. (B) Synapse silent due to presynaptic molecular deficiency. (C) Conditionally silent mammalian synapse (NMDA receptors only). (D) Synapse silent due to postsynaptic deficiency (receptors nonfunctional). (E) Synapse in transition (postsynaptic components being assembled). Adapted from Atwood and Wojtowicz (1999).

attention aside from the pre-versus-post debate. Specifically, these experiments have focused new attention on the importance of considering the postsynaptic compartment in a cell-biological context. Mechanisms of receptor insertion, trafficking, and turnover that had been studied in nonneuronal cells are now beginning to get the attention they deserve in neurons as well. Like retrograde signaling, experiments arising from investigating mechanisms for activation of silent synapses have led to important "spin-off" studies that are important independent of the precipitating issue of pre-versus-post.

Given the variety of evidence described so far, should one conclude that LTP at Schaffer collateral synapses resides presynaptically, postsynaptically, or both? Although there is not yet an unambiguous consensus in the pre-versus-post debate, overall the available literature indicates that changes are occurring in both the presynaptic and postsynaptic compartments. First, a number of experiments using sophisticated imaging techniques have found LTP to be associated with presynaptic changes such as increased vesicle recycling and increased presynaptic membrane turnover (see, e.g., Malgaroli *et al.*, 1995; Zakharenko *et al.*, 2001). Also, direct biochemical measurements of the phosphorylation of proteins selectively localized to the presynaptic compartment have shown LTP-associated changes. Conceptually similar experiments looking at phosphorylation of postsynaptic proteins have found the same thing. Thus, imaging and biochemistry studies have fairly clearly illustrated that sustained biochemical changes are occurring in both the presynaptic and postsynaptic cell.

This conclusion, and indeed all the pre-versus-post experiments, have a very important caveat to keep in mind. In trying to reach a consensus conclusion, comparisons are being made across a wide spectrum of different types of experiments and different preparations (e.g., results with cultured cells versus hippocampal slices), between different types of LTP (e.g., pairing versus tetanic stimulation protocols), different stages of LTP, and material from different developmental stages in the animal. These considerations are a good reason to exercise caution in interpreting the experiments at this point, and indeed these issues may be contributing greatly to the apparent incompatibility of the results obtained in different laboratories.

LTP Can Be Divided into Phases

Contemporary models divide very long-lasting LTP (i.e., LTP lasting in the range of 5–6 h) into at least three phases. LTP comprising all three phases can be induced with repeated trains of high-frequency stimulation in area CA1, and the phases are expressed sequentially over time to constitute what we call "LTP." Late LTP (L-LTP) is hypothesized to be dependent for its induction on changes in gene expression, and this phase of LTP lasts many hours (see Winder *et al.*, 1998). Early LTP (E-LTP) is likely subserved by persistently activated protein kinases, starts at around 30 minutes or less post tetanus, and is over by about 2–3 h. The first stage of LTP, generally referred to as *short-term potentiation* (STP), is independent of protein kinase activity for its induction and

lasts about 30 minutes. The mechanisms for STP are essentially a complete mystery at present.

The Induction, Maintenance, and Expression of LTP

Mechanistically speaking, LTP can also be divided into three component parts: induction, maintenance, and expression. *Induction* refers to the transient events serving to trigger the formation of LTP. *Maintenance*, or more specifically a maintenance mechanism, refers to the persisting biochemical signal that lasts in the cell. This persisting biochemical signal acts upon an effector, for example, a glutamate receptor or the presynaptic release machinery, resulting in the *expression* of LTP.

It is important to keep in mind that, depending on the design of the experiment, induction, maintenance, and expression can be differentially inhibited (Fig. 19.17). The simplest type of experiment does not do this. For example, imagine if one applies an enzyme inhibitor (or knocks out a gene) before, during, and after the period of LTP-inducing high-frequency stimulation; this manipulation may block LTP. However, this does not distinguish whether the missing activity is required for the induction, expression, or maintenance of LTP. To distinguish among these possibilities, imagine instead applying the inhibitor selectively at different time points during the experiment. If inhibitor is applied only during the tetanus and then washed out, and it blocks the generation of LTP, one can conclude that the enzyme is necessary for LTP induction. If the inhibitor is applied after the tetanus and reverses the potentiation, it may be blocking either the maintenance or expression of LTP, as was nicely illustrated in an early experiment by Malinow, Madison, and Tsien

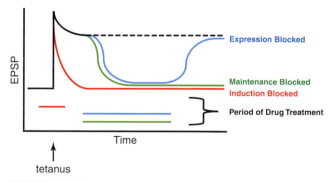

FIGURE 19.17 Induction, maintenance, and expression of LTP. This schematic illustrates the different experimental approaches to dissecting effects on the biochemical mechanisms subserving LTP induction, maintenance, or expression. See text for additional details.

(Malinow *et al.*, 1988), where they applied a protein kinase inhibitor after LTP induction. In this experiment, transient application of a kinase inhibitor after tetanus blocked synaptic potentiation, but the potentiation recovered after removal of the inhibitor. This is a blockade of LTP *expression*. However, if the kinase inhibitor had caused the potentiation to be lost irreversibly, the inhibitor would then by definition have blocked the *maintenance* of LTP.

Finally, it is important to synthesize the concepts of induction, maintenance, and expression with the concept of phases. Simply stated, three phases of LTP (STP, E-LTP, and L-LTP) times three distinct underlying mechanisms for each phase (induction, maintenance, and expression) gives nine separate categories into which any particular molecular mechanism contributing to LTP may fit.

LTP OUTSIDE THE HIPPOCAMPUS

The abundance of literature dedicated to studying LTP in the hippocampus might lead a newcomer to the field to suppose that LTP is somehow restricted to these synapses. However, plasticity of synaptic function, including phenomena such as LTP and LTD, is the rule rather than the exception for most forebrain synapses. LTP outside the hippocampus has been mostly studied in the cerebral cortex and the amygdala. The likely functional roles for LTP at these other sites are quite diverse, but two specific examples are worth highlighting. LTP-like processes in the cerebral cortex play a role in activity-dependent development of the visual system and other sensory systems. LTP in the amygdala has received prominent attention as a mechanism contributing to cued fear conditioning. The role of LTP in amygdala-dependent fear conditioning in fact is the area for which the strongest case can be made for a direct demonstration of a behavioral role for LTP. This example of learning will be discussed in more detail in a following section of the chapter.

MODULATION OF LTP INDUCTION

In one sense the hippocampal slice is a denervated preparation. In the intact animal, the hippocampus receives numerous input fibers that provide modulatory inputs of the neurotransmitters dopamine (DA), norepinephrine (NE), serotonin (5-HT), and acetylcholine (ACh). Functionally these inputs are largely lost as a necessity of physically preparing the hippocampal slice for the experiment. However, these lost modulatory inputs can be partially reconstituted by directly applying the neurotransmitters (or, more commonly, pharmacologic substitutes) to the slice preparation *in vitro*. This approach has been used quite successfully to gain insights into the physiologic mechanisms and functional roles of these inputs in the intact brain.

Norepinephrine, DA, and ACh-mimicking compounds can all modulate the induction of LTP at Schaffer collateral synapses. Specifically, agents acting at various subtypes of receptors for these compounds can increase the likelihood of LTP induction and the magnitude of LTP that is induced. Several examples of this type of modulation experiment are shown in Figure 19.18. In one example (Fig. 19.18, panel A), 5 Hz stimulation of Schaffer collateral synapses for 3 minutes gives essentially no potentiation. Coapplication of isoproterenol, a beta-adrenergic receptor agonist that mimics endogenous NE, converts a nonpotentiating signal into a potentiating one (Thomas *et al.*, 1996). Under other conditions beta-adrenergic agonists can augment the *magnitude* of LTP induced as well, if different physiologic stimulation protocols are used that evoke modest LTP. Similar types of effects can be observed for activation of various subtypes of receptors for ACh and DA (see Yuan *et al.*, 2002).

One known site of action of neuromodulators is regulation of back-propagating action potentials in pyramidal neuron dendrites. All these agents that modulate LTP induction can modulate the magnitude of back-propagating action potentials (Fig. 19.18, panel B). The augmentation of back-propagating action potentials is a means by which these neurotransmitters can enhance membrane depolarization and thereby enhance NMDA receptor opening.

The growth factor brain-derived neurotrophic factor (BDNF) can also modulate the induction of LTP by a number of mechanisms, at least one of which is presynaptic (Gottschalk *et al.*, 1998; Lu and Chow, 1999; Xu *et al.*, 2000; and Fig. 19.18, panel C). BDNF, acting through its cell-surface receptor TrkB, acts on presynaptic terminals to selectively facilitate neurotransmitter release during high-frequency stimulation. This is an interesting example of modulation of LTP induction that is activity-dependent but localized to the presynaptic compartment. The mechanisms controlling the levels of BDNF in the adult hippocampus are not entirely clear at this point, but it is fairly well established that hippocampal BDNF levels can be regulated by a variety of neuronal activity-dependent processes and indeed in response to environmental signals impinging upon the behaving animal.

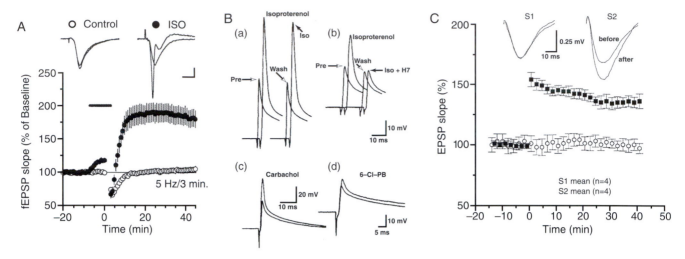

FIGURE 19.18 Neuromodulation of LTP induction. (A) Modulation of LTP induction by the beta-adrenergic agonist isoproterenol. Activity-dependent β-adrenergic modulation of low-frequency stimulation-induced LTP in the hippocampus CA1 region. In control experiments (no ISO), 3 min of 5 Hz stimulation (delivered at time = 0, open symbols, n = 26) had no lasting effect on synaptic transmission (45 min after 5 Hz stimulation, fEPSPs were not significantly different from pre–5 Hz baseline, $t_{(25)} = 1.01$). However, 3 min of 5 Hz stimulation delivered at the end of a 10 min application of 1.0 mM ISO (indicated by the bar) induced LTP (closed symbols, p < 0.01 compared with baseline). The traces are fEPSPs recorded during baseline and 45 min after 5 Hz stimulation in the presence and absence (control) of ISO. Calibration bars are 2.0 mV and 5.0 ms. Reproduced from Thomas *et al.* (1996). (B) One potential mechanism for neuromodulation is regulation of back-propagating action potentials in CA1 dendrites. The data shown illustrate amplification of dendritic action potentials by isoproterenol (a) and its susceptibility to inhibition by the protein kinase inhibitor H7. The traces shown are from dendritic patch-clamp recordings from hippocampal pyramidal neurons. Muscarinic agonist (carbachol, lower traces) and the dopamine receptor agonist 8-Cl-PB also can give various degrees of action potentail modulation. (a) Bath application of 1 μM isoproterenol resulted in a 104% increase in amplitude, from 41 mV ("Pre") to 84 mV, of an antidromically initiated action potential recorded 220 μm from the soma. Washout of isoproterenol amplitude (38 mV; "Wash"). With a second application of isoproterenol (dark arrow labeled "Iso"), the amplitude again increased two-fold to 80 mV. (b) In a different recording 300 μM H-7, a generic kinase inhibitor, was included in the control saline during the washout of isoproterenol. The subsequent second application of isoproterenol failed to lead to a second increase in amplitude (dark arrow labeled "Iso + H7"). (c) In a distal recording (300 μM), 1 μM carbachol increased the action potential amplitude by 81%. In the carbachol experiments, cells were held hyperpolarized to −80 mV to remove Na$^+$ channel inactivation. (d) One of the 6 out of 10 recordings where 6-Cl-PB led to an increase in amplitude. In a recording 220 μm from the soma, 10 μM 6-Cl-PB increased dendritic action potential amplitude by 26%, from 21 to 26.5 mV. The cells were held at −70 mV in all 6-Cl-PB experiments. Adapted from Johnston *et al.* (1999). (C) BDNF also modulates LTP induction in response to theta-frequency-type stimulation. Two stimulating electrodes were positioned on either side of a single recording electrode to stimulate two different groups of afferents converging in the same dendritic field in CA1. Stimulation was applied to Schaffer collaterals alternately at low frequency (1 per min). After a period of baseline recording, LTP was induced with a theta burst stimulation applied at time 0 only to one pathway (S1, filled squares). Simultaneous recording of an independent pathway (S2, open circles) showed no change in its synaptic strength after the theta burst was delivered to S1. BDNF (closed squares) selectively facilitates the induction of LTP in the tetanized pathway without affecting the synaptic efficacy of the untetanized pathway. EPSPs were recorded in the CA1 area of BDNF-treated slices. Synaptic efficacy (initial slope of field EPSPs) is expressed as a percentage of baseline value recorded during the 20 min before the tetanus. Representative traces of field EPSPs from S1 and S2 pathways were taken 10 min before and 40 min after the theta burst stimulation. Adapted from Gottschalk *et al.* (1998).

BIOCHEMICAL MECHANISM FOR NMDAR-DEPENDENT LTP

Sweatt has described the complexity of the biochemical mechanisms that underlie LTP and summarized the roles of hundreds of different molecules that have been implicated in LTP induction and expression (Sweatt, 2003). Here we consider only a few of these that have attracted widespread interest. It is known that glutamate receptors and signaling molecules that modulate synaptic transmission are closely associated with the postsynaptic density (PSD) of hippocampal

neurons. Ca^{2+}/calmodulin-dependent protein kinase II (CaMKII), a major constituent of the PSD (see Chapter 12), is a large multimeric enzyme that consists of 10–12 subunits. In its basal state it is inactive due to an autoinhibitory domain that blocks substrate binding. It has several interesting biochemical properties that make it particularly well suited to be a transducer of Ca^{2+} signals and potentially a molecular information-storage device.

CaMKII can autophosphorylate, generating a constitutively active kinase that remains active long after $[Ca^{2+}]_i$ has returned to baseline levels. Phosphorylation disables the autoinhibitory domain by not allowing it to block the active site even after CaM dissociates.

Autophosphorylation, which first occurs on Thr286, allows a transient elevation in Ca^{2+} concentration to be transduced into prolonged kinase activity that persists until the appropriate protein phosphatase dephosphorylates Thr286. Alcino Silva and colleagues explored the potential importance of this site in transgenic mice with a Thr286 point mutation, finding that brain slices from these mice were deficient with respect to LTP and memory formation (Giese et al., 1998).

Although CaMKII can phosphorylate numerous protein substrates, one particularly important to LTP is the AMPAR, which appears to be altered during LTP. The AMPAR is a heteromeric complex composed of various combinations of subunits, GluR1 to GluR4, each complex consisting of four or five such subunits. In the CA1 region of the hippocampus, the AMPAR is composed of GluR1 and GluR2. When in its active state, CaMKII phosphorylates Ser831 on GluR1 with no analogous site on the GluR2 subunit. Interestingly, a recently developed knockout mouse lacking GluR1 shows an LTP deficit in region CA1 (Zamanillo et al., 1999). Phosphorylation of the AMPAR by CaMKII increases the AMPAR-mediated synaptic current. The GluR1 subunit can adopt multiple conductance states ranging from 9 to 28 pS (Derkach et al., 1999). Phosphorylation of the AMPAR is suggested to increase the single-channel conductance from its lower basal conductance state to a higher conductance state. An additional target of kinase regulation in LTP is the insertion of new functional AMPA receptors into the postsynaptic membrane and PSD, increasing synaptic conductance by this mechanism as well. CaMKII has been implicated in controlling this process as well.

Several other protein kinases, including extracellular signal-regulated kinase–mitogen-activated protein kinase (ERK MAPK), PKC, PKA, and the SRC family of tyrosine kinases, have also been demonstrated to contribute to LTP and memory formation using both pharmacologic and genetic engineering approaches (Sweatt, 2003). PKC can phosphorylate Ser831 on GluR1, whereas PKA can phosphorylate the adjacent Ser845. PKC has been suggested to play a role analogous to CaMKII, primarily because increasing PKC activity can enhance synaptic transmission (Carroll et al., 1998), and certain PKC isoforms have been shown to be persistently activated in LTP and memory (Sweatt, 2003). PKC and ERK MAPK have also been demonstrated to control glutamate receptor trafficking in LTP as well, as an additional target for increasing synaptic strength.

Phosphorylation of Ser845 on GluR1 appears to occur under basal conditions, which suggests that PKA may not play a primary role in LTP induction. However, PKA can indirectly support LTP by boosting CaMKII activity. By decreasing competing protein phosphorylation activity by means of phosphorylation of inhibitor 1, an endogenous protein phosphatase inhibitor, PKA can help sustain CaMKII activity (Blitzer et al., 1998). A recently developed mouse model genetically inhibited calcineurin, a protein phosphatase, resulting in increased LTP both in vitro and in vivo (Malleret et al., 2001). This facilitation was PKA dependent, lending support to the importance of a balance between protein kinases and phosphatases in the intact brain. The facilitation of LTP was accompanied by an improvement in learning in hippocampus-dependent tasks, and the improvement was reversed by suppression of transgene expression.

NMDAR function is also altered during LTP due to phosphorylation by tyrosine kinases. The NMDAR is a heteromeric complex consisting of the NR1 subunit in combination with the NR2A–D subunits. The NR2A and NR2B subunits are subject to phosphorylation of the intracellular C-terminal tyrosine residues, thereby relieving a basal zinc inhibition of the NMDAR. Phosphorylation of NR2A or NR2B thus potentiates the current through the NMDAR complex. Consequently, induction of LTP can activate the SRC family of kinases, resulting in phosphorylation and enhancement of the NMDAR-mediated current. The increase in Ca^{2+} influx can then trigger autophosphorylation of CaMKII and persistent activation of PKC, which in due course leads to potentiation of the AMPAR-mediated current, as discussed previously.

Neurons also contain many different protein phosphatases, including protein phosphatases 1, 2A, and 2B (PP1, PP2A, and PP2B or calcineurin). There are significant levels of PP1 and PP2A within the PSD, both of which are effective at dephosphorylating CaMKII, although PP1 is generally thought to be primarily responsible. PP1 normally is suppressed from dephosphorylating CaMKII during LTP via an inhibitor protein called inhibitor 1 (I1). I1 becomes active only once phosphorylated by PKA. A plausible scenario is that, in the initial stages of LTP induction, PKA activation causes it to phosphorylate I1, which in turn inhibits PP1, with the end result of prolonging CaMKII activity.

Hippocampal neurons contain a variety of voltage-dependent ion channels, some of which are expressed nonuniformly across the neuron. A transient K^+ channel whose density increases from the soma outward to the distal dendrites has a dampening effect on the BPAP. Activation of this K^+ current attenuates the BPAP amplitude, decreases the amplitude of EPSPs, and increases the threshold for dendritic spike initiation. Reducing this K^+ current therefore can be expected to have a significant effect on LTP induction. Johnston and coworkers have already identified two mechanisms that can control this K^+ current. First, activation of several different kinases (PKA, PKC, MAPK) can reduce this current (Hoffman and Johnston, 1998). Second, the current can be partly inactivated by trains of summating EPSPs. If a BPAP occurs within 15–20 ms of the onset of an EPSP train, its amplitude is larger in the dendrites. BPAP enlargement due to kinase activation and/or prior EPSP trains could enhance LTP induction through the increased dendritic depolarization (which should facilitate removal of the Mg^{2+} block in the NMDARs) and through enhanced Ca^{2+} influx (through VDCCs), resulting in an increase in dendritic $[Ca^{2+}]_i$ in or near the head of the dendritic spine (Johnston et al., 2000).

Gene Expression and Protein Synthesis

Long-lasting forms of synaptic plasticity are associated with modifications in gene expression and protein synthesis. Although transcription and translational control mechanisms are immensely complex, one central signal transduction mechanism for controlling gene expression and protein translation is the ERK MAPK cascade. ERK MAPK operates to control a critical transcription factor, CREB, and to control the protein synthesis machinery in synaptic plasticity and memory formation in the hippocampus (Sweatt, 2003).

Several specific candidate genes are known to be targets for altered transcription in LTP and memory. For instance, synaptic stimulation that induces LTP also promotes the expression of the immediate early gene activity-regulated cytoskeleton-associated protein (*Arc*) (Steward and Worley, 2001a). Newly synthesized *Arc* mRNA is targeted to synapses that have recently undergone specific forms of synaptic activity where it is locally translated (Steward and Worley, 2001b). Kelly and Deadwyler (2002) have also recently shown an experience-dependent upregulation of *Arc* mRNA in several temporal lobe structures following acquisition of an operant lever-pressing task. The targeting of *Arc* mRNA to specific synapses is interrupted by application of NMDA receptor antagonists (MK-801 or APV), suggesting its dependence on the NMDA receptor (Steward and Worley, 2001b). In addition to mRNA, an accumulation of *Arc* protein is observed in activated synapses. Using an *Arc*-specific antibody, Steward and coworkers (Steward et al., 1998) uncovered a band of newly synthesized *Arc* protein in the same dendritic laminae in which *Arc* mRNA was concentrated.

A large and old literature has consistently pointed to the importance of protein synthesis in certain aspects of learning and/or memory (Matthies, 1989). Application of protein synthesis inhibitors can interfere with the formation and retention of memories. One *in vivo* study suggested that a late stage of LTP, which can last for weeks, depends on protein synthesis (Krug et al., 1984). As described previously, hippocampal LTP experiments using brain slice experiments demonstrate that an early stage of LTP, which lasts for about 1 h and is not dependent on protein synthesis, is followed by a later stage of LTP, which does depend on protein synthesis (Nguyen et al., 1994). Finally, recent work on the role of protein synthesis in memory reconsolidation, or the need for memories to be restabilized after recall, casts a whole new light on the role of protein synthesis in various forms of learning, memory, and synaptic plasticity (Kida et al., 2002; Nader et al., 2000).

The Problem of Targeting Newly Synthesized Proteins

The involvement of protein synthesis in LTP immediately raises the question of how the synthesized proteins are targeted to just those synapses whose enhanced efficacy is to be maintained. This targeting is required because LTP can be remarkably "input specific" (Kelso and Brown, 1986). If neural activity ultimately affects gene expression in the nucleus, then the proteins produced in the soma could in principle be transported to synapses at any location on the dendritic arbor. One possible solution is that a synapse-specific molecular marker or tag—set by neuronal activity during tetanization—can sequester plasticity-related proteins being transported within the dendrites (Frey and Morris, 1998). The idea of an activity-dependent synaptic marker is a very appealing hypothesis, and the mechanisms for this synaptic tagging mechanism are an active area of research at present.

DEPOTENTIATION AND LTD

If synapses can be potentiated and this potentiation is very long lasting, over time the synapses will be driven to their maximum synaptic strength. In this condition there is no longer synaptic plasticity and no further capacity for that synapse to participate in synaptic plasticity–dependent processes. Worse yet, over the lifetime of an animal synapses will by random chance experience LTP-inducing conditions (presynaptic activity coincident with a postsynaptic action potential, e.g.) numerous times. If LTP is irreversible, ultimately every synapse will be maximally potentiated—obviously not a desirable condition vis-à-vis memory storage.

Consideration of this conundrum raises two implications. First, synapses that are involved in lifelong memory storage must be rendered essentially aplastic. In order to have good fidelity of memory storage over the lifetime of an animal a synapse involved in permanent memory storage must be rendered immutable to a change in synaptic strength due to the random occurrence of what would normally be LTP-inducing stimulation. But what about synapses like those in the hippocampus that are not sites of memory storage, but rather whose plasticity is part of the active processing of forming new long-term memories? In order to retain their plasticity and hence their capacity to contribute to memory formation, their potentiation must be reversible. Schaffer collateral synapses can undergo activity-dependent reversal of LTP; a phenomenon termed *depotentiation* (Fig. 19.19). Another activity-dependent way to decrease synaptic strength is long-term depression (LTD), the mirror image of LTP. LTD is a long-lasting decrease of synaptic strength below baseline. Using a logic similar to that of the first paragraph of this section, the phenomenon of dedepression of synaptic transmission is implied, although this has not been widely studied at this point.

As a practical matter it is often difficult to separate depotentiation from LTD experimentally. For example, a "baseline" response in hippocampal slices or *in vivo* likely is a mixture of basal synaptic activity and activity at previously potentiated synapses. Moreover, for the most part the stimulation protocols used to induce depotentiation are variations of the protocols used to induce LTD. Nevertheless, mechanistic investigations have made clear that depotentiation and LTD use different mechanisms (Lee *et al.*, 1998, 2000) and thus must be considered as distinct processes.

Physiologic LTD (and depotentiation) induction protocols generally involve variations of repetitive 1 Hz stimulation (Kemp *et al.*, 2000; Lee *et al.*, 1998). A common protocol is to deliver 900 stimuli at 1 Hz, but there also are LTD protocols that use random small variations in frequency in the 1 Hz region, and variations that use paired-pulse stimuli delivered at 1 Hz. Synaptic depression appears to be fairly robust *in vivo* but is quite difficult to induce in hippocampal slices from adult animals. LTD *in vitro* is almost always studied using slices from immature animals, or cultured immature neurons, and it is possible that LTD as it is currently studied *in vitro* is largely a manifestation of what is normally a developmental mechanism.

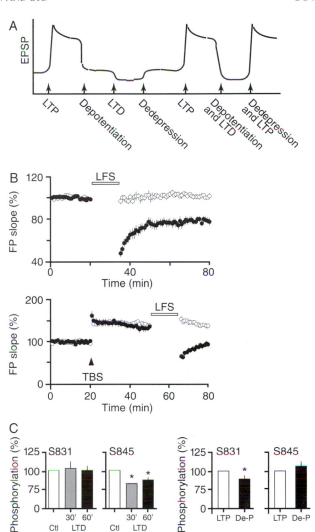

FIGURE 19.19 Depotentiation and LTD. (A): Schematic illustrating LTP, depotentiation, LTD, dedepression, and combinations of them. (B): LTD and depotentiation in hippocampal neurons. Simultaneous recording of slices receiving baseline stimulation (control, open circles) and 1 Hz stimulation (closed circles). FP, field potential. Regulation of distinct AMPA receptor phosphorylation sites during bidirectional synaptic plasticity. (C): Homosynaptic LTD in CA1 is associated with dephosphorylation of GluR1 at a PKA site (Ser845). Depotentiation gives dephosphorylation at a CaMKII–PKC site (Ser831). Adapted with permission from Macmillan Publishers Ltd. Lee, H.K., Barbarosie, M., Kameyama, K., Bear, M.F., and Huganir, R.L. (2000). Regulation of distinct AMPA receptor phosphorylation sites during bidirectional synaptic plasticity. *Nature*, **405**, 955–959.

One ironic aspect of the LTP–LTD story is that both phenomena at Schaffer collateral synapses can be blocked by NMDA receptor antagonists. This suggests that calcium influx triggers both processes, and indeed current models of LTD induction hypothesize that LTD is caused by an influx of calcium that achieves a lower level than that needed for LTP induction. This lower level of calcium is hypothesized to selectively activate protein phosphatases, and by this mechanism lower synaptic efficacy through AMPA receptor dephosphorylation and AMPA receptor internalization (Figs. 19.8 and 19.19).

Another very different type of LTD is cerebellar LTD. Cerebellar LTD occurs at synapses onto Purkinje neurons in the cerebellar cortex. Cerebellar LTD is a very interesting phenomenon because its behavioral role is much better understood than the hippocampal plasticity phenomena we are discussing throughout this book. Among other things, cerebellar LTD is involved in associative eyeblink conditioning, a cerebellum-dependent classical conditioning paradigm. Considerable progress has been made in investigating the roles and mechanisms of cerebellar LTD, as will be described in a later section of this chapter.

SUMMARY

Like learning, LTP and LTD are long-lasting changes in synaptic efficacy that occur in response to a transient stimulus. In many cases the persistence of these effects has been demonstrated to extend many hours *in vitro* and several weeks *in vivo*. We do not know how LTP and LTD relate to memory, especially in the hippocampus, although there is a considerable amount of evidence available supporting the hypothesis that hippocampal LTP is involved in memory formation. Regardless, LTP and LTD are the best understood examples of long-lasting synaptic plasticity in the mammalian CNS, and they are cellular models for how long-lasting memory-associated changes are likely to occur in the CNS. One premise underlying many studies of LTP and LTD is that understanding these phenomena will yield valid insights into the mechanisms of plasticity that underlie learning and memory in the brain. The bona fide changes in neuronal connections that occur *in vivo* may or may not be identical to LTP and LTD as they are presently studied in the laboratory, but this does not diminish their utility as cellular model systems for studying lasting neuronal change in the mammalian CNS. Finally, in two notable cases, cerebellar LTD and amygdalar LTP, these plasticity phenomena have been strongly implicated as directly underlying memory-associated behavioral changes. These two specific examples will be addressed in more detail in subsequent sections of this chapter.

PARADIGMS HAVE BEEN DEVELOPED TO STUDY ASSOCIATIVE AND NONASSOCIATIVE LEARNING

Associative Learning

Associative learning is a broad category that includes much of our daily learning activities that involve the formation of associations among stimuli and/or responses. It is usually subdivided into classical conditioning and instrumental conditioning. Classical (or Pavlovian) conditioning is induced by a procedure in which a generally neutral stimulus, termed a *conditioned stimulus* (CS), is paired with a stimulus that generally elicits a response, termed an *unconditioned stimulus* (US). Two examples of unconditioned stimuli are food that elicits salivation and a shock to the foot that elicits limb withdrawal.

Instrumental (or *operant*) *conditioning* is a process by which an organism learns to associate consequences with its own behavior. In an operant conditioning paradigm, the delivery of a reinforcing stimulus is contingent on the expression of a designated behavior. The probability that this behavior will be expressed is then altered. This chapter focuses on classical conditioning because it is mechanistically the best-understood type of associative conditioning.

An astute observation by Ivan Pavlov, a Russian physiologist who had been studying digestion in dogs, led to his discovery of classical conditioning. He first noticed that the mere sight of the food dish caused dogs to salivate. He continued the experiments to see if dogs would also salivate in response to a bell rung at feeding time. Pavlov trained dogs to stand in a harness and, after the sound of a bell, fed them meat powder (Fig. 19.20) (Rachlin, 1991). At first, the bell by itself did not elicit any response, but the meat powder elicited reflex salivation, which was termed the *unconditioned response*. He noted that after a few pairings of the bell and meat powder the dogs began to salivate when the bell rang, before they received the meat powder. This response is termed the *conditioned response*. This type of conditioning came to be called *reward* or *appetitive classical conditioning*. If the bell or another stimulus was followed by an unpleasant event, such as an electric shock, then a variety of autonomic responses became conditioned. This type of conditioning is often termed *aversive* or *fear conditioning*. Skeletal muscle movements appropriate to deal with the US (e.g., leg flexion after a shock delivered to a paw) are also learned in aversive classical conditioning.

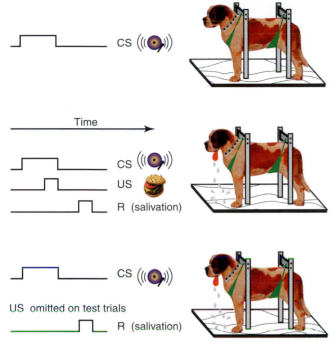

FIGURE 19.20 Classical conditioning. In the procedure introduced by Pavlov, the production of saliva is monitored continuously. Presentation of meat powder reliably leads to salivation, whereas some "neutral" stimulus such as a bell initially does not. With repeated pairings of the bell and meat powder, the animal learns that the bell predicts the food and salivates in response to the bell alone. Modified from Rachlin (1991).

Traditionally, classical or Pavlovian conditioning is an operation that pairs one stimulus, the conditioned stimulus, with a second stimulus, the unconditioned stimulus, as noted earlier. The US reliably elicits a response termed the unconditioned response (UR). Repeated pairings of the CS and US result in the CS eliciting a response, which is defined as the conditioned response (CR). The CR is often similar to the UR—for example, in Pavlov's experiment both were salivation—but this is not always the case; for example, in fear conditioning in rat (see later) the initial UR to foot shock is increased activity but the CR is freezing (Fanselow and Poulos, 2005). Conditioning procedures in which the CS and US overlap in time are called *delay conditioning,* whereas in trace conditioning a short interval is interposed between the CS and the US. Although the traditional view of Pavlovian conditioning emphasized the contiguity of the CS and US, a more general and contemporary view of Pavlovian conditioning emphasizes the informational relationship between the CS and the US. In other words, the information that the CS provides about the occurrence of the US, the *contingency* between occurrence of the CS and the US, is the critical feature for learning (Rescorla, 1988).

Nonassociative Learning

Three examples of nonassociative learning have received the most experimental attention: habituation, dishabituation, and sensitization. *Habituation* is defined as a reduction in the response to a stimulus that is repeatedly delivered. *Dishabituation* refers to the restoration or recovery of a habituated response due to the presentation of another, typically strong, stimulus to the animal. *Sensitization* is an enhancement or augmentation of a response produced by the presentation of a strong stimulus. The following sections introduce the *Aplysia* and focus on the neural and molecular mechanisms of sensitization.

INVERTEBRATE STUDIES: KEY INSIGHTS FROM *APLYSIA* INTO BASIC MECHANISMS OF LEARNING

Since the mid-1960s, the marine mollusk *Aplysia* has proven to be an extremely useful model system for gaining insights into the neural and molecular mechanisms of simple forms of memory. Indeed, the pioneering discoveries of Eric Kandel using this animal were recognized by his receipt of the Nobel Prize in Physiology or Medicine in 2000. A number of characteristics make *Aplysia* well suited for examination of the molecular, cellular, morphological, and network mechanisms underlying neuronal modifications (plasticity) and learning and memory. The animal has a relatively simple nervous system with large, individually identifiable neurons that are accessible for detailed anatomical, biophysical, biochemical, and molecular studies. Neurons and neural circuits that mediate many behaviors in *Aplysia* have been identified precisely. In several cases, these behaviors have been shown to be modified by learning. Moreover, specific loci within neural circuits at which modifications occur during learning have been identified, and aspects of the cellular mechanisms underlying these modifications have been analyzed (Byrne *et al.*, 1993; Byrne and Kandel, 1996; Hawkins *et al.*, 1993).

The Siphon–Gill and Tail–Siphon Withdrawal Reflexes of *Aplysia*

Within the mantle cavity of *Aplysia* lies the respiratory organ of the animal, the gill, and protruding from the mantle cavity is the siphon (Fig. 19.21). The siphon–gill withdrawal reflex is elicited when a tactile or electrical stimulus is delivered to the siphon; the stimulus causes withdrawal of the siphon and gill (Fig. 19.21A). A second behavior that has been examined extensively is the tail–siphon withdrawal reflex. Tactile or electrical stimulation of the tail elicits a coordinated set of defensive responses composed of a reflex withdrawal of the tail and the siphon (Fig. 19.21B).

These two defensive reflexes in *Aplysia* can exhibit three forms of nonassociative learning: habituation, dishabituation, and sensitization. A single sensitizing stimulus, such as a brief several-second-duration electric shock, can produce a reflex enhancement that lasts minutes (short-term sensitization), whereas prolonged

training (e.g., multiple stimuli over an hour or longer) produces an enhancement that lasts from days to weeks (long-term sensitization). *Aplysia* also exhibit several forms of associative learning, including classical conditioning (see later) and operant conditioning (Botzer *et al.*, 1998; Brembs *et al.*, 2002; Cook and Carew, 1989).

A prerequisite for successful analysis of the neural and molecular bases of these different forms of learning is an understanding of the neural circuit that controls the behavior. The afferent limb of the siphon–gill withdrawal reflex consists of a population of approximately 24 sensory neurons with somata in the abdominal ganglion. The siphon sensory neurons (SN) monosynaptically excite a population of approximately 13 gill and siphon motor neurons (MN) that are also located in the abdominal ganglion (Fig. 19.22A). Activation of the gill and siphon motor neurons leads to contraction of the gill and siphon. Excitatory, inhibitory, and modulatory interneurons

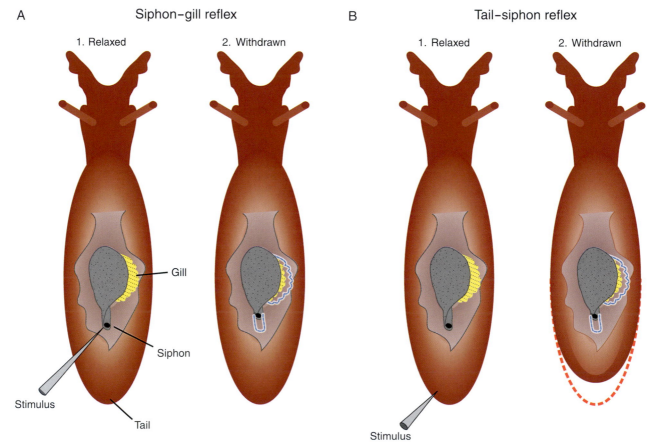

FIGURE 19.21 Siphon–gill and tail–siphon withdrawal reflexes of *Aplysia*. (A) Siphon–gill withdrawal. Dorsal view of *Aplysia*: (1) relaxed position; (2) a stimulus (e.g., a water jet, brief touch, or weak electric shock) applied to the siphon causes the siphon and the gill to withdraw into the mantle cavity. (B) Tail–siphon withdrawal reflex: (1) relaxed position; (2) a stimulus applied to the tail elicits a reflex withdrawal of the tail, the siphon, and the gill.

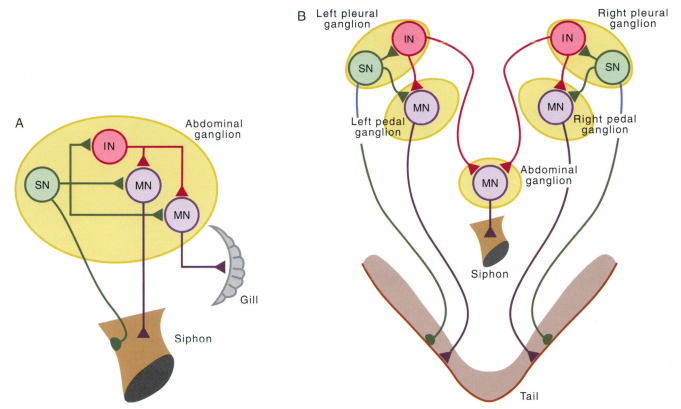

FIGURE 19.22 Simplified circuit diagrams of the siphon–gill (A) and tail–siphon (B) withdrawal reflexes. Stimuli activate the afferent terminals of mechanoreceptor sensory neurons (SN), the somata of which are located in central ganglia (abdominal, pedal, and pleural). The sensory neurons make excitatory synaptic connections (triangles) with interneurons (IN) and motor neurons (MN). The excitatory interneurons provide a parallel pathway for excitation of the motor neurons. Action potentials elicited in the motor neurons, triggered by the combined input from the SNs and INs, propagate out peripheral nerves to activate muscle cells and produce the subsequent reflex withdrawal of the organs. Modulatory neurons (not shown here, but see Fig. 19.23A1), such as those containing serotonin (5-HT), regulate the properties of the circuit elements and, consequently, the strength of the behavioral responses.

(IN) in the withdrawal circuit have also been identified, although only excitatory interneurons are illustrated in Figure 19.22. The afferent limb of the tail–siphon withdrawal reflex consists of a bilaterally symmetrical cluster of approximately 200 sensory neurons located in the left and right pleural ganglia. These sensory neurons make monosynaptic excitatory connections with at least three motor neurons in the adjacent pedal ganglion, which produce withdrawal of the tail (Fig. 19.22B). In addition, the tail sensory neurons form synapses with various identified excitatory and inhibitory interneurons. Some of these interneurons activate motor neurons in the abdominal ganglion, which controls reflex withdrawal of the siphon. Moreover, several additional neurons modulate the tail–siphon withdrawal reflex (Cleary *et al.,* 1995) (Fig. 19.23A1).

The sensory neurons for both the siphon–gill and tail–siphon withdrawal reflexes are similar and appear to be important, although probably not exclusive (e.g., Cleary *et al.,* 1998), sites of plasticity in their respective neural circuits. Changes in their membrane properties and the strength of their synaptic connections (synaptic efficacy) are associated with sensitization.

Multiple Cellular Processes and Short- and Long-Term Sensitization in *Aplysia*

Short-Term Sensitization

Short-term (minutes) sensitization is induced when a single brief train of shocks to the body wall results in the release of modulatory transmitters, such as serotonin (5-HT), from a separate class of interneurons (INs) referred to as *facilitatory neurons* (Fig. 19.23A1). These facilitatory neurons regulate the properties of the sensory neurons and the strength of their connections with postsynaptic interneurons and motor neurons (MNs) through a process called *heterosynaptic facilitation* (Byrne and Kandel, 1996) (Figs. 19.23A2 and A3). The molecular mechanisms contributing

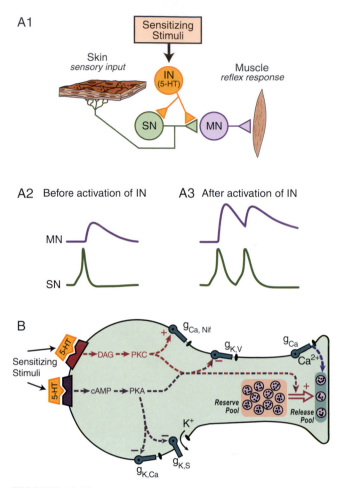

FIGURE 19.23 Model of short-term heterosynaptic facilitation of the sensorimotor connection that contributes to short- and long-term sensitization in *Aplysia*. (A1) Sensitizing stimuli activate facilitatory interneurons (IN) that release modulatory transmitters, one of which is 5-HT. The modulator leads to an alteration of the properties of the sensory neuron (SN). (A2, A3) An action potential in SN after the sensitizing stimulus results in greater transmitter release and, hence, a larger postsynaptic potential in the motor neuron (MN, A3) than an action potential before the sensitizing stimulus (A2). For short-term sensitization, the enhancement of transmitter release is due, at least in part, to broadening of the action potential and an enhanced flow of Ca^{2+} (I_{Ca}) into the sensory neuron. (B) Model of a sensory neuron that depicts the multiple processes for short-term facilitation that contribute to short-term sensitization. 5-HT released from facilitatory neurons binds to at least two distinct classes of receptors on the outer surface of the membrane and leads to the transient activation of two intracellular second messengers, DAG and cAMP, and their respective kinases (PKC and PKA). These two kinases affect multiple cellular processes, the combined effects of which lead to enhanced transmitter release when subsequent action potentials are fired in the sensory neuron (see text for additional details). Modified from Byrne and Kandel (1996).

to heterosynaptic facilitation are illustrated in Figure 19.23B. The first step is the binding of 5-HT to one class of receptors on the outer surface of the membrane of the sensory neurons. This leads to the activation of

adenylyl cyclase, which in turn leads to an elevation of the intracellular level of the second-messenger cyclic AMP (cAMP) in sensory neurons. When cAMP binds to the regulatory subunit of cAMP-dependent protein kinase (protein kinase A, or PKA), the catalytic subunit is released and can now add phosphate groups to specific substrate proteins, thereby altering their functional properties. One consequence of this protein phosphorylation is an alteration in the properties of membrane channels. Specifically, the increased levels of cAMP lead to a decrease in the serotonin-sensitive potassium conductance [S-K^+ conductance ($g_{K,S}$)], a component of the calcium-activated K^+ conductance ($g_{K,Ca}$) and the delayed K^+ conductance ($g_{K,V}$). (See Chapters 5 and 6 for more information on these channel types.) These changes in membrane currents lead to depolarization of the membrane potential, enhanced excitability, and an increase in the duration of the action potential (i.e., spike broadening). Reflections of enhanced excitability include an increase in the number of action potentials elicited in a sensory neuron by a fixed extrinsic current injected into the cell or by a fixed stimulus to the skin.

Cyclic AMP also appears to activate a facilitatory process that is independent of membrane potential and spike duration. This process is represented in Figure 19.23B (large arrow) as the translocation or mobilization of transmitter vesicles from a reserve pool to a releasable pool. The translocation makes more transmitter-containing vesicles available for release with subsequent action potentials in the sensory neuron. The overall effect is a short-term cAMP-dependent enhancement of transmitter release. One substrate for cAMP–PKA-dependent phosphorylation is synapsin (Angers *et al.*, 2002). Synapsin is a synaptic vesicle-associated protein that tethers synaptic vesicles to cytoskeletal elements and thus helps control the reserve pool of vesicles in synaptic terminals (see Chapter 8). Phosphorylation of synapsin would allow vesicles in the reserve pool to migrate to the releasable pool and thus contribute to enhanced transmitter release (Fioravante *et al.*, 2007).

Serotonin also acts through another class of receptors to increase the level of the second messenger DAG. DAG activates PKC, which, like PKA, contributes to facilitation that is independent of spike duration (e.g., mobilization of vesicles). In addition, PKC regulates a nifedipine-sensitive Ca^{2+} channel ($g_{Ca, Nif}$) and the delayed K^+ channel ($g_{K,V}$). Thus, the delayed K^+ channel ($g_{K,V}$) is dually regulated by PKC and PKA. The modulation of $g_{K,V}$ contributes importantly to the increase in duration of the action potential (Fig. 19.23A3). Because of its small magnitude, the modulation of $g_{Ca, Nif}$ appears to play a minor role in the facilitatory process.

Prolonged treatment of 5-HT (1.5 h) also activates MAPK (Martin *et al.*, 1997). However, this pathway was originally suggested to be important only for the induction of longer-term processes (see later). Of general significance is the observation that a single modulatory transmitter (i.e., 5-HT) activates multiple kinase systems. The involvement of multiple second-messenger systems in synaptic plasticity also appears to be a theme emerging from mammalian studies. For example, the induction of LTP in the CA1 area of the hippocampus (see section on LTP–LTD) appears to involve MAPK, PKC, CaMKII, and tyrosine kinase (reviewed in Dineley *et al.*, 2001).

The consequences of activating these multiple second-messenger systems and modulating these various cellular processes are expressed when test stimuli elicit action potentials in the sensory neuron at various times after the presentation of the sensitizing stimuli (Fig. 19.23A3). More transmitter is available for release as a result of the mobilization process, and each action potential is broader, allowing a larger influx of Ca^{2+} to trigger release of the available transmitter. The combined effects of mobilization and spike broadening lead to facilitation of transmitter release from the sensory neuron and, consequently, to a larger postsynaptic potential in the motor neuron. Larger postsynaptic potentials lead to enhanced activation of interneurons and motor neurons and, thus, to an enhanced behavioral response. The maintenance of short-term sensitization is dependent on the persistence of the PKA-, PKC-induced phosphorylations of the various substrate proteins.

Long-Term Sensitization

Sensitization also exists in a long-term form, which persists for at least 24 h. Whereas short-term sensitization can be produced by a single brief stimulus, the induction of long-term sensitization requires a more extensive training period of an hour or longer.

Both short- and long-term sensitization share some common cellular pathways during their induction. For example, both forms activate the cAMP–PKA cascade (Fig. 19.24). However, in the long-term form, unlike the short-term form, activation of the cAMP–PKA cascade induces gene transcription and new protein synthesis (Byrne *et al.*, 1993; Hawkins *et al.*, 1993). Repeated training leads to a translocation of PKA to the nucleus where it phosphorylates the transcriptional activator CREB1 (cAMP-responsive element-binding protein). CREB1 binds to a regulatory region of genes known as *CRE* (cAMP-responsive element). Next this bound and phosphorylated form of

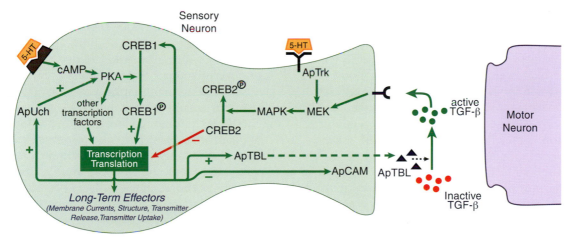

FIGURE 19.24 Simplified scheme of the mechanisms in sensory neurons that contribute to long-term sensitization and some aspects of short-term sensitization. Sensitization training leads to the release of 5-HT and the cAMP-dependent regulation of short-term effectors (see Fig. 19.23 for details) and phosphorylation of CREB1. 5-HT, either by acting through cAMP or though activation of a tyrosine receptor kinase-like molecule (ApTrk), also leads to the activation of MAPK, which regulates the repressor CREB2. The combined effects of activation of CREB1 and derepression of CREB2 lead to regulation of the synthesis of at least 10 proteins, only four of which (ApTBL, ApCAM, ApUch, and CREB1) are shown. Three of these proteins (ApTBL, ApUch, and CREB1) appear to be components of positive feedback loops. ApTBL is believed to activate latent forms of TGF-β, which can then bind to receptors on the sensory neuron. TGF-β activates MAPK, which can have both acute and long-term effects. One of its acute effects is the regulation of transmitter release. MAPK may also act by initiating a second round of gene regulation by affecting CREB2-dependent pathways. Increased synthesis of ApUch leads to enhanced degradation of the regulatory subunit of PKA, leading to enhanced activation of the catalytic subunit and increased phosphorylation of CREB1. Synthesis of CREB1 itself is increased, further boosting CRE-dependent transcriptional regulation. The fourth protein, ApCAM, is down-regulated. Down-regulation of ApCAM is involved in regulating growth processes associated with long-term facilitation.

CREB1 leads to increased transcription. Serotonin, either by acting through cAMP (Martin *et al.*, 1997) or though activation of a tyrosine receptor kinase-like molecule (ApTrk) (Ormond *et al.*, 2004) leads to the activation of MAPK, which phosphorylates the transcriptional repressor CREB2. Phosphorylation of CREB2 by MAPK leads to a derepression of CREB2 and therefore promotes CREB1-mediated transcriptional activation (Bartsch *et al.*, 1995).

The role of transcription factors in long-term memory formation is not limited to the induction phase but may also extend to the consolidation phase, where consolidation is defined as the time window during which RNA and protein synthesis are required for converting short- to long-term memory. For example, treatment of ganglia with five pulses of 5-HT over a 1.5 h period to mimic sensitization training leads to the binding of CREB1 to the promoter of its own gene and induces CREB1 synthesis (Liu *et al.*, 2008; Mohamed *et al.*, 2005). This observation agrees well with earlier findings that the requirement for gene expression is not limited to the induction phase. The necessity of prolonged transcription and translation for LTF observed at 24 h persists for at least 7–9 h after induction.

The combined effects of activation of CREB1 and derepression of CREB2 lead to changes in the synthesis of specific proteins. So far, more than 10 gene products that are regulated by sensitization training have been identified, and others are likely to be found in the future. These results indicate that as was the case with LTP, there is not a single memory gene or protein, but that multiple genes are regulated, and they act in a coordinated way to alter neuronal properties and synaptic strength. In the following section, four regulated proteins of particular significance are discussed.

The down-regulation of a homologue of a neuronal cell adhesion molecule (NCAM), ApCAM, plays a key role in long-term facilitation. This down-regulation has two components. First, the synthesis of ApCAM is reduced. Second, preexisting ApCAM is internalized via increased endocytosis. The internalization and degradation of ApCAM allow for the restructuring of the axon arbor (Bailey *et al.*, 1992). The sensory neuron can now form additional connections with the same postsynaptic target or make new connections with other cells. Another protein whose synthesis is regulated is *Aplysia* tolloid–BMP-like protein (ApTBL–1). Tolloid and the related molecule BMP–1 appear to function as secreted Zn^{2+} proteases. In some preparations, they activate members of the transforming growth factor β (TGF-β) family. Indeed, in sensory neurons, TGF-β mimics the effects of 5-HT in that it produces long-term increases in synaptic strength of the

sensory neurons (Zhang *et al.*, 1997). Interestingly, TGF-β activates MAPK in the sensory neurons and induces its translocation to the nucleus. Thus, TGF-β could be part of an *extracellular* positive feedback loop possibly leading to another round of protein synthesis (Fig. 19.24) to further consolidate the memory (Zhang *et al.*, 1997). A third important protein, *Aplysia* ubiquitin hydrolase (ApUch), appears to be involved in an *intracellular* positive feedback loop. During induction of long-term facilitation, ApUch levels in sensory neurons are increased, possibly via CREB phosphorylation and a consequent increase in ApUch transcription. The increased levels of ApUch increase the rate of degradation, via the ubiquitin–proteosome pathway, of proteins including the regulatory subunit of PKA (Chain *et al.*, 1999). The catalytic subunit of PKA, when freed from the regulatory subunit, is highly active. Thus, increased ApUch will lead to an increase in PKA activity and more protracted phosphorylation of CREB1 (Liu *et al.*, 2008). This phosphorylated CREB may act to further prolong ApUch expression, thereby closing a positive feedback loop. Moreover, the consequent activation of PKA can phosphorylate other substrate proteins. A fouth important protein is CREB1, which as mentioned previously, shows a CRE-dependent increase in expression with LTF. The combination of its increased expression post induction and its persistent phosphorlation closes another intracellular positive feedback loop that is necessary for memory consolidation (Liu *et al.*, 2008).

One simplifying hypothesis is that the mechanisms underlying the expression of short- and long-term sensitization are the same but extended in time for long-term sensitization. Some evidence supports this hypothesis. For example, long-term sensitization, like short-term sensitization, is associated with an enhancement of sensorimotor connections. In addition, K^+ currents and the excitability of sensory neurons are modified by long-term sensitization (Cleary *et al.*, 1998). However, the mechanisms underlying the *expression* of short- and long-term sensitization differ in four fundamental ways. First, long-term sensitization training leads to a *decrease* in the duration of the sensory neuron action potential (Antzoulatos and Byrne, 2007). Second, structural changes such as neurite outgrowth and active zone remodeling are associated with the expression of long-term but not with short-term sensitization (Bailey and Kandel, 1993). Interestingly, the structural changes take time to develop. Long-term synaptic facilitation measured 1 day after 4 days of training is associated with structural changes, whereas long-term facilitation measured 1 day after a single training session is not (Wainwright *et al.*, 2004). Third, long-term sensitization is associated with an increase in high-affinity glutamate uptake (Levenson *et al.,* 2000). A change in glutamate uptake could potentially exert a significant effect on synaptic efficacy by regulating the amount of transmitter available for release, the rate of clearance from the cleft, and thereby the duration of the EPSP and the degree of receptor desensitization. Finally, long-term sensitization has been correlated with changes in the postsynaptic cell (i.e., the motor neuron; Cleary *et al.,* 1998). Thus, as with other examples of memory, multiple sites of plasticity exist even within this simple reflex system.

Other Temporal Domains for the Memory of Sensitization

Operationally, memory has frequently been divided into two temporal domains: short term and long term. It has become increasingly clear from studies of a number of memory systems that this distinction is overly restrictive. For example, in *Aplysia,* Carew and his colleagues (Sutton *et al.,* 2001, 2002) and Kandel and his colleagues (Ghirardi *et al.,* 1995) discovered an intermediate phase of memory that has distinctive temporal characteristics and a unique molecular signature. The intermediate-phase memory for sensitization is expressed at times approximately 30 min to 3 h after the beginning of training. Like long-term sensitization, its induction requires protein synthesis, but unlike long-term memory it does not require mRNA synthesis. The expression of the intermediate-phase memory requires the persistent activation of PKA.

In addition to intermediate-phase memory, it is likely that *Aplysia* has different phases of long-term memory. For example, at 24 h after sensitization training, there is increased synthesis of a number of proteins, some of which are different from those whose synthesis is increased during and immediately after training. These results suggest that the memory for sensitization that persists for times longer than 24 h may be dependent on the synthesis of proteins occurring at 24 h and may have a different molecular signature from the 24 h memory.

MECHANISMS UNDERLYING ASSOCIATIVE LEARNING IN *APLYSIA*

Classical Conditioning of Withdrawal Reflexes in *Aplysia*

The withdrawal reflexes of *Aplysia* are subject to classical conditioning (Byrne *et al.,* 1993; Hawkins *et al.,* 1993). The short-term classical conditioning observed at the behavioral level reflects, at least in part, a cellular mechanism called *activity-dependent neuromodulation.* A diagram of the general scheme is presented in Figure 19.25. The US pathway is activated by a shock to the animal, which elicits a withdrawal response (the UR). When a CS is consistently paired with the US, the animal develops an enhanced withdrawal response (CR) to the CS. Activity-dependent neuromodulation is proposed as the mechanism for this pairing-specific effect. The US activates both a motor neuron (UR) and a modulatory system. The modulatory system delivers the neurotransmitter serotonin (5-HT) to all the sensory neurons (parts of the various CS pathways), which leads to a nonspecific enhancement of transmitter release from the sensory neurons. This nonspecific enhancement contributes to short-term sensitization (see prior discussion). Sensory neurons whose activity is temporally contiguous with the US-mediated reinforcement are additionally modulated. Spiking in a sensory neuron during the presence of 5-HT leads to changes in that cell relative to other sensory neurons whose activity was not paired with the US. Thus, a subsequent CS will lead to an enhanced activation of the reflex (Fig. 19.25B).

Figure 19.26 illustrates a more detailed model of the proposed cellular mechanisms responsible for this example of classical conditioning. The modulator (US) acts by increasing the activity of a Type I adenylyl cyclase (AC), which in turn increases the levels of cAMP in the sensory neuron. Spiking in the sensory neurons (CS) leads to increased levels of intracellular calcium, which greatly enhances the action of the modulator to increase the cAMP cascade. This system determines CS–US contiguity by a method of coincidence detection at the presynaptic terminal.

Now consider the postsynaptic side of the synapse. The postsynaptic region contains NMDA-type receptors (see section on LTP–LTD and Chapters 9 and 11). These receptors need concurrent delivery of glutamate and depolarization to allow calcium to enter. The glutamate is provided by the activated sensory neuron (CS), and the depolarization is provided by the US (Antonov *et al.,* 2003; Bao *et al.,* 1998; Murphy and Glanzman, 1997; Roberts and Glanzman, 2003) (for review see Lechner and Byrne, 1998). Thus, the postsynaptic neuron provides another example of coincidence detection. This aspect of plasticity at the sensorimotor synapse is similar to the phenomenon of LTP in the CA1 region of hippocampus (see section on LTP–LTD). The increase in intracellular calcium putatively causes a retrograde signal to be released from the postsynaptic to the presynaptic terminal, ultimately acting to further enhance the cAMP cascade in the sensory neuron. The overall amplification

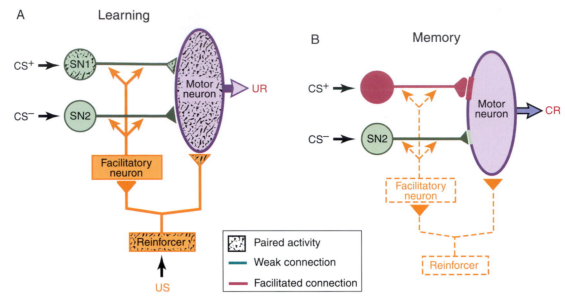

FIGURE 19.25 Model of classical conditioning of a withdrawal reflex in *Aplysia*. (A) Activity in a sensory neuron (SN1) along the CS⁺ (paired) pathway is coincident with activity in neurons along the reinforcement pathway (US). However, activity in the sensory neuron (SN2) along the CS⁻ (unpaired) pathway is not coincident with activity in neurons along the US pathway. The US directly activates the motor neuron, producing the UR. The US also activates a modulatory system in the form of the facilitatory neuron, resulting in the delivery of a neuromodulatory transmitter to the two sensory neurons. The pairing of activity in SN1 with the delivery of the neuromodulator yields the associative modifications. (B) After the paired activity in (A), the synapse from SN1 to the motor neuron is selectively enhanced. Thus, it is more likely to activate the motor neuron and produce the conditioned response (CR) in the absence of US input. Modified from Lechner and Byrne (1998).

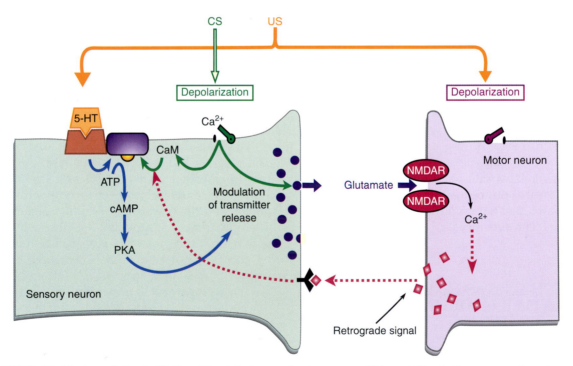

FIGURE 19.26 Model of associative facilitation at the *Aplysia* sensorimotor synapse. This model has both a presynaptic and a postsynaptic detector for the coincidence of the CS and the US. Furthermore, a putative retrograde signal allows for the integration of these two detection systems at the presynaptic level. The CS leads to activity in the sensory neuron, yielding presynaptic calcium influx, which enhances the US-induced cAMP cascade. The CS also induces glutamate release, which results in postsynaptic calcium influx through NMDA receptors if paired with the US-induced depolarization of the postsynaptic neuron. The postsynaptic calcium influx putatively induces a retrograde signal, which further enhances the presynaptic cAMP cascade. The end result of the cAMP cascade is to modulate transmitter release and enhance the strength of the synapse. Modified from Lechner and Byrne (1998).

of the cAMP cascade acts to raise the level of PKA, which in turn leads to the modulation of transmitter release. These activity-dependent changes enhance synaptic efficacy between the specific sensory neuron of the CS pathway and the motor neuron. Thus, the sensory neuron along the CS pathway is better able to activate the motor neuron and produce the CR.

Operant Conditioning of Feeding Behavior in *Aplysia*

Although this section has focused on the use of withdrawal reflexes of *Aplysia* as a model system to study sensitization and classical conditioning, the study of feeding behavior of *Aplysia* has recently begun to reveal the mechanisms underlying operant conditioning. Feeding behavior in *Aplysia* exhibits several features that make it amenable to the study of learning. For example, the behavior occurs in an all-or-nothing manner and is therefore easily quantified, and the neural circuitry underlying the generation of the behavior, termed the *central pattern generator* (CPG), is well characterized to the extent that many of the key individual neurons responsible for the generation of feeding movements have been identified. As with the study of classical conditioning of the defensive withdrawal reflex, researchers are exploiting the advantages of *Aplysia* and identifying loci of plasticity and changes in membrane properties in the key neurons of the CPG that occur during operant conditioning.

In an operant conditioning paradigm, the delivery of a reinforcing stimulus is contingent on the expression of a designated behavior. In the case of feeding behavior, the operant behavior is ingestive or biting movements and the reinforcement is a stimulus to the esophageal nerve (En), which, based on previous work, is a neural pathway enriched in dopamine processes and which signals the presence of food in the mouth (Kabotyanski, *et al.*, 1998; Lechner *et al.*, 2000). A training protocol was developed in which, over a 10 min training period, the En was stimulated each time the animal performed a spontaneous biting movement (contingent reinforcement) (Fig. 19.27) (Brembs *et al.*, 2002). Control animals received the same number of stimulations over the 10 min period; however, they were explicitly unpaired. Learning was assessed by measuring the number of spontaneous bites during a test period 1 and 24 h following training. The group of animals that had received paired training showed a significantly larger number of spontaneous bites both 1 and 24 h after training compared with control animals.

The development of behavioral protocols for operant conditioning allowed for further investigations into the mechanisms underlying the learning. Neural

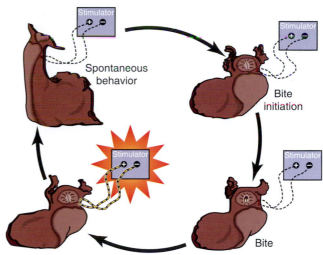

FIGURE 19.27 Operant conditioning of feeding behavior. Throughout the experiment, the animal was observed and all bites were recorded. In the contingent reinforcement group a bite was immediately followed by a brief electric stimulation of the esophageal nerve. A control group received the same sequence of stimulations as the contingent group, but the stimulation was uncorrelated with the animal's behavior. Experimental sessions consisted of a 5 min pretest, 10 min of training, and a final test period. In each period, the number of bites was recorded. The final test period was either 1 or 24 h after training. The group of animals that had received contingent reinforcement showed a significantly larger number of spontaneous bites both 1 and 24 h after training compared with control animals. Modified from Brembs *et al.* (2002).

correlates of operant conditioning were identified by removing the buccal ganglia, where the CPG underlying feeding behavior is located, from recently trained animals and studying the change in cellular properties of key cells that are essential for producing the neural activity underlying biting movements. The CPG in the buccal ganglia produces activity termed *buccal motor programs* (BMPs), which underlie the various types of feeding movements. These BMPs have been well characterized (Church and Lloyd, 1994; Morton and Chiel, 1993a,b), and one cell that has proven to be essential in the generation of BMPs underlying biting movements (ingestive-like BMPs) is cell B51 (Nargeot *et al.*, 1999b,c). Therefore, changes in the cellular properties of B51 were measured in ganglia taken from trained and untrained animals. This analysis revealed that the input resistance was significantly higher and the burst threshold significantly lower in B51 of ganglia from trained animals compared with untrained animals. These types of changes would serve to increase the probability that B51 would become active and would therefore facilitate the generation of the neural activity underlying biting movements in the trained animals.

Because the ganglia of *Aplysia* can be removed and studied for extended periods in isolation, *in vitro* and single-cell analogues of operant conditioning can be developed and used not only to further probe the network, cellular, and membrane properties supporting training but also to validate the identified mechanisms underlying learning at multiple levels of analysis. An *in vitro* analogue of operant conditioning has been developed in which buccal ganglia from naïve animals are removed and placed in a chamber (Nargeot *et al.*, 1997, 1999b,c). Over a 10 min training period, every ingestion-like BMP that occurred was reinforced by stimulation of the En (paired or contingent reinforcement). Controls consisted of ganglia that received explicitly unpaired (noncontingent) delivery of the same number of En stimulations. "Learning" was assessed by measuring the number of ingestion-like BMPs generated 1 h following training in both groups. The ganglia from the contingent reinforcement group generated significantly more ingestion-like BMPs than the control group. In addition, similar to what was seen in the neural correlates in *in vivo* operant conditioning, in ganglia that had received contingent reinforcement, the burst threshold of cell B51 was significantly lower and the input resistance was significantly higher compared with control ganglia (Nargeot *et al.*, 1999b).

Taken together, the studies described previously suggested that changes in the properties of B51 are correlated with learning-induced changes that occur during operant conditioning. However, they did not indicate whether the cellular mechanisms underlying this associative plasticity were intrinsic to B51 or whether they were due to extrinsic factors such as plasticity that may occur in other cells within the CPG that exert their influence on B51. This issue was addressed in a series of experiments in which a single-cell analogue of operant conditioning that consisted of cell B51 isolated in culture was developed (Brembs *et al.*, 2002). In this procedure, an individual B51 cell was removed from a naïve ganglion and maintained in culture. Because a series of previous experiments had provided strong evidence that dopamine (DA) was a critical neurotransmitter in the reinforcement pathway (Kabotyanski *et al.*, 1998; Nargeot *et al.*, 1999a), reinforcement was mimicked by the application of a brief "puff" of DA onto the cell. In addition, previous experiments had revealed that B51 exhibits a plateau potential during each ingestion-like BMP (Nargeot *et al.*, 1999b,c). Therefore, DA reinforcement was made contingent on a plateau potential that was elicited in B51 by injection of a brief depolarizing current pulse. Training consisted of delivering seven depolarizations, which elicited plateau potentials, over a 10 min interval. Each plateau potential was immediately followed by a DA puff onto the cell. Controls consisted of cells that received the DA puff 40 s after the plateau potential. Training produced a significant increase in input resistance and significant decrease in burst threshold (Fig. 19.28), similar to the changes observed in the neural correlates (Brembs *et al.*, 2002) and *in vitro* analogues (Nargeot *et al.*, 1999b) of operant conditioning. These data suggest B51 is an important locus of plasticity in operant conditioning of feeding behavior and that intrinsic cellwide plasticity may be one important mechanism underlying this type of learning. The continued analysis will provide insights into the molecular mechanisms of operant conditioning as well as into the mechanistic relationship between operant and classical conditioning.

A model based on recent work of Lorenzetti et al. (2008) of the molecular mechanisms underlying appetitive operant conditioning is shown in Fig. 19.28C. The CPG that mediates feeding behavior produces synaptic input to neuron B51. When this input is suprathreshold, it triggers an all-or-nothing sustained several second burst of spikes (plateau potential) in B51, which is critical for the expression of ingestive behavior. A secondary consequence of the plateau potential is to produce an accumulation of Ca^{2+} in B51, which leads to the activation of PKC. The activated PKC then weakly activates and primes a type II adenylyl cyclase. Reinforcement (reward) activates the dopaminergic (DA) modulatory system. DA binds to a D1-like receptor, but the DA-induced activation of the cAMP cascade is weak and insufficient to modulate downstream effectors (e.g., membrane channels regulating input resistance and burst threshold). However, if the ingestive behavior just precedes the delivery of the reward, as occurs during operant conditioning, the adenylyl cyclase will have been primed by PKC, and due to the synergistic interaction of the two pathways, the levels of cAMP will be significantly greater than that produced by either behavior (activity in B51) alone or reinforcement (DA) alone. After a sufficient number of contingent reinforcements, the increased levels of cAMP activate PKA sufficiently to produce the increase in input resistance and excitability of B51. Consequently, subsequent CPG-driven synaptic input to B51 is more likely to fire the cell and lead to the increase in ingestive behavior associated with operant conditioning.

Interestingly, the mechanisms of activity-dependent neuromodulation for this appetitive form of operant conditioning appear to be very similar to a form of aversive classical conditioning observed in sensory neurons of *Aplysia*, which mediate withdrawal reflexes (Fig. 19.26). In the sensory neuron, the coincidence detection involves, at least in part, a synergistic interaction between a Ca^{2+}/calmodulin-

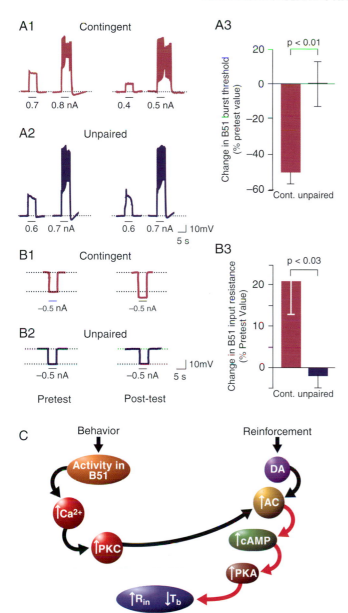

FIGURE 19.28 Changes in B51 produced by single-cell *in vitro* analogue of operant conditioning. Contingent-dependent changes in burst threshold and input resistance in cultured B51. (A) Burst threshold. (A1, A2) Intracellular recordings from a pair of contingently reinforced and unpaired neurons. Depolarizing current pulses were injected into B51 before (pretest) and after (post-test) training. In this example, contingent reinforcement led to a decrease in burst threshold from 0.8 to 0.5 nA (A1), whereas it remained at 0.7 nA in the corresponding unpaired cell (A2). (A3) Summary data. The contingently reinforced cells had significantly decreased burst thresholds. (B) Input resistance. (B1, B2) Intracellular recordings from a pair of contingently reinforced and unpaired control neurons. Hyperpolarizing current pulses were injected into B51 before (pretest) and after (post-test) training. In this example, contingent reinforcement led to an increased deflection of the B51 membrane potential in response to the current pulse (B1), whereas the deflection remained constant in the corresponding unpaired cell (B2). (B3) Summary data. The contingently reinforced cells had significantly increased input resistances. Modified from Brembs *et al.* (2002). (C) Model of the molecular mechanisms underlying operant conditioning. See text for details. Modified from Lorenzetti *et al.* (2008).

sensitive adenylyl cyclase (Type I) and a serotonin-activated cAMP cascade (Ocorr *et al.*, 1985; Yovell *et al.*, 1992). Also, in *Drosophila*, a type I adenylyl cyclase that is necessary for classical conditioning does not appear to be necessary for operant conditioning (Brembs and Plendl, 2008). Although the specific isoform of adenylyl cyclase appears to differ (Type I for classical conditioning and Type II for operant conditioning), adenylyl cyclase appears to serve as the molecule of convergence in both forms of learning. This finding is striking, but does not imply that adenylyl cyclase is a universal coincidence detector for all examples of classical and operant conditioning. Indeed, the membrane properties of B51 were also modulated by classical conditioning of feeding behavior (Lorenzetti *et al.*, 2006). However, following classical conditioning of feeding behavior, the burst threshold of B51 increases rather than decreases as it does following operant conditioning and the input resistance does not change, indicating that the mechanisms underlying the modulation of B51 by classical conditioning involve a different as yet unidentified coincidence detector from that for operant conditioning. In addition, and as described earlier in the chapter, the properties of the NMDA receptor serve as a coincidence detector in many CNS circuits. Further characterization of the signaling cascades upstream and downstream of the various cellular coincidence detectors will provide important insights into the molecular logic of operant and classical conditioning.

Growing evidence indicates that many of the same molecular pathways that mediate operant conditioning in *Aplysia* are also involved in vertebrate reward learning in the striatum. For example, *in vivo* operant conditioning was blocked by infusions into the nucleus accumbens of antagonists for D1 dopamine receptors and NMDA receptors (Smith-Roe and Kelley, 2000) and also by inhibitors of cAMP/PKA (Baldwin *et al.*, 2002). Furthermore, D1 dopamine receptors are necessary for potentiation of cortico-striatal synapses in an analogue of reward learning where induced electrical activity in striatal spiny neurons was paired with reward inducing stimulation of the substantia nigra (Reynolds *et al.* 2001). In addition to synaptic plasticity, striatal spiny neurons can display an increased level of intrinsic excitability known as the up-state. Increasing CREB levels in the nucleus accumbens increased the intrinsic excitability of spiny neurons mimicking the up-state and blocking CREB decreased the excitability (Dong *et al.*, 2006). Also, a recent study in *Drosophila* has shown tha PKC may play a role operant conditioning (Brembs and Plendl, 2008).

Summary

Certain invertebrates display an enormous capacity for learning and offer particular experimental advantages for analyzing the cellular and molecular mechanisms of learning (see Box 19.1). For example, behaviors in *Aplysia* are mediated by relatively simple neural circuits, which can be analyzed with conventional electroanatomical approaches. Once the circuit is specified, the neural locus for the particular example of learning can be found, and biophysical,

---

BOX 19.1

SOME INVERTEBRATES THAT ARE USEFUL FOR PROVIDING INSIGHTS INTO THE MECHANISMS UNDERLYING LEARNING

Gastropod Mollusks

Aplysia
See text.

Hermissenda
The gastropod mollusk *Hermissenda* exhibits associative learning of light-elicited locomotion and foot length (CRs). The conditioning procedure consists of pairing visual stimuli (light, unconditioned stimulus, CS) with vestibular stimuli (high-speed rotation, US). After conditioning, the CS suppresses normal light-elicited locomotion and elicits foot shortening. The associative memory can be retained from days to weeks depending on the number of conditioning trials administered during initial acquisition. The type A and B photoreceptors mediating the CS pathway have been identified as critical sites of plasticity for associative learning. Cellular correlates of conditioning have identified changes in membrane properties and in several K^+ conductances of the photoreceptors, and have identified several second-messenger systems that mediate the plasticity supporting learning (Crow, 2003).

Pleurobranchaea
The opisthobranch *Pleurobranchaea* is a voracious marine carnivore. When exposed to food, the animal exhibits a characteristic bite–strike response. After pairing of a food stimulus (CS) with a strong electric shock to the oral veil (US), the CS, instead of eliciting a bite–strike response, elicits a withdrawal and suppression of feeding responses (CR). The CR is acquired within a few trials and retained for up to 4 weeks. Neural correlates of associative learning have been analyzed by examining responses of various identified neurons in the circuit to chemosensory inputs in animals that have been conditioned. One correlate is an enhanced inhibition of command neurons for feeding (Gillette, 1992).

Tritonia diomedea
To escape a noxious stimulus, the opisthobranch *Tritonia diomedea* initiates stereotypical rhythmic swimming. This response exhibits both habituation and sensitization and involves changes in many different components of swim behavior in each case (Frost *et al.*, 1996). The neural circuit consists of sensory neurons, precentral pattern-generating (CPG) neurons, and motor neurons. Habituation appears to involve plasticity at multiple loci, including decrement at the first afferent synapse. Sensitization appears to involve enhanced excitability and synaptic strength in one of the CPG interneurons (Frost, 2003). Modulation of interneurons can be mediated by 5-HT, which has diverse effects on multiple loci of the circuit (Sakurai *et al.*, 2007).

Pond Snail (Lymnaea stagnalis)
The pulmonate *Lymnaea stagnalis* exhibits fairly rapid nonaversive conditioning of feeding behavior. A neutral chemical or mechanical stimulus (CS) applied to the lips is paired with a strong stimulant of feeding such as sucrose (US) (Kemenes and Benjamin, 1994). Greater levels of rasping, a component of the feeding behavior, can be produced by a single trial, and this response can persist for at least 19 days. The circuit consists of a network of three types of CPG neurons, 10 types of motor neurons, and a variety of modulatory interneurons (Vavoulis *et al.*, 2007). An analog of the behavioral response occurs in the isolated central nervous system. The enhancement of the feeding motor program appears to be due to facilitation of the motor neurons, the modulatory neurons, and presumably the CPG neurons, resulting in increased activation of the CPG cells by mechanosensory inputs from the lips (Benjamin *et al.*, 2000; Kemenes *et al.*, 2006).

Land Snail (Helix)
Land snails withdraw in response to weak tactile stimulation. The withdrawal behavior is mediated by a neuronal circuit involving four groups of nerve cells: sensory neurons, motor neurons, modulatory neurons, and command neurons (Balaban, 2002). This withdrawal can be habituated or sensitized, depending on the

BOX 19.1 *(continued)*

intensity of stimulation. Habituation of the withdrawal behavior emerges from depletion of neurotransmitter at sensory cell synapses as well as heterosynaptic inhibition mediated by FMRFa-containing neurons (Balaban *et al.*, 1991). Sensitization appears to be mediated by serotonergic modulatory cells whose spiking frequency increases following noxious stimulation (Balaban, 2002). These serotonergic cells are electrically coupled so that they get recruited and fire synchronously in response to strong excitatory input. One gene that is upregulated by external noxious input is the *Helix* Command Specific #2 (HCS2) (Balaban *et al.*, 2001). The HCS2 gene encodes a precursor protein whose processed products may function as neuromodulators or neurotransmitters mediating the withdrawal reactions of the snail (Korshunova *et al.*, 2006). Application of neurotransmitters and second messengers known to be involved in withdrawal behavior result in upregulation of HCS2 gene (Balaban, 2002).

The mechanisms underlying habituation and sensitization in the *Helix* can be further investigated by reconstructing behaviorally relevant synapses in culture. Using this approach, mechanosensory neuron-withdrawal interneuron synapses were found to display several forms of short-term synaptic plasticity such as facilitation, augmentation, and post-tetanic potentiation (Fiumara *et al.*, 2005).

Limax

The pulmonate *Limax* is an herbivore that locomotes toward desirable food odors. This behavior makes it well suited to food-avoidance conditioning. The slug's normal attraction to a preferred food odor (CS) is significantly reduced when the preferred odor is paired with a bitter taste (US). In addition to this example of classical conditioning, food avoidance in *Limax* exhibits higher-order features of classical conditioning, such as blocking and second-order conditioning. An analog of taste-aversion learning occurs in the isolated central nervous system, facilitating subsequent cellular analyses of learning in *Limax*. The procerebral lobe in the cerebral ganglion processes olfactory information and is a likely site for the plasticity. The procerebral lobe contains several types of neuropeptides, and FMRFamide as well as SCPB have been shown to affect activity of the rhythmic motor network underlying the *Limax* feeding system, making these neuropeptides attractive candidates for modulating plasticity within the network. In addition, NO synthase is present in the procerebral lobe, and NO has been shown to affect oscillations within the procerebral lobe. Thus, NO, acting through the cGMP signaling pathway, may

modulate plasticity via long-term changes in cells within the procerebral lobe (Gelperin, 2003; Watanabe *et al.*, 2008).

Arthropods

Cockroach (*Periplaneta americana*) and Locust (*Schistocerca gregaria*)

Learned modifications of leg positioning in the cockroach and locust may be useful in the cellular analysis of operant conditioning. When the animal is suspended over a dish containing a fluid, initially it makes many movements, including those that cause the leg to come in contact with the liquid surface. When contact with the fluid is paired with an electric shock, the insect rapidly learns to hold its foot away from the fluid. Neural correlates of the conditioning have been observed in somata of the leg motor neurons. These correlates include changes in intrinsic firing rate and membrane conductance (Eisenstein and Carlson, 1994).

Crayfish (*Procambarus clarkii*)

A crayfish escapes from noxious stimuli by flipping its tail. A key component of the tail-flip circuit is a pair of large neurons called the *lateral giants* (LGs), which run the length of the animal's nerve cord. The LGs are the decision and command cells for the tail-flip. The crayfish tail-flip response exhibits habituation (Wine *et al.*, 1975) and sensitization (Krasne and Glanzman, 1986). Plastic changes induced during learning involve modulation of the strength of synaptic input driving the LGs (Edwards *et al.*, 1999). A diminution of transmitter release with repeated activation of afferents is thought to underlie habituation (Krasne and Roberts, 1967; Zucker, 1972). An inhibitory pathway was also identified that can tonically inhibit the LGs (Krasne and Wine, 1975; Vu and Krasne, 1992; Vu *et al.*, 1993). This putatively GABAergic (Vu and Krasne, 1993, but see Heitler *et al.*, 2001) inhibitory pathway also plays a major role in habituation (Krasne and Teshiba, 1995). In addition to the regulation of synaptic strength, habituation also results in decreased excitability of LGs (Araki and Nagayama, 2005). Bath application of the endogenous neuromodulators 5-HT and octopamine decrease the rate of LG habituation to repetitive sensory stimulation (Araki *et al.*, 2005). Octopamine is also thought to at least partly mediate sensitization, because it mimics the sensitizing effects of strong stimulation on the tail-flip (Glanzman and Krasne, 1986; Krasne and Glanzman, 1986).

BOX 19.1   *(continued)*

Honeybee (*Apis mellifera*)

Honeybees, like other insects, are superb at learning. For example, sensitization of the antenna reflex of *Apis mellifera* is produced as a result of presenting gustatory stimuli to the antennae. Classical conditioning of feeding behavior can be produced by pairing a visual or olfactory CS with sugar solution (US) to the antennae. The small size of bee neurons is an obstacle in pursuing detailed cellular analyses of these behavioral modifications. Nevertheless, regions of the brain necessary for associative learning have been identified. In particular, intracellular recordings have revealed that one identified cell, the ventral unpaired median (VUM) neuron that is putatively octopaminergic, mediates reinforcement during olfactory conditioning and represents the neural correlate of the US. The learning has been dissected into several phases of memory, including short term, midterm, and long term, and the cellular and molecular mechanisms supporting the various phases are beginning to be unraveled. For example, numerous studies have revealed that, as in other species, the molecular mechanisms underlying memory formation in the honeybee involve upregulation of the cAMP pathway and activation of PKA, resulting in CREB-mediated transcription of downstream genes (Menzel, 2001; Menzel *et al.*, 2006).

Drosophila

Because the neural circuitry in the fruit fly is both complex and inaccessible, the fly might seem to be an unpromising subject for studying the neural basis of learning. However, the ease with which genetic studies are performed compensates for the difficulty in performing electrophysiological studies (DeZazzo and Tully, 1995). A frequently used protocol employs a two-stage differential odor–shock avoidance procedure, which is performed on large groups of animals simultaneously rather than on individual animals. Animals learn to avoid odors paired (CS+) with shock but not odors explicitly unpaired (CS–). This learning is typically retained for 4–6 h, but retention for 24 h to 1 week can be produced by a spaced training procedure. Several mutants deficient in learning have been identified. Analysis of the affected genes has revealed elements of the cAMP signaling pathway as key in learning and memory. It is now known that the formation of long-term memory requires activation of the cAMP signaling pathway, which, in turn, activates members of the CREB transcription family. These proteins appear to be the key molecular switch that enables expression of genes necessary for the formation of long-term memories. Further, the expression pattern of these genes, as well as other mutational analyses, has revealed that brain structures called *mushroom bodies* are crucial sites for olfactory learning. These transcription factors are also important for long-term memory in *Aplysia* and in vertebrates (Davis, 2005; Dudai and Tully, 2003; Waddel and Quinn, 2001).

Annelids

Leech

In the leech *Hirudo medicinalis*, nonassociative learning has been studied in several well-characterized behaviors: movements in response to light and water currents (Ratner, 1972), bending (Lockery and Kristan, 1991), shortening reflex to repeated light (Lockery *et al.*, 1985) or tactile stimulation (Belardetti *et al.*, 1982; Boulis and Sahley, 1988; Sahley *et al.*, 1994), and swimming (Catarsi *et al.*, 1993; Zaccardi *et al.*, 2001).

In the shortening reflex of the leech, the neuronal changes underlying habituation and sensitization occur in the pathway from mechanosensory neurons to electrically coupled neurons, the S cells (Bagnoli and Magni, 1975; Sahley *et al.*, 1994). Habituation of this reflex can reach asymptotic levels after 20 training trials and correlates with decreased S-cell excitability (Burrell *et al.*, 2001). The reflex can be restored following application of a single noxious stimulus (dishabituation) (Boulis and Sahley, 1988). The potentiation of the shortening reflex observed during sensitization requires the S neurons, as their ablation disrupts sensitization (Sahley *et al.*, 1994). This potentiation is mediated by 5-HT through an increase of cAMP (Belardetti *et al.*, 1982), which also increases S-cell excitability (Burrell *et al.*, 2001). Depletion of 5-HT disrupts sensitization (Sahley *et al.*, 1994). Interestingly, ablation of the S cells only partly disrupts dishabituation, indicating that separate processes contribute to dishabituation and sensitization (Ehrlich *et al.*, 1992; Sahley *et al.*, 1994).

An additional mechanism that could potentially contribute to habituation of the shortening reflex involves depression of the synapses of touch (T) sensory neurons onto their follower target neurons. This synaptic depression has been associated with an increase in the amplitude of the T-cell afterhyperpolarizing potential (AHP) that follows their discharge (Brunelli *et al.*, 1997; Scuri *et al.*, 2002). The lasting increase in AHP amplitude, following low-frequency stimulation of T cells, has been attributed, in turn, to increased activity of the electrogenic

---

BOX 19.1 *(continued)*

Na$^+$ pump, and requires activation of phospholipase A2 and the downstream arachidonic acid metabolites (Scuri *et al.*, 2005).

Nematoda

Caenorhabditis elegans

C. elegans is a valuable model system for cellular and molecular studies of learning. Its principal advantages are three-fold. First, its nervous system is extremely simple. It has a total of 302 neurons, the anatomical connectivity of which has been described at the electron microscopy level. Second, the developmental lineage of each neuron is completely specified. Third, its entire genome has been sequenced, making it highly amenable to a number of genetic and molecular manipulations. *C. elegans* respond to a vibratory stimulus applied to the medium in which they locomote by swimming backward. This reaction, known as the *tap withdrawal reflex*, exhibits habituation, dishabituation, sensitization, and long-term (24 h) retention of habituation training (Rankin *et al.*, 1990). Laser ablation studies have been used to elucidate the neural circuitry supporting the tap withdrawal reflex and to identify likely sites of plasticity

within the network. Plastic changes during habituation appear to occur at the chemical synapses between presynaptic sensory neurons and postsynaptic command interneurons (Wicks and Rankin, 1997). Analysis of several *C. elegans* mutants has revealed that synapses at the locus of plasticity in the network may be glutamatergic (Rose and Rankin, 2001). Mutation of the gene coding for the brain-specific inorganic phosphate transporter *eat-4* results in more rapid habituation compared to wild-type worms and slower recovery (Rankin and Wicks, 2000). The protein coded by *eat-4* is involved in the regulation of glutamatergic transmission and is homologous to the mammalian vesicular glutamate transporter VGLUT1 (Bellocchio *et al.*, 2000). *Eat-4* worms also do not display dishabituation, suggesting that neurotransmitter regulation plays a role in habituation and dishabituation (Rankin and Wicks, 2000). Moreover, worms that carry a mutation in glr-1, an excitatory glutamate receptor expressed in postsynaptic command interneurons, do not display long-term memory for habituation (Rose *et al.*, 2003). In general, the study of behavioral genetics in the worm has provided significant insights into the ways in which genes regulate behavior (Rankin, 2002).

---

biochemical, and molecular approaches can then be used to identify mechanisms underlying the change. The relatively large size of some of these cells allows these analyses to take place at the level of individually identified neurons. Individual neurons can be surgically removed and assayed for changes in the levels of second messengers, protein phosphorylation, and RNA and protein synthesis. Moreover, peptides and nucleotides can be injected into individual neurons. This chapter has focused exclusively on *Aplysia*, but many other invertebrates have proven to be valuable model systems for the cellular and molecular analysis of learning and memory (for review, see Byrne, 1987). Each has its own unique advantages. For example, *Aplysia* is excellent for applying cell biological approaches to the analysis of learning and memory mechanisms. Other invertebrate model systems such as *Drosophila* and *Caenorhabditis elegans* are not well suited for cell biological approaches because of their small neurons but offer tremendous advantages for obtaining insights into mechanisms of learning and memory through the application of genetic approaches. (See Box 19.1

and Carew, 2000, for a review of several selected invertebrate model systems that have contributed significantly to the understanding of memory mechanisms.)

CLASSICAL CONDITIONING IN VERTEBRATES: DISCRETE RESPONSES AND FEAR REACTIONS AS MODELS OF ASSOCIATIVE LEARNING

When animals, including humans, are faced with an aversive or threatening situation, at least two complementary processes of learning occur. Learned fear or arousal reactions develop very rapidly, often in one trial. Subsequently, the organism learns to make adaptive behavioral motor responses to deal with the situation. These observations led to so-called two-process theories of learning: development of an initial learned fear or arousal, followed by slower learning of discrete behavioral responses (Rescorla and

Solomon, 1967). As the latter form of learning develops, fear subsides. It is now believed that, at least in mammals, a third process typically takes place in which declarative memory for the events and their relationships develops. This section focuses on the learning of discrete responses, using eyeblink conditioning as the model system, and on conditioned fear.

Eyeblink Is a Model System for Studying the Conditioning of Discrete Behavioral Responses in Vertebrates

A vast amount of research has used Pavlovian conditioning of the eyeblink response in humans and other mammals in various types of investigations (Gormezano *et al.*, 1983). The eyeblink response exhibits all the basic laws and properties of Pavlovian conditioning equally in humans and other mammals. The basic procedure is to present a neutral CS, such as a tone or a light, followed approximately a quarter of a second later (while the CS is still present) by a puff of air to the eye or by a periorbital (around the eye) shock (US). This is an example of what is known as a *delay procedure*. Initially, there is no response to the CS and a reflex eyeblink to the US. After a number of such trials, the eyelid begins to close in response to the CS before the US occurs. In a well-trained subject, the eyelid closure (CR) becomes very precisely timed so that it is maximally closed about the time that the air puff or shock (US) onset occurs (Fig. 19.29). This very adaptive timing of the eyeblink CR develops over the range of CS–US onset intervals in which learning occurs, which is about 100 ms to 1 s. Thus, the conditioned eyeblink response is a very precisely timed and elementary learned motor skill. The same is true of other discrete behavioral responses learned to deal with aversive stimuli (e.g., the forelimb or hindlimb flexion response and the head turn).

Classical conditioning consists of two basic procedures: delay (see earlier) and trace. Pavlov (1927) was the first to describe trace classical conditioning. He stressed that the organism must maintain a "trace" of the CS in the brain in order for the CS and the US to become associated. In eyeblink conditioning in animals, a typical trace interval between CS offset and US onset is 500 ms. The trace eyeblink procedure is much more difficult to learn than the delay procedure.

Two brain systems, the hippocampus and the cerebellum, become massively engaged in eyeblink conditioning (Thompson and Kim, 1996). The cerebellum is essential for learning and retention of both delay and trace conditioning and the hippocampus is necessary for learning and initial retention of trace, but not

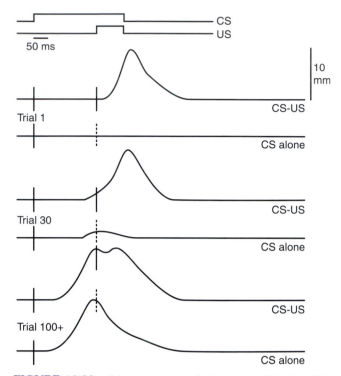

FIGURE 19.29 Adaptive nature of classical conditioning. This example shows the development of the conditioned eyeblink response over the trials of training. The CS is typically a "neutral" light or tone; the US here is a puff of air to the cornea. The eyelid closure response is indicated by upward movement of the tracing. The first marker is tone CS onset; the second is air puff US onset. In trial 1 the eyeblink does not move to the CS but closes (blinks) following onset of the US. The conditioned response (CR) is any measurable degree of eyelid closure prior to the onset of the US. Note that after learning, the CR peaks at the onset of the US, that is, maximum eyelid closure at air puff onset. If the CS–US onset interval were longer (e.g., 500 ms), the CR would now peak at the onset of the US, 500 ms after CS onset. The conditioned response is adaptive. For this type of learning, a period (ISI) of about 250 ms between CS onset and US onset (shown here) yields the best learning. This best learning time varies widely depending on the type of response (e.g., for fear learning, several seconds is best).

delay, conditioning (see later). The amygdala also plays a role when the US is sufficiently aversive to elicit learned fear. In an experimental design using a click CS and glabellar (forehead) tap US in restrained cats, a very short latency (<20 ms) eyeblink muscle EMG (electrical response recorded from muscles around the eye) CR involving the motor cortex develops (Woody *et al.*, 1974; Woody and Yarowsky, 1972). However, bilateral removal of the motor cortex does not appear to affect either learning or expression of the standard longer-latency adaptive delay or trace CRs (Ivkovich and Thompson, 1997). This short-latency EMG response may be a component of the startle response elicited by a sudden acoustic stimulus (Davis, 1984).

Hippocampus and Classical Conditioning

In eyeblink conditioning, neuronal unit activity in hippocampal fields CA1 and CA3 increases very rapidly in paired (tone CS–corneal air puff US) training trials, shifts forward in time as learning develops, and forms a predictive "temporal model" of the learned behavioral response both within trials and during repeated trials of training (Berger *et al.*, 1976; Berger and Thompson, 1978) (Fig. 19.30). To summarize a large body of research on this topic, the growth of the hippocampal unit response, under normal conditions, invariably and strongly predicts subsequent behavioral learning (Berger *et al.*, 1986). This increase in neuronal activity in the hippocampus becomes significant

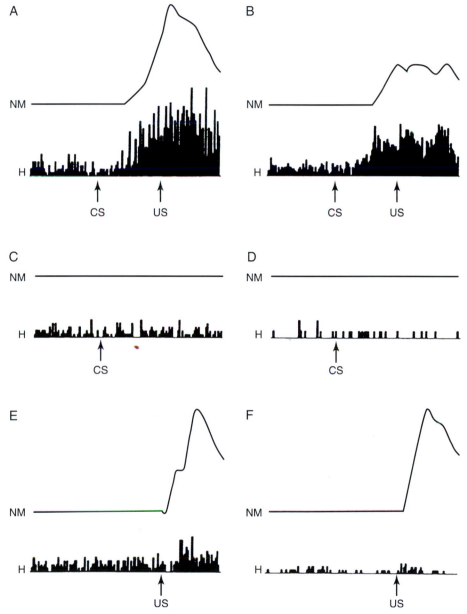

FIGURE 19.30 Engagement of hippocampal neurons in eyeblink conditioning. Responses of identified pyramidal neurons during paired (A, B) and unpaired (C–F) presentations of tone and corneal air puff. The upper traces show the averaged nictitating membrane (NM, a component of the eyeblink) response for all trials during which a given cell was recorded. The bottom traces show the response of the recorded neuron in the form of a peristimulus time histogram. The total length of both NM responses and histograms was 750 ms. Arrows occurring early in the trial period indicate tone onset; arrows occurring late in the trial period indicate air puff onset. H, hippocampus. In this particular figure, A and B show examples of responses of two pyramidal neurons recorded from two different animals during delay conditioning. The results in C and E show the response of a pyramidal neuron recorded from an animal given unpaired tone alone (C) and air puff alone (E) presentations. (D, F) Same for a different pyramidal cell recorded from a different control animal. From Berger *et al.* (1986).

by the second or third trial of training, long before behavioral signs of learning develop. Neurons in the hippocampus become engaged in many other types of learning as well (Isaacson and Pribram, 1986).

In eyeblink conditioning, many neurons identified as pyramidal neurons in fields CA1 and CA3 show learning-related increases in discharge frequency during the trial period (Berger *et al.*, 1983) (Fig. 19.30). Typically, a given neuron models only some limited period of the trial, although some pyramidal neurons model the entire learned behavioral response, as shown in Figures 19.30A and B. Thus, the pyramidal neuron representation of the behavioral learned response is distributed over both space and time in the hippocampus (Thompson, 1990). The results just described were obtained using the basic delay procedure, in which hippocampal lesions do not impair simple response acquisition in rabbits. Similarly, humans with hippocampal–temporal lobe anterograde amnesia are able to learn simple acquisition of the eyeblink conditioned response but cannot recall the learning experience (Weiskrantz and Warrington, 1979). The involvement of the hippocampus depends on the difficulty of the task. For example, such amnesic humans are massively impaired on conditional discriminations in eyeblink conditioning (e.g., blink to tone only if preceded by light) but not on simple discriminations (Daum *et al.*, 1991). Bilateral hippocampal lesions in rabbits markedly impair subsequent acquisition of the trace CR (Moyer *et al.*, 1990; Solomon *et al.*, 1986). Interestingly, when the hippocampal lesion is made immediately after trace learning in rabbits, the CR is abolished; when the lesion is made a month after training, the CR is not impaired (Kim *et al.*, 1995) (Fig. 19.31).

These results are striking in light of reports of declarative memory deficit following damage to the temporal lobe of the hippocampal system in humans and monkeys. These deficits have two key temporal characteristics: (1) profound and permanent anterograde amnesia and (2) profound but clearly time-limited retrograde amnesia. Subjects have great difficulty learning new declarative tasks and/or information (anterograde amnesia) and have substantial memory loss for events for some period preceding brain damage (retrograde amnesia) but relatively intact memory for earlier events (Zola-Morgan and Squire, 1990).

One of the hallmarks of declarative memory in humans is awareness. In the case of eyeblink conditioning, *awareness* refers to the ability to describe the experience and the contingencies between the CS and the US. Patients with temporal lobe–hippocampal amnesia can learn the delay eyeblink procedure at a normal rate, as noted previously, but are unable to

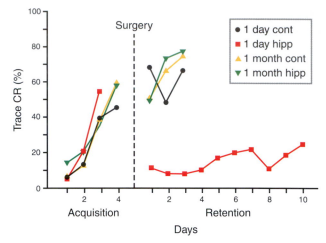

FIGURE 19.31 Effects of hippocampal lesions on retention of trace CRs. Shown are the mean percentage of CRs during initial training and following postoperative training: 1 day cont, controls given cortical or sham lesions 1 day after training; 1 month cont, controls given lesions 1 month after training; 1 day hipp, bilateral hippocampal lesions made 1 day after training; 1 month hipp, hippocampal lesions made 1 month after training. Only the hippocampal lesions made immediately after training abolished the trace CR. From Kim *et al.* (1995).

acquire the trace eyeblink procedure (Clark and Squire, 1998; McGlinchey-Berroth *et al.*, 1997). For normal humans, awareness of the situation is unrelated to successful learning in the delay paradigm (procedural learning) but is a prerequisite for successful trace conditioning. Trace conditioning is temporal lobe–hippocampus-dependent because, as in other tasks of declarative memory, conscious knowledge must be acquired across training sessions. Trace conditioning may thus provide a simple means for studying awareness in nonhuman animals in the context of current ideas about multiple memory systems in the brain (Clark *et al.*, 2001; Clark and Squire, 1998). Therefore, even in simple memory tasks such as eyeblink conditioning, hippocampus-dependent "declarative" memory processes can develop. Consistent with this is the striking observation that extreme deficits in ability to learn eyeblink conditioning occur in Alzheimer patients; indeed, marked deficits in normal elderly humans may be diagnostic of the subsequent development of Alzheimer disease (Woodruff-Pak *et al.*, 1990). In recent work, Disterhoft and associates have focused on the role of cerebral cortex as well as hippocampus in trace eyeblink conditioning (Weiss *et al.*, 2006). In brief, the anterior cingulated cortex appears critical for trace learning, as is the hippocampus, and these systems play on the essential cerebellar circuit (see Fig. 19.32).

What are the mechanisms of the changes in the hippocampus? The process of LTP is widely considered

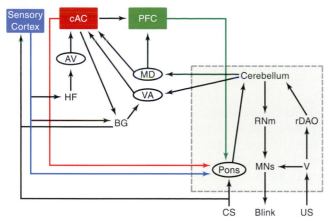

FIGURE 19.32 A schematic of the connections that might mediate trace eyeblink conditioning (arrows do not necessarily indicate monosynaptic connections). The circuit shows a forebrain circuit that converges upon the cerebellar circuit. The structures mediating delay conditioning are enclosed by the dashed box. AV, anteroventral thalamus; BG, basal ganglia; cAC, caudal anterior cingulate; PFC, prefrontal cortex; HF, hippocampal formation; rDAO, rostral dorsal accessory olive; MD, dorsomedial thalamus; MNs, motor neurons; PFC, prefrontal cortex; RNm, magnocellular red nucleus; V, trigeminal nucleus; VA ventral anterior thalamus. From Weiss et al. (2006).

to be the most likely mechanism of memory storage in the hippocampal system (see section on LTP–LTD). In the case of classical eyeblink conditioning, a number of parallels exist between the properties of LTP and the properties of the learning-induced increase in neuronal activity in the hippocampus (Baudry et al., 1993; Berger et al., 1986; Maren et al., 1993; Thompson et al.,

1992). Both LTP and the learning-induced increase in hippocampal neuron activity are associated with pyramidal neurons, and both begin to develop after very brief periods (e.g., 100 Hz for 1 s for LTP; one to three trials of training in eyeblink conditioning). Also, both approach a limit asymptotically over a period of many minutes, show the same magnitude of increase, and are developed only with very specific parameters of stimulation. Furthermore, there is a persistent increase in the extracellularly recorded monosynaptic population spike in the dentate gyrus in response to stimulation of the perforant path as a result of eyeblink conditioning, just as occurs when LTP is induced by tetanus of the perforant path (Weisz et al., 1984).

There are strikingly parallel and persistent increases in AMPA receptor binding on hippocampal membranes in both eyeblink conditioning (well-trained animals) and in vivo expression of LTP by stimulation of the perforant path projection to the hippocampal dentate gyrus. The pattern of increased binding is similar in both (Baudry et al., 1993; Maren et al., 1993; Tocco et al., 1992). NMDA receptors play a critical role in induction of LTP (at least in the dentate gyrus and CA1) and also appear to be involved in acquisition of the trace eyeblink CR (Thompson et al., 1992). Mechanisms of LTP are discussed at length in the section above.

In addition to alterations in synaptic strength, as in LTP, changes occur in the intrinsic excitability of pyramidal neurons in the hippocampus following eyeblink conditioning (Disterhoft et al., 1986) (see also Box 19.2). Specifically, the calcium-dependent slow after-hyperpolarization (AHP), mediated by a

---

BOX 19.2

MEMORY BEYOND THE SYNAPSE

The search for the biological basis of learning and memory has led many of the twentieth century's leading neuroscientists to direct their efforts to investigating the synapse. The focus of most recent work on LTP (and indeed the bulk of this chapter) has been to elucidate the mechanisms underlying changes in synaptic strength. Although changes in synaptic strength are certainly ubiquitous, they are not the exclusive means for the expression of neuronal plasticity associated with learning and memory. Both short-term and long-term sensitization and classical conditioning of defensive reflexes in Aplysia are associated with an enhancement of excitability of the sensory neurons in addition to changes in synaptic strength (Fig. 19.23). Classical and operant conditioning of feeding

behavior in Aplysia also leads to changes in excitability of neuron B51 (Fig. 19.28). Interestingly, in this case there is a bidirectional control. The excitability of B51 is increased by operant conditioning, whereas it is decreased by classical conditioning. Changes in excitability of sensory neurons in the mollusk Hermissenda are produced by classical conditioning. In vertebrates, classical conditioning of eyeblink reflexes produces changes in excitability of hippocampal and cortical neurons and changes in the spike after-potential of hippocampal pyramidal neurons (see Fig. 19.33). Finally, as described in their original report on LTP, Bliss and Lomo (1973) found that the expression of LTP was also associated with an apparent enhanced excitability.

voltage-gated potassium conductance, is markedly reduced following learning but not in control animals given unpaired CS and US trials (de Jonge *et al.*, 1990). These observations were first made in the delay paradigm, where pyramidal neurons in the hippocampus markedly increase their responsiveness as a result of learning (Fig. 19.30). A marked decrease in the AHP and accommodation in hippocampal pyramidal neurons is seen following trace eyeblink conditioning (Moyer *et al.*, 1996). Importantly, this trace learning–induced decrease in the AHP and accommodation is time limited, being maximum 1 to 24 h after training and returning toward baseline after a week, consistent with the time-dependent effect of hippocampal lesions on the memory for trace conditioning (see Fig 19.33 and Fig. 19.31) (Thompson *et al.*, 1996). Interestingly, both AHP and accommodation are enhanced in aged subjects (Disterhoft and Oh, 2006).

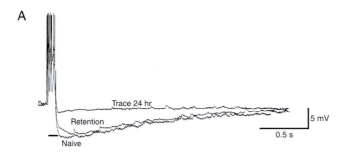

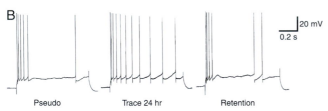

FIGURE 19.33 Hippocampal CA1 pyramidal cells (intracellular recordings) exhibit a transient increase in excitability following acquisition of trace eyeblink conditioning. (A) Overlapping traces of AHPs in CA1 neurons following injection of a 100 msec depolarizing current (horizontal bar). Current injection was the minimum required to elicit burst of four action potentials and did not differ significantly between groups. Amplitude and duration of AHP is similar in recordings from animals not exposed to training stimuli (naïve) and recordings from animals 14 d after acquisition (retention). Recordings from pyramidal cells 24 h following acquisition show a significant decrease in the duration and amplitude of the AHP (trace 24 h). (B) Spike accommodation in response to 800 msec depolarizing current is significantly reduced in CA1 pyramidal neurons 24 h after acquisition (trace 24 h) as compared with pyramidal cells from animals receiving explicitly unpaired stimuli presentations (pseudo) or animals trained 14 d prior to recording. From Moyer *et al.* (1996).

The Cerebellar System and Classic Conditioning of Discrete Responses

Since publication of the classic papers of Marr (1969) and Albus (1971), the cerebellum has been favored as a structure for modeling neuronal learning. Figure 19.34 is a highly simplified diagram of a current qualitative working model of the role of the cerebellum in basic classic (delay) conditioning of eyeblink and other discrete responses (Thompson, 1986; Thompson and Krupa, 1994). Laterality is not addressed in Figure 19.34; the critical region of the cerebellum is ipsilateral to the trained eye (or limb), whereas the critical regions of the pontine nuclei, red nucleus, and inferior olive are contralateral. For a more realistic representation, see Figure 19.37. In this section, the data refer to the basic delay eyeblink CR, unless otherwise noted (Lavond *et al.*, 1993; Thompson and Krupa, 1994; Yeo, 1991).

In brief, the reflex eyeblink response pathways activated by corneal air puff (or periorbital shock) include the trigeminal nucleus, direct projections to the relevant motor nuclei (mostly the seventh and accessory sixth), and indirect projections to the motor nuclei via the brain stem reticular formation (Fig. 19.34). Analysis of response latencies rules out any direct role of the cerebellum in the reflex response. The tone (and light) CS pathways project to the cerebellum as mossy fibers, mostly relaying through the pontine nuclei. The mossy fibers, in turn, activate granule cells, which project to Purkinje cells via parallel fibers. The US pathway projects from the trigeminal nucleus to the inferior olive and then as climbing fibers to the cerebellum. The CS-activated mossy fiber–parallel fiber pathway and the US-activated climbing fiber pathway converge and make synaptic connections on Purkinje neurons in the cerebellar cortex (parallel fiber–climbing fiber) and on neurons in the interpositus nucleus (mossy fiber–climbing fiber). The CR pathway projects from the interpositus nucleus of the cerebellum via the superior cerebellar peduncle to the red nucleus, and from there to the premotor and motor nuclei (mostly seventh and accessory sixth), controlling the eyeblink response.

This circuitry has been identified using a number of methods, including lesion studies, electrophysiological recordings, electrical microstimulation, and anatomical characterization of projection pathways. For example, in animals, neurons in the cerebellar cortex and interpositus nucleus respond to the CS and US before training and develop amplitude–time course models (e.g., Fig. 19.35) of the learned behavioral response. These models precede and predict the occurrence and form of the CR within trials and over the trials of training (McCormick and

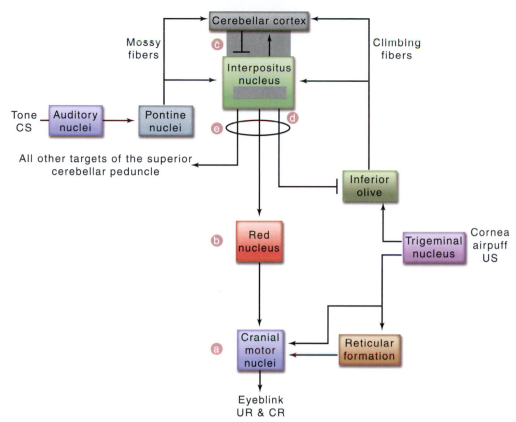

FIGURE 19.34 Simplified schematic of the essential brain circuitry involved in standard delay classical conditioning of discrete responses (e.g., eyeblink response). Shadowed boxes represent areas that have been reversibly inactivated during training. (a) Inactivation of motor nuclei including facial (seventh) and accessory sixth. (b) Inactivation of magnocellular red nucleus. (c) Inactivation of dorsal aspect of the anterior interpositus and overlying cerebellar cortex. (d) Inactivation of ventral anterior interpositus nucleus and associated white matter. (e) Complete inactivation of the superior cerebellar peduncle (scp), essentially all output from the cerebellar hemisphere. Inactivation of each of these regions in trained rabbits abolishes performance of the CR. Significantly, inactivation of the motor nuclei (a), the red nucleus (b), the superior cerebellar peduncle (e), and the output of the interpositus nucleus (d) during training do not prevent learning at all, but inactivation of a localized region of the anterior interpositus nucleus and overlying cortex (c) during training completely prevents learning. From Thompson and Krupa (1994).

Thompson, 1984) (Fig. 19.35). By inference from PET analysis, the same process is thought to occur in humans (Logan and Grafton, 1995) (Fig. 19.36). In animals, appropriate lesions of the anterior interpositus nucleus completely and permanently prevent learning. Similar lesions inflicted after training permanently abolish the CR but are without effect on the UR (McCormick and Thompson, 1984; Steinmetz *et al.*, 1992; Yeo *et al.*, 1985a,b). In the same way, in humans, cerebellar lesions can completely prevent learning of the CR but again are without effect on the UR (Daum *et al.*, 1993). Interestingly, when the interpositus lesion (in rabbits) is incomplete, resulting in a marked impairment in the CR but not complete abolition, the attenuated CR does not recover with further training (Clark *et al.*, 1984; Welsh and Harvey, 1989).

Appropriate lesions of the pontine nuclei (the CS pathway) can selectively abolish the CR to one

modality of CS, and stimulation of the pontine nuclei serves as a supernormal CS, yielding learning faster than peripheral CSs (Steinmetz *et al.*, 1986, 1987). Finally, lesions of the appropriate region of the inferior olive completely prevent learning if they are made before training. Lesions made at the same location after training result in extinction and abolition of the CR (McCormick, *et al.*, 1985; Voneida *et al.*, 1990; Yeo *et al.*, 1986). Electrical microstimulation of this same region elicits discrete movements, and the exact movements elicited can be trained to occur in response to any neutral stimulus (Mauk *et al.*, 1986). Note that the inferior olive–climbing fiber system is the only system in the brain other than reflex afferents where this type of training can take place (Thompson, 1989).

These results constitute a verification of the theories initially developed in the classic papers of Marr (1969) and Albus (1971) and elaborated on by Eccles

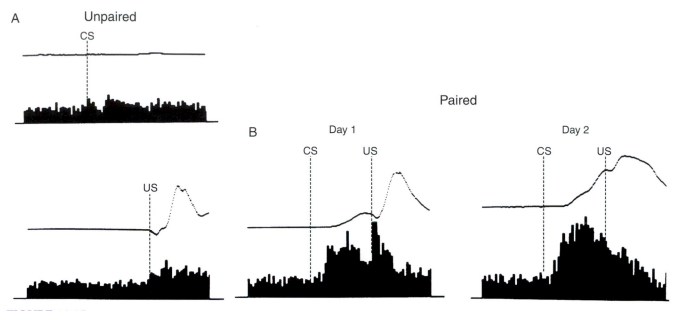

FIGURE 19.35 Engagement of neurons in the cerebellar interpositus nucleus in eyeblink conditioning. Histograms of a unit cluster recording from the anterior interpositus nucleus over the course of training are shown. The eyeblink response (here nictitating membrane extension) is shown on the tracing above each histogram. (A) Results of a day of unpaired CS and US presentations. There is some activity to the US. However, when paired training (B) is given (days 1 and 2), as behavioral learning develops (eyelid closure prior to US onset) there is a massive increase in neuronal discharges in the CS period that precedes and correlates with performance of the conditioned eyeblink response. Total trace duration, 750 ms; CS–US onset interval, 250 ms. Each trace and histogram is the average or cumulation of 1 day of training (120 trials). From McCormick and Thompson (1984).

(1977), Ito (1984), and Thach *et al.* (1992). These theories proposed that the cerebellum was a neuronal learning system in which there was a convergence of mossy fibers–parallel fibers that conveyed information about stimuli and movement contexts (CSs) and the climbing fibers that conveyed information about specific movement errors and aversive events (USs). This convergence of fibers might occur on Purkinje neurons in the cerebellar cortex to alter the synaptic efficacy of the parallel fiber synapses on Purkinje dendrites. A similar convergence of mossy and climbing fibers exists on neurons in the interpositus nucleus (Fig. 19.34 and Fig. 19.37).

The Cerebellum: The Locus of the Long-Term Memory Trace

The results described so far demonstrate that the cerebellum is necessary for learning, retention, and expression of classical conditioning of the eyeblink and other discrete responses. The next and more

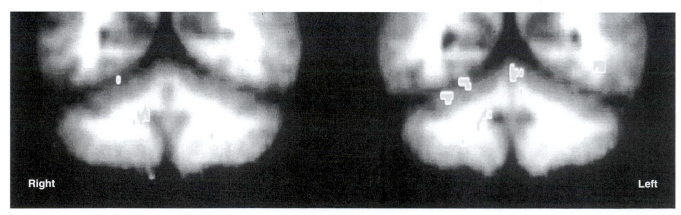

FIGURE 19.36 Functional localization of brain cerebellar activity (PET scan) in human eyeblink conditioning. Regions showing bright yellow are the regions where activation correlated significantly with degree of learning. The cerebellar anterior interpositus nucleus and several regions of cerebellar cortex show highly significant increases in activation with learning. From Logan and Grafton (1995).

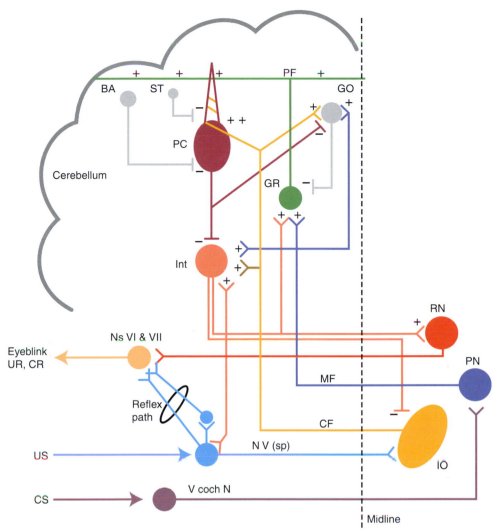

FIGURE 19.37 Commonly accepted eyeblink conditioning circuit based on experimental findings and anatomy of the cerebellum and the brain stem. The conditioned stimulus (CS) pathway consists of excitatory (+) mossy fiber (MF) projections primarily from the pontine nuclei (PN) to the interpositus nucleus (Int) and to the cerebellar cortex. In the cortex, the mossy fibers form synapses with granule cells (GR), which in turn send excitatory parallel fibers to the Purkinje cells (PC). Purkinje cells are the exclusive output neurons from the cortex, and they send inhibitory (−) fibers to deep nuclei such as the interpositus. The unconditioned stimulus (US) pathway consists of excitatory climbing fiber (CF) projections from the inferior olive to the interpositus nucleus and to the Purkinje cells in the cerebellar cortex. Within the cerebellar cortex, Golgi (GO), stellate (ST), and basket (BA) cells exert inhibitory actions on their respective target neurons. The efferent conditioned response (CR) pathway projects from the interpositus nucleus to the red nucleus (RN) and via the descending rubral pathway to act ultimately on the motor neurons generating the eyeblink response. V Coch N, ventral cochlear nucleus; N V (sp), spinal fifth cranial nucleus; N VI, sixth and accessory sixth cranial nuclei; N VII, seventh cranial nucleus; UR, unconditioned response. Note that the reflex pathways do not involve the cerebellar circuitry. From Kim and Thompson (1997).

critical issue concerns the locus of the memory traces. Considerable evidence has accrued pointing to the cerebellum as the location where long-term memory traces for this type of learning are formed and stored (Christian and Thompson, 2003).

Reversible inactivation has proven to be a powerful tool in the localization of sites of memory storage in systems where the essential circuitry is known, such as eyeblink conditioning (Figs. 19.34 and 19.37). In brief, if inactivation of a structure abolishes the

learned response, the structure is considered to be part of the circuitry necessary for expression of the learned response. If the structure is inactivated during training and the animal immediately shows complete learning when the inactivation is subsequently removed, then the structure is not involved in acquiring the learned response but lies on the efferent path from the memory trace. However, if the animal shows no evidence of learning following inactivation training, then either the memory trace is normally located in the structure or the structure is a necessary afferent to the trace. Reversible inactivation can be produced by local cooling using a cold probe or by infusion of a drug. A variety of drugs can produce reversible inactivation, including muscimol, a GABA agonist that inactivates only neuron somata and not axons, and tetrodotoxin (TTX), a sodium channel blocker that blocks both neuron somata and axons.

Several parts of the circuit shown in Fig. 19.34 have been reversibly inactivated during training in naïve animals (Clark and Lavond, 1993; Hardiman *et al.*, 1996; Krupa *et al.*, 1993; Lavond *et al.*, 1993; Thompson and Kim, 1996; Thompson and Krupa, 1994). In brief, inactivation of the motor nuclei (Fig. 19.34a), red nucleus (Fig. 19.34b), superior cerebellar peduncle (Figs. 19.34d and e), or a localized region of the anterior interpositus nucleus and overlying cerebellar cortex (Fig. 19.34c) prevents expression of the CR (only motor nucleus inactivation also prevents expression of the UR). When tested without inactivation after training with inactivation of motor nuclei, the red nucleus, or the superior cerebellar peduncle, the CR is found to be fully learned as soon as the inactivation has ceased (Figs. 19.34a,b,d,e). However, localized cerebellar inactivation (Fig. 19.34c) completely prevents learning; animals must learn from scratch as if completely untrained. These results argue strongly for cerebellar localization of the memory trace. Indeed, when muscimol inactivation is limited to the interpositus nucleus (Krupa and Thompson, 1997) learning is completely prevented and then occurs normally with no savings after inactivation is removed. Note that here, the mossy and climbing fiber projections to cerebellar cortex are not impaired, that is, inputs to cerebellar cortex are normal, yet no learning at all occurs during inactivation of the interpositus. This hypothesis is supported by the observation that inhibition of protein synthesis in the cerebellar interpositus nucleus appears to prevent long-term retention of the conditioned eyeblink response (Bracha *et al.*, 1998; Chen and Steinmetz, 2000; Gomi *et al.*, 1999).

Experimentally, it has proven extremely difficult to determine the relative roles of the cerebellar cortex and interpositus nucleus in eyeblink conditioning using the lesion method (Lavond *et al.*, 1987; Yeo *et al.*, 1985a,b). The overlying cortex is not much more than a mm above the dorsal aspect of the interpositus nucleus. One solution to this problem has been the use of the mutant Purkinje cell degeneration (pcd) mouse strain (Chen *et al.*, 1996). In this mutant, Purkinje neurons (and all other neurons studied) are normal throughout pre- and perinatal development. Approximately 2–4 weeks after birth, the Purkinje neurons degenerate and disappear. For the next 2 months or so, other neuronal structures appear relatively normal, at least at the light microscopic level. Thus, during the period of young adulthood, the animals have a complete, selective functional decortication of the cerebellum (Goldowitz and Eisenman, 1992). The pcd mice learned very slowly, poorly, and at a much lower level than wild-type controls but did show substantial and significant learning and showed extinction with subsequent training to the CS alone. Thus, the cerebellar cortex plays a critically important role in normal learning (of discrete behavioral responses), but some degree of learning is possible without the cerebellar cortex. Appropriate lesions of the interpositus nucleus in pcd mice completely prevent eyeblink conditioning (Chen *et al.*, 1999), demonstrating that the residual learning in nonlesioned pcd mice requires the interpositus.

Considerable evidence exists that the cerebellar cortex plays a key role in adaptive timing of the learned response (McCormick and Thompson, 1984; Perrett *et al.*, 1993). It also plays a role in the normal rate of learning. The Purkinje neurons in the cerebellar cortex were counted (in representative sections) in a group of rabbits aged 3–50 months. The correlation between trials to criterion and number of Purkinje neurons was –0.79 (Woodruff-Pak *et al.*, 2000). A consistent finding in normal humans is a marked and virtually linear increase in difficulty of learning with increasing age (Solomon *et al.*, 1989; Woodruff-Pak *et al.*, 1996; Woodruff-Pak and Thompson, 1988). It seems clear from the preceding that processes of neuronal plasticity known as "memory traces" develop in both the cerebellar cortex and the interpositus nucleus as a result of classical conditioning of the eyeblink and other discrete responses. In our view, the overwhelming bulk of evidence argues strongly that the *essential* memory trace is in the interpositus nucleus (see Bao *et al.*, 2002; Chen *et al.*, 1996; Christian and Thompson, 2003, 2005; Freeman and Nicholson, 2000; Kleim *et al.*, 2002; Steinmetz and Woodruff-Pak, 2000). However, some other groups still favor the cerebellar cortex as the essential structure; the debate continues (see Attwell *et al.*, 2002; Garcia *et al.*, 1999a; Gruart and Yeo, 1995; Jirenhed *et al.*, 2007).

Putative Mechanisms of Memory Storage in the Cerebellum

Classic theories of the cerebellum as a learning machine (see earlier) proposed that conjoint activation of Purkinje neurons by parallel fibers and climbing fibers would lead to alterations in synaptic strength of the parallel fiber synapses. Ito (1982) discovered that such conjoint activation leads to a long-lasting depression in the efficacy of parallel fiber synapses to Purkinje neuron dendrites. This process is known as cerebellar LTD (see section on LTP–LTD). Ito and associates showed that such a process plays a key role in adaptation of the vestibulo-ocular reflex (VOR) (duLac et al., 1995; Ito, 1984, 1989, 1993).

In eyeblink conditioning, many of the Purkinje neurons that exhibit learning-related changes show decreases in simple spike responses in the CS period (Christian, 2004; Thompson, 1990) that are consistent with LTD. The current view at the molecular level is that LTD is caused by a persistent decrease in AMPA receptor function at parallel fiber synapses on Purkinje neuron dendrites (Ito, 1993; Linden et al., 1991; Linden and Connor, 1995). This decrease in AMPA receptor function is, in turn, the result of glutamate activation of AMPA and metabotropic receptors on Purkinje neuron dendrites, together with increased levels of intracellular calcium, which is normally caused by climbing fiber activation.

Because of the wide range of mutant and transgenic mice exhibiting cerebellar abnormalities, the mouse has become a major player in the study of neural substrates of eyeblink conditioning (e.g., Chen et al., 1996). Confusion has recently been introduced in a study arguing that the standard measure of eyeblink behavior in mouse (and rat), EMG from the obicularis oculi, is not a good measure of eyelid movement (Koekkoek et al., 2002). Actually, these authors recorded EMG from the muscle ventral to the orbit, where the measure is contaminated by other muscle actions. In all other studies in mouse and rat, EMG was recorded from the muscle dorsal to the orbit, where no such contamination occurs (Skelton, 1988; Stanton et al., 1992). Indeed, the correlations between the *dorsal* EMG and lid movements range from 0.98 to 0.99 (McCormick et al., 1982). A much more serious problem with mouse, unlike rabbit, is the marked increase in startle responses and spontaneous responses with "CR" temporal properties to higher frequency tone CSs (e.g., 10 KHz) independent of training (Lee, 2007). Thus, Koekkoek et al. (2003), using a 10 KHz CS, claimed that mice with interpositus lesions still showed CRs but the onset latency of their "CR" was 20 msec, a startle response; the minimal onset latency of actual CRs is never less than about 60 msec. Indeed, studies of eyeblink conditioning in mice using tone CSs in the range of 10 KHz cannot be interpreted. Use of much lower tone CS frequencies, 1–2 KHz, obviates this problem.

Both cerebellar cortical LTD and eyeblink conditioning have been studied in a number of mutant and KO mice. The consistent result is a high correlation between impairments in LTD and in eyeblink learning (Aiba et al., 1994; Chen et al., 1995; De Zeeuw et al., 1998; Ichise et al., 2000; Kashiwabuchi et al., 1995; Kim et al., 1996; Kishimoto et al., 2002; Kishimoto and Kano, 2006; Miyata et al., 2001; Shibuki et al., 1996). Cerebellar cortical abnormalities powerfully modulate eyeblink conditioning.

The memories formed in classical conditioning of discrete responses are long-term, relatively permanent memories. As with other forms of long-term memories (see section on LTP–LTD and *Aplysia*), these memories appear to require protein synthesis. Thus, infusion of anisomycin and protein kinase inhibitor H7 into the interpositus markedly impairs learning, and infusion of actinomycin D completely prevents learning (Bracha et al., 1998; Chen and Steinmetz, 2000; Gomi et al., 1999). However, when learning occurs in normal circumstances, subsequent infusion of these substances in the interpositus does not impair performance of the learned response. Thus, it is apparent that protein synthesis is necessary for formation of the memory trace but not for expression of the memory once it is formed.

In extremely important observations, Kleim et al. (2002) showed that following eyeblink conditioning in rats, a marked increase is found in the number of excitatory synaptic terminals in the interpositus (presumed mossy fiber terminals) but no increase in number of inhibitory synaptic terminals (presumed Purkinje neuron terminals). This finding is the most direct evidence to date for the structural basis of a long-term associative memory in mammals. Gene expression in the cerebellar interpositus nucleus resulting from eyeblink conditioning has been reported in rabbit and mouse. Over the course of learning there are two separate patterns of expression; one set of genes are expressed early in training bilaterally and another set are expressed later in training ipsilateral to the trained eye (mouse; Park et al., 2006), consistent with two-process theories of learning (see earlier). In rabbit, eyeblink conditioning is also associated with expression of a cell division cycle-type protein kinase (Gomi et al., 1999).

This section has focused on the essential role of the cerebellum in the classical conditioning of discrete behavioral responses, which is a basic form of associative learning and memory. To date, this is perhaps the clearest and most decisive example of evidence for the localization of a memory trace to a particular brain region (the cerebellum) in mammals. The cerebellum has also been pinpointed as the location where complex, multijoint movements are learned and stored (Thach et al., 1992). In addition, a growing body of evidence suggests that the cerebellum is critically involved in many other forms of learning and memory, including cardiovascular conditioning (Supple and Leaton, 1990), discrete response instrumental avoidance learning (Steinmetz *et al.*, 1993), maze learning (Pelligrino and Altman, 1979), spatial learning and memory (Goodlett *et al.*, 1992; Lalonde and Botez, 1990), and adaptive timing (Keele and Ivry, 1990). A growing literature also implicates the cerebellum in complex cognitive processes (Schmahmann, 1997).

The type of learning exemplified by the cerebellar circuitry underlying classical conditioning of discrete responses has been termed *supervised learning* (Knudsen, 1994). Information from one network of neurons acts as an instructive signal (US) to influence the pattern of connectivity in another network (e.g., the CS); other examples include adaptation of the VOR and calibration of the auditory space map in the barn owl. In eyeblink conditioning, the neutral CSs (e.g., tone or light) only weakly influence the activity of neurons in the cerebellum and do not yield the behavioral response. The strong connections established between networks of neurons as a result of training are not functionally coupled prior to learning. That is, diffuse cerebellar mossy and/or parallel fibers activated by the CS develop sufficient strength of their synaptic connections to successfully signal the specific circuit initially formed by the very localized climbing fiber projections to the cerebellum activated by the corneal air puff US. In summary, weak and ineffective anatomical connections become powerful and effective through learning. However, note that the connections do indeed exist before training. This may be considered a general principle in all aspects of learning and memory.

Fear Conditioning Is a Model System for Investigating the Neural Substrates of Emotional Memory

Significant progress has been made in identifying the neural substrates of emotional memory processing. Many of the advances have involved studies of classical fear conditioning. This model system has been used extensively in animal studies and, more recently, in human experiments as well.

What Is Fear Conditioning?

Classical fear conditioning involves repeated temporal pairings of an emotionally neutral stimulus (CS) with an aversive unconditioned stimulus (US) (Fig. 19.38). The US elicits a multitude of physiological and behavioral unconditioned responses, and over one or more conditioning trials, conditioned responses

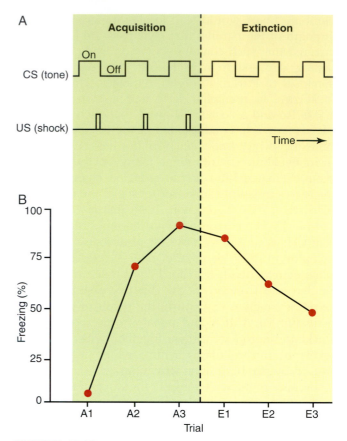

FIGURE 19.38 Fear conditioning paradigm: (A) Typical parametric arrangement of stimuli during the acquisition and extinction phases of a simple delay conditioning task. CS, conditioned stimulus; US, unconditioned stimulus. (B) Hypothetical (but realistic) acquisition and extinction learning curves for defense (freezing) responses conditioned to the CS. Note the rapid increase of freezing during the CS during acquisition trials (A1–A3) and the decline of freezing during extinction trials (E1–E3).

develop in reaction to the CS itself. This procedure was first devloped by Pavlov, hence the term *Pavlovian fear conditioing*, although Pavlov referred to it as *defense conditioning* (Pavlov, 1927). In contemporary work, the subject is typically presented with a tone followed by a brief electric shock. After several tone–shock pairings, the tone becomes aversive to the subject and begins to elicit a set of responses (the CRs) characteristic of a state of fear: freezing; autonomic responses (such as changes in heart rate, blood pressure, skin conductance, or pupillary dilation); endocrine responses (such as the release of so-called *stress hormones*, especially ACTH, adrenal steroids, and adrenal catecholamines); and potentiation of somatic reflexes (like eyeblink and startle) (see Choi *et al.*, 2001; Davis, 1998; Fendt and Fanselow, 1999; Kapp *et al.*, 1992; LeDoux, 2000; Maren 2001). The CRs thus form a set of quantifiable indices that can be used to gauge emotional learning and memory as the organism acquires the association between the CS and the US.

Conditioned fear responess are reactions that are automatically elicited by the CS. In addition, organisms often perform instrumental responses that cope with the CS and the US it warns about. These are called *avoidance responses*. As noted, while Pavlov had developed defense conditioning, and Watson (1925) used it in his famous study of Little Albert, the behaviorist movement in psychology later emphasized instrumental learning as the key to understanding complex human behaviors, including fear-related behaviors (Hull, 1943; Skinner, 1938). Thus, starting in the 1940s, instrumental avoidance conditioning became the primary behavioral approach to study learned fear (Miller, 1948, 1951; Mowrer, 1947; Mowrer and Lamoreaux, 1946). Subsequently, much research attempted to understand the neural basis of avoidance (Gabriel *et al.,* 1983; Goddard, 1964; Isaacson, 1982; Sarter and Markowitsch, 1985; Weiskrantz, 1956). This research generated complex results that were not easily integrated into a coherent view of the brain mechanisms of avoidance, much less fear (Cain and LeDoux, 2007; LeDoux, 1996). In the early 1970s researchers turned to Pavlovian conditioning and other simple forms of learning in order to facilitate the search for neural mechanisms of learning and memory in invertebrates (see earlier). Stimulated by the success of this approach, mammalian researchers began to use Pavlovian over instrumental approaches (Cohen, 1974; Thompson, 1976). By the late 1970s and early 1980s, the neural mechanisms of fear learning in mammals were much more likely to be studied using Pavlovian fear conditioning than instrumental avoidance.

As discussed in the previous section, memory researchers often distinguish between explicit or declarative (conscious) and implicit or nondeclarative (unconscious) memory. Fear conditioning falls into the latter category because it occurs normally in humans who have a loss of explicit memory due to brain damage (Bechara *et al.,* 1995). Similarly, brain damage disrupting fear conditioning fails to interfere with explicit memory of the conditioning experience (Bechara *et al.,* 1995; LaBar *et al.,* 1995). Thus, in an emotionally arousing situation in normal life, one brain system will form implicit (unconscious) emotional memories (expressed as conditioned fear responses) and another brain system will form explicit (conscious) memories about the experience. These have been called *emotional memories* and *memories about emotions,* respectively (LeDoux, 1996). The following section focuses on the neural circuits involved in the implicit formation of emotional memories

through Pavlovian defense or fear conditioning. The circuits involved in the formation of conscious memories about emotions are the same circuits involved in any other kind of explicit memory. It should also be noted that the two kinds of memories can interact. In other words, conscious explicit memories about an emotional situation can be modulated (enhanced or diminished) by the concurrent activation of an implicitly functioning emotional system (Cahill and McGaugh, 1998).

Specific Neural Circuits Within the Amygdala Are Involved in Fear Conditioning

Across various paradigms, species, and response measures, the amygdala has consistently emerged as a brain structure essential to the acquisition and expression of conditioned fear (see Davis, 1998; Fendt and Fanselow, 1999; Kapp *et al.,* 1992; LeDoux, 2000; Lee *et al.,* 2001; Maren 2001; Rodrigues *et al.,* 2004). Pretraining lesions of the amygdala prevent the development of CRs, whereas post-training lesions of the amygdala disrupt the expression of CRs that have already been learned, even after extensive overtraining (Maren, 1998).

In combination with lesion data, neuroanatomical tract-tracing studies have begun to elucidate the afferent and efferent connections of the amygdala to sensory and motor areas involved in transmitting information about the CS and US and generating emotional responses (for reviews see LeDoux, 2000; McDonald, 1998; Pitkanen *et al.,* 1997). Most of the work on CS pathways has involved auditory stimuli, although visual CS pathways have also been described (Shi and Davis, 2001). Information regarding an auditory CS reaches the amygdala by way of two neural routes: a direct thalamo–amygdala pathway from the auditory thalamus (medial portion of the medial geniculate nucleus and posterior intralaminar nucleus) to the lateral nucleus of the amygdala, and an indirect thalamo–cortico–amygdala pathway linking the auditory thalamus with the lateral nucleus of the amygdala by way of connections within the auditory cortex (Fig. 19.39) (LeDoux *et al.,* 1991). The direct thalamo–amygdala pathway is more rapid but provides a relatively crude representation of the incoming sensory stimulus compared with the thalamo–cortico–amygdala pathway (LeDoux, 1986). The direct pathway is thought to function in two ways: first as a quick route for simple stimulus features to evoke defensive emotional responses; and second as a method of priming the amygdala to set up appropriate emotional responses to incoming stimuli that are more highly processed by the indirect pathway. Lesion studies have shown that either of

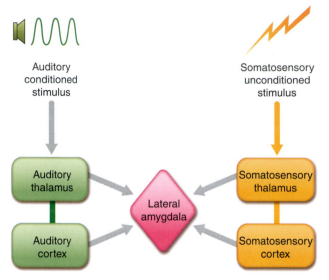

FIGURE 19.39 CS and US transmission pathways in fear conditioning. The auditory conditioned stimulus and somatosensory unconditioned stimulus are transmitted through brain stem to the thalamic stations in each pathway (the line is broken to reflect that the pathways include stations in the brain stem that are not shown). The thalamic regions give rise to connections to cortical receiving areas. Components of both the thalamic and cortical processing regions in each pathway then connect with the lateral nucleus of the amygdala. Brain lesion studies suggest that the CS and US can reach the amygdala through either thalamic or cortical areas.

these routes in isolation is sufficient to mediate responding in simple conditioning protocols involving one CS (Romanski and LeDoux, 1992). However, in situations involving more complex stimulus processing it is likely that the cortical projections to the amygdala will be involved (Jarrell *et al.*, 1987; Lindquist *et al.*, 2004), but the exact conditions requiring the auditory cortex are poorly understood (Armony *et al.*, 1997).

The two incoming CS pathways converge in the lateral nucleus of the amygdala (LeDoux *et al.*, 1991; Li *et al.*, 1996), which functions as the sensory gateway to the amygdala (Amaral *et al.*, 1992; Pitkanen *et al.*, 1997). Indeed, damage to the lateral amygdala (Campeau and Davis, 1995; LeDoux *et al.*, 1990) or inactivation of this region and adjacent areas (Muller *et al.*, 1997; Wilensky *et al.*, 1999) prior to conditioning prevents fear learning from taking place. In addition, cells in the lateral nucleus that receive incoming acoustic CS information also respond to somatosensory signals elicited by footshock. Although the exact sources of US inputs to the amygdala are still not fully understood, it appears that stimulation of the posterior intralaminar nucleus can serve as an effective US in fear conditioning (Cruikshank *et al.*, 1992). However, some evidence exists that the US pathway, like the CS pathway, involves both direct thalamo–amygdala and indirect thalamo–cortico–amygdala circuits that converge in the lateral amygdala (Shi and Davis, 1999). Therefore, it appears that the lateral nucleus of the amygdala is a likely region of integration, both within CS and US systems and between them. At the same time, plasticity in the acoustic thalamus and auditory cortex (Weinberger, 1995) may also play an important role in auditory fear conditioning (Apergis-Schoute *et al.*, 2005), but the amygdala appears critical for the development of these plastic changes in the wider fear network (Maren *et al.*, 2001).

Incoming sensory information reaching the lateral nucleus is then projected by intra-amygdala circuitry to several targets within the amygdala, including the basal, accessory basal, medial, intercalated, and central nuclei (LeDoux, 2007; Pare *et al.*, 1995; Paré *et al.*, 1995; Pitkanen *et al.*, 1997). For simple fear conditioning, transmission from the lateral to the central, either directly or by way of intermediate synapses, in other amygdala areas is required. The central nucleus, in turn, mediates emotional responses through efferent connections with motor and autonomic centers (see Choi *et al.*, 2001; Davis, 1998; Fendt and Fanselow, 1999; Kapp *et al.*, 1992; LeDoux, 2000). Interestingly, lesions of the target structures or transection of fibers projecting to these target areas selectively disrupt CRs expressed in specific response modalities, leaving other CR measures intact. For example, lesions of the central gray region disrupt freezing to the CS, but increases in blood pressure elicited by the same stimulus remain evident. These results, along with others, suggest that brain stem projections of the central amygdala are involved in the expression of conditioned responses in different modalities. Therefore, along the neural routes involved in fear conditioning, the amygdala appears to function as the key station where emotional significance is assessed. This assessment takes place independently of the manner in which the stimulus enters the brain and independently of the way the responses are expressed.

The contribution of amygdala areas to appetitive Pavlovian conditioning has also been studied (Balleine and Killcross, 2006; Cardinal and Everitt, 2004). Although some of the same nuclei are involved, their contributions to appetitive conditioning appear to differ from fear conditioning, which is not surprising given the different behavioral demands and corresponding neural requirements of the two forms of conditioning.

Cellular Mechanisms of Fear Conditioning

The fact that CS and US pathways converge in the lateral amygdala suggests that this sensory interface of the amygdala may be the site of plasticity in the fear conditioning circuit. Indeed, single-unit recording studies have shown that cells in the lateral amygdala have short-latency responses to acoustic stimuli (Bordi and LeDoux, 1992) and that the firing rate of these cells increases dramatically during training (Collins and Pare, 2000; Maren, 2000; Quirk *et al.*, 1995, 1997; Repa *et al.*, 2001). Recent studies have shown that such changes predict the acquisition of conditioned fear behavior and that the cellular responses in at least some of the cells persist during extinction until the behavioral response has been eliminated by presentations of the CS without the US (Repa *et al.*, 2001). Although conditioned neural responses have also been observed in the basal (Maren *et al.*, 1991) and central (Pascoe and Kapp, 1985) nuclei, the latency of these is longer than that in the lateral nucleus, suggesting the changes in the lateral amygdala may play a primary or driving role in amygdala plasticity.

The change in the responsiveness of lateral amygdala cells during and after fear conditioning suggests that alterations in excitatory synaptic transmission in the lateral amygdala might be critical for fear conditioning. Accordingly, many of the recent studies that have examined the cellular basis of fear conditioning have focused on the potential link between LTP in the lateral amygdala and fear memory formation. A number of studies have shown, for example, that LTP exists in each of the major auditory input pathways to the lateral amygdala, including the thalamic and cortical auditory pathways (Chapman *et al.*, 1990; Clugnet and LeDoux, 1990; Dityatev and Bolshakov, 2005; Huang and Kandel, 1998; Humeau *et al.*, 2003; Pape and Stork, 2003; Rogan and LeDoux, 1995; Weisskopf, *et al.*, 1999). Furthermore, induction of LTP in CS pathways alters the processing of auditory stimuli (Rogan and LeDoux, 1995), which is similar to the way that fear conditioning alters the processing of auditory stimuli in the lateral amygdala (Rogan *et al.*, 1997). In addition, fear conditioning *in vivo* alters the responses of amygdala cells to afferent stimulation *in vitro* (McKernan and Shinnick-Gallagher, 1997). These various findings have strengthened the hypothesis that meaningful plasticity occurs in the lateral amygdala during fear conditioning, and also highlight the potential of fear conditioning as an approach to understanding the relationship of LTP to memory formation.

Molecular Basis of Fear Conditioning

Following the discovery that long-term potentiation in the hippocampus depends on NMDA receptors (see section on LTP–LTD), researchers pursued the possibility that fear conditioning involves an NMDA-dependent form of synaptic plasticity in the

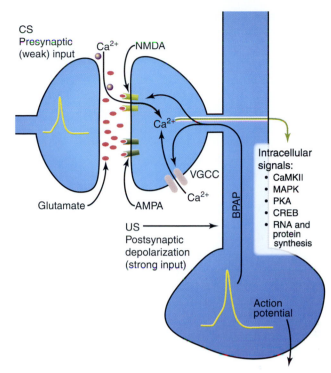

FIGURE 19.40 Model of synaptic plasticity during fear conditioning. The CS provides weak glutamate-mediated activation of presynaptic inputs to lateral amygdala cells. The glutamate activates AMPA receptors but cannot activate NMDA receptors in the absence of strong, depolarizing inputs to the cell. If the weak presynaptic input from the CS arrives at about the same time that the postsynaptic cell is also depolarized by a strong input, such as by the US, postsynaptic depolarization occurs, allowing calcium to enter through NMDA receptors and voltage-gated calcium channels (VGCCs). This combined calcium signal then activates a variety of signaling pathways, including MAPK, PKA, and CREB, and initiates RNA and protein synthesis. The proteins then act at the CS input synapses to strengthen and stabilize the connection.

amygdala (Fig. 19.40). Several studies have shown that conditioning is disrupted by a blockade of NMDA receptors in the lateral (and basal) amygdala prior to fear conditioning (Gerwitz and Davis, 1997; Lee *et al.*, 1998; Maren *et al.*, 1996; Miserendino *et al.*, 1990; Walker and Davis, 2000). Other recent studies have specifically implicated the NR2B subunit of the NMDA receptor in fear conditioning (Rodrigues *et al.*, 2001; Tang *et al.*, 1999). In these later studies, manipulation of NR2B impaired both short-term memory (STM) and long-term memory (LTM) of fear conditioning, suggesting that NMDA receptors play an essential role in the initial synaptic plasticity underlying fear learning (Rodrigues *et al.*, 2001; Tang *et al.*, 1999; Walker and Davis, 2000).

Other studies have found that LTP in the lateral amygdala is NMDA receptor dependent, which is consistent with the behavioral findings (Bauer *et al.*, 2002; Gean *et al.*, 1993; Huang and Kandel, 1998). However, under some conditions, associative and synapse-specific LTP in the lateral amygdala requires calcium influx in the postsynaptic cell through L-type voltage-gated calcium channels (VGCCs) rather than NMDA receptors (Bauer *et al.*, 2002; Weisskopf *et al.*, 1999). Thus, as in the hippocampus (Magee and Johnston, 1997), both NMDA receptors and L-type VGCCs are involved in aspects of synaptic plasticity in the amygdala (see section on LTP–LTD), suggesting that fear conditioning may require some combination of these mechanisms at the cellular level (Blair *et al.*, 2001). This hypothesis is supported by studies showing that blockade of L-type VGCCs in the lateral amygdala disrupts fear conditioning (Bauer *et al.*, 2002).

A great deal of progress has also been made in pursuing the molecular basis of fear conditioning in the amygdala. As we have seen in previous sections, one of the immediate downstream consequences of NMDAR-mediated activity-dependent increases in Ca^{2+} at the time of LTP induction is the activation of Ca^{2+}/calmodulin (CaM)-dependent protein kinase II (CaMKII). Anatomical studies have shown that αCaMKII is robustly expressed in lateral amygdala pyramidal neurons (McDonald *et al.*, 2002), where it coexists with NR2B in spines postsynaptic to terminals that originate in the auditory thalamus (Rodrigues *et al.*, 2004). Fear conditioning leads to increases in the autophosphorylated form of αCaMKII at Thr^{286} in spines of lateral amygdala neurons. Further, intra-amygdala infusion or bath application of an inhibitor of CaMKII (KN-62) impairs acquisition and STM formation of fear conditioning and LTP at thalamic inputs to LA neurons, respectively. This latter finding is consistent with molecular genetic experiments indicating that induced overexpression of active αCaMKII by a transgene that replaces Thr^{286} with an aspartate residue in the amygdala and striatum results in a reversible deficit in fear conditioning (Mayford *et al.*, 1996). Although the mechanism by which αCaMKII promotes plasticity and STM formation in the lateral amygdala is at present unknown, it likely involves modulation of glutamatergic receptors at lateral amygdala synapses. As we have seen, evidence has accumulated indicating that new AMPA receptors are trafficked and inserted into synapses during and after LTP (Grosshans *et al.*, 2002; Malenka, 2003; Malinow, 2003). Fear conditioning is also known to lead to enhanced surface expression of the GluR1 subunit (Yeh *et al.*, 2006), and intra-amygdala infusion of a viral vector that encodes for a transgene that prevents GluR1 from being inserted into synaptic sites impairs fear acquisition and synaptic plasticity in the lateral amygdala (Rumpel *et al.*, 2005). Importantly, the insertion of the AMPA receptor subunit GluR1 into synapses has been shown to be αCaMKII dependent, and blockade of αCaMKII-mediated synaptic delivery of GluR1 prevents LTP (Hayashi *et al.*, 2000). Thus, while additional studies are needed, these findings collectively suggest that activation of αCaMKII during fear acquisition may regulate the insertion of AMPARs at LA synapses and thereby contribute, in part, to the formation and maintenance of STM.

The mechanisms underlying fear memory consolidation, or the conversion of STM to LTM, have also been extensively studied. Findings have shown, for example, that LTM, but not STM, of fear conditioning requires the synthesis of mRNA and new proteins in the lateral amygdala, as well as the activity of several kinases, including MAPK and PKA (Bailey *et al.*, 1999; Schafe *et al.*, 1999, 2000; Schafe and LeDoux, 2000). These kinases have also been implicated in LTP in the lateral amygdala (Huang *et al.*, 2000; Huang and Kandel, 1998; Schafe *et al.*, 2000). Both MAPK and PKA phosphorylate the transcription factor CREB. Mice deficient in CREB have impaired fear memory consolidation (Bourtchouladze *et al.*, 1994). Further, upregulation of CREB in the amygdala enhances the consolidation of fear conditioning (Josselyn *et al.*, 2001) and influences the probability that individual lateral amygdala neurons are recruited into a fear memory trace (Han *et al.*, 2007). These findings are consistent with the general roles of CREB, PKA, and MAPK discovered in studies as diverse as synaptic facilitation in *Aplysia* (see section on Invertebrate Studies), conditioning in *Drosophila* (Yin *et al.*, 1994, 1995), hippocampal LTP (English and Sweatt, 1997; Frey *et al.*, 1993; Huang *et al.*, 1994; Impey *et al.*, 1996), and hippocampus-dependent spatial learning (Abel *et al.*, 1997; Blum *et al.*, 1999; Bourtchouladze *et al.*, 1994; Guzowski and McGaugh, 1997). The conservation of these molecular contributions across species and training conditions suggests that the uniqueness of memory may not be due to the underlying molecules so much as to the circuit in which these molecules act. It is important to note that the discovery that protein synthesis in the lateral amygdala is also required for the stabilization of memories after retrieval (so-called *reconsolidation*) is stimulating new research and raising questions about classic notions of consolidation (Nader *et al.*, 2000).

In addition to Ca^{2+}-mediated signaling, neurotrophins have been widely implicated in driving protein kinase signaling pathways necessary for long-term synaptic plasticity and memory formation, including fear conditioning. A recent study has shown that BDNF-mediated signaling in the amygdala is critical to fear learning (Rattiner *et al.*, 2004, 2005). In that study, fear conditioning led to increases in both TrkB receptor phosphorylation and decreases in TrkB receptor immunoreactivity in the lateral amygdala during the consolidation period, which is typically indicative of bound BDNF. Further, disruption of TrkB receptor signaling in the amygdala using either a Trk receptor antagonist or lentiviral overexpression of a dominant negative TrkB isoform impaired fear memory formation (Rattiner et al., 2004). Although this study did not distinguish between acquisition and consolidation phases of fear learning, the assumption is that BDNF signaling in the lateral amygdala plays a critical role in the establishment of long-term fear memories, possibly by promoting the activation and nuclear translocation of protein kinases such as MAPK (Patterson et al., 2001; Ying et al., 2002). Additional experiments will be necessary to define the signaling pathways through which BDNF acts during fear learning.

Although CRE-mediated transcription clearly supports the development of long-term plasticity and memory, the downstream targets of CREB in the lateral amygdala, like in other memory systems, have remained largely unknown. However, a number of studies have shown that fear conditioning induces the expression of both immediate-early (Beck and Fibiger 1995; Malkani and Rosen 2000; Reijmers et al., 2007; Rosen et al., 1998; Scicli et al., 2004) and downstream late response genes (Ressler et al., 2002; Stork et al., 2001) in the LA. Although the specific contributions of many of these genes to fear conditioning is still unclear, it is widely believed that learning-induced gene expression ultimately contributes to changes in cell (especially synaptic) structure that stabilizes memory (Bailey and Kandel 1993; Rampon and Tsien 2000; Woolf 1998), presumably by altering the actin cytoskeleton underlying synaptic organization (Kasai et al., 2003; Matus, 2000; van Rossum and Hanisch, 1999). Indeed, fear conditioning leads to alterations in cytoskeletal proteins and to new spine formation in the lateral amygdala. Fear conditioning, for example, leads to the transcription of genes involved in cytoskeletal remodeling, including the CRE-mediated gene NF-1 (Ressler et al., 2002). Further, interference with molecular pathways known to be involved in structural plasticity during early development, such as the Rho-GAP signaling pathway, disrupts memory formation (Lamprecht et al., 2002), and fear conditioning drives actin cytoskeleton–regulatory proteins, such as profilin, into amygdala spines shortly after training (Lamprecht et al., 2006). Finally, a recent morphological study has suggested that fear conditioning leads to an increase in spinophilin-immunoreactive dendritic spines in the

lateral amygdala (Radley et al., 2006). Although much remains to be done in this area, the fear memory system clearly holds great promise in linking molecular signaling pathways and synaptic plasticity to memory formation.

Human Fear Conditioning

Fear conditioning has been readily demonstrated in human subjects, and many experimental preparations for measuring fear in animals can be adapted for use with human populations. The brain structures that mediate aspects of fear learning in animals appear to perform similar functions in humans, although the cellular and molecular mechanisms are not as well understood. In humans, autonomic indices of fear are commonly investigated, such as the skin conductance response (SCR), a measure of sympathetic arousal taken from the sweat glands located on the palms of the hand and bottom of the feet. Studies of neurologic patients with medial temporal lobe damage, including the amygdala, indicate that they fail to acquire conditioned fear reactions, as measured by SCR (Bechara et al., 1995; LaBar et al., 1995; Hamann et al., 2002; Phelps et al., 1998; Peper et al., 2001). Importantly, if the lesion extent does not affect the hippocampus bilaterally, these patients are able to verbally report the correct CS–US contingency, demonstrating intact declarative knowledge of the training parameters. In contrast, amnesic patients with selective bilateral hippocampal damage show the opposite dissociation: they have intact fear responses to the CS on simple tasks but fail to acquire knowledge about the reinforcement contingency (Bechara et al., 1995; LaBar and Disterhoft, 1998; LaBar and Phelps, 2005). Amnesic patients also fail to associate conditioned cues with the environmental contexts in which conditioning takes place (LaBar and Phelps, 2005), which is necessary to form a declarative memory for the conditioning episode. The hippocampus contributes to other complex forms of fear conditioning as well, such as trace conditioning (Buchel et al., 1999; Knight et al., 2004). Thus, the human research supports distinctive contributions of the amygdala and hippocampus to fear learning, as predicted by the animal literature (Fig. 19.41).

Because organic lesions in humans rarely affect the amygdala selectively, it is important to corroborate the human lesion results with neuroimaging techniques that have a better ability to discern the functional contributions of neighboring structures in the healthy human brain. Moreover, functional magnetic resonance imaging (fMRI) can reveal how activity in the relevant neural circuitry changes across training trials. During conditioned fear learning, activity in

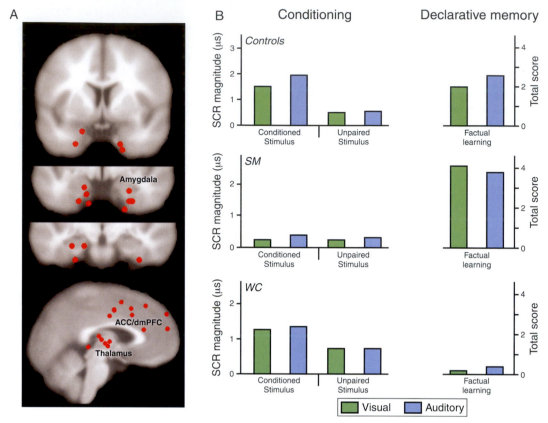

FIGURE 19.41 Neural circuits for conditioned fear learning in humans. (A) A summary of nine studies showing hemodynamic changes in the amygdala and adjacent periamygdaloid cortex, thalamus, and anterior cingulate–dorsomedial prefrontal cortex of healthy human adults during conditioned fear learning. (B) Damage to the amygdala versus hippocampus yields dissociable deficits during conditioned fear learning. Whereas a patient with selective bilateral amygdala damage (SM) has intact declarative memory for a conditioning episode, she fails to acquire conditioned skin conductance responses (SCR) to visual or auditory conditioned stimuli. In contrast, a patient with selective bilateral hippocampal damage (WC) shows the reverse pattern. Healthy controls acquire both conditioned SCRs and declarative knowledge, presumably via the distinct influences of these brain regions. μS = microsiemens. Adapted with permission from Macmillan Publishers Ltd. LaBar, K.S., and Cabeza, R. (2006). Cognitive neuroscience of emotional memory. *Nature Rev. Neurosci.* **7**, 54–64.

the amygdala, periamygdaloid cortex, thalamus, and dorsal anterior cingulate–medial prefrontal cortex is consistently reported (reviewed in Buchel and Dolan, 2000; LaBar and Cabeza, 2006) (Fig. 19.41). Some imaging studies have further elucidated the temporal dynamics of the amygdala's response. Its activity appears to be strongest during the early portion of acquisition training, diminishes over time when the CS–US association is well-learned, and then transiently re-emerges during extinction (Buchel *et al.*, 1998; LaBar *et al.*, 1998; Morris *et al.*, 2001; Phelps *et al.*, 2004). This pattern suggests that the amygdala is predominantly engaged when the emotional salience of the CS changes, which is indicative of a learning and/or attentional function. During extinction or reversal learning, the amygdala interacts with the ventral prefrontal cortex and ventral anterior cingulate, which putatively provide executive signals for fear suppression (Morris and Dolan, 2004; Phelps *et al.*, 2004). Such observations dovetail with other studies of emotion regulation in humans

(Schaefer *et al.*, 2002) and converge with electrophysiological evidence for prefrontal-amygdala interactions during extinction learning in nonhuman animals (Milad and Quirk, 2002).

The degree of amygdala activation is tightly coupled with the generation of conditioned SCRs, both within subjects (on a trial-by-trial basis) and between subjects (reflecting individual differences in conditionability) (Cheng *et al.*, 2003; Furmark *et al.*, 1997; LaBar *et al.*, 1998). This functional relationship is stronger for the amygdala than for other brain regions that contribute to conditioned fear acquisition, which suggests a special relationship between amygdala activity and peripheral indices of fear learning and expression. Studies in other related domains, such as facial affect perception (Adolphs *et al.*, 1994; Breiter *et al.*, 1996; Morris *et al.*, 1996; Whalen *et al.*, 1998) and anticipatory anxiety (Phelps *et al.*, 2001), confirm a central role for the amygdala in detecting and responding to potentially threatening signals in the environment.

Interestingly, the amygdala may interact with different afferent pathways depending on whether conditioned fear learning proceeds consciously or not. Morris and colleagues (1999) used functional connectivity modeling to compare how the amygdala's response correlated with other brain regions when participants were conditioned subliminally versus supraliminally. Subliminal conditioning was accomplished by masking the visual CS so that it was rendered imperceptible to the participants. In the subliminal case, the amygdala exhibited greater functional connectivity with subcortical regions, including the thalamus and superior colliculus, whereas in the supraliminal case, the amygdala interacted with cortical regions to a greater extent.

Fear conditioning–induced plasticity in sensory processing regions has also been implicated by neuroimaging studies. Morris and colleagues (1998) identified regions of auditory cortex that responded preferentially to 2000 Hz or 800 Hz tones and then investigated how these regions responded following aversive conditioning to one of the tone frequencies. A conditioning-induced, frequency-specific modulation of auditory cortex activity was found that covaried with activity in the amygdala and orbitofrontal cortex. Auditory cortex activity to an acoustic CS is further altered by the presence of a visual context that indicates greater probability of US occurrence (Armony and Dolan, 2001). In sum, the existing human studies suggest a marked conservation of function in the relevant fear pathways.

A strong similarity exists between the clinical symptoms of anxiety in humans and measures of conditioned fear in animals, and drugs with anxiolytic effects in animal models of fear have been applied to treat human anxiety (Davis, 1992). Some researchers have thus proposed using fear conditioning as a model for studying human affective disorders (Charney et al., 1993; Wolpe and Rowan, 1988). For example, one clinical marker for post-traumatic stress disorder (PTSD) is an increase in startle. PTSD patients often have increased fear-potentiated startle and conditioned autonomic responses relative to control subjects (Orr et al., 2000). They also exhibit an imbalanced pattern of corticolimbic activity during fear conditioning in which the amygdala is over-responsive to the CS during fear acquisition and the anterior cingulate is under-responsive to the CS during fear extinction (Bremner et al., 2005). Psychopaths are characterized by the opposite pattern, with reduced amygdala activity during fear conditioning and a lack of behavioral expression of fear learning (Birbaumer et al., 2005). Therefore, fear conditioning paradigms can be useful to help differentiate imbalances in emotional learning circuits across affective disorders. There is a current emphasis on linking individual differences in fear conditioning with genetic markers of neurotransmitter function in an attempt to identify potential biomarkers for individuals at risk for developing anxiety disorders (Garpenstrand et al., 2001). These advances, in combination with the incorporation of classical conditioning principles in cognitive-behavioral therapies (Marks and Tobena, 1990), are an excellent example of how animal models of plasticity can be potentially translated into effective clinical treatments for disorders of emotion and memory. For additional information about the relation of fear conditioning and the amygdala to anxiety disorders see Lang et al. (2000), Bremner (2007), Rauch et al. (2003), and Yehuda and LeDoux (2007).

HOW DOES A CHANGE IN SYNAPTIC STRENGTH STORE A COMPLEX MEMORY?

The relationship between the synaptic changes and the behavior of conditioned reflexes can be straightforward, because the locus for the plastic change is part of the mediating circuit. Thus, the change in the strength of the sensorimotor synapse in *Aplysia* can be related to the memory for sensitization (e.g., Fig. 19.23A). However, the idea that an increase in synaptic strength leads to an enhanced behavioral response, and a decrease in synaptic strength leads to a decreased behavioral response, can be misleading. For example, a decrease in synaptic strength in a postsynaptic neuron that exerts an inhibitory action can be translated into an enhanced behavioral response. Indeed, in the parallel fiber–to–Purkinje cell connection in the cerebellum, such a disinhibition is precisely the mechanism that has been proposed to mediate classical conditioning of the eyeblink reflex. Nevertheless, for relatively simple reflex systems in which the circuit is well understood, it is possible to directly relate a change in synaptic strength to learning. However, in most other examples of memory, it is considerably less clear how the synaptic changes are induced and, once induced, how the information is retrieved. This is especially true in memory systems that involve the storage of information for patterns, facts, and events. Neurobiologists have turned to artificial neural circuits to gain insights into these issues.

A simple network that can store and "recognize" patterns is illustrated in Figure 19.42. The network is artificial but nevertheless is inspired by actual circuitry in the CA3 region of the hippocampus. In this example, six different input projections make

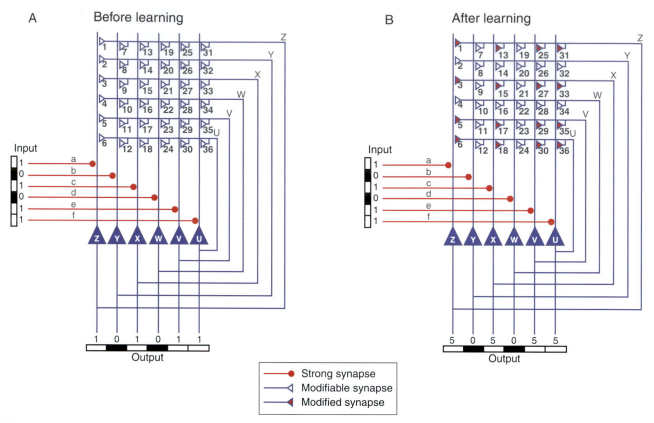

FIGURE 19.42   Autoassociation network for recognition memory. The artificial circuit consists of six input pathways that make strong connections to each of six output neurons. The output neurons have axon collaterals that make synaptic connections with each of the output cells. (A) A pattern represented by activity in the input lines or axons (a, b, c, d, e, f) is presented to the network. A *1* represents an active axon (e.g., a spike) whereas a *0* represents an inactive axon. The input pathways make strong synapses (filled circles) with the postsynaptic output cells. Thus, the output cells (u, v, w, x, y, z) generate a pattern that is a replica of the input pattern. The collateral synapses were initially weak and do not contribute to the output. Nevertheless, the activity in the collaterals that occurred in conjunction (assume minimal delays within the circuit) with the input pattern led to a strengthening of a subset of the 36 synapses. (B) A second presentation of the input produces an output pattern that is an amplified but an otherwise intact replica of the input. An incomplete input pattern can be used as a cue to retrieve the complete pattern.

synaptic connections with the dendrites of each of six postsynaptic neurons (Fig. 19.42A). The postsynaptic neurons serve as the output of the network. Input projections can carry multiple types of patterned information, and these patterns can be complex. To simplify the present discussion, consider that the particular input pathway in Figure 19.42A carries information regarding the pattern of neural activity induced by a single brief flash of a spatial pattern of light. For example, activity in the top pathway (line *a*) might represent light falling on the temporal region of the retina, whereas activity in the pathway on the bottom (line *f*) might represent light falling on the nasal region of the retina. Thus, the spatial pattern of an image falling on the retina could be reconstructed from the pattern of neuronal activity over the *n* (in this case, 6) input projections to the network.

Three aspects of the circuit endow it with the ability to store and retrieve patterns. First, each of the input

lines makes a sufficiently strong connection with its corresponding postsynaptic cell to activate it reliably. Second, each output cell (z to u) sends an axon collateral that makes an excitatory connection with itself as well as the other five output cells. This pattern of synaptic connectivity leads to a network of 36 synapses (a total of 42, including the 6 input synapses). Third, each of the 36 synaptic connections are modifiable through an LTP-like mechanism (see earlier). Specifically, the strength of a particular synaptic connection is initially weak, but it will increase if the presynaptic *and* postsynaptic neurons are active at the same time. The circuit configuration with the embedded synaptic "learning rule" leads to an autoassociation or autocorrelation matrix. The autoassociation is derived from the fact that the output is fed back to the input where it associates with itself.

Now consider the consequences of presenting the patterned input to the network of Figure 19.42A.

The input pattern will activate the six postsynaptic cells in such a way as to produce an output pattern that will be a replica of the input pattern. In addition, the pattern will induce changes in the synaptic strength of the active synapses in the network. For example, synapse 3 will be strengthened because the postsynaptic cell, cell z, and the presynaptic cell, cell x, will be active at the same time. Note also that synapses 1, 5, and 6 will be strengthened as well. This occurs because these input pathways to cell z are also active. Thus, all synapses that are active at the same time as cell z will be strengthened. When the pattern is presented again as in Figure 19.42B, the output of the cell will be governed not only by the input but also by the feedback connections, a subset of which were strengthened (Fig. 19.42B, filled synapses) by the initial presentation of the stimulus. Thus, for output cell z a component of its activity will be derived from input a, but components will also come from synapses 1, 3, 5, and 6. If each of the initially strong and newly modified synapses is assumed to contribute equally to the firing of output cell z, the activity will be five times greater than the activity produced by input a before the learning. After learning, the output is an amplified version of the input but the basic features of the pattern are preserved.

Note that the "memory" for the pattern does not reside in any one synapse or in any one cell. Rather, it is *distributed* throughout the network at multiple sites. The properties of these types of autoassociation networks have been examined by James Anderson, Teuvo Kohonen, David Marr, Edmond Rolls, David Wilshaw, and their colleagues (see Kohonen, 1989) and found to exhibit a number of phenomena that would be desirable for a biological recognition memory system. For example, such networks exhibit pattern completion. If a partial input pattern is presented, the autoassociation network can complete the pattern in the sense that it can produce an output that is approximately what is expected for the full input pattern. Thus, any part of the stored pattern can be used as a cue to retrieve the complete pattern. For the example of Figure 19.42, the input pattern was {101011}. This pattern led to an output pattern of {505055}. If the input pattern was degraded to {101000}, the output pattern would be {303022}. Some change in the strength of firing occurs but the basic pattern is preserved. Autoassociation networks also exhibit a phenomenon known as *graceful degradation*, which means that the network can still function even if some of the input connections or postsynaptic cells are lost. This property arises from the distributed representation of the memory within the circuit.

SUMMARY

The search for the biological basis of learning and memory has occupied the research of many of the twentieth century's leading neuroscientists. This work has ranged from studies on the behavior of whole organisms down to the molecular changes occurring at individual synapses. Within the past two decades, great strides have been made in increasing our understanding of learning and memory. For example, the neural and molecular mechanisms of short- and long-term sensitization and classical and operant conditioning in *Aplysia* have been elucidated. The search for the engram underlying classical eyeblink conditioning has been continually narrowed and refined. The anatomical and physiological properties subserving Pavlovian fear conditioning have also been elucidated, at least in general terms. Finally, although much work remains to be done, neuroscientists have garnered a greater understanding of the subcellular changes underlying synaptic plasticity, particularly within the hippocampal formation and cerebellum.

In the near future, a major experimental question to be answered is the extent to which the mechanisms for learning are common both within any one animal and between different species. Although many common features are emerging, there seem to be some differences. Thus, it will be important to understand the extent to which specific mechanisms are used selectively for one type of learning and not another. Irrespective of the particular example of learning and memory that is analyzed, whether it be simple or complex, it will be important to pay attention to three major details: the details of the circuit interactions, the details of the learning rule, and the details of the intrinsic biophysical properties of the neurons within the circuit.

References

Suggested Readings

Aggleton, J. P. (2000). "The Amygdala: A Functional Analysis," 2nd ed. Oxford Univ. Press, New York.

Andersen, P., Morris, R., Amaral, D., Bliss, T., and O'Keefe, J., Eds. (2007). "The Hippocampus Book." Oxford Univ. Press, New York.

Bear, M. F., and Malenka, R. C. (1994). Synaptic plasticity: LTP and LTD. *Curr. Opin. Neurobiol.* **4**, 389–399.

Bliss, T. V. P., and Collingridge, G. L. (1993). A synaptic model of memory: Long-term potentiation in the hippocampus. *Nature (London)* **361**, 31–39.

Byrne, J. H. (1987). Cellular analysis of associative learning. *Physiol. Rev.* **67**, 329–439.

Carew, T. J. (2000). "Behavioral Neurobiology." Sinauer Associates, Sunderland, MA.

Fendt, M., and Fanselow, M. S. (1999). The neuroanatomical and neurochemical basis of conditioned fear. *Neurosci. Biobehav. Rev.* **23**, 743–760.

Ito, M. (1989). Long-term depression. *Annu. Rev. Neurosci.* **12**, 85–102.

Kandel, E. R. (2001). The molecular biology of memory storage: A dialogue between genes and synapses. *Science* **294**, 1030–1038.

Malenka, R. C., and Nicoll, R. A. (1999). Long-term potentiation—a decade of progress? *Science* **285**, 1870–18744.

Martin, S. J., Grimwood, P. D., and Morris, R. G. (2000). Synaptic plasticity and memory: An evaluation of the hypothesis. *Annu. Rev. Neurosci.* **23**, 649–711.

Squire, L. R., and Kandel, E. R. (1999). "Memory: From Mind to Molecules." Freeman, New York.

Sweatt, J. D. (2003). "Mechanisms of Memory." Elsevier, London.

Cited

Abel, T., Nguyen, P. V., Barad, M., Deuel, T. A., Kandel, E. R., and Bourtchouladze, R. (1997). Genetic demonstration of a role for PKA in the late phase of LTP and in hippocampus-based long-term memory. *Cell* **88**, 615–626.

Adolphs, R., Tranel, D., Damasio, H., and Damasio, A. R. (1994). Impaired recognition of emotion in facial expressions following bilateral damage to the human amygdala. *Nature, 372, 669–672.*

Aiba, A., Kano, M., Chen, C., Stanton, M. E., Fox, G. D., Herrup, K., Zwingman, T. A., and Tonegawa, S. (1994). Deficient cerebellar long-term depression and impaired motor learning in mGluR1 mutant mice. *Cell* **79**, 377–388.

Albus, J. S. (1971). A theory of cerebellar functions. *Math. Biosci.* **10**, 25–61.

Amaral, D. G., Price, J. L., Pitkanen, A., and Carmichael, S. T. (1992). Anatomical organization of the primate amygdaloid complex. *In* "The Amygdala: Neurobiological Aspects of Emotion, Memory, and Mental Dysfunction (J. Aggleton, Ed.), pp. 1–66. Wiley–Liss, New York.

Andersen, P., Sundberg, S. H., Sveen, O., Wigstrom, H. (1977). Specific long-lasting potentiation of synaptic transmission in hippocampal slices. *Nature* **266**, 736–737.

Angers, A., Fioravante, D., Chin, J., Cleary, L. J., Bean, A. J., and Byrne, J. H. (2002). Serotonin stimulates phosphorylation of *Aplysia* synapsin and alters its subcellular distribution in sensory neurons. *J. Neurosci.* **22**, 5412–5422.

Antonov, I., Antonova, I., Kandel, E. R., and Hawkins, R. D. (2003). Activity-dependent presynaptic facilitation and Hebbian LTP are both required and interact during classical conditioning in *Aplysia. Neuron* **37**, 135–147.

Antzoulatos, E. G., and Byrne, J. H. (2007). Long-term sensitization training produces spike narrowing in *Aplysia* sensory neurons. *J. Neurosci.* **27**, 676–683.

Apergis-Schoute, A. M., Debiec, J., Doyere, V., LeDoux, J. E., and Schafe, G. E. (2005). Auditory fear conditioning and long-term potentiation in the lateral amygdala require ERK/MAP kinase signaling in the auditory thalamus: A role for presynaptic plasticity in the fear system. *J. Neurosci.* **25**(24), 5730–5739.

Araki, M., and Nagayama, T. (2005). Decrease in excitability of LG following habituation of the crayfish escape reaction. *J. Comp. Physiol. A. Neuroethol. Sens. Neural. Behav. Physiol.* **191**, 481–489.

Araki, M., Nagayama, T., and Sprayberry, J. (2005). Cyclic AMP mediates serotonin-induced synaptic enhancement of lateral giant interneuron of the crayfish. *J. Neurophysiol.* **94**, 2644–2652.

Armony, J. L., and Dolan, R. J. (2001). Modulation of auditory neural responses by a visual context in human fear conditioning. *Neuroreport.* **12**(15), 3407–3411.

Armony, J. L., Servan-Schreiber, D., Romanski, L. M., Cohen, J. D., and LeDoux, J. E. (1997). Stimulus generalization of fear responses: Effects of auditory cortex lesions in a computational model and in rats. *Cerebral Cortex* **7**(2), 157–165.

Attwell, P. J., Cooke, S. F., and Yeo, C. H. (2002). Cerebellar function in consolidation of a motor memory. *Neuron* **34**, 1011–1020.

Atwood, H. L., and Wojtowicz, J. M. (1999). Silent synapses in neural plasticity: Current evidence. *Learning & Memory* **6**(6), 542–571.

Bagnoli, P., and Magni, F. (1975). Synaptic inputs to Retzius' cells in the leech. *Brain. Res.* **96**, 147–152.

Bailey, C. H., Chen, M., Keller, F., and Kandel, E. R. (1992). Serotonin-mediated endocytosis of apCAM: An early step of learning-related synaptic growth in *Aplysia. Science* **256**, 645–649.

Bailey, C. H., and Kandel, E. R. (1993). Structural changes accompanying memory storage. *Annu. Rev. Physiol.* **55**, 397–426.

Bailey, D. J., Kim, J. J., Sun, W., Thompson, R. F., and Helmstetter, F. J. (1999). Acquisition of fear conditioning in rats requires the synthesis of mRNA in the amygdala. *Behav. Neurosci.* **113**, 276–282.

Bailey, H. C., and Kandel, E. R. (1993). Structural changes accompanying memory storage. *Ann. Rev. Physiol.* **55**, 397–426.

Balaban, P. M. (2002). Cellular mechanisms of behavioral plasticity in terrestrial snail. *Neurosci. Biobehav. Rev.* **26**, 597–630.

Balaban, P. M., Bravarenko, N. I., and Zakharov, I. S. (1991). [The neurochemical basis of recurrent inhibition in the reflex arch of the defensive reaction]. *Zh. Vyssh. Nerv. Deiat. Im. I. P. Pavlova* **41**, 1033–1038.

Balaban, P. M., Maksimova, O. A., and Bravarenko, H. I. (1994). Behavioral plasticity in a snail and its neural mechanisms. *Neuros. Behav. Physiol.* **24**, 97–104.

Balaban, P. M., Poteryaev, D. A., Zakharov, I. S., Uvarov, P., Malyshev, A., Belyavsky, A. V. (2001). Up- and down-regulation of *Helix* command-specific 2 (HCS2) gene expression in the nervous system of terrestrial snail *Helix lucorum. Neuroscience* **103**, 551–559.

Baldwin, A. E., Sadeghian, K., Holahan, M. R., and Kelley, A. E. (2002). Appetitive instrumental learning is impaired by inhibition of cAMP-dependent protein kinase within the nucleus accumbens. *Neurobiol. Learn. Mem.* **77**, 44–62.

Balleine, B. W., and Killcross, S. (2006). Parallel incentive processing: An integrated view of amygdala function. *Trends Neurosci.* **29**(5), 272–279.

Bao, J. X., Kandel, E. R., and Hawkins, R. D. (1998). Involvement of presynaptic and postsynaptic mechanisms in a cellular analog of classical conditioning at *Aplysia* sensory–motor neuron synapses in isolated cell culture. *J. Neurosci.* **18**, 458–466.

Bao, S., Chen, L., Kim, J. J., and Thompson, R. F. (2002). Cerebellar cortical inhibition and classical eyeblink conditioning. *Proc. Natl. Acad. Sci.* **99**, 1592–1597.

Barrionuevo, G., and Brown, T. H. (1983). Associative long-term potentiation in hippocampal slices. *Proc. Natl. Acad. Sci. USA* **80**, 7347–7351.

Barrionuevo, G., Kelso, S., Johnston, D., and Brown, T. H. (1986). Conductance mechanism responsible for long-term potentiation in monosynaptic and isolated excitatory synaptic inputs to the hippocampus. *J. Neurophysiol.* **55**, 540–550.

Bartsch, D., Ghirardi, M., Skehel, P. A., Karl, K. A., Herder, S. P., Chen, M., Bailey, C. H., and Kandel, E. R. (1995). *Aplysia* CREB2 represses long-term facilitation: Relief of repression converts transient facilitation into long-term functional and structural change. *Cell* **83**, 979–992.

Baudry, M., Davis, J. L., and Thompson, R. F. (Eds.) (1993). "Synaptic Plasticity: Molecular and Functional Aspects." MIT Press, Cambridge, MA.

Bauer, E. P., Schafe, G. E., and LeDoux, J. E. (2002). NMDA receptors and L-type voltage-gated calcium channels contribute to long-term potentiation and different components of fear memory formation in the lateral amygdala. *J. Neurosci.* **22**, 5239–5249.

Bechara, A., Tranel, D., Damasio, H., Adolphs, R., Rockland, C., and Damasio, A. R. (1995). Double dissociation of conditioning and declarative knowledge relative to the amygdala and hippocampus in humans. *Science* **269**, 1115–1118.

Beck, C. H., and Fibiger, H. C. 1995. Conditioned fear-induced changes in behavior and in the expression of the immediate early gene c-fos: With and without diazepam pretreatment. *J. Neurosci.* **15**(1 Pt 2), 709–720.

Bekkers, J. M., Stevens, C. F. (1990). Presynaptic mechanism for long-term potentiation in the hippocampus. *Nature* **346**, 724–729.

Belardetti, F., Biondi, C., Colombaioni, L., Brunelli, M., and Trevisani, A. (1982). Role of serotonin and cyclic AMP on facilitation of the fast conducting system activity in the leech *Hirudo medicinalis*. *Brain. Res.* **246**, 89–103.

Bellocchio, E. E., Reimer, R. J., Fremeau, R. T., Jr., and Edwards, R. H. (2000). Uptake of glutamate into synaptic vesicles by an inorganic phosphate transporter. *Science* **289**, 957–960.

Benjamin, P. R., Staras, K., and Kemenes, G. (2000). A systems approach to the cellular analysis of associative learning in the pond snail *Lymnaea*. *Learn. Memory* **7**, 124–131.

Berger, T. W., Alger, B. E., and Thompson, R. F. (1976). Neuronal substrate of classical conditioning in the hippocampus. *Science* **192**, 483–485.

Berger, T. W., Berry, S. D., and Thompson, R. F. (1986). Role of the hippocampus in classical conditioning of aversive and appetitive behaviors. *In* "The Hippocampus" (R. L. Isaacson and K. H. Pribram, Eds.), pp. 203–239. Plenum, New York.

Berger, T. W., Rinaldi, P. C., Weisz, D. J., and Thompson, R. F. (1983). Single-unit analysis of different hippocampal cell types during classical conditioning of the rabbit nictitating membrane response. *J. Neurophysiol.* **50**, 1197–1219.

Berger, T. W., and Thompson, R. F. (1978). Identification of pyramidal cells as the critical elements in hippocampal neuronal plasticity during learning. *Proc. Natl. Acad. Sci. USA* **75**, 1572–1576.

Bi, G. Q., Poo, M. M. (1998). Synaptic modifications in cultured hippocampal neurons: Dependence on spike timing, synaptic strength, and postsynaptic cell type. *J. Neurosci.* **18**, 10464–10472.

Birbaumer, N., Veit, R., Lotze, M., Erb, M., Hermann, C., Grodd, W., and Flor, H. (2005). Deficient fear conditioning in psychopathy: A functional magnetic resonance imaging study. *Arch. Gen. Psychiatry* **62**, 799–805.

Blair, H. T., Schafe, G. E., Bauer, E. P., Rodrigues, S. M., and LeDoux, J. E. (2001). Synaptic plasticity in the lateral amygdala: A cellular hypothesis of fear conditioning. *Learn. Memory* **8**, 229–242.

Bliss, T. V. P., and Lomo, T. (1973). Long-lasting potentiation of synaptic transmission in the dentate area of the anaesthetized rabbit following stimulation of the perforant path. *J. Physiol.* **232**, 331–356.

Blitzer, R. D., Connor, J. H., Brown, G. P., Wong, T., Shenolikar, S., Iyengar, R., and Landau, E. M. (1998). Gating of CaMKII by cAMP-regulated protein phosphatase activity during LTP. *Science* **280**, 1940–1942.

Blum, S. A., Moore, N., Adams, F., and Dash, P. K. (1999). A mitogen-activated protein kinase cascade in the CA1/CA2 subfield of the dorsal hippocampus is essential for long-term spatial memory. *J. Neurosci.* **19**, 3535–3544.

Bordi, F., and LeDoux, J. (1992). Sensory tuning beyond the sensory system: An initial analysis of auditory response properties of neurons in the lateral amygdaloid nucleus and overlying areas of striatum. *J. Neurosci.* **12**, 2493–2503.

Botzer, D., Markovich, S., and Susswein, A. J. (1998). Multiple memory processes following training that a food is inedible in *Aplysia*. *Learn. Memory* **5**, 204–219.

Boulis, N. M., and Sahley, C. L. (1988). A behavioral analysis of habituation and sensitization of shortening in the semi-intact leech. *J. Neurosci.* **8**, 4621–4627.

Bourtchouladze, R., Frenguelli, B., Blendy, D., Cioffi, D., Schultz, G., and Silva, A. J. (1994). Deficient long-term memory in mice with a targeted mutation of the cAMP-responsive element-binding protein. *Cell* **79**, 59–68.

Bracha, V., Irwin, K. B., Webster, M. L., Wunderlich, D. A., Stachowiak, M. K., and Bloedel, J. R. (1998). Microinjections of anisomycin into the intermediate cerebellum during learning affect the acquisition of classically conditioned responses in the rabbit. *Brain Research* **788**, 169–178.

Breiter, H. C., Etcoff, N. L., Whalen, P. J., Kennedy, W. A., Rauch, S. L., Buchner, R. L., Strauss, M. M., Hyman, S. E., and Rosen, B. R. (1996). Response and habituation of the human amygdala during visual processing of facial expression. *Neuron* **17**, 875–877.

Brembs, B., Lorenzetti, F. D., Reyes, F. D., Baxter, D., and Byrne, J. H. (2002). Operant reward learning in *Aplysia*: Neuronal correlates and mechanisms. *Science* **296**, 1706–1709.

Brembs, B., and Plendl, W. (2008). Double dissociation of protein-kinase C and adenylyl cyclase manipulations on operant and classical learning in *Drosophila*. *Curr. Biol.* In press.

Bremner, J. D. (2007). Functional neuroimaging in post-traumatic stress disorder. *Expert. Rev. Neurother.* **7**, 393–405.

Bremner, J. D., Vermetten, E., Schmahl, C., Vaccarino, V., Vythilingam, M., Afzal, N., Grillon, C., and Charney, D. S. (2005). Positron emission tomographic imaging of neural correlates of a fear acquisition and extinction paradigm in women with childhood sexual-abuse-related post-traumatic stress disorder. *Psychol. Med.* **35**, 791–806.

Brown, T. H., Ganong, A. H., Kairiss, E. W., and Keenan, C. L. (1990). Hebbian synapses: Biophysical mechanisms and algorithms. *Annu. Rev. Neurosci.* **13**, 475–512.

Brown, T. H., and Johnston, D. (1983). Voltage-clamp analysis of mossy-fiber synaptic input to hippocampal neurons. *J. Physiol.* **50**, 487–507.

Brown, T. H., Perkel, D. H., and Feldman, M. W. (1976). Evoked neurotransmitter release: Statistical effects of nonuniformity and nonstationarity. *Proc. Nat. Acad. Sci.* **73**, 2913–2917.

Brown, T. H., and Zador, A. M. (1990). The hippocampus. *In* "Synaptic Organization of the Brain" (G. Shepherd, Ed.), pp. 346–388. Oxford Univ. Press, New York.

Brunelli, M., Garcia-Gil, M., Mozzachiodi, R., Scuri, R., Zaccardi, M. L. (1997). Neurobiological principles of learning and memory. *Arch. Ital. Biol.* **135**, 15–36.

Buchel, C., and Dolan, R. J. (2000). Classical fear conditioning in functional neuroimaging. *Curr. Opin. Neurobiol.* **10**(2), 219–223.

Buchel, C., Dolan, R. J., Armony, J. L., and Friston, K. J. (1999). Amygdala-hippocampal involvement in human aversive trace conditioning revealed through event-related functional magnetic resonance imaging. *J. Neurosci.* **19**(24), 10869–10876.

Buchel, C., Morris, J., Dolan, R. J., and Friston, K. J. (1998). Brain systems mediating aversive conditioning: An event-related fMRI study. *Neuron* **20**, 947–957.

Burrell, B. D., Sahley, C. L., and Muller, K. J. (2001). Non-associative learning and serotonin induce similar bi-directional changes in excitability of a neuron critical for learning in the medicinal leech. *J. Neurosci.* **21**, 1401–1412.

Byrne, J. H. (1987). Cellular analysis of associative learning. *Physiol. Rev.* **67**, 329–439.

Byrne, J. H., and Kandel, E. R. (1996). Presynaptic facilitation revisited: State- and time-dependence. *J. Neurosci.* **16**, 425–435.

Byrne, J. H., Zwartjes, R., Homayouni, R., Critz, S., and Eskin, A. (1993). Roles of second messenger pathways in neuronal plasticity and in learning and memory: Insights gained from *Aplysia*. *Adv. Second Messenger Phosphoprotein Res.* **27**, 47–108.

Cahill, L., and McGaugh, J. L. (1998). Mechanisms of emotional arousal and lasting declarative memory. *Trends Neurosci.* **21**, 294–299.

Cahill, L., Weinberger, N. M., Roozendaal, B., and McGaugh, J. L. (1999). Is the amygdala a locus of "conditioned fear"? Some questions and caveats. *Neuron* **23**, 227–278.

Cain, C. K., and LeDoux, J. E. (2007). Escape from fear: A detailed behavioral analysis of two atypical responses reinforced by CS termination. *J. Exp. Psychol. Anim. Behav. Process* **33**, 451–463.

Campeau, S., and Davis, M. (1995). Involvement of subcortical and cortical afferents to the lateral nucleus of the amygdala in fear conditioning measured with fear-potentiated startle in rats trained concurrently with auditory and visual conditioned stimuli. *J. Neurosci.* **15**, 2312–2327.

Cardinal, R. N., and Everitt, B. J. (2004). Neural and psychological mechanisms underlying appetitive learning: Links to drug addiction. *Curr. Opin. Neurobiol.* **14**(2), 156–162.

Carew, T. J. (2000). "Behavioral Neurobiology." Sinauer Associates, Sunderland, MA.

Carroll, R. C., Nicoll, R. A., and Malenka, R. C. (1998). Effects of PKA and PKC on miniature excitatory postsynaptic currents in CA1 pyramidal cells. *J. Neurophysiol.* **80**, 2797–2800.

Cavus, I., and Teyler, T. J. (1996). Two forms of long-term potentiation in area CA1 activate different signal transduction pathways. *J. Neurophysiol.* **76**, 3038–3047.

Chain, D. G., Casadio, A., Schacher, S., Hegde, A. N., Valbrun, M., Yamamoto, N., Goldberg, A. L., Bartsch, D., Kandel, E. R., and Schwartz, J. H. (1999). Mechanisms for generating the autonomous cAMP-dependent protein kinase required for long-term facilitation in *Aplysia*. *Neuron* **22**, 147–156.

Chapman, P. F., Kairiss, E. W., Keenan, C. L., and Brown, T. H. (1990). Long-term synaptic potentiation in the amygdala. *Synapse* **6**, 271–278.

Charney, D. S., Deutch, A. V., Krystal, J. H., Southwick, A. M., and Davis, M. (1993). Psychobiologic mechanisms of posttraumatic stress disorder. *Arch. Gen. Psychiatry* **50**, 294–305.

Chen, C., Masanobu, K., Abeliovich, A., Chen, L., Bao, S., Kim, J. J., Hashimoto, K., Thompson, R. F., and Tonegawa, S. (1995). Impaired motor coordination correlates with persistent multiple climbing fiber innervation in PKCγ mutant mice. *Cell* **83**, 1233–1242.

Chen, G., and Steinmetz, J. E. (2000). Microinfusion of protein kinase inhibitor H7 into the cerebellum impairs the acquisition but not the retention of classical eyeblink conditioning in rabbits. *Brain Res.* **856**, 193–201.

Chen, L., Bao, S., Lockard, J. M., Kim, J. J., and Thompson, R. F. (1996). Impaired classical eyeblink conditioning in cerebellar lesioned and Purkinje cell degeneration (pcd) mutant mice. *J. Neurosci.* **16**, 2829–2838.

Chen, L., Bao, S., and Thompson, R. F. (1999). Bilateral lesions of the interpositus nucleus completely prevent eyeblink conditioning in Purkinje cell-degeneration mutant mice. *Behav. Neurosci.* **113**, 204–216.

Cheng, D. T., Knight, D. C., Smith, C. N., Stein, E. A., and Helmstetter, F. J. (2003). Functional MRI of human amygdala activity during Pavlovian fear conditioning: Stimulus processing versus response expression. *Behav. Neurosci.* **117**: 3–10.

Chittajallu R., Alford S., Collingridge G. L. (1998). Ca2+ and synaptic plasticity. *Cell Calcium* **24**, 377–385.

Choi, J.-S., Lindquist, D. H., and Brown, T. H. (2001). Amygdala lesions prevent conditioned enhancement of the rat eyeblink reflex. *Behav. Neurosci.* **115**, 764–775.

Christian, K. M. (2004). Acquisition, consolidation and storage of an associative memory in the cerebellum. Doctoral dissertation, University of Southern California.

Christian, K. M., and Thompson, R. F. (2003). Neural substrates of eyeblink conditioning: Acquisition and retention. *Learn. Mem.* **10**, 427–455.

Christian, K. M., and Thompson, R. F. (2005). Long-term storage of an associative memory trace in the cerebellum. *Behav. Neurosci.* **119**, 526–537.

Church, P., and Lloyd, P. (1994). Activity of multiple identified motor neurons recorded intracellulary during evoked feeding like motor programs in *Aplysia*. *J. Neurophysiol.* **72**, 1794–1809.

Clark, G. A., McCormick, D. A., Lavond, D. G., and Thompson, R. F. (1984). Effect of lesions of cerebellar nuclei on conditioned behavioral and hippocampal neuronal responses. *Brain Res.* **291**, 125–136.

Clark, R. E., and Lavond, D. G. (1993). Reversible lesions of the red nucleus during acquisition and retention of a classically conditioned behavior in rabbits. *Behav. Neurosci.* **107**, 264–270.

Clark, R. E., and Squire, L. R. (1998). Classical conditioning and brain systems: The role of awareness. *Science* **280**, 77–81.

Clark, R. E., West, A. N., Zola, S. M., and Squire, L. R. (2001). Rats with lesions of the hippocampus are impaired on the delayed nonmatching-to-sample task. *Hippocampus* **11**, 176–186.

Cleary, L. J., Byrne, J. H., and Frost, W. N. (1995). Role of interneurons in defensive withdrawal reflexes in *Aplysia*. *Learn. Memory* **2**, 133–151.

Cleary, L. J., Lee, W. L., and Byrne, J. H. (1998). Cellular correlates of long-term sensitization in *Aplysia*. *J. Neurosci.* **18**, 5988–5998.

Clugnet, M. C., and LeDoux, J. E. (1990). Synaptic plasticity in fear conditioning circuits: Induction of LTP in the lateral nucleus of the amygdala by stimulation of the medial geniculate body. *J. Neurosci.* **10**, 2818–2824.

Cohen, D. H. (1974). The neural pathways and informational flow mediating a conditioned autonomic response. *In* "Limbic and Autonomic Nervous System Research" (L. V. Di Cara, Ed). Plenum Press, New York.

Collingridge G. L., Kehl, S. J., McLennan, H. (1983). Excitatory amino acids in synaptic transmission in the Schaffer collateral-commissural pathway of the rat hippocampus. *J. Physiol.* **334**, 33–46.

Collins, D. R., and Pare, D. (2000). Differential fear conditioning induces reciprocal changes in the sensory responses of lateral amygdala neurons to the CS(+) and CS(−). *Learn. Memory* **7**, 97–103.

Cook, D. G., and Carew, T. J. (1989). Operant conditioning of head-waving in *Aplysia*. III. Cellular analysis of possible reinforcement pathways. *J. Neurosci.* **9**, 3115–3122.

Crow, T. (2003). Invertebrate learning: Associative learning in *Hermissenda*. *In* "Learning and Memory," 2nd ed. (J. H. Byrne, Ed.), pp. 277–281. Macmillan, New York.

Cruikshank, S. J., Edeline, J. M., and Weinberger, N. M. (1992). Stimulation at a site of auditory–somatosensory convergence in the medial geniculate nucleus is an effective unconditioned stimulus for fear conditioning. *Behav. Neurosci.* **106**, 471–483.

Daum, I., Channon, S., Polkey, C. E., and Gray, J. A. (1991). Classical conditioning after temporal lobe lesions in man: Impairment in conditional discrimination. *Behav. Neurosci.* **105**, 396–408.

Daum, I., Schugens, M. M., Ackerman, H., Lutzenberger, W., Dichgans, J., and Birbaumer, N. (1993). Classical conditioning after cerebellar lesions in human. *Behav. Neurosci.* **107**, 748–756.

Davis, R. L. (2005). Olfactory memory formation in *Drosophila*: From molecular to systems neuroscience. *Annu Rev Neurosci.* **28**, 275–302.

Davis, M. (1984). Mammalian startle response. In "Neural Mechanisms of Startle Behavior" (R. C. Eaton, Ed.), pp. 287–351. Plenum, New York.

Davis, M. (1992). The role of the amygdala in fear and anxiety. *Annu. Rev. Neurosci.* **15**, 353–375.

Davis, M. (1998). Are different parts of the extended amygdala involved in fear versus anxiety? *Biol. Psychiatry* **44**, 1239–1247.

de Jonge, M. C., Black, J., Deyo, R. A., and Disterhoft, J. F. (1990). Learning-induced afterhyperpolarization reductions in hippocampus are specific for cell type and potassium conductance. *Exp. Brain Res.* **80**, 456–462.

Derkach, V., Barria, A., and Soderling, T. R. (1999). Ca^{2+}/calmodulin-kinase II enhances channel conductance of alpha-amino-3-hydroxy-5-methyl-4-isoxazolepropionate type glutamate receptors. *Proc. Natl. Acad. Sci. USA* **96**, 3269–3274.

DeZazzo, J., and Tully, T. (1995). Dissection of memory formation: From behavioral pharmacology to molecular genetics. *Trends Neurosci.* **18**, 212–218.

De Zeeuw, C. I., Hansel, C., Bian, F., Koekkoek, S. K. E., van Alphen, A. M., Linden, D. J., and Oberdick, J. (1998). Expression of a protein kinase C inhibitor in Purkinje cells blocks cerebellar LTD and adaptation of the vestibulo-ocular reflex. *Neuron.* **20**, 495–508.

Dineley, K. T., Weeber, E. J., Atkins, C., Adams, J. P., Anderson, A. E., and Sweatt, J. D. (2001). Leitmotifs in the biochemistry of LTP induction: Amplification, integration and coordination. *J. Neurochem.* **77**, 961–971.

Disterhoft, J. F., Coulter, D. A., and Alkon, D. L. (1986). Conditioning-specific membrane changes of rabbit hippocampal neurons measured *in vitro*. *Proc. Natl. Acad. Sci.* **83**, 2733–2737.

Disterhoft, J. F., and Oh, M. (2006). Learning, aging and intrinsic neuronal plasticity. *TINS.* **29**, 587–599.

Dityatev, A. E., and Bolshakov, V. Y. (2005). Amygdala, long-term potentiation, and fear conditioning. *Neuroscientist* **11**(1), 75–88.

Dolphin, A. C., Errington, M. L., and Bliss, T. V. (1982). Long-term potentiation of the perforant path *in vivo* is associated with increased glutamate release. *Nature* **297**, 496–498.

Dudai, Y., and Tully, T. (2003). Invertebrate learning: Neurogenetics of memory in *Drosophila*. In "Learning and Memory," 2nd ed. (J. H. Byrne, Ed.), pp. 292–296. Macmillan, New York.

duLac, S., Raymond, J. L., Sejnowski, T. J., and Lisberger, S. G. (1995). Learning and memory in the vestibulo-ocular reflex. *Annu. Rev. Neurosci.* **18**, 409–441.

Dyer, J. R., Manseau, F., Castellucci, V. F., Sossin, W. S. (2003). Serotonin persistently activates the extracellular signal-related kinase in sensory neurons of Aplysia independently of cAMP or protein kinase C. *Neuroscience* **116**, 13–17.

Eccles, J. C. (1977). An instruction-selection theory of learning in the cerebellar cortex. *Brain Res.* **127**, 327–352.

Edwards, D. H., Heitler, W. J., and Krasne, F. B. (1999). Fifty years of a command neuron: The neurobiology of escape behavior in the crayfish. *Trends Neurosci.* **22**, 153–161.

Ehrlich, J. S., Boulis, N. M., Karrer, T., and Sahley, C. L. (1992). Differential effects of serotonin depletion on sensitization and dishabituation in the leech, *Hirudo medicinalis*. *J. Neurobiol.* **23**, 270–279.

Eichenbaum, H. (1999). Conscious awareness, memory and the hippocampus. *Nat. Neurosci.* **2**, 775–776.

Eisenstein, E., and Carlson, A. (1994). Leg position learning in the cockroach nerve cord using an analog technique. *Physiol. Behav.* **56**, 687–691.

English, J. D., and Sweatt, J. D. (1997). A requirement for the mitogen-activated protein kinase cascade in hippocampal long-term potentiation. *J. Biol. Chem.* **272**, 19103–19106.

Fanselow, M. S., and LeDoux, J. E. (1999). Why we think plasticity underlying Pavlovian fear conditioning occurs in the basolateral amygdala. *Neuron* **23**, 2239–2242.

Fanselow, M. S., and Poulos, A. M. (2005). The neuroscience of mammalian associative learning. *Annu. Rev. Psychol.* **56**, 207–234.

Fendt, M., and Fanselow, M. S. (1999). The neuroanatomical and neurochemical basis of conditioned fear. *Neurosci. Biobehav. Rev.* **23**, 743–760.

Fioravante, D., Liu, R. Y., Netek, A., Cleary, L. J., and Byrne, J. H. (2007). Synapsin regulates basal synaptic strength, synaptic depression and serotonin-induced facilitation of sensorimotor synapses in *Aplysia*. *J. Neurophysiol.* **98**, 3568–3580.

Fiumara, F., Leitinger, G., Milanese, C., Montarolo, P. G., and Ghirardi, M. (2005). In vitro formation and activity-dependent plasticity of synapses between *Helix* neurons involved in the neural control of feeding and withdrawal behaviors. *Neuroscience* **134**, 1133–1151.

Frankland, P. W., Cestari, V., Filipkowski, R. K., McDonald, R. J., and Silva, A. J. (1998). The dorsal hippocampus is essential for context discrimination but not for contextual conditioning. *Behav. Neurosci.* **112**, 863–874.

Freeman, J. H., and Nicholson, D. A. (2000). Developmental changes in eye-blink conditioning and neuronal activity in the cerebellar interpositus nucleus. *J. Neurosci.* **20**, 813–819.

Frey, U., Huang, Y. Y., and Kandel, E. R. (1993). Effects of cAMP stimulate a late stage of LTP in hippocampal CA1 neurons. *Science* **260**, 1661–1664.

Frey, U., and Morris, R. G. M. (1998). Synaptic tagging: Implications for late maintenance of hippocampal long-term potentiation. *Trends Neurosci.* **21**, 181–188.

Frost, W. N. (2003). Invertebrate learning: Habituation and sensitization in *Tritonia*. In "Learning and Memory," 2nd ed. (J. H. Byrne, Ed.), pp. 291–292. Macmillan, New York.

Frost, W. N., Brown, G. D., and Getting, P. A. (1996). Parametric features of habituation of swim cycle in the marine mollusc *Tritonia diomedea*. *Neurobiol. Learn. Memory* **65**, 125–135.

Furmark, T., Fischer, H., Wik, G., Larsson, M., and Fredrikson, M. (1997). The amygdala and individual differences in human fear conditioning. *Neuroreport* **8**, 3957–3960.

Gabriel, M., Foster, K., Orona, E., Saltwick, S. E., and Stanton, M. (1980). Neuronal activity of cingulate cortex, anteroventral thalamus, and hippocampal formation in discriminative conditioning: Encoding and extraction of the significance of conditional stimuli. In "Progress in Psychobiology and Physiological Psychology," Vol. 9 (J. M. Sprague and A. N. Epstein, Eds.). New York, Academic Press.

Garcia, K. S., Steele, P. M., and Mauk, M. D. (1999a). Cerebellar cortex lesions prevent acquisition of conditioned eyelid responses. *J. Neurosci.* **19**, 10940–10947.

Garcia, R., Vouimba, R. M., Baudry, M., and Thompson, R. F. (1999b). The amygdala modulates prefrontal cortex activity relative to conditioned fear. *Nature* **402**, 294–296.

Garpenstrand, H., Annas, P., Ekblom, J., Oreland, L., and Fredrikson, M. (2001). Human fear conditioning is related to dopaminergic and serotonergic biological markers. *Behav. Neurosci.* **115**, 358–364.

Gean, P. W., Chang, F. C., Huang, C. C., Lin, J. H., and Way, L. J. (1993). Long-term enhancement of EPSP and NMDA receptor mediated synaptic transmission in the amygdala. *Brain Res. Bull.* **31**, 7–11.

Gelperin, A. (2003). Invertebrate learning: Associative learning in *Limax*. In "Learning and Memory," 2nd ed. (J. H. Byrne, Ed.), pp. 281–287. Macmillan, New York.

Gerwitz, J. C., and Davis, M. (1997). Second-order fear conditioning prevented by blocking NMDA receptors in amygdala. *Nature* **388**, 471–474.

Gerwitz, J. C., Falls, W. A., and Davis, M. (1997). Normal conditioned inhibition and extinction of freezing and fear-potentiated startle following electrolytic lesions of medial prefrontal cortex in rats. *Behav. Neurosci.* **111**, 712–726.

Ghirardi, M., Montarolo, P. G., and Kandel, E. R. (1995). A novel intermediate stage in the transition between short- and long-term facilitation in the sensory to motor neuron synapse of *Aplysia*. *Neuron* **14**, 413–420.

Giese, K. P., Fedorov, N. B., Filipkowski, R. K., and Silva, A. J. (1998). Autophosphorylation at the Thr^{286} of the alpha calcium-calmodulin kinase II in LTP and learning. *Science* **279**, 870–873.

Gillette, R. (1992). Invertebrate learning: Associative learning in *Pleurobranchaea*. In "Encyclopedia of Learning and Memory" (L. R. Squire, Ed.), pp. 302–305. Macmillan, New York.

Glanzman, D. L., Krasne, F. B. (1986). 5,7-Dihydroxytryptamine lesions of crayfish serotonin-containing neurons: effect on the lateral giant escape reaction. *J. Neurosci.* **6**, 1560–1569.

Goddard, G. (1964). Functions of the amygdala. *Psychol. Rev.* **62**, 89–109.

Goddard, G. (1982). Hippocampal long-term potentiation: Mechanisms and implications for memory. *Neurosci. Res. Program Bull.* **20**, 676–680.

Goldowitz, D., and Eisenman, L. M. (1992). Genetic mutations affecting murine cerebellar structure and function. In "Genetically Defined Animal Models of Neurobehavioral Dysfunctions" (P. Driscoll, Ed.), pp. 66–88. Birkhauser, Boston.

Gomi, H., Sun, W., Finch, C. E., Itohara, S., Yoshimi, K., and Thompson, R. F. (1999). Learning induces a CDC2-related protein kinase, KKIAMRE. *J. Neurosci.* **19**, 9530–9537.

Goodlett, C. R., Hamre, K. M., and West, J. R. (1992). Dissociation of spatial navigation and visual guidance performance in Purkinje cell degeneration (*pcd*) mutant mice. *Behav. Brain Res.* **47**, 129–144.

Gormezano, I., Kehoe, E. J., and Marshall-Goodell, B. S. (1983). Twenty years of classical conditioning research with the rabbit. In "Progress in Physiological Psychology" (J. M. Sprague and A. N. Epstein, Eds.), pp. 197–275. Academic Press, New York.

Gottschalk W., Pozzo-Miller L. D., Figurov A., Lu B. (1998). Presynaptic modulation of synaptic transmission and plasticity by brain-derived neurotrophic factor in the developing hippocampus. *J. Neurosci.* **18**, 6830–6839.

Grosshans, D. R., Clayton, D. A., Coultrap, S. J., and Browning, M. D. (2002). LTP leads to rapid surface expression of NMDA but not AMPA receptors in adult rat CA1. *Nat. Neurosci.* **5**(1), 27–33.

Gruart, A., and Yeo, C. H. (1995). Cerebellar cortex and eyeblink conditioning: Bilateral regulation of conditioned responses. *Exp. Brain Res.* **104**, 431–448.

Grover, L. M., Teyler, T. J. (1990). Two components of long-term potentiation induced by different patterns of afferent activation. *Nature* **347**, 477–479.

Guzowski, J. F., and McGaugh, J. L. (1997). Antisense oligodeoxynucleotide-mediated disruption of hippocampal cAMP response element binding protein levels impairs consolidation of memory for water maze training. *Proc. Natl. Acad. Sci. USA* **94**, 2693–2698.

Hamann, S., Monarch, E. S., and Goldstein, F. C. (2002). Impaired fear conditioning in Alzheimer's disease. *Neuropsychologia* **40**(8), 1187–1195.

Han, J. H., Kushner, S. A., Yiu, A. P., Cole, C. J., Matynia, A., Brown, R. A., Neve, R. L., Guzowski, J. F., Silva, A. J., and Josselyn, S. A. (2007). Neuronal competition and selection during memory formation. *Science* **316**(5823), 457–460.

Hardiman, M. J., Ramnani, N., and Yeo, C. H. (1996). Reversible inactivations of the cerebellum with muscimol prevent the acquisition and extinction of conditioned nictitating membrane responses in the rabbit. *Exp. Brain Res.* **110**, 235–247.

Harris E. W., Cotman C. W. (1986). Long-term potentiation of guinea pig mossy fiber responses is not blocked by N-methyl D-aspartate antagonists. *Neurosci. Lett.* **70**, 132–137.

Hawkins, R. D., Kandel, E. R., and Siegelbaum, S. (1993). Learning to modulate transmitter release: Themes and variations in synaptic plasticity. *Annu. Rev. Neurosci.* **16**, 625–665.

Hayashi, Y., Shi, S. H., Esteban, J. A., Piccini, A., Poncer, J. C., and Malinow, R. (2000). Driving AMPA receptors into synapses by LTP and CaMKII: Requirement for GluR1 and PDZ domain interaction. *Science* **287**(5461), 2262–2267.

Hebb, D. O. (1949). "The Organization of Behavior." Wiley– Interscience, New York.

Heitler, W. J., Watson, A. H., Falconer, S. W., and Powell, B. (2001). Glutamate is a transmitter that mediates inhibition at the rectifying electrical motor giant synapse in the crayfish. *J. Comp. Neurol.* **430**, 12–26.

Hoffman, D. A., and Johnston, D. (1998). Downregulation of transient K$^+$ channels in dendrites of hippocampal CA1 pyramidal neurons by activation of PKA and PKC. *J. Neurosci.* **18**, 3521–3528.

Huang, Y.-Y., and Kandel, E. R. (1998). Postsynaptic induction and PKA-dependent expression of LTP in the lateral amygdala. *Neuron* **21**, 169–178.

Huang, Y. Y., Li, X. C., and Kandel, E. R. (1994). cAMP contributes to mossy fiber LTP by initiating both a covalently mediated early phase and macromolecular synthesis-dependent late phase. *Cell* **79**, 69–79.

Huang, Y. Y., Martin, K. C., and Kandel, E. R. (2000). Both protein kinase A and mitogen-activated protein kinase are required in the amygdala for the macromolecular synthesis-dependent late phase of long-term potentiation. *J. Neurosci.* **20**, 6317–6325.

Hull, C. L. (1943). "Principles of behavior." Appleton-Century-Crofts, New York.

Humeau, Y., Shaban, H., Bissiere, S., and Luthi, A. (2003). Presynaptic induction of heterosynaptic associative plasticity in the mammalian brain. *Nature* **426**(6968), 841–845.

Ichise, T., Kano, M., Hashimoto, K., Yanagihara. D., Nakao, K., Shigemoto, R., Katsuki, M., and Aiba, A. (2000). mGluR1 in cerebellar Purkinje cells essential for long-term depression, synapse elimination, and motor coordination. *Science* **288**, 1832–1835.

Impey, S., Mark, M., Villacres, E. C., Poser, C., Chavkin, C., and Storm, D. R. (1996). Induction of CRE-mediated gene expression by stimuli that generate long-lasting LTP in area CA1 of the hippocampus. *Neuron* **16**, 973–982.

Isaacson, R. L. (1982). "The Limbic System." Plenum Press, New York.

Isaacson, R. L., and Pribram, K. H. (Eds.) (1986). "The Hippocampus," **Vol. 4**. Plenum, New York.

Ito, M. (1984). "The Cerebellum and Neural Control." Appleton– Century–Crofts, New York.

Ito, M. (1989). Long-term depression. *Annu. Rev. Neurosci.* **12**, 85–102.

Ito, M. (1993). Cerebellar mechanisms of long-term depression. In "Synaptic Plasticity: Molecular and Functional Aspects" (M. Baudry, J. L. Davis, and R. F. Thompson, Eds.), pp. 117–146. MIT Press, Cambridge, MA.

Ito, M., Sukurai, M., and Tongroach, P. (1982). Climbing fibre induced depression of both mossy fibre responsiveness and glutamate sensitivity of cerebellar Purkinje cells. *J. Physiol.* **324**, 113–134.

Ivkovich, D., and Thompson, R. F. (1997). Motor cortex lesions do not affect learning or performance of the eyeblink response in rabbits. *Behav. Neurosci.* **111**, 727–738.

Jarrell, T. W., Gentile, C. G., Romanski, L. M., McCabe, P. M., and Schneiderman, N. (1987). Involvement of cortical and thalamic auditory regions in retention of differential bradycardiac conditioning to acoustic conditioned stimuli in rabbits. *Brain Res.* **412**, 285–294.

Jirenhed, D.-A., Bengtsson, F., Hesslow, G. (2007). Acquisition, extinction, and reaquisition of a cerebellar cortical memory trace. *J. Neurosci.* **27**, 2493–2502.

Johnston D., and Amaral D. G. (1998). Hippocampus. *In* "The Synaptic Organization of the Brain," 4th ed. (G. M. Shepherd, Ed.), pp. 417–458. Oxford Univ. Press, New York.

Johnston, D., Hoffman, D. A., Colbert, C. M., and Magee, J. C. (1999). Regulation of back-propagating action potentials in hippocampal neurons. *Curr. Opin. Neurobiol.* **9**, 288–292.

Johnston, D., Hoffman, D. A., Magee, J. C., Poolos, N. P., Watanabe, S., Colbert, C. M., and Migliore, M. (2000). Dendritic potassium channels in hippocampal pyramidal neurons. *J. Physiol.* **525**, 75–81.

Johnston, D., Williams, D., Jaffe, D., and Gray, R. (1992). NMDA-receptor-independent long-term potentiation. *Annu. Rev. Physiol.* **54**, 489–505.

Josselyn, S. A., Shi, C., Carlezon, W. A., , Jr., Neve, R. L., Nestler, E. J., and Davis, M. (2001). Long-term memory is facilitated by cAMP response element-binding protein overexpression in the amygdala. *J. Neurosci.* **21**, 2402–2412.

Kabotyanski, E., Baxter, D., and Byrne, J. (1998). Identification and characterization of catecholaminergic neuron B65, which initiates and modifies patterned activity in the buccal ganglia of *Aplysia*. *J. Neurophysiol.* **79**, 605–621.

Kamondi A., Acsady L., and Buzsaki, G. (1998). Dendritic spikes are enhanced by cooperative network activity in the intact hippocampus. *J. Neurosci.* **18**, 3919–3928.

Kapp, B. S., Whalen, P. J., Supple, W. F., and Pascoe, J. P. (1992). Amygdaloid contributions to conditioned arousal and sensory information processing. *In* "The Amygdala: Neurobiological Aspects of Emotion, Memory, and Mental Dysfunction" (J. Aggleton, Ed.), pp. 229–254. Wiley–Liss, New York.

Kapur A., Yeckel M. F., Gray R., and Johnston, D. (1998). L-type calcium channels are required for one form of hippocampal mossy fiber LTP. *J. Neurophysiol.* **79**, 2181–2190.

Kasai, H., Matsuzaki, M., Noguchi, J., Yasumatsu, N., and Nakahara, H. (2003). Structure-stability-function relationships of dendritic spines. *Trends Neurosci.* **26**(7), 360–368.

Kashiwabuchi, N., Ikeda, K., Araki, K., Hirano, T., Shibuki, K., Takayama, C., Inoue, Y., Kutsuwada, T., Yagi, T., Kang, Y., Aizawa, S., and Mishina, M. (1995). Impairment of motor coordination, Purkinje cell synapse formation, and cerebellar long-term depression in GluR2 mutant mice. *Cell* **81**, 245–252.

Keele, S. W., and Ivry, R. B. (1990). Does the cerebellum provide a common computation for diverse tasks: A timing hypothesis. *In* "The Development and Neural bases of Higher Cognitive Functions" (A. Diamond, Ed.), pp. 179–211. Academic Press, New York.

Kelly, M. P., and Deadwyler, S. A. (2002). Acquisition of a novel behavior induces higher levels of Arc mRNA than does overtrained performance. *Neurosci.* **110**, 617–626.

Kelso, S. R., and Brown, T. H. (1986). Differential conditioning of associative synaptic enhancement in hippocampal brain slices. *Science* **232**, 85–87.

Kelso, S. R., Ganong, A. H., and Brown, T. H. (1986). Hebbian synapses in hippocampus. *Proc. Natl. Acad. Sci.* **83**, 5326–5330.

Kemenes, G., and Benjamin, P. R. (1994). Training in a novel environment improves the appetitive learning performance of the snail, *Lymnaea stagnalis*. *Behav. Neurosci.* **61**, 139–149.

Kemenes, I., Straub, V. A., Nikitin, E. S., Staras, K., O'Shea, M., Kemenes, G., and Benjamin, P. R. (2006). Role of delayed nonsynaptic neuronal plasticity in long-term associative memory. *Curr. Biol.* **16**(13), 1269–7129.

Kemp, N., McQueen, J., Faulkes, S., and Bashir, Z. I. (2000). Different forms of LTD in the CA1 region of the hippocampus: Role of age and stimulus protocol. *Eur. J. Neurosci.* **12**, 360–366.

Kida, S., Josselyn, S. A., de Ortiz, S. P., Kogan, J. H., Chevere, I., Masushige, S., and Silva, A. J. (2002). CREB required for the stability of new and reactivated fear memories. *Nat. Neurosci.* **5**, 348–355.

Kim, J. J., Chen, L., Bao, S., Sun, W., and Thompson, R. F. (1996). Genetic dissections of the cerebellar circuitry involved in classical eyeblink conditioning. *In* "Gene Targeting and New Developments in Neurobiology" (S. Nakanishi, A. J. Silva, S. Aizawa, and M. Katsuki, Eds.), pp. 3–15. Japan Scientific Societies Press, Tokyo.

Kim, J. J., Clark, R. E., and Thompson, R. F. (1995). Hippocampectomy impairs the memory of recently, but not remotely, acquired trace eyeblink conditioned responses. *Behav. Neurosci.* **109**, 195–203.

Kim, J. J., and Thompson, R. F. (1997). Cerebellar circuits and synaptic mechanisms involved in classical eyeblink conditioning. *Trends Neurosci.* **20**, 177–181.

Kishimoto, Y., and Kano, M. (2006). Endogenous cannabinoid signaling through the CB1 receptor is essential for cerebellum-dependent discrete motor learning. *J. Neurosci.* **26**, 8829–8837.

Kishimoto, Y., Fujimichi, R., Araishi, K., Kawahara, S., Kano, M., Aiba, A., and Kirino, Y. (2002). mGluR1 in cerebellar Purkinje cells is required for normal association of temporally contiguous stimuli in classical conditioning. *Eur. J. Neurosci.* **16**, 2416–2424.

Kleim, J. A., Freeman, J. H., , Jr., Bruneau, R., Nolan, B. C., Cooper, N. R., Zook, A., and Walters, D. (2002). Synapse formation is associated with memory storage in the cerebellum. *Proc. Natl. Acad. Sci.* **99** 13228–13231.

Knight, D. C., Cheng, D. T., Smith, C. N., Stein, E. A., and Helmstetter, F. J. (2004). Neural substrates mediating human delay and trace fear conditioning. *J. Neurosci.* **24**, 218–228.

Knudsen, E. I. (1994). Supervised learning in the brain. *J. Neurosci.* **14**, 3985–3997.

Koekkoek, S. K. E., Den Ouden, W. L., Perry, G., Highstein, S. M., and De Zeeuw, C. I. (2002). Monitoring kinetic and frequency-domain properties of eyelid responses in mice with magnetic distance measurement technique. *J. Neurophysiol.* **88**, 2124–2133.

Koekkoek, S. K. E., Hulscher, H. C., Dortland, B. R., Hensbroek, R. A., Elgersma, Y., Ruigrok, T. J., and De Zeeuw, C. I. (2003). Cerebellar LTD and learning-dependent timing of conditioned eyelid responses. *Science* **301**, 1736–1739.

Kohonen, T. (1989). "Self-Organization and Associative Memory," 3rd ed. Springer-Verlag, Heidelberg.

Korshunova, T. A., Malyshev, A. Y., Zakharov, I. S., Ierusalimskii, V. N., and Balaban, P. M. (2006). Functions of peptide CNP4, encoded by the HCS2 gene, in the nervous system of *Helix lucorum*. *Neurosci. Behav. Physiol.* **36**, 253–260.

Krasne, F. B., and Roberts, A. (1967). Habituation of the crayfish escape response during release from inhibition induced by picrotoxin. *Nature* **215**, 769–770.

Krasne, F. B., and Wine, J. J. (1975). Extrinsic modulation of crayfish escape behaviour. *J. Exp. Biol.* **63**, 433–450.

Krasne, F. B., and Glanzman, D. L. (1986). Sensitization of the crayfish lateral giant escape reaction. *J. Neurosci.* **6**, 1013–1020.

Krasne, F. B., and Teshiba, T. M. (1995). Habituation of an invertebrate escape reflex due to modulation by higher centers rather than local events. *Proc. Nat. Acad. Sci. USA* **92**, 3362-3366.

Krug, M., Loessner, B., and Otto, T. (1984). Anisomycin blocks the late phase of long-term potentiation in the dentate gyrus of freely moving rats. *Brain Res. Bull.* **13**, 39–42.

Krupa, D. J., Thompson, J. K., and Thompson, R. F. (1993). Localization of a memory trace in the mammalian brain. *Science* **260**, 989–991.

Krupa, D. J., and Thompson, R. F. (1997). Reversible inactivation of the cerebellar interpositus nucleus completely prevents acquisition of the classically conditioned eye-blink response. *Learn. Mem.* **3**, 545–556.

LaBar, K. S., and Cabeza, R. (2006). Cognitive neuroscience of emotional memory. *Nature Rev. Neurosci.* **7: 54–64.**

LaBar, K. S., and Disterhoft, J. F. (1998). Conditioning, awareness, and the hippocampus. *Hippocampus* **8**, 620–626.

LaBar, K. S., Gatenby, J. C., Gore, J. C., LeDoux, J. E., and Phelps, E. A. (1998). Human amygdala activation during conditioned fear acquisition and extinction: A mixed-trial fMRI study. *Neuron* **20**, 937–945.

LaBar, K. S., LeDoux, J. E., Spencer, D. D., and Phelps, E. A. (1995). Impaired fear conditioning following unilateral temporal lobectomy in humans. *J. Neurosci.* **15**, 6846–6855.

LaBar, K. S., and Phelps, E. A. (2005). Reinstatement of conditioned fear in humans is context dependent and impaired in amnesia. *Behav. Neurosci.* **119: 677–686.**

Lalonde, R., and Botez, M. I. (1990). The cerebellum and learning processes in animals. *Brain Res. Rev.* **15**, 325–332.

Lamprecht, R., Farb, C. R., and LeDoux, J. E. (2002). Fear memory formation involves p190 RhoGAP and ROCK proteins through a GRB2-mediated complex. *Neuron* **36**(4), 727–738.

Lamprecht, R., Farb, C. R., Rodrigues, S. M., and LeDoux, J. E. (2006). Fear conditioning drives profilin into amygdala dendritic spines. *Nat. Neurosci.* **9**(4), 481–483.

Lang, P.J., Davis, M., and Ohman, A. (2000). Fear and anxiety: Animal models and human cognitive psychophysiology. *J. Affect. Disord.* **61**, 137–159.

Lavond, D. G., Kim, J. J., and Thompson, R. F. (1993). Mammalian brain substrates of aversive classical conditioning. *Annu. Rev. Psychol.* **44**, 317–342.

Lavond, D. G., Steinmetz, J. E., Yokaitis, M. H., and Thompson, R. F. (1987). Reacquisition of classical conditioning after removal of cerebellar cortex. *Exp. Brain Res.* **67**, 569–593.

Lechner, H. A., and Byrne, J. H. (1998). New perspectives on classical conditioning: A synthesis of Hebbian and non-Hebbian mechanisms. *Neuron* **20**, 355–358.

Lechner, H., Baxter, D., and Byrne, J. (2000). Classical conditioning of feeding in *Aplysia*. I. Behavioral analysis. *J. Neurosci.* **20**, 3369–3376.

LeDoux, J. (2007). The amygdala. *Curr. Biol.* **17**, R868–R874.

LeDoux, J. E. (1986). Sensory systems and emotion. *Integrative Psychiatry* **4**, 237–248.

LeDoux, J. E. (1996). "The Emotional Brain: The Mysterious Underpinnings of Emotional Life." Simon and Schuster, New York.

LeDoux, J. E. (2000). Emotion circuits in the brain. *Annu. Rev. Neurosci.* **23**, 155–184.

LeDoux, J. E., Cicchetti, P., Xagoraris, A., and Romanski, L. M. (1990). The lateral amygdaloid nucleus: Sensory interface of the amygdala in fear conditioning. *J. Neurosci.* **10**, 1062–1069.

LeDoux, J. E., Farb, C. R., and Romanski, L. M. (1991). Overlapping projections to the amygdala and striatum from auditory processing areas of the thalamus and cortex. *Neurosci. Lett.* **134**, 139–144.

Lee, K. H. (2007). Eyeblink conditioning in mutant mice. Doctoral dissertation, University of Southern California.

Lee, H. K., Barbarosie M., Kameyama, K., Bear, M. F., and Huganir, R. L. (2000). Regulation of distinct AMPA receptor phosphorylation sites during bidirectional synaptic plasticity. *Nature* **405**, 955–959.

Lee, H. J., Choi, J.-S., Brown, T. H., and Kim, J. J. (2001). Amygdalar N-methyl-D-aspartate (NMDA) receptors are critical for the expression of multiple conditioned fear responses. *J. Neurosci.* **21**, 4116–4124.

Lee, H.-K., Kameyama, K., Huganir, R. L., and Bear, M. F. (1998). NMDA induces long-term synaptic depression and dephosphorylation of the GluR1 subunit of AMPA receptors in hippocampus. *Neuron* **21**, 1151–1162.

Levenson, J., Endo, S., Kategaya, L. S., Fernandez, R. I., Brabham, D. G., Chin, J., Byrne, J. H., and Eskin, A. (2000). Long-term regulation of neuronal high-affinity glutamate and glutamine uptake in *Aplysia*. *Proc. Natl. Acad. Sci. USA* **97**, 12858–12863.

Li, X. F., Armony, J. L., and LeDoux, J. E. (1996). GABA$_A$ and GABA$_B$ receptors differentially regulate synaptic transmission in the auditory thalamo-amygdala pathway: An *in vivo* micro-iontophoretic study and a model. *Synapse* **24**, 115–124.

Linden, D. J. (1999). The return of the spike: Postsynaptic action potentials and the induction of LTP and LTD. *Neuron* **22**, 661–666.

Linden, D. J., and Connor, J. A. (1995). Long-term synaptic depression. *Annu. Rev. Neurosci.* **18**, 319–335.

Linden, D. J., Dickinson, M. H., Smeyne, M., and Connor, J. A. (1991). A long-term depression of AMPA currents in cultured cerebellar Purkinje neurons. *Neuron* **7**, 81–89.

Lindquist, D. H., Jarrard, L. E., and Brown, T. H. (2004). Perirhinal cortex supports delay fear conditioning to rat ultrasonic social signals. *J. Neurosci.* **24**(14), 3610–3617.

Liu, R.Y., Fioravante, D., Shah, S., and Byrne, J. H. (2008). CREB1 feedback loop is necessary for consolidation of long-term synaptic facilitation in *Aplysia*. *J. Neurosci.* **28**(8), 1970–1976.

Lockery, S. R., and Kristan, W. B., Jr, (1991). Two forms of sensitization of the local bending reflex of the medicinal leech. *J. Comp. Physiol.* [A] **168**, 165–177.

Lockery, S. R., Rawlins, J. N., and Gray, J. A. (1985). Habituation of the shortening reflex in the medicinal leech. *Behav. Neurosci.* **99**, 333–341.

Logan, C. G., and Grafton, S. T. (1995). Functional anatomy of human eyeblink conditioning determined with regional cerebral glucose metabolism and positron-emission tomography. *Proc. Natl. Acad. Sci. USA* **92**, 7500–7504.

Lorenzetti, F. D., Baxter, D. A., Byrne, J. H. (2008). Molecular mechanisms underlying a cellular analogue of operant reward learning. *Neuron* **59**, 815–828.

Lu, B., and Chow, A. (1999). Neurotrophins and hippocampal synaptic transmission and plasticity. *J. Neurosci. Res.* **58**, 76–87.

Lynch, G., Larson, J., Kelso, S., Barrionuevo, G., and Schottler, F. (1983). Intracellular injections of EGTA block induction of hippocampal long-term potentiation. *Nature* **305**, 719–721.

Magee, J., and Johnston, D. (1997). A synaptically controlled, associative signal for hebbian plasticity in hippocampal neurons. *Science* **275**, 209–213.

Malenka, R. C. (2003). Synaptic plasticity and AMPA receptor trafficking. *Ann. NY Acad. Sci.* **1003**, 1–11.

Malgaroli, A., Ting, A. E., Wendland, B., Bergamaschi, A., Villa, A., Tsien, R. W., Scheller, R. H. (1995). Presynaptic component of long-term potentiation visualized at individual hippocampal synapses. *Science* **268**, 1624–1628.

Malinow, R. (2003). AMPA receptor trafficking and long-term potentiation. *Philos. Trans. R. Soc. Lond. B. Biol. Sci.* **358**(1432), 707–714.

Malinow, R. (1991). Transmission between pairs of hippocampal slice neurons: Quantal levels, oscillations, and LTP. *Science* **252**, 722–724.

Malinow, R., Madison, D.V., and Tsien, R. W. (1988). Persistent protein kinase activity underlying long-term potentiation. *Nature* **335**, 820–824.

Malinow, R., Tsien, R. W. (1990). Presynaptic enhancement shown by whole-cell recordings of long-term potentiation in hippocampal slices. *Nature* **346**, 177–180.

Malkani, S., and Rosen, J. B. (2000). Specific induction of early growth response gene 1 in the lateral nucleus of the amygdala following contextual fear conditioning in rats. *Neurosci.* **97**(4), 693–702.

Malleret, G., Haditsch, U., Genoux, D., Jones, M. W., Bliss, V. P., Vanhoose, A. M., Weitlauf, C., Kandel, E. R., Winder, D. G., and Mansuy, I. M. (2001). Inducible and reversible enhancement of learning, memory and long-term potentiation by genetic inhibition of calcineurin. *Cell* **104**, 675–686.

Maren, S. (1998). Overtraining does not mitigate contextual fear conditioning deficits produced by neurotoxic lesions of the basolateral amygdala. *J. Neurosci.* **18**, 3088–3097.

Maren, S. (2000). Auditory fear conditioning increases CS-elicited spike firing in lateral amygdala neurons even after extensive overtraining. *Eur. J. Neurosci.* **12**, 4047–4054.

Maren, S. (2001). Neurobiology of Pavlovian fear conditioning. *Annu. Rev. Neurosci.* **24**, 897–931.

Maren, S., Aharonov, G., Stote, D. L., and Fanselow, M. S. (1996). N-Methyl-D-aspartate receptors in the basolateral amygdala are required for both acquisition and expression of conditional fear in rats. *Behav. Neurosci.* **110**, 1365–1374.

Maren, S., Poremba, A., and Gabriel, M. (1991). Basolateral amygdaloid multi-unit neuronal correlates of discriminative avoidance learning in rabbits. *Brain Res.* **549**, 311–316.

Maren, S., Tocco, G., Standley, S., Baudry, M., and Thompson, R. F. (1993). Postsynaptic factors in the expression of long-term potentiation (LTP): Increased glutamate receptor binding following LTP induction *in vivo. Proc. Natl. Acad. Sci. USA* **90**, 9654–9658.

Maren, S., Yap, S. A., and Goosens, K. A. (2001). The amygdala is essential for the development of neuronal plasticity in the medial geniculate nucleus during auditory fear conditioning in rats. *J. Neurosci.* **21**(6), RC135.

Marks, I., and Tobena, A. (1990). Learning and unlearning fear: A clinical and evolutionary perspective. *Neurosci. Biobehav. Rev.* **14**, 365–384.

Marr, D. (1969). A theory of cerebellar cortex. *J. Physiol.* **202**, 437–470.

Martin, K. C., Michael, D., Rose, J. C., Barad, M., Casadio, A., Zhu, H., and Kandel, E. R. (1997). MAP kinase translocates into the nucleus of the presynaptic cell and is required for long-term facilitation in *Aplysia. Neuron* **18**, 899–912.

Matthies, H. (1989). In search of cellular mechanisms of memory. *Prog. Neurobiol.* **32**, 277–349.

Matus, A. 2000. Actin-based plasticity in dendritic spines. *Science* **290**(5492), 754–758.

Mauk, M. D., Steinmetz, J. E., and Thompson, R. F. (1986). Classical conditioning using stimulation of the inferior olive as the unconditioned stimulus. *Proc. Natl. Acad. Sci.* **83**, 5349–5353.

Mayford, M., Bach, M. E., Huang, Y. Y., Wang, L., Hawkins, R. D., and Kandel, E. R. (1996). Control of memory formation through regulated expression of a CaMKII transgene. *Science* **274**(5293), 1678–1683.

McCormick, D. A., Lavond, D. G., and Thompson, R. F. (1982). Concomitant classical conditioning of the rabbit nictitating membrane and eyelid responses: Correlations and implications. *Physiol. & Behav.* **28**, 769–775.

McCormick, D. A., Steinmetz, J. E., and Thompson, R. F. (1985). Lesions of the inferior olivary complex cause extinction of the classically conditioned eyeblink response. *Brain Res.* **359**, 120–130.

McCormick, D. A., and Thompson, R. F. (1984). Cerebellum: Essential involvement in the classically conditioned eyelid response. *Science* **223**, 296–299.

McDonald, A. J. (1998). Cortical pathways to the mammalian amygdala. *Prog. Neurobiol.* **55**, 257–332.

McDonald, A. J., Muller, J. F., and Mascagni, F. (2002). GABAergic innervation of alpha type II calcium/calmodulin-dependent protein kinase immunoreactive pyramidal neurons in the rat basolateral amygdala. *J. Comp. Neurol.* **446**(3), 199–218.

McGlinchey-Berroth, R., Carrillo, M. C., Gabrieli, J. D., Brawn, C. M., and Disterhoft, J. F. (1997). Impaired trace eyeblink conditioning in bilateral, medial-temporal lobe amnesia. *Behav. Neurosci.* **111**, 873–882.

McKernan, M. G., and Shinnick-Gallagher, P. (1997). Fear conditioning induces a lasting potentiation of synaptic currents *in vitro. Nature* **390**, 607–611.

McNaughton, B. L., Douglass, R. M., and Goddard, G. V. (1978). Synaptic enhancements in fascia dentata: Cooperativity among coactive efferents. *Brain Res.* **157**, 277–293.

Menzel, R. (2001). Searching for the memory trace in a mini-brain: The honeybee. *Learn. Memory* **8**, 53–62.

Menzel, R., Leboulle, G., and Eisenhardt, D. (2006). Small brains, bright minds. *Cell.* **124**(2), 237–239.

Milad, M. R., and Quirk, G. J. (2002). Neurons in medial prefrontal cortex signal memory for fear extinction. *Nature* **420**(6911), 70–74.

Miller, N. E. (1948). Studies of fear as an acquirable drive: I. Fear as motivation and fear reduction as reinforcement in the learning of new responses. *J. Experimen. Psychol.* **38**, 89–101.

Miller, N. E. (1951). Learnable drives and rewards. In "Handbook of Experimental Psychology" (S. S. Stevens, Ed), pp 435–472. Wiley, New York.

Miserendino, M. J. D., Sananes, C. B., Melia, K. R., and Davis, M. (1990). Blocking of acquisition but not expression of conditioned fear-potentiated startle by NMDA antagonists in the amygdala. *Nature* **345**, 716–718.

Miyata, M., Kim, H.-T., Hashimoto, K., Lee, T.-K., Cho, S.-Y., Jiang, H., Wu, Y., Jun, K., Wu, D., Kano, M., and Shin, H.-S. (2001). Deficient long-term synaptic depression in the rostral cerebellum correlated with impaired motor learning in phospholipase C4 mutant mice. *Eur. J. Neurosci.* **13**, 1945–1954.

Mohamed, H. A., Yao, W., Fioravante, D., Smolen, P. D., and Byrne, J. H. (2005). cAMP-response elements in *Aplysia* creb1, creb2, and Apuch promoters: Implications for feedback loops modulating long term memory. *J. Biol. Chem.* **280**, 27035–27043.

Morris, J. S., Buchel, C., and Dolan, R. J. (2001). Parallel neural responses in amygdala subregions and sensory cortex during implicit fear conditioning. *Neuroimage* **13**(6 Pt 1), 1044–1052.

Morris, J. S., and Dolan, R. J. (2004). Dissociable amygdala and orbitofrontal responses during reversal fear conditioning. *Neuroimage* **22**(1), 372–380.

Morris, J. S., Friston, K. J., and Dolan, R. J. (1998). Experience-dependent modulation of tonotopic neural responses in human auditory cortex. *Proc. Biol. Sci.* **265**(1397), 649–657.

Morris, J. S., Frith, C. D., Perret, D. I., Rowland, D., Young, A. W., Calder, A. J., and Dolan, R. J. (1996). A differential neural response in the human amygdala to fearful and happy facial expressions. *Nature* **383**, 812–815.

Morris, J. S., Ohman, A., and Dolan, R. J. (1999). A subcortical pathway to the right amygdala mediating "unseen" fear. *Proc. Natl. Acad. Sci. USA* **96**, 1680–1685.

Morton, D. W., and Chiel, H. J. (1993a). *In vivo* buccal nerve activity that distinguishes ingestion from rejection can be used to predict behavioral transitions in *Aplysia*. *J. Comp. Physiol. A* **172**, 17–32.

Morton, D., and Chiel, H. (1993b). The timing of activity in motor neurons that produce radula movements distinguishes ingestion from rejection in *Aplysia*. *J. Comp. Physiol. A* **173**, 519–536.

Mowrer, O. H. (1947). On the dual nature of learning: A reinterpretation of "conditioning" and "problem solving." *Harvard Educational Review* **17**, 102–148.

Mowrer, O. H., and Lamoreaux, R. R. (1946). Fear as an intervening variable in avoidance conditioning. *J. Compar. Psychol.* **39**, 29–50.

Moyer, J. R., Jr., Deyo, R. A., and Disterhoft, J. F. (1990). Hippocampectomy disrupts trace eye-blink conditioning in rabbits. *Behav. Neurosci.* **104**, 243–252.

Moyer, J. R., Thompson, L. T., and Disterhoft, J. F. (1996). Trace eye-blink conditioning increases CA1 excitability in a transient and learning-specific manner. *J. Neurosci.* **16**, 5536–5546.

Muller, J., Corodimas, K. P., Fridel, Z., and LeDoux, J. E. (1997). Functional inactivation of the lateral and basal nuclei of the amygdala by muscimol infusion prevents fear conditioning to an explicit conditioned stimulus and to contextual stimuli. *Behav. Neurosci.* **111**, 863–891.

Murphy, G. G., and Glanzman, D. L. (1997). Mediation of classical conditioning in *Aplysia californica* by long-term potentiation of sensorimotor synapses. *Science* **278**, 467–471.

Naber, P. A., and Witter, M. P. (1998). Subicular efferents are organized mostly as parallel projections: A double-labeling, retrograde-tracing study in the rat. *J. Comp. Neurol.* **393**, 284–297.

Nader, K., Schafe, G. E., and LeDoux, J. E. (2000). Fear memories require protein synthesis in the amygdala for reconsolidation after retrieval. *Nature* **406**, 722–726.

Nargeot, R., Baxter, D., and Byrne, J. (1997). Contingent-dependent enhancement of rhythmic motor patterns: An *in vitro* analog of operant conditioning. *J. Neurosci.* **17**, 8093–8105.

Nargeot, R., Baxter, D., and Byrne, J. (1999a). Dopaminergic synapses mediate neuronal changes in an analogue of operant conditioning. *J. Neurophysiol.* **81**, 1983–1987.

Nargeot, R., Baxter, D., and Byrne, J. (1999b). *In vitro* analog of operant conditioning in *Aplysia*. I. Contingent reinforcement modifies the functional dynamics of an identified neuron. *J. Neurosci.* **15**, 2247–2260.

Nargeot, R., Baxter, D., and Byrne, J. (1999c). *In vitro* analog of operant conditioning in *Aplysia*. II. Modifications of the functional dynamics of an identified neuron contribute to motor pattern selection. *J. Neurosci.* **19**, 2261–2272.

Nguyen, P. V., Abel, T., and Kandel, E. R. (1994). Requirement of a critical period of transcription for induction of a late phase of LTP. *Science* **265**, 1104–1107.

Nicoll, R. A., and Malenka, R. C. (1995). Contrasting properties of two forms of long-term potentiation in the hippocampus. *Nature* **377**, 115–118.

Ocorr, K., Walters, E. T., and Byrne, J. H. (1985). Associative conditioning analog selectively increases cAMP levels of tail sensory neurons in *Aplysia*. *Proc. Natl. Acad. Sci. USA* **82**, 2548–2552.

Ormond, J., Hislop, J., Zhao, Y., Webb, N., Vaillaincourt, F., Dyer, J. R., Ferraro, G., Barker, P., Martin, K. C., and Sossin, W. S. (2004). ApTrkl, a Trk-like receptor, mediates serotonin-dependent ERK activation and long-term facilitation in *Aplysia* sensory neurons. *Neuron* **44**, 715–728.

Orr, S. P., Metzger, L. J., Lasko, N. B., Macklin, M. L., Peri, T., and Pitman, R. K. (2000). De novo conditioning in trauma-exposed individuals with and without posttraumatic stress disorder. *J. Abnormal Psychol.* **109**, 290–298.

Pape, H. C., and Stork, O. (2003). Genes and mechanisms in the amygdala involved in the formation of fear memory. *Ann. NY Acad. Sci.* **985**, 92–105.

Paré, D., Quirk, G. J., and Ledoux, J. E. (2004). New vistas on amygdala networks in conditioned fear. *J. Neurophysiol.* **92**(1), 1–9.

Pare, D., Smith, Y., and Pare, J. F. (1995). Intra-amygdaloid projections of the basolateral and basomedial nuclei in the cat: *Phaseolus vulgaris*-leucoagglutinin anterograde tracing at the light and electron microscope level. *Neurosci.* **69**, 567–583.

Park, J.-S., Onodera, T., Nishimura, S., Thompson, R. F., and Itohara, S. (2006). Molecular evidence for two-stage learning and partial laterality in eyeblink conditioning of mice. *Proc. Natl. Acad. Sci.* **103**, 5549–5554.

Pascoe, J. P., and Kapp, B. S. (1985). Electrophysiological characteristics of amygdaloid central nucleus neurons in the awake rabbit. *Brain Res. Bull.* **14**, 331–338.

Patterson, S. L., Pittenger, C., Morozov, A., Martin, K. C., Scanlin, H., Drake, C., and Kandel, E. R. (2001). Some forms of cAMP-mediated long-lasting potentiation are associated with release of BDNF and nuclear translocation of phospho-MAP kinase. *Neuron* **32**(1), 123–140.

Pavlov, I. P. (1927). "Conditioned Reflexes." Dover, New York.

Pavlov, I. P. (1927). "Conditioned Reflexes" (G. V. Anrep, Tran.). Oxford Univ. Press, London.

Pelligrino, L. J., and Altman, J. (1979). Effects of differential interference with postnatal cerebellar neurogenesis on motor performance, activity level and maze learning of rats: A developmental study. *J. Comp. Physiol. Psychol.* **93**, 1–33.

Peper, M., Karcher, S., Wohlfarth, R., Reinshagen, G., and LeDoux, J. E. (2001). Aversive learning in patients with unilateral lesions of the amygdala and hippocampus. *Biol. Psychol.* **58**(1), 1–23.

Perrett, S. P., Ruiz, B. P., and Mauk, M. D. (1993). Cerebellar cortex lesions disrupt learning-dependent timing of conditioned eyelid responses. *J. Neurosci.* **13**, 1708–1718.

Phelps, E. A., Delgado, M. R., Nearing, K. I., and LeDoux, J. E. (2004). Extinction learning in humans: Role of the amygdala and vmPFC. *Neuron* **43**, 897–905.

Phelps, E. A., LaBar, K. S., Andersen, A. K., O'Connor, K. J., and Fulbright, R. K. (1998). Specifying the contributions of the human amygdala to emotional memory: A case study. *NeuroCase* **4**, 527–540.

Phelps, E. A., O'Connor, K. J., Gatenby, J. C., Gore, J. C., Grillon, C., and Davis, M. (2001). Activation of the left amygdala to a cognitive representation of fear. *Nat. Neurosci.* **4**(4), 437–441.

Pitkanen, A., Savander, V., and LeDoux, J. E. (1997). Organization of intra-amygdaloid circuitries in the rat: An emerging framework for understanding functions of the amygdala. *Trends Neurosci.* **20**, 517–523.

Quirk, G. J., Armony, J. L., and LeDoux, J. E. (1997). Fear conditioning enhances different temporal components of tone-evoked spike trains in auditory cortex and lateral amygdala. *Neuron* **19**, 613–624.

Quirk, G. J., Repa, J. C., and LeDoux, J. E. (1995). Fear conditioning enhances short-latency auditory responses of lateral amygdala neurons: Parallel recordings in the freely behaving rat. *Neuron* **15**, 1029–1039.

Rachlin, H. (1991). "Introduction to Modern Behaviorism," 3rd ed. Freeman, New York.

Radley, J. J., Johnson, L. R., Janssen, W. G., Martino, J., Lamprecht, R., Hof, P. R., Ledoux, J. E., and Morrison, J. H. (2006). Associative Pavlovian conditioning leads to an increase in spinophilin-immunoreactive dendritic spines in the lateral amygdala. *Eur. J. Neurosci.* **24**(3), 876–884.

Rampon, C., and Tsien, J. Z. (2000). Genetic analysis of learning behavior-induced structural plasticity. *Hippocampus* **10**(5), 605–609.

Rankin, C. H. (2002). From gene to identified neuron to behaviour in Caenorhabditis elegans. *Nat. Rev. Genet.* **3**, 622–630.

Rankin, C. H., and Wicks, S. R. (2000). Mutations of the caenorhabditis elegans brain-specific inorganic phosphate transporter eat-4 affect habituation of the tap-withdrawal response without affect the response itself. *J. Neurosci.* **20**, 4337–4344.

Rankin, C. H., Beck, C. D., and Chiba, C. M. (1990). Caenorhabditis elegans: a new model system for the study of learning and memory. *Behav. Brain Res.* **37**, 89–92.

Ratner, S. C., (1972). Habituation and retention of habituation in the leech (Macrobdella decora). *J. Comp. Physiol. Psychol.* **81**, 115–121.

Rattiner, L. M., Davis, M., French, C. T., and Ressler, K. J. (2004). Brain-derived neurotrophic factor and tyrosine kinase receptor B involvement in amygdala-dependent fear conditioning. *J. Neurosci.* **24**(20), 4796–4806.

Rattiner, L. M., Davis, M. and Ressler, K. J. (2005). Brain-derived neurotrophic factor in amygdala-dependent learning. *Neuroscientist* **11**(4), 323–333.

Rauch, S. L., Shin, L. M., and Wright, C. I. (2003). Neuroimaging studies of amygdala function in anxiety disorders. *Ann NY Acad. Sci.* **985**, 389–410.

Reijmers, L. G., Perkins, B. L., Matsuo, N., and Mayford, M. (2007). Localization of a stable neural correlate of associative memory. *Science* **317**(5842), 1230–1233.

Repa, J. C., Muller, J., Apergis, J., Desrochers, T. M., Zhou, Y., and LeDoux, J. E. (2001). Two different lateral amygdala cell populations contribute to the initiation and storage of memory. *Nat. Neurosci.* **4**, 724–731.

Rescorla, R. A. (1988). Behavioral studies of pavlovian conditioning. *Annu. Rev. Neurosci.* **11**, 329–352.

Rescorla, R. A., and Solomon, R. L. (1967). Two process learning theory: Relationships between pavlovian conditioning and instrumental learning. *Psychol. Rev.* **55**, 151–182.

Ressler, K. J., Paschall, G., Zhou, X. L., and Davis, M. (2002). Regulation of synaptic plasticity genes during consolidation of fear conditioning. *J. Neurosci.* **22**(18), 7892–7902.

Rickert, E. J., Bennett, T. L., Lane, P., and French, J. (1978). Hippocampectomy and the attenuation of blocking. *Behav. Biol.* **22**, 147–160.

Roberts, A.C., and Glanzman, D.L. (2003). Learning in *Aplysia*: Looking at synaptic plasticity from both sides. *Trends in Neurosci.* **26**, 662–670.

Rodrigues, S. M., Farb, C. R., Bauer, E. P., LeDoux, J. E., and Schafe, G. E. (2004). Pavlovian fear conditioning regulates Thr286 autophosphorylation of Ca^{2+}/calmodulin-dependent protein kinase II at lateral amygdala synapses. *J. Neurosci.* **24**(13), 3281–3288.

Rodrigues, S. M., Schafe, G. E., and LeDoux, J. E. (2001). Intra-amygdala blockade of the NR2B subunit of the NMDA receptor disrupts the acquisition but not the expression of fear conditioning. *J. Neurosci.* **21**, 6889–6896.

Rodrigues, S. M., Schafe, G. E., and LeDoux, J. E. (2004). Molecular mechanisms underlying emotional learning and memory in the lateral amygdala. *Neuron* **44**(1), 75–91.

Rogan, M. T., and LeDoux, J. E. (1995). LTP is accompanied by commensurate enhancement of auditory-evoked responses in a fear conditioning circuit. *Neuron* **15**, 127–136.

Rogan, M. T., Staubli, U. V., and LeDoux, J. E. (1997). Fear conditioning induces associative long-term potentiation in the amygdala. *Nature* **390**, 604–607.

Romanski, L. M., and LeDoux, J. E. (1992). Equipotentiality of thalamo-amygdala and thalamo-cortico-amygdala circuits in auditory fear conditioning. *J. Neurosci.* **12**, 4501–4509.

Rose, J. K., Kaun, K. R., Chen, S. H., and Rankin, C. H. (2003). GLR-1, a non-NMDA glutamate receptor homolog, is critical for long-term memory in Caenorhabditis elegans. *J. Neurosci.* **23**, 9595–9599.

Rose, J. K., and Rankin, C. H. (2001). Analysis of habituation in Caenorhabditis elegans. *Learn. Memory* **8**, 63–69.

Rosen, J. B., Fanselow, M. S., Young, S. L., Sitcoske, M., and Maren, S. (1998). Immediate-early gene expression in the amygdala following footshock stress and contextual fear conditioning. *Brain Res.* **796**(1–2), 132–142.

Rumpel, S., LeDoux, J., Zador, A., and Malinow, R. (2005). Postsynaptic receptor trafficking underlying a form of associative learning. *Science* **308**(5718), 83–88.

Sahley, C. L., Modney, B. K., Boulis, N. M., and Muller, K. J. (1994). The S cell: an interneuron essential for sensitization and full dishabituation of leech shortening. *J. Neurosci.* **14**, 6715–6721.

Sakurai, A., Calin-Jageman, R. J., and Katz, P. S. (2007). Potentiation phase of spike timing-dependent neuromodulation by a serotonergic interneuron involves an increase in the fraction of transmitter release. *J. Neurophysiol.* **98**(4), 1975–1987.

Sarter, M. F., and Markowitsch, H. J. (1985). Involvement of the amygdala in learning and memory: A critical review, with emphasis on anatomical relations. *Behavioral Neuroscience* **99**, 342–380.

Schafe, G. E., Atkins, C. M., Swank, M. W., Bauer, E. P., Sweatt, J. D., and LeDoux, J. E. (2000). Activation of ERK/MAP kinase in the amygdala is required for memory consolidation of pavlovian fear conditioning. *J. Neurosci.* **20**, 8177–8187.

Schafe, G. E., and LeDoux, J. E. (2000). Memory consolidation of auditory pavlovian fear conditioning requires protein synthesis and protein kinase A in the amygdala. *J. Neurosci.* **20**, RC96.

Schafe, G. E., Nadel, N. V., Sullivan, G. M., Harris, A., and LeDoux, J. E. (1999). Memory consolidation for contextual and auditory fear conditioning is dependent on protein synthesis, PKA, and MAP kinase. *Learn. Memory* **6**, 97–110.

Schaefer, S. M., Jackson, D. C., Davidson, R. J., Aguirre, G. K., Kimberg, D. Y., and Thompson-Schill, S. L. (2002). Modulation of amygdalar activity by the conscious regulation of negative emotion. *J. Cogn. Neurosci.* **14**(6), 913–921.

Schmahmann, J. D. (Ed.). (1997). "International Review of Neurobiology," Vol. 41. Academic Press, San Diego.

Scicli, A. P., Petrovich, G. D., Swanson, L. W., and Thompson, R. F. (2004). Contextual fear conditioning is associated with lateralized expression of the immediate early gene c-fos in the central and basolateral amygdalar nuclei. *Behav. Neurosci.* **118**(1), 5–14.

Scuri, R., Mozzachiodi, R., and Brunelli, M. (2002). Activity-dependent increase of the AHP amplitude in T sensory neurons of the leech. *J. Neurophysiol.* **88**, 2490–2500.

Scuri, R., Mozzachiodi, R., and Brunelli, M. (2005). Role for calcium signaling and arachidonic acid metabolites in the activity-dependent increase of AHP amplitude in leech T sensory neurons. *J. Neurophysiol.* **94**, 1066–1073.

Shi, C., and Davis, M. (1999). Pain pathways involved in fear conditioning measured with fear-potentiated startle: Lesion studies. *J. Neurosci.* **19**, 420–430.

Shi, C., and Davis, M. (2001). Visual pathways involved in fear conditioning measured with fear-potentiated startle: Behavioral and anatomic studies. *J. Neurosci.* **21**(24), 9844–9855.

Shibuki, K., Gomi, H., Chen, C., Bao, S., Kim, J. J., Wakatsuki, H., Fujisaki, T., Fujimoto, K., Katoh, A., Ikeda, T., Chen, C., Thompson, R. F., and Itohara, S. (1996). Deficient cerebellar long-term depression, impaired eyeblink conditioning and normal motor coordination in GFAP mutant mice. *Neuron* **16**, 587–599.

Skelton, R.W. (1988). Bilateral cerebellar lesions disrupt conditioned eyelid responses in unrestrained rats. *Behav. Neurosci.* **102**, 586–590.

Skinner, B. F. (1938). "The behavior of organisms: An experimental analysis." Appleton-Century-Crofts, New York.

Smith-Roe, S. L. and Kelley, A. E. (2000). Coincident activation of NMDA and dopamine D1 receptors within the nucleus accumbens core is required for appetitive instrumental learning. *J. Neurosci.* **20**, 7737–7742.

Solomon, P. R., Pomerleau, D., Bennett, L., James, J., and Morse, D. L. (1989). Acquisition of the classically conditioned eyeblink responses in humans over the life span. *Psychol. Aging* **4**, 34–41.

Solomon, P. R., Vander Schaaf, E. R., Thompson, R. F., and Weisz, D. J. (1986). Hippocampus and trace conditioning of the rabbit's classically conditioned nictitating membrane response. *Behav. Neurosci.* **100**, 729–744.

Stanton, M. E., Freeman, J. H., , Jr., and Skelton, R. W. (1992). Eyeblink conditioning in the developing rat. *Behav. Neurosci.* **106**, 657–665.

Steinmetz, J. E., Lavond, D. G., Ivkovich, D., Logan, C. G., and Thompson, R. F. (1992). Disruption of classical eyelid conditioning after cerebellar lesions: Damage to a memory trace system or a simple performance deficit? *J. Neurosci.* **12**, 4403–4426.

Steinmetz, J. E., Logan, C. G., Rosen, D. J., Thompson, J. K., Lavond, D. G., and Thompson, R. F. (1987). Initial localization of the acoustic conditioned stimulus projection system to the cerebellum essential for classical eyelid conditioning. *Proc. Natl. Acad. Sci.* **84**, 3531–3535.

Steinmetz, J. E., Logue, S. F., and Miller, D. P. (1993). Using signaled barpressing tasks to study the neural substrates of appetitive and aversive learning in rats: Behavioral manipulations and cerebellar lesions. *Behav. Neurosci.* **107**, 941–954.

Steinmetz, J. E., Rosen, D. J., Chapman, P. F., Lavond, D. G., and Thompson, R. F. (1986). Classical conditioning of the rabbit eyelid response with a mossy-fiber stimulation CS. I. Pontine nuclei and middle cerebellar peduncle stimulation. *Behav. Neurosci.* **100**, 878–887.

Steinmetz, J. E., and Woodruff-Pak, D. S. (2000). Animal models in eyeblink classical conditioning. *In* "Eyeblink classical conditioning: Animal models," Vol. 2 (D. S. Woodruff-Pak and J. E. Steinmetz, Eds.), pp. 81–103. Kluwer Academic Publishers, Boston, MA.

Steward, O., Wallace, C. S., Lyford, G. L., and Worley, P. F. (1998). Synaptic activation causes the mRNA for the IEG Arc to localize selectively near activated postsynaptic sites on dendrites. *Neuron* **21**, 741–751.

Steward, O., and Worley, P. F. (2001a). A cellular mechanism for targeting newly synthesized mRNAs to synaptic sites on dendrites. *Proc. Nat. Acad. Sci. USA* **98**, 7062–7068.

Steward, O., and Worley, P. F. (2001b). Selective targeting of newly synthesized Arc mRNA to activated synapses requires NMDA receptor activation. *Neuron* **30**, 227–240.

Stork, O., Stork, S., Pape, H. C., and Obata, K. (2001). Identification of genes expressed in the amygdala during the formation of fear memory. *Learn. Mem.* **8**(4), 209–219.

Supple, W. F., Jr., and Leaton, R. N. (1990). Lesions of the cerebellar vermis and cerebellar hemispheres: Effects on heart rate conditioning in rats. *Behav. Neurosci.* **104**, 934–947.

Sutton, M. A., Ide, J., Masters, S. E., and Carew, T. J. (2002). Interaction between amount and pattern of training in the induction of intermediate- and long-term memory for sensitization in *Aplysia*. *Learn. Mem.* **9**, 29–40.

Sutton, M. A., Masters, S. E., Bagnall, M. W., and Carew, T. J. (2001). Molecular mechanisms underlying a unique intermediate phase of memory in *Aplysia*. *Neuron* **31**, 143–154.

Sweatt, J. D. (2003). "Mechanisms of Memory." Elsevier, London.

Tang, Y.-P., Shimizu, E., Dube, G. R., Rampon, C., Kerchner, G. A., Zhuo, M., Liu, G., and Tsien, Z. (1999). Genetic enhancement of learning and memory in mice. *Nature* **401**, 63–69.

Thach, W. T., Goodkin, H. G., and Keating, J. G. (1992). The cerebellum and the adaptive coordination of movement. *Annu. Rev. Neurosci.* **15**, 403–442.

Thomas, M. J., Moody, T. D., Makhinson, M., and O'Dell, T. J. (1996). Activity-dependent beta-adrenergic modulation of low frequency stimulation induced LTP in the hippocampal CA1 region. *Neuron* **17**, 475–482.

Thompson, L. T., Deyo, R. A., and Disterhoft, J. F. (1992). Hippocampus-dependent learning facilitated by a monoclonal antibody or D-cycloserine. *Nature* **359**, 838–841.

Thompson, L. T., Moyer, J. R., and Disterhoft, J. F. (1996). Transient changes in excitability of rabbit CA3 neurons with a time course appropriate to support memory consolidation. *J. Neurophysiol.* **76**, 1836–1849.

Thompson, R. F. (1976). The search for the engram. *Am. Psychol.* **31**, 209–227.

Thompson, R. F. (1986). The neurobiology of learning and memory. *Science* **233**, 941–947.

Thompson, R. F. (1989). Role of inferior olive in classical conditioning. *In* "The Olivecerebellar System in Motor Control" (P. Strata, Ed.), pp. 347–362. Springer-Verlag, New York.

Thompson, R. F. (1990). Neural mechanisms of classical conditioning in mammals. *Philos. Trans. R. Soc. London Ser. B* **329**, 161–170.

Thompson, R. F., Bao, S., Chen, L., Cipriano, B. D., Grethe, J. S., Kim, J. J., Thompson, J. K., Tracy, J. A., Weninger, M. S., and Krupa, D. J. (1997). Associative learning. *Int. Rev. Neurosci.* **41**, 151–189.

Thompson, R. F., and Kim, J. J. (1996). Memory systems in the brain and localization of a memory. *Proc. Nat. Acad. Sci. USA* **93**, 13438–13444.

Thompson, R. F., and Krupa, D. J. (1994). Organization of memory traces in the mammalian brain. *Annu. Rev. Neurosci.* **17**, 519–549.

Tocco, G., Annala, A. J., Baudry, M., and Thompson, R. F. (1992). Learning of a hippocampal-dependent conditioning task changes the binding properties of AMPA receptors in rabbit hippocampus. *Behav. Neural. Biol.* **58**, 222–231.

van Groen, T., Wyss, J. M. (1990). Extrinsic projections from area CA1 of the rat hippocampus: Olfactory, cortical, subcortical, and bilateral hippocampal formation projections. *J. Comp. Neurol.* **302**, 515–528.

van Rossum, D., and Hanisch, U. K. (1999). Cytoskeletal dynamics in dendritic spines: Direct modulation by glutamate receptors? *Trends Neurosci.* **22**(7), 290–295.

Vavoulis, D. V, Straub, V. A., Kemenes, I., Kemenes, G., Feng, J., Benjamin, P. R. (2007). Dynamic control of a central pattern generator circuit: A computational model of the snail feeding network. *Eur. J. Neurosci.* **25**(9), 2805–2818.

Voneida, T., Christie, D., Bogdanski, R., and Chopko, B. (1990). Changes in instrumentally and classically conditioned limb-flexion responses following inferior olivary lesions and olivocerebellar tractotomy in the cat. *J. Neurosci.* **10**, 3583–3593.

Vu, E. T., Krasne, F. B. (1992). Evidence for a computational distinction between proximal and distal neuronal inhibition. *Science* **255**, 1710–1712.

Vu, E. T., and Krasne, F. B. (1993). Crayfish tonic inhibition: prolonged modulation of behavioral excitability by classical GABAergic inhibition. *J. Neurosci.* **13**, 4394–4402.

Vu, E. T., Lee, S. C., and Krasne, F. B., (1993). The mechanism of tonic inhibition of crayfish escape behavior: distal inhibition and its functional significance. *J. Neurosci.* **13**, 4379–4393.

Waddel, S., and Quinn, W. G. (2001). Flies, genes and learning. *Annu. Rev. Neurosci.* **24**, 1283–1309.

Wainwright, M.L., Byrne, J.H., and Cleary, L.J. (2004). Dissociation of morphological and physiological changes associated with long-term memory in Aplysia. *J. Neurophysiol.* **92**, 2628–2632.

Walker, D. L., and Davis, M. (2000). Involvement of NMDA receptors within the amygdala in short- versus long-term memory for fear conditioning as assessed with fear-potentiated startle. *Behav. Neurosci.* **114**, 1019–1033.

Watanabe, S., Kirino, Y., Gelperin, A. (2008). Neural and molecular mechanisms of microcognition in Limax. *Learn. Mem.* **15,** 633–642.

Watson, J. B. (1925). Behaviourism. New York: WW. Norton.

Weinberger, N. M. (1995). Retuning the brain by fear conditioning. In "The Cognitive Neurosciences" (M. S. Gazzaniga, Ed.), pp. 1071–1090. MIT Press, Cambridge, MA.

Weiskrantz, L. (1956). Behavioral changes associated with ablation of the amygdaloid complex in monkeys. *J. Com. Physiol. Psychol.* **49**, 381–391.

Weiskrantz, L., and Warrington, E. K. (1979). Conditioning in amnesic patients. *Neuropsychologia* **17**, 187–194.

Weiss, C., Weible, A. P., Galvez, R., and Disterhoft, J. P. (2006). Forebrain-cerebellar interactions during learning. *Cellsci. Rev.* **3**, 200–230.

Weisskopf, M. G., Bauer, E. P., and LeDoux, J. E. (1999). L-Type voltage gated calcium channels mediate NMDA-independent associative long-term potentiation at thalamic input synapses to the amygdala. *J. Neurosci.* **19**, 10512–10519.

Weisz, D. J., Clark, G. A., and Thompson, R. F. (1984). Increased activity of dentate granule cells during nictitating membrane response conditioning in rabbits. *Behav. Brain Res.* **12**, 145–154.

Welsh, J. P., and Harvey, J. A. (1989). Cerebellar lesions and the nictitating membrane reflex: Performance deficits of the conditioned and unconditioned response. *J. Neurosci.* **9**, 299–311.

Whalen, P. J., Rauch, S. L., Etcoff, N. L., McInerney, S. C., Lee, M. B., and Jenike, M. A. (1998). Masked presentation of emotional facial expressions modulate amygdala activity without explicit knowledge. *J. Neurosci.* **18**, 411–418.

Wicks, S. R., and Rankin, C. H. (1997). Effects of tap withdrawal response habituation on other withdrawal behaviors: the localization of habituation in the nematode Caenorhabditis elegans. *Behav. Neurosci.* **111**, 342–353.

Wigstrom, H., and Gustafsson, B. (1986). Postsynaptic control of hippocampal long-term potentiation. *J. Physiol. (Paris)* **81**, 228–236.

Wilensky, A. E., Schafe, G. E., and LeDoux, J. E. (1999). Functional inactivation of the amygdala before but not after auditory fear conditioning prevents memory formation. *J. Neurosci.* **19**(RC48), 1–5.

Winder, D. G., Mansuy, I. M., Osman, M., Moallem, T. M., and Kandel, E. R. (1998). Genetic and pharmacological evidence for a novel, intermediate phase of long-term potentiation suppressed by calcineurin. *Cell* **92**, 25–37.

Wine, J. J., Krasne, F. B., and Chen, L. (1975). Habituation and inhibition of the crayfish lateral giant fibre escape response. *J. Exp. Biol.* **62**, 771–782.

Wolpe, J., and Rowan, V. C. (1988). Panic disorder: A product of classical conditioning. *Behav. Res. Ther.* **26**, 441–450.

Woodruff-Pak, D. S., Finkbiner, R. G., and Sasse, D. K. (1990). Eyeblink conditioning discriminates Alzheimer's patients from nondemented aged. *NeuroReport* **1**, 45–48.

Woodruff-Pak, D. S., Goldenberg, G., Downey-Lamb, M. M., Boyko, O. B., and Lemieux, S. K. (2000). Cerebellar volume in humans related to magnitude of classical conditioning. *NeuroReport* **11**, 609–615.

Woodruff-Pak, D. S., Romano, S., and Papka, M. (1996). Training to criterion in eyeblink classical conditioning in Alzheimer's disease, Down's syndrome with Alzheimer's disease, and healthy elderly. *Behav. Neurosci.* **110**, 22–29.

Woodruff-Pak, D. S., and Thompson, R. F. (1988). Classical conditioning of the eyeblink response in the delay paradigm in adults aged 18–83 years. *Psychol. Aging* **3**, 219–229.

Woodworth, R. S. (1921). "Psychology: A Study of Mental Life." Holt, New York.

Woody, C. D., Yarowsky, P., Owens, J., Black-Cleworth, P., and Crow, T. (1974). Effect lesions of cortical motor areas on acquisition of conditioned eye blink in the cat. *J. Neurophysiol.* **37**, 385–394.

Woody, C. D., and Yarowsky, P. J. (1972). Conditioned eyeblink using electrical stimulation of coronal-precruciate cortex of cats. *J. Physiol.* **35**, 242–252.

Woolf, N. J. (1998). A structural basis for memory storage in mammals. *Prog. Neurobiol.* **55**(1), 59–77.

Xu, B., Gottschalk, W., Chow, A., Wilson, R. I., Schnell, E., Zang, K., Wang, D., Nicoll, R. A., Lu, B., and Reichardt, L. F. (2000). The role of brain-derived neurotrophic factor receptors in the mature hippocampus: Modulation of long-term potentiation through a presynaptic mechanism involving TrkB. *J. Neurosci.* **20**, 6888–6897.

Yeckel, M. F., Kapur, A., and Johnston, D. (1999). Multiple forms of LTP in hippocampal CA3 neurons use a common postsynaptic mechanism. *Nat. Neurosci.* **2**, 625–633.

Yeh, S. H., Mao, S. C., Lin, H. C., and Gean, P. W. (2006). Synaptic expression of glutamate receptor after encoding of fear memory in the rat amygdala. *Mol. Pharmacol.* **69**(1), 299–308.

Yehuda, R., and LeDoux, J. E. (2007). Response variation following trauma. A translational neuroscience approach to understanding PTSD. *Neuron* **56**, 19–32.

Yeo, C. H. (1991). Cerebellum and classical conditioning of motor responses. *Ann. NY Acad. Sci.* **627**, 292–304.

Yeo, C. H., Hardiman, M. J., and Glickstein, M. (1985a). Classical conditioning of the nictitating membrane response of the rabbit. I. Lesions of the cerebellar nuclei. *Exp. Brain Res.* **60**, 87–98.

Yeo, C. H., Hardiman, M. J., and Glickstein, M. (1985b). Classical conditioning of the nictitating membrane response of the rabbit. II. Lesions of the cerebellar cortex. *Exp. Brain Res.* **60**, 99–113.

Yeo, C. H., Hardiman, M. J., and Glickstein, M. (1986). Classical conditioning of the nictitating membrane response of the rabbit. IV. Lesions of the inferior olive. *Exp. Brain Res.* **63**, 81–92.

Yin, J. C., Del Vecchio, M., Zhou, H., and Tully, T. (1995). CREB as a memory modulator: Induced expression of a dCREB2 activator isoform enhances long-term memory in Drosophila. *Cell* **81**, 107–115.

Yin, J. C., Wallach, J. S., Del Vecchio, M., Wilder, E. L., Zhou, H., Quinn, W. G., and Tully, T. (1994). Induction of a dominant negative CREB transgene specifically blocks long-term memory in Drosophila. *Cell* **79**, 49–58.

Ying, S. W., Futter, M., Rosenblum, K., Webber, M. J., Hunt, S. P., Bliss, T. V., and Bramham, C. R. (2002). Brain-derived neurotrophic factor induces long-term potentiation in intact adult hippocampus: Requirement for ERK activation coupled to CREB and upregulation of Arc synthesis. *J. Neurosci.* **22**(5), 1532–1540.

Yovell, Y., Kandel, E. R., Dudai, Y., and Abrams, T. W. (1992). A quantitative study of the Ca^{2+}/calmodulin sensitivity of adenylyl cyclase in Aplysia, Drosophila, and rat. *J. Neurochem.* **59**, 1736–1744.

Yuan, L. L., Adams, J. P., Swank, M., Sweatt, J. D., and Johnston, D. (2002). Protein kinase modulation of dendritic K+ channels in hippocampus involves a mitogen-activated protein kinase pathway. *J. Neurosci.* **22**, 4860–4868.

Zaccardi, M. L., Traina, G., Cataldo, E., and Brunelli, M. (2001). Nonassociative learning in the leech *Hirudo medicinalis. Behav. Brain Res.* **126**, 81–92.

Zakharenko, S. S., Zablow, L., and Siegelbaum, S. A. (2001). Visualization of changes in presynaptic function during long-term synaptic plasticity. *Nat. Neurosci.* **4**, 711–717.

Zalutsky, R. A., and Nicoll, R. A. (1990). Comparison of two forms of long-term potentiation in single hippocampal neurons. *Science* **248**, 1619–1624.

Zamanillo, D., Sprengel, R., Hvalby, O., Jensen, V., Burnashev, N., Rozov, A., Kaiser, K. M., Koster, H. J., Borchardt, T., Worley, P., Lubke, J., Frotscher, M., Kelly, P. H., Sommer, B., Andersen, P., Seeburg, P. H., and Sakmann, B. (1999). Importance of AMPA receptors for hippocampal synaptic plasticity but not for spatial learning. *Science* **284**, 1805–1811.

Zhang, F., Endo, S., Cleary, L. J., Eskin, A., and Byrne, J. H. (1997). Role of transforming growth factor-beta in long-term facilitation in *Aplysia. Science* **275**, 1318–1320.

Zola-Morgan, S. M., and Squire, L. R. (1990). The primate hippocampal formation: Evidence for a time-limited role in memory storage. *Science* **250**, 288–290.

Zucker, R. S. (1972). Crayfish escape behavior and central synapses. II. Physiological mechanisms underlying behavioral habituation. *J. Neurophysiol.* **35**, 621–637.

Molecular and Cellular Mechanisms of Neurodegenerative Disease

Mark R. Cookson, Randy Strong, P. John Hart,

and James L. Roberts

INTRODUCTION

The adult-onset neurodegenerative diseases are a set of disorders that have certain shared characteristics:

- They are all associated with a dramatic loss of neurons in some, but not all, parts of the brain.
- They are all associated with age, older people being more likely to develop them.
- They are all progressive, with minor subtle symptoms developing into disabling and, in some cases, fatal diseases.

What doesn't define neurodegeneration is the actual symptoms, although there are overlaps. At a basic level, this is because each disease affects different regions of the brain (Fig. 20.1). Thus, damage to neurons in the spinal cord leads to motor neuron disease, exemplified by ALS; loss of cells in the basal ganglia underlies movement disorders including Parkinson disease; and diminished function in the cerebral cortex gives rise to personality and memory impairments in the dementing illnesses, which include Alzheimer disease. Because all these diseases have neuronal loss, albeit in different brain regions and with different neurotransmitter groups affected, one could argue that there must be common aspects to the degenerative process. While these three disorders are by no means the only ones associated with the nervous system, they do serve to demonstrate the common underlying molecular and cellular mechanisms as examples of how the material outlined in this textbook leads to a better understanding of disorders of the nervous system.

Neurodegenerative diseases are also often defined by their pathology (Table 20.1). At autopsy, brain samples from patients with neurodegenerative diseases have deposits of aggregated proteins in the few neurons that survive. While neuronal loss is common to Alzheimer, Parkinson, or ALS diseases, protein deposition varies between them and each has a characteristic pathology. Alzheimer disease has plaques, composed of the amyloid-beta peptide, and tangles, made up of tau. The neurons in the midbrain remaining at the end of Parkinson disease have very distinctive inclusions called *Lewy bodies,* composed principally of alpha-synuclein. The characteristic protein deposition in ALS is less well understood, although recent data suggest that a protein called *TDP43* may be present in many cases (Neumann *et al.,* 2006). Because protein deposition is somewhat distinctive, this leads to the proposal that each has a distinct underlying cause, or etiology.

Discussion of the root etiology of neurodegeneration is very difficult because, by definition in some of these diseases, we typically exclude cases where we know the cause. This seems counterintuitive, but there is a good reason for it. For example, many drugs used to treat epilepsy can damage the basal ganglia and lead to a clinical syndrome that has aspects of Parkinson disease. Often, we use the term *parkinsonism* to separate out these cases. The problem with this definitional approach is that we are left with the concept that

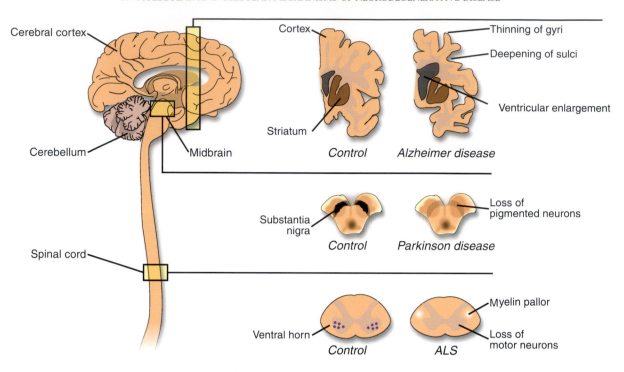

FIGURE 20.1 The different brain areas affected by Alzheimer disease (AD), Parkinson disease (PD), and amyotrophic lateral sclerosis (ALS). The cortex and underlying hippocampus represent major sites of neurodegeneration in Alzheimer disease, resulting in a widening of sulci and enlargement of cerebral ventricles. In PD, loss of the pigmented, dopamine-secreting neurons in the substantia nigra pars compacta depletes dopamine from the substantia nigra and striatum. Degeneration of acetylcholine-secreting motor neurons in the ventral horn of the spinal cord results in loss of innervation of skeletal muscles in ALS. Each of these diseases is associated with abnormally folded proteins: tau and amyloid β (Aβ) in AD; α-synuclein in PD; and copper–zinc superoxide dismutase (SOD1) in ALS.

TABLE 20.1 Genes That Cause Neurodegenerative Diseases

Disease	Genes			Proteins
	Dominant	**Recessive**	**Risk Factors**	
Alzheimer disease	APP	?	ApoE	Aβ
	Presenilin 1–2			Tau
FTDP-17[1]	MAPT[2]	-	?	Tau
Parkinson disease or parkinsonism	SNCA[3]	Parkin	?	α-synuclein
	LRRK2	PINK1		
		DJ-1		
		ATP13A2		
ALS and other motor neuron diseases	SOD1[4]	Alsin[5]	?	
	Senataxin[5]			
	VAPB[5]			

[1]FTDP-17 is frontotemporal dementia with parkinsonism linked to chromosome 17.
[2]MAPT is the gene encoding for the tau protein.
[3]SNCA is the gene name for α-synuclein.
[4]SOD1 mutations can also have recessive inheritance.
[5]These genes cause mixed ALS-like phenotypes.

neurodegenerative diseases are only *sporadic,* a term that means there is no known cause. Although this might be true, and Alzheimer and other diseases might be simple bad luck, it is rather unhelpful in trying to understand the disease in terms that might allow us to develop therapies based on anything other than trying to minimize the effects of symptoms for the patient. To give a historical example, Jenner wouldn't have developed smallpox vaccines unless there was an idea that the causative agent was related to cowpox.

One approach that has had some utility is the identification of genetic forms of common neurodegenerative diseases. Perhaps the best example is Alzheimer, where families with early-onset Alzheimer were studied and found to have two genes that together define a common biochemical pathway for plaque deposition (Hardy and Selkoe, 2002). The genetics of ALS and Parkinson are more complex, and the relationships between familial and sporadic diseases have not been fully resolved. But the point for this chapter is that genetics supports separateness over similarity. Alzheimer has different genes from Parkinson, or from ALS, although there are some fuzzy boundaries with cases with unusual phenotypes that blur clear distinctions.

Here, we will cover a few well-understood diseases to give an idea how neuroscientists approach the problem of neurodegenerative diseases at several levels. This multilayer thinking is helpful because the brain is an inherently complex organ with hierarchical organization. Therefore, we can think about neurodegenerative diseases in terms of genes, at the level of biochemical processes, with respect to the process of cellular dysfunction or framed by brain regions and circuits. Dealing progressively with each of these ways of organizing the available data is an important way to get an overall picture of the disease and may help us clarify some of the issues around similarity and dissimilarity of neurodegeneration. We will first reiterate some of the key clinical aspects of the disease and then discuss genetic, biochemical, and cellular data for Alzheimer, Parkinson, and ALS diseases.

ALZHEIMER DISEASE

The clinical symptoms of Alzheimer disease result from the deterioration of selective cognitive domains, particularly those related to memory. Memory decline initially manifests as a loss of episodic memory, which is considered a subcategory of declarative memory. The dysfunction in episodic memory impedes recollection of recent events, including autobiographical activities. Elucidating the underlying molecular determinants that trigger the disruption of recent episodic memory, and eventually the decline in the other cognitive domains, is among the most crucial unanswered questions in the AD field. In 1907, Alois Alzheimer described two pathological alterations in the brain of a female patient suffering from dementia. These two lesions represent the hallmark pathognomic features of the disease, and their observation during postmortem examination is still required for a diagnosis of AD (Fig. 20.2). Alzheimer described a "peculiar substance" occurring as extracellular deposits in specific brain regions, which are now referred to as *amyloid plaques.* Only in the mid-1980s was it discovered that the plaques consist primarily of aggregates of a small peptide called *amyloid-β* (Aβ). The second lesion described by Alzheimer, neurofibrillary tangles (NFTs), occurs intraneuronally. Shortly after, it was discovered that NFTs are composed of aggregates of the tau protein, which becomes abnormally hyperphosphorylated. Although plaques and NFTs are pathognomic, it would be misleading to create the impression that these are the only significant pathological changes occurring in the AD brain. In fact, numerous other structural and functional alterations ensue, including inflammatory responses and oxidative stress. The combined consequences of all the pathological changes, including the effects of the Aβ and tau pathologies, is severe neuronal and synaptic dysfunction and loss; at the time of death, the brain of a patient with AD may weigh one-third less than the brain of an age-matched, nondemented individual. Understanding the molecular pathways by which the various pathological alterations compromise neuronal function and integrity and lead to clinical symptoms has been a long-standing goal of AD research. Success in developing mouse models that

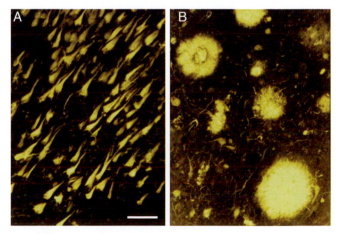

FIGURE 20.2 Photomicrographs of neurofibrillary tangles (A) and amyloid-containing plaques (B) in the hippocampal formation of a patient who died with late-stage Alzheimer disease. Note that the size and morphological characteristics of plaques in B vary widely. Scale bar: 50 É m. Courtesy of P. Hof, Mount Sinai School of Medicine.

mimic various facets of the disease process has greatly facilitated this effort. One recurring theme that has emerged from the study of some, albeit not all, AD mouse models is the occurrence of the Aβ peptide within neurons.

The Genetic Basis of Alzheimer and Related Dementias

A recurrent theme in this chapter is that while most cases are apparently sporadic (i.e., no known cause), the rare families where we can identify a gene defect are extremely important for identifying underlying biochemical pathways.

The genetics of AD are a little simpler than some of the other diseases we will discuss in that the Mendelian variants identified to date are all dominant and highly penetrant. Because these are dominant traits, a single chromosome containing a gene defect is sufficient to pass on the disease from an affected parent to their sons or daughters. Because a parent has two copies of each chromosome, this gives each sibling in a given generation an approximately 50% chance of developing the disease. We will discuss some examples later where this is not true, but for the Alzheimer genes, the proportion of people with a given mutation who develop the disease is high, that is, the trait is almost completely penetrant. The one caveat is that the penetrance is age dependent. For the families discussed here, the disease onset is between 30 and 50 years of age. The age, and the fact that the disease is inherited, is the major distinction between these forms of familial Alzheimer and the sporadic disease. Prior to this age, there are no obvious signs of dementia, and by middle age, essentially all mutation carriers will have developed disease.

The first reported gene for Alzheimer disease was APP, the gene for the amyloid precursor protein (Chartier-Harlin et al., 1991). APP is a type of transmembrane protein at the surface of many cells, including neurons, that is cleaved to form a number of proteoloytic fragments, which will be discussed later in more detail. Many of the early mutations were named after the place where they were found, such as the London or Swedish mutations. Since the initial discoveries, several more APP mutations have been found in different families. APP mutations cluster in the region of the protein around the plasma membrane where many of these cleavage events occur. Thus an immediate hypothesis (Hardy and Higgins, 1992) is that APP mutations alter how the protein is processed, which, as we will see, is very likely to be true. Most mutations are single amino-acid changing point mutations.

The APP gene is on the distant end of chromosome 22, within the region that is famously duplicated in Down syndrome (the so-called Down syndrome critical region). Interestingly, Down syndrome patients have a similar amyloid deposition to that found in Alzheimer (Ikeda et al., 1989). Recently, duplication mutations in APP have been found that also cause Alzheimer disease or vascular dementia (Rovelet-Lecrux et al., 2006). This reinforces the apparently strange observations that Down syndrome and Alzheimer might have some common pathologies—too much APP can lead to dementia in both cases. This is starting to be a general principle in neurodegeneration as a similar dosage phenomena has recently been reported for Parkinson disease.

A few years after APP mutations had been identified, mutations in a second gene were found to cause a similar early-onset familial Alzheimer disease. Because the gene was uncharacterized, the group that discovered it named it presenilin1 (gene name PS1) for the presenile dementing illness (Sherrington et al., 1995). Mutations in PS1 are more numerous than APP mutations. There is one homologous gene, PS2, in the human genome, and mutations in PS2 are a very rare cause of early-onset Alzheimer (Rogaev et al., 1995). Therefore, there are three clearly defined mutations (APP, PS1–2) that cause Alzheimer disease. Like APP mutations, PS1–2 mutations are generally single-point mutations that change amino acids, although there is a very interesting mutation that removes exon 9 of the gene (△Ex9) that is associated with unusual pathology and clinical presentation (Crook et al., 1998).

At this point, we should discuss mutations in another gene, MAPT (Hutton et al., 1998). MAPT codes for a protein, tau, that was suspected to be involved in Alzheimer as its part of the defining pathology of the illness, namely the formation of tangles in neurons. But, strictly speaking, MAPT–tau mutations do not cause Alzheimer. The disease associated with MAPT mutations was named by committee, which is perhaps why it was given the unwieldy but accurate nomenclature of "frontotemporal dementia with parkinsonism linked to chromosome 17," or FTDP17. FTD is a different "flavor" of dementia, with more involvement of the frontal lobes than is normally seen in Alzheimer. Because the frontal lobes are involved in personality and suppression of inappropriate behaviors, a patient with FTD will often have more extreme behavioral problems and fewer memory problems than a person with Alzheimer disease. Parkinsonism, which will be defined more precisely in the section "Parkinson Disease," is a prominent feature of many cases. This discussion brings up the complicated relationship between genes and phenotype, but as we will discuss later, the biochemical relationship between the amyloid processing mutations (APP and PS1–2) and MAPT is very interesting and probably critical for understanding Alzheimer disease (reviewed in Golde *et al.*, 2006). Again, most MAPT mutations are point mutations, but there are also a relatively large number of mutations around intron–exon splice junctions that do not change amino acid sequence but instead alter the inclusion of exon 10 of the gene (reviewed in D'Souza and Schellenberg, 2005).

At a few points in this discussion we have skirted the issue of the relationship between genetic and "sporadic" forms of Alzheimer disease. To re-emphasize, *sporadic* in this context means we have no idea about etiology. Ignorance is often reinterpreted in review articles as evidence for a "gene–environment" interaction, even though a truly sporadic disease is also logically compatible with simple bad luck. For Alzheimer, we have one strong piece of evidence that the "sporadic" disease is not truly sporadic, as there are genetic factors that clearly contribute to the lifetime risk of getting the disease.

Before discussing what these genes are, it is worth discussing what genetic risk factors are. In a family where there is a disease or other phenotype (red hair, for example), we can be sure that there is a gene, because inheritance follows either a dominant or recessive pattern, examples of which are given in this text. Therefore, these genes pass one of the two acceptable tests for whether a gene causes disease, segregation for the gene with phenotype. The other test for causation of a gene variant with disease is association. Under an association model, the gene variant does not need to be present in each case of the disease or absent in all unaffected persons, but rather works at a population level. That is to say, there are more or less affected people with a given allele (a gene variant) than unaffected people. Associated genes can either act as risk alleles or be protective, depending on whether possessing a specific variant increases or decreases the chance that a person gets the disease. The problem with assessing association is that people are genetically diverse, and therefore, by chance, out of millions of bases of DNA, each person has thousands of variants that are *not* associated with any disease or other phenotypes. Worse, human populations with distinct heritages differ between each other at random, and some populations have more or less genetic diversity than others; the African populations are more diverse, in part because they are more ancient, than peoples on isolated islands like Sardinia or Iceland (Li *et al.,* 2008). What we are ideally looking for in association studies is an allele that is found in many more cases than controls within a given population but the allele also has a similar effect in many populations.

Many association studies fail these criteria, but an interesting story here is the ApoE effect in Alzheimer disease. ApoE is a lipid metabolism protein, and there are three common variants in most populations: the ε2, ε3, and ε4 alleles, with ε3 being the most common. Many studies have, in aggregate, shown the ε2 is a protective allele, while ε4 is a risk allele for Alzheimer (reviewed in Cedazo-Mínguez, 2007). ApoE ε4–ε4 homozygotes have the highest risk of developing AD because they have inherited two risk alleles. The strength of association can be measured by estimating the odds ratio: the increased risk for a disease if one inherits a specific allele relative to the general population. ApoE ε4 has an odds ratio of about 4 in most populations, which is a relatively strong effect. This has two practical implications. First, because it is a strong effect, it is relatively easy to detect even with modest numbers of cases and controls analyzed, and most studies of the risk of ApoE have noted a positive association. Second, it means that someone with an ε4 allele is at increased risk of developing Alzheimer disease, but not all will—the odds ratio is high but not infinite.

Therefore, in an apparently "sporadic" disease we have one strong risk factor that contributes to risk across the population, outside of the rare families with inherited disease. Are there other genes that make similar contributions? The answer comes from two sources. Since the discovery of ApoE, many groups have suggested genes that might be associated with lifetime risk. Many of these are not well replicated across additional studies, either because the initial report was wrong or because the effect is rather weak and hard to detect without assaying large numbers of samples or combining the results across studies. Some good resources are now available for this type of "meta-analysis"; a great one for Alzheimer disease is the Alzgene database (http://www.alzgene.org). Looking across many studies, small risk factors emerge for some of the genes involved in APP processing, that is, APP and PS1–2, along with some of the genes thought to be involved in downstream processing of APP fragments (Bertram *et al.,* 2007).

If genetics is the kingdom of the blind, having no preset goal other than finding that a gene may be there, it isn't clear if the one-eyed person with a hypothesis is really king. Maybe a gene will be associated with disease, or maybe not, and if most risk factors are rather modest (odds ratios less than 1.5), then finding them requires large numbers of samples. A recent trend in human genetics has been to develop tactics for whole-genome association studies. In this approach, one takes randomly placed variants (usually single nucleotide polymorphisms, or SNPs) along the genome and assays all of them simultaneously in a given set of cases and controls. Such approaches are statistically fraught, but the basic principle is that the larger the case series, the smaller the odds ratio detectable, with calculations suggesting that thousands, perhaps tens of thousands, of samples are needed to identify samples with an odds ratio below 1.5. This type of approach has been performed in Alzheimer and shows that the major risk factor is perhaps four times as strong as any other gene variant (Coon *et al.,* 2007). Therefore, there is one strong genetic basis for sporadic Alzheimer and perhaps a number of weaker ones.

Biochemical–Cellular Basis of Alzheimer Disease

The genetics of the familial forms of a neurodegenerative disease often help us to understand the biochemical and cellular basis underpinning the disease. With Alzheimer, we have seen that several different, seemingly unrelated, proteins can lead to the disorder as a result of mutations in the genome. However, the commonality is that aggregates of one form or another, be they of hyperphosphorylated tau or aggregated amyloid form that seemingly cannot be cleared from the cell, lead to a breakdown of cellular function and eventual apoptosis. Hyperphosphorylated tau leads to aggregates that result in what are referred to as *tangles,* and aggregated β-amyloid is a constituent of the amyloid plaque (Fig. 20.2)

The primary constituents of amyloid plaques are the protein β-amyloid proteins (Aβ), which are composed of proteolytic fragments of APP, which is a widely expressed transmembrane protein in brain. These are Aβ1-40, 1-42, 11-40, and 11-42, which are generated by cleavages of two aspartyl proteases, β- and γ-secretase within the interior of the APP (Cai *et al.,* 2001; Iwatsubo *et al.,* 1994). Part of the sequence of these short peptides is derived from the transmembrane spanning region of the APP, which by its nature is highly hydrophobic and leads to its predisposition to aggregate and form insoluble complexes, as shown in Figure 20.3. These peptides can be found both intracellularly and extracellularly (Fig. 20.4). When located extracellularly, they can aggregate and form the basis of an amyloid plaque, which will grow as damage to surrounding neuropil grows and cells die releasing other aggregated proteins. Intracellularly, they can interfere with numerous cellular functions, particularly when present in smaller aggregates, referred to as *Aβ oligomers.* They can enhance the formation of tangles—the other hallmark of Alzheimer—cause mitochondrial stress, disrupt synaptic function, and damage proteasome function, the main intracellular system for clearing damaged–misfolded protein from the cell. The proteasome also plays a major role in the pathology of

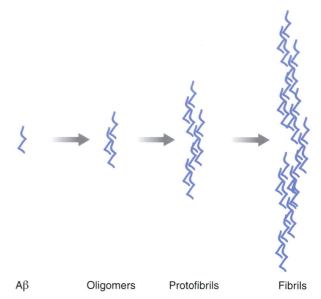

| Aβ | Oligomers | Protofibrils | Fibrils |

FIGURE 20.3 The Aβ fragment derived from the APP protein forms progressively larger aggregates with time. It is believed that the intermediate oligomer form is the most toxic to the cell.

Parkinson disease, as will be discussed later in this chapter.

The brain circuitry that is disrupted by the death of neurons is complex and involves many circuits important in learning, memory, and cognitive performance. These include the hippocampus, entorhinal cortex, limbic cortex, basal forebrain cholinergic

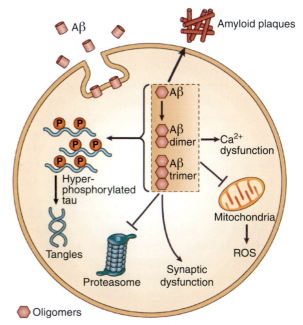

FIGURE 20.4 The toxic nature of the Aβ oligomers derives from numerous cellular functions disrupted, including mitochondria, resulting in reactive oxygen species (ROS) production, disruption of synaptic function, and inhibition of proteasome function.

system, and neocortex (Braak *et al.*, 1999). Thus, differences in severity and the progression of the memory loss occur depending upon which circuits become lost and in which order, ultimately resulting in complete long-term memory loss.

PARKINSON DISEASE

Parkinson disease is a progressive neurodegenerative disease primarily characterized by the loss of neurons that use dopamine as a neurotransmitter. Their cell bodies are located in the substantia nigra pars compacta, and their axon terminals form synapses on neurons in the caudate and putamen, collectively called the *striatum* (Fig. 20.1). Although Parkinson disease was described as early as 4500 to 1000 BC in the ancient Ayurvedic medical literature native to the Indian subcontinent, direct mention of the disease was sparse, referring mainly to related symptoms including tremors, which went by several names. In that literature it is most commonly called *Kampavata*. It wasn't until 1817 that a comprehensive picture of the disease was described in a monograph entitled "An Essay on the Shaking Palsy" by James Parkinson, an English physician, paleontologist, and geologist. The three cardinal signs of the disease include tremor, rigidity, and bradykinesia. *Tremor* refers to a shaking of the extremities, often manifesting as a "pill-rolling tremor," in which the forefinger and thumb are rubbed together in a movement reminiscent of the motion made by an early pharmacist compounding a remedy into a pill. *Bradykinesia* refers to a slowness of movement characterized by difficulty in initiating walking or reaching for an object. *Rigidity* is manifested as a resistance of the limbs to being manipulated by the clinician. At least two of these three symptoms are required for a diagnosis of Parkinson disease. Other signs of the disease that are not necessarily diagnostic include expressionless or "masked" face, difficulty in speaking above a whisper, difficulty in swallowing, writing in very small letters or micrographia, flexed posture, and shuffling gait.

The symptoms of the disease are associated with marked depletion of dopamine from the substantia nigra and from the caudate and putamen (together called the *neostriatum*) (see Fig. 20.1), which together with the globus pallidus, the subthalamic nucleus, the pedunculopontine nucleus, and the pars reticulata of the substantia nigra are collectively called the *basal ganglia*. At death, the loss of dopamine in the neostriatum is greater than 90%. Less than a 50% depletion of dopamine is asymptomatic, and symptoms are manifested when around 70% of dopamine is lost

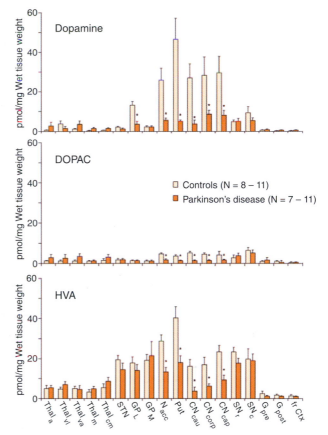

FIGURE 20.5 Content of dopamine and its metabolites in affected and unaffected regions of the brain. Note that the loss of dopamine is greatest in the caudate and putamen. Also note that the degree of the loss of dopamine in the caudate and putamen is greater than that of its HVA metabolite.

from the striatum (Hornykiewicz, 1998) (Fig. 20.5). The fact that at least 50% of the dopamine is lost before symptoms appear is related to the remarkable ability of the surviving neurons to increase the synthesis and turnover of dopamine in response to the loss of their neighbors, as indicated by an increased ratio of the dopamine metabolites DOPAC and HVA to dopamine (Fig. 20.5 and Fig. 20.6) (Hornykiewicz, 1998). The severity of the loss of dopamine correlates best with bradykinesia. The deficiency of dopamine is symptomatic regardless of whether the disease is from known or unknown causes. Replacement of dopamine by treatment with L-dopa (Fig. 20.6) is effective early in the disease, but it does not treat the underlying disease process and eventually the treatment fails.

To understand how the loss of dopamine produces the motor symptoms of Parkinson disease, it is necessary to understand the connectivity of the basal ganglia. The connectivity of the basal ganglia is illustrated

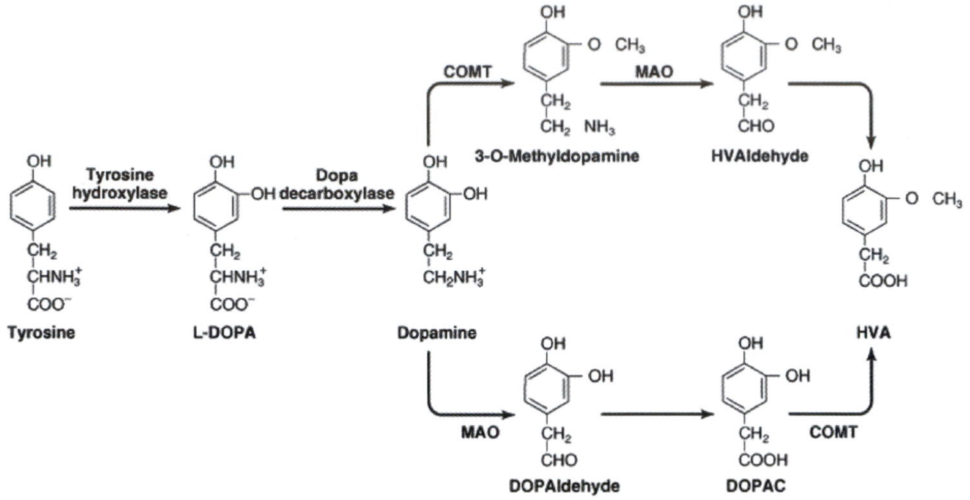

FIGURE 20.6   The biosynthetic pathway for dopamine and its metabolites.

in Figure 20.7. The neostriatum is the main input portion of the basal ganglia receiving input from virtually all portions of the cortex, but mainly from the motor and prefrontal cortices. The output circuitry of the basal ganglia is often divided into two major pathways, the direct pathway and the indirect pathway. The neurons in the striatum that make up the different pathways have two different types of receptors, although each releases GABA as a neurotransmitter. The neurons that contribute to the direct pathway express the D_1 receptor subtype, which responds to dopamine stimulation by increasing their activity, while the neurons that contribute to the indirect pathway respond to stimulation of the D_2 receptor subtype by inhibiting their activity. As shown in Figure 20.5, the loss of dopamine results in an imbalance in the output of the two pathways, in the end causing decreased stimulation of neurons in the motor cortex and ultimately the symptoms of Parkinson disease.

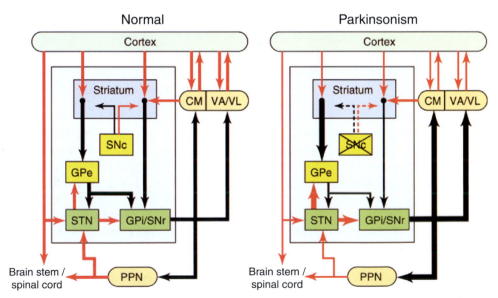

FIGURE 20.7   The role of the direct and indirect pathways in the symptoms of Parkinson disease. Red lines represent the influence of neurotransmitters acting at excitatory neurotransmitter receptors, and black lines represent the effect of neurotransmission at inhibitory receptors. Loss of dopamine in the striatum reduces neurotransmission at excitatory and inhibitory dopamine receptors on neurons contributing to direct and indirect pathways, respectively. This results in both increased excitatory and decreased inhibitory influences on the inhibitory GABA neurons in the internal segment of the globus pallidus. The result is a greater inhibitory influence of the basal ganglia on excitatory thalamic neurons that project to motor regions of the neocortex.

About 85% of the cases of Parkinson disease are of unknown origin. The known causes include manganese poisoning, encephalitis, carbon monoxide poisoning, and recurrent head trauma. Due to improvements in public health, no new cases of postencephalitic Parkinson disease have occurred since the 1960s. The rest of the cases are referred to as *idiopathic,* in keeping with their unknown origin. Although several gene mutations have been identified in families in which Parkinson disease is inherited as discussed elsewhere in this chapter, they account for relatively few cases worldwide. Most cases are sporadic and most likely related to a combination of environmental exposure and genetic susceptibility. As discussed earlier in this chapter, the most definite risk factor for Parkinson disease is old age. The disease is rare before the age of 30 and only 4–10% and of the cases occurs before age 40. The mean age of onset of Parkinson's disease is 62.4 years. Other risk factors include herbicides, pesticides, heavy metals, proximity to industry, rural residence, well water, and repeated head trauma. Alcohol and nicotine are sometimes cited as factors that reduce the risk of Parkinson disease because in some epidemiological studies patients with Parkinson disease are less likely to drink or smoke. However, the tendency not to engage in these behaviors is more likely related to the loss of dopamine from the mesolimbic dopamine system, which is involved in feelings of pleasure or reward from smoking and drinking.

There is growing agreement among scientists that both genetic susceptibility and environmental agents may play a role in the disease. It has been recognized that idiopathic Parkinson disease is associated with impaired mitochondrial function (Greenamyre et al., 2001; Haas et al., 1995). Most notably, diminished NADH dehydrogenase (complex I) activity is observed in various cell types throughout the body (Greenamyre et al., 2001b; Haas et al., 1995b). The reduction in complex I activity may be seen early in Parkinson disease and was detected in double-blind studies that controlled for drug treatment (Haas et al., 1995c). In the early 1980s, young drug users began developing Parkinson disease after self-administration of 1-methyl-4-phenyl-1, 2,5,6-tetrahydropyridine (MPTP), a by-product of the illicit manufacture of the narcotic meperidine. Studies of acute or short-term administration of this complex I inhibitor provided many clues to the potential mechanisms underlying environmental exposure in PD. However, hallmark manifestations of PD such as the presence of Lewy bodies were notably absent. Therefore, the significance of complex I dysfunction in PD was unclear, except that it might contribute to increased oxidative stress. A significant advancement in understanding the role of complex I deficiency in Parkinson disease was made when Greenamyre and coworkers showed that the complex I inhibitor rotenone, administered systemically to rats, reproduced hallmark features of the disease including degeneration of substantia nigra dopamine neurons and Lewy body–like inclusions staining for ubiquitin and α-synuclein in surviving dopamine neurons (Betarbet et al., 2000). Recently, chronic systemic administration of MPTP for 4 weeks by osmotic minipump was reported to produce parkinsonian pathology in mice, like that produced by rotenone, including Lewy body–like inclusions staining positively for α-synuclein and ubiquitin in substantia nigra and locus ceruleus (Fornai et al., 2005). More recently, oral administration of rotenone to mice for 30 days produced similar PD pathology, including inclusions staining positively for α-synuclein (Inden et al., 2007). Thus, chronic complex I inhibition, with structurally dissimilar inhibitors in two different species, recapitulates characteristic behavioral and pathological manifestations of Parkinson disease.

The pathway by which chronic complex I inhibition leads to the characteristic pathology of Parkinson disease remains unclear; however, several hypotheses have emerged based on data from animal and cellular models that implicate increased oxidative stress. Complex I inhibitors increase hydrogen peroxide (H_2O_2) production; glutathione peroxidase removes H_2O_2 using glutathione (GSH) as a substrate. Perry et al. (1982) reported that in presymptomatic Lewy body–positive control brains, the substantia nigra showed depletion of glutathione content. Other laboratories have reported that glutathione was 30% to 47% lower in patients with PD (Fitzmaurice et al., 2003; Jenner et al., 1992; Perry et al., 1982). Furthermore, it has been reported that glutathione peroxidase 1 (Gpx1) plays a crucial role in protecting cells against oxidative damage by reducing the level of H_2O_2 and therefore decreasing the production of the reactive hydroxyl radicals (Halliwell, 1992). More recently, Ridet and coworkers (2006) reported that the lentivirus-mediated expression of Gpx1 in nigral dopaminergic neurons caused a significant neuroprotection after an intrastriatal injection of 6-hydroxydopamine (6-OHDA) in mice. Recently, it was reported that the expression of glutathione peroxidase 1 is notably reduced in the SNpc of PD patients (Duke et al., 2006).

Aside from free radicals, the most destructive sources of oxidative stress are biogenic aldehydes. Unlike free radicals, aldehydes possess half-lives ranging from a few hours to days, allowing the aldehyde products to accumulate at the site of injury or disease, injuring neighboring healthy cells and slowly but progressively enlarging the lesion. Complex I inhibitors also reduce the ability of aldehyde dehydrogenases to remove toxic aldehydes by reducing the availability of NAD+, a cofactor required for aldehyde dehydrogenase activity. It has been reported that Parkinson disease is associated with elevated levels of 3,4-dihydroxyphenylacetaldehyde (DOPAL), a neurotoxic product of dopamine metabolism by monoamine oxidase (MAO; Fig. 20.6) (Burke *et al.*, 2003; Castellani *et al.*, 2002b; Galter *et al.*, 2003; Mandel *et al.*, 2005), and 4-hydroxynonenal (4-HNE), the major end-product of lipid peroxidation. The available evidence suggests that the elevation of biogenic aldehydes in PD results, at least in part, from deficiency in ALDH1. There are three lines of evidence that implicate ALDH1 in the pathogenesis of PD. First, ALDH1, a cytosolic enzyme, has been found to be specifically and strongly expressed by dopaminergic neurons in the substantia nigra and ventral tegmental area of PD patients (Galter *et al.*, 2003). Second, high expression of mRNA for Aldh1a1, the rodent homolog of human ALDH1, is found in developing mesencephalic dopamine neurons of both C57BL/6 mice and Sprague-Dawley rats (Westerlund *et al.*, 2005), suggesting a role in aldehyde metabolism in these neurons. Third, Galter *et al.* (2003) reported significantly decreased levels of ALDH1 expression in surviving dopamine cells of the substantia nigra of PD patients. The selective expression of ALDH1 in midbrain dopamine neurons and the loss of its expression and activity may help explain the selective degeneration of dopamine neurons in PD.

Lewy bodies are cytoplasmic inclusions found in surviving dopamine neurons and are a pathological hallmark of PD, although they are also found in other neurological diseases (Fig. 20.8). A major constituent of Lewy bodies is α-synuclein, a protein that is believed to be involved in dopamine release. Formation of α-synuclein aggregates is proposed to be a crucial event in the pathogenesis of Parkinson disease. Soluble oligomers of α-synuclein have been identified and proposed to be intermediates during fibril formation, and these, or related aggregates, may constitute the toxic element that triggers neurodegeneration (Castellani *et al.*, 2002a; Lashuel *et al.*, 2002; Mytilineou *et al.*, 2004). Inhibition of proteasome activity sensitizes dopamine neurons to these protein alterations and increases their susceptibility to oxidative stress (Mytilineou *et al.*, 2004). Oxidatively modified α-synuclein has been isolated from cells after complex I inhibition by rotenone (Mirzaei *et al.*, 2006). Recent studies show that lipid peroxidation products, 4-HNE and malondialdehyde, and the dopamine metabolite, DOPAL, promote oligomerization of α-synuclein (Trostchansky *et al.*, 2006; Dalfo *et al.*, 2005; Qin *et al.*, 2007; Galvin, 2006). Analysis of postmortem human tissue of patients with diffuse Lewy body disease identified malondialdehyde and 4-HNE adducts of α-synuclein (Dalfo *et al.*, 2005). Others report that 4-HNE modifies α-synuclein at a sequence that includes His[50], a site important in copper-mediated fibril formation (Trostchansky *et al.*, 2006). Moreover, 4-HNE-modified oligomers are neurotoxic to cells in culture (Qin *et al.*, 2007). In addition, DOPAL was recently reported to trigger α-synuclein aggregation in living dopaminergic cells, suggesting a mechanism for the selective death of dopamine neurons in PD (Galvin, 2006).

Genetic Basis: The PARK Loci, PD, and Parkinsonism

If the genetics of Alzheimer disease seemed straightforward, whether PD is really a genetic illness is a more complex question. In fact, PD was considered to be definitively *not* genetic until about 1995, when the first gene for a Mendelian disorder related to PD was cloned (Polymeropoulos *et al.*, 1997). The gene in question is NACP, which codes for a small protein expressed at high levels in neurons, α-synuclein. Mutations in NACP–α-synuclein are rare, with three point mutations being reported. A53T (Polymeropoulos *et al.*, 1997) and E46K (Zarranz *et al.*, 2004) are definitively pathogenic, but A30P (Krüger *et al.*, 1998) is found in one small German family, and the evidence for segregation is a little weak. As for APP, multiplication mutations around the NACP genomic region are found, either triplication (Singleton *et al.*, 2003; i.e., twice the normal gene dosage) or duplication (Chartier-Harlin *et al.*, 2004; 1.5 times normal gene dosage) of a normally sequenced gene.

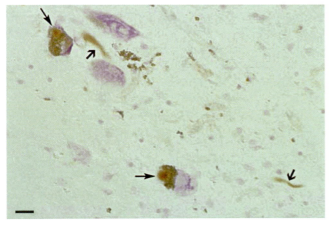

FIGURE 20.8 Photomicrograph of dopamine neurons containing Lewy bodies in the brain of a patient with Parkinson disease. The Lewy bodies are darkly stained intracellular inclusions indicated by the arrows.

Like APP, α-synuclein is important for our understanding of PD in general because it represents the major deposited protein in the characteristic pathology of PD, the Lewy body (Spillantini *et al.*, 1997). What is particularly interesting here is that the disease defined by NACP mutations is not "simply" Parkinson disease. In fact, many people with mutations that affect α-synuclein have a broader disease that involves cortical areas and might be better described as diffuse Lewy body disease (DLBD). Therefore, α-synuclein defines a range of diseases with Lewy body deposition in a number of brain regions from the basal ganglia to the cerebral cortex. Interestingly, the genetic forms suggest the basal ganglia are affected at lower expression levels than the cerebral cortex, comparing duplication and triplication cases. This mirrors the proposed distribution of Lewy bodies over the time course of the sporadic disease proposed by Braak and coworkers (Braak *et al.*, 2003). About a third of all sporadic Parkinson disease patients eventually become demented, presumably because of later cortical involvement.

The other dominant gene for Parkinson disease is LRRK2, which codes for the protein leucine-rich repeat kinase 2 (Paisán-Ruíz *et al.*, 2004; Zimprich *et al.*, 2004). The LRRK2 protein has a complicated domain structure, with several protein–protein interaction domains, including the leucine-rich repeats for which it is named and two enzymatic domains, a GTPase and a kinase. Mutations are spread out throughout the domain structure, with a particularly common mutation in the kinase domain (G2019S) that is present in 1–30% of all apparently sporadic Parkinson depending on the population (reviewed in Cookson *et al.*, 2005). Depending on which mutation is examined, LRRK2 mutations can have a reduced penetrance; thus there are cases that are clinically unaffected and there are cases that (by chance) have some other unusual pathology. The disease caused by LRRK2 mutations usually have Lewy bodies, but some cases have neuronal loss in the absence of obvious protein deposition, and there are cases with unusual inclusion body pathology (Funuyama *et al.*, 2005; Zimprich *et al.*, 2004). These observations tell us two things. Because many cases have Lewy body pathology, there is likely to be a relationship between LRRK2 and synuclein. But there isn't a required connection between neuronal loss in the nigra and the deposition of synuclein into Lewy bodies, which would be a surprising conclusion if we had only NACP mutations to interpret.

Next come three recessive genes associated with an early-onset form of "Parkinson disease." Here, inheritance of disease requires the presence of two mutant alleles, which generally code for nonfunctional versions of the proteins. Parkin was the first discovered of these and is the numerically most common recessive gene for PD known to date (Kitada *et al.*, 1998). PINK1 mutations (Valente *et al.*, 2004) are a little less common, and DJ-1 mutations (Bonifati *et al.*, 2003) are vanishingly rare. Although there are point mutations in all three genes, there are also genomic deletions and truncating or other mutations that disrupt the sequence of the gene. In fact, some mutations in DJ-1 and PINK1 work in a rather trivial way, by causing the protein to become very unstable in the cell, thus in these cases there is no protein (Beilina *et al.*, 2005; Miller *et al.*, 2003). One way or another, all the convincing recessive mutations known to date cause a loss of protein function.

Several aspects distinguish recessive gene mutations from the dominant counterparts. Recessive mutations produce an early-onset disease, exceptionally from the age of midteens but rarely over the age of 50, that is very mild in comparison to either dominant genes or sporadic Parkinson. Recessive cases tend to respond very well to dopamine-replacement therapy and can have features like foot dystonia that would be unusual in the setting of sporadic Parkinson disease. Recessive cases only rarely develop dementia, in contrast to this being a common event in NACP mutations and unfortunately frequent for sporadic PD. Finally, although autopsy studies are limited, cases with parkin mutations are not generally associated with Lewy body formation. Because of these differences, the term *recessive parkinsonism* is often preferred to *Parkinson disease* for these cases.

It isn't clear, therefore, if the recessive genes tell us much about sporadic Parkinson, although the common phenotype of loss of nigral neurons suggests that understanding the functions of the recessive genes might help us understand why it is some neurons are more vulnerable than others to specific stressors. In fact, given that neuronal damage is *more* specific in recessive parkinsonism, with no cortical involvement identified to date, one can argue that these genes are more informative about why some types of cells are so readily lost.

There are several rare genes that have been suggested to be causal for PD but which, for various reasons, are not clear-cut. Heterozygous mutations in UCHL1 were reported in a pair of siblings, but no other mutations in the same gene have ever been found, making them uncertain. Further, if we assume the mutations are dominant, there must be one parent who carries the same mutation and neither was reported to be affected, implying either the mutations are not real or they have very low penetrance (Leroy *et al.,* 1998). Similarly, mutations in SNCAIP1 (Marx *et al.,* 2003) and NR4A2 (Le *et al.,* 2003) have been reported, but these do not show convincing segregation with phenotype and the number of variants is low, leaving these two genes ambiguous. Mutations in ATP13A2 were originally described as associated with Kufor–Rakeb syndrome, an unusual recessive phenotype where parkinsonism is a small part of the clinical spectrum (Ramirez *et al.,* 2006). However, some cases with a more typical presentation of recessive early-onset parkinsonism have been identified (Di Fonzo *et al.,* 2007), suggesting that this may actually be a gene of interest for understanding parkinsonism.

Finally, it is worth discussing genetic risk factors for apparently sporadic Parkinson. There are simply too many candidates to discuss here, with many studies proposing gene variants that have a moderate association with PD. None reported to date has as large an effect as ApoE in Alzheimer and many have not been reproducible. As for Alzheimer, whole genome association studies have the potential to answer this question by looking in a relatively unbiased fashion at all gene variants in a large series of cases and controls. Recent studies that have done this have suggested several possible SNPs that might be associated with Parkinson disease (Fung *et al.,* 2006; Maraganore *et al.,* 2005), but these still need confirmation. What is perhaps more interesting is what these studies have not found: there are no single genetic risk factors of the same magnitude as the ApoE effect in Alzheimer disease.

AMYOTROPHIC LATERAL SCLEROSIS

Amyotrophic lateral sclerosis (ALS, Lou Gehrig's disease, motor neuron disease) is the most common adult motor neuron disease. First described in the late nineteenth century by Jean-Martin Charcot (Charcot and Joffroy, 1869), the hallmark of ALS is the progressive death of the neurons leading to spasticity, hyperreflexia, muscle atrophy, and paralysis (Mulder *et al.,* 1986). Death generally occurs between 1 and 5 years of symptom onset, most often from respiratory failure. The majority (~90%) of ALS cases are sporadic, with no family history of the disease, while the remaining cases are familial, with a genetic lesion passed from generation to generation (reviewed in Bruijn *et al.,* 2004, and in Valentine *et al.,* 2005). Clinically, sporadic and familial forms of ALS are nearly indistinguishable, suggesting that the underlying molecular causes for the two forms of the disease may be related.

Genes Causal for ALS and Other Motor Neuron Diseases

The genetics of inherited motor neuron disease, exemplified by ALS, are more like Parkinson than Alzheimer disease, with several genes shown or proposed to be causative but with different modes of inheritance and some uncertainty about how we classify these diseases.

Mutations in Cu–Zn superoxide dismutase (SOD1), a major cytosolic antioxidant protein (Fig. 20.9), were

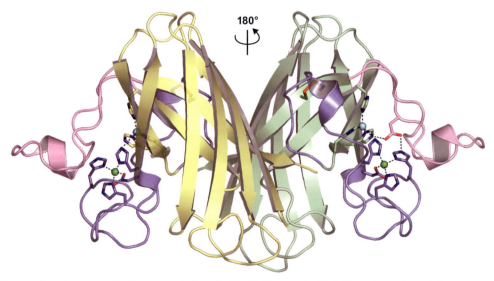

FIGURE 20.9 Human Cu–Zn superoxide dismutase (pdb code 1AZV; Hart *et al.,* 1998). The relationship of the two monomers is indicated. Intrasubunit disulfide bonds are shown as orange sticks; the metal-binding loops (loop IV and VI) are shown in blue and pink, respectively. Copper and zinc ions are shown as cyan and green spheres, respectively.

the first reported autosomal-dominant cause of ALS (Rosen *et al.,* 1993). SOD1 is a small but abundant protein, and mutations have now been found throughout the protein. Most are point mutations, although a few frame shift, and truncating mutations have been found. The disease is similar to sporadic ALS, with the exception that some families have earlier onset than would be typical for sporadic ALS, although this is variable.

One of the very interesting things about SOD1 mutations is that they are not always dominantly inherited. For example, the D90A variant is associated with recessive inheritance in Scandinavian families but is dominant in other populations (Jonsson *et al.,* 2002). Why this should be is simply not clear, but one hypothesis is that a nearby gene acts as a protective factor in some populations (Al-Chalabi *et al.,* 1998). Such a hypothesized gene has not been found.

Two further dominant genes, Senataxin and VAPB, cause diseases where ALS is part of the clinical picture. Senataxin genes can cause an autosomal-dominant juvenile disease with both upper and lower motor neuron damage (Chen *et al.,* 2004). However, in other families, Senataxin mutations cause autosomal recessive ataxia, that is, damage to other neuronal groups with a different mode of inheritance (Moreira *et al.,* 2004). Mutations in VAPB are all dominant but the clinical picture is quite variable, with some families showing a severe ALS-like phenotype, with both upper and lower motor neurons affected, but others have upper motor neuron dysfunction only (Dierick *et al.,* 2008; Nishimura *et al.,* 2004). Therefore, VAPB mutations cause damage to motor neurons, but the actual presentation can be variable, affecting different groups of neurons. Although, because the rarity of some of the families may raise questions about whether the mutations are really pathogenic, at face value these examples illustrate how variable clinical presentation can be for apparently simple dominant mutations. Similar considerations probably also apply to parkinsonism, which can be caused by mutations in genes more typically associated with cerebellar damage such as spinocerebellar ataxia genes 2 and 3 (Gwinn-Hardy *et al.,* 2000, 2001).

Mutations in ALS2–alsin are associated with recessive motor neuron disorders (Hadano *et al.,* 2001; Yang *et al.,* 2001). Mutations generally truncate the protein and cause loss of normal alsin function. Several families have now been studied, and the disease is quite variable in terms of whether upper or lower motor neurons are involved, thus leading to a number of terms being used, including *juvenile ALS, juvenile primary lateral sclerosis,* or *infantile ascending spastic paraparesis.* If we include alsin in the "ALS" genes, we

might therefore also consider genes that cause lower (e.g., Dynactin; Puls *et al.,* 2003) or upper (e.g., Spastin, Atlastin, Paraplegin; reviewed in James and Talbot, 2006) motor neuron diseases. Most of these are clinically distinct from sporadic ALS in number and can express as very different phenotypes closer to Charcot-Marie-Tooth than ALS, although the same can be true for alsin. What is probably most interesting about this set of genes is what they say in aggregate about the processes that maintain either upper or lower motor neurons, which will be discussed later.

As for Parkinson disease, the genetic risk factors for "sporadic" ALS are currently unclear. Many association studies based on hypotheses about what the disease process might be have been performed, and some have found risk factors with small odds ratios. Whole genome approaches are in their infancy, but as for PD, there do not seem to be ApoE-like risk factors in sporadic ALS. Overall, while there are clearly genes for familial ALS, how much genetics contributes to sporadic ALS is less clear (reviewed in Schymick *et al.,* 2007).

SOD1-Linked ALS Is Not a Loss-of-Function Disease

In the early 1990s, 11 families with histories of ALS were discovered to have dominant mutations in the gene encoding copper–zinc superoxide dismutase (SOD1) (Deng *et al.,* 1993; Rosen *et al.,* 1993), a ubiquitous 32 kDa homodimeric enzyme that detoxifies reactive superoxide anion (O_2^-), a normal by-product of respiration and fatty acid oxidation, to molecular oxygen and water [$2O_2^- + 2H^+ \rightarrow H_2O_2 + O_2$] (Fridovich, 1989). SOD1 is estimated to comprise between 0.1 and 2.0% of the detergent-soluble protein in spinal cord and brain (Marklund, 1984; Pardo *et al.,* 1995), and this abundance presumably reflects the plentiful superoxide generated by these metabolically active (respiring) tissues. In the years subsequent to the discovery of the SOD1–ALS link, the number of distinct ALS–SOD1 mutations–families has risen to ~115, falling at about 65 of 153 positions in the primary amino acid sequence (Fig. 20.10; reviewed in Hart, 2006).

Given the critical role of SOD1 as an antioxidant protein and its abundance in neural tissue, the initial (and logical) hypothesis was that the fALS mutations resulted in an enzyme that could not detoxify reactive oxygen species, and over time, this deficiency would lead to oxidative damage and death of neural cells. However, it is now well established that SOD1-linked ALS is not a loss-of-function disease. For example, mice with their SOD1 genes deleted do not develop symptoms akin to ALS (Reaume *et al.,* 1996), and mice

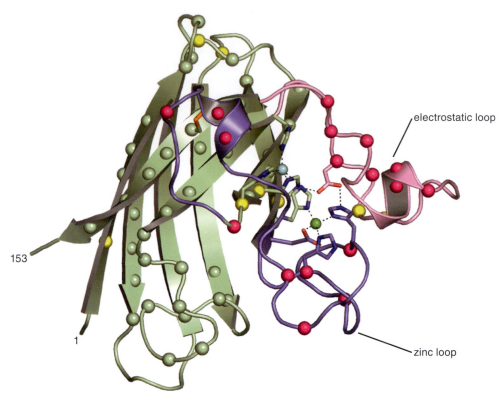

153

1

electrostatic loop

zinc loop

FIGURE 20.10 The spatial distribution of the known pathogenic SOD1 mutations. A monomer of SOD1 is shown in approximately the same orientation as the right-most subunit in Figure 20.9. The α-carbon positions of fALS mutations falling in the β-barrel and in the metal-binding loop elements are shown as green and hot pink spheres, respectively. The α-carbon positions of pathogenic SOD1 mutants for which there are mouse models are shown as yellow spheres.

that express human SOD1–ALS mutant proteins, in addition to their own normal mouse SOD1 (and therefore possess full superoxide dismutase activity), develop paralytic symptoms strikingly similar to what is observed in human patients (Bruijn *et al.*, 1997; Gurney *et al.*, 1994; Wang *et al.*, 2005). Thus, toxicity appears to be linked only with the presence of the mutant polypeptide and not to enzymatic activity, and this has led researchers to suggest that pathogenic SOD1 proteins have acquired a property toxic to motor neurons.

Aggregation of Mutant SOD1 in fALS

Examination of the neural tissue of patients and transgenic mice expressing human fALS–SOD1 proteins reveals prominent proteinaceous inclusions that are enriched for SOD1 (Bruijn *et al.*, 1997; Gurney *et al.*, 1994; Wang *et al.*, 2005). In general, aggregates are derived from assemblies of protein that attain relatively high molecular weight and include fibrous aggregates as well as somewhat smaller oligomeric structures. Pathologic protein aggregates often resist dissociation in detergent, and larger aggregates can be separated from tissue homogenates by ultracentrifugation and/or size exclusion chromatography. In transgenic

mouse models of the disease in which the mice overexpress human pathogenic SOD1, the appearance of these inclusions tends to coincide with the onset of paralytic symptoms. An example of these inclusions is shown in Figure 20.11. The acquired toxic property of the mutant SOD1 may therefore come from the tendency of these proteins to assemble into higher-order structures that somehow interfere with the neuronal cellular machinery. Importantly, this propensity to form higher-order oligomers is not a feature of the normal protein, which does not aggregate. In humans, the most consistently reported pathologic structures are hyaline or Lewy body–like inclusions that are immunoreactive to SOD1 antibodies (Wang *et al.*, 2003). Mutant SOD1 aggregates with similar features are also formed in cultured cells that express high levels of mutant protein (Karch and Borchelt, 2008; Wang *et al.*, 2003). Thus, aggregation of the mutant protein seems to be a consequence of fALS mutation, and it is now generally accepted that SOD1-linked ALS is protein misfolding and aggregation disorder, although it remains to be clarified whether the observed inclusions are causal or symptomatic of motor neuron dysfunction.

Although there is a paucity of detail regarding the precise structure of mutant SOD1 aggregates that form

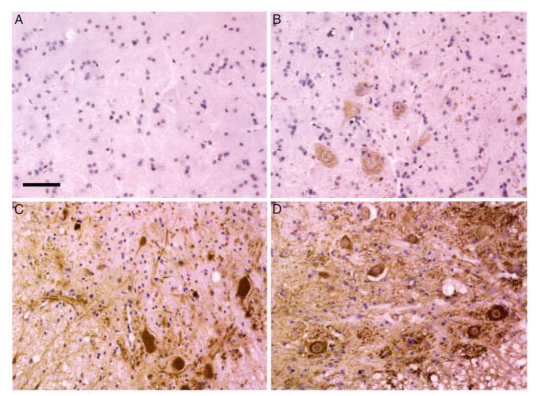

FIGURE 20.11   Accumulation of SOD1–L126 truncation variant (L126Z) in somatodendritic compartments of spinal motor neurons. Tissue sections embedded in paraffin were deparaffinized and immunostained with hSOD1 antiserum at a dilution of 1:500. (A) Nontransgenic littermate 9 months old. (B) Representative image from 3.5-month-old L126Z mice. (C) Image from a 7-month-old symptomatic SOD1–L126Z mouse shows longitudinal profiles of dendrites and motor neuron corpses filled with immunoreactivity. (D) Image from 9-month-old symptomatic SOD1–L126Z mouse shows intensifying of motor neuron soma and circular profiles resembling dendritic cross sections. Scale bar = 50 mm. Adapted from Wang (2005).

in vivo, metal-depleted mutant forms of SOD1 (but not metal-replete forms) can engage in nonnative SOD1–SOD1 protein–protein interactions to form amyloid-like filaments and "amyloid pores" *in vitro* (Elam *et al.*, 2003) (Fig. 20.12). Amyloid pores have been observed in other neurodegenerative diseases in which protein aggregation is a characteristic feature (Lashuel *et al.*, 2002). In most, but not all, fALS mouse models, there is evidence of the accumulation of amyloid-like material in the form of thioflavin-S–positive fibrillar structures (Wang *et al.*, 2003, 2005). At present, however, there is a lack of sufficient understanding of the role of specific SOD1 aggregate structures in disease pathogenesis to predict whether disruption of mutant SOD1 aggregation would be beneficial or detrimental, and there is a continuing debate as to whether the insoluble aggregates or their soluble precursors are the toxic entities.

Immature Pathogenic SOD1 and Toxicity

An additional potential player in SOD1-linked ALS is the copper chaperone for SOD1 (CCS), a helper protein that confers two critical stabilizing

post-translational modifications on newly synthesized SOD1: the insertion of the catalytic copper ion (Culotta *et al.*, 1997) and the oxidation of the SOD1 intrasubunit disulfide bond (Brown *et al.*, 2004). The presence of a disulfide bond is rare for cytosolic proteins given the strong reducing environment of the cytosol, and recent studies suggest that CCS-mediated oxidation of this disulfide bond occurs concomitant with copper delivery and is oxygen (or superoxide) dependent (Brown *et al.*, 2004). At the protein level, the ratio of SOD1 to CCS in the cytosol ranges between 15- and 30-fold (Rothstein *et al.*, 1999), and because of this, CCS must cycle through the nascent SOD1 pool to activate these molecules (Furukawa and O'Halloran, 2006). The metal-replete, disulfide-oxidized SOD1 holoenzyme is remarkable in that it is one of the most stable protein molecules known, with a melting point of approximately 93°C (Rodriguez *et al.*, 2005). In contrast, the metal-free, disulfide-reduced wild-type SOD1 protein melts at approximately 42°C, and metal-free, disulfide-reduced SOD1 proteins with substitutions located in the β-barrel exhibit no endothermic transition, suggesting they are at best only partially structured (Rodriguez *et al.*, 2005). It is important to note

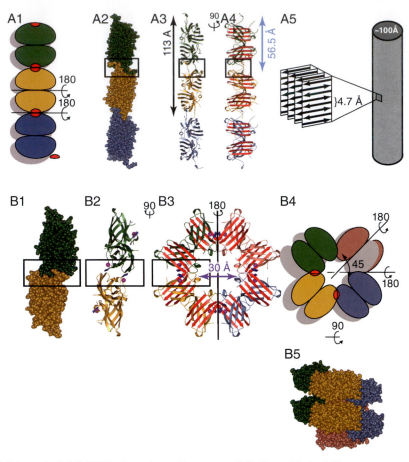

FIGURE 20.12 Metal-deficiency in fALS–SOD1 gives rise to linear cross-β fibrils and helical filamentous arrays. (A) Linear, amyloid-like filaments formed by three dimers shown from top to bottom in green, gold, and blue. Nonnative SOD1–SOD1 interactions are shown as red patches in 1 and are boxed in 2–4. The "cross-β" structure observed in amyloid fibrils is shown schematically in 5. (B) Metal-deficiency in pathogenic SOD1 also gives rise to water-filled helical filamentous arrays. (1) One-half of one turn of the helical filament is represented by the two dimers shown from top to bottom in green and gold. (2–3) Ribbon representation. The blue arrow indicates the diameter of the central cavity. The nonnative interactions between SOD1 dimers are boxed. (4) Schematic view of the helical filamentous array shown in 3, with the new interdimer contacts shown as red patches. 5 This view of the helical filament is rotated 90° around a horizontal axis relative to the view in 3 and 4. Successive Zn–H46R dimers (green, yellow, blue, and red) comprise one turn of the helical filament with a pitch of <35 Å. Adapted from Elam (2003).

that the newly translated (nascent) SOD1 molecules are also metal free, disulfide reduced, and therefore expected to be similarly destabilized relative to the holoenzyme, at least until they can be acted upon by CCS or bind metal ions adventitiously.

Recent studies in transgenic mice reveal that a significant fraction of the pathogenic SOD1 proteins in the soluble fraction are metal deficient and/or disulfide reduced (Jonsson *et al.,* 2006; Zetterstrom *et al.,* 2007). These observations are consistent with the possibility that certain pathogenic SOD1 mutations might interfere with nascent SOD1–CCS protein–protein interactions or other aspects of CCS action, resulting in a pool of immature molecules (Cao *et al.,* 2008). Such molecules are essentially off-pathway folding intermediates, and recent studies strongly suggest that

these immature SOD1 molecules eventually end up in the aggregates observed in humans and in transgenic mouse models (Jonsson *et al.,* 2006; Wang *et al.,* 2005). For example, the H46R and H48Q substitutions replace copper ligands, and these variants are known to be unable to properly bind copper (Wang *et al.,* 2007). Although CCS can attempt to load copper ion into nascent H46R SOD1, because the copper-binding site is ablated, it will never be able to attain the highly stable state exhibited by the wild-type enzyme. In addition, it is likely that pathogenic SOD1 variants that contain substitutions in the metal-binding loop IV and loop VI elements will have similar problems being post-translationally modified by CCS. Given its critical role in SOD1 maturation, studies of CCS action on pathogenic SOD1 variants are intensifying.

GLIA AND NEURODEGENERATIVE DISEASE

The interplay between cells of the nervous system plays a crucial role in the survival or demise of a particular neuron. In the past decade it has become clear that glia play a crucial role in the neurodegenerative process, both positive and negative. However, there were implications of this involvement over a hundred years ago when Ramon-y-Cajal described activated glia in the brains of demented patients whom Alzheimer had described (García-Marín *et al.*, 2007). The axis formed between neurons and glial cells appears to constitute a basic unit in brain function, which together with the blood vessels present a set of interacting elements that define the functional capacity of the brain. The relationships between glial and neuronal cells involve both growth–survival promoting factors and proinflammatory cytokines. Often it is the balance and/or timing between these two opposing elements that determines whether a neuron will live or die.

Growth factors like BDNF or GDNF and certain cytokines IL-1β form the positive element of this neuronal unit, positively influencing neuronal health and mediating protection from the everyday stress of neuronal function and/or damaging stresses on the neurons due to genetic or environmental influences. Chapter 10 dealt with the growth factors and cytokines in detail, but in this context, it is helpful to point out that glia produce significant levels of the trophic factors in addition to neurons. Upon stress to a neuron, glia are known to become activated, with astroglia swelling in size with process ramification, and microglia begin to migrate into the stress area. This activation results in an increased metabolism of the glia, helping to remove toxins from the neuropil and buffer the extracellular ionic environment, protecting the neuron from further stress. In addition, they begin to release trophic factors, which further supports the survival of the neuron.

On the other hand, inflammation is a process that has been actively associated with the onset of several neurodegenerative disorders, including AD. However, the precise implications of the inflammatory response for neurodegeneration have not been elucidated. A current hypothesis considers that an extracellular insult to neurons could trigger the production of inflammatory cytokines by astrocytes and microglia. These cytokines, namely, IL-1β, TNF-α, and IL-6, could affect the normal behavior of neuronal cells. Therefore, dysfunction at this core level may lead to abnormalities such as neurofibrillary degeneration in AD. The brain has the ability to activate inflammatory reactions in response to an immune challenge in the periphery. Various cytokines are implicated, such as IL-1β, which responds with a widespread pattern of activities within the brain, and interleukin-6, whose overproduction has been considered as a step in the pathway to neuronal degeneration. The well-known TNF-α constitutes an important cytokine that shares these patterns of expression. In spite of these studies, the precise role of cytokines in neurodegenerative processes is not fully understood. In this context, it has been described that cytokines secreted by microglial cells, astrocytes, and/or neuronal cells may induce synthesis of certain acute-phase proteins, including the amyloid precursor protein APP. On the other hand, the amyloid-β peptide (Aβ) itself can induce the expression of IL-1β, TNF-α, and IL-6 in astrocytes and microglial cells in culture. IL-6 has been found at an early stage of senile plaque formation. It was also observed that IL-6 secretion by peripheral blood mononuclear cells was increased in patients with AD as opposed to normal subjects or those suffering from other brain disorders such as vascular dementia. In addition to these findings, overexpression of IL-6 in the brain of transgenic mice that overproduce this cytokine is associated with a variety of neuropathological findings, including gliosis and disruption of cholinergic neurotransmission in the hippocampus. Thus, a direct correlation has been established between the Aβ-induced neurotoxicity in neurodegenerative conditions and cytokine production.

SUMMARY

The adult-onset neurodegenerative diseases are a set of disorders that have certain shared characteristics: a dramatic loss of neurons in some, but not all, parts of the brain, a progressive development, with minor subtle symptoms developing into disabling and, in some cases, fatal diseases getting worse as the person gets older. While there are differing cellular and molecular mechanisms underlying each disease, they do share some common characteristics. One is the accumulation of insoluble material in the cell that increases with age and that ultimately causes the cell function to collapse resulting in death. In part, this has been borne out by the data produced from the genetic analysis of the familial forms of these diseases, which points a finger to the disruption of the machinery responsible for clearing damaged proteins or the production of mutated forms of a protein, which is then highly susceptible to aggregation if not cleared. Despite the fact that the manifestation of the disease is through the loss of specific neurons and the circuits they subserve, it is apparent that glia, both astroglia and microglia, play an important role in either protecting from, or in some cases exacerbating, the neurodegeneration.

References

Al-Chalabi A., Andersen P.M., Chioza B., Shaw C., Sham P.C., Robberecht W., Matthijs G., Camu W., Marklund S.L., Forsgren L., Rouleau G., Laing N.G., Hurse P.V., Siddique T., Leigh P.N., and Powell J.F. (1998). Recessive amyotrophic lateral sclerosis families with the D90A SOD1 mutation share a common founder: evidence for a linked protective factor. Hum. Mol. Genet. 7(13), 2045-2050.

Beilina A., Van Der Brug M., Ahmad R., Kesavapany S., Miller D.W., Petsko G.A., and Cookson M.R. (2005). Mutations in PTEN-induced putative kinase 1 associated with recessive parkinsonism have differential effects on protein stability. Proc. Natl. Acad. Sci. USA 102(16), 5703-5708.

Bertram L., McQueen M.B., Mullin K., Blacker D., and Tanzi R.E. (2007). Systematic meta-analyses of Alzheimer disease genetic association studies: the AlzGene database. Nat. Genet. 39(1), 17-23.

Betarbet R., Sherer T.B., MacKenzie G., Garcia-Osuna M., Panov A.V., and Greenamyre J.T. (2000). Chronic systemic pesticide exposure reproduces features of Parkinson's disease. Nat. Neurosci. 3, 1301-1306.

Bonifati V., Rizzu P., van Baren M.J., Schaap O., Breedveld G.J., Krieger E., Dekker M.C., Squitieri F., Ibanez P., Joosse M., van Dongen J.W., Vanacore N., van Swieten J.C., Brice A., Meco G., van Duijn C.M., Oostra B.A., and Heutink P. (2003). Mutations in the DJ-1 gene associated with autosomal recessive early-onset parkinsonism. Science 299(5604), 256-259.

Braak H., Del Tredici K., Rüb U., de Vos R.A., Jansen Steur E.N., and Braak E. (2003). Staging of brain pathology related to sporadic Parkinson's disease. Neurobiol. Aging 24(2), 197-211.

Braak E., Griffing K., Arai K., Bohl J., Bratzke H., and Braak H. (1999). Neuropathology of Alzheimer's disease: what is new since A. Alzheimer? Eur. Arch. Psychiatry Clin. Neurosci. 249 Suppl. 3, 14-22.

Brown N.M., Torres A.S., Doan P.E., and O'Halloran T.V. (2004). Oxygen and the copper chaperone CCS regulate posttranslational activation of Cu,Zn superoxide dismutase. Proc. Natl. Acad. Sci. USA 101(15), 5518-5523.

Bruijn L.I., Miller T.M., and Cleveland D.W. (2004). Unraveling the mechanisms involved in motor neuron degeneration in ALS. Annu. Rev. Neurosci. 27, 723-749.

Bruijn L.I., Becher M.W., Lee M.K., Anderson K.L., Jenkins N.A., Copeland N.G., Sisodia S.S., Rothstein J.D., Borchelt D.R., Price D.L., and Cleveland D.W. (1997). ALS-linked SOD1 mutant G85R mediates damage to astrocytes and promotes rapidly progressive disease with SOD1-containing inclusions. Neuron 18(2), 327-338.

Burke W.J., Li S.W., Williams E.A., Nonneman R., and Zahm D.S. (2003). 3,4-Dihydroxyphenylacetaldehyde is the toxic dopamine metabolite in vivo: implications for Parkinson's disease pathogenesis. Brain Res. 989, 205-213.

Cai H., Wang Y., McCarthy D., Wen H., Borchelt D.R., Price D.L., and Wong P.C. (2001). BACE1 is the major beta-secretase for generation of Abeta peptides by neurons. Nat. Neurosci. 4, 233-234.

Cao X., Antonyuk S.V., Seetharaman S.V., Whitson L.J., Taylor A.B., Holloway S.P., Strange R.W., Doucette P.A., Valentine J.S., Tiwari A., Hayward L.J., Padua S., Cohlberg J.A., Hansain S.S.,

and Hart P.J. (2008). Structures of the G85R variant of SOD1 in familial amyotrophic lateral sclerosis. J. Biol. Chem. 283(23), 16169-16177.

Castellani R.J., Perry G., Siedlak S.L., Nunomura A., Shimohama S., Zhang J., Montine T., Sayre L.M., and Smith M.A. (2002a). Hydroxynonenal adducts indicate a role for lipid peroxidation in neocortical and brainstem Lewy bodies in humans. Neurosci. Lett. 319, 25-28.

Castellani R.J., Perry G., Siedlak S.L., Nunomura A., Shimohama S., Zhang J., Montine T., Sayre L.M., and Smith M.A. (2002b). Hydroxynonenal adducts indicate a role for lipid peroxidation in neocortical and brainstem Lewy bodies in humans. Neurosci. Lett. 319, 25-28.

Cedazo-Mínguez A. (2007). Apolipoprotein E and Alzheimer's disease: molecular mechanisms and therapeutic opportunities. J. Cell. Mol. Med. 11(6), 1227-1238.

Charcot J.M. and Joffroy A. (1869). Deux cas d'atrophie musculaire progressive avec lesions de la substance grise et de faisceaux anterolateraux de la moelle epiniere. Arch. Physiol. Norm. Pathol. 1, 354-367.

Chartier-Harlin M.C., Crawford F., Houlden H., Warren A., Hughes D., Fidani L., Goate A., Rossor M., Roques P., Hardy J., and Mullan M. (1991). Early-onset Alzheimer's disease caused by mutations at codon 717 of the beta-amyloid precursor protein gene. Nature 353, 844-846.

Chartier-Harlin M.C., Kachergus J., Roumier C., Mouroux V., Douay X., Lincoln S., Levecque C., Larvor L., Andrieux J., Hulihan M., Waucquier N., Defebvre L., Amouyel P., Farrer M., and Destée A. (2004). Alpha-synuclein locus duplication as a cause of familial Parkinson's disease. Lancet 364, 1167-1169.

Chen Y.Z., Bennett C.L., Huynh H.M., Blair I.P., Puls I., Irobi J., Dierick I., Abel A., Kennerson M.L., Rabin B.A., Nicholson G.A., Auer-Grumbach M., Wagner K., De Jonghe P., Griffin J.W., Fischbeck K.H., Timmerman V., Cornblath D.R., and Chance P.F. (2004). DNA/RNA helicase gene mutations in a form of juvenile amyotrophic lateral sclerosis (ALS4). Am. J. Hum. Genet. 74(6), 1128-1135.

Cookson M.R., Xiromerisiou G., and Singleton A. (2005). How genetics research in Parkinson's disease is enhancing understanding of the common idiopathic forms of the disease. Curr. Opin. Neurol. 18(6), 706-711.

Coon K.D., Myers A.J., Craig D.W., Webster J.A., Pearson J.V., Lince D.H., Zismann V.L., Beach T.G., Leung D., Bryden L., Halperin R.F., Marlowe L., Kaleem M., Walker D.G., Ravid R., Heward C.B., Rogers J., Papassotiropoulos A., Reiman E.M., Hardy J., and Stephan D.A. (2007). A high-density whole-genome association study reveals that APOE is the major susceptibility gene for sporadic late-onset Alzheimer's disease. J. Clin. Psychiatry 68(4), 613-618.

Crook R., Verkkoniemi A., Perez-Tur J., Mehta N., Baker M., Houlden H., Farrer M., Hutton M., Lincoln S., Hardy J., Gwinn K., Somer M., Paetau A., Kalimo H., Ylikoski R., Pöyhönen M., Kucera S., and Haltia M. (1998). A variant of Alzheimer's disease with spastic paraparesis and unusual plaques due to deletion of exon 9 of presenilin 1. Nat. Med. 4(4), 452-455.

Culotta V.C., Klomp L.W., Strain J., Casareno R.L., Krems B., and Gitlin J.D. (1997). The copper chaperone for superoxide dismutase. J. Biol. Chem. 272(38), 23469-23472.

Dalfo E., Portero-Otin M., Ayala V., Martinez A., Pamplona R., and Ferrer I. (2005). Evidence of oxidative stress in the neocortex in incidental Lewy body disease. J. Neuropathol. Exp. Neurol. 64, 816-830.

Deng H.X., Hentati A., Tainer J.A., Iqbal Z., Cayabyab A., Hung W.Y., Getzoff E.D., Hu P., Herzfeldt B., and Roos R.P. (1993). Amyotrophic lateral sclerosis and structural defects in Cu,Zn superoxide dismutase. Science 261(5124), 1047-1051.

Dierick I., Baets J., Irobi J., Jacobs A., De Vriendt E., Deconinck T., Merlini L., Van den Bergh P., Rasic V.M., Robberecht W., Fischer D., Morales R.J., Mitrovic Z., Seeman P., Mazanec R., Kochanski A., Jordanova A., Auer-Grumbach M., Helderman-van den Enden A.T., Wokke J.H., Nelis E., De Jonghe P., and Timmerman V. (2008). Relative contribution of mutations in genes for autosomal dominant distal hereditary motor neuropathies: a genotype-phenotype correlation study. Brain 131(Pt. 5), 1217-1227.

Di Fonzo A., Chien H.F., Socal M., Giraudo S., Tassorelli C., Iliceto G., Fabbrini G., Marconi R., Fincati E., Abbruzzese G., Marini P., Squitieri F., Horstink M.W., Montagna P., Libera A.D., Stocchi F., Goldwurm S., Ferreira J.J., Meco G., Martignoni E., Lopiano L., Jardim L.B., Oostra B.A., and Barbosa E.R. (2007). The Italian Parkinson Genetics Network, Bonifati V ATP13A2 missense mutations in juvenile parkinsonism and young onset Parkinson disease. Neurology 68(19), 1557-1562.

D'Souza I. and Schellenberg G.D. (2005). Regulation of tau isoform expression and dementia. Biochim. Biophys. Acta. 1739(2-3), 104-115.

Duke D.C., Moran L.B., Kalaitzakis M.E., Deprez M., Dexter D.T., Pearce R.K., Graeber M.B. (2006). Transcriptome analysis reveals link between proteasomal and mitochondrial pathways in Parkinson's disease. Neurogenetics 7(3), 139-148.

Elam J.S., Taylor A.B., Strange R., Antonyuk S., Doucette P.A., Rodriguez J.A., Hasnain S.S., Hayward L.J., Valentine J.S., Yeates T.O., and Hart P.J. (2003). Amyloid-like filaments and water-filled nanotubes formed by SOD1 mutant proteins linked to familial ALS. Nat. Struct. Biol. 10(6), 461-467.

Fitzmaurice P.S., Ang L., Guttman M., Rajput A., Furukawa Y., and Kish S.J. (2003). Nigral glutathione deficiency is not specific for idiopathic Parkinson's disease. Movement Disorders 18, 969-976.

Fornai F., Schluter O.M., Lenzi P., Gesi M., Ruffoli R., Ferrucci M., Lazzeri G., Busceti C.L., Pontarelli F., Battaglia G., Pellegrini A., Nicoletti F., Ruggieri S., Paparelli A., and Sudhof T.C. (2005). Parkinson-like syndrome induced by continuous MPTP infusion: convergent roles of the ubiquitin-proteasome system and alpha-synuclein. Proc. Natl. Acad. Sci. USA 102, 3413-3418.

Fridovich I. (1989). Superoxide dismutases: an adaptation to a paramagnetic gas. J. Biol. Chem. 264(14), 7761-7764.

Funayama M., Hasegawa K., Ohta E., Kawashima N., Komiyama M., Kowa H., Tsuji S., and Obata F. (2005). An LRRK2 mutation as a cause for the parkinsonism in the original PARK8 family. Ann. Neurol. 57(6), 918-921.

Fung H.C., Scholz S., Matarin M., Simón-Sánchez J., Hernandez D., Britton A., Gibbs J.R., Langefeld C., Stiegert M.L., Schymick J., Okun M.S., Mandel R.J., Fernandez H.H., Foote K.D., Rodríguez R.L., Peckham E., De Vrieze F.W., Gwinn-Hardy K., Hardy J.A., and Singleton A. (2006). Genome-wide genotyping in Parkinson's disease and neurologically normal controls: first stage analysis and public release of data. Lancet Neurol. 5(11), 911-916.

Furukawa Y. and O'Halloran T.V. (2006). Posttranslational modifications in Cu,Zn-superoxide dismutase and mutations associated with amyotrophic lateral sclerosis. Antioxid. Redox. Signal. 8(5-6), 847-867.

Galter D., Buervenich S., Carmine A., Anvret M., and Olson L. (2003). ALDH1 mRNA: presence in human dopamine neurons and decreases in substantia nigra in Parkinson's disease and in the ventral tegmental area in schizophrenia. Neurobiol. Dis. 14, 637-647.

Galvin J.E. (2006). Interaction of alpha-synuclein and dopamine metabolites in the pathogenesis of Parkinson's disease: a case for the selective vulnerability of the substantia nigra. Acta. Neuropathol. (Berl) 112, 115-126.

García-Marín V., García-Lopez P., and Freire M. (2007). Cajal's contributions to the study of Alzheimer's disease. J. Alzheimer's Dis. 12, 161-174.

Golde T.E., Dickson D., and Hutton M. (2006). Filling the gaps in the abeta cascade hypothesis of Alzheimer's disease. Curr. Alzheimer Res. 3(5), 421-430.

Greenamyre J.T., Sherer T.B., Betarbet R., and Panov A.V. (2001). Complex I and Parkinson's disease. IUBMB. Life 52, 135-141.

Gurney M.E., Pu H., Chiu A.Y., Dal Canto M.C., Polchow C.Y., Alexander D.D., Caliendo J., Hentati A., Kwon Y.W., and Deng H.X. (1994). Motor neuron degeneration in mice that express a human Cu,Zn superoxide dismutase mutation. Science 264(5166), 1772-1775.

Gwinn-Hardy K., Singleton A., O'Suilleabhain P., Boss M., Nicholl D., Adam A., Hussey J., Critchley P., Hardy J., and Farrer M. (2001). Spinocerebellar ataxia type 3 phenotypically resembling parkinson disease in a black family. Arch. Neurol. 58(2), 296-299.

Gwinn-Hardy K., Chen J.Y., Liu H.C., Liu T.Y., Boss M., Seltzer W., Adam A., Singleton A., Koroshetz W., Waters C., Hardy J., and Farrer M. (2000). Spinocerebellar ataxia type 2 with parkinsonism in ethnic Chinese. Neurology 55(6), 800-805.

Haas R.H., Nasirian F., Nakano K., Ward D., Pay M., Hill R., and Shults C.W. (1995). Low platelet mitochondrial complex I and complex II/III activity in early untreated Parkinson's disease. Ann. Neurol. 37, 714-722.

Hadano S., Hand C.K., Osuga H., Yanagisawa Y., Otomo A., Devon R.S., Miyamoto N., Showguchi-Miyata J., Okada Y., Singaraja R., Figlewicz D.A., Kwiatkowski T., Hosler B.A., Sagie T., Skaug J., Nasir J., Brown R.H., Jr., Scherer S.W., Rouleau G.A., Hayden M. R., and Ikeda J.E. (2001). A gene encoding a putative GTPase regulator is mutated in familial amyotrophic lateral sclerosis 2. Nat. Genet. 29(2), 166-173.

Halliwell B. (1992). Reactive oxygen species and the central nervous system. J Neurochem. 59, 1609-1623.

Hardy J.A. and Higgins G.A. (1992). Alzheimer's disease: the amyloid cascade hypothesis. Science 256:184-185.

Hardy J. and Selkoe D.J. (2002). The amyloid hypothesis of Alzheimer's disease: progress and problems on the road to therapeutics. Science 297, 353-356.

Hart P.J. (2006). Pathogenic superoxide dismutase structure, folding, aggregation and turnover. Curr. Opin. Chem. Biol. 10(2), 131-138.

Hart P.J., Liu H., Pellegrini M., Nersissian A.M., Gralla E.B., Valentine J.S., and Eisenberg D. (1998). Subunit asymmetry in the three-dimensional structure of a human CuZnSOD mutant found in familial amyotrophic lateral sclerosis. Protein Sci. 7(3), 545-555.

Hornykiewicz O. (1998). Biochemical aspects of Parkinson's disease. Neurol. 51(2, Suppl. 2), S2-S9.

Hutton M., Lendon C.L., Rizzu P., Baker M., Froelich S., Houlden H., Pickering-Brown S., Chakraverty S., Isaacs A., Grover A., Hackett J., Adamson J., Lincoln S., Dickson D., Davies P., Petersen R.C., Stevens M., de Graaff E., Wauters E., van Baren J., Hillebrand M., Joosse M., Kwon J.M., Nowotny P., Che L.K., Norton J., Morris J.C., Reed L.A., Trojanowski J., Basun H., Lannfelt L., Neystat M., Fahn S., Dark F., Tannenberg T., Dodd P. R., Hayward N., Kwok J.B., Schofield P.R., Andreadis A., Snowden J., Craufurd D., Neary D., Owen F., Oostra B.A., Hardy J., Goate A., van Swieten J., Mann D., Lynch T., and Heutink P. (1998). Association of missense and 5'-splice-site mutations in tau with the inherited dementia FTDP-17. Nature 393, 702-705.

Ikeda S., Allsop D., and Glenner G.G. (1989). A study of the morphology and distribution of amyloid beta protein immunoreactive plaque and related lesions in the brains of Alzheimer's disease and adult Down's syndrome. Prog. Clin. Biol. Res. 317, 313-323.

Inden M., Kitamura Y., Takeuchi H., Yanagida T., Takata K., Kobayashi Y., Taniguchi T., Yoshimoto K., Kaneko M., Okuma Y., Taira T., Ariga H., and Shimohama S. (2007). Neurodegeneration of mouse nigrostriatal dopaminergic system induced by repeated oral administration of rotenone is prevented by 4-phenylbutyrate, a chemical chaperone. J. Neurochem. 101(6), 1491-1504.

Iwatsubo T., Odaka A., Suzuki N., Mizusawa H., Nukina N., and Ihara Y. (1994). Visualization of A beta 42(43) and A beta 40 in senile plaques with end-specific A beta monoclonals: evidence that an initially deposited species is A beta 42(43). Neuron 13(1), 45-53.

James P.A. and Talbot K. (2006). The molecular genetics of non-ALS motor neuron diseases. Biochim. Biophys. Acta. 1762(11-12), 986-1000.

Jenner P., Schapira A.H., and Marsden C.D. (1992). New insights into the cause of Parkinson's disease. Neurology 42, 2241-2250.

Jonsson P.A., Graffmo K.S., Andersen P.M., Brännström T., Lindberg M., Oliveberg M., and Marklund S.L. (2006). Disulphide-reduced superoxide dismutase-1 in CNS of transgenic amyotrophic lateral sclerosis models. Brain 129(Pt. 2), 451-464.

Jonsson P.A., Bäckstrand A., Andersen P.M., Jacobsson J., Parton M., Shaw C., Swingler R., Shaw P.J., Robberecht W., Ludolph A.C., Siddique T., Skvortsova V.I., and Marklund S.L. (2002). CuZn-superoxide dismutase in D90A heterozygotes from recessive and dominant ALS pedigrees. Neurobiol. Dis. 10(3), 327-333.

Karch, C.M. and Borchelt D.R. (2008). A limited role for disulfide cross-linking in the aggregation of mutant SOD1 linked to familial amyotrophic lateral sclerosis. J. Biol. Chem. 283, 13528-13537.

Kitada T., Asakawa S., Hattori N., Matsumine H., Yamamura Y., Minoshima S., Yokochi M., Mizuno Y., and Shimizu N. (1998). Mutations in the parkin gene cause autosomal recessive juvenile parkinsonism. Nature 392(6676), 605-608.

Krüger R., Kuhn W., Müller T., Woitalla D., Graeber M., Kösel S., Przuntek H., Epplen J.T., Schöls L., and Riess O. (1998). Ala30-Pro mutation in the gene encoding alpha-synuclein in Parkinson's disease. Nat. Genet. 18(2), 106-108.

Lashuel H.A., Hartley D., Petre B.M., Walz T., and Lansbury P.T. Jr. (2002). Neurodegenerative disease: amyloid pores from pathogenic mutations. Nature 418(6895), 291.

Leroy E., Boyer R., Auburger G., Leube B., Ulm G., Mezey E., Harta G., Brownstein M.J., Jonnalagada S., Chernova T., Dehejia A., Lavedan C., Gasser T., Steinbach P.J., Wilkinson K.D., and Polymeropoulos M.H. (1998). The ubiquitin pathway in Parkinson's disease. Nature 395(6701), 451-452.

Le W.D., Xu P., Jankovic J., Jiang H., Appel S.H., Smith R.G., and Vassilatis D.K. (2003). Mutations in NR4A2 associated with familial Parkinson disease. Nat. Genet. 33(1), 85-89.

Li J.Z., Absher D.M., Tang H., Southwick A.M., Casto A.M., Ramachandran S., Cann H.M., Barsh G.S., Feldman M., Cavalli-Sforza L.L., and Myers R.M. (2008). Worldwide human relationships inferred from genome-wide patterns of variation. Science 319(5866), 1100-1104.

Mandel S., Grunblatt E., Riederer P., Amariglio N., Jacob-Hirsch J., Rechavi G., and Youdim M.B. (2005). Gene expression profiling of sporadic Parkinson's disease substantia nigra pars compacta reveals impairment of ubiquitin-proteasome subunits, SKP1A, aldehyde dehydrogenase, and chaperone HSC-70. Ann. N. Y. Acad. Sci. 1053, 356-375.

Maraganore D.M., de Andrade M., Lesnick T.G., Strain K.J., Farrer M.J., Rocca W.A., Pant P.V., Frazer K.A., Cox D.R., and Ballinger D.G. (2005). High-resolution whole-genome association study of Parkinson disease. Am. J. Hum. Genet. 77(5), 685-693.

Marklund S.L. (1984). Extracellular superoxide dismutase in human tissues and human cell lines. J. Clin. Invest. 74, 1398-1403.

Marx F.P., Holzmann C., Strauss K.M., Li L., Eberhardt O., Gerhardt E., Cookson M.R., Hernandez D., Farrer M.J., Kachergus J., Engelender S., Ross C.A., Berger K., Schöls L., Schulz J.B., Riess O., and Krüger R. (2003). Identification and functional characterization of a novel R621C mutation in the synphilin-1 gene in Parkinson's disease. Hum. Mol. Genet. 12(11), 1223-1231.

Miller D.W., Ahmad R., Hague S., Baptista M.J., Canet-Aviles R., McLendon C., Carter D.M., Zhu P.P., Stadler J., Chandran J., Klinefelter G.R., Blackstone C., and Cookson M.R. (2003). L166P mutant DJ-1, causative for recessive Parkinson's disease, is degraded through the ubiquitin-proteasome system. J. Biol. Chem. 278, 36588-36595.

Mirzaei H., Schieler J.L., Rochet J.C., and Regnier F. (2006). Identification of rotenone-induced modifications in alpha-synuclein using affinity pull-down and tandem mass spectrometry. Anal. Chem. 78, 2422-2431.

Moreira M.C., Klur S., Watanabe M., Németh A.H., Le Ber I., Moniz J.C., Tranchant C., Aubourg P., Tazir M., Schöls L., Pandolfo M., Schulz J.B., Pouget J., Calvas P., Shizuka-Ikeda M., Shoji M., Tanaka M., Izatt L., Shaw C.E., M'Zahem A., Dunne E., Bomont P., Benhassine T., Bouslam N., Stevanin G., Brice A., Guimarães J., Mendonça P., Barbot C., Coutinho P., Sequeiros J., Dürr A., Warter J.M., and Koenig M. (2004). Senataxin, the ortholog of a yeast RNA helicase, is mutant in ataxia-ocular apraxia 2. Nat. Genet. 36(3), 225-227.

Mytilineou C., McNaught K.S., Shashidharan P., Yabut J., Baptiste R.J., Parnandi A., and Olanow C.W. (2004). Inhibition of proteasome activity sensitizes dopamine neurons to protein alterations and oxidative stress. J. Neural Transm. 111, 1237-1251.

Mulder D.W., Kurland L.T., Offord K.P., and Beard C.M. (1986). Familial adult motor neuron disease: amyotrophic lateral sclerosis. Neurology 36(4), 511-517.

Neumann M., Sampathu D.M., Kwong L.K., Truax A.C., Micsenyi M.C., Chou T.T., Bruce J., Schuck T., Grossman M., Clark C. M., McCluskey L.F., Miller B.L., Masliah E., Mackenzie I.R., Feldman H., Feiden W., Kretzschmar H.A., Trojanowski J.Q., and Lee V.M. (2006). Ubiquitinated TDP-43 in frontotemporal lobar degeneration and amyotrophic lateral sclerosis. Science 314, 130-133.

Nishimura A.L., Mitne-Neto M., Silva H.C., Richieri-Costa A., Middleton S., Cascio D., Kok F., Oliveira J.R., Gillingwater T., Webb J., Skehel P., and Zatz M. (2004). A mutation in the vesicle-trafficking protein VAPB causes late-onset spinal muscular atrophy and amyotrophic lateral sclerosis. Am. J. Hum. Genet. 75(5), 822-831.

Paisán-Ruíz C., Jain S., Evans E.W., Gilks W.P., Simón J., van der Brug M., López de Munain A., Aparicio S., Gil A.M., Khan N., Johnson J., Martinez J.R., Nicholl D., Carrera I.M., Pena A.S., de Silva R., Lees A., Martí-Massó J.F., Pérez-Tur J., Wood N. W., and Singleton A.B. (2004). Cloning of the gene containing mutations that cause PARK8-linked Parkinson's disease. Neuron 44(4), 595-600.

Pardo C.A., Xu Z., Borchelt D.R., Price D.L., Sisodia S.S., and Cleveland D.W. (1995). Superoxide dismutase is an abundant component in cell bodies, dendrites, and axons of motor neurons and in a subset of other neurons. Proc. Natl. Acad. Sci. USA 92(4), 954-958.

Perry T.L., Godin D.V., and Hansen S. (1982). Parkinson's disease: A disorder due to nigral glutathione deficiency. Neurosci. Lett. 33, 305-310.

Polymeropoulos M.H., Lavedan C., Leroy E., Ide S.E., Dehejia A., Dutra A., Pike B., Root H., Rubenstein J., Boyer R., Stenroos E.S., Chandrasekharappa S., Athanassiadou A., Papapetropoulos T., Johnson W.G., Lazzarini A.M., Duvoisin R.C., Di Iorio G., Golbe L.I., and Nussbaum R.L. (1997). Mutation in the alpha-synuclein gene identified in families with Parkinson's disease. Science 276(5321), 2045-2047.

Puls I., Jonnakuty C., LaMonte B.H., Holzbaur E.L., Tokito M., Mann E., Floeter M.K., Bidus K., Drayna D., Oh S.J., Brown R.H., Jr., Ludlow C.L., and Fischbeck K.H. (2003). Mutant dynactin in motor neuron disease. Nat. Genet. 33(4), 455-456.

Qin Z., Hu D., Han S., Reaney S.H., Di Monte D.A., and Fink A.L. (2007). Effect of 4-hydroxy-2-nonenal modification on alpha-synuclein aggregation. J. Biol. Chem. 282, 5862-5870.

Ramirez A., Heimbach A., Gründemann J., Stiller B., Hampshire D., Cid L.P., Goebel I., Mubaidin A.F., Wriekat A.L., Roeper J., Al-Din A., Hillmer A.M., Karsak M., Liss B., Woods C.G., Behrens M.I., and Kubisch C. (2006). Hereditary parkinsonism with dementia is caused by mutations in ATP13A2, encoding a lysosomal type 5 P-type ATPase. Nat. Genet. 38 (10), 1184-1191.

Reaume A.G., Elliott J.L., Hoffman E.K., Kowall N.W., Ferrante R.J., Siwek D.F., Wilcox H.M., Flood D.G., Beal M.F., Brown R.H. Jr, Scott R.W., and Snider W.D. (1996). Motor neurons in Cu/Zn superoxide dismutase-deficient mice develop normally but exhibit enhanced cell death after axonal injury. Nat. Genet. 13(1), 43-47.

Ridet J.L., Bensadoum J.C., Deglon N., Aebischer P., and Zurn A.D. (2006). Lentivirus-mediated expression of glutatione peroxidase: neuroprotection in murine models of Parkinson's disease. Neurobiol. Dis. 21, 29-34.

Rodriguez J.A., Shaw B.F., Durazo A., Sohn S.H., Doucette P.A., Nersissian A.M., Faull K.F., Eggers D.K., Tiwari A., Hayward L.J., and Valentine J.S. (2005). Destabilization of apoprotein is insufficient to explain Cu,Zn-superoxide dismutase-linked ALS pathogenesis. Proc. Natl. Acad. Sci. USA 102(30), 10516-10521.

Rogaev E.I., Sherrington R., Rogaeva E.A., Levesque G., Ikeda M., Liang Y., Chi H., Lin C., Holman K., Tsuda T., Mar L., Sorbi S., Nacmias B., Piacentini S., Amaducci L., Chumakov I., Cohen D., Lannfelt L., Fraser P.E., Rommens J.M., and St George-Hyslop P.H. (1995). Familial Alzheimer's disease in kindreds with missense mutations in a gene on chromosome 1 related to the Alzheimer's disease type 3 gene. Nature 376, 775-778.

Rosen D.R., Siddique T., Patterson D., Figlewicz D.A., Sapp P., Hentati A., Donaldson D., Goto J., O'Regan J.P., and Deng H.X. (1993). Mutations in Cu/Zn superoxide dismutase gene are associated with familial amyotrophic lateral sclerosis. Nature 362(6415), 59-62.

Rothstein J.D., Dykes-Hoberg M., Corson L.B., Becker M., Cleveland D.W., Price D.L., Culotta V.C., and Wong P.C. (1999). The copper chaperone CCS is abundant in neurons and astrocytes in human and rodent brain. J. Neurochem. 72(1), 422-429.

Rovelet-Lecrux A., Hannequin D., Raux G., Le Meur N., Laquerrière A., Vital A., Dumanchin C., Feuillette S., Brice A., Vercelletto M., Dubas F., Frebourg T., and Campion D. (2006). APP locus duplication causes autosomal dominant early-onset Alzheimer disease with cerebral amyloid angiopathy. Nat. Genet. 38(1), 24-26.

Schymick J.C., Talbot K., and Traynor B.J. (2007). Genetics of sporadic amyotrophic lateral sclerosis. Hum. Mol. Genet. 15(16 Spec. No. 2), R233-R242.

Sherrington R., Rogaev E.I., Liang Y., Rogaeva E.A., Levesque G., Ikeda M., Chi H., Lin C., Li G., Holman K., Tsuda T., Mar L., Foncin J.-F., Bruni A.C., Montesi M.P., Sorbi S., Rainero I., Pinessi L., Neestar L., Chumakov I., Pollen D., Brookes A., Sanseau P., Polinsky R.J., Wasco W., Da Silva H.A.R., Haines J.L., Pericak-Vance M.A., Tanzi R.E., Roses A.D., Fraser P.E., Rommens J.M., and St George-Hyslop P.H. (1995). Cloning of a gene bearing missense mutations in early-onset familial Alzheimer's disease. Nature 375, 754-760.

Singleton A.B., Farrer M., Johnson J., Singleton A., Hague S., Kachergus J., Hulihan M., Peuralinna T., Dutra A., Nussbaum R., Lincoln S., Crawley A., Hanson M., Maraganore D., Adler C., Cookson M.R., Muenter M., Baptista M., Miller D., Blancato J., Hardy J., and Gwinn-Hardy K. (2003). Alpha-Synuclein locus triplication causes Parkinson's disease. Science 302, 841.

Spillantini M.G., Schmidt M.L., Lee V.M., Trojanowski J.Q., Jakes R., and Goedert M. (1997). Alpha-synuclein in Lewy bodies. Nature 388(6645), 839-840.

Trostchansky A., Lind S., Hodara R., Oe T., Blair I.A., Ischiropoulos H., Rubbo H., and Souza J.M. (2006). Interaction with phospholipids modulates alpha-synuclein nitration and lipid-protein adduct formation. Biochem. J. 393, 343-349.

Valente E.M., Abou-Sleiman P.M., Caputo V., Muqit M.M., Harvey K., Gispert S., Ali Z., Del Turco D., Bentivoglio A.R., Healy D.G., Albanese A., Nussbaum R., González-Maldonado R., Deller T., Salvi S., Cortelli P., Gilks W.P., Latchman D.S., Harvey R.J., Dallapiccola B., Auburger G., and Wood N.W. (2004). Hereditary early-onset Parkinson's disease caused by mutations in PINK1. Science 304(5674), 1158-1160.

Valentine J.S., Doucette P.A., and Zittin-Potter S. (2005). Copper-zinc superoxide dismutase and amyotrophic lateral sclerosis. Annu. Rev. Biochem. 74, 563-593.

Wang J., Caruano-Yzermans A., Rodriguez A., Scheurmann J.P., Slunt H.H., Cao X., Gitlin J., Hart P.J., and Borchelt D.R. (2007). Disease-associated mutations at copper ligand histidine residues of superoxide dismutase 1 diminish the binding of copper and compromise dimer stability. J. Biol. Chem. 282(1), 345-352.

Wang J., et al. (2005). Somatodendritic accumulation of misfolded SOD1-L126Z in motor neurons mediates degeneration: alphaB-crystallin modulates aggregation. Hum. Mol. Genet. 14(16), 2335-2347.

Wang J., et al. (2003). Copper-binding-site-null SOD1 causes ALS in transgenic mice: aggregates of non-native SOD1 delineate a common feature. Hum. Mol. Genet. 12(21), 2753-2764.

Westerlund M., Galter D., Carmine A., Olson L. (2005). Tissue- and species-specific expression patterns of class I, III, and IV Adh and A/dh 1 mRNAs in rodent embryos. Cell Tissue Res. 322, 227-236.

Yang Y., Hentati A., Deng H.X., Dabbagh O., Sasaki T., Hirano M., Hung W.Y., Ouahchi K., Yan J., Azim A.C., Cole N., Gascon G., Yagmour A., Ben-Hamida M., Pericak-Vance M., Hentati F., and Siddique T. (2001). The gene encoding alsin, a protein with three guanine-nucleotide exchange factor domains, is mutated in a form of recessive amyotrophic lateral sclerosis. Nat. Genet. 29(2), 160-165.

Zarranz J.J., Alegre J., Gómez-Esteban J.C., Lezcano E., Ros R., Ampuero I., Vidal L., Hoenicka J., Rodriguez O., Atarés B., Llorens V., Gomez Tortosa E., del Ser T., Muñoz D.G., and de Yebenes J.G. (2004). The new mutation, E46K, of alpha-synuclein causes Parkinson and Lewy body dementia. Ann. Neurol. 55(2), 164-173.

Zetterstrom P., Stewart H.G., Bergemalm D., Jonson P.A., Graffmo K.S., Andersen P.M., Brännström T., Oliveberg M., and Marklund S.L. (2007). Soluble misfolded subfractions of mutant superoxide dismutase-1s are enriched in spinal cords throughout life in murine ALS models. Proc. Natl. Acad. Sci. USA 104(35), 14157-14162.

Zimprich A., Biskup S., Leitner P., Lichtner P., Farrer M., Lincoln S., Kachergus J., Hulihan M., Uitti R.J., Calne D.B., Stoessl A.J., Pfeiffer R.F., Patenge N., Carbajal I.C., Vieregge P., Asmus F., Müller-Myhsok B., Dickson D.W., Meitinger T., Strom T.M., Wszolek Z.K., and Gasser T. (2004). Mutations in LRRK2 cause autosomal-dominant parkinsonism with pleomorphic pathology. Neuron 44(4), 601-607.

Index

Note: Page numbers followed by "f" indicate figures, "t" indicate tables, and "b" indicate boxes.